XXIX R-P-S Seminar 2020

IOP Conference Series: Materials Science and Engineering
Volume 1015

Wroclaw, Poland
30 November – 4 December 2020

Editors:

Zbigniew Wójcicki
Jacek Grose
Monika Podwórna

ISBN: 978-1-7138-4549-2
ISSN: 1757-8981

Printed from e-media with permission by:

Curran Associates, Inc.
57 Morehouse Lane
Red Hook, NY 12571

CURRAN ASSOCIATES INC. proceedings .com

Some format issues inherent in the e-media version may also appear in this print version.

Additional copies of this publication are available from:

Curran Associates, Inc.
57 Morehouse Lane
Red Hook, NY 12571 USA
Phone: 845-758-0400
Fax: 845-758-2633
Email: curran@proceedings.com
Web: www.proceedings.com

TABLE OF CONTENTS

IOP Conf. Series: Materials Science and Engineering 1015 (2021) 011001 doi:10.1088/1757-899X/1015/1/011001

XXIX Russian-Polish-Slovak *Seminary Theoretical Foundation of Civil Engineering* (RPS Seminar TFoCE),

30 November – 4 December 2020 on-line meetings

Edited by:

Zbigniew WÓJCICKI – Wrocław University of Science and Technology

Jacek GROSEL – Wrocław University of Science and Technology

Monika PODWÓRNA – Wrocław University of Science and Technology

Dear Friends

The phrase *Dear Friends* is a typical salutation currently commonly used in electronic communications. For participants of the **Russian-Polish-Slovak *Seminary Theoretical Foundation of Civil Engineering* (RPS Seminar TFoCE)**, this phrase has a much deeper dimension than the typical *Dear Friends*. A group of scientists from three neighbouring Slavic countries (occasionally from other countries also) has been systematically meeting annually for nearly 30 years, alternately in Russia, Slovakia or Poland, on the occasion of the exchange of information in the scientific field of Civil Engineering. During such a long series of meetings, seminar participants will get to know each other much better than in typical sporadic conference meetings. This is the basic specific feature of RPS Seminar TFoCE. The participants got to know each other, but by the way they got to know the countries that they briefly but systematically visited. They could exchange scientific information taking into account the geopolitical specificities of the three countries, of which Russia, due to its size and ethnic diversity, was perhaps the most important. In my opinion, it is not possible to exchange scientific knowledge only. Seminar participants also exchanged socio-political views and information. They presented their points of view on scientific and other topics, including those related to the "vagaries of history" in the informal discussions. In total, I think at least 1,000 scientists have participated in the Seminary over these 30 years. But a certain group of people, the backbone of the conference, has been basically unchanged for many years. There have been times when some of them disappeared for some time for various reasons, such as health care or excess of responsibilities resulting from the performance of organizational functions at home universities. After a while, however, most returned. Those who retired – unfortunately not. Nevertheless, as you can see from today's perspective, a slightly "nebulous" group of scientists has formed the "core" of the seminar. And this group of scientists is, in my opinion, the seminar's greatest achievement. Those who have known each other for so many years and remained faithful participants of the Seminary, because they wanted to, really became friends with a capital "F". That's why, for the RPS Seminar TFoCE, the phrase Dear Friends has such a completely different, much deeper and literal meaning!

IOP Conf. Series: Materials Science and Engineering 1015 (2021) 011001 doi:10.1088/1757-899X/1015/1/011001

Dear Friends!

In 2020, the XXIX RPS Seminar TFoCE was organized in Poland. According to previous determinations, the organizer was to be the Wrocław University of Science and Technology. From the beginning, there were troubles. Practically a revolution in the organization of science in Poland. The new law 2.0, turning upside down almost everything related to the science in the *strict sense* of its organization, but above all to the financing of science in the already hugely underfunded Poland. All this posed a threat that the Seminary would not take place. It was disturbing and frustrating. But the worst was yet to come. Pandemic! Coronavirus! Covid-19! SARS-CoV-2.

What to do?!

After many discussions and deliberations, including with colleagues from the Warsaw University of Technology supporting us, we decided! We posted a message on the conference page:

> *What should we do about the pandemic? We don't interrupt our meetings. This year we will meet virtually (online). The articles will be published as last year in the IOP Conference Series: Materials Science and Engineering. If you have any suggestions for our virtual meeting, please write to rsp2020@pwr.edu.pl.*

After a short time we received information from Žilina in Slovakia, from Moscow and from Samara in Russia. Everyone supported this decision!

And today we finished developing conference materials – articles for publication in the *IOP Conference Series: Materials Science and Engineering.*

In the end, 105 articles were sent to the editor of the magazine divided into 5 Sections. In Section 1. – Mechanics there are 23 articles, in Section 2. – Structures there are 23 articles, in Section 3 – Geodesy there are 13 articles, in Section 4. – Materials there are 36 articles and in the last Section 5. – Transport is 10 articles.

The seminar will take place online from 30.11-04.12.2020 and then it will be possible to perform a final scientific evaluation of the entire conference.

On behalf of the Organizers from Wrocław, I also encourage you to participate in the next jubilee XXX RPS Seminar TFoCE. Participants from other countries are also welcome.

Wrocław, 17.11.2020 Prof. Zbigniew Wójcicki

LOCAL ORGANIZING COMMITTEE

XXIX R P S

Dr. Jacek Grosel, WrUST - **chair**
Prof. Zbigniew Wójcicki, WrUST - **co-chair**
Dr. Monika Podwórna, WrUST - **co-chair**
Prof. Wojciech Glabisz, WrUST
Prof. Piotr Ruta, WrUST
Dr. Wojciech Pakos, ,WrUST
Dr. Krzysztof Majcher, WrUST
Dr. Wojciech Sawicki, WrUST
Dr. Irena Bagińska, WrUST
Prof. Vladimir Andreev MGSU
Assoc. prof. Michał Krzemiński, WUT
Assoc. prof. Peter Koteš, UNIZA
Dr. Vaddim Alpatov, SamSTU

INTERNATIONAL SCIENTIFIC COMMITTEE

- **Prof. Zbigniew Wójcicki, PhD.**, WrUST
- **Prof. Wojciech Glabisz, PhD.**, WrUST
- **Prof. Bożena Hoła, PhD.**, WrUST
- **Prof. Jerzy Hoła, PhD.**, WrUST
- **Prof. Wojciech Kostecki, PhD.**, WrUST
- **Prof. Ryszard Kutyłowski, PhD.**, WrUST
- **Dr. Jacek Grosel, PhD.**, WrUST
- **Dr. Monika Podwórna, PhD.**, WrUST

Politechnika Wrocławska

- **Prof. Vladimir Andreev, PhD.**, MGSU
- **Prof. Pavel Akimov, PhD.**, MGSU
- **Prof. Tatiana Matseevich, PhD.**, MGSU
- **Prof. Zaven Ter-Martirosyan, PhD.**, MGSU

UNIVERSITY OF CIVIL ENGINEERING

- **Prof. Ján Bujňák, PhD.**,UNIZA
- **Prof. Ján Mikolaj PhD.**,UNIZA
- **Assoc. prof. Daniela Durčanská, PhD.**,UNIZA
- **Assoc. prof. Janka Ižvoltová, PhD.**,UNIZA
- **Assoc. prof. Koteš, PhD.** UNIZA
- **Assoc. prof. Mária Kudelčíková, PhD.** UNIZA
- **Assoc. prof. Daniela Kucharová, PhD.**,UNIZA
- **Assoc. prof. Dana Sitanyiová, PhD.**,UNIZA

UNIVERSITY OF ŽILINA

- **Prof. Dmitriy Shlyakhin, PhD.**,SamSTU
- **Dr. Vaddim Alpatov, PhD**, SamSTU
- **Dr. Alexander Pishchulev, PhD**, SamSTU
- **Dr. Aleksey Solov'yev, PhD**, SamSTU

SAMARA POLYTECH Flagship University

- **Prof. Wojciech Gilewski, PhD.**,WUT
- **Prof. Stanisław Jemioło, PhD.**,WUT
- **Prof. Tomasz Lewiński, PhD.**,WUT
- **Prof. Anna Siemińska-Lewandowska, PhD.**,WUT
- **Prof. Artur Zbiciak, PhD.**,WUT
- **Assoc. prof. Michał Krzemiński, PhD.**,WUT
- **Assoc. prof. Marcin Gajewski, PhD.**, WUT
- **Assoc. prof. Nabi Ibadov, PhD.**,WUT
- **Assoc. prof. Monika Mitew-Czajewska, , PhD.**, WUT
- **Assoc. prof. Piotr Woyciechowski, PhD.**, WUT
- **Dr. Hubert Anysz, PhD.**, WUT

Politechnika Warszawska

IOP Conf. Series: Materials Science and Engineering 1015 (2021) 011002 doi:10.1088/1757-899X/1015/1/011002

Peer review declaration

XXIX Russian-Polish-Slovak Seminar *Theoretical Foundation of Civil Engineering*

All papers published in this volume of IOP Conference Series: Materials Science and Engineering have been peer reviewed through processes administered by the Editors. Reviews were conducted by expert referees to the professional and scientific standards expected of a proceedings journal published by IOP Publishing.

- **Type of peer review: Double-blind**
- **Describe criteria used by Reviewers when accepting/declining papers. Was there the opportunity to resubmit articles after revisions? Articles were rejected if they were not within the scope of the conference and if, according to reviewers or editors, their quality was low. There was no opportunity to resubmit articles after declining decision and there was obligation to resubmit corrected articles if reviewers decision was minor or major revision.**
- **Conference submission management system: e-mail**
- **Number of submissions received: 134**
- **Number of submissions sent for review: 119**
- **Number of submissions accepted: 106**
- **Acceptance Rate (Number of Submissions Accepted / Number of Submissions Received X 100): 79,1%**
- **Average number of reviews per paper: 2**
- **Total number of reviewers involved: 86**
- **Any additional info on review process (ie plagiarism check system): -each reviewer obtained a review form to fill up.**
- **Contact person for queries: Jacek Grosel, Wrocław University of Science and Technology, jacek.grosel@pwr.edu.pl**

IOP Conf. Series: Materials Science and Engineering 1015 (2021) 012001 doi:10.1088/1757-899X/1015/1/012001

First and second order sensitivity analysis of discontinuous parametric periodic systems

Z Wójcicki

Wroclaw University of Science and Technology, 27 Wybrzeże Wyspiańskiego st., 50-370 Wrocław, Poland

E-mail address: zbigniew.wojcicki@pwr.edu.pl

Abstract The paper concerns sensitivity analysis of discontinuous parametric periodic systems. The eigenderivatives of monodromy matrix (Floquet transient matrix) were the first calculated. Then the first and second derivatives of multiplier, which are the complex eigenvalues of monodromy matrix, were obtained. Then, by solving the sensitivity equation obtained in this way, to evaluate the first and second derivative of monodromy matrix and finally the first and second derivatives of multipliers. The theory has been expanded to cover a class of discontinuous parametric periodic excitations, including a situations when: the period of parametric excitation depends on the design parameters, the position of discontinuity points depends on the design parameters, and both discontinuity points and parametric constraints are a function of the design parameter. In analysis of discontinuous parametric systems the definition of distributional derivatives was applied.

1. Introduction

A generalized first and second order sensitivity analysis method of a continuous parametric periodic system was presented in paper [1]. The sensitivity analysis method presented there is much easier than the method presented in [2-3], where only the first order sensitivity analysis was done.
In the current paper, a generalized first and second order sensitivity analysis method of a discontinuous parametric periodic system is presented. Discontinuous dynamical systems are a wide class of structures which have a significant practical importance e.g. [4-8]. Despite this, a discontinuous parametric periodic systems sensitivity analysis analyses are relatively seldom realised. However, the resonant vibration in unstable parametric systems are very dangerous because the amplitude of a parametric resonant vibration growing grows exponentially in time, even when a damping exists in the system exists. Thus, from the a practical point of view, the stabilization of an unstable parametric systems [9] is usually the most important problem.

The basic theory of the first order sensitivity analysis of discontinuous parametric periodic systems was very shortly presented in [10]. An innovation of the method is the idea to find the sensitivity equation by evaluating the derivative of the parametric equation of motion with respect to design parameter of the system firstly. Then, by solving the differential equation of sensitivity obtained in this way, to evaluate the derivative of monodromy matrix of the system and finally the derivatives of its multipliers. The advantage of the presented method is then the fact that method does not require knowledge of the analytical form of the solutions, which practically do not exist in parametric systems.

The current paper presents the generalized method not only of the first but also the second order sensitivity analysis of discontinuous parametric periodic systems (with discontinuous parametric

IOP Conf. Series: Materials Science and Engineering 1015 (2021) 012001 doi:10.1088/1757-899X/1015/1/012001

periodic excitations), and an example of the method's implementation. The generalisation of the method algorithm allows to correctly determine the derivatives of multipliers not only with respect to those system parameters on which the parametric excitation period depends but also when parametric excitation discontinuity points depend on the design parameter. Thanks to this approach, one can carry out sensitivity analyses of any system whose analytical solution cannot be found – in this case, of a system with parametric periodic discontinuous excitations. The results of this analysis can be useful in resonance vibration stabilization algorithms of this type of systems. In analysis of discontinuous parametric systems, the definition of distributional derivatives was applied. It remains possible to treat the discontinuous parametric periodic excitation in a similar way as the continuous parametric systems in paper [1].

This approach was also tested theoretically and experimentally also in a nonparametric system [11], where the sensitivity analysis of an cable stayed bridge eigenproblem formulated in accordance with second order theory was realized. Implementation of this researches is Polish patent No 226397.

This paper also analyses the problem of how general the formulas presented in paper [1] are. Can they be used to calculate the derivative of the multipliers correctly if the function describing the parametric excitation is discontinuous? The answer to this question is not clear. It was found that the formulas derived earlier are still correct in some cases, but not always. The purpose of this paper is to generalize the presented method of eigenproblem sensitivity analysis to the case of discrete parametric excitation and to clearly define the scope of applicability of the above generalisation.

2. The first and the second derivatives of multipliers

A linear first order homogeneous periodic parametric system can be written in form

$$\dot{\mathbf{x}}(t) = \mathbf{A}(t)\,\mathbf{x}(t) \tag{1}$$

where

$\mathbf{x}(t)$ is a vector of state variables and

$\mathbf{A}(t)$ is a system matrix periodic with period T.

On the basis of Floquet theory [12] the steady monodromy matrix $\mathbf{D}$ of a homogenous system (1) is defined as the value of the transition matrix $\mathbf{X}(t)$ (a standard fundamental solution matrix normalized at zero) at time point $t = T$ i.e. $\mathbf{D} = \mathbf{X}(T)$. Multipliers ρ_k and the left- $\mathbf{l}^{\mathrm{T}}$ and right-side $\mathbf{r}^{\mathrm{T}}$ eigenvectors are the solutions of eigenproblem of the monodromy matrix $\mathbf{D}$. In the work [1] formulas for the first and for the second derivatives of multipliers were derived, i.e.

$$\rho'_p = \mathbf{l}^{\mathrm{T}}\mathbf{D}'_p\mathbf{r} \tag{2}$$

$$\rho''_{p_i p_j} = \mathbf{l}^{\mathrm{T}}\mathbf{D}''_{p_i p_j}\mathbf{r} + \mathbf{l}^{\mathrm{T}}\mathbf{D}'_{p_i}\mathbf{r}'_{p_j} + \mathbf{l}^{\mathrm{T}}\mathbf{D}'_{p_j}\mathbf{r}'_{p_i} \tag{3}$$

where vectors $\mathbf{r}'_{p_i}$ and $\mathbf{r}'_{p_j}$ can be found from formula (proof in [1])

$$\mathbf{r}'_{p_i} = -\mathbf{C}^{-1}\left(\mathbf{D}'_{p_i} - \rho'_{p_i}\mathbf{I}\right)\mathbf{r} \tag{4}$$

where matrices $\mathbf{C} = (\mathbf{D} - \rho\mathbf{I} - \mathbf{\Omega})$, $\mathbf{\Omega} = \mathbf{V}\,\mathbf{V}^*$.

Matrix $\mathbf{V} = [\bar{\mathbf{l}}_1, \ldots, \bar{\mathbf{l}}_d]$ contains d columns which are conjugated in complex sense with vectors of being combined with the left-handed eigenvectors of the monodromy matrix corresponding to the multiplier ρ of the algebraic multiplicity d. The matrix $\mathbf{V}^* = \bar{\mathbf{V}}^{\mathrm{T}}$ means conjugate transpose of matrix $\mathbf{V}$.

3. Discontinuous parametric excitation

3.1. Generalized formulas for system matrix and its derivatives

Now are examined a generalization of formulas for system matrix $\mathbf{A}(t)$ which refer to cases where parametric excitation is described by a periodic function that is not continuous one. It is assumed that a discontinuity of the parametric excitation fulfils the first two conditions of Dirichlet ([13]), i.e. in the interval $(0,T)$ the function is (intervally) monotone and has a finite number of discontinuity points t_k of the first type, i.e., has discontinuity jumps with values

$$h_k = f(t_k+0) - f(t_k-0) = f(t_k^+) - f(t_k^-) \tag{5}$$

Function

$$\varphi(t) = f(t) - \sum_{k=1}^{n} h_k 1(t-t_k) \tag{6}$$

is a continuous function, where $1(t)$ is the Heaviside step function ([13]). Function $f(t)$ can be written in the form

$$f(t) = \langle f(t) \rangle = \varphi(t) + \sum_{k=1}^{n} h_k 1(t-t_k) \tag{7}$$

Then the first distributional derivative of the function (7) can be obtained in form

$$f'_{dystr}(t) = \langle f(t) \rangle' = f'(t) + \sum_{k=1}^{n} h_k \delta(t-t_k) \tag{8}$$

in which $f'(t)$ is derived in the usual sense, but $\delta(t)$ is the Dirac delta function (pseudofunction) [13].

Assuming that the periodic matrix $\mathbf{A}(t)$ of the system (1), with period T, is the matrix function of the class $C^1(0,T)$, which satisfies the equation (1) everywhere alternatively except the points of discontinuity t_k. Therefore matrix $\mathbf{A}(t)$ can be presented in the form

$$\langle \mathbf{A}(t) \rangle = \sum_{k=1}^{n} \mathbf{A}_k(t)\,[1(t-t_{k-1}^+) - 1(t-t_k^-)] \tag{9}$$

or

$$\langle \mathbf{A}(t) \rangle = \mathbf{A}(t)\,[1(t-0^+) - 1(t-T^-)] + \sum_{k=1}^{n-1} \Delta\mathbf{A}_k(t_k)\, 1(t-t_k) \tag{10}$$

where

$$\Delta\mathbf{A}_k(t_k) = \Delta\mathbf{A}_k(t_k^+) - \Delta\mathbf{A}_{k-1}(t_k^-) \tag{11}$$

The symbol t_k^+ indicates an infinitely close position to the right side of the discontinuity point t_k, and the symbol t_k^- is infinitely close to its left side. It was further assumed that

$$k = 1,2,\ldots,n \quad t_o = 0 \quad t_n = T \tag{12}$$

where n is the number of continuity intervals of the function matrix $\mathbf{A}(t)$. The form of the function matrix $\langle \mathbf{A}(t) \rangle$ written in (10) is analogous to (8). Therefore, by calculating the derivative of the differential equation (1), with respect to the parameter p it must be taken into account that the distribution derivative of the matrix $\mathbf{A}(t)$ should, in this case, be computed in accordance with formula (8). It is assumed that time points t_k, in which a change in the coefficients of system matrix $\mathbf{A}(t)$ occurs, depends on the parameter with respect to which the sensitivity is analysed, i.e. $t_k = t_k(p)$.

For a system (1) in which functional system matrix $\mathbf{A}(t)$ fulfils the conditions for discontinuity (9) or (10) a practical method of calculating a matrix derivative is then use formula

$$\langle \mathbf{A}(t) \rangle'_p = \sum_{k=1}^{n} \mathbf{A}'_{k,p}(t)[1(t-t_{k-1}^+) - 1(t-t_k^-)] - \sum_{k-1}^{n-1} \Delta\mathbf{A}_k(t_k)\,\delta(t-t_k)t'_{k,p} + \mathbf{A}_n(T^-)\,\delta(t-T^-)T'_p \tag{13}$$

where

$$\mathbf{A}'_{k,p}(t) = \frac{\partial}{\partial p}\mathbf{A}_k(t)$$

The first sum to the right side of equation (13) can be represented as

$$\sum_{k=1}^{n}\mathbf{A}'_{k,p}(t)[1(t-t^{+}_{k-1})-1(t-t^{-}_{k})] = \mathbf{A}'_{p}(t)[1(t-0^{+})-1(t-T^{-})] + \sum_{k=1}^{n-1}\Delta\mathbf{A}'_{k,p}(t)1(t-t_k) \tag{14}$$

where

$$\Delta\mathbf{A}'_{k,p}(t) = \frac{\partial}{\partial p}\Delta\mathbf{A}_k(t_k)$$

Using a similar convention of writing and marking as before, the second distributional derivative may be presented using the formula

$$\begin{aligned}\langle\mathbf{A}(t)\rangle''_{p_i p_j} &= \sum_{k=1}^{n}\mathbf{A}''_{k,p_i p_j}(t)[1(t-t^{+}_{k-1})-1(t-t^{-}_{k})] - \\ &-\sum_{k-1}^{n-1}\Delta\mathbf{A}'_{k,p_i}(t_k)\,\delta(t-t_k)t'_{k,p_j} + \mathbf{A}'_{n,p_i}(T^-)\,\delta(t-T^-)T'_{p_i} - \\ &-\sum_{k-1}^{n-1}\Delta\mathbf{A}'_{k,p_j}(t_k)\,\delta(t-t_k)t'_{k,p_i} + \mathbf{A}'_{n,p_j}(T^-)\,\delta(t-T^-)T'_{p_i} + \\ &+\sum_{k-1}^{n-1}\Delta\mathbf{A}_k(t_k)\,\dot{\delta}(t-t_k)t'_{p_j}t'_{k,p_j} + \mathbf{A}'_{n,p_i}(T^-)\,\dot{\delta}(t-T^-)T'_{p_i}T'_{p_j} - \\ &-\sum_{k-1}^{n-1}\Delta\mathbf{A}_k(t_k)\,\delta(t-t_k)\,t''_{k,p_i p_j} + \mathbf{A}'_n(T^-)\,\delta(t-T^-)T''_{p_i p_j}\end{aligned} \tag{15}$$

As in the case of equations (9) and (13), the first sum to the right side of equation (16) can be represented as

$$\sum_{k=1}^{n}\mathbf{A}''_{k,p_i p_j}(t)[1(t-t^{+}_{k-1})-1(t-t^{-}_{k})] = \mathbf{A}''_{p_i p_j}(t)[1(t-0^{+})-1(t-T^{-})] + \sum_{k=1}^{n-1}\Delta\mathbf{A}''_{k,p_i p_j}(t)1(t-t_k) \tag{16}$$

where

$$\Delta\mathbf{A}'_{k,p_i p_j}(t) = \frac{\partial^2}{\partial p_i \partial p_j}\Delta\mathbf{A}_k(t)$$

3.2. Generalized formulas for derivatives of a monodromy matrix

This paper presents a generalization of the formulas obtained in paper [1]. Below are the formulas for the calculation of derivatives of monodromy matrix refers to cases where parametric excitation is described by a periodic function that is not continuous.

$$\mathbf{D}'_p = \mathbf{X}'_p(T) = \mathbf{D}\int_0^T \mathbf{X}^{-1}(t)\langle\mathbf{A}(t)\rangle'_p\,\mathbf{X}(t)\,\mathrm{d}t \tag{17}$$

Subsequent insertion (13) to (17) is achieved

$$\tilde{\mathbf{D}}'_p = \mathbf{D}'_p + \bar{\mathbf{D}}'_p \tag{18}$$

where

$$\mathbf{D}'_p = \mathbf{D}\sum_{k=1}^{n}\int_{t_{k-1}}^{t_k}\mathbf{X}_k^{-1}(t)\mathbf{A}'_{k,p}(t)\mathbf{X}_k(t)\,\mathrm{d}t \tag{19}$$

$$\bar{\mathbf{D}}'_p = \mathbf{A}_n(T^-)\mathbf{D}T'_p - \mathbf{D}\sum_{k=1}^{n-1}\mathbf{X}_k^{-1}(t)\Delta\mathbf{A}_k(t_k)\mathbf{X}_k(t)t'_{k,p} \tag{20}$$

The formula (18) with (19) and (20) is a generalisation of formulas in paper [1]. It takes into account:

- the case when the function is not continuous parametric excitation;
- the case of the dependence of the discontinuity points t_k position on design parameters.

If the location of the discontinuity points t_k is not a function of the design parameter p, the formula (19) is equivalent to formula in paper [1], the first component in formula (20) stays equivalent to formula in paper [1] and the last component in the formula (20) disappears and finally a generalised formula in paper [1] is obtained. If also the period of parametric excitation does not depend on the parameter p, then the proper formula in paper [1] is obtained.

Calculating of the second derivative of the monodromy matrix is much more complicated. The number of mathematical transformations increases dramatically and to make sure the formulas are correct, the calculations are done in two independent ways. First, the derivative of the first derivative is calculated with accordance to the formulas (18), (19) and (20) with respect to another design parameter, taking into account the formula for the derivative of the integral with respect to parameter. Secondly, the compounds (13) and (16) were added to the modified formula formulated in paper [1]. The modified formula has then the form

$$\begin{aligned}\mathbf{D}''_{p_ip_j} = {} & \mathbf{D}\int_0^T \mathbf{X}^{-1}(t)\langle\mathbf{A}(t)\rangle'_{p_j}\,\mathbf{X}(t)\int_0^t \mathbf{X}^{-1}(\tau)\langle\mathbf{A}(\tau)\rangle'_{p_i}\,\mathbf{X}(\tau)\,d\tau\,dt + \\ & + \mathbf{D}\int_0^T\int_t^T \mathbf{X}^{-1}(\tau)\langle\mathbf{A}(\tau)\rangle'_{p_j}\,\mathbf{X}(\tau)\,d\tau\,\mathbf{X}^{-1}(t)\langle\mathbf{A}(t)\rangle'_{p_i}\,\mathbf{X}(t)\,dt + \\ & + \mathbf{D}\int_0^T \mathbf{X}^{-1}(t)\langle\mathbf{A}''(t)\rangle''_{p_ip_j}\,\mathbf{X}(t)\,dt\end{aligned} \tag{21}$$

Using the filtering properties of the Dirac delta function (distribution) and its derivatives [13,14], the formula for the second derivative of the monodromy matrix was obtained. When the results obtained for these two independent methods were consistent, it was found that these formulas were correctly derived. The formulas thus obtained are tested on examples. Finally, $\mathbf{D}''_{p_ip_j}$ was obtained in the form of

$$\begin{aligned}\mathbf{D}''_{p_i p_j} = {} & \mathbf{D}\sum_{k=1}^{n}\int_{t_{k-1}}^{t_k}\mathbf{X}_k^{-1}(t)\mathbf{A}''_{k,p_ip_j}(t)\mathbf{X}_k(t)\,dt + \\ & + \mathbf{D}\sum_{k=1}^{n}\left\{\int_{t_{k-1}}^{t_k}\mathbf{X}_k^{-1}(t)\mathbf{A}'_{k,p_i}(t)\mathbf{X}_k(t)\left[\sum_{l=1}^{k-1}\int_{t_{l-1}}^{t_l}\mathbf{X}_l^{-1}(\tau)\mathbf{A}'_{l,p_j}(\tau)\mathbf{X}_l(\tau)\,d\tau\right]dt\right\} + \\ & + \mathbf{D}\sum_{k=1}^{n}\left\{\int_{t_{k-1}}^{t_k}\mathbf{X}_k^{-1}(t)\mathbf{A}'_{k,p_j}(t)\mathbf{X}_k(t)\left[\sum_{l=1}^{k-1}\int_{t_{l-1}}^{t_l}\mathbf{X}_l^{-1}(\tau)\mathbf{A}'_{l,p_i}(\tau)\mathbf{X}_l(\tau)\,d\tau\right]dt\right\} + \\ & + \mathbf{D}\sum_{k=1}^{n}\left\{\int_{t_{k-1}}^{t_k}\mathbf{X}_k^{-1}(t)\mathbf{A}'_{k,p_i}(t)\mathbf{X}_k(t)\int_{t_{k-1}}^{t}\mathbf{X}_k^{-1}(\tau)\mathbf{A}'_{k,p_j}(\tau)\mathbf{X}_k(\tau)\,d\tau dt\right\} + \\ & + \mathbf{D}\sum_{k=1}^{n}\left\{\int_{t_{k-1}}^{t_k}\mathbf{X}_k^{-1}(t)\mathbf{A}'_{k,p_j}(t)\mathbf{X}_k(t)\int_{t_{k-1}}^{t}\mathbf{X}_k^{-1}(\tau)\mathbf{A}'_{k,p_i}(\tau)\mathbf{X}_k(\tau)\,d\tau dt\right\}\end{aligned} \tag{22}$$

It is a matrix formula that allows to calculate the second derivative of the monodromy matrix with respect to design parameters, where the parametric excitation function is continuous, but is the result of "gluing together" several (e.g. n) different functions. It is the equivalent of the formula in paper [1] but is more general and in the case of $n = 1$ it becomes identical with it. The formula (22) is sufficient only if the position of the discontinuity points and the parametric excitation are not functions of the design parameters with respect to which the sensitivity is analysed. If the position of discontinuity points depends on the design parameters, the formula (22) needs to be supplemented by adding a matrix $\mathbf{D}''_{t,p_i p_j}$ that can be calculated from the formula

$$
\begin{aligned}
\mathbf{D}''_{t,p_i p_j} = & -\mathbf{D}\sum_{k=1}^{n-1}\left\{\mathbf{X}_k^{-1}(t)\Delta\mathbf{A}_k(t)\mathbf{X}_k(t)\left[\sum_{l=1}^{k}\int_{t_{l-1}}^{t_l}\mathbf{X}_l^{-1}(t)\mathbf{A}'_{l,p_j}(t)\mathbf{X}_l(t)\,dt\right]t'_{k,p_i}\right\}- \\
& -\mathbf{D}\sum_{k=1}^{n-1}\left\{\mathbf{X}_k^{-1}(t)\Delta\mathbf{A}_k(t)\mathbf{X}_k(t)\left[\sum_{l=1}^{k}\int_{t_{l-1}}^{t_l}\mathbf{X}_l^{-1}(t)\mathbf{A}'_{l,p_i}(t)\mathbf{X}_l(t)\,dt\right]t'_{k,p_j}\right\}- \\
& -\mathbf{D}\sum_{k=1}^{n}\left\{\int_{t_{k-1}}^{t_k}\mathbf{X}_k^{-1}(t)\mathbf{A}'_{k,p_i}(t)\mathbf{X}_k(t)dt\left[\sum_{l=1}^{k-1}\mathbf{X}_l^{-1}(\tau)\Delta\mathbf{A}_l(\tau)\mathbf{X}_l(\tau)t'_{l,p_j}\right]\right\}- \\
& -\mathbf{D}\sum_{k=1}^{n}\left\{\int_{t_{k-1}}^{t_k}\mathbf{X}_k^{-1}(t)\mathbf{A}'_{k,p_j}(t)\mathbf{X}_k(t)dt\left[\sum_{l=1}^{k-1}\mathbf{X}_l^{-1}(\tau)\Delta\mathbf{A}_l(\tau)\mathbf{X}_l(\tau)t'_{l,p_i}\right]\right\}- \\
& -\mathbf{D}\sum_{k=1}^{n-1}\mathbf{X}_k^{-1}(t_k)\Delta\mathbf{A}'_{k,p_j}(t_k)\mathbf{X}_k(t_k)\,t'_{k,p_i}-\mathbf{D}\sum_{k=1}^{n-1}\mathbf{X}_k^{-1}(t_k)\Delta\mathbf{A}'_{k,p_i}(t_k)\mathbf{X}_k(t_k)\,t'_{k,p_j}+ \\
& +\mathbf{D}\sum_{k=1}^{n-1}\left\{\mathbf{X}_k^{-1}(t_k)\Delta\mathbf{A}_k(t_k)\mathbf{X}_k(t_k)\left[\sum_{l=1}^{k-1}\mathbf{X}_l^{-1}(t_l)\Delta\mathbf{A}_l(t_l)\mathbf{X}_l(t_l)t'_{l,p_j}\right]t'_{k,p_i}\right\} \\
& +\mathbf{D}\sum_{k=1}^{n-1}\left\{\mathbf{X}_k^{-1}(t_k)\Delta\mathbf{A}_k(t_k)\mathbf{X}_k(t_k)\left[\sum_{l=1}^{k-1}\mathbf{X}_l^{-1}(t_l)\Delta\mathbf{A}_l(t_l)\mathbf{X}_l(t_l)t'_{l,p_i}\right]t'_{k,p_j}\right\}
\end{aligned}
\tag{23}
$$

If the period of parametric excitation depends on the design parameters, for which sensitivity is analysed, the matrix formula (22) should be also additionally supplemented by adding the matrix $\mathbf{D}''_{T,p_i p_j}$ specified by the formula

$$
\begin{aligned}
\mathbf{D}''_{T,p_i p_j} = & \,\mathbf{A}_n(T^-)\mathbf{D}\left[\sum_{k=1}^{n}\int_{t_{k-1}}^{t_k}\mathbf{X}_k^{-1}(t)\mathbf{A}'_{k,p_j}(t)\mathbf{X}_k(t)\,dt\right]T'_{p_i}+ \\
& +\mathbf{A}_n(T^-)\mathbf{D}\left[\sum_{k=1}^{n}\int_{t_{k-1}}^{t_k}\mathbf{X}_k^{-1}(t)\mathbf{A}'_{k,p_i}(t)\mathbf{X}_k(t)\,dt\right]T'_{p_j}+ \\
& +\mathbf{A}'_{n,p_j}(T^-)\,\mathbf{D}T'_{p_i}+\mathbf{A}'_{n,p_i}(T^-)\,\mathbf{D}T'_{p_j}+ \\
& +\mathbf{A}_n^2(T^-)\,\mathbf{D}T'_{p_i}T'_{p_i}+\mathbf{A}_n(T^-)\,\mathbf{D}T''_{p_i p_j}
\end{aligned}
\tag{24}
$$

If both discontinuity points and parametric constraints are a function of the design parameter, then one more matrix component $\mathbf{D}''_{tT,p_i p_j}$ defined by the following formula must be added

$$
\begin{aligned}
\mathbf{D}''_{tT,p_i p_j} = & -\mathbf{A}_n(T^-)\mathbf{D}\left[\sum_{k=1}^{n-1}\mathbf{X}_k^{-1}(t_k)\Delta\mathbf{A}_k(t_k)\mathbf{X}_k(t_k)\,t'_{k,p_j}\right]T'_{p_i}- \\
& -\mathbf{A}_n(T^-)\mathbf{D}\left[\sum_{k=1}^{n-1}\mathbf{X}_k^{-1}(t_k)\Delta\mathbf{A}_k(t_k)\mathbf{X}_k(t_k)\,t'_{k,p_i}\right]T'_{p_j}
\end{aligned}
\tag{25}
$$

In the most general case, the complete solution is the sum of the formulas (22), (23), (24)

$$
\tilde{\mathbf{D}}''_{p_i p_j} = \mathbf{D}''_{p_i p_j} + \bar{\mathbf{D}}''_{p_i p_j} \tag{26}
$$

where $\mathbf{D}''_{p_ip_j}$ is described by formula (22) but $\bar{\mathbf{D}}''_{p_ip_j}$ by formula

$$\bar{\mathbf{D}}''_{p_ip_j} = \mathbf{D}''_{t,p_ip_j} + \mathbf{D}''_{T,p_ip_j} + \mathbf{D}''_{tT,p_ip_j} \tag{27}$$

The first and the second order sensitivity analysis of discontinuous parametric systems algorithm can be realised now. The proper formulas from (19) to (26) may by putted to the formulas for the first (2) and for the second (3) derivatives of multipliers to exam the sensitivity of the discontinuous parametric systems on a little change its parameters having influence on stability/instability of the system (1).
Finally in general formulas (2) and (3) for first and second order derivatives of multipliers instead of matrices $\mathbf{D}'_p$ and $\mathbf{D}''_{p_ip_j}$, matrices $\tilde{\mathbf{D}}'_p$ and $\tilde{\mathbf{D}}''_{p_ip_j}$ ought to be substituted, which differ from the matrices $\mathbf{D}'_p$ and $\mathbf{D}''_{p_ip_j}$ in matrix components $\bar{\mathbf{D}}'_p$ and $\bar{\mathbf{D}}''_{p_ip_j}$ described by formulas (22) and (27) respectively.

4. Example

The example shown in paper [1] allowed to validate the operation of the generalized sensitivity algorithm for calculating the first and second derivatives of multipliers in continuous parametric systems. That example was simple enough to obtain solutions implementing three independent ways of calculation: starting from direct analytical calculating derivatives of multipliers presented as an explicit function of the design parameters, through a symbolic calculation with the use of *Mathematica* system procedures in accordance with the algorithms described in paper [1], to numerical calculations using the same algorithms. On the other hand, the example was too simple to fully test the algorithm, especially in the situation of discontinuous parametric excitation which is presented here. The element of that example which most simplified the course of proceedings within the presented generalized variant of the method was that both left- and right-hand modal matrices which were identity matrices in that example in paper [1]. This is a great advantage, from the point of view of the analytical testing capabilities of the algorithm, but also a great drawback from the point of view of testing all the paths, variants and nuances of the algorithm. So there was a need to look for an example that would allow for a full check of the algorithm, but at the same time would be simple enough to allow analytical control in key points. The example presented in work [12] seems to fulfill the above assumptions.
A differential equation is considered in form

$$\ddot{\mathbf{x}} + p(t)\mathbf{x} = \mathbf{0} \tag{28}$$

where (figure 1)

$$p(t) = \begin{cases} \alpha^2 & \text{dla} \quad 0 < t < c \\ \beta^2 & \text{dla} \quad c < t < T \\ (\alpha > 0, & \beta > 0) \end{cases} \tag{29}$$

and

$$p(t+T) = p(t) \tag{30}$$

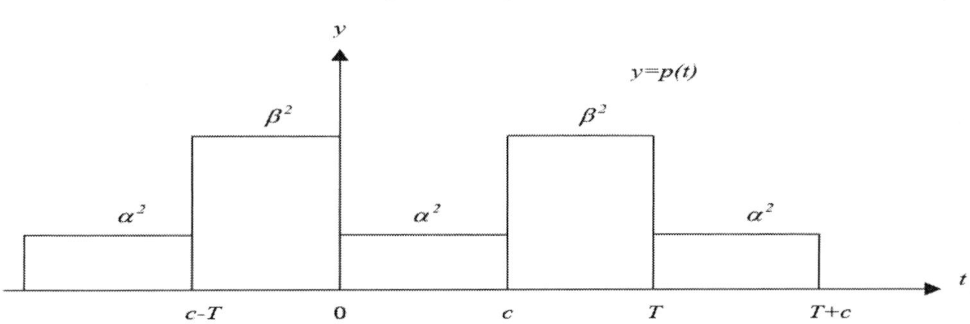

Figure 1. Parametric excitation function.

The solutions of equation (28) are "*functions of a* $C^1(-\infty,+\infty)$ *class, satisfying the equation everywhere excluding, possibly, discontinuity points* kT, $kT+c$ $(k=0,\pm1,\pm2,\ldots)$ *of the coefficient* $p(t)$ " [12].

Replacing the equation (28) by the system of first order equations is obtained

$$\begin{cases} \dfrac{dx}{dt} = y \\ \dfrac{dy}{dt} = -p(t)\,x \end{cases} \qquad \mathbf{x}(t) = \begin{bmatrix} x(t) \\ y(t) \end{bmatrix} \qquad \mathbf{A}(t) = \begin{bmatrix} 0 & 1 \\ -p(t) & 0 \end{bmatrix} \tag{31}$$

This system, in matrix notation, takes the form of a equation (1). The normalized ($\mathbf{X}(0)=\mathbf{I}$) standard fundamental solution matrix of this system can be written in analytical form, [12], where for $0 \le t \le c$

$$\mathbf{X}(t) = \begin{bmatrix} \cos\alpha t & \dfrac{\sin\alpha t}{\alpha} \\ -\alpha\sin\alpha t & \cos\alpha t \end{bmatrix} \tag{32}$$

and for $c \le t \le T$

$$\mathbf{X}(t)=\begin{bmatrix} \cos\alpha c\cos\beta(t-c)-\dfrac{\alpha\sin\alpha c\sin\beta(t-c)}{\beta} & \dfrac{\sin\alpha c\cos\beta(t-c)}{\alpha}+\dfrac{\cos\alpha c\sin\beta(t-c)}{\beta} \\ -\beta\cos\alpha c\sin\beta(t-c)-\alpha\sin\alpha c\cos\beta(t-c) & -\dfrac{\beta\sin\alpha c\sin\beta(t-c)}{\alpha}+\cos\alpha t\cos\beta(t-c) \end{bmatrix} \tag{33}$$

The monodromy matrix shows the formula ([12])

$$\mathbf{D}=\begin{bmatrix} \cos\alpha c\cos\beta(T-c)-\dfrac{\alpha\sin\alpha c\sin\beta(T-c)}{\beta} & \dfrac{\sin\alpha c\cos\beta(T-c)}{\alpha}+\dfrac{\cos\alpha c\sin\beta(T-c)}{\beta} \\ -\beta\cos\alpha c\sin\beta(T-c)-\alpha\sin\alpha c\cos\beta(T-c) & -\dfrac{\beta\sin\alpha c\sin\beta(T-c)}{\alpha}+\cos\alpha t\cos\beta(T-c) \end{bmatrix} \tag{34}$$

In control calculations, several functional variants of the dependence of the period of parametric excitation T on the design parameters α and β are assumed. Among others, it was assumed that $T=\pi/\alpha$, $T=\pi/\beta$, $T=\pi/(\alpha\beta)$. In addition, the cases were considered when this period is not dependent on any design parameter, and when the period itself is a design parameter for which the sensitivity of the system is analysed. Further, some of the results obtained in the most general calculation variant are presented, i.e., when $T=\pi/(\alpha\beta)$.

After the analytical solution of the eigenvalue problem for the monodromy matrix, the multipliers were obtained (written in the vector ρ) and the right-hand(side) R and left-hand(side) L modal matrices in form below

$$\boldsymbol{\rho}=\begin{bmatrix} \dfrac{a-(a^2-b^2)^{1/2}}{b} \\ \dfrac{a+(a^2-b^2)^{1/2}}{b} \end{bmatrix} \quad \mathbf{R}=\begin{bmatrix} \dfrac{g+(a^2-b^2)^{1/2}}{b\cdot u} & \dfrac{g-(a^2-b^2)^{1/2}}{b\cdot u} \\ 1 & 1 \end{bmatrix} \quad \mathbf{L}=\begin{bmatrix} \dfrac{b\cdot u}{2(a^2-b^2)^{1/2}} & \dfrac{-m\cdot ss+(a^2-b^2)^{1/2}}{2(a^2-b^2)^{1/2}} \\ -\dfrac{b\cdot u}{2(a^2-b^2)^{1/2}} & \dfrac{m\cdot ss+(a^2-b^2)^{1/2}}{2(a^2-b^2)^{1/2}} \end{bmatrix} \tag{35}$$

where

$a=g-d$, $b=2\alpha\beta$, $h=\alpha^2+\beta^2$, $g=h\,ss$, $d=b\,cc$,

$ss=\sin\alpha c\sin(\pi/\alpha-\beta c)$, $cc=\cos c\alpha\cos(\pi/\alpha-\beta c)$,

$m=\alpha^2-\beta^2$, $sr=\sin\pi/\alpha-\beta c$, $cr=\cos(\pi/\alpha-\beta c)$,

$u=\alpha\cdot sc+\beta\cdot cs$, $cs=\cos c\alpha\sin(\pi/\alpha-\beta c)$, $sc=\sin c\alpha\cos(\pi/\alpha-\beta c)$.

The Jordanian normal form of the monodromy matrix was also checked to see if there were multiple eigenvalues corresponding to Jordan blocks of order greater than one. Next, the derivatives of multiplies were calculated analytically with respect to the parameters α and β. The following was obtained

$$\boldsymbol{\rho}'_\alpha = \begin{bmatrix} \dfrac{a+(a^2-b^2)^{1/2}}{2b\cdot\alpha} + ba - aa \\ \dfrac{a-(a^2-b^2)^{1/2}}{2b\cdot\alpha} + ba + aa \end{bmatrix} \qquad \boldsymbol{\rho}'_\beta = \begin{bmatrix} \dfrac{a+(a^2-b^2)^{1/2}}{2b\cdot\alpha} + bb - ab \\ \dfrac{a-(a^2-b^2)^{1/2}}{2b\cdot\alpha} + bb + ab \end{bmatrix} \tag{36}$$

where

$$aa = \frac{2a\cdot so((c\alpha^2 b - \pi h)cr - 2\alpha^3 sr) - \alpha\cdot co(b\cdot cr - (c\cdot \alpha h - 2\pi\beta)sr) - 2b^2\alpha}{2b\cdot\alpha^2(a^2-b^2)^{1/2}}, \quad so = \sin c\alpha, \quad co = \cos c\alpha,$$

$$bb = \frac{2\alpha\cdot cc + c\cdot h\cdot sc + c\cdot b\cdot cs - 2\beta\cdot ss}{b}, \quad ab = \frac{2a\cdot so(2\beta\cdot sr - c\cdot h\cdot cr) - 2\alpha\, cc - c\cdot b\cdot cs) - 4b\cdot\alpha}{2b(a^2-b^2)^{1/2}}$$

The second derivatives of multipliers were calculated analytically with respect to design parameters α and β. Unfortunately, the presentation of these results in an analytical form is meaningless. These formulas are too complex, and despite the use of the *FullSimplify* function of *Mathematica*, which simplifies and condenses the symbolic notation, the simplest formula would encompass more than a dozen pages. However, these formulas were saved in the memory of *Mathematica* and were control formulas. The results of the symbolic and numerical calculations later performed on the basis of the algorithms presented in this paper must have been consistent with them.

Below are presented the results of the numerical calculations, made with the following data: $p_1 = \alpha = 1$, $p_2 = \beta = 2$, $c = 1$. As a consequence of these assumptions, the period of parametric excitation is $T = \pi/(\alpha\beta) = \pi/2$. Calculations were performed in two ways: on the basis of analytical formulas and according to the algorithm presented above. All possible control tests have been performed, although they are not presented in the paper. Part of the intermediate and final results obtained is presented below.

Multipliers are the complex coordinates of the vector

$$\boldsymbol{\rho} = \begin{bmatrix} -0.731589 - 0.681746i \\ -0.731589 + 0.681746i \end{bmatrix} \tag{37}$$

The right-hand and the left-hand modal matrices are

$$\mathbf{R} = \begin{bmatrix} -0.430578 + 0.511527i & -0.430578 - 0.511527i \\ 1 & 1 \end{bmatrix} \quad \mathbf{L} = \begin{bmatrix} 0.000000 - 0.977466i & 0.5 - 0.420876i \\ 0.00000 + 0.977466i & 0.5 + 0.420876i \end{bmatrix} \tag{38}$$

Derivatives of a monodromy $\tilde{\mathbf{D}}'_\alpha$ matrix with respect to parameter α and $\tilde{\mathbf{D}}'_\beta$ matrix with respect to parameter β calculated according to formula (18) are

$$\tilde{\mathbf{D}}'_\alpha = \begin{bmatrix} 1.11511 & 1.54269 \\ -0.0357638 & 3.9412 \end{bmatrix} \qquad \tilde{\mathbf{D}}'_\beta = \begin{bmatrix} 0.85767 & 0.529901 \\ -0.806753 & 0.426499 \end{bmatrix} \tag{39}$$

Derivatives of multipliers with respect to parameters α and β calculated according to formulas (2) are

$$\boldsymbol{\rho}'_\alpha = \begin{bmatrix} 2.52815 - 2.71299i \\ 2.52815 + 2.71299i \end{bmatrix} \qquad \boldsymbol{\rho}'_\beta = \begin{bmatrix} 0.642085 - 0.689028i \\ 0.642085 + 0.689028i \end{bmatrix} \tag{40}$$

Calculations for the second mixed derivative with respect to the parameters α and β were also performed. The matrices have the form

$$\mathbf{D}''_{\alpha\beta}=\begin{bmatrix}0.586744 & 0.247661\\ 1.88208 & 0.853671\end{bmatrix}\quad \bar{\mathbf{D}}''_{\alpha\beta}=\begin{bmatrix}-0.529779 & -1.85038\\ 1.81966 & 3.60567\end{bmatrix}\quad \tilde{\mathbf{D}}''_{\alpha\beta}=\begin{bmatrix}0.0569655 & -1.60272\\ 3.70173 & 4.45934\end{bmatrix} \tag{41}$$

Derivatives of the right-hand modal matrix with respect to parameters α and β were calculated on the basis of formulas (3), using the formula (4) and using the procedure described in paper [1]. For given data received

$$\mathbf{R}'_{\alpha}=\begin{bmatrix}1.38685+0.559851i & 1.38685-0.559851i\\ -1.97632+1.04764i & -1.97632-1.04764i\end{bmatrix}\quad \mathbf{R}'_{\beta}=\begin{bmatrix}0{,}13671+0{,}0620598i & 0{,}13671-0{,}0620598i\\ -0{,}20268+0{,}0966525i & -0{,}20268-0{,}0966525i\end{bmatrix} \tag{42}$$

Finally, the second mixed derivative of multipliers can be calculated

$$\boldsymbol{\rho}''_{\alpha\beta}=\begin{bmatrix}2.52815+2.69979i\\ 2.52815-2.69979i\end{bmatrix} \tag{43}$$

The above results were obtained by performing calculations using all the calculation methods. These results are identical to the accuracy limited by the capabilities of CPU and the *Mathematica* system.

A different result is obtained by applying the algorithm presented in [9]. The second derived of multipliers vector in the case of discontinuous parametric excitation is obtained in the form of

$$\boldsymbol{\rho}''_{\alpha\beta}=\begin{bmatrix}0.720208+1.83646i\\ 0.720208-1.83646i\end{bmatrix} \tag{44}$$

This is a wrong result, and the cause of the error is the forbidden omission in this algorithm of the non-zero matrix $\bar{\mathbf{D}}''_{p_i p_j}$.

5. Conclusions

- The use of the generalized algorithm presented in this paper guarantees the correctness of the calculation also in the case of discontinuous parametric induction.
- The results obtained calculating the first and second derivatives of multipliers in the case of discontinuous parametric excitation are consistent with those obtained analytically.
- The analysis of the examples shows an additional conclusion that is related to the specifics of systems that are not continuous:
  - if the matrix $\mathbf{A}(t)$ of a system is periodic, with period T, but is discontinuous in points $t=0$ and $t=T$, then the use of the periodicity condition $\mathbf{A}(0)=\mathbf{A}(T)$ in integral formulas on derivatives of the monodromy matrix is an error. Consequently, assuming that, the additional component (19) in formula (20) in paper [1] on the derivative of the monodromy matrix takes the form $\bar{\mathbf{D}}'_p=\mathbf{A}(T)\,\mathbf{D}\,T'_p=\mathbf{A}(0)\,\mathbf{D}\,T'_p$. The substitution of an $\mathbf{A}(T)=\mathbf{A}(0)$ generates an error. This error was made in [9], although it should be noted that that paper only analysed the case of continuous parametric excitation.
  - in these formulas there are the values of the system matrix in points $t=0^+$ and $t=T^-$, i.e. $\mathbf{A}(0^+)$ and $\mathbf{A}(T^-)$. Therefore, although $\mathbf{A}(t)$ is a periodic function $\mathbf{A}(0^+)\neq\mathbf{A}(T^-)$ due to the discontinuities in these points. The analysis of the example shows that in the $\mathbf{A}(T)=\mathbf{A}(T^-)$ should be inserted in formulas from (17) to (27).

References

[1] Wójcicki Z 2019 Eigenproblem sensitivity analysis of continuous parametric periodic systems, *IOP Conf. Series: Materials Science and Engineering*, 661 012042 (XXVIII R-P-S Seminar, doi:10.1088/1757-899X/661/1/012042)

[2] Lu Y and Murthy V R 1992 Sensitivity analysis of discrete periodic systems with applications to helicopter rotor dynamics, *AIAA Journal* **30** (8) pp 1962-69

[3] Gu Y, Chen L and Wang W 1999 Further research for sensitivity analyses of discrete periodic systems, *AIAA Journal* **3** pp 653-6

[4] Aizerman M A and Pyatnitskii E S 1974 Foundation of a theory of discontinuous systems-1 *Autom. Remote Control* **35** pp 1066–79

[5] Luo A C J 2009 *Discontinuous dynamical systems on time-varying domains* (Berlin: HEP-Springer)

[6] Luo A C J 2012 *Discontinuous dynamical systems* (Berlin: HEP-Springer)

[7] Huang J and Luo A C J 2017 Complex dynamics of bouncing motions at boundaries and corners in a discontinuous dynamical system, *J. of Computational and Nonlinear Dynamics* **12** 6

[8] Babitsky V I and Krupenin V L 2001 *Vibration of strongly nonlinear discontinuous systems, foundations of engineering mechanics* (Berlin-Heidelberg: Springer-Verlag)

[9] Seyranian A P, Solem F and Pedersen P 2000 Multi-parameter linear periodic systems: sensitivity analysis and applications *Journal of Sound and Vibrations* **229** pp 89-111

[10] Wójcicki and Z Śniady P 2005 Sensitivity analysis of discontinuous parametric periodic systems *PAMM Proc. Appl. Math. Mech.* **5** p 189–190 (DOI 10.10 02/pamm.200510072)

[11] Pakos W M, Wójcicki Z, Grosel J, Majcher K P and Sawicki W 2016 Experimental research of cable tension tuning of a scaled model of cable stayed bridge. *Archives of Civil and Mechanical Engineering*, **16** (1) pp 41-52

[12] Demidowicz B P 1972 *Mathematical theory of stability* (Warsaw: WNT in Polish)

[13] Fichtenholz G M 1981 *Differential and integral calculus* (Warsaw: WNT in Polish)

IOP Conf. Series: Materials Science and Engineering 1015 (2021) 012002 doi:10.1088/1757-899X/1015/1/012002

Calculation of an inhomogeneous polymer thick-walled cylindrical shell taking into account creep under the action of temperature load

V I Andreev and S A Sereda

National Research Moscow State University of Civil Engineering, 129337 Yaroslavskoye shosse, 26, Moscow, Russia

E-mail address: asv@mgsu.ru

Abstract. The article presents the calculation of a polymer thick-walled cylindrical shell taking into account creep under the action of an uneven temperature field. The relevance of the work lies in the widespread introduction of polymer pipes in the repair and construction of heating pipelines, sewerage systems, water supply systems. The problem is solved in an axisymmetric setting under conditions of plane deformation. The calculation is based on the nonlinear Maxwell - Gurevich equation, which is widely used in the calculations of polymer structures.

1. Introduction

In the problem under consideration, the cylindrical shell is in a plane deformed state. Temperature is considered to be a function of radius and time. Relaxation and elastic characteristics, which are strongly dependent on temperature, will be functions of the coordinate and time on the inner and outer surfaces of the shell, and temperatures In the problem under consideration, the cylindrical shell is in a plane deformed state. Temperature is also considered to be a function of radius and time. The relaxation and elastic characteristics, which are strongly dependent on temperature, will be functions of the coordinate and time on the inner and outer surfaces of the shell, where temperatures $T(a)=T_a$ and $T(b)=T_b$ act. Here a and b are, respectively, the radii of the inner and outer surfaces of the shell. Taking into account the axial symmetry, the problem in the geometric formulation is one-dimensional (all functions depend only on r), and with regard to the influence of temperature, which in some cases can change in time, it is considered as quasi-stationary.

2. Physical relations for a viscoelastic material

The problems of viscoelasticity of continuously inhomogeneous bodies considered in this article are related to problems of polymer mechanics. Below are physical relations in differential form that are valid for polymers and composites [1], which can be considered quite general and in particular cases applicable to other materials [2, 3].

Consider the equations describing the proper viscoelastic behavior of the material, i.e. defining term ε^*_{jk} . Due to the micro inhomogeneity of the structure at the molecular and supramolecular level, the polymer can have a spectrum of relaxation times; therefore, highly elastic deformations in the general case are the sums of individual components, each of which corresponds to a certain member of the spectrum:

$$\varepsilon^*_{jk} = \sum_s (\varepsilon^*_{jk})_s \quad (1)$$

Usually, when solving problems of mechanics in relation to rigid polymers in quasi-static regimes, we will consider it sufficient to take into account one or two components of highly elastic deformation. For the rate of the components of highly elastic deformation, there is an unambiguous dependence on the parameters of the deformation process. The generalized nonlinear Maxwell equation is valid as analytical expressions for these dependences (constraint equations). This equation was derived in [4] based on molecular concepts of the behavior of materials. Later, it was repeatedly tested in solving various problems in the mechanics of polymers and composites [2]

$$\frac{\partial(\varepsilon^*_{jk})_s}{\partial t} = \left[\frac{3}{2}(\sigma_{jk} - \sigma_{cp}\delta_{jk}) - E_{\infty s}(\varepsilon^*_{jk})_s\right]\frac{1}{\eta^*_s}; \quad (j,k = \xi,\eta,\zeta) \quad (2)$$

The function η^*_s in (2) is the relaxation viscosity coefficient, invariant with respect to the coordinate system, has the form

$$\eta^*_s = \eta^*_{0s} \exp\left\{-\frac{1}{m^*_s}\left[\gamma^*_s\sigma_{cp} + \left|\frac{3}{2}(\sigma_{ll} - \sigma_{cp}) - E_{\infty s}(\varepsilon^*_{ll})_s\right|_{\max}\right]\right\} \quad (3)$$

In (2) and (3) the following designations are accepted: $E_{\infty s}$, – modulus of high elasticity $(\sigma = E_{\infty s} \cdot \varepsilon_{\infty s})$; m^*_s – deformation rate modulus; γ^*_s – is the volumetric coefficient; η^*_{0s} – is the coefficient of the initial relaxation viscosity of the s-th component of the highly elastic deformation.
The index ll in (3) denotes the principal stresses and strains for which the equalities are valid

$$3\sigma_{cp} = \sum_{l=1}^{3}\sigma_{ll}; \quad 3\varepsilon_{cp} = \sum_{l=1}^{3}\varepsilon_{ll} \quad (4)$$

The relaxation viscosity coefficient η^*_s, determined by equality (3), is related to the relaxation time T^*_s by the relation

$$\eta^*_s = 3G\left(1 + \frac{G_\infty}{G}\right)T^*_s \quad (5)$$

2. Derivation of resolving equations

As in the problems of the theory of elasticity, the problems of the theory of creep can be solved in stresses and in displacements. Below is the way to deal with the displacements. The basic formulas of the theory of creep formally coincide with the equations of the theory of elasticity. For deformations we have

$$\varepsilon_z = \varepsilon^0_z + \varepsilon^*_z + \alpha T$$

Assuming from the plane deformation condition $\varepsilon_z = 0$, we obtain

$$\varepsilon^0_z = -\varepsilon^*_z - \alpha_T T$$

Then the elastic component of the average deformation will be equal to

$$\varepsilon^0_{cp} = \frac{1}{3}(\varepsilon^0_r + \varepsilon^0_\theta + \varepsilon^0_z) = \frac{1}{3}\left[\frac{\partial u}{\partial r} + \frac{u}{r} - (\varepsilon^*_r + \varepsilon^*_\theta + \varepsilon^*_z) - 3\alpha_T T\right] = \frac{1}{3}\left(\frac{\partial u}{\partial r} + \frac{u}{r}\right) - \alpha_T T \quad (6)$$

The last transformation is made on the basis of the hypothesis that the volumetric creep strain is zero.

Using the Cauchy relations and Hooke's law for elastic deformations

$$\varepsilon_r^0 = \frac{1}{E}(\sigma_r - \nu\sigma_\theta); \qquad \varepsilon_\theta^0 = \frac{1}{E}(\sigma_\theta - \nu\sigma_r)$$

we get expressions:

$$\varepsilon_r^0 = \frac{\partial u}{\partial r} - \varepsilon_r^* - \alpha_T T; \qquad \varepsilon_\theta^0 = \frac{u}{r} - \varepsilon_\theta^* - \alpha_T T \tag{7}$$

Substituting (6) and (7) into Hooke's law in the Lamé form, we obtain

$$\begin{aligned} \sigma_r &= \lambda\left(\frac{\partial u}{\partial r} + \frac{u}{r}\right) + 2\mu\frac{\partial u}{\partial r} - 2\mu\varepsilon_r^* - 3K\alpha_T T \\ \sigma_\theta &= \lambda\left(\frac{\partial u}{\partial r} + \frac{u}{r}\right) + 2\mu\frac{u}{r} - 2\mu\varepsilon_\theta^* - 3K\alpha_T T \end{aligned} \tag{8}$$

We obtain the resolving equation in displacements by substituting the expression for stresses (8) into the equilibrium equation

$$\frac{\partial^2 u}{\partial r^2} + \varphi(r,t)\frac{\partial u}{\partial r} + \psi(r,t)\frac{u}{r} = f(r,t) \tag{9}$$

In this equation

$$\varphi(r,t) = \frac{1}{r} + \frac{1}{\lambda + 2\mu}\frac{\partial(\lambda + 2\mu)}{\partial r}; \qquad \psi(r,t) = \frac{1}{r^2} + \frac{1}{r(\lambda + 2\mu)}\frac{\partial \lambda}{\partial r}$$

$$f(r,t) = \frac{1}{\lambda + 2\mu}\left[3\alpha_T\frac{\partial(KT)}{\partial r} + 2\mu\left(\frac{\partial \varepsilon_r^*}{\partial r} + \frac{\varepsilon_r^* - \varepsilon_\theta^*}{r}\right) + 2\frac{\partial \mu}{\partial r}\varepsilon_r^*\right] \tag{10}$$

3. Solution method

Resolving equation (9) is a second order partial differential equation with variable coefficients. Due to the nonlinearity and complexity of the coefficients, an analytical solution of these equations cannot be found even with significant simplifications. If we assume that the temperature field and force loads change slowly in time, then the creep problems can be considered as quasi-stationary. One of the first works in which a "layer by layer" method was proposed for solving quasistationary creep problems was [5]. Later this method was used in [6, 7] (such methods for solving creep problems are also called stepwise, incremental, etc).

Let us explain the essence of the "layer-by-layer integration" method by the example of solving equation (9) under the action of a temperature field. At the zero stage (at $t = 0$), the problem of determining the temperature field $T = T_0(r)$ is first solved and the dependences of the elastic and relaxation parameters of the material on temperature and coordinates are found

$$E = E_0[T_0(r)], \quad \nu = \nu_0[T_0(r)], \quad E_{\infty s} = E_{\infty s,0}[T_0(r)], \ldots$$

If the loading is considered instantaneous, then at time $t = 0$ the initial conditions will be valid:

$$t = 0; \qquad \varepsilon_{r,0}^* = \varepsilon_{\theta,0}^* = 0 \tag{11}$$

Thus, at the zero stage, we arrive at an elastic problem. In this case, in equation (9), the partial derivatives with respect to the radius can be replaced by ordinary ones (in the subsequent formulas, a prime is used to denote such derivatives):

$$u_0'' + \varphi(r,0)u_0' + \psi(r,0)u_0 = f(r,0) \tag{12}$$

where

$$\varphi(r,0) = \frac{1}{r} + \frac{\lambda_0' + 2\mu_0'}{\lambda_0 + 2\mu_0}; \qquad \psi(r,0) = \frac{1}{r^2} - \frac{\lambda_0'}{r(\lambda_0' + 2\mu_0)}$$

$$f(r,0) = \frac{3\alpha_T}{\lambda_0' + 2\mu_0}(K_0 T_0)' \tag{13}$$

Here, the dependences of the mechanical characteristics on the radius are due to the initial (in this case, temperature) inhomogeneity of the material.
Equation (12) with boundary conditions

$$t = 0; \quad \varepsilon^*_{r,0} = \varepsilon^*_{\theta,0} = 0$$

is a two-point boundary value problem, which, due to the complexity of the coefficients of the equation, must be solved numerically. One of the effective methods for solving such boundary value problems is the run-through method [8, 9].
Having determined at the zero stage all the necessary quantities (displacements, deformations and stresses), from the corresponding physical equations, for example (2), one can find the creep strain rates

$$\left(\frac{\partial \varepsilon^*_r}{\partial t}\right)_0, \left(\frac{\partial \varepsilon^*_\theta}{\partial t}\right)_0$$

Assuming that the time step Δt can be arbitrarily small, one can carry out a linear approximation in time and calculate the creep deformations at the next "time layer" $t = \Delta t$:

$$t = t_1 = \Delta t; \quad \varepsilon^*_{r,1} = \left(\frac{\partial \varepsilon^*_r}{\partial t}\right)_0 \cdot \Delta t; \quad \varepsilon^*_{\theta,1} = \left(\frac{\partial \varepsilon^*_\theta}{\partial t}\right)_0 \cdot \Delta t$$

By numerical differentiation, the derivatives of the creep deformations along the radius are also found in the right-hand side of Eq. (9). In the case of a non-stationary thermal process, one should also analytically or by linear approximation in time to determine a new temperature distribution and new dependences of mechanical characteristics on temperature, and, consequently, on radius. Having thus formed the right-hand side of equation (9), we again come to the elastic problem with a new function $f(r)$ and, in the general case, with new coefficients $\varphi(r)$ and $\psi(r)$. Also solving this problem numerically, we obtain a solution at the first stage. Continuing the process to an arbitrary point in time, it is possible to determine stresses, deformations and displacements at any point in the body.
In step-by-step methods, the question of their convergence remains open. One way to analyse convergence is to compare the results obtained at different values Δt. For long-term processes, when the creep rate decreases, an uneven (increasing) time step can be used. There are some other possibilities for analysing and improving the convergence of the method, based, for example, on comparing the results at two successive stages of the iterative process with the subsequent adjustment of the previous step in time.

4. Creep of an unevenly heated cylinder

A characteristic feature of temperature problems for structural elements made of polymeric materials is a strong dependence of all mechanical characteristics on temperature. Thus, even with relatively small gradients of the temperature field, it is necessary to solve the problem of mechanics taking into account the inhomogeneity of the material.
Consider the problem of calculating a thick-walled cylinder in an axisymmetric temperature field determined by the following boundary and initial conditions

$$\begin{aligned} &t = 0, && T(r,0) = T_0 = const \\ &0 < t < t_1, && T(a,t) = T_0 + \beta t; \quad T(b,t) = T_0 \\ &t > t_1, && T(a,t) = T_0 + \beta t_1; \quad T(b,t) = T_0 \end{aligned} \tag{14}$$

In accordance with (14), a constant temperature T_0 is maintained on the outer surface of the cylinder throughout the entire time, and the inner surface is first heated (over time t_1), and then its temperature is also maintained constant. If you set the final temperature T_1 of the inner surface of the cylinder, then the heating time can be determined by the formula $t_1 = (T_1 - T_0)/\beta$, where β is the heating rate.

The process is unsteady until a certain moment of time $t_2 > t_1$, until the final distribution of temperature along the radius is established. In Figure 1 shows the results of the numerical solution of the equation

$$\lambda \nabla^2 T - c\rho \frac{\partial T}{\partial t} + W = 0$$

obtained under the assumption of constancy of the temperature conductivity coefficient with the following initial data: $a = 8$mm; $b = 28$mm; $T_0 = 28^\circ$C; $T_1 = 100^\circ$C; $\beta = 60^\circ$ C/h. From the given dependences, it can be concluded that the stabilization of the temperature field occurs at a time of 3.6 hours.

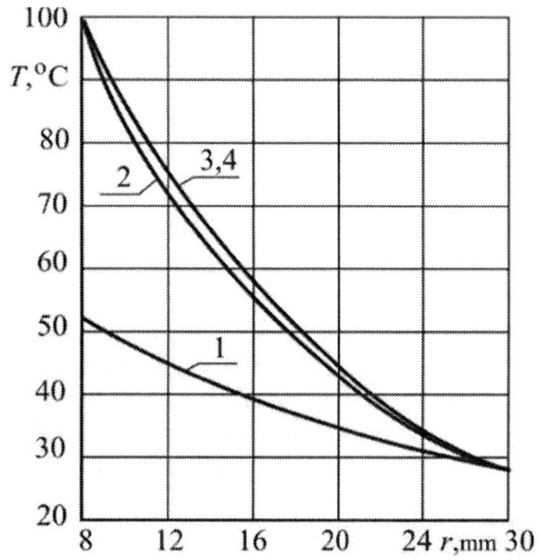

Figure 1. Temperature distribution incylinder at different times: 1 - 0.4 hours; 2 - 1.2 hours; 3 - 3.6 hours; 4 - 100 hours.

Assuming the cylinder is sufficiently long, we will assume that a plane deformed state occurs in it, for which the resolving equation in displacements (9) is valid. Considering a short-term process (up to 100 hours), we will restrict ourselves in calculations only to the "senior" component of highly elastic deformation.

Below are the results of solving the problem of creep for a cylinder made of EDT-10 epoxy resin. For this material, the dependences of mechanical characteristics on temperature were studied in the fundamental work [10]. Taking into account the insignificant change in the considered temperature range of Poisson's ratio ν and coefficient of linear thermal expansion α, we will assume that they are constant and equal to $\nu = 0.3$; $\alpha = 8 \cdot 10^{-5}$ 1/deg. In the aforementioned work, for EDT-10, the following empirical dependences on the temperature of the elastic modulus and relaxation characteristics corresponding to the leading component of the highly elastic deformation are given:

$$E = (8302 - 1{,}75T_K) \ [\text{MPa}]; \qquad E_{\infty 1} = (11340 - 30T_K) \ [\text{MPa}];$$

$$m_1^* = (7{,}75 - 0{,}011T_K) \ [\text{MPa}]; \quad \eta_{01}^* = 685 \cdot 10^8 \exp(-0{,}0275T_K) \ [\text{MPa} \cdot \text{sec}].$$

Here T_K is the temperature in Kalvin scale, K.

Figures 2 and 3 show the variation along the radius of these characteristics for some points in time. Numerical designations in these and subsequent figures are the same as in figure 1. It should be noted that, with the exception of the modulus of elasticity, all other characteristics of the material stabilize already by the end of heating of the inner surface of the cylinder at t = 1.2 hours.

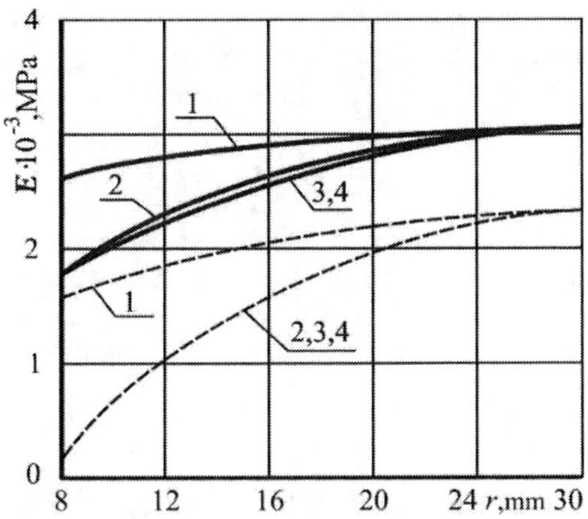

Figure 2. Modifying a module elasticity E (——) and high elasticity modulus E_∞ (- - -) in cylinder for various moments in time.

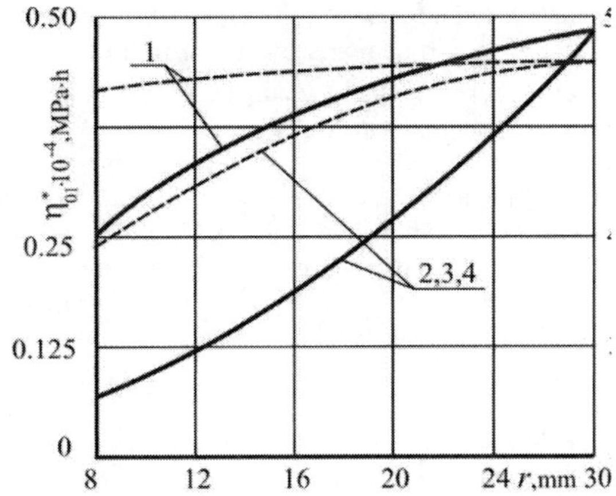

Figure 3. Change in the coefficient of the initial relaxation viscosity η_{01}^* (——) and the modulus of velocity m* (- - -) in the cylinder for different moments of time.

As will be shown below, it is essential to take into account in the calculations a relatively short heating time compared to the entire process, since the increase in temperature occurring during this period significantly accelerates the creep process from the very beginning. The calculation was carried out by the numerical method described in Section 3.

Figure 4 shows the stress plots A for some points in time.

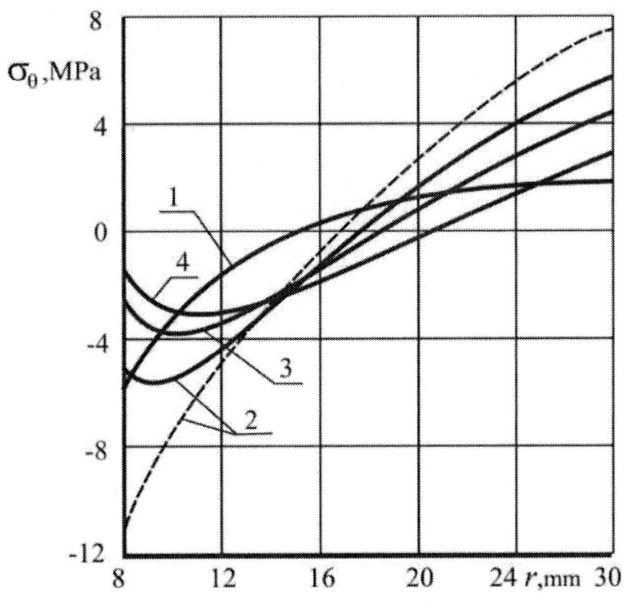

Figure 4. Distribution of stresses in the cylinder for different points in time; - - - elastic solution.

For comparison, the same figure shows the stress diagram obtained from the elastic solution and corresponding to the heating end time ($t = 1.2$ hours). The following results should be noted. In the initial period, during the heating process, the stresses increase, this is natural, since the temperature loads increase. Then, during the creep of the cylinder, a significant relaxation of stresses occurs in both the stretched and compressed zones. In this case, if at $t > 3.6$ hours the temperature distribution along the radius remains unchanged (see figure 1), then the relaxation process continues, which leads to an even greater decrease in stresses.

Conclusion

The article shows that the solution of viscoelasticity problems based on the nonlinear Maxwell - Gurevich equation using differential equations is quite simple. Several papers on a related topic have been published [11 - 13]. Note that the method of sequential loading was used to solve the nonlinear problem in the article. You can also use the method of successive approximations [14], which may be a development of the topic under consideration.

References

[1] Rabinovich A L 1970 *Vvedeniye v mehaniku armirovannyh polimerov* (Moscow: Nauka) p 482 (in Russian)
[2] Rabinovich A L, Shtarkov M G and Dmitriyeva E I 1959 *O nekotoryh zakonomernostyah odnorodnogo deformirivaniya metallov, obladayushchih uprochneniyem* (Moscow: Oborongiz Pub. MFTI) (in Russian)
[3] Treloar L 1979 *Fizika uprugosti kauchuka* (Moscow: Nauka) p 560 (in Russian)
[4] Gurevich G I 1974 *Deformiruyemost' sred I rasprostraneniye seysmicheskih voln.* (Moscow: Nauka) p 482 (in Russian)
[5] Andreev V I 1968 Ob ustoychivosti polimernyh sterzhney pri polzuchesti *Mehanika polimerov* **1** pp 22-28 (in Russian)
[6] Tsybin N Yu, Andreev V I and Turusov R A 2016 Comparison of creep in free polymer *Procedia engineering* **153** pp 51–8
[7] Yazyyev B M, Andreev V I and Turusov R A 2009 Nekonorye zadachi I metody mehaniki makroneodnorodnoy vyazkouprugoj sredy *Rostov-on-Don, Pub. RGSU* p 208 (in Russian)
[8] Samarskij A A 1982 *Vvedeniye v chislennyye metody* (Moscow: Nauka) p 269 (in Russian)
[9] Pirumov U G 1998 *Chislennyye metody* (Moscow: Pub. MAI) p 224 (in Russian)
[10] Babich V F 1966 *Issledovaniye vliyaniya temperatury na mehanicheskie harakteristiki polimerov* (Moscow: Diss. of the Cand.. of Technical Sciences) (in Russian)
[11] Yazyyev B M 1989 Nelineynaya polzuchest' neodnorodnyh cilindrov *XI Miedzynar. Symp. Nauk. Ziel. Gora* (in Russian)
[12] Andreev V I, Ter-Semenov A, Turebayeva R D and Frolova I I *1989* Uprugost' I polzuchest' neodnorodnoj poloy sfery *Vsecoyuznaya Konf. "Fundamt. issledovaniya I novyje tehnilogii v stroitel'stve". Tezisy doklada, Belgorod. P* 1 (in Russian)
[13] Tsybin N Yu, Andreev V I and Turusov R A 2018 Issledovaniye polzuchesti polimerov v razlichnyh usloviyah deformirovaniya *Stroitel'naya mehanika I raschet sooruzhenij* **3** pp 30-5 (in Russian)
[14] Polyakova L S and Andreev V I 2018 Calculation of a nonlinearly elastic three-layer cylindrical shell taking into account the continuous inhomogeneity caused by the temperature field *IOP Conference Series: Materials Science and Engineering* **456** 012124

IOP Conf. Series: Materials Science and Engineering 1015 (2021) 012003 doi:10.1088/1757-899X/1015/1/012003

Absorbers impact on the reliability of structures subjected to random vibrations

M Podwórna[1], J Grosel[1] and P Śniady[2]

[1] Faculty of Civil Engineering, Wrocław University of Science and Technology, Wyb. Wyspiańskiego 27, 50-370 Wrocław, Poland
[2] Faculty of Environmental Engineering and Geodesy, Wrocław University of Environmental and Life Science, Norwida St. 25, 50-375 Wrocław, Poland

E-mail address: monika.podworna@pwr.edu.pl

Abstract. The paper considers the problem of optimizing the absorber parameters when random vibrations of the structure create a filtered Poisson process. A new parameter optimization criterion was proposed, consisting in minimizing the probability of the maximum vibrations of the structure exceeding a certain permissible value of displacements. This criterion means maximizing the reliability of the vibrating structure with regard to the maximum vibration amplitudes.

1. Introduction

Reliability analyses can help the designer to determine acceptable system's tolerances and determine construction's parameters for safe structures. The problem of reliability analysis of structures subjected to vibration is extremely important from the engineer's point of view. It is essential to develop vibration reduction for structures - an interesting option is using dynamic vibration absorbers. A Tuned Mass Damper (TMD) or a Dynamic Vibration Absorber (DVA) is a passive vibration device used to reduce resonant vibration. The different kinds of vibration absorbers are applied for reducing the vibration level in various engineering structures such as buildings, towers, bridges, etc.

The topic of a dynamic vibration absorber is well established in the literature. The first absorber was invented in the form of a spring supported mass in 1911 by Frahm [1]. On the other hand, the first damped vibration absorber was created in 1928 by Ormondroyd and Den Hartog [2]. Since Den Hartog [3] found an analytical optimal solution for an undamped primary system, many researches have tried to optimize Dynamic Vibration Absorber (DVA) or Tuned Mass Damper (TMD). The absorber optimization criteria proposed so far can be classified into three categories [4]: H_∞ optimization, H_2 optimization, and stability maximization. The H_∞ optimization is the earliest proposed criterion [2] and aims to minimize the maximum amplitude response. The H_2 optimization criterion was devised to minimize the kinetic energy over the entire frequency range of the primary system subjected to random excitation [5]. The stochastic vibrations were analysed in the field of correlation theory and spectral density analysis.

In the case of stochastic vibrations of structures, an important issue is the assessment of the impact of absorbers on the assessment of the reliability of this structure. As a criterion for assessing the effectiveness of the absorber in reducing vibrations, the study proposes the assessment of the reliability

IOP Conf. Series: Materials Science and Engineering 1015 (2021) 012003 doi:10.1088/1757-899X/1015/1/012003

of the structure with regard to the established criterion. In stochastic structure dynamics, the problem of reliability is treated as the first-crossing problem [6,7].

In this paper, vibrations of the system (absorber and structure) caused by road traffic which is modelled by filtered Poisson process are considered. The very issue of structure reliability is analysed in the database of random variables, the infinite sequence (series) of which creates Poisson "white noise". The probability of not exceeding a certain value was adopted as a measure of the reliability evaluation of the vibrating system. This value may indicate an acceptable level of vibrations due to people or may result from the reliability of the fixed structural element of the object. It can be deterministically defined or assume random values. In the paper it was assumed that all random variables have a log-normal distribution.

2. The problem

Let us consider vibration of a two degree freedom system caused by kinematic excitation (figure 1) where $y_s(t)$ represent vibration of a primary system and $y_a(t)$ of a dynamic vibration absorber.

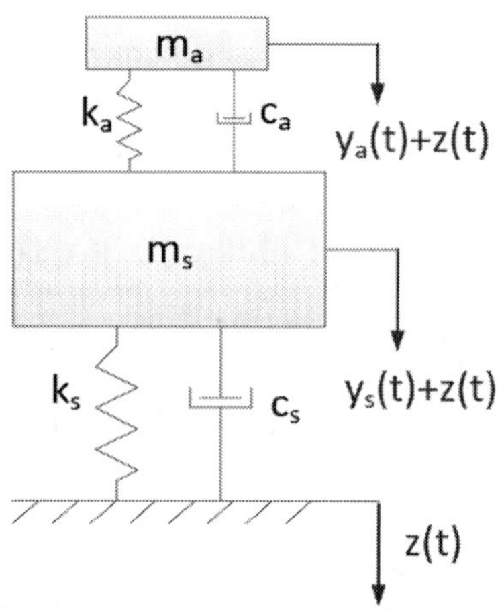

Figure 1. Two degree freedom system.

Vibration of this system is described by system two differential equations:

$$\frac{d^2y_a(t)}{dt^2}+2\xi_a\omega_a[\frac{dy_a(t)}{dt}-\frac{dy_s(t)}{dt}]+\omega_a^2[y_a(t)-y_s(t)]=-\frac{d^2z(t)}{dt^2}=f(t) \quad (1)$$

$$\frac{d^2y_s(t)}{dt^2}+2\xi_s\omega_s\frac{dy_s(t)}{dt}+\omega_s^2y_s(t)+2\xi_a\omega_a\mu[\frac{dy_s(t)}{dt}-\frac{dy_a(t)}{dt}]+\omega_a^2\mu[y_s(t)-y_a(t)]=-\frac{d^2z(t)}{dt^2}=f(t) \quad (2)$$

where $\omega_s^2=k_s/m_s$, $\omega_a^2=k_a/m_a$, $\alpha_s=c_s/2m_s=\xi_s\omega_s$, $\alpha_a=c_a/2m_a=\xi_a\omega_a$, $\mu=m_a/m_s$ and excitation process $f(t)$ is a stochastic process, while m_s, c_s and k_s are the mass, damping coefficient and stiffness of the primary system, respectively, m_a, c_a and k_a are the mass, damping coefficient and stiffness of the absorber system, respectively. The function $z(t)$ represents ground motion (kinematic excitation).

IOP Conf. Series: Materials Science and Engineering 1015 (2021) 012003 doi:10.1088/1757-899X/1015/1/012003

The stochastic process is assumed to be a series of pulses with random amplitudes occurring at random times. The excitation process has the form of filtered Poisson process:

$$f(t)=\sum_{n=1}^{N(t)} A_n\varphi(t-t_n)=\sum_{n=1}^{N(t)} A_n\cdot(t-t_n)\cdot e^{-\alpha_f(t-t_n)}\sin\Omega_f(t-t_n) \tag{3}$$

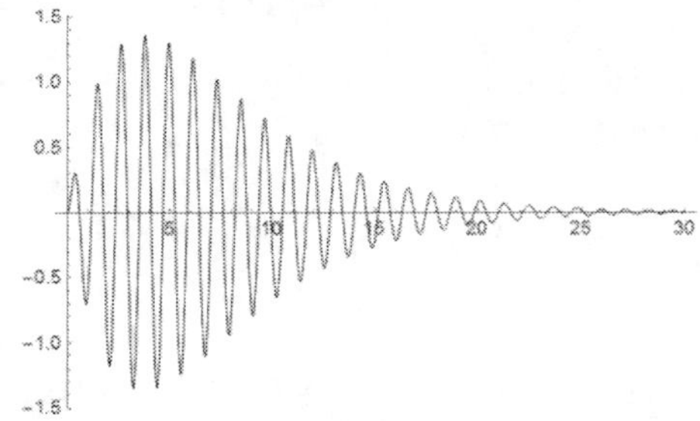

Figure 2. An example of the single excitation process.

whereas amplitudes are assumed to be random variables which are both mutually independent and independent of the instants t_n. It is assumed that the expected values $E[A_n]=E[A]=const$, $E[A_n^2]=E[A^2]=const$ are known and does not depend on n.

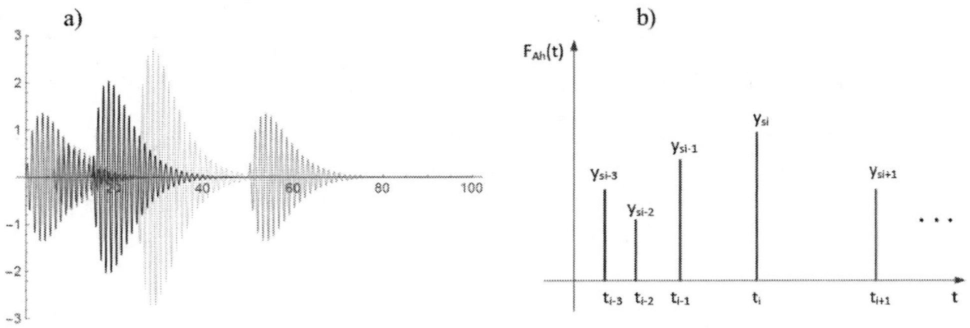

Figure 3. a) Response of the structure with overlapping effect.
b) Random series of maximum response of the structure.

The excitation process, equation (3), is a sum of the pulses (figure 2) loading the system during observation. For this reason the response of the system is also a sum of the responses to pulses. Let introduce the dynamic influence functions $H_s(t-t_n)$ and $H_a(t-t_a)$ which are response of the system caused by a single impulse occurring in time t_n $(A_n=1)$. Using dynamics response functions the displacement of system can be written in the form of a Stieltjes integral as

$$y_s(t_i)=AH_s(t_i)+\int_{t_0}^{t_i} A(\tau)H_s(t_i-\tau)dN(\tau)=y_s^*(t_i)+y_s^{**}(t_i) \tag{4}$$

$$y_a(t_i)=AH_a(t_i)+\int_{t_0}^{t_i} A(\tau)H_a(t_i-\tau)\mathrm{d}N(\tau) \tag{5}$$

where t_i is the moment when one of the impulses is located in the time-step in which the response of the system has the maximum value. [8] If $t_0=0$ then one considers transient vibration, and for $t_0=-\infty$ one considers steady-state vibration case.

The symbol $dN(t)$ denotes increment of the process $N(t)$ in the time interval $(\tau, \tau + d\tau)$ and has the following properties

$$E[dN(t)] = \gamma dt, \quad E[dN^{n}(t)] = \gamma dt \tag{6}$$

$E[dN(t_1)dN(t_2)] = \gamma^2 dt_1 dt_2$ for $t_1 \neq t_2$, where γ is an intensity of the Poisson process $N(t)$.

The expected value and the variance of the system response can be obtained , after taking into account equations (4), (5) and (6). This yields

$$E[y_s(\infty)] = E[y_{si}] = E[A]H_s(t_i) + E[A]\gamma \int_0^\infty H_s(\tau)d\tau \tag{7}$$

$$E[y_a(\infty)] = E[y_{ai}] = E[A]H_a(t_i) + E[A]\gamma \int_0^\infty H_a(\tau)d\tau \tag{8}$$

$$\sigma_{y_s}^2(\infty) = \sigma_{y_s}^2(y_{si}) = \sigma_A^2 H_s^2(t_i) + E[A^2]\gamma \int_0^\infty H_s^2(\tau)d\tau \tag{9}$$

$$\sigma_{y_a}^2(\infty) = \sigma_{y_a}^2(y_{ai}) = \sigma_A^2 H_a^2(t_i) + E[A^2]\gamma \int_0^\infty H_a^2(\tau)d\tau \tag{10}$$

where $\sigma_A^2 = E[A^2]\dfrac{1}{1+v_A^2}$, while $v_A = 0.3$.

Let assume $F_Y(t)$ denote the log-normal distribution function of the random variables y_i, which expected value is given by equation (7) and variance by equation (9).

Let function $F(y,t)_{\max}$ determine the distribution of structure response maxima in a given period (0, t). This function is given by the formulae

$$F_{\max}(y,t) = \sum_{i=1}^{\infty} P\{N(t)=n\} \prod_{i=1}^{n} \{P(y_i \le y)\} = \sum_{n=1}^{\infty} p_n(t) F_Y^n(y) \tag{11}$$

where $p_n(t)$ is probability of n impulses in the time (0, t).

Considering that the $N(t)$ process is a Poisson process, the above formula can be presented in the form

$$F_{\max}(y,t) = \exp\{-\gamma t[1 - F_Y(t)]\} \tag{12}$$

Above solutions can be used in optimization the parameters of the absorber. As the first criterion we assume minimum probability exceeding the permissible level Δ by a single pulse. This probability p_f is given by formulae

$$p_f = 1 - p_r = 1 - \Phi\left[\frac{\ln(\Delta) - E[\ln y]}{\sigma_{lny}^2}\right] \tag{13}$$

where $\Phi(.)$ - standard normal distribution.

The next criterion we assume the probability that the maximum vibrations in the period (0, T) will not exceed Δ level. This probability is equal

$$p_r(T) = 1 - p_f(T) = \exp\{-\gamma T[1 - F_Y(\Delta)]\} \tag{14}$$

where

$$F_Y(\Delta) = \Phi\left[\frac{\ln(\Delta) - E[\ln y]}{\sigma^2_{lnY}}\right] \tag{15}$$

Let's consider the situation that the vibrations of the system caused by one shock are so dominant that the effect from the other shocks can be avoided, the question arises as to the distribution of structure response maxima in a given time interval (figure 4). Let h be the maximum vibration amplitude caused by a single shock (pulse) with unit amplitude $A=1$:

$$h = \max H_s(t), \quad 0 \le t \le \infty \tag{16}$$

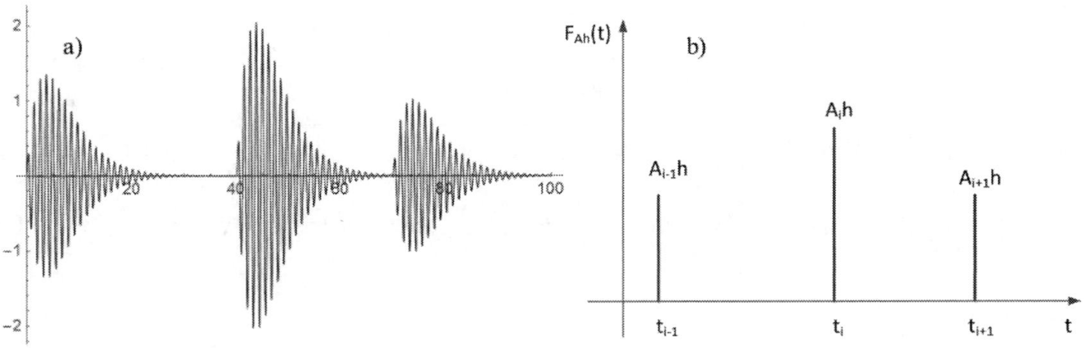

Figure 4. A simplified model of structure vibrations

If the system is loaded with an A pulse, then the maximum displacement of the system is equal Ah. The excitation process (equation 3) is a pulse (figure 2) loading the system during observation. Equation (4) comes down to:

$$y_s(t_i) = y_s^*(t_i) = AH_s(t_i) \tag{17}$$

In above model the equations (11)-(15) need modification. Let function $F_{max}(y,t)$ determine the distribution of structure response maxima in a given period $(0,t)$.

$$F_{max}(x,t) = \sum_{n=1}^{\infty} P\{N(t)=n\} \prod_{i=1}^{n} P\{A_i h \le x\} = \sum_{n=1}^{\infty} p_n(t) F_A^n\left(\frac{x}{h}\right) \tag{18}$$

Considering that the $N(t)$ process is a Poisson process, the above formula can be presented in the form

$$F_{max}(x,t) = \exp\left\{-\gamma t\left[1 - F_A\left(\frac{x}{h}\right)\right]\right\} \tag{19}$$

where $F_A(x)$ is log-normal distribution of the amplitudes A.
In this case the probability p_f is given by formulae

$$p_f = 1 - \Phi\left[\frac{\ln\left(\frac{\Delta}{h}\right) - E[\ln A]}{\sigma^2_{lnA}}\right] \tag{20}$$

The probability the Δ level will be exceeded in the period $(0, T)$ is equal

$$p_f(T) = \exp\left\{-\gamma t\left[1 - F_A\left(\frac{\Delta}{h}\right)\right]\right\} \tag{21}$$

where

$$F_A\left(\frac{\Delta}{h}\right) = \Phi\left[\frac{\ln\left(\frac{\Delta}{h}\right) - E[\ln A]}{\sigma_{\ln A}^2}\right] \tag{22}$$

3. The numerical example

In the article, the calculations were made for two cases:

1) there is a lot of traffic, so vibrations overlap (figure 3),
2) the movement is so small that excitations can be considered separately (figure 4).

Moreover, two excitation variants were considered: one was close to the resonance frequency $\lambda_1 = \Omega_{f1} / \omega_s$, and the second variant of the excitation was far from the resonance frequency - $\lambda_2 = \Omega_{f2} / \omega_s$. All the graphs below present the results for the data $\kappa = \omega_a / \omega_s = \langle 0,3 \rangle$, $\mu = m_a / m_s = \langle 0,0.8 \rangle$ and the excitation $\lambda_1 = \Omega_{f1} / \omega_s = 1.1$ or $\lambda_2 = \Omega_{f2} / \omega_s = 2.2$. One can see results with a single impulse $y_s^*(t)$ (equation 17) and with series of impulses $y_s^{**}(t)$ or the sum $y_s(t)$ (equation 4) The graphs on the left below show the results when a single excitation with frequency $\lambda_1 = 1.1$, while on the right the graph shows results when $\lambda_2 = 2.2$

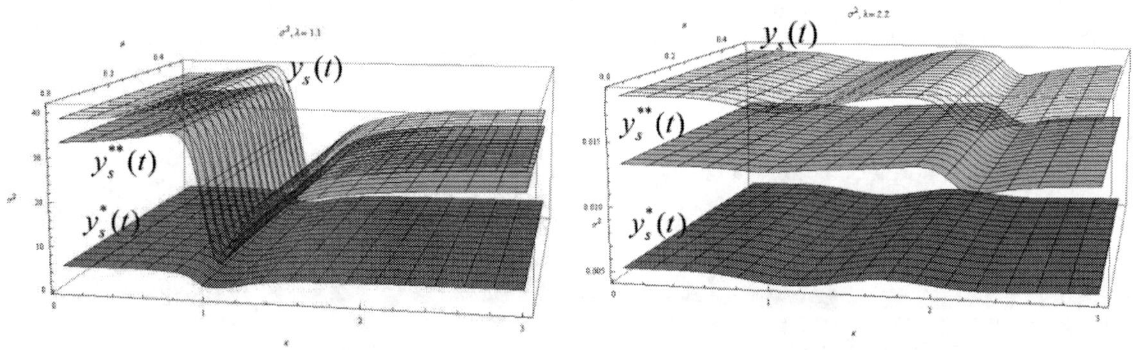

Figure 5. The variance of the system response for $\lambda_1 = 1.1$ and $\lambda_2 = 2.2$.

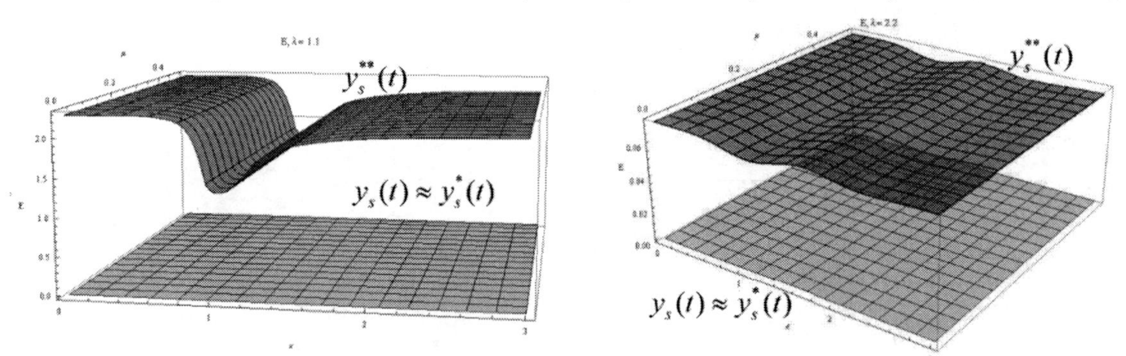

Figure 6. The expected value of the system response for $\lambda_1 = 1.1$ and $\lambda_2 = 2.2$.

Figures 5-6 illustrates results with a single impulse $y_s^*(t)$ - blue plane - and with series of impulses $y_s^{**}(t)$ - red plane - or the sum $y_s(t) = y_s^*(t) + y_s^{**}(t)$ – pale yellow surface. Comparing those figures, the following conclusions were obtained:

1. The level of the function of the variance of the system response $\sigma_{y_s}^2$ is mostly stabled (figure 5). In the case of $\lambda_1 = 1.1$ (when the excitation is close to a resonance) results are more various - one can see a very significant decrease in the level of the function $\sigma_{y_s}^2$ for $\kappa = \omega_a / \omega_s \approx 1.1$. In the second case, $\lambda_2 = 2.2$ (when the excitation is far from a resonance) is similar situation for $\kappa = \omega_a / \omega_s \approx 2.3$ but not so significant as in the first case.

2. The level of the function of the expected value of the system response $E[y_s]$ is mostly stabled (figure 6). Moreover results with a single impulse $y_s^*(t)$ are almost the same as the sum $y_s(t)$ - the planes are coincided. Similarly to the function of the variance of the system response, in the case of $\lambda_1 = 1.1$ - there is a significant decrease in the level of the function $E[y_s]$ for $\kappa = \omega_a / \omega_s \approx 1.1$ but only for the sum $y_s(t)$. In the second case, $\lambda_2 = 2.2$ - the planes are almost flat .

4. Conclusions

The article considers the problem of optimizing the parameters of an absorber in a situation where random vibrations of the structure create a filtered Poisson process, e.g. vehicles passing by a building. A parameter optimization criterion was proposed, consisting in minimizing the probability of the maximum vibrations of the structure, i.e. the maximum reliability of the structure in relation to the maximum vibration amplitudes.

Two load models were analysed – a frequent vehicle traffic and a single vehicle passing. Numerical simulations indicate that both the variance and the mean value depend on the ratio of the frequency of the absorber to the structure $\kappa = \omega_a / \omega_s$ in both the first and second load models. However, it seems that they hardly depend on the ratio of the absorber mass to the structure mass $\mu = m_a / m_s$. There is significant decrease in the level of the variance function of the excitation when a frequency of the absorber is close to a frequency of the structure ($\kappa = \omega_a / \omega_s \approx 1.1$).

Probability criterion allows objectify criterion limits the vibration of the structure.

References

[1] Frahm H 1911 Device for damping vibrations of bodies United States Patent p 3576-3580
[2] Ormondroyd J, Den Hartog J P 1928 The theory of the dynamic vibration absorber, *Transactions of ASME, J. of Applied Mechanics* **50** (7) pp 9–22
[3] Den Hartog J P 1985 *Mechanical Vibrations* (New York: Dover 4th ed.)
[4] Asami T, Nishihara O and Baz A M 2002 Analytical Solutions to H∞ and H2 Optimization of Dynamic Vibration Absorbers Attached to Damped Linear Systems *ASME J. of Vibration and Acoustics* **124** (2) pp 284-95
[5] Crandall S H, Mark W D 1963 *Random Vibration in Mechanical Systems* (Academic Press)
[6] Soong T T, Grigoriu M 1993 *Random vibration of mechanical and structural systems* (PTR Prentice-Hall, Inc.)
[7] Melchers R E1999 *Structural Reliability analysis and prediction* (John Wiley & Sons)
[8] Rystwej A, Sniady P 2007 Dynamic response of an infinite beam and plate to a stochastic train of moving forces *J. of Sound and Vibration* **299** pp 1033-48

IOP Conf. Series: Materials Science and Engineering 1015 (2021) 012004 doi:10.1088/1757-899X/1015/1/012004

A slightly compressible hyperelastic material model with the Mullins effect

S Jemioło and A Franus

Faculty of Civil Engineering, Warsaw University of Technology, Armii Ludowej 16, 00-637 Warsaw, Poland

E-mail address: s.jemiolo@il.pw.edu.pl

Abstract. The paper presents a hyperelastic material model which is intended to describe slightly compressible materials with encountered the Mullins effect. It is assumed the volumetric-isochoric stored energy function split regarding the primary response of a material. Constitutive equations based on the concept of pseudo-elasticity are derived. The problems of a simple uniaxial compression/tension involving one finite element and a compression of a ball are solved using ABAQUS. The material model is defined with the help of user subroutine UHYPER in terms of the primary material response. Its main advantage is a physically correct description of volumetric changes while the polynomial model in ABAQUS does not meet basic growth conditions. In order to include the Mullins effect, it is combined with available in the software library the Mullins effect material model which is based on Ogden-Roxburgh proposition of stress softening function. Experimental data on synthetic rubber neoprene published in the literature is utilized.

1. Introduction

Rubber-like materials show a significant change in their mechanical response resulting from the first deformation [1]. The stress softening phenomenon, observed in filled and non-filled rubber, has been investigated intensively by Mullins and his co-workers [2] and consequently is called the Mullins effect. More precisely, the Mullins effect refers to a strain-induced stress softening of the material that takes place almost exclusively after the first loading path of loading cycles. It is often assumed that a damage causing the stress softening only depends on the maximum strain energy attained previously by the material.

There are a large number of works concerning the Mullins effect [3]. Many authors use the thermodynamical framework of continuum damage mechanics to develop constitutive equations. Others derive a large strain viscoelastic constitutive model with damage for rubber-like materials. A different approach is based on the network theory which states that parts of the network are reformed during loading.

Ogden and Roxburgh [4,5] proposed a pseudo-elastic model which differs from the mentioned theories because the stress softening function - that depends on the maximum strain energy - is activated only on unloading paths. A single continuous damage parameter is incorporated into the model in order to modify the stored energy function so that the material responses differently on unloading paths. The approach greatly simplifies the identification of material parameters. Its main advantage is a convenient description of the stress-strain relationship in a cyclic loading as well as their numerical implementation for the finite element method [6,7]. Based on the concept of pseudo-elasticity, Naumann and Ihlemann

[8] proposed an integral form of the specific energy function to derive the constitutive equations. However, the Mullins effect described using the approach should be treated as a quasi-static phenomenon.

In the paper we derive constitutive equations based on the concept of pseudo-elasticity for a slightly compressible hyperelastic material model [10-12]. Example boundary value problems, which concern a simple compression/tension and a compression of a ball, are solved using ABAQUS/Standard [6,7]. The material model is defined via user subroutine UHYPER in terms of the primary material response. It is combined with available in the software library the Mullins effect material model which is based on the Ogden-Roxburgh proposition.

2. Pseudo-hyperelasicity

2.1. An interal variable model

The local equations of thermodynamics yield the Clausius-Duhem [9] inequality for purely mechanical material models

$$D=\frac{1}{2}\mathbf{T}\cdot\dot{\mathbf{C}}-\dot{\Psi}\geq 0 \tag{1}$$

where D denotes the mechanical dissipation, $\mathbf{T}$ the second Piola-Kirchhoff stress tensor, $\mathbf{C}=\mathbf{F}^T\mathbf{F}$ the right Cauchy-Green tensor, and Ψ the specific free energy. Here $\mathbf{F}$ denotes the deformation gradient with $J=\det\mathbf{F}>0$. We assume that Ψ is a function of $\mathbf{C}$ and an internal variable g, i.e. $\Psi=\Psi(\mathbf{C},g)$. Hence, the inequality (1) can be written in the form

$$D=\left(\frac{1}{2}\mathbf{T}-\left.\frac{\partial\Psi}{\partial\mathbf{C}}\right|_{\mathbf{C}=\mathbf{C}^T}\right)\cdot\dot{\mathbf{C}}+P\dot{g}\geq 0,\quad P=-\frac{\partial\Psi}{\partial g} \tag{2}$$

where P a thermodynamic force associated with a damage variable g. In the case of pure hyperelasticity, we have $g=0$ and $D=0$ (process is fully reversible). A constitutive relation can be established as

$$\mathbf{T}=2\left.\frac{\partial\Psi}{\partial\mathbf{C}}\right|_{\mathbf{C}=\mathbf{C}^T} \tag{3}$$

The inequality (2) contains two parts describing the dissipation such that $D=D_1+D_2$, which are functions of $\dot{\mathbf{C}}$ and $\dot{g}$. In the article we assume that the specific internal energy [9] is of the form

$$\Psi=\Psi(\mathbf{C},g)=\phi_1(g)W_0(\mathbf{C})+\phi_2(g) \tag{4}$$

where $W_0(\mathbf{C})$ denotes the stored energy function of the virgin material model (it is assumed that in the undeformed configuration $W_0(\mathbf{I})=0$). Therefore, the inequality (2.2) yields

$$\begin{aligned} D&=\left(\frac{1}{2}\mathbf{T}-\phi_1\left.\frac{\partial W_0}{\partial\mathbf{C}}\right|_{\mathbf{C}=\mathbf{C}^T}\right)\cdot\dot{\mathbf{C}}+P\dot{g}=D_1+D_2\geq 0\\ P&=-\frac{\partial\Psi(\mathbf{C},g)}{\partial g}=-\left(\frac{\partial\phi_1}{\partial g}W_0+\frac{\partial\phi_2}{\partial g}\right)=\dot{g}\alpha \end{aligned} \tag{5}$$

Here, both $\dot{g}\geq 0$ and $\dot{g}\leq 0$ are possible. If $\alpha>0$ and $D_1=0$, equations (5) give the constitutive equation for stress tensor and the evolution equation for internal variable

$$\mathbf{T} = 2\phi_1 \left.\frac{\partial W_0}{\partial \mathbf{C}}\right|_{\mathbf{C}=\mathbf{C}^T}, \qquad \dot{g} = \frac{P}{\alpha} = -\frac{1}{\alpha}\left(\frac{\partial \phi_1}{\partial g} W_0 + \frac{\partial \phi_2}{\partial g}\right) \tag{6}$$

Equations (6) also state basic continuum damage mechanics models [3]. In the case of a typical damage modelling, the internal variable g often describes only an irreversible process, so that only $\dot{g} \geq 0$ is possible.

2.2. The Ogden-Roxburgh model

Assumption $\dot{g} = 0$ yields the pseudo-elastic Ogden–Roxburgh model [4]

$$\frac{\partial \phi_1}{\partial g} W_0 + \frac{\partial \phi_2}{\partial g} = 0 \tag{7}$$

In such case it holds $D_2 = 0$. Therefore, the function Ψ, which meets condition (7), is called a pseudo-elastic potential (it is not a stored elastic energy potential) and ϕ_1 is a damage (softening) function. In the work [4] it is assumed that $\phi_1 \equiv \eta$ and $\phi_2 \equiv \phi(\eta)$, so that $\partial\phi / \partial\eta = -W_0$. The damage variable η is given by

$$\eta = 1 - \frac{1}{c_2}\operatorname{Erf}\left[\frac{1}{c_1}\left(W_0 - W_{\max}\right)\right] \tag{8}$$

where c_1, c_2 are positive constants and $W_{\max}(t) \equiv \max\{W_0(\tau), \tau \leq t\}$. Thus, the damage of material is controlled by the maximum energy state $W_{\max}$ attained.

2.3. An integral form of pseudo-hyperelastic potential

In the approach, the internal energy dissipation occurs during loading and unloading cycles of the material. It is convenient to define a pseudo-hyperelastic potential in an integral form [8]

$$\Psi(W_0, W_{\max}) \equiv \int_0^{W_0} \eta(\xi, W_{\max})\,\mathrm{d}\xi \tag{9}$$

where $\eta = \eta(\xi, W_{\max})$ is the softening function. The definition (9) directly leads to $\Psi(W_0(\mathbf{I}) = 0, W_{\max}) = 0$. Material derivative of Ψ yields

$$\dot{\Psi} = \frac{\partial \Psi(W_0, W_{\max})}{\partial W_0}\dot{W}_0 + \frac{\partial \Psi(W_0, W_{\max})}{\partial W_{\max}}\dot{W}_{\max} = \eta(W_0, W_{\max})\dot{W}_0 + \frac{\partial \Psi(W_0, W_{\max})}{\partial W_{\max}}\dot{W}_{\max} \tag{10}$$

so that from (1) we have

$$D = \left(\frac{1}{2}\mathbf{T} - \eta \left.\frac{\partial W_0}{\partial \mathbf{C}}\right|_{\mathbf{C}=\mathbf{C}^T}\right)\cdot\dot{\mathbf{C}} - \frac{\partial \Psi(W_0, W_{\max})}{\partial W_{\max}}\dot{W}_{\max} \geq 0 \tag{11}$$

Similarly to the previous case, it holds

$$\mathbf{T} = 2\eta \left.\frac{\partial W_0(\mathbf{C})}{\partial \mathbf{C}}\right|_{\mathbf{C}=\mathbf{C}^T}, \qquad \frac{\partial \Psi(W_0, W_{\max})}{\partial W_{\max}}\dot{W}_{\max} \leq 0 \tag{12}$$

By definition $\dot{W}_{\max}$ is always non-negative. Then, a sufficient condition for thermomechanical consistency is

$$\frac{\partial \eta(W_0, W_{max})}{\partial W_{max}} \leq 0 \tag{13}$$

Hence, every softening function which monotonically decreases with respect to W_{max} leads to a positive dissipation. Taking into consideration (8), the specific energy function may be explicitly stated as

$$\Psi(W_0, W_{max}) = \frac{c_1}{c_2\sqrt{\pi}} \mathrm{Exp}\left[-\frac{1}{c_1^2}(\xi - W_{max})^2\right] + \xi - \frac{1}{c_2}(\xi - W_{max})\mathrm{Erf}\left[\frac{1}{c_1}(\xi - W_{max})\right]\Bigg|_0^{W_0} \tag{14}$$

3. A slightly compressible pseudo-hyperelastic material model

3.1. Slightly compressible hyperelastic models

Models of isotropic slightly compressible materials are a generalisation of models describing incompressible materials. In the case we assume that no coupling between the stored energy function of isochoric $\bar{W}$ and volumetric W_{vol} deformations occurs [10-12]. Therefore, it is convenient to express stored energy function in the form

$$\tilde{W}(\bar{I}_1, \bar{I}_2, J) = \bar{W}(\bar{I}_1, \bar{I}_2) + W_{vol}(J) \tag{15}$$

where the modified invariants are given by $\bar{I}_1 = \mathrm{tr}\bar{\mathbf{B}}$, $\bar{I}_2 = \mathrm{tr}\bar{\mathbf{B}}^{-1}$ and $\bar{\mathbf{B}} = \bar{\mathbf{F}}\bar{\mathbf{F}}^T$, $\bar{\mathbf{F}} = J^{-1/3}\mathbf{F}$. These can be explicitly written as

$$\bar{I}_1 = J^{-\frac{2}{3}} I_1, \quad \bar{I}_2 = J^{-\frac{4}{3}} I_2 \tag{16}$$

Due to the form of stored energy function (15), the constitutive relation in the current configuration yields

$$\boldsymbol{\sigma} = \frac{1}{J}\mathbf{F}\mathbf{T}\mathbf{F}^T = \frac{\partial W_{vol}}{\partial J}\mathbf{I} + \frac{2}{J}\left(\frac{\partial \bar{W}}{\partial \bar{I}_1}\bar{\mathbf{B}}_D - \frac{\partial \bar{W}}{\partial \bar{I}_2}\bar{\mathbf{B}}_D^{-1}\right) \tag{17}$$

where $\boldsymbol{\sigma}$ is the Cauchy stress tensor and $\bar{\mathbf{B}}_D$ stands for deviatoric part of the modified left Cauchy-Green tensor $\bar{\mathbf{B}}$ as follows

$$\bar{\mathbf{B}}_D = \bar{\mathbf{B}} - \frac{1}{3}\bar{I}_1\mathbf{I}, \quad \bar{\mathbf{B}}_D^{-1} = \bar{\mathbf{B}}^{-1} - \frac{1}{3}\bar{I}_2\mathbf{I} \tag{18}$$

3.2. The pseudo-hyperelastic model

As mentioned in the previous section, Ogden and Roxburgh proposed a phenomenological pseudo-elastic model for the Mullins effect in filled rubber which concerns an incompressible material [4,11]. In the case of slightly compressible material, the pseudo-hyperelastic potential is modified such that

$$\Psi = \Psi(\bar{I}_1, \bar{I}_2, \eta) = \eta\bar{W}(\bar{I}_1, \bar{I}_2) + \phi(\eta) + W_{vol}(J) \tag{19}$$

The consequence of the above form of the function is that the Mullins effect is associated only with the deviatoric part of the deformation. The pressure stress of the augmented response is the same as that of the primary response [6]. Here, the inequality (1)

IOP Conf. Series: Materials Science and Engineering 1015 (2021) 012004 doi:10.1088/1757-899X/1015/1/012004

$$D=\left(\frac{1}{2}\mathbf{T}-\eta\left.\frac{\partial\overline{W}(\overline{I}_1,\overline{I}_2)}{\partial\mathbf{C}}\right|_{\mathbf{C}=\mathbf{C}^T}-\frac{J}{2}\frac{\partial W_{vol}(J)}{\partial J}\mathbf{C}^{-1}\right)\cdot\dot{\mathbf{C}}+P\dot{g}\geq 0 \tag{20}$$

yields the constitutive relation of the form

$$\mathbf{T}=J\frac{\partial W_{vol}(J)}{\partial J}\mathbf{C}^{-1}+2\eta\left.\frac{\partial\overline{W}(\overline{I}_1,\overline{I}_2)}{\partial\mathbf{C}}\right|_{\mathbf{C}=\mathbf{C}^T} \tag{21}$$

or in the spatial description

$$\boldsymbol{\sigma}=\frac{1}{J}\mathbf{FTF}^T=\frac{\partial W_{vol}}{\partial J}\mathbf{I}+\frac{2}{J}\eta\left(\frac{\partial\overline{W}}{\partial\overline{I}_1}\overline{\mathbf{B}}_D-\frac{\partial\overline{W}}{\partial\overline{I}_2}\overline{\mathbf{B}}_D^{-1}\right) \tag{22}$$

In the article we consider so called MCMV five-parameter model which deviatoric part is defined by the stored energy function of the form

$$\overline{W}(\overline{I}_1,\overline{I}_2)=\frac{1}{2}\left[a_1(\overline{I}_1-3)+\frac{1}{2}a_2(\overline{I}_1^2-9)+\frac{1}{3}a_3(\overline{I}_1^3-27)+a_4(\overline{I}_2-3)+a_5(\overline{I}_1\overline{I}_2-9)\right] \tag{23}$$

It is worth mentioning that the function (23) is a consistent third order approximation in terms of deformation tensor $\overline{\mathbf{C}}$ norm such that

$$\left|W(\overline{I}_1,\overline{I}_2)-\overline{W}(\overline{I}_1,\overline{I}_2)\right|=\mathcal{O}\left(\left\|\overline{\mathbf{C}}\right\|_2^4\right) \tag{24}$$

For more details regarding the model we refer the reader to [10,12].

In the literature, there are a large number of propositions of the volumetric stored energy function [12]. Such function needs to meet certain growth conditions. Here, we define

$$W_{vol}(J)=\mathrm{K}_0\left[\frac{1}{4}(J^2-1)-\frac{1}{2}\ln J\right] \tag{25}$$

where is the K_0 initial bulk modulus, which is several orders of magnitude greater than the initial shear modulus in the case of rubber-like materials. We emphasize that available in the ABAQUS standard material library all hyperelastic models include the volumetric stored energy function which do not meet basic growth conditions. It may lead to unrealistic response of the material model.

4. Example boundary value problems

In this section we present solutions of two boundary value problems which illustrates application of the discussed model. The example problems are solved using ABAQUS/Standard which is a finite element program designed for general use in nonlinear problems [6,7]. The MCMV model is defined with the help of user subroutine UHYPER in terms of the primary material response. It is combined with available in the software library the Mullins effect material model which is based on the Ogden-Roxburgh model. The damage variable is given by

$$\eta=1-\frac{1}{c_2}\mathrm{Erf}\left[\frac{W_{\max}-W_0}{c_1+\beta W_{\max}}\right] \tag{26}$$

which generalizes the function (8) by β parameter. We use values of parameters given in table 1 for the problems presented in the following part of the article. The initial bulk modulus is step up to be $\mathrm{K}_0=1000\mu_0=340.74$ MPa.

Table 1. Parameters of MCMV for neoprene [13,14].

Model	a_1 [MPa]	a_2 [MPa]	a_3 [MPa]	a_4 [MPa]	a_5 [MPa]	c_1	c_2	β
MCMV	$3.402\ 10^{-1}$	$-9.644\ 10^{-3}$	$1.685\ 10^{-4}$	$2.902\ 10^{-2}$	$-1.762\ 10^{-4}$	2.2	3.0	0

4.1. A simple compression/tension test

In order to illustrate basic properties of the model, a simple compression/tension test involving one finite element (C3D8H) is considered. The loading is realised by prescribed displacement over one face according to figure 1.

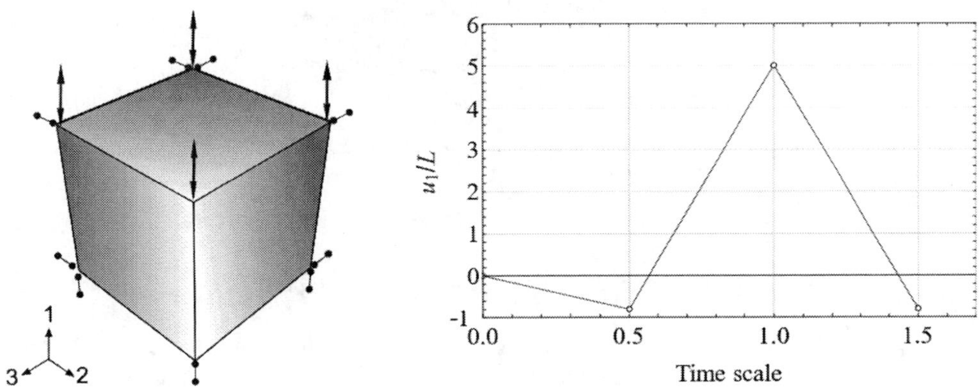

Figure 1. Uniaxial tension/compression test.

Results in the form of the Cauchy stress versus principal stretch are shown in figure 2. The first part of loading process concerns compression – path marked as '1'. Unloading path '2' clearly shows predicted stress softening. One can notice that the path in the range of tension is not a smooth curve. After marked in the plot filled point a primary response of the material occurs. It means that the stored energy function $\bar{W}(\bar{I}_1,\bar{I}_2)$ attains greater value than at maximum compression. Following unloading path '3' is indeed a smooth one.

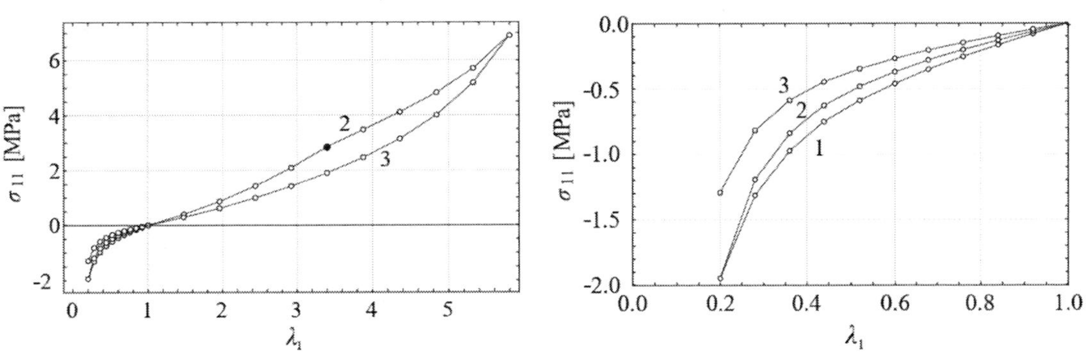

Figure 2. The Cauchy stress vs principal stretch of uniaxial compression/tension.

4.2. Compression of a ball

The next discussed problem is a boundary value problem of compression of a ball with radius $R = 1$, see figure 3. The force is applied by a contact with non-deformable membranes with prescribed traction on one of them. The computational model consists of 9572 finite elements: type C3D8H with linear interpolation functions for the ball and type R3D4 for the membranes.

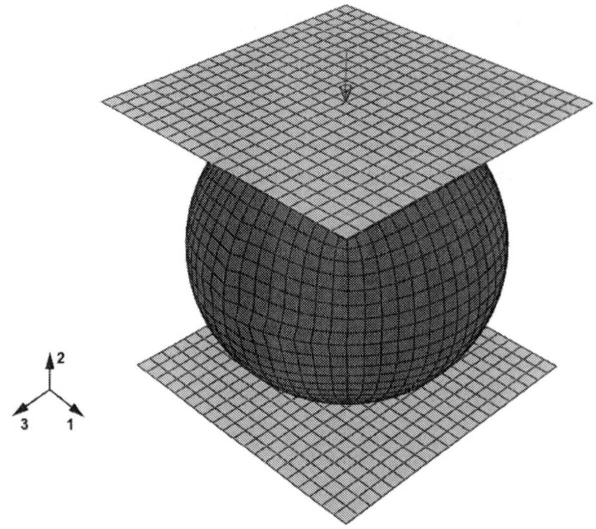

Figure 3. Finite element mesh of the ball.

In order to show the Mullins effect, the ball is compressed and then unloaded. figure 4 presents two curves of the Cauchy stress versus displacement of the ball's top point. Values of the stress are obtained from four top, central elements on the edge. On the other hand, values of internal energy are given in the figure 5 for the whole model (it does not concern membranes as they are fully rigid). In the example the damage energy (ALLDMD) is much smaller than the recoverable (elastic) one (ALLSE). In this case the internal energy denoted in ABAQUS as ALLIE is a sum of ALLSE and ALLDMD.

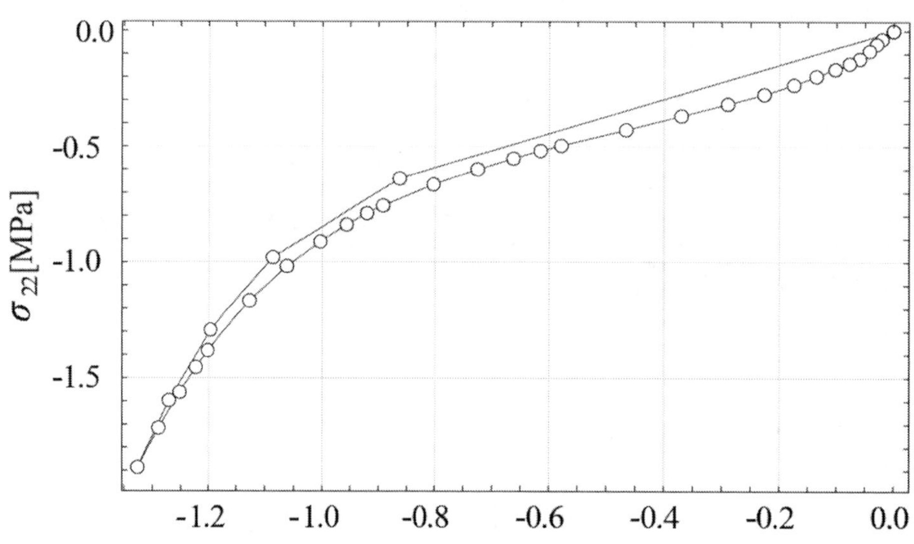

Figure 4. The Cauchy stress versus normalized displacement of the ball's top point.

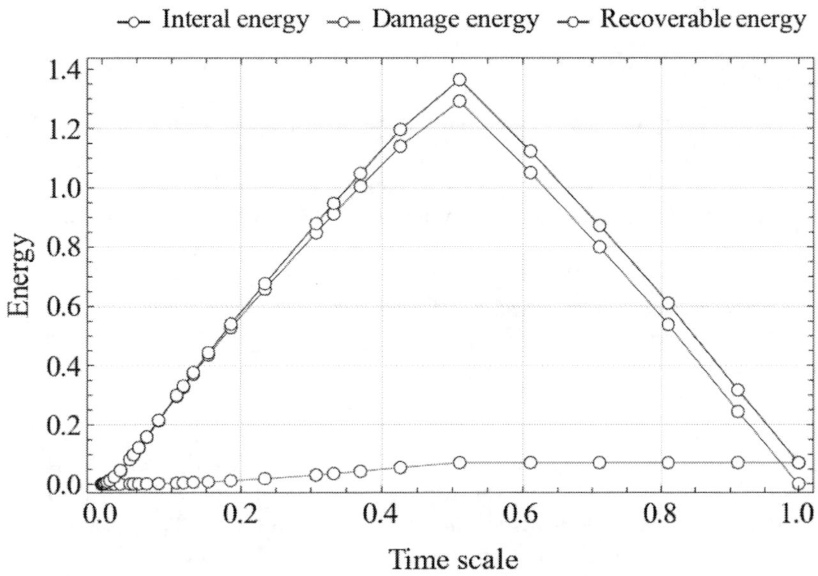

Figure 5. Finite element mesh for uniaxial compression of a rectangular rod.

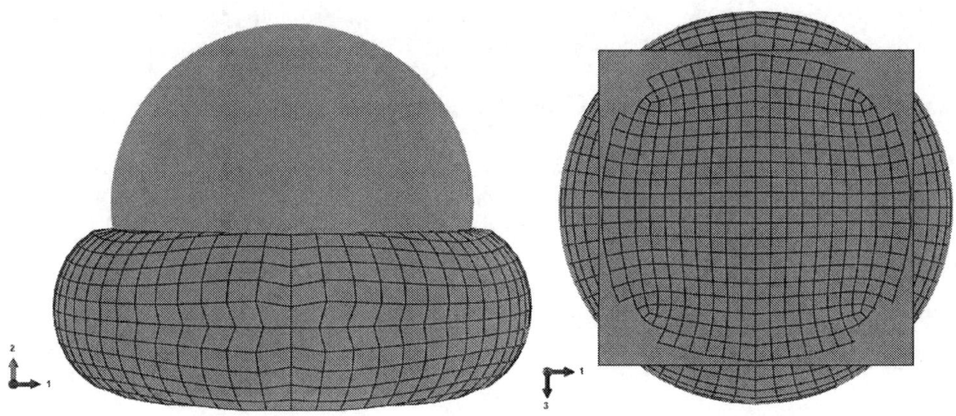

Figure 6. Maximum compression deformation of the ball.

5. Conclusions

The paper concerns a phenomenological, constitutive modelling of the Mullins effect. We derive constitutive equations based on the concept of pseudo-elasticity for a slightly compressible hyperelastic material model. The approach was firstly proposed by Ogden and Roxburgh [4] where an incompressible material model is discussed. In this case the Mullins effect should be treated as a quasi-static phenomenon as the theory does not predict any rate-dependent effects.

In order to demonstrate the Mullins effect occurring in a cycle loading, example boundary value problems are solved using ABAQUS/Standard. The proposed material model is defined via user subroutine UHYPER which is suitable for definition of isotropic material models. The primary hyperelastic response is combined with available in the software library the Mullins effect material model which is based on the Ogden-Roxburgh model with stress softening function. The main advantage of the proposed model is the physically correct description of volumetric changes while the polynomial model in ABAQUS does not meet basic growth conditions.

IOP Conf. Series: Materials Science and Engineering 1015 (2021) 012004 doi:10.1088/1757-899X/1015/1/012004

References

[1] Diani J, Fayolle B and Gilormini P 2009 A review on the Mullins effect. *Eur. Polym. J.* **45** pp 601–12

[2] Mullins L and Tobin N R 1957 Theoretical model for the elastic behavior of filler-reinforced vulcanized rubbers, *Rubber Chem. Technol* **30** (2) pp 555-71

[3] Chagnon G, Verron E, Gornet L, Marckmann G and Charrier P 2004 On the relevance of Continuum Damage Mechanics as applied to the Mullins effect in elastomers *J. Mech. Phys. Solids* **52** pp 1627–50

[4] Ogden R W and Roxburgh D G 1999 A pseudo-elastic model for the Mullins effect in filled rubber, *Proc. R. Soc. Lond. A* **455** pp 2861–77

[5] Jemioło S 2001 Pseudo-Hyperelastic Model of Elastomers. In *Theoretical Foundations of Civil Enginierings*, Rymsza B. [ed], Proc. The X Russian-Polish Semminar pp 89-98 [in Polish], *OWPW* Warsaw.

[6] Abaqus 2016, Theory guide, *Dassault Systèmes*

[7] Abaqus 2016, Analysis user's guide. Volume III: Materials, *Dassault Systèmes*

[8] Naumann C and Ihlemann J 2015 On the thermodynamics of pseudo-elastic material models which reproduce the Mullins effect, *Int. J. Solids Struct.* **69–70** pp 360–9

[9] Berjamin H, Favrie N, Lombard B and Chiavassa G 2017 Nonlinear waves in solids with slow dynamics: an internal-variable model *Proc. R. Soc. A Math. Phys. Eng. Sci.* **473** 20170024,.

[10] Jemioło S 2002 *Studium hipersprężystych własności materiałów izotropowych* (Warsaw: OWPW)

[11] Suchocki C and Jemioło S 2019 On finite element implementation of polyconvex incompressible hyperelasticity: theory, coding and applications, *Int. J. Comput. Methods*,**1950049**

[12] Franus A, Jemioło S and Antoni M 2020 A slightly compressible hyperelastic material model implementation in ABAQUS, *Engineering Solid Mechanics*, **8** (4) pp 365–80

[13] Alexander H 1968 A constitutive relation for rubber-like materials *Int. J. Eng. Sci.* **6** (9) pp 549-63

[14] Jemioło S and Franus A 2019 *Izotropowe modele konstytutywne hipersprężytości. Wyznaczanie parametrów materiałowych i implementacja numeryczna* (Warsaw: OWPW)

IOP Conf. Series: Materials Science and Engineering 1015 (2021) 012005 doi:10.1088/1757-899X/1015/1/012005

Third formulation of the space-time finite element method

Z Kacprzyk

The Faculty of Civil Engineering, Warsaw University of Technology, al. Armii Ludowej 16, 00-637 Warsaw, Poland

E-mail address: z.kacprzyk@il.pw.edu.pl

Abstract. The paper explains differences existing between the space-time finite elements, which fundamentals were presented by the Kączkowski in 1975 and 1976, and its modification. The replacement of load with concentrated forces and moments (which are to satisfy equilibrium conditions at nodes) was given up, namely in the latter one; instead, a requirement was made that functions approximating the solution would satisfy differential equation in inter-node areas in a specified integral way. This leads to a system of algebraic equations, in which kinematic parameters existing in the solution approximating functions are the unknowns. Displacements and their derivatives (of any order) at nodes are those parameters.

1. Introduction

The space-time finite element method is an adaptation of the classical finite element method [4,17,18] to a time-dependent problem. The first formulation of the space-time finite element method, the so-called stationary presented by Kączkowski in the works [8-10]. The stationary formulation of the method was generalized in the doctoral dissertation by Kacprzyk [6]. In 1983, Kączkowski presented the second formulation of the method in which he applied the non-stationary division of space-time [11]. This approach was developed by Bajer in the works [1-3]. The method was used and developed by many researchers [7,14-16].

Below we present a modified space-time element method, in which we will give up replacing loads acting on an element with a system of loads concentrated at nodes. Functions approximating the solution, i.e. satisfying approximately a system of differential equations in the finite element area, we will make dependent on any number of kinematic node parameters. Displacement values and any number of consecutive derivatives of functions approximating the solution will be those parameters. Each increase in those parameters number (with-out changing the nodes grid) results in a significant increase in the solution accuracy

2. General relationships

Let us assume that in the space-time area Ω the motion of a body is described by a differential matrix equation

$$\underset{n\times n}{\mathbf{L}}\;\underset{n\times 1}{\mathbf{u}}\left(\mathbf{x},t\right)=\underset{n\times 1}{\mathbf{p}}\left(\mathbf{x},t\right) \tag{1}$$

in which $\mathbf{L}$ is a matrix composed of differential operators multiplied by coefficients dependent on physical parameters characterising the material, $\mathbf{u}$ is the displacement vector, and $\mathbf{p}$ - the vector of external load. Number n depends on the number of kinematic parameters describing the motion of the

IOP Conf. Series: Materials Science and Engineering 1015 (2021) 012005 doi:10.1088/1757-899X/1015/1/012005

considered body. Equation (1) multiplied by the volume of infinitely small element $d\Omega$ expresses the equilibrium condition for impulses acting on this element.

Equation (1) may be also considered as a mathematical notation of the Newton's third law; the right side of equation (1) means all loads acting on the considered body from outside, while the left side - an equal but oppositely directed action (reaction) of the body.

After transferring load $\mathbf{p}$ to the left side of the equal sign, equation (1) gets the following form:

$$\mathbf{Lu}-\mathbf{p}=\mathbf{0} \tag{2}$$

If vector **u** is an exact solution of equation (1), then the product of any functional matrix $\bar{\mathbf{u}}^T(\mathbf{x},t)$ and equation (2) as well as an integral of this product in area Ω are as an identity equal to zero:

$$\int_{\Omega} \bar{\mathbf{u}}^T [\mathbf{Lu}-\mathbf{p}] d\Omega \equiv 0 \tag{3}$$

Components of matrix $\bar{\mathbf{u}}$ may be treated as virtual increments in displacements arranged in matrix **u**, and equation (3) - as an equation of virtual *work-time*

Let us present an approximate solution of differential equation (1) in the following form

$$\mathbf{v}(\mathbf{x},t)=\mathbf{N}(\mathbf{x},t)\mathbf{C} \tag{4}$$

So-called shape functions, satisfying specific geometrical conditions, which we will discuss below, are components of matrix **N**. Sub-matrices $\mathbf{C}_j\ (j=1,2,3,...,k)$ are components of one-column matrix $\mathbf{C}$, each of them consists of geometrical parameters expressing values of function $\mathbf{v}$ and its derivatives of the first and possibly higher orders at node j of coordinates

$$\mathbf{x},t=\mathbf{x}_j,t_j, \qquad j=1,2,3,\ldots,k \tag{5}$$

One of components of function matrix $\mathbf{N}$ is assigned to each of constants forming vector $\mathbf{C}$. If elements featuring the shape of a rectangle or cuboid are used, the shape functions usually get the shape of basic functions, each of which depends only on one space or time variable and on the approximation order. Table 1 specifies examples of basic functions of orders *I, II, III,* and *IV*. These functions are components of the following one-row matrices

$$\mathbf{N}^I=[\rho],\quad \mathbf{N}^{II}=[\rho,\omega],\quad \mathbf{N}^{III}=[\rho,\omega,\chi],\quad \mathbf{N}^{IV}=[\rho,\omega,\chi,\lambda] \tag{6}$$

The following notation has been used in the table 1

$$\vartheta=\vartheta(\xi)=\frac{1-\xi^2}{4},\quad \xi_i=\pm 1,\quad \xi\in(-1,+1) \tag{7}$$

Taking into consideration the fact that function ϑ for both extreme values of ξ is equal zero, and function $\frac{1+\xi_i\xi}{2}$ only at $\xi=\xi_i$ takes the value of 1, based on table 1 it is easy to determine boundary conditions, satisfied by individual components of matrices specified in formulae (6).

Table 1. Basic shape functions

Order		f
I	ρ	$\frac{1+\xi_i\xi}{2}$
II	ρ	$\frac{1+\xi_i\xi}{2}+\xi_i\xi\vartheta$
	ω	$-(\xi_i+\xi)\vartheta$
III	ρ	$\frac{1+\xi_i\xi}{2}+\xi_i\xi(\vartheta+3\vartheta^2)$
	ω	$-(\xi_i+\xi)\vartheta-(\xi_i+3\xi)\vartheta^2$
	χ	$(1+\xi_i\xi)\vartheta^2$
IV	ρ	$\frac{1+\xi_i\xi}{2}+\xi_i\xi(\vartheta+\vartheta^2)+10\xi_i\xi\vartheta^3$
	ω	$-(\xi_i+\xi)\vartheta-(\xi_i+3\xi)\vartheta^2-2(\xi_i+5\xi)\vartheta^3$
	χ	$(1+\xi_i\xi)\vartheta^2+(2+4\xi_i\xi)\vartheta^3$
	λ	$-\frac{2}{3}(\xi_i+\xi)\vartheta^3$

Having substituted an approximate solution (4) to differential equation (2) this equation will be satisfied only approximately

$$\mathbf{Lv}(\mathbf{x},t)-\mathbf{p}(\mathbf{x},t)=\mathbf{LN}(\mathbf{x},t)\mathbf{C}-\mathbf{p}(\mathbf{x},t)=\varepsilon(\mathbf{x},t)\approx\mathbf{0} \tag{8}$$

Let us multiply equation (25) by a function matrix

$$\bar{\mathbf{v}}^T(\mathbf{x},t)=\bar{\mathbf{C}}^T\mathbf{N}^T(\mathbf{x},t) \tag{9}$$

i.e. by the function matrix consisting of shape functions **N**, defined above, multiplied by any coefficients being components of vector $\bar{\mathbf{C}}$. Although the product of function (9) and equation (8) is different from zero, however, we can demand this product integral within boundaries of area Ω to be equal to zero

$$\bar{\mathbf{C}}^T[\mathbf{KC}-\mathbf{P}]=\mathbf{0} \tag{10}$$

In this equation symbols **K** and **P** stand for the following quantities

$$\mathbf{K}=\int_\Omega \mathbf{N}^T\mathbf{LN}d\Omega,\quad \mathbf{P}=\int_\Omega \mathbf{N}^T\,d\Omega \tag{11}$$

To satisfy equation (10) irrespective of the number of matrix $\bar{\mathbf{C}}$ components, the expression in brackets must be equal to zero

$$\mathbf{KC}-\mathbf{P}=\mathbf{0} \tag{12}$$

In this way we have received a system consisting of m equations depending on m parameters making vector **C**. Part of those parameters is known from boundary and initial conditions, and also from possible kinematic conditions at points situated within area Ω Having split vector **C** into two sub-vectors: ${}^u\mathbf{C}$ composed of unknown parameters and ${}^k\mathbf{C}$ - of known parameters, and after an appropriate regrouping of matrix **K** columns and rows, the system of equations (12) gets the following form:

$$\begin{bmatrix} \mathbf{K}_{uu} & \mathbf{K}_{uk} \\ \mathbf{K}_{ku} & \mathbf{K}_{kk} \end{bmatrix} \begin{Bmatrix} {}^{u}\mathbf{C} \\ {}^{k}\mathbf{C} \end{Bmatrix} - \begin{Bmatrix} {}^{u}\mathbf{P} \\ {}^{k}\mathbf{P} \end{Bmatrix} = \begin{Bmatrix} \mathbf{0} \\ \mathbf{0} \end{Bmatrix} \tag{13}$$

Hence unknown parameters ${}^{u}\mathbf{C}$ will be determined from the following formula:

$${}^{u}\mathbf{C} = -\mathbf{K}_{uu}^{-1}\mathbf{K}_{uk}\,{}^{k}\mathbf{C} + \mathbf{K}_{uu}^{-1}\,{}^{u}\mathbf{P} \tag{14}$$

and after substituting them and components of vector ${}^{k}\mathbf{C}$ to formula (4) we obtain an approximate solution of the problem. The presented procedure may be applied not only in the space-time element method but also in other numerical methods used in mechanics.
Below, we will test the derived formulae on a problem from dynamics.

3. Example - A Free Motion of Mass Concentrated in the Gravitational Field

We will follow the procedure in the modified space-time element method using the simplest example of mass m motion in a free-fall. In the space-time a concentrated mass is a one-dimensional object, referred to as the mass life line. As long as the mass is in rest, its life line is a straight line, parallel to the time axis. The life line of a mass moving under the influence of the force of gravity is a curve described by a differential equation

$$m\frac{d^2u}{dt^2} = mg \tag{15}$$

After dividing the mass life line into elements $2h$ long (measured in time units) and after introducing dimensionless quantities

$$\tau = \frac{t}{h}, \quad v = \frac{u}{gh^2} \tag{16}$$

equation (15) gets the form

$$\ddot{v} = 1, \quad where \quad \ddot{v} = \frac{d^2v}{d\tau^2} \tag{17}$$

We look for the integral of this equation in the area of element comprised between points *a (ante)* and *p (post)* (see figure 1) in the form of

$$v(\tau) = \begin{bmatrix} \mathbf{N}_a(\tau) & \mathbf{N}_p(\tau) \end{bmatrix} \begin{Bmatrix} \mathbf{C}_a \\ \mathbf{C}_p \end{Bmatrix} \tag{18}$$

where

$$\mathbf{N}_i(\tau) = \begin{bmatrix} \rho_i(\tau) & \omega_i(\tau) \end{bmatrix}, \quad \mathbf{C}_i = \begin{Bmatrix} v_i \\ \dot{v}_i \end{Bmatrix}, \quad i = a, p \tag{19}$$

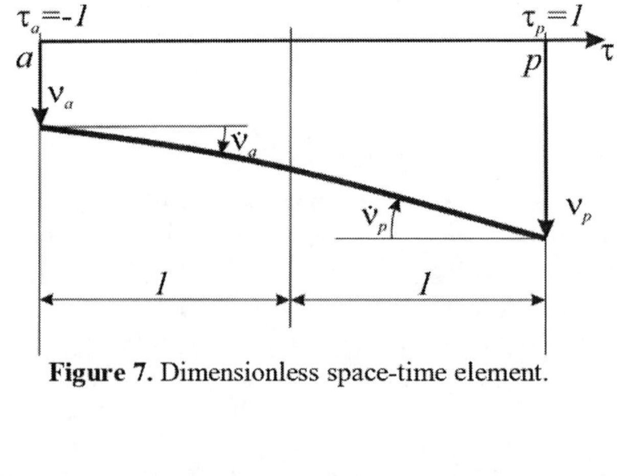

Figure 7. Dimensionless space-time element.

IOP Conf. Series: Materials Science and Engineering 1015 (2021) 012005 doi:10.1088/1757-899X/1015/1/012005

We assumed that order *II* functions are components of the shape matrix (cf. table 1). The application of order *I* shape function is impossible, because the substitution of a linear function to an inhomogeneous differential equation (17) would result in a contradiction.
With the symbols used here the components of matrices **K** and **P** (cf. formulae (11) and (13) have the following structure

$$\mathbf{K}_{ij} = \int_{-1}^{1} \mathbf{N}_i^T \ddot{\mathbf{N}}_j d\tau = \frac{1}{30}\begin{bmatrix} -18\tau_i\tau_j & 18\tau_i + 15\tau_j \\ 3\tau_j & -3-5\tau_i\tau_j \end{bmatrix}, \quad \mathbf{P}_i = \int_{-1}^{1} \mathbf{N}_i^T d\tau = \frac{1}{3}\begin{Bmatrix} 3 \\ -\tau_i \end{Bmatrix} \tag{20}$$

where $i, j = a, p$.

In particular it is

$$\mathbf{K}_{pp} = \frac{1}{30}\begin{bmatrix} -18 & 33 \\ 3 & -8 \end{bmatrix}, \quad \mathbf{K}_{pa} = \frac{1}{30}\begin{bmatrix} 18 & 3 \\ -3 & 2 \end{bmatrix}, \quad \mathbf{P}_p = \frac{1}{3}\begin{Bmatrix} 3 \\ -1 \end{Bmatrix} \tag{21}$$

Then we calculate successively

$$\mathbf{K}_{pp}^{-1} = -\frac{2}{3}\begin{bmatrix} 8 & 33 \\ 3 & 18 \end{bmatrix}, \quad -\mathbf{K}_{pp}^{-1}\mathbf{K}_{pa} = \begin{bmatrix} 1 & 2 \\ 0 & 1 \end{bmatrix}, \quad \mathbf{K}_{pp}^{-1}\mathbf{P}_p = \begin{Bmatrix} 2 \\ 2 \end{Bmatrix} \tag{22}$$

Hence we will find the next vectors of dimensionless displacements based on a recurrent formula (14)

$$\begin{Bmatrix} v_i \\ \dot{v}_i \end{Bmatrix} = \begin{bmatrix} 1 & 2 \\ 0 & 1 \end{bmatrix}\begin{Bmatrix} v_{i-1} \\ \dot{v}_{i-1} \end{Bmatrix} + \begin{Bmatrix} 2 \\ 2 \end{Bmatrix} = \begin{bmatrix} 1 & 2 & 2 \\ 0 & 1 & 2 \end{bmatrix}\begin{Bmatrix} v_{i-1} \\ \dot{v}_{i-1} \\ 1 \end{Bmatrix} \tag{23}$$

and - having returned to denominated quantities - we will obtain

$$\begin{Bmatrix} u_i \\ \dot{u}_i \end{Bmatrix} = \begin{bmatrix} gh^2 & 0 \\ 0 & gh \end{bmatrix}\begin{Bmatrix} v_i \\ \dot{v}_i \end{Bmatrix} + \begin{Bmatrix} 2gh^2 \\ 2gh \end{Bmatrix} = \begin{bmatrix} gh^2 & 0 \\ 0 & gh \end{bmatrix}\begin{bmatrix} 1 & 2 & 2 \\ 0 & 1 & 2 \end{bmatrix}\begin{Bmatrix} v_{i-1} \\ \dot{v}_{i-1} \\ 1 \end{Bmatrix} \tag{24}$$

Let us notice that - irrespective of the element dimension - we obtain exact results. If we assume that at moment $t = 0$ the displacement u_0 and the velocity of mass motion $\dot{u}_0$ are equal to zero, then at moments $t = 2h$ and $t = 4h$, based on formula (24), we will obtain

$$u_1 = 2gh^2 = \frac{g(2h)^2}{2}, \quad \dot{u}_1 = g(2h); \quad u_2 = 8gh^2 = \frac{g(4h)^2}{2}, \quad \dot{u}_2 = g(4h) \tag{25}$$

Also it is easy to check that not only at nodes but also at any point between nodes a and p the displacement:

$$u = gh^2\left[v_a\rho_{-1}(\tau) + \dot{v}_a\omega_{-1}(\tau) + v_p\rho_{-1}(\tau) + \dot{v}_p\omega_{-1}(\tau)\right] \tag{26}$$

exactly satisfies the differential equation of the problem.

4. Conclusions

The basic difference between the finite element method and the modification presented above consists in the fact, that the replacement of load with concentrated forces and moments (which are to satisfy equilibrium conditions at nodes) was given up, instead, a requirement was made that functions approximating the solution would satisfy differential equations in inter-node areas in a specified integral way. This leads to a system of algebraic equations, in which kinematic parameters existing in the solution approximating functions are the unknowns. Displacements and their derivatives (of any order) at nodes are those parameters. The modified space-time element method was tested on a few experimental numerical examples - in this article, we present one simple example.

References

[1] Bajer C 1986 Triangular and tetrahedral space-time fnite elements in vibration analysis, *Int. Jou. for Num. Methods in Eng.* **23** (11) pp 2031-2048

[2] Bajer C 2009 *The space-time fnite elements method in dynamic structures calculation* (Warszawa: IPPT PAN) (in Polish)

[3] Bajer C 1997 Numerical space-time modeling of dynamic contact problem. *Report IPPT* **5** Warszawa (in Polish)

[4] Bathe K J and Wilson E 1976 *Numerical methods in fnite element analysis* (New Jersey: Prentice-Hall)

[5] Kacprzyk Z 1981 Vibration analysis of a chimney under seismic load *Arch. Inż. Ląd.* **27** pp 507-516 (in Polish)

[6] Kacprzyk Z 1984 On the application of weighting functions in the space-time fnite element method *Zeszyty Naukowe Politechniki Warszawskiej s. Budownictwo* **85** pp 83-95 (in Polish)

[7] Kacprzyk Z 2016 A Stationary Formulation of the Space-time Finite Element Method in: *Proc. Engineering* **153** pp 248-255

[8] Kączkowski Z 1975 The method of fnite space-time elements in dynamics of structures *J. Tech. Phys.* **16** (1) pp 69-84.

[9] Kączkowski Z 1976 The method of space-time fnite elements *Arch. Inż. Ląd.* **22** (3) pp 365-378 (in Polish)

[10] Kączkowski Z 1979 General formulation of the stifness matrix for the space-time fnite elements *Arch. Inż. Ląd.* **25** (3) pp 351-357

[11] Kączkowski Z 1983 On applications of non-rectangular space-time elements *Journal of Theoretical and Applied Mechanics* **21** (4) pp 531-542 (in Polish)

[12] Kączkowski Z 1986 Unconstrained systems of equations in the method of space-time fnite elements *Arch. Inż. Ląd.* **32** pp 39-50 (in Polish).

[13] Kączkowski Z 1998 *Space-time element metod in: Handbook of Computational Solid Mechanics,* Kleiber M (ed.) (Berlin Heidelberg: Springer Verlag) p 149-199

[14] Pelc J 1984 Non-linear shape functions in the space-time element method *Arch. Inż. Ląd.* **30** (1) p 53-63 (in Polish).

[15] Podhorecki A 1986 The viscoelastic space-time element *Computers & Structures* **23** (4) p 535-544

[16] Witkowski M 1983 *On the time space in structural dynamics* Prace Nauk PW Budownictwo **80** (in Polish)

[17] Zienkiewicz O C Taylor R L and Zhu J Z 2013 *The Finite Element Method: Its Basis and Fundamentals*, seven edition (Elsevier)

[18] Zienkiewicz O C, Taylor R L and Fox D D 2013 *The Finite Element Method for Solid and Structural Mechanics* seven edition (Elsevier)

IOP Conf. Series: Materials Science and Engineering 1015 (2021) 012006 doi:10.1088/1757-899X/1015/1/012006

Analysis of a reinforced concrete dome

P Czumaj, S Dudziak and Z Kacprzyk

The Faculty of Civil Engineering, Warsaw University of Technology, al. Armii Ludowej 16, 00-637 Warsaw, Poland

E-mail address: p.czumaj@il.pw.edu.pl

Abstract. FEM models of axi-symmetrical reinforced concrete dome with two rings have been analysed. Different complexity level of computational models (2D and 3D), geometry simplifications and FEM codes (Abaqus, FEAS, ARSAP) have been compared. Assessment of building structure deflections has been performed with several approaches, which gave opportunity to confront them and estimate mistakes of most commonly used models.

1. Introduction

Nowadays, static analysis of building structures is most often performed with the Finite Element Method (FEM) and 3D models consisting of shells and bars [1]. During phase of geometry preparation for such models, structural designers have to make many decisions concerning e.g. relative position of neutral axis and mid-surfaces (resulting in eccentricities). It is obvious that every designer adopts various simplifications [2–4]. Hence, modern design codes take into account the uncertainty of models in safety factors [5]. However, some simplifications can lead to major mistakes [6], therefore, in case of complex systems, it is worthy to perform validation and verification of models before final static analysis.

Axial symmetry of geometry is rarely used in static analyses due to unsymmetrical variable actions (especially wind or thermal actions). On the other hand, axi-symmetry allows significant reduction of model variables, which means, that it is possible to prepare 2D models based on the theory of elasticity, which ensure almost exact results for symmetrical loads [7]. Such models can be used in the verification and validation procedure for more complex 3D models consisting of shells and bars.

The main objective for this paper is to compare models of different level of complexity, concerning axial-symmetrical dome with two circular flanges, shown in figure 1.

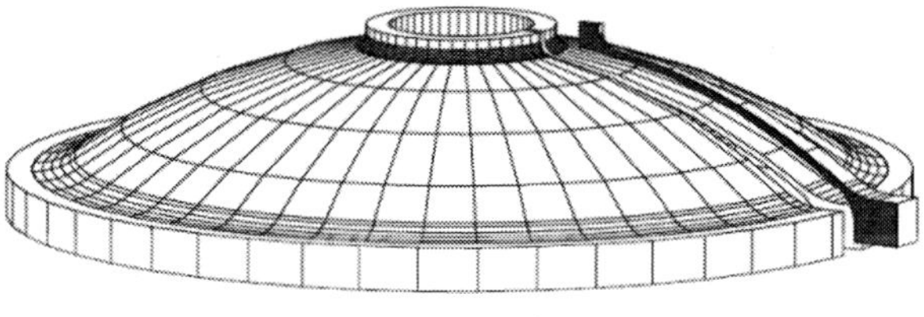

Figure 1. Concrete dome with two circular flanges.

Three groups of models have been analyzed:

- **P1**: 2D axial-symmetrical problem of theory of elasticity (this model will be treated as a reference for further 3D models),
- **P2**: 3D task with axial-symmetrical cylindrical finite elements,

- **P3**: 3D models consisting of bar and shell type of finite elements.

Each group contained at least two models created with different FEM codes:

- Abaqus FEA (Abaqus) [8]
- Finite Element Analysis System (FEAS) [9]
- Autodesk Robot Structural Analysis Professional 2020 (ARSAP) [10]

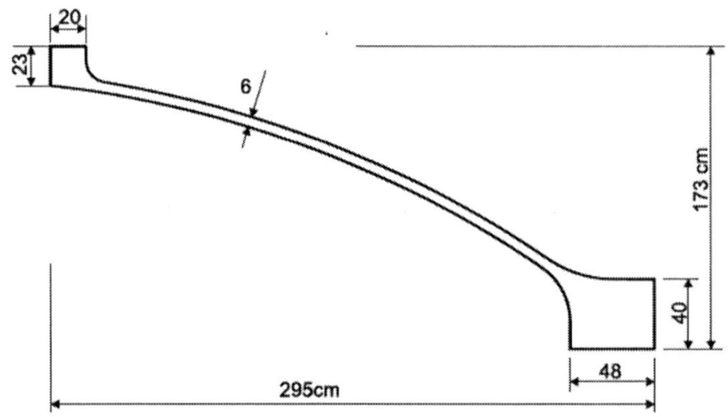

Figure 2. Dimensions of the dome.

Each software was operated by a different person, so personal influences in modelling process were also included.

2. Geometry and material parameters

Geometry of the dome has been shown in figure 2. However, it is not always possible to model accurate geometry. Important issue of modelling structures is approximation of geometry. In first group of models, it could have been almost accurate, while in further models, volumetric structural components have been represented with bar and shell finite elements. On the left side of figure 3, exemplary approaches to geometry simplifications have been presented respectively: in the first model, mid-surface of shell has been assumed on the top surface of the dome and its span has been extended to centroids of both flanges. In the second model, surface in the middle of dome thickness has been taken as shell mid-surface, while its span has been assumed in the same way as in the first model. The third model takes mid surface in same way as the second one, though the span is cut by flange faces.

2.1. Material parameters

Mechanical properties for concrete have been adopted in the analysis. Parameters have been taken, as follows:

- Young modulus $E = 27027$ MPa,
- Poisson ration $\nu = 1/6$,
- unit weight $\rho = 27{,}5$ kN/m^3.

2.2. Loads

Two load cases has been analysed: dead load of the structure and linear force of 500 kN/m over top ring perimeter. Since top flange radius equals 0,751 m, total force applied is 2359,34 kN.

2.3. Boundary conditions

In this example only bottom flange has been supported in central point of its bottom surface. One degree of freedom has been taken: vertical translation in one node, or more precisely: over circuit (see figure 4)

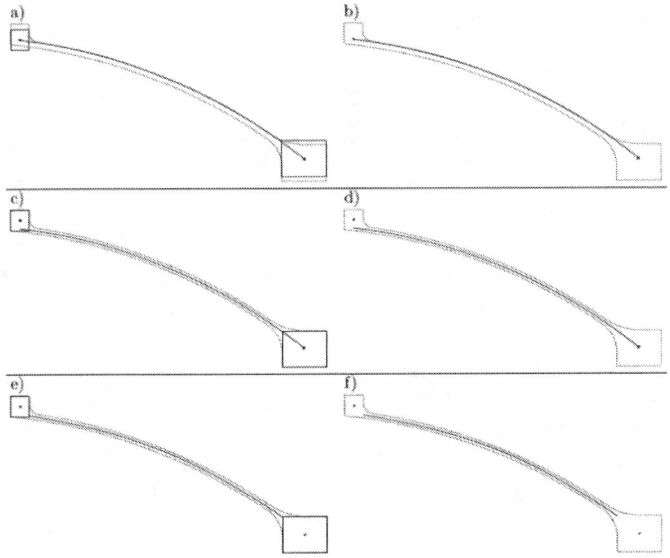

Figure 3. Dome models: a) top face adjust, b) computational model, c) middle surface adjustment, d) computational model, e) middle surface adjustment and cut with bar sections, f) computational model.

3. Analysis

3.1. Model p1. Two-dimensional problem of theory of elasticity: axial symmetry

2D Calculations have been performed with all aforementioned three FEM codes. With each, at least two different size of finite elements have been checked, to ensure displacement convergence. In FEAS and ARSAP models, triangle finite elements with three nodes have been used, whereas triangle elements with six nodes (AX6) have been adopted in Abaqus.

Figures 4÷6 depicts final finite elements meshes for each software.

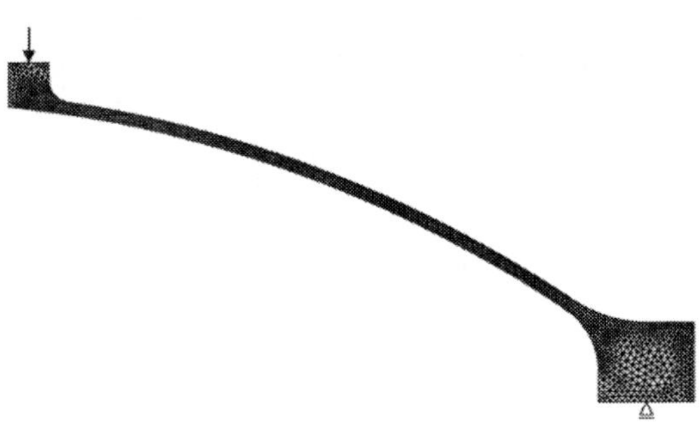

Figure 4. FEAS, 1482 nodes, 2486 elements.

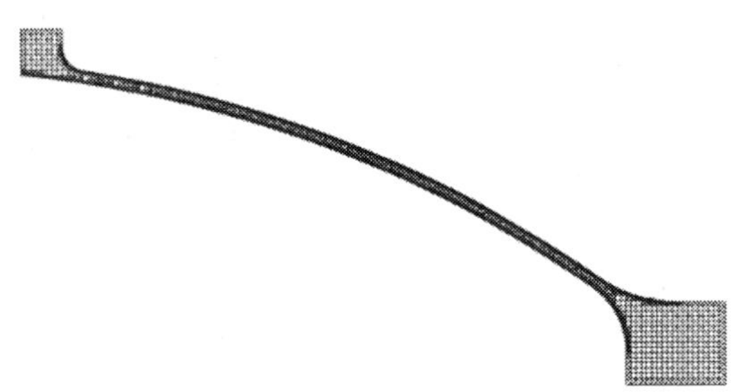

Figure 5. ARSAP, 3345 nodes, 4974 elements.

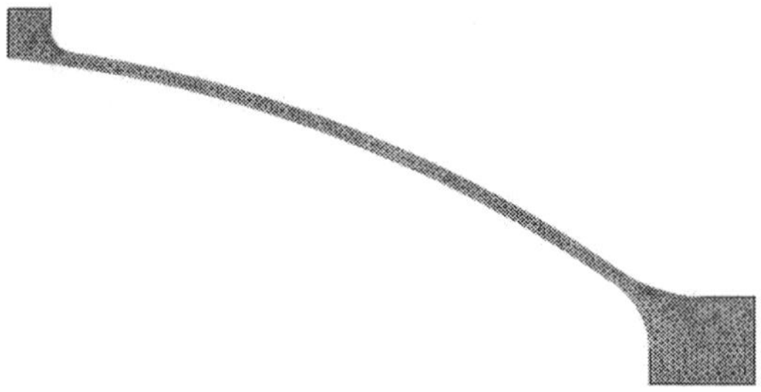

Figure 6. Abaqus, 3642 nodes, 1571 elements.

In ARSAP, finite elements generator for curvilinear geometry structures is bound with geometry definition. High accuracy of geometry imposes large number of elements (see figure 5), although high mesh density is not always desired. In this example, the amount of elements is rather too high.

3.2. Model p2. 3D task with axi-symmetrical cylindrical finite elements

In this point, the dome has been modelled with rarely used cylindrical finite elements, which are not implemented in ARSAP, nonetheless are available in FEAS and Abaqus. These elements can be considered as one-dimensional as was shown in figure 7 (mid-surface is parametrised by one natural coordinate s). Such elements are quite similar to beam elements since each node has three degrees of freedom – two translations and one rotation angle.

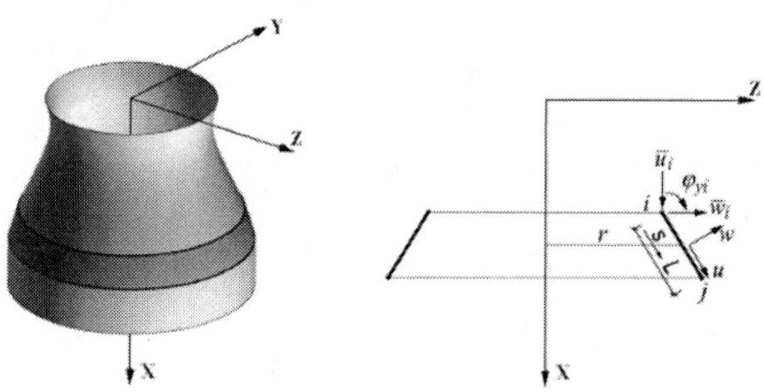

Figure 7. Cylindrical finite element.

3.2.1. Abaqus

With Abaqus, 17 three-node shell elements (SAX2), which gives total of 35 nodes, have been used. Two-parameter springs, with rotation (k_φ) and translation (k_u) stiffness as in equations (1) and (2), have substituted flanges. Both stiffness were acquired from [11], though eccentricities between shell and rings have been neglected.

$$k_u = \frac{EA}{R^2} \tag{1}$$

$$k_\varphi = \frac{EJ}{R^2} \tag{2}$$

Horizontal translation and rotation described with (1) have been added to boundary conditions from previous models. with parameters from equations (1) and (2).

3.2.2. FEAS

Model created with FEAS consists of 120 nodes, 119 two-node shell elements and 2 single-node cylindrical elements for top and bottom flange. It is worth pointing out that initial mesh studies revealed that 59 elements, 60 nodes and 2 do not guarantee correct results.

3.3. Model p3. 3D models consisting of bar and shell type of finite elements

This group contains models from each of three presented FEM codes. Table 1 gathers all models and its main properties.

Table 1. Analysed computational models.

no	shell element	bar element	flange concentricity	FEM code
P3-1	8-nodal curvilinear	3-nodal curvilinear	no	Abaqus
P3-2	8-nodal curvilinear	3-nodal curvilinear	yes	Abaqus
P3-3	3-nodal plain	2-nodal	yes	FEAS
P3-4	3-nodal plain	2-nodal	yes	ARSAP

Figure 8. Abaqus, computational model P3-1, 3312 nodes, 1008 shell elements, 64 bar elements.

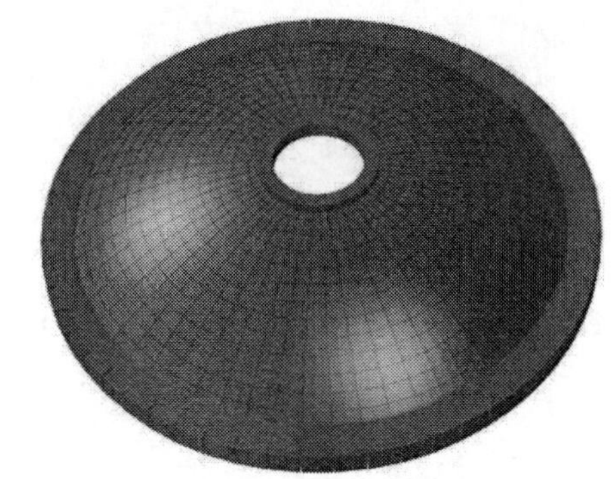

Figure 9. Abaqus, computational model P3-2, 2900 nodes, 812 shell elements, 116 bar elements.

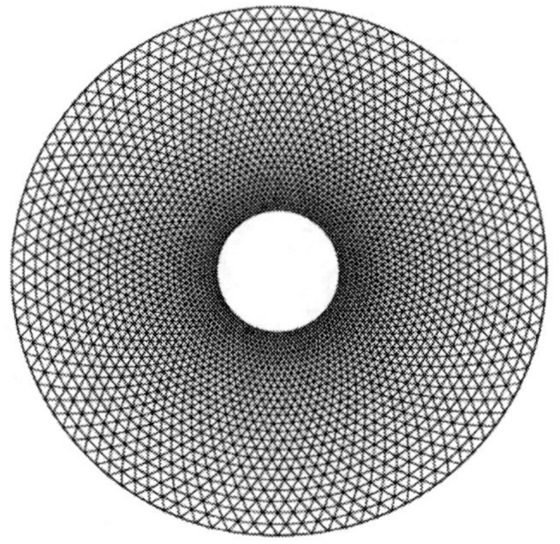

Figure 10. FEAS, computational model P3-3 (plan), 2800 nodes, 5400 shell elements, 200 bar elements

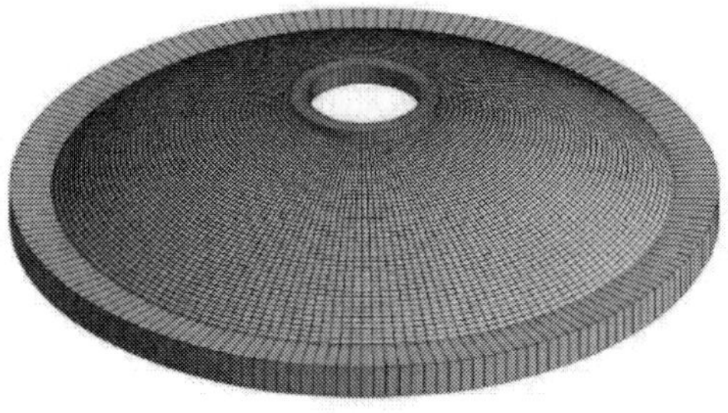

Figure 11. ARSAP, computational model P3-4, 9190 nodes, 3836 shell elements, 360 bar elements

IOP Conf. Series: Materials Science and Engineering 1015 (2021) 012006 doi:10.1088/1757-899X/1015/1/012006

Two Abaqus models with same types of elements (S8R and B32) have been created, as shown in figures 8 and 9. They differed with approach to flange position: model P3-1 represents exact geometry, therefore flange concentricity has been preserved. Model P3-2 neglected eccentricity of flanges, according to figure 3f.

In FEAS (figure 10) and ARSAP (figure 11) models, similar finite elements have been used. In the FEAS model, finite elements mesh has been obtained with use of geometry primitives generator. It allows obtaining regular elements, according to requested parameters.

4. Results summary

In tables 2 and 3, vertical displacements of top ring over all models have been presented for two analysed load cases. Ratios between results obtained from model P1 (considered as reference – the most accurate model) and other models were shown in brackets. For comparable models which were analysed in all three FEM codes mean, standard deviation and coefficient of variation (COV) were calculated (see tables 4 and 5).

Table 2. Vertical displacement of the top ring – dead weight [$m \cdot 10^{-5}$].

	P1	P2	P3
Abaqus	5,05	5,75 (1,14)	5,39 (1,07) 5,75 (1,14)
FEAS	5,34	5,37 (1,01)	5,43 (1,02)
ARSAP	5,32	-	5,64 (1,06)

Table 3. Vertical displacement of the top ring – line load [cm].

	P1	P2	P3
Abaqus	0,968	1,27 (1,31)	1,06 (1,06) 1,27 (1,31)
FEAS	0,948	1,01 (1,07)	1,15 (1,21)
ARSAP	0,953	-	1,30 (1,36)

Table 4. Vertical displacement of the top ring – dead weight – comparison of results [$m \cdot 10^{-5}$].

	P1	P3
Abaqus	5,05	5,75
FEAS	5,34	5,43
ARSAP	5,32	5,64
mean	5,24	5,61
standard deviation	0,16	0,16
COV	3,09%	2,90%

IOP Conf. Series: Materials Science and Engineering 1015 (2021) 012006 doi:10.1088/1757-899X/1015/1/012006

Table 5. Vertical displacement of the top ring – line load – comparison of results [cm].

	P1	P3
Abaqus	0,968	1,27
FEAS	0,948	1,15
ARSAP	0,953	1,30
mean	0,956	1,24
standard deviation	0,010	0,08
COV	1,09%	6,40%

Sample deformations of models loaded with dead weight are presented in figures 12÷15. Figures 16÷20 shows deformations of models loaded with linear load over top ring.

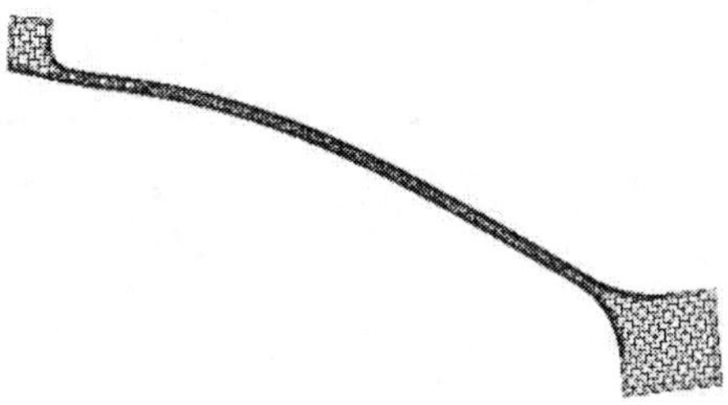

Figure 12. P1 ARSAP, dead weight.

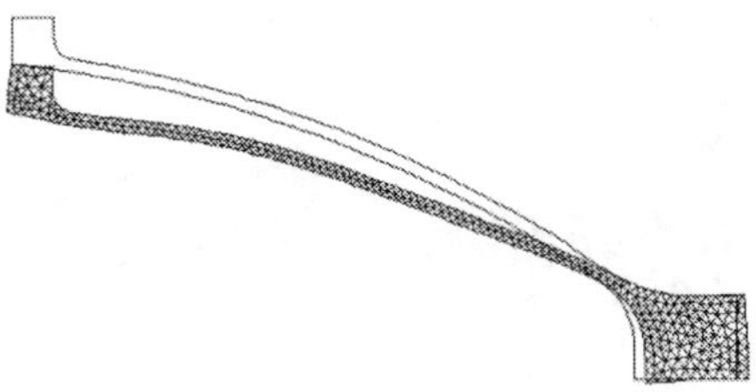

Figure 13. P1 FEAS, dead weight.

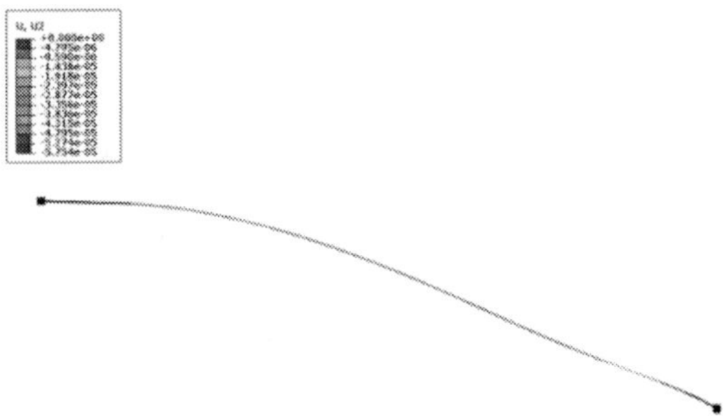

Figure 14. P2 Abaqus, dead weight.

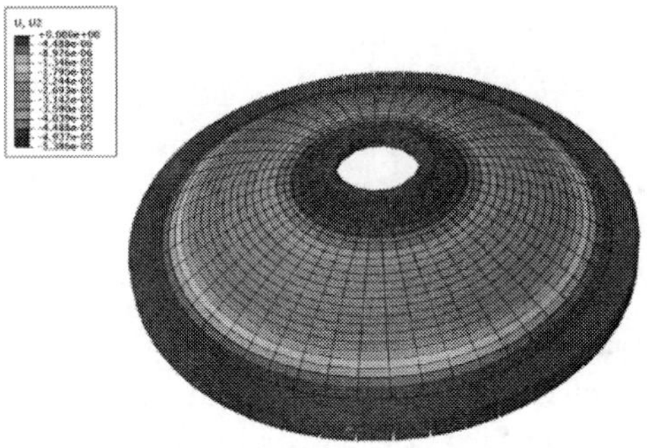

Figure 15. P3 Abaqus, dead weight.

Figure 16. P1 ARSAP, line load.

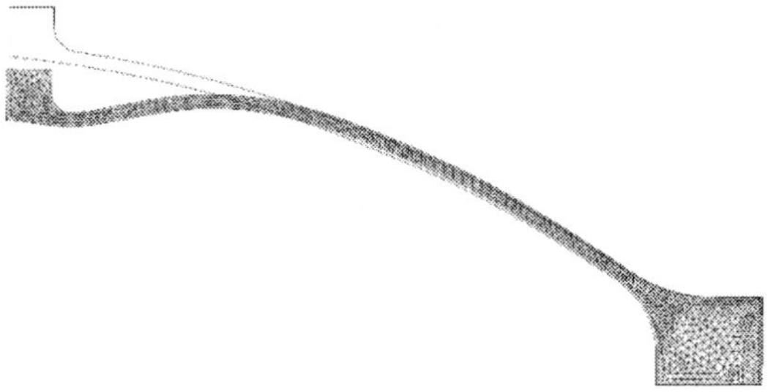

Figure 17. P1 FEAS, line load.

IOP Conf. Series: Materials Science and Engineering 1015 (2021) 012006 doi:10.1088/1757-899X/1015/1/012006

Figure 18. P2 Abaqus, line load.

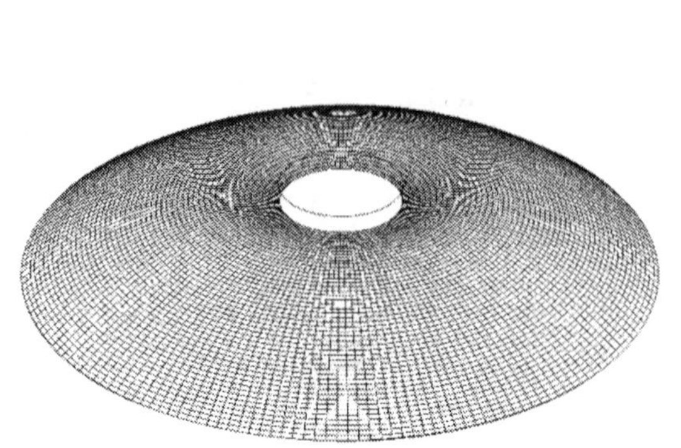

Figure 19. P3 ARSAP, line load.

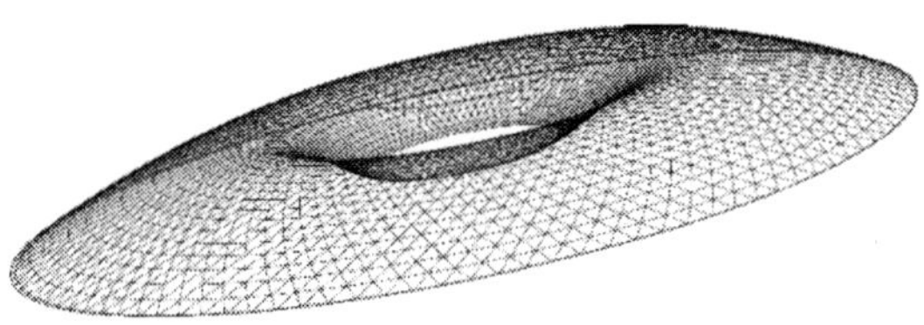

Figure 20. P3 FEAS, line load.

5. Result discussion and conclusions

Differences between reference P1 model and models consisting of structural elements – P2 and P3 are significant (up to 30%). On the other hand, deflections predicted by these models are overestimated, therefore from engineering point of view, they are "on the safe side". However, using models which overestimate deflections so significantly, can lead to uneconomical design. In P3 model family, the best agreement between accurate model and 3D model, was obtained for Abaqus model taking into account the eccentrics between the shell and the rings (7% for dead weight and 6% for line load). For programs analyzed separately the smallest differences for different class of models occurs for FEAS (1-2% for dead weight and to 21% for line load).

Dispersion between results obtained with models prepared in different programs by different users are smaller than for models of different classes. The smaller COV value (which is measure of result relative dispersion) was obtained in case of P1 model for line load, while the greatest was in case of P3 model and line load as well. The small dispersion of results from different software, indicate that users have not made major mistakes.

The paper presents the verification procedure of the concrete dome with two rings. The dome was modelled with different class models (2D and 3D) in three different FEM codes. The 2D axi-symmetrical models were used to verify the complex 3D shell-bar model, which later could be applicate to the static analysis covering unsymmetrical load cases also.

References

[1] Perelmuter A V and Slivker V I 2003 *Numerical Structural Analysis: Methods, Models and Pitfalls* (Berlin Heidelberg: Springer)

[2] Zobel H, Zbiciak A, Oleszek R, Michalczyk R and Mossakowski P 2014 Numeryczna identyfikacja cech dynamicznych stalowo-betonowego mostu kolejowego Roads Bridg. -

Drog. i Most. **13** pp 275–301
[3] Czumaj P, Dudziak S and Kacprzyk Z 2018 Computational models of reinforced concrete ribbed floor *MATEC Web Conf.* **196** pp 1–8
[4] Kacprzyk Z and Czumaj P 2019 Modelling of Multi-Storey Frame Interacting with Rigid Core of the Building *IOP Conf. Ser. Mater. Sci. Eng.* **661** pp 0–8
[5] EN 1990 Eurocode - Basis of structural desing
[6] Sidorov V N and Nowak K 2018 Numerical investigation of the long-term work of arches under material creep conditions *IOP Conf. Ser. Mater. Sci. Eng.* **456**
[7] Lewiński P M and Dudziak S 2018 Nonlinear interaction analysis of RC cylindrical tank with subsoil by adopting two kinds of constitutive models for ground and structure AIP *Conference Proceedings* **1922** (AIP Publishing LLC) p 130007
[8] Abaqus Theory Manual (6.12)
[9] Kacprzyk Z and Postek E 1993 System FEAS Isoparametric Finite Element Library *Comput. Methods Civ.* Eng. **3** pp 89–108
[10] Robot Structural Analysis Professional 2020 - manual
[11] Mazurkiewicz Z 2004 *Cienkie powłoki sprężyste: teoria liniowa* (Warszawa: Oficyna Wydawnicza Politechniki Warszawskiej)

IOP Conf. Series: Materials Science and Engineering 1015 (2021) 012007 doi:10.1088/1757-899X/1015/1/012007

Dynamic and static response of a slightly compressible hyperelastic solid

A Franus and A Marek

Faculty of Civil Engineering, Warsaw University of Technology, Armii Ludowej 16, 00-637 Warsaw, Poland

E-mail address: aleksander.franus.dokt@pw.edu.pl

Abstract. The paper presents dynamic and static response of a slightly compressible hyperelastic solid. Material models under consideration are obtained by extending incompressible material models which are consistent second and third order approximation of existing stored energy function in terms of a deformation measure, namely MCMV and MCIZ. The problems of a uniaxial compression of a rod and a twisting column are solved using ABAQUS which is a powerful finite element program designed for general use in nonlinear problems. The models are defined with a help of user subroutine UHYPER. In the case of including dynamic effects, an implicit integration scheme available in the software is used. For the twisting column problem, results are compared against available in the ABAQUS library polynomial model where the volumetric stored energy function does not meet basic growth conditions. Experimental data on synthetic rubber neoprene published in the literature is utilized.

1. Introduction

The aim of this paper is to present dynamic and static response of slightly compressible hyperelastic material models [1]. Discussed models are obtained by extending incompressible material models which are consistent second and third order approximations of existing stored energy function in terms of a deformation measure. These are widely applied to model rubber-like materials, e.g. elastomers which are characterized by the volumetric compressibility modulus, couple orders of magnitude larger than the shear modulus, $K_0 >> \mu_0$ [2]. Therefore, the stored energy function (SEF) is assumed to be sum of two functions: the function of the isochoric deformation $\bar{W}$ and the volumetric one W_{vol} [3]. It means that volumetric and isochoric parts of the energy are decoupled. Such approach leads to a clear physical interpretation of basic experimental tests when determining material parameters.

We focus on material models, namely MCMV and MCIZ, that were proposed in [4,5]. Example initial-value problems, which concern a simple compression of a rod and a twisting column, are solved using ABAQUS/Standard which is a finite element program designed for general use in nonlinear problems [6,7]. In the case of including dynamic effects, we use an implicit integration scheme available in the software. Since the models are not available in ABAQUS library, we utilize a user subroutine UHYPER which is suitable for definition of isotropic material model. Moreover, in the second example we compare results against available in the library polynomial model where the volumetric stored energy function does not meet basic growth conditions [8]. We use experimental data on synthetic rubber neoprene published in the literature [9,10].

2. Basic equations of hyperelasticty

A description of finite deformations in the framework of continuum mechanics requires to distinct configurations of a deformed and undeformed body. As a consequence, different tensors appear in equilibrium equations, i.e. the first Piola-Kirchhoff stress tensor $\mathbf{S}$ (unsymmetric) and the Cauchy stress tensor $\boldsymbol{\sigma}$ (symmetric) in the reference configuration and in the spatial one, respectively. The equations of elastodynamics, which may be written in the Lagrangian (reference) and Eulerian (spatial) coordinates, with absence of body forces have the form

$$\rho_0 \frac{\partial^2 \mathbf{u}}{\partial t^2} = \operatorname{Div}\mathbf{S}, \quad \rho\ddot{\mathbf{u}} = \operatorname{div}\boldsymbol{\sigma} \tag{1}$$

where ρ_0, ρ are densities in the reference and actual configurations, $\mathbf{u} = \mathbf{u}(t, \mathbf{X})$ is the displacement vector. 'Div' and 'div' denote the divergence operator in the Lagrangian and Eulerian coordinates, respectively [8]. A deformation of a body is described by so called deformation gradient given by

$$\mathbf{F} = \frac{\partial \chi(\mathbf{X},t)}{\partial \mathbf{X}}, \quad \dot{\mathbf{F}} = \mathbf{L}\mathbf{F}, \quad \mathbf{L} = \frac{\partial \mathbf{v}}{\partial \mathbf{x}} = \frac{\partial \dot{\mathbf{u}}}{\partial \mathbf{x}} \tag{2}$$

with the relation between the Eulerian and the Lagrangian coordinates systems $\mathbf{u}(t,\mathbf{X}) = \mathbf{x} - \mathbf{X}$. The velocity gradient tensor is denoted by $\mathbf{L}$. Since $J = \det\mathbf{F} > 0$, the mapping $\mathbf{x} = \chi(\mathbf{X},t)$ is orientation-preserving and locally invertible. The polar decomposition states $\mathbf{F} = \mathbf{R}\mathbf{U} = \mathbf{V}\mathbf{R}$, which means that any $\mathbf{F}$ can be multiplicatively decomposed into a rotation tensor $\mathbf{R} \in SO(3)$ and $\mathbf{U} = \sqrt{\mathbf{F}^T\mathbf{F}}$, $\mathbf{V} = \sqrt{\mathbf{F}\mathbf{F}^T}$ the stretch tensors. Assumption of existence of a stored energy function $W = W(\mathbf{F})$ such that

$$\mathbf{S} = \frac{\partial W(\mathbf{F})}{\partial \mathbf{F}} \tag{3}$$

leads to the hyperelastic material model. The energy is defined per unit volume in the reference configuration. In order to avoid interpenetration of matter, it is necessary to put some restrictions on the stored energy function

$$W(\mathbf{F}) \to +\infty \quad \text{as } J \to 0^+, \quad W(\mathbf{F}) \to +\infty \quad \text{as } \|\mathbf{F}\| \to +\infty \tag{4}$$

Here, we also assume that stored-energy function is isotropic, i.e.

$$W(\mathbf{F}) = \breve{W}(\mathbf{C}) = \breve{W}(\mathbf{B}) = \tilde{W}(I_1, I_2, I_3), \tag{5}$$

where $I_1 = \operatorname{tr}\mathbf{C} = \operatorname{tr}\mathbf{B}$, $I_2 = \operatorname{tr}\operatorname{cof}\mathbf{C} = \operatorname{tr}\operatorname{cof}\mathbf{B}$ and $I_3 = J^2$ are the invariants of each of the left and right Cauchy-Green deformation tensors $\mathbf{C} = \mathbf{F}^T\mathbf{F}$ and $\mathbf{B} = \mathbf{F}\mathbf{F}^T$. There exists a solution for a wide class of boundary values problems associated with hyperelastic model when W has the form $W(\mathbf{F}) = \mathcal{W}(\mathbf{F}, \operatorname{cof}\mathbf{F}, \det\mathbf{F})$ for some convex $\mathcal{W}$. Such functions are said to be polyconvex [11]. We emphasize that $\mathbf{F}, \operatorname{cof}\mathbf{F}, \det\mathbf{F}$ describe deformation of fibres, surface and volume of a body, respectively. Since $\mathbf{S}$ is non-symmetric, it is convenient to introduce the second Piola-Kirchhoff stress tensor

$$\begin{gathered}\mathbf{T} = \left.\frac{\partial \hat{W}(\mathbf{E})}{\partial \mathbf{E}}\right|_{\mathbf{E}=\mathbf{E}^T} = 2\left.\frac{\partial \breve{W}(\mathbf{C})}{\partial \mathbf{C}}\right|_{\mathbf{C}=\mathbf{C}^T} = 2\left(\gamma_1\mathbf{I} + \gamma_2\mathbf{C} + \gamma_3\mathbf{C}^2\right) \\ \gamma_1 = \frac{\partial \breve{W}}{\partial I_1} + \frac{\partial \breve{W}}{\partial I_2} I_1 + \frac{\partial \breve{W}}{\partial I_3} I_2, \quad \gamma_2 = -\left(\frac{\partial \breve{W}}{\partial I_2} + \frac{\partial \breve{W}}{\partial I_3} I_1\right), \quad \gamma_3 = \frac{\partial \breve{W}}{\partial I_3}\end{gathered} \tag{6}$$

where $\mathbf{E}=(\mathbf{C}-\mathbf{I})/2$ denotes the Lagrangian strain tensor. The relationship between the Cauchy stress tensor $\boldsymbol{\sigma}$, the first and second Piola-Kirchhoff ones, namely $J\boldsymbol{\sigma}=\mathbf{S}\mathbf{F}^T=\mathbf{F}\mathbf{T}\mathbf{F}^T$, implies a constitutive relation in the spatial description such that

$$\begin{gathered} J\boldsymbol{\sigma}=2\mathbf{B}\left.\frac{\partial \breve{W}(\mathbf{B})}{\partial \mathbf{B}}\right|_{\mathbf{B}=\mathbf{B}^T}=2\left(\beta_1\mathbf{I}+\beta_2\mathbf{B}+\beta_3\mathbf{B}^2\right) \\ \beta_1=\frac{\partial \breve{W}}{\partial I_3}I_3,\quad \beta_2=\frac{\partial \breve{W}}{\partial I_1}+\frac{\partial \breve{W}}{\partial I_2}I_1,\quad \beta_3=-\frac{\partial \breve{W}}{\partial I_3} \end{gathered} \tag{7}$$

From the principle of conservation of mechanical energy and the principle of conservation of mass the following work conjugations arise

$$\frac{1}{J}\dot{W}=\boldsymbol{\sigma}\cdot\mathbf{D}=\frac{1}{J}\mathbf{S}\cdot\dot{\mathbf{F}}=\frac{1}{2J}\mathbf{T}\cdot\dot{\mathbf{C}}=\frac{1}{J}\mathbf{T}\cdot\dot{\mathbf{E}} \tag{8}$$

A symmetric part of the velocity gradient tensor is denoted by $\mathbf{D}$. It should be noted that $\dot{\mathbf{C}}=2\mathbf{F}^T\mathbf{D}\mathbf{F}=2\dot{\mathbf{E}}$.

3. Slightly compressible hyperelastic material models

3.1. *Constitutive relations of slightly compressible materials*

Models of isotropic slightly compressible materials are a simple generalisation of models describing incompressible materials. In the case we assume that no coupling between the stored energy function of isochoric $\bar{W}$ and volumetric W_{vol} deformations occurs [1]. Therefore, it is convenient to express stored energy function in the form

$$\tilde{W}(\bar{I}_1,\bar{I}_2,J)=\bar{W}(\bar{I}_1,\bar{I}_2)+W_{vol}(J) \tag{9}$$

where the modified invariants are given by $\bar{I}_1=\mathrm{tr}\bar{\mathbf{B}}$, $\bar{I}_2=\mathrm{tr}\bar{\mathbf{B}}^{-1}$ and $\bar{\mathbf{B}}=\bar{\mathbf{F}}\bar{\mathbf{F}}^T$, $\bar{\mathbf{F}}=J^{-1/3}\mathbf{F}$. These can be explicitly written as

$$\bar{I}_1=J^{-\frac{2}{3}}I_1,\quad \bar{I}_2=J^{-\frac{4}{3}}I_2 \tag{10}$$

Due to the form of SEF (9), the constitutive relationship of a slightly compressible material in the current configuration yields

$$\boldsymbol{\sigma}=\frac{\partial W_{vol}}{\partial J}\mathbf{I}+\frac{2}{J}\left(\frac{\partial \bar{W}}{\partial \bar{I}_1}\bar{\mathbf{B}}_D-\frac{\partial \bar{W}}{\partial \bar{I}_2}\bar{\mathbf{B}}_D^{-1}\right) \tag{11}$$

Tensor $\bar{\mathbf{B}}_D$ stands for deviatoric part of the modified left Cauchy-Green tensor $\bar{\mathbf{B}}$ as follows

$$\bar{\mathbf{B}}_D=\bar{\mathbf{B}}-\frac{1}{3}\bar{I}_1\mathbf{I},\quad \bar{\mathbf{B}}_D^{-1}=\bar{\mathbf{B}}^{-1}-\frac{1}{3}\bar{I}_2\mathbf{I} \tag{12}$$

3.2. *The MCMV material model*

In the article we consider so called MCMV model [4] which isochoric part is defined by the stored energy function of the form

$$\breve{W}\left(\bar{\mathbf{C}}\right)=b+c\,\mathrm{tr}\bar{\mathbf{C}}+d_1\left(\mathrm{tr}\bar{\mathbf{C}}\right)^2+d_2\,\mathrm{tr}\bar{\mathbf{C}}^2+e_1\left(\mathrm{tr}\bar{\mathbf{C}}\right)^3+e_2\,\mathrm{tr}\bar{\mathbf{C}}\,\mathrm{tr}\bar{\mathbf{C}}^2 \tag{13}$$

The function is a consistent third order approximation in terms of deformation tensor $\bar{\mathbf{C}}$ norm such that

$$\left|W\left(\bar{I}_1,\bar{I}_2\right)-\bar{W}\left(\bar{I}_1,\bar{I}_2\right)\right|=\mathcal{O}\left(\left\|\bar{\mathbf{C}}\right\|_2^4\right) \tag{14}$$

Assuming the existence of natural state

$$\breve{W}\left(\mathbf{I}\right)=b+3c+9d_1+9d_2+27e_1+27e_2=0 \tag{15}$$

we eliminate one of the constants obtaining

$$\begin{gathered}\breve{W}\left(\bar{\mathbf{C}}\right)=c\left(\mathrm{tr}\bar{\mathbf{C}}-3\right)+d_1\left[\left(\mathrm{tr}\bar{\mathbf{C}}\right)^2-9\right]+d_2\left(\mathrm{tr}\bar{\mathbf{C}}^2-9\right)+\\+e_1\left(\left(\mathrm{tr}\bar{\mathbf{C}}\right)^3-27\right)+e_2\left(\mathrm{tr}\bar{\mathbf{C}}\,\mathrm{tr}\bar{\mathbf{C}}^2-27\right)\end{gathered} \tag{16}$$

Substituting definitions of the invariants into (16) yields five parameters model

$$\begin{gathered}\breve{W}\left(\bar{\mathbf{C}}\right)=\bar{W}\left(\bar{I}_1,\bar{I}_2\right)=\\\frac{1}{2}\left[a_1\left(\bar{I}_1-3\right)+\frac{1}{2}a_2\left(\bar{I}_1^2-9\right)+\frac{1}{3}a_3\left(\bar{I}_1^3-27\right)+a_4\left(\bar{I}_2-3\right)+a_5\left(\bar{I}_1\bar{I}_2-9\right)\right]\end{gathered} \tag{17}$$

Similarly, as a special case of (17) we obtain isochoric part of SEF of the Ishihara-Zahorski (MCIZ) model [5]

$$\bar{W}\left(\bar{I}_1,\bar{I}_2\right)=\frac{1}{2}\left[a_1\left(\bar{I}_1-3\right)+\frac{1}{2}a_2\left(\bar{I}_1^2-9\right)+a_4\left(\bar{I}_2-3\right)\right] \tag{18}$$

which states the second order approximation.

The part of SEF describing volumetric changes should meet certain mathematical conditions. Firstly, we assume that $W_{vol}\left(J\right)$ is differentiable with respect to J and $W_{vol}\left(1\right)=0$. Based on assumption of existence of natural state and convexity of the function we have

$$\left.\frac{\partial W_{vol}\left(J\right)}{\partial J}\right|_{J=1}=0,\quad \frac{\partial^2 W_{vol}\left(J\right)}{\partial J^2}\geq 0 \tag{19}$$

To ensure proper growth, the conditions should hold

$$\begin{gathered}W_{vol}\left(J\to 0\right)=+\infty,\quad \left.\frac{\partial W_{vol}\left(J\right)}{\partial J}\right|_{J\to 0}=-\infty\\W_{vol}\left(J\to +\infty\right)=+\infty,\quad \left.\frac{\partial W_{vol}\left(J\right)}{\partial J}\right|_{J\to +\infty}=+\infty\end{gathered} \tag{20}$$

For such models the initial bulk modulus is given by

$$\mathrm{K}_0=\left.\frac{\partial^2 W_{vol}}{\partial J^2}\right|_{J=1} \tag{21}$$

In the case of rubber-like materials, the initial bulk modulus is several orders of magnitude greater than the initial shear modulus. Therefore, these are typically described taking into consideration the

IOP Conf. Series: Materials Science and Engineering 1015 (2021) 012007 doi:10.1088/1757-899X/1015/1/012007

isochoric-volumetric split. The approach significantly simplifies interpretation of basic experimental results for such materials.

Figure 1 presents nominal stress vs stretch plots for synthetic rubber neoprene [9] in the case of incompressibility. Parameter values are summarized in table 1. For more details we refer the reader to [10].

Table 1. Parameters of MCMV and MCIZ for neoprene [9,10].

Model	a_1 [MPa]	a_2 [MPa]	a_3 [MPa]	a_4 [MPa]	a_5 [MPa]
MCIZ	$2.486\ 10^{-1}$	$3.054\ 10^{-4}$	-	$1.839\ 10^{-2}$	-
MCMV	$3.402\ 10^{-1}$	$-9.644\ 10^{-3}$	$1.685\ 10^{-4}$	$2.902\ 10^{-2}$	$-1.762\ 10^{-4}$

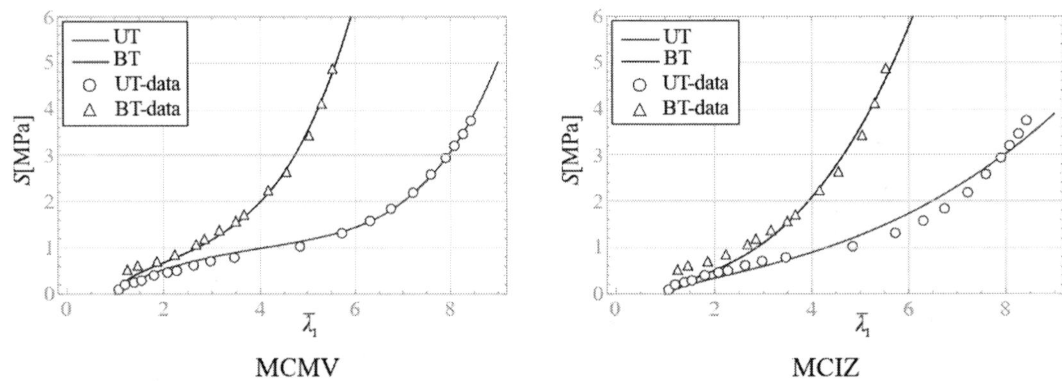

Figure 1. Nominal stress vs principal stretch for uniaxial (UT) and biaxial (BT) tensile test for neoprene in the case of incompressibility.

In the literature, there are a large number of propositions of volumetric stored energy function that meet the mathematical conditions. For examples presented in the following part of the article we choose function of the form

$$W_{vol}(J) = \mathrm{K}_0\left[\frac{1}{4}\left(J^2-1\right)-\frac{1}{2}\ln J\right] \tag{22}$$

We leave it to the reader to verify that the function meets discussed conditions.

4. Example initial-boundary value problems

In this section we present solutions of two initial-boundary value problems which illustrates application of the discussed models. The example problems are solved using ABAQUS/Standard which is a finite element program designed for general use in nonlinear problems [6,7]. In case of including dynamic effects, we use an implicit integration scheme available in the software. The MCMV and MCIZ models are defined with a help of user subroutine UHYPER.

4.1. Uniaxial compression of a rod

The first problem illustrating the behaviour of the MCMV material model is a simple uniaxial compression of a rectangular rod with dimensions $L \times L/20 \times L/20$. The FEM model consists of 450 C3D8H (hybrid formulation) elements. The initial bulk modulus is step up to be

$K_0 = 1000\mu_0 = 340.74$ MPa and density is assumed to be $\rho = 1000$ kg/m^3 in the case of initial-boundary value problem. Values of the parameters are given in table 1.

Over the marked face A, the prescribed displacement is applied $u_3 = -L/2$. On the face B $\mathbf{u} = \mathbf{0}$ is imposed and zero traction is prescribed over the other faces. In the case of dynamic problem, zero initial velocity is given which means that dynamic effect is due to the inertial forces.

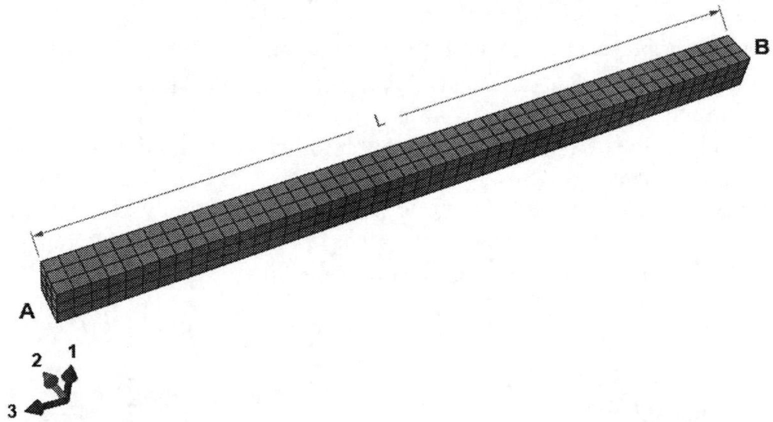

Figure 2. Finite element mesh for uniaxial compression of a rectangular rod.

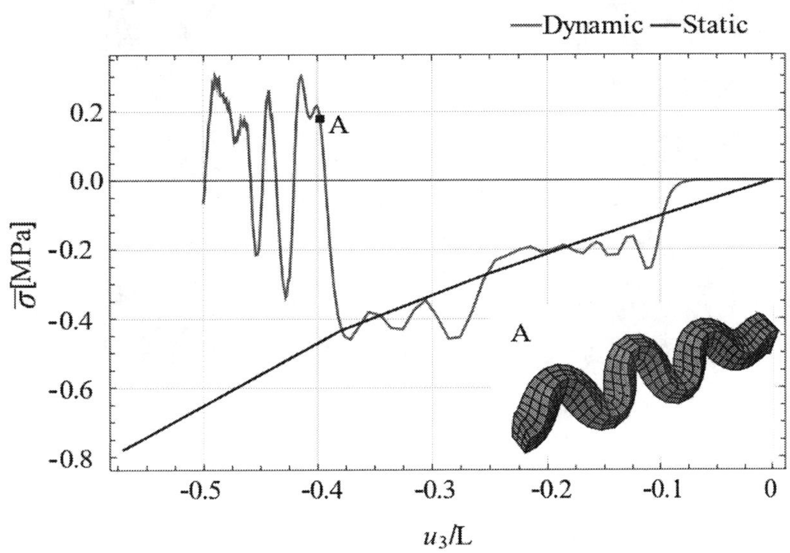

Figure 3. Plot of averaged stress $\overline{\sigma}$ as a function normalized displacement u_3 / L.

Figure 3 shows plots of averaged stress $\overline{\sigma}$ as a function of displacement of the face A. The averaged stress values are obtained by averaging values of the component σ_{33} from the Gauss points of finite elements at face B. In the case of the static problem, we obtain solution which results with a simple uniaxial deformation. This is not unique solution because of nonlinear nature of the problem. Taking into consideration dynamics effects, we observe a loss of stability at $u_3 / L = -0.4$ after uniaxial deformation. In other to obtain the solution in the static case an initial imperfection should be applied.

IOP Conf. Series: Materials Science and Engineering 1015 (2021) 012007 doi:10.1088/1757-899X/1015/1/012007

4.2. Twisting column

The second example concerns a twisting column 100x100x600 mm clamped on its bottom face [12]. An initial sinusoidal rotational velocity field relative to the origin is given by

$$\omega_0 = \left[0, 0, \Omega \sin\left(\frac{\pi Z}{200}\right)\right]^{\mathrm{T}} \tag{23}$$

where $\Omega = 300$ rad/s represents the magnitude of initial velocity. Finite elements C3D8H with hybrid formulation are used for the problem. Material parameters and density are assumed to be the same as for the previous example.

Figure 5 presents change of a height of the column for different models. We observe that there is no significant difference between proposed volumetric stored energy function and the one available in ABAQUS. Results of MCMV and MCIZ do not align, because mainly isochoric deformation occurs. However, there a significant difference between values of total energy when comparing the models. The initial bulk modulus K_0 has high impact on the total energy, see figure 7.

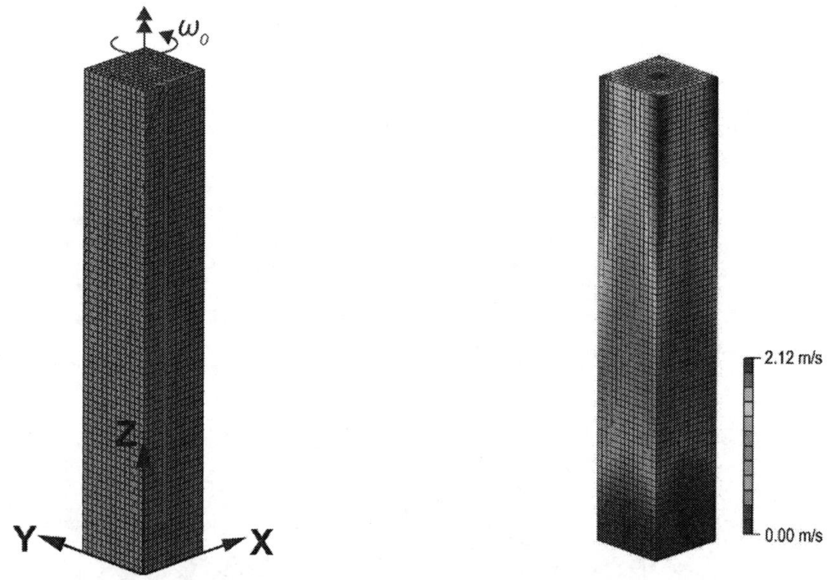

Figure 4. Twisting column – mesh and initial velocity.

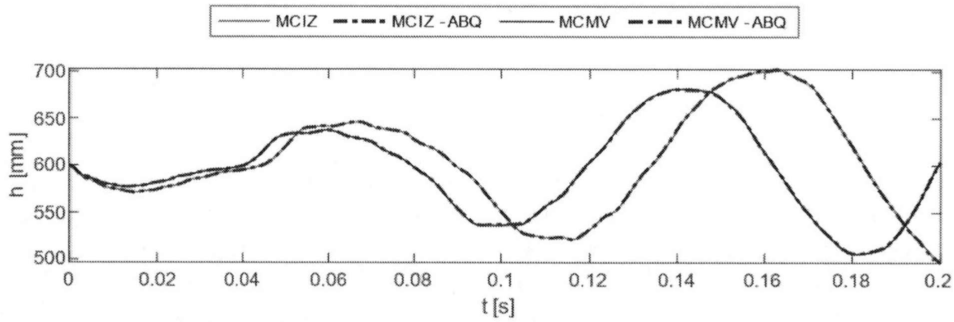

Figure 5. Twisting column: change of a height of the column for different models.

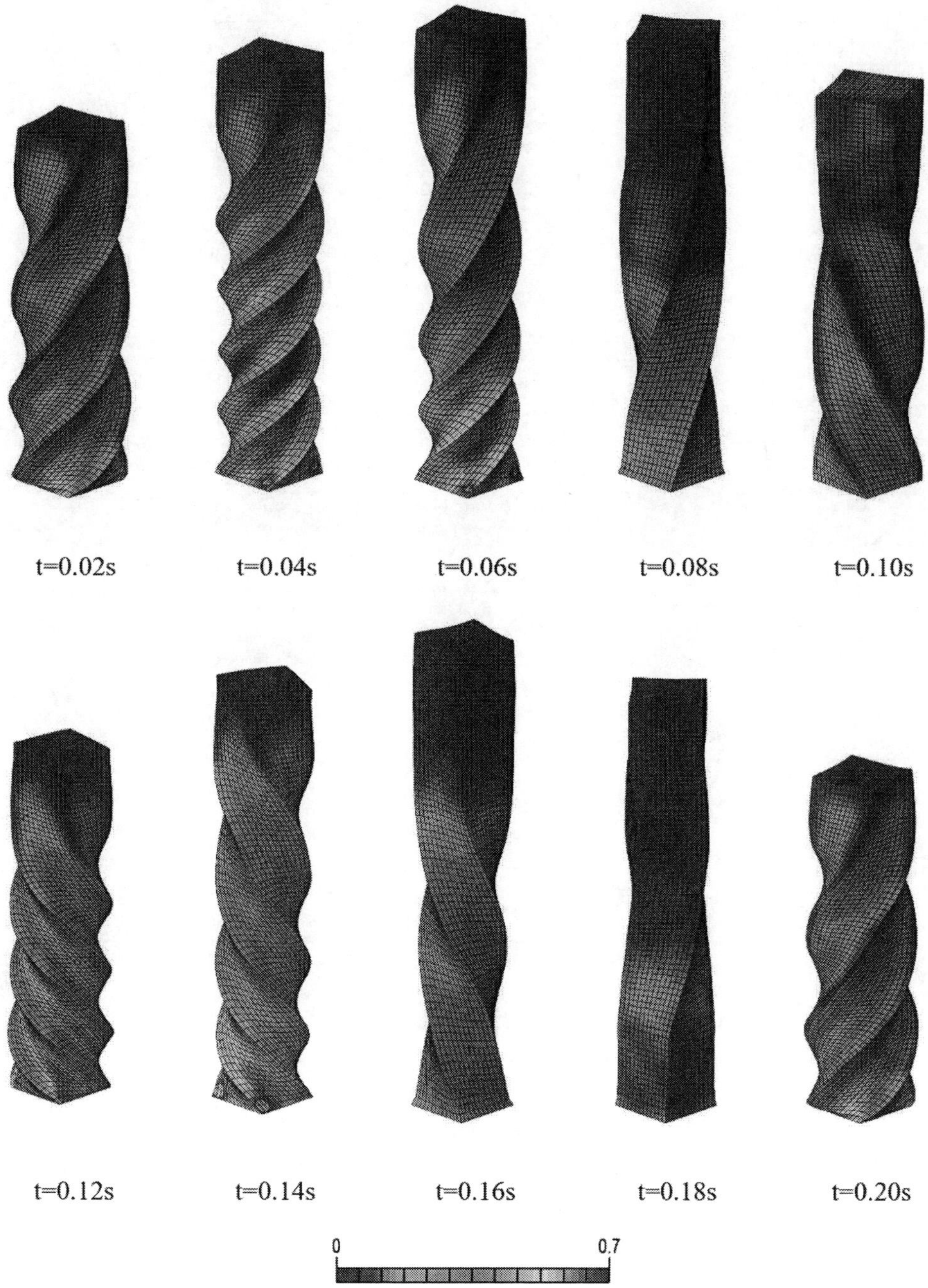

Figure 6. Twisting column: a sequence of deformations with Huber-Mieses stress map [MPa].

IOP Conf. Series: Materials Science and Engineering 1015 (2021) 012007 doi:10.1088/1757-899X/1015/1/012007

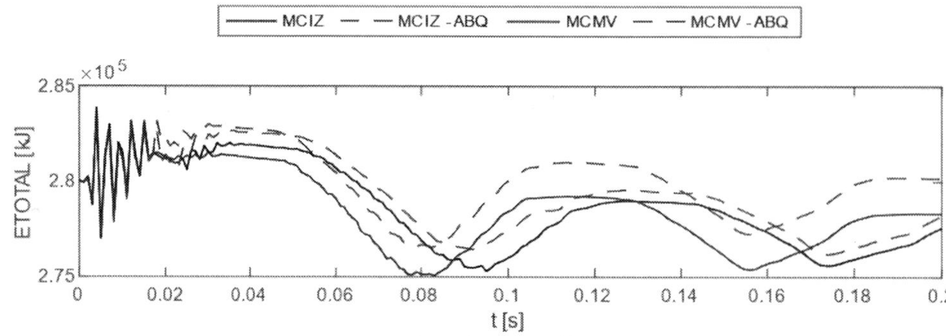

a) $K_0 = 10MPa$

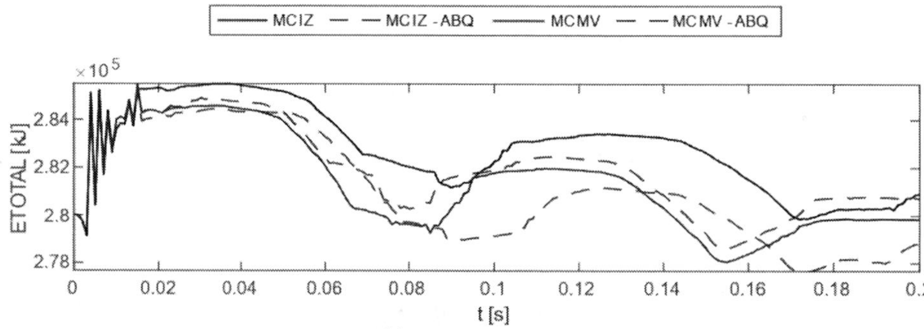

b) $K_0 = 100MPa$

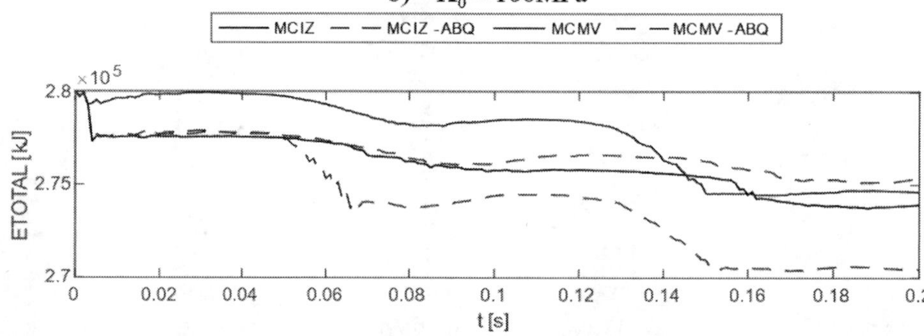

c) $K_0 = 1000MPa$

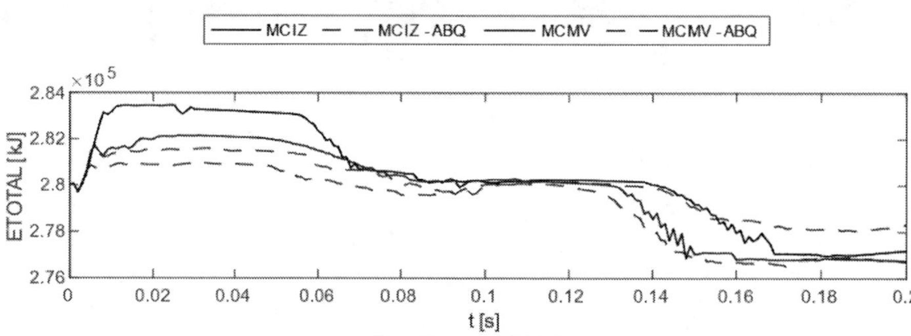

d) $K_0 = 5000MPa$

Figure 7. Twisting column: plots of total energy.

5. Conclusions

The article is intended as an attempt to present a dynamic and static response of certain slightly compressible hyperelastic models. These are obtained by extending incompressible material models which are consistent second and third order approximations of existing stored energy function in terms of a deformation measure, namely MCMV and MCIZ models. The approach leads to clear physical interpretation of a response of a material, i.e. volumetric and isochoric parts of deformation are decoupled. The description is especially suitable for rubber-like material, because of their nearly incompressibility. Moreover, the approach significantly simplifies interpretation of experimental results for such materials.

Example initial-value problems, which concern a simple compression of a rod and a twisting column, are solved using ABAQUS/Standard. In the case of including dynamic effects, we use an implicit integration scheme available in the software. Since the models are not available in ABAQUS library, we utilize a user subroutine UHYPER which is suitable for definition of isotropic material model. Moreover, in the second example we compare results against available in the library polynomial model where the volumetric stored energy function does not meet basic growth conditions.

In the case of the static uniaxial compression of a rod, we obtain solution which results with a simple uniaxial deformation. This is not unique solution because of nonlinear nature of the problem. Taking into consideration dynamics effects, we observe a loss of global stability after uniaxial deformation. In order to obtain the solution in the static case an initial imperfection should be applied. In the case of dynamically twisted columns we observe that there is no significant difference between discussed models when observing change of a height of the column. Results of MCMV and MCIZ do not align, because mainly isochoric deformation occurs. However, there a significant difference between values of total energy when comparing the models. The initial bulk modulus has high impact on the total energy.

References

[1] Franus A, Jemioło S and Antoni M 2020 A slightly compressible hyperelastic material model implementation in ABAQUS *Engineering Solid Mechanics* **8** (4) pp 365–80

[2] Chadwick P 1974 Thermo-mechanics of rubberlike materials *Philos. Trans. A Math. Phys. Eng. Sci.* **276** (1260) pp 371–403

[3] Suchocki C and Jemioło S 2019 On finite element implementation of polyconvex incompressible hyperelasticity: theory, coding, and applications *Int. J. Comput. Methods* 1950049

[4] Jemioło S 2002 *Studium hipersprężystych własności materiałów izotropowych* (Warsaw: OWPW)

[5] Isihara A, Hashitsume N and Tatibana M 1951 Statistical theory of rubber-like elasticity 4 (Two-Dimensional Stretching) *J. Chem. Phys* **19** pp 1508–12

[6] Abaqus 2016 Theory guide *Dassault Systèmes*

[7] Abaqus 2016 Analysis user's guide. Volume III: Materials, *Dassault Systèmes*

[8] Marsden J E 1983 *Mathematical foundation of elasticity* (Englewood Cliffs N.J: Prentice-Hall)

[9] Alexander H 1968 A constitutive relation for rubber-like materials *Int. J. Eng. Sci.* **6** (9) pp 549-63

[10] Jemioło S and Franus A 2019 *Izotropowe modele konstytutywne hipersprężytości. Wyznaczanie parametrów materiałowych i implementacja numeryczna* (Warsaw: OWPW)

[11] Ball J M 1976 Convexity conditions and existence theorems in nonlinear elasticity, *Arch. Ration. Mech. Anal.* **63** (4) pp 337–403

[12] Marek A M 2020 Model konstytutywny hipersprężytości elastomerów – implementacja numeryczna w programie ABAQUS z przykładami zastosowań (Warsaw: University of Technology Master Degree Thesis)

IOP Conf. Series: Materials Science and Engineering 1015 (2021) 012008 doi:10.1088/1757-899X/1015/1/012008

Modern lightweight deployable engineering structures

A Al Sabouni-Zawadzka[1], A Zawadzki[2] and W Gilewski[1]

[1]Faculty of Civil Engineering, Warsaw University of Technology, Al. Armii Ludowej 16, 00-637 Warsaw, Poland
[2]Faculty of Automotive and Construction Machinery Engineering, Warsaw University of Technology, Narbutta 84, 02-524 Warsaw, Poland

E-mail address: adam.zawadzki@pw.edu.pl

Abstract. The paper concerns the formulations and areas of possible scientific research in the field of modern deployable structures. The key problem is to provide a precise definition of modern deployable structures and to define the inherent properties that such structures should have. The objective was accomplished by specifying how the words deployable and lightweight are understood and assigning the constructions the necessary smart features. The defined features are fulfilled by tensegrity structures whose characteristics, from the point of view of deployability, are given in the paper.

1. Introduction

The work is the result of the authors' considerations in the area of scientific research in the field of modern deployable structures. The key aspect seems to be giving a precise definition and specifying the research area - studies of the literature indicate a very wide spectrum of structures that can be considered. Even greater variety of studies concerns techniques and algorithms used in deployable structures, the computational methods used and potential implementation areas. Detailed discussion of the extensive literature on the subject exceeds the scope of this work. However, we will mention two particularly valuable papers [1,2] in which interesting classifications of deployable structures and their systematics can be found.

The extensive literature on the subject relates to a large extent of theoretical issues. There are fewer experimental works or works describing realized projects, and if they are, they are usually cycles of papers, describing research, algorithms and their implementation in the long term (even several years). This means that deployable structures is a difficult scientific and technological challenge.

An important aspect of the research is the scientific discipline in which we would place deployable structures. Most often, deployable structures are related to: mechanical engineering, robotics, aerospace engineering and civil engineering. Robotics and aerospace engineering are beyond the scope of the paper. The authors' considerations also focus on lightweight structures in the field of modern solutions that are dated, due to the development of technologies and materials, to the 21st century.

2. About definitions

The issues that are considered in the paper are differently defined or not defined at all in the literature, referring to the colloquial understanding of words or the reader's intuition. It seems, however, that the precise definition of certain concepts is most appropriate and even necessary when we want to deal with them for scientific purposes.

IOP Conf. Series: Materials Science and Engineering 1015 (2021) 012008 doi:10.1088/1757-899X/1015/1/012008

The most commonly used term is "deployable structure". According to English dictionaries, the word "deployable" means "able to be moved to a place where it can be used when it is needed". In turn, "deployable structure is a structure that can change shape so as to significantly change its size". It seems that when defining the term "deployable structure" with precision, three important features should be emphasized:

a. The structure should be assembled/disassembled by itself, based on built-in mechanisms, with limited external factors (human or assembly team), if possible.
b. After reaching the target configuration, the structure should be ready for implementation for the predefined purpose.
c. The structure should significantly change its shape during erection.

In fact, in the above-mentioned definitions, there is no information about two-way operation, i.e. there is no requirement that the structure, after unfolding, may also be folded back to its original or other form. Deployable is understood to mean joining together. Considering the assembly and possible disassembly of the structure should be defined more precisely,. The above considerations look a bit like splitting hair, but they can be important in formulating algorithms and procedures for the practical functioning of this type of structures.

Other terms used in the literature are (following the dictionary of the English language):
"Erectable", "capable of being erected or raised up",
"Temporary" - "not lasting or needed for very long",
"Mobile" - "moving or walking around freely" or "able to be moved from one place to another",
"Demountable" - "able to be dismantled or removed from its setting and readily reassembled or repositioned",
"Retractable" - "that can be pulled back or in".
The terms "inflatable", "articulated" and "foldable" are used less frequently. According to the authors' opinion, these terms are narrower than "deployable", although they are related to specific and important features of this class of structures. When defining the area of scientific work it is necessary to use some of the above-mentioned terms additionally to the term "deployable".

Smart structures are the structures with the ability to sense and respond adaptively to changes in their environment. This feature distinguishes them from the conventional ones. Whereas the main purpose of the traditional structures is to provide strength and carry loads acting on them, the smart ones adapt in a pre-designed manner to a functional need, modifying their shape, stiffness or damping characteristics in order to minimize deflection and possible damage.

Further considerations are dedicated to structures that meet the a-c features mentioned above, and are also lightweight and smart.

3. Heavyweight versus lightweight structures

According to Meriam-Webster dictionary heavy is something having great weight and lightweight is something having less than average weight. When it comes to distinguishing these features between erectable engineering structures a problem of a limit appears. Another need for a limit definition comes with ultralightweight structures.

It is not obvious where to set these boarders because they should be defined for specific groups of structures, not for structures in general. Authors propose aforementioned limits in terms of different aspects of analyzing deployable structures (Tab. 1). Most of the entries do not require additional comment since they are explained plainly in the table. Line no. 4 should be interpreted so that heavy structures meet all the requirements set for standard (not deployable) realizations. Lightweight structures in comparison to heavy ones are designed for limited, safe conditions of use. Ultralightweight structures are designed to be used in extreme conditions but their behavior differs significantly from heavy structures (e.g. high deformations, low lifecycle, resonance at unsafe frequencies for civil engineering etc.).

The table does not include amount of the load acting on the structure since the difference between deployable applications is vast e.g. between a bridge and a supporting column. This parameter could be

IOP Conf. Series: Materials Science and Engineering 1015 (2021) 012008 doi:10.1088/1757-899X/1015/1/012008

compared indirectly by introducing a dimensionless parameter of load carrying capacity (LCC) that is a ratio of a load capacity and a weight of the structure.

Table 1. Classification of structures according to their weight

No.	Category	Heavyweight	Lightweight	Ultralightweight
1.	Density of the structure	>100 kg/m^3	30÷100 kg/m^3	<30 kg/m^3
2.	Equipment needed for deployment	Deployable using equipment with >1kW power	Deployable using equipment with ≤1kW power	Deployable with hands
3.	Ability to be deployed with hands	Not possible	More than one person needed	One person needed
4.	Sensitivity to external loads	Insensible	Not designed for extreme conditions	Typical limits: e.g. deformation do not apply also in standard conditions
5.	Dimensionless load carrying capacity (LCC)	LCC<0.04	LCC=0.04÷0.2	LCC>0.2

Various comparison tests are proposed in the literature. A good example is [3], where authors compared pantographic structures with membranes between struts that were to be used as temporary roof structures. They defined following parameters for comparison:

- the minimum dimensions of the boundary edges of the unfolded structure (when it lays flat on a ground),
- effective floor space range that the structure can assure once it is deployed and acceptable height is reached,
- the minimum folding angle understood as the smallest angle between two adjacent surfaces during the deployment – authors point out that it is important because it determines the minimum rotation angle which the joints(hinges) have to assure without limiting the adaptability,
- variety of triangles that is a characteristic feature of compared structures and influences the complexity of the structure – it affects the transportability and deployment procedures,
- the minimum number of hinges for a real structure assuming minimum two hinges per internal fold line,
- estimation of the weight.

The above information (first of all table 1) allow to define the lightweight structure sufficiently.

4. Selection of the concepts

The most important feature of a deployable structure is the deploying mechanism which characterizes the behavior of the system and its mechanical properties. Due to this concepts from the literature can be divided into groups that focus on separate mechanisms. These groups are: pantographic mechanisms (scissor-like mechanisms, Bennett mechanism, Myard linkage, coilable structures etc.), tensegrity mechanisms, mechanisms based on air-filled elements, origami based mechanisms, mechanisms based on shape memory materials. Since these structures are morphing from time to time, there are analyses in terms of utilized control algorithms that were also reviewed by the authors.

Promising type of structures which allow to build deployable structures are tensegrities [4]. Tensegrity structure can be defined as a pin-jointed system with a particular configuration of cables and struts that form a statically indeterminate structure in a stable equilibrium. Tensegrities consist of a discontinuous set of compressed elements inside a continuous net of tensioned members, which have no compressive stiffness. Infinitesimal kinematic mechanisms, that occur in tensegrity structures, are balanced with self-stress states [4]. Occurrence of a self-stress state in a structure indicates that there is a certain set of internal forces in structural members, which are independent from external loading and boundary conditions because they are in self-equilibrium.

To major advantages of tensegrity systems belong: large stiffness-to-mass ratio, deployability, reliability and controllability [5]. Moreover, tensegrities have some unique features that result from infinitesimal mechanisms, which are stabilized by self-equilibrated normal forces. It is possible to control their static and dynamic properties by adjusting the pre-stressing forces [5]. It is also to identify and control of mechanical properties of tensegrity systems [6].

As it was presented in [7] there are some particular features of tensegrity structures following which one can classify them as smart structures. There are: self-control, self-diagnosis, self-repair and self-adjustment (active control) with the use of self-stress as well as geometrical properties of the structure.

From the point of view of deployable structures, it is proposed to classify tensegrity structures as:

- 0D - this class includes all regular tensegrity modules; if the module is a "pure tensegrity" (according to the definitions [4,5,7]) it is deployable in a natural way with controllability by a single cable or strut.
- 1D - structures made of many modules connected along one straight line or curve; we can distinguish here: tensegrity beams, columns and arches, flat or spatial; the class of 1D structures can be extended to structures that are not made of modules but have tensegrity features.
- 2D - structures made of many modules connected on a plane or curved surface; we can distinguish between tensegrity plates and shells; non-modular structures are also allowed in this class.
- 3D - structures made of many modules connected in three directions; non-modular structures are also allowed in this class.

5. Conclusions

The analysis of the literature on the subject and the authors' preliminary computational simulations [8] indicate the great potential that lies in the application of the tensegrity concept to the construction and operation of deployable structures. Applications in civil/mechanical engineering is a challenging task.

The structure referred to as "modern lightweight deployable engineering structure" should have the following inherent properties:

- Should be assembled by itself, based on built-in mechanisms, without external factors.
- After reaching the target configuration, the structure should be ready for implementation for the predefined purpose.
- Should significantly change its shape during erection.
- Should be lightweight or ultralightweight in the sense of table 1.
- Should be smart in the sense of ability to: self-control, self-diagnosis, self-repair and self-adjustment (active control).

References

[1] Fenci G E and Currie N G R 2017 Deployable structures classification: a review, *International Journal of Space Structures* **32** (2) pp 112-30

[2] De Temmerman N 2007 *Design and analysis of deployable bar structures for mobile architectural applications* (Vrije Universitat Brussel, Faculty of Engineering)

[3] Van Knippenberg R, Habraken A and Teuffel P 2016 Deployable structures using non-singular rigid foldable patterns, *Procedia Engineering* **155** pp 388-97

[4] Tensegrity M R 2003 *Structural Systems for the Future* (London: Kogan Page Science)

[5] Al Sabouni-Zawadzka A and Gilewski 2018 W Inherent properties of tensegrity structures controlled by self-stress *Applied Sciences* – Basel, **8** (5) p 787

[6] Al Sabouni-Zawadzka A., Gilewski W 2016 On orthotropic properties of tensegrity structures, *Procedia Engineering* **153** pp 887-894

[7] Gilewski W and Al Sabouni-Zawadzka A 2015 On possible applications of smart structures controlled by self-stress *Archives of Civil and Mechanical Engineering* **15** (2) pp 469-78

[8] Zawadzki A and Al Sabouni-Zawadzka A 2020 Simulation of a deployable tensegrity column based on the finite element modeling and multibody dynamics simulations *Archives of Civil Engineering* **66** (4) (in print)

IOP Conf. Series: Materials Science and Engineering 1015 (2021) 012009 doi:10.1088/1757-899X/1015/1/012009

The associated non-stationary thermal elasticity problem for a two-layer plate

D A Shlyakhin and Zh M Kusaeva

[1] Samara State Technical University, 244, Molodogvardeyskaya St., 443100, Samara, Russian Federation

E-mail address: zhkusaeva@mail.ru

Abstract. A new analytical solution of a non-stationary thermal elasticity problem for a rigidly fixed round two-layer plate was constructed. The case of temperature change on its upper face surface is considered and the heat exchange of the lower face surface with the environment is taken into account. When constructing a General solution, the mathematical apparatus for separating variables in the form of finite integral Fourier–Bessel transformations and generalized biorthogonal transformation is used. The solution peculiarity is the application of control and measuring device. The calculated relations make it possible to determine the stress-strain state and the distribution nature of the temperature field in a rigid round two-layer plate under any external temperature influence.

1. Introduction.

Uneven non-stationary temperature influence on the structure leads to the formation of thermal deformations and stresses, which are to be taken into account when determining the strength characteristics of elastic systems [1-4].

The problem of thermoelasticity becomes more complex when studying multilayer structures. The investigations often consider an infinitely long cylinder. Thus, in [5,6], the Laplace transform in time with numerical implementation of the calculated relations in the field of function images was used for the effect of non-stationary thermal and mechanical loads on the structure.

In this paper, the object of research is a multilayer rigidly fixed round plate in the case of non-stationary axisymmetric temperature action on its upper front surface (boundary conditions of the 1st kind). At the same time, the initial calculation ratios do not take into account the inertial characteristics of the elastic system, which is acceptable when studying extremely thin plates [7] (appendix).

2. Statement of the problem.

Let a round multilayer rigidly fixed plate occupy the area Ω: $\{0 \leq \gamma_* \leq b,\ 0 \leq \theta \leq 2\pi,\ 0 \leq z_* \leq h^*\}$ in the cylindrical coordinate system: $(r_* \leq \theta, z_*)$.

The case of temperature $\omega_1^*(r_*, t_*)$ change on its upper $(z_* = 0)$, and on the lower $(z_* = h)$ plane the temperature of the external environment is set ϑ^*. The number of layers in the plate is not limited, but for certainty, we assume it to be two-layer (figure1).

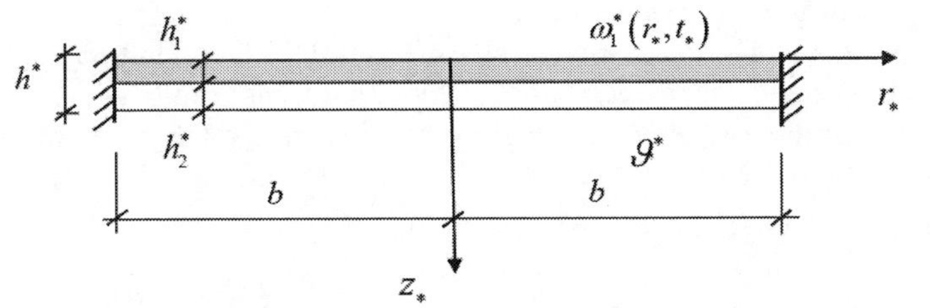

Figure 1. Calculation scheme.

The resolving system of differential axisymmetric equations of equilibrium and thermal conductivity of the classical (CTE) theory of thermoelasticity of an isotropic medium j – layer, as well as the boundary conditions of the problem under consideration in a dimensionless form have the form [8,9]:

$$\frac{\partial}{\partial r}\nabla U + a_1^{(j)}\frac{\partial^2 U}{\partial z^2} + a_2^{(j)}\frac{\partial^2 W}{\partial r \partial z} - a_3^{(j)}\frac{\partial^2 T}{\partial r} = 0 \quad (1)$$

$$a_1^{(j)}\nabla\frac{\partial W}{\partial r} + \frac{\partial^2 W}{\partial z^2} + a_2^{(j)}\frac{\partial}{\partial z}\nabla U - a_3^{(j)}\frac{\partial T}{\partial z} = 0$$

$$\nabla\frac{\partial T}{\partial r} + \frac{\partial^2 T}{\partial z^2} - a_4^{(j)}\frac{\partial T}{\partial t} - a_5^{(j)}\frac{\partial}{\partial t}\left(\nabla U + \frac{\partial W}{\partial z}\right) = 0$$

$r = 1$
$$\frac{\partial T}{\partial r}_{|r=1|} = 0, \qquad U = W = 0 \quad (2)$$

$r = 0$
$$\{U, W, T\} < \infty \quad (3)$$

$z = 0$
$$a_6^{(1)}\nabla U + \frac{\partial W}{\partial z} - T = 0, \quad \frac{\partial W}{\partial r} + \frac{\partial U}{\partial z} = 0, \; T = \varpi_1 \quad (4)$$

$z = h_1$
$$\{U, W, T\}|_{-z} = \{U, W, T\}|_{+z}, \; -\Lambda^{(1)}\frac{\partial T}{\partial z}\bigg|_{-z} = -\Lambda^{(2)}\frac{\partial T}{\partial z}\bigg|_{+z} \quad (5)$$

$$\left(\frac{\partial W}{\partial r} + \frac{\partial U}{\partial z}\right)_{|-z} = a_3\left(\frac{\partial W}{\partial r} + \frac{\partial U}{\partial z}\right)_{|+z}, \quad \left[a_6^{(1)}\nabla U + \frac{\partial W}{\partial z} - T\right]_{|-z} = \frac{a_7}{a_3^{(2)}}\left[a_6^{(2)}\nabla U + \frac{\partial W}{\partial z} - a_3^{(2)}T\right]_{|+z}$$

$z = h$
$$a_6^{(2)}\nabla U + \frac{\partial W}{\partial z} - a_3^{(2)}T = 0, \quad \frac{\partial W}{\partial r} + \frac{\partial U}{\partial z} = 0, \; -\Lambda^{(2)}\frac{\partial T}{\partial z} = \alpha(T - \vartheta) \quad (6)$$

$t = 0$
$$U = W = T = 0 \quad (7)$$

where $\{U^*, W^*\} = b\{U, W\}$, $\{T, \omega_1, \vartheta\} = a_9^{(1)}\{[T^*, \omega_1^*, \vartheta^*] - T_0\}$, $t = t_*\frac{k^{(1)}}{b^2}$, $a_1^{(j)} = \frac{(1 - 2\nu^{(j)})}{2(1 - 2\nu^{(j)})}$

$$a_2^{(j)} = \frac{1}{2(1 - 2\nu^{(j)})}, \; a_3^{(j)} = \frac{a_9^{(j)}}{a_9^{(1)}}, \; a_4^{(j)} = \frac{k^{(1)}}{k^{(j)}}, \; a_5^{(j)} = \frac{\gamma^{(j)}T_0}{\Lambda^{(j)}}a_9^{(1)}k^{(1)}, \; a_6^{(j)} = \nu^{(j)}(1 - \nu^{(j)})^{-1}, \; a_7 = \frac{\gamma^{(2)}}{\gamma^{(1)}}$$

$$a_8 = \frac{E^{(2)}(1 + \nu^{(1)})}{E^{(1)}(1 + \nu^{(2)})}, \; a_9^{(j)} = \frac{\gamma^{(j)}(1 + \nu^{(j)})(1 - 2\nu^{(j)})}{E^{(j)}(1 - \nu^{(j)})}, \; \gamma^{(j)} = \frac{E^{(j)}}{(1 - 2\nu^{(j)})}a_t^{(j)}, \; k^{(j)} = \frac{\Lambda^{(j)}}{c_\varepsilon^{(j)}}, \; \nabla = \frac{\partial}{\partial r} + \frac{1}{r}$$

$U^*(r_*, z_*, t_*)$, $W^*(r_*, z_*, t_*)$, $T^*(r_*, z_*, t_*)$, $T_0(r_*, z_*, t_*)$ – components of the displacement vector, temperature, and absolute temperature of the initial undeformed state of the body in dimensional form; t_* – time in measurable form; $E^{(j)}, \nu^{(j)}$ – elastic modulus and Poisson's ratio of an isotropic material of the j – layer; $\Lambda^{(j)}, \alpha_t^{(j)}, c_\varepsilon^{(j)}$ – thermal conductivity coefficient, linear thermal expansion and bulk heat capacity of the j – layer material; α – heat transfer coefficient.

3. The construction of the General solution.

Initial-boundary problem (1)-(7) is solved by sequential use of finite integral Fourier-Bessel [8] transform by variable r and generalized finite biorthogonal transform (FIT) [9] by coordinate z.

Considering that the Fourier-Bessel transform applied below allows to investigate problems with mixed homogeneous boundary conditions [8], we replace the last equality (2) with the condition of the presence of tangent stresses $P_1(z,t)$ at $r=1$:

$$\sigma_{rz|r=1} = \frac{E^{(j)}}{2\left(1+\nu^{(j)}\right)}\frac{\partial W}{\partial r} + \frac{\partial U}{\partial z} = P_1 \tag{8}$$

and enter a new function $w(r,z,t)$:

$$W(r,z,t) = \frac{1+\nu^{(j)}}{E^{(1)}} r^2 P_1(z,t) + W_1(t) + w(r,z,t) \tag{9}$$

Expressions for P_1, W_1 are defined in the process of solving the problem when the condition is met $W(1,z,t)=0$.

Substituting the expression (9) into (1) - (8) converts the problem relative to functions U,w,T, in which the differential equations (1), the first (4), (6) and last (5) boundary conditions, as well as the initial condition relative to the vertical component of the displacement vector become inhomogeneous with the right parts $R_1...R_3$, $B_1...B_4{}^*$:

$$\{R_1,R_2,R_3\} = -\left[a_{10}^{(j)}\right]^{-1}\left\{2a_2^{(j)} r\frac{\partial}{\partial z}, \left(4a_1^{(j)} + r^2\frac{\partial^2}{\partial z^2}\right), a_5^{(j)} r^2\frac{\partial^2}{\partial z \partial t}\right\} P_1$$

$$B_4 = -\left[a_{10}^{(j)}\right]^{-1} r^2 P_1(z,0) - W_1(0)$$

$$\{B_1,B_2,B_3\} = -\left[a_{10}^{(j)}\right]^{-1} r^2\left\{\frac{\partial P_1}{\partial z}_{|z=0}, \frac{\partial P_1}{\partial z}_{|z=h}, \left(\frac{\partial P_1}{\partial z}_{|-h_1} - \frac{a_7}{a_3^{(2)}}\frac{\partial P_1}{\partial z}_{|+h_1}\right)\right\}$$

$$a_{10}^{(j)} = E^{(j)}\left(1+\nu^{(j)}\right)^{-1}$$

and the boundary conditions on a cylindrical surface take the form: ($r=1$)

$$U(1,z,t)=0, \quad \frac{\partial w}{\partial r}_{|r=1|} = \frac{\partial T}{\partial r}_{|r=1|} = 0 \tag{10}$$

At the first stage, we apply the Fourier-Bessel transform to the boundary value problem using the following transformants:

$$u_H(n,z,t) = \int_0^1 U(r,z,t) r J_1(j_n r)dr \quad \{w_H(n,z,t),\phi_H(n,z,t)\} = \int_0^1 \{w(r,z,t),T(r,z,t)\} r J_0(j_n r)dr \tag{11}$$

and conversion formulas:

$$U(r,z,t) = 2\sum_{n=1}^{\infty}\frac{u_H(n,z,t)}{J_0(j_n)^2}J_1(j_n r), \quad \{w(r,z,t),T(r,z,t)\} = 2\sum_{n=0}^{\infty}\frac{\{w_H(n,z,t),\phi_H(n,z,t)\}}{J_0(j_n)^2}J_0(j_n r) \tag{12}$$

and in the second stage, biorthogonal finite integral transformation [9] is applied with unknown components of eigenvector-functions of transform nuclei $K_1(\lambda_{in},z)..K_3(\lambda_{in},z)$, $N_1(\mu_{in},z)..N_3(\mu_{in},z)$:

$$G(\lambda_{in},n,t) = \int_0^h \left[a_5^{(j)} j_n U_H(n,z,t) + a_4^{(j)} L_H(n,z,t) - a_5^{(j)} W_H(n,z,t)\frac{d}{dz}\right] K_3(\lambda_{in},z)dz \tag{13}$$

$$\{U_H(n,z,t),W_H(n,z,t),L_H(n,z,t)\} = \sum_{i=1}^{\infty} G(\lambda_{in},n,t)\{N_1(\mu_{in},z),N_2(\mu_{in},z),N_3(\mu_{in},z)\}\|K_{in}\|^{-2} \tag{14}$$

$$\|K_{in}\|^2 = \int_0^h K_3(\lambda_{in}, z) N_3(\mu_{in}, z) dz;$$

where j_n – are the positive zeros of the function $J_1(j_n)$ $(n = \overline{0,\infty};\ j_0 = 0)$, λ_{in}, μ_{in} – eigenvalues of corresponding homogeneous linear boundary value problems with respect to conjugate $K_k(\lambda_{in}, z)$ and invariant $N_k(\mu_{in}, z)$ components of vector functions of kernels FIT ($k = 1,2,3$).

As a result we get function expressions $U(r,z,t), W(r,z,t), T(r,z,t)$

$$U(r,z,t) = 2\sum_{n=1}^{\infty} \frac{J_1(j_n r)}{J_0(j_n)^2}\left[H_1(n,z,t) + \sum_{n=1}^{\infty} G(\lambda_{in}, n, t) N_1(\mu_{in}, z)\|K_{in}\|^{-2}\right] \tag{15}$$

$$W(r,z,t) = \frac{1+\nu^{(j)}}{E^{(j)}} r^2 P_1(z,t) + W_1(t) + 2\sum_{n=0}^{\infty} \frac{J_0(j_n r)}{J_0(j_n)^2}\left[H_2(n,z,t) + \sum_{i=1}^{\infty} G(\lambda_{in}, n, t) N_2(\mu_{in}, z)\|K_{in}\|^{-2}\right]$$

$$T(r,z,t) = 2\sum_{n=0}^{\infty} \frac{J_0(j_n r)}{J_0(j_n)^2}\left[H_3(n,z,t) + \sum_{i=1}^{\infty} G(\lambda_{in}, n, t) N_3(\mu_{in}, z)\|K_{in}\|^{-2}\right]$$

The final stage of the study is to determine the functions $W_1(t), P_1(z,t), H_1(n,z,t)..H_3(n,z,t)$. The case of temperature load only is initially considered ω_1, ϑ ($P_1 = 0$). Then $H_1 ... H_3$ they are determined from the condition of simplifications of the right parts of equations:

$$j_n^2 H_1 - a_1^{(j)} \frac{\partial^2 H_1}{\partial z^2} + a_2^{(j)} j_n \frac{\partial H_2}{\partial z} - a_3^{(j)} j_n H_3 = 0$$

$$a_1^{(j)} j_n^2 H_2 - \frac{\partial^2 H_2}{\partial z^2} - a_2^{(j)} j_n \frac{\partial H_1}{\partial z} + a_3^{(j)} \frac{\partial H_3}{\partial z} = 0, \quad j_n^2 H_3 - \frac{\partial^2 H_3}{\partial z^2} = 0$$

As a result, systems of equations are formed with respect to $f_1(z), f_2(z), f_{11}(z)$ and $f_3(z), f_4(z), f_{12}(z)$, the solution of which, taking into account the conditions (15), allows us to determine $H_1 .. H_3$.

The function $W_1(t)$ is defined from a condition $W(1,h,t) = 0$

$$W_1(t) = -2\sum_{n=0}^{\infty} J_0(j_n)^{-1}\left[H_2(n,h,t) + \sum_{i=1}^{\infty} G(\lambda_{in}, n, t) N_2^{(2)}(\mu_{in}, h)\|K_{in}\|^{-2}\right]$$

In the next step of the solution, tangent stresses are applied to the cylindrical surface of the plate $P_1(z,t)$ $(\omega_1 = \vartheta = 0)$.Considering the balance conditions of the plate, twin and inseparability of tangent stresses

$$\int_0^h P_1(z,t) dz = 0, P_{1|z=0,h_1,h} = 0$$

we present it as the following dependency:

$$P_1(z,t) = S_0 W_1(t)\left\{\frac{h-h_1}{h_1}\sin\left(\frac{\pi}{h_1} z\right) H(h_1 - z) - \sin\left[\frac{\pi}{h-h_1}(z-h_1)\right] H(z-h_1)\right\}$$

Functions H_1, H_2 are also calculated from the condition of simplifications of the right parts of differential equations

$$j_n^2 H_1 - a_1^{(j)} \frac{\partial^2 H_1}{\partial z^2} + a_2^{(j)} j_n \frac{\partial H_2}{\partial z} = 0, \quad a_1^{(j)} j_n^2 H_2 - \frac{\partial^2 H_2}{\partial z^2} - a_2^{(j)} j_n \frac{\partial^2 H_1}{\partial z} = 0.$$

In this case, systems of differential equations with respect to functions are formed $f_5(z), f_6(z), f_7(z), f_8(z)$ and $f_9(z), f_{10}(z)$, which are solved by satisfying the boundary conditions (19). The sum of the two calculation results allows us to determine the constant S_0 from the condition of integral equality to zero of vertical displacements of the cylindrical surface of the plate $\int_0^h W(1,z,t_{max})dz = 0$.

4. Numerical analysis of results. Conclusions.

Let's consider the case of acting on the upper face ($z_* = 0$) of the elastic system of temperature load

$$\omega_1^*(r_*,t_*) = (1-r_*)T_{max}\left[\sin\left(\frac{\pi}{2t_{max}}t_*\right)H(t_{max}^* - t_*) + H(t_* - t_{max}^*)\right]$$

at a known ambient temperature of $\vartheta^* = 293\,K$ ($20\,^0C$) – heat transfer coefficient between the plate surface and the air $\alpha = 8.7$ (W/(m²K)).

Here $T_{max} = T_{max}^* - T_0$; T_{max}^*, t_{max}^* – the maximum value of the temperature load and the time at which the thermal external influence reaches the greatest value in the dimensional form ($T_{max}^* = 373$ K ($100\,^0C$) , $T_0 = 293$ K, ($20\,^0C$), $t_{max}^* = 10$ s). The two-layer plate ($b = 1$ m) is made of steel ($j = 1$) and plastic ($j = 2$), which has the following physical and mechanical characteristics of materials:

$$\{E^{(1)}, E^{(2)}\} = \{20, 0.33\} \times 10^{10}\,\text{Pa},\ \{\Lambda^{(1)}, \Lambda^{(2)}\} = \{50, 0.2\}\ \text{W/(mK)},$$

$$\{c_\varepsilon^{(1)}, c_\varepsilon^{(2)}\} = \{3.8, 0.23\} \times 10^6\ \text{J/(m}^3\text{K)},\ \{\nu^{(1)}, \nu^{(2)}\} = \{0.28, .0.33\},\ \{\alpha_t^{(1)}, \alpha_t^{(2)}\} = \{1.2, 8\} \times 10^{-5}\ \text{1/K}.$$

Figures 2,3 show graphs of changes in temperature field $T^*(0,z,t)$ and $W^*(0,z,t)$ movements and components of stress tensor $\sigma_{rr}(r,z,t)$ over spatial variables at various times t. The analysis of the numerical results of the calculation makes it possible to draw the following conclusions:

1. When the maximum temperature load is reached $t = t_{max}$, due to the high thermal conductivity of the metal, a sufficiently high temperature is observed on the inner surface of the steel part of the plate (figure 2, graphs 1) ($h_1 = 0.1 - T^*(0,h_1,t_{max}) = 63^\circ C$, $h_1 = 0.05 - T^*(0,h_1,t_{max}) = 71^\circ C$). At the same time, the thickness of the plate has a significant effect on the time of complete heating of the structure and on the temperature of its lower ($z = h$) face. For a thick ($h = 0.2$) plate $t = 1100t_{max}$, $T^*(0,h,t) = 60\,^0C$, and for a thinner one ($h = 0.1$) design $t = 300t_{max}$, $T^*(0,h,t) = 64\,^0C$ (figure 2, graphs 3);

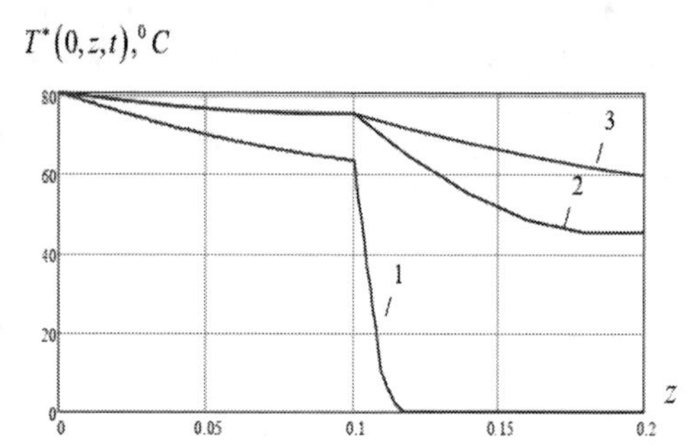

$a) h_1 = h_2 = 0.1 (1 - t = t_{max}$, $1 - t = 550t_{max}, 3 - 1100t_{max})$

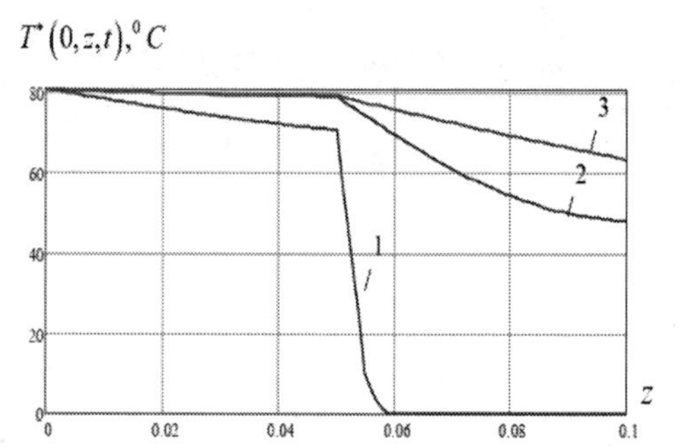

$b) h_1 = h_2 = 0.05 (1 - t = t_{max}$, $2 - t = 150t_{max}, 3 - 300t_{max})$

Figure 2. Graphs of temperature $T(0,z,t)$ change by axial coordinate z at different times.

2. The coupling of thermoelastic fields leads to a decrease in the temperature of the plate (figure 3,a), an increase in displacements (figure 3, b), which leads to an increase in the radial component of the stress tensor (figure 3,c) $\sigma_{rr}(r,0,t)$ up to 10% during its heating. The greatest effect of field connectivity is observed in the center of the plate when the maximum temperature load is reached $t = t_{max}$ (figure 3,c, graphs 1); When the temperature load reaches the maximum values ($t = t_{max}$), the upper face surface of the plate ($z = 0$) on the section $0 \le r < 0.55$ experiences stretches in the radial plane, and on the remaining section - compression. However, during the heating of the structure, the compression area increases and with steady thermal mode, this plane $t = 1000t_{max}$ will be compressed, and the stresses $\sigma_{rr}(r,z,t)$ are constant (figure 3, c, graphs 2).

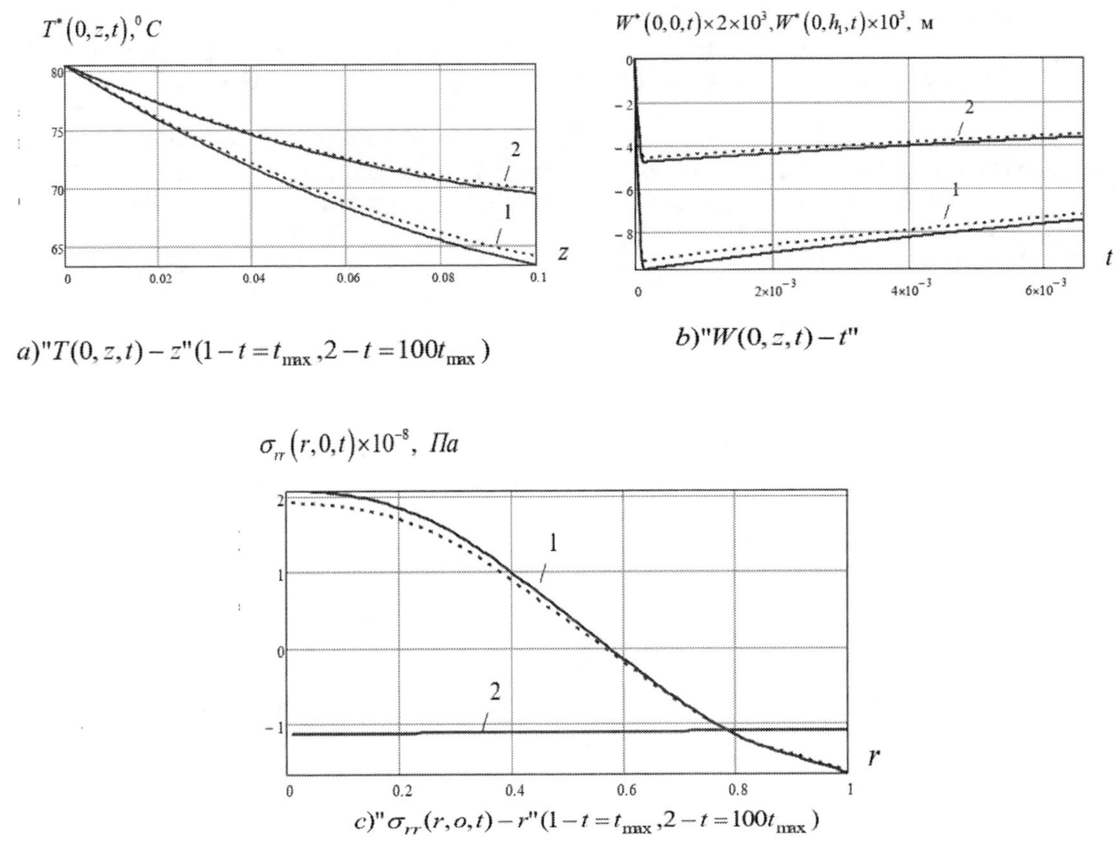

Figure 3. Graphs of temperature, displacement, and voltage changes with (solid line) and without (dotted line) connectedness of thermoelastic fields ($h = 0.2$).

In conclusion, it should be noted that the constructed calculation algorithm also allows to investigate initial-edge thermoelasticity problems for the multilayer plate in case of force load action and kinematic action on its face surfaces, as well as to take into account the effect of internal heat sources located in the area of layers connection.

Acknowledgments
The reported study was funded by RFBR according to the research project № 20-31-90042.

References

[1] Podstrigach Ya S, Lomakin V A and Kolyano Yu M 1984 Thermal elasticity of bodies of non-uniform structure (Moscow) p 368
[2] Pain B and Wayner J 1964 The theory of temperature stresses (Moscow) p 520
[3] Novatsky V 1970 Dynamic problems of thermoelast. (Moscow) p 256
[4] Lychev S A, Manzhirov A V and Uuber S V 2010 Closed solutions of boundary value problems of coupledthermoelasticity *Mechanics of Solids* **45** (4) pp 138–154
[5] Fu J W, Chen Z T and Qian L F 2015 Coupled thermoelastic analysis of a multi-layered hollow cylinder based on the C–T theory and its application on functionally graded materials *Composite Structures* **131** pp 139−150
[6] Gennaro V and Gennady M 2016 Analysis of residual stresses in thermoelastic multilayer cylinders *J. of the European Ceramic Society* **36** pp 2411−17
[7] Shlyakhin D A and Dauletmuratova Zh M 2018 Non-stationary coupled axisymmetric thermoelasticity problem for a rigidly fixed round plate *Eng. J.: science and innovations* **5** (77) (doi.1018698 / 2308-6033-2018-5)
[8] Sneddon I N 1955 Fourier Transforms *Foreign Literature Publishing House* (Moscow) p 668
[9] Senitsky Yu E 1996 Biorthogonal multicomponent finite integral transformation and its application to boundary value problems of mechanics *Izv. vuzov. Mathematics* **8** pp 71–81

IOP Conf. Series: Materials Science and Engineering 1015 (2021) 012010 doi:10.1088/1757-899X/1015/1/012010

Related thermoelastic elastomeric task for cylinder

D A Shlyakhin and M A Kalmova

Samara State Technical University, 244, Molodogvardeyskaya St., 443100, Samara, Russian Federation

E-mail address: kalmova@inbox.ru

Abstract. A new closed solution of the coupled non-stationary problem of thermoelectroelasticity for a long piezoceramic radially polarized cylinder is constructed when the boundary conditions of the 1st kind of thermal conductivity are satisfied on its front surfaces. The case is considered when the rate of change of the temperature field does not affect the inertial characteristics of the elastic system, which makes it possible to include in the initial calculated relations of the problem under consideration the linear equations of equilibrium, electrostatics and thermal conductivity with respect to the radial component of the displacement vector, electric potential, as well as the function of changing the temperature field. The calculations use the classical Fourier law of thermal conductivity. To solve the problem, the mathematical apparatus of incomplete separation of variables is used in the form of a generalized biorthogonal finite integral transformation (KIP), based on a multicomponent relation of eigenvector-functions of two homogeneous boundary value problems. An important point in the procedure of the structural algorithm of this method is the selection of the adjoint operator, without which it is impossible to solve non-self-adjoint linear problems of mathematical physics.

1. Introduction

Currently, piezoelectric temperature sensors of the resonant and non-resonant classes are widely used in measuring devices. Their work is based on the dependence of the electric field induced in the piezoceramic element on the magnitude of the external temperature effect [1–4]. To expand the functionality of sensors of this type, there is a need for in-depth analysis of non-stationary processes, which makes it possible to understand the effect of the interaction of mechanical, temperature and electric fields.

The mathematical formulation of the initial – boundary value problems describing the connectivity of thermoelectric fields includes a system of non-self-adjoint differential equations, the study of which has recently been carried out using numerical methods [5, 6]. However, rather weak effects of the interaction of fields of different physical nature can be analyzed only with the help of closed analytical solutions. In this case, the problem of integrating the initial calculated relations and the construction of a general solution leads to calculations in a simplified setting, namely, unrelated problems are investigated [4, 7] or infinitely long bodies are analyzed [8, 9].

The aim of this work is to solve a coupled non-stationary problem of thermoelectroelasticity for a long hollow piezoceramic cylinder under the action of a temperature load on its surfaces (boundary conditions of the 1st kind) [10]. The case is considered when the rate of change in the load is significantly less than the speed of propagation of elastic waves, which makes it possible not to take into account the inertial properties of the structure and use the equilibrium equations in calculations [11, 12].

2. Formulation of the problem

Let a long, hollow piezoceramic cylinder occupy the region in the cylindrical coordinate system (r_*, θ, z) the area $\Omega:\{a \le r_* \le b, 0 \le \theta \le 2\pi, -\infty < z < \infty\}$. Let us consider the case of action on its cylindrical surfaces of an unsteady load in the form of functions of temperature change with time t_* – $\omega_1^*(t_*)$ ($r_* = a$), $\omega_2^*(t_*)$ ($r_* = b$) (boundary conditions of the 1st kind). The front electroded surfaces of the element are short-circuited. In the general case, the differential equations of equilibrium, electrostatics, and heat balance based on the classical Fourier law of thermal conductivity in dimensionless form have the form [13–15]

$$\nabla\frac{\partial U}{\partial r} - a_1\frac{U}{r^2} + \nabla\frac{\partial \phi}{\partial r} - a_2\frac{1}{r}\frac{\partial \phi}{\partial r} - \nabla\Theta + a_3\frac{\Theta}{r} = 0 \tag{1}$$

$$-\nabla\frac{\partial \phi}{\partial r} + a_4\nabla\frac{\partial U}{\partial r} + a_5\frac{1}{r}\frac{\partial U}{\partial r} + a_6\nabla\Theta = 0$$

$$\nabla\frac{\partial \Theta}{\partial r} - \frac{\partial}{\partial t}\left(a_7\nabla U + \Theta - a_8\frac{\partial \phi}{\partial r}\right) = 0$$

$$r = R,1 \quad \frac{\partial U}{\partial r} + a_9\frac{U}{r} + \frac{\partial \phi}{\partial r} - \Theta = 0,\ \phi = 0,\ \Theta|_{r=R} = \omega_1,\ \Theta|_{r=1} = \omega_2 \tag{2}$$

$$t = 0 \quad U = \phi = \Theta = 0 \tag{3}$$

where $\{U, r, R\} = \{U^*, r_*, a\}/b$, $\phi = \frac{e_{33}}{C_{33}b}\phi^*$, $\{\Theta, \omega_1, \omega_2\} = \frac{\gamma_{33}}{C_{33}}\{\Theta^*, (\omega_1^* - T_0), (\omega_2^* - T_0)\}$

$$t = \frac{\Lambda}{kb^2}t_*,\quad a_1 = \frac{C_{11}}{C_{33}},\quad a_2 = \frac{e_{31}}{e_{33}},\quad a_3 = \frac{\gamma_{11}}{\gamma_{33}},\quad a_4 = \frac{e_{33}^2}{C_{33}\varepsilon_{33}},\quad a_5 = \frac{e_{31}e_{33}}{C_{33}\varepsilon_{33}},\quad a_6 = \frac{g_3 e_{33}}{\gamma_{33}\varepsilon_{33}},\quad a_7 = \frac{\gamma_{33}^2}{C_{33}k}T_0$$

$$a_8 = \frac{\gamma_{33}g_3}{e_{33}k}T_0,\quad a_9 = \frac{C_{13}}{C_{33}},\quad \nabla = \frac{\partial}{\partial r} + \frac{1}{r}$$

3. General solution construction

At the first stage of the solution, the procedure of reducing the inhomogeneous boundary conditions (2) to homogeneous is performed, which allows further use of the biorthogonal method of finite integral transformations to solve the problem under consideration [16]. For this, new functions are introduced $u(r,t)$, $\chi(r,t)$ associated with $U(r,t)$, $\Theta(r,t)$ in the following way

$$U(r,t) = H_1(r,t) + u(r,t),\quad \Theta(r,t) = H_2(r,t) + \chi(r,t) \tag{4}$$

where $H_1(r,t) = f_1(r)\omega_1 + f_2(r)\omega_2$, $H_2(r,t) = f_3(r)\omega_1 + f_4(r)\omega_2$.

Substitution of (4) into (1) - (3), under the conditions

$$f_1(R)' + a_9\frac{f_1(R)}{R} = 1,\ f_1(1)' + a_9 f_1(1) = 0,\ f_2(R)' + a_9\frac{f_2(R)}{R} = 0,\ f_2(1)' + a_9 f_2(1) = 1 \tag{5}$$

$$f_3(R) = f_4(1) = 1,\ f_3(1) = f_4(R) = 0$$

IOP Conf. Series: Materials Science and Engineering 1015 (2021) 012010 doi:10.1088/1757-899X/1015/1/012010

allows to obtain a new initial-boundary value problem for the functions $u(r,t)$, $\phi(r,t)$, $\chi(r,t)$ we solve it using the structural algorithm of biorthogonal finite integral transformation (KIP) [16]. For this, on the segment [R, 1], we introduce the KIP with unknown components of the eigenvector-functions of the transformation kernels $K_1(\lambda_i,r)...K_3(\lambda_i,r)$, $N_1(\mu_i,r)...N_3(\mu_i,r)$

$$G(\lambda_i,t)=\int_R^1\left(a_7\nabla u(r,t)+\chi(r,t)-a_8\frac{\partial\phi(r,t)}{\partial r}\right)rK_3(\lambda_i,r)dr \tag{6}$$

$$\{u(r,t),\phi(r,t),\chi(r,t)\}=\sum_{i=1}^{\infty}G(\lambda_i,t)\{N_1(\mu_i,r),N_2(\mu_i,r),N_3(\mu_i,r)\}\|K_i\|^2$$

$$\|K_i\|^2=\int_R^1 K_3(\lambda_i,r)N_3(\lambda_i,r)rdr$$

where λ_i, μ_i - eigenvalues of the corresponding homogeneous linear boundary value problems with respect to the conjugate $K_k(\lambda_i,r)$ and invariant $N_k(\mu_{ir},r)$ the components of the vector-functions of the KIP kernels ($k=1,2,3$).

As a result of using the KIP algorithm [16], we obtain a countable set of Cauchy problems for the transformant $G(\lambda_i,t)$

$$\frac{\partial G_i}{\partial t}+\lambda_i G_i=-F_H\ , \tag{7}$$

$$t=0 \qquad G_0=-\int_R^1\left[H_1(r,0)K_1(\lambda_i,r)+H_2(r,0)K_3(\lambda_i,r)\right]rdr\ , \tag{8}$$

whose solutions have the form

$$G_i=G_0\exp(-\lambda_i t)-\int_0^t F_H(\tau)\exp\lambda_i(\tau-t)d\tau\ ,\ F_H=\int_R^1\left(R_1K_1+R_2K_2+R_3K_3\right)rdr\ , \tag{9}$$

as well as two systems of differential equations and boundary conditions for the unknown components of the transformations $K_1(\lambda_i,r)...K_3(\lambda_i,r)$

$$\nabla\frac{dK_1}{dr}-a_1\frac{K_1}{r^2}-a_4\nabla\frac{dK_2}{dr}+\frac{a_5}{r}\frac{dK_2}{dr}-\lambda_i a_7\frac{dK_3}{dr}=0 \tag{10}$$

$$\nabla\frac{dK_1}{dr}+a_2\frac{1}{r}\frac{dK_1}{dr}+\nabla\frac{dK_2}{dr}+\lambda_i a_8\nabla K_3=0$$

$$\frac{dK_1}{dr}+a_3\frac{K_1}{r}+a_6\frac{dK_2}{dr}+\nabla\frac{dK_3}{dr}+\lambda_i K_3=0$$

$$r=R,1 \quad \frac{dK_1}{dr}+a_9\frac{K_1}{r}-a_4\frac{dK_2}{dr}=0,\ K_2=K_3=0 \tag{11}$$

and $N_1(\mu_i,r)...N_3(\mu_i,r)$

$$\nabla\frac{dN_1}{dr}-a_1\frac{N_1}{r^2}+\nabla\frac{dN_2}{dr}-\frac{a_2}{r}\frac{dN_2}{dr}-\nabla N_3+a_3\frac{N_1}{r}=0 \tag{12}$$

$$\nabla\frac{dN_2}{dr}-a_4\nabla\frac{dN_1}{dr}-\frac{a_5}{r}\frac{dN_1}{dr}-a_6\nabla N_3=0$$

$$\nabla\frac{dN_3}{dr}+\mu_i\left[a_7\nabla N_1+N_3-a_8\frac{dN_2}{dr}\right]=0$$

$$r=R,1 \quad \frac{dN_1}{dr}+a_9\frac{N_1}{r}+\frac{dN_2}{dr}=0,\ N_2=N_3=0 \tag{13}$$

Systems (10), (12) when using the condition $\{a_7, a_8\} \ll \{a_1 ... a_6\}$, valid for piezoceramic materials, as well as the introduction of new functions $\{K_4, N_4\} = r\frac{\mathrm{d}\{K_2, N_2\}}{\mathrm{d}r}$, $\{K_5, N_5\} = r\{K_3, N_3\}$, are reduced to the following resolving equations

$$\left(\frac{\mathrm{d}^2}{\mathrm{d}r^2}+\frac{1}{r}\frac{\mathrm{d}}{\mathrm{d}r}-\frac{1}{r^2}\right)\left[\frac{\mathrm{d}^2}{\mathrm{d}r^2}-\frac{1}{r}\frac{\mathrm{d}}{\mathrm{d}r}+\left(\frac{1}{r^2}+\lambda_i\right)\right]K_5 = D_{1i}\frac{p_1}{r^2} \quad (14)$$

$$\left(\frac{\mathrm{d}}{\mathrm{d}r}+\frac{2}{r}\right)\left(\frac{\mathrm{d}^2}{\mathrm{d}r^2}-\frac{1}{r}\frac{\mathrm{d}}{\mathrm{d}r}-\mu_i\right)\left(\frac{\mathrm{d}}{\mathrm{d}r}-\frac{1}{r}\right)N_5 = E_{1i}\frac{p_2}{r^2} \quad (15)$$

where

$$p_1 = a_6 + \frac{a_5(a_3 - a_2 a_6)}{(1+a_4)}, \quad p_2 = \frac{\mu_i a_8 (b_1^2 - b_2^2) b_1 + b_2 (a_2 b_1^2 - a_8 b_2)}{b_1^3}, \quad b_1 = (a_4 a_8 - a_7), \quad b_2 = (a_5 a_8 - a_7)$$

General solutions of differential equations (14), (15) have the form

$$K_5 = \sum_{n=1}^{5} D_{mi} F_n(\lambda_i, r), \quad N_5 = \sum_{n=1}^{5} E_{ni} P_n(\mu_i, r) \quad (16)$$

where $F_1 = -p_1 \frac{\pi}{2} r\left[Y_0(B_i r)\int J_0(B_i r)dr - J_0(B_i r)\int Y_0(B_i r)dr\right]$, $\{F_2, F_3\} = r\{J_0(B_i r), Y_0(B_i r)\}$

$$F_4 = \frac{\pi B}{2}\left[rY_0(B_i r)\int J_0(B_i r)r^{-1}dr - rJ_0(B_i r)\int Y_0(B_i r)r^{-1}dr\right], \quad F_5 = \frac{r}{\lambda_i}, \quad B_i = \sqrt{\lambda_i}$$

$$P_1 = p_2 r\left[\int V_1(r) I_1(A_i r)dr - \int V_2(r)\tilde{K}_1(A_i r)dr\right], \quad \{P_3, P_4\} = r\{I_0(A_i r), \tilde{K}_0(A_i r)\}$$

$$P_2 = r\left[\int V_3(r) I_1(A_i r)dr - \int V_4(r)\tilde{K}_1(A_i r)dr\right], \quad P_5 = r, \quad V_1(r) = \int r^{-1}\tilde{K}_1(A_i r)dr$$

$$V_2(r) = \int r^{-1} I_1(A_i r)dr, \quad V_3(r) = \int r^{-2}\tilde{K}_1(A_i r)dr, \quad V_4(r) = \int r^{-2} I_1(A_i r)dr, \quad A_i = \sqrt{\mu_i}$$

In equalities (14) - (16) $D_{1i} ... D_{5i}$, $E_{1i} ... E_{5i}$ the integration constants, $J_v(...), Y_v(...)$, $I_v(...), \tilde{K}_v(...)$ - ordinary and modified Bessel functions of the 1st and 2nd kind of order v.

Using the differential dependences obtained in the process of reducing (10), (12) to (14), (15) we obtain expressions for the functions $K_1(\lambda_i, r) ... K_3(\lambda_i, r)$, $N_1(\mu_i, r) ... N_3(\mu_i, r)$.

Substitution $K_1(\lambda_i, r) ... K_3(\lambda_i, r)$, $N_1(\mu_i, r) ... N_3(\mu_i, r)$ into the corresponding boundary conditions (11), (13), allows one to form two systems of algebraic equations, the solution of which allows determining the integration constants $D_{1i} ... D_{6i}$, $E_{1i} ... E_{6i}$ and eigenvalues λ_i, μ_i.

Final function expressions $U(r,t)$, $\phi(r,t)$, $\Theta(r,t)$ we obtain by applying inversion formulas (6) to transform (9). As a result, taking into account (4), we have

$$\{U(r,t), \Theta(r,t)\} = \{H_1(r,t), H_2(r,t)\} + \sum_{i=1}^{\infty} G(\lambda_i, t)\{N_1(\mu_i, r), N_3(\mu_i, r)\}\|K_i\|^{-2} \quad (17)$$

$$\phi(r,t) = \sum_{i=1}^{\infty} G(\lambda_i, t) N_2(\mu_i, r)\|K_i\|^{-2}$$

Functions $H_1(r,t)$, $H_2(r,t)$ are determined by solving the following differential equations

$$\nabla \frac{\partial H_m}{\partial r} = 0, \ (m = 1, 2) \quad (18)$$

As a result of solving (18), when conditions (8) are satisfied, we obtain expressions for $f_1(r) ... f_4(r)$.

4. Conclusion

As an example, we consider a radially polarized piezoceramic cylinder ($R = 0.8$) of composition PZT-4, which has the following physical characteristics: $\{C_{11}, C_{33}, C_{13}\} = \{13.9,\ 11.5,\ 7.43,\}\times 10^{10}$ N/m² $\varepsilon_{33} = 5.62\times10^{-9}$ F/m, $\{e_{31}, e_{33}\} = \{-5.2,\ 15.1\}$ Kl/m², $\rho = 7500$ kg/m³, $\{\gamma_{11}, \gamma_{33}\} = \{4.6,\ 3.9\}\times 10^{5}$ N/(m²K), $g_3 = 2\times10^{-4}$ Kl/(m²K), $k = 3\times10^{6}$ J/(m³K), $\Lambda = 1.6\ t/(\mathrm{mK})$, $\alpha = 5.6\ Vt/(\mathrm{m^2K})$, in the case of action on the inner surface ($r_* = a$) of a temperature load in the form of

$$\omega_1^*(t_*) = T_{\max}\left[\sin\left(\frac{\pi}{2t_{\max}^*}t_*\right)H\left(t_{\max}^* - t_*\right) + H\left(t_* - t_{\max}^*\right)\right],\ \omega_2^*(t_*) = 0 \qquad (19)$$

where $H(\tilde{t})$ - unit Heaviside function ($H(\tilde{t} = 1)$ at $\tilde{t} \geq 0$), $H(\tilde{t}) = 0$ at $\tilde{t} < 0$), $T_{\max} = T_{\max}^* - T_0$, $T_{\max}^*$, $t_{\max}^*$ - the maximum value of the external temperature effect and the corresponding time in dimensional form ($T_{\max}^* = 373$ K$(100\ ^\circ\mathrm{C})$, $T_0 = 293$ K$(20\ ^\circ\mathrm{C})$, $t_{\max}^* = 1$ s).

Figure 1 shows graphs of changes in functions $\Theta^*(r,t)$, $\phi(r,t)$, $U(r,t)$ radial coordinate r at different times t ($b = 0.02$ m). The numbers 1–3 respectively denote the results for $t = t_{\max}$, $3t_{\max}$, $15t_{\max}$.

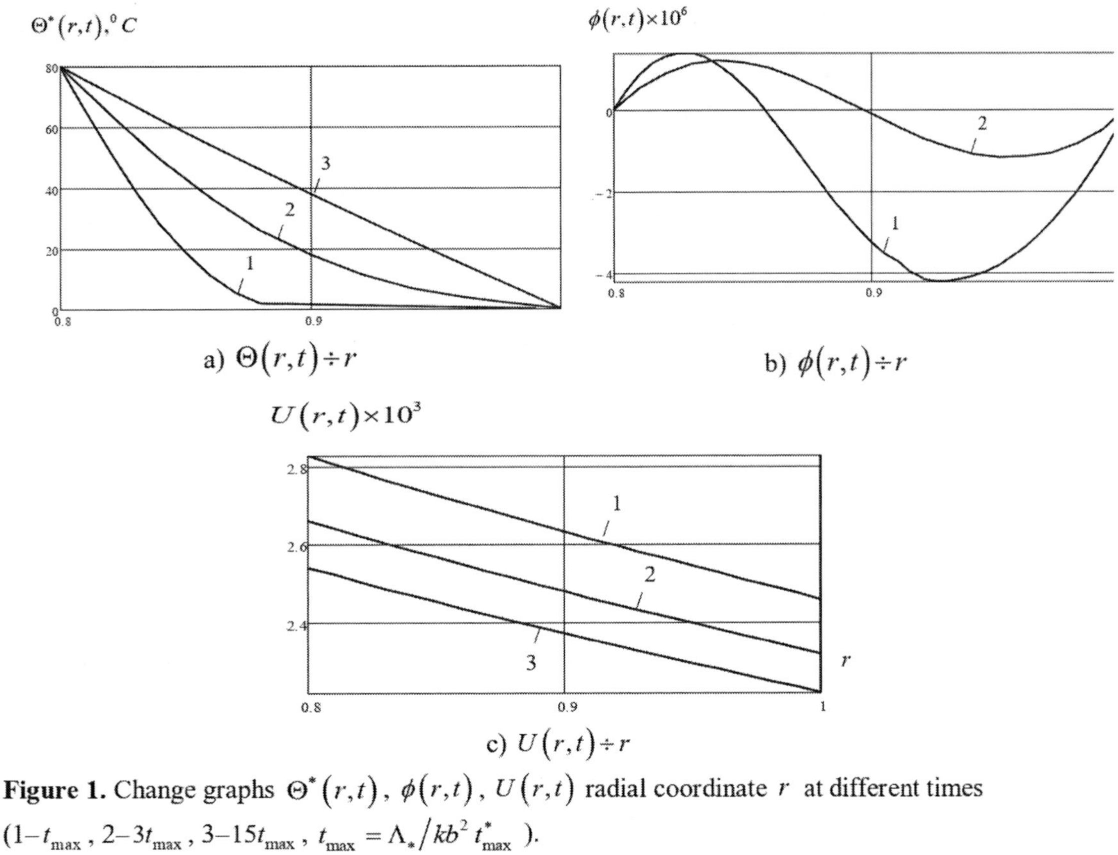

Figure 1. Change graphs $\Theta^*(r,t)$, $\phi(r,t)$, $U(r,t)$ radial coordinate r at different times ($1 - t_{\max}$, $2 - 3t_{\max}$, $3 - 15t_{\max}$, $t_{\max} = \Lambda_* / kb^2\, t_{\max}^*$).

Analysis of the graphs presented allows us to draw the following conclusions
1. When the temperature load reaches the maximum values $t = t_{max}$ (figure 1a, curve 1) the temperature field changes in the region close to the front heated surface ($r = R$).Subsequently, the temperature field of the cylinder increases (figure 1a, 2), and complete heating of the piezoceramic structure is observed at $t = 15t_{max}$ (figure 1a, 3).
2. Under the action of a temperature load (19), the piezoceramic cylinder is compressed in the radial plane (figure 1 c), which leads to the formation of an electric field. In this case, the electric potential $\phi(r,t)$ and moving $U(r,t)$ have a maximum value at $t = t_{max}$ (figure 1.b, curve 1; figure 1.c, 1). Further, the numerical values $\phi(r,t)$ and $U(r,t)$ decrease (figure 1, b, 2, figure 1, c, 2, 3), and the electric field disappears at a steady temperature $t = 15t_{max}$.

Figure 2 shows graphs of changes in the radial component of the electric field strength $E_r = \partial\phi/\partial r$ radial coordinate r . The numbers 1–3 respectively denote the results for a cylinder with a radius $b = 0.01$, 0.02, 0.03 m. The calculation results allow us to conclude that an increase in the cylinder wall thickness, i.e. a decrease in the rigidity of the structure, leads to an increase in the electric field strength E_r .

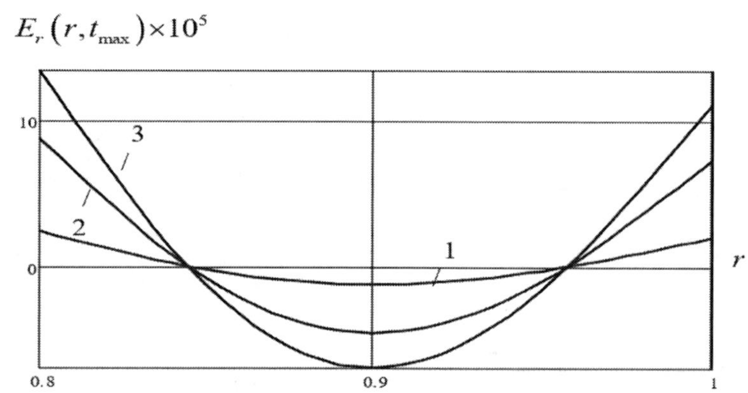

Figure 2. Graphs of changes in electric field strength $E_r(r,t_{max})$ radial coordinate r (1–$b = 0.01$ m , 2–$b = 0.02$ m, 3–$b = 0.03$ m).

When studying the initial-boundary value problems of thermoelectroelasticity for a long hollow piezoceramic radially polarized cylinder in the case of an unsteady temperature load, the coupling of electric and temperature fields must be taken into account in thick-walled elements $R \leq 0.01$. In turn, taking into account the rate of change in the volume of the electroelastic body ($\frac{\partial}{\partial t}\nabla U$) does not lead to a redistribution of temperature in it. This dependence is observed in structures made of a material with a significantly higher coefficient of linear thermal expansion, in particular, in polyvinyl compounds [11].

References

[1] Kozlov V L 2005 *Optoelectronic sensors* (Minsk) p 116
[2] Kulchin Yu N 2001 Distributed fiber optic measurement systems (Moscow) p 272
[3] Dmitriev S A and N N Slepov 2010 *Fiber optic technology: current state and new perspectives: collection of articles* (Technosphere) p 608
[4] Pankov A A 2018 *Bulletin of the Perm National Research Polytechnic University* **2** pp 72-82

[5] Abbas I A and Youssef H M 2009 *Arch Appl Mech.* **79** pp 917-925
[6] He T 2005 *European J. of Mechanics A/Solids* **24** pp 349-359
[7] Abbas I A andA Zenkour M 2013 *Composite Structures* **96** pp 89-96
[8] Vatulyan A O, Kiryutenko A Y and Nasedkin A V 1996 *PMTF* **37** (5) pp 135-142
[9] Vatulyan A O and Nesterov S A 2017 *Computational mechanics of continuous media* **10** (2) pp 117-126
[10] Kovalenko A D 1965 *Introduction to thermoelasticity* (Dumka) p 204
[11] Shlyakhin D A and Dauletmuratova Zh M 2018 *Engineering journal: science and innovations* **5** (77) (doi.1018698 / 2308-6033-2018-5)
[12] Shlyakhin D and Dauletmuratova Zh M 2019 *Bulletin of PNRPU Mechanics* **4** pp 191-200 (doi: 10.15593/perm.mech/2019.4.18)
[13] Lychev S A, Manzhirov A V and Uuber S V 2010 *Mechanics of Solids* **4** pp 138–154
[14] Radaev Yu N and Taranova M V 2011 *Bulletin of the Samara State techn. un–ty. Ser. Phys. - mat. science* **2** (23) pp 53−61
[15] Grinchenko V T, Ulitko A F and Shulga N A *The mechanics of related fields in structural members* (Dumka) p 279
[16] Senitsky Yu E 1996 *Izv. vuzov. Mathematics* (8) pp 71–81
[17] Kobzar V N and Filshtinsky L A 2008 *Applied Math. and Mech.* **72** (5) pp 842−851

IOP Conf. Series: Materials Science and Engineering 1015 (2021) 012011 doi:10.1088/1757-899X/1015/1/012011

Hybrid cross-tie system with passive dampers study

M Pustovalova, A Konovalov and Y Selivanov

Structural Mechanics and Strength of Materials Department, Northern (Arctic) Federal University named after M.V. Lomonosov, Arhangelsk, Russia

E-mail address: m.pustovalova@narfu.ru

Abstract. The authors have investigated the use of passive vibration dampers through the implementation of concentrated masses in the hybrid cross-tie system of a cable-stayed bridge. A 3D model of a cable-stayed bridge was built in ANSYS suite. The authors have studied the stress-strain behavior and dynamic characteristics of the model under the gravity load. The study has shown that the most effective model appears to be the one with passive dampers implemented by means of six concentrated masses. The increased stiffness of links of the concentrated masses had stronger effect on the model with passive dampers implemented by means of ten concentrated masses, however the model where passive dampers were implemented by six concentrated masses displayed comparable values of natural vibration frequencies with deflections in the spans being 3% lower, displacements of the towers being 8% lower, and longitudinal forces in the main cable being 6% lower. The results obtained by the authors allow for the conclusion that all models of a hybrid cross-tie system with passive dampers reviewed have high aerodynamic stability.

1. Introduction

Hybrid systems in which the roadway comprises a girder with tower supports and staying system are widely used in bridge engineering. This kind of structures provide high load-bearing capacity to cable-stayed bridges, are lightweight, and allow for covering long bridge spans without having to install additional supports. Yet, such systems are sensitive to dynamic loading. Different dampening devices are used for the prevention of wind-induced vibrations in bridge structures. This study is intended as the study of stress-strain behavior, evaluation of wind-induced vibration critical velocities and aerodynamic stability of the structures when passive dampers are used in the form of additional concentrated masses attached by elastic constraint. To that end, a 3D model of a cable-stayed bridge was built in ANSYS suite. The dimensions of the model were assigned as per the contemplated design of a bridge across the Kuznechikha River in Arkhangelsk [1].

The influence of passive vibration dampers on the strain-stress behavior and dynamic characteristics of the structure were studied for a cable-stayed bridge with the end spans each 200 m long and 450 m central span, 90 m high towers, and the bridge height of 30 m. Stiffeners, dimensions and configuration of a hybrid system meets the model described in ref. [2].

2. Stress-Strain Analysis and Modal Analysis

Three cases of passive vibration dampers modeling were analyzed: ten, six and two concentrated masses fixed at the ends of the stiffening truss in the central span on elastic elements by means of constraint $EI = 2.46 \cdot 10^{11}$ Nm^2, $EA = 2.51 \cdot 10^8$ N. Passive dampers implemented by means of ten concentrated masses are shown on figure 1. The distance from the tower to the nearest concentrated mass is 80 meters; the

IOP Conf. Series: Materials Science and Engineering 1015 (2021) 012011 doi:10.1088/1757-899X/1015/1/012011

distance between the masses along the stiffening truss is 60 meters, while the distance between the masses across the stiffening truss is 30 meters.

Figure 1. Passive dampers implemented by means of concentrated masses.

Initially, the effect of the weight of concentrated masses on the stress-strain behavior of the model was analyzed. The variation was between 0 and 8% of the bridge structure weight without dampers. The calculations showed that increase of the weight of 10 passive dampers to 8% of the weight of the original bridge structure will increase the deflections at midspan by 35%. A 40% increase was noted in the displacement of the tops of towers.

The modal analysis of a cable-stayed bridge model with different concentrated masses has shown that the use of passive dampers has no significant effect on the variation of natural torsional frequencies and natural transverse frequencies of a cable-stayed bridge. The transverse frequency has changed by 1.25%, and the torsional frequency has changed by 2.9 %. The frequency of vertical motions has decreased by 6%. The relationships between the variation of natural frequencies and the value of additional masses are shown on figure 2. The mod spacing is not consistent with the original model in which the first harmonic vibrational modes were located in the 3rd, 4th and 5th positions in the spectrum. Transverse, torsional and vertical vibrations of the model with dampers are located respectively in the 3rd, 24th and 35th positions in the spectrum (figure 3). Intermediate positions in the spectrum are occupied by vibrations of additional masses.

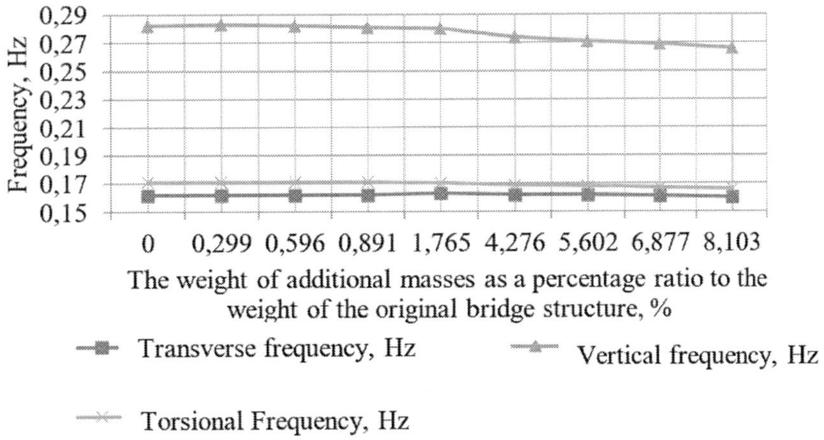

Figure 2. The Relationship of Natural Vibrations and the Weight of Additional Masses.

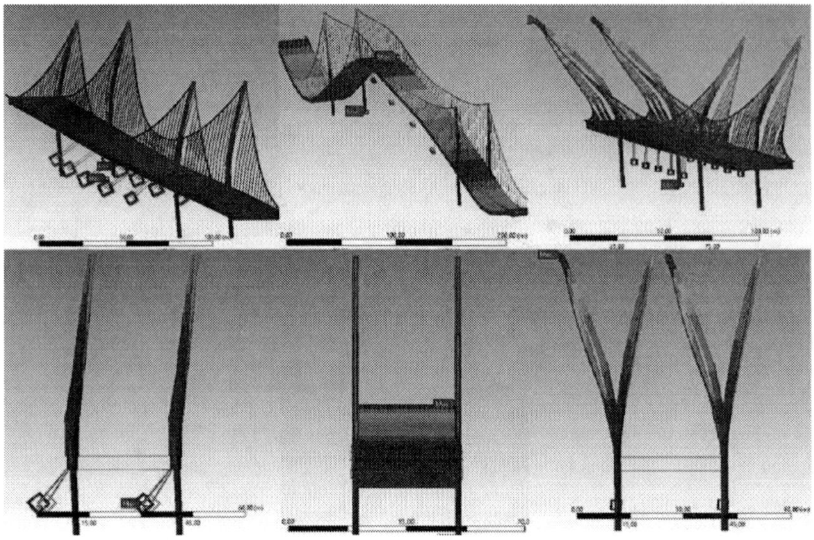

Figure 3. The first harmonic vibrational modes with passive dampers: transverse, vertical and torsional.

At the next phase of the study, the value of additional concentrated masses was assumed to be constant and equal to 8% of the original weight of the bridge structure. The modal analysis of a cable-stayed bridge with varying stiffness of links of the concentrated masses has shown that the increased stiffness of links of additional masses has resulted in 60% increase of transverse vibrational frequency. The changes in the stiffness of links of the concentrated masses have no effect on the frequencies of vertical and torsional vibrations (figure 4).

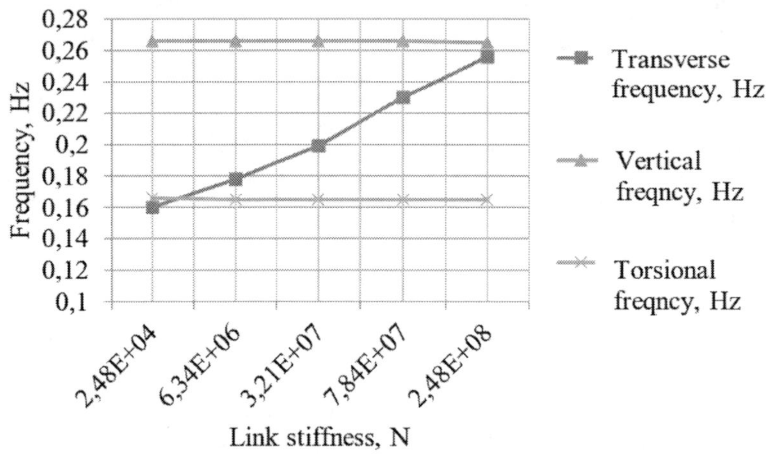

Figure 4. The Effect of Additional Masses Link Stiffness on the Natural Frequency.

Next, the study focused on the influence caused by the number of concentrated masses used in the cable-stayed bridge structure on its stress-strain behavior and dynamic characteristics. In all cases, the displacements and internal forces in the structural elements were confirmed to be tending to increase as an effect of the increased load by means of additional concentrated masses. In contrast, the frequencies of natural vibrations of the structure showed very small change. In all cases reviewed using 2, 6 and 10 additional concentrated masses, considerable effect on transverse frequency of the structure was caused only by the increase in the stiffness of links that attached these masses to the structure (Figure 4).

On the basis of the analysis completed on the models, the most efficient model appears to be the model with passive dampers implemented by means of six concentrated masses. As compared to the models having passive dampers implemented by two and ten concentrated masses, the model with dampers represented by six concentrated masses has lower deflections in the spans, lower displacements of the towers, lower stress in the stiffening truss, and lower longitudinal forces in the main cable (table 1).

Table 1. Characteristics of a Cable-Stayed Bridge with Passive Vibration Dampers.

Parameter	The Number of Concentrated Masses With the Total Weight Equal to 8% of the Bridge Weight			
	0	2	6	10
Maximal stress in the stiffening truss at midspan, MPa	100.1	135	133.17	133.2
Maximal stress in the stiffening truss at the end span, MPa	67.34	125	111.4	115.5
Longitudinal forces in the main cable, $\times 10^7$, N	9.17	11.5	10.5	11.2
Deflection in the middle of the central span, m	1.819	2.725	2.46	2.548
Deflection in the middle of the end span, m	0.68	1.16	1.08	1.113
Displacement of the tower top, m	0.443	0.646	0.607	0.621

3. Aerodynamic Stability

The change of aerodynamic characteristics of the structure has an effect on its aerodynamic stability [3]. Aerodynamic stability condition is written as [4, 5]

$$v_{cr} >> v_{WND} \quad or \quad v_{cr} > 1.5 v_{WND} \tag{1}$$

where v_{WND} is the design wind speed, i.e. the maximum wind speed that can be expected in this bridge construction area (normally, $v_{WND} = 25 \ldots 35$ m/s); 1.5 is specific safety factor [6, para. 2.24].

In the practice of cable-stayed bridges structural analysis, aerodynamic stability is evaluated using various methods and criteria that are based on the natural vibrational frequency of the bridge structures.

A.A. Petropavlovskij states the relationship between the flow velocity v, characteristic effective size h of the structure towards the cross flow, and Karman vortex shedding frequency Θ_W [6, p.183]

$$S_H = \Theta_W h v^{-1} \tag{2}$$

where Θ_W is the circular frequency of Karman vortex generation; S_N is the Strouhal number, dependent on the shape of the structure; for bridges, $S_N = 0.12 \div 0.18$, $S_N = 0.15$ is assumed. Wind-induced vibrations are understood to be the build-up of amplitudes of self-exited vibrations directed across the air flow that occurs when the Karman vortex shedding frequency coincides with one of the natural vibrational frequencies of the structure. For the estimation of aerodynamic stability by criterion of critical velocity for wind-induced vibrations the following formula is used:

$$v_{cr} = w \cdot 2\pi \cdot (h_B \cdot B)^{1/2} \cdot S_N^{-1} \tag{3}$$

where h_B is the stiffening truss height, (m); B is the bridge breadth, (m); ω is the natural vibrational frequency of the structure, (Hz). Table 2 is the summary of results obtained by the calculation of wind-induced vibrations critical velocities using the formula (3) for the models reviewed in this study.

Table2. Results Obtained by the Calculation of Wind-Induced Vibrations Critical Velocities

Vibration frequency w, Hz / Critical velocity, v_{cr}, m/s	Cable-stayed bridge model without passive vibration dampers	Model with passive vibration dampers which total weight is 8% of the bridge weight		
		The number of concentrated masses		
		2	6	10
transverse	$\frac{0.162}{74.3}$	$\frac{0.158}{72.5}$	$\frac{0.161}{73.8}$	$\frac{0.160}{73.4}$
vertical	$\frac{0.282}{129.3}$	$\frac{0.261}{119.7}$	$\frac{0.269}{123.3}$	$\frac{0.266}{122.0}$
torsional	$\frac{0.171}{78.4}$	$\frac{0.165}{75.7}$	$\frac{0.167}{76.6}$	$\frac{0.166}{76.1}$

3. Conclusions

In contrast to the models of passive dampers implemented by means of two and ten concentrated masses, the model of passive dampers implemented by six concentrated masses has less deflection in the spans, lower displacements of the towers, lower stress in the stiffening truss, and lower longitudinal forces in the main cable (see table 1). However, the greatest effect on the frequencies of natural vibrations was seen in the model with passive dampers implemented by means of two concentrated masses (table 2). The increased stiffness of links of the concentrated masses had stronger effect on the model with passive dampers implemented by ten concentrated masses, but the model where passive dampers are represented by six concentrated masses also displays comparable values of natural vibration frequencies with deflections in the spans being 3% lower, displacements of the towers being 8% lower, and longitudinal forces in the main cable being 6% lower. Aerodynamic stability conditions are met for all modeled cases. Based on the analysis of the models described above, the model with passive dampers represented by six concentrated masses can be considered as the most efficient model.

References

[1] http://www.ects.ru/images/552/File/jakovleva_l.g._bolshie_mosty_uchebnoe_posobie_2012.pdf / online: 03/2019.
[2] Konovalov A, Pustovalova M 2017 00083 E *26th RSP MATEC Web Conf.* **117**
[3] Dorogan A 2011 Cable-stayed bridges of Increased stiffness. *Str. Mech. and An. of Constr.* E4
[4] Baranovskij A 2005 *Bridges of large spans. Design of suspended and cable-stayed bridges* (PGUPS)
[5] Kachurin V 1971 *Design of suspended and cable-stayed bridges* (Moscow)
[6] Petropavlovskij A 1985 *Cable-stayed bridges* (Moscow) p 183

IOP Conf. Series: Materials Science and Engineering 1015 (2021) 012012 doi:10.1088/1757-899X/1015/1/012012

On the problem of oscillation control of viscoelastic mechanical systems with integral aftereffect

T Bobyleva[1], V Chernik[2] and A Shamaev[2,3]

[1] Moscow State University of Civil Engineering, Yaroslavskoye Shosse, 26, Moscow, 129337, Russia
[2] Ishlinsky Institute for Problems in Mechanics of the Russian Academy of Sciences, Pr. Vernadskogo, 101-1, Moscow, 119526, Russia
[3] Lomonosov Moscow State University, GSP-1, Leninskie Gory, Moscow, 119991, Russia

Email address: tatyana2211@outlook.com

Abstract. The article investigates the problems of control of dynamical systems with nonlocal terms of the convolution type using a force distributed over the area in which the motion is considered. The conditions are established under which it is possible to bring the system to complete rest with the help of a distributed force limited in absolute value. The case of one spatial variable is considered, but the proposed spectral method of research can be generalized to dimensions two and three. A connection is established between the behavior of the Fourier coefficients of the initial condition and the solvability of the control problem. The method can be used for mechanical systems with hereditary properties that arise in structural mechanics.

1. Introduction

In this paper, we develop a method for controlling viscoelastic mechanical systems with an integral delay based on the spectral method. The application of the method is demonstrated in four examples. This method can be used for various mechanical systems with hereditary properties arising, in particular, in structural mechanics.

A very large number of works by Russian and foreign authors are devoted to the problems of controlling systems with distributed parameters. This work is devoted to the study of the controllability of some systems with an integral aftereffect using a distributed force limited in absolute value.

For an equation of the form (3) (see below, which is sometimes called the Gurtin-Pipkin equation), it is established that there is controllability with the help of a force distributed over the entire region with a restriction in absolute value. Previously, control problems for this equation were investigated in [1-3]. Controllability is lost if a force is applied to a part of the domain, this was proved in [4].

Equation (1) (see below) was studied in [5]. It is shown that control over a part of the region is impossible, but the controllability problem becomes solvable if the control force is applied to the moving subdomain. At the same time, in these works, restrictions on the control action are not imposed.

In the present work, it is shown that in problem (1) controllability by means of a force applied to the complete region (in this case, to a segment) takes place. In this case, the modulus force can be limited to a small value.

The same statement is valid for equations (2) - (4) (see below), describing in a simplified form the oscillations of liquids with viscosity in integral form. The correctness of these models was investigated, for example, in [6].

In this paper, we solve the problem of damping oscillations for systems (1) - (4) using the spectral method that is common for all four models. For this, it is sufficient to use a force of small magnitude applied to the full segment. The kernel of the convolution, which determines the viscosity in integral form, contains only one exponent.

In [7], the problems of control of systems of differential equations with the help of a force bounded in absolute value distributed over the domain were investigated. The authors of this work showed that it is possible to completely stop the motion if the initial conditions are sufficiently smooth and satisfy some additional boundary conditions.

In the present work, a connection is established between the decrease in the velocity of the Fourier coefficients of the function specifying the initial condition and the ability to bring the system to rest with the help of a force limited in absolute value. The required velocity of decrease of the Fourier coefficients can also be a consequence of the smoothness of the initial conditions and additional boundary conditions for them. A similar method was applied in [8].

2. Problem specification and decision

Consider the following four control problems for dynamical systems with nonlocal terms of the convolution type using a force limited in absolute value and distributed over the domain in which the motion is considered. Namely, on an interval $[0,\pi]$ in the spatial variable x and on an interval $(0,+\infty)$ in time t, the following control problems are considered

$$\begin{cases} \dot{u} - c^2 u''_{xx} + K\int_0^t e^{-\lambda(t-\tau)} u(\tau,x)d\tau = f(t,x) \\ u(t,x) = \varphi(x),\ u(t,0) = u(t,\pi) = 0, \\ |f(t,x)| \le \varepsilon,\ f(t,x) \text{ is a control function} \end{cases} \tag{1}$$

$$\begin{cases} \dot{u} - c^2 u_{xx} + K\int_0^t e^{-\lambda(t-\tau)} u''(\tau,x)d\tau = f(t,x) \\ u(t,x) = \varphi(x),\ u(t,0) = u(t,\pi) = 0, \\ |f(t,x)| \le \varepsilon,\ f(t,x) \text{ is a control function} \end{cases} \tag{2}$$

$$\begin{cases} \dot{u} + K\int_0^t e^{-\lambda(t-\tau)} u''(\tau,x)d\tau = f(t,x) \\ u(t,x) = \varphi(x),\ u(t,0) = u(t,\pi) = 0, \\ |f(t,x)| \le \varepsilon,\ f(t,x) \text{ is a control function} \end{cases} \tag{3}$$

$$\begin{cases} \dot{u} - c^2 \dot{u}''_{xx} + K\int_0^t e^{-\lambda(t-\tau)} u''(\tau,x)d\tau = f(t,x) \\ u(t,x) = \varphi(x),\ u(t,0) = u(t,\pi) = 0, \\ |f(t,x)| \le \varepsilon,\ f(t,x) \text{ is a control function} \end{cases} \tag{4}$$

In (1) - (4) u is the displacement, $f(t,x)$ is a mass force, $\lambda > 0$.

System (1) corresponds, in particular, to a simplified model of the dynamics of the delayed neutron density in the reactor, system (2) describes the fluid oscillation control problem in the Oldroyd model, system (3) is the Maxwell model for fluids, as well as it is the heat conduction model with distributed

IOP Conf. Series: Materials Science and Engineering 1015 (2021) 012012 doi:10.1088/1757-899X/1015/1/012012

delay, and system (4) represents the Kelvin-Voight fluid model. For all four systems, the problem of choosing a control function $f(t,x)$ is posed so that for problems (1) - (4) $u(t,x) \equiv 0$ and $f(t,x) \equiv 0$ for all t, such that $t > T$, T is some positive constant.

In this paper, the existence $T > 0$ and the corresponding control function $f(t,x)$ is proved by a unified method. This method consists in representing the sought control function in the form of Fourier series in a system of functions $\{\sin nx\}_{n=1}^{\infty}$ at each value $t > 0$. This approach leads to sequences of control problems for oscillatory systems with dissipation, such as pendulum systems with friction. An external force, which depends on time, is here the controlling impact.

An important element is the proof of the possibility of choosing the stopping time $T > 0$, which is the same for the entire infinite set of the indicated simplest oscillatory systems. Some spectra values of problems (1) - (4) are presented in figures 1 - 4, respectively, μ_n are eigenvalues, $n = 1, 2, \ldots$. In each of the cases (1) - (4), the spectrum is a combination of $n = 1, 2, \ldots$ pairs of points on the complex plane, which are the spectra of the above-mentioned simplest oscillatory systems.

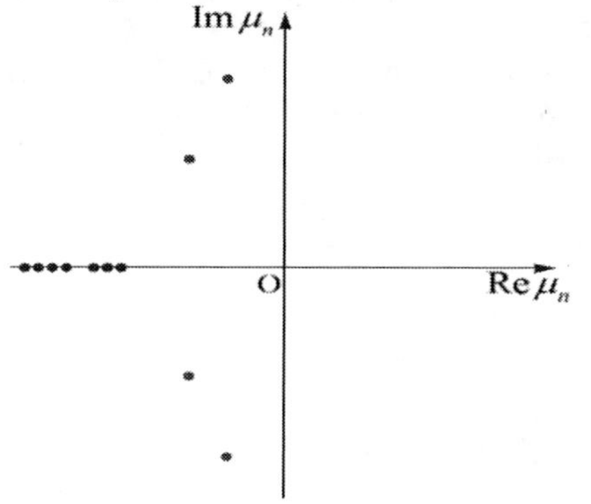

Figure 1a. The case of a finite number of spectrum points corresponding to damped oscillations is possible.

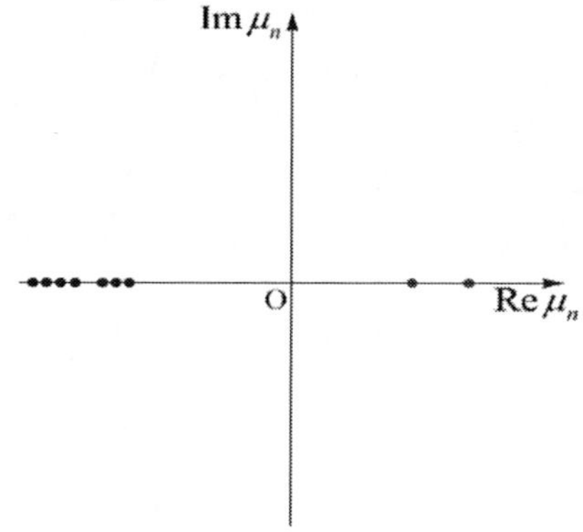

Figure 1b. The case of a finite number of points of the spectrum corresponding to unstable motion is possible, including unstable points of the spectrum may be absent, which corresponds to the stable motion of the system.

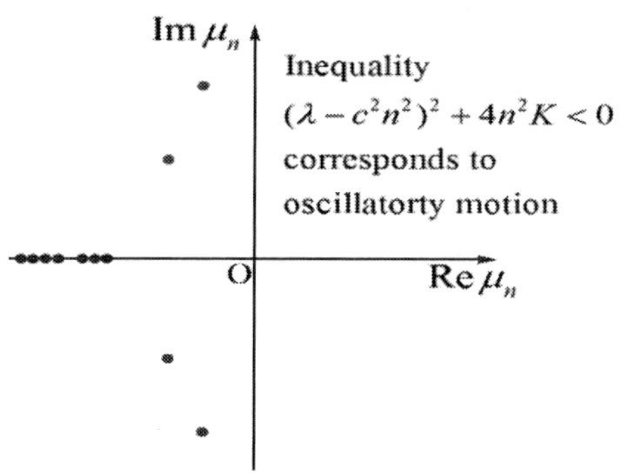

Figure 2. A finite number of spectrum points corresponding to unstable motion are possible. In this case, there are no eigenvalues corresponding to oscillations. There can be only a finite number of eigenvalues corresponding to oscillations.

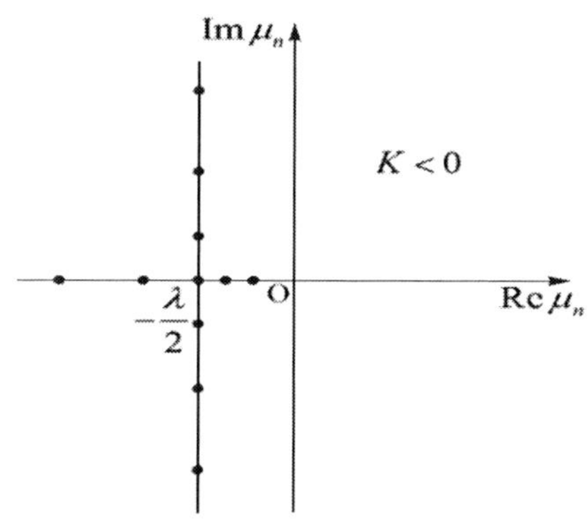

Figure 3. A finite number of spectrum points are possible, corresponding to attenuation without oscillations. Oscillations correspond to an infinite number of points in the spectrum.

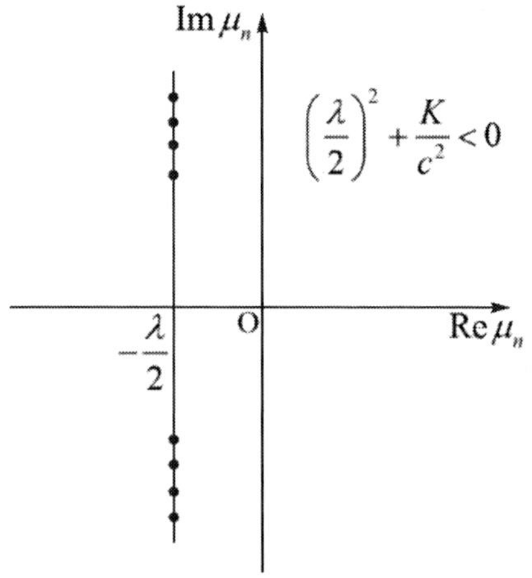

Figure 4a. All points of the spectrum correspond to damped oscillations.

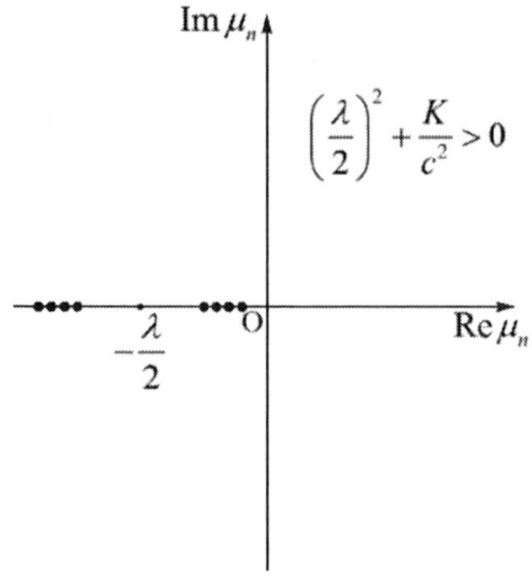

Figure 4b. All points of the spectrum correspond to decaying motions without oscillations.

The following statement holds.

Statement 1.

Consider the next sequence of systems

$$\begin{cases} \ddot{u}_n + A_n^2\dot{u}_n + B_n^2 u_n = g_n(t), \ n=1,2,\ldots \\ u_n(0)=a_n, \quad \dot{u}_n(0)=b_n \\ |g_n(t)|\le \varepsilon_n, \quad \sum_{n=1}^{\infty}\varepsilon_n \le \varepsilon \end{cases} \tag{5}$$

Here $g_n(t)$ is control function, ε is an arbitrary positive number. Let be $A_n \sim c_1 n,\ B_n \sim c_2 n,\ a_n \le c_3,\ b_n \le c_4$, c_1,c_2,c_3,c_4 are some constants. Then there are a constant $T>0$ independent of n and such a control function $g_n(t)$, such that $u_n(t)\equiv 0$ at $t\ge T$ and $g_n(t)\equiv 0$ if $t\ge T$ $n=1,2,\ldots$.

Proof.

The general solution of (5) has the form

$$u_n(t)=a_n e^{\alpha_n t}\cos\beta_n t+\frac{b_n-\alpha_n a_n}{\beta_n}e^{\alpha_n t}\sin\beta_n t+\frac{1}{\beta_n}\int_0^t e^{\alpha_n(t-\tau)}\sin\beta_n(t-\tau)g_n(\tau)d\tau \tag{6}$$

where $\alpha_n \pm i\beta_n$ are the roots of the characteristic equation.

We will look for a control function $g_n(t)$ in the form

$$g_n(t)=C_1^n(T)e^{-\alpha_n t}\cos\beta_n t+C_2^n(T)e^{-\alpha_n t}\sin\beta_n t \tag{7}$$

where $C_1^n(T), C_2^n(T)$ are to be determined from the conditions $u_n(T)=\dot{u}_n(T)=0$. Calculations show that the following inequality holds: $|C_1^n(T)|+|C_2^n(T)|\le C\exp(\gamma_n)$, $\gamma_n<0$, $\frac{|\gamma_n|}{|\alpha_n|}>1, n=1,2,\ldots$.

For the two real roots of the characteristic equation, control function $g_n(t)$ should also be sought in the form

$$g_n(t) = K_1^n(T)e^{-\lambda_n^{(1)}t} + K_2^n(T)e^{-\lambda_n^{(2)}t} \tag{8}$$

where $\lambda_n^{(1)} \le \lambda_n^{(2)} < 0$ are two negative real roots of the characteristic equation. In this case, it is also easy to show that $|K_1^n(T)| + |K_2^n(T)| \le C\exp(\gamma_n)$, $\gamma_n < 0$, $\frac{|\gamma_n|}{|\lambda_n|} > 1, n = 1,2,\ldots$ Hence it can be seen that for a sufficiently large T there will be $|g_n(\tau)| \le \varepsilon$ if $\tau < T$, $\varepsilon > 0$ is given small number, and more $T > 0$ can be chosen independent of $n = 1,2,\ldots$.

Consequence. We will consider the function $g(t,x) = \sum_{n=1}^{\infty} g_n(t)\sin nx$, which is determined by the condition $|g_n(t)| \le \varepsilon_n$, $\sum_{n=1}^{\infty} \varepsilon_n \le \varepsilon$. Then $g(t,x)$ is a control function that satisfies the constraint $|g(t)| \le \varepsilon$ and $g(t,x)$ is leading to "rest" the next system

$$\begin{cases} \ddot{u}_n + \lambda \dot{u}_n = \lambda c^2 u_{xx} + c^2 \dot{u}_{xx} + ku + g(t,x) \\ u_n(0,x) = \varphi(x) = \sum_{n=1}^{\infty} a_n \sin nx \\ \dot{u}_n(0,x) = \psi(x) = \sum_{n=1}^{\infty} b_n \sin nx \\ u(t,0) = u(t,\pi) = 0, \ t > 0 \end{cases} \tag{9}$$

in a finite time $T \ge 0$, if $A_n^2 = \lambda + c^2 n^2, B_n^2 = \lambda c^2 n^2 - K$.

Proof. The function $u(t,x)$ can be represented as a series $u(t,x) = \sum_{n=1}^{\infty} u_n(t)\sin nx$, where $u_n(t)$ is a solution to problem (5). According to Statement 1, there is such an n-independent value $T > 0$, that $u_n(t) \equiv 0$ if $n = 1,2,\ldots$. The series representing the function $u(t,x)$, is convergent due to the conditions on a_n, b_n, A_n, B_n. This is proved in the same way as in the justification of the classical Fourier method. Consequently, the control function $g(t,x)$ is indeed a solution to problem (5).

Now consider the control problem with a nonlocal term of the convolution type

$$\begin{cases} \dot{u} = c^2 u_{xx} + K\int_0^t e^{-\lambda(t-\tau)} u(\tau,x)d\tau + f(t,x) \\ u(t,0) = u(t,\pi) = 0, \quad u(t,0) = \varphi(x) \end{cases} \tag{10}$$

$f(t,x)$ is a control function. Problem (10) can be easily reduced to problem (9), in which there is no integral term. To do this, we differentiate equation (10) with respect to t, multiply this equation by λ and add the two equations obtained in this way. It is easy to see that we get problem (9) for which $g(t,x) = \lambda f(t,x) + \dot{f}(t,x)$. If the function $g(t,x)$ satisfies the condition $|g(t,x)| < \varepsilon$, then the function $f(t,x)$ can be obtained as a solution to a linear inhomogeneous first order equation and has the following form:

$$f(t,x) = \int_0^t e^{-\lambda(t-\tau)} g(\tau,x)d\tau \tag{11}$$

In this case, a positive constant $\varepsilon > 0$ can obviously be chosen so small that $|f(t,x)| \le \varepsilon_1$, $\varepsilon_1 > 0$ is a given small value. Choosing a control function $f(t,x)$ in accordance with formula (11), we obtain a solution to the control problem.

Statements similar to Statement 1 are easy to prove for the equations of oscillatory systems that arise when the Fourier method is applied to the Gurtin-Pipkin equations, as well as for equations similar to the equations describing the Oldroyd and Kelvin-Voight fluids, respectively:

$$\ddot{u}_n + \lambda \dot{u}_n - Kn^2 u_n = g_n(t) \tag{12}$$

$$\ddot{u}_n + (\lambda + c^2 n^2)\dot{u}_n + n^2(\lambda c^2 - K)u_n = g_n(t) \tag{13}$$

$$\ddot{u}_n + \lambda \dot{u}_n - \left(\frac{Kn^2}{1+c^2n^2}\right)u_n = \frac{g_n(t)}{1+c^2n^2} \tag{14}$$

Statement 2. The proof of the possibility of choosing a quantity $T > 0$ independently of $n = 1,2,\ldots$ is completely analogous to the proof of Statement 1, if in case (12) we replace in Statement 1 the conditions for a_n, b_n by the condition $a_n, b_n \sim Cn^{-(1+\alpha)}$, in case (14) by the condition $a_n, b_n \sim Cn^{-(3+\alpha)}$.

Hence, using the method of expansion in a Fourier series, it is easy to obtain statements about the controllability of the corresponding problems for partial differential equations.

3. Conclusions

The main result of this work is the proof of the possibility of bringing to rest a number of mechanical systems with an integral aftereffect using a distributed force applied to the complete area and limited in absolute value. Attention should be paid to the interesting fact of loss of controllability in the transition from the complete area to a part of the area as the area of application of the control action. In some cases, controllability can be maintained by applying a control action to a moving subdomain [5].

Acknowledgments

This work was supported by the Russian Science Foundation (project No. 16-11-10343).

References

[1] Pandofi L 2005 The controllability of the Gurtin-Pipkin equations: a cosine operator approach *Appl. Math. Optim.* **52** pp 143–165

[2] Ivanov S and Pandofi L 2009 Heat equations with memory: lack of controllability to rest *J. Math. Analysis and Appl.* **355** (1) pp 1–11

[3] Romanov I V and Shamaev A S 2018 Some problems of distributed and boundary control for systems with integral aftereffect *J. Math. Sci.* **234** (4) pp 470–484

[4] Shamaev A S and Romanov I V 2016 Noncontrollability to rest of the two-dimensional distributed system governed by the integrodifferential equation *J.Optim. Theory App.* **170** (3) pp 772–782

[5] Chaves- Silvia F W, Zhang X and Zuazua E 2017 Controbillity of evolution equations with memory *arXiv:* 1705.07683v1 [*math.OC*] 22 May 2017

[6] Oskolkov A P 1989 Initial boundary value problems for the equations of motion of Kelvin – Voigt fluids and Oldroyd fluids *Proc. Steklov Inst. Math.* **179** pp 137–182

[7] Chernousko F L 1992 Bounded control in distributed-parameter systems *J. Appl.Math. and Mech.* **56** (5) pp 707–723

[8] Bobyleva T and Shamaev A 2020 Problem of damping oscillations of a mechanical system with integral memory *IOP Conf. Ser.: Mater. Sci. Eng.* **869** p 022011

IOP Conf. Series: Materials Science and Engineering 1015 (2021) 012013 doi:10.1088/1757-899X/1015/1/012013

Experimental - theoretical study of the stress-strain state of cable-stayed-rod systems of the dome type

A D Akhmedov, I V Loseva and M A Kalmova

Samara State Technical University, 244, Molodogvardeyskaya St., 443100, Samara, Russian Federation

E-mail address: kalmova@inbox.ru

Abstract. The problem of determining forces and displacements in a spatial hinge-rod system with missing links under static loading is considered. Equilibrium equations and geometric relationships are written in matrix form. The calculated equations are not linear. To solve the equations, a technique is proposed that is not associated with assumptions regarding the structure of the system or the nature of the load, based on the fact that the displacements obtained by solving the linear part of the equations can be significantly improved by interpreting the nonlinear terms as a load, under which the system can be considered as linear. To test the theoretical assumptions and preconditions, an experimental study was carried out on a model made at a scale of 1:30 of a natural design. The analysis of the results obtained is presented.

1. Introduction

The cable-stayed-rod system of the dome type [1] is a spatial structure consisting of radially located, bending, rigid ribs and a pre-stressed cable-stayed network, limited by a cable-selection (flexible contour). The design scheme of the cable-stayed-rod system is a hinge-rod model of a variable, in particular instant-rigid type [4].

2. Formulation of the problem

The problem of determining the forces and displacements arising in a spatial hinge-rod system with missing (up to geometrically unchangeable) connections when the acting load changes is considered. The displacements are counted from a certain stable equilibrium state, called the initial one, which is considered given if the load and the corresponding system configuration and the forces in its elements (rods) are known. Any equilibrium state can be taken as the initial one: the state of prestressing, or loading with a constant load, or one of the limiting states. It is assumed that the relative deformations of the rods are small compared to unity, and for the material of the rods a linear relationship between stresses and deformations is valid. These provisions with a sufficient degree of accuracy reflect the nature of the work of the materials used in cable-stayed structures.

All further reasoning is reduced to a prestressed spatial hinge-rod system, which is understood as a system of material points (hinge nodes), interconnected by linear elastic ties (rods). In order to define a hinge-rod system, it is necessary, first of all, to provide information about the incidence of its nodes and rods. The specified information about the nature of the links reflects the topological structure of the system. A graph is a visual way to describe the structure of a system. The incidence matrix of this graph is called a rectangular table A, which uniquely defines a graph with numbered vertices and edges, and, consequently, the structure of the hinge-rod system [3].

3. General solution construction

The equilibrium of a hinge-rod system of instant rigid type under the action of a load applied to it is described by the system of equations

$$\mathbf{A} \cdot \mathbf{X} \cdot \mathbf{A}^{\mathrm{T}} \cdot \mathbf{Z} = \mathbf{F} \tag{1}$$

where $\mathbf{A}$ - incidence matrix [3]; $\mathbf{X}$, $\Delta\mathrm{X}$ - respectively, the diagonal matrix of linear forces and their increments in the rods of the system; $\mathbf{Z}$, $\mathbf{\Delta Z}$, $\mathbf{\Delta F}$ - three-column matrices, respectively, of the coordinates of the nodes, displacements of the nodes and the disturbing load along the coordinate axes; $\mathbf{C}$ - equilibrium matrix; $\mathbf{B}$ - quasi-diagonal matrix formed by three blocks of the form $\mathbf{AXA}^{|}$ (the dash denotes the transportation operation, the dash denotes the vector corresponding to the replacement of three-column matrices with a triple-length vector or a diagonal matrix by a vector consisting of elements of the main diagonal).

Moving on to considering the deformed state of the system, we transform equation (1), which describes the initial state, taking into account the change in the configuration and the efforts in the system caused by the increment in the load. Denoting by the symbol Δ the increment of the corresponding quantities, we write the equilibrium equation

$$\mathbf{A}(\mathbf{X} + \mathbf{\Delta X})\mathbf{A}^{|}(\mathbf{Z} + \mathbf{\Delta Z}) = \mathbf{F} + \mathbf{\Delta F} \tag{2}$$

Taking into account equation (1), we have

$$\mathbf{A\Delta XA}^{|}\mathbf{Z} + \mathbf{AXA}^{|}\mathbf{\Delta Z} + \mathbf{A\Delta XA}^{|}\mathbf{\Delta Z} = \mathbf{\Delta F} \tag{3}$$

Equation (3) is nonlinear, the solution of which requires individual analysis to develop a rational solution method. In this regard, let us consider the question of linearization of equation (3). Since for a given cable-stayed-rod system of the dome type the possible displacements caused by the load are quite small, we can neglect the values of the nonlinear terms in (3), then we obtain a linear equation of the form:

$$\mathbf{A\Delta XA}^{|}\mathbf{Z} + \mathbf{AXA}^{|}\mathbf{\Delta Z} = \mathbf{\Delta F} \tag{4}$$

or in a more compact form

$$\mathbf{C\Delta\bar{X}} + \mathbf{B\Delta\bar{Z}} = \mathbf{\Delta\bar{F}} \tag{5}$$

In equations (1), (2), the values of the corresponding quantities ($\mathbf{X}$, $\mathbf{Z}$) in the initial state of equilibrium under load $\mathbf{F}$. Calculation of the system according to the linear theory in accordance with the above-mentioned assumptions is possible only in the case when the rank of the matrix ($\mathbf{C}$, $\mathbf{B}$) is equal to the dimension of the load space. The linearization of equation (3) is based on the predominant use of the assumption that the initial form remains unchanged. Accordingly, after multiplying equations (5) by the matrix $(\mathbf{C}_1, \mathbf{D})^{|}$ ($\mathbf{D}$ - matrix orthogonal to $\mathbf{C}_1$) equilibrium conditions are reduced to the equations:

$$\mathbf{C}_1^{'}\mathbf{C\Delta\bar{X}} = \mathbf{C}_1^{'}\mathbf{\Delta\bar{F}} \tag{6}$$

$$\mathbf{D}^{'}\mathbf{B\Delta\bar{Z}} = \mathbf{D}^{'}\mathbf{\Delta\bar{F}} \tag{7}$$

where $\mathbf{C} = \mathbf{C}_1\mathbf{T}$, $\mathbf{C}_1, \mathbf{T}$ - matrices whose rank is equal to the rank of matrix $\mathbf{C}$,

$$\mathbf{\Delta X}_1 = \mathbf{T\Delta X} \tag{8}$$

$$\mathbf{D}^{'}\mathbf{C}_1 = \mathbf{0} \tag{9}$$

From (6) the solution for $\mathbf{\Delta X}_1$. In this case, the equilibrium part of the load

$$\mathbf{\bar{F}}_1 = \mathbf{C}_1(\mathbf{C}_1^{'}\mathbf{C}_1)^{'}\mathbf{C}_1^{'}\mathbf{\Delta\bar{F}} \tag{10}$$

As a solution to (7), the normal solution of the corresponding underdetermined system of equations

$$\bar{\mathbf{Z}}_2 = \mathbf{D}(\mathbf{D}'\mathbf{B}\mathbf{D})^{-1}\mathbf{D}'\Delta\bar{\mathbf{F}} \tag{11}$$

In this case $\bar{\mathbf{Z}}_2$ - displacements generated by the nonequilibrium part of the load. Displacements caused by the action of the equilibrium part of the load

$$\Delta\bar{\mathbf{X}} = \mathbf{G}\mathbf{C}'^{1}\bar{\mathbf{Z}}_1 \tag{12}$$

where $\mathbf{G}$ - diagonal stiffness matrix of bars.
According to (5) and (8)

$$\bar{\mathbf{Z}}_1 = \mathbf{C}_1(\mathbf{C}_1'\mathbf{C}_1)^{-1}(\mathbf{T}\mathbf{G}\mathbf{T}')\Delta\bar{\mathbf{X}} \tag{13}$$

where in

$$\Delta\bar{\mathbf{X}} = \mathbf{G}\mathbf{T}'(\mathbf{T}\mathbf{G}\mathbf{T}')^{-1}\Delta\bar{\mathbf{X}}_1 \tag{14}$$

The results are proved to be independent of the specific choice of matrices $\mathbf{C}_1$, $\mathbf{D}$. Work of the balanced part of the load $\mathbf{F}_1$ on non-equilibrium movements $\mathbf{Z}_2$ and work of the non-equilibrium part of the load $(\Delta\mathbf{F}-\mathbf{F}_1)$ on equilibrium displacements $\mathbf{Z}_1$ is equal to zero (Betty's theorems). Non-equilibrium movements do not cause changes in linear forces $\Delta\mathbf{X}$. A partial refinement of the described linear approach is obtained by taking into account in the matrix B instead $\mathbf{X}$, $\mathbf{X}+\Delta\mathbf{X}$, where $\Delta\mathbf{X}$ obtained by (14), which corresponds to the quasilinear approach to solving the problem [2], [5], [7].

Figure 1. General view of the model.

The study of the developed algorithm and program showed the acceptability of the described linear approach for calculating a cable-stayed-rod system of a dome type in a fairly wide range of initial states and disturbing loads. Due to the physical content, the proposed method of bringing the load to equilibrium (MPNR) can be effectively used when choosing the main parameters of cable-stayed rod systems [6], [8]. An experimental study of the cable-stayed rod system was carried out to test the

theoretical assumptions and premises of the MPNR. The overall dimensions of the model, made on a scale of 1:30 natural construction, in the plan were $D = 2.36$ m (figure 1).
The span was modeled together with rigid and support contours. The section of the model was selected in accordance with the similarity theorem, taking into account the physical and mechanical properties, the materials used and the acting loads. The model, representing 8 cyclically symmetric sectors, consisted of a rigid contour, a cable-stayed network limited by a pick-up cable and a supporting contour in the form of tables. The rigid contour was a spatial rod system of meridianly located stiffeners made of channels № 5 connected to the central ring. Between the meridian stiffeners, a cable-stayed network was suspended, consisting of circular and radial cables made of rope wire ($d = 0.8$ mm). The selection rope is made of class A-1 reinforcing steel ($d = 4$ mm, SP 63.13330: M, 2012). The cable-stayed network, consisting of 35 rods, formed 12 internal and 16 contour nodes in each sector (figure 2).

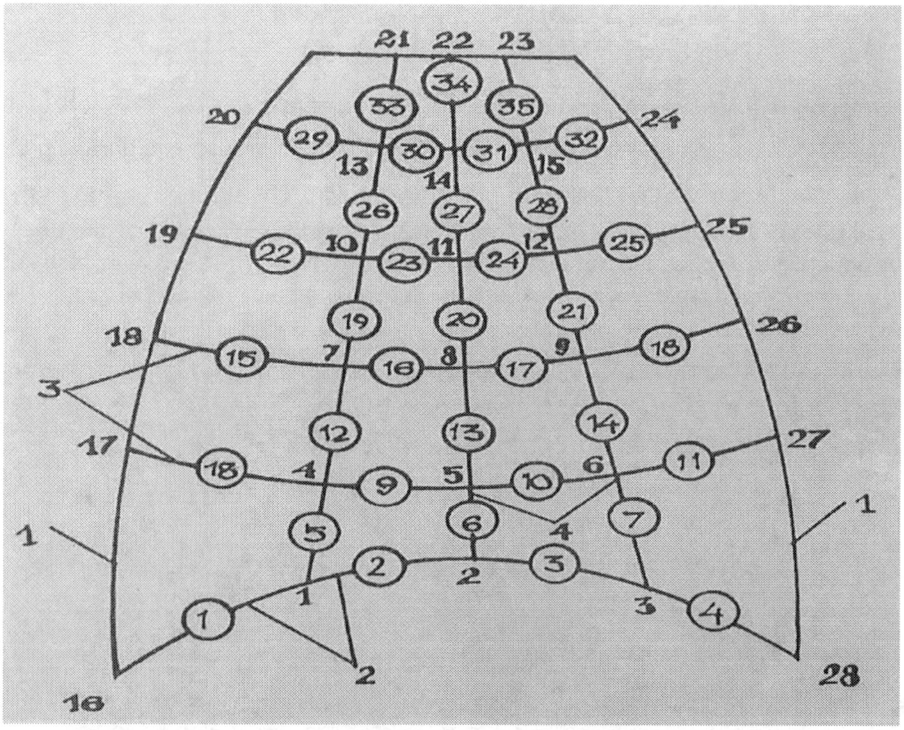

Figure 2. Plan of the sector of the model.
1 - stiffeners; 2 - rope selection; 3 - ring cables; 4 - radial cables.

According to the calculation, a preliminary voltage was created in the network of each sector of the model. After prestressing, the shape of the cable-stayed network surface became close to a hyperbolic paraboloid. The study of the model was carried out on the static effect of the load, the model was loaded according to four schemes of temporary and constant loads. The design of the model allowed only angular load transfer. In this regard, the loading was carried out with concentrated loads suspended from the internal nodes of the cable-stayed set with the help of straps. For all loading schemes, floor loading was carried out up to a load of 24.5 N (five stages of 4.9 N each) in each node, as well as stage-by-stage unloading. To determine the deformations in the rope-selection, resistance strain gauges and an automatic strain gauge AID-1M were used. displacements of five stiffeners were measured by deflection meters PAO - 6, displacements of nodes - with a leveling compensator "KONI - 007", efforts in the cables - by a deformer [9, 10]. Tables 1, 2, 3 show the experimental data, as well as the results of theoretical studies obtained using the developed method of bringing the load to equilibrium (MPNR) and the finite element method (MKE). The load is applied in all sectors.

Table 1. Efforts in the middle ring cables (N).

Load (N)	Node numbers											
	1			2			3			4		
	Exp	MPNR	FEM	Exp	MPNR	FEM	Exp	MPNR	FEM	Exp	MPNR	FEM
4.9	143.4	156.4	178.8	124.5	135.9	164.2	143.3	153.0	178.7	145.4	158.1	182.7
9.8	188.3	205.3	239.4	134.4	146.5	167.5	149.6	160.1	193.7	190.3	209.1	247.1
14.7	234.2	255.3	304.2	154.8	168.7	198.6	172.8	187.5	233.5	236.2	247.8	318.1
19.6	284.8	310.4	367.9	183.3	199.8	239.7	196.0	215.6	246.9	294.8	324.2	388.0
24.5	302.6	329.8	395.6	198.7	216.6	267.3	215.1	236.5	279.5	303.6	330.9	372.7

Table 2. Efforts in the middle radial cables (N).

Load, (N)	Node numbers											
	1			2			3			4		
	Exp	MPNR	FEM	Exp	MPNR	FEM	Exp	MPNR	FEM	Exp	MPNR	FEM
4.9	94.1	102.4	121.8	90.2	98.9	117.2	91.1	100.0	111.7	93.1	102.1	122.1
9.8	97.2	105.7	130.1	98.0	106.5	127.4	98.0	106.8	129.8	97.6	105.1	122.9
1.7	100.2	109.3	132.2	100.5	108.7	130.6	102.5	115.5	133.5	107.1	117.8	139.1
19.6	103.5	113.4	133.9	105.3	115.8	131.2	108.2	117.7	146.9	110.2	119.2	143.0
24.5	128.8	140.8	165.6	114.7	124.2	148.1	116.7	127.5	150.8	124.4	136.9	161.2

Table 3. Moving middle nodes in a circular direction (mm).

Load, (N)	Node numbers								
	1			2			3		
	Exp	MPNR	FEM	Exp	MPNR	FEM	Exp	MPNR	FEM
4.9	0.151	0.164	0.196	0.181	0.199	0.238	0.161	0.177	0.209
9.8	0.233	0.254	0.303	0.262	0.288	0.346	0.243	0.267	0.316
14.7	0.412	0.453	0.536	0.483	0.532	0.638	0.422	0.464	0.548
19.6	0.643	0.702	0.836	0.653	0.718	0.862	0.664	0.731	0.863
24.5	0.987	1.105	1.283	1.402	1.146	1.375	0.993	1.093	1.291

Under the action of the temporary load of the 1st and 2nd stages, the movements of the nodes in the unloaded sectors are directed in the opposite direction of the line of action of the load. The forces in the ring cables exceed the forces in the radial cables by almost 2.5 times in the loaded sectors by several times.

4. Conclusions

The proposed method (iterative process) converges in a wide range of system parameters, as evidenced by the performed calculations. The convergence has a quadratic character, as a result of which, with the required degree of accuracy, two or three iterations are sufficient for practical calculations. The application of a temporary load to any sector causes a change in forces and displacements in all sectors, which confirms the spatial work of the structure. Under the action of temporary loads in the dome system, the forces are redistributed due to changes in geometry in individual sectors. The efforts in the cable-stayed ones vary according to the parabolic law.

Analysis of the results shows the discrepancy between the experiment and the developed method (MPNR) within 9-12%, comparison of the experiment with the MKE gives a discrepancy of 28-32%.

References

[1] Gershanok R A and Rasha K K 1974 *Publ. in bull. Discoveries. Inventions. Industrial goods. Trademarks* (Moscow: PPP Patent) **31**
[2] Gantmakher F R 2002 *Matrix theory* (Moscow: Nauka) p 572
[3] Zykov A A 2004 *Fundamentals of graph theory* (Moscow: University book) p 664
[4] I M Rabinovich, 1962 *Instantly rigid systems, their properties and calculation bases. Hanging cover* p 247
[5] A D Akhmedov 2014 Almanac of modern science and education. **5-6** (84) pp 27-35
[6] A D Akhmedov 2014 Don's Engineering Herald. **4** pp 27-35
[7] A D Akhmedov, A Litikov 2014 Published monthey **5**(14) pp 97-101
[8] A D Akhmedov 2018 Urban planning and architecture **8** (2) pp 4-8
[9] A D Pitluk 1971 *Test of building structures on models*. (Leningrad, publishing house of literature on construction) p 152
[10] D A Shlyakhin, Zh M Dayletmyratova, M A Kalmova 2020 IOP Conf. Series: Materials Science and Engineering **775** 012137 doi:10.1088/1757-899X/775/1/012137

IOP Conf. Series: Materials Science and Engineering 1015 (2021) 012014 doi:10.1088/1757-899X/1015/1/012014

Parametric study of the stress state of beams with round holes in pure bending

S Saiyan and A Paushkin

Moscow State University of Civil Engineering, National Research University, 26 Yaroslavskoe shosse, Moscow, 129337, Russia

E-mail address: berformert@gmail.com

Abstract. A parametric study of the stress state of a rectangular cross-section beam with round holes in pure bending is performed by the finite element method implemented in the ANSYS Mechanical APDL software package. Correction coefficients were obtained for normal stresses determined by strength of materials methods. The coefficients of stress concentration during bending were determined depending on the size of the holes, as well as the distance between them. The research results obtained can be used in practical problems of designing perforated beams.

1. Introduction

Perforated beams are often used as load-bearing structures that work on bending. Such beams work well under static loads, while it is possible to save material and lighten the weight of building structures [1]. Existing methods for calculating perforated beams [2-7] are not able to consider the stress distribution across the section of perforated beams and do not answer the question of how exactly the stresses are distributed taking into account the holes. In this work, the objective was determine the correction factors of stress concentration for results obtained from the engineering methods of strength of materials. For reliable determination of the stress state of perforated beams with circular cutouts subjected to pure bending, depending on the size of the perforation and the perforation step, was conducted the parametric computer simulation of beams using the software ANSYS Mechanical APDL.

2. Research

To conduct the numerical study, we considered a rigidly pinched rectangular cross-section beam with a perforation in the form of round notches, the radius of which changes parametrically to the height of the beam. The cross-section of a detail of the perforation in the wall beams for parametric studies of the stress state shown in figure 1. For the parametric study varied the radius of the holes R, and the distance between them (pitch) L. The value of the beam height was taken $H = 0.9$ m, a thickness of the beam T=0.1 m. The beam material is steel with an elastic modulus $E = 2.1 \cdot 10^5$ MPa and a Poisson's ratio ν=0.3. The beam's own weight was not taken into account. The value of the bending moment $M = 900$ kN·m. The material model is selected as linear-elastic.

IOP Conf. Series: Materials Science and Engineering 1015 (2021) 012014 doi:10.1088/1757-899X/1015/1/012014

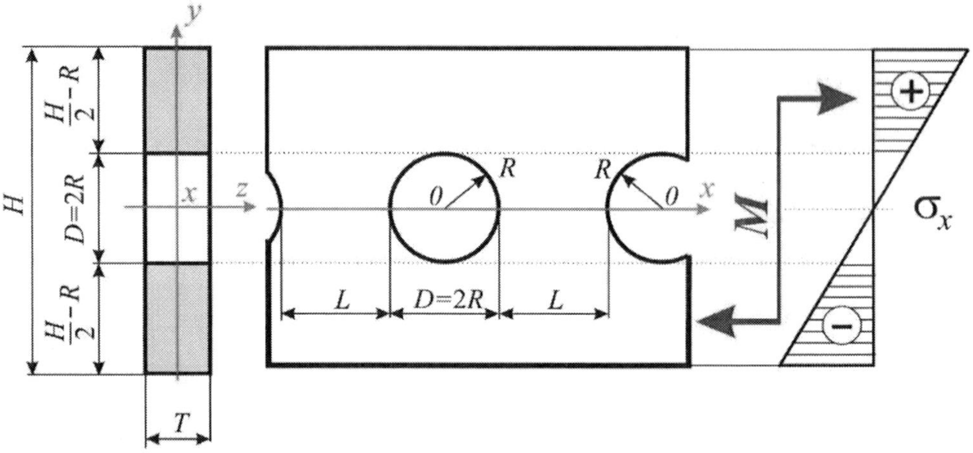

Figure 1. The cross section (left), a fragment of a beam perforation (center) and a normal stress σ_x diagram using strength of materials formulas (right).

As an analytical solution, we will use a solution using strength of materials formulas that gives acceptable accuracy results for solid beams when bending, but does not take into account the perforation effects of stress concentration near the holes. The finite element calculation was performed in a two-dimensional setting for a generalized plane stress state. The size of the end element varies depending on the perforation parameters and is determined based on the number of end elements with local thickening in the hole areas. Due to the constancy of the bending moment along the beam, normal stresses along the perforated beam (away from kinematic and static boundary conditions) [8-13] will only change due to the influence of holes, which allows us to study the influence of perforation on the distribution of normal stresses in local areas near the holes.

Based on the considered scheme in figure 1, we present the Isofields of normal cross-section stresses at $R = H/6$ and $L = 5R$, determined by the finite element method in the ANSYS Mechanical APDL software package (figure 2).

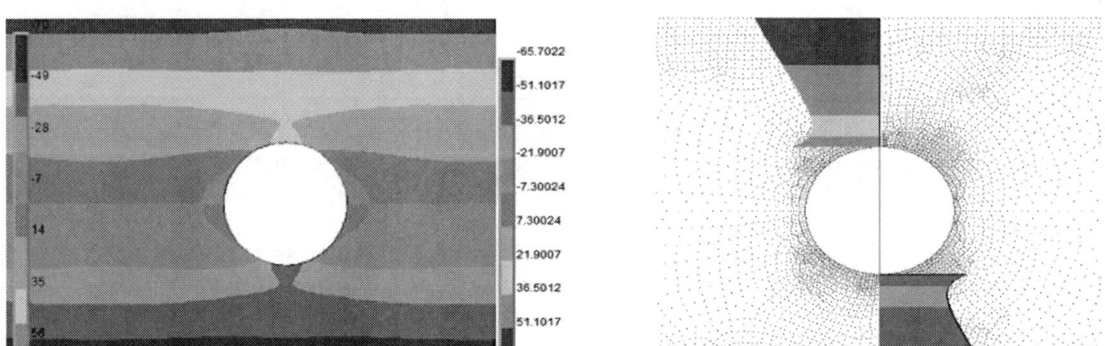

Figure 2. The normal stress σ_x field near the hole (left) and the normal stress σ_x diagram at the center of the hole in the ANSYS Mechanical APDL software package (right) at $R = H/6$ and $L = 5R$, MPa.

As you can see, if there is a perforation, there is a stress concentration that occurs at the border of the cutout. Moreover, the maximum stress values occur directly in the section at the center of the hole. This effect is characterized by a non-linear distribution of normal stresses across the cross-section (figure 2), while the linear law is preserved by methods of strength of materials (figure 1).

Having analyzed on the basis of parametric studies the changes in normal stresses at different radii of holes R, we present graphs of changes in normal stresses at the level of the border of the holes (figure 3) and at the level of the extreme fibers (figure 4), depending on the radius of the holes R at a step between them $L = 5R$. These graphs show the stress values determined numerically by the finite element method in the ANSYS Mechanical APDL software package, as well as analytically by the strength of materials formulas.

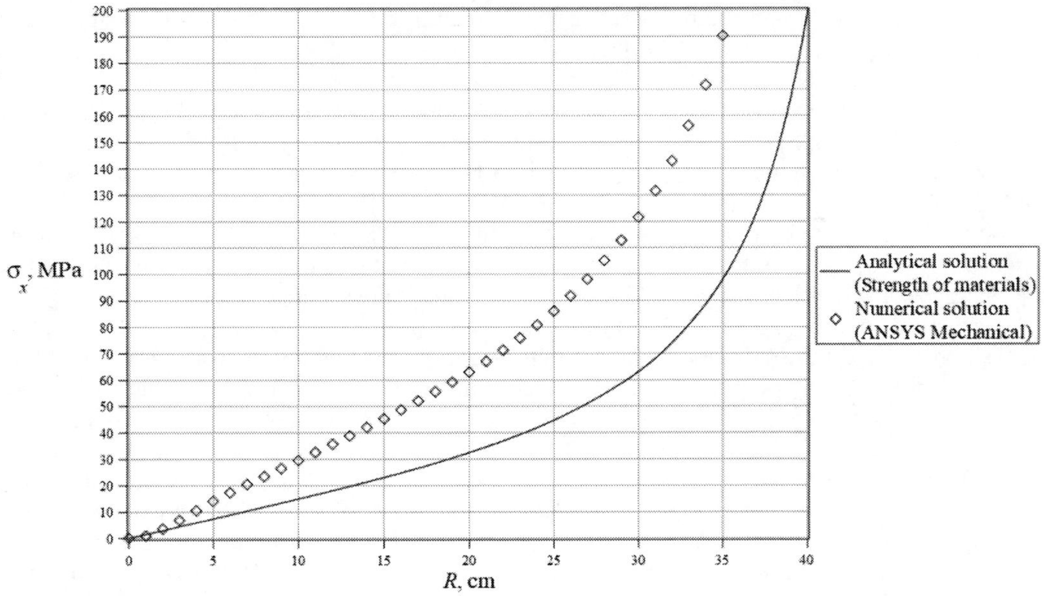

Figure 3. Graph of changes in normal stresses at the level of the hole boundary depending on their radius R at $L = 5R$ for the beam shown in figure 1.

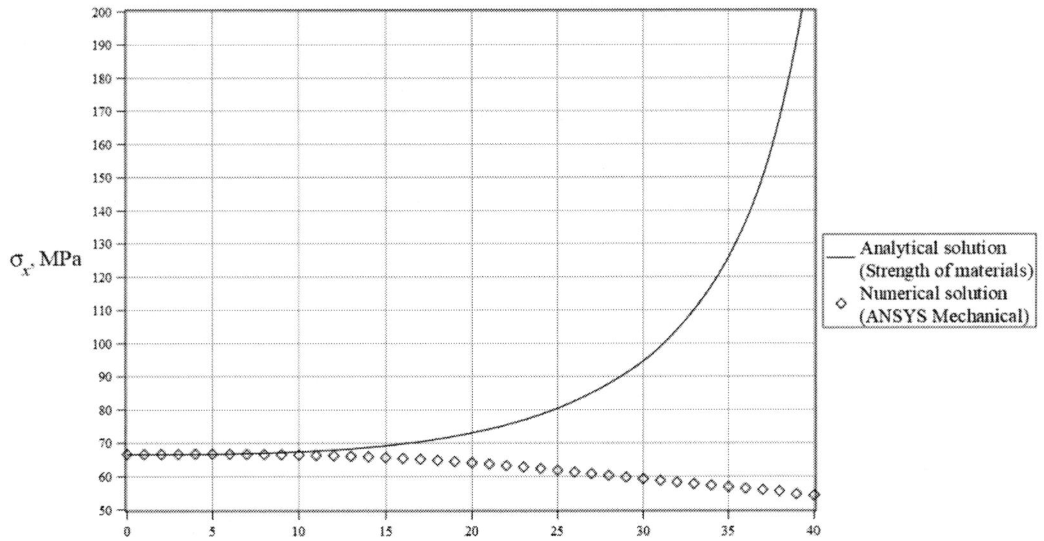

Figure 4. Graph of changes in normal stresses at the level of the extreme fibers depending on the radius of the holes R at $L = 5R$ for the beam shown in figure 1.

Let's consider some features of changing the stress state when changing the parameters of the perforation:

1. The normal stresses σ_x defined in the cross section along the center of the holes in the finite element solution are distributed according to a nonlinear law, while in the solution by methods of strength of materials, the law of variation of normal stresses is linear.
2. At the level of the hole boundary, the stress concentration coefficient at the step $L = 5R$ for different hole sizes practically does not change and corresponds to $K \approx 2$ (figure 3).
3. As in the case with the solution by methods of strength of materials, in the numerical solution the value of the normal stress at the level of the border of the holes tends to infinity $\sigma_x \rightarrow \infty$ with increasing diameter (figure 3).
4. Normal stresses σ_x at the level of extreme fibers almost completely coincide when solving by methods of strength of materials and in a finite element solution up to the value of the ratio of the hole radius to the height of the section equal to 1/6, after which large discrepancies are observed (figure 4).
5. At the level of the extreme fibers when the ratio of the radius of the holes to the height of the section more than 1/6, have seen a significant increase of the normal stresses in the solution by methods of strength of materials ($\sigma_x \rightarrow \infty$), while in the finite-element solution is observed a slight decrease of the stress level (figure 4).
6. When the ratio of the radius of the holes to the height of the section is approximately 1/4, the effect of equal stresses is observed in the finite element solution, at which the values of the normal stresses σ_x at the level of the boundary of the holes and at the level of the extreme fibers are equal to each other.
7. When the ratio of the hole radius to the cross-section height is approximately 5/18 or higher, the values of normal stresses σ_x in the finite element solution at the level of the hole boundary become higher than at the level of the extreme fibers in the solution by strength of materials methods, which indicates that such perforated structures should be designed from the strength condition for normal stresses with a radius-to-height ratio greater than 5/18, taking into account the influence of the stress concentration coefficient for stresses.

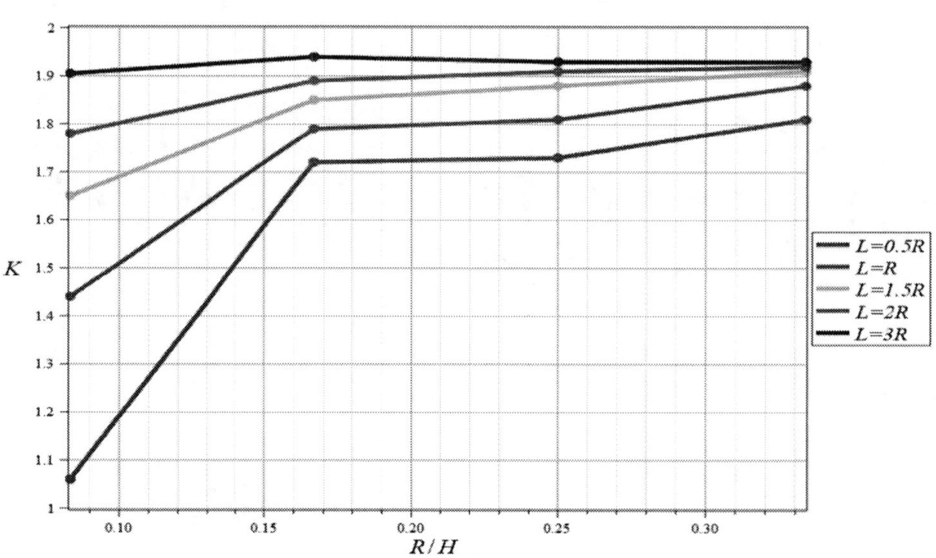

Figure 5. Graph of the dependence of the concentration coefficient K of normal stresses σ_x at the level of the hole boundary on the ratio of the radius R to the height of the beam H at a different step L.

By examining these graphs, the values of normal stresses σ_x determined by strength of materials methods can be correlated through correction values, which can be called stress concentration coefficients K.

After analyzing some types of perforation sizes for different perforation steps, we have plotted the dependence of stress concentration coefficients K for normal stresses at the level of the hole border (figure 5) and at the level of the extreme fibers (figure 6) depending on the radius of the holes R and the step between them L.

The ratio of the hole radius to the cross-section height 1/12, 1/6, 1/4, 1/3 at steps $L = 0.5R$, $L = R$, $L=1.5R$, $L = 2R$, $L = 3R$ was considered as the standard sizes.

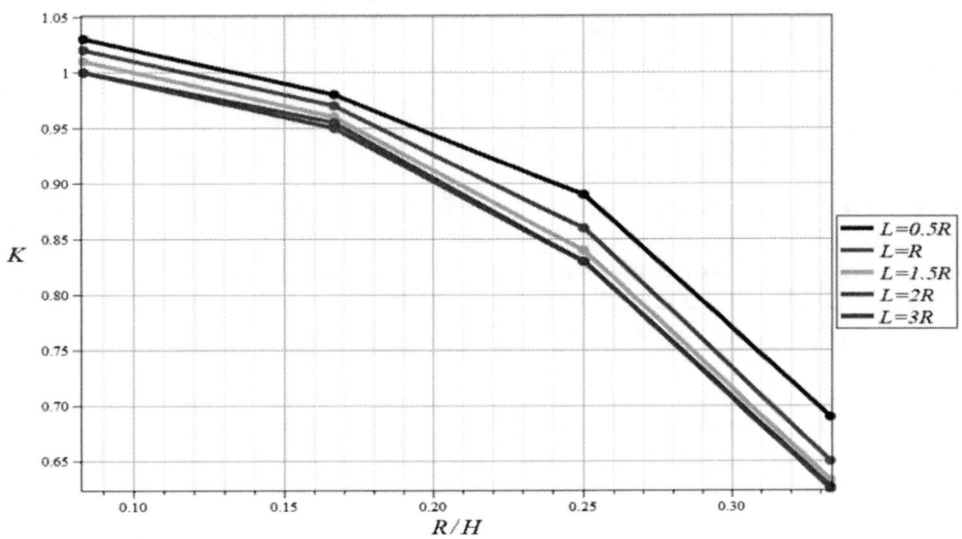

Figure 6. Graph of the dependence of the concentration coefficient K of normal stresses σ_x at the level of the extreme fibers on the ratio of the radius R to the height of the beam H at a different step L between the holes.

Let's do some analysis of these graphs:

1. Normal stresses σ_x at the level of the hole boundary are most sensitive to changes in the perforation parameters. For almost all values of the step L, the stress concentration coefficient K increases with increasing hole sizes. Almost independent of the size of the holes, the concentration coefficient $K \approx 2$ occurs at $L = 5R$ (figure 3).
2. The influence of stress concentration at the level of the extreme fibers practically does not depend on the parameters of perforation, but when the size of the holes increases, the normal stress at the level of the extreme fibers decreases (figure 6).
3. The greater the step L between the holes, the more flat the graph of stress concentration K becomes at the level of the border of the holes, which indicates that the value of stress concentration ceases to depend on their size (figure 5).
4. Verification studies of intermediate values of perforation parameters have shown that the use of linear interpolation based on the data of graphs of stress concentration coefficients shows fairly reliable results in comparison with the finite element solution.

3. Conclusions

1. The presence of a perforation changes the character of normal stress σ_x diagrams to a non-linear one with a stress concentration near the boundaries of the holes.

2. The normal stress concentration factor K depends on the ratio of the hole radius to the cross-section height R/H, as well as on the pitch L.

3. Based on the finite element solution, when the ratio of the radius of the holes R to the height of the section becomes equal 1/4, an effect is observed in which the values of normal stresses σ_x at the level of the border of the holes and at the level of the extreme fibers are equal to each other.

4. The use of correction coefficients of stress concentration K in determining stresses by methods of strength of materials gives acceptable results in terms of accuracy when performing engineering calculations.

5. Further research can be devoted to the study of the behavior of shear stresses during transverse bending in beams with holes.

References

[1] Saiyan S and Paushkin A 2019 Fatigue cracks in castellated I-beams under cyclic loads *J. of Physics: Conference Series* **1425** 012163

[2] Drobachev V M and Litvinov E V 2003Analytical determination of the stress-strain state of the wall-bridge of a perforated beam *University news. Construction* **5** pp 128 –33

[3] Filatov V V 2013 Calculation of through beams based on the theory of composite rods by A. R. Rzhanitsyn *MGSU Bulletin* **9** pp 23 – 31

[4] Bondarenko V M, Zaitsev P I and Lyubimov A A 1963 Calculation of steel beams from split rolled I-beams with holes in the wall *Collection of Scientific Papers Kharkov Civil Engineering Institute* (Kharkov: Kharkov Civil Engineering Institute) **16** pp 19 – 25

[5] Besedin M T 1962 Beams from developed rolled I-beams with holes in the wall *Scientific works of the Kharkov Civil Engineering Institute* (Kharkov: Kharkov Civil Engineering Institute) **19** pp 15 – 19

[6] Hahnemann G A and Kikot A A 2017 Analysis of metal beams with a perforated wall *Polzunovsky Almanac* **4** (2) pp 49 – 52

[7] Al Khetari A A 2018 Features of work and calculation of beams with a perforated wall *Symbol of science* **6** pp 11 – 4

[8] Saiyan S G and Paushkin A G 2018 Investigation of shear stresses in a straight rod near the load application zone *BST: Byulleten' Stroitel'noj Tehniki* **1** pp 56 –9

[9] Sayan S and Paushkin A 2017 Numerical study of the applicability of the Saint-Venant principle *MATEC Web of Conferences* **117** 00134

[10] Sayan S and Paushkin A 2018 Investigation of shear stress attenuation in a straight rod near the load application point *IOP Conference Series: Materials Science and Engineering* **365** 042026

[11] Sayan S and Paushkin A 2018 Investigation of perturbations of normal stresses in an I-beam rod in the load application zone *MATEC Web of Conferences* **251** 04025

[12] Sayan S and Paushkin A 2020 Saint-Venant principle for kinematic boundary conditions *Journal of Physics: Conference Series* **1425** 012198

[13] Sayan S and Paushkin A 2019 Numerical study of the shear stress distribution in an i-beam in the load application zone *Material Science Forum* **974** pp 659 – 64

IOP Conf. Series: Materials Science and Engineering 1015 (2021) 012015 doi:10.1088/1757-899X/1015/1/012015

Numerical study of the stress-strain state of the vertical four-compartment cylindrical tank

S Saiyan and A Paushkin

Moscow State University of Civil Engineering, National Research University, 26 Yaroslavskoe shosse, Moscow, 129337, Russia

E-mail address: berformert@gmail.com

Abstract. A numerical study of the stress-strain state of the construction of the vertical four-compartment cylindrical tank at the stage of operation for storing various substances has been carried out. The purpose of the finite element calculation is to determine the optimal value of the thickness of its load-bearing walls and partitions, which ensure the mechanical safety of the construction under different modes of filling the containers. The reliability of the calculation results, which are carried out both in linear and geometrically nonlinear formulations, is provided by calculations in two software packages SCAD and ANSYS Mechanical APDL with comparison of solutions.

1. Introduction

Thin-walled cylindrical shells are widely used in a wide variety of areas of modern technology: in industrial and civil construction, mechanical engineering, shipbuilding, aircraft and rocketry, etc. Such systems are used in the form of silo buildings for the storage and processing of bulk materials, in port hydraulic construction, in construction such as reactors, etc. [1]. Multi-chamber tank systems provide the flexibility to store bulk solids substance, gases or liquids by allowing multiple different substances to be stored simultaneously in a single tank.

There are various types of intersection of construction elements and other constructional features, leading to the fact that calculations of this kind are presented due to the presence of local disturbances of the stress-strain state [2-5]. The resulting local stresses and deformations can reach significant values, so a thorough study of the behavior of multi-compartment constructions obtained by joining shells into a single construction is a very difficult task.

Modern analytical methods of calculation do not allow calculations the stress-strain state of multi-compartment constructions due to the presence of geometric features (changes in the cross-section of structural elements, hatches, load-bearing partitions, holes, etc.) which affect stresses and deformations in local zones [6]. The finite element method makes it possible to determine the stress-strain state of the elements of cylindrical tanks of various configurations, allowing us to take into account all local effects [7-9].

2. Research

2.1. Research task

Within the framework of this article, the calculations of the stress-strain state of a four-compartment vertical cylindrical tank under the action of hydrostatic pressure under various operating modes are carried out. The calculations were carried out in a linear and geometrically nonlinear formulation.

The calculated four-chamber tank is 9.5 m high, 5.5 m in diameter, in which the liquid is poured up to the 8.5 m mark. The tank has the following operating modes: filling one, two, three and all sections of the tank (figure 1).

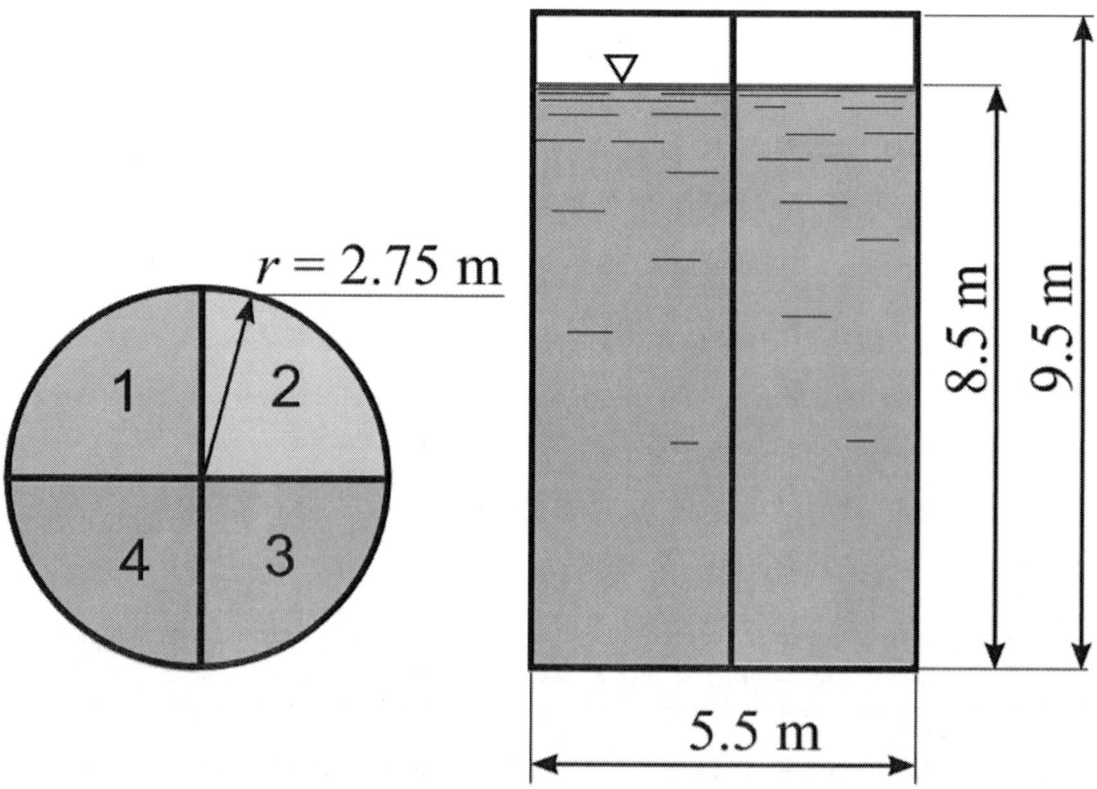

Figure 1. Geometric model of a cylindrical vertical tank with internal walls for liquid storage.

The purpose of the finite element calculation is to determine the optimal value of the thickness of the bearing walls and partitions of the cylindrical tank, which, together with the rest of the characteristics of the construction, provide mechanical safety and tightness of the construction.

2.2. Finite element models

The cylindrical vertical tank with inner walls for storing liquids is the system of outer walls and inner partitions that create sealed compartments for storing substances. The developed finite element models of a cylindrical reservoir are shown in figure 2. The calculations used the finite element software packages ANSYS Mechanical APLD and SCAD 21.1.

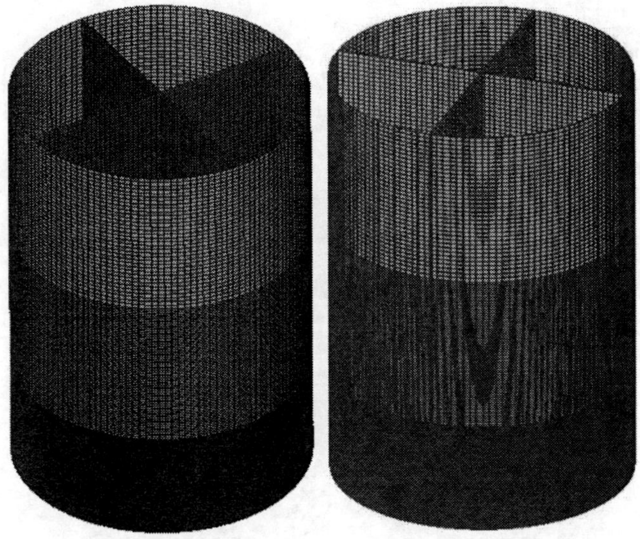

Figure 2. Finite element models of a cylindrical vertical tank with internal partitions for storing liquids in the software packages ANSYS Mechanical APDL (left) and SCAD (right).

The calculations used finite elements of the shell type (Type 44 in the SCAD and SHELL181 in the ANSYS Mechanical APDL). The element nodes have six degrees of freedom each-UX, UY, UW, ROTX, ROTY, and ROTZ (three linear movements along and three rotation angles around the coordinate axes). Degrees of freedom UX, UY correspond to membrane, and UZ, ROTX, ROTY — bending deformations. The finite element meshes developed in the ANSYS Mechanical APDL and SCAD software packages are identical in terms of the number of nodes and elements that adequately reflect geometric, inertial, and load characteristics. Depending on the assumed gradient of change in the stress-strain state, the finite element mesh was divided into three zones. Finite element meshes broken down in the meridional direction with a step of 0.025 m at a height Z from the bottom, from 0.0 m to 2.5 m, with a step of 0.050 m at a height Z from the bottom with 2.5 m to 6.0 m in increments of 0.100 m at a height Z from the bottom 6.0 m to 9.5 m and in the circumferential direction in increments of 2.5° (144 element). This partitioning into finite elements allows us to refine the solution in the areas of high gradients of changes in the desired functions and is chosen based on the optimal calculation time and the accuracy of results based on preliminary test calculations.

When carrying out calculations, the allowable stress was taken for a given steel grade of 168 MPa. The value of the yield stress was taken to be 252 MPa, and the ultimate strength was 520 MPa. The value of the elastic modulus was taken to be $1.96 \cdot 10^5$ MPa and a Poisson's ratio 0.29, and the density of steel 7.9 t/m^3. The material model is selected linearly elastic.

The boundary conditions are assigned by prohibiting all linear and angular displacements of all lower nodes of the model. Static boundary conditions are set hydrostatic pressure caused by the liquid filling density 1 t/m^3 based on the conditions of the assumed operating modes of the cylindrical tank.

2.3. Linear calculation results

The thickness selection was based on the third (maximum shear stress) and fourth (distortion energy) strength hypotheses, which are in good agreement with experimental data for plastic materials such as steel [10]. Based on preliminary linear calculation studies, the minimum required thicknesses for wall and partition elements were determined, which is defined as the greatest thickness when selecting according to two strength theories. The wall thickness is equal to t_w=1.0 cm, and the partition thickness is t_p =4.5 cm.

Let us present the results of linear calculation in two software packages for the most unfavorable operating mode (mode No. 3), shown in figure 3 and figure 4. As the parameters to be compared, consider the displacement (in the cartesian coordinate system), as well as the values of the largest equivalent (according to the fourth theory of strength) stresses.

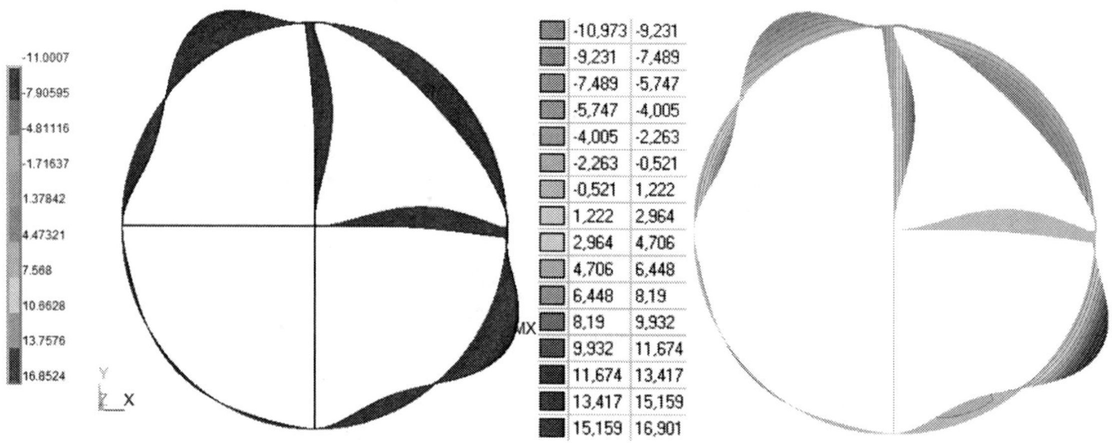

Figure 3. Isofields of displacements u_x in the Cartesian coordinate system according to the results of linear calculations in the ANSYS (left) and SCAD (right) software packages for mode No. 3, mm.

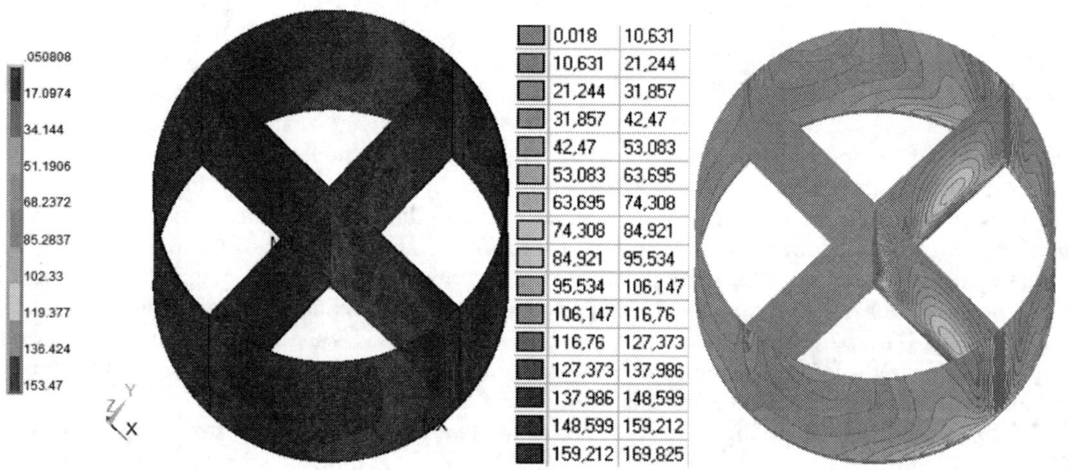

Figure 4. Isofields of equivalent stresses on the bottom layer of the shell based on the results of linear calculations in the ANSYS (left) and SCAD (right) software packages for mode No. 3, MPa.

As seen from figures 3 and 4 results of linear calculation in two software packages are well correlated with each other both in qualitative and quantitative characteristics.

Considering the results of the calculation during normal operation with the stress level not exceeding the maximum for the material, it is worth noting the most dangerous zones of the structure in terms of strength based on linear calculation. For the outer wall of a cylindrical tank, the maximum stresses are localized in the supporting part, however, it is worth noting the zones of high stresses along the height of the interface between the wall and the partition. For internal partitions, the greatest stresses are localized in the central zones of intersecting partitions, as well as in the interface with the outer wall, by analogy with a plate rigidly supported along the contour.

2.4. Nonlinear calculation results

To take into account the influence of membrane forces on bending effects, the interaction between bending effects and longitudinal forces (analogous to the phenomenon of transverse-longitudinal bending in beams) was considered in the form of geometric nonlinearity, when the work of an elastic system is associated with the need to take into account the change in the geometry of the system during its deformations under load. Geometric nonlinearity can be taken into account through equations relating displacements to deformations or using equilibrium equations. Calculations taking into account geometric nonlinearity are often referred to as deformed calculations.

In the SCAD software package, to take into account geometric nonlinearity, it is possible to take into account only the "Karman approximation", when it is assumed that the squares of the angles of rotation of the elements of the considered design scheme are quantities of the same order of smallness as the relative elongations in the material, which in turn are considered small in comparison with unit [11]. Nonlinear calculations are performed using a stepwise method, the idea of which is based on tracking the behavior of the system at relatively small load increments. In this case, at each step, a linearized system of resolving equations is solved for the current increment of the nodal load vector generated for the load under consideration. The calculation is focused on solving nonlinear problems in several modifications of the step method: a simple step method, a step method with refinements, a step-iterative method [12].

The ANSYS Mechanical APDL software package provides a Newton-Raphson approach to account for nonlinearities, in which the entire load is divided into a number of increments of this load. Before each solution, the Newton-Raphson method evaluates the unbalanced load vector, which is the difference between restoring forces (loads corresponding to element stresses) and applied loads. The program then performs a linear solution using unbalanced loads and verifies convergence. If the convergence criteria are not met, the unbalanced load vector is reevaluated, the stiffness matrix is recalculated and a new solution is obtained. This iterative procedure continues until the problem reaches convergence according to user-specified criteria [13].

Based on the thicknesses obtained from the linear calculation, a refined nonlinear numerical study of the stress-strain state was carried out taking into account the membrane forces on the bending component, the results of which are shown in figure 5 and figure 6.

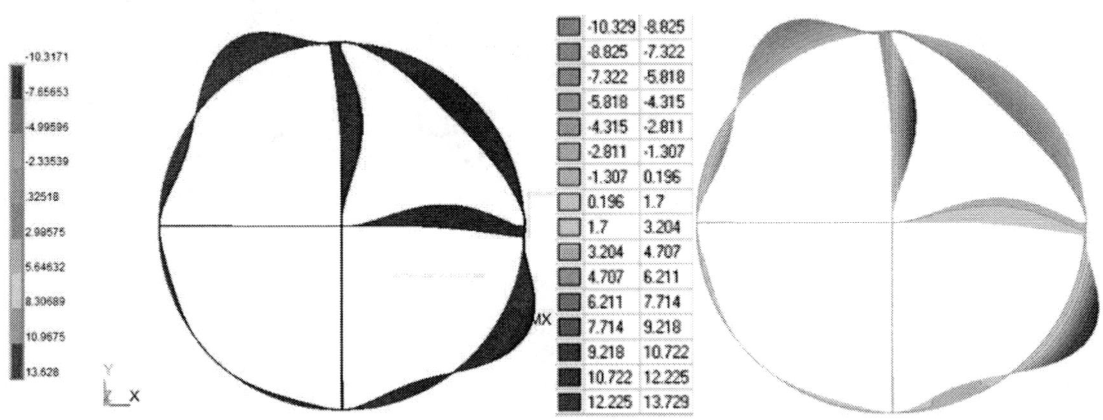

Figure 5. Isofields of displacements u_x in the Cartesian coordinate system according to the results of nonlinear calculations in the ANSYS (left) and SCAD (right) software packages for mode No. 3, mm.

As seen from figure 5 and 6, the results of nonlinear calculation in two software packages are well correlated with each other qualitatively and quantitatively.

In comparison with the results of linear calculations, the overall stress level slightly decreased, however, in the local areas of intersection of elements with each other and in the zone of the support

IOP Conf. Series: Materials Science and Engineering 1015 (2021) 012015 doi:10.1088/1757-899X/1015/1/012015

part, the stress state underwent only a slight change, since here the stresses depend little on geometric nonlinearity.

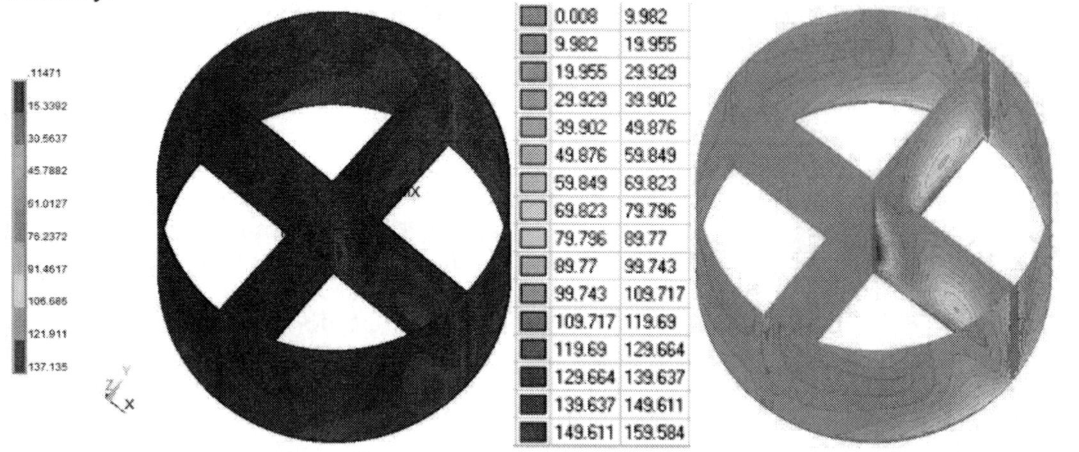

Figure 6. Isofields of equivalent stresses on the inner layer of the shell according to the results of nonlinear calculations in software packages ANSYS (left) and SCAD (right) for mode No. 3, MPa.

2.5. Comparison of calculation results

Comparative analysis of computational studies using alternative software packages is an integral part of the computational process associated with the verification of a numerical solution.

Comparative analysis of computational studies (table 1) presents the results of calculations in software packages with each other and a comparison of solutions in various settings (linear and nonlinear).

Table 1. Comparison of the results of computational studies

Calculation type	Compared parameters	ANSYS Mechanical APDL	SCAD	Δ, %
Linear solution	u_{max}, mm	16.85	16.90	0.30
	σ_{max}, MPa	153.47	169.83	10.66
Nonlinear solution	Method	Newton-Raphson	Step	
	u_{max}, mm	13.63	13.73	0.73
	σ_{max}, MPa	137.14	159.58	16.36
δ, %	δ_u, %	23.62	23.01	
	δ_σ, %	11.91	6.41	

3. Conclusions

As a result of the performed computational studies of the stress-strain state of an atypical construction of a vertical four-compartment cylindrical tank with inner partitions for storing liquid under various operating modes, the following conclusions and recommendations can be formulated:

1. The finite element models of the supporting structures of the atypical vertical four-compartment cylindrical tank were developed in the software packages ANSYS Mechanical APDL and SCAD. The results of the comparative analysis show similar resultant stress-strain states in both linear and nonlinear formulations.

2. Based on the performed computational studies in linear and nonlinear formulations, it was determined that taking into account the membrane forces on the bending component of the considered atypical structure does not change the qualitative characteristics, but changes the quantitative characteristics of the stress-strain state: the difference in the maximum displacements δ_u is about 20% to a lesser side, the difference in the maximum stresses δ_σ of the order of 6-12% is also downward. This fact indicates that when calculating such structures, it is allowed to use a linear relationship between

forces and displacements, since such a formulation of the calculation goes into a margin of safety. However, it should be noted that this conclusion is valid for the considered tank design. With an increase in the size of the tank, as well as the spans of the partitions inside, the geometric nonlinearity can make a greater contribution to the stress-strain state.

3. Based on the calculated studies, it was found that the use of a continuous steel sheet as a partition of the structure is impractical, because according to the results of calculations, the thickness of the steel sheets turned out to be significant (up to 4.5 cm).

4. Calculations in a geometrically nonlinear formulation, taking into account the influence of membrane forces on bending effects in order to justify a decrease in pre-assigned thicknesses from a linear calculation, does not lead to a significant change in the stress-strain state and reasoning to a decrease in the thicknesses of steel sheets for the considered construction, since the greatest stresses are concentrated in local areas of interface shells and are less sensitive to geometric nonlinearity.

5. The considered disadvantages allow us to conclude that this structural scheme of the tank with continuous steel partitions is impractical due to their significant thicknesses, and therefore requires the development of a design diagram of the partition, which provides mechanical safety and tightness in conjunction with economic indicators and technological features of a construction.

References

[1] Shagivaleev K F and Burov S N 2011 Calculation of thin-walled spatial systems consisting of interconnected closed cylindrical shells using the force method *Building mechanics of engineering structures and structures* **1** pp 8–12

[2] Kositsyn S B and Tran Xuan Linh 2013 Analysis of the stress-strain state of intersecting cylindrical shells under elastic-plastic deformations taking into account geometric nonlinearity *Structural mechanics of engineering structures and structures. Review-analytical and scientific-technical journal* (Moscow) **1** pp 3–9

[3] Kositsyn S B and Tran Xuan Linh 2014 Comparative analysis of various models of material as applied to the calculations of the shell of a railway tank taking into account elastoplastic deformations and geometric nonlinearity *Mechanization of construction* **838** (4) pp 12 – 5

[4] Kositsyn S B and Tran Xuan Linh 2014 Numerical analysis of the stress-strain state of orthogonally intersecting cylindrical shells without and taking into account their one-way interaction with the surrounding soil mass *International Journal for Computational Civil and Structural Engineering* **10** (1) pp 72–8

[5] Manuilov G A and Kositsyn S B 2006 Investigation of the loss of stability of the railway tank shell *Computational mechanics of a deformable solid - Proceedings of the international scientific and technical conference. In two volumes* (Moscow: Russian University of transport) pp 236 – 240

[6] Sherstobitova R T 2008 *Improving the safety of vertical steel tanks operated for a long time at low temperatures* (Ufa: Ufa State Petroleum Technological University) p 131

[7] Rozin L A 2000 Finite element method *Soros Educational Journal* **6** (4) pp 120 – 7

[8] Buryaka V A, Fokin V G, Soldusova E A, Glazunova N A and Adeyanov I E 2010 Engineering Analysis in ANSYS Worckbench *Textbook manual. SamSTU* p 271

[9] Chigarev A V, Kravchuk A S and Smalyuk A F 2004 ANSYS for engineers *Mechanical Engineering* (Moscow) p 511

[10] Vardanian G S, Andreev V I, Atarov N M and Gorshkov A A 2011 Strength of materials with the fundamentals of theory of elasticity and plasticity *ASV* (Moscow) p 568

[11] Perelmuter A V and Slivker V I 2007 Computational models of structures and the possibility of their analysis *DMK Press* (Moscow) p 600

[12] Karpilovsky V S, Kriksunov E Z, Malyarenko A A, Mikitarenko M A, Perelmuter A V and Perelmuter M A 2015 SCAD Office. *Computing complex SCAD ++ ASV* (Moscow) p 848

[13] ANSYS, Inc 2018 Structural Analysis Guide. Release 19.2

IOP Conf. Series: Materials Science and Engineering 1015 (2021) 012016 doi:10.1088/1757-899X/1015/1/012016

The regions of dynamic instability of a plate, the material of which obeys the hereditary low of viscoelasticity, at the second resonance

E Kosheleva

Moscow State University of Civil Engineering (National Research University)
Yaroslavskoe Shosse 26, Moscow, Russia

E-mail address: elleok@mail.ru

Abstract. The problem of determining the side regions of dynamic instability of a rectangular plate, the material of which obeys the hereditary law of deformation, is considered. The solution of the integro-differential equation of vibrations of a plate loaded in its plane with constant and variable loads is considered. The solution used is a series with separated variables. An equation is obtained to determine the critical frequencies of even regions of instability. The influence of the parameters of the nucleus on the position of the second region of dynamic instability is investigated.

1. Introduction

In [1], the main regions of dynamic instability of a rectangular plate loaded in its plane by forces uniformly distributed along the edges were investigated

$$N_x = N_{x0} + N_{xt}\cos\theta t, \quad N_y = N_{y0} + N_{yt}\cos\theta t \tag{1}$$

The viscoelastic properties of the material will obey the hereditary Boltzmann-Volterra law with a weakly singular creep kernel [2]

$$R(t-s) = A\exp\left[-\beta(t-s)\right](t-s)^{\alpha-1}\left(0<\alpha<1\right) \tag{2}$$

Three kernel parameters allow you to describe the behavior of many polymeric materials. This paper considers the solution to the problem of determining even domains of dynamic instability and investigates the influence of the kernel parameters on them.

2. Problem setting

We will consider a rectangular plate loaded with forces (1). The viscoelastic properties of the material will obey the hereditary law (2). Taking ν constant, we have the relationship between stresses and deformations

$$\sigma_x = \frac{E}{1-\nu^2}\left(\varepsilon_x + \nu\varepsilon_y\right) - \frac{E}{1-\nu^2}\int_0^t R(t-\theta)\left(\varepsilon_x + \nu\varepsilon_y\right)\mathrm{d}\theta$$

$$\sigma_y = \frac{E}{1-\nu^2}\left(\varepsilon_y + \nu\varepsilon_x\right) - \frac{E}{1-\nu^2}\int_0^t R(t-\theta)\left(\varepsilon_y + \nu\varepsilon_x\right)\mathrm{d}\theta \tag{3}$$

$$\tau_{xy} = \frac{E}{2(1+\nu)}\gamma_{xy} - \frac{E}{2(1+\nu)}\int_0^t R(t-\tau)\gamma_{xy}\,d\theta$$

To describe the transverse vibrations of the plate, we will use the equation [3]

$$D\left(\nabla^4 w - \int_0^t R(t-s)\nabla^4 w\,ds\right) + (N_{x0} + N_{xt}\cos\theta t)\frac{\partial^2 w}{\partial x^2} + (N_{y0} + N_{yt}\cos\theta t)\frac{\partial^2 w}{\partial y^2} + m\frac{\partial^2 w}{\partial t^2} = 0 \quad (4)$$

where $D = \dfrac{Eh^3}{12(1-\nu^2)}$ - cylindrical stiffness, E - instant modulus of elasticity.

The article [1] considered the problem of finding the boundaries of the regions of instability, for which equation (4) has periodic solutions with a period of $2T$, including the main regions of dynamic instability. In this paper, we consider the problem of finding solutions with period T. This will allow us to construct even regions of dynamic instability.

3. Solution technique

The solution to equation (4) will be sought by the Bubnov-Galerkin method, separating the variables in the form [1]

$$w = \sum f_i(t) X_i(x) Y_i(y) \quad (5)$$

We have the equation in dimensionless form:

$$f'' - \varepsilon\varphi\int_0^\tau R(\tau - s) f(s)\,ds + \varphi(1 - 2\mu\cos\tau) f = 0 \quad (6)$$

where the notation is used [3]
dimensionless time $\tau = \theta t$, ω - natural frequency of unloaded plate,

$$\varphi = \frac{\Omega^2}{\theta^2},\ \varepsilon = \frac{\omega^2}{\Omega^2},\ \Omega^2 = \omega^2\left(1 - \frac{N_{x0}N_{2*} + N_{y0}N_{1*}}{N_{1*}N_{2*}}\right),\ \mu = \frac{1}{2}\frac{N_{xt}N_{2*} + N_{yt}N_{1*}}{N_{1*}N_{2*} - N_{x0}N_{2*} - N_{y0}N_{1*}},$$

Ω - the frequency of natural vibrations of the plate loaded N_{x0} and N_{y0}, μ - excitation factor, N_{1*}, N_{2*} - critical values of efforts N_{x0} and N_{y0} at their independent static action,

$$I_1 = \int_0^a\int_0^b\left(\frac{d^4X}{dx^4}Y + 2\frac{d^2X}{dx^2}\frac{dY^2}{dy^2} + \frac{dY^4}{dy^4}X\right)XY\,dx\,dy,\ I_2 = \int_0^a\int_0^b X^2Y^2\,dx\,dy$$

$$I_3 = -\int_0^a\int_0^b X\frac{d^2X}{dx^2}Y^2\,dx\,dy,\ I_4 = -\int_0^a\int_0^b X^2\frac{d^2Y}{dy^2}Y\,dx\,dy,\ \omega^2 = \frac{DI_1}{mI_2},\ N_{1*} = \frac{I_1}{I_3}D,\ N_{2*} = \frac{I_1}{I_4}D$$

As is known [4], domains of unboundedly increasing solutions are separated from stability domains by periodic solutions with periods T and $2T$. Moreover, two solutions of the same period limit the region of instability, two solutions of different periods - the region of stability. We seek periodic solutions with period T in the form of a series:

$$f(t) = \sum_{k=2,4,6}^{\infty}\left(a_k\sin\frac{k\tau}{2} + b_k\cos\frac{k\tau}{2}\right) + b_0 \quad (7)$$

Substitute series (7) into equation (6):

$$\sum_{k=2,4}^{\infty}\left(-a_k\frac{k^2}{4}\sin\frac{k\tau}{2}-b_k\frac{k^2}{4}\cos\frac{k\tau}{2}\right)-\varepsilon\phi\sum_{k=2,4}^{\infty}\left[\sin\frac{k\tau}{2}(a_kB_k+b_kA_k)+\cos\frac{k\tau}{2}(b_kB_k-a_kA_k)\right]$$
$$-\varepsilon\phi b_0R_0+\varepsilon\phi(1-2\mu\cos\tau)+\phi\sum_{k=2,4}^{\infty}\left(a_k\sin\frac{k\tau}{2}+b_k\cos\frac{k\tau}{2}\right)- \tag{8}$$
$$\mu\phi\sum_{k=2,4}^{\infty}a_k\left(\sin\frac{k-2}{2}\tau+\sin\frac{k+2}{2}\tau\right)-\mu\phi\sum_{k=2,4}^{\infty}b_k\left(\cos\frac{k-2}{2}\tau+\cos\frac{k+2}{2}\tau\right)=0$$

where the following notation is introduced

$$R_0=\int_0^\tau R(x)\,\mathrm{d}x \quad A_k=\int_0^\tau R(x)\sin\frac{kx}{2}\,\mathrm{d}x, \quad B_k=\int_0^\tau R(x)\cos\frac{kx}{2}\,\mathrm{d}x, \quad x=\tau-s \tag{9}$$

Let us equate the coefficients at the same $\sin k\tau/2$ and $\cos k\tau/2$ equations (8). As a result, we obtain a system of linear algebraic equations with respect to a_k and b_k with constant coefficients

$$\begin{aligned}
&\varepsilon\phi b_0R_0-b_0\phi+\mu\phi b_0=0\\
&a_2+\varepsilon\phi a_2B_2+\varepsilon\phi b_2A_2-\phi a_2+\mu\phi a_2=0\\
&b_2+\varepsilon\phi b_2B_2-\varepsilon\phi a_2A_2+2\mu b_0\phi-\phi b_2+\mu\phi b_4=0\\
&\frac{k^2}{4}a_k+\varepsilon\phi a_kB_k+\varepsilon\phi b_kA_k-\phi a_k+\phi\mu(a_{k+2}+a_{k-2})=0\\
&\frac{k^2}{4}b_k+\varepsilon\phi b_kB_k-\varepsilon\phi a_kA_k-\phi b_k+\phi\mu(b_{k+2}+b_{k-2})=0\\
&(k=2,4,6...)
\end{aligned} \tag{10}$$

We divide the equations of system (10) by φ and perform the replacement $1/\varphi=\theta^2/\Omega^2=4p^2$. Let's compose the determinant of the system and equate it to zero. We obtain the equation of critical frequencies

$$\begin{vmatrix}
\varepsilon R_0 & 0 & \mu & 0 & 0 & \dots\\
0 & 4p^2 & \varepsilon A_2 & \mu & 0 & \dots\\
2\mu & -\varepsilon A_2 & 4p^2+\varepsilon B_2-1 & 0 & \mu & \dots\\
0 & \mu & 0 & 16p^2+\varepsilon B_4-1 & \varepsilon A_4 & \dots\\
0 & 0 & \mu & -\varepsilon A_4 & 16p^2+\varepsilon B_4-1 & \dots\\
\dots & \dots & \dots & \dots & \dots & \dots
\end{vmatrix}=0 \tag{11}$$

Solving equation (11), we find the boundaries of even domains of dynamic instability. To calculate in the first approximation the second domain of dynamic instability, we will store in the determinant nine terms

$$\begin{vmatrix}
\varepsilon R_0 & 0 & \mu\\
0 & 4p^2+\varepsilon A_2 & \varepsilon A_2\\
2\mu & -\varepsilon A_2 & 4p^2+\varepsilon B_2-1
\end{vmatrix}=0 \tag{12}$$

After disclosing the determinant from the obtained equation, we obtain the formula for determining the critical frequencies

IOP Conf. Series: Materials Science and Engineering 1015 (2021) 012016 doi:10.1088/1757-899X/1015/1/012016

$$p = 0.5\left[1-\varepsilon B_2 + \left(\mu^2 \pm \left(\mu^4 - \varepsilon^2 A_2^2 (\varepsilon R_0 - 1)^2\right)^{1/2}\right)(\varepsilon R_0 - 1)^{-1}\right]^{1/2} \quad (13)$$

As long as the expression under the inner radical is positive, this expression gives two real roots for the value of p, corresponding to the two boundaries of the second instability region. In the extreme case

$$\mu^4 - \varepsilon^2 A_2^2 (\varepsilon R_0 - 1)^2 = 0 \quad (14)$$

We determine from (14) the minimum value of the excitation coefficient μ_*, at which the occurrence of sustained oscillations is still possible, and the corresponding value p_*

$$\mu_* = \left[\left(\varepsilon A_2 (\varepsilon R_0 - 1)\right)^2\right]^{1/4}, \quad p_* = 0.5\left[1 - \varepsilon B_2 + \mu_* (\varepsilon R_0 - 1)^{-1}\right]^{1/2} \quad (15)$$

It should be noted that the coefficients of the equation of critical frequencies (11) are variable, since they depend on time. As time changes, they quickly approach the values at which the integrals have an infinite upper limit. In this case, the regions do not change their position, starting from some time. To calculate R_0, B_k, A_k you can use the relations

$$R_0 = A\beta^{-\alpha}\Gamma(\alpha),$$
$$A_k = A\Gamma(\alpha)\left(\beta^2 + \frac{k^2}{4}\right)^{-\alpha/2} \sin\left(\alpha \operatorname{arctg}\frac{k}{2\beta}\right) \quad (16)$$
$$B_k = A\Gamma(\alpha)\left(\beta^2 + \frac{k^2}{4}\right)^{-\alpha/2} \cos\left(\alpha \operatorname{arctg}\frac{k}{2\beta}\right)$$

4. Results

Let us investigate the influence of the parameters of the nucleus on the position of the even regions of instability. Consider the change in the values of the excitation coefficient μ_* and p_* in the case of constant kernel parameters α, β, and variable A. We will use formulas (15) obtained in the first approximation. Let's accept the following values of the kernel parameters [1]: α =0.15, β =0.05, A in the range from 0.019 to 0.103, ε =1.25. To calculate the coefficients, we use formulas (16), substituting the kernel parameters and $k = 2$:

$$R_0 = A\beta^{-\alpha}\Gamma(\alpha) = A(0.05)^{-0.15}\Gamma(015) = 3.968\,A$$
$$A_2 = A\Gamma(\alpha)(\beta^2+1)^{-\alpha/2} \sin\left(\alpha \operatorname{arctg}\frac{1}{\beta}\right) = 1.4067A \quad (17)$$
$$B_2 = A\Gamma(\alpha)(\beta^2+1)^{-\alpha/2} \cos\left(\alpha \operatorname{arctg}\frac{1}{\beta}\right) = 6.058A$$

The calculation results are shown in table 1.

Table 1. Influence of A on the values μ_* and p_* at α =0.15, β =0.05

A	0.019	0.032	0.051	0.064	0.083	0.103
μ_*	0.174	0.217	0.259	0.277	0.293	0.294
p_*	0.453	0.419	0.362	0.317	0.237	0.104

An increase in A leads to an increase in the value of the excitation coefficient μ_*. This indicates an increase in the amplitudes of the periodic components of the forces, which can cause a loss of dynamic stability. On the plane μ, p the boundaries of the dynamic instability region will move towards increasing values to the right. The results presented in the bottom row of table 1 show a decrease p_* with increasing A. This allows us to conclude that the frequency of the periodic component of the load decreases. In this case, the boundary of the dynamic instability region on the plane μ, p will shift downward toward the axis μ. Consider the stability regions at constant α=0.25 and variables β and A. Retaining twenty-five elements in equation (11), we obtain the equation for the second approximation. The calculation results are shown in figure 1. At a constant value α, an increase in β and A leads to a decrease in the frequency of the periodic load, causing a loss of dynamic stability, with first an increasing and then decreasing minimum excitation coefficient μ_* and amplitudes of forces

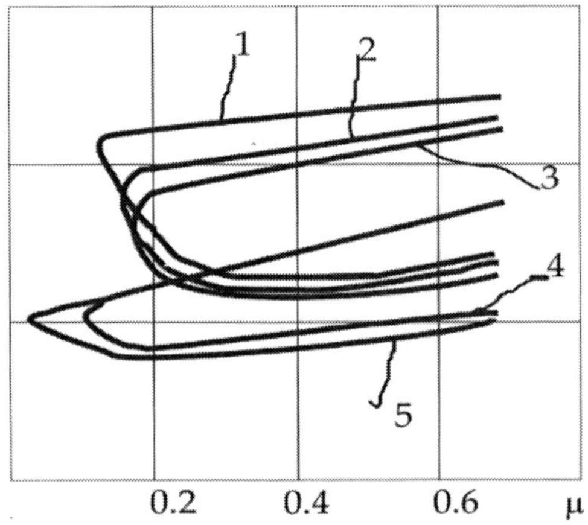

Figure 1. Dynamic area boundaries instability of the second order at α =0.25:
1 - β =0.0054, A= 0.05,
2 - β =0.046, A=0.085,
3 - β =0.086, A=0.1,
4 - β =3.37, A=0.25,
5 - β =54, A=0. 5.

5. Conclusions

In this work, an equation of critical frequencies for even regions of dynamic instability is obtained and the influence of the parameters of the nucleus on the position of the second region of dynamic instability. It was revealed that an increase in parameter A with constant other parameters leads to an increase in the amplitudes and a decrease in the frequencies of the periodic components of the forces, which can cause a loss of dynamic stability. It should be noted that a simultaneous increase in A and β with unchanged α, also leads to a decrease in the frequency of the periodic component of the load at first an increasing and then decreasing minimum excitation coefficient. In [1], similar results were obtained for the main region of instability.

References

[1] Kosheleva E L 2018 *2nd Int. Conf. on Civil, Arch. and Env. (IFCAE 2018) Advances in Eng. Research* **163** (Atlantis Press Zhuhai China)
[2] Rzhanitsyn A R 1968 *Creep theory* (Stroyizdat)
[3] Troyanovsky I E 1970 *Bulletin of the Moscow State University* **6** pp 67-72
[4] Bolotin V V 1956 *Dynamic Stability of Elastic Systems* (GITL)
[5] Koltunov M A 1966 *Mechanics of polymers* **4** pp 483-97

IOP Conf. Series: Materials Science and Engineering 1015 (2021) 012017 doi:10.1088/1757-899X/1015/1/012017

Reliability of a reinforced concrete frame building under earthquake

O Mkrtychev and S Bulushev

Moscow State University of Civil Engineering (National Research University), Research center "Safety and Seismic Resistance of Structures", 129337 Moscow, Russia

E-mail address: sergey.bulushev@gmail.com

Abstract. The research objective is performing probabilistic nonlinear dynamic seismic resistance analysis for the reinforced concrete building frame. The building design is compliant with SP 14.13330.2018 Construction Code. The LS-DYNA software was used for nonlinear dynamic analysis. A set of two-component accelerograms to be analyzed was generated. The set consists of 20 random seismic events with the specified expected values for the predominant frequency and the intensity. The structural response as random processes and values was studied. The resulting experimental response value distributions were fitted to distribution laws. The results showed that the K_1 factor accounting for acceptable damage to buildings and structures should be much larger than specified in the regulatory codes. It indicates that the seismic resistance of the building is insufficient in terms of the near-collapse failure criterion.

1. Introduction

By its nature, the seismic action represented by the accelerogram is a multidimensional random process [1]. Generally, it has six components. A deterministic analysis does not always adequately estimate the structural response. A probabilistic analysis provides a more adequate representation of the structural response and correctly evaluates the building seismic resistance for the specified safety.

When exposed to strong earthquakes, structures go beyond the elastic range. Still, when buildings and structures are designed to the seismic engineering principles [2], the response spectrum method considers only the elastic range. The K_1 factor accounting for acceptable damage to buildings and structures (inverse of the reduction factor R [3]) represents the nonlinear component.

The studies [4–8] and the actual earthquake consequences show that in most cases steel structures have some bearing capacity margins under earthquakes. The reason is that steel structures can dissipate a lot of energy through plastic response. It is different for reinforced concrete structures. Concrete is not plastic, while the reinforcement ratio in concrete structures is too low to dissipate a lot of energy. The K_1 factor for reinforced concrete structures is nearly equal to that for structural steel. It indicates that the bearing capacity of reinforced concrete structures under strong earthquakes [4, 5, 9, 10] may be insufficient.

2. Problem statement

The research objective is assessing the seismic resistance of a 5-storey reinforced concrete frame building with the near-collapse failure criterion using probabilistic nonlinear dynamic analysis concerning the variable nature of the seismic impact. Refer to [7] for the detailed analysis strategy. The building is designed to the existing seismic engineering regulations [2] with the PC LIRA SAPR 2016 software. The designed spectrum of the 9-point earthquake was used. The K_0 factor (reliability factor for the responsibility of the calculated building and structure) was assumed to be 1, K_ψ (factor taking into account the ability of buildings and structures to dissipate energy) was equal to 1, K_1 was equal to 0.35 [2]. The column and beam cross-sections are 40×40 cm. As specified in [10], the concrete's compressive strength grade was assumed to be B25 (C20). The longitudinal and transverse rebars are - A500.

LS-DYNA was used for deterministic estimations as required for the probabilistic analysis. The analysis covered both physical and geometric nonlinearities. The K_1 factor was assumed to be 1, K_0 was equal to 1. The damping parameter ξ was 5% of the critical value. Explicit integration of the motion equations was used. The software can directly simulate reinforced concrete structures. The structural model is shown in figure 1. The concrete was represented with solid finite elements and the Continuous Surface Cap Model (MAT_159 CSCM) [12–17]. The rebars were represented with 1D finite elements and the material's elasto-plastic response (figure 2). The building first natural mode frequency is 1.04 Hz.

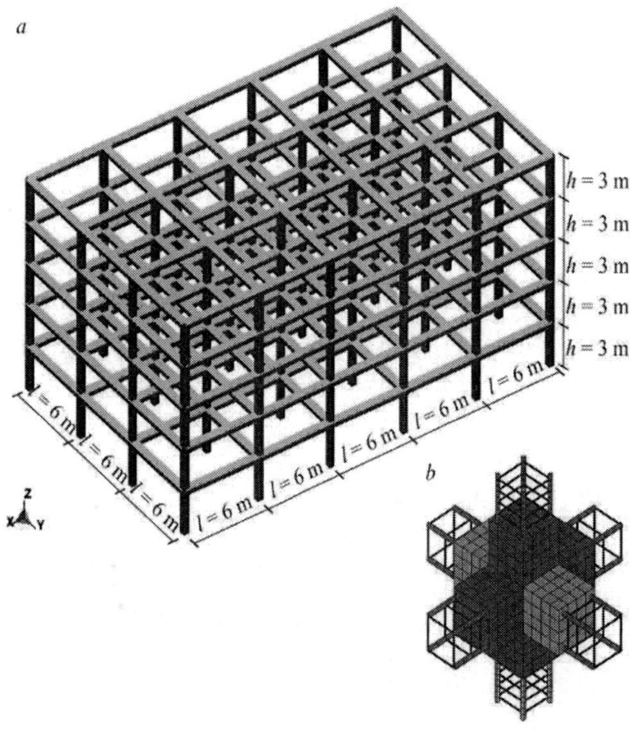

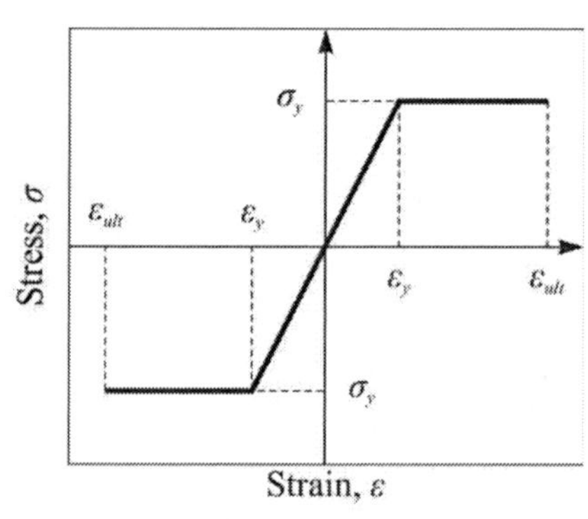

Figure 1. Calculation scheme with direct reinforcement: *a* - general view; *b* – fragment.

Figure 2. General view of rebars deformation diagram.

To perform the analysis in the time domain, a set of 20 random seismic events was generated. The impacts contained two components along the X and Y axis. The vertical component was disregarded. The impact intensity was 9 points. The predominant frequency of accelerograms was a random value. The expected value for the predominant frequency of a set was 1.04 Hz that matched the building first natural mode frequency. The modeling strategy was as specified in [18–19]. Figures 3–4 show one case in each direction. The envelope was formed on the basis of the paper [19] (figure 5). Duration of the oscillations intense phase depends on the magnitude of earthquake.

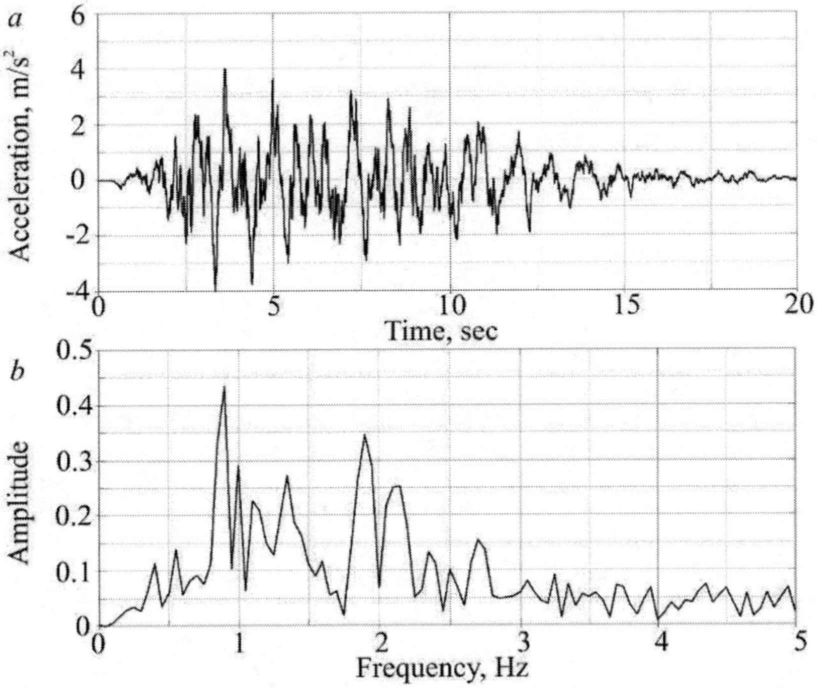

Figure 3. One case of random seismic action along the X axis: *a* - accelerogram; *b* - spectral composition.

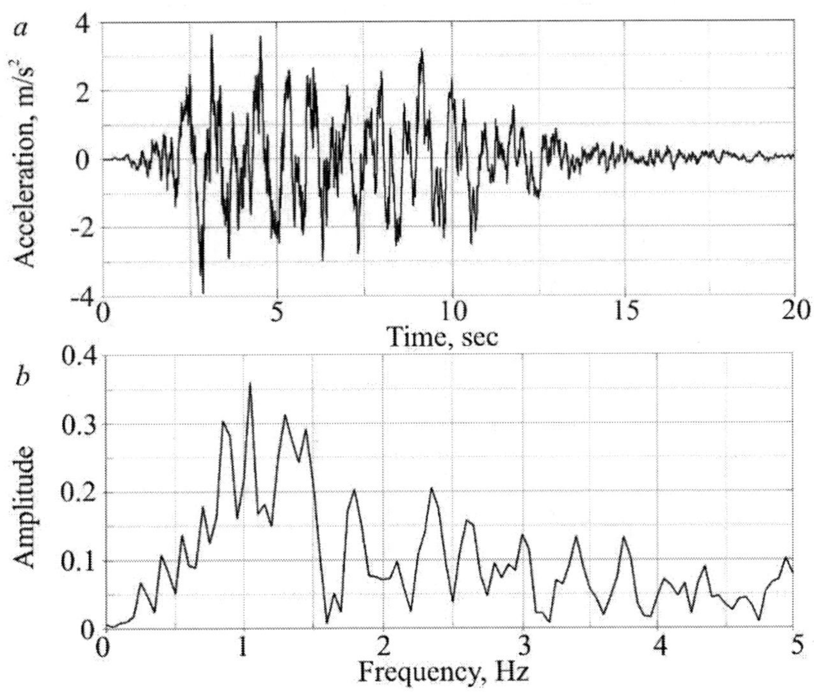

Figure 4. One case of random seismic action along the Y axis: *a* - accelerogram; *b* - spectral composition.

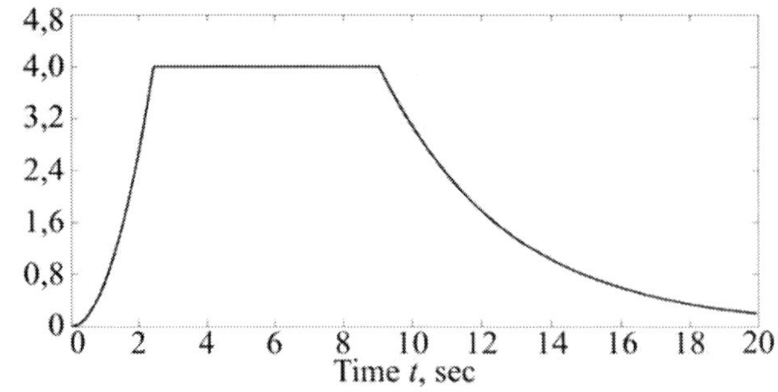

Figure 5. The envelope of accelerograms.

3. Results and discussions

The probabilistic analysis produced the K_c bearing capacity margin factor and the K_1 factor of the reinforced concrete framed building for the random load cases (refer to tables 1, 2). K_1 was estimated by dividing the rated value by the K_c for each load case. The average bearing capacity margin factor K_c for these load cases was 0.570, and the K_1 factor was 0.637.

Table 1. The K_c bearing capacity margin factor

№ of case	1	2	3	4	5	6	7	8	9	10
The K_c factor	0.600	0.500	0.700	0.500	0.600	0.400	0.600	0.600	0.800	0.700
№ of case	11	12	13	14	15	16	17	18	19	20
The K_c factor	0.500	0.600	0.600	0.500	0.700	0.600	0.500	0.600	0.400	0.500

Table 2. The K_1 factor

№ of case	1	2	3	4	5	6	7	8	9	10
The K_1 factor	0.583	0.700	0.500	0.700	0.583	0.875	0.583	0.583	0.438	0.500
№ of case	11	12	13	14	15	16	17	18	19	20
The K_1 factor	0.875	0.700	0.583	0.583	0.700	0.500	0.583	0.700	0.583	0.875

The resulting K_c series was fitted to distribution laws. We considered three distribution laws [7]: the normal distribution, the lognormal distribution, and the double exponential Gumbel distribution (for maximums). Figure 6 shows the K_c empirical distribution density and the respective theoretical distribution density curves.

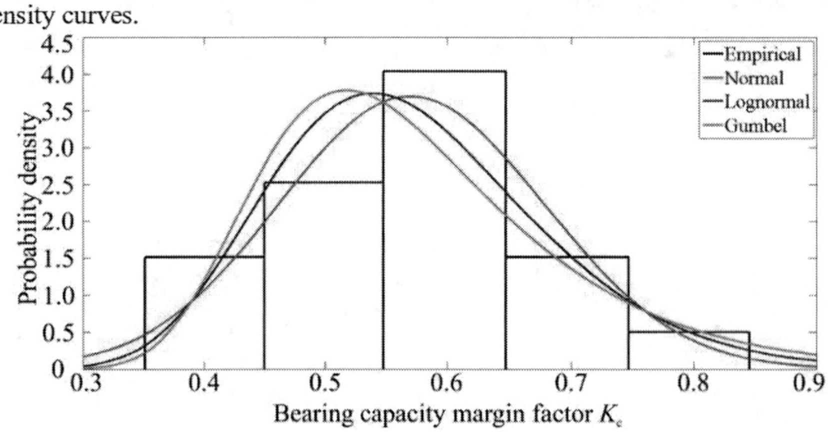

Figure 6. The empirical and the respective theoretical distribution density curves for K_c.

With Pearson's chi-square test it was found that the normal distribution best fitted the dataset. For the normal distribution, the actual bearing capacity margin factor K_c was 0.392 with the 0.95 reliability. A more precise K_1 value was 0.893. The figure 7 shows the empirical distribution curve with 95% confidence bounds and the normal distribution curve.

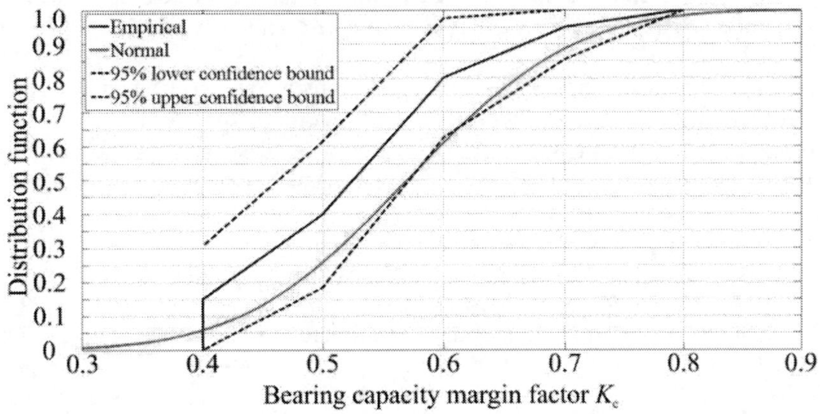

Figure 7. The empirical and the normal distribution curve with 95% confidence bounds for K_c.

Whether the K_1 values for various buildings and structures are valid or not is extensively discussed in the research community [4–10, 20, 21]. Still, the research in this area is insufficient. The studies [4, 5] present a nonlinear dynamic analysis of reinforced concrete structures. It is shown that the K_1 factor for these structures shall be higher than specified in the Construction Code. It means that the seismic resistance is insufficient. Similar conclusions were made in papers [9, 10]

4. Conclusions

The statistical simulation for the 5-storey reinforced concrete framed building produced the K_1 values for the random load cases within the 0.438 to 0.875 range. The actual bearing capacity margin factor K_c was 0.392 with the 0.95 reliability. A more precise K_1 value was 0.893.

The study showed that a reinforced concrete framed building designed to the seismic engineering principles with $K_1 = 0.35$ would have insufficient seismic resistance as estimated with the near-collapse failure criterion.

The more precise K_1 value is larger than the rated one. Therefore, nonlinear dynamic methods should be used to assess the actual seismic resistance of buildings and structures. Using the response spectrum method without prior estimating a more precise K_1 value may lead to disastrous consequences. The proposed strategy based on the probabilistic nonlinear dynamic analysis provides a more accurate assessment of building/structure seismic resistance and ensures their safety.

References

[1] Bolotin V V Radin V P and Chirkov V P 1999 Modeling Dynamic Processes in the Elements of Building Structures in Case of Earthquakes *News of Higher Educational Institutions. Construction* **5** pp 17-21

[2] Standard SP 14.13330.2018 2018 Construction in seismic regions. The updated edition of SNiP II-7-81* (Moscow) p 208

[3] Standard EC8 2004 Eurocode 8: Design of structures for earthquake resistance. - Part 1: General Rules, Seismic Actions and Rules for Buildings. EN1998-1:2003, CEN (Brussels) p 229

[4] Mkrtychev O V and Dzhinchvelashvili G A 2012 Accounting problems of nonlinear seismic stability in the theory hypothesis and errors (Moscow)

[5] Mkrtychev O V Dzhinchvelashvili G A and Busalova M S 2016 Normative approaches to

structural design calculations in a non-linear framework *MATEC Web of Conferences* **86** p 01018
[6] Mkrtychev O V and Bulushev S V 2019 Probabilistic estimation seismic resistance of plain steel frame *IOP Conf. Series: Materials Science and Engineering* **661** p 012016
[7] Mkrtychev O V and Bulushev S V 2020 Probabilistic estimation seismic resistance of spatial steel frame under earthquake *Structural Mechanics of Engineering Constructions and Buildings* **16** (2) pp 87-94
[8] Mkrtychev O V and Bulushev S V 2020 A probabilistic analysis of operation of a flat steel frame under seismic action *Industrial and Civil Engineering* **5** pp 45-50
[9] Kabantsev O V and Umarov K I 2020 Features of elastic-plastic deformation of reinforced concrete shear-wall structures under earthquake excitations *Earthquake engineering. Constructions safety* **1** pp 18-28
[10] Sosnin A V 2017 About refinement of the seismic-force-reduction factor (K1) and its coherence with the concept of seismic response modification in formulation of the spectrum method (in order of discussion) *Bulletin of Civil Engineers* **60** (1) pp 92-116
[11] SP 63.13330.2018 2018 Concrete and reinforced concrete structures. General provisions (Moscow) pp 96
[12] Mkrtychev O V Sidorov D S and Bulushev S V 2017 Comparative analysis of results from experimental and numerical studies on concrete strength *MATEC Web of Conferences* **117** p 00123
[13] Mkrtychev O V and Andreev M I 2018 Verification of the reinforced concrete beam model based on the results of a full-scale experimental study *MATEC Web of Conferences* **196** p 01029
[14] Andreev M I Bulushev S V and Dudareva M S 2018 Verification of the eccentrically com-pressed reinforced concrete column calculation model based on the results of a full-scale experimental study *MATEC Web of Conferences* **251** p 04013
[15] Hallquist, J O 2006 LS-DYNA Theory Manual. Livermore Software Technology Corporation (LSTC)
[16] Murray Y D 2007 Users Manual for LS-DYNA Concrete Material Model 159. Report No. FHWA-HRT-05-062. Federal Highway Administration (McLean) p 77
[17] US Department of Transportation 2007 Federal Highway Administration. Evaluation of LS-DYNA Concrete. Material Model 159. Publication NO. FHWA-HRT-05-063 (McLean) p 190
[18] Mkrtychev O V and Reshetov A A 2015 Method for determining baseline characteristics most unfavorable accelerograms for linear systems with finite number of degrees of freedom *Vestnik MGSU* **8** pp 80-91
[19] Mkrtychev O V and Reshetov A A 2017 A representative set of earthquake accelerograms for calculation of buildings and structures on seismic effects *Vestnik MGSU* **12** (7) pp 754-760
[20] Mkrtychev O V and Dzhinchvelashvili G A 2016 Fundamental principles of earthquake resistance calculation to be reflected in the next generation regulations *MATEC Web of Conferences* **86** p 01017
[21] Mkrtychev O V and Bulushev S V 2018 Actual problems of earthquake engineering, *"Loleyt readings-150". Modern methods of calculation of reinforced concrete and stone structures by limit states* pp 270-8

IOP Conf. Series: Materials Science and Engineering 1015 (2021) 012018 doi:10.1088/1757-899X/1015/1/012018

Natural frequencies of plates of medium thickness

A N Leontiev[1] **T N Gorbunova**[2] **and M R Missurin**[1]

[1] Department of Strength of Materials, National Research Moscow State University of Civil Engineering, 26, Yaroslavskoye Shosse, Moscow, 129337, Russia
[2] Department of Applied Mathematics, National Research Moscow State University of Civil Engineering, 26, Yaroslavskoye Shosse, Moscow, 129337, Russia

E-mail address: an_leontiev@mail.ru

Abstract. A comparison is made of the frequencies obtained by the classical theory of thin plates and the refined theory of plates of medium thickness. Considered a rectangular plate with articulation along the contour. To determine the fundamental frequency, the Bubnov-Galerkin variational method is used. It is shown that with a relative plate thickness $h/a = 0.2$, the calculation refinement is approximately 15-20% for the values of the fundamental frequency of natural vibrations of a rectangular plate.

1. Introduction

A widespread element of building structures is rectangular plates with various fixings of the contour, the most important dynamic of which are the frequencies of natural vibrations.

The technical theory of plate bending, proposed by G. Kirchhoff, describes well the stress-strain state of fairly thin plates, however, with an increase in thickness, the results may not be entirely correct, and their work under load is much more accurately described by the theory of plates of medium thickness.

This area includes the work of a number of scientists (E. Reisner and others [1 – 7]), including the theory developed by B.F. Vlasov [8] and successfully implemented in a number of works [9, 10].

2. State of the problem

In accordance with the B.F. Vlasov, the stress-strain state of a rectangular plate of average thickness under the action of a transverse load $p(x,y)$ is reduced to the determination of two sought functions $f(x,y)$ and $\varphi(x, y)$ from the solution of a system of two differential equations

$$\begin{cases} D\nabla^2\nabla^2 f = p(x,y), \\ \nabla^2\varphi = \dfrac{2Gh}{D(1-\nu)}\varphi. \end{cases} \tag{1}$$

Here $D = Eh^3(12(1-\nu^2))^{-1}$ – is the cylindrical stiffness of the plate, E, G are the elastic and shear moduli of the plate material, h – is its thickness, ν – is the Poisson's ratio of the material, and ∇^2 – is the Laplace operator.

The deflection of the plate $w(x,y)$ is related to the desired function $f(x, y)$ as follows

$$w = f - \frac{D}{Gh}\nabla^2 f \tag{2}$$

When considering natural vibrations, in accordance with the d'Alembert principle, inertial forces are taken as the external load, equal to, where m is the distributed mass of the plate, and the points above the function w denote differentiation in time

In this case, the first equation of system (1) takes the form

$$D\nabla^2\nabla^2 f + m\ddot{w} = 0 \tag{3}$$

Taking into account relation (2), equation (3) takes the form:

$$D\nabla^2\nabla^2 f + m\left(\ddot{f} - \frac{D}{Gh}\nabla^2\ddot{f}\right) = 0 \tag{4}$$

In the considered setting, the function f depends both on coordinates and on time: $f(x,y,t)$, therefore, in accordance with the Fourier method, we represent it as a product of two functions, one of which depends only on coordinates, and the other on time:

$$f(x,y,t) = W(x,y)\cdot T(t) \tag{5}$$

Substituting expression (5) into equation (4), we obtain

$$D\nabla^2\nabla^2 W\cdot T + m\left(W - \frac{D}{Gh}\nabla^2 W\right)\cdot\ddot{T} = 0 \tag{6}$$

The separation of variables, traditional for the Fourier method, leads to the following relations

$$\frac{D}{m}\frac{\nabla^2\nabla^2 W}{W - \frac{D}{Gh}\nabla^2 W} = -\frac{\ddot{T}}{T} = \tilde{\omega}^2$$

from which for the function $T(t)$ we obtain the equation of harmonic oscillations:

$$\ddot{T}(t) + \tilde{\omega}^2 T(t) = 0$$

and for the function $W(x,y)$ – the equation

$$\frac{D}{m}\nabla^2\nabla^2 W - \tilde{\omega}^2\left(W - \frac{D}{Gh}\nabla^2 W\right) = 0 \tag{7}$$

with which, depending on the type of boundary conditions, it is possible to determine the values of the frequencies of natural vibrations $\tilde{\omega}^2$.

3. Analytical method of solving

Consider a rectangular plate hinged on all sides.

In this case, the deflection $W(x,y)$ can be represented as a double trigonometric series satisfying all boundary conditions

$$W(x,y) = \sum_{i=1}^{m}\sum_{j=1}^{n} C_{ij}\sin\frac{i\pi x}{a}\sin\frac{j\pi y}{b} \tag{8}$$

Here a and b – are the dimensions of the plate in plan.

Substituting series (8) into equation (7), we obtain for each pair of numbers (i, j) an equation of the form

$$\frac{D\pi^4}{m}\left[\left(\frac{i}{a}\right)^2 + \left(\frac{j}{b}\right)^2\right]^2 - \tilde{\omega}^2\left\{1 + \frac{D\pi^2}{Gh}\left[\left(\frac{i}{a}\right)^2 + \left(\frac{j}{b}\right)^2\right]\right\} = 0 \tag{9}$$

from which it is possible to obtain the value of the corresponding frequency

$$\tilde{\omega}^2 = \frac{\dfrac{D\pi^4}{m}\left[\left(\dfrac{i}{a}\right)^2 + \left(\dfrac{j}{b}\right)^2\right]^2}{1 + \dfrac{D\pi^2}{Gh}\left[\left(\dfrac{i}{a}\right)^2 + \left(\dfrac{j}{b}\right)^2\right]} \tag{10}$$

Considering that $\dfrac{D}{Gh} = \dfrac{Eh^3}{12(1-\nu^2)}\dfrac{2(1+\nu)}{Eh} = \dfrac{h^2}{6(1-\nu)}$ and taking out of the brackets one of the dimensions of the plate, for example, a, we get

$$\tilde{\omega}^2 = \frac{\dfrac{E}{ma}\dfrac{\pi^4}{12(1-\nu^2)}\left(\dfrac{h}{a}\right)^3\left[i^2 + (\alpha j)^2\right]^2}{1 + \dfrac{\pi^2}{6(1-\nu)}\left(\dfrac{h}{a}\right)^2\left[i^2 + (\alpha j)^2\right]} \tag{11}$$

where $\alpha = a/b$.

For comparison, we present the formula by which the frequencies of rectangular plates are determined in accordance with the technical theory of G. Kirchhoff

$$\omega^2 = \frac{D\pi^4}{m}\left[\left(\frac{i}{a}\right)^2 + \left(\frac{j}{b}\right)^2\right]^2 = \frac{E}{ma}\frac{\pi^4}{12(1-\nu^2)}\left(\frac{h}{a}\right)^3\left[i^2 + (\alpha j)^2\right]^2 \tag{12}$$

4. Examples of calculation

Let us determine for two types of plates the values of the basic dimensionless frequency ($i = j = 1$) in accordance with formulas (11) and (12)

$$\frac{ma}{E}\tilde{\omega}^2 = \frac{\dfrac{\pi^4}{12(1-\nu^2)}\left(\dfrac{h}{a}\right)^3\left(1+\alpha^2\right)^2}{1 + \dfrac{\pi^2}{6(1-\nu)}\left(\dfrac{h}{a}\right)^2\left(1+\alpha^2\right)} \tag{13}$$

$$\frac{ma}{E}\omega^2 = \frac{\pi^4}{12(1-\nu^2)}\left(\frac{h}{a}\right)^3\left(1+\alpha^2\right)^2 \tag{14}$$

In accordance with formulas (13) and (14), figure 1 shows the graphs of the dependence of the values of dimensionless frequencies $\dfrac{ma}{E}\tilde{\omega}^2$ and $\dfrac{ma}{E}\omega^2$ on the parameter h/a for square plates ($\alpha = 1$).

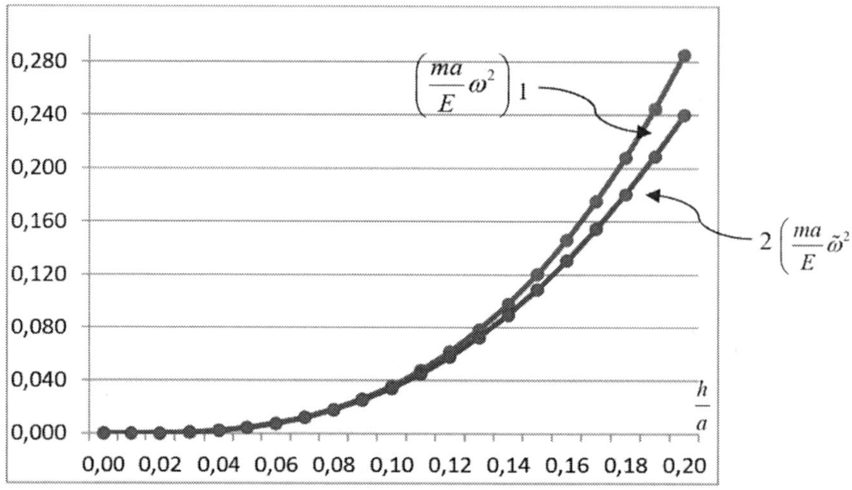

Figure 1. Dependence of the value of natural vibration frequencies on h/a for square plates according to two theories:
1. on the technical theory of G. Kirchhoff,
2. according to B.F. Vlasov's theory of plates of medium thickness.

Can be seen that the results obtained on the technical theory of thin plates with increasing relative thickness are overestimated.

To estimate the value of the resulting error, we compose the ratio of the squares of the frequencies obtained by the two theories under consideration

$$\frac{\omega^2}{\tilde{\omega}^2}=1+\frac{D\pi^2}{Gh}\left[\left(\frac{i}{a}\right)^2+\left(\frac{j}{b}\right)^2\right]=1+\frac{D\pi^2}{Gh}\left[\left(\frac{i}{a}\right)^2+\left(\frac{j}{b}\right)^2\right]$$

which for the fundamental frequencies (for $i=j=1$) leads to the following result

$$\frac{\omega^2}{\tilde{\omega}^2}=1+\frac{\pi^2}{6(1-\nu)}\left(\frac{h}{a}\right)^2\left(1+\alpha^2\right) \tag{15}$$

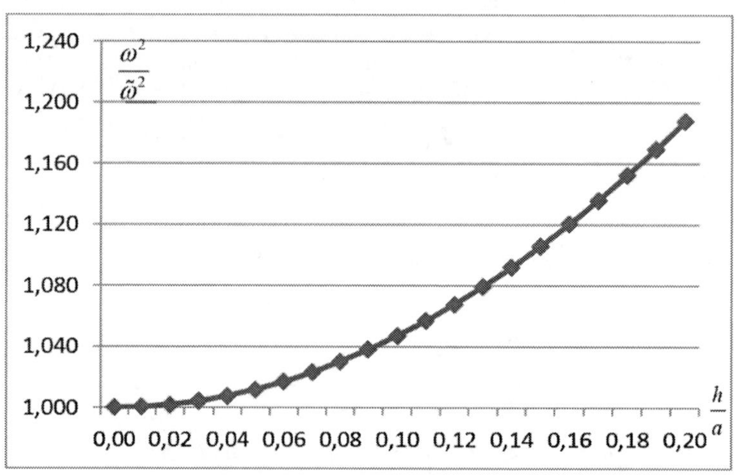

Figure 2. Ratio of the squares of natural frequencies obtained by two theories of bending of square plates at different values of the parameter h/a.

IOP Conf. Series: Materials Science and Engineering 1015 (2021) 012018 doi:10.1088/1757-899X/1015/1/012018

Figure 2 shows a graph constructed in accordance with formula (15) and showing the ratio of the squares of the natural vibration frequencies obtained according to two bending theories for square plates (at $\alpha = 1$) at different values of the parameter h/a.

5. Conclusions

Comparison of the results obtained according to the classical theory of thin plates and the refined theory of plates of average thickness, showed that with a relative plate thickness $h/a = 0.2$, the refinement of the calculation is about 15-20% for the values of the fundamental frequency of natural vibrations.

Acknowledgments

The authors would like to express their gratitude for the help in the preparation of the article from the head of the Department of Strength of Materials Andreev V

References

[1] Reissner E 1945 The effect of transverse shear deformation on the bending of elastic plates *J Appl Mech* **12** pp 68–77
[2] Green A E 1949 On Reissner's theory of bending of elastic plates *Quart Appl Math* **7** (2) pp 223–8
[3] Reissner E 1975 On transverse bending of plates, including the effect of transverse shear deformation *Int J Solids Struct* **11** (5) pp 569–73
[4] Rychter Z 1986 An improved bound on the error in Reissner's theory of plates *Arch Mech* (Warszawa) **38** (1 2) pp 209–213
[5] Huang C S, Chang M and Leissa A 2006 Vibrations of mindlin sectorial plates using the Ritz method considering stress singularities *J of Vibration and Control* **12** pp 635–57
[6] Gabbasov R F and Hoang T A 2014 Dynamic load calculation of a bending plate of average thickness using general equations of finite differences method. *Vestnik MGSU* **10** pp 16–23
[7] Blevins R D 2016 Natural frequency of plates and shells *Formulas for Dynamics, Acoustics and Vibration* pp 203–59
[8] Vlasov B F 1989 Bending equations for Medium thickness plates *Proc. of the Univ. of Civil Eng.* p 107-16
[9] Mardji M and Vlasov B F 1989 Calculation of rectangular plates on an elastic foundation according to the refined theory *Proc. of the Univ. of Civil Eng.* pp 116-20
[10] Vlasov B F and Papuch A V 1989 *Proc. of the Univ. of Civil Eng.* Pp 121-7

IOP Conf. Series: Materials Science and Engineering 1015 (2021) 012019 doi:10.1088/1757-899X/1015/1/012019

Analysis of the impact of shallow subway train traffic on urban development

V A Mitroshin and V L Mondrus

National Research Moscow State Civil Engineering University. 26 Yaroslavskoye av., Moscow, 129337 Russian Federation

E-mail address: mitroshin.vasiliy@yandex.ru

Abstract. The article discusses the process of propagation of vibrations in a soil massif from a moving subway train. With the help of previously performed measurements of ground surface vibrations above a shallow metro line, it was found that the electrical insulating joint of rail strings is a source of increased vibrations when a train enters the station from a tunnel, which can lead to an excess of vibration acceleration levels in residential and industrial buildings limited by sanitary standards. The main types of wheel and rail defects are considered. The propagation of vibrations in unlimited space from a moving subway train in a tunnel, represented by an absolutely rigid body, loaded with dynamic action, taking into account the shock impact at the rail joint, has been studied. The dependences of the parameters of the wave field in infinite space and the displacement of the points of the tunnel on the frequency of the forcing action are found. It is noted that the parameters of the wave field decrease with an increase in the frequency of the external forcing action. The results obtained can be used to solve the problem of the propagation of vibrations in a confined space from moving subway trains.

1. Introduction

The metro is an integral part of the transport network of any modern metropolis. Ease of movement between different parts of the city, bypassing traffic jams on public roads, cannot be overestimated. The location of future metro lines is closely related to the development of new territories and is necessarily reflected in the master plan for the city's development. Subway tunnels, depending on the method of construction and hydrogeological conditions, can be located at different depths, practically independent of the building on the day surface. The underground is a complex movable mechanical system, which can be simplified to the following constituent elements: rolling stock - railroad bed - ballast - tunnel lining. Each of these components in the process of manufacturing and operation acquires a certain random number of various defects and damages, the interaction of which leads to the formation of an increased vibration background. The tunnel, thus, becomes the emitter of a wave field propagating radially in the soil massif, which, attenuating with distance, collides with the nearby building structures of buildings and structures. A number of works have been devoted to the propagation of vibrations from the subway in domestic science, they have in common the presentation of a moving load on a railroad bed in the form of a certain periodically repeating average weight of a car. This approach allows you to simplify mathematical calculations and, with a certain accuracy, estimate the vibration parameters of a tunnel or soil mass. However, both domestic and foreign researchers note several main factors affecting the formation of vibration from a moving train, namely: defects in wheels and rails, as well as a wheel hitting an obstacle - a rupture of rail strings (the so-called electrical insulating joint) in one level or vertically offset to form a step. Figure 1 shows an example of a junction of rails.

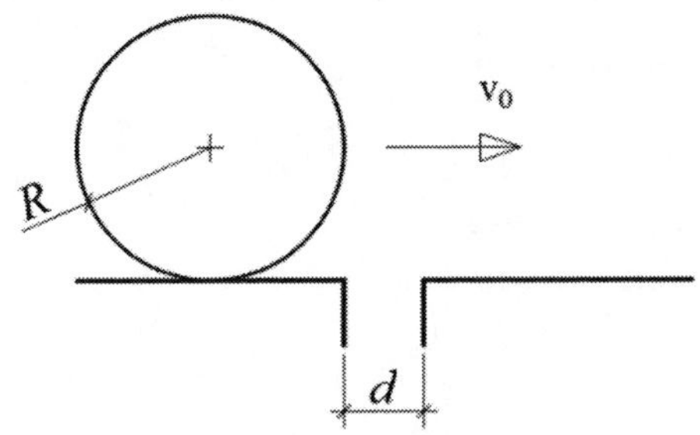

Figure 1. An example of junctions of rails.

The following works [1, 2, 3] are devoted to the issues of numerical simulation of the interaction of a wheel and a rail during their impact. The authors describe the physics of the process in sufficient detail and determine with sufficient accuracy the arising internal forces both in the rail and in the wheel. For example, [4] propose, by solving a linear partial differential equation of beam vibrations on a viscoelastic basis, to determine the significant components of the stress-strain state at a specific point of the rail for a given law of dynamic action.

A number of studies are devoted to solving problems of the propagation of an oscillatory front from a metro tunnel as a "radiator", among which the work of M.A. Dashevsky [5], describing in detail the application of the method of successive wave approximations together with the method of compensating loads. This method makes it possible to find a solution for a half-plane with a free boundary, simulating a day surface, using a combination of certain solutions for an unbounded domain in a flat formulation of the problem. In [6], the process of propagation of a wave field in a soil massif under the action of a periodic load on a tunnel is considered, and the wave interaction of the tunnel structures with the surrounding soil environment is also taken into account. As an external influence, the author adopted a loading scheme obtained by replacing concentrated forces from car bogies with a uniformly distributed averaged load on a section bounded by a railroad bed. In the article [7] it is shown that the impact of the composition on the electrical insulating joint leads to the formation of increased vibrations

2. Determination of wave field parameters

Following the logic of works [5, 6], consider a rigid inclusion in the form of a washer, "soldered" into an elastic space, loaded with an average dynamic shock load.

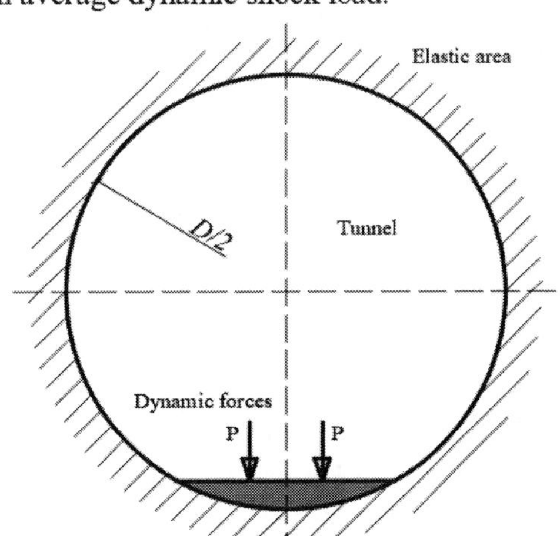

Figure 2. Diagram of load application to the tunnel.

The force of the impact interaction of the wheel and the rail can be written as follows [8]

$$P = P_0 \sin \frac{\pi}{\tau} t \tag{1}$$

where P_0 is the amplitude value of the impact force.

Impact impulse

$$I_y = m\omega d \tag{2}$$

where d is the width of the insulating joint.

Using the Hertz theorem describing contact interactions [9] and also [8], we write

$$P_0 = k^{2/5} \left(\frac{5 I_y^2}{4m} \right)^{3/5} \tag{3}$$

where $k = \frac{4}{3\pi} \frac{\sqrt{R}}{\delta_1 + \delta_2}$, R - radius of the wheel rolling circle, δ_1, δ_2 - material characteristics.

For this and subsequent problems, the main dependencies are derived from the Lamé equation of motion of points of a continuous elastic medium

$$\mu \cdot \Delta \bar{u} + (\lambda + \mu) \cdot grad(div\bar{u}) = \rho \frac{\partial^2 \bar{u}}{\partial t^2} \tag{4}$$

where u is the displacement vector, λ, μ are the Lamé constants, ρ is the density of the elastic medium.

According to the Helmholtz theorem [10], almost any vector field can be represented as a gradient of some scalar φ and a curl of some vector ψ, the divergence of which is equal to zero

$$\bar{u} = \mathrm{grad}\varphi + \mathrm{rot}\,\psi, \ \mathrm{div}\,\psi = 0 \tag{5}$$

where φ is the scalar potential ψ is the vector potential.

Thus, the problem of determining the displacement field is reduced to solving the following set of equations for unknown potentials

$$\left[\begin{array}{l} \Delta\varphi = \frac{1}{a^2} \cdot \frac{\partial^2 \varphi}{\partial t^2} \\ \Delta\psi = \frac{1}{b^2} \cdot \frac{\partial^2 \psi}{\partial t^2} \end{array} \right. \tag{6}$$

where $a^2 = \frac{\lambda + 2\mu}{\rho}$, $b^2 = \frac{\mu}{\rho}$ are the velocities of longitudinal and transverse waves in an elastic medium.

In this and subsequent problems, along with the Cartesian one, a polar coordinate system is used, the origin of which coincides with the center of an inclusion soldered into the elastic space.

In accordance with [11], elastic displacements of points can be written as follows

$$\begin{array}{l} u_r = \frac{\partial \varphi}{\partial r} + \frac{1}{r} \cdot \frac{\partial \psi}{\partial \theta} \\ u_\theta = \frac{1}{r} \cdot \frac{\partial \varphi}{\partial \theta} - \frac{\partial \psi}{\partial r} \end{array} \tag{7}$$

Using the known relations for the components of the linear theory of elasticity [12] and the results [6], we write, with some refinements, the expressions for stresses, expressed in terms of potentials in a polar coordinate system

$$\sigma_r = -\rho a^2 k_1^2 \varphi - \frac{2\rho b^2}{r}\left(\frac{\partial\varphi}{\partial r} + \frac{1}{r^2}\frac{\partial^2\varphi}{\partial\theta^2} + \frac{1}{r}\frac{\partial\psi}{\partial\theta} - \frac{\partial^2\psi}{\partial r\partial\theta}\right),$$
$$\sigma_\theta = -\rho a^2 k_1^2 \varphi + \frac{2\rho b^2}{r}\left(r\frac{\partial^2\varphi}{\partial r^2} + \frac{1}{r}\frac{\partial\psi}{\partial\theta} - \frac{\partial^2\psi}{\partial r\partial\theta}\right), \quad (8)$$
$$\sigma_{r\theta} = \tau_{r\theta} = \frac{4\rho b^2}{r}\left(\frac{\partial^2\varphi}{\partial r\partial\theta} - \frac{1}{r}\frac{\partial\varphi}{\partial\theta} + \frac{r}{2}k_2^2\psi + \frac{\partial\psi}{\partial r} + \frac{1}{r}\frac{\partial^2\psi}{\partial\theta^2}\right),$$

where: $k_1^2 = \dfrac{\omega^2}{a^2} = \dfrac{\rho\omega^2}{\lambda+2\mu}$, $k_2^2 = \dfrac{\omega^2}{b^2} = \dfrac{\rho\omega^2}{\mu}$.

In general form, solutions of the Helmholtz equations for retarded potentials [11] can be written in the form

$$\varphi = \varphi_1\cos\theta = L_1 H_1^{(2)}(k_1 r)\cos\theta$$
$$\psi = \psi_1\sin\theta = Q_1 H_1^{(2)}(k_2 r)\sin\theta \quad (9)$$

where: L_1, Q_1 are the required coefficients that fully characterize the wave field in unlimited space.

Using the boundary conditions for the plane problem of radiation of cylindrical waves into an unbounded plane [5], we can write the following system of equations for the coefficients L_1, Q_1.

$$L_1\left(\alpha H_0^{(2)}(k_1R) - \frac{\alpha}{k_1R}H_1^{(2)}(k_1R) - \frac{\beta\omega^2}{R}H_2^{(2)}(k_1R)\right) + Q_1\left(\frac{\alpha}{R}H_1^{(2)}(k_1R) + \frac{\beta\omega^2}{R}H_1^{(2)}(k_1R)\right) + 2P_0 = 0 \quad (10)$$
$$L_1\left(\frac{1}{R}H_1^{(2)}(k_1R) + H_0^{(2)}(k_1R) - \frac{1}{k_1R}H_1^{(2)}(k_1R)\right) + Q_1\left(\frac{1}{R}H_1^{(2)}(k_2R) + H_0^{(2)}(k_2R) - \frac{1}{k_2R}H_2^{(2)}(k_1R)\right) = 0$$

3. Results and discussions

Having solved the system of equations (10) in the mathematical software with respect to L_1, Q_1, we will construct graphs of the dependence of the parameters of the wave field in infinite space and the radial displacement of the tunnel on the change in the frequency of the external influence (figures 3, 4).

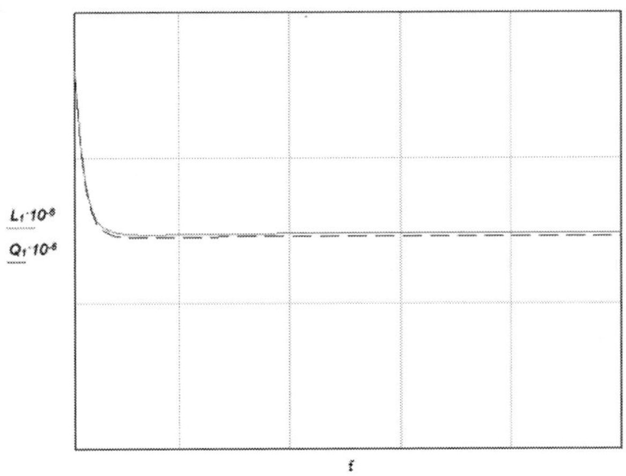

Figure 3. Dependence of the parameters of the wave field on the frequency of the external influence.

Note that figure 3 shows that the intensity of the wave field decreases with increasing frequency of the external influence.

4. Conclusions

The studies carried out within the framework of the set task allow us to formulate the following conclusions:

- the impact at the junction of the rails induces all modes of vibration, but the largest are vibrations with frequencies close to 31.5 Hz and 63 Hz;
- vibration levels, measured in octaves 31.5 and: 63 Hz, at points located above the metro line, significantly exceed the permissible according to the Sanitary Standards of the Russian Federation;
- with an increase in the frequency of external influence, the intensity of the wave field in infinite space decreases;
- the obtained coefficients L_l, Q_l, characterizing the wave field in unbounded space, will be further used to solve the problem of the propagation of oscillations induced by a tunnel in a semi-infinite space.

References

[1] Kuznecov A V 2000 Udarnoe vzaimodejstvie kolesa i rel'sa. *Dissertaciya na soiskanie uchyonoj stepeni kandidata tekhnicheskih nauk (special'nost' 05.23.17 – Stroitel'naya mekhanika)* (MGUPS (MIIT) Moskva) p 88

[2] Kouroussis G, Connolly D, Alexandrou G, Vogiatzis K, Verlinden O 2015 Modelling the singular rail and wheel surface defect for predicting railway ground vibration. *Proceedings of the 22nd International Congress on Sound and Vibration* (Florence Italy) p 8

[3] Ivanov V V 2011 Snizhenie udarnogo vozdejstviya na koleso gruzovogo vagona pri prohozhdenii rel'sovogo styka. *Dissertaciya na soiskanie uchyonoj stepeni kandidata tekhnicheskih nauk (special'nost' 05.22.07 – Podvizhnoj sostav zheleznyh dorog, tyaga poezdov i elektrifikaciya)* (Moskva: OmGUPS) p 170

[4] Kogan A YA, Pejch YU L 2002 Raschyot nestacionarnogo napryazhyonno-deformirovannogo sostoyaniya elementov konstrukcii puti v zone styka rel'sov *Vestnik VNIIZHT* 2 pp 31-39

[5] Dashevskij M A 1991 Zashchita zdanij ot vibracii, vozbuzhdaemoj dvizheniem poezdov metropolitena *Dissertaciya na soiskanie stepeni doktora tekhnicheskih nauk. (special'nost' 05.23.47* – (Moscow: Stroitel'naya mekhanika. CNIISK im. V. A. Kucherenk*o)* pp 270 – 324

[6] Kolotovichev YU A 2010 Zadachi prognoza kolebanij poverhnosti grunta pri dvizhenii poezdov metropolitena v tonnelyah neglubokogo zalozheniya *Dissertaciya na soiskanie uchyonoj stepeni kandidata tekhnicheskih nauk special'nost' 05.23.17* (Moscow: Stroitel'naya mekhanika. Moscow State Uiversity of Civil Engineering) p 215

[7] Mitroshin V, Mondrus, Sizov V D 2019 Analysis of ground vibrations caused by the shallow subway. *XXVIII RPS Seminar 2019. IOP Conf. Series: Materials Science and Engineering.* (661 doi:10.1088/1757-899X/661/1/012022).

[8] Manashkin L A, Myamlin S V Prizod'ko V I 2008 Ocenka sily udarnogo vzaimodejstviya kolesa i rel'sa na styke dvuh rel'sov. *Vestnik Dnepropetrovskogo nacional'nogo universiteta zheleznodorozhnogo transporta* **22** pp 36-9

[9] Gol'dsmit V 1965 *Udar. Teoriya i fizicheskie svojstva soudaryaemyh tel*. (Moskva: Strojizdat) p 448

[10] Mors F M, Fershbah G 1958 *Metody teoreticheskoj fiziki* (Moscow: Izdatel'stvo inostrannoj literatury) **1** p 930

[11] Kupradze V D 1935 *Osnovnye zadachi matematicheskoj teorii difrakcii (ustanovivshiesya processy)* (Moskva: Glavnaya redakciya obshchetekhnicheskoj literatury) p 111

[12] Novozhilov V V 1958 *Teoriya uprugosti* (Leningrad: SUDPROMGIZ) p 371

IOP Conf. Series: Materials Science and Engineering 1015 (2021) 012020 doi:10.1088/1757-899X/1015/1/012020

Determination of dynamic parameters of the elastically supported beam system model by optical and numerical methods

S Gridnev, I Ravodi and V Volko

VSTU, Structural Mechanics Department, 394026, Voronezh, Russia

E-mail address: iravodin@yandex.ru

Abstract. The article describes the design of a test facility that designed and manufactured by the authors for parametric studies of the features of free and forced vibrations of elastically supported dynamical systems. The results of a full-scale experiment are presented in the article, during which a frequency analysis of a beam with an intermediate elastic support is performed. Measurements are made with the use of the optical method of determining the movements and inductive sensors. A finite element model of the test facility was built in the Midas NFX environment with the aim to verify and evaluate the results. The results of the modal analysis performed for the finite element model of the test facility are compared with the natural frequency spectra obtained as a result of experiments. Conclusions are drawn about the prospects of using the test facility, the accuracy of the measuring equipment used, and about the way the finite element model built in the Midas NFX program environment corresponds to the full-scale model of the dynamic system in terms of its dynamic characteristics.

1. Introduction

Currently, when designing buildings and structures, a significant number of calculations related to the determination of the stress-strain state of structures are performed using software systems based on the finite element method (FEM). The accuracy of such calculations depends to varying degrees on the numerical approximation of the model, accepted assumptions and hypotheses. Often, errors caused by FEM inaccuracy can be estimated and minimized by selecting the optimal type of finite elements (FE) used in solving the problem, the size of the grid FE, the settings of the computational algorithm, etc. It is more difficult to estimate errors caused by a number of simplifications, such as idealization of construction materials, geometry, loads, etc.

In [1], notes a number of problems that arise during the transition from the physical model to its FE approximation. From the point of view of design reliability, the simplified representation of the design scheme is compensated by using a group of coefficients (reliability coefficient, operating conditions coefficient, etc.). However, for example, in nonlinear dynamical systems and in systems with feedback, it is quite difficult to provide a certain confidence interval for the results. There is a need to consider several calculation situations, which ultimately leads to less cost-effective solutions.

According to the recommendations given in [2] and [3] in dynamic calculations implemented in modern software systems, it is necessary to use mainly refined calculation schemes. In such schemes should be most considered features of the design (rigidity, pliability of the supports and the base, the spatial work of the construction, etc.). While in [2], the authors point out the possible error in the determination of the natural frequencies of buildings and load-bearing elements. For approximate schemes, it varies in the range from 15 to 35 %, depending on the type of construction. In refined schemes, the error value can be 10-15%. Given the multidimensional nature of the practical problems being solved today, and as a result, the variety of forms of vibrations of structures, the results of calculations at this level of confidence in the dynamic parameters of the system become difficult to analyse.

One of the means of developing a verified FE model is an experiment. Experimental data can be obtained from a full-scale or scale model using various types of sensors. If necessary, the design as a whole or its separate element is considered. To determine the characteristics of linear design schemes, it is sufficient to consider the free vibrations of the structure.

In [4], a full-scale model of a girder bridge and a vehicle was created on a geometric scale of 1 to 10 in order to work out and test the proposed numerical models. The theoretical calculation scheme is performed in the Ansys program and consists of solid and shell finite elements. The monolithic reinforced concrete slab and diaphragms are modelled by the shell63 element, while the beams are modelled by the solsh190 element. The rigidity of the asphalt pavement of the bridge and the restrictive curb is ignored in the calculation. As a result, the difference between the measured and theoretical first frequency of natural vibrations of the superstructure was 4%.

The authors in [5] developed a finite element model of a three-axle truck EQ3166 in the LS-DYNA program for dynamic calculations of a road bridge with a mobile load. The car model consists of a total of 49 parts, 9 types of materials, 106 multi-node connections, 7008 nodes, 369 solid elements, 7959 shell elements and 237 beam elements. The nonlinear character of wheel deformation is modelled using the AIRBAG_SIMPLE_AIRBAG_MODEL option in LS-DYNA. The measured base frequency of the full-scale truck was 1.680 Hz, while the first frequency of the developed finite element model was 1.634 Hz. Thus, the error of the numerical model does not exceed 3 %.

The necessary experimental data about the investigated object can be obtained using mechanical, electromechanical, optical, electronic devices and their combinations. However, the use of mechanical devices is limited, especially in dynamic systems, where the recorded values from the sensors are functions of time. V. V. Volkov and S. Yu. Gridnev in [6] suggest using an optical method to analyse the movement of road bridge nodes from a mobile load. The authors use an algorithm that allows to determine changes in the coordinate of the investigated point by comparing several images of the superstructure during loading.

The issues of modelling dynamic mobile load on a span structure in modern finite element complexes are discussed in [7]. The authors investigate the possibility of using contact interaction mechanisms in the Midas NFX program to describe joint vibrations of a beam and a moving load.

The article [8] presents a FE model of the rubber-metal support part (RMSP), developed in order to clarify the dynamic behaviour of road bridges. The numerical modelling of the composite structure of the support material, the formation of interaction features of rubber and steel plates is of interest in this problem. In addition to the numerical implementation, in [9] describes an experimental analysis of the vibrations of the span structure of a steel-reinforced concrete bridge with the use of RMSP. Registration and processing of values of dynamic displacements of supports and deflections is performed by an optical device supplemented with special programs. The authors note the importance of taking into account the shear and vertical deformations of the RMSP in practical calculations.

Thanks to the spread of electronic gadgets with the ability to record photos and videos in high resolution, a wide range of researchers are becoming available to new opportunities in the experimental analysis of building structures. So in the article [10] to determine the movement of span structures or parts of them, it is proposed to use a smartphone with a pre-installed program D-Viewer. The program algorithm is based on tracking the movement of a light spot on the projection screen from a laser fixed to the investigated object. In order to test the proposed method, the authors measure the deflections of the beams of a reduced copy of the suspension bridge using iPhone 6 and iPhone 6 Plus smartphones and classic arrow indicators. The peak value of the oscillation amplitude measured by the two types of sensors differs by 1.71%.

The article describes the developed facility for studying the features of dynamic behaviour of elastically supported systems. Using the readings of inductive sensors and the results of optical measurements, the natural frequencies of vibrations of a beam with an intermediate elastic support are determined. To verify the results, the values of the frequencies and forms of natural vibrations of the investigated dynamic system are determined in the Midas NFX program environment. The results presented in the article complete the first stage of research in the field of experimental and theoretical analysis of the behaviour of elastically supported dynamical systems under the action of harmonic stationary and mobile loads.

2. Description of the design scheme of the investigated dynamic system

In mechanical engineering there is a whole class of dynamic machines and aggregates that have not rigid, but pliable junctions with the base or structural elements. In this case, the boundary conditions are represented by a system of elastic connections and rigid limiters, which are included in the operation when the set gap is selected. In construction practice, elastic connections are widely used in modelling pontoons of floating bridges.

Figure 1 shows the design scheme of the investigated dynamic system. The beam is pivotally supported at the ends and has one intermediate elastic connection in the middle.

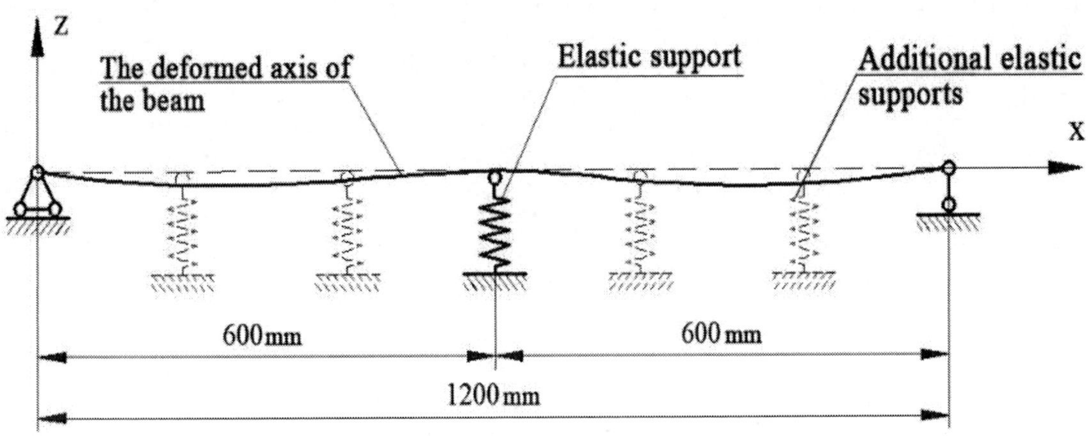

Figure 1. Design scheme of the investigated dynamic system.

Horizontal longitudinal and transverse horizontal deformations of the beam are not taken into account due to their smallness. A linear deformation diagram was used for the material. The initial equilibrium state of the system was selected in such a way that after applying the gravitational load, the upper part of the supports was at the same level.

3. Development and production of a facility for obtaining experimental data

The experimental stand was made by the authors of the article in the framework of research on the analysis of the deformed state of elastically supported systems under the action of a moving load. Experimental data obtained on the test facility are more reliable than data taken from measuring equipment on a real structure, since in an enclosed space, undesirable effects of external factors are minimized, and the same loading can be implemented repeatedly. The General view of the manufactured installation and its main components are shown in figure 2.

IOP Conf. Series: Materials Science and Engineering 1015 (2021) 012020 doi:10.1088/1757-899X/1015/1/012020

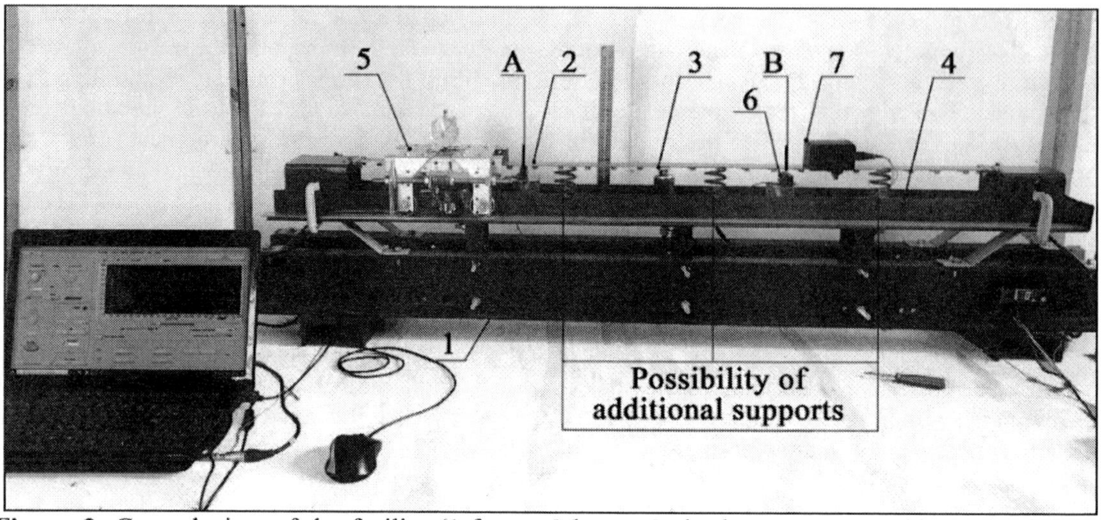

Figure 2. General view of the facility (1-frame, 2-beam, 3-elastic support, 4-guides, 5-bogie with micrometer, 6-inductive sensors, 7-eccentric with adjustable speed).

The frame 1 is made of rolled paired channels that provide stability and rigidity of the structure. The frame is used for fixing the supports of the studied beam, measuring sensors and guides. The main parts and components are attached to the frame through split connections (bolts and screw clamps), which allows to change the parameters of the dynamic system if necessary.

Beam 2 at the ends has one fixed and one movable attachment with mating sections for the entrance and exit of the mobile vehicle model. The stiffness and weight of the beam may vary during experiments. To obtain the desired frequency range of the dynamic system, lead weights with a total mass of 168.6 g were suspended from the steel band.

At this stage, the beam has one intermediate elastic support 3 in the middle of the span, the rigidity of which was determined as a result of static tests. However, in General, the number and location of elastic supports can be changed if this is required during the experiment. The facility can also be supplemented with restrictive supports at the ends, which are included in the work when selecting the set gap, which makes it possible to study the structural nonlinear vibrations of the beam.

For checking the vertical marks of the beam before testing, as well as for determining static deflections, guides 4 and a bogie with micrometer 5 are used. The bogie can move freely along the beam, while in other directions its position is fixed by a system of roller bearings. The accuracy of the micrometer is 0.1 mm. The trolley can also be used to simulate the impact of a moving load.

The displacements of the studied dynamic system are recorded by means of optical control and subsequently processed using special software. The capabilities of optical measurement tools are limited by the frame rate when it comes to short-term processes or dynamic effects resulting from the influence of higher harmonics. To obtain a detailed spectrum of the system's natural frequencies, the test facility is equipped with inductive sensors 6. A real-time signal is recorded from inductive sensors, whose amplitude is proportional to the speed of the beam movement.

The article presents the results of the analysis of only free vibrations of the beam, but the versatility of the manufactured facility allows us to study forced vibrations when the load is moving. The authors performed a number of experiments aimed at studying the dynamic behaviour of elastically supported systems under the influence of harmonic stationary and moving loads. For this purpose, an eccentric motor 7 can optionally be installed on the beam, which exerts a harmonic force effect of a given amplitude. The load frequency can be changed using a single-channel pulse width modulator (PWM) power. The controller display shows the percentage of the output equivalent voltage to the rated voltage.

IOP Conf. Series: Materials Science and Engineering 1015 (2021) 012020 doi:10.1088/1757-899X/1015/1/012020

4. Optical method of measuring the movements of a dynamic system

In the optical method for determining spatial coordinates, which is used for analysing the deformed state of the installation, two stages can be distinguished.

At the first stage, a digital video camera is used to record beam vibrations at a certain time interval. In this case, the optical images projected on the light-sensitive matrix are converted to digital form and then recorded on an external storage medium.

At the second stage, the coordinates of the studied cross sections of the beam are calculated from the resulting video recording. To do this, we use the analysis of equidistant points in space and the correlation of their brightness images (figure 3). To determine the absolute movement of points, entered a scale factor that shows the number of pixels contained in a unit of length. It can be determined using a ruler or other template equidistant from the camera along with the system's calculated points (figure 4).

As you can see in the picture, the accuracy of the measurement results directly depends on the quality of the shooting. The higher the resolution of the recording device, the lower the error caused by the discretization of the image. So, for current research, the following value of the scale coefficient is accepted: λ=8.8. Then the error in determination of displacements is 0.06 mm.

Measurement errors may also occur as a result of vibrations in the recording device itself. In order to avoid this effect, the camera tripod was installed at a sufficient distance from the sources of vibration on an isolated monolithic reinforced concrete structure.

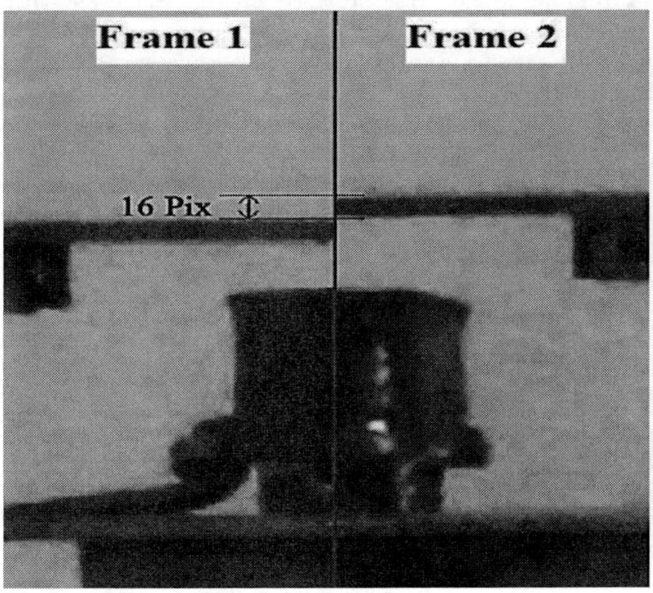

Figure 3. Correlation of system points.

Figure 4. Determining the scale factor.

A camera with a frequency of 30 frames per second was used in order to determine the dynamic movements of the beam. If we take into account Kotelnikov's theorem in the field of digital signal processing, we can say that the camera's capabilities allow us to analyse vibrations occurring with a frequency not exceeding 15 Hz. Often, the first natural frequencies of buildings and structures are below this value. However, in some cases, for example, if the building uses centrifuges, vibrating sieves or other dynamic equipment, the frequency of forced vibrations of structures may be higher and this will require the use of high-speed cameras for movement analysis. In this paper, inductive sensors are used to obtain the frequency spectrum of higher harmonics of an elastically supported beam.

5. Determination of free oscillation frequencies of a dynamic system using inductive sensors

The principle of operation of inductive sensors is based on changing the parameters of the magnetic field created by a permanent magnet when a metal object is introduced into the core. The inductor coil tends to keep its magnetic flux unchanged, so when the magnetized beam oscillates, aт electric current occurs in the coil, resulting in a voltage at its ends. The value of the output signal is proportional to the product of the beam oscillation amplitude and frequency.

During the experiments, two inductive sensors were placed on the facility, as shown in figure 2. The sensor leads are connected to a 6.35 mm TRS jack for transmitting analog audio signal. Recording from the sensors was performed in WAV format, with two channels, with a sampling rate of 44.1 kHz.

In order to obtain the natural frequency spectrum, in the program Audacity performed a fast Fourier transform for the signal obtained when recording free vibrations of the beam. The best frequency resolution can be obtained when selecting Hanning window functions for fast conversion.

It should be noted that it is difficult to use inductive sensors to determine the speed and amplitude of the beam vibration, as the frequency of vibrations increases and the value of the voltage increases. In other words, the ratio of harmonics in the output signal will differ from the real one. However, inductive sensors can be successfully used to determine the frequencies of both free and forced vibrations of dynamic systems.

6. Description of the FE model of the experimental facility and results of modal analysis

The finite element model of the studied beam in the Midas NFX program environment is shown in figure 5. The beam in the calculation was modeled by 4-node rectangular finite elements of the shell, which have 5 degrees of freedom in the node. The shape and size of the finite element grid depended on the position of the supports and stability of the calculation. The concentrated masses from the additional weights and the elastic support were set using scalar elements that express the strain energy or kinetic energy when the selected node moves relative to the fixed support.

In order to create an accurate model of the dynamic system, the actual mass of the structure, the mass of additional weights, the geometric dimensions of the beam and other necessary characteristics of the system were determined using measuring instruments.

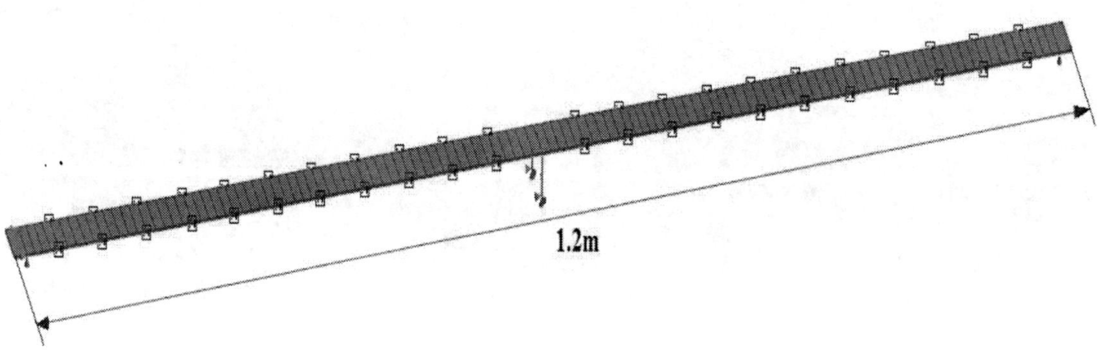

Figure 5. Finite element model of an elastically supported beam.

The stiffness of the elastic support was obtained as a result of a series of static tests, during which the deformations of the support were determined using the optical method.

The cross section of the beam is 37.5 × 1.8 mm. The beam material is steel with an elastic modulus $E = 2.14 \times 10^5$ MPa. Since the maximum stresses in the beam obtained during the experiments are significantly less than the proportionality limit, a linear diagram of steel deformation was used.

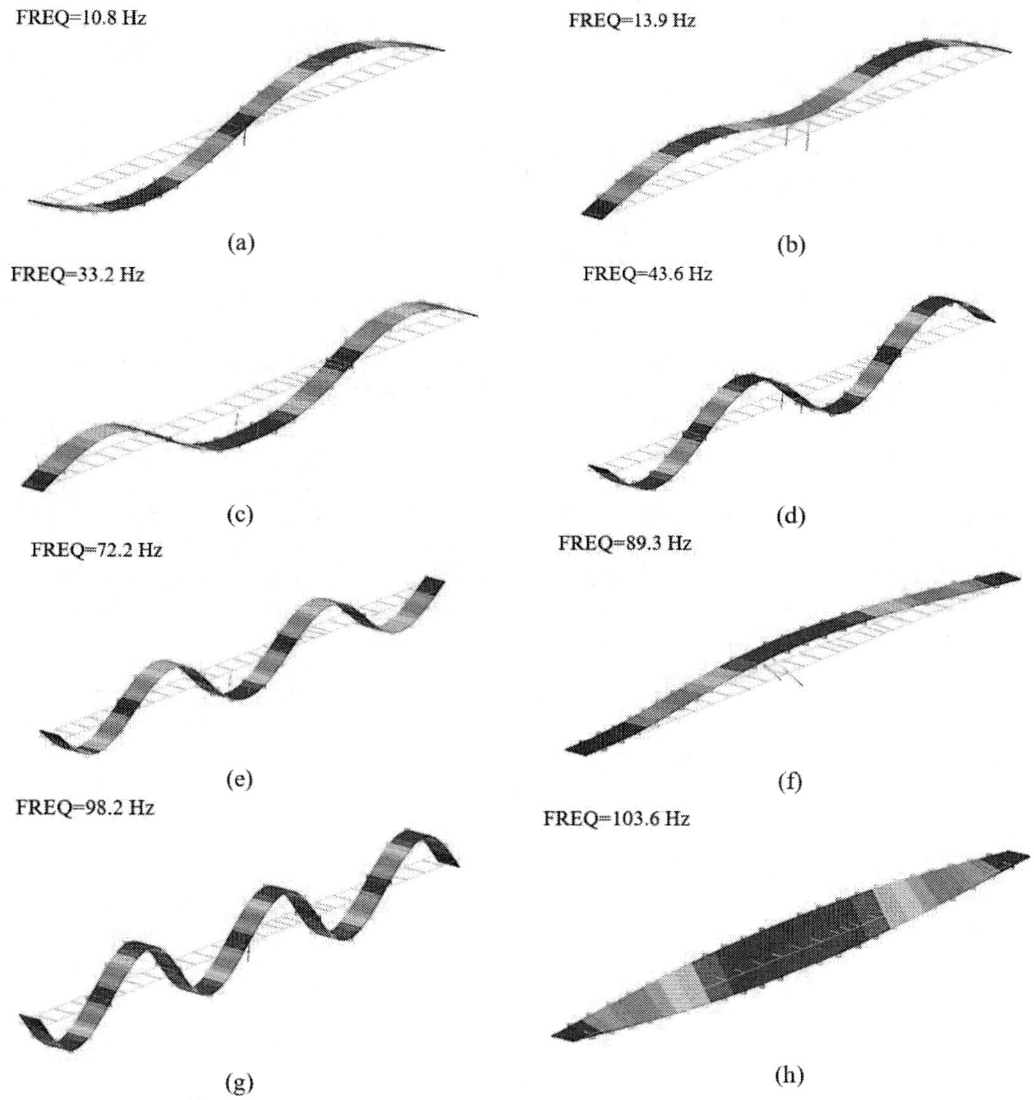

Figure 6. Forms of vibrations of a beam with an elastic support in the middle of the span.

The Midas NFX program contains functionality that allows performing modal and harmonic analysis of a FE model. The problem of determining the natural frequencies and forms of vibrations of the system is the first step to performing other, more complex types of dynamic calculations.

It should be noted that when determining the vibration frequencies of the beam in the Midas NFX FE complex, the influence of damping forces arising under the conditions of a full-scale experiment was not taken into account. This approach is due to the fact that the value of the damping coefficient δ, established during the analysis of the oscillatory process, is significantly less than the first circular frequency of free vibrations of the dynamic system in the absence of friction ω_0.

For the constructed FE model of an elastically supported beam, a modal analysis is performed and the first 8 natural frequencies and corresponding oscillation forms are obtained, which are shown in figure 6.

The figure shows that the elastic connection in the middle of the beam span affects the appearance of certain forms of free vibrations. Traditionally, the first form with a single half-wave is modified in this case (figure 6 b), and its frequency grows with increasing elastic connection stiffness. As a result of modal analysis, it is established that the 6th form of beam oscillation occurs in a horizontal plane, and the 8th is a torsional form.

In General, the difference between the natural frequencies of the presented forms from the first to the eighth is significant and allows to analyze the capabilities of control and measuring equipment in a wide range when working with a full-scale model.

7. Analysis of the obtained experimental data

The purpose of the experiment was to determine the natural frequencies of vibrations of an elastically supported beam. Measurements were made by optical means of movement control, as well as inductive sensors.

At the first stage of the experiment, 10 seconds of background were recorded (the beam was at rest) in order to determine the measurement error caused by the vibration of the installation or the influence of external electromagnetic fields on the electrical circuit. Then, after deflecting the beam from the static equilibrium position, free damping vibrations occurred, which were recorded by inductive sensors and a video camera.

The positions of the cross-sections of the beam, which movements were analysed during the experiment, were selected taking into account the obtained forms of vibrations as a result of modal analysis. One of the inductive sensors was installed in the middle of the first span (section A), the second – in the first third of the second span (section B). Figure 7 shows the vertical movement of the cross section of the A beam with free vibrations.

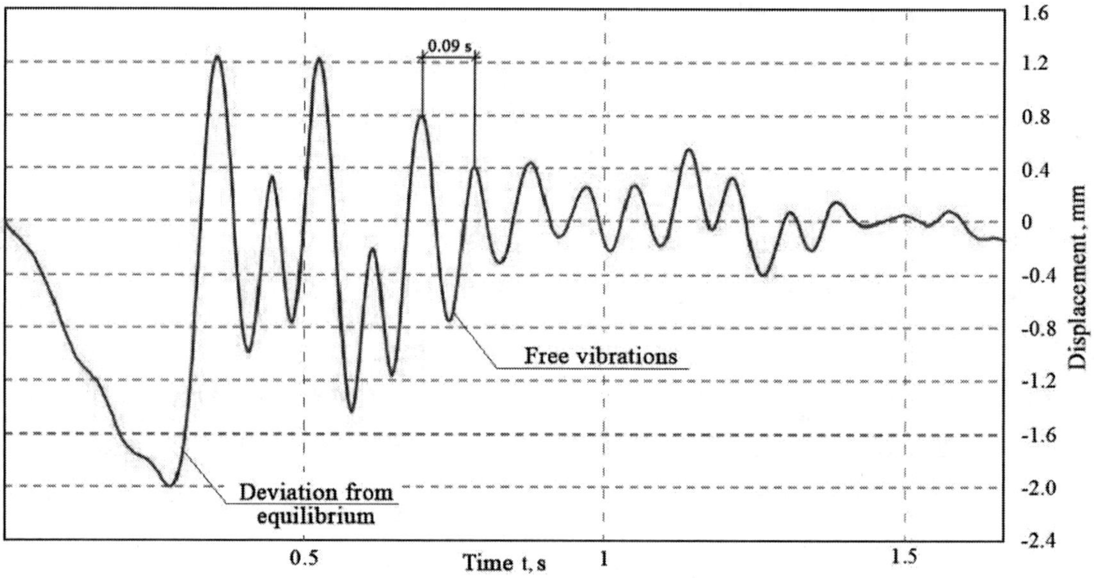

Figure 7. Vertical movements of the beam section A.

As mentioned above, the capabilities of the video camera used theoretically allow you to record vibrations occurring at a frequency of no more than 15 Hz. For rice.7 you can see that the main frequency of the system is 11 Hz, while the frequency of the higher forms of the graph cannot be determined. It can also be noted that the amplitude of the beam vibrations in the section under consideration does not exceed 2 mm, so the displacements and deformations remain small, which allows us to use linear physical and geometric dependencies in the FEM equations.

Figure 8 shows the results of a fast Fourier transform performed using the Audacity program for the recorded signal with free vibrations of the beam.

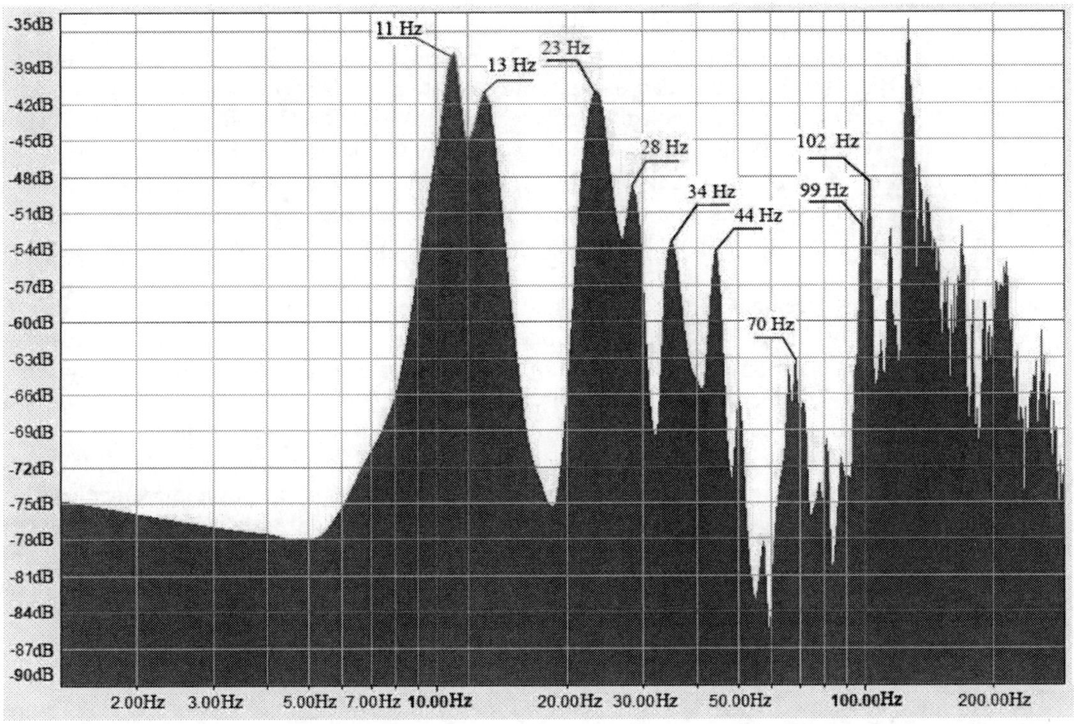

Figure 8. Graph of the spectrum of free vibrations of the beam.

The spectrum graph shows several distinct peaks that correspond to the prevailing harmonics in the processed signal. The results obtained are in good agreement with the numerical calculation data. The natural oscillation frequencies of the investigated beam differ from the theoretical ones by no more than 2-3%. The comparison results are shown in table 1.

Table 1. The frequency of free oscillations.

Form №	1	2	3	4	5	6	7	8
FEM	10.8 Hz	13.9 Hz	33.2 Hz	43.6 Hz	72.2 Hz	89.3 Hz	98.2 Hz	103.6 Hz
Experiment	11 Hz	13 Hz	34 Hz	44 Hz	70 Hz	-	99 Hz	102 Hz

Graph figure 8 lacks the 6th oscillation frequency, the theoretical value of which is 89.3 Hz. This suggests that the design and position of inductive sensors in this experiment does not allow us to analyse fluctuations occurring in the horizontal plane effectively enough. For rice.8 you can see pronounced peaks corresponding to the frequencies of 23 Hz and 28 Hz, which are not present in the results of modal analysis. The appearance of these harmonics on the graph can be explained by non-linearity in the electrical circuit, since they are multiples of the first two oscillation frequencies of the dynamical system.

8. Conclusions

The test facility was developed and manufactured in order to study the features of the dynamic behaviour of elastically supported beams. At the current stage, the research was limited to the study of free vibrations of the system. During the experiments, the natural forms and frequencies of vibrations were determined, and the results obtained were compared with the results of modal analysis performed in the Midas NFX program environment. On the basis of the conducted research we can draw the following conclusions and recommendations:

IOP Conf. Series: Materials Science and Engineering 1015 (2021) 012020 doi:10.1088/1757-899X/1015/1/012020

- The test facility made by the authors of the article, due to its universal design, can be used to solve a number of scientific and practical problems. By varying the number of elastic supports, beam stiffness, boundary conditions, and other system parameters, a large amount of empirical material can be collected to analyse the behaviour of elastically supported structures.
- Using the optical method of measuring displacements and electromagnetic sensors, it was found that the dynamic parameters of the FE model in the Midas NFX program correspond to its physical prototype. Based on these data, the next step is supposed to perform an analysis of an elastically supported dynamic system under the influence of harmonic stationary and mobile loads.
- The optical displacement measuring method can be used to analyse vibrations of a wide range of dynamic systems due to its accessibility and clarity of results. The accuracy of the data obtained depends largely on the resolution and frame rate of the video recording system. In addition, when conducting research, it is necessary to exclude the movement of the video camera itself during recording, or to control them with additional equipment.
- The use of inductive sensors and subsequent signal processing allowed us to obtain the natural frequency spectrum of the elastically supported beam, which is in good agreement with the results of modal analysis of the FE model. Inductive sensors do not allow to get information about the actual movement of the structure and forms of vibrations. Therefore, it is advisable to use this approach in conjunction with theoretical calculations or when there are already some ideas about the dynamic behaviour of the structure.

References

[1] Perelmuter A V and Slivker V I 2011 *Calculation Models of Structures and the Possibility of their Analysis* (Moscow: DMK Press) p 709
[2] Kucherenko V A 1970 *Instruction for Calculating the Load-bearing Structures of Industrial Buildings and Structures on Dynamic Loads* (Moscow: Stroizdat) p 288
[3] Standard SP 413.1325800.2018 *The Buildings and Structures under Dynamic Actions. Design Rules* (Moscow: Standardinform) p 39
[4] Gui Shuirong, Wan Shui and Chen Shuisheng 2016 *ICACIE* (Hong Kong /Atlantis Press) pp 0116-21
[5] Gui Shuirong, Liu Lu, Chen Shuisheng and Zhao Hui 2014 *J. of Highway and Transportation Research and Development.* **8** pp 76-82
[6] Method for Determining the displacements of road bridge beams from the moving load by an optical meter (Pat. 2010107650 Russian Federation: IPC G01B 11/03)
[7] Gridnev S Y and Ravodin I V 2018 *XXVII Seminar R-S-P TFoCE* (Rostov-on-Don: EDP Sciences) p 8
[8] Gridnev S Y 2008 *Scientific Herald of VSTU.* **17** pp 112-120
[9] Gridnev S Y and Ravodin I V 2008 *XI Int. Sc. and Tech. Conf.* (Tula: Publishing Tulgu) p 2
[10] Zhao X, Zhao Q, Yu Y, Chen Y et al. 2016 *J. Perform. Constr. Facil.* **31** p 13

Stress-stain state in elastic spherical shells under the action some asymmetrical of circular loads

A Astakhova

Moscow State University of Civil Engineering, National Research University, 26 Yaroslavskoe shosse, Moscow, 129337, Russia

E-mail address: ast965@yandex.ru

Abstract. The paper presents the basic differential equations for calculating shells of rotation. It is considered that the material of the shell works beyond the elastic limit. The conclusion of the resolving system of differential equations is based on the linear shell theory and the Kirchhoff-Love hypothesis. The physical correlations are based on the small elastic-plastic deformation theory. With the construction of the solution of the load function, also unknown resolving functions expand to the Fourier series along the circumferential coordinate. In the present work, the action of the asymmetrical distribution along the parallel loads is considered. The solution is approached by the elastic decisions method. The study of stress-stain state in the isotropic spherical shells is presented in the work. The results of the calculation are presented in the example of a spherical shell. The work of the material beyond the elastic limit is taken into account using assuming linear hardening of the material. The shell is loaded with ring distributed loads, and they are applied on a certain segment in a parallel circle. The diagrams of bending moments and areas of the appearance of plastic deformations along the thickness of the shell are presented. The areas of plastic deformations for ring loads distributed along the parallel relative to the zero meridian in the range of 45, 90,135 degrees are shown.

1. Introduction

The work experience of the sheet structures demonstrated that the material of the constructions in the sphere of boundary and local effects usually works beyond the elastic limit. In the present work thin isotropic rotation shells with constant thickness regarding the operation of the material beyond the elastic limit are considered. The result of the resolving system of the differential equations is based on the linear shell theory on the Kirchhoff-Love hypothesis. The theories of small elastic-plastic deformations are accepted as a basement of physical correlations. The thin shells calculation theory is based on the scientific works by V Z Vlasov, A L Goldenveizer, A A Ilyushin, I S Tsurkov [1-3]. Inner forces tensor and additional terms tensor expressed physical nonlinearity of the problem. The numerical realization of the solution is based on the system of differential equations proposing by Y M Grigorenko for elastic rotational shells [4]. In the present work the main resolvent equations for rotation shells in the orthogonal coordinate system (s, θ, ζ) are presented. Where s – meridian arc length, θ – azimuthal angle in the horizontal plane, ζ – normal to the mid-surface. It is considered that the generatrix is quite smooth curve, the curvature of which is smaller than $1/h$.

2. The method of solution

The whole differential equations system is led to eight ordinary differential equations of the first order according to the coefficients of Fourier series of resolvent functions, loads and functions considering physical nonlinearity of the material (the indexes of the coefficients of Fourier series k are omitted). The length of the arc forming "s" is selected as an independent variable [5]. The solving system of equations is represented as

$$\frac{d\vec{N}}{ds} = A(s)\vec{N} + \vec{F}(s) \tag{1}$$

$$A(s) = a_{ij} \qquad (i, j = 1,2,3\ldots8) \tag{2}$$

$$\begin{gathered}
a_{11} = (1-\nu)\frac{\cos\alpha}{r}, \quad a_{12} = \nu\frac{\sin\alpha}{r}, \quad a_{13} = k\frac{\cos\alpha}{r}, \quad a_{14} = k^2h\nu\frac{\sin\alpha}{r^2}, \quad a_{15} = \frac{h}{r^2}\left(1+\frac{k^4h^2\sin^2\alpha}{12r^2}\right) \\
a_{16} = \frac{k^4h^3\sin\alpha\cos\alpha}{12r^4}, \quad a_{17} = \frac{kh}{r^2}\left(1+\frac{k^2h^2\sin^2\alpha}{12r^2}\right), \quad a_{18} = -\frac{k^2h^2\sin\alpha\cos\alpha}{12r^3}, \quad a_{21} = 0, a_{22} = \frac{\cos\alpha}{r} \\
a_{23} = -\frac{k\sin\alpha}{rp_0}\left(1-\frac{h^2\cos^2\alpha}{3r^2}\right), \quad a_{24} = \frac{k^2h\nu\cos\alpha}{r^2}, \quad a_{25} = -\frac{k^4h^3\sin\alpha\cos\alpha}{12r^4} \\
a_{26} = \frac{k^2h^3}{12r^4}\left(\frac{2}{(1+\nu)p_0}+k^2\cos^2\alpha\right), \quad a_{27} = \frac{k^3h^3\sin\alpha\cos\alpha}{12r^4}, \quad a_{28} = -\frac{k^2h^2}{12r^3}\left(\frac{2}{(1+\nu)p_0}+\cos^2\alpha\right) \\
a_{31} = -k\nu\frac{\cos\alpha}{r}, \quad a_{32} = k\nu\frac{\sin\alpha}{r}, \quad a_{33} = \frac{2\cos\alpha}{r}, \quad a_{34} = kh\nu\frac{\sin\alpha}{r^2}, \quad a_{35} = \frac{kh}{r^2}\left(1+\frac{k^2h^2\sin^2\alpha}{12r^2}\right) \\
a_{36} = \frac{k^3h^3\sin\alpha\cos\alpha}{12r^4}, \quad a_{37} = \frac{k^2h}{r^2}\left(1+\frac{h^2\sin^2\alpha}{12r^2}\right), \quad a_{38} = \frac{kh^2\sin\alpha\cos\alpha}{12r^3}, \quad a_{41} = \frac{\sin\alpha}{h} \\
a_{42} = \frac{\cos\alpha}{h}, \quad a_{43} = -\frac{kh\sin\alpha}{3r^2p_0}, \quad a_{44} = (1-\nu)\frac{\cos\alpha}{r}, \quad a_{45} = \frac{k^2h^2\sin\alpha\cos\alpha}{12r^3} \\
a_{46} = -\frac{k^2h^2}{12r^3}\left(\frac{2}{(1+\nu)p_0}+\cos^2\alpha\right), \quad a_{47} = \frac{kh^2\sin\alpha\cos\alpha}{12r^3}, \quad a_{48} = \frac{h}{12r^2}\left(\frac{2k^2}{(1+\nu)p_0}+\cos^2\alpha\right) \\
a_{51} = (1-\nu^2)\frac{\cos^2\alpha}{h}, \quad a_{52} = (1-\nu^2)\frac{\sin\alpha\cos\alpha}{h}, \quad a_{53} = a_{54} = 0, \quad a_{55} = \nu\frac{\cos\alpha}{r}, a_{56} = 0 \\
a_{57} = k\nu\frac{\cos\alpha}{r}, \quad a_{58} = -\frac{\sin\alpha}{h}, \quad a_{61} = (1-\nu^2)\frac{\sin\alpha\cos\alpha}{h}, \quad a_{62} = (1-\nu^2)\frac{\sin^2\alpha}{h}, \quad a_{63} = a_{64} = 0 \\
a_{65} = -\frac{\nu\sin\alpha}{r}, \quad a_{66} = 0, \quad a_{67} = -\frac{k\nu\sin\alpha}{r}, \quad a_{68} = -\frac{\cos\alpha}{h}, \quad a_{71} = a_{72} = 0, \quad a_{73} = \frac{2(1+\nu)}{p_0h}, \quad a_{74} = 0 \\
a_{75} = -\frac{k\cos\alpha}{r}, \quad a_{76} = \frac{k\sin\alpha}{p_0r}\left(1-\frac{h^2\cos^2\alpha}{3r^2}\right), \quad a_{77} = -\frac{\cos\alpha}{r}, \quad a_{78} = \frac{kh\sin\alpha}{3r^2p_0} \\
a_{81} = a_{82} = a_{83} = 0, \quad a_{84} = \frac{12(1-\nu^2)}{h}, \quad a_{85} = -\frac{k^2h\nu\sin\alpha}{r^2}, \quad a_{86} = -\frac{k^2h\nu\cos\alpha}{r^2} \\
a_{87} = -kh\nu\frac{\sin\alpha}{r^2}, a_{88} = \nu\frac{\cos\alpha}{r}, \quad p_0 = 1+\frac{h^2\sin^2\alpha}{3r^2}
\end{gathered} \tag{3}$$

Where r – radial coordinate, h – shell thickness, k – number of the coefficients of Fourier series, $\nu =$ 0.5 – Poisson ratio

$$\vec{F}(s) = \vec{f}(s) + \vec{\psi}(s) \tag{4}$$

$$\vec{f}(s)=(f_1,f_2\dots f_8),\ \vec{\psi}(s)=(\psi_1,\psi_2,\dots\psi_8)$$
$$f_1=-q_r,f_2=-q_z,f_3=-q_\theta,f_4=f_5=f_6=f_7=f_8=0 \tag{5}$$
$$\psi_1=\frac{1}{r}(0.5\Delta N_s-\Delta N_\theta)+\frac{k^2h\sin^2\alpha}{r^2}(0.5\Delta M_s-\Delta M_\theta)$$
$$\psi_2=\frac{2kh}{r^2p_0}\left(\frac{h\sin\alpha}{6r}\Delta S-\Delta H\right)+\frac{k^2h\cos\alpha}{r^2}(0.5\Delta M_s-\Delta M_\theta)$$
$$\psi_3=\frac{k}{r}(0.5\Delta N_s-\Delta N_\theta)+\frac{kh\sin\alpha}{r^2}(0.5\Delta M_s-\Delta M_\theta) \tag{6}$$
$$\psi_4=-\frac{4k}{rp_0}\left(\frac{h\sin\alpha}{6r}\Delta S-\Delta H\right)-\frac{\cos\alpha}{r}(0.5\Delta M_s-\Delta M_\theta)$$
$$\psi_5=-\frac{3\cos\alpha}{4h}\Delta N_s,\psi_6=\frac{3\sin\alpha}{4h}\Delta N_s,\psi_7=\frac{1}{p_0}(\frac{3}{h}\Delta S+\frac{6\sin\alpha}{r}\Delta H),\psi_8=\frac{9}{h}\Delta M_s$$

The differential equations system is formulated according to the following dimensionless functions $N_r,N_z,S,M_s,u_r,u_z,\upsilon,\vartheta_s$ (1). That's why the coefficients of Fourier series of the efforts N_r,N_z,S are refered to the composition of the modulus of elasticity to the thickness of the shell Eh, meridian bending moment M_s – to the value Eh^2, the coefficients of Fourier series of the movements u_r,u_z,υ are refered to the thickness of the shell h, ϑ_s – the angle of the normal rotation in the radians. The coefficients of Fourier series of the components of the load q_r,q_z,q_θ are refered to E. The coefficients of Fourier series corrections to the efforts $\Delta N_s,\Delta N_\theta,\Delta S$ are refered to Eh, but the corrections to the moments $\Delta M_s,\Delta M_\theta,\Delta H$ – to Eh^2, $\varepsilon_s,\varepsilon_\theta$ – the coefficients of Fourier series of the linear deformation and $\varepsilon_{s\theta}$ – of the shear deformations, $\varkappa_s,\varkappa_\theta$ – the coefficients of Fourier series of the deformation of the bend and $\varkappa_{s\theta}$ – the torsion of median surface. Horizontal, vertical, tangential and meridian components are marked with the indexes r, z, θ, s. The boundary value problem in the left and right ends of the generatrix is solved. Thus, it is essential to define eight boundary terms.

The obtained values of the resolvent functions allow us to determine the coefficients of the Furier series of deformations using the following formulas

$$\varepsilon_s=\frac{Eh}{D_N}(-N_r\cos\alpha+N_z\sin\alpha+\Delta N_s)-\frac{\nu h}{r}(u_r+k\upsilon),\varepsilon_\theta=\frac{h}{r}(u_r+k\upsilon)$$
$$\varepsilon_{s\theta}=\frac{2Eh}{(1-\upsilon)D_Np_0}\left(S+\Delta S+\frac{2h\sin\alpha}{r}\Delta H\right)-\frac{4kD_M\sin\alpha}{r^2p_0D_N}\left(\frac{h}{r}u_z-\vartheta_s\right)$$
$$\varkappa_s=\frac{Eh^2}{D_M}(M_s+\Delta M_s)-\nu\left(\frac{k^2h\sin\alpha}{r^2}u_r+\frac{k^2h\cos\alpha}{r^2}u_z+\frac{kh\sin\alpha}{r^2}\upsilon-\frac{\cos\alpha}{r}\vartheta_s\right)$$
$$\varkappa_\theta=\frac{kh\sin\alpha}{r^2}(ku_r+\upsilon)+\frac{k^2h\cos\alpha}{r^2}u_z-\frac{\cos\alpha}{r}\vartheta_s \tag{7}$$
$$\varkappa_{s\theta}=\frac{2Eh\sin\alpha}{(1-\nu)rD_Np_0}\left(S+\Delta S+\frac{2h\sin\alpha}{r}\Delta H\right)+\frac{k}{rp_0}\left(\frac{h}{r}u_z-\vartheta_s\right),D_N=\frac{Eh}{(1-\nu^2)},D_M=\frac{Eh^3}{12(1-\nu^2)}$$

Based on the elastic decisions method in the first approximation the coefficients of Fourier series of the functions considered physical nonlinearity of the material $\vec{\psi}(s)$ are considered equal to zero. Their value for each subsequent of approximation is defined by the functions of deformation, using the following formulas:

$$\begin{aligned}
\Delta N_s &= \frac{1}{(1-\nu^2)}(\varepsilon_s + \nu\varepsilon_\theta) - \frac{4}{3Eh}\left[(\varepsilon_s + 0.5\varepsilon_\theta)J_1 + (\varkappa_s + 0.5\varkappa_\theta)J_2\right] \\
\Delta N_\theta &= \frac{1}{(1-\nu^2)}(\varepsilon_\theta + \nu\varepsilon_s) - \frac{4}{3Eh}\left[(\varepsilon_\theta + 0.5\varepsilon_s)J_1 + (\varkappa_\theta + 0.5\varkappa_s)J_2\right] \\
\Delta S &= \frac{1}{2(1+\nu)}\varepsilon_{s\theta} - \frac{1}{3Eh}\left[(\varepsilon_{s\theta}J_1 + 2\varkappa_{s\theta}J_2) - \frac{1}{R_s}(\varepsilon_{s\theta}J_2 + 2\varkappa_{s\theta}J_3)\right] \\
\Delta M_s &= \frac{h}{12(1-\nu^2)}(\varkappa_s + \nu\varkappa_\theta) - \frac{4}{3Eh^2}\left[(\varepsilon_s + 0.5\varepsilon_\theta)J_2 + (\varkappa_s + 0.5\varkappa_\theta)J_3\right] \\
\Delta M_\theta &= \frac{h}{12(1-\nu^2)}(\varkappa_\theta + \nu\varkappa_s) - \frac{4}{3Eh^2}\left[(\varepsilon_\theta + 0.5\varepsilon_s)J_2 + (\varkappa_\theta + 0.5\varkappa_s)J_3\right] \\
\Delta H &= \frac{h}{12(1+\nu)}\varkappa_{s\theta} - \frac{1}{3Eh^2}(\varepsilon_{s\theta}J_2 + 2\varkappa_{s\theta}J_3)
\end{aligned} \tag{8}$$

The functions J_1, J_2, J_3 – the variables along the median surface of the stiffness of the shell [7].

$J_1 = \int_{-h/2}^{h/2} \bar{\psi} dz$, $J_2 = \int_{-h/2}^{h/2} \bar{\psi} z dz$, $J_3 = \int_{-h/2}^{h/2} \bar{\psi} z^2 dz$

The function $\bar{\psi}$ is equal to the material elasticity module $\bar{\psi} = E$ in the field of elastic deformations, and the stiffness of the shell are $J_1 = Eh, J_2 = 0, J_3 = Eh^3 / 12$ for the elastic area, where

$\sigma_i = E\,\varepsilon_i$

Beyond the elastic limit, the relation between tensity σ_i, and deformation intensity ε_i must be set [6], [7]. The values of the functions $\Delta N_s, \Delta N_\theta, \Delta S, \Delta M_s, \Delta M_\theta, \Delta H$ from the relations (8) are substituted with the right part of the system (1) for the following approximation of the elastic decisions method.

3. Results

The example of the nonsymmetrical application of inner forces is presented. The circular spherical shell at the edges is supported by a rigid support $\delta = 5°$ and $\alpha_0 = 160°$. The radius ratio to the thickness R/h = 100. The steel uniaxial tension test is approximated by the broken line with the hardening module λ = 0.95 is presented:

$$\text{for } \varepsilon_i > \varepsilon_{i_0}\ \sigma_i = E\,\varepsilon_i\left[1 - \lambda\left(1 - \frac{\varepsilon_{i_0}}{\varepsilon_i}\right)\right]$$

The intensity of deformations corresponding to the beginning of yielding is equal to $\varepsilon_{i_0} = 0.001$. The elastic modulus $E = 2.1 \cdot 10^5$ MPa, the loads equal $P/Eh = 0.143 \cdot 10^{-3}$.The shell is loaded by three ring loads P along the parallel, figure 1. The angles in the parallel circle have the boundaries of the load action, its equal $45°, 90°, 135°$. The epure of the bending moment M_s at the zero meridian for the studying cases of the load is presented in the figure 2.

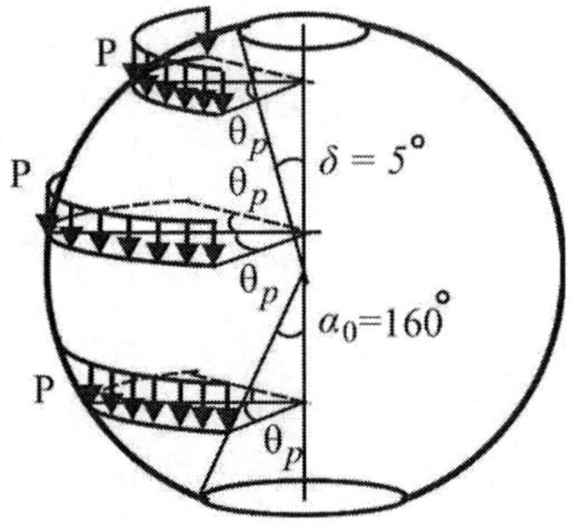

Figure 1. The shell with loads

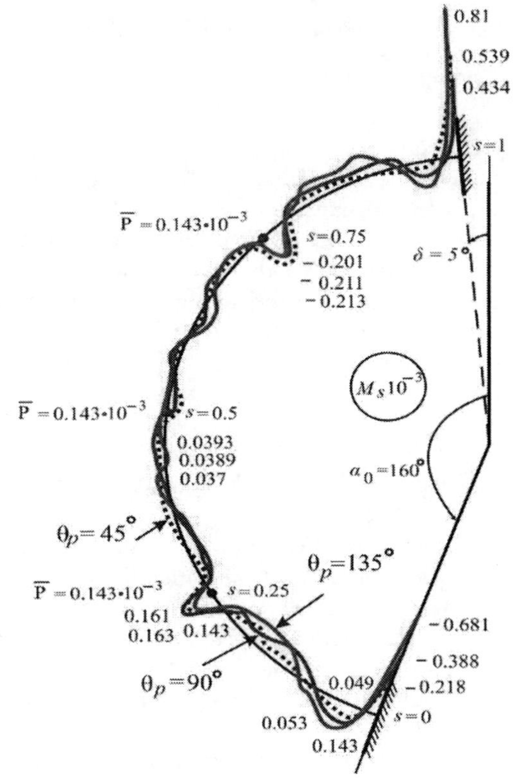

Figure 2. The diagram of the bending moment M_s at the zero meridian

The boundary effects are brightly expressed near the rigid supports. The moment M_s near the supports is 4-7 times higher than the maximum moments at the middle points. It should be noticed that beneath the point of the application of the middle load for the studying cases of the load are approximately equal. The areas of occurrence of plastic deformations are shown in the figure 3 for the range of action of ring loads in a parallel circle relative to the zero meridian $\theta_p = 45°, 90°, 135°$.

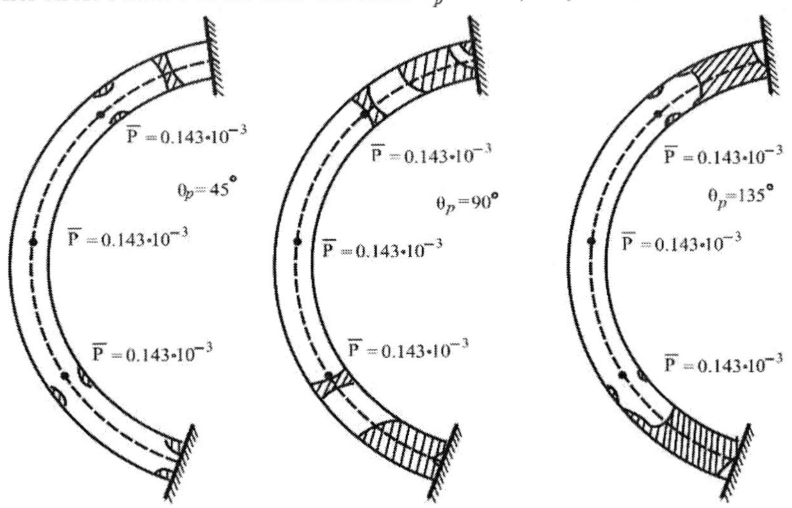

Figure 3. The areas of occurrence of plastic deformations.

Very small areas of plastic deformation appear in the shell at $\theta_p = 45°$ and significant areas are located near the supports at $\theta_p = 90°$ and $\theta_p = 135°$, they are approximately $0.12s$ and $0.2s$.

4. Conclusions

The paper presents the main relations for calculating thin shells of rotation beyond the elastic limit. In the calculation example, a material with linear hardening is considered. The results of calculations allow to determine the propagation of plastic deformations with increasing the range of distributed loads, to compare the values of the internal forces, deformations, displacements without considering the occurrence of plastic deformations at specific points and taking into account the formation of plastic deformations, to analyze the redistribution of the internal forces resulting from the appearance of plastic deformations in the thin shell.

References

[1] Goldenveizer A L 1976 *Theory of elastic thin shells* (Moscow: Nauka)
[2] Vlasov V Z 1949 *Shell theory and its appllication to the technics* (Moscow-Leningrad: Gostekhizdat)
[3] Tsurkov I S 1965 Towards the integration of the equations for thin non-elastic shells *Collected papers of MISI named after V V Kuibyshev* (Moscow)
[4] Grigorenko Y M and Vasilenko F T 1968 Unsymmetrical deformation of isotropic and nonisotropic rotational shell *Apllied mechanics* **3** (Moscow)
[5] Astakhova A Y 1985 *Calculation of elastoplastic shells under fixed load. MTT, 4* (Moscow)
[6] Astakhova A 2016 Plastic deformations in thin rotational shells *Published online, 28 November 2016, Numer of page(s) 7. Matec Web Conf.* **8** 601016, *5th International Scientific Conference: "Integration, Partnership and Innovation in Construction Science and Education", Section 1 Structural Mechanics.*
[7] Astakhova A 2018 Calculation of thin isotopic shells beyond the elastic solutions, , *Published on line, 03 September 2018, Number of page(s) 8. Matec Web Conf.* **19** 601014, *XXVII R-S-P Seminar, 'Theoretical Foundation of Civil Engineering', Section Structural Mechanics.*

IOP Conf. Series: Materials Science and Engineering 1015 (2021) 012022 doi:10.1088/1757-899X/1015/1/012022

Different models of connection flexibility in dynamic analysis of plane frame systems

T T V Tran and T H Tran

Hanoi Architectural University, Hanoi, Viet Nam

E-mail address: ttthvan.hau@gmail.com

Abstract: Nowadays, in Vietnamese structural design standards for steel frames and semi-precast reinforced concrete structural frame systems, the connections of beams and columns are commonly considered as perfectly hinged or rigidly fixed. This is adopted because of the simplicity in design and analysis procedure but it is not consistent with the condition taking into account actual behaviour in practice of structural systems. The connections between columns and beams have partial restraint depending on the type of used connections. The presence of connection flexibility tends to reduce frame stiffness in comparison with the case of fully fixed connection and hence increases the frame's natural period of vibration. However, it is also more time required because of complexity of analysis process with consideration of connection flexibility in the case of lack of suitable analysis method. The paper presents the dynamic analysis of plane frame systems with different models of semi-rigid connections using finite element method. The used models in dynamic analysis are rotational spring and rotational-lateral spring. The analysis procedure was implemented using MathCad programming software and the results of this study will obtained by conducting of numerical examples. From there, the conclusions and recommendation of dynamic analysis for steel and semi-precast concrete frames with models of connection flexibility and the differences between used models will be proposed.

1. Introduction

Semi-rigid frames are considered frames, in which the beam-column connections are neither perfectly pinned nor rigid. In practical engineering structures, all frames are semi-rigid in nature because truly pinned or perfectly rigid connections do not exist. The design instructions for these types of structures have been mentioned in many foreign construction standards with specific calculation recommendations considering the connection flexibility. However, in Vietnam, there is still no specific domestic calculation instructions in design of these structures considering the actual behavior of the connections. When designing these structures, it is either used foreign standards or considered the assumption that the connections between beams and columns are perfectly hinged or rigidly fixed. This assumption does not reflect the actual behavior of practical structures. The calculation with consideration of connection flexibility is more constituent with the behavior of the work of the practical structures. Thereof, if the calculation takes into account the connection flexibility, the analysis results will be more accurate and closer to the actual behavior of the real structures. If the influence of flexibility of semi-rigid connections is underestimated, and they are treated in the design as pinned, it has a negative impact on cost of a structure. Thus, in the analysis and design of steel frames the connection between beams and columns should be modeled as semi-rigid connections.

The connection flexibility of steel and semi-precast concrete frame structures have been concerned by many scientists from over the worlds and many studies on this issue are published with different problems [1-8]. In most of these studies the connections between beams and columns of semi-rigid frames were modeled by rotational springs. In fact, frame elements have semi-rigid behavior through the axial direction of themselves. In [9], the behaviors of beam-column connections are designed through either rotational or lateral effects. The additional effect of lateral springs already discussed in [9] on statically analysis. In this paper, the dynamic analysis is further discussed with the whole range of the values rotational and lateral springs which correspond with the end-fixity factor. The end-fixity factor is analyzed from a value greater than 0 to near 1 (0 is the value of end-fixity factor that corresponds with the case of pinned connection and 1-the case of rigidly fixed).

2. Dynamic analysis of plane frame systems with consideration of different models of semi-rigid connection

The flexibility of beam-column connections of frames is modelled by rotational springs and lateral springs. Each model has its characteristics. In this section, it is introduced about their properties and element stiffness matrix in finite element analysis.

2.1. Rotational spring model

In rotational spring model, the connection is represented by rotational springs near to beam-column joints. The presence of rotational springs will restrain the rotations θ_{rA} and θ_{rB} at the ends of the beam AB [9].

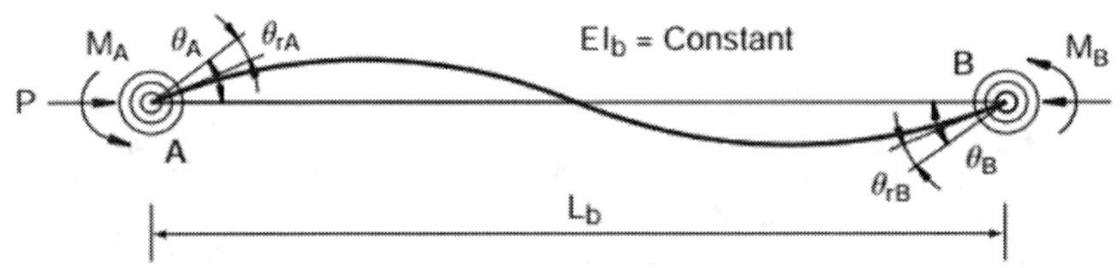

Figure 1. Beam element with semi-rigid connections.

The unit of stiffness of rotational spring is kN·m/rad. In the analysis of semi-rigid plane frames this model is illustrated in figure 2.

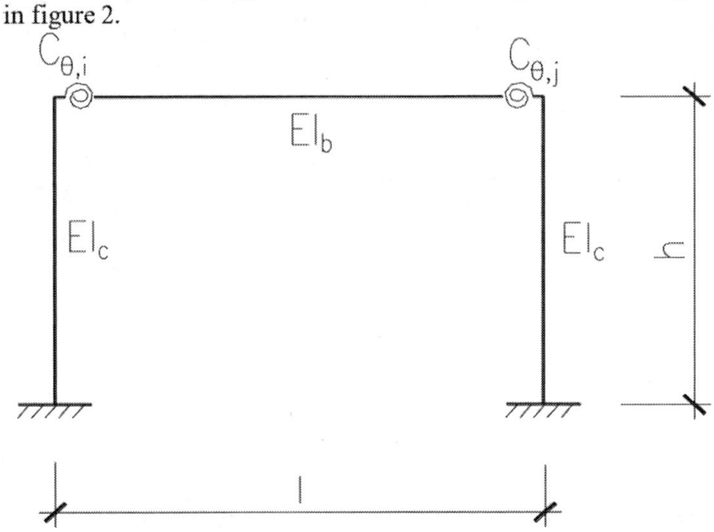

Figure 2. The rotational spring model for beam-column connections.

The important step of analysis using finite element method is determination of global stiffness matrix of structure system. The element stiffness matrices for beams and columns can be written as following The beam stiffness matrix [10]

$$\mathbf{K}_{bi\_r}=\frac{E\cdot I_b}{l}\begin{bmatrix} \frac{A}{I_b} & 0 & 0 & -\frac{A}{I_b} & 0 & 0 \\ & \frac{\left(s_{ii}^*+2s_{ij}^*+s_{jj}^*\right)}{l^2} & \frac{\left(s_{ii}^*+s_{ij}^*\right)}{l} & 0 & -\frac{\left(s_{ii}^*+2s_{ij}^*+s_{jj}^*\right)}{l^2} & \frac{\left(s_{ii}^*+s_{ij}^*\right)}{l} \\ & & s_{ii}^* & 0 & -\frac{\left(s_{ii}^*+s_{ij}^*\right)}{l} & s_{ij}^* \\ & & & \frac{A}{I_b} & 0 & 0 \\ & & & & \frac{\left(s_{ii}^*+2s_{ij}^*+s_{jj}^*\right)}{l^2} & -\frac{\left(s_{ij}^*+s_{jj}^*\right)}{l} \\ & & & & & s_{jj}^* \end{bmatrix} \tag{1}$$

In which

$$\begin{aligned} s_{ii}^* &= \left(4+\frac{12E\cdot I_b}{l\cdot R_{kiB}}\right)\Bigg/\left(\left(1+\frac{4E\cdot I_b}{l\cdot R_{kiA}}\right)\cdot\left(1+\frac{4E\cdot I_b}{l\cdot R_{kiB}}\right)-\left(\frac{E\cdot I_b}{l}\right)^2\cdot\frac{4}{R_{kiA}\cdot R_{kiB}}\right) \\ s_{jj}^* &= \left(4+\frac{12E\cdot I_b}{l\cdot R_{kiA}}\right)\Bigg/\left(\left(1+\frac{4E\cdot I_b}{l\cdot R_{kiA}}\right)\cdot\left(1+\frac{4E\cdot I_b}{l\cdot R_{kiB}}\right)-\left(\frac{E\cdot I_b}{l}\right)^2\cdot\frac{4}{R_{kiA}\cdot R_{kiB}}\right) \\ s_{ij}^* &= s_{ji}^* = 2\Bigg/\left(\left(1+\frac{4E\cdot I_b}{l\cdot R_{kiA}}\right)\cdot\left(1+\frac{4E\cdot I_b}{l\cdot R_{kiB}}\right)-\left(\frac{E\cdot I_b}{l}\right)^2\cdot\frac{4}{R_{kiA}\cdot R_{kiB}}\right) \end{aligned} \tag{2}$$

R_{kiA}, R_{kiB} – initial stiffness rotational spring (in this study it is used the symbol $C_{\theta i}= R_{kiA}$, $C_{\theta j}= R_{kiB}$)

E, I_b, A_b, l are respectively elastic modulus, moment of inertia, area of cross section, length of the beams.

The column stiffness matrix [10]

$$\mathbf{K}_{ci\_r}=\frac{E\cdot I_c}{l_c}\cdot\begin{bmatrix} \frac{12}{l_c^2}\cdot\gamma_1 & 0 & \frac{6}{l_c}\cdot\gamma_2 & -\frac{12}{l_c^2}\cdot\gamma_1 & 0 & \frac{6}{l_c}\cdot\gamma_2 \\ & \frac{A_c}{I_c} & 0 & 0 & -\frac{A_c}{I_c} & 0 \\ & & 4\gamma_3 & -\frac{6}{l_c}\cdot\gamma_2 & 0 & 2\gamma_4 \\ & & & \frac{12}{l_c^2}\cdot\gamma_1 & 0 & -\frac{6}{l_c}\cdot\gamma_2 \\ & & & & \frac{A_c}{I_c} & 0 \\ & & & & & 4\gamma_3 \end{bmatrix} \tag{3}$$

In which the coefficients are determined by following formula

$$\gamma_1 = \frac{(kl_c)^3 \cdot \sinh(kl_c)}{12 \cdot \left(2 - 2\cosh(kl_c) \pm kl_c \cdot \sinh(kl_c)\right)};\ \gamma_2 = \frac{(kl_c)^2 \cdot \left(\cosh(kl_c) - 1\right)}{6 \cdot \left(2 - 2\cosh(kl_c) \pm kl_c \cdot \sinh(kl_c)\right)};$$
$$\gamma_3 = \frac{(kl_c) \cdot \left(kl_c \cdot \cosh(kl_c) - \sinh(kl_c)\right)}{4 \cdot \left(2 - 2\cosh(kl_c) \pm kl_c \cdot \sinh(kl_c)\right)};\ \gamma_4 = \frac{(kl_c) \cdot \left(\sinh(kl_c) - kl_c\right)}{\left(2 - 2\cosh(kl_c) \pm kl_c \cdot \sinh(kl_c)\right)} \quad (4)$$

Where $k = (P / EI_c)^{1/2}$, "+" sign corresponds with tensile axial force and "-" sign – compressive axial force. In the case of that the axial force is small, $\gamma_1 = \gamma_2 = \gamma_3 = \gamma_4 = 1$.

2.2. Rotational-lateral spring model

This model is represented by rotational and lateral springs near beam-column connections. The presence of these spring not only restrain the rotations θ_i and θ_k, but also the lateral displacements δ_i and δ_k at the ends of beams. This model is illustrated in figure 3.

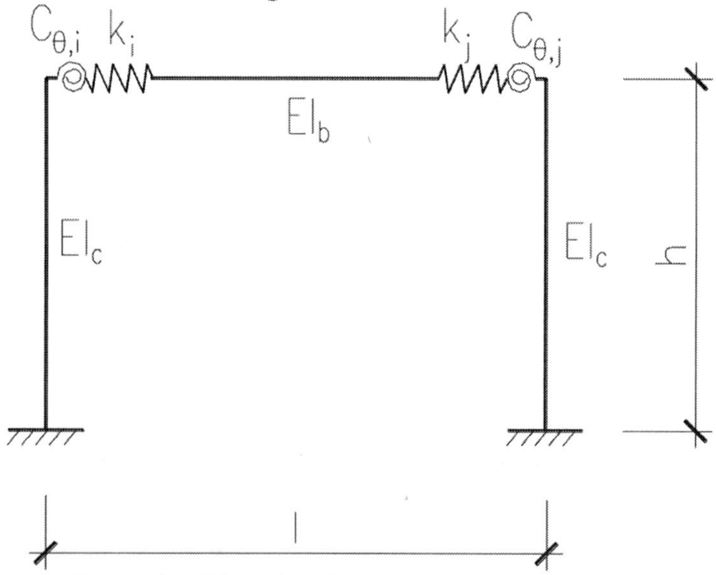

Figure 3. The rotational-lateral spring model in semi-rigid frame structures.

The beam stiffness matrix can be written as followings:

$$\mathbf{K}_{bi\_rl} = \frac{E \cdot I_b}{l} \begin{bmatrix} \frac{A_b}{I_b} + R_{bi} & 0 & 0 & -\frac{A_b}{I_b} - R_{bi} & 0 & 0 \\ & \frac{\left(s_{ii}^* + 2s_{ij}^* + s_{jj}^*\right)}{l^2} & \frac{\left(s_{ii}^* + s_{ij}^*\right)}{l} & 0 & -\frac{\left(s_{ii}^* + 2s_{ij}^* + s_{jj}^*\right)}{l^2} & \frac{\left(s_{ii}^* + s_{ij}^*\right)}{l} \\ & & s_{ii}^* & 0 & -\frac{\left(s_{ii}^* + s_{ij}^*\right)}{l} & s_{ij}^* \\ & & & \frac{A_b}{I_b} + R_{bi} & 0 & 0 \\ & & & & \frac{\left(s_{ii}^* + 2s_{ij}^* + s_{jj}^*\right)}{l^2} & -\frac{\left(s_{ij}^* + s_{jj}^*\right)}{l} \\ & & & & & s_{jj}^* \end{bmatrix} \quad (5)$$

in which

$$R_{bi} = \frac{l}{I_b} \cdot \frac{A_b \cdot (k_i \cdot k_j)}{l \cdot (k_i \cdot k_j) + E \cdot A_b \cdot (k_i \cdot k_j)} \tag{6}$$

where, k_i, k_j – lateral stiffness of the second model. The unit of lateral stiffness is kN/m;

2.3. Dynamic analysis of plane frames with consideration of two models of connection flexibility using finite element method

This study is to investigate the dynamic characteristics of semi-rigid frames and the influences of connection flexibility on the dynamic characteristics. The equation of motion for a semi-rigid frames in the case of free vibration is written as followings:

$$\mathbf{M} \cdot \ddot{\mathbf{v}} + \mathbf{K} \cdot \mathbf{v} = 0 \tag{7}$$

Where $\ddot{\mathbf{v}}$ and $\mathbf{v}$ are acceleration and displacement vectors of a structure respectively.

M and **K** are respectively mass and element stiffness matrices of semi-rigid plane frames.

In this study, the dynamic characteristics of semi-rigid frames are determined by modal analysis. The frequency $\boldsymbol{\omega_i}$ and period $\mathbf{T_i}$ of a vibration will be investigated. The influences of connection flexibility will be discussed. It is solved the eigenvalue problem, that means the determinant of the system has to be equal to zero. From equation (8) it is found corresponding frequency $\boldsymbol{\omega_i}$.

$$\left|\mathbf{K} - \omega^2 \cdot \mathbf{M}\right| = 0 \tag{8}$$

The effects of connection flexibility are modeled using rotational springs having stiffness $C_{\theta i,j}$ and rotational-lateral springs having stiffness $k_{i,j}$ at the ends of the beams. To reflect the relative stiffness of the beams and the rotational end-spring connections, an end-fixity factor is used [10]. It is defined as:

$$\eta = \frac{\alpha}{\phi} = \frac{1}{1 + \frac{3}{\overline{R}_k}} \tag{9}$$

Where α is the member end rotation, ϕ is the combined rotation of the member and the connection due to a unit end moment, $\overline{R}_k = (C_{\theta i,j} \cdot l_b) \cdot (EI_b)^{-1}$. Note that η falls between zero and one, it defines the stiffness of each end connection relative to the adjoining beam (η =0 corresponds with the case of perfectly pinned connection, η =1 – fully fixed connection).

3. Numerical example

In this paper, two-storey semi-rigid frames with two above mentioned models as shown in figure 4 and figure 5. All frames have same geometrical dimensions, cross-section and material properties for comparison of influences of connection flexibility on dynamic characteristics. Similarly, in this study it is clarified the different effects of used different models of flexibility connection on dynamic characteristics.

Given material properties and geometrical sizes as followings: elastic modulus E=20·10^7 kN/m^2, cross-section of beams C150×70, cross-section of columns I160×81×5.0, l=4m, h=3.5m.

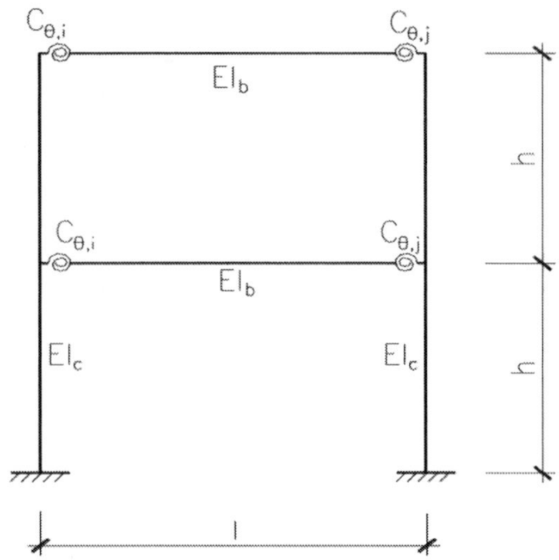

Figure 4. Two-storey semi-rigid frame with rotational spring model in dynamic analysis.

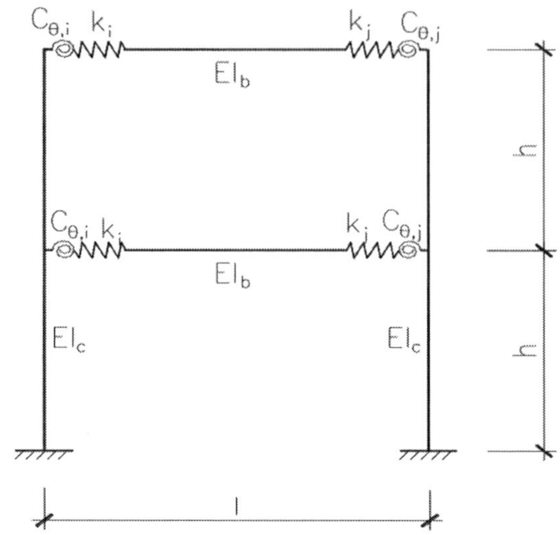

Figure 5. Two-storey semi-rigid frame with rotational -lateral spring model in dynamic analysis.

Table 1. Results of dynamic analysis with rotational spring model

End-fixity factor η	Frequency ω_i (rad/s)			Period T_i (s)		
	1-st mode	2-nd mode	3-rd mode	1-st mode	2-nd mode	3-rd mode
1 (rigid)	6.532	6.795	34.52	0.9617	0.9245	0.182
0.8	6.532	6.795	30.03	0.96175	0.9246	0.209
0.6	6.532	6.795	26.6	0.9618	0.9246	0.236
0.4	6.5317	6.794	23.48	0.96185	0.92465	0.267
0.3	6.5315	6.793	21.82	0.96185	0.92467	0.287
0.1	6.5316	6.792	17.66	0.9619	0.9247	0.355
0 (pinned)	6.531	6.663	14.44	0.962	0.943	0.435

Table 2. Results of dynamic analysis with rotational-lateral spring model

End-fixity factor η	Lateral spring stiffness k_i (kN/m)	Frequency ω_i (rad/s)			Period T_i (s)		
		1-st mode	2-nd mode	3-rd mode	1-st mode	2-nd mode	3-rd mode
1 (rigid)	10^7	6.532	6.795	34.52	0.9617	0.9245	0.182
0.8	10^6	6.5317	6.795	30.0	0.9618	0.92454	0.209
0.6	10^5	0.65314	6.794	26.59	0.96184	0.9246	0.2362
0.4	10^4	0.6531	6.794	23.4	0.96191	0.92466	0.268
0.3	10^3	6.531	6.79	21.6	0.96198	0.9248	0.289
0.1	10^2	0.65308	6.78	17.55	0.96199	0.9249	0.358
0 (pinned)	0	6.5305	6.662	14.43	0.962	0.943	0.435

From the results of calculation, it can be shown that the different models of semi-rigid connection effects differently on the dynamic analysis.

4. Conclusions

The flexibility of beam-column connections is presented by rotational and rotational-lateral springs models that partially reflect actual behaviour of real structures. The stiffness matrices are introduced for dynamical analysis by finite element method. In this study, the subroutine is written for dynamical analysis using finite element method by MathCad programming software that helps to perform any geometrically dimensions of structures easily.

The effects of connection flexibility are investigated. The different types of models of connection flexibility effect differently on dynamic behaviour of structures. The study results show that the connection flexibility tends to increase periods, whereas it tends to decrease the frequency. The first and the second frequencies tend to increase with very little changes, but the third frequency changes significantly corresponding with the change of the value of fixity factor. The frequencies of plane frames with rotational spring model are greater in comparison with the case with rotational-lateral spring models, whereas the periods of the same analysis are reversed.

References

[1] Monforton G R 1962 Matrix analysis of frames with semi-rigid connections *Electronic Theses and Dissertations* (Windsor, Ontario, Canada)

[2] Csebfalvi A and Csebfalvi G *Effect of semi-rigid connection in optimal design of frame structures* (6th World Congresses of Structural and Multidisciplinary Optimization)

[3] Ozturk A U 2000 Dynamic analysis of semi-rigid connected frames *Ms.Thesis* (Izmir Dokuz EyluI University)

[4] Salatic R and Sekilovic M 2001 Nonlinear analysis of frames with flexible connections, *Computers and Structures* **79** pp 1097-1107

[5] Fu Z, Ohi K, Takanashi K and Lin X 1998 Seismic behavior of steel frames with semi-rigid connections and braces, *Journal of Constructional Steel Research* **46** pp 440-61

[6] Kattner M and Crisinel M 2000 Finite element modeling of semirigid composite joints, *J. Computers & Structures* **78** pp 341-53

[7] Kim S E and Choi S H 2001 Practical advanced analysis for semirigid space frames, *J. of Solids and Structures* 38 pp 9111-31

[8] Lui E M and Lopes A 1997 Dynamic analysis and response of semi rigid frames, *Journal of Engineering Structures* 19 pp 644-654

[9] Ozturk A U and Yeslce Y 2005 *The statical analysis of semi-rigid frames by different connection type* (Izmir Dokuz EyluI University)

[10] Chen W F, Kishi N, Komuro M 2011 *Semi-rigid connections handbook* (J. Ross Publishing Inc)

[11] Folic R 1988, Links and connections of precast reinforced buildings, *Precast civil engineering structures* (Ekonomika, Belgrade in Serbian) pp 117-67

IOP Conf. Series: Materials Science and Engineering 1015 (2021) 012023 doi:10.1088/1757-899X/1015/1/012023

Mixed finite element formulation for the nonlinear analysis of buckling behavior of truss under effects of temperature variation

V T B Quyen, D N Tien and T M Khuyen

Faculty of Civil Engineering, Hanoi Architectural University, Vietnam

E-mail address: bquyen1312@gmail.com; quyenvtb@hau.edu.vn

Abstract. This paper proposes a novel approach to consider the effects of temperature variability in the nonlinear finite element analysis of buckling behavior of truss system subjected to initial mechanical load. Generally, using displacement-based finite formulation the procedure for solving geometrically nonlinear problem of the truss system subjected to temperature load is required to implement the temperature deformation constraint depending on the incremental element length to the master stiffness equation. For escaping the mathematical difficulties of treating the additional constraint of temperature deformation this research proposes to formulate this problem based on mixed finite element formulation. The balanced mixed equation of truss considering temperature loading effect is constructed based on the principle virtual work with establishing the mixed matrix of truss element, which consists the element temperature deformation, considering large displacement. For investigating the effects of temperature variation to the nonlinear buckling behavior of truss this research develops the incremental-iterative algorithm for solving the nonlinear equation with incremental displacement and temperature by using Newton-Raphson technique. Based on proposed algorithm the calculation program is written for numerical investing the effects of temperature variation the buckling behavior of plan truss.

1. Introduction

In many practical cases, the buckling behavior of truss need to be considered for design purposes when the truss is subjected to temperature load. However, the problem of elastic buckling of truss subjected to the mechanical temperature load is substantially different. In fact not as many research works have been published regarding temperature load variation. Especially, the little of study for nonlinear analysis of buckling behavior of truss under temperature variation have been found [1,2]. In geometrical nonlinear finite element analysis, the temperature deformation of truss element is dependent on the incremental element length. In geometrical linear finite element analysis the temperature loads usually are calculated using the restraining method suggested by J. M. C. Duhamel (1838) by adding the thermal loads to the nodal external force vector as equivalent loads [3-5]. Generally, in geometrical nonlinear finite element analysis based on displacement-based formulation, the solving a geometrically nonlinear problem of system under thermal loading is required to implement the temperature deformation constraint depending on the incremental element length to the master stiffness equation by using mathematical methods for constrained optimization such as penalty augmentation method or Lagrange multiplier adjunction method [6-9]. The implementation of temperature deformation constraint considerably increases the difficulty in finite element formulation of nonlinear buckling analysis of truss.

Based on mixed finite formulation, this research introduces novel approach to formulate the geometrical nonlinear buckling problem of truss under the effects of temperature load without difficulties of imposing the temperature deformation constraint. The mixed matrix of truss element, consisting element temperature deformation, is constructed based on large displacement assumption. Based on the formulated equation this research develops the incremental-iterative algorithm for solving the nonlinear equations with incremental displacement and temperature by using Newton-Raphson technique. Based on proposed algorithm, the calculation program is written for investigating the effects of temperature variation to the nonlinear buckling behavior of truss. The numerical results are presented to verify the efficiency of the proposed method.

2. Problem formulation

2.1. Balanced equation of mixed truss element considering temperature loading effect

Consider the homogeneous truss bar subjected to uniform thermal load due to a constant temperature change shown in figure 1. The axial deformation of truss element is defined as

$$\Delta L = L - L_T = L - L_0 - \Delta L_T \tag{1}$$

Where: L_0 and L - distances between two end nodes of the truss element before and after deformation due to mechanical load; L_T – the length of truss element after deformation considering temperature deformation; $\Delta L_T = \alpha.\Delta T.L_0$ - the change in length due to temperature deformation [10] in case of temperature change ΔT with linear thermal expansion coefficient α

Using the relationship of node displacements in geometry, the length of the truss element is computed as follows

$$L = \sqrt{(L_0.\sin\alpha_0 + u_4 - u_2)^2 + (L_0.\cos\alpha_0 + u_3 - u_1)^2} \tag{2}$$

where u_1, u_2, u_3, u_4 nodal displacements in a global coordinate system

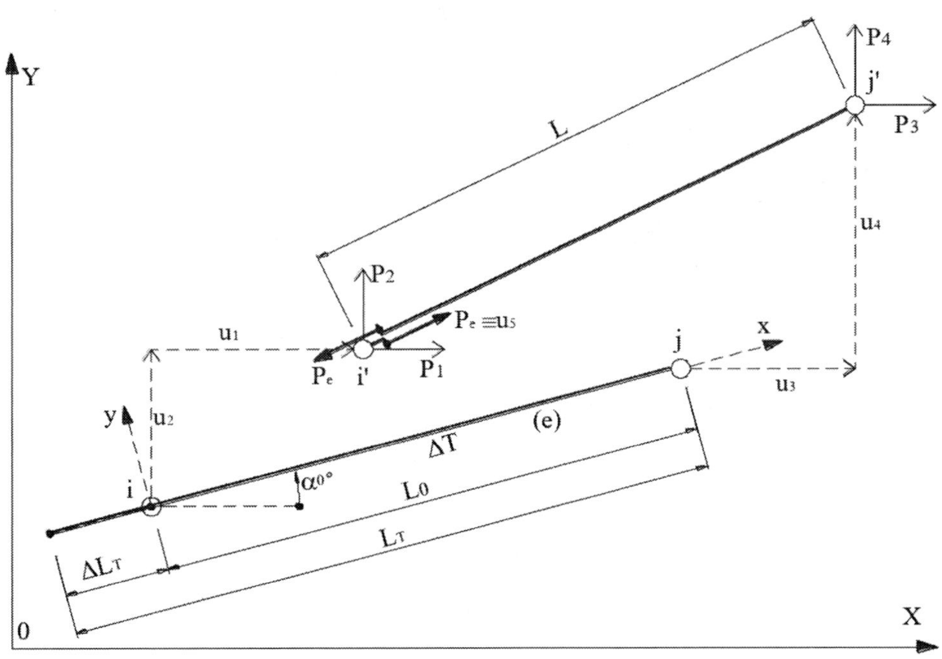

Figure 1. Mixed truss element cosidering thermal loading effect.

Work done by internal forces can be computed by

$$\begin{aligned}\delta V &= -\int \sigma_x \delta\varepsilon_x \mathrm{d}V = -\int_A \sigma_x dA \int_0^{L_0} \delta\varepsilon_x \mathrm{d}x = -N\int_0^{L_0} \frac{\delta(\Delta \mathrm{d}x)}{\mathrm{d}x}\mathrm{d}x = -N\int_0^{\Delta L} \delta(\Delta dx) = \\ &= -N\cdot\delta\left(\int_0^{\Delta L} \Delta \mathrm{d}x\right) = -N\cdot\delta\Delta L = -N\cdot\left\{\sum_{i=1}^{4}\frac{\partial \Delta L}{\partial u_i}\delta u_i + \frac{\partial \Delta L}{\partial \Delta L_T}\delta\Delta L_T\right\}\end{aligned} \tag{3}$$

where: N - internal normal force of truss element.

The virtual external work can be defined as

$$\delta\overline{V} = P_1.\delta u_1 + P_2.\delta u_2 + P_3.\delta u_3 + P_4.\delta u_4 - P_e.\delta\Delta L_T = \sum_{i=1}^{4} P_i.\delta u_i - P_e.\delta\Delta L_T \tag{4}$$

where: P_1, P_2, P_3, P_4 - nodal forces in a global coordinate system.

The total work done by internal and external forces is obtained by summing equation (3) and equation (4), having

$$\begin{aligned}\delta V + \delta\overline{V} &= -N\left\{\sum_{i=1}^{4}\frac{\partial \Delta L}{\partial u_i}\delta u_i + \frac{\partial \Delta L}{\partial \Delta L_T}\delta\Delta L_T\right\} + \left\{\sum_{i=1}^{4} P_i\,\delta u_i - P_e\delta\Delta L_T\right\} = \\ &= \sum_{i=1}^{4}\left\{-N\frac{\partial \Delta L}{\partial u_i} + P_i\right\}\delta u_i + \left\{-N\frac{\partial \Delta L}{\partial \Delta L_T} - P_e\right\}\delta\Delta L_T = 0\end{aligned} \tag{5}$$

Where P_e - resultant external force at the ith cross section after deformation.

Based on the principle of virtual work, in equilibrium the virtual work of the forces applied to a system is zero, getting

$$\begin{cases} -N\dfrac{\partial \Delta L}{\partial u_i} + P_i = 0 \quad (i = 1,2,3,4) \\ -N\dfrac{\partial \Delta L}{\partial \Delta L_T} - P_e = 0 \end{cases} \tag{6}$$

Replacing the axial deformation of truss in equation (6) with deformation in equation (1), then expressing axial force through deformation, getting

$$\begin{cases} \dfrac{EA}{L_0}.(L - L_0 - \Delta L_T).\dfrac{\partial(L - L_0 - \Delta L_T)}{\partial u_i} = P_i \quad (i = 1,2,3,4) \\ -\dfrac{EA}{L_0}.(L - L_0 - \Delta L_T).\dfrac{\partial(L - L_0 - \Delta L_T)}{\partial \Delta L_T} - P_e = 0 \end{cases} \tag{7}$$

Combining equation (7) with equation (2), having

$$\begin{cases} \dfrac{EA}{L_0}.(L - L_0 - \Delta L_T).\dfrac{\partial L}{\partial u_i} = P_i \quad (i = 1,2,3,4) \\ \dfrac{EA}{L_0}.(L - L_0 - \Delta L_T) - P_e = 0 \end{cases} \tag{8}$$

Designate $\begin{cases} q_i^{(e)}(u) = \dfrac{EA}{L_0}.(L - L_0 - \Delta L_T).\dfrac{\partial L}{\partial u_i} \quad (i=1,2,3,4) \\ q_5^{(e)}(u) = \dfrac{EA}{L_0}.(L - L_0 - \Delta L_T) - P_e \end{cases}$; $\begin{cases} P_i^{(e)} = P_i \quad (i=1,2,3,4) \\ P_5^{(e)} = 0 \end{cases}$

The equation (8) can be compactly written as

$$q_k^{(e)}(u) = P_k^{(e)} \quad k=1,2,..,5 \tag{9}$$

Where

$\mathbf{u} \equiv \{u_1, u_2, u_3, u_4, u_5\}^T$ is the vector consists of nodal variables (primary unknowns) in the global coordinate system, including displacement unknowns and external force unknown at the ith cross section $u_5 \equiv P_e = N$

Based on mixed finite element formulation, the balanced equation truss element can be written as

$$\mathbf{q}^{(e)}(\mathbf{u}, \Delta L_T) = \mathbf{P}^{(e)} \tag{10}$$

Designating $\begin{cases} \mathbf{q}^{(e)}(\mathbf{u}, \Delta L_T) \equiv \{q_1^{(e)}(\mathbf{u}, \Delta L_T), q_2^{(e)}(\mathbf{u}, \Delta L_T), q_3^{(e)}(\mathbf{u}, \Delta L_T), q_4^{(e)}(\mathbf{u}, \Delta L_T), q_5^{(e)}(\mathbf{u}, \Delta L_T)\}^T \\ \mathbf{P}^{(e)} \equiv \{P_1^{(e)}, P_2^{(e)}, P_3^{(e)}, P_4^{(e)}, P_5^{(e)}\}^T \end{cases}$

Input incremental loading into the Eq. (10), getting

$$\mathbf{q}^{(e)}(\mathbf{u}, \Delta L_T) + \Delta\mathbf{q}^{(e)}(\mathbf{u}, \Delta L_T) = \mathbf{P}^{(e)} + \Delta\mathbf{P}^{(e)} \tag{11}$$

$\Delta\mathbf{q}^{(e)}(\mathbf{u})$ is second order infinitesimal and it can be negligible, the equation (11) is becoming

$$\mathbf{q}^{(e)}(\mathbf{u}, \Delta L_T) + \frac{\partial \mathbf{q}^{(e)}(\mathbf{u}, \Delta L_T)}{\partial \mathbf{u}} \delta\mathbf{u} = \mathbf{P}^{(e)} + \Delta\mathbf{P}^{(e)} \tag{12}$$

Setting $\mathbf{k}^{(e)}(\mathbf{u}, \Delta L_T) = \dfrac{\partial \mathbf{q}^{(e)}(\mathbf{u}, \Delta L_T)}{\partial \mathbf{u}}$

The equation (12) can be compactly written as

$$\mathbf{k}^{(e)}(\mathbf{u}, \Delta L_T)\delta\mathbf{u} = (\mathbf{P}^{(e)} + \Delta\mathbf{P}^{(e)}) - \mathbf{q}^{(e)}(\mathbf{u}, \Delta L_T) \tag{13}$$

In the equation (13), is a mixed matrix of truss element considering the temperature loading effect and large displacement, is given by

$$\mathbf{k}^{(e)}(\mathbf{u}, \Delta L_T) = \begin{bmatrix} k_{11} & k_{12} \dots k_{15} \\ k_{21} & k_{22} \dots k_{25} \\ \dots & \dots \quad \dots \\ k_{51} & k_{52} \dots k_{55} \end{bmatrix}, k_{ij} = \frac{\partial q_i^{(e)}(\mathbf{u}, \Delta L_T)}{\partial u_j}, (i,j=1,2,...,5) \tag{14}$$

where

$$k_{11}=\frac{EA}{L_0}\left\{\begin{array}{l}\dfrac{\left[\sqrt{(L_0.\sin\alpha_0+u_4-u_2)^2+(L_0.\cos\alpha_0+u_3-u_1)^2}-L_0-\Delta L_T\right]}{\sqrt{(L_0.\sin\alpha_0+u_4-u_2)^2+(L_0.\cos\alpha_0+u_3-u_1)^2}}\\+\dfrac{-\left[\sqrt{(L_0.\sin\alpha_0+u_4-u_2)^2+(L_0.\cos\alpha_0+u_3-u_1)^2}-L_0-\Delta L_T\right]}{[(L_0.\sin\alpha_0+u_4-u_2)^2+(L_0.\cos\alpha_0+u_3-u_1)^2]^{\frac{3}{2}}}\\.(L_0\cos\alpha_0-u_1+u_3)^2\\+\dfrac{(L_0\cos\alpha_0-u_1+u_3)^2}{(L_0.\sin\alpha_0+u_4-u_2)^2+(L_0.\cos\alpha_0+u_3-u_1)^2}\end{array}\right\}$$

$$k_{12}=\frac{EA}{L_0}\left\{\begin{array}{l}\dfrac{(-L_0\cos\alpha_0+u_1-u_3)(-L_0\sin\alpha_0+u_2-u_4)}{(L_0.\sin\alpha_0+u_4-u_2)^2+(L_0.\cos\alpha_0+u_3-u_1)^2}\\+\dfrac{-\left[\sqrt{(L_0.\sin\alpha_0+u_4-u_2)^2+(L_0.\cos\alpha_0+u_3-u_1)^2}-L_0-\Delta L_T\right]}{[(L_0.\sin\alpha_0+u_4-u_2)^2+(L_0.\cos\alpha_0+u_3-u_1)^2]^{\frac{3}{2}}}\\.(-L_0\cos\alpha_0+u_1-u_3)(-L_0\sin\alpha_0+u_2-u_4)\end{array}\right\}$$

$$k_{22}=\frac{EA}{L_0}\left\{\begin{array}{l}\dfrac{\left[\sqrt{(L_0.\sin\alpha_0+u_4-u_2)^2+(L_0.\cos\alpha_0+u_3-u_1)^2}-L_0-\Delta L_T\right]}{\sqrt{(L_0.\sin\alpha_0+u_4-u_2)^2+(L_0.\cos\alpha_0+u_3-u_1)^2}}\\+\dfrac{-\left[\sqrt{(L_0.\sin\alpha_0+u_4-u_2)^2+(L_0.\cos\alpha_0+u_3-u_1)^2}-L_0-\Delta L_T\right]}{[(L_0.\sin\alpha_0+u_4-u_2)^2+(L_0.\cos\alpha_0+u_3-u_1)^2]^{\frac{3}{2}}}\\.(L_0\sin\alpha_0-u_2+u_4)^2\\+\dfrac{(L_0\sin\alpha_0-u_2+u_4)^2}{(L_0.\sin\alpha_0+u_4-u_2)^2+(L_0.\cos\alpha_0+u_3-u_1)^2}\end{array}\right\}$$

$$k_{51}=\frac{EA}{L_0}\left\{\frac{-(L_0\cos\alpha_0+u_3-u_1)}{\sqrt{(L_0.\sin\alpha_0+u_4-u_2)^2+(L_0.\cos\alpha_0+u_3-u_1)^2}}\right\}$$

$$k_{52}=\frac{EA}{L_0}\left\{\frac{-(L_0.\sin\alpha_0+u_4-u_2)}{\sqrt{(L_0.\sin\alpha_0+u_4-u_2)^2+(L_0.\cos\alpha_0+u_3-u_1)^2}}\right\}$$

$$k_{13}=-k_{11},k_{14}=-k_{12},k_{15}=0,k_{21}=k_{12},k_{23}=-k_{12},k_{24}=-k_{22},k_{25}=0,k_{31}=-k_{11},k_{32}=-k_{12},k_{33}=k_{11}$$

$$k_{34}=k_{12},k_{35}=0,k_{41}=-k_{12},k_{42}=-k_{22},k_{43}=k_{12},k_{44}=k_{22},k_{45}=0,k_{53}=-k_{51},k_{54}=-k_{52},k_{55}=-1$$

2.2. *Balanced equation of truss system*

The global equation of truss system can be established by assembling elements . Based on the equation of (10), the equilibrium condition of the system can be expressed as

$$\mathbf{q}(\mathbf{u},\Delta\mathbf{L}_T)=\mathbf{P} \tag{15}$$

where:

$$\begin{cases} \mathbf{u}\equiv\{u_1,u_2,...,u_n\}^T;\ \mathbf{q}(\mathbf{u},\Delta\mathbf{L}_T)\equiv\{\mathbf{q}_1(\mathbf{u},\Delta\mathbf{L}_T),\mathbf{q}_2(\mathbf{u},\Delta\mathbf{L}_T),...,\mathbf{q}_n(\mathbf{u},\Delta\mathbf{L}_T)\}^T \\ \mathbf{P}\equiv\{\mathbf{P}_1,\mathbf{P}_2,...,\mathbf{P}_n\}^T;\ \mathbf{P}_i=\sum_{e=1}^{m}\mathbf{P}_i^{(e)};\mathbf{q}_i(\mathbf{u},\Delta\mathbf{L}_T)=\sum_{e=1}^{m}\mathbf{q}_i^{(e)}(\mathbf{u},\Delta L_T);\ (i=1,2,..,n) \\ \Delta\mathbf{L}_T=\{\alpha_1\Delta T_1L_0^{(1)},\alpha_2\Delta T_2L_0^{(2)},...,\alpha_m\Delta T_mL_0^{(m)}\}^T \end{cases}$$

"m" is a number of truss elements and "n" is number of unknowns

The equation (15) is a nonlinear system, it can be expressed in incremental form as

$$\mathbf{q}(\mathbf{u},\Delta\mathbf{L}_T)+\frac{\partial\mathbf{q}(\mathbf{u},\Delta\mathbf{L}_T)}{\partial\mathbf{u}}\delta\mathbf{u}=\mathbf{P}+\Delta\mathbf{P} \tag{16}$$

Setting a global mixed matrix, developed by assembling mixed matrix of truss elements

$$\mathbf{K}(\mathbf{u},\Delta\mathbf{L}_T)=\frac{\partial\mathbf{q}(\mathbf{u},\Delta\mathbf{L}_T)}{\partial\mathbf{u}} \tag{17}$$

The equations (16)&(17) can be compactly written as

$$\mathbf{K}(\mathbf{u},\Delta\mathbf{L}_T)\delta\mathbf{u}=\mathbf{P}+\Delta\mathbf{P}-\mathbf{q}(\mathbf{u},\Delta\mathbf{L}_T) \tag{18}$$

$$\mathbf{K}_{i,j}(\mathbf{u},\Delta\mathbf{L}_T)=\sum_{e=1}^{m}\mathbf{k}_{i,j}^{(e)}(\mathbf{u},\Delta L_T),\ (i,j=1,2,...,n) \tag{19}$$

For geometrically nonlinear analysis of system subjected to thermal load increasingly putting temperature parameter $\boldsymbol{\lambda}_{\Delta T}$ and vector of element deformation due to the thermal load $\overline{\Delta\mathbf{L}_T}$ (the initial value is guessed in solving process using incremental-iterative solution techniques). The initial thermal expansion vector can be expressed as

$$\Delta\mathbf{L}_T=\{\alpha_1\Delta T_1L_0^{(1)},\alpha_2\Delta T_2L_0^{(2)},...,\alpha_m\Delta T_mL_0^{(m)}\}^T=\boldsymbol{\lambda}_{\Delta T}\overline{\Delta\mathbf{L}_T} \tag{20}$$

When the mechanical load remains unchanged, the equation (18) is becoming

$$\mathbf{K}(\mathbf{u},\boldsymbol{\lambda}_{\Delta T}\overline{\Delta\mathbf{L}_T})\delta\mathbf{u}=\mathbf{P}-\mathbf{q}(\mathbf{u},\boldsymbol{\lambda}_{\Delta T}\overline{\Delta\mathbf{L}_T}) \tag{21}$$

3. Algorithm for solving problem based on Newton-Raphson technique

Based on Newton-Raphson method, in this research a nonlinear finite element analysis performed using an incremental formulation, in which the variables including displacement and temperature are updated incrementally corresponding to successive load steps in order to trace out the complete solution path (equilibrium path).

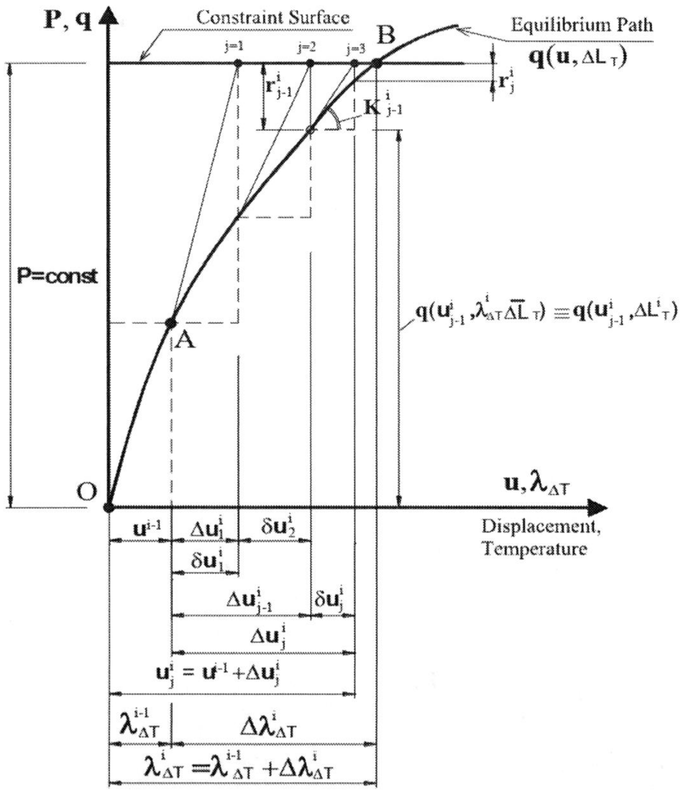

Figure 2. Newton Raphson technique for solving nonlinear truss problem

Using Newton-Raphson technique [11-13], external loads are computed at the first iteration of each incremental step and held constant throughout the remaining iterations in the step, as illustrated in figure 2.

The following iterative algorithm of jth iterative step at the ith thermal load step is proposed:

- Computing the axial deformation of truss element $\mathbf{\Delta L}_T^i$ corresponding temperature parameter $\lambda_{\Delta T}^i$: $\mathbf{\Delta L}_T^i \leftarrow \lambda_{\Delta T}^i \overline{\mathbf{\Delta L}_T}$
- Computing the tangent mixed matrix at the (j-1)th iterative step: $\mathbf{K}_{j-1}^i \leftarrow \mathbf{K}(\mathbf{u}_{j-1}^i, \mathbf{\Delta L}_T^i)$
- Computing the incremental displacement due to the residual load $\mathbf{r}_{j-1}^i$: $\delta \mathbf{u}_j^i \leftarrow \left[\mathbf{K}_{j-1}^i\right]^{-1} \mathbf{r}_{j-1}^i$
- Computing the $\mathbf{u}_j^i$: $\mathbf{u}_j^i \leftarrow \mathbf{u}_{j-1}^i + \delta \mathbf{u}_j^i$
- Computing corresponding load in equilibrium path: $\mathbf{q}(\mathbf{u}_j^i, \mathbf{\Delta L}_T^i)$
- Computing the residual load: $\mathbf{r}_j^i = \mathbf{P} - \mathbf{q}(\mathbf{u}_j^i, \mathbf{\Delta L}_T^i)$
- Comparing the residual load to convergence tolerance ε :

  + If $\left\|\mathbf{r}_j^i\right\| \leq \varepsilon$, go to the next thermal load step: $i \leftarrow i+1, \left(\lambda_{\Delta T}^i \leftarrow \lambda_{\Delta T}^{i-1} + \Delta\lambda_{\Delta T}^i ; \mathbf{\Delta L}_T^i \leftarrow \lambda_{\Delta T}^i \overline{\mathbf{\Delta L}_T}\right)$;

  + If $\left\|\mathbf{r}_j^i\right\| \leq \varepsilon$ is not satisfied, go to next iterative step: $j \leftarrow j+1$.

Based on the proposed iterative algorithm, the block diagram of algorithm for solving the nonlinear system of equations is established (shown in figure 3).

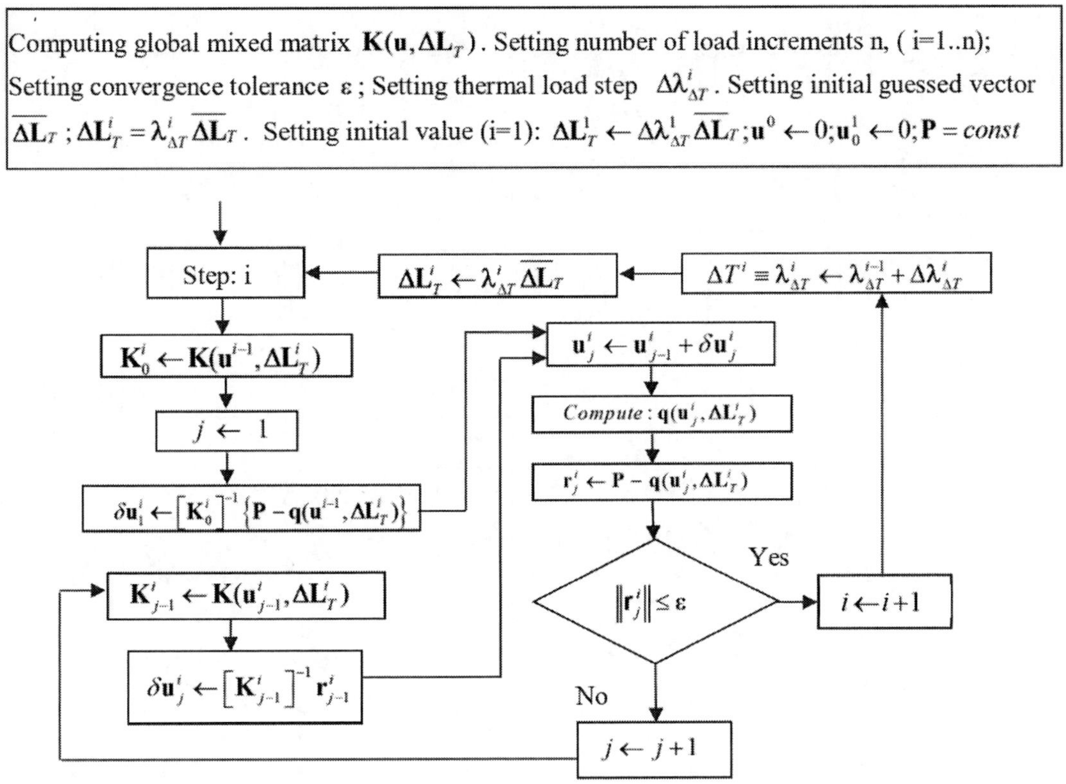

Figure 3. Block diagram of algorithm for solving nonlinear system of equations based on Newton Raphson technique

4. Numerical investigation

Based on proposed incremental-iterative algorithm, the calculation program to solve the above truss problem is written using Matlab software.

4.1. Example formulation

Investigate the nonlinear buckling behavior of truss system subjected to constant mechanical load and varied temperature load ΔT shown in figure 4. All of the truss bars made of the same material and have the same cross-sectional area. The parameters are given:

$$E=2.10^4\,kN/cm^2,\quad A=4cm^2,\quad \alpha=11.10^{-6}\,(^\circ C)^{-1},\quad P=36kN,\quad \Delta T=\left[0^\circ C,\,-50^\circ C\right]$$

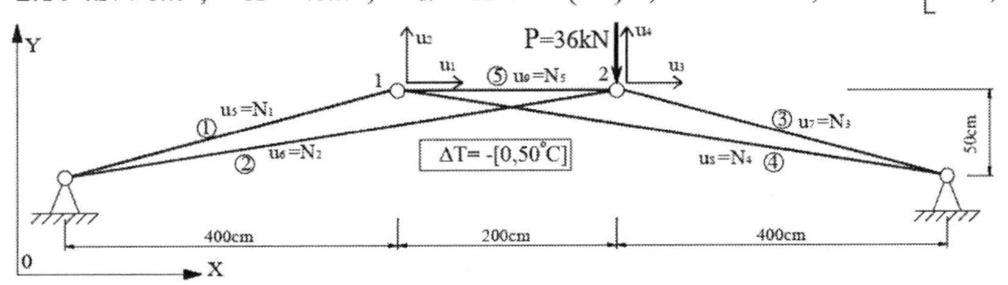

Figure 4. Examined system, designating unknowns of the system in the mixed formulation

The unknowns of the problem are designated as shown in figure 4, including (u_1,u_2,u_3,u_4) - nodal displacement unknowns and (u_5,u_6,u_7,u_8,u_9) - axial force unknowns.

IOP Conf. Series: Materials Science and Engineering 1015 (2021) 012023 doi:10.1088/1757-899X/1015/1/012023

4.2. Numerical results

The calculating results are temperature load-displacement and temperature load-internal force equilibrium path in two cases of solving, shown in figures 5-6.

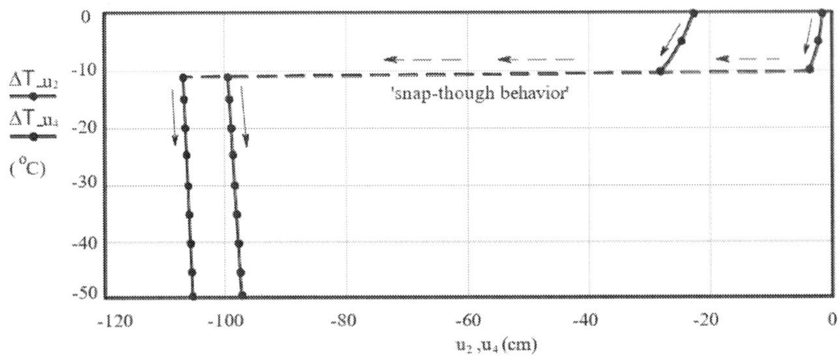

Figure 5. Temperature load-displacement equilibrium path $\Delta T - u_2, \Delta T - u_4$ in cases of temperature variation $\Delta T = \left[0^oC, -50^oC\right]$

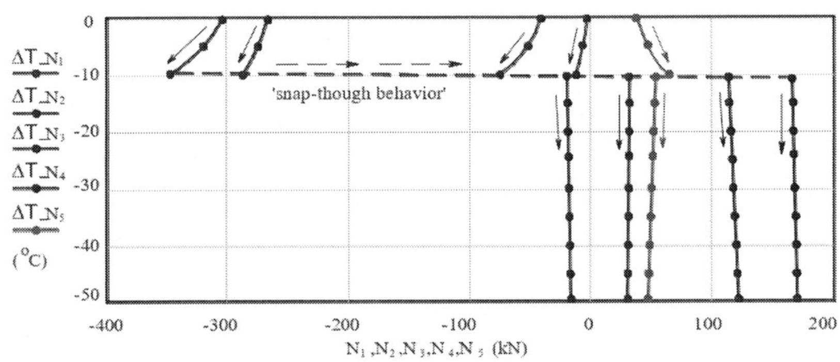

Figure 6. Temperature load-internal force equilibrium path $\Delta T - N_{1,2,3,4,5}$ in cases of temperature variation $\Delta T = \left[0^oC, -50^oC\right]$

4.3. Comments

The temperature variation significantly influence to the nonlinear buckling behavior of truss system. Due to the effects of temperature variation the snap-through behavior is appearing in the equilibrium path. The limit point of temperature variation $\Delta T = \Delta T_{Limit} = -10^oC$ can be found in the graphics in figure 5 and figure 6.

5. Conclusions

The mixed based formulation mathematical model for solving the geometrically nonlinear buckling problem of truss under temperature loading effects has significant advantage over displacement based formulation model. The proposed above algorithm for solving a geometrically nonlinear problem of system under thermal loading had been built without treating the temperature deformation constraint depending on the incremental element length. Taking mixed unknowns gives possibility to insert the length-dependent temperature deformation into the mixed matrix of truss element and to simplify the solving algorithm. The proposed method and established algorithm can be effectively used to investigate the nonlinear buckling and snap-through behaviour of truss, to define the equilibrium path and limit point of temperature variation.

IOP Conf. Series: Materials Science and Engineering 1015 (2021) 012023 doi:10.1088/1757-899X/1015/1/012023

References

[1] Ibrahimbegovic A, Hajdo E and Dolarevic S 2013 Linear instability or buckling problems for mechanical and coupled thermomechanical extreme conditions *Coupled Systems Mechanics* **2** (4) pp 349-374

[2] Gebrail B, Rasim T and Yusuf C T 2014 Analyses of truss structures under thermal effects *Recent Advances in Civil Engineering and Mechanics* (Italy, November 22-24, WSEAS Press) pp 235-239

[3] Bathe K J 2016 *Finite Element Procedures* (Prentice Hall)

[4] Zienkiewicz O C and Taylor R L 2014 *The Finite Element Method for Solid and Structural Mechanics* (Butterworth-Heinemann, 7th edn)

[5] Radeş M 2006 *Finite Element Analysis* (Editura Printech)

[6] Strodiot J J 2002 *Numerical Methods in Optimization* (Belgium: Namur)

[7] Sun W and Yuan Y -X 2006 *Optimization Theory and Methods - Nonlinear Programming* (Springer)

[8] Bertsekas D P 1982 *Constrained optimization and Lagrange multiplier method* (Academic Press Inc)

[9] Oden J T 1981 Penalty-finite element methods for constrained problems in elasticity, *Symposium on finite element method* (Hefei Politecnical University, China, May 19-23)

[10] Naotake N, Hetnarski R B and Yoshinobu T 2003 *Thermal stresses* (Taylor & Francis Group)

[11] Crisfield M A 1997 *Nonlinear Finite Element Analysis of Solids and Structures* (John Wiley & Sons Ltd.)

[12] Crisfield M A 1981 A Fast Incremental/Iterative Solution Procedure That Handles Snap-Through *Comput. Struct.* **13** (1–3) pp 55–62

[13] Riks, E. 1972 The Application of Newton's Method to the Problem of Elastic Stability *ASME J. Appl. Mech.* **39**(4) pp 1060–6

IOP Conf. Series: Materials Science and Engineering 1015 (2021) 012024 doi:10.1088/1757-899X/1015/1/012024

Using COSMOSWORKS abilities in construction glass strength research

V Alpatov and O Y Veremeenko

Samara State Technical University, Academy of Architecture and Civil Engineering, Molodogvardeyskaya St., 194, Samara, 443001, Russia

E-mail address: avu75@mail.ru

Abstract. Considered the issues of definition maximum acceptable load on supporting construction, made of glass. Object of research is a complex construction – insulated glass unit. Glass unit has two loaded layers of three in triplex and central layer – air chamber. Air chamber is formed by rubber layer, connecting external layers of triplex by perimeter. Glass unit loading is single-sided, only on top layer. Glass unit fastening was made only in four corner points. Researches were made in program complex COSMOSWORKS. Maximum acceptable load is searched by iterative calculation with step-by-step increase of calculation load on model. Established, that covering deformation and the stress distribution in them changes linearly. The value of maximum acceptable load for glass unit with special geometrical and physical characteristics was calculated. In this article the optical, thermal and other glass unit properties were not considered. The article presents the results of glass unit analysis when only mechanical action is applied to them. Temperature effect on glass units was not considered in this article.

1. Introduction

Glass is a very interesting material. For a long time this material was used only in manufacturing of enclosing structures. It was not presented with increased requirements for strength. As a result, methods for assessing the strength of glass structures nearly did not exist [1]. But recently situation has changed. Interest in transparent constructions, that are used with significant force effects has increased [2]. One of the factors increasing this interest is the massive construction of high-rise buildings in dense urban development. Requirements for the strength of glass structures when exposed to heavy loads led to the need to assess their reliability. As a result, the need to develop methods for calculation of glass structures appeared. The concept of "glass supporting structures" appeared [3]. Glass structures began to be used not only as enclosing structures, but also as load-bearing elements in the frames of buildings and structures. Glass started to be considered as structural material, which means a material for the manufacture of a structure, parts, elements that receive load.

We should notice, that today there are no methods of glass constructions calculation, that are set out as generally accepted design standards [1]. In each situation justification of their using possibility is required. Today, the evidence justification is numerical modeling в in calculation complexes, followed by necessary testing of experimental constructions in laboratories.

Numerical researches of constructions can be made in modern calculation complexes, such as Ansys, ABAQUS, Nastran, Autodesk Simulation, COSMOS, RFEM, APM, SCAD, Lira and others [4-6].

Authors made a numerical research of glass unit with dimensions 1.16×1.99 m (figure 1). Glass unit consists of three layers: triplex t=16 mm, air chamber t=18 mm, triplex t=16 mm. The air chamber is formed by a rubber layer, that connects layers of triplex around the perimeter. Triplex – Is A Multilayer Tempered Glass, Connected With Polymer Membrane. In This Glass Unit a Three-Layer triplex of tempered glass 8 mm + PVB + tempered glass 8 mm is used. Glass unit connection to the building frame is supposed at four points. For fastening the glass unit, it provides steel fastening elements for the system Spider d=60 mm. Impact of snow and wind on glass unit during exploitation is supposed.

The described glass unit is the object of research. The target of the research is to determine the maximum permissible load on a glass unit.

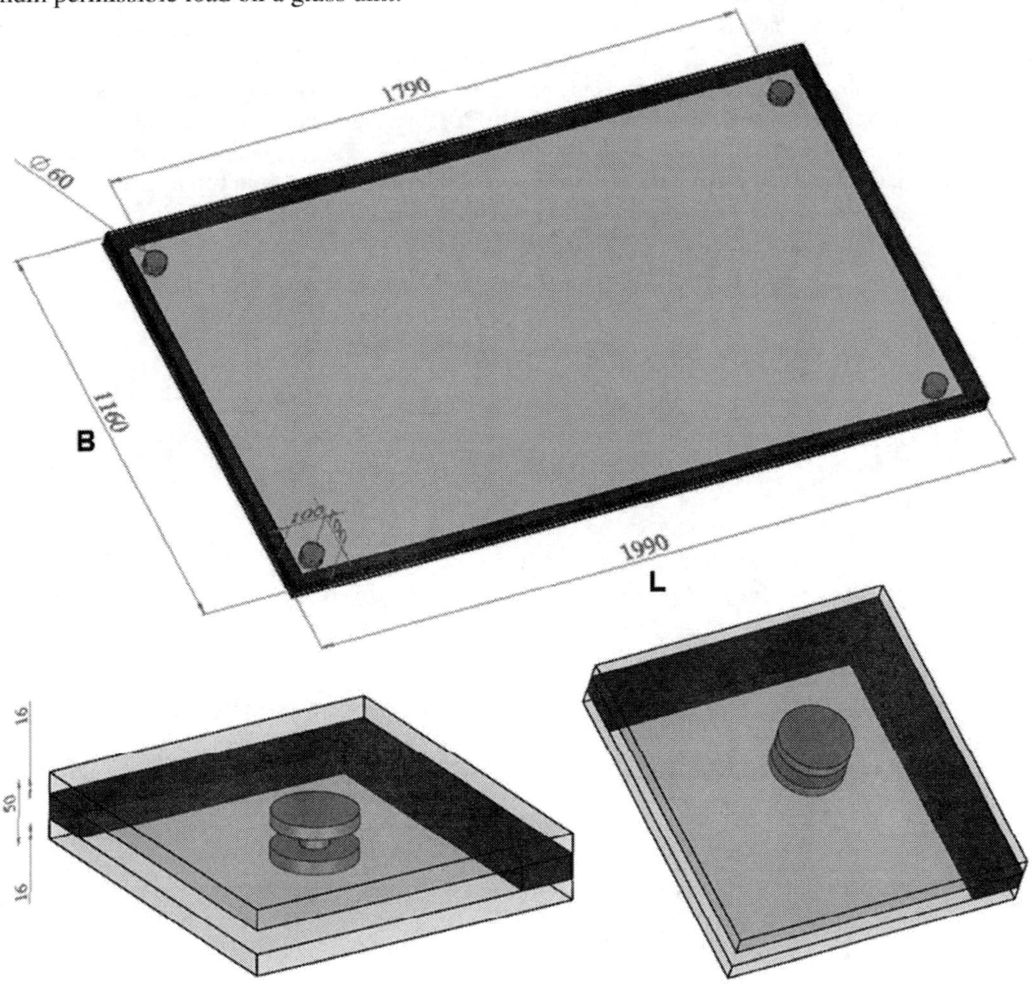

Figure 1. Researched glass unit.

2. Materials and methods

The task of determining the maximum permissible load was formulated as a numerical research. Method of research is iterative calculations of virtual model with step-by-step load increase. The COSMOSWORKS calculation complex was chosen as the research tool. The algorithm for solving the task is to perform the following sequence of actions:

1. Creating calculation model;
2. Adding load to calculation model;
3. Structural analysis of calculation model;
4. Assessment of load acceptability.

Calculation model was creates in SolidWorks program complex (figure 2). Load is added to the upper surface of the model. Calculations are performed only for a uniformly distributed load. The load is applied in steps, that is, it increases in each next calculation.

As a result of calculation model structural analysis its deformations and load in its elements are calculated. For further analysis amounts of calculation model maximum vertical displacement and maximum equivalent stresses in model elements are recorded.

The criterion for the admissibility of the applied load is taken to simultaneously ensure the strength and rigidity requirements of the glass unit. Ensuring the strength of a glass unit is assessed by comparing the calculated equivalent stresses in the elements of a glass unit with the reference value of strength limit of the material. Ensuring the rigidity of the glass unit is evaluated by comparing the calculated maximum vertical displacement with the allowable value. Acceptable amount of glass unit vertical displacement was set by B/300=3.87 mm according to the requirements of norms [7]. The amounts of materials strength limits are taken according to reference books [8, 9].

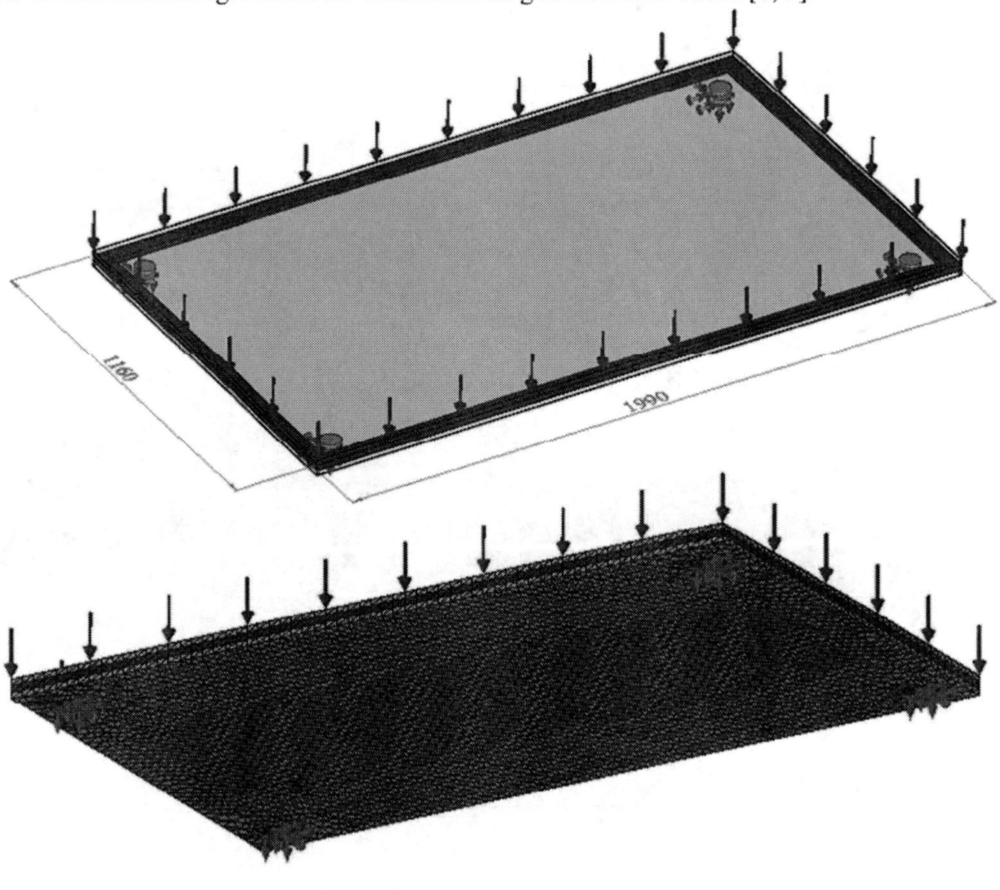

Figure 2. Calculation model

3. Results and discussions

Load increase step during calculation the stresses, was taken equal to 2 kPa. The step of increasing the load during calculating deformations was 1 kPa. Smaller step during вычислении calculating deformations was chosen because maximum deformations are achieved much faster than limit of strength (figures 2, 3).

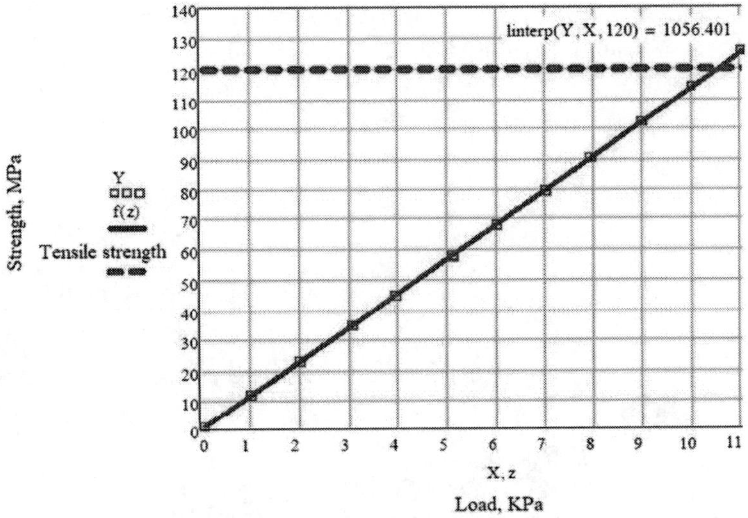

Figure 3. Calculated strength.

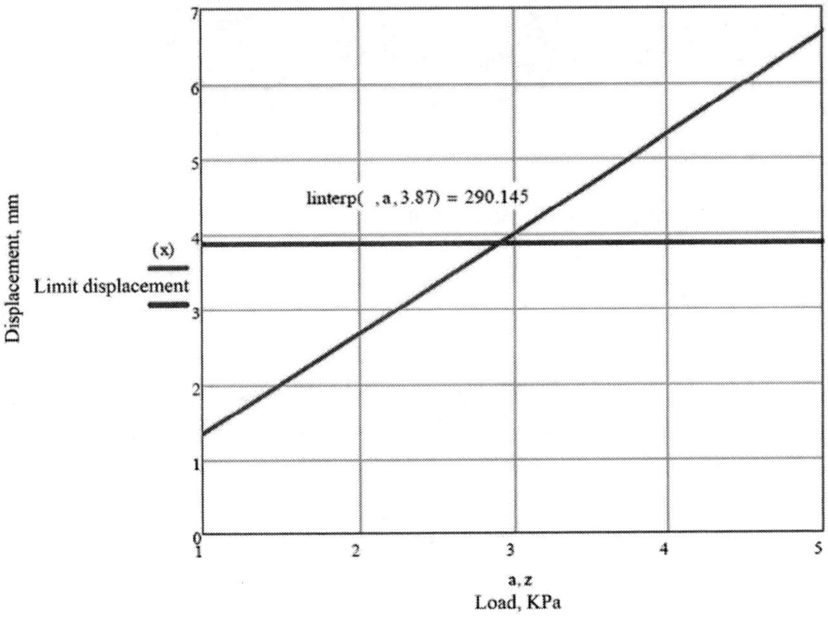

Figure 4. Calculated displacements.

As a result of performing a series of calculations with a step-by-step change in the load, stresses and displacements were obtained in the calculation model. Stresses in the rubber layer and in steel elements were not analyzed. Loads in triplex were defined by Maxwell–Huber–Hencky–von Mises theory. The calculation results are shown in tables 1, 2. On figures 3, 4 graphs of stress and displacement versus load are shown. The graphs are constructed by linear approximation of the calculated in program complex COSMOSWORKS amounts of stresses and displacements. The calculation results were processed by program complex MathCAD. On figures 5-7 most illustrative graphical results of model calculation with load equal 11 kPa are shown.

Table 1. – Stress calculation results.

Load, (kPa)	Calculated strength (MPa)	Tensile strength, (MPa)
1	11.6	
3	34.2	
5	55.9	
7	79.6	
9	102.3	
11	125.0	120

Table 2. Displacement calculation results.

Load, (kPa)	Displacement, (mm)	Limit displacement (mm)
1	1.362	
2	2.702	
3	4.107	3.87
4	5.430	
5	6.798	
11	14.912	

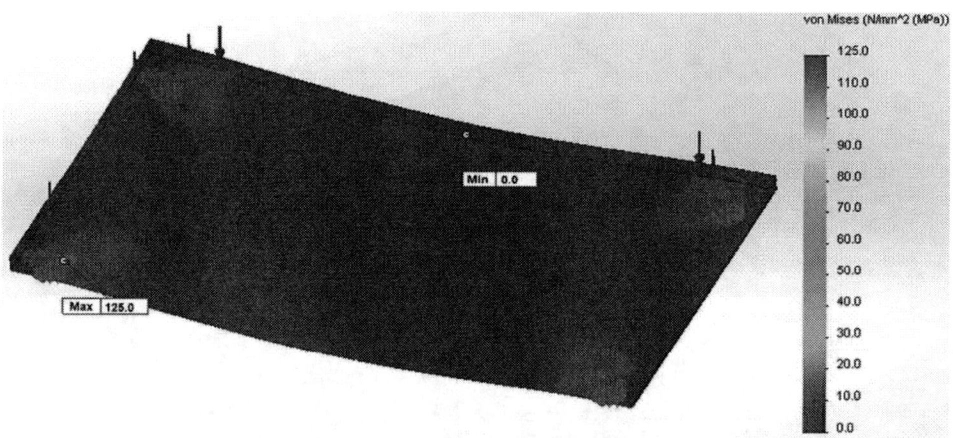

Figure 5. Stress distribution in the calculation model.

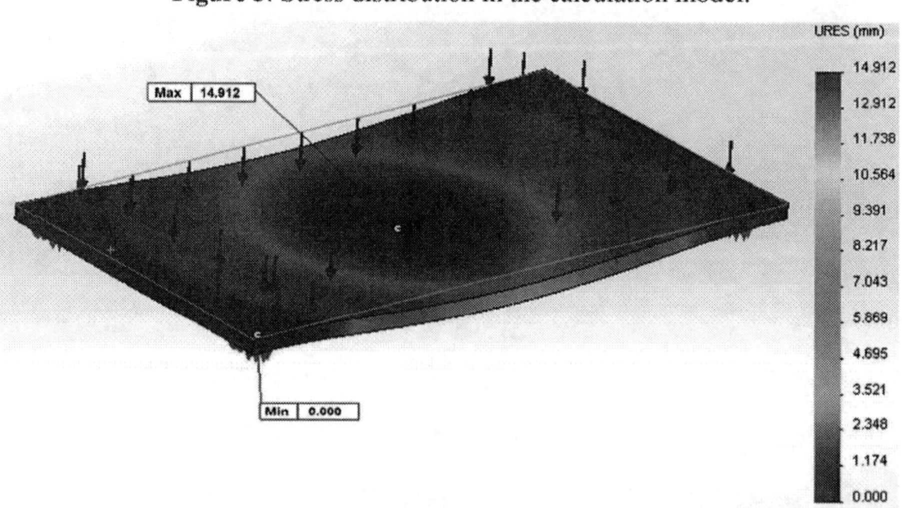

Figure 6. Calculation model deformation. View on upper triplex.

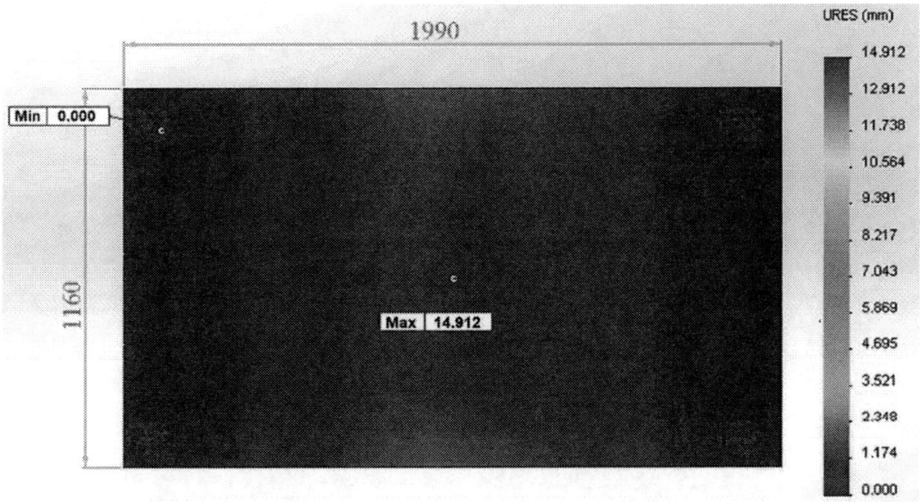

Figure 7. Calculation model deformation. View on lower triplex.

4. Conclusions

Numerical experiment established:

1. Stresses in external layers of glass unit (triplex) change linearly until reaching the glass strength limit.
2. Glass unit external layers deformations change linearly until reaching the glass strength limit. Deformation character of upper (loaded) and lower (not loaded) layers are different. Upper layer deformates as a plate, supported by four points (figure 6). Lower layer deformates as a plate, supported by two sides (figure 7).
3. Maximum equivalent stresses in glass are seen in zone of spider connection, here we should expect initial cracking.
4. We consider maximum acceptable load as load, that causes appearance of loads in glass, that are equal to strength limit. These are the peak stresses seen in areas of glass connection to a metal element spider (figure 5). Amount of accessible load is 10.9 kPa.

References

[1] Zubkov V and Kondratieva N 2013 *Flexural strength of sheet glass* (Samara: SamLuxPrint)
[2] Zubkov V and Kondratieva N 2018 *MATEC Web Conf.* **196** p 02015
[3] Zubkov V and Kondratieva N 2008 Characteristics of calculation of flat glass in translucent structures. Glass performance days. New Delhi pp 27-29
[4] Alpatov V 2018 *IOP Conf. Ser.: Mater. Sci. Eng.* **463** pp 032103
[5] Alpatov V et al 2018 *IOP Conf. Ser.: Mater. Sci. Eng.* **456** 012010
[6] Balzannikov M et al 2016 *MATEC Web Conf.* **73** p 01012
[7] Standard SP 20.13330.2016 Loads and impacts, http://docs.cntd.ru/document/456044318/
[8] Standard GOST 111-2014 Colorless sheet glass, http://docs.cntd.ru/document/1200119804
[9] Handbook TROSIFOL 2005 (Troisdorf: Kuraray Specialities Europe GmbH) 53840

IOP Conf. Series: Materials Science and Engineering 1015 (2021) 012025 doi:10.1088/1757-899X/1015/1/012025

Plate tensegrity structures controlled with self-stress

K Martyniuk-Sienkiewicz and A Al Sabouni-Zawadzka

Warsaw University of Technology, Faculty of Civil Engineering, Al. Armii Ludowej 16, 00-637 Warsaw, Poland

E-mail address: a.sabouni@il.pw.edu.pl

Abstract. The present paper focuses on the analysis of plate tensegrity structures based on the 3-strut simplex module. Two structural systems are considered: a single plate tensegrity module and a multi-module segment. It is proved that the implementation of plate elements in the basic tensegrity module not only opens new application perspectives, but also increases its stiffness. The analyses are performed with the use of the finite element method (FEM), using the second order theory and the geometrically nonlinear analysis.

1. Introduction

Nowadays, tensegrity structures are becoming more and more popular – not only because of their modern appearance in real structures, but mostly because of their undeniable advantages such as a high stiffness-to-mass ratio, economic use of materials, increased capacity under the increasing external load or possibility to construct modular systems. The development of tensegrity has brought a new type of structural system – plate tensegrity structures.

Generally, tensegrity is a term that characterizes a construction rule, in which a structure consists of compressed elements (struts) joined by tensile members (cables) [1, 2]. These structural elements stabilize each other. In tensegrity structures there are infinitesimal mechanisms [3] that are eliminated by self-stress states, which guarantee structural stiffness. Self-stress forces are independent of supports and loads. Tensegrities are classified as a specific type of trusses. To their important features belongs adaptiveness to the environment due to self-stress [4-7].

Plate tensegrity structures have all previously mentioned properties and additionally, an integrated plate element which is a working part of the system, not an additional part attached to it. Such structures are developed by replacing some of the cables with plate elements. It allows us to change the designing approach – not only the construction stiffness is increased but also it opens new application possibilities.

Through years of evolution of tensegrity systems, plates have been repeatedly implemented into the structure, but always as an additional element, which only increased its functionality. The first real integration of plate and bars was made by Andreas Falk [8, 9]. In his consideration, Falk created two plate tensegrity modules – one based on a triangular plate and the second based on a rectangular plate. Plates are perforated in the middle and a strut goes through the opening. Both ends of the strut are joined to the plate corners by cables. The module is prestressed to eliminate an infinitesimal mechanism. As a result of prestressing, struts and plates are compressed and cables are tensed. The buckling length of the strut is halved by a plate which increases its capacity.

Using those modules Falk and Kirkegaard analyzed various slab structures [9] – from flat to arched. Each of them is created by joining plate tensegrity modules on plate level and both strut ends.

Plates transmit each other's parallel forces – compression and shear. Struts are joined by cables. As a result, a three-level structure is obtained with external cable nets and an internal plate surface. The structure is in equilibrium when every element is prestressed with the set of forces attuned to modules.

Falk stated two ways of creating slabs – using one large plate or joining smaller modules. The first one creates a stiffer structure and decreases material usage. The second one allows us to create more flexible system – arched or with specific shapes. By stabilizing the systems with cable nets some of the modules may be removed and non-structural elements such as transparent plates or plates from lighter material may be applied. What is more, openings could be left in the slabs.

During their analysis, Falk and Kirkegaard noted that cables in nets and modules have different directions and therefore, prestressing of these cables could cause some deformation. To avoid the possibility of deformation they proposed a special geometrical ratio based on prestressing forces in cables. In order to adapt the structure to various load cases, self stress needs to be adjusted a few times during use. In his analysis, Falk consciously resigned from compression in tension rule. On this account, plate capacity can be bigger and it also helps to reapply prestressed elements.

In his analysis, Falk also mentions that plate tensegrity structures are stronger than tensegrity systems in case of cable damage. It resulted from the fact that damaging one cable provides variations in the prestressing forces set. The plate has a greater capacity limit and by manipulating its thickness and level of integration with cables, the structure is prepared for different load cases.

This feature confirms that plate tensegrity structures, just as traditional tensegrity systems, can be regarded as smart structures [2, 10]. Their smartness results from infinitesimal mechanisms that occur in the modules. There are four major inherent properties of smart structures [2]: self-control, self-diagnosis, self-repair, and self-adjustment (active control). Self-control refers to self-stiffening structures with the increase of applied load. Self-diagnosis refers to identification of damages by noting changes in the internal forces set. Damage of active members changes the level of self-stress in the entire structure. Self-repair comes directly from infinitesimal mechanisms stabilized by self-stress. Proper change of prestressing compensates damages in the structure and restores the initial state of displacements. Self-adjustment refers to the ability of adjusting self stress in the whole structure or separate parts of it. Each of these features is also present in plate tensegrity structures.

In present paper an analysis of a plate tensegrity module based on the simplex module is presented. The analysis involves comparing calculations in two types of theories – a second-order analysis which considers the prestressing forces and a geometrically nonlinear analysis. Moreover, an analysis of a multi-module plate tensegrity structure is presented.

2. Plate tensegrity module

A plate tensegrity structure is based on simplex (figure 1) – a basic tensegrity unit, which is composed of three struts and nine cables. It has the shape of a regular triangular prism. Struts – compressed elements – are diagonals of the prism sides and cables – tensed elements – are edges of the block. The final form of the module is created by rotating the upper and lower base of the prism in opposite directions creating a 150° angle. In plate tensegrity modules the upper and lower cables are replaced with panels.

The analyzed module is composed of two triangular plates with 1 m sides for the lower element and with 0.5768 m sides for the upper element. Both panels are 40 mm thick. For struts and cables, catalog cross-sections are applied. Compressed elements are steel tubes with 60 mm diameter and 2.9 mm wall thickness. Tensed elements are steel rods with 10 mm diameter.

For plate elements cross laminated timber (CLT) is applied. The production process eliminates anisotropy by gluing wooden boards in altering cross-wise layers to obtain two-dimensional form stability. There is always an uneven number of layers and an angle between two next sheets is suited to the specific construction. CLT has also profitable weight to stiffness ratio which is the desired feature for tensegrity structures. Moreover, the production process is environmentally friendly and suited to the case.

Structural steel S355JR is used as a material for cables and struts. Material is commonly used in cable structures, not only because of high ultimate tensile strength and compression strength but also due to its plasticity and ease of shaping.

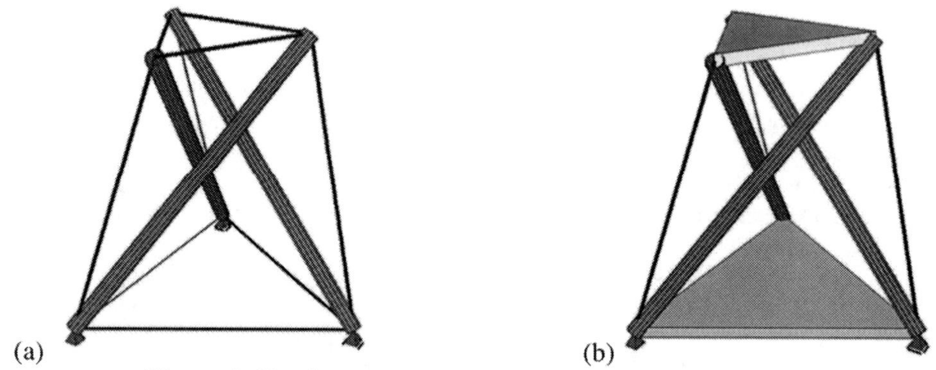

Figure 1. Simplex module: (a) tensegrity, (b) plate tensegrity.

The module is fixed by pin support in three points of the lower plate (figure 2). In node 1 movement in *X*, *Y*, *Z* direction is blocked, in node 2 *Z* motion is blocked and in node 3 *X* and *Z* movement is blocked.

The module was prestressed before applying external load to remove the infinitesimal mechanism encountered in tensegrity structures. Struts and cables are compressed with different forces: $S = S_0 \cdot a$, where S_0 – initial force value, a – force multiplier value (table 1).

Table 1. Force multiplier values for struts and cables.

Element	Force multiplier value
Strut	-0.421637
Cable	0.3333333

Table 2 presents the values of eigenfrequencies obtained for the module before and after the application of self-stress. It can be notice the prestressing forces eliminate the infinitesimal mechanism.

Table 2. Eigenfrequencies in the module.

	Eigenfrequencies	
	Before prestressing (Hz)	After prestressing (Hz)
1	0.0	24.14
2	104.86	104.53
3	155.78	148.53
4	213.71	207.41
5	319.29	314.96
6	325.37	324.72
7	513.74	511.50
8	623.59	616.06
9	805.90	807.28
10	865.54	862.19

For the analysis, SOFiSTiK software was used. The analysis was performed for two load cases (figure 2). In the first case, three vertical loads equal to 10 kN each were applied to upper plate nodes 4, 5, 6. In the second one, the horizontal force equal to 10 kN in the global *X* direction was applied to node 5.

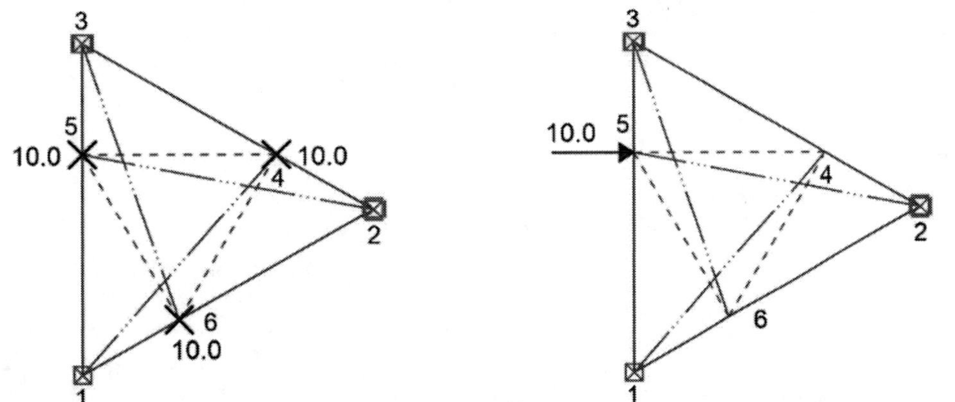

(a) (b)

Figure 2. Load cases: (a) vertical loads applied to upper plate nodes, (b) horizontal load applied to node 5.

The module was analyzed with the use of the finite element method (FEM) in two types of theories – a second-order analysis which considers the stability of the structure and geometrically nonlinear analysis.

In the compressed element, buckling could occur. Intending to reduce that adverse mechanism, the applied prestress load needs to be limited. The critical load was calculated using Euler buckling formula. Its value is 247.81 kN. Analysis was limited by the critical load, the prestress applied to the module was in a range from 0 to 600 kN.

During the analysis the displacement of node 5 in the global Z direction for vertical load case (figure 3a) and global X direction for horizontal load case (figure 3b) was noted. Additionally, the behavior of the analyzed plate tensegrity module was compared to the characteristics of the basic 3-strut simplex module (figure 4). Displacement values are given in mm and force values in kN.

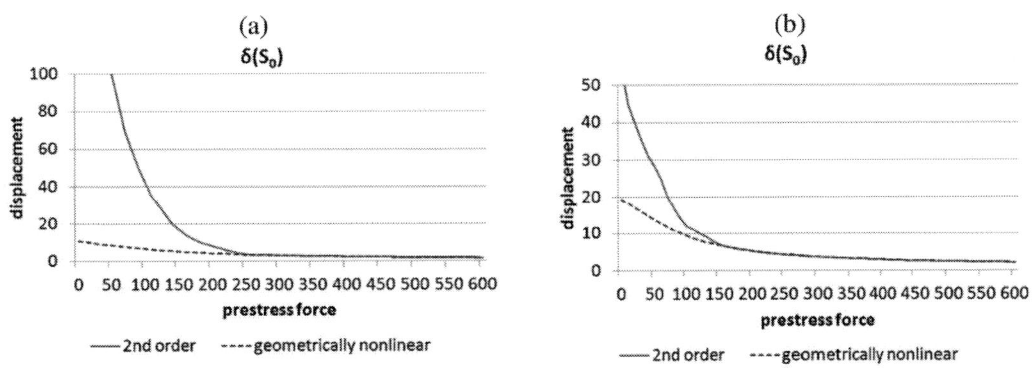

Figure 3. Influence of self-stress on the displacement of node 5: (a) vertical load case, b) horizontal load case.

Finite element analysis according to the second order theory allows to consider prestressing forces through geometric stiffness matrix and to obtain structural response for external loading. Fully nonlinear analysis leads to big differences within the scope of small prestressing forces, which disappear along with an increase of force values (figure 3).

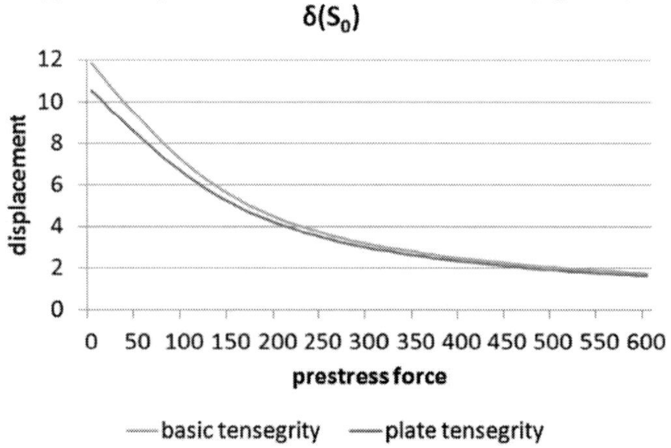

Figure 4. Influence of self-stress on the displacement of node 5 – comparison of the 3-strut simplex module and the simplex-based plate tensegrity.

Moreover, it can be noticed that the implementation of the plate element in the basic tensegrity module increases its stiffness (figure 4). The greatest difference is noted for small values of prestressing forces and slowly disappears with an increase of self-stress.

3. Multi-module plate tensegrity structure

Natural applications for multi-module plate tensegrity structures are slabs. Plate elements, which are integral part of the structure, could be an overlay as well as a construction surface. What is more, slabs can be created with single repetitive modules, what could simplify the erection process. Two major slab segment shapes – hexagon and elongated hexagon could be created by using simplex plate tensegrity module. The first segment is used in round or skew slabs, the second in rectangle ones.

Hexagonal segment is assembled from six simplex modules joined in one node (figure 5). Modules need to be arranged clockwise and counter-clockwise in rotation with the aim of stiffening the structure. The segment is fixed in four nodes.

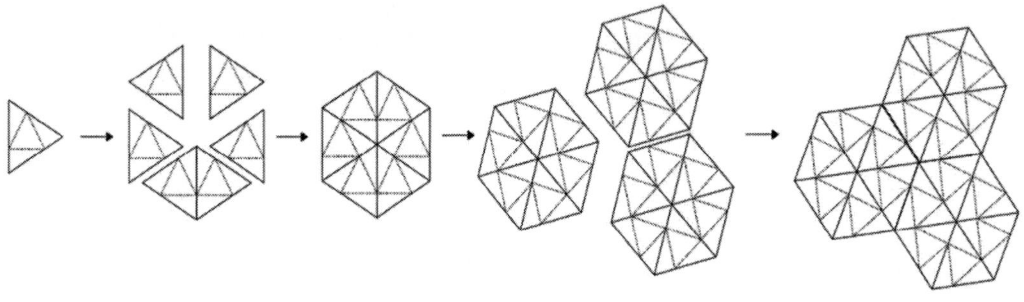

Figure 5. Scheme of creating a basic hexagonal segment and a slab.

The elongated hexagonal segment could be assembled with any amount of modules arranged clockwise and counter-clockwise, alternately (figure 6). Analogically to the hexagonal segment, in the elongated hexagonal segment the infinitesimal mechanism is compensated with the same prestress as in a single module.

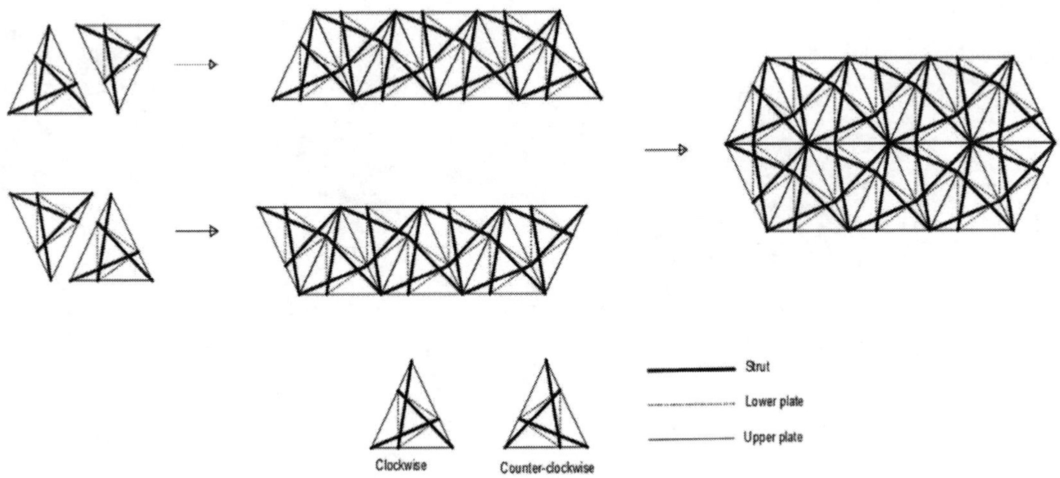

Figure 6. Scheme of creating a basic elongated hexagonal segment.

Slab dimensions are limited only by a segment dimensions. Hexagonal segments are arranged concentrically around the chosen point in the case of round slabs or next to each other in rows for skew slabs. Elongated hexagonal segment creates a basic slab with a 4 m width and unlimited length. Slab with endless dimensions could be formed by joining more of those basic slabs.

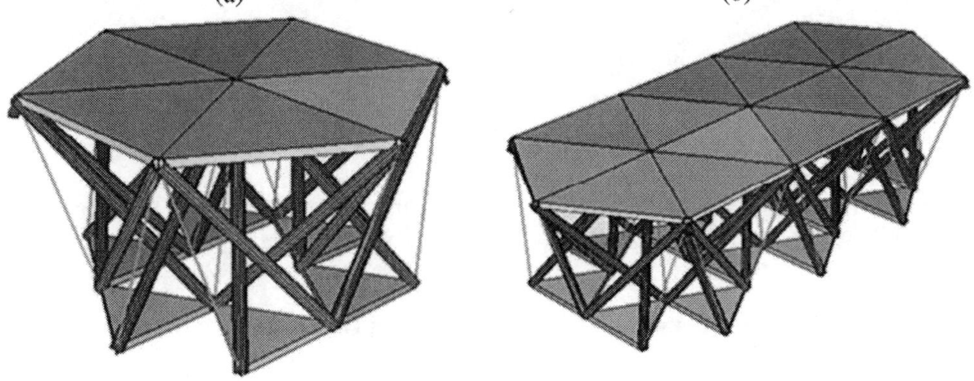

Figure 7. 3D views of plate tensegrity segments: (a) hexagonal segment, (b) elongated segmen

The analyzed slab segment was created with fourteen modules of the same type – seven clockwise and seven counter-clockwise oriented (figure 8). The module used here is the single module analyzed before. The module is 1 m high. To create a continuous walking surface and avoid increasing the slab weight by adding more plate elements, modules need to be turned upside down.

As was mentioned before, modules are joined in rotation clockwise and counterclockwise oriented to block mutual movements. Modules are assembled in rows, seven in every and then both rows are joined together. Some of the struts are removed to avoid duplication in nodes. Finally, the structure is 1.73 meter wide, 4 m long and 1 m high.

The segment is fixed in four nodes on the lower surface by pin supports (figure 8). In node 1 movement in *X*, *Y*, *Z* direction is blocked, in node 7 *X* and *Z* motion is blocked, in node 23 *Y* and *Z* movement is blocked and in node 25 *Z* motion is blocked. The structure is loaded with vertical forces of 10 kN applied to the upper nodes (figure 8).

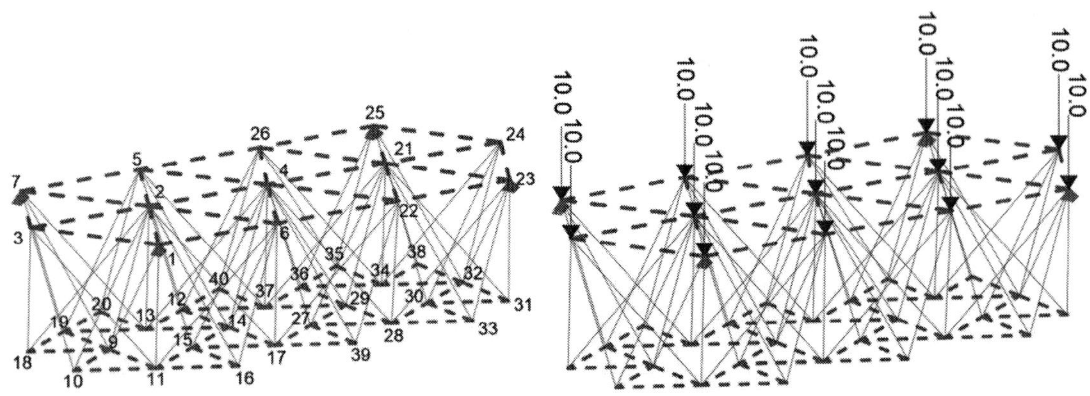

Figure 8. The elongated hexagonal segment with nodes numbering and loading.

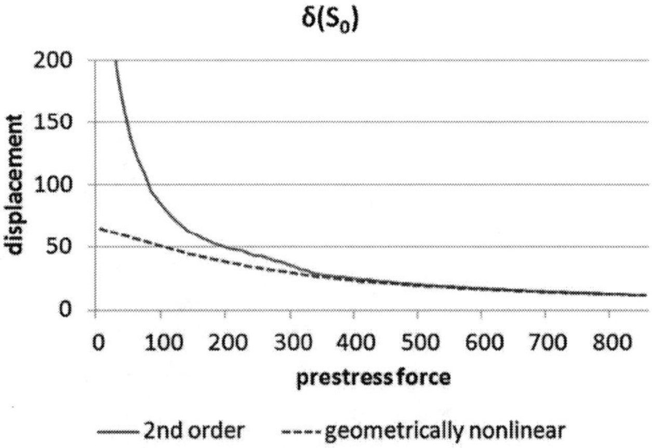

Figure 9. Influence of self-stress on the vertical displacement of node.

It can be noticed that the behavior of the elongated segment is very similar to the characteristics of the single module, just as in the case of traditional tensegrity structures [11].

4. Conclusions

In present paper analyses of plate tensegrity structures based on the 3-strut simplex module are presented. Two types of theories are compared – a second-order analysis which considers the prestressing forces and a geometrically nonlinear analysis. Finite element analysis according to the second order theory allows to consider prestressing forces through geometric stiffness matrix and to obtain structural response for external loading. Fully nonlinear analysis leads to big differences within the scope of small prestressing forces, which disappear along with an increase of force values.

Two structural systems are analyzed: a single plate tensegrity module and a plate segment that consists of 14 modules connected in common nodes.

The performed analyses show that the implementation of plate element in the basic tensegrity module increases its stiffness and, in the same time, opens new application perspectives. The greatest difference is noted for small values of prestressing forces and slowly disappears with an increase of self-stress.

In the future studies the authors plan to analyze other structures based on the module considered in this paper, focusing on various support schemes and real loads acting in different directions. Moreover, new solutions of tensegrity modules will be searched for, where the cable elements may be replaced by rigid plates.

References

[1] Motro R 2003 *Tensegrity: Structural Systems for the Future* (London: Kogan Page Science)
[2] Al Sabouni-Zawadzka A and Gilewski W 2018 Inherent properties of smart tensegrity structures *Applied Sciences* **8** (5)
[3] Calladine C R and Pellegrino S 1991 First-order infinitesimal mechanisms *International Journal of Solids and Structures* **27** pp 505-15
[4] Wroldsen A S 2007 *Modelling and control of tensegrity structures* (PhD Thesis, Department of Marine Technology, Norwegian University of Science and Technology)
[5] Adam B and Smith I F C 2007 Self-diagnosis and self-repair of an active tensegrity structure *Journal of Structural Engineering* pp 1752-1761
[6] Bel Hadj Ali N. and Smith I F C 2010 Dynamic behavior and vibration control of a tensegrity structure *International Journal of Solids and Structures* **47** pp 1285-96
[7] Fest E, Shea K and Smith I F C 2004 Active tensegrity structure *J. of Struct. Eng.* pp 1454-65
[8] Falk A 2006 Architectural and structural developement of plate tensegrity *Int. Association for Shell and Spatial Structures Symp.* (Pekin China)
[9] Falk A and Kirkegaard P H 2012 Pre-stressing Timber-Based Plate Tensegrity Structures. IASS-APCS 2012 *Spatial structures to space structures* (Seoul Korea)
[10] Gilewski W and Al Sabouni-Zawadzka A 2015 On possible applications of smart structures controlled by self-stress *Archives of Civil and Mechanical Engineering* **15** pp 469-478
[11] Al Sabouni-Zawadzka A Active control of smart tensegrity structures *Archives of Civil Engineering* **60** (4) pp 517-34

IOP Conf. Series: Materials Science and Engineering 1015 (2021) 012026 doi:10.1088/1757-899X/1015/1/012026

Design and modeling of glulam beams with holes

J Pełczyński[1], P Brodniewicz[2], A Al Sabouni-Zawadzka[1] and W Gilewski[1]

[1] Warsaw University of Technology, Faculty of Civil Engineering,
al. Armii Ludowej 16, 00-637 Warsaw, Poland
[2] Andrewex sp. z o.o., ul. Sosnowa 14, 87-167 Cierpice,
Konstimber – Zespół Projektowy, ul. Toruńska 69/4, 87-800 Włocławek

E-mail address: j.pelczynski@il.pw.edu.pl

Abstract. The paper presents a sketch of design recommendations and numerical simulations of stress distributions in perforated double-tapered beams made of glued laminated timber, taking into account the reinforcement used. Conscious design requires, according to the authors opinion, calculations performed using the finite element method. The analysis presented in the paper concerns three possible positions of the opening in the apex zone. It is recommended to position the opening halfway up, as it causes the least stress disturbance in the beams. Bottom position is definitely not recommended, as stress disturbances are large and cannot be levelled with transverse reinforcement. Reinforcement allows for the reduction of stresses in the range from several to several dozen percent.

1. Introduction

The construction of a building structure is always subordinated to the overriding goal, which is the functionality. The structure is not designed for itself, but in order to form a specific spatial context: to create places where users will spend time, build routes on which people will move during their daily duties and exceptional situations.

People involved in wooden construction are particularly sensitive to the fact that such structures are always designed to be something more than a simple structure. Their shape is supposed to add value to the aesthetics of the whole building. Furthermore, wooden parts increase the comfort of use and have a positive effect on the user's health. Therefore, when the location of installation parts is planned, such as sanitary and electrical installations, they have to be weaved together with the main structure, so they coexist in harmony.

Hence, the designer often adjusts the shape of the structure to the installations serving the facility, e.g. by the design of the openings in structural elements where the electric cables, water and air pipes will be placed. Many possible cases of the modifications may be indicated. The static scheme of a structural element, its geometry and connections may vary. In addition, the required holes can have various locations (near the supports, in the span), sizes (from several to several dozen centimeters) or shapes (most often round, less often rectangular). The preparation of the hole can become a considerable challenge for the designer. Undoubtedly, such a case is the opening in the glued laminated timber girders placed in an extremely sensitive place – the apex zone.

The present paper indicates the sources of stresses in specified holes locations and in their immediate vicinity. Furthermore, the outline of recommendations for their design, resulting from the available literature, is presented. In addition, a possible method of analyzing a complex case is proposed. Application of the finite element method for the analyze of the openings in the apex zone with possible reinforcement is justified and recommended.

2. Holes in glulam beams

To enable the proper functioning of the building there is a requirement to include the elements of the building infrastructure (electrical, air conditioning, sanitary). Often, those parts are located in the areas usually dedicated to the structural elements. Meeting not only functional but also economical or aesthetical expectances, this need may cause a necessity of creating openings in the structural elements, which become a gateway for linear installation members.

The locations and reasons for placing apertures vary. For example, the drainage of a double sloping roof, which includes a set of pipes below roofing, is limited by parapets and requires a series of round holes with a diameter up to 35 cm. The pipes are situated in the area, with the lowest roof, usually near the supports of the structural elements. On the other hand, the air vents are usually larger, rectangular and located in a place depending only on their functionality, namely in the support area as well as in the middle of the span of a wooden beam.

Designing openings is an exceptional challenge when it comes to timber structural elements. Every hole is a weakening of the cross-section. Since timber is an anisotropic material, the holes are a source of compressive and tensile stresses perpendicular to grain. Tensile stress $\sigma_{t,90}$, with the value of only 0.4-0.5 MPa of strength [1, 2], is a unique threat for timber elements, because it is a potential source of cracks or delamination.

The stress distribution in the vicinity of a hole depends on its shape, size and position as well as on the ratio of the section forces. The openings located in the areas where the shear force and the bending moment interact, cause tension perpendicular to grain on two diagonally opposite edges (see figure 1a). When the area of a timber member is subjected to pure or almost pure bending, the concentration occurs on two upper edges of the opening (see figure 1b).

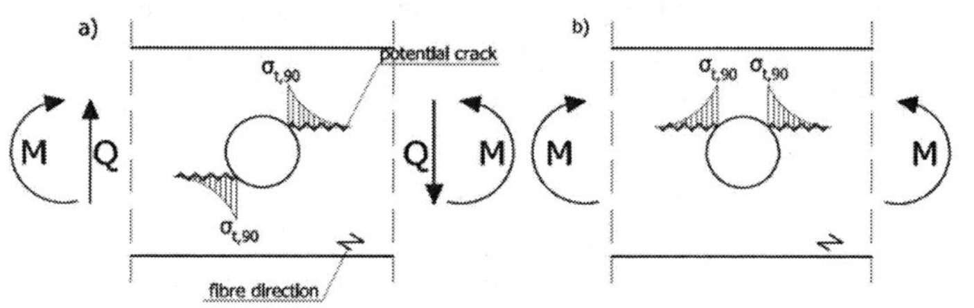

Figure 1. Tension perpendicular to grain in areas of openings.

The openings have to be designed with special care and the vulnerable zones often have to be reinforced. Even more precision is needed for beam zones that are already under stress, which can dangerously combine with stress concentrations induced by the opening. That case is characteristic for the locations exposed to shear stresses, like support areas and zones where tensile stress perpendicular to grain is already present. The most popular example is the apex zone of the elements typical for glued laminated timber roof structures: double tapered (see figure 2a), pitched cambered (figure 2b) and curved beams (figure 2c).

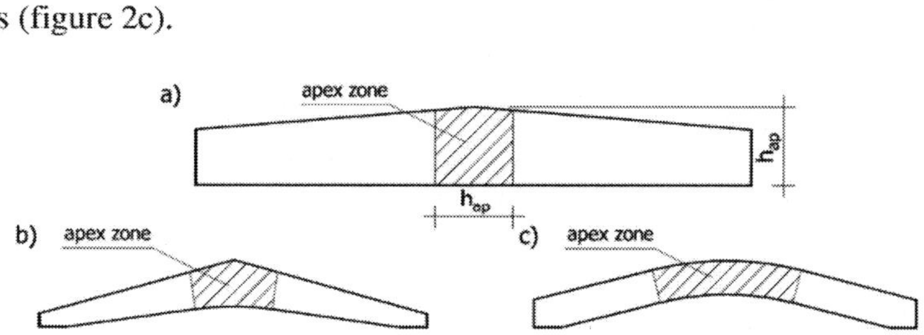

Figure 2. Apex zone of beams with fiber direction parallel to the lower edge of the beam.

IOP Conf. Series: Materials Science and Engineering 1015 (2021) 012026 doi:10.1088/1757-899X/1015/1/012026

Tensile stress perpendicular to grain appears when the bending moment in the apex zone stretches lower grains of the cross-section and compresses upper ones. The stress is caused by a difference between the directions of tensile and compressive forces and also an effect of straightening the curvature of the glulam member by the bending moment. In this case laminations try to move apart, creating tensile radial stresses perpendicular to grain between them.

3. Design practice – an international view

The basic source for designing timber structures, Eurocode [3], does not touch upon a problem of openings. However, the National Annex to the German standard [4] provides fundamental rules for holes placement in glulam elements and the necessary reinforcement. The literature allows only two types of holes: circular and rectangular with corner radius. Moreover, they have to be considered as weakening of a cross-section and taken into account in calculations.

The German standard points out the source of tensional force as a result of the additive influence of bending and shear forces, calculated as

$$F_{\mathrm{t,V,d}} = \frac{V_{\mathrm{d}} h_{\mathrm{d}} \left(3 - \frac{h_{\mathrm{d}}^2}{h^2}\right)}{4h} \tag{1}$$

$$F_{\mathrm{t,M,d}} = \frac{0.008 M_{\mathrm{d}}}{h_{\mathrm{r}}} \tag{2}$$

where $F_{\mathrm{t,V,d}}$ and $F_{\mathrm{t,M,d}}$ are tensional forces perpendicular to grain caused by the shear force and the bending moment respectively, V_{d} is the value of the shear force, h_{d} is the height of the opening, h is the height of the beam member, M_{d} is the value of the bending moment and h_{r} is the distance between the hypothetical crack and the edge of the beam. Notably, timber strength that the force above is compared to, is modified by the volume factor $k_{\mathrm{t,90}}$.

Drilling holes is a subject of the AITC Guidelines [5]. Similarly, it recommends calculating cross-sections with openings as weakened, but adapts a procedure proposed by Aicher and Höfflin [6] to evaluate stress concentrations, given by the formula

$$f_{\mathrm{t,\perp,u}} = 3V \frac{1.23 + \frac{0.82h}{d}}{2bd} + \frac{0.6M}{bd^2} \cdot \frac{h}{d} \tag{3}$$

where $f_{\mathrm{t,\perp,u}}$ is the tensile stress perpendicular to grain, V is the shear force, b, d are the beam width and height respectively, h is the diameter of a hole and M is the bending moment. Equation (3) is compared with the value of the design shear strength of a beam multiplied by 1.6 and modified by the following factors: the shear reduction C_{vr}, the load duration C_{d}, the temperature C_{t}, the wet-use C_{M} and the volume factor for stress concentrations at holes C_{vol}.

Both of the approaches mentioned above include an important volume factor ($k_{\mathrm{t,90}}$ [4], C_{vol} [5] – relying on a Weibull weak link serial system). Load capacity as well as stress magnitude are both dependent on the size of an element.

Moreover, both standards [4, 5] show simplified rules of proper location of openings. Additionally, AITC presents critical zones in the beam (due to the moment, shear and bearing), where holes have to be approved by an engineer or an architect qualified in timber design.

The German standard recommends that openings in zones subjected to shear and perpendicular to grain tensile stresses should be avoided. The rule also applies to load-bearing structural elements in the 3rd service class. If that is impossible, reinforcement should be applied.

Another tool, helpful for structural engineers, the Swedish Limtrahändbok [7] recommends rules presented in the German Annex [4] with an addition of a significant note, which states that not only the tensile stresses should be considered on the edges of openings, but also the shear stresses. The fact

IOP Conf. Series: Materials Science and Engineering 1015 (2021) 012026 doi:10.1088/1757-899X/1015/1/012026

can be evaluated with the method provided by Blaß and Bejtka [8] and should be checked especially for the rectangular holes.

Another approach, temporarily included in some drafts of Eurocode 5, uses an assumed similarity of the mechanical problem of holes with end-notched beams in a shape currently included in the European standard [3]. Nevertheless, it has been evaluated by Aichner and Höfflin [6] as convincing for rectangular holes and doubtful for the round ones. Moreover, Aichner and Höfflin [9] emphasize the importance of considering timber as an anisotropic material dependent on Weibull theory rules and show that using shape, size and moment to shear force ratio factors is appropriate only with an additional calibration factor.

The surveys of the experimental research results in the improvement of calculation methods, provided by Danielsson [10] and Jeleč, Varevac and Zovkić [11]. The works show that the simplification consisting in considering timber as homogeneous and linear elastic material and abandoning the Weibull theory often leads to the insufficient or inaccurate results. One can find more conclusions useful for designing of openings: the influence of a slope of fibers is important, there are advantages of placing an opening near the neutral axis rather than eccentrically, affection of material strength of the homogeneous or combined glulam beams is negligible. The last two cases have been also brought up by Aichner and Tapia [12].

Recently, two design procedures have been developed. Both are based on the German National Annex approach [4]. The first is mentioned above [12] and the other is proposed by Danzer, Dietsch and Winter [13], highlighting a possible eccentricity of holes and considering placing them in groups. Both have been compared and evaluated by Aichner and Tapia [15] and may be considered as a valuable, less conservative, closer to experimental results alternatively to the official regulation.

It has to be emphasized that the location of openings in the apex zone, which is the area where tensile stress perpendicular to grain is already present, is not a subject of any design recommendations. Eurocode 5 [3] provides a method for the evaluation of this stress as a result of a bending moment, given by the formula

$$\sigma_{t,90,d} = \frac{k_p 6 M_{ap,d}}{b h_{ap}^2} \tag{4}$$

where $M_{ap,d}$ is the bending moment in the apex zone, h_{ap} and b are the height and the width of a beam in the apex zone, respectively. The factor k_p rises together with an increase of the angle of the upper edge of the beam. In this case the timber design strength is modified by the factor dependent on the volume of the zone.

The area subjected to the concentration of tensional stress perpendicular to grain may be strengthened by additional structural elements in a way that ensures bearing all of the crucial actions. The reinforcement is assumed to work in a fully cracked cross-section. According to DIN-EN [4], the reinforcement can be made using wood panels glued to lateral surfaces of a beam, steel rods glued into a timber member or self-tapping screws. These rules have been adapted by producers of various fasteners and other building materials and included in the European Technical Approvals, becoming a basic source of the calculation procedures for structural engineers.

A number of literature positions offer more possible reinforcement materials. One can find numerical and experimental analysis prepared to help with the designing process and to show the efficiency of self-tapping screws with consideration of holes' eccentricity [15] or a comparison of plywood and glass fiber reinforcements [16], where a failure mode is less brittle. Especially useful for engineers are the comparisons between the research results and the standard equations [17] as well as an introduction to some promising new approaches and calculation methods [18].

Undoubtedly, one recommendation from the German National Annex [4] is very significant in the case of openings in the apex zone. According to it openings in the areas under shear and tensile perpendicular to grain stresses should be avoided and, if that is impossible, vulnerable places have to be reinforced. Moreover, the studies in which the case has already been researched are not numerous

[19]. In that cases the numerical calculations, including the finite element method, may be the right aid.

4. Finite element modelling

In order to assess quantitatively and qualitatively the effect of openings and reinforcement on the stress distributions in double tapered beams, the girder shown in figure 3 is modelled with the use of the finite element method [20]. The plane stress state within the linear theory of elasticity is adopted.

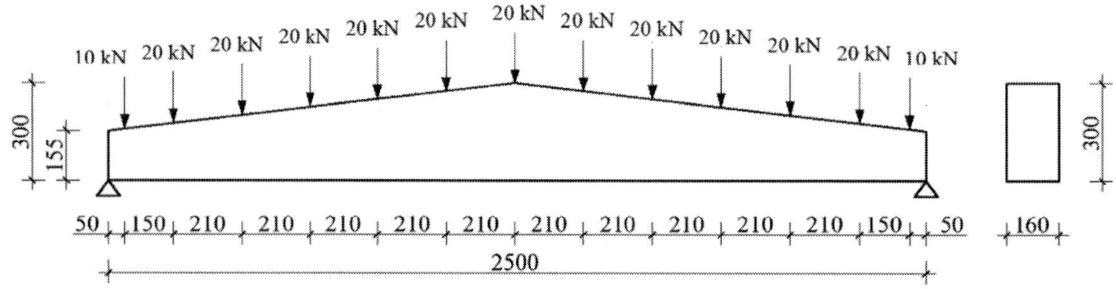

Figure 3. Double tapered beam: geometry, boundary conditions and loads.

The actual (design) load values are assumed for a beam with a span of 25 m and height above the supports equal to 1.55 m. Two heights in the apex zone are analysed: 2 m (called beam 1) and 3 m (called beam 2), with angles of inclination of the upper edges 2.1°, 6.6°, respectively. Moreover, three locations of single holes with a diameter of 30 cm each, positioned in the middle section of the beam (figure 4), and reinforcement with four bars in three different configurations (figure 5) are taken into account.

The height of 3 m in the apex zone is proposed for comparative purposes, although due to the production technology it is not found in real engineering. The height in the apex zone is important as a dimension relative to the diameter of the openings. It also generates a greater slope of the upper edge, which is important from the point of view of the standard calculation of delaminating stresses.

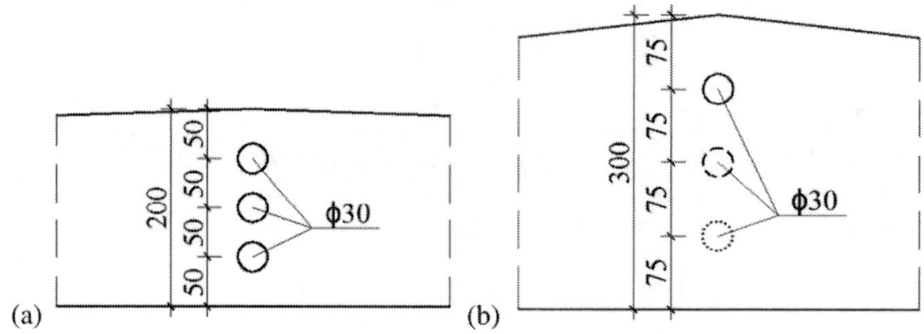

Figure 4. Placement of holes: (a) beam 1, (b) beam 2.

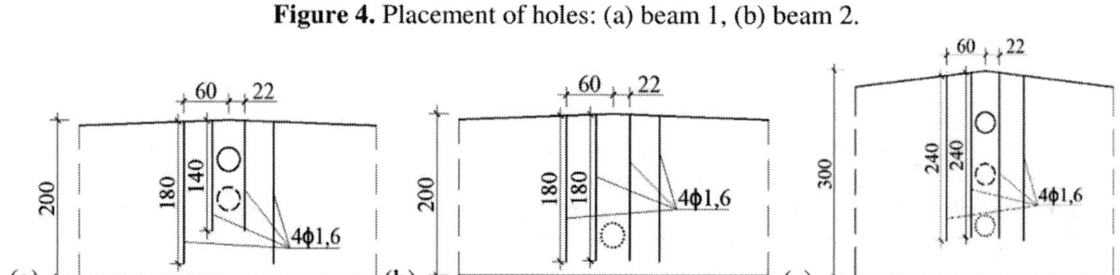

Figure 5. Placement of reinforcement: (a) beam 1: upper and middle hole; (b) beam 1: lower hole; (c) beam 2: all hole cases.

The material properties of the orthotropy with the parameters given in the European standard [1] for GL28h class wood are considered, which means E_1=12600 MPa, E_2=300 MPa, ν_{12}=0.35, G_{12}=650 MPa. The calculations are performed using the finite element method [20] in the Abaqus environment [21]. Two-dimensional eight-node finite elements with square shape functions (CPS8R) are used. In the adopted finite elements the so-called reduced integration is assumed. After analysing the convergence of the solution, a division grid with a mesh size of approx. 0.1 m was selected. A broad parametric analysis of beams without openings can be found in [22-24]. The selected results are presented in figures 6-11 and in tables 1 and 2.

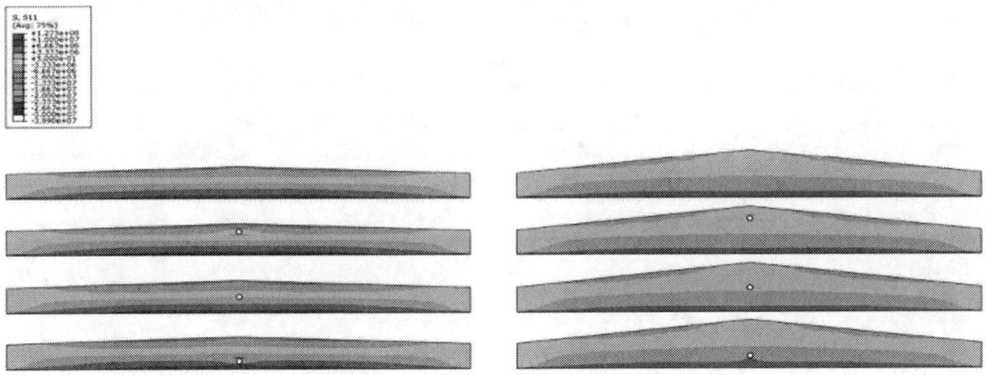

Figure 6. Normal stresses along the wood grain without reinforcement.

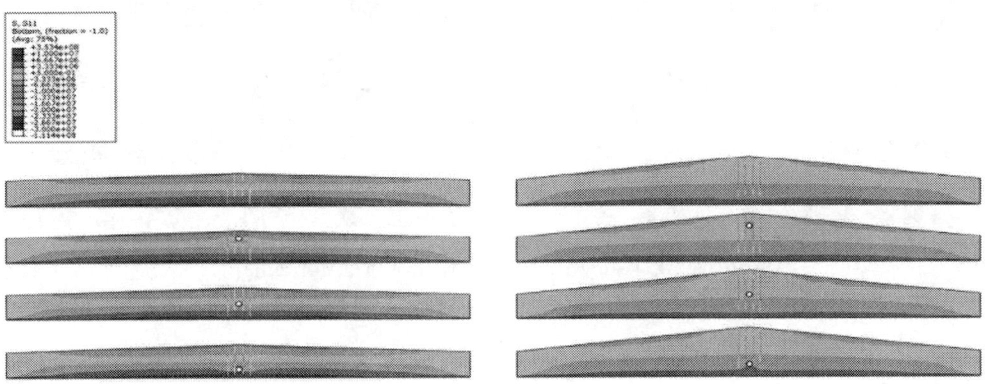

Figure 7. Normal stresses along the wood grain with reinforcement.

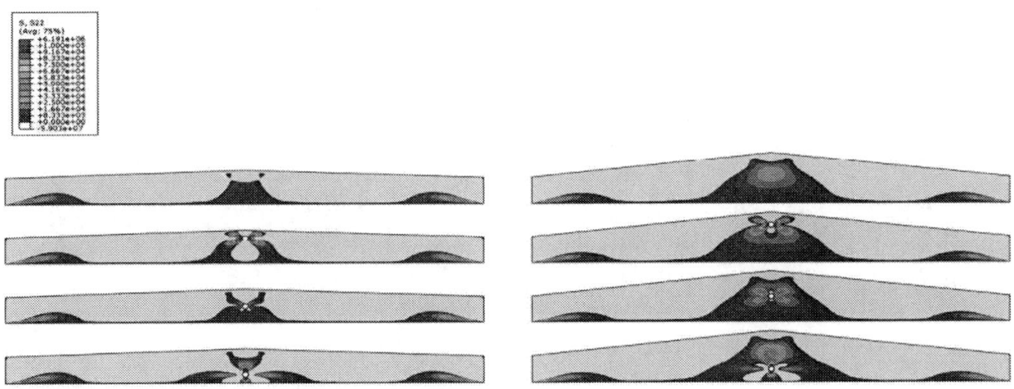

Figure 8. Delamination stresses without reinforcement.

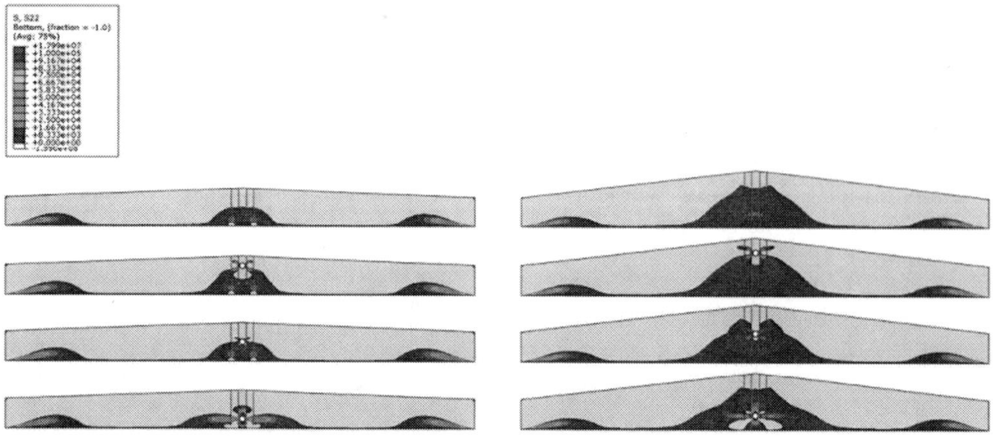

Figure 9. Delamination stresses with reinforcement.

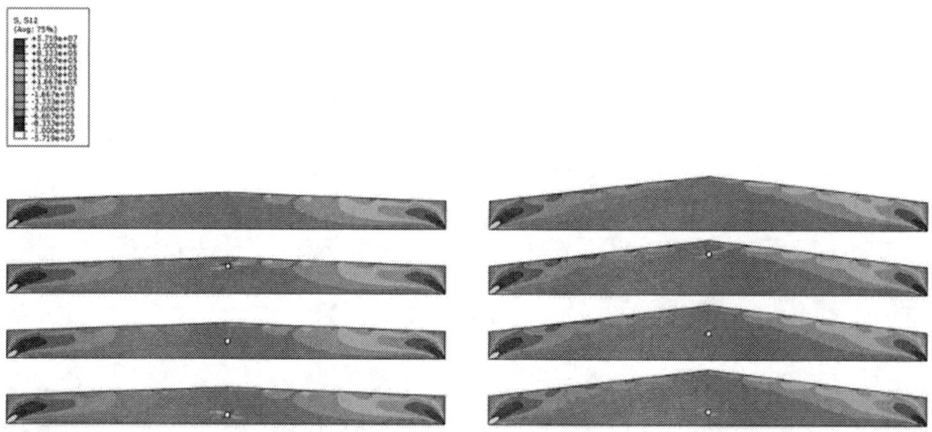

Figure 10. Tangential stress without reinforcement.

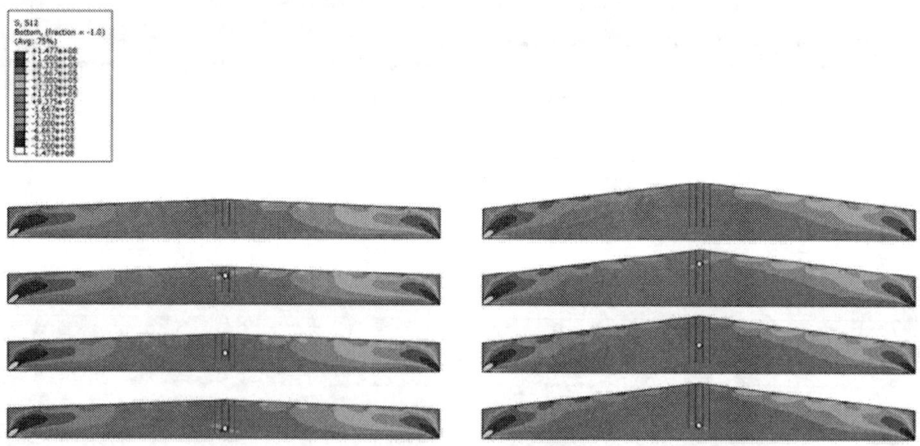

Figure 11. Tangential stresses with reinforcement.

IOP Conf. Series: Materials Science and Engineering 1015 (2021) 012026 doi:10.1088/1757-899X/1015/1/012026

Table 1. Selected stresses [MPa] – beam 1.

	No holes		Upper hole		Middle hole		Lower hole	
	Reinf.	No reinf.	Reinf.	No reinf.	Reinf.	No reinf.	Reinf.	No reinf.
$S_{22-max}^{positive}$	0.006	0.041	0.708	0.434	0.142	0.102	0.807	0.700
S_{11-max}	8.427	8.395	8.615	8.555	8.429	8.389	12.279	10.108
S_{12-max}	0.827	0.827	0.827	0.827	0.827	0.827	0.827	0.827

Table 2. Selected stresses [MPa] – beam 2.

	No holes		Upper hole		Middle hole		Lower hole	
	Reinf.	No reinf.	Reinf.	No reinf.	Reinf.	No reinf.	Reinf.	No reinf.
$S_{22-max}^{positive}$	0.030	0.019	0.423	0.240	0.168	0.094	0.384	0.319
S_{11-max}	4.544	4.501	4.602	4.552	4.550	4.500	7.206	6.330
S_{12-max}	0.760	0.760	0.760	0.760	0.760	0.760	0.760	0.760

The distribution of normal stresses along the wood grain depends to a small extent on the presence of openings at the top and in the middle (see table 1 and table 2) of the beam. Only a slight disturbance in the vicinity of the holes is observed. Reinforcement minimally reduces the values of these stresses. It is different when the hole is at the bottom. In this case, normal stress along the fibres increases by a little over 31-37% depending on the height at the ridge. The introduction of reinforcement reduces this increase by 17-12%. Local concentrations of negative stresses occur near the openings in the upper part of the beam. In the beam with a lower angle of inclination of the upper edge (low ridge) the stress value of -2.662 MPa can be reduced to -10.194 MPa by the application of reinforcement. In the beam with a greater angle of inclination of the upper edge (high ridge) the corresponding values are -8.177 MPa and -6.952 MPa.

The distribution of the maximum tangential stresses do not depend on the presence of holes in principle (see table 1 and table 2). In the beams with openings slight disturbances of tangential stresses are observed in the area of openings placed at the top and at the bottom. These disturbances are more visible in low apex beams.

The presence of openings in double trapezoidal beams significantly changes the distribution of delaminating stresses (figure 8 and figure 9). Beams with a low ridge, in which these stresses are trace, should be analysed slightly differently. More attention should be paid to high ridge beams, where these stresses are much greater and dangerous. The location of the holes essentially affects the distribution and values of delaminating stresses. The characteristics of quantitative changes can be presented as follows:

1) in the beams with a low ridge (table 1):
   - opening at the top: stresses increase 118 times, reinforcement reduces them by 38%,
   - center hole: stresses increase 24 times, reinforcement reduces them by 28%,
   - bottom opening: stresses increase 135 times, reinforcement reduces them by 13%;
2) in the beams with a high ridge (table 2):
   - opening at the top: stresses increase 141 times, reinforcement reduces them by 43.26%,
   - center hole: stresses increase by 5.6 times, reinforcement reduces them by 44.05%,
   - bottom opening : stresses increase 13 times, reinforcement reduces them by 16.93%.

5. Conclusions

The analysis presented in the present paper concerns a design problem that occurs in engineering practice. The possibility of numerical analysis allows for the discussion of hypothetical situations that may arise in the design.

The following aspects of design and numerical modelling are considered:

- assessment of the practically possible diameter of the holes,
- possibility of placing openings in other places of the beam,
- analysis of the reinforcement arrangement in the vicinity of the holes, including the diameter and length of the reinforcing bars and the spacing between them,
- possibility and advisability of using reinforcing rings around holes.

According to the authors, conscious design requires calculations performed using the finite element method. As stress concentrations occur in the beams with openings, it is recommended to use recognized calculation programs and much caution in interpreting the results, both qualitatively and quantitatively. The analysis presented in the present paper concerns three possible positions of the opening in the apex zone. It is recommended to position the opening halfway up, as it causes the least stress disturbance in the beams. Bottom position is definitely not recommended, as stress disturbances are large and cannot be leveled with transverse reinforcement. Reinforcement allows for the reduction of stresses in the range from several to several dozen percent.

References

[1] Standard EN 14080:2013-07 - Timber structures - Glued laminated timber and glued solid timber – Requirements

[2] Standard EN 14081-1+A1:2019-11 - Timber structures -- Strength graded structural timber with rectangular cross section -- Part 1: General requirements

[3] Standard EN 1995-1-1:2010 - Eurocode 5: Design of timber structures - Part 1-1: General – 1 Common rules and rules for buildings

[4] Standard DIN-EN 1995-1-1/NA:2010-12: Nationaler Anhang- National festgelegte Parameter – Eurocode 5: Bemessung und Konstruktion von Holzbauten – Teil 1-1: Allgemeines – Allgemeine Regeln und Regeln für den Hochbau

[5] July 2012, AITC Technical Note 19: Guidelines for evaluation of drilled holes and notches in structural glued laminated timber beams *American Institute of Timber Construction.*

[6] Aicher S and Hofflin L 2004 New Design Model for Round Holes in Glulam Beams *Proceedings of the 8th World Conference on Timber Engineering (WCTE 2004)* **1** pp 67-72

[7] Crocetti R, Danielsson H, et al 2016 *Limträhandbok. Del 2.Projektering av limträkonstruktioner*

[8] Blaß H J and Bejtka I 2003 *Querzugverstärkungen in gefährdeten Bereichen mit selbstbohrenden Holzschraube, Forschungsbericht, Versuchanstalt für Stahl, Holz und Steine*

[9] Aicher S and Hofflin L 2008 Fracture behaviour and design of glulam beams with round holes, *10th World Conference on Timber Engineering 2008* **1** pp 132-9

[10] Danielsson H 2009 *The strenght of glulam beams wigh holes. A survey of tests and calculation methods*

[11] Jeleč M, Varevac D and Zovkić J 2014 *Glulam beams with holes*

[12] Aicher S and Tapia C 2017 Improved design equations for the resultant tensile forces in glulam beams with holes *International Network on Timber Engineering Research*

[13] Danzer M, Dietsch P and Winter S 2017 Round holes in glulam beams arranged eccentrically or in groups *International Network on Timber Engineering Research*

[14] Aicher S and Tapia C 2019 Evaluation of design concepts for holes in glulam beams – comparison with test results *Otto-Graf-Journal* **18** pp 329-42

[15] Dröscher J, Augustin M andSchickhofer G 2016 Comparison of verification and reinforcement concepts for timer beams with large round holes *World Conf. on Timber Eng.* pp 4497-506

[16] Danzer M, Dietsch P and Winter S 2016 Reinforcement of round holes in glulam beams arranged eccentrically or in group *World Conf. on Timber Eng.* pp 2772-80.
[17] Uthman R and Othman R 2009 *Holes in glulam beams – possible methods of reinforcements*
[18] Aicher S 2012 Glulam beams with internally and externally reinforced holes – tests, detailing and design, *CIB Working Commission W18*
[19] Al. Sabouni-Zawadzka A, Brodniewicz P, Gilewski W and Pełczyński J 2020 Projektowanie drewnianych belek dwutrapezowych z otworami w strefie kalenicy, *Inżynieria i budownictwo* **6** pp 293-95
[20] Zienkiewicz O C and Taylor R L 2000 *The finite element method. Vol. 1. The basis* (New Jersey, USA: Butterworth-Heinemann)
[21] ABAQUS Documentation 2017
[22] Al Sabouni-Zawadzka A, Gilewski W and Pełczyński J 2017 Some considerations on perpendicular to grain stress in double-tapered glulam beams *Proc. of XXVI R-S-P Seminar 2017, Theoretical Foundations of Civil Engineering*
[23] Al Sabouni-Zawadzka A, Gilewski W and Pełczyński J 2018 Perpendicular to grain stress concentrations in glulam beams of irregular shape – finite element modelling in the context of standard design *Int. Wood Products J.* **9** (4) pp 157-63
[24] Gilewski W and Pełczyński J 2017 The influence of orthotropy level for perpendicular to grain stresses in glulam double tapered beams, *23rd Int. Conf. Engineering Mechanics 2017*

IOP Conf. Series: Materials Science and Engineering 1015 (2021) 012027 doi:10.1088/1757-899X/1015/1/012027

Derivation of statistical parameters for flexure and shear resistance of reinforced concrete beams

Z Abdullah and M Lutomirska

The Faculty of Civil Engineering, Warsaw University of Technology,
Al. Armii Ludowej 16, 00-637 Warsaw, Poland

Email address: m.lutomirska@il.pw.edu.pl

Abstract. The objective of this paper is a study on factors influencing the derivation of statistical parameters of flexural resistance and shear resistance for reinforced concrete beams. A variety of RC beams under bending and shear forces are designed and analyzed, taking into consideration different cross-sectional sizes, compressive concrete strengths, reinforcement ratios, and statistical distributions. The latest version of the American Building Code ACI-318-19 and ASCE/SEI 7-16 are used. The coefficient of variation (V) and bias factor (λ) for the selected models of reinforced concrete beams in flexure and shear are determined using the Monte Carlo simulation technique.

1. Introduction

Reinforced concrete beams can be designed with vast range of properties. The statistical parameters of resistance available in the literature cover the basic design cases. The research conducted in this paper is supposed to extend the available database and identify factors having influence on the flexural and shear resistance parameters for reinforced concrete beams. The results could be dependent on sizes of cross-sections, concrete compressive strengths, reinforcement ratios, and assumed statistical distributions. Therefore, a variety of design cases are selected and analyzed. The nominal resistance for flexure and shear for RC beams are calculated based on the latest version of the American Building Code ACI-318-19. ASCE/SEI 7-16 Standards are used for the load and load combination factors. For the purpose of the generation of the statistical parameters of resistance, the coefficient of variation (V) and bias factor (λ), the Monte Carlo simulation technique is applied.

The currently available statistical parameters of resistance are developed from a normal distribution for all the parameters [1]. Since some of the researchers use a normal distribution for material strengths, while others suggest lognormal distribution [2]-[3], in this study the simulations of statistical parameters of resistance are performed for both types of distributions.

2. Statistical parameters of resistance

2.1. Resistance model

During the design process the resistance of any structural member "R" is considered as a deterministic value. In reality, the resistance "R" is a random variable due to natural and human causes. There are three sources of uncertainty in the load carrying capacity (resistance): material properties, fabrication, and analysis [4].

IOP Conf. Series: Materials Science and Engineering 1015 (2021) 012027 doi:10.1088/1757-899X/1015/1/012027

$$R = R_n \cdot M \cdot F \cdot P \quad (1)$$

where:

R_n – nominal (design) value of resistance,
M – materials factor,
F – fabrication factor,
P – professional factor.

The material factor (M) represents the variability of material properties in strength and modulus of elasticity [4]. The material factors for concrete compressive strength and steel yield strength are based on Nowak et al. 2005 and 2011 [1, 6]. The values of bias factors are between 1.31 and 1.08, while the values of the coefficient of variation are between 0.17 and 0.11, correspondingly for concretes from 10.7 MPa (3 ksi) to 82.7 MPa (12 ksi). Both bias factor and coefficient of variation decrease for higher concrete classes. The statistical parameters for yield strength of reinforcing steel vary with the bar diameter. For reinforcing steel 414 MPa (grade 60) the bias factor varies from 1.18 to 1.12 and the coefficient of variation varies from 0.04 to 0.02.

The statistical parameters for Fabrication factor (F) are based on the data from Ellingwood et al. 1980. The statistical parameters for dimensions of the reinforced concrete beam are listed in table 1. For steel reinforcement the bias factor is λ=1.0 and coefficient of variation V=0.01 for rebar diameter, and λ=1.0 and V=0.015 for total steel reinforcement area A_S.

Table 1. Fabrication factor for reinforced concrete beams.

Bar size	Bias factor λ	Coefficient of variation V
Width of beam, cast-in-place.	1.01	0.04
Effective depth of a reinforced concrete beam.	0.99	0.04

Professional (analysis) factor (P) points out the variation in the actual resistance ratio and what can be expected analytically using the strength of materials and exact dimensional values [4]. The values for professional factors were taken from Ellingwood *et al.* [2].

Table 2. Professional factor for reinforced concrete beams.

Bar size	Bias factor λ	Coefficient of variation V
Beam, flexure	1.020	0.06
Beam, shear	1.075	0.10

2.2. Selection of Design Cases

The design cases considered in the study are selected in such way to cover all the statistical data that was found for RC statically determinate beam components and be close to realistic values. First, three different dimensions for beam cross section are chosen. Secondly, for each cross section a different rebar diameter and stirrup spacing are selected. Thirdly, according to the design code limitations different flexural and shear reinforcement ratios are taken into consideration. For flexure it is maximum and minimum reinforcement area. For shear it is maximum, minimum and no shear reinforcement. A cross-section without shear reinforcement means that the concrete cross-section is carrying half of the shear. Then, for each model, thirteen different compressive strengths of concrete are used. In total 78 design cases for flexure and 117 design cases for shear are identified and analyzed to obtain resistance parameters. Moreover, for each case two types of distribution (normal and lognormal) for material strength are considered. For simulation of the dimensions a normal distribution is assumed [5]. All design cases are listed in table 3.

Table 3. Selected design cases.

Resistance	Beam	Dimensions mm (in)	Main bar diameter (ø)	Stirrup diameter (ø)	Reinforcement area	f_c' MPa (ksi)	f_y MPa (ksi)
flexure	I	508×406 (20×16)	19.05mm (#6)	9.52mm (#3)	maximum minimum	20.7 to 82.7 (3 to 12)	414 (60)
	II	813×508 (32×20)	25.4mm (#8)	9.52mm (#3)	maximum minimum		
	III	406×305 (16×12)	32.25mm (#10)	9.52mm (#3)	maximum minimum		
shear	I	508×406 (20×16)	19.05mm (#6)	9.52mm (#3)	maximum minimum no shear reinforcement	20.7 to 82.7 (3 to 12)	414 (60)
	II	813×508 (32×20)	25.4mm (#8)	9.52mm (#3)	maximum minimum no shear reinforcement		
	III	406×305 (16×12)	32.25mm (#10)	9.52mm (#3)	maximum minimum no shear reinforcement		

2.3. Flexural Resistance of RC Beams

To calculate the nominal flexural resistance for the reinforced concrete beam R_n the design formula (2) is used [10]. Based on the Monte Carlo simulation technique statistical parameters of resistance for all the selected design cases are determined. First, a normal distribution is used for all parameters such as A_s, f_y, d, b, and f_c'. Then, a lognormal distribution is used for material strength f_y and f_c', while other parameters remained normally distributed.

$$R_n = A_s \cdot f_y \cdot \left(d - \frac{a}{2}\right) \tag{2}$$

Where equivalent compression zone is:

$$a = \frac{A_s \cdot f_y}{0.85 \cdot f_c' \cdot b} \tag{3}$$

Statistical parameters for compressive strength of concrete are taken from Nowak *et al.* 2005 and 2011 [1, 6] based on cylindrical test data. The value of f_c' obtained from cylinder tests are reduced by 10% for calculating nominal moment resistance because the average in-place strength is lower than a standard cylinder test [1, 6]. The reinforcement ratios for all models are taken from minimum and maximum allowed by the ACI requirements. For computation of minimum reinforcement equations (4) and (5) are used. Values of maximum reinforcement are taken from allowable design phase. A reinforcement limitation is changed for different models by using a different material strength:

$$\rho_{\min} = \frac{0.25\sqrt{f_c'}}{f_y} \tag{4}$$

$$\rho_{\min} = \frac{1.4}{f_y} \tag{5}$$

2.4. *Shear Resistance of RC Beams*

Nominal shear resistance for the reinforced concrete beam (R_n) with shear reinforcement can be calculated by formula (6). Monte Carlo simulations are used to determine resistance of components for all selected design cases. First, a normal distribution for all components (A_V, f_y, f_c', S, b and d) is assumed. Then, a lognormal distribution is used to calculate strength of materials such as (f_y and f_c'), while other parameters remained normally distributed.

$$R_n = V_n = V_C + V_S \quad (6)$$

where, the shear resistance provided by concrete is calculated by following formula (7)

$$V_c = \frac{1}{6}\sqrt{f_C'} \cdot b \cdot d \quad (7)$$

and shear resistance provided by shear reinforcement (8)

$$V_S = \frac{A_V \cdot f_y \cdot d}{S} \quad (8)$$

Parameters for compressive strength of concrete are assumed from Nowak et al. 2005 [6]. Concrete compressive strengths f_c' are reduced 5% for calculating nominal shear resistance for the reinforced concrete beam [7]. Three design situations are taken into consideration for amount of shear reinforcement: maximum shear reinforcement, minimum shear reinforcement, and no shear reinforcement (the cross-section carries half of shear). Based on the ACI requirement for stirrup spacing, three equations (9), (10), and (11) are used to calculate maximum and minimum shear reinforcement area.

Maximum shear reinforcement

$$A_{V_{max}} = \frac{4}{6} \cdot \sqrt{f_C'} \cdot \frac{b}{f_y} \cdot d \cdot S \quad (9)$$

$$A_{Vmin} = 0.062 \cdot \sqrt{f_C'} \cdot \frac{b \cdot S}{f_y} \quad (10)$$

But not less than

$$A_{Vmin} = 0.35 \cdot \frac{b \cdot S}{f_y} \quad (11)$$

The stirrup spacing is assumed to vary between 200 mm to 305 mm (6 in to 12 in), according to beam dimension and materials. For the cross-sectional dimension, compressive strength and yield strength components are taken to be the same as flexural resistance.

3. Results of statistical parameters

3.1. *Monte Carlo Simulations*

Based on Monte Carlo simulations, the resistance of reinforced concrete beams for flexure and shear are computed. In order to derive statistical parameters of flexural resistance, one thousand simulations are performed for each of the selected design cases.

3.2. *Results for flexure resistance parameters (beam I, II and III)*

The variability of the results for flexural resistance can be observed in figure 1 to figure 8 plotted below. The statistical parameters for flexural resistance are gathered in table 4 and are presented in accordance to reinforcement ratios.

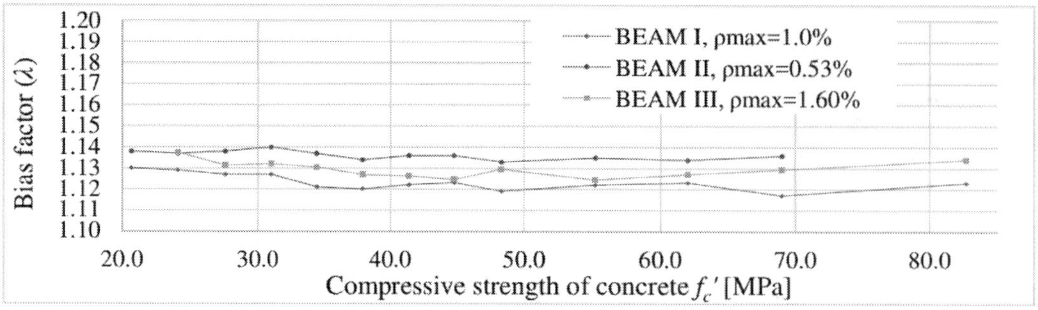

Figure 1. Bias factor vs. material strength for flexure, normal distribution.

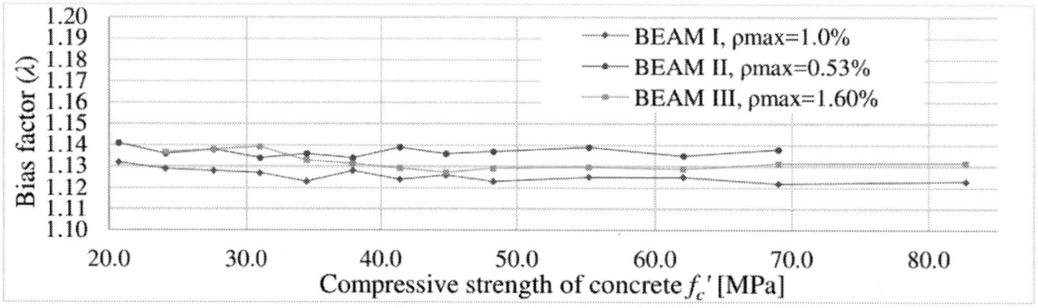

Figure 2. Bias factor vs. material strength for flexure, lognormal distribution.

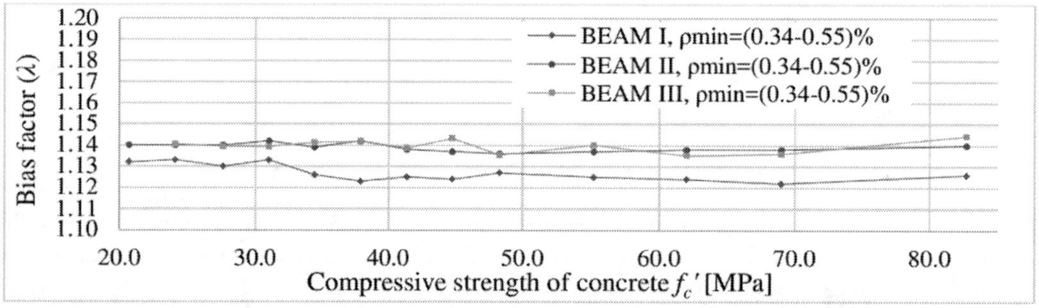

Figure 3. Bias factor vs. material strength for flexure, normal distribution.

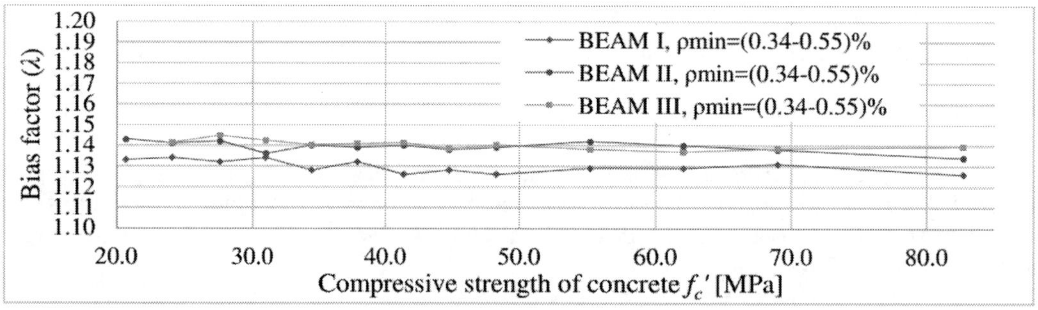

Figure 4. Bias factor vs. material strength for flexure, lognormal distribution.

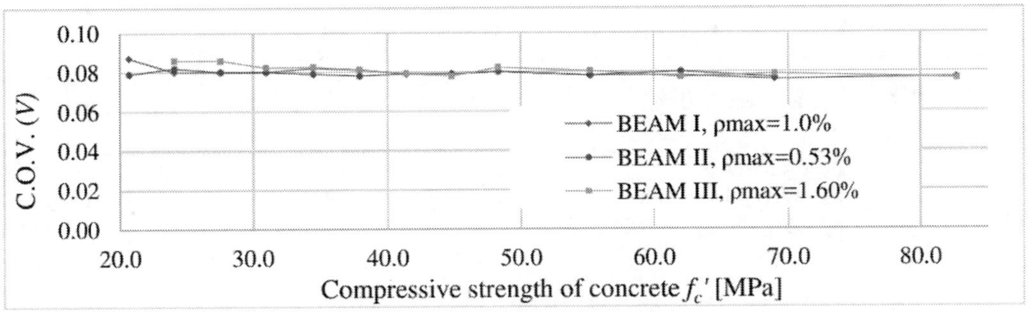

Figure 5. C.O.V. vs. material strength for flexure, normal distribution.

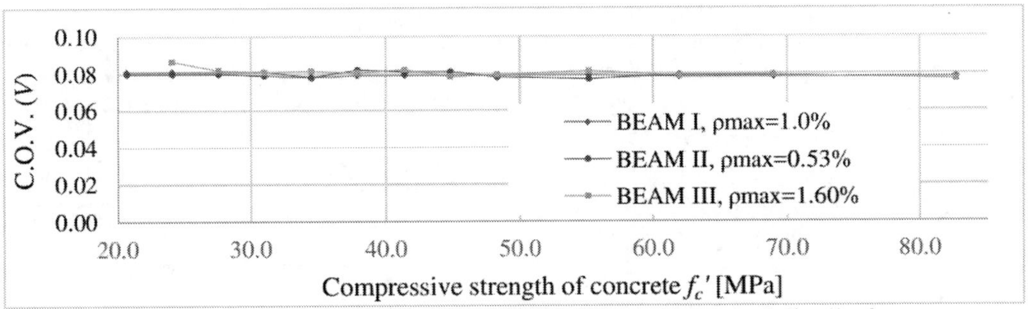

Figure 6. C.O.V. vs. material strength for flexure, lognormal distribution.

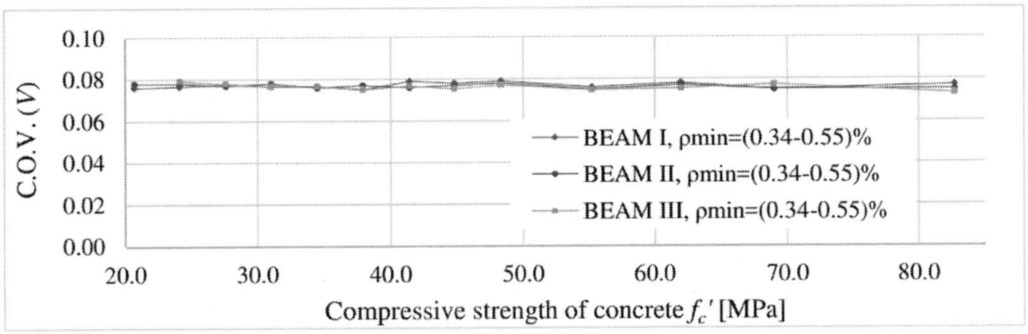

Figure 7. C.O.V. vs. material strength for flexure, normal distribution.

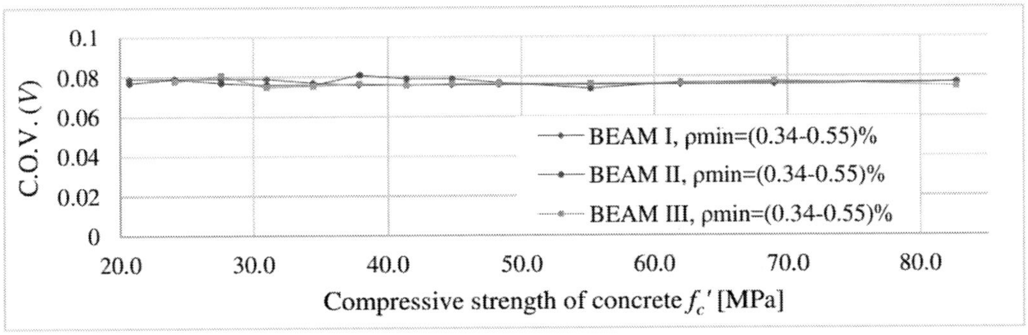

Figure 8. C.O.V. vs material strength for flexure, lognormal distribution.

Table 4. Statistical parameters for flexural resistance.

Flexural Resistance		Normal Distribution		Strength Lognormal Distribution	
		λ	V	λ	V
beam I	ρ_{max} =1.0%	1.123	0.080	1.126	0.079
	ρ_{min}= (0.34-0.55) %	1.127	0.077	1.130	0.077
beam II	ρ_{max} =0.53%	1.136	0.079	1.137	0.080
	ρ_{min}= (0.34-0.55) %	1.139	0.077	1.139	0.077
beam III	ρ_{max} =1.6%	1.130	0.081	1.132	0.081
	ρ_{min}= (0.34-0.55) %	1.140	0.076	1.140	0.077

3.3. Results for shear (beam I, II and III)

The variability of the results for shear resistance can be observed in figure 9 to figure 12 plotted below. The statistical parameters for shear resistance are presented in table 5 according to concrete compressive strength and shear reinforcement due to their straight impact on shear resistance [9].

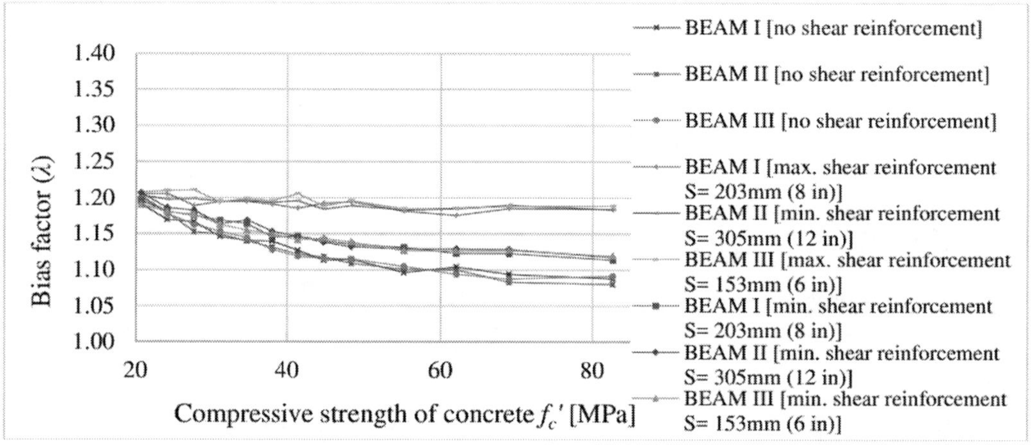

Figure 9. Bias factor vs. material strength for shear resistance, normal distribution.

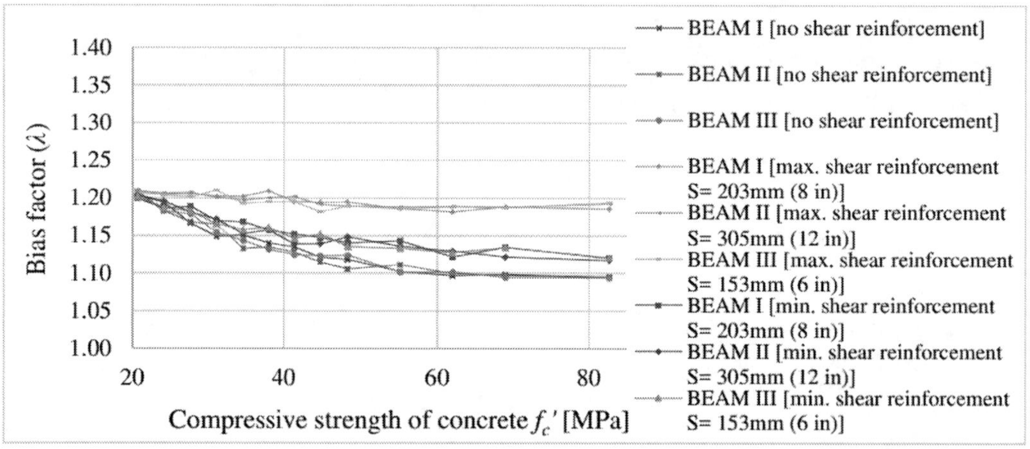

Figure 10. Bias factor vs. material strength for shear resistance, lognormal distribution.

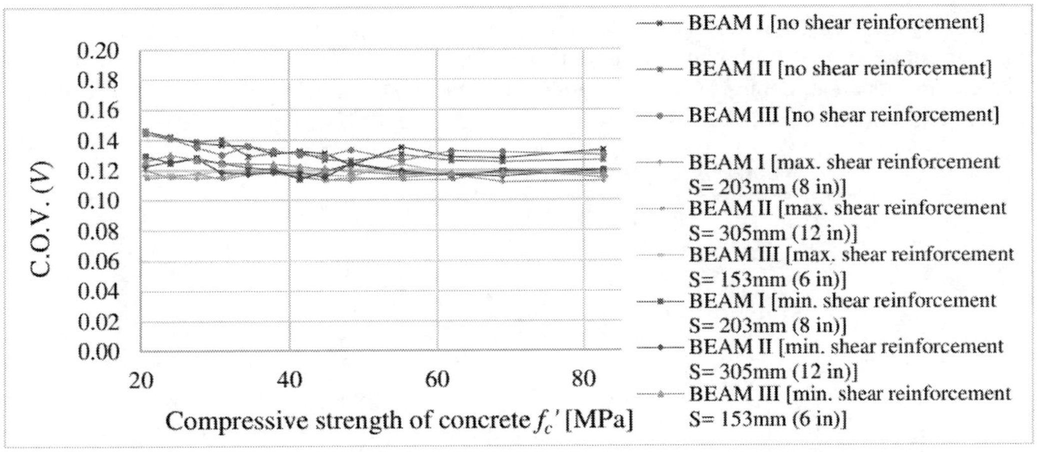

Figure 11. C.O.V. vs. material strength for shear resistance, normal distribution.

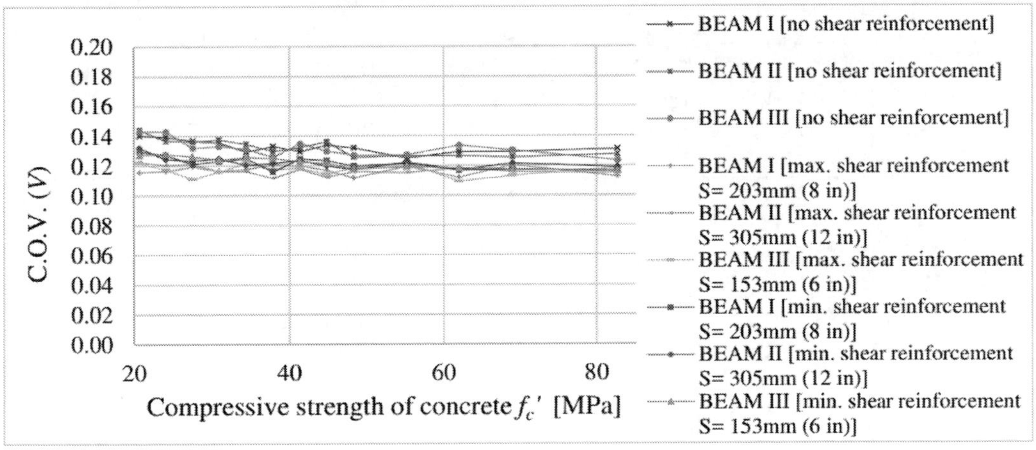

Figure 12. C.O.V. vs. material strength for shear resistance, lognormal distribution.

4. Analysis of results

It is recognized that flexural resistance is carried out by reinforcement [8]. Therefore, the reinforcement ratio can control statistical parameters for resistance more than other components. The flexural resistance parameters were calculated for the maximum and minimum reinforcement ratio. From table 4, it was observed that the bias factor (λ) decreases for higher reinforcement ratios. The change in bias factor is not significant, about 1% for beam I and beam III, while for beam II the change was lower than 1%. It is because the maximum reinforcement ratio obtained from the design phase is close to the minimum reinforcement ratio provided by ACI, or sometimes less than the minimum. Figure 5 and figure 6 show a slightly higher coefficient of variation (V) for higher reinforcement ratios for beams I, II and III. However, the differences in the coefficients of variation are very small and they can be assumed to be equal (table 4).

The shear reinforcement ratio can have a large impact on the statistical parameters for shear resistance. The statistical parameters that are calculated for maximum shear reinforcement are significantly different from those without shear reinforcement (up to 10% in high strength concrete). Figure 9 and figure 10 show a higher value of bias factor for maximum shear reinforcement. The coefficient of variation decreases for maximum shear reinforcement (figure 11 and figure 12). Another observation is that higher compressive strength of concrete results in decreased statistical parameters of shear resistance, bias and coefficient of variation (table 5). It was proven that statistical parameters for shear resistance can be affected by the shear reinforcement ratio more than compressive strength of concrete, when a higher shear reinforcement ratio is used. But for minimum and no shear reinforcement concrete compressive strength can control statistical parameters rather than shear reinforcement ratio. This fact was also proven by Nowak et al. (2011) [1].

Table 5. Statistical parameters for flexural resistance

f_c'	Normal distribution		Strength lognormal distr.	
	λ	V	λ	V
Maximum shear reinforcement				
20.7 MPa (3 ksi)	1.205	0.117	1.208	0.120
24.1 MPa (3.5 ksi)	1.204	0.116	1.205	0.118
27.6 MPa (4 ksi)	1.199	0.117	1.205	0.117
31.0 MPa (4.5 ksi)	1.195	0.117	1.204	0.116
34.5 MPa (5 ksi)	1.197	0.118	1.197	0.118
37.9 MPa (5.5 ksi)	1.193	0.119	1.201	0.117
41.4 MPa (6 ksi)	1.196	0.117	1.197	0.119
44.8 MPa (6.5 ksi)	1.189	0.114	1.189	0.114
48.3 MPa (7 ksi)	1.193	0.115	1.191	0.115
55.2 MPa (8 ksi)	1.183	0.118	1.187	0.119
62.0 MPa (9 ksi)	1.182	0.116	1.186	0.113
69.0 MPa (10 ksi)	1.188	0.117	1.188	0.116
82.7 MPa (12 ksi)	1.186	0.115	1.191	0.115
Minimum shear reinforcement				
20.7 MPa (3 ksi)	1.198	0.126	1.201	0.130
24.1 MPa (3.5 ksi)	1.182	0.127	1.190	0.126
27.6 MPa (4 ksi)	1.178	0.128	1.184	0.124
31.0 MPa (4.5 ksi)	1.165	0.123	1.169	0.124
34.5 MPa (5 ksi)	1.162	0.121	1.160	0.124
37.9 MPa (5.5 ksi)	1.149	0.121	1.159	0.121
41.4 MPa (6 ksi)	1.144	0.118	1.146	0.123
44.8 MPa (6.5 ksi)	1.141	0.119	1.146	0.122
48.3 MPa (7 ksi)	1.135	0.123	1.141	0.118
55.2 MPa (8 ksi)	1.129	0.118	1.138	0.121
62.0 MPa (9 ksi)	1.126	0.117	1.126	0.117
69.0 MPa (10 ksi)	1.126	0.118	1.130	0.119
82.7 MPa (12 ksi)	1.117	0.119	1.120	0.117
No shear reinforcement				
20.7 MPa (3 ksi)	1.198	0.146	1.206	0.142
24.1 MPa (3.5 ksi)	1.175	0.141	1.186	0.140
27.6 MPa (4 ksi)	1.160	0.138	1.171	0.135
31.0 MPa (4.5 ksi)	1.149	0.136	1.156	0.135
34.5 MPa (5 ksi)	1.141	0.134	1.142	0.132
37.9 MPa (5.5 ksi)	1.133	0.132	1.136	0.129
41.4 MPa (6 ksi)	1.122	0.132	1.129	0.132
44.8 MPa (6.5 ksi)	1.115	0.129	1.120	0.133
48.3 MPa (7 ksi)	1.113	0.128	1.116	0.128
55.2 MPa (8 ksi)	1.101	0.131	1.105	0.126
62.0 MPa (9 ksi)	1.099	0.129	1.099	0.130
69.0 MPa (10 ksi)	1.088	0.129	1.096	0.128
82.7 MPa (12 ksi)	1.086	0.130	1.095	0.127

By comparison of results for different beam sizes (figure 1 to 12), it can be concluded that changes in the beam cross-section did not have significant influence on statistical parameters for flexure and shear resistance. This confirms the assumption by Ellingwood et al. 1980 [2], and Nowak et al. 2011 [1]. It was also observed that the type of the assumed distribution of the material strength does not have significant influence on the results.

More detailed explanation of the results can be found in Zheer 2020 [11].

5. Conclusions

The objective of this paper was to extend the available database and identify factors influencing the statistical parameters of resistance (bias factor and coefficient of variation) for rectangular reinforced concrete beams under flexure and shear. The study was performed for a variety of design cases with different cross sections, concrete strengths, reinforcement ratios, and types of statistical distribution for material strength. ACI 318-19 Code and ASCE/SEI 7-16 Standards were used. The simulation of statistical parameters was based on the Monte Carlo technique. It is concluded that reinforcement ratio and bar diameter causes a small effect on the statistical parameters for flexural resistance, and they affect mostly the bias factor. The statistical parameters for shear resistance are governed by the shear reinforcement ratio and concrete compressive strength. Generally, the coefficient of variation of shear resistance is larger than the coefficient of variation of flexural resistance, because shear failure is much harder to predict. It was found that the element size and probability distribution type of the material strength does not significantly influence the results.

References

[1] Nowak A S, Rakoczy A M and Szeliga E K 2011 Revised statistical resistance models for R/C structural components *American concrete institute* **SP-284-6** pp 1-16

[2] Ellingwood B, Galambos T V, MacGregor J G, Cornell C A 1980 Development of a probability based load criterion for american national standard A58 *NBS Special Publication* p 577

[3] Sakai T, Nakajima M, Tokaji K and Hasegawa N H 1997 Statistical distribution patterns in mechanical and fatigue properties of metallic materials *Materials science research international* **3** pp 63-74

[4] Nowak A S and Collins K R 2013 *Reliability of Structures* (CRC Press Taylor and Francis Group Boca Raton)

[5] Mirza S A and MacGregor J G 1979 Variations in dimensions of reinforced concrete members *Journal of The Structural Division* **ST4** pp 751-66

[6] Nowak A S, Szerszeń M M, Szeliga E K, Szwed A and Podhorecki P J 2005 Reliability-based calibration for structural concrete *Portland Cement Association and Precast/Prestressed Concrete Institute*

[7] Bartlett Ml F and MacGregor 1996 Statistical analysis of the compressive strength of concrete in structures *ACI Materials Journal* **93** pp 158-68

[8] Wright J K and MacGregor J G 2012 *Reinforced concrete Mechanics and design* New Jersey Pearson Prentice and Hall Upper Saddle River

[9] Mirza S A and MacGregor J G 1979 Statistical study of shear strength of reinforced concrete slender beams *ACI Journal* **76** pp 1159-78

[10] ACI 318-19, Building Code Requirements for Structural Concrete and Commentary 2019 *American Concrete Institute* Farmington Hills

[11] Abdullah Z M 2020 Reliability model for flexural and shear resisatnce of renfocred concrete beams *Warsaw University of Technology* Warsaw

IOP Conf. Series: Materials Science and Engineering 1015 (2021) 012028 doi:10.1088/1757-899X/1015/1/012028

Investigation of shear connection between new and old concrete layers

P Koteš and M Vavruš

University of Žilina, Faculty of Civil Engineering, Univerzitná 8215/1 10 26 Zilina, Slovakia

E-mail address: peter.kotes@uniza.sk

Abstract. The structure and all its elements must be load-bearing, safe and serviceable. Due to time and external factors such as wind, thawing and freezing cycles, seismic activity, the structure of the concrete is disturbed and this leads to a reduction in the load-bearing capacity of the element. For vertical elements (columns), this usually means that the member´ strengthening must be used. The most common way of strengthening a concrete column is to use a new reinforced layer of concrete. In this case, great emphasis is placed on the contact between the original and the new layer of concrete. This article deals with two ways of surface treatment, such as notches and indents using different types of strengthening.

1. Introduction

Due to the aggressiveness of the environment, the materials itself degrades over time in concrete structures and also reduces the load-bearing capacity. We can extend the life-time of the structure by regular inspections and maintenance. Early detection of the causes of structural damage and subsequent proper selection of strengthening can restore the structure's ability to satisfying operational requirements [1], [2], [3].

In the case of corrupted elements, or the entire structure, it is needed to take into account how to proceed with strengthening. In order to achieve the required level of reliability, the correct amplification method must be chosen. Nowadays and with today's technologies, we have a huge variability of strengthening. Adding concrete reinforcement to the new enlargement cross-section is one of the most traditional ways of reinforcement (figures 1 a, b). If this method is used, there must be sufficient space around the column where the cross-section is increased, which is not possible in many cases. The recommended minimum thickness of the reinforced concrete layer according to [4] is 30 to 60 mm.

Using this method, it must be ensured that the original and new concrete do not slip, so it is considered the full co-operation. A state where the tensile force is transmitted to the pressed concrete [5] by shear forces is considered to be a full co-operation. It follows that the resistance depends on the concrete-concrete connection (contact between new and old layers). With such a joint, it is assumed that the newly added layer will has the same concrete quality or higher.

a) b)

Figure 1. Traditional ways of strengthening by using new layer of concrete.

Shear resistance [6] in the shear joint is able to transmit shear forces to a certain level of resistance. After this level, the cross-sectional resistance is ensured by the action of several factors such as the strength of individual materials, technological processes or constructional arrangements. In the case of concrete structures, 3 factors contribute to the shear resistance when using concrete reinforcement [7], [8]:

- cohesion,
- friction,
- shear reinforcement.

These factors act simultaneously, only after overcoming a certain level of stress does cohesion cease to work. After overcoming cohesion, there is no collapse of the element, but a reduction of stresses in the cross section, which means that the element can no longer be considered monolithic [9], [10]. The standard STN EN 1992-1-1 [11] states the values of the coefficients of cohesion and friction for various surface treatments from very smooth to roughed. These factors are essential for calculating the load-bearing capacity of a composite element.

In practice, we often encounter the fact that the new additional reinforcement is anchored with hooks, which is quite laborious [12], [13]. Therefore, the article deals only with surface treatment for notches and indents by means of fibre concrete concreting. It follows that there is no need for concrete reinforcement and also no anchoring. The relevance of cohesion will also be demonstrated by inverse calculation from experimental measurements for notches and also the cohesion for notches will be determined which is not given in [11].

2. Experiments

Push tests consisted of a central column and an encasement. The cross-section of the column was chosen with floor plan dimensions of 0.16×0.16 m and a height of 0.35 m. The column (core) was reinforced with 4Ø10 mm, the length of the bars is 0.3 m. The stirrups of diameter 8 mm were placed at an axial distance of 85 mm (figure 2a). The concrete grade was used C 16/20. Samples after concreting were treated for 28 days. Subsequently, the samples were modified according to figure 2b and c. The samples were encasement with a layer of 35 mm thickness.

Temporary polystyrene was placed on the samples on the lower part (the polystyrene was removed after the encasement hardened). Subsequently, the samples were placed in the formwork where they were centered with spacer screws (figure 2d). Prior to concreting, the samples were cleaned, moistened and then the samples were concreted with fiber concrete. The composition of the concrete mix was based on previous research [14], where they compared the amount of fibers added to the concrete mix and their effect on the flexural tensile strength (table 1).

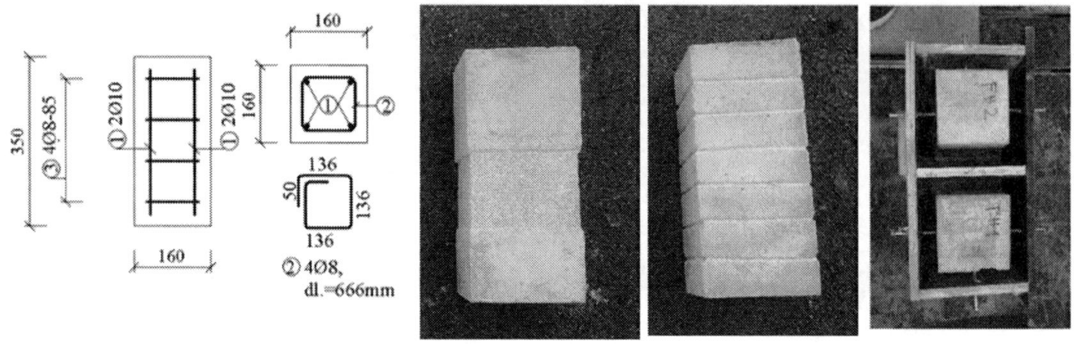

a) b) c) d)

Figure 2. Reinforcement of concrete column and modification of surface.

Table 1. Results of the modify materials characteristics.

Concrete mixture	1 kg/m^3
CEM II/B-S 32,5R (Ladce)	400
Aggregate 0/8 mm	910
Aggregate 8/16 mm	685
Fly ash (USS Košice)	80
Water	200
Fibers DRAMIX 3D	40

Cubes with an edge length of 150 mm were cast to determine the compressive strength of the concrete. The average cube strength of fiber-reinforced concrete was 45.6 MPa.

The samples were loaded with axial force. The deformation between the core and the encasement was measured. Trajectory sensors were used for sliding. It was loaded by deformation at a speed of 0.5 mm/min. Theoretical load capacities of samples were calculated according to STN EN 1992-1-1 [11] and they are shown in table 2.

The results of measurements showed shear resistance for individual types of surface treatments. The six samples (three in two series) were tested and the test results are shown in figure 3 for indent and figure 4 for notches.

After analysis of the indent tests (ZZ), it was clearly visible that the indent in the core was filled with the concrete mixture and were slipped after reaching the shear resistance (figures 5a and d). Using fiber concrete, one or two major longitudinal cracks were formed, which occurred at the time of shear resistance. After reaching the maximum shear force, cracks were formed and opened (figures 5b and c). The average value transferred by the surface treatment to the indent was 262.17 kN.

The notches (ZR) samples showed the high shear resistance. In the analysis of the cladding, it was visible that the formation of cracks was situated more on the corners of the core, but the effectiveness of the notch was also clearly visible (figures 6 a, b, c). The mechanism of the violation was the same as for the indent. The average value transferred by the surface treatment to the notches was 250.58 kN.

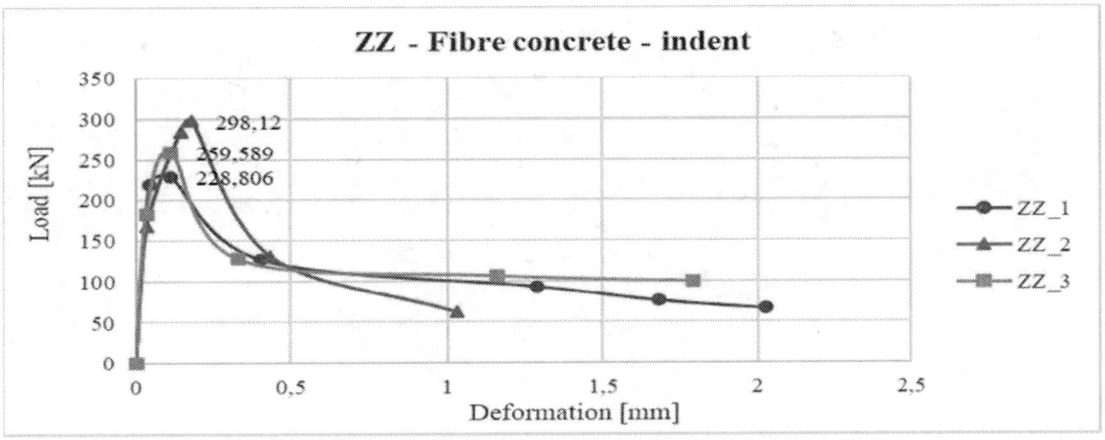

Figure 3. Load-deformation, fibre concrete – indent.

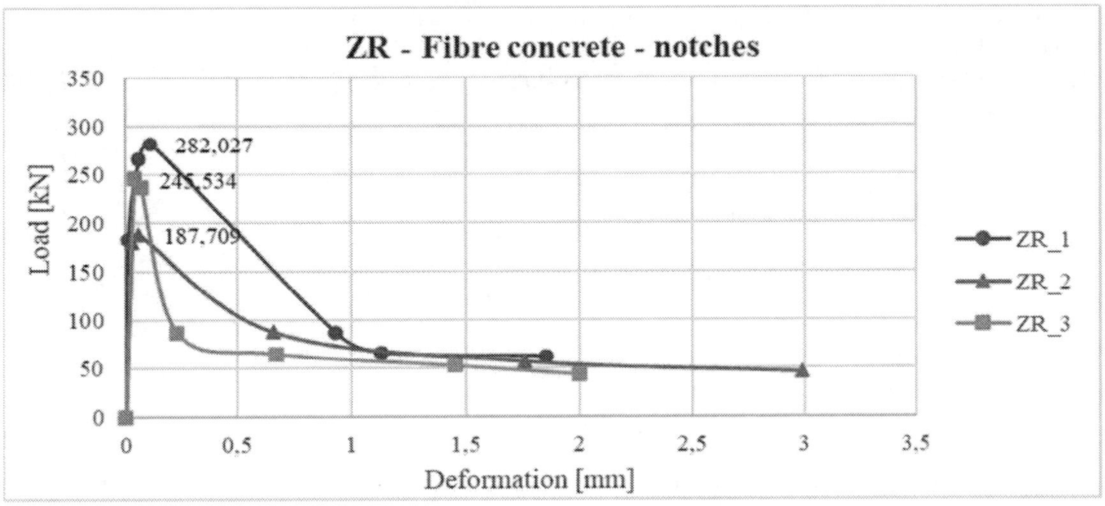

Figure 4. Load-deformation, fibre concrete – notches.

a) b) c) d)

Figure 5. Sample after load – indent.

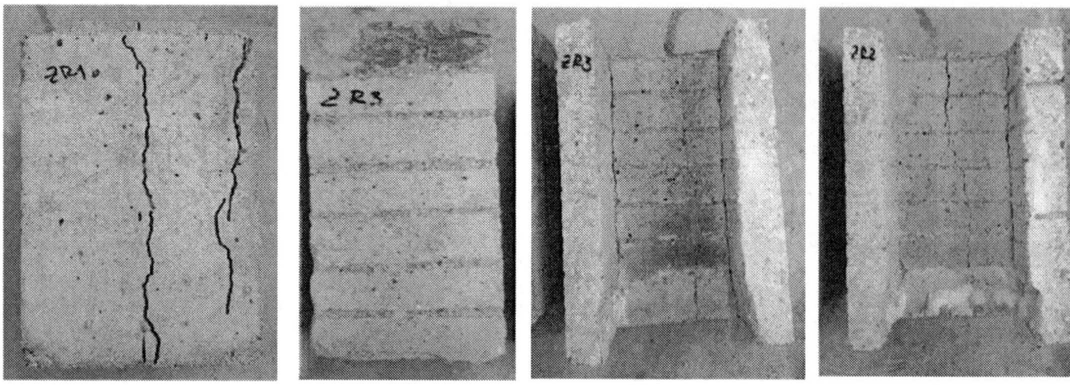

a) b) c) d)

Figure 6. Sample after load – notches

For comparison, the characteristic values of shear strength τ [11] were calculated according to equation (1 and 2). For the indent, the shear resistance was calculated at 105.0 kN. A coefficient of friction of 0.45 was used for the notches and the value of shear calculated according to STN EN 1992-1-1 [11] was 94.5 kN.

$$\tau_n = c + \mu\sigma_n \tag{1}$$

$$\tau = \tau_n \cdot S \tag{2}$$

where:

- τ_n is the shear stress [MPa],
- τ is the shear resistance [kN],
- σ_n is the normal stress [MPa],
- S is the contact shear surface [m^2].

Table 2. Determination of the coefficient of cohesion.

Series	Average shear resistance from experiment τ [kN]	Shear resistance standard τ [kN]	Shear stress from experiment τ_n [MPa]	Calculated cohesion c [MPa]	Standard cohesion c [MPa]	Cohesion difference calculated-norm [%]
ZZ	262.18	105.0	1.248	1.25	0.50	40.05
ZR	250.58	94.5	1.193	1.20	0.45	37.50

From the obtained results (table 2) the calculations from equations (1) and (2) were confirmed, where the values of the tests in some cases exceeded the calculated results for indent and notch many times. As the standard [11] does not specify the coefficients of friction and cohesion for the notches surface treatments, it was necessary to calculate these values by back-calculation and compare them with the standard values.

However, it should be noted that the results came from large differences between the individual samples. The diversity of the results is largely due to the concrete mix and the air cavities at the contact surfaces of the encasement.

IOP Conf. Series: Materials Science and Engineering 1015 (2021) 012028 doi:10.1088/1757-899X/1015/1/012028

3. Conclusions

If is necessary to strengthen the vertical load-bearing elements by using a new layer concrete, the emphasis is placed on the slippage between the original and the new concrete and also on the load-bearing capacity of this new layer. This contact is influenced by the type of surface from which cohesion depends. The mechanism ensures the cohesion of the cement composite through porosity and also by surface treatment. The article proves that the values of cohesion are several times on the side of the security side compared to the measured and calculated values of up to 40.05% for indent and 37.50% for notches. They also point out that by surface treatment we achieve higher shear resistance. In our case, the modification of notch indicates a higher shear capacity than notches but smaller by only 11.6 kN.

Acknowledgments

This research is supported by the Slovak Research and Development Agency under contract No. APVV-14-0772, and by Research Project No. 1/0413/18 and Project No. 1/0045/19.

References

[1] Koteš P. 2012 *Reliability of existing bridge structures and possibilities of its increase* (Žilina)
[2] Vičan J and Koteš P 2018 *Evaluation of existing bridge structures* (Žilina: EDIS)
[3] Tomica V, Sokolík A, Zemko Š 1992 *Maintenance and reconstruction of bridges* (Bratislava)
[4] Bilčík J and Dohnálek J 2003 *Rehabilitation of concrete structures* (Bratislava: JAGA)
[5] Priganc S and Bahleda F 2007 *Reinforcement of concrete structures* (Košice)
[6] Sadowski L 2019 *Adhesion in Layered Cement Composites* (Switzerland: Springer)
[7] Wall J S and Shrive N G 1988 Factors affecting bond between new and old concrete *ACI Materials* **95** pp 117-125
[8] Czarnecki L and Chmielewska B 1993 Factors affecting adhesion on building joints *Proc. of Fifth Int. Conf. on Structural Faults and Repair*
[9] Vavruš M, Koteš P and Bahleda F 2019 Analysis of shear contact between wrapped layout of fiber reinforced concrete *POLLACK PERIODICA Int. J. for Eng. and Information Sciences*
[10] Bakhsh K N 2010 *Evaluating of Bond Strength between Overlay and Substrate in Concrete Repairs* (Master of Thesis, Stockholm)
[11] Standard 2004 STN EN 1992-1-1: Eurocode 2. Design of concrete structures-Part1-1: General rules and rules for buildings
[12] Čítek D, Kolísko J, Řeháček S and Mandlík T 2015 Concrete cover effect on bond behaviour of UHPC *22nd Int. Conf. Concrete Days* (Litomysl; Czech Republic) **249** pp 273-7
[13] Čítek D, Huňka P, Řeháček S, Mandlík T and Kolísko J 2014 Testing of bond behavior of UHPC *11th Int. Conf. on Special Concrete and Composites* **1054** pp 95-98
[14] Vítek J L and Smiřinský S 2010 Interaction of classical and diffuse reinforcement *BETON*

IOP Conf. Series: Materials Science and Engineering 1015 (2021) 012029 doi:10.1088/1757-899X/1015/1/012029

Investigation of the influence of heat protection characteristics of protecting designs on residential buildings heat storage capacity

Yu S Vytchikov, A A Chulkov and V A Golikov

Samara State Technical University, Molodogvardeiskaya street, 244, Samara, Russian Federation 443100

E-mail address: vladislavgol@rambler.ru

Abstract. The article presents the methodology of determining the heat storage factor of residential buildings heated by the central heating system. This calculation method is considered on the example of one-room corner apartment on the ground floor of an apartment building. The results of heat accumulation factor calculation are presented and the parameter changes depending on the value of normative multiplicity of air exchange and external air temperature are shown. The calculation of indoor air temperature after switching off the heating system was performed. The results presented in the article allow to draw a conclusion about the significant impact of natural ventilation during an accident on the heat network. The increase in air exchange in this case reduces the internal air temperature significantly. The obtained results should be used when choosing measures to ensure energy security of buildings and structures.

1. Introduction

According to the technical regulations “safety of buildings and structures”, the parameters of the microclimate of premises should be harmful to human health.

During the operation of residential buildings that are heated from centralized heat supply systems accidents occur periodically. They are associated with the pipeline wear and tear and resulting poor coolant circulation. Therefore, the designed buildings are subject to regulatory requirements for the heat storage capacity of residential premises. Before emergency situations are eliminated, there is a significant decrease in the temperature of the internal air in buildings and the heat carrier in heating and hot water systems. The process of premises cooling occurs in conditions of non-stationary heat transfer between internal and external air, which complicates the method of solving the problem under consideration significantly. In works [1-10] the results of thermal engineering calculation of enclosing structures operated under conditions of changing internal air temperature are presented. In [2, 6] the results of the study of the influence of the non-stationary thermal regime of enclosing structures on the heating system of buildings are presented. Works [1, 3-10] are devoted to solving the conjugate problem of heat exchange in the process of premises heating and cooling.

2. Methodology for calculating room storage capacity

The heat storage capacity of buildings according to [4] is characterized by the heat storage coefficient β, measured in hours. To determine it, there are experimental and computational methods described in detail in [1, 4].

IOP Conf. Series: Materials Science and Engineering 1015 (2021) 012029 doi:10.1088/1757-899X/1015/1/012029

The indoor air temperature in living quarters t_{in} when the heating system is turned off is determined according to [4] by the formula

$$t_{\text{in}} = t_{\text{out}} + \left(t_{\text{in}_0} - t_{\text{out}}\right) e^{-\frac{\tau}{\beta}} \tag{1}$$

where t_{out} - outdoor temperature, (°C); t_{in_0} - internal air temperature before the heating system is turned off, (°C); τ - room cooling time, (h).

The value of the heat storage coefficient is significantly influenced by the heat-protective characteristics of enclosing structures, their heat capacity, as well as the multiplicity of air exchange. An approximate analytical method for determining the above coefficient was proposed by Yu. V. Kononovich in [7].

According to [7], the value of the coefficient β is calculated using the formula

$$\beta = \frac{C_{\text{protecting design}}}{3.6\, Q_{\text{sp}}} \tag{2}$$

where $C_{\text{protecting design}}$ - heat capacity of the building protecting design, (kJ/°C); Q_{sp} - specific heat loss of the room, (W/°C).

Value $C_{\text{protecting design}}$ external protecting design and internal partitions is calculated by the formula

$$C_{\text{ptotecting design}} = K_t \sum_{i=1}^{n} \delta_i c_i \rho_i \frac{F_i}{2} \tag{3}$$

where δ_i, c_i, ρ_i - thickness, specific heat and density of the layer material of protecting design; F_i - protecting design square, (m^2); $K_t = \frac{\tau_{\text{R}} - t_{\text{out}}}{t_{\text{in}_0} - t_{\text{out}}}$ - dimensionless coefficient; τ_{R} - radiation temperature of the protecting design, (°C).

The radiation temperature of the building protecting design to [7] is determined by the formula

$$\tau_i = t_{in_0} - \frac{t_{in_0} - t_{\text{out}}}{R_{\text{o}}^{\text{pr}} \alpha_{\text{in}}} \tag{4}$$

where R_{o}^{pr} - resistance to heat transfer of the protecting design, determined according to SP 50.13330.2012 "Thermal protection of buildings", (m$^{2}\cdot$°C)/W; α_{in} - heat transfer coefficient from the side of the inner surface of the protecting design, W/(m$^{2}\cdot$°C).

The value of specific heat loss according to [7] is calculated by the formula

$$Q_{\text{sp}} = \sum_{i=1}^{n} k_i F_i + L c_{\text{in}} \rho_{\text{in}} \tag{5}$$

where k_i - coefficient of heat transfer of the external protecting design W /(m$^{2}\cdot$°C); L - the flow of infiltrating outdoor air through the enclosing structures, the standard value of which for residential buildings is 3 m^3/h per 1 m^2 of residential premises of the apartment.

3. Results of calculating the heat storage capacity of the room

As an example, we considered the determination of the heat storage capacity of a residential apartment located on the first floor of an apartment building.

Figure 1 shows the plan of the corner room in the apartment under consideration. Thermal engineering indicators of the apartment's enclosing structures are shown in table 1.

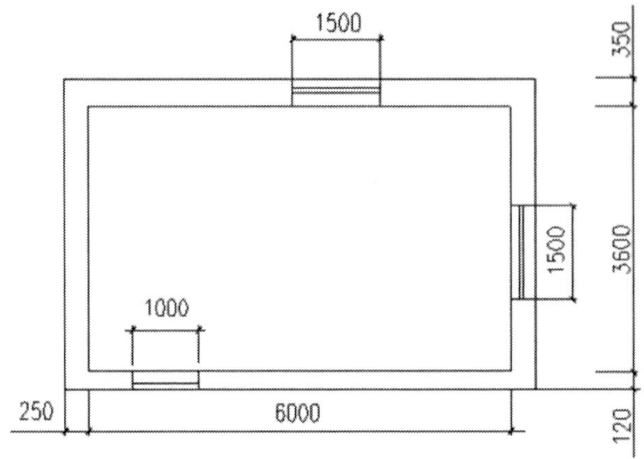

Figure 1. Plan of the corner living room.

Table 1. Thermal characteristics of structures and materials used in the calculation.

Protecting design type	Area of the fence F_i m^2	Thickness δ_i m	Coefficient of thermal conductivity λ_i W/(m·°C)	Specific heat capacity c_i kJ/kg °C	Density ρ_i kg/m^3	Heat transfer resistance of the protection R_0 (m^2·°C)/W
			External protecting design			
External walls:	20.94	-	-	-	-	3.24
- expanded clay concrete panels	-	0.350	0.330	0.84	1000	-
basalt mineral wool	-	0.100	0.042	0.84	120	-
- textured layer of the facade system	-	0.0035	0.700	0.84	1600	-
Window blocks with double-glazed Windows in bindings	4.5	-	-	-	-	0.55
The ceiling above the unheated basement:	21.6	-	-	-	-	1.75
the linoleum on mastic	-	0.0015	0.29	1.47	1600	-
- cement-sand solution	-	0.03	0.76	0.84	1800	-
- PENOPLEX	-	0.04	0.031	1.34	35	-
- cement-sand solution	-	0.02	0.76	0.84	1800	-
- reinforced concrete hollow floor slab	-	0.22	1.74	0.84	2500	-
			Internal protecting design			
Interroom partition:	13.9	-	-	-	-	-
- laying of ceramic bricks on a cement-sand solution	-	0.12	0.70	0.88	1800	-
- lime-sand plaster	-	0.04	0.70	0.84	1700	-
Interroom partition:	9.54	-	-	-	-	-
- laying of ceramic bricks on a cement-sand solution	-	0.25	0.7	0.88	1800	-
- lime-sand plaster	-	0.04	0.7	0.84	1700	-
Floor:	21.6	-	-	-	-	-
- heat and sound insulation linoleum	-	0.0035	0.04	1.47	1000	-
- cement-sand solution	-	0.03	0.76	0.84	1800	-
- PENOPLEX	-	0.03	0.031	1.34	35	-
- reinforced concrete floor slab	-	0.22	1.74	0.84	2500	-

The value of the specific heat loss of a living room was determined by the formula (6), the heat capacity of protecting design by the formula (3). The results of calculating the coefficient of heat storage at different outdoor temperatures are shown in table 2.

Table 2. Results of calculating the accumulating capacity of a residential apartment.

Outdoor temperature t_{out} °C	Outdoor air density ρ_{out} kg/m^3	Mass flow rate of infiltrating air G_{in} kg /h	Specific heat loss Q_{sp}, W/°C	Radiant temperature τ_{out} °C	K_i	Heat capacity $C_{\text{protecting design}}$ kJ/°C	β h
0	1.29	83.6	50.17	19.4	0.97	7955.3	44.05
-10	1.34	86.8	51.07	19.21	0.974	7988.2	43.45
-20	1.39	90.1	52.0	19.01	0.975	7996.3	42.72
-30	1.45	93.98	53.07	18.8	0.976	8004.5	41.9

From the data presented in table 2, it follows that the coefficient of heat storage decreases slightly when the outdoor temperature decreases. The average value of the coefficient β was 43 hours.

When operating residential buildings, the multiplicity of air exchange varies from 0.25 to 1.0. The minimum value corresponds to the duty mode of the ventilation system in the absence of people.

Table 3 shows the results of calculating the heat storage coefficient for different values of air exchange multiplicity.

Table 3. Heat storage capacity of a residential apartment at different values of the multiplicity of air exchange.

Outdoor temperature t_{out} °C	Ventilation rate n h^{-1}	Specific heat loss Q_{sp} W/°C	Heat capacity $C_{\text{protecting design}}$ kJ/°C	β h
0	0.25	32.78	7955.3	67.41
	0.50	36.58		60.41
	0.75	44.39		49.78
	1.00	50.17		44.05
-10	0.25	33.00	7988.2	67.24
	0.50	39.03		56.85
	0.75	45.05		49.26
	1.00	51.07		43.45
-20	0.25	33.23	7996.3	66.84
	0.50	39.48		56.26
	0.75	45.74		48.56
	1.00	52.00		42.72
-30	0.25	33.50	8004.5	66.37
	0.50	40.02		55.56
	0.75	46.55		47.77
	1.00	53.07		41.90

From the data presented in table 3, it follows that the coefficient of heat accumulation decreases with an increase in the multiplicity of air exchange.

The term from formula (1) will be denoted as the dimensionless temperature of the internal air

$$\Theta_{in} = \frac{t_{\text{in}} - t_{\text{out}}}{t_{\text{in}_0} - t_{\text{out}}} \qquad (6)$$

The temperature of the internal air during the cooling of the living room was determined by the formula (6)

$$\Theta_{in} = e^{-\frac{\tau}{\beta}} \quad (7)$$

To calculate the dimensionless temperature Θ_{in}, it is proposed to use table 4, which shows the results of calculating Θ_{in} for different values of the heat storage coefficient.

Table 4. Results of calculating the process of cooling residential premises.

Cooling time τ, h	Dimensionless internal air temperature at values β, h			
	40	50	60	70
0	1.000	1.000	1.000	1.000
5	0.880	0.904	0.920	0.931
10	0.778	0.818	0.845	0.866
15	0.686	0.740	0.778	0.806
20	0.605	0.669	0.716	0.750
25	0.533	0.605	0.658	0.699
30	0.471	0.547	0.605	0.650

Using table 4, it is possible to determine the value of the dimensionless temperature of indoor air in residential premises in emergency situations.

4. Discussion

The results of calculation of accumulation capacity of the considered room show, that the heat storage factor changes insignificantly relative to its average value. The data presented in table 2 were obtained with a standard air exchange equal to one rate.

Further investigation of room storage capacity at air exchange multiplicity from 0.25 to 1, has shown changes in average value of heat accumulation factor from 67 to 43 hours. It is obvious that in order to increase room storage capacity during emergencies on heat networks it is necessary to reduce specific heat losses during this period of time. According to the data presented in table 3, it is necessary to reduce air exchange in the given room by closing vents, transoms, window vestibules or breathing valves.

Thus, the air exchange rate is of great importance for the room storage capacity. It should be noted that these studies were conducted for a residential building with a natural ventilation system. However, not only the air exchange but also the heat protection characteristics of the protecting designs should be taken into account when switching off the heating system.

5. Conclusions

The results of calculation of heat accumulation factor of an apartment building showed that the work of natural ventilation system has a significant influence on the degree of heat accumulation in a room at violation of circulation of the coolant in the heating system. Normative air exchange decrease allows maintaining the maximum allowed temperature of internal air at switching-off of heating system. To increase the cooling time of the room, it is necessary to reduce air exchange to 0.25 times at the time of stopping heat supply, however, when restoring heat supply, it is recommended to increase air exchange to standard values. The results obtained should be taken into account when developing energy saving measures and measures to ensure buildings energy security.

References

[1] Borodin A I, Tsvetkov N A and Ivanova E A 2007 *Bulletin of the TSUACE* **2** p 161
[2] Malyavina E G and Asatov R R 2010 *Academia. Architecture and construction* **3** p 324
[3] Matyukhov D V, Nizovtsev M I and Terekhov V I 2002 *University news. Construction* **7** p 72
[4] Sokolov E Ya 2002 *District heating and heat networks* (Moscow: Moscow power engineering Institute publishing house) p 472

[5] Manukovskaya T G and Sterligov V A 2011 Questions of modern science and practice. *V. I. Vernadsky University* **3** (34) p 384
[6] Malyavina E G and Petrov D Yu 2012 *Housing building* **6** p 66
[7] Kononovich Yu V 1986 *Thermal regime of mass construction buildings* (Moscow: Stroyizdat) p 157
[8] Datsyuk T A, Ivlev Yu P and Pukhkal V A 2014 *Modern problems of science and education* **5** p 145
[9] Vytchikov Yu S, Saparev M E and Belyakov I G 2016 *International research journal* **6–2** (48) p 42
[10] Gorshkov A S, Rymkevich P P and Vatin N N 2014 *Engineering and Construction J.* **8** p 38

IOP Conf. Series: Materials Science and Engineering 1015 (2021) 012030 doi:10.1088/1757-899X/1015/1/012030

On the need to take into account the effect of wind on the frame of thin-walled steel structures during installation

A V Soloviev and O N Solovieva

Department of metal and wooden structures Samara State Technical University.
Samara State Technical University, Molodogvardeyskaya 244, 443001 Samara, Russia

E-mail address: savsmr@rambler.ru

Abstract. Recently, the construction of buildings and structures using elements of light steel thin-walled structures has been increasing in Russia. The use of lightweight galvanized profiles provides a number of advantages in the manufacture, transportation and installation of frames. In the practice of designing structures of this type, the calculation of frames, as a rule, is carried out for the loads acting on the structure during its operation. In many cases, analysis of possible wind effects on frames during installation is not performed. This article examines the effect of wind on the frame of a 120-meter-long single-span building made of light steel thin-walled structures. Such structures, having a low mass, under certain parameters, can experience dynamic vibrations caused by the effect of wind. Most of the aeroelastic processes are self-oscillating in nature, occurring during energy exchange between the flow and the streamlined body. The main phenomena of aeroelasticity are flutter (stall flutter) buffing. If we consider the issues of aeroelasticity from the point of view of the effect on the structure, then such concepts as vortex excitation, galloping across the air flow, divergence and parametric resonance should be considered. If, during the design process, the analysis of possible wind effects on the frame is not carried out and measures are not taken to reduce the negative effect of wind, this can lead to deformation of the frame elements or their destruction.

1. Introduction

Calculation of building structures for the effects of wind is a very urgent problem at the present time [1-4]. This becomes especially important when designing frames made of light steel profiles of rather long buildings. As a rule, projects provide for the erection of a supporting frame, and then the installation of a light wall and roof fence. It should be noted that the erected frames are exposed to wind influences before the installation of the wall and roof fencing. Depending on the type of terrain and wind speed, impacts can significantly contribute to the performance of the framing elements. Both the entire frame and its individual elements can fall into the zone of resonant aerodynamic vibrations. This is especially dangerous for elements of square or rectangular cross-section. When the wind stream flows around the elements of the building frame, the so-called aerodynamic instability of the aerodynamic parameters of the environment occurs. There is a sharp change in pressure associated with the separation of the air flow from the structure. Vortex formation processes appear. The emerging aerodynamic forces lead to additional loading of the frame elements. The resulting forces must be taken into account in strength calculations. The main part of aeroelastic processes is self-oscillating in nature, occurring during energy exchange between the flow and the streamlined body. The main phenomena of aeroelasticity are flutter, buffeting. If we consider the issues of aeroelasticity from the point of view of the effect on the structure, then one should analyze such phenomena as vortex excitation, galloping across the air flow, divergence and parametric resonance. Flutter - non-damping elastic vibrations of a structure (parts of a structure) when an air flow is running at a certain (critical) speed. Flutter is a type of self-oscillation, where the source of vibration energy is the wind flow, and the feedback is realized due to the elasticity of the building structure (or building element). At certain angles of attack, a flow stall can occur. It should be noted that stall flutter is usually realized at low flow rates. Buffeting - forced vibrations of the entire structure or its parts under the action of non-stationary aerodynamic forces when the flow is disrupted from the surface of an elastic object at high angles of attack (as a rule, from poorly streamlined surfaces). When buffering, the stall flow from one part of the structure acts on another part of the structure. If the phenomenon of flutter can be studied by highlighting a certain "significantly elastic" part of the structure "locally", then during buffeting it is necessary to consider the entire building, even its "relatively inelastic" ("rigid") elements. The vortex excitation mechanism is similar to the flutter mechanism (stall flatter) and is associated with the fact that when the flow is running, the bluff body drops vortices alternating in a checkerboard pattern with a frequency depending on the flow velocity, size and shape of the streamlined body. Galloping across the air flow typically occurs in flexible structures and features with special cross-sectional shapes, such as power lines, cable-stayed or suspended structures. This phenomenon is a self-oscillation of the body, occurring almost perpendicular to the direction of the incident flow. Divergence is primarily associated with the twisting of a body of small transverse thickness in an air flow directed along the longitudinal axis of the structure's cross section. The phenomenon is characterized by torsional self-oscillations of the structure increasing in time (causing an increase in the drag of the structure profile) and leading to the destruction of the structure. It is also a problem of air resistance. Divergence usually occurs in bridge spans and structures with flat cross sections. Parametric resonance is characterized by a complex nature of the interaction between the structure and the incident flow and, in the general case, is associated with a change in the parameters of the dynamic system over time, leading to an increase in the amplitude of oscillations. As a rule, the natural vibration of the structure or the damping coefficient undergo a change. As an example, we can cite changes in the tension force of the cables of a suspension bridge or the tension force of the link elements and, as a consequence, the excitation of vibrations of structures.

This paper deals with the issues of aerodynamic wind effects on building frames during installation. An analysis of the causes of deformation of the frame of a single-span building and the destruction of its elements at the stage of installation due to the effect of wind is also presented.

2. Materials and methods

In accordance with the design standards for buildings and structures, as well as their elements, it is necessary to take into account the main type of wind load, resonant vortex excitation and aerodynamically unstable oscillations such as galloping, divergence and flutter.

The main type of wind load and peak wind loads are associated with the direct effect on buildings and structures of hurricane winds, maximum for the construction site, and must be taken into account when designing all structures. Resonant vortex excitation and aerodynamic unstable vibrations must be taken into account for buildings, solid-wall structures or their individual sections that have a rectilinear central axis, as well as unchanging or smoothly changing shapes and sizes of the cross-section, for which the condition λ_e>20 is met, where λ_e is determined depending on geometric parameters of a structure or its element. Criteria for the excitation of aerodynamically unstable vibrations are established in the design standards.

So the critical wind speeds V [m/s], at which resonant vortex excitation occurs according to the i-th natural mode of vibration, are determined by the formula:

$$V_{cr,i} = \frac{k_v \cdot f_i \cdot d}{St} \tag{1}$$

here:

f_i, (Hz) - is the natural frequency of vibrations in the i-th flexural natural form;

IOP Conf. Series: Materials Science and Engineering 1015 (2021) 012030 doi:10.1088/1757-899X/1015/1/012030

d, (m) - is the transverse dimension of the structure or element;

St - is the Strouhal number of the cross section, determined experimentally or by reference data (for rectangular sections $St = 0.11$, for square sections $St = 0.12$).

k_v - is the coefficient that takes into account the effect of trapping the natural frequency of oscillations, from the range $0.9 < k_v < 1.1$ from the condition for the implementation of the worst case loading.

Resonant vortex excitation does not occur if:

$$V_{cr,i} > V_{\max}(z_h) \qquad (2)$$

where

$V_{max}(z_h)$ - maximum wind speed at z_h level, (m/s)

$$V_{\max}(z_h) = 1.5\sqrt{w_0 k(z_h)} \qquad (3)$$

here:

w_0, (Pa) - the standard value of the wind pressure taken in accordance with the wind area;

$k(z_h)$ - coefficient taking into account the change in wind pressure for the height z_h.

When the wind speed reaches a critical value, there is a likelihood that the frame elements, during installation, into the resonant vortex excitation. This can lead to the accumulation of fatigue damage and in some cases to failure.

2.1. Destruction of building frame elements from wind loads at the installation stage

Let's consider the case of destruction of elements of a light steel frame at the stage of installation. As an example, a single-span building with a length of 120 meters is considered. The building frame diagram and the structural design of the main elements are shown in figures 1, 2.

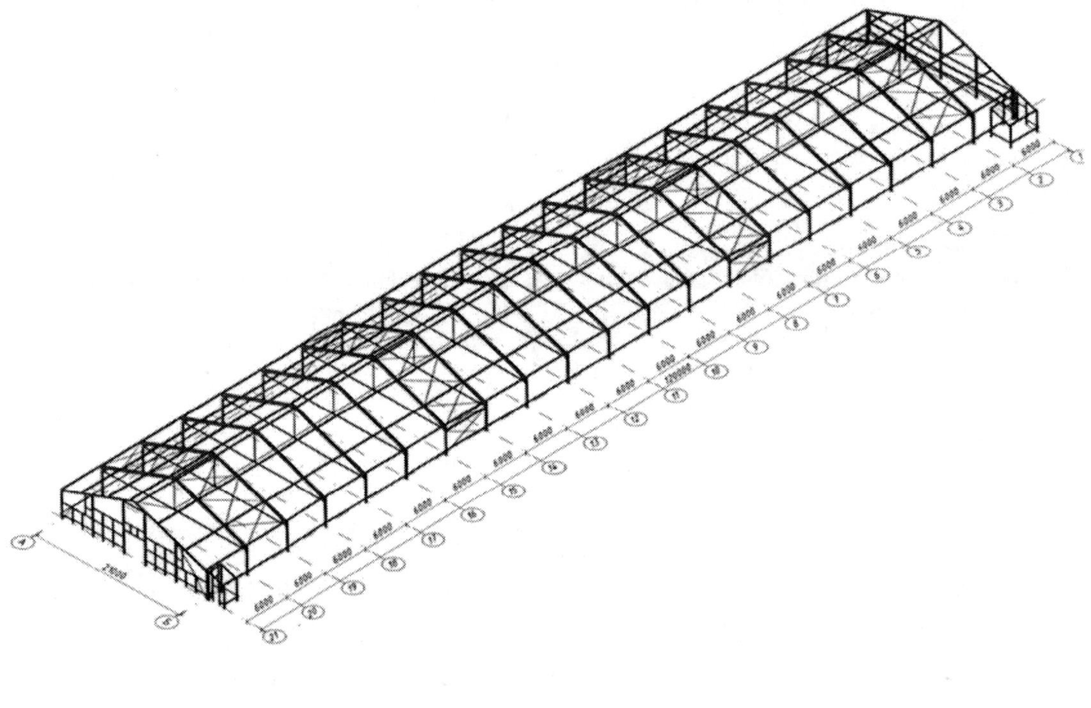

Figure 1. Scheme of the building frame.

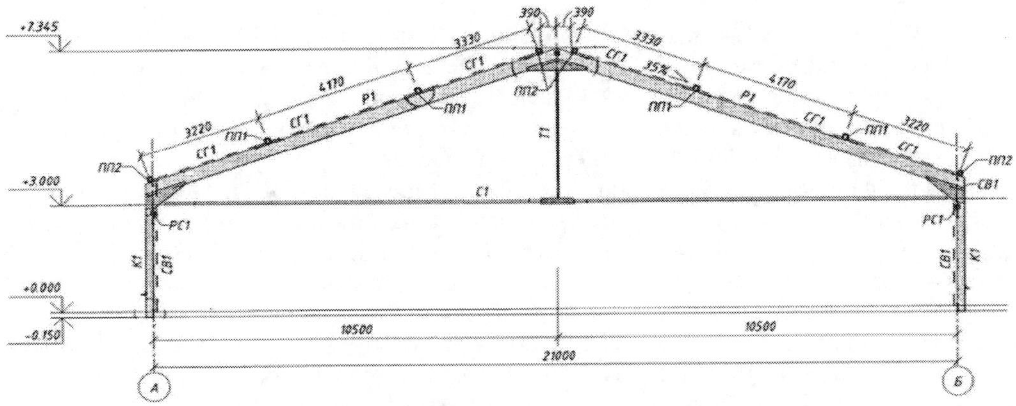

Figure 2. Transverse frame of the frame along the axis 7.
K1 - frame column (□ 200×120×4); P1 - beam (□ 350×150×5);
CB1 - vertical connection (○ 16); □□1 - spacer (□ 100×100×3);
СГ1 - horizontal connection (○ 12); ПП1, ПП2 - cover runs (□ 160×120×4);
C1 - tightening (□ 80×80×3); T1 - suspension (○ 12).

The project for the production of work envisaged first the installation of the frame elements, and then the installation of the elements of the wall and roof fencing. In total, 18 building frames were installed. For technical and economic reasons, the installation of wall and roof panels was postponed for 3 months. When the installation work was resumed, damage to the frame was found in the form of: destruction of the nodal elements of fastening the puffs C1; the occurrence of cracks in the nodal element of fastening the tightening C1 to the column K1; breakage of suspensions T1 at the points of adjoining to the tightening of C1; sagging of stiffening elements and suspensions. The nature of the damage is shown in figure 3.

a)

b)

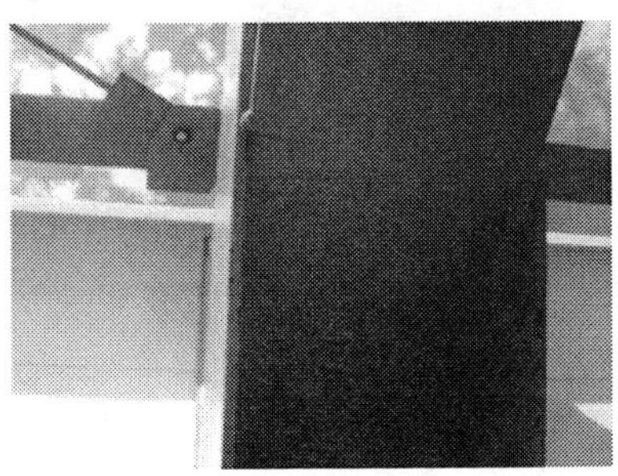

c)

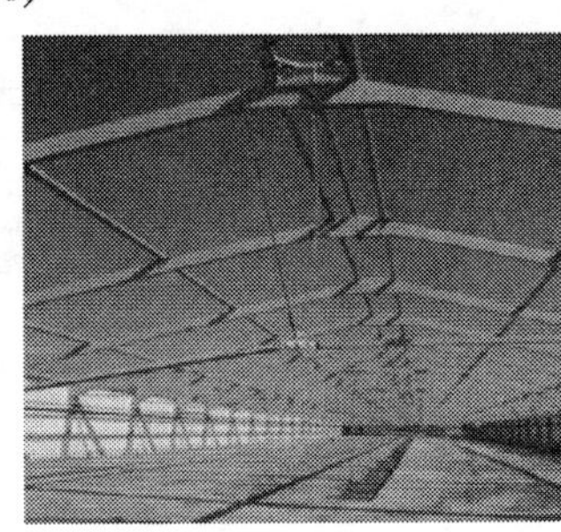

d)

Figure 3. Destruction of frame elements
a) destruction of the nodal elements of fastening the guy wires C1 to the column;
b) the appearance of cracks in the nodal element of fastening the guy wires C1 to the column K1;
c) breakage of the tightening hangers;
d) sagging of tie and suspension elements.

IOP Conf. Series: Materials Science and Engineering 1015 (2021) 012030 doi:10.1088/1757-899X/1015/1/012030

2.2. Analysis of the causes of destruction from wind load at the stage of installation of the frame

Analysis of the causes of the destruction of the frame elements at the installation stage shows that the destruction occurred due to the effect of wind. Obviously, the wind was blowing longitudinally along the building. At the same time, the wind speed reached such a value at which the building frames fell into parametric resonance, and the puff was galloped across the air flow.

Let's analyze the parameters of the structure and the wind impact. The estimated area of the wind impact on the end of the building, in the presence of wall and roof fencing, is 115 m^2. In the absence of wall and roof fencing, the wind load is transferred to the building frames. The total area of the frame elements on the windward side is 230 m^2. It is obvious that doubling the impact area leads to overstressing of vertical and horizontal structures of connections. In addition, prestressed bond elements accompany the transition of the entire frame to parametric resonance when the wind speed reaches critical values. Residual deformations (sagging) of the bonds show that the elements have undergone such a load, which led to the formation of irreversible plastic deformations. Tightening of the building frames when the wind reaches critical speeds turned into galloping (figure 4).

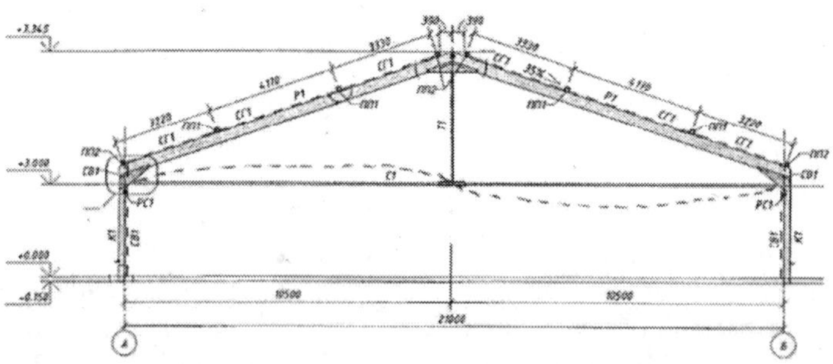

Figure 4. Schema of galloping of the frame tightening element.

Since the tightening in the middle is fixed by the suspension, the galloping occurs along the second form of the natural frequency. All mounted frames are located in an open area. The maximum wind speed at a height of 3 meters (3) will be

$$V_{\max}(z_h) = 1.5\sqrt{380 \cdot 0,75} = 25.3 \text{ m/s} \tag{4}$$

The frequency of natural vibrations of the tightening according to the $i = 2$ form will be:

$$f_i = \left(\frac{\pi \cdot i}{l}\right)^2 \cdot \sqrt{\frac{EI}{m}} = \left(\frac{3.14 \cdot 2}{21}\right)^2 \cdot \sqrt{\frac{210 \cdot 10^8 \cdot 91.4 \cdot 10^{-8}}{0.74}} = 14.4\text{Hz} \tag{5}$$

here:

l - puff length in meters;

E – the modulus of elasticity of the material;

I - the moment of inertia of the section;

m - the mass of a running meter of the tightening element.

The critical wind speed at which the resonant vortex excitation of the puff occurs along $i = 2$ of the natural vibration mode in accordance with (1) will be:

$$V_{cr,i} = \frac{k_v \cdot f_i \cdot d}{St} = \frac{1 \cdot 14.4 \cdot 0.08}{0.12} = 9.6 \text{ m/s} \tag{6}$$

here:

d - the height of the tightening section in the plane perpendicular to the wind flow.

Condition (2) is not met. This means that at a wind speed of about 10 m/s, vortex excitation can occur and puff oscillations can begin (figure 4). In this case, alternating loads will arise on the nodal elements of fastening the puffs. These alternating loads led to fatigue fractures of some elements of the nodal gussets (figure 3, a, b) and breakage of suspensions (figure 3, c).

3. Conclusions

The work done on the analysis of wind effects on structures under construction showed that at certain wind speeds and angle of attack, aerodynamic vibrations of the frame and its individual elements can appear. These vibrations can lead to deformation of structures and their destruction. When designing structures, such architectural and structural solutions should be used that exclude the excitation of aerodynamically unstable vibrations.

References

[1] Barshtein M F 1978 *Manual for the calculation of buildings and structures on the action of the wind* (M.F. Barshtein.-M.: Stroyizdat) p 216

[2] Birbraer A N 2009 *Extreme impacts on structures* (SPb. Publishing house of Polytechnic. University) p 594

[3] Mendis P, Ngo T, Haritos N and Hira A 2007 Wind loading on Tall Buildings *EJSE* (Special Issue: Loading on Structures) pp 41-54

[4] Simiu E and Scanlan R 1984 *Impact of wind on buildings and structures* (Moscow: Stroyizdat) p 360

IOP Conf. Series: Materials Science and Engineering 1015 (2021) 012031 doi:10.1088/1757-899X/1015/1/012031

Numerical verification of the saw-cut method

J Kraľovanec and M Moravčík

Department of Structures and Bridges, Faculty of Civil Engineering, University of Žilina, Univerzitná 8215/1, 010 26 Žilina, Slovakia

E-mail address: jakub.kralovanec@uniza.sk

Abstract. In Slovakia and other countries, a contemporary challenge for engineers and academics is an assessment of the state of existing prestressed concrete structures. The main components of assessment include regular inspections, maintenance, and diagnostics. For defining the load-carrying capacity and remaining service life, more reliable results of diagnostics are required. A need for detailed knowledge of the actual value of prestressing force is becoming a crucial aspect of civil engineering. There are many prestressed concrete structures built in the previous century which are about reaching their limit of service life. These structures are about sixty years old and many of them are exhibiting signs of deterioration due to the design and construction deficiencies, such as overloaded vehicles, environmental distress, and inadequate maintenance. In this paper, dozen concrete specimens are numerically analysed using the indirect non-destructive Saw-cut method. Ten of them are loaded by different level of compressive force which initiates compressive stress from 2.0 MPa to 12.0 MPa. Performed numerical analysis is an important basis for the experiment. The influence of temperature change caused by sawing will be later experimentally investigated on the remaining two concrete specimens. For numerical analysis, a 2D finite element model with the assumption of nonlinear material behaviour is performed. Subsequently, the correlation between the distance and depth of the saw-cuts, and amount of released strain or stress, as a result of numerical analysis, is outlined. Finally, the outputs obtained from numerical analysis are discussed and recommendations for the experiment are summarized.

1. Introduction

Nowadays, we are facing the problems related to precast prestressed concrete structures designed and constructed in the middle part of the 20th century. It is evident that this fact results in a growing need for new in-situ methods to define their load-carrying capacity and remaining service life. Replacement of existing structure by new, technologically advanced structure should be used only as a last resort. The rapid development of prestressing technology increased in the early 1980s. The length of girder increased to 30 m. The maximum length of post-tensioned girders made from three members is 42.0 m [1]. In the case of prestressed structures, knowledge of the state of prestressing force is required. Reliable methods for determining the level of prestressing force are essential for assessment.

The importance of this issue can be presented on the first generation of precast prestressed concrete bridges situated in Northern Slovakia (figure 1). Since 2015, three bridges of same building technology were classified as the state of emergency. Their construction problems can be summed up to two main groups of leading causes of failure. First, the problems related to improper design such as the absence of conventional reinforcement. The girders have been acting as the plain concrete members with partially bonded or even unbonded prestressing. Second, insufficient inspections and inadequate maintenance result in corrosion of prestressing wires due to absence of proper grouting of the ducts. Moreover, the anchors were installed without grouting or any other relevant protection. Half of them have not been protected against the influence of water. On the other hand, concrete of the beams was in surprisingly good condition.

Figure 1. The collapse of a precast prestressed concrete bridge in Trstená, Slovakia.

The methods for determining the state of prestressing force can be classified according to their influence on the structure or approach to the calculation of the level of prestressing. In relation to the effect of the performed technique on the investigated structure, we distinguish destructive and non-destructive methods. Classification of some techniques is debatable because the line between destructive and non-destructive approach is extremely thin and depends on the author's judgment. Generally, methods with none or only minimal impact on the structure are considered as non-destructive. These methods include, for example, a method based on the exposure of strand, Drilling method, and Saw-cut method, which is the pivotal object of this paper. In the case of using a destructive approach, we are not able to avoid a significant effect of our actions on the investigated structure. This group involve a method based on the initiation of crack or crack reopening in concrete member, and the Strand cutting method. Another classification divides the techniques into direct and indirect methods. As the name suggests, direct methods enable us to determine the level of prestressing directly from the results of the measurement. Unfortunately, these methods suffer from a plethora of pitfalls. In practice, we mostly use the Magnetoelastic method which is applied typically in new structures. The benefit of this method is the possible real-time monitoring of the prestressing force. Application of direct methods in existing structures is very limited. Indirect methods require measurement of physical or other quantities such as deformation, strain, or width of the crack. Subsequently, the value of the prestressing force is calculated using the results obtained from the abovementioned measurement.

The principle of the Saw-cut method is stress relief initiated by sawing, which fully isolates the concrete block from the acting forces. Stress relief is monitored in the area adjacent to performed saw-cuts. The concrete block can be considered as fully isolated if the increasing depth of saw-cuts no longer causes any significant change of measured strain or stress. Eventually, the results gained from the strain or stress relief measurement enable us to calculate the actual state of the prestressing force in the investigated structure. This method has only a negligible impact on the concrete structure because saw-cuts can be properly fixed. Therefore, the Saw-cut method is usually considered as a non-destructive indirect method.

2. Description of the experiment

Twelve concrete push-out specimens with the label from CS1 to CS12 will be tested in the laboratory. For the experiment, standard concrete testing specimens (figure 2) with a length of 700.0 mm and cross-section dimensions of 150.0 mm will be prepared. Beams will be produced from low-strength concrete of grade C30/37. Since only compressive stress in specimens is considered, no reinforcement will be used. In addition to specimens, nine cylinders (150.0 × 300.0 mm) for mechanical properties testing will be produced to determine compressive strength and modulus of elasticity of concrete.

Figure 2. Preparation of concrete specimens in the laboratory.

To simulate different conditions in the real prestressed concrete structures, non-identical level of compressive stress will be initiated in each pair of specimens. Specifically, the range of compressive stress in specimens will be from 2.0 MPa to 12.0 MPa. The saw-cuts application will be performed in the central area of the specimens, while the 5 cm long foil strain gage will be placed in the middle of the axial distance between adjacent saw-cuts. Figure 3. presents the testing scheme with the labelling of saw-cuts applied in the experimental specimen.

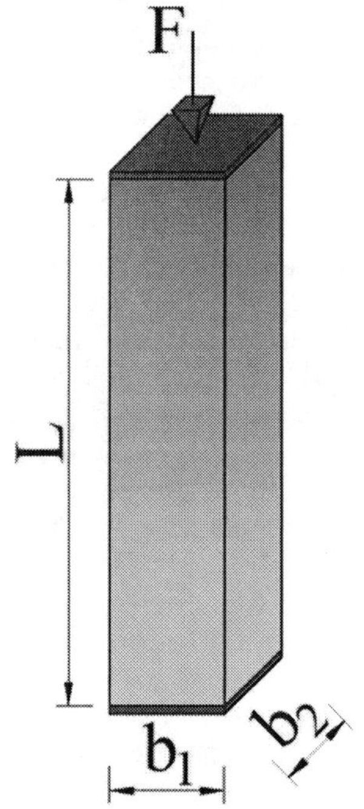

Figure 3. Scheme of the concrete specimen.

The realization of the Saw-cut method will be performed after compressive stress in the concrete specimen reaches the intended value. The parameters of specimens, external load, level of compressive stress during push-out testing, and applied saw-cuts are presented in table 1.

Table 1. Parameters of the specimens and saw-cuts.

Specimen	Specimen dimensions			F [kN]	σ_c [MPa]	Saw-cut parameters	
	b_1 [mm]	b_2 [mm]	L [mm]			d [mm]	h [mm]
CS1, CS2	150.0	150.0	700.0	45.00	2.00	100.0	35.0
CS3, CS4	150.0	150.0	700.0	90.00	4.00	100.0	35.0
CS5, CS6	150.0	150.0	700.0	135.00	6.00	100.0	35.0
CS7, CS8	150.0	150.0	700.0	180.00	8.00	100.0	35.0
CS9, CS10	150.0	150.0	700.0	270.00	12.00	100.0	35.0
CS11, CS12*	150.0	150.0	700.0	-	-	100.0	35.0

*Verification of the influence of temperature change caused by sawing.

3. Numerical analysis

In order to analyse the stress change in concrete specimens, a 2D numerical model with the assumption of nonlinear material behaviour in ATENA Software was performed. Quadrilateral finite element mesh with a defined element size of 0.01 m was generated automatically (figure 4) for all used macro-elements. Dimensions of concrete specimens and saw-cuts were used in accordance with table 1, while the thickness of steel plates placed on the edges of the specimens was 1.0 cm. The contact between the concrete specimen and steel plates was considered as rigid. Material properties of macro-elements are presented in table 2.

Table 2. Material properties of macro-elements used in a 2D numerical model.

Macro-element	Description	Material type	E [GPa]	Other properties
1	Concrete specimen	SBeta	28.06	$f_{cu} = 25.00$ MPa $f_c = 21.25$ MPa $f_t = 2.052$ MPa $\mu = 0.20$
2 and 3	Steel plates	Plane Stress Elastic	210.00	$\mu = 0.30$
4 and 5	Saw-cuts	Isotropic	1.00×10^{-6}	$\mu = 0.30$

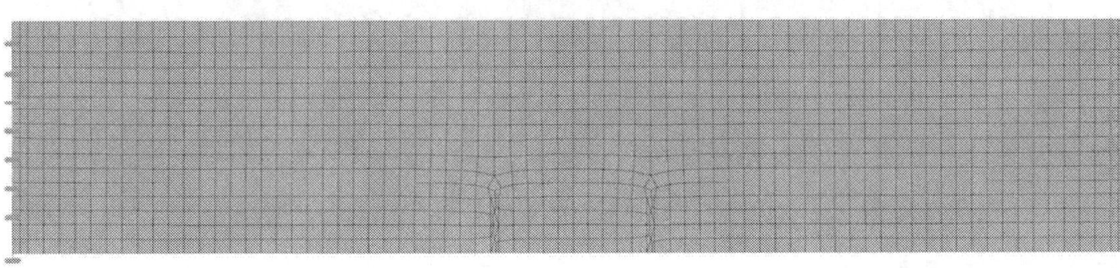

Figure 4. A 2D numerical model of the push-out specimen in ATENA Software.

The analysis using the Newton – Raphson method was divided into two construction stages (figure 5.). First, the specimen is supported on the upper and lower edge on the steel plates, while the load is applied on the upper steel plate longitudinally to the axis of the specimen. Second, the load remains constant and the saw-cuts are performed. The application of the Saw-cut method is exhibited by the reduction of modulus of elasticity of macro-elements which represent saw-cuts. Stress monitoring point is placed in the middle of the axial distance between adjacent saw-cuts.

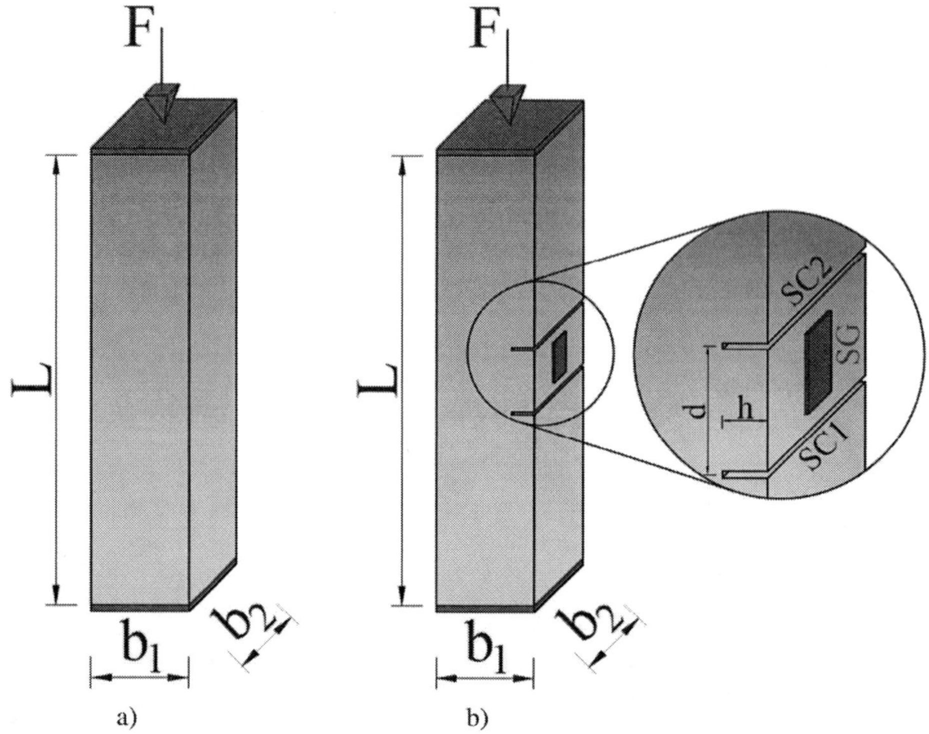

Figure 5. Construction stages of numerical analysis – a) first stage; b) second stage.

Since the initial stress in the concrete structure in practice is unknown, the stress (or strain) relief is presented as percentage change which well describes the rate of isolation of concrete block. The distribution of compressive stress in the concrete specimen with the initial stress level of 2.0 MPa before and after application of saw-cuts is shown in figures 6, 7 8.

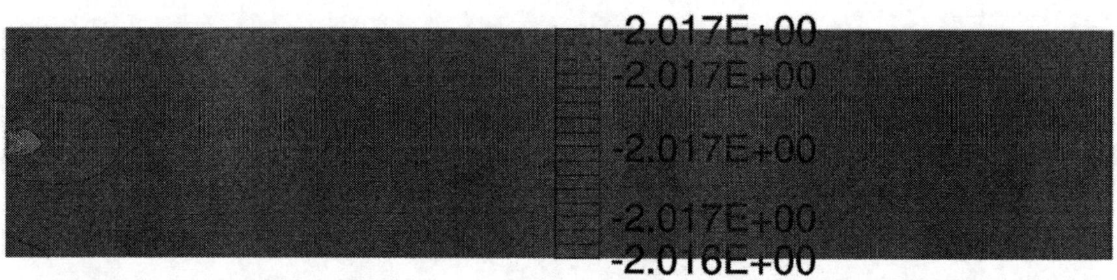

Figure 6. Stress [MPa] in concrete push-out specimen CS1/CS2 before application of saw-cuts.

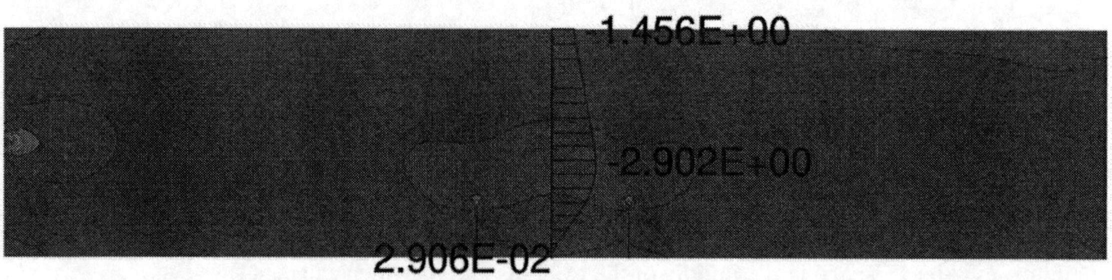

Figure 7. Stress [MPa] in concrete push-out specimen CS1/CS2 after application of saw-cuts.

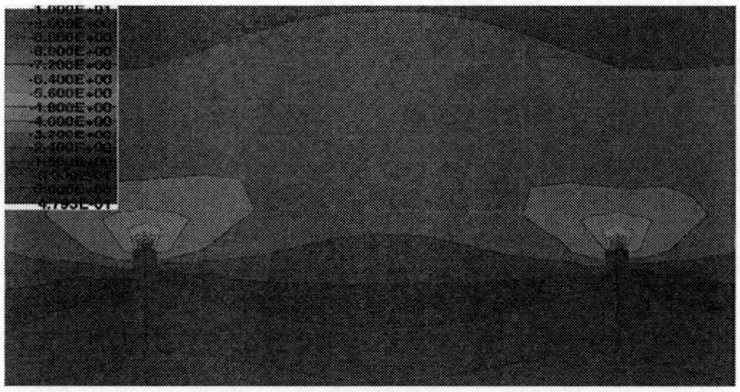

Figure 8. Stress [MPa] in concrete push-out specimen CS1/CS2 after application of saw-cuts (detail).

4. Conclusions

Saw-cut parameters for push-out testing are based on the results of numerical analysis. The goal was to find the relation between depth and axial distance of saw-cuts which would result in full stress relief in monitored point. The numerical analysis indicates that the level of initial stress does not significantly influence the percentage change of stress or strain after application of saw-cuts (figure 9 and table 3). This fact is promising for the planned experimental push-out testing and application of this technique in real structures in practice. Monitoring of influence of the temperature change will be very important for the description of measurement accuracy in real structures. In case of its significant effect, the consideration of a modification of saw-cuts parameters, especially the axial distance, will be necessary.

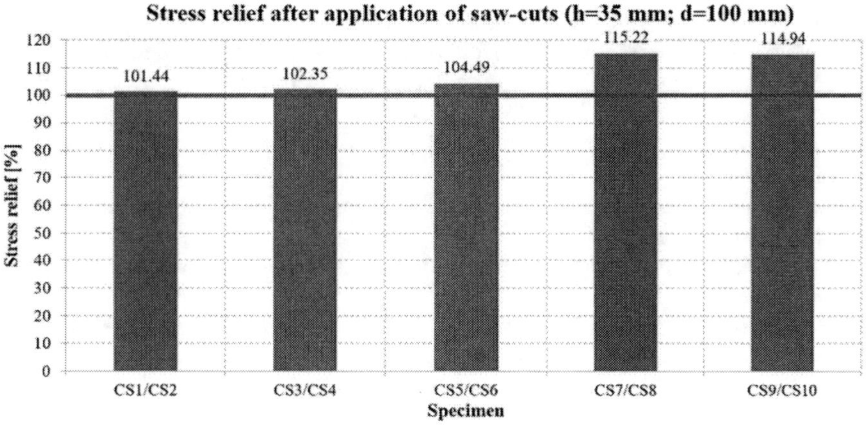

Figure 9. Stress relief in concrete push-out specimens after application of saw-cuts.

Table 3. Stress relief results obtained from a 2D numerical model.

Specimen	Saw-cut parameters		Stress		Stress relief
	h [mm]	d [mm]	σ_{c1} [MPa]	σ_{c2} [MPa]	$\Delta\sigma_c$ [%]
CS1, CS2	35.0	100.0	-2.02	0.03	101.44
CS3, CS4	35.0	100.0	-4.02	0.09	102.35
CS5, CS6	35.0	100.0	-6.02	0.27	104.49
CS7, CS8	35.0	100.0	-8.01	1.22	115.22
CS9, CS10	35.0	100.0	-12.00	1.79	114.94
CS11, CS12 *	35.0	100.0	-	-	-

*Verification of the influence of temperature change caused by sawing.

Acknowledgements
This research project was supported by the Slovak Grant Agency under contract No. 1/0045/19.

References

[1] Bujňáková P and Strieška M 2017 Development of Precast Concrete Bridges During the Last 50 Years in Slovakia *International scientific conf. on sustainable, modern and safe transport TRANSCOM 2017* Procedia Engineering, **192** pp 75-79 DOI:10.1016/j.proeng.2017.06.013

[2] Bagge N, Nilimaa J, Blanksvärd T and Elfgren L 2014 Instrumentation and Full-Scale Test of a Post-Tensioned Concrete Bridge *Nordic Concrete Research* **51** pp 63-83

[3] Bagge N, Nilimaa J and Elfgren L In-situ 2017 Methods to Determine Residual Prestress Forces in Concrete Bridges *Engineering Structures*

[4] fib, Comité Euro – International du Béton 1998 Strategies for Testing and Assessment of Concrete Structures *Guidance Report* **243** fib Bulletin May 1998

[5] Ciolko A and Tabatabai H 2010 Nondestructive Methods for Condition Evaluation of Prestressing Steel Strands in Concrete Bridges *Structures Congress NCHRP* (Web Document 23 Project 10-53 1999)

[6] Moravčík M and Bujňáková P 2017 Analysis of Brittle Failure of Prestressed Bridge in the North of Slovakia *ROCZNIKI INŻYNIERII BUDOWLANEJ* **17**

[7] Moravčík M Bujňáková P and Bahleda F 2019 Conceptual Problems of First Generation of Precast Bridges. *International fib Symposium on Conceptual Design of Structures* (Instituto de Ciencias de la Construcción Eduardo Torroja Madrid Spain September 26-28)

[8] Moravčík M, Bujňáková P and Bahleda F 2017 Príčiny havárie a rekonštrukcia predpätého mosta v Podbieli. *BETON* **4** (in Slovak)

[9] Moravčík M, Bahleda F and Jošt J 2018 Využitie nepriamych metód stanovenia úrovne predpätia v konštrukcii *Betonárske dni* 2018 *Zborník príspevkov* November 8-9 2018 (in Slovak)

[10] Klier T, Míčka T, Polák M, Plachý T Hedbávný M, Jelínek M and Bláha F 2018 Application of the Modified Magnetoelastic Method and an Analysis of the Magnetic Field. *Acta Polytechnica CTU Proceedings* **15** 46-50

[11] Červenka V, Jendele L and Červenka J 2018 ATENA Program Documentation – Part 1. Theory (Prague January 26)

[12] Červenka V and Červenka J 2015 ATENA Program Documentation – Part 2-1. User's Manual ATENA 2D (Prague May)

[13] Červenka J 2015 ATENA Program Documentation – Part 4-1. Tutorial for Program ATENA 2D (Prague June 2)

[14] Janda Z and Červenka J 2009 ATENA Program Documentation – Part 4-3. Tutorial for Construction Process Modelling in ATENA 2D (Prague March 9)

Results of the study of vibrations of concrete structures of the hydropower plant building

S V Evdokimov, V A Seliverstov and A A Orlova

Samara State Technical University, Molodogvardeyskaya st., 244, Samara, Russia, 443100

E-mail address: sali5@mail.ru

Abstract. The paper considers the results of studying vibrations of certain components of the hydropower plant building and their influence on the structure of the building as a whole. Vibration loads on various elements of the hydropower plant building which can cause damage to the building, thereby reducing the operational reliability of the whole structure and doing potential harm to the environment and human life support, is analyzed. The necessity for vibration monitoring of structures which should be conducted constantly or periodically to determine the present condition of structures is reasoned. Vibrations of the section concrete parts and vibrations of hydroelectric power plant units are considered. To detect the vibrational movements of the section as a solid body on an elastic base Section № 2of the hydropower plant building has been chosen. The recording of vibrations has been conducted for all three components: Z (vertical), H II (horizontal, parallel to the flow) and H ⊥ (horizontal, perpendicular to the flow) and the obtained data are given in this paper. The intervals of observed and dominant periods of vibrations are estimated. The research provides the data on amplitudes and accelerations of vibrations, as well as on the maximum recorded peak-to-peak swing. Based on the analysis of the received vibrograms, it is stated that the recorded vibrations are non-periodic and have a relatively narrow range of periods. In conclusion it is underlined that in order to determine the maximum possible amplitudes of section vibrations as well as to calculate the corresponding dynamic stresses in ceiling trusses it is necessary to observe vibrations when hydraulic units are operated simultaneously and the flow passes through the fully open bottom spillways of the inspected section.

1. Introduction

The most important challenge when operating hydraulic structures is to ensure their safe performance. This area of the research is of high priority for scientists of the Samara State Technical University [1-5]. In this regard, the detailed study of the condition of objects that have been in direct contact with the pressure waterfront for a long period and (or) subject to constant dynamic impacts is particularly relevant and important [6-10].

A typical example of such structures is the building of a hydropower plant (HPP), structural elements of which are dynamically affected by the rotating masses of hydraulic units and the water flow.The building of the Zhigulevskaya HPP has been in operation for a long time and is subject to various vibration loads, both natural and man-caused. Moreover, vibration loads on various elements of the hydropower plant building can cause damage to the building structure, thereby reducing the operational reliability of the HPP building itself and doing potential harm to the environment and human life support. Loss in operational reliability can appear in the form of cracks, damaged elements,

and other structural changes. Vibration monitoring of structure condition should be conducted constantly or periodically to estimate the present state of structures [11].
Recent investigations lack sufficient data that allow determining a correspondence between the vibration severity grade and the damage caused by it.

2. Materials and methods

To study the vibrations of the Zhigulevskaya HPP building, it is proposed to estimate the influence of certain parts of the hydropower plant building on the structure of the building as a whole. Let us consider the vibrations of concrete parts of the Zhigulevskaya HPP section. To detect the vibrational movements of the section acting as a solid body on an elastic base, section № 2 of the Zhigulevskaya HPP building has been chosen. The devices were installed in the following point locations of the section:

1- power house, at level 41.30, the middle of the section;
2- at level 37.25, the middle of the section;
3- at level 37.25, the left bank face of the section;
4- at level 37.25, the right bank face of the section;
5-crane beam over the ground gate, at level 41.50 m, the middle of the section.

Vibrations of the first (counting from the right bank) support of the high-voltage line tower of this section have been also recorded (point location 6). Vibrations have been recorded in all three components: Z (vertical), H II (horizontal, parallel to the flow) and H ⊥ (horizontal, perpendicular to the flow). Sample recording of section vibrations is shown in figure 1.

Based on the analysis of vibrograms obtained for point locations 1-5 the following facts can be stated: recorded vibrations are non-periodic, but they have a relatively narrow range of periods. Based on the research results, histograms of vibration periods for all three components discussed above have been constructed. The intervals of observed periods have been put on the *X*-axis, and the probability of the period occurrence of the corresponding interval - on the *Y*-axis. The time intervals between two adjacent humps or depressions have been assumed to be the periods of vibrations when processing oscillograms. Histograms allowed determining the intervals of observed and dominant periods (table 1).

Due to the fact that the recorded vibrations are not periodic, and the rotation period of hydraulic units (0.88 s) lies outside the dominant periods of section vibrations, the observed vibration of the HPP section cannot be explained by the unbalance of the units' rotors.

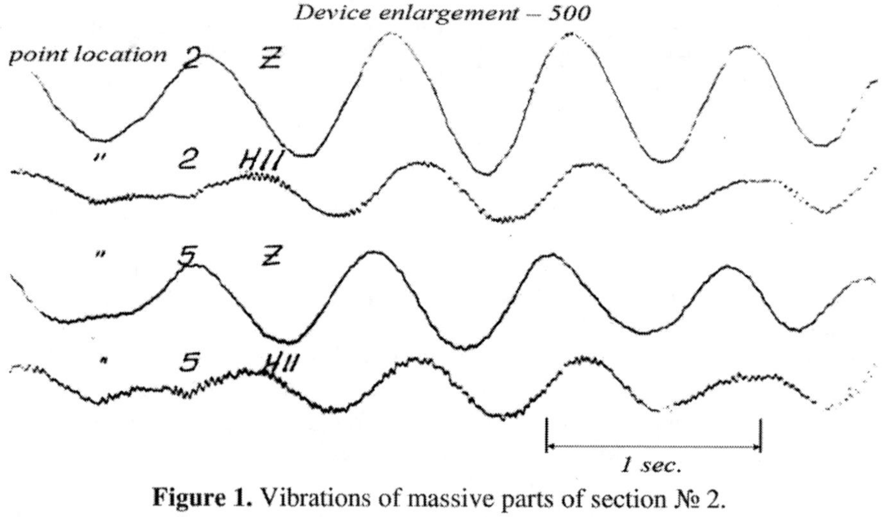

Figure 1. Vibrations of massive parts of section № 2.

IOP Conf. Series: Materials Science and Engineering 1015 (2021) 012032 doi:10.1088/1757-899X/1015/1/012032

Table 1. Intervals of the observed and dominant periods of vibrations.

Location of devices	Component	Dominant periods of vibrations, (s)	Range of observed periods of vibrations, (s)
Section №2, point locations 1-5; levels 37.25-41.50	Z	0.6÷0.8	0.3÷1.0
	H II	0.6÷0.8	0.3÷1.1
	H ⊥	0.6÷0.8	0.3÷1.1

At the same time the vibration records, obtained at closed spillways, show that in this case the sections of the HPP experience periodic vibrations with the period that lasts 0.88 seconds and coincides with the period of hydraulic units rotation and with an amplitude of no more than 3-4μm.

3. Results and discussions

Having compared the results of observations at operating and non-operating bottom spillways, it can be concluded that the vibration of HPP sections during the operation of hydroelectric units alone is rather insignificant, and the main vibration source is the pressure pulsation that occurs when flow passes through the bottom water outlets.

Data on amplitudes and accelerations of vibrations are given in table 2. The third column of the table shows the maximum observed peak-to-peak swings (double amplitudes) of vibrations $2A_{max}$, the fourth column shows the average quadratic amplitude A_δ which has been calculated in the standard way based on the records, the fifth one shows the amplitudes maximum for this hydraulic mode $A_{3\delta}$, which have been calculated using the "three-sigma rule". The last column gives the values of conditional accelerations which have been calculated using the formula of the sinusoidal vibration law and correspond to the maximum amplitudes $A_{3\delta}$and dominant periods. Thus, vertical vibrations in amplitude are the largest, and vibrations in direction H ⊥ are the smallest.

The records of tower support' vibration (point location 6) indicate the following:

- vibrations in directions Z and H⊥ are identical in periods and amplitudes to vibrations measured at point locations 1-5;
- for vibrations H II, the interval of dominant periods is 0.4÷0.6 s, and amplitudes also increase.

Table 2. Data on amplitudes and accelerations of vibrations.

Location of devices	Component	Maximum observed peak-to-peak swing $2A_{max}$ (μm)	Average square amplitude A_δ (μm)	Maximum amplitude $A_{3\delta}$ (μm)	Maximum acceleration $Y_{3\delta}$(cm/s^2)
Section № 2 point locations 1–5 levels 37.25-41.50	Z	87	20	60	0.66
	H II	54	14	14.8	0.50
	H ⊥	26.6	-	≈20	≈0.50

Data on dominant periods, amplitudes and accelerations are shown in table 3.

IOP Conf. Series: Materials Science and Engineering 1015 (2021) 012032 doi:10.1088/1757-899X/1015/1/012032

Table 3. Data on dominant periods, amplitudes and accelerations.

Location of devices	Component	Dominant periods T (s)	Maximum observed peak-to-peak swing $2A_{max}$ (μm)	Average square amplitude A_{δ} (μm)	Maximum amplitude $A_{3\delta}$ (μm)	Maximum acceleration $Y_{3\delta}$(cm/s^2)
Section № 2	Z	0.6÷0.8	47	-	≈60	≈0.66
point location 6	H II	0.6÷0.8	122	34.2	103	2.54
tower support	H ⊥	0.6÷0.8	37	-	≈20	≈0.50

In some recording areas H II angular points can be observed that indicate abrupt changes in the velocity of vibration motion at certain time intervals. For the component H II (tower support) higher accelerations can be expected than it is indicated in table №3. The increase in vibrational amplitudes of the tower support under the component H II can hardly account completely for the movement of the section as a solid body (the increase of the horizontal component with height). It is most likely due to bending vibrations of the power house wall in the direction H II.

Taking into account that at some areas of recording vibrations of the section and the power support there can be observed vibration groups with approximately the same period, it is recommended to make a verification calculation of periods of the high-voltage line tower' free vibrations. Their occurrence within the range of 0.3-1.0 seconds is undesirable due to the possible coincidence of forced and natural frequencies and the related large dynamic effect.

To measure vibrations of hydraulic units the vibrographs have been installed with the approval of the maintenance service at the following point locations:

1- on generator bearings №№ 1, 2, 3- respectively point locations 12, 22, 32;
2 - on covers of hydraulic unit turbines №№ 1, 2, 3- respectively point locations 13, 23, 33;
3 - on upper generator brackets №№1, 3, 4- respectively point locations 14, 34, 44;
4 - at the remote end of the stiffener plate of upper generator bracket № 3- point location 35.

Besides the devices located on the parts of hydraulic units in indicated places, the sensors have been also installed on the concrete floor of the power and turbine houses (at levels 41.3m and 37.25m). All sensors have been installed from the upstream side respect to the axis of the unit. Vibrations have been recorded for all three elements: vertical Z, horizontal in the direction of the flow H II, and horizontal across the flow H ⊥.

Analysis of the obtained vibrograms shows the following:

1. Vertical vibrographs mounted on the turbine cover record steady almost periodic vibrations (Fig. 2), which maximum recorded swings (double amplitudes) and periods are shown in table. № 4;
2. As may be supposed, it is possible to correlate the vibration occurrence either with free vibrations of individual elements of the units, or with the turbine rotation, as the period of the turbine blades move coincides with recorded values;
3. All devices record non-periodic vibrations with a peak-to-peak swing of about several tens of microns and dominant "periods" of 0.6÷0.8 seconds. These vibrations are nothing more than vibrations of the hydropower plant building, which are transmitted to the hydraulic units installed in the section.
4. Vibrations of hydroelectric units related to the unbalance of the rotors in most cases turned out to be less in amplitudes than the HPP building vibrations, and therefore it is impossible to distinguish them from the vibration of the building.

In some cases, when the level of hydroturbine vibration of the units exceeded the level of vibrations of the HPP building, vibrations with a constant period coinciding with the rotation period of the unit (0.88 s) have been recorded on the vibrograms. Their average and largest observed double amplitudes are shown in table 4.

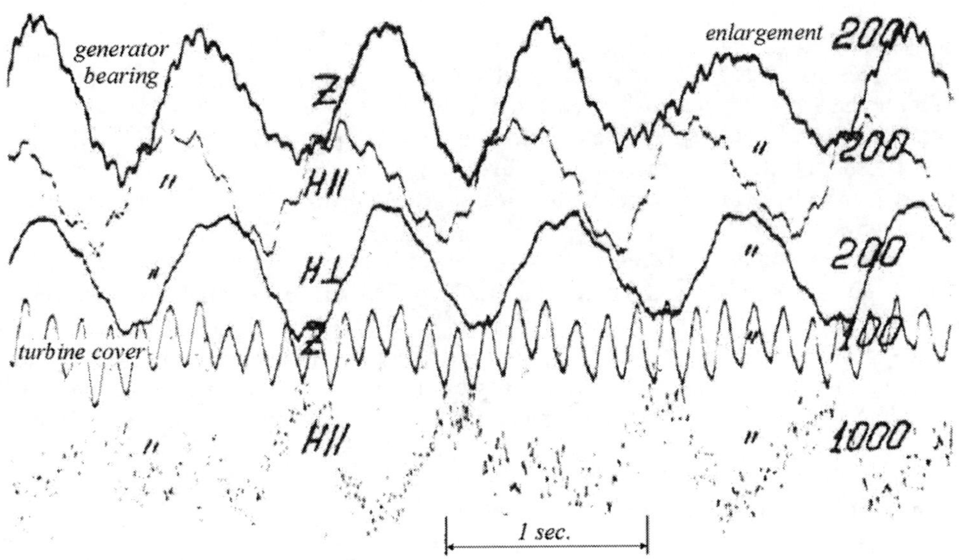

Figure 2. Vibrations of unit № 2.

Table 4. Data on the maximum recorded swings.

Measurable value	Unit №1	Unit № 2	Unit № 3
Vibration period (s)	0.146	0.147	0.143÷0.145
The largest recorded swing (μm)	118	196	100
Acceleration (cm/s^2)	10.9	18.0	9.5

In all other cases of measuring it was impossible to distinguish vibrations of hydraulic units with a period of 0.88 seconds from the hydropower plant vibration, therefore it can be assumed that the swing of vibrations with a period of 0.88 seconds in these other cases of measuring was less than 90-100μm. Vibrations with a period of 0.88 seconds are caused by the unbalance of the rotor or insufficient shaft alignment. The average amplitude of these vibrations reaches 67.5μm (180μm is acceptable).

Besides vibrations with a period of 0.88 seconds and swinging related to the movement of the section, devices H II and H ⊥, mounted on bearings of units and upper brackets, have recorded unsteady horizontal vibrations with the dominant period of 0.14-0.15 seconds and average peak-to-peak swing of about 30-40μm, which are likely to be free vibrations of the upper brackets.

Due to the presence of a relatively large background noise, which, in that case, meant vibrations of the building, it turned out to be impossible to monitor the nature of transmitting vibrations with periods of 0.88 seconds and 0.14 seconds from units to concrete parts of the hydropower plant.

As indicated above, vibrations of the building caused by the pulsation of hydrodynamic pressure in the flow are also transmitted to the hydraulic units installed in the sections.

4. Conclusions

Based on the results of studying vibrations of concrete structures in section № 2 of the Zhigulevskaya HPP building the following conclusions can be drawn:

1) according to the results of study, the observed vibration of section 2 of the HPP building cannot be explained by the unbalance of the rotors of hydraulic units. A number of factors suggest that this vibration is caused by a pulsation of hydrodynamic pressure in the water flow passing through the bottom spillways;

2) the maximum amplitudes of vertical vibrations of section № 2 have been observed under hydraulic conditions - up to 60μm, the maximum amplitudes of horizontal vibrations of the section - up to 45μm.

3) to establish the maximum possible amplitudes of section vibrations, as well as to calculate the corresponding dynamic stresses in the ceiling trusses, it is necessary to conduct vibration observations while simultaneously operating hydraulic units and passing the flow through the fully open bottom spillways of the section being studied;

4) vibrations of the power house' roof, caused by vibrations of section № 2 at the Zhigulevskaya HPP, have been recorded. But it needs more scope of study.

Acknowledgment

The authors of the paper express their appreciation to the Head of the Environmental and Hydrotechnical Engineering Department of Samara State Technical University and the team of the department for their comprehensive scientific advice on the subject of this publication. Special gratitude is expressed to the reviewers of this paper for their work.

References

[1] Standard GOST 52892-2007 Vibration and shock. Vibration of buildings. Measurement of vibration and evaluation of its effects on structure.
[2] Balzannikov M I, Zubkov V A, Kondratieva N V and Khurtin V A 2013 Complex survey of technical condition of the Zhigulevskaya HPP structures, *Hydrotechnical Constr.* **6** pp 21-27
[3] Balzannikov M I, Ivanov B G and Mikhasek A A 2012 Hydraulic structures controllingsystem, *Vestnik MGSU* **7** pp 119-124
[4] Balzannikov M I, Rodionov M V, Seliverstov V A 2011 Improvement of ecological safety of the operated soil hydraulic engineering structures,*Vestnik SGASU. Urban planning and architecture* **1** pp 100-105
[5] Leonov O V, Romanov A A and Evdokimov S V 2011 Analysis of the seismic conditions of the Zhigulevskaya hydroelectric power plant location, *Bull., Urban planning and architecture* **2** pp 109-114
[6] Romanov A A, Ivanov B G and Evdokimov S V 2012 Assessment of seismic resistance of main hydraulic structures of the Zhigulevskaya HPP *Environmental Engineering* **5** pp 49-53
[7] Romanov A A, Leonov O V and Evdokimov S V 2011 Seismic resistance of foundation soils and structures of the main Zhigulevskaya HPP buildings *Vestnik SGASU, Urban planning and architecture* **4** pp 66-72
[8] Balzannikov M I and Seliverstov V A 2010 Onidentification of main parameters for designinga water outlet unit of the sectional type at a large pumping station *Industrial and Civil Engineering* **8** pp 17-19
[9] Balzannikov M I and Kurmanaev A L 2014 Using Flow Guiding Devices in the Rainwater Construction of the Hydropower Installation *Eastern European Scientific Journal (Gesellschaftswissenschaften)* (Germany Düsseldorf: Auris Verlag) **2** pp 282-286
[10] Balzannikov M I and Elistratov V V 2008 *Renewable energy sources. Aspects of integrated use*, (Samara: Samara State Univiversity of Architecture and Civil Engineering) p 331
[11] Romanov A A 2010 *Zhigulevskaya HPP. Hydraulic structuresoperation* (Samara: Publishing house "Agni") **1** p 360

IOP Conf. Series: Materials Science and Engineering 1015 (2021) 012033 doi:10.1088/1757-899X/1015/1/012033

To the optimization model of flexible reinforced concrete elements in monolithic frames of buildings

A G Tamrazyan

Department of Reinforced Concrete and Stone Structures, Moscow State University of Civil Engineering, Moscow, Yaroslavskoe shosse, 26, Russia

E-mail address: tamrazian@mail.ru

Abstract. This article examines a three-span 8-storey frame building with the aim of optimizing the girders based on heuristic algorithms. A comparative analysis of modern heuristic algorithms for optimizing bending elements as applied to reinforced concrete frames of monolithic multi-storey buildings for the possibility of their use in practical engineering calculations. A mathematical model of the adapted algorithm has been formed that meets the requirements of Russian standards. The target function for cost has been determined, including the cost of materials (concrete and reinforcement), formwork, construction with the corresponding rise in price coefficients. In this case, the girder is optimized, since, despite the fact that their number in the building frame is 24, the cost is about 65% of the total cost, including reinforced concrete columns in the amount of 32.5 constant parameters and 19 variable parameters are selected, a set of limitations of the optimization algorithm was developed, including regulatory requirements for the calculation of limit states, design requirements for the reinforcement of the girder and its geometric characteristics, including the concrete cover, the step of the rods, anchoring, etc. The calculation of girders as part of a reinforced concrete monolithic frame of an 8-storey building was carried out using the developed methodology and the comparison of the results obtained with numerical calculations and solutions obtained by other methods. It is shown that the results of numerical calculations have a sufficiently high efficiency of the developed optimization technique. The final cost is after 100.000 iterations. Savings in terms of the main design characteristics range from 3 to 13%, depending on the complexity of the problem and the accepted constraints.

1. Introduction

Design of reinforced concrete elements is a complex task of ensuring the safety and manufacturability of a structure with minimal costs [1]. As a rule, the design of reinforced concrete elements includes two main stages: the assignment of the assumed geometric parameters of the element cross-section and the selection of the required reinforcement for this section. When the criteria for safety or manufacturability in design are met, the economic efficiency of reinforced concrete elements is significantly reduced [2]. When designing any structure, one should not only strive for maximum efficiency of the structure, but also to ensure the strength, rigidity and stability of its elements. The big problem is the number of variable parameters. If it is large, then the number of possible solutions and combinations in the problem becomes so large that it becomes irrational to enumerate them all, the computational cost of the problem grows exponentially. To solve such problems of optimizing design solutions for reinforced concrete elements, modern heuristic algorithms are used that significantly speed up the process of finding the objective function [3].

The bending elements of multi-storey reinforced concrete frames adopted as the object of research are usually considered either as separate structures within various engineering structures, or as a component of the load-bearing frame of multi-storey frame-type buildings [4]. Among the most often used heuristic algorithms for optimizing concrete frames are the simulated annealing (SA), BB-BC, colliding particles (CBO), harmony search (HS), particle swarm (PSO) algorithm [5, 6]. A great contribution to the research of such algorithms for use in the optimization and calculation of reinforced concrete frames was made by the scientists of X. S. Yang and A. Kaveh. Also noteworthy are the studies by A. Akin and M. P. Saka, I. Paya and V. Yepez, Y. C. Toklu, J. Kim, S. Rajeev and C. S. Krishnamoorthy.

2. Optimization model of girders of reinforced concrete frames

The formation of the target function is made necessarily taking into account the following criteria: efficiency, manufacturability and safety [2]. In most works on the study of heuristic algorithms for the optimization of reinforced concrete structures, the function of the element cost is taken as the target optimization function, which includes the totality of costs for concrete and reinforcement, for formwork and the construction of a structure. In general terms, the target function of the design cost of the girder C_r conventional unit (c.u.) can be represented as

$$C_r = C_b + C_{rf} + C_f \tag{1}$$

where C_b -is the cost of the material to be reinforced (c.u.);
C_{rf} - cost of the reinforcing material (c.u.);
C_f - the cost of the formwork and the cost of the work (c.u.).
Here C_b can be represented as

$$C_b = b_r \times h_r \times L_r \times C_c \tag{2}$$

where b_r - cross-sectional width of the girder (m);
h_r - height of the cross-section of the girder (m);
L_r - girder span (m);
C_c - the cost of 1 m^3 of concrete "in business" (c.u. / m^3).

The value of reinforcement C_{rf} is determined as the total value of all reinforcing bars in the girder using the formula

$$C_{rf} = 0.25\pi C_s \gamma_s (\sum_{i=1}^{m} n_i d_i^2 l_i + k d_s^2 l_s) \tag{3}$$

where C_s - the cost of 1 ton of reinforcing steel "in business", (c.u. / t);
γ_s - specific weight of reinforcing steel (t / m^3);
m -is the number of rows of longitudinal reinforcement bars;
n_i - the number of longitudinal reinforcement bars of the i-th row of bars;
d_i - diameter of longitudinal reinforcement bars (m);
l_i - the length of the bars of the i-th row of reinforcement (m);
k - is the number of transverse reinforcement bars;
d_s - diameter of transverse reinforcement bars (m);
l_s - length of cross-reinforcement stirrups (m).

The cost of reinforcement C_{rf} may include the costs of an additional increase in the consumption of reinforcement in the amount of $5 \div 10\%$ of the total longitudinal reinforcement volume, which are used for the manufacture of embedded parts, docking devices, etc.

The C_f indicator is calculated by the formula

$$C_f = C_w L_r (b_r + 2h_r) \tag{4}$$

where C_w is the cost of 1 m^2 of formwork, provided that the upper edge of the girder is open for the production of concrete works (c.u./m^2). C_w also includes the costs of construction production: performance of reinforcement, formwork, concreting and auxiliary works (bending, cutting, welding of reinforcing bars, etc. in the amount of $5 \div 10\%$ of the main work). Further, this parameter is usually called the "cost of formwork", omitting the additional costs of production.

If necessary, the target function of the cost of the girders C_r can be multiplied by increasing factors that take into account the rise in the cost of work in the winter, as well as changes in the cost of materials and work due to inflation.

Thus, based on the above, the target function of optimizing the girder as part of the frame can be described by the formula

$$C_r = b_r h_r L_r C_c + 0.25\pi C_s \gamma_s (\sum_{i=1}^{m} n_i d_i^2 l_i + k d_s^2 l_s) + C_w L_r (b_r + 2h_r) \quad (5)$$

Determination of the main constant and variable parameters of the algorithm

Of the fourteen parameters of the girders optimization objective function proposed above, five are designated as constant and nine - as variables. It should be noted that the varied parameters must necessarily be discrete in order to be able to implement them in real design conditions.

For the remaining nine variable parameters, the range of acceptable values is set. The range of permissible values is determined based on design considerations and the requirements of the manufacturability of the design. In this case, for the geometric dimensions of the section $b_r \times h_r$, the following restrictions are adopted, reflecting the requirements [7]:

- for section width b_r: values from 250 to 500 mm with a step of 50 mm;
- for section height h_r: values from 400 to 700 mm with a step of 50 mm.

The number of longitudinal reinforcement bars of the bottom row: n_1=3÷5; additional bottom row: n_2=2÷4; top row: n_3=2÷4.

The range of used diameters of reinforcement bars is limited to the range of the most common bars in the design of real structures, regardless of the number of reinforcement: d_i=12÷32 mm.

The number of rows of longitudinal reinforcement bars m is ultimately determined when calculating the actual bearing capacity of the section with the adopted reinforcement option and section dimensions. For the problems under consideration, the values m=2 or 3 are taken.

The length of the longitudinal reinforcement bars l_i generally depends on the arrangement row m, and for the span reinforcement of the lower row m=1 the length value $l_1=L_r+40\,d_i$, where 40 d_i - is the anchorage length of the bars of the lower row. For the spanning reinforcement of the second row $l_2=0.8\,L_r$, for the upper row $l_2=L_r+40\,d_i$.

The diameter of the transverse reinforcement d_s is taken in the range d_s=6÷10mm. The number of stirrups of transverse reinforcement k, depending on their pitch in the span and on the supporting sections, is determined by calculation. In order to unify the step of the cross-reinforcement stirrups in the girders near the support sections ($0.25L_r$) is taken equal to 150 mm, in the span - 300 mm. As a result, the parameter of their number k can be approximately determined, while rounding the resulting value up

$$K = \frac{0.5L_r}{0.3m} + 2\frac{0.25L_r}{0.15m} \approx ceil(L_r \times 5\frac{pcs}{m}) \quad (6)$$

The length of the stirrups of the transverse reinforcement l_s directly depends on the geometric parameters of the width and height of the section and is found by the formula

$$L_s \approx 2(b_r + h_r) \times 0.9 \quad (7)$$

The main variable parameters of the girder cost objective function and some design constraints are shown in figure 1 and are summarized in table 1, the penult column of which indicates the range of values of the variable parameters.

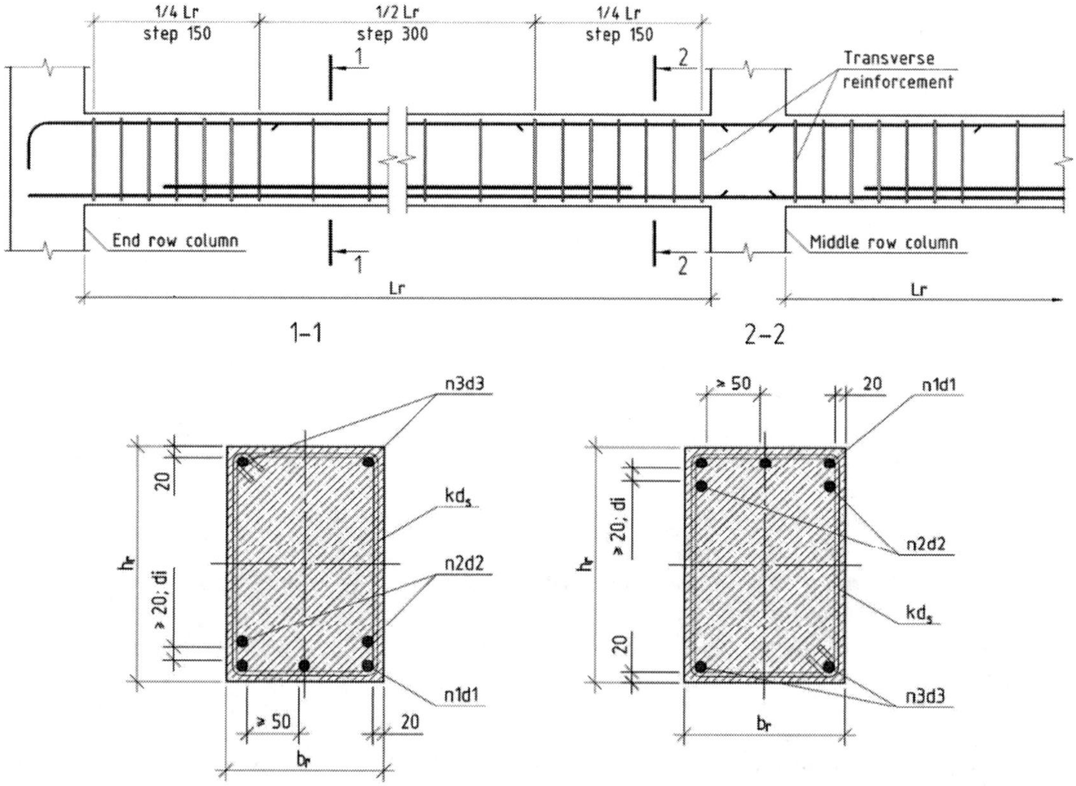

Figure 1. Main variable parameters of the girder cost objective function (the number and diameter of reinforcement bars are shown conditionally).

Table 1. Main variable parameters of the girder cost objective function.

Denomination	Designation	Unit measurements	Value / step	Number of possible values
Section width	br	m	250÷500/50	6
Section height	hr	m	400÷700/50	7
Number of rods, bottom row1	n_1	pcs.	3÷5/1	3
Number of rods, bottom row 2	n_2	pcs.	2÷4/1	3
Number of rods, top	n_3	pcs.	2÷4/1	3
Number of rows	m	pcs.	2÷3/1	2
Diameter of longitudinal bars	d_i	mm	12÷32	9
Stirrups diameter	d_s	mm	6÷10	3
Length of longitudinal bars	l_i	m	depending by m	-
Stirrups length	l_s	m	by formula 7	-
Number of stirrups	k	pcs.	by formula 6	-

As additional variable parameters of the algorithm, it is possible to introduce classes of concrete and reinforcement. In this case, for the class of concrete, values from B25 to B40 are accepted as the most common in practice among structures of this type. The reinforcement classes are limited to those used in non-stressed structures: A400, A500C and A600.
The main characteristics of these reinforcing bars are presented in table 2.

Table 2. Mechanical properties of reinforcing steels

Mechanical properties of reinforcement (MPa)	A400	A500C	A600
Temporary tensile strength	590	600	885
Yield point	390	500	590
Design tensile strength	365	450	520
Elongation after rupture δ_s, %	14	14	6

2.1. Algorithm conditions and limitations

The main limitations of the g_i algorithm include strict compliance with the requirements of the strength and bearing capacity of elements, structural requirements for reinforcement, including the minimum percentage of reinforcement, requirements for the minimum size of the concrete cover, etc. The main regulatory document for the introduction of all restrictions are the 1st and 2nd groups of limiting states and design restrictions [8]. The performance of these limitations of the objective function ensures the safety and reliability of the structure during its normal operation.

With these restrictions, the minimum concrete cover thickness is assumed to be 20 mm, as well as not less than the diameter of the working rods. The minimum clear distance between the rods is at least 50 mm, between the rows of tensile reinforcement - at least 20 mm and more than the maximum accepted diameter of the reinforcement rods.

2.2. Heuristic optimization algorithm

This paper provides for the use of the heuristic HS algorithm in order to solve the optimization problem. The HS algorithm was chosen as considered due to the fact that it has a sufficiently high convergence rate to the global optimum of the objective function [9], contains a small number of specific parameters for its adjustment, and is also widely used in solving optimization problems of various building structures [10].

For such an algorithm, first of all, the design constants of the objective function, the ranges of variation of the main variable parameters (indicated in table 1), as well as the specific parameters of the algorithm are set.

According to [9, 11] and [10], the variables HMCR = 0.85 are written into the specific parameters of the algorithm; PAR = 0.17 and HMS = 50. Such values of these variables allow achieving good performance when optimizing the objective function according to [11].

The concrete class is B30, the class of reinforcement is A500C.

The initial matrix of variable parameters of the construction h_r, b_r, n_i, d_i, d_s, m, m is randomly generated and written into the memory of the algorithm, then the conditions and constraints g_i set for the algorithm are checked, and the objective function C_r is calculated for each of the possible solutions in the initial matrix.

Next, a new solution is generated by changing the varied parameters in one direction or another with a certain probability (i.e., a new solution is not generated randomly). The new solution is written into the memory of the algorithm over the previous one - the most expensive in comparison with the rest. Then the algorithm is repeated until the moment when the optimization goal is achieved.

The stop criterion of the algorithm (100.000 iterations) is the moment when the maximum difference between the cost of the last five variants of one iteration of the algorithm is less than 2%. The stage of generating new solutions includes checks of all design constraints of the g_i algorithm.

In general, the girder optimization problem can be represented as

$$\text{Minimize } f(x_i) = C_r \quad \text{provided: } x \in X_i \quad i=1,2,\dots, N \qquad g_j \le 0 \quad j=1,2,\dots,M \tag{8}$$

where

X_i - set of design parameters for variable i;
N - is the number of variables in the algorithm;
g_j - constraints of the objective cost function;
M - is the number of constraints on the objective cost function.

The algorithm memory matrix (the so-called HM matrix) is composed of a set of design parameters for the algorithm. The pool of design parameters is initialized at the first step of the algorithm from a discrete set of possible values for each variable parameter. The algorithm memory matrix is written in the following form:

$$[H]=\left[\begin{pmatrix} x_1^1 & x_2^1 & \dots & x_{N-1}^1 & x_N^1 \\ x_1^2 & x_2^2 & \dots & x_{N-1}^2 & x_N^2 \\ \dots & \dots & \dots & \dots & \dots \\ x_1^{HMS-1} & x_2^{HMS-1} & \dots & x_{N-1}^{HMS-1} & x_N^{HMS-1} \\ x_1^{HMS} & x_2^{HMS} & \dots & x_{N-1}^{HMS} & x_N^{HMS} \end{pmatrix}\right] \tag{9}$$

Each row in this matrix corresponds to a vector of a possible solution to the problem, i.e. optimal solution candidate and determines a set of variable parameters for its cost function. Moreover, each vector of a possible solution necessarily satisfies all the constraints of the algorithm, and a cost function is calculated for it. The number of columns of the matrix HM corresponds to the number of variable parameters of the algorithm, and the number of rows is limited by the amount of memory of the HMS algorithm. A prerequisite for the third step is to sort the candidates in ascending order of the value of the cost function.

The synthesis of a new solution vector at the fourth step is performed in accordance with the following rules:

1) by setting the probability of accessing the values already written in the memory matrix using the parameter HMCR=0.7÷0.95. The value of the first variable parameter $x_1^{'}$ with a given probability is chosen and the corresponding (first) column of the matrix HM: $\{x_1^1, x_1^2, x_1^3, \dots, x_1^{HMS-1}, x_1^{HMS}\}^T$. The values of the remaining variable parameters are selected in a similar way from the corresponding columns. With a probability of 1-HMCR, the values of the variable parameters can be selected at random from the pool of design solutions for the problem. This rule can be represented as:

$$[H]=\begin{cases} x_i^{'} \in \{x_1^1, x_1^2, \dots, x_1^{HMS}\}^T \text{with probability HCMR} \\ x_i^{'} \in X_I \qquad\qquad \text{with probability 1 - HCMR} \end{cases} \tag{10}$$

2) each variable of the new vector, if it was selected from the memory matrix of the algorithm, is estimated through the parameter of the probability of additional change in the variable parameter PAR=0.15÷0.20. With a given probability, the selected parameter $x_i^{\prime}$, which is the k-th element of the pool of values X_i, is replaced with the adjacent parameter x_i (k±1) from this pool. This rule was introduced into the algorithm in order to prevent stagnation of variable variables and to diversify the memory matrix, which ultimately increases the chance of finding the global optimum of the objective cost function.

As noted above, the vector of the new solution at the fifth step of the algorithm is checked for all the established constraints. The cost of the synthesized solution is calculated. This vector has the right to overwrite the least economical solution contained in the HM memory only if the cost is lower in comparison with it and all the constraints of the algorithm are checked.

3. Optimization model of girders of reinforced concrete frames

The algorithm for optimizing reinforced concrete structures is proposed to be implemented using the MATLAB software package from The MathWorks. The package uses an interpreted programming language M-language, focused primarily on working with matrices and data vectors, as well as having a wide range of capabilities for implementing various kinds of algorithms, data visualization and interaction with.

As the output of the program, a file of the main properties of the model is generated in the form of coordinates of nodes, geometry, material properties and other initially entered data; it also contains the results of strength analysis - components of the SSS system: mainly, efforts in the rods, displacements of specific nodes, a journal with the history of calculation etc.

In order to make it possible to carry out a comparative analysis, as well as to verify the results obtained, as test problems in this work, it is proposed to use the design scheme of a three-span, eight-story frame (figure 2), described in the works of A Kaveh, O Sabzi [12] and S Gholizadeh, V Aligolizadeh [13].

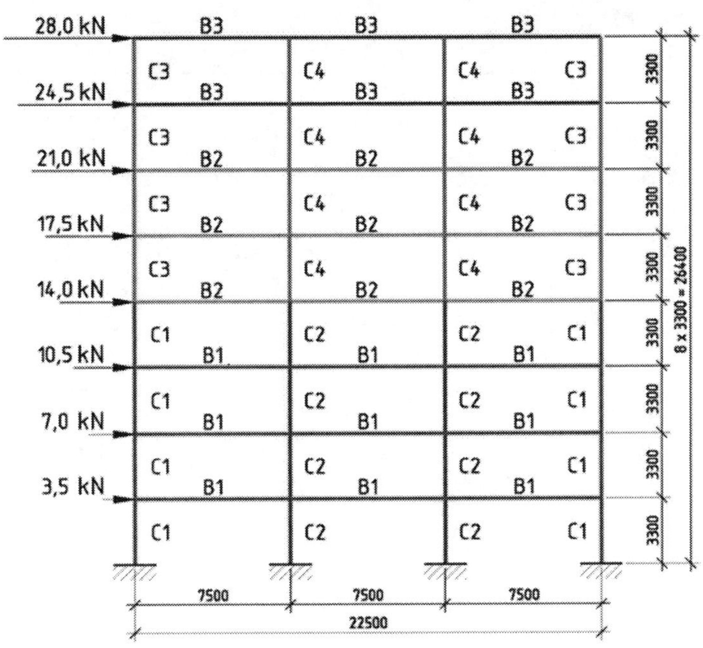

Figure 2. Design scheme of a multi-storey frame.

The height of the floor is taken as constant 3.3 m, and the spans are 7.5 m.

In the design model of loads, several load cases are considered in accordance with the requirements of the standards.

The design scheme and design operating conditions also remain constant, not considering possible technogenic loads [14,15,16].

This diagram contains 24 girders, 32 columns. The memory size for this problem is 50 initial solution vectors. The total cost after 100,000 iterations expired was 29,000 (c.u.).

Figure 3 shows the history of the convergence of the crossbar cost objective function to this value. Table 3 compares the main output parameters of the calculation with the results of studies using the BB-BC-algorithm [12].

It should be noted that the constant parameters of the cost of the main structural materials and formwork for all tasks are taken the same and equal:

- the cost of concrete: C_C=105 c.u./m^3;
- the cost of reinforcing steel: C_S=900 c.u./t;
- the cost of formwork: C_W=92 c.u./m^2.

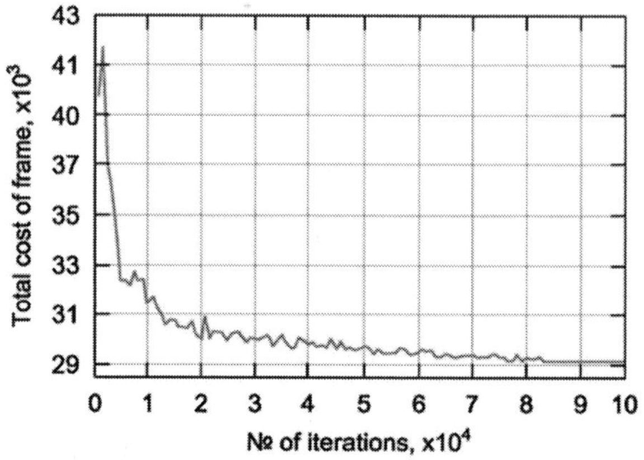

Figure 3. Graph of convergence of the girder cost objective function.

Table 3. Girders optimization results and comparison with BB-BC algorithm.

Elements		HS- algorithm			BB-BC- algorithm [12]		
		Section sizes mm		The main	Section sizes, mm		The main
Type	Group	Width	Height	reinforcement	Width	Height	reinforcement
	B1	300	500	5Ø22	300	500	6Ø22
Girders	B2	300	500	5Ø22	300	500	6Ø22
	B3	300	500	4Ø22	300	500	5Ø22
Cost, c.u.			29000			29730	

The final cost difference between the synthesized solution and the one obtained in [12] using the BB-BC heuristic algorithm is 2.52% in favor of the HS-algorithm.

4. Conclusions

1. The spatial frame system made of reinforced concrete is quite complex in mathematical optimization. Therefore, as an object of research, a reinforced concrete girder was adopted as a component of the supporting frame of a three-span, eight-story frame-type frame.
2. The structure of the objective function of a monolithic reinforced concrete girder with 5 constant and 19 variable parameters, including its reinforcement, geometric characteristics, concrete cover, rod pitch, anchoring, etc., as well as an optimization algorithm for given constraints, taking into account the cost criterion.
3. The results of numerical calculations showed a sufficiently high efficiency of the developed optimization technique. Savings on the main design characteristics range from 3 to 13%, depending on the complexity of the task.

References

[1] Tamrazyan A and Filimonova E 2014 Searching method of optimization of bending reinforced concrete slabs with simultaneous assessment of criterion function and the boundary conditions *Applied mechanics and materials* **467** pp 404-409

[2] Tamrazyan A and Filimonova E 2013 The structure of the objective function during optimization reinforced concrete slabs, taking into account structural safety *Industrial new and civil engineering* **9** pp 14-15

[3] Xin-She Y 2010 *Engineering Optimization: an introduction with metaheuristic appications* (New Jersey: John Wiley& Sons Inc.) ISBN 978-0-470-58246-6

[4] Standard SP 52-103-2007 2007 Reinforced concrete monolithic structures of buildings (Introduction Research Center Construction)

[5] Fakhimi AH, Masoonmeh J and Cengiz TY 2017 The most applicable metaheuristic optimization methods on concrete structures *3rd Iran Int. Conf. on Structural Eng.*

[6] Tamrazyan A and Alekseytsev A 2019 Evolutionary optimization of reinforced concrete beams, taking into account design reliability, safety and risks during the emergency loss of supports *E3S Web of Conferences* **97** 4005

[7] Standard SP 430.132580.2018 2018 Monolinyye konstruktivnyye sistemy *Pravila proyektirovaniya Vved* (Minstroy Rossii)

[8] Standard SP 63.13330.2018 2018 Concrete and reinforced concrete structures Basic provisions Updated edition of SNiP 52-01-2003 (Ministry of Construction of Russia)

[9] Zong Woo Geem 2009 *Music Inspired Harmony Search Algorithm Theory and Application* (Springer)

[10] Xin-She Y, Bekdas G, Nigdeli S M 2016 Metaheuristics and Optimization in Civil engineering *Modelling and optimization in science and technologies* **7**

[11] Akin A and Saka MP 2014 Harmony search algorithm based optimum detailed design of reinforced concrete plane frames subject to ACI 318-05 provisions *Computers and Structures* **147** pp 79-95

[12] Kaveh A, Sabzi OA 2011 Comparative study of two metaheuristic algorithms for optimum design of reinforced concrete frames *International Journal of Civil Engineering* **9** (3) pp 483-497

[13] Gholizadeh S and Aligholizadeh V 2013 Optimum design of reinforced concrete frames using bat metaheuristic algorithm International Journal of Optimization in Civil Engineering **3** (3) pp 483-497

[14] Kabantsev OV, Tamrazyan AG 2014 Allowing for changes in the calculated scheme during the analysis of structural behavior *Magazine of civil engineering* **49** (5) pp 15-26

[15] Tamrazyan AG, Avetisyan LA 2018 Behavior of compressed reinforced concrete columns under thermodynamic influences taking into account increased concrete deformability *IOP Conference Series: Materials Science and Engineering* **365** (5) p 52034 (DOI 10.1088 /1757-899X/365/5/052034)

[16] Tamrazyan A, Avetisyan L 2016 Comparative analysis of analytical and experimental results of the strength of compressed reiforced concrete columns under special combinations of loads *MATEC Web of Conference* **86** p 1029 (DOI 10.1051/matecconf/20168601029)

IOP Conf. Series: Materials Science and Engineering 1015 (2021) 012034 doi:10.1088/1757-899X/1015/1/012034

Geometric and physio-mechanical characteristics of the contact layer area of multilayer reinforced concrete structures

E A Korol[1] **and Vu Dinh Tho**[1,2]

[1] Moscow State University of Civil Engineering, 129377, 26, Yaroslavskoe Shosse, Moscow, Russian
[2] University of Transport Technology (UTT), Thanh Xuân, Ha Noi, Viet Nam

E-mail address: professorkorol@mail.ru

Abstract. In the practice of modern construction, multi-layer building envelopes made of concrete of various physio-mechanical characteristics has been wide applied. During successive layer-by-layer laying of various concretes, contact layers are formed between them, and their physical and mechanical characteristics are different from the adjacent layers. Technological and time factors for the manufacture of this structure affect the formation, physical and mechanical characteristics of the contact layer between the inner and outer structural layers, and effecting to the working of the multi-layer structure. In this article, the geometric and physio-mechanical characteristics of the contact layer zone of multi-layer reinforced concrete structures are researched. The main content of study was conducted on samples of three-layer structural blocks with using strong structural concrete B25 in the external layers, and low-strength concrete in the middle layer. The geometric and physio-mechanical characteristics of the contact layers were researched by the experimental method. Based on the results of experiments, the geometric and physio-mechanical characteristics of the contact layer of multilayer reinforced concrete structures are shown, taking into account the influence of technological and time factors for the manufacture of this structure. Moreover, the results of the study allow us to determine rational parameters in design of multilayer envelopes with reinforced concrete layers of various strengths, interconnected by a monolithic contact layer. According to the results, modification and calculation method for multilayer reinforced concrete building envelopes has been evaluated and applied into the production process of these constructions.

1. Introduction

In the practice of the present construction, multi-layer reinforced concrete enclosing structures are widely applied by using a middle heat-insulating layer of lightweight concretes, and their characteristics are light density and low thermal conductivity.

One of the most promising and effective applications of these structures is a multi-layer reinforced concrete structure when apply the polystyrene concrete in the internal layer and structural concrete in the external layers [1-4]. According to the results of research [5-9], various methods for calculating multilayer reinforced concrete structures, which using different materials in the internal and external layers were carried out.

Experimental studies have been proved the possibility of forming a monolithic contact zone in the manufacture of multilayer structures made by structural concrete and polystyrene concrete [10-15]. The main parameters of the contact zone were determined and the strength and deformability of the

contact zone of various concretes of multilayer structures was carried out in the series of experimental [16-20]. As it was known by research, during the manufacture of multilayer reinforced concrete structures, a contact zone is formed in boundary layer between adjacent layers. Therefore, a dense large and small aggregate of structural concrete have been penetrated into the adjacent layer of lightweight concrete, and, conversely, a light aggregate of lightweight concrete into the structural layer. Based on research [21-24], methods for calculating multilayer reinforced concrete structures with consideration for the contact layer zone between adjacent layers was carried out. To ensure the joint operation of layers and the monolithic cross-section of multilayer reinforced concrete structures, the authors of the research evaluated the influence of technological parameters of manufacturing on the formation of connections between different layers of concrete [25-26]. However, these studies have not considered the effects of factors of production technology on the geometrical and mechanical properties of the contact zone between outer and inner layers of multilayer reinforced concrete structures by different materials.
This article also presents a method for determining the geometric and physico-mechanical characteristics of the contact layer zone of multilayer reinforced concrete structures.

2. Materials and methods of research

In the study two different types of concrete were used, include the normal concrete (NC) and the lightweight concrete (LWC) – polystyrene concrete. Raw materials of the manufacture of normal concrete of strength class B25 and polystyrene concrete D350 are used in the following composition [15, 16] and showed in table 1.

Table 1. Composition by mass of mixes.

	Density (kg/m^3)	Cement M400 (kg)	Water (l)	Limestone 0.5-1cm (m^3)	Sand 0-2mm (m^3)	EPS 2-5mm (kg/m^3)	Chemical additives SilkRoad SR-5000F (g)
D350	338	330	105	-	-	6.9	710
B25	2325	358	205	0.783	0.483	-	-

The results of experimental about the determination of density, compressive strength and tensile strength of heavy concrete and polystyrene concrete in different ages of hardening has been presented in table 2.

Table 2. Properties of heavy concrete and polystyrene concrete.

	Average density, (kg/m^3)	The compressive strength in different ages (MPa)					The tensile strength in hardening at the age of 28 days (MPa)
		1 day	3 days	7 days	14 days	28 days	
D350	338	0.37	0.45	0.63	0.75	0.83	0.29
B25	2325	12.7	20.8	25.2	29.7	38.6	3.51

The experiments were performed on composite cubic samples with dimensions 150×150×150 mm. At first, concrete B25 with thickness 40mm and lightweight concrete D350 with thickness 70mm were placed in the forms in sequence. After that, the external layer by B25 concrete with thickness 40mm were proceeded. Finally, samples were made in inventory forms by layer-by-layer laying of concrete mixtures with varying time intervals between laying layers of concrete of different densities from 30 minutes to 4 hours in steps of 30 minutes. The contact surfaces were prepared after 28 days of curing in laboratory conditions at a temperature of 20°C and a relative humidity of 95%.
Microscopic examination of the structure of the contact zone between layers of different materials evaluated the adhesion between different layers by the instrument - scanning electron microscope

Quanta 250 and microscope stereoscopic MBS10. Experiment has been analysed on composite three-layer cubic specimens has been showed in figure 1 and 2.

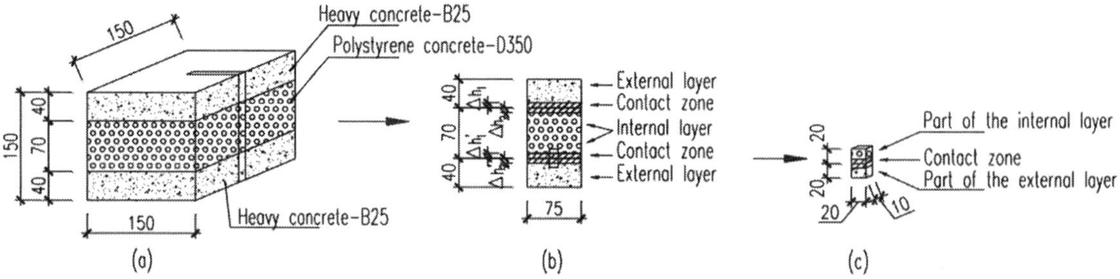

Figure 1. Sample of three-layer concrete for observing the structure of the contact zone between layers.

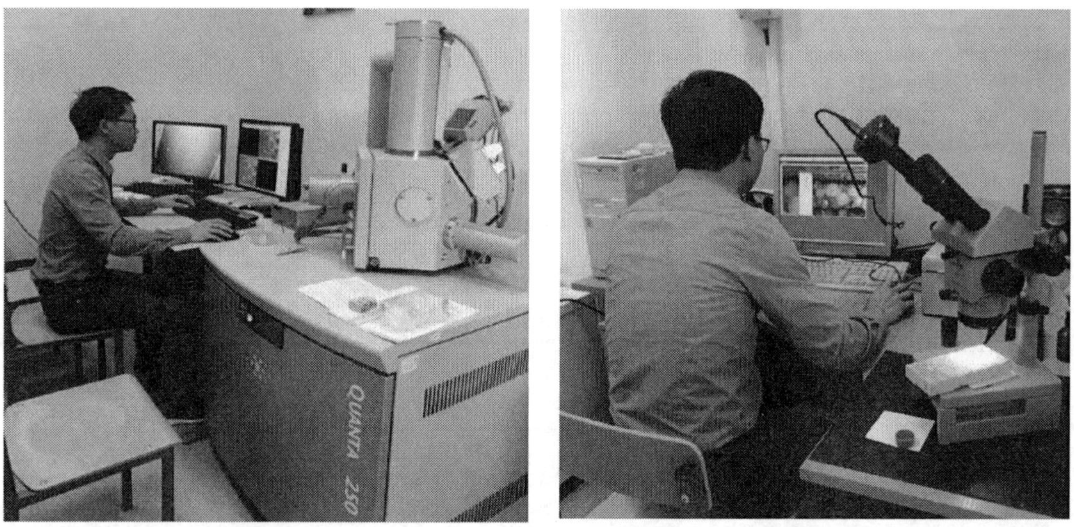

(a) (b)

Figure 2. General view of the structure of the contact zone of layers between the outer and inner layer of three-layer reinforced concrete structures using Microscope Quanta 250 (a) and MBS10 (b).

To determine the adhesive strength of the layers between the layers of different concrete tensile contact layer ridiculous layers of three-layer reinforced concrete structures with considering the influence of technological parameters of manufacture of which designs were carried out three-layer cylindrical samples measuring 150×300MM with an outer layer of structural concrete B25 thickness 50 mm and inner layer of polystyrene D350 with thickness 200 mm (figure 3). During the production of experimental samples, that were made in inventory forms by layering concrete mixes with different time intervals between layers of laying concrete of different densities from 0 to 4 hours in increments of 30 minutes. After 28 days of curing in the laboratory at a temperature of 20°C and a relative humidity of 95%, the contact surfaces were prepared. Tests for axial tension were carried out according to the method in accordance with the requirements [13]. Tests of each series of samples at the age of 28 days were carried out. Anchors were used as grippers for fixing the sample in the breaking machine. Loading was performed continuously at a speed that increases the calculated stress in the sample to its complete destruction in the range of 0.05...0.02 MPa/s. The destroyed sample was subjected to visual inspection, while the test log noted the characteristics and place of destruction

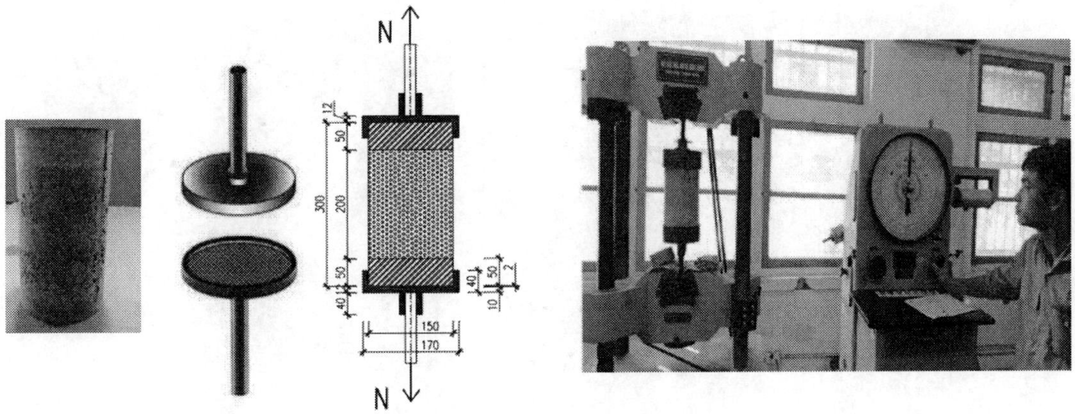

Figure 3. The samples and tensile test scheme of three-layer blocks.

3. Results and discussions

3.1. Experimental study of the structure of the contact zone between layers using the microscopy instrument

There were tested 24 samples for observation of the structure of the contact layers between the outer layer and inner layer of three-layer reinforced concrete structures was obtained forces and the structure of the contact zone of layers with the influence of technological and time factors make this structure. Depending on the time elapsed between concreting two adjacent layers, all tests can be divided into 8 times groups, from 0 hours to 4 hours. The results of the experiments are shown in table 3 and figure 4 and 5.

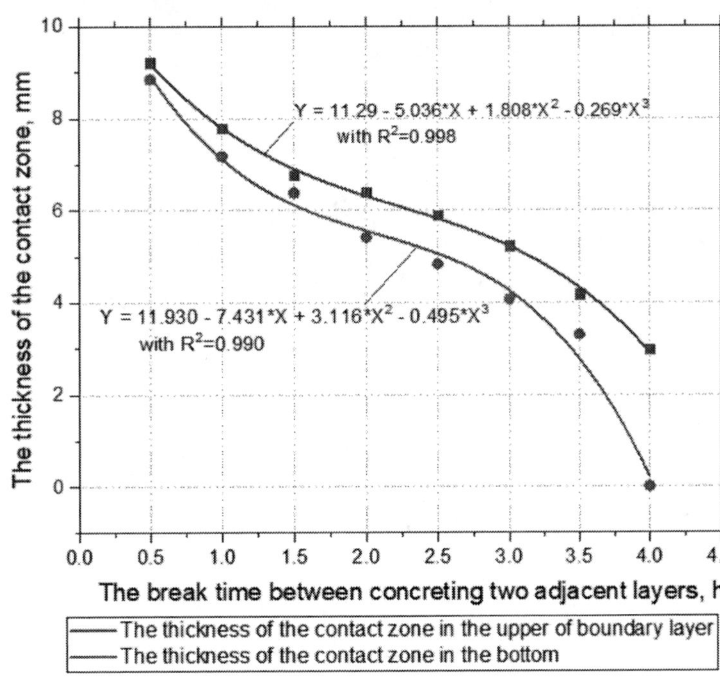

Figure 4. Dependence of the adhesive strength of layers on the time interval between layers.

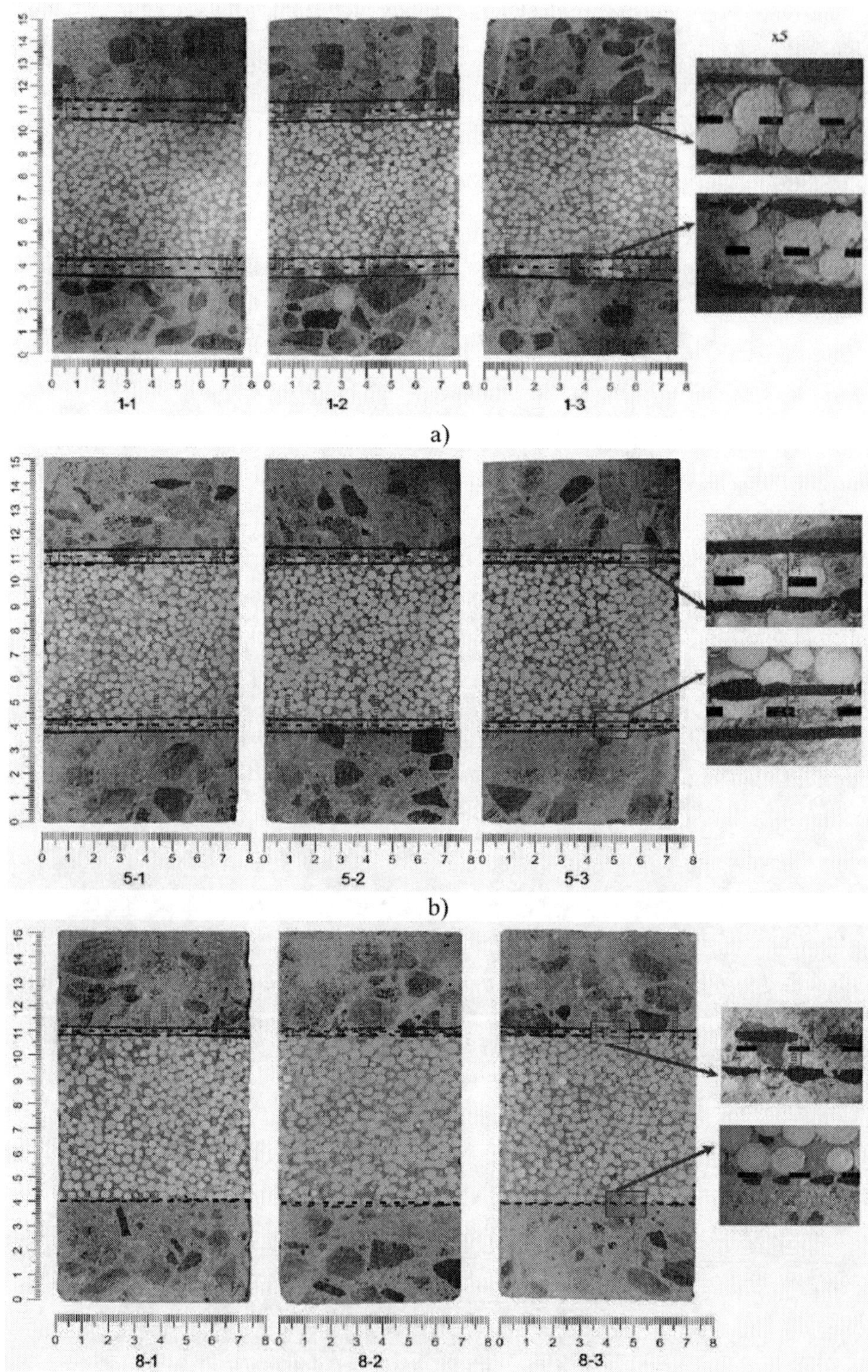

Figure 5. Observe the microstructure of the contact seams between the layers of blocks with a break time of 1h (5a), 2.5h (5b) and 4.0 h (5c).

IOP Conf. Series: Materials Science and Engineering 1015 (2021) 012034 doi:10.1088/1757-899X/1015/1/012034

Table 3. Three-layer test device microscopy.

The series of samples	The position of the contact zone		The thickness of the contact zone section of the samples									$^{a}\overline{\Delta h_i}$	$^{b}h^*$
			Samples-1			Samples-2			Samples-3				
0.5	In the upper	Δh_1	4.5	5.0	5.5	4.5	4.7	4.9	4.0	4.5	4.5	4.7	9.3
		Δh_2	4.5	4.9	4.7	4.5	4.4	4.1	5.0	4.5	4.5	4.6	
	In the bottom	Δh_1	3.6	4.1	5.0	4.8	4.5	4.2	4.3	5.3	5.3	4.6	8.8
		Δh_2	4.4	4.4	4.1	3.9	4.3	4.1	4.3	4.5	4.7	4.3	
1	In the upper	Δh_1	3.6	3.4	3.2	3.9	4.0	4.0	4.2	3.5	2.9	3.6	7.8
		Δh_2	3.9	4.2	4.3	3.9	4.0	4.0	4.2	4.5	4.5	4.2	
	In the bottom	Δh_1	4.3	4.0	3.7	3.3	3.0	2.8	3.8	4.2	4.2	3.7	7.2
		Δh_2	3.1	3.1	3.0	3.9	4.0	4.1	3.5	3.3	3.4	3.5	
1.5	In the upper	Δh_1	3.5	3.0	2.9	3.4	3.6	3.4	3.3	3.2	3.1	3.3	6.8
		Δh_2	3.8	4.0	3.9	3.3	3.4	3.6	3.3	3.2	3.1	3.5	
	In the bottom	Δh_1	3.4	3.1	3.7	3.3	3.3	3.9	3.1	3.4	3.3	3.4	6.4
		Δh_2	2.8	3.1	2.5	3.3	3.3	2.9	3.1	2.9	3.2	3.0	
2	In the upper	Δh_1	3.0	3.1	3.3	3.3	3.2	3.3	3.4	2.8	2.2	3.0	6.4
		Δh_2	3.0	3.1	2.9	3.1	3.2	3.3	3.4	4.0	4.5	3.4	
	In the bottom	Δh_1	2.8	2.7	2.5	2.5	2.4	2.7	2.6	2.7	2.8	2.6	5.4
		Δh_2	2.3	2.7	2.9	2.3	2.2	2.0	3.3	3.6	3.7	2.8	
2.5	In the upper	Δh_1	2.6	2.9	3.4	3.3	3.3	3.3	2.8	2.9	3.3	3.1	5.9
		Δh_2	3.4	2.9	2.5	3.7	3.3	3.0	2.8	2.3	1.5	2.8	
	In the bottom	Δh_1	2.2	2.4	2.9	2.5	2.4	2.5	1.8	2.0	2.3	2.3	4.8
		Δh_2	3.5	2.8	2.3	2.9	2.8	2.3	2.1	2.0	1.8	2.5	
3	In the upper	Δh_1	2.2	2.5	2.9	2.6	2.6	2.3	2.4	2.4	2.2	2.4	5.2
		Δh_2	3.4	3.2	2.4	1.9	2.6	3.1	2.7	2.6	3.0	2.8	
	In the bottom	Δh_1	1.3	1.7	2.3	2.0	1.7	1.7	1.7	1.9	2.1	1.8	4.1
		Δh_2	2.6	2.3	1.5	2.0	2.2	2.5	2.5	2.3	2.3	2.3	
3.5	In the upper	Δh_1	2.2	1.9	1.6	2.0	1.8	2.0	1.9	1.8	1.8	1.9	4.2
		Δh_2	1.5	2.0	2.1	2.4	2.7	2.3	2.3	2.5	2.9	2.3	
	In the bottom	Δh_1	1.7	1.5	1.8	1.4	1.4	1.8	1.3	1.2	1.3	1.5	3.3
		Δh_2	1.8	2.1	2.4	1.8	1.9	2.0	1.8	1.6	1.2	1.8	
4	In the upper	Δh_1	1.6	1.3	1.0	1.1	1.3	1.4	1.7	1.0	0.2	1.2	3.0
		Δh_2	1.5	1.8	2.1	1.8	1.6	1.4	1.7	1.9	2.5	1.8	
	In the bottom	Δh_1	-	-	-	-	-	-	-	-	-	-	-
		Δh_2	-	-	-	-	-	-	-	-	-	-	

Note:

[a] Average thickness of the contact zone's section: $\overline{\Delta h_i} = \frac{1}{n}\sum_{n} \Delta h_i$, mm;

[b] Average thickness of the contact zone: $h^* = \Delta h_{1(1')} + \Delta h_{2(2')}$, mm.

The results in table 3 and figure 4 show that when successive layers of heavy concrete and polystyrene concrete are laid with the same adhesive material, the type of cement is used. The cross-section of the block between layers of different materials is a monolithic section.

IOP Conf. Series: Materials Science and Engineering 1015 (2021) 012034 doi:10.1088/1757-899X/1015/1/012034

The thickness of the concrete bonding zone between the layers is from 0 cm to 0.93 cm, which depends on the break time for laying layers of heavy concrete and polystyrene concrete in series. The shorter the break time of laying layers between different layers of concrete, the more thickness of the zone connection between the layers exists. Because when the concrete of the previous layer has not solidified, the fresh concrete part of the next layer will be fed into the concrete zone of the previous layer and conversely. When the break time of laying different layers increases, the concrete part of the previous filling layer solidifies and hardens, the penetration of the next layer of concrete into the previous concrete being filled becomes less. Therefore, the thickness of the bonding area is reduced.

3.2. Tests for determining the tensile strength of the bound zone of the layers of three-layer structures

To determine the adhesion strength of the contact layer of composite layers of three-layer reinforced concrete structures, accounting the influence of technological parameters of manufacturing. The destroyed sample was subjected to visual inspection, and the nature and location of the destruction were noted in the test report.

The tensile strength of the adhesion between layers at the junction of the layers for each sample is determined by the formula:

$$f_t = \frac{P}{A}$$

P - maximum load, [N] and A - the working cross-sectional area of the test piece, mm^2.

A total of 24 samples were tested to determine the tensile strength of the adhesion of the layers and 24 values of the breaking load were obtained. Depending on the time elapsed between concreting two adjacent layers, all tests can be divided into 8 groups by time, from 0 hours to 4 hours. The results of the experiments are shown in table 4 and figures 6 and 7.

a) b) c)

Figure 6. Testing the tensile strength of layers with a break time 0.5h (6a), 2.5h (6b) and 4.0h (6c)

According to the test results (table 4), the dependence of the strength of the contact zone of the layers on the slice on the holding time of the previous layer before laying the next one was established (figure 6 and 7).

When the value of the time intervals between the laying of the layers is less than 1.5h, the destruction during the adhesion strength between laying layers occurred as in the least durable material, i.e. polystyrene concrete, and in the contact area. For values of the time interval exceeding 1.5h, the fraction of samples destroyed directly in the contact zone increased significantly, i.e. the formation of a monolithic bond between the layers did not occur.

Table 4. Three-layer tensile tests

	№ Sample	Break time (h)	Ultimate load (kg)	Tensile strength, f_t (MPa)	Average tensile strength (MPa)	The nature and location of destruction
1	1-1	0.5	525	0.293		Destruction of polystyrene concrete
2	1-2	0.5	530	0.297	0.294	with a break in part of the granules and
3	1-3	0.5	525	0.293		a break in part of the middle layer
4	2-1	1.0	520	0.282		Destruction of polystyrene concrete
5	2-2	1.0	515	0.278	0.281	with a break in part of the granules and
6	2-3	1.0	525	0.282		a break in part of the middle layer
7	3-1	1.5	490	0.269		Destruction of polystyrene concrete
8	3-2	1.5	490	0.268	0.271	with a break in part of the granules and
9	3-3	1.5	495	0.275		a break in part of the middle layer
10	4-1	2.0	420	0.236		Destruction along the adhesion plane
11	4-2	2.0	400	0.223	0.227	without destruction of the polystyrene
12	4-3	2.0	405	0.224		concrete layer
13	5-1	2.5	350	0.197		Destruction along the adhesion plane
14	5-2	2.5	335	0.186	0.191	without destruction of the polystyrene
15	5-3	2.5	340	0.189		concrete layer
16	6-1	3.0	310	0.173		Destruction along the adhesion plane
17	6-2	3.0	310	0.173	0.172	without destruction of the polystyrene
18	6-3	3.0	305	0.169		concrete layer
19	7-1	3.5	285	0.160		Destruction along the adhesion plane
20	7-2	3.5	295	0.163	0.162	without destruction of the polystyrene
21	7-3	3.5	295	0.163		concrete layer
22	8-1	4.0	275	0.152		Destruction along the adhesion plane
23	8-2	4.0	255	0.143	0.147	without destruction of the polystyrene
24	8-3	4.0	260	0.147		concrete layer

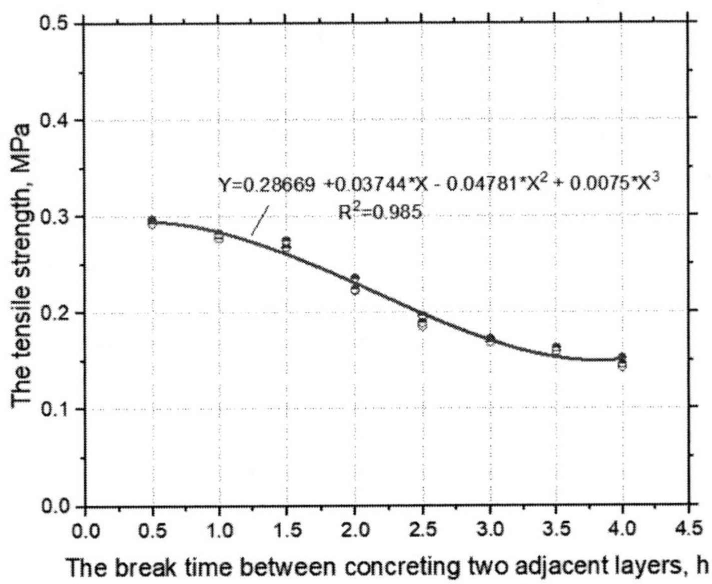

Figure 7. Dependence of tensile strength of the adhesive of the layers on the time interval between layers.

For samples whose production intervals were more than 2.5 hours, failure occurred only along the adhesion plane. So, with a break of 4.0 hours, the adhesion strength of the layers decreases to 50% of the original value.
The calculated dependences for assessing the adhesion strength of the layers are differentiated depending on the time interval between laying layers:

$$f_r = 0.28669 + 0.03744\,X - 0.04781\,X^2 + 0.0075\,X^3 \text{ with } R^2 = 0.985$$

4. Conclusions

Based on the experiments pertaining to the development of the adhesion strength between the layers in composite components of three-layer concrete with different strengths, the following conclusions are drawn:
- The influence of some factors and manufacturing time parameters on the physical-mechanical characteristics of the monolithic bond of concrete layers in three-layer structures made of concrete of various densities is considered and investigated.
- During the manufacture of reinforced concrete structures from concrete of various strengths, a contact layer is formed between the external and internal layers. This layer has the property of intermediate strength between the strength of the concrete of the outer and inner layer. The thickness of the contact layer depends on the technological parameters, from the preparation, and the fractional composition used for the manufacture of the beam.
- According to the results of experiments, the dependences of the strength of the monolithic bond of the layers on the temporal parameters of the manufacture of structural elements after manufacture are established. The optimal value of aging duration of exposure of each of the layers of the three-layer structure is from 0.5 to 1.5 hour.

References

[1] Andreev V I, Turusov R A and Tsybin N Yu 2016 Determination of the stress-strain state of a three-layer beam using the contact layer method *Vestnik MGSU* (Moscow) **4** pp 17–26
[2] Andreev V I, Turusov R A, Tsybin N Yu 2016 Application of the Contact Layer in the Solution of the Problem of Bending the Multilayer Beam, *Elsevier Engineering* **153** pp 59 – 65
[3] Kolchunov V I 2001 The model of resistance of the boundary layer between different concretes and the method of numerical and experimental studies, *Third All-Russian Conference "New in Architecture, Design of Building Structures and Structures (Cheboksary)* pp 245-250
[4] Turusov R A, Andreev VI and Tsybin N Yu 2017 A general solution to the problem of the bending of a multilayer beam in the Fourier series *Structural Mechanics of Engineering Structures and Structures* **4** pp 34–42
[5] Gara F, Ragni L, Roia D and Dezi L 2012 Experimental behaviour and numerical analysis of floor sandwich panels *Engineering Structures* **36** pp 258–269
[6] Vu Dinh Tho and Korol EA 2019 Influence of geometrical parameters of the cross section, strength and deformability of the materials used on stressstrain state of three-layered reinforced concrete *IOP Conf. Series: Materials Science and Engineering* **661** 012121
[7] Trussoni M, Hays C D and Zollo R F 2013 Fracture properties of concrete containing expanded polystyrene aggregate replacement *ACI Materials Journal* **110** (5) pp 549-558
[8] Benayoune A, Samad A, Trikha D N, Ali A A and Ellinna S H M 2008 Flexural behaviour of pre-cast concrete sandwich composite panel – Experimental and theoretical investigations *Constr. and Building Materials* **22** (4) pp 580–592 doi.10.1016/j.conbuildmat. 2006.11.023
[9] Korol E and Berlinova M 2018 Calculation of multilayer enclosing structures with middle layer of polystyrene concrete *MATEC Web of Conf.* **193** doi.10.1051/matecconf/201819303020
[10] Yue Z and Xiao H 2002 Generalized Kelvin solution based boundary element method for crack problems in multilayered solids *Eng. Analysis with Boundary Elements* **26** pp 691–705 doi.10.1016/s0955-7997(02)00038-3

[11] Marčiukaitis G and Juknevičius L 2002 Influence of the internal layer cracks on the cracking of flexural three-layer concrete members *Journal of Civil Engineering and Management* **8** (3) pp 153-158 doi.10.1080/13923730.2002.10531270
[12] Andreev V I, Turusov R A and Tsybin N Yu 2017 The contact layer method in calculating of the shear compounds *MATEC Web of Conferences* **117** *26th R-S-P Seminar Theoretical Foundation of Civil Engineering* doi.org/10.1051/matecconf/201711700008
[13] Korol E A, Pugach E M and Kharkin Yu A 2014 The influence of technological factors on the formation of the connection of layers of a multilayer building envelope *Vestnik MGSU* **3** pp 67-75
[14] Gara F, Ragni L, Roia D and Dezi L 2012 Experimental tests and numerical modelling of wall sandwich panels *Engineering Structures* **37** pp 193–204
[15] Lander V and Shauman Z 2005 Microstructure research of contact zone aggregate hardened cement binder in filtering concretes *The 111 Int. Simposium on Silicate Chemistry* (Brno)
[16] Andreev V, Turusov R and Tsybin N 2018 Layered composite and contact layer. Normal separation and transversal strength *MATEC Web of Conferences* **251** (6) 04066
[17] Sayadi A A, Tapia J V, Neitzert T R and Clifton G C 2016 Effects of expanded polystyrene (EPS) particles on fire resistance, thermal conductivity and compressive strength of foamed concrete, *Construction and Building Materials* **112** pp 716–724
[18] Lam T V, Tho V D, Dien V K, Bulgakov B I and Korol E A 2018 Properties and thermal insulation performance of lightweight concrete *Magazine of Civil Eng.* **84** (8) pp 173–191
[19] Xu Y, Jiang L, Xu J, Chu H and Li Y 2015 Prediction of compressive strength and elastic modulus of expanded polystyrene lightweight concrete *Magazine of Concrete Research* **67** (17) pp 954–962 doi.10.1680/macr.14.00375
[20] Chen B and Liu N 2013 A novel lightweight concrete-fabrication and its thermal and mechanical properties *Construction and Building Materials* **44** pp 691–698
[21] Pshenichny G N 2006 To the question of the contact zone of concrete and reinforced concrete *Concrete and reinforced concrete* **2** pp 7-9
[22] Kolchunov Vl I and Shoemakers P V 2004 Deformability and crack resistance of the contact zone of multilayer concrete and reinforced concrete structures *Construction transport* **2** p 5
[23] Tsybin N, Turusov R, Andreev V and Kolesnikov A 2018 Stress-strain state of a three-layer rod. Comparison of the results of analytical and numerical calculations with the experiment *MATEC Web of Conferences* **196** 01057 doi.org/10.1051/matecconf/ 201819601057
[24] Pugach E M and Korol O A 2011 Experimental research of three-layer structure with middle layer of concrete with low thermal conductivity in nonstationary heat and humidity mode *Vestnik MGSU* **3** pp 154-162
[25] Korol E, Vu Dinh Tho and Nguyen Huy Hoang 2018 Analysis the effects of lightweight concrete in the middle layer of multi-layered reinforced concrete structures on the stress-strain state using the finite element method *MATEC Web of Conferences* **196**, 02022 doi.org/10.1051/matecconf/201819602022
[26] Korol E A and Kharkin YA 2010 On the choice of software for simulation of stress-strained state of sandwich concrete elements and structures with solid coupling layers *Vestnik MGSU* **3** pp 156-163

IOP Conf. Series: Materials Science and Engineering 1015 (2021) 012035 doi:10.1088/1757-899X/1015/1/012035

Investigation of the humidity regime of multilayer enclosing structures of buildings and structures

Yu S Vytchikov [1], M E Saparev [1] and A B Kostuganov [2]

[1] Samara State Technical University, Construction and Architecture Academy, Molodogvardeyskaya street, 244, Russian Federation
[2] Orenburg State University, Pobeda avenue, 13, Russian Federation

E-mail address: msx072007@yandex.ru

Abstract. The article describes an approximate analytical method for calculating the moisture regime of multilayer building envelope structures. The novelty of the proposed method lies in the use an analytical dependence for determination the position of the flatness possible condensation, obtained as a result of studying the function which is the difference between the partial pressures of water vapor in the considered section the enclosing structure and the partial pressure of saturated water vapor at the extremum. The article presents the results of calculating the moisture regime of the outer wall of a residential building under construction in Samara using monolithic sandless expanded clay blocks with a density equal to 600 kg/m3. The calculation showed that during the heating period the coordinate of flatness of possible condensation in sandless expanded clay concrete, measured from the inner surface of the wall moves from 0.444 m to 0.5 m. The made assessment of moisture accumulation in the considered outer wall, both for the annual period of operation of the building, and for the period of months with negative temperatures, showed compliance with the regulatory requirements for heated rooms.

1. Introduction

One of the most important conditions for a person's stay in buildings and structures is the absence of moisture on the inner surfaces of the enclosing structures. The accumulation of moisture in external walls, coatings and ceilings can lead to the formation of molds, which are very dangerous for humans, as well as to the destruction of building structures. The main reasons for the accumulation of moisture in the outer walls are considered in work [1]. In works [2, 3] is considered the application of the theory of moisture potential in the study of the moisture regime of enclosing structures. Experimental studies of this problem are reflected in the works as well as domestic [4, 5] and foreign [6, 7] scientists. In works [3, 8, 9] are considered the methods of determining the plane of possible condensation.

In work [9] is considered an approximate analytical method for determining the plane's position of possible condensation, convenient for manual counting.

By developing the above method was used the dependence of the partial pressure of saturated water vapor from temperature, given in the reference book edited by I.G. Staroverov. [10].

At present, according to the current standards, the protection calculation against water logging of building enclosing structures is made for months with negative outside temperatures. The average temperature and partial pressure of water vapor for the cold period are taken as the calculation parameters of the outside air.

IOP Conf. Series: Materials Science and Engineering 1015 (2021) 012035 doi:10.1088/1757-899X/1015/1/012035

2. Method for calculating the moisture regime of enclosing structures

By actualization the author's method for calculating the moisture regime of multilayer enclosing structures, described in detail in work [8], is suggested a more accurate dependence of the partial pressure of saturated water vapor E from the temperature τ_i of the form [°C]

$$E = 611\left(1+\frac{t}{100}\right)^{7.6} \tag{1}$$

Figure 1 shows a schematic diagram of the heat transfer process through a multilayer outer wall. Temperature values on the outer surface of the i-th layer are determined according to [7] by the formula [°C]

$$\tau_i = t_{\text{in}} - \frac{t_{\text{in}} - t_{\text{out}}}{R_0^{con}}\left(\frac{1}{\alpha_{in}} + \sum_{i=1}^{i} R_i\right) \tag{2}$$

where t_{in}, t_{out} - the temperature of the indoor and outdoor air, respectively, [°C]; R_0^{con} - heat transfer resistance of the surface of the outer wall, [m^2·°C/W]; α_{in} – heat transfer coefficient from the inner and outer wall [W/(m^2·°C)]; thermal resistance of the i-th layer, [m^2·°C/W]; δ_i , λ_i - thickness, [m], and thermal conductivity of the i-th layer, respectively, [W/(m·°C)].

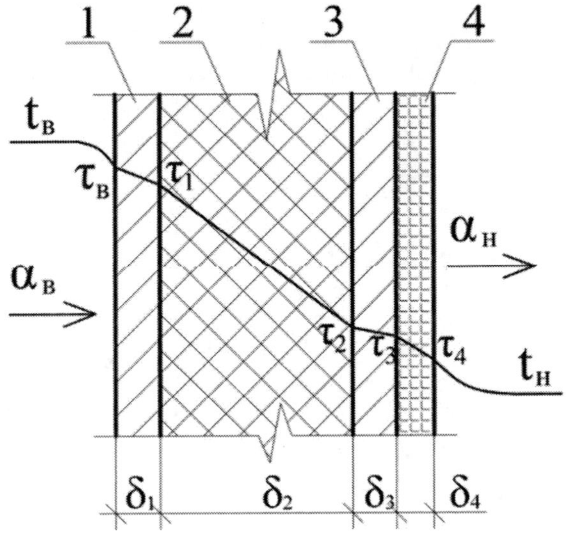

Figure 1. Schematization of the heat transfer process in a multilayer outer wall.

The partial pressure of water vapor contained in humid air passing through the outer wall is determined by the formula

$$e_i = e_{\text{in}} - \frac{e_{\text{in}} - e_{\text{out}}}{R_{p_0}} \sum_{i=1}^{i} R_{p_i} \tag{3}$$

where e_{in}, e_{out} is the partial pressure of water vapor inside the heated room and outside air, respectively, Pa; R_{p0} - resistance to vapor permeation of the outer wall [m^2·h·Pa/mg]; $R_{ni}=\delta_i/\mu_i$ - resistance to vapor permeation of the i-th layer [m^2·h·Pa/mg]; μ_i –is the vapor permeability coefficient of the material of the i-th layer [mg/(m·h·Pa)].

To determine the coordinate x_i of the plane of possible condensation, it is necessary to investigate the function $\Phi = e_i - E_i$ for an extremum. For this, the first derivative of the function Φ is equated to zero. Then we get

$$\frac{d\Phi}{dx_i} = \frac{d\left(e_{x_i} - E_{x_i}\right)}{dx_i} = 0 \qquad (4)$$

After transformation, the expression for finding the coordinate of the plane of possible condensation will take the form [m]

$$x_i = \lambda_i \left\{ \frac{\left[t_{\text{in}} - 55.9 \left(\frac{e_{\text{in}} - e_{\text{out}}}{t_{\text{in}} - t_{\text{out}}} \frac{R_0^{con}}{R_{p_0}} \frac{\lambda_i}{\mu_i} \right)^{0.1515} + 100 \right] R_0^{con}}{t_{\text{in}} - t_{\text{out}}} - \frac{1}{\alpha_{\text{in}}} - \sum_{i=1}^{i} R_i \right\} \qquad (5)$$

Substitution of values xi in the expression of the second derivative of the function Φ leads to a positive value, which indicates its maximum value in the plane of possible condensation. If the value x_i exceeds the layer thickness δ_i, then the outer surface of the calculated layer is taken as the plane of possible condensation, i.e. $x_i = \delta_i$.

3. Calculating's results the moisture regime of a multilayer enclosing structure

According to the above methodology, the calculation the humidity regime of the outer wall of a residential building under construction in Samara was carried out. The considered enclosing structure is shown in figure 2.

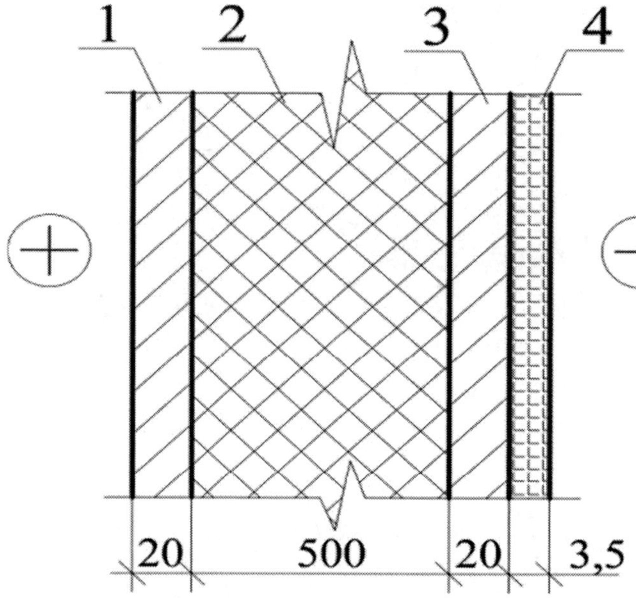

Figure 2. Outer wall of a residential building.

Table 1. Thermal characteristics of the outer wall materials.

Material names	Layer thickness δ_i. m	Thermophysical characteristics of the layer material: Density ρ_i. kg/m^3	Thermal conductivity coefficient λ_i. W/(m·°C)	vapor permeability coefficient μ_i. mg/(m·h·Pa)
Lime-sand mortar	0.02	1600	0.7	0.12
Monolithic expanded clay concrete	0.5	600	0.14	0.155
Cement-sand mortar	0.02	1800	0.76	0.09
The textured layer of the facade system	0.0035	1600	0.7	0.05

By calculating the moisture regime of the outer wall shown in figure 2 the following initial data were taken:

1. Construction area of Samara
2. Temperature of the coldest five-day period t_{out} = -30°C
3. Average temperature for the heating period t_h = - 4.7°C
4. Duration of the heating period $Z_{h.p}$= 197 days
5. Air temperature inside the building t_{in} = 20°C
6. Relative humidity inside the building φ_{in}=50%.

The calculation to find the plane of possible condensation was carried out in the following sequence:

1. The resistance to heat transfer the surface of the outer wall was determined by the formula [m^2·°C/W]

$$R_0^{con} = \frac{1}{\alpha_{in}} + \sum_{i=1}^{i} R_i + \frac{1}{\alpha_{out}} \quad (6)$$

2. The resistance to vapor permeation of the outer wall was found by the formula [m^2·h·Pa/mg]

$$R_{p_0} = \frac{\delta_1}{\mu_1} + \frac{\delta_2}{\mu_2} + \frac{\delta_3}{\mu_3} + \frac{\delta_4}{\mu_4} = \frac{0.02}{0.12} + \frac{0.5}{0.155} + \frac{0.02}{0.09} + \frac{0.0035}{0.05} = 3.69 \quad (7)$$

3. By formulas (2) and (3) were determined the values of temperature τ_i and partial pressure of water vaporei on the outer surface of each layer of the outer wall. The calculation results are summarized in table 2
4. The obtained values of τ_i were used to calculate the values of the partial pressure of saturated water vapor according to the following formulas given in the reference book [10].

$$E_i = 2.88\left(1.098 + \frac{\tau}{100}\right)^{8.02}, \tau \geq 0 \quad (8)$$

$$E_i = 4.688\left(1.486 + \frac{\tau}{100}\right)^{12.3}, \tau \leq 0 \quad (9)$$

5. By comparing the values of e_i and E_i for months with negative outside temperatures and the entire cold period was found that condensation of water vapor occurs in sandless expanded clay concrete from December to February. Slight condensation in the third layer was noted in December.
6. According to the formula (4) were determined the values of the coordinates of the plane of possible condensation in sandless expanded clay concrete.

From the data presented in table 2 follows that when the outside temperature drops, the plane of possible condensation moves from the outer to the inner surface of the wall. The value x_2 varies from 0.444 to 0.5 m. The calculation of moisture accumulation in the outer wall for the annual period of operation of the building and the period of months with negative outside temperatures, presented in Table 3 showed conformity with the regulatory documents for heated buildings. Based on the calculated data was plotted a graphical dependence of the distribution of the elasticity of water vapor E_x and e_x in the thickness of the outer wall in January, shown in figure 3.

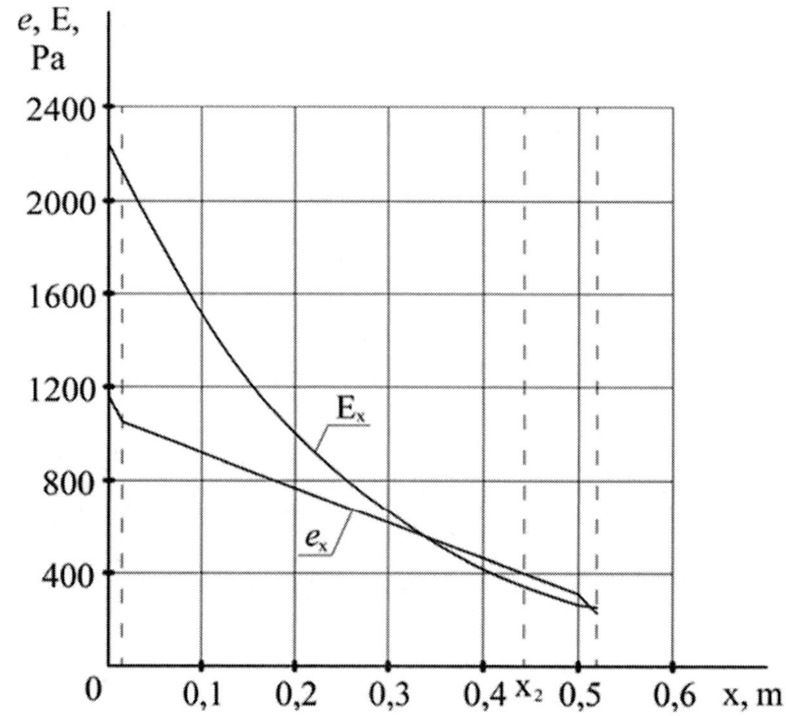

Figure 3. Distribution of water vapor elasticities E_x and ex in the thickness of the outer wall in January: E_x - the value of the elasticity of saturated water vapor, e_x- the calculated value the elasticity of water vapor in the layers of the outer wall, x – the current coordinate of the outer wall layer.

Table 2. Estimated and required values of resistance to vapor permeation of the outer wall.

Coordinate of the plane of possible condensation x_2, m	Resistance to vapor permeation. $m^2 \cdot h \cdot Pa/mg$		Required resistance to vapor permeation. $m^2 \cdot h \cdot Pa/mg$	
	$R^*_{p_{out}}$	R^*_p	$R^{*req}_{p_1}$	$R^{*req}_{p_2}$
0.5	0.292	3.39	0.0405	0.116
0.444	0.653	3.03	-0.034	0.145

From the data presented in table 3, it can be concluded that the resistance to vapor permeation R^*_p significantly exceeds the values $R^{*req}_{p_1}$ and $R^{*req}_{p_2}$ obtained by the known regulatory method. Consequently, moisture accumulation in the wall is unlikely. Taking this into account, we can conclude that the accumulation of moisture in the wall does not occur over the year of the building's operation.

Table 3a. Results of calculating the values of temperature τ_i and partial pressure of water vaporei on the outer surface of each layer of the outer wall.

Months	$t_н$, °C	$e_н$, Pa	The temperature of the outer surface layer, ° C				Coordinates of the plane possible condensation
			τ_1	τ_2	τ_3	τ_4	x_2, m
November	-2.1	450	19.16	-1.66	-1.82	-1.85	-
December	-8.3	300	18.93	-7.74	-7.94	-7.98	0.489
January	-11.2	220	18.82	-10.58	-10.80	-10.84	0.444
February	-10.6	220	18.84	-10.00	-10.21	-10.25	0.448
March	-3.7	360	19.10	-3.23	-3.40	-3.43	-
months with negative temperatures	-7.18	310	19.00	-6.66	-6.85	-6.88	0.500

Table 3b. Results of calculating the values of temperature τi and partial pressure of water vaporei on the outer surface of each layer of the outer wall.

Months	$t_н$, °C	$e_н$, Pa	The partial pressure of saturated water vapor, Pa				The partial pressure of water vapor on the outer surface of the layer, Pa			
			E_1	E_2	E_3	E_4	e_1	e_2	e_3	e_4
November	-2.1	450	2218.80	532.70	526.00	524.70	1136.50	507.90	464.60	451
December	-8.3	300	2187.02	316.97	311.20	310.20	1129.70	369.90	317.70	301
January	-11.2	220	2169.60	246.30	241.70	240.70	1126.00	296.40	239.30	221
February	-10.6	220	2175.18	259.60	254.80	253.90	1126.00	296.40	293.30	221
March	-3.7	360	2211.00	466.70	460.10	458.90	1132.00	425.10	376.50	361
months with negative temperatures	-7.18	310	2197.00	348.20	342.60	1.70	1130.00	379.10	327.40	311

Analyzing the graphical dependencies shown in figure 3, it can be concluded that the condensation zone is located in sandless expanded clay concrete at a distance from 0Y to 0.52 m from the inner surface of the wall.

4. Discussions

Comparative data on the partial pressure of saturated water vapor, taken from various sources, are presented in table 4.

Comparison with the experimental values given in the reference literature showed that by negative outside temperatures, the error in determining the partial pressure of saturated water vapor ranges from 5% to 16% when the temperature changes from -5°C to -20°C. When developing an approximate analytical method, the outside air temperature and partial pressure were determined according to climatic data for the coldest month.

According to the current requirements for thermal protection of buildings and in work [9] is used an auxiliary function to find the plane of possible condensation, the numerical values of which are given in a special table, by interpolation is determined the temperature value in the plane of possible condensation, which is used then to find the coordinate of the plane of possible condensation. According to formula (5), it is possible to obtain an approximate solution to the problem of determining the position of the plane of possible condensation in the form of an analytical dependence convenient for manual calculation.

Table 4. Values of the partial pressure of saturated water vapor *E*, Pa, for temperatures from -30°C to +30°C.

The source of information	Temperature, t, °C (the denominator shows the error)						
	-30	-20	-10	0	+10	+20	+25
Experimental value	38	103	260	611	1228	2338	3168
$E = 10^{2.125+\frac{156+8.12t}{236+t}}$, according to [11]	50.1 / 31.8	124.5 / 26.9	285.8 / 9.9	610.9 / 0	1227.4 / 0	2339 / 0	3162 / 0.18
$E = 1.84 \cdot 10^{11} \exp\left(-\frac{5370}{273+t}\right)$ (regulatory requirement)	55.3 / 45.5	130.8 / 27.0	291.3 / 12.0	611.7 / 0	1221 / −0.6	2319.4 / −0.8	3146 / 0.7
$E = 4.688\left(1.486 + \frac{t}{100}\right)^{12.3}$ for $t \le 0$ °C $E = 288.58\left(1.098 + \frac{t}{100}\right)^{8.02}$ for $t \ge 0$ °C, according [12]	38.2 / 0	103 / 0	260 / 0	611 / 0	1228 / 0	2340 / 0	3169 / 0
$E = 611\left(1 + \frac{t}{100}\right)^{7.6}$ (data of the authors of the article)	40.6 / 6.8	112 / 8.7	274.3 / 5.5	611 / 0	1260.7 / 2.66	2442 / 4.4	3331 / 5.1

From the above it follows that the technique proposed by the authors of the article significantly simplifies the solution of the problem of finding the plane of possible condensation with a higher accuracy.

The analytical method proposed by the authors of the article for determining the plane of possible condensation due to the simplicity of implementation can be recommended both for manual counting and when using personal computers. Compared with the existing method for calculating the humidity regime of enclosing structures, set out in the regulatory requirements, this method more accurately determines the temperature of the plane of possible condensation at low outdoor temperatures typical for northern regions.

5. Conclusions

1. An analytical dependence to determine the position of the plane of possible condensation in a multilayer building envelope is obtained. In comparison with the existing regulatory methodology, the obtained dependence allows us to obtain more accurate results for construction areas with low negative outdoor temperatures.

2. The results of calculating the coordinates of the plane of possible condensation for the outer wall of a residential building under construction in Samara are presented.

3. An assessment of moisture accumulation in the outer wall for the annual period the building's exploitation and for the period of months with negative temperatures was made.

References

[1] Frolov M V 2017 Problems of energy saving in industrial and housing and communal complexes *Proceedings of the XVIII International scientific and practical conference* (Penza: Penza state University of architecture and construction) pp185-189
[2] Gagarin V G and Zubarev K P 2019 *Bulletin of the MSUCE* **4** pp 485-495
[3] Bogoslovskiy V N 1979 *Thermal mode of the building* (Moscow: Stroyizdat)
[4] Petrov A S and Kupriyanov V N 2018 *IOP Conf. Series Materials Science and Engineering* **463**(2) doi: 10.1088/1757-899X/463/2/022064
[5] Korniyenko S V 2012 *Magazine of Civil Engineering* **2** pp 18-23

[6] Künzel H M 1995 *Simultaneous heat and moisture transport in building components: One- and two-dimensional calculation using simple parameters* (Stuttgart: IRBVerl.) p 65
[7] Rode C and Burch D M 1995 Thermal Performance of the Exterior Envelopes of Building VI **4–8** pp 283–295
[8] Il'inskiy V M 1974 *Construction thermophysics* (Moscow: Vysshaya shkola) p 319
[9] Vytchikov Yu S and Beljakov I G 1998 News of higher educational institutions *Construction* **3** pp 76-79
[10] Staroverov I G and Shiller I Yu 1990 *Internal sanitary-technical devices: Designer's reference* **1** (Moscow: Stroyizdat)
[11] Vytchikov Yu S 2006 *Building material* **4** pp 92-94
[12] Blazi V 2004 *Designer's reference book. Building physics* (Moscow: Technosphera) p 480

IOP Conf. Series: Materials Science and Engineering 1015 (2021) 012036 doi:10.1088/1757-899X/1015/1/012036

Degradation of wooden products used in non-ventilated flat roofs

P Juras

Faculty of Civil Engineering, University of Zilina, Univerzitna 8215/1, 010 26 Zilina, Slovakia

E-mail address: peter.juras@uniza.sk

Abstract. In the recent years, the amount of flat roofs with load bearing structure from wood in Slovakia increases. This is caused by the costs and the effort to maximize the speed of built. Using the wood rafter in the flat roofs cause problem of using an organic material between two vapor tight layers, from interior a vapor barrier and from outside the roofing membrane. Additional, also oriented strand boards are also used instead of wooden boards. These are used either in condensations zones of the roofs or direct under the roofing membrane. This creates degradation of the material because of use in the humid environment in very short time. This paper briefly describes the state-of-art situation in Slovakia with some examples from the buildings and the numerical evaluation of the problem.

1. Introduction

Use of wood in different products increases in the building sector in Slovakia every year. It is caused by various factors, such as the speed of the construction, or the energy efficient buildings, where the timber framed buildings have a more subtle building envelope in comparison to the traditional masonry wall with ETICS or masonry from insulated bricks.

To avoid a wet process, some civil engineers decide to use the lightweight load carrying structure made of wood, instead of the concrete slab. This causes several problems, because the hydrothermal regime of flat roof differs from the regime of pitched roof and the same composition cannot be just used in both types.

Figure 1. Flat roof with wooden rafters and OSB under the PVC roofing membrane. Total degradation of OSB caused by the condensed vapor [3].

Figure 2. Wet OSB (caused by the condensed vapor under PVC membrane) [3].

In this paper, some examples and the hydrothermal regime of selected roof, which is nowadays unfortunately used in more variants in Slovakia, are shown.

2. State of the art and Oriented strand boards

Use of wooden products in the non-ventilated flat roof is more obvious because of using the oriented strand board (OSB) in these roofs as a cheap alternative instead of other materials. OSB is a material with favorable mechanical properties which make it particularly suitable for load-bearing applications in constructions [1]. It is now more popular than plywood, commanding 66% of the structural panel market [1]. OSB is a type of engineered wood similar to particle board, formed by adding adhesives and then compressing layers of wood strands (flakes) in specific orientations. It was invented by Armin Elmendorf in California in 1963 [2]. OSB may have a rough and variegated surface with the individual strips of around 2.5 cm×15 cm, lying unevenly across each other, and it is produced in a variety of types and thicknesses. There are also five classes for use in different environment. Most used is the OSB 3 type, which should be suitable for use in load-bearing boards for use in humid conditions.

OSB boards are usually used in two positions depending on the type of the flat roofs:

- as a layer under the roof membrane (figure 1 and 2),
- as a layer on the top of rafters, with thermal insulation on top of each other (figure 3 and 4).

In both positions, its use is not recommended, because in the first case, the roofing membrane has usually a high vapor resistance, and in case of daily temperature changes, the water vapor condenses at the bottom of membrane and the surface of OSB starts to rot (figure 1, 2). In a short time period, during the normal maintenance of roof, the worker can break through into the thermal insulation.

In second case, the roof composition consists of two layers of thermal insulation, one is above the OSB, second is bellow, between the rafters. The idea, in which the roof is constructed is that in the first stage, normal flat roof composition is realized on the OSB, starting with the vapor barrier, thermal insulation and ending with roofing membrane. In second stage, in order to achieve the higher thermal resistance, additional thermal insulation is added from bottom between the rafters and another vapor barrier is added with the ceiling from gypsum board (figure 5). Usually, for lower thermal insulation is used the cheapest option, which is a spray applied polyurethane foam (PUR). This has high vapor resistance and also high water retention. Insulation also in short time becomes wet and the water starts to flow in the interior of building between the joint of the gypsum boards (figure 5 and 6). This caused the total deterioration of the OSB (figure 7 and 8) and also the rafters, but they last longer, because of their bigger dimensions and also the fact, that the massive wood has higher resistance than the OSB. High vapor amount and change of the wet/dry state increases the speed of deterioration of the boards. The glue cannot hold the strands together.

Figure 3. Wet upper part of the OSB board used within the composition of the flat roof.

Figure 4. OSB board as a base layer under the vapor barrier.

IOP Conf. Series: Materials Science and Engineering 1015 (2021) 012036 doi:10.1088/1757-899X/1015/1/012036

Figures 1, 2, 5, 6 are collected by the worker for the company producing the roofing membranes as results of the warranty claims, that the roof is not waterproof and has a leak. The reality showed almost every time a problem within the roof composition and the membrane was waterproof.

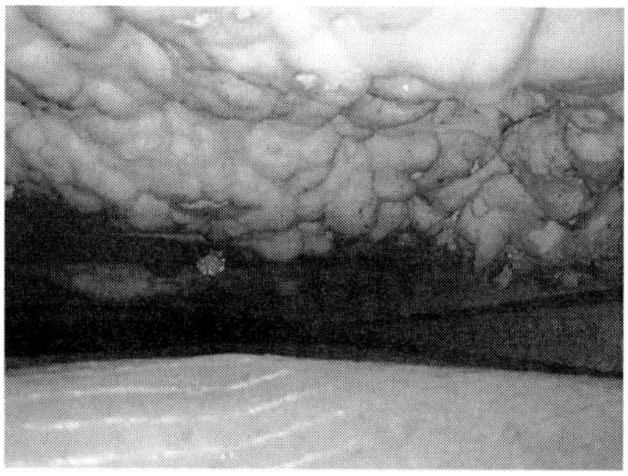

Figure 5. Wet surface of the polyurethane foam. Water drops on the vapor barrier [3].

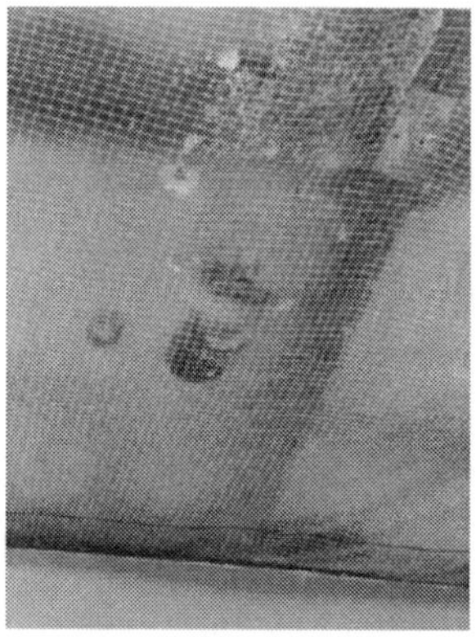

Figure 6. Collected water from the condense in foam [3].

3. Case study of the roof

To investigate the hydrothermal regime of such kind of roof, one representative roof for this article was chosen. The composition of this roof is shown in figure 9, and pictures from its observation are presented in figures 3 -5 and 8. The roof survey was conducted almost after 2 years of construction of the building. Building was 3 family, two story house. Construction of the roof was as described before, in two stages, starting with wooden rafters and OSB on it. Usual problem is the fact that sometimes when the roofing workers started the work, the OSB was wet from the rain. In this case, the vapor barrier was used from foil and not the bitumen type, which should be melted to the surface and the board will be dried.

Figure 7. View at oriented strand board after removing the sprayed thermal insulation. Growth mold and rot degradation.

Figure 8. Totally degraded OSB, removed with bare hands after nearly 2 years after construction.

If the roof was left with the first stage, it would work flawlessly as a simple flat roof with vapor barrier. Using the foil/plastic vapor barrier has problematic seams, so the real vapor resistance of the layer can vary. During conducting the roof survey, samples of materials for gravimetric measurement were taken. Measured water contents are stated in table 1.

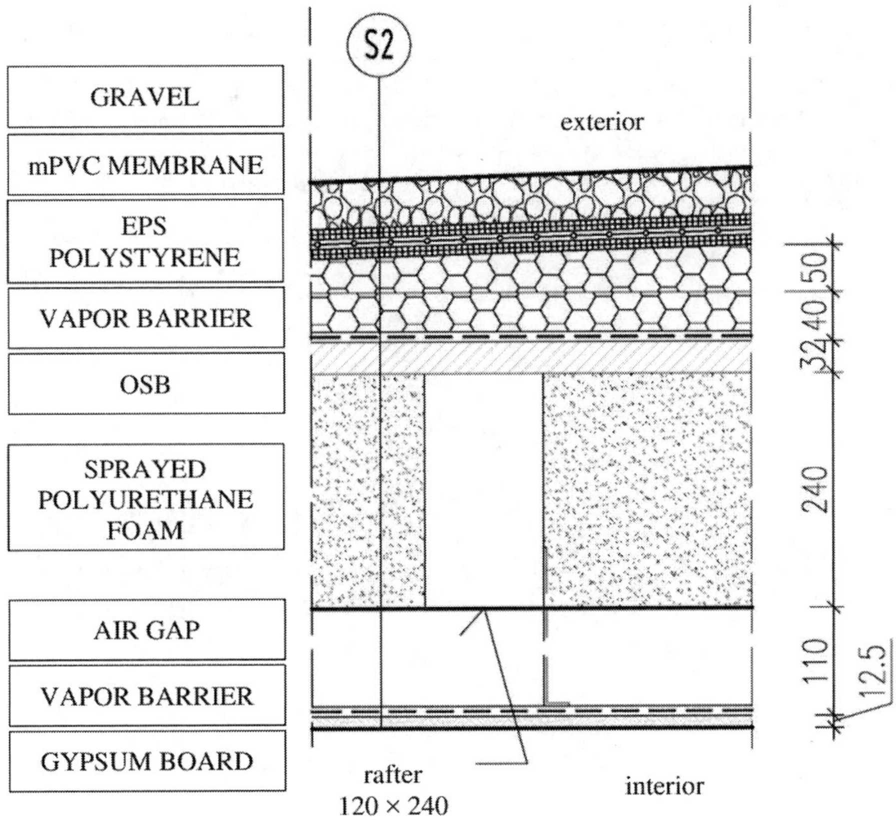

Figure 9. Composition of flat roof used for the numerical calculation.

Very high water content was measured in almost all layers, but especially in the PU foam. During the survey it was found out that the roof membrane was watertight, so the measured water content was not caused by the leak of rainwater.

Table 1. Gravimetric measurement results.

Material	water content [%]
Geotextile	45.9
EPS polystyrene	18.9
EPS polystyrene	1.1
OSB	40.25
Polyurethane foam	380
Gypsum board	14.1

4. Calculation according to the Slovak standard in the software Teplo

In this paper, only a mandatory calculation of water vapor condensation for the roof was made, in order to find out, if this theoretical model based on the Glaser model shows potential problem of this kind of roof, or the more advanced method should be used. For the condensation calculation, two methods can be used, either the one described in the STN 73 0540-2 [4], which is season made, or the one from STN EN ISO 13788 [5], which is monthly based. There are two conditions to be fulfilled:

amount of condensed vapour ($M_{,condensed}$) is less than 0.1 kg/(m^2·a) and the difference between condensed and evaporated ($M_{,evaporated}$) water vapor is positive.

Calculations were made in the Svoboda s software Teplo in version 2017.3 [6]. Monthly calculation of the roof is in figure 10. During the cold season of the year, water condenses in the roof, during summer some of this water content dries, but there is more condensed water than evaporated, so the roof does not fulfil the conditions and has to be considered as wrong.

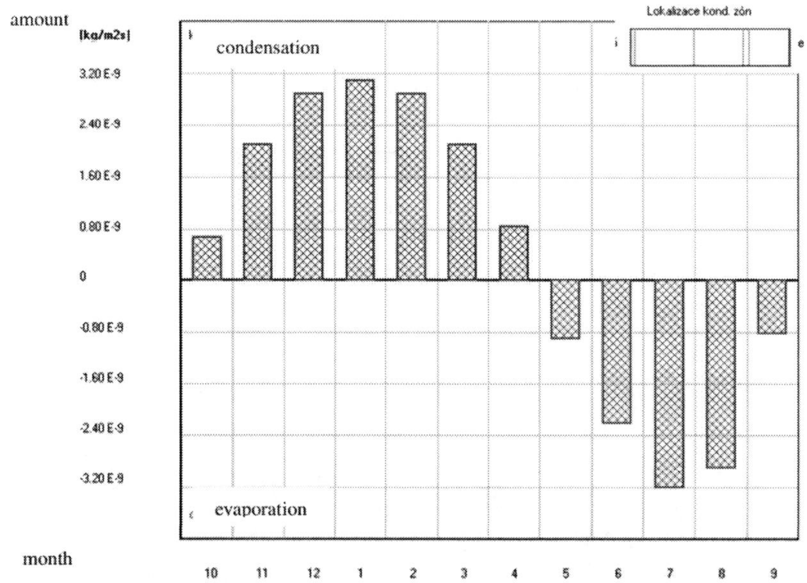

Figure 10. Composition of flat roof used for the numerical calculation.

Parameters of used materials in the roof are summarized in the table 2. Because of various boundary conditions and some uncertain material properties, more cases were calculated: higher indoor relative humidity (60 instead of 50%), without middle or bottom vapor barrier, possible perforation of vapor barriers (lowered vapor diffusion resistance factor), more efficient bottom vapor layer and wet OSB layer as starting condition. All variants and results of the calculations are summarized in table 3.

Because this calculation of condensation does not reflect the ability of some organic materials to deteriorate caused by the moisture, there is analyze table (table 4), which shows number of days of the year in which the material was in some area of relative air humidity. Each material, if is exposed long enough to some specific relative humidity, is going to reach equilibrium with its sorption curve. Especially for wooden materials, high water content can cause the mold growth, rot etc.

5. Results and discussion

Results of the gravimetric measurement in table 1 show very high water content in the OSB board and the foam thermal insulation. Such a high water content caused the total degradation of the board and the polyurethane foam was completely wet.

Calculated hydrothermal cases are summarized in table 3. Simple method of calculation according to the STN 73 0540-2 showed good results in all cases, the flat roof compositions works. According to the monthly method from STN EN ISO 13788, some considered cases show a negative balance, but in some cases neither standard calculation showed possible problems.

The probable risks with non-ventilated flat roofs showed the result table 4.

Table 2. Material properties.

Layer	Thickness d [m]	Thermal conductivity λ [W/(m·K)]	Vapor diffusion resistance factor μ [-]
Gypsum board	0.0125	0.270	17.0
Vapor barrier lower	0.0002	0.350	600 000.0
Air gap	0.120	0.750	0.08
Polyurethane foam	0.250	0.025	72.5
OSB	0.032	0.130	50.0
Vapor barrier middle	0.0002	0.300	500 000.0
EPS 150	0.090	0.035	50.0
mPVC membrane	0.0015	0.350	24 000.0

Table 3. Summary of calculated cases.

	STN 73 0540-2, Appendix B			STN EN ISO 13788	
case	$M_{,condensated}$	$M_{,evaporated}$	result	$M_{,condensated}$	result
real composition	0.0209	0.0592	pass	0.0344	positive
increased indoor humidity	0.0209	0.0592	pass	0.0728	wet
without middle vapor barrier	0.0396	0.0966	pass	0.0603	wet
without bottom vapor barrier	0.0237	0.0600	pass	0.0389	positive
perforated bottom vapor barrier	0.0233	0.0599	pass	0.0383	positive
perforated middle vapor barrier	0.0392	0.0961	pass	0.0596	wet
better bottom vapor barrier	0.0006	0.0534	pass	0.0001	positive
wet OSB 0.5kg/m^2	0.0209	0.0592	pass	0.5344	wet

Table 4. Number of days and relative humidity's for individual materials.

material	< 60%	60-70%	70-80%	80-90%	> 90%
Gypsum board	182	121	62	---	---
Vapor barrier lower	182	121	62	---	---
Air gap	273	92	---	---	---
Polyurethane foam	---	---	153	31	181
OSB	---	---	153	31	181
Vapor barrier middle	---	---	153	31	181
EPS 150	---	92	212	61	---
EPS 150	---	62	91	61	151
mPVC membrane	---	62	91	61	151

The oriented strand board and the polyurethane foam are whole calculated year in the area of relative humidity above 70 %, more than a half of the year above 90 %. This represents around 150 kg/m^3 of water in the material. The results correlate with the gravimetric measurement results. OSB and polyurethane foam have a very high water content. Problem of such roof composition is a wrong placement of the vapor barrier, which creates two condensation zones. One is under the roofing membrane and the other one is at the bottom of the middle one. The second one creates condense, which moistens the OSB surface (figure 3). Even if the composition was without the middle vapor barrier, the yearly balance is wet. The calculation will be even worse with use of roofing membrane with higher vapor diffusion resistance, such as FPO (flexible polyolefin) or bitumen membranes.

6. Conclusions

In this paper, problems with use of wooden materials in non-ventilated flat roofs were described, which is unfortunately very often in Slovakia nowadays. Flat roof composition, which was analyzed is based on the real roof, where was conducted roof survey. Results from this survey showed serious problem with this roof in very short time after construction. This is caused by the wrong design of roof and use of wooden layers in composition, where is a very small possibility of drying, because the wood is between two, in this case even among three vapor barriers.

These results were supported by the analysis of water content in selected materials during the annual calculation. In such composition, there are two condensation zones, one under the badly placed middle vapor barrier and second under the roofing membrane. Based on the survey and this analysis it is highly recommended to avoid use of wooden layers, such as OSB in the composition with minimal possibility of drying.

Acknowledgment

The research is supported by the grant project VEGA No. 1/0673/20 and KEGA 032ŽU-4/2018.

References

[1] web <https://en.wikipedia.org/wiki/Oriented_strand_board>
[2] Document US000003164511A (PDF). DEPATISnet. Deutsches Patent- und Markenamt Retrieved 2 May 2016.
[3] Mr. Pavol Krajcovic photo archive from Roof surveys, Fatra Izolfa
[4] Standard STN 730540-2+Z1+Z2: 2019: Thermal protection of buildings. Thermal performance of buildings and components. Part 2. Functional requirements Change 1 and 2
[5] Standard STN EN ISO 13788:2013: Hygrothermal performance of building components and building elements. Internal surface temperature to avoid critical surface humidity and interstitial condensation. Calculation methods
[6] Svoboda software Teplo 2017.3

IOP Conf. Series: Materials Science and Engineering 1015 (2021) 012037 doi:10.1088/1757-899X/1015/1/012037

The use of included stiffeners to strengthen the frames of one-story industrial buildings

O Yu Veremeenko and V Yu Alpatov

Samara State Technical University, Molodogvardeyskaya str. 244, Samara, Russia 443100

E-mail address: overemeenko@yandex.ru

Abstract. The article considers issues of industrial buildings frames strengthening by the introduction of stiffening elements at the change of calculated seismicity. The principle of materials concentration in the form of stiffness elements with included connections is used. The degree of their influence on the work of building frame is estimated. In this case, the work of coating disk as an element distributing horizontal forces changes. The main dependencies for gap definition determined by the reinforced concrete frame movements, at which moments in the lower section of the columns reach the limit values, are determined. Thus, in the process of influence the seismic force will remain approximately constant, as the total stiffness will be slightly higher than the initial columns stiffness. Introduction of rigidity elements into the frame should be accompanied by corresponding measures for strengthening of foundations for steel supports.

1. Introduction

The authors propose to use the linking scheme of the frame with stiffening elements, developed by CNIIPromzdanie, PI-1 and NIIZHB to strengthen one-story industrial buildings [1-6]. In such frames, the stiffeners take up the bulk of the horizontal loads, and the row columns and the corresponding foundations work almost as centrally compressed.

In the longitudinal direction of building frames, there are steel cross or portal connectors, which should be considered as stiffeners. The possibility of placing the connectors in the transverse direction along the ends of the building is determined by the sufficient bearing capacity of the coating disk. When the stiffeners are introduced in the transverse direction, the forces in the coating disc change. These efforts become greater when the distance between the stiffeners and the stiffness ratio of the stiffeners and row columns become the greater.

2. Main provisions

2.1. Increased horizontal loads

To determine the stiffness characteristics of stiffening elements and row columns, a well-known method of calculation of interconnecting frames is used [2]. Rigid characteristics of stiffening elements are determined using the dependence of the frame deflection on the current horizontal load and the total back pressure of the frame

$$f = \frac{W}{\sum n_i \cdot \dfrac{N_i}{\dfrac{tg\, k_i \cdot l}{k_i} - l}} \tag{1}$$

where: W - horizontal load,
N_i - vertical load on the column,
l - calculated column length,

and

$$k_i = \sqrt{\frac{N_i}{B_i}} \tag{2}$$

where B_i – rigidity of i column.

We obtain the moments in the design sections using the dependence

$$M_i = P_i \cdot l + N_i \cdot f \tag{3}$$

where

$$P_i = \frac{N_i}{\dfrac{tg\, k_i \cdot l}{k_i} - l} \quad - \mathit{backpressure}\ i - \mathit{column}$$

With the reinforcement associated with the planned increase in horizontal loads, it is sufficient to assign the ratio BRC/BCO in such a way that the forces in the design sections of the row columns remain at the same level as they were before the load increase.

Writing the dependence in the form

$$W = f \cdot \sum n_i \cdot \frac{N_i}{\dfrac{tg\, k_i \cdot l}{l} - l} \tag{4}$$

and analyzing it, it can be seen that the value of the bending moment in row columns remains unchanged at a constant deflection value. By fixing the value and increasing the stiffness of the stiffeners, it is possible to determine how much the horizontal loads can increase with the given stiffeners parameters.

2.2. Increase in vertical loads

During the reconstruction associated with an increase in vertical loads on the frame, the condition of maintaining the bearing capacity of the columns was used. According to the graph of the bearing capacity, it was determined by how much it was necessary to reduce the bending moment in the design section of the columns to maintain the bearing capacity. Then, placing the stiffeners and using dependencies (2) and (3), their stiffness characteristics were determined, at which the necessary reduction in bending moments occurs.

3. Change in the calculated seismicity

Strengthening the frames of one-story industrial buildings using stiffeners when changing the design seismicity has its own specifics. Simple introduction of stiffeners leads to an increase in seismic forces due to an increase in the rigidity of the frame [7, 8]. Therefore, the structures of stiffening elements included in the work were developed in the process of exposure to seismic forces at the stage when the stiffness of the columns decreases. Thus, at the moment of seismic forces the rigidity of the frame is equal to the initial stiffness before strengthening. At the end of the impact, the rigidity may increase slightly.

3.1. Structures of included stiffeners

The stiffeners are closed steel supports 2, surrounding two or four columns 1 (see figure 1).

Frames are rigidly fixed in a reinforced concrete clip fortification of foundations. Disc springs 4 are fixed to the frame at column header, through which horizontal seismic forces should be transmitted to the stiffening elements.

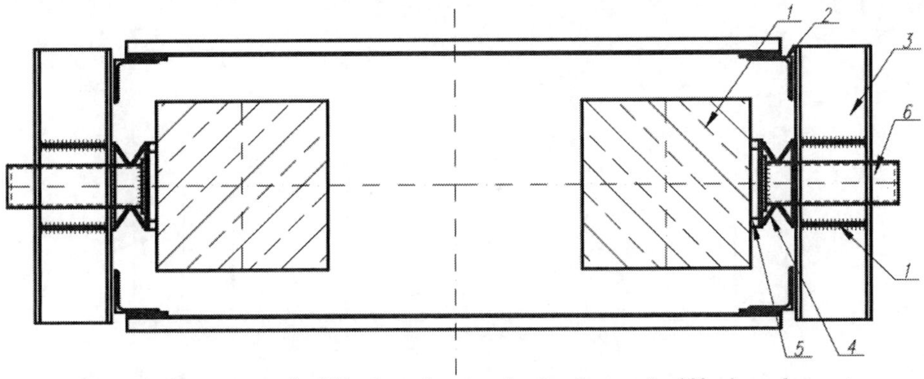

Figure1. Structure of stiffening element in the form of stiffening elements, surrounding reinforced concrete columns (*1* - reinforced concrete column; *2* - steel support; *3* - steel stop; *4* - disc spring; *5* - foam band; *6* - steel guide, pipe).

The springs 5 are installed in such a way that there is a gap between them and the column surface. Therefore, the elements of rigidity are not included in the work immediately, but when cracks in the columns are formed and the column's stiffness falls. The gap size is determined by the movement of the reinforced concrete frame, at which the moments in the lower section of the columns reach the limit values. In this case, the total displacement of reinforced concrete columns, consisting of the displacement of the reinforced concrete frame before the inclusion in the work of the steel ring and after its inclusion, excludes the achievement in the concrete reinforcement column ultimate deformation and possible damage. The gap value was determined from the condition:

$$\delta = \frac{M_u \cdot l^2}{3 \cdot B} \tag{5}$$

M_u - the maximum allowable moment,

B - bending stiffness of the column.

The stiffness of steel supports is selected so that the total stiffness of the reinforced concrete frame and the total stiffness of steel supports are of the same order. This ensures that the total stiffness of the frame is always the same. At the initial moment, the rigidity of the framework is determined by the stiffness of reinforced concrete columns and after reaching the moment of limiting values in the columns and the formation of plastic joints in the lower sections of the columns with fixed forces, the stiffness of the framework will be determined by the total stiffness of steel supports and reinforced concrete columns with plastic joints. As a result, in the process of influence the seismic force will remain approximately constant, because the total stiffness will slightly exceed the initial stiffness of the columns. However, the gap size is determined from the condition of achieving the limit forces in the calculated sections using the calculated characteristics of the material. The probability of plastic hinge appearance is extremely low and, therefore, after the stiffness elements inclusion in the work, rigidity of the frame should increase. In this case the seismic load on stiffening elements will also increase. That is why steel supports should be calculated with some reserve.

Introduction of stiffening elements into the frame should be accompanied by appropriate measures to strengthen foundations for steel supports.

4. Stiffness elements arrangement

When choosing the stiffeners placement scheme they should be arranged in such a way as to minimize the impact on existing technology in the reinforced buildings and to strengthen without stopping production. For this purpose, it is advisable to place the stiffening elements on the temperature joints and external building axes (see figure 2). Moreover, the placement of stiffening elements at temperature joints makes it possible to create stiffening elements by combining the neighboring columns this or that way.

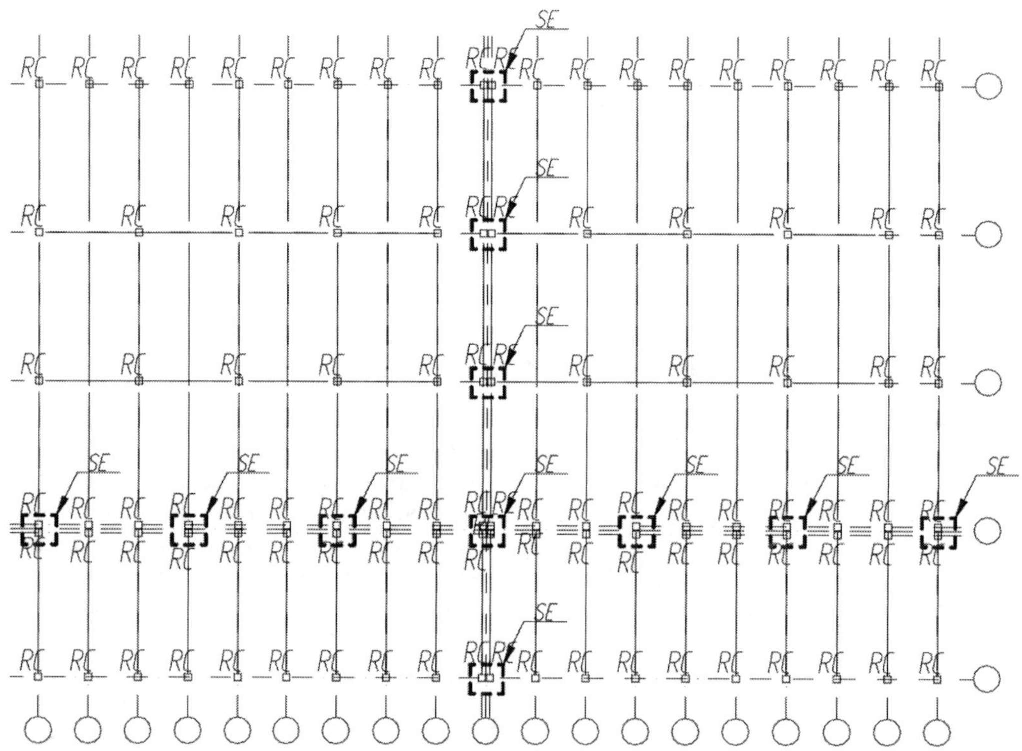

Figure 2. The scheme of the stiffeners in the building plan: *KC* - reinforced concrete columns (*RC*); *SE* – stiffeners elements.

5. Conclusions

1. When placing the included stiffening elements in the plan of the building it is necessary to take into account the peculiarities of the existing production technology, as well as the possibility of strengthening without stopping production. Moreover, it is recommended to place the stiffness elements close to the building axes of symmetry, which has a positive impact on the work of the building coating disk, which is the main distribution element of horizontal forces under seismic loads.
2. The parameters and dimensions of steel through supports of stiffening elements should be determined taking into account the influence of support lattice on given support flexibility. Support stiffness should be approximately equal to the stiffness of reinforced concrete column.
3. For smooth seismic forces transfer to the supports, it is recommended to use disc springs reducing the dimensions of steel supports.
4. The size of the gap between the reinforced concrete column and the stiffness element is determined by the stiffness of the reinforced concrete column and its maximum load-bearing capacity, taking into account the formation of cracks in dangerous sections of the column and is 40-50 mm.

References

[1] Vasiliev B F and Rosenblum A Ya 1974 Reinforced concrete columns of one-story industrial buildings (Stroyizdat) p 195

[2] Gershanok R A and Ivanov A V 1988 New constructive solution for a one-storey industrial building *Design and calculation of building structures* pp 16-20

[3] Veremeenko O Yu 1990 Determination of the stiffness characteristics of stiffeners when reinforcing the reinforced concrete frames of one-story industrial buildings by introducing connections at the ends *Methods and methods of strengthening building structures at reconstructed enterprises* pp 11-15

[4] Veremeenko O Yu 1990 Determination of the bearing capacity of the coating disc when reinforcing the frames of one-story industrial buildings by introducing stiffening elements *Methods of strengthening building structures at reconstructed enterprises* pp 8-11

[5] Veremeenko O Yu 1991 *Strengthening the frames of one-story industrial buildings by introducing stiffening elements* (Stroyizdat) pp 243-244

[6] Klevtsov V A 2004 The work of frame structures for one-story industrial buildings and their strengthening *All-Russian Research Institute of Problems of Scientific and Technological Progress and Information in Construction* p 40

[7] Standard SP14.13330. 2014 Amendment 1 with Crimea - SNiP II-7-81- *Construction in seismic areas.*

[8] Sakharov O A 2004 On the issue of setting the seismic impact in the multi-level design of seismic structures *Earthquake-resistant construction Safety of structures* **4** pp 7-9

IOP Conf. Series: Materials Science and Engineering 1015 (2021) 012038 doi:10.1088/1757-899X/1015/1/012038

Accumulation of damage in reinforced concrete elements under cyclic loads

O Mkrtychev and E Lokhova

Moscow State University of Civil Engineering, National Research University, 26 Yaroslavskoe shosse, Moscow, 129337, Russia

E-mail address: elm97@mail.ru

Abstract. The article presents the results of research on the behavior of a frame structure in a reinforced concrete building with a full rigid frame under cyclic load typical for strong earthquakes. When creating a finite element model for the research, the models of materials were used that take into account the nonlinear nature of the structure behavior, as well as the accumulation of damage under cyclic loads. The calculation was carried out by a direct dynamic method using explicit schemes for integrating the equations of motion in the multipurpose finite element software package LS-DYNA.

1. Introduction

One of the main objectives of earthquake engineering is the development of effective seismic protection systems and methods for calculating buildings and structures designed with these systems, which allow to adequately assess the ability of a structure to resist seismic impacts. A considerable number of modern buildings and structures designed for seismic areas are being constructed using reinforced concrete (figure 1).

Figure 1. The use of reinforced concrete in the construction of buildings.

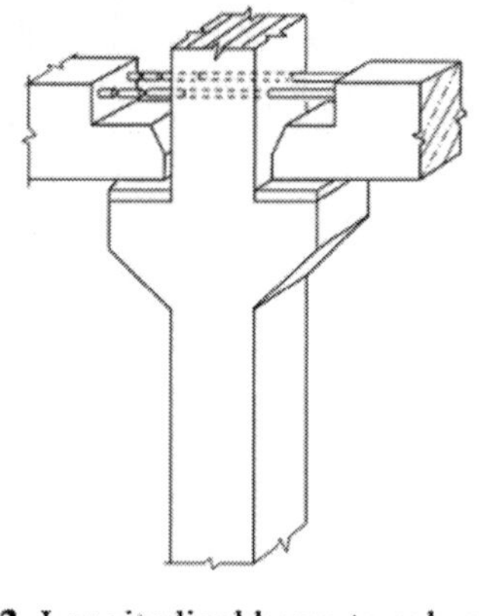

Figure 2. Longitudinal beam to column joint.

At present, the longitudinal beam to column joints (figure 2) in the calculation schemes of frame structures are modeled, as a rule, either absolutely rigid or hinged. However, the analysis of their behavior shows that such modeling results in a significant error in the calculation for seismic impacts. Research shows that the joints of reinforced concrete elements are mobile and have the ability to dissipate energy during cyclical operation during an earthquake.[1], [3].

In this regard, one of the actual objectives of earthquake engineering is the development of methods for calculation of reinforced concrete structures that take into account the mobility and dissipative properties of the joints. In this work the behavior of reinforced concrete frame under cyclic load typical for intensive seismic impacts was studied. The main objective of this work was to determine the dissipative reserves in the behavior of a reinforced concrete structure joint. The research was carried out using the LS-DYNA software package. This package uses explicit integration schemes. The explicit method is the most effective, and it is advisable to use it for calculation of structures with dynamic loads.

2. Research

2.1. Description of the finite element model

A single-span reinforced concrete frame was chosen as the object of research. When developing the design model, the following sections were adopted: for the column – a square section of 0.4×0.4 m, for the longitudinal beam – a rectangular section of 0.4×0.5 m. The span of the longitudinal beam is 6m, the height of the columns is 4 m. The frame material is B25(C20) concrete (the strength of this concrete is 25 MPa, and the density is 2300-2400 kg/m^3 [4, 5]). The longitudinal beam and the columns are reinforced with four reinforcing bars with a diameter of 28 mm (used reinforcement A400, yield point not less than 400 MPa, [6]), as well as ties with a diameter of 12 mm (used reinforcement A240, yield point not less than 235 MPa [6]), with a spacing of 200 mm.

The model used is shown in figure 3, two types of material models were used to create it (for modeling of reinforcement – Piecewise Linear Isotropic Plasticity material, for modeling of concrete – Continuous Surface Cap Model). The main feature of the mathematical model of concrete is that this model allows you to take into account the degradation of stiffness and strength, as well as the accumulation of damage under cyclic loads. Figure 4 shows a closed model of the Continuous Surface Cap Model material developed by the Federal Highway Administration of the United States Department of Transportation [1], it shows the yield surface of primary stresses in space in accordance with this model of concrete [2].

Figure 3. Used finite element model.

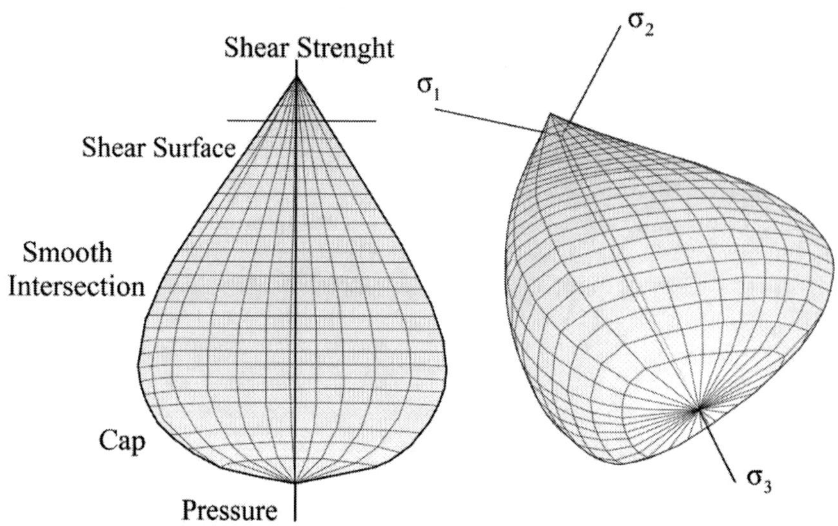

Figure 4. The mathematical model of concrete (CSCM – Continuous Surface Cap Model)[2]

The yield surface for this model is expressed in terms of stress invariants:

$$J_1 = 3P, \quad J_2' = \frac{S_{ij}S_{ij}}{2}, \quad J_3' = \frac{S_{ij}S_{ik}S_{ki}}{3}, \tag{1}$$

where J_1 – stands for the first invariant of the stress tensor, J_2 – second invariant of the stress deviator, J_3 – third invariant of the stress deviator, S_{ij} – stress deviator, P – pressure.

The model consists of 52920 nodes and 4792 elements. The model contains 2 types of elements: beam and solid. Element length 0.04 m. Due to the small size of the grid, the element was integrated at one point. A horizontal distributed load was applied to the upper surface of the longitudinal beam, two concentrated vertical forces were applied to the columns, as well as one horizontal force, whose change curve is shown in figure 5. The impact amplitude varied from 10 kN to 200 kN. The load period was determined so that significant dynamic effects did not arise, that is, so that the load period did not exceed the period of oscillation of the structure (also not allowed resonance). This load is close to intense seismic impact.

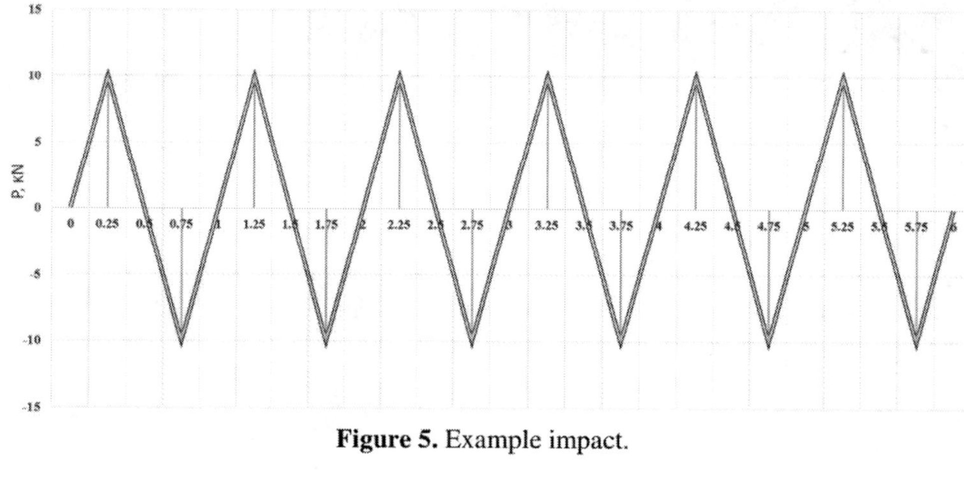

Figure 5. Example impact.

2.2. Destruction criterion. Accumulation of damage

In this work, a special material model was used, which takes into account the accumulation of damage under cyclic loads. Damage accumulation is based upon two distinct formulations, which we call brittle damage and ductile damage.

Brittle damage builds up when stretched. Accumulation of brittle damage depends on the maximum principal deformation ε_{max}:

$$\tau_b = \sqrt{E\varepsilon_{max}^2} \tag{2}$$

Brittle damage accumulates when the pressure is tensile and an energy-type term, τ_b exceeds the damage threshold r_{0b}.

Plastic damage builds up during compression. It does not accumulate when stretched. The accumulation of plastic damage depends on the components of the total deformation ε_{ij}:

$$\tau_d = \sqrt{\frac{1}{2}\sigma_{ij}\varepsilon_{ij}} \tag{3}$$

Plastic fracture begins when τ_d exceeds the initial threshold r_{0d} [1].

2.3. Results obtained

The model takes into account both compression and tension damage.Figure 6-9 show deformed schemes with isofields of stress intensity for reinforcement and concrete, isofields of plastic flow intensity for reinforcement, and isofields of changes in the function of accumulation of concrete damage before the onset of structural failure when applying an impact with an amplitude of 130 kN. The damage accumulation function varies from 0 to 1 (0 corresponds to no damage, 1 – complete destruction of the material)

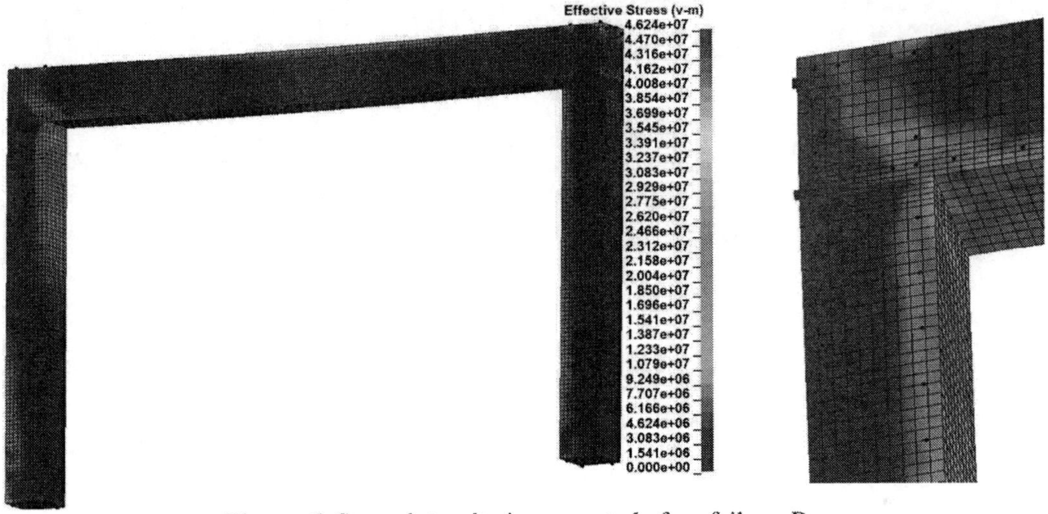

Figure 6. Stress intensity in concrete before failure, Pa

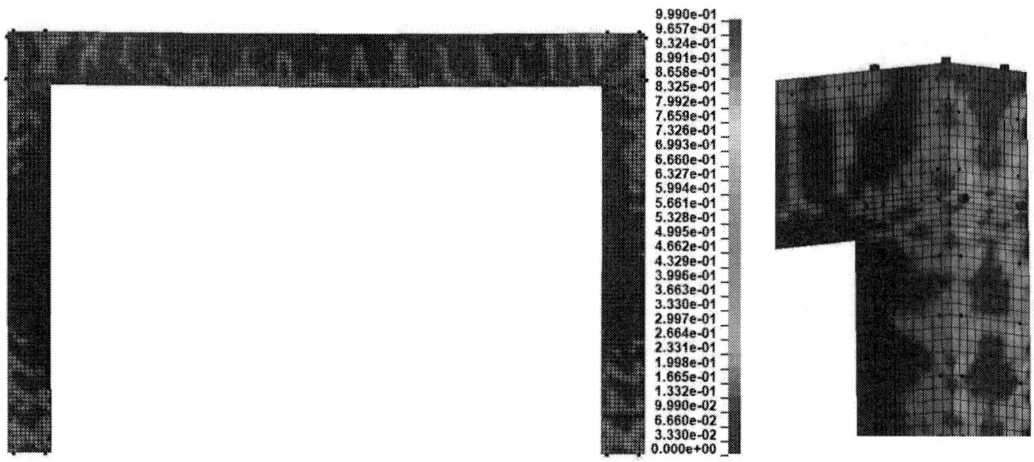

Figure 7. Function of damage accumulation in concrete before failure.

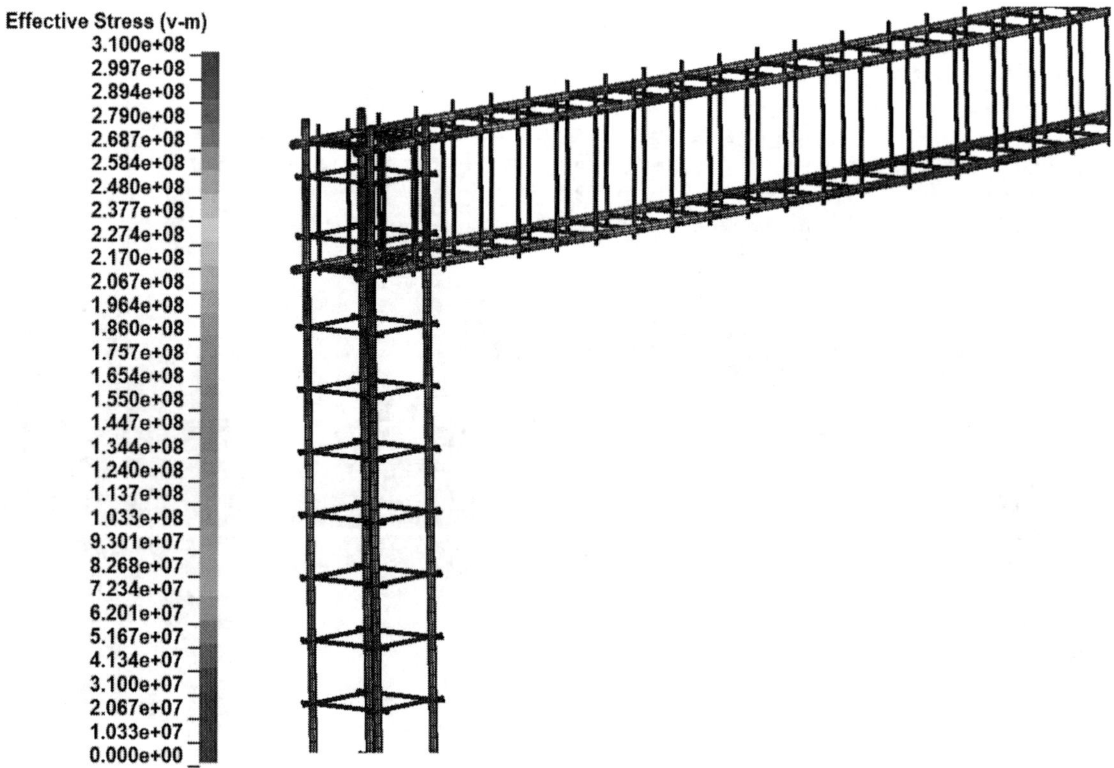

Figure 8. Stress intensity in reinforcement before failure, Pa

IOP Conf. Series: Materials Science and Engineering 1015 (2021) 012038 doi:10.1088/1757-899X/1015/1/012038

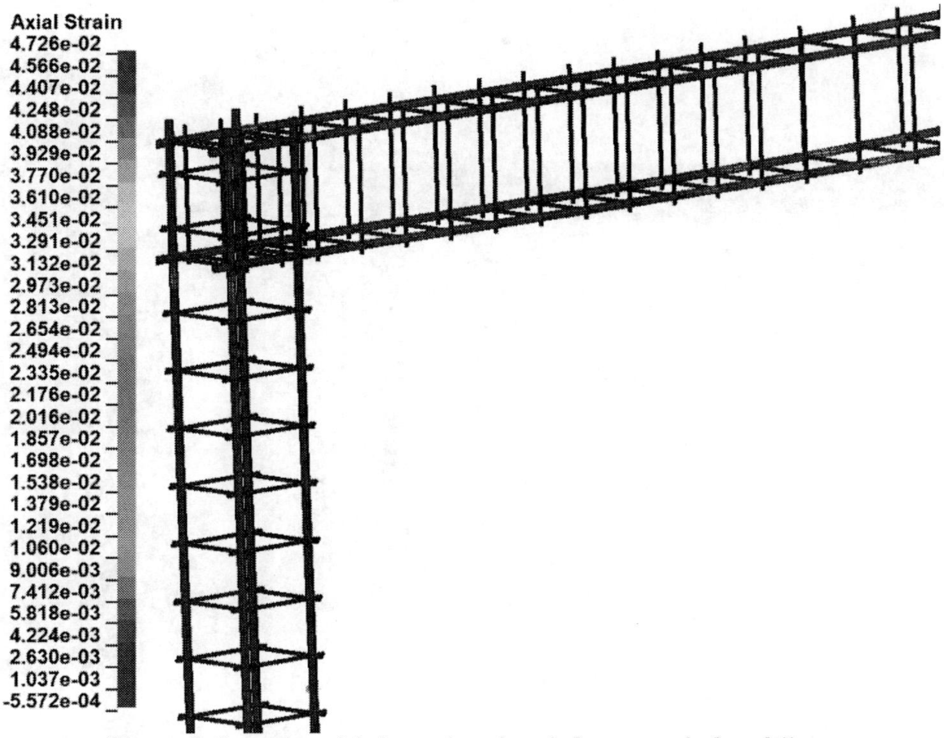

Figure 9. Intensity of deformations in reinforcement before failure.

Figure 10-13 show deformed schemes with isofields of stress intensity for reinforcement and concrete, isofields of plastic flow intensity for reinforcement, and isofields of changes in the function of accumulation of concrete damage after the structural failure when applying an impact with an amplitude of 130 kN.

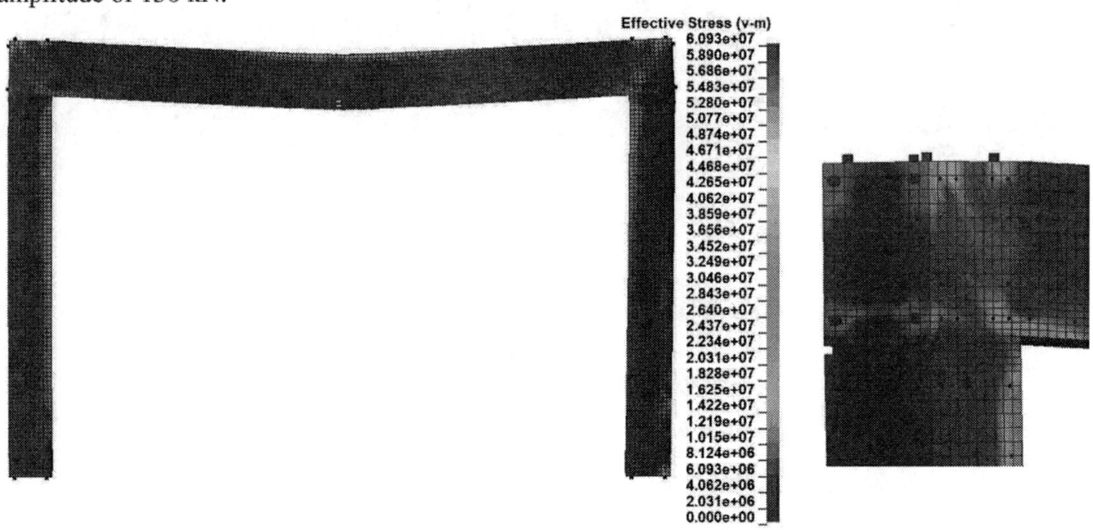

Figure 10. Stress intensity in concrete after failure, Pa

IOP Conf. Series: Materials Science and Engineering 1015 (2021) 012038 doi:10.1088/1757-899X/1015/1/012038

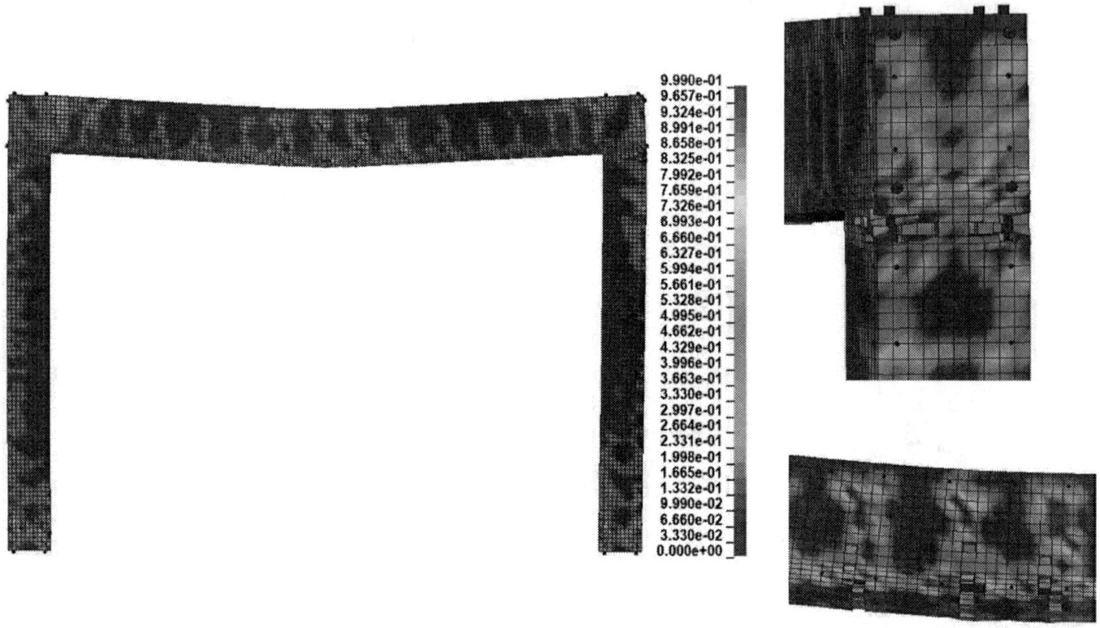

Figure 11. Function of damage accumulation in concrete after failure.

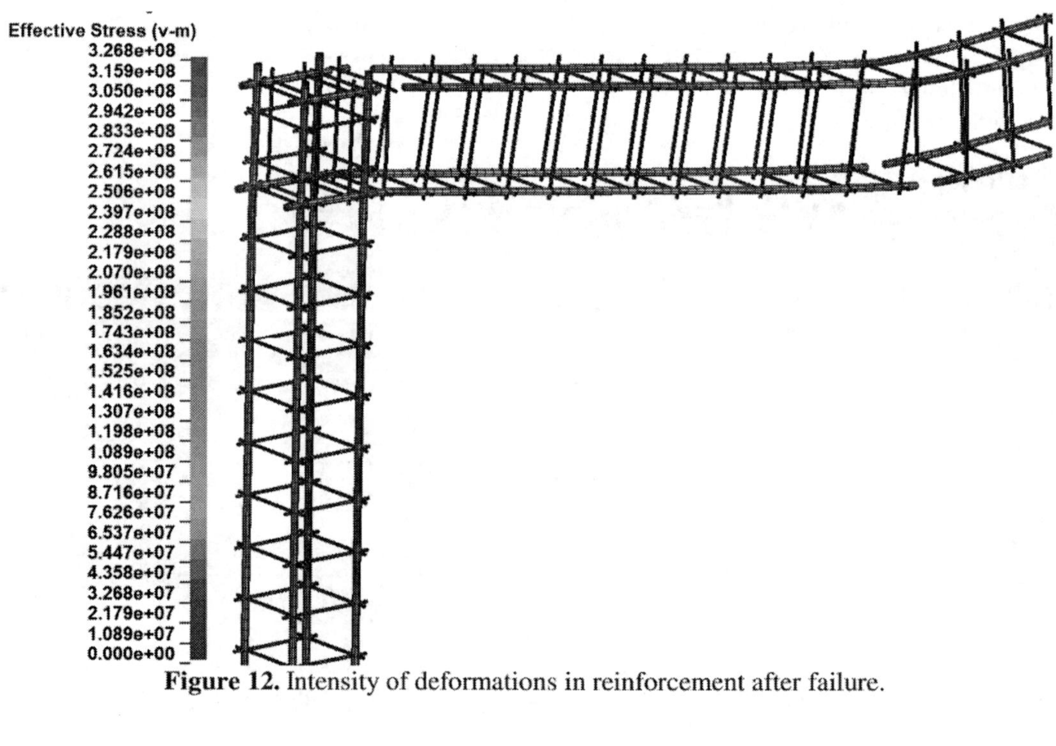

Figure 12. Intensity of deformations in reinforcement after failure.

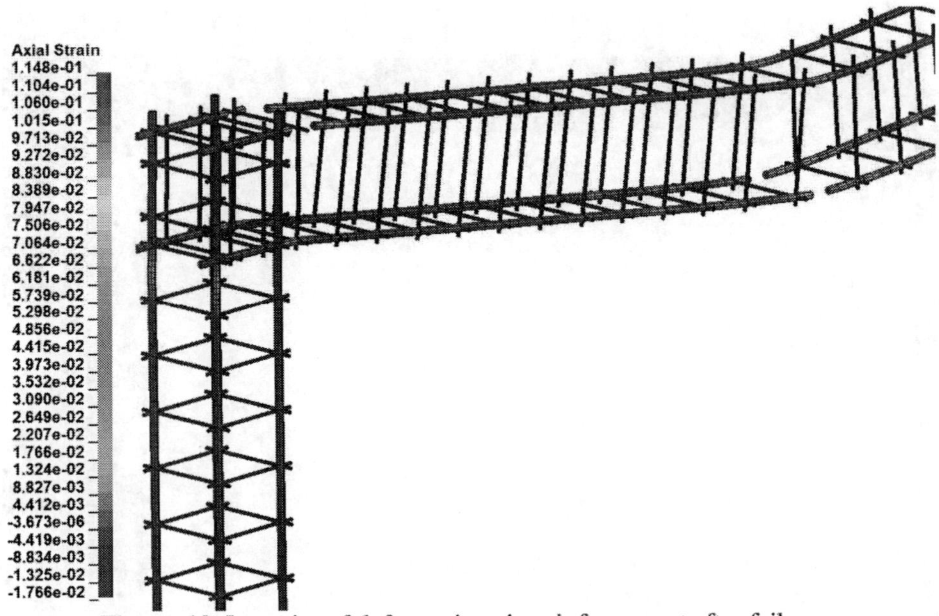

Figure 13. Intensity of deformations in reinforcement after failure.

Figure 14-16 show the isofields of damage accumulation at different amplitudes of cyclical impact before the structural failure.

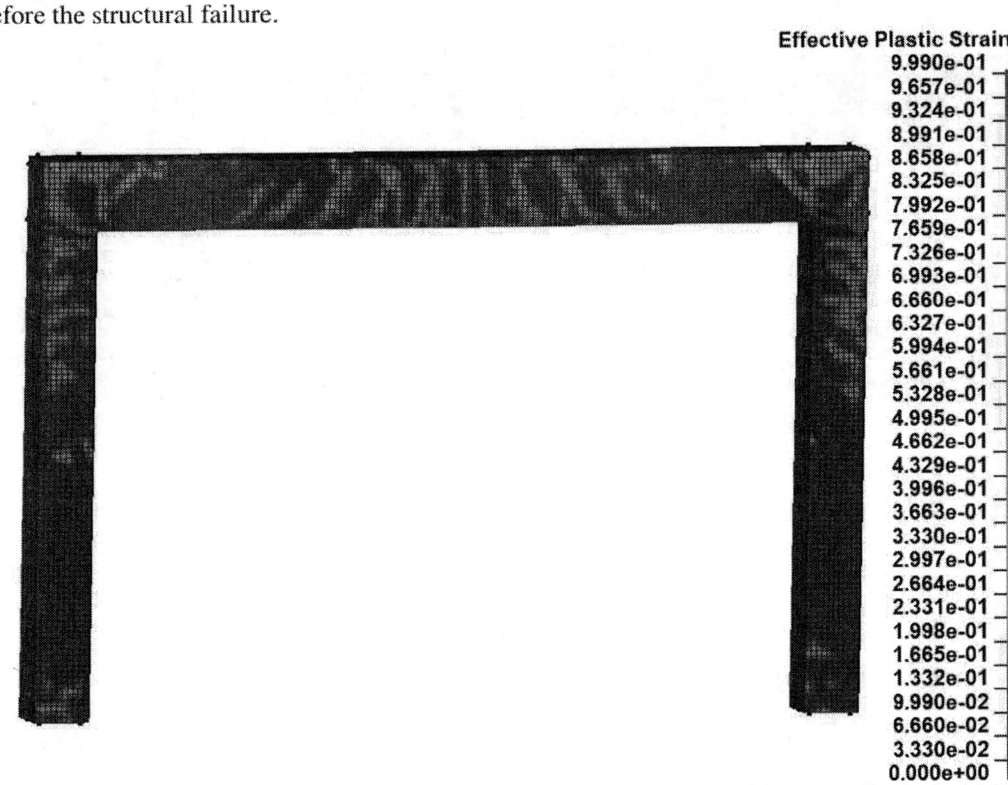

Figure 14. Function of damage accumulation applying an impact with an amplitude P=30 kN.

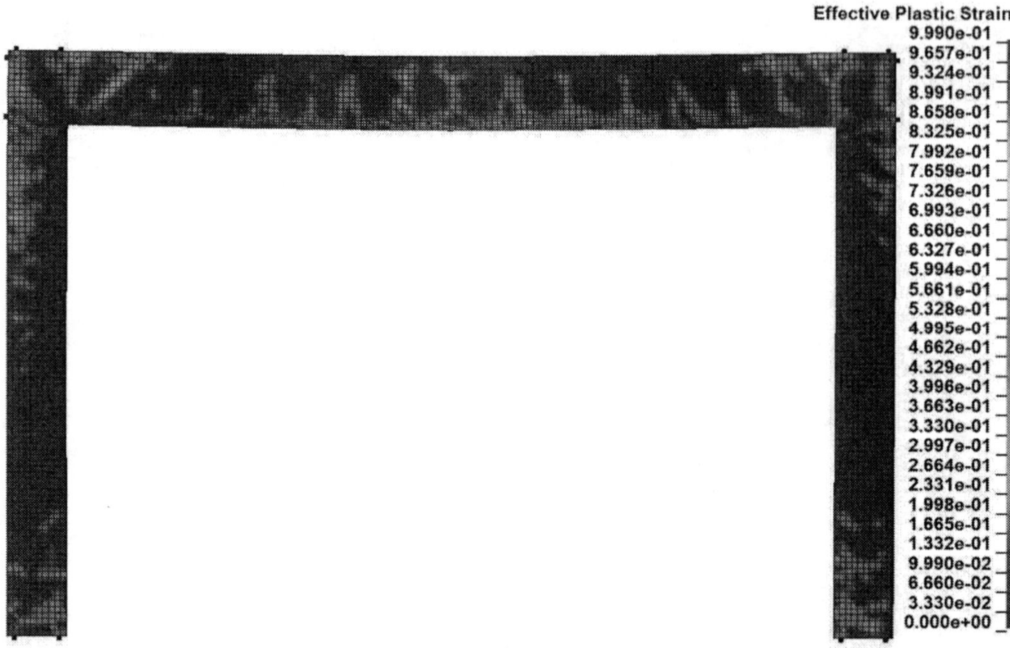

Figure 15. Function of damage accumulation applying an impact with an amplitude P=100 kN.

Figure 16. Function of damage accumulation applying an impact with an amplitude P=180 kN

After analyzing the nature of failure for cases with different amplitudes of external impact, the dependence of the number of cycles before the failure of the reinforced concrete joint on the amplitude of the applied cyclic load (figure 17).

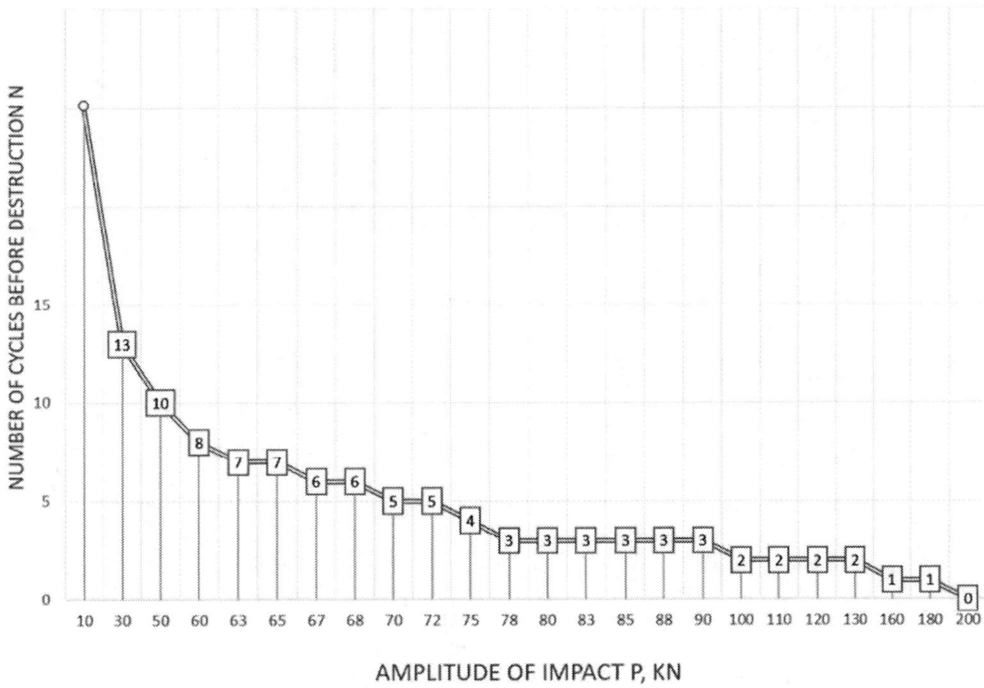

Figure 17. The dependence of the number of cycles before the destruction of a frame reinforced concrete connection on the amplitude of the applied cyclic load.

3. Conclusions

1. The isofields of stress and deformation intensity, as well as isofields of changes in the damage accumulation function for the structure in question at given time points were drawn.

2. The dependence of the number of cycles before the failure of reinforced concrete frame on the amplitude of the applied cyclic load is drawn (from 10 to 200 kN).

3. A method for modeling a reinforced concrete frame has been developed, which allows taking into account degradation stiffness and strength, as well as the accumulation of damage under cyclic loads. And there were also identified determine the dissipative reserves in the behavior of a reinforced concrete structure joint.

4. The proposed approach makes it possible to evaluate the seismic stability of frame joints of reinforced concrete buildings and structures in a nonlinear dynamic setting.

References

[1] US Department of Transportation. Federal Highway Administration. Evaluation of LSDYNA Concrete. Material Model 159. FHWA-HRT-05-063, 190 (2007)

[2] Murray YD 2007 Users Manual for LS-DYNA Concrete Material Model 159, FHWAHRT-05-062

[3] Aoyoma H and Naguchi H 1979 Mechanical propernies oof concrete under load cycles idealiving seinnieactions *Comit Euto-International du beton. Bulletin dinfor-mation* **131** (Rome)

[4] Standard GOST 26633-2015 Heavy and fine-grained concrete

[5] Standard Eurocode 2: Design of concrete structures

[6] Standard GOST 5781-82 Hot-rolled steel for reinforcing reinforced concrete structures

IOP Conf. Series: Materials Science and Engineering 1015 (2021) 012039 doi:10.1088/1757-899X/1015/1/012039

Theoretical-experimental analysis of selected passive timber house wall

D Michálková and P Ďurica

Department of Building Engineering and Urban Planning, Faculty of Civil Engineering, University of Žilina, Univerzitná 1, Žilina 010 26, Slovak Republic

E-mail address: danielamichalkova33@gmail.com

Abstract. The advancing times and legislation require constant research of new progressive building envelope structures. Walls based on wooden material are perspective in this regard. These are capable of elimination the wall thickness while providing the required thermal resistance. The pavilion research of KPSU UNIZA laboratory includes this kind of wall structures intended for passive wooden houses. This paper deals with one of the researched wall assemblies in the annual operation. The measurements show that such a construction is suitable for near to zero energy buildings. It also points out that the correct design can eliminate thermal bridges in the area of the load-bearing wooden studs.

1. Introduction

Reducing the energy consumption of the buildings is a current topic since the global oil crisis in 1973 [1]. The "global warming" and the associated need to change our behaviour also contribute to developing new ways of thinking.

The current legislation STN 73 0540-2+Z1+Z2 [2] determine the main requirements for passive buildings. It defines the heat transfer coefficient U W/(m^2·K) of envelope structures, while also respecting the values for specific heat demand for heating $Q_{H,nd,N}$ according to the building shape factor.

Besides, the airtightness of the whole building can be maximal $n_N \leq 0.6$ h^{-1} and must be verified according to STN EN ISO 9972 (730576) [3] by the so-called blower-door test [2].

It is possible to meet these demanding requirements by the increase of thermal insulation thickness or by the development of new materials with higher thermal resistance. The possible compromise is buildings with a wooden load-bearing structure that enables decreasing the wall thickness while also increasing the thermal resistance by placing the insulation within the wall studs. However, such structural system is flammable, therefore the fire safety restrictions must be taken into account [4].

2. Analysed construction – theoretical base

The selected wall assembly is shown in figure 1. It is placed within the pavilion research of wooden passive house wall structures of KPSU in Žilina. This laboratory was built in 2011. A more detailed description of it is in the literature [5].

This structure is located in a south-facing room with monitoring of the temperature and relative humidity of indoor air. Within the assembly are placed probes between individual materials and on the studs in three high levels.

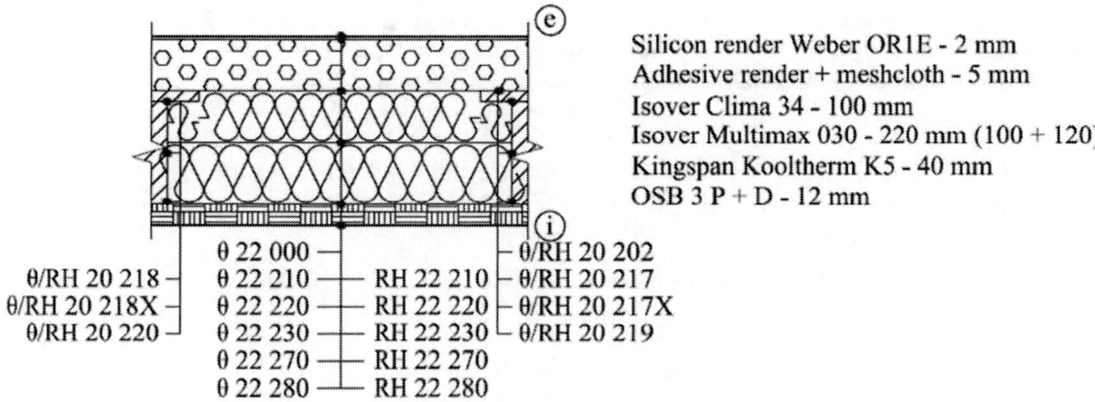

Figure 1. The selected structure with the probes placement.

This wall structure consists of a wooden load-bearing system filled with thermal insulation of glass fibres and phenolic board. Each material and its main characteristics summarize table 1.

Table 1. The main material characteristics.

Material	d [m]	ρ [kg/m^3]	λ [W/(m·K)]	μ [-]	c [J/(kg·K)]
Silicon render	0.002	1600	0.860	130	920
Adhesive render + mesh cloth	0.005	1660	0.900	20	900
Isover Clima 34 (glass fibres)	0.100	148	0.034	1	940
Isover Multimax 030 (glass fibres)	0.220	64	0.030	1	940
Kingspan Kooltherm (phenolic foam)	0.040	35	0.021	35	1400
OSB 3 P+D	0.012	650	0.130	50	1700

Note: d – thickness, ρ – bulk density, λ – thermal conductivity, μ – water vapor diffusion resistance factor, c – specific heat capacity.

It is a diffusely open structure provided with the exterior silicone render. table 2 comprises main properties of the assembly from a stationary point of view – the resulting thermal resistance and heat transfer coefficient – as well as from a non-stationary point of view – thermal damping factor and phase shift of thermal oscillation.

Table 2. The main properties of the structure.

Name	Thermal resistance R_T [(m^2·K)/W]	Heat transfer coefficient U [W/(m^2·K)]	Thermal damping factor ν [-]	Phase shift of thermal oscillation ψ [h]
J4	10.75	0.093	206.2	9.1

3. Experimental part

The outdoor climatic data were collected by a weather station on the laboratory roof. It records the outside temperature, relative humidity, air pressure, solar radiation (direct and diffuse), wind speed and direction, and rain. All of them are discussed in the literature [6]. This article further shows only the temperature and rain intensity of the year 2019.

The temperature and relative humidity of the indoor air records the Fluke Hydra III multimeter. The NTC thermistors scan the temperature within the structure with the accuracy of ±0.2 °C. The capacity probes scan the relative humidity with an accuracy of ±2 %.

3.1. Outside and inside boundary conditions

Figure 2 graphically shows the mean outside temperature and rain intensity of the reference year 2019. The maximum temperatures are visible in June, while the maximum total rain was recorded in May.

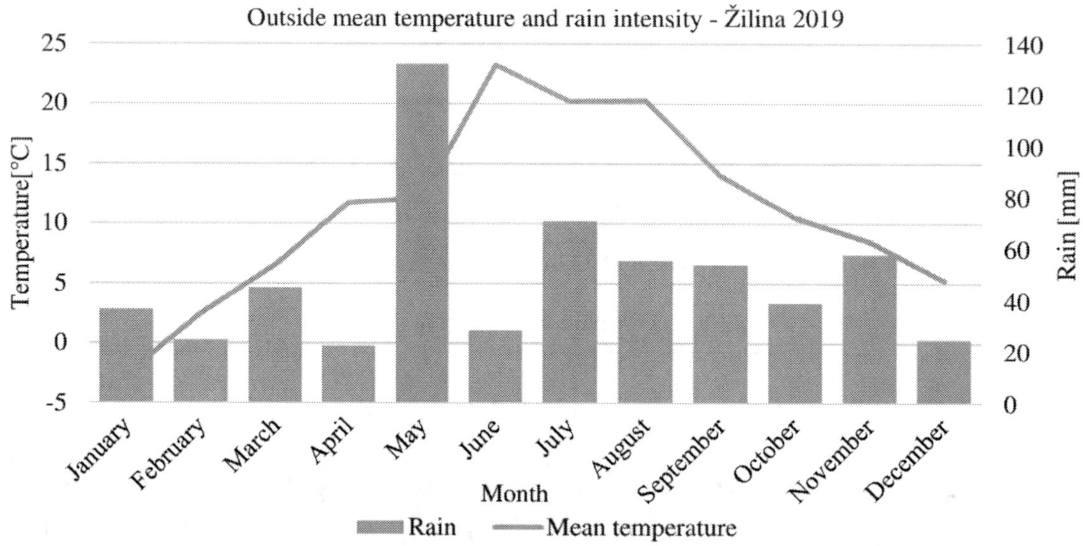

Figure 2. Outside mean temperature and intensity in Žilina (2019).

Figure 3 shows the temperature and relative humidity of indoor air in the south-facing room. To increase clarity, the data in the following graphs are presented as moving-diameter trendlines with a period of 150.

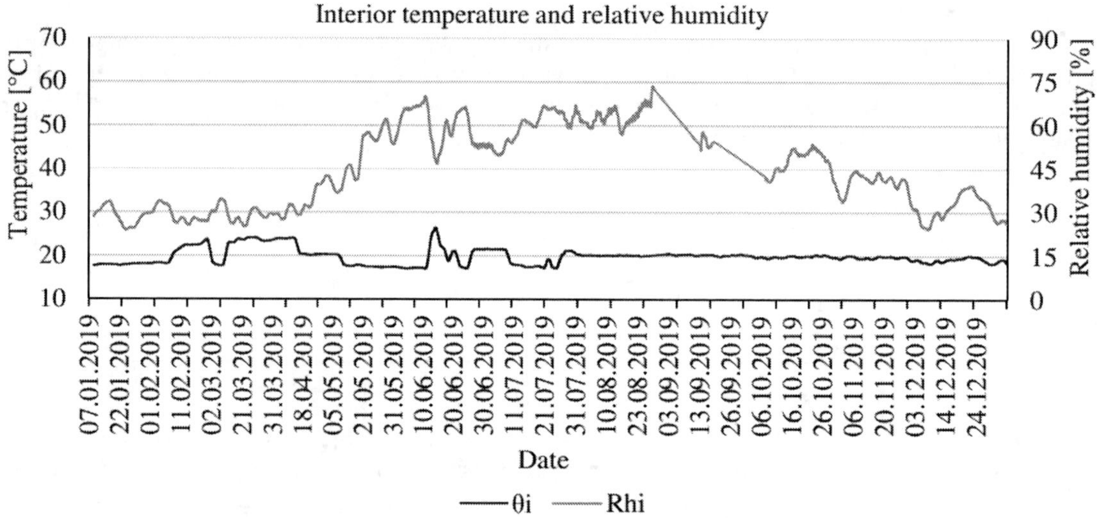

Figure 3. Interior temperature and relative humidity (2019).

IOP Conf. Series: Materials Science and Engineering 1015 (2021) 012039 doi:10.1088/1757-899X/1015/1/012039

3.2. The temperature and relative humidity of the measured structure

The analysed assembly is 380 mm thick. It consists of the silicone render made on an adhesive render with a mesh cloth from the inside. Towards the interior is glass fibre thermal insulation Isover Clima 034 with a thickness of 100 mm. This layer is followed by another glass fibre insulation Isover Multimax 030 100 mm and 120 mm thick. To the interior follows phenolic board Kingspan Kooltherm K5 of 40 mm. The last layer is a 12 mm thick OSB board.
Figure 4 displays the temperature within the construction throughout the year 2019 between individual layers with the signatures as shows figure 1.

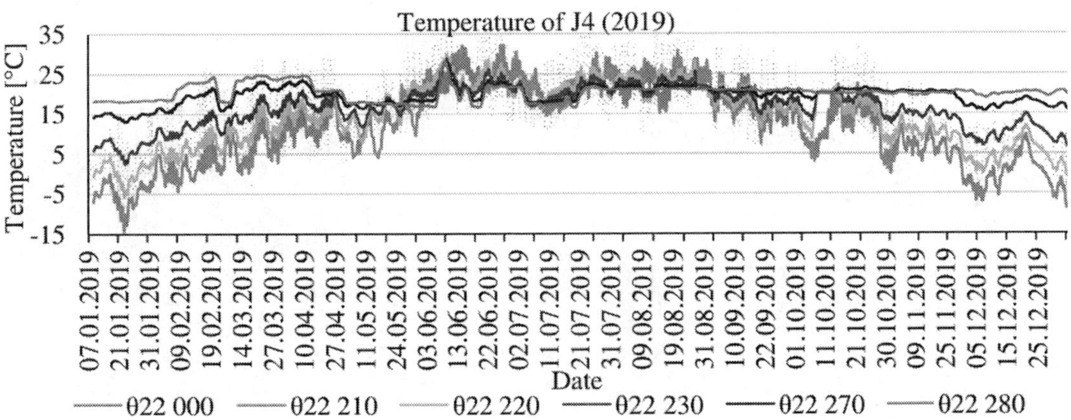

Figure 4. The temperature of the southern wall assembly 4 (J4) in 2019.

The external surface temperature coincides with the temperature between render system and the first Isover Clima thermal insulation so it's barely visible. The differences between the individual layers increase significantly towards the interior. The internal temperature essentially copy the temperature of the indoor air.

The temperature within the structure rises by 5°C after the first layer from the exterior. The interior surface temperature reaches 18 °C even during the lowest outdoor temperature of -17 °C on January 23, 2019. In summer, on the other hand, it provides a temperature reduction of 15 °C on the day with the maximum outdoor temperature on June 14, 2019, when the outdoor temperature was 37 °C. The internal surface temperature was a maximum of 21.7 °C. Figure 5 presents the relative humidity of the assembly between each layer, similar to the temperature.

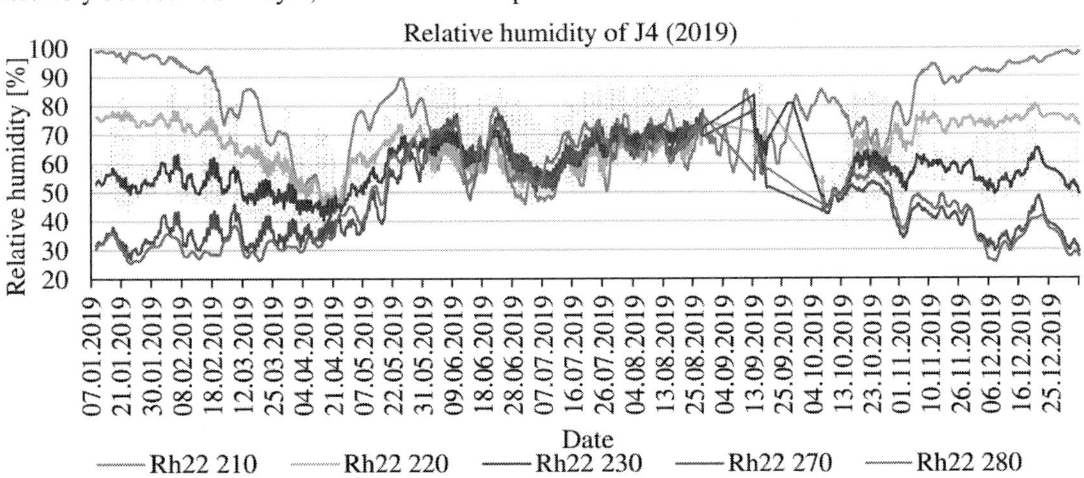

Figure 5. The relative humidity of the southern wall assembly 4 (J4) in 2019.

Significant differences among the relative humidity across the structure are in the winter of 2019. On the interior side, the relative humidity in the range 25-45 %. It increases towards the exterior until the relative humidity reaches almost 100 % under the plaster system. The differences in the summer are not so significant and the whole structure achieves essentially the same relative humidity.

To compare the temperature and relative humidity across the whole structure are further discussed three probes placements on the left stud, in the thermal insulation and on the right stud. These probes are at three stud levels that are identical to the three sensor locations within the insulation – between Isover Clima and Isover Multimax (e), between the two layers of Isover Multimax (middle), and between Isover Multimax and Kingspan Kooltherm (i) (figure 6 and figure 7).

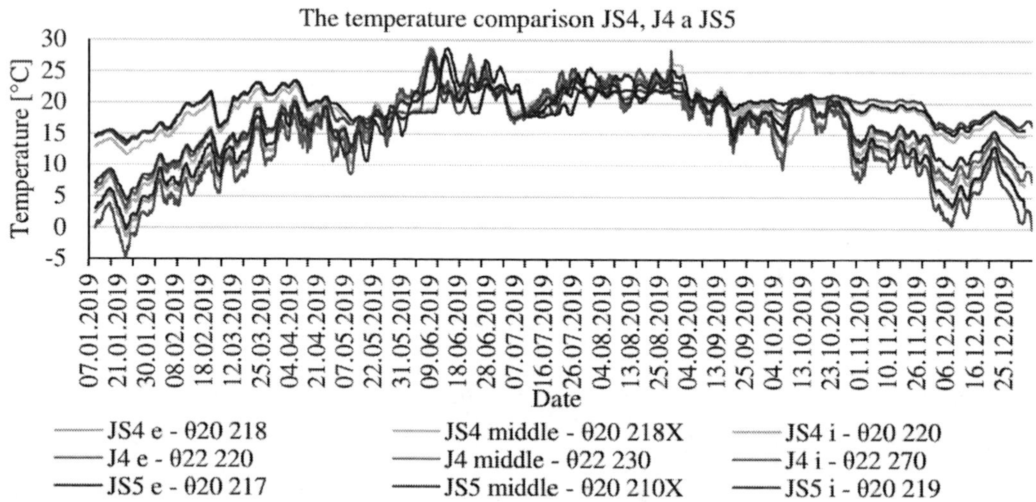

Figure 6. The temperature of the southern stud 4 (JS4), southern wall assembly 4 (J4) and southern stud 5 (JS5).

Based on the temperature curves (figure 6), it is apparent that the temperature is evenly distributed between the studs and the assembly itself. In winter, the temperature in the middle of the wall is even lower than on the studs.

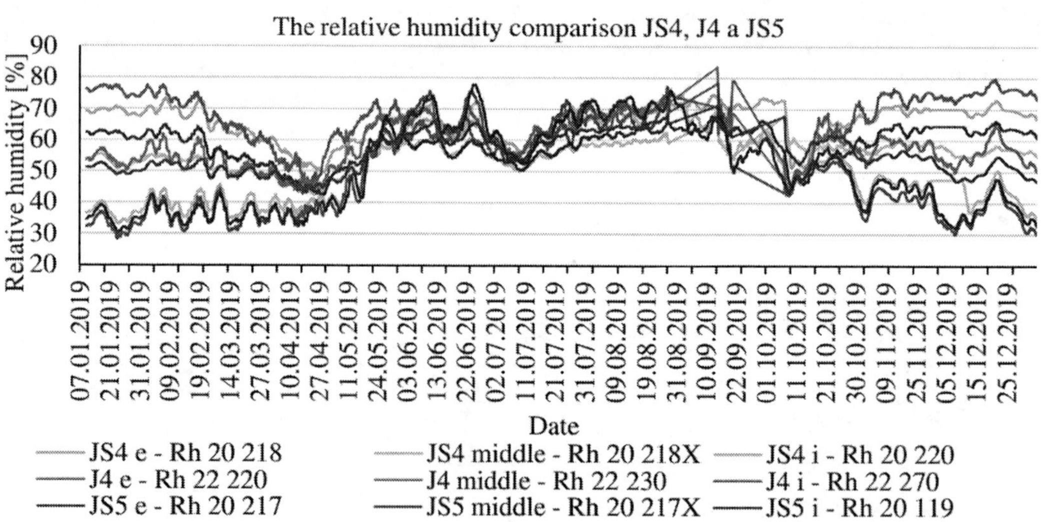

Figure 7. The relative humidity of the southern stud 4 (JS4), southern wall assembly 4 (J4) and southern stud 5 (JS5).

The sensors mounted on the load-bearing studs recorded less relative humidity than in the middle of the thermal insulation. This is explainable by the properties of wood as a material, as it is a hygroscopic material that can absorb and release moisture based on the surrounding environment.

4. Conclusions

The analysed assembly of a light-weight wooden wall is highly effective in terms of its hygro-thermal performance in year-round operation. The wall is installed since 2011, so the temperature and relative humidity are without the influence of built-in moisture.

The temperature within the structure in 2019 was in a range from -17 °C to 37 °C. The analysed composition showed significant protection against these maxima. In the winter it provided the temperature of 18 °C on the interior surface – 35 °C higher temperature than in the exterior. In summer, it provided a temperature of 15 °C lower on this surface.

The relative humidity during the whole period reached values from 25 % to 98 % in the exterior part of the assembly. Towards the interior, this humidity gradually decreased. The water management of the structure was evident in these discrepancies.

When comparing the temperature and relative humidity between the middle of the structure and the studs, no significant differences occurred. In some cases was the temperature by the studs even higher, which proves the elimination of thermal bridges. The relative humidity was lower on the studs, which indicates the hygroscopic properties of the wooden material.

Acknowledgments

The research presented in this paper was possible thanks to the support from the VEGA grant project Nr. 1/0673/20: Theoretical and experimental analysis of energy effective and environmental friendly building envelopes.

References

[1] Hens H 2018 Energy efficiency in residential buildings, less straight forward then presumed STP1813155H

[2] Standard STN 73 0540-2+Z1+Z2 2019 *Tepelná ochrana budov. Tepelnotechnické vlastnosti stavebných konštrukcií a budov. Časť 2: Funkčné požiadavky. Konsolidované znenie* (Thermal protection of buildings. Thermal performance of buildings and components. Part 2: Functional requirements)

[3] Standard STN EN ISO 9972 (730576) 2016 *Tepelnotechnické vlastnosti budov. Stanovenie vzduchovej priepustnosti budov. Metóda pretlaku pomocou ventilátora* (Thermal performance of buildings. Determination of air permeability of buildings. Fan pressurization method)

[4] Standard STN 92 0201-1 2000 *Požiarna bezpečnosť stavieb. Spoločné ustanovenia. Časť 1: Požiarne riziko, veľkosť požiarneho úseku* (Structural fire protection. Common regulations. Part 1: Firerisk, fire compartment area)

[5] Ďurica P, Iringová A, Ponechal R, Rybárik J, and Vertaľ M 2017 Energetické a environmentálne navrhovanie a hodnotenie budov (Žilina/ Edis)

[6] Michálková D 2020 Teoreticko-experimentálna analýza a vývoj ľahkých obvodových plášťov pasívnych drevodomov (Theoretical-experimental analysis and development of lightweight envelope structures of passive wooden houses) (Dissertation thesis Žilina)

IOP Conf. Series: Materials Science and Engineering 1015 (2021) 012040 doi:10.1088/1757-899X/1015/1/012040

Assessment features of bearing capacity of large-span covering constructions in the reconstruction of religious Orthodox buildings

A P Shepelev and R R Ibatullin

Samara State Technical University, Molodogvardeyskaya st., 244, Samara, Russia, 443100

E-mail address: shepelevap@mail.ru

Abstract. The research object is the middle part and the refectory overlap of the temple in the honor of the Holy Kosma and Domian Asiysky, located in the village of Ssezzhe of Bogatovsky district of Samara region. At the time of the works, the object was undergoing a large-scale reconstruction. Based on the data of the detailed technical condition survey of the building structures of the temple overlap, the authors made calculations in a physically nonlinear setting, allowing to fully assess the stress-strain state of the monolithic reinforced concrete structure of the middle part and refectory, which is a complex unified system of thin-walled spatial structures of different contours of the middle surfaces. The recommendations on the completion of the overlap construction and its further operation were developed on the basis of the calculations results analysis.

1. Introduction

The change in political and social realities in Russia in the late 1980s gave rise to the church revival in the country. As the number of parishioners grew, it became necessary to increase the number of parishes. This period can be marked as the beginning of active restoration and construction of religious buildings of the Russian Orthodox Church. In 1988, the Russian Orthodox Church had 6.5 thousand cathedrals; in 2020 there are 40 thousand cathedrals. In the last 10 years the number of parishes of the Russian Orthodox Church has increased by 10 thousand. Most buildings for religious purposes are not new construction objects, but the objects transferred from state property back to the jurisdiction of the Russian Orthodox Church. The transferred religious buildings were erected in the pre-revolutionary period. During the Soviet period some of them were either destroyed or rebuilt for various purposes: in rural areas the buildings were often used as granaries, mechanical workshops, and clubs; in the urban areas they were mainly turned into public buildings, production workshops and even apartment buildings. Of course, when the functional purpose of religious buildings was changed, architectural, volumetric-planning and structural solutions were significantly changed too. First of all the brightest religious “signs” were dismantled: architectural decor, belfry, quadrangles, octagons, drums, domes, etc.

When re-profiling the building back to the needs of the Russian Orthodox Church, there is a need for restoration work. Over the long period of operation the buildings didn’t have often timely repairs. Some temple facilities were abandoned for decades without any reservation measures, the building structures have significant amount of damage and their technical condition deteriorated greatly. Thus, the restoration work implies a whole complex of reconstruction measures.

Now there is a practice when the restoration of half-ruined cult constructions is carried out not with working out of necessary volume of the design-budget documentation, but in an economic way. Most often this happens in rural parishes when all cares on restoration of a church are taken by the rector. At the same time, at the reconstruction, very laborious and technically complex tasks are being solved, including the restoration of the lost large-span coverings (domes, vaults, and apses), multi-tiered bell towers, and domes. It should be noted that temple overlap is often the most important element of architectural, volumetric and structural solutions of the building. Thus, it is very difficult to control the restoration work quality, including the design stage, and there is a need to determine the actual technical condition and bearing capacity of already erected or reinforced structures. Given the specifics of these circumstances, often the most common engineering methods are not suitable for the task of identifying the actual carrying capacity of structures. It is necessary to use more labour-consuming and more exact methods with modern software and computer complexes application. It allows to reveal the unaccounted reserves of bearing ability and thus to minimize financial expenses for restoration.

Members of Construction Structures department of Samara State Technical University have participated in scientific and technical support of reconstruction of buildings of religious purpose of Russian Orthodox Church. This work reflects the results of the survey of the technical condition of the middle part and church refectory, located in the village of Ssezzheye Bogatovsky district of Samara region of the Russian Federation. The purpose of the set of measures for the technical inspection was to assess the stress-strain state of the overlap structure with the subsequent recommendations development for the completion of construction and installation work related to the construction of the church overlap.

The building of the temple was erected in 1827. The temple was illuminated in honor of the saints Cosma and Damian. In the 1880s, the temple was reconstructed, which was aimed at its expansion. Until the autumn of 1938, the building was used in accordance with its functional purpose, after which it was equipped with a granary. In the 1990s, the vaulted stone structures of the building envelope collapsed (figure 1). Until 2005, the building was in an abandoned condition. Since 2005, a long reconstruction process has been underway. At present the building has been restored: the lost fragments of stone walls have been reconstructed, a monolithic reinforced concrete covering has been arranged, a stone drum above the middle part has been arranged, the drum and the bell tower have been crowned with the dome.

The considered overlap is made of monolithic reinforced concrete and is a complex continuous system of thin-walled spatial structures with different outlines of the middle surfaces. The survey work was carried out in accordance with the current regulatory documents [1-5], taking into account the methods developed by the members of Construction Structures department of Samara State Technical University, partially set out in [6-8].

2. Construction planning and design scheme

A stone cross-domed one-storey one-domed church in the plan has the shape of a cross with maximum dimensions in the axes of 31.2×19.7 m. The compositional solution is axially symmetrical. The schematic plan and general view of the building are shown in figures 2 and 3, the coverage plan is shown in figure 4.

The plan structure is four-part plan; it consists of the following parts connected by a "ship": an altar in axes 4-6, a church (middle part) in axes 3-4, a refectory in axes 2-3, and a vestibule with a built-on bell tower in axes 1-2.

The covering of the refectory is a monolithic reinforced concrete on a rectangular plan composite smooth shell formed of two shells of zero Gaussian curvature intersecting in mutually perpendicular directions: in the right and left naves, the longitudinal axes of the vaults are oriented in the North-South direction; the longitudinal axis of the middle nave is in the East-West direction.

Figure 1. View of the southern facade and covering of the temple (as of 2005)

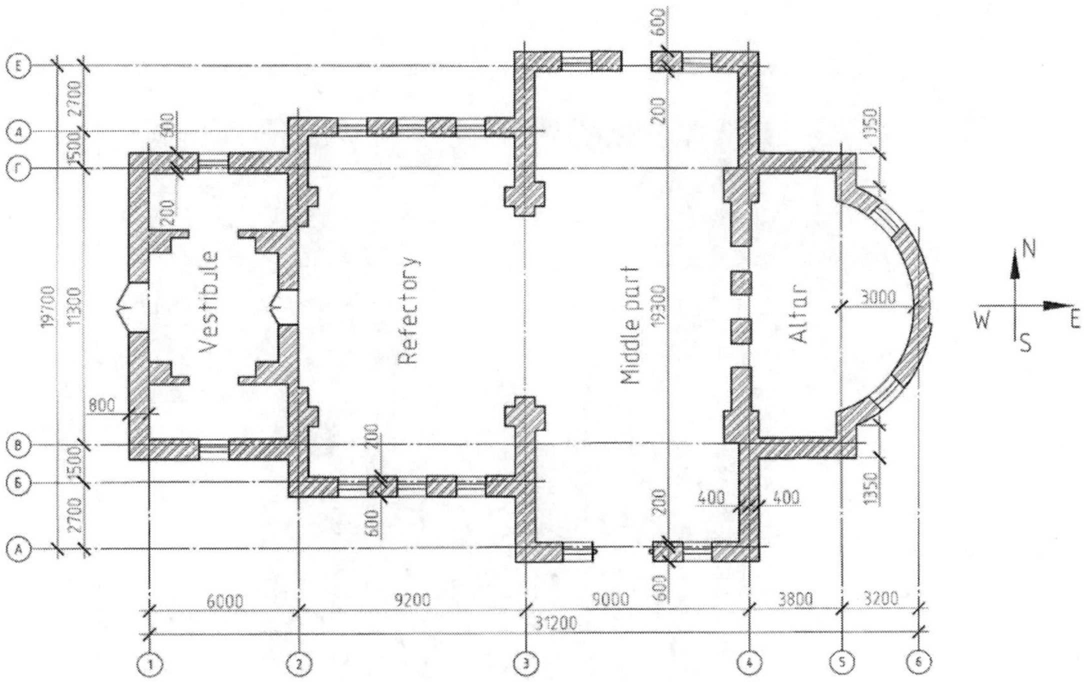

Figure 2. Schematic plan of the temple building

Figure 3. General view of the building (http://kosma-i-damian.cerkov.ru/)

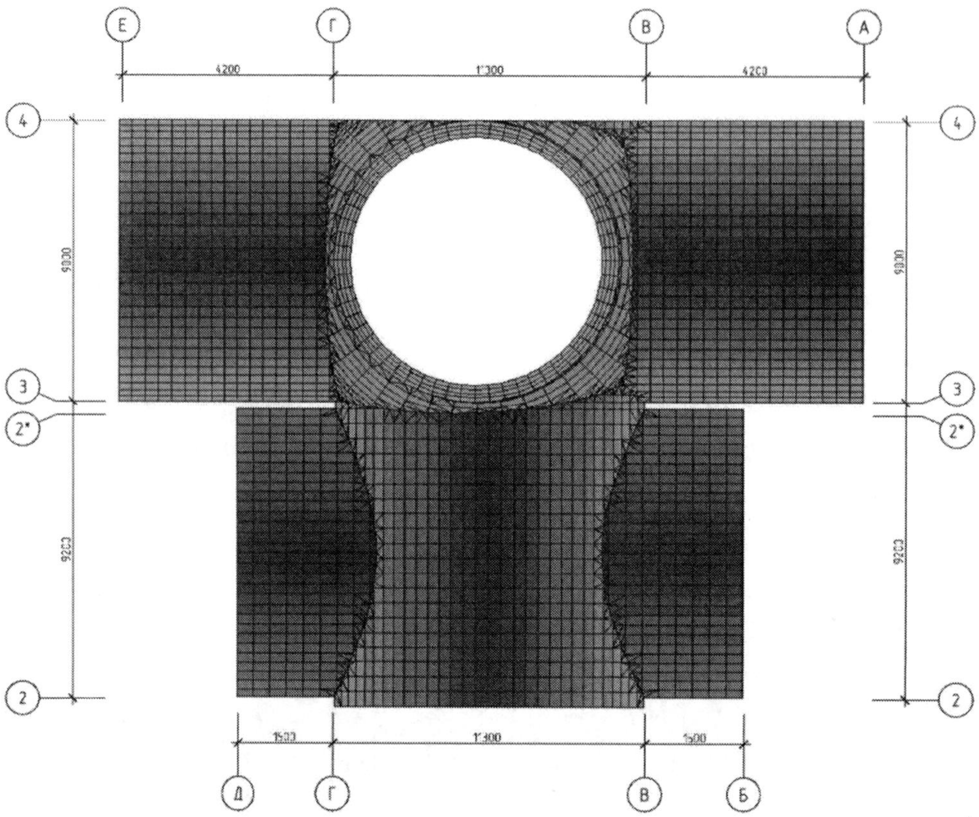

Figure 4. Plan of covering the middle part and refectory

The cover of the middle part is formed as follows: the outer naves are covered with cylindrical vaults (the longitudinal axes are oriented in the North-South direction); a conical drum is arranged in the middle nave; the drum, vaults of the outer naves of the middle part and the vault of the middle nave of the refectory are united by monolithic shells in the form of concave spherical triangles (scoinson arches).

The thickness of the cylindrical vaults and scoinson arches is 100 mm, the conical drum is 500 mm thick. In axes B/3-4; Г/3-4 and B-Г/2 cylindrical vaults are thickened (with development in the lower direction) up to 300 mm, forming stiffeners 900 mm wide.

The composite shell is supported by horizontal supporting edges on the wall fence of the temple through a monolithic reinforced concrete stiffening belt to which it is monolithically connected. The dimensions of the belt cross-section are 500×200 mm. Curved lines of cylindrical vaults freely rest on the arch of the wall fence, excluding the western refectory wall.

3. Calculation model

The bearing capacity and deformability estimation of a covering design is made by a method of finite elements with use of program complex "Lira-SAPR".

The computational model is a composite shell with different contours of the middle surfaces. Discretization of the continuum system is achieved by means of four- and triangular finite elements of the shell and two-node finite elements.

The finite element model of the shell is shown in figure 5.

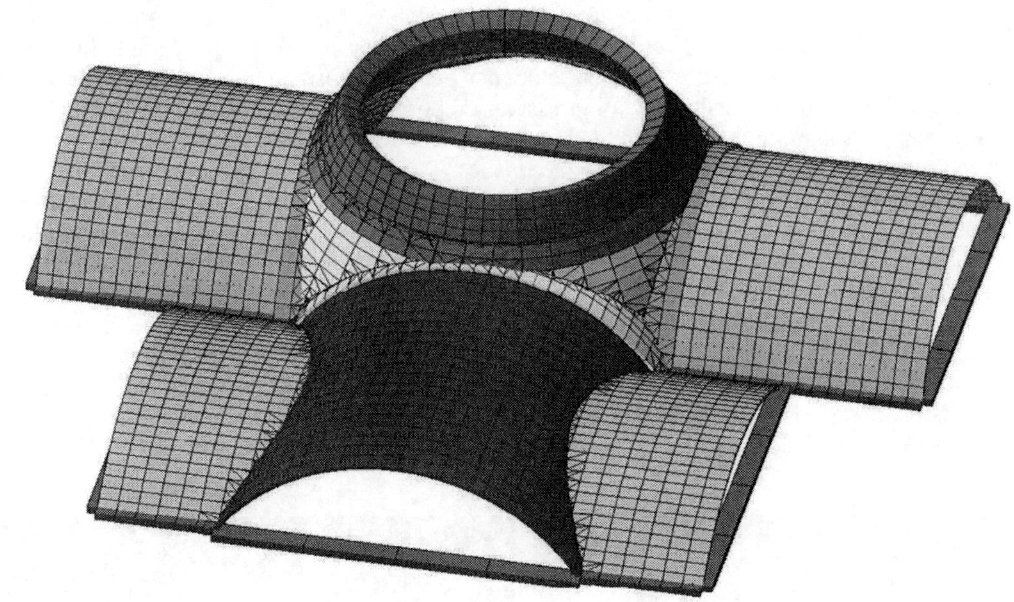

Figure 5. Spatial finite element overlap model

The calculation of the system is based on the physical non-linearity of materials. When forming the finite element model we used physically nonlinear universal finite elements of the shell (№№241, 242, 244).

Geometric invariability of the system has been achieved by superimposing external relations: along the global Z axis in all units when resting on walls, excluding the cylindrical arch in the Г-В/2 axes. On one node of axes Д-Г/2* superimposed bonds *X*, *Y*, *Z*; on one node in axes Б-В/2* of bonds *Y*, *Z*.

Modules of elasticity of concrete in the overlap final elements are assigned in accordance with the class of concrete determined during natural examination by non-destructive methods.

The law of nonlinear deformation of concrete - № 35 is exponential (calculated strength). The law of nonlinear deformation of reinforcing material - № 11 is exponential. Geniev theory of strength is for reinforced concrete. Type of reinforcing inclusions is physical equivalent of a grid. The method of calculation of physically nonlinear problem is simple stepwise with the factor to loading on steps makes 0.05.

Calculation is made on the action of the calculated load from the own weight, including the dome and the drum which are not erected at the time of examination.

The analysis of the material state of the model shows that at the first step of loading the elements are mainly formed by lower cracks on the main sites. Concrete parts of the elements are destroyed under tension.

At the second step of loading, lower and upper cracks are formed in the elements along the main areas. In the single elements there was the destruction of the main material (concrete) in compression and crushing of compressed concrete on the upper and lower surfaces. On the fourth step of loading along with the increasing number of elements with lower and upper cracks on the main platforms and the destruction of concrete elements under tension and compression (in single elements); crushing of compressed concrete on the upper and lower surfaces (in single elements), the formation of plastic joints is observed. The tension in reinforcing material has reached the yield point (in single elements). Further the loading effect is characterized by an increase in the number of elements with stress-strain state described in the fourth loading step. In this case, the formation of plastic joints is observed only in some elements. This is due to the fact that longitudinal forces prevail in the stressed state. The bending moments in the structural elements are negligible compared to the longitudinal forces. Destruction of the structure is fixed when the system has been transformed into geometrically variable.

Longitudinal forces, both tensile and compressive, predominate in the end elements of cylindrical arches. The highest longitudinal forces (compressive forces) in combination with bending moments occur at the rib supports in the direction of curved lines. The greatest bending moments occur in the final elements of the conical drum. Scoinson arches are mainly compressed in the vertical direction.

The analysis of the tensely deformed state of the shell shows that the end elements of cylindrical vaults are destroyed by tensile stresses (excluding the support parts of the stiffening edges, single elements at the supports and the joints of parts of the structure of the coating); the elements of the scoinson arches are destructed from compressive stress; in the elements of the conical drum destruction occurs mainly from tensile stresses, and in some areas destruction occurs from compressive stress. State of materials in the model at each step of loading is shown in figure 6 (red color and its shades - the destruction of elements under tension; blue color and its shades - the destruction of elements under compression).

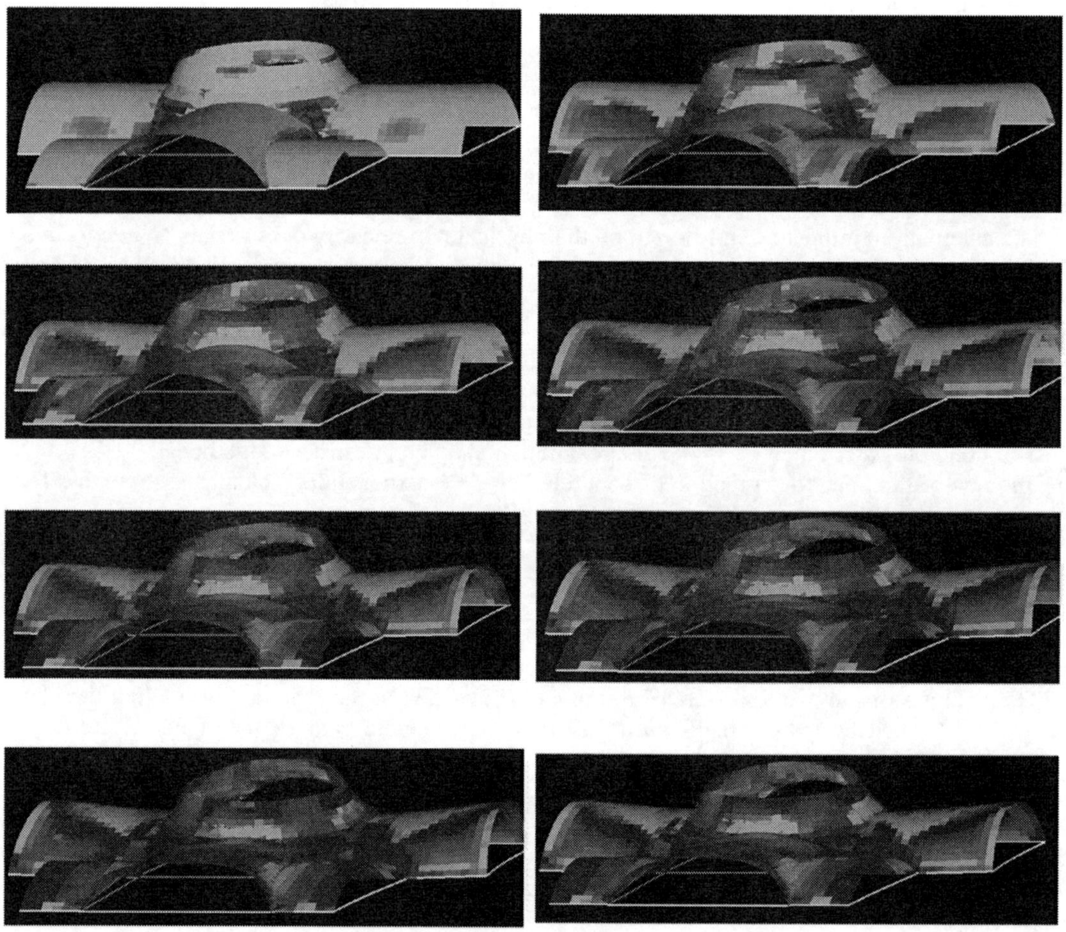

Figure 5. The materials state in the model at each loading step

4. Conclusions

The end elements of cylindrical vaults are dominated by longitudinal forces, both tensile and compressive. The highest longitudinal forces (compressive forces) in combination with bending moments occur at the rib supports in the direction of curved lines. The greatest bending moments occur in the final elements of the conical drum. Scoinson arches are mainly compressed in the vertical direction.

The analysis of the tensely deformed state of the shell shows that the end elements of cylindrical vaults are destroyed by tensile stresses (excluding the support parts of the stiffening edges, single elements at the supports and the joints of parts of the structure of the coating); the elements of the scoinson arches - from compressive; in the elements of the conical drum destruction occurs mainly from tensile stresses, in some areas - from compressive. State of materials in the model at each step of loading is shown in figure 5 (red color and its shades - the destruction of elements under tension; blue color and its shades - the destruction of elements under compression).

The performed calculations of a complex spatial system in a physically nonlinear setting taking into account real deformation properties of concrete and reinforcement allowed to determine the stress-strain state and bearing capacity of the overlap structure.

The recommendations on the final structure of the overlap were developed on the basis of calculations results.

References

[1] Standard GOST 31937-2011 Buildings and constructions. Rules of inspection and monitoring of the technical condition 2014 IV, p 54

[2] Standard GOST 27751-2014 Надежность строительных конструкций и оснований. Основные положения - Reliability for constructions and foundations. General principles (Москва: Стандартинформ)

[3] Standard SP 13-102-2003 Rules for inspection of load-bearing structures of buildings and structures p. 60

[4] Standard SP 63.13330.2012. Concrete and reinforced concrete structures: updated version of SNiP 52-01-03. Main provisions: official publication: approved by the order of the Ministry of regional development of the Russian Federation dated December 29, 2011 г. № 635/8/пр: 2012.

[5] Standard SP 20.13330.2016. Loads and actions: updated version of SNiP 2.01.07-85

[6] Shepelev A P, Ibatullin R R and Buzovskaya Y A 2019 Features of technical condition examination of historical buildings on the example Mansion of merchant M.M. Diakova (hostel № 3 of Samara State Technical University) *Traditions and innovations in construction and architecture. Construction*

[7] Ilyin N A, Pishchulev A A, Slavkin P N, Shepelev A P and Ibatullin R R 2013 Restoration of compressed reinforced concrete structures of the building *Vestnik SGASU Urban planning and architecture* **4** (12) pp 62-67

[8] Shepelev A P, Panfilov D A and Ibatullin R R 2017 Operational damage to the walls of multi-storey stone buildings as a result of the difference in their deformations *Scientific review* **11** pp 25–30

IOP Conf. Series: Materials Science and Engineering 1015 (2021) 012041 doi:10.1088/1757-899X/1015/1/012041

Hydraulic aspects of the layout of head structures during water intake from lowland rivers

D Bazarov[1], I Markova[2], B Norkulov[1] and O Vokhidov[1]

[1] Tashkent Institute of Irrigation and Agricultural Mechanization Engineers, Tashkent, 100000, Uzbekistan
[2] National Research Moscow State University of Civil Engineering (NRU MGSU), Moscow, Russia

E-mail address: dr.bazarov@mail.ru

Abstract. The article discusses various types of new improved dam water intake hydroelectric systems, sedimentation facilities, sediment control devices on rivers and canals with a rational layout. The results of the analysis of studies on the use of the theory of transverse circulation and methods of artificial, active regulation of the structure of the water intake flow are presented, which has become widespread in world practice and is the main one for creating an effective design of water intake headworks for pumping stations and hydropower facilities that provide them with reliable protection from sediments. The author of the article, on the basis of numerous experimental data obtained in various countries of the world, makes a conclusion about the effectiveness and prospects of using front-type structures with more than 50% water intake. The design of water intakes of this type has been significantly improved; they can be successfully used in various conditions, including on Mountain Rivers. The author draws attention to the development of a new design of the pioneer ditch, which ensures the efficient passage of bottom sediments into the downstream of the structure, which prevents intensive silting of the reservoir basin by bottom sediments and improves the efficiency of the dam water intake.

1. Introduction

For a country located in a region with a hot climate, with a large irrigated area, guaranteed provision of water intake facilities of pumping stations, operational energy paths of hydropower facilities with the required volume of water, satisfying the need for water resources and the generation of the required capacity of hydropower facilities through the construction and reconstruction of water management and reclamation systems , is among the highest priority tasks for the national economy. In the practice of hydraulic engineering and hydropower construction, very often water intake is carried out by a dam method [1]. In cases where the household levels of the river are insufficient for the intake of water or the creation of the necessary pressure, dam water intake hydroelectric systems are installed to the power structure by gravity flow to the consumer, where the dams create a backwater level and provide a guaranteed water intake or the required demanded pressure to the structure [2–5]. The development of new and improvement of existing layout diagrams and individual design solutions for elements of water intake hydrosystem, guaranteeing the selection of design water flow rates with the least possible entrainment of riverbed sediments into the supply canals of pumping stations, becomes more urgent. The object of the study is the existing water intake waterworks in Uzbekistan. Usually, a number of structures and devices are included in the structure of dam water intake hydroelectric systems: a dam (blind and spillway or only spillway); upstream and downstream regulatory structures; washing devices; head water intake regulating structure and sedimentation tanks [6–9].

Basically, the operational services, together with a guaranteed water intake to the head structure, are faced with the task of removing clarified water and preventing siltation of the water intake area in the head structure. Many scientists have dealt with the issues of research and solution of the problem of the operation of hydroelectric facilities, which represent operational efficiency [10–14]. Many studies devoted to solving water intake problems provide recommendations for engineering solutions to combat sediment in water intake facilities built in India, the USA, Turkey, Egypt, Japan, Iran and other countries of the world [13, 15]. Despite the abundance of works devoted to solving the problem of clarified water in the head hydrotechnical and hydropower structure at the dam water intake, the issue of its effective operation has not been brought to its logical conclusion. Based on the above, the main goal of this work and the research objectives are determined. The main goal of the work is to select the most effective type of water intake structure that minimizes the flow of bottom and suspended sediments to the maximum extent, as well as reduces the intensity of siltation and inflow of reservoirs. To solve this goal, the following research objectives have been identified:

1. Consideration of various types of dam water intake hydroelectric systems used in practice by scientific research, design, construction and water management organizations in many countries of the world;

2. Generalization of the results of studies devoted to the use of the theory of transverse circulation and methods of artificial, active regulation of the flow structure to create an effective design for water intake structures of canals, providing them with reliable protection from sediments;

3. Determination of the efficiency and prospects of the use of water intake facilities and installation of the type of this structure.

2. Method

Analysis of the advantages and disadvantages of the existing water intake waterworks during operation, development of a new structural element contributing to the improvement of the operational characteristics of the water intake structure.

3. Results

As you know, water intake structures by design features belong to the following types of water intake waterworks: lateral, frontal, bottom-lattice and goby. Each type of hydrosystem has various modifications aimed at improving its design [16]. Approximate conditions for the use of various types of water intake waterworks are shown in table 1.

Table 1. Recommendations for the use of various types of water intake waterworks.

Water intake type	River sections	Supply canal configuration	Water intake coefficient, R not more	Recomm. water intake flow rate (m^3/s)	Comment
Side: unilateral bilateral	Foothill and flat, less often mountainous	curvilinear straightforward	0.5	no limit	At $R > 0.5$, the accumulation of entrained sediments in the headwater and periodic flushing are required.
Frontal: unilateral	Predominantly foothill	curvilinear	0.8	also	It is allowed to transfer part of the water flow (up to 30%) to the other side
Bilateral	Mainly flat, foothills are acceptable	straightforward	0.7	also	Not recommended for rivers with a lot of writhing and fin
Donno - lattice	Mountain	straightforward	0.4	< 20	Maximum design river discharge up to 300 m^3/s

In many water intake waterworks one of the main problems is the prevention of siltation and the introduction of the headwater where the intake structure is located. Silting up of the water intake area occurs due to bottom sediments. Especially, the problem of siltation of the water intake area due to bottom sediments occurs intensively in Mountain Rivers (for example, such as Pskem). The existing dam water intake waterworks in Central Asia, based on the principle of preventing the flow of bottom sediments into water intake facilities, can be divided into three main types:

1) a water intake unit with a frontal location of the dam and a lateral location of a water intake (of the "calm backwater" type - Indian type, figure 1).

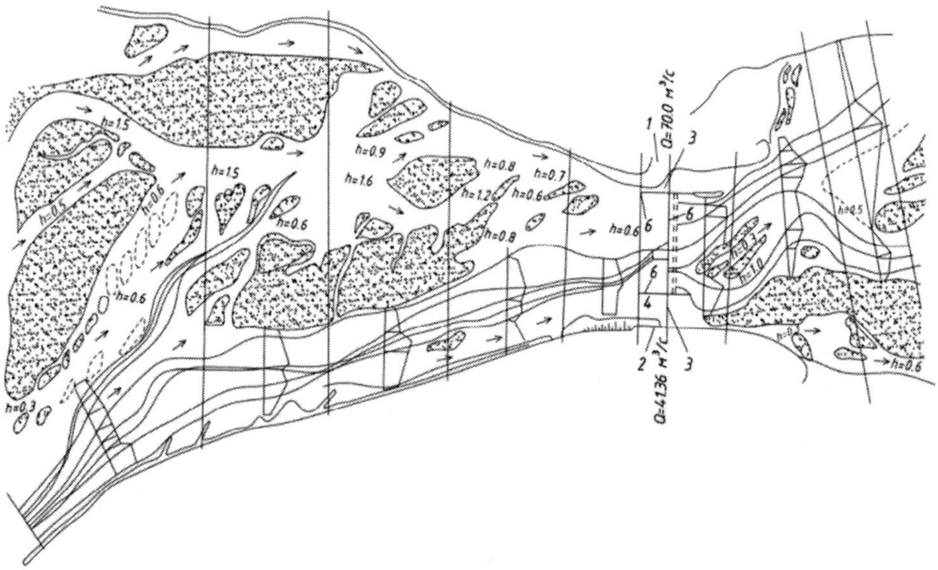

Figure 1. Water intake unit of Indian type: 1-left bank water intake; 2-right bank water intake; 3-washing sluices; 4 pockets; 5-spillway dam; 6-wall pocket.

2) a water intake unit using transverse circulation (Fergana type, figure 2)

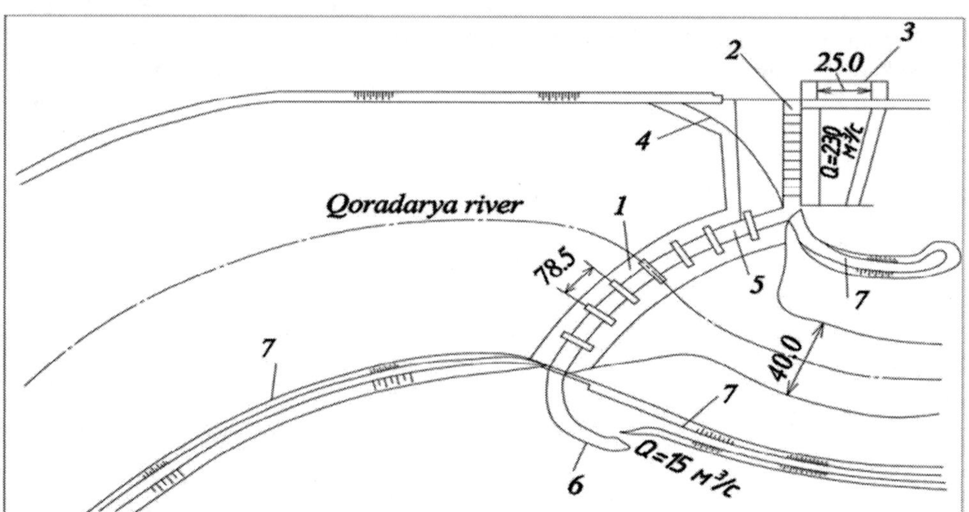

Figure 2. Fergana-type water intake node: 1-harvest dam; 2-water receiver; Canal 3; 4-curvilineary threshold water receiver; 5-duke; canal 6; 7-stringing dams.

3) a water intake unit built on the principle of vertical flow stratification (European type, figure 3)

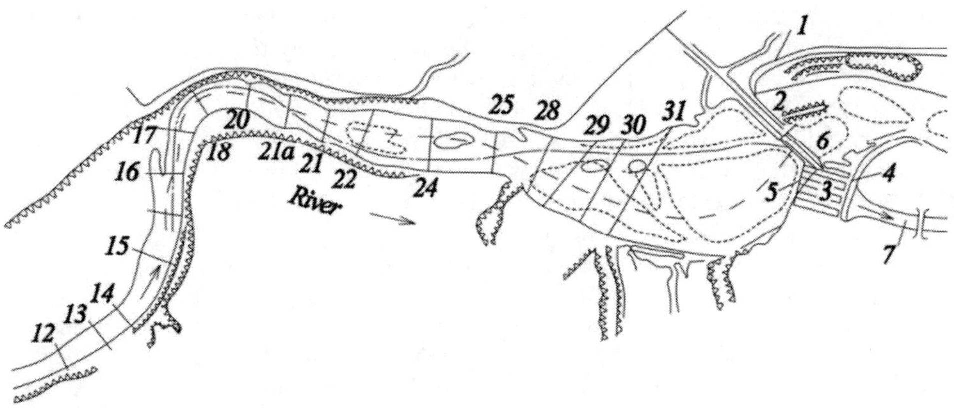

Figure 3. Water intake unit of European type: 1-foldable part of the dam; 2-blind part of the dam; 3-settling tank; 4-mud flap; 5-water intake; 6 wash galleries; 7-canal (12-31 diameters).

To control sediment in the layouts of the dam hydroelectric systems, the properties of the flow are used: its characteristic features with a lateral outlet from the river; natural and artificial transverse circulation in curved sections of a river or canal; reactive bottom currents excited by artificial barriers and screens; layer-by-layer water drainage and others.

Dam hydroelectric complex of the first type (Indian type), built in 1929 on the foothill section of the Zarafshan River (Pervomayskaya dam), which consists of a blind and spillway dam, two wash locks, two (left-bank and right-bank) head regulators of main canals, upstream right-bank stream-guide dam and right-bank water retention dam. This type has serious drawbacks: unregulated backed level, unstable approach of the flow to the points of water intake, ineffectiveness of washing devices with a large loss of water, significant ingress of sediment into canals and interruptions of water intake during washing. After 20 years of operation, the project proposals for the assembly of the unit did not come true. Such intakes were reconstructed.

The arrangement of water intake units of the second type (Fergana type) has proved itself well in practice and successfully protects the canals from the ingress of large bottom sediments. In such hydrosystems, transverse circulation is used, for the occurrence of which the supply canal is given a curvilinear outline in the plan, and its width is taken equal to the width of the stable canal. The intake structure is located on a concave bank. The unit includes a curvilinear supply canal; spillway dam with gates, water intake flow of various designs, head regulators. However, in structural and operational terms, units of this type have the following disadvantages: increased length of the dam due to its lateral or oblique location; non-smooth flow of water through the dam; unsatisfactory hydraulic conditions for the flow of waste flows through the dam; very poor washing conditions of the narrow and curvilinear in terms of the headwater from sediments and the introduction of the downstream by sediments. The indicated disadvantages (inherent to some extent and other types of structures) in the designed nodes of this type are easily eliminated. The water intake unit of the third type (European type), built in 1940 (Gazalkent hydroelectric complex), depends on the relative position of the water intake, bottom and waste structures, as well as on the regulation of the river canal leading to the water intake. With this arrangement of the unit, the water intake is satisfactory only under a certain operating mode of the structures (the ratio of the flow rates discharged through the dam, taken into the water intake and passed through the bottom holes). The arrangement of these types requires a significant head at the waterworks. A common disadvantage in the layout of the constructed hydroelectric complexes is the unregulated river canals and a very wide front in front of the junction (dam and water intake). On the other hand, the variable discharge regime of the river presents a great difficulty in the layout of the water intake. A certain arrangement of the unit can justify itself at certain limits of fluctuations in river flow rates, while at other flows it remains unsatisfactory.

The gradual narrowing of the supply canal with the help of a dam with a series of side discharge holes, adapts the operation of the water intake to the river regimes. In this direction, new schemes for the layout of hydraulic structures of the second type, proposed by many scientists [16, 17] make it possible to create, at all river flows, a stable frontal and concentrated river flow to the water intake, which provides a one-way water intake 4 *B*, with double-sided water intake (5-6) *B*. In the downstream, the length of the regulated canal depends on the layout of the unit and the share of water intake. It should not be less than the width of the stable canal $L_{n,b} = B$. The radii of curvature of the regulated canal are assumed to be variable. Smallest radius of curvature $R_1 = 3.5\ B$ and the greatest $R_2 = 7\ B$. The backwater curves are connected smoothly.

Under conditions of pronounced variable flow regimes in a river with water intake fluctuations from 15 - 20% during floods and up to 100% during low water periods, the most positive effect of preventing sediment into canals during water intake is achieved with the combined use of two principles - transverse circulation and stratification of the flow with canal regulation [10, 15].

Experience in the operation and design of water intakes, as well as research and development work in this area show that the application of principles using cross-circulation and stratification separately and in combination, in certain conditions, should be limited. The best results can be achieved on the foothill sections of the river with a water intake of up to 60 - 70%. With a water intake exceeding this indicator, a combination of two principles is advisable - transverse circulation and vertical stratification of the flow with canal regulation. [18, 19, 20].

To combat sediment, the authors of this article recommend a pioneer digging device on the reservoir bowl, preferably in the old riverbed. The hydraulic parameters and geometrical dimensions of the ditch must correspond to the parameters of the canal, which provides the everyday hydrological regime of the river before the construction of the dam. The slope of the canal bottom should correspond to the domestic regime. The lining of the bottom and side walls of the pit is made with concrete lining of increased smoothness with a roughness coefficient of no more than n = 0.012-14. The pioneer pit will communicate with the reservoir bank with a slope of at least m = 3.5. The length of the pioneer dug is equal to the length of the canal from the dam to the alignment, where the violation of the river's daily routine begins. For mountain reservoirs, it can be determined from the graph plotted as a result of the analytical solution of the reservoir equation.

$$l_i = \frac{h_0}{i}\{[\eta_1 - (1-\overline{j})\Phi(\eta_1)] - [\eta_2 - (1-\overline{j})\Phi(\eta_2)]\} \tag{1}$$

here: l_i – length for each section; $\overline{j}$ – average value; η – the relative depth of their function $\Phi(\eta)$ look [21], pp.530-533; h_0 – normal depth to rivers, h_0=5; i – slope of the Pskem river bottom, i=0.0095; m – inclined; m=3.5.

$$\overline{j} = \frac{\alpha \cdot i \cdot \overline{C}^2 \cdot B}{g\overline{\chi}} \tag{2}$$

$\overline{B}$ – average canal width in each section

$$\overline{B} = b + 2m\overline{h} \quad [\text{m}] \tag{3}$$

$\overline{\chi}$ – average wetted perimeter of the river in each section

$$\overline{\chi} = b + 2\overline{h}\sqrt{1+m^2} \quad [\text{m}] \tag{4}$$

$\overline{\omega}$ – average cross-sectional area of the river in each section

$$\overline{\omega} = (b + m\overline{h})\overline{h} \quad [\text{m}^2] \tag{5}$$

$\overline{R}$ – average hydraulic radius in each section:

$$\overline{R} = \frac{\overline{\omega}}{\overline{\chi}} \quad [\text{m}] \tag{6}$$

Shezy coefficient:

$$C = \frac{1}{n} R^{1/6} \qquad [\mathrm{m}^{0.5}/\mathrm{s}] \qquad (7)$$

here: n - canal roughness, $n = 0.02$.

Then we proceed as follows. We divide the given canal into a number of sections, find out within what limits the required depths should lie. We set, in accordance with these limits, the depths h at the beginning and at the end of each selected section of the canal (one depth at the end or at the beginning of our canal is not specified for the calculation).

After that, knowing the depths at the beginning and at the end of each section, we calculate the lengths l of all sections using equation (1). All calculations are summarized in table 2.

Table 2. An example of calculating the construction of a curved free surface in the Pskem reservoir

Plot numbers	h (m)	$\bar{h}$ (m)	$\bar{B}$ (m)	$\bar{\chi}$ (m)	$\bar{\omega}$ (m)	$\bar{R}$ (m)	$\bar{C}$ ($\mathrm{m}^{0.5}/\mathrm{s}$)	$\bar{j}$	η	$\Phi(\eta)$	$\eta-(1-\bar{j})\Phi(\eta)$	l_i (m)
5	50	45	330	342.6	7762.5	22.6	84.1	7.25	10	0	10	1051
	40								8	0.0002	8.003	
4	40	35	260	269.8	4812.5	17.8	80.8	6.95	8	0.0002	8.002	1037
	30								6	0.0005	6.03	
3	30	25	190	197	2562.5	13.0	76.7	6.04	6	0.0005	6.03	1063
	20								4	0.002	4.01	
2	20	15	120	124.2	1012.5	8.1	70.9	5.17	4	0.002	4.008	1042
	10								2	0.025	2.10	
1	10	7.5	65	69.6	309.4	4.4	64.1	4.08	2	0.025	2.077	1638
	5								1	1.362	5.19	
											Σl_i	5831

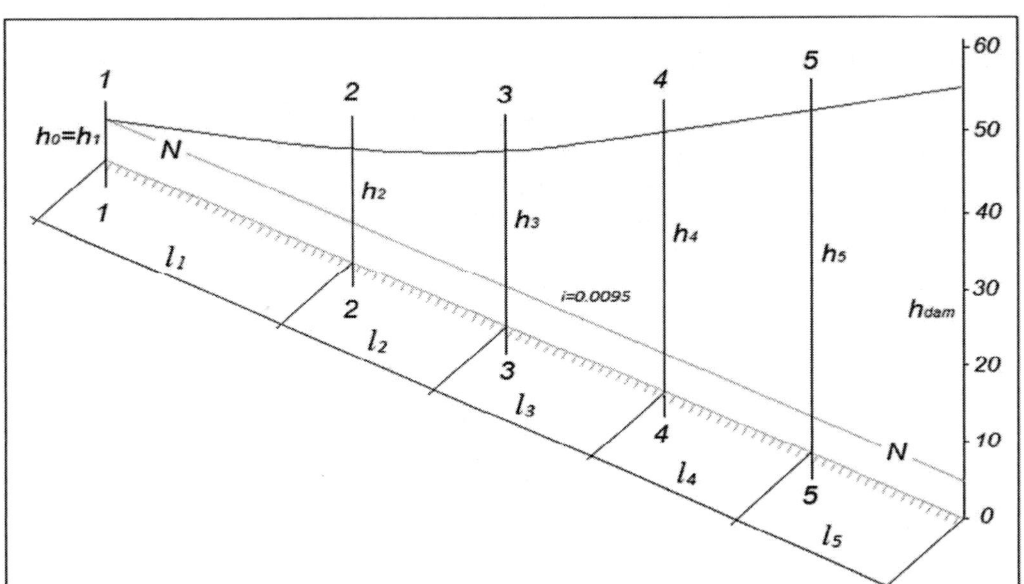

Figure 5. Graph of determination of pioneer digging in the upper reaches of reservoirs: N-N – normal water depth; h_1, h_2, h_3, h_4, h_5 – depth of water at each section; l_1, l_2, l_3, l_4, l_5 – section length in each section.

The proposed design allows periodic passage of bottom sediments into the downstream of the dam. Determination of the characteristic dimensions of the pit, its type (open or closed) and the study of the effect of bottom pumping on the energy absorbers of the downstream of the structure is the task of further research.

4. Conclusions

Analysis of studies related to the issue of sediment control is of undoubted interest for specialists in water management and scientific organizations. The results of the above studies allow us to draw the following conclusions:

1. The various types of dam water intake hydroelectric systems discussed above have shown that research, design, construction and water management organizations in many countries of the world have done significant work to create rational layouts of new improved structures of existing types of water intake and settling facilities, as well as sediment control devices on rivers and canals.

2. The research results presented in the reports of the IX International Congress show that the use of the theory of transverse circulation and methods of artificial, active regulation of the flow structure for the purpose of water intake has become widespread in world practice and is currently the main one for creating effective structures for water intake structures of canals that ensure their reliable protection from sediment.

3. Based on numerous experimental data obtained in various countries of the world when studying the operation of individual objects and the results of theoretical studies, it is possible to draw a conclusion about the effectiveness and prospects of using frontal-type structures with more than 50% water intake. The design of water intakes of this type has been significantly improved, they can be successfully used in various conditions, including on Mountain Rivers;

4. A new design of the pioneer ditch has been developed, which ensures efficient passage of bottom sediments into the downstream of the structure, which prevents intensive silting of the reservoir basin by bottom sediments and improves the efficiency of the dam water intake.

References

[1] Mukhamedov A M 1982 The first waterworks on the Amudarya river in Central Asia *Hydraulic engineering* **1** (12) pp 30–33

[2] Makhmudov E Zh and Paluanov D T 2010 The problem of the bearing capacity of the foundations of a low-pressure dam. *Mechanics problems* **1** pp 27–31

[3] Temirova G V 1975 To the question of the influence of the flow approach on the operation of the supply canal in the Fergana type of water intake *Collection of sci. papers* **147** pp 33–47

[4] Rumyantsev N S and Cromer R K 1981 Forecast of entering the headwater of the water intakehydroelectric complex *Hydraulic engineering and land reclamation* **8** pp 29–31

[5] Krutov A, Norkulov B, Artikbekova F and Nurmatov P 2020 Optimal location of an intake at a reservoir prone to salt diffusion *IOP Conf. Ser.: Mater. Sci. Eng.* **869** 072020 DOI:10.1088/1757-899x/869/7/072020

[6] Obidov B, Choriev R, Vokhidov O and Rajabov M 2020 Experimental studies of horizontal flow effects in the presence of cavitation on erosion – free dampers *IOP Conf. Ser.: Mater. Sci. Eng.* **883** 012051 DOI:10.1088/1757-899x/883/1/012051

[7] Bazarov D, Markova I, Raimova I and Sultanov S 2020 Water flow motion in the vehicle of main canals. IOP Conference Series: *Materials Science and Engineering* **883** p 012001 DOI:10.1088/1757-899x/883/1/012001

[8] Bazarov D, Uralov B, Matyakubov B and Vokhidov O 2020 The effects of morphometric elementsof the canal on hydraulic resistance of machine canals of pumping stations. *Mater. Sci. Eng.* **869** (072014). DOI:10.1088/1757-899X/869/7/072015

[9] Bazarov D, Shaazizov F and Erjigitov S 2020 Transfer of Amudarya flowing part to increase the supportability of the Uzbekistan southern regions. *IOP Conf. Ser.: Mater. Sci. Eng.* **883** p 012068 DOI:10.1088/1757-899x/883/1/012068
[10] Frog R, Brashiera H A 1975 River bed stabilization for irrigation water transport and sediment control *Proc. of the IX Int. Congress on Irrigation and Drainage* **30** pp 234–237
[11] Khachatryan A G, Shapiro K S and Sharova Z I 1966 *Silting and flushing of irrigation sedimentation tanks and reservoirs* (Moscow)
[12] Shapiro K S 1967 Model study of sediment loading and washing in the upper pool of the Kyzyl-Ayak hydroelectric complex on the Amudarya river *Hydraulic engineering* **36** pp 121–130
[13] Tohru K 1975 Irrigation water intake design (Japan) *Proc. of the IX Int. Congress on Irrigation and Drainage* pp 260–279
[14] Fattach F A and Gasser M M 1975 Problems of water intake into canals and reduction of sediment inflow into them (Egypt) *Proc. of the IX Int. Congress on Irrigation and Drainage* **30** pp 123–138
[15] Sharma K D and Astkhana B N 1975 Canal sediment management challenges in India. *Proc. of the IX Int. Congress on Irrigation and Drainage* **30** pp 50–83
[16] Yangiev A, Eshe, Panjie S and Rakhimov A 2020 Calculation of sediment flow in canals taking into account passing and counter wind waves *IOP Conf. Ser.: Mater. Sci. Eng.* **883** doi.org/10.1088/1757-899X/883/1/012036.
[17] Eshev S S, Khazratov A N, O'Gli Rahimov A R and O`Gli Latipov S A 2020 Influence of wind waves on the flow in flowing reservoirs. *IIUM Eng. J.* **21** pp 125–132 https://doi.org/ 10.31436/iiumej.v21i2.1329.
[18] Obidov B, Vokhidov O, Shodiev B, Ashirov B and Sapaeva M 2020 Hydrodynamic loads on a water drain with cavitation quenchers *IOP Conf. Ser.: Mater. Sci. Eng.* **883** p 012011. DOI:10.1088/1757-899x/883/1/012011
[19] Khidirov S, Norkulov B, Ishankulov Z, Nurmatov P and Gayur A 2020 Linked pools culverts facilities *IOP Conf. Ser.: Mater. Sci. Eng.* **883** 012004 DOI:10.1088/1757-899x/ 883/1/ 012004.
[20] Pastukhov S M 2010 Calculation of curves of the free flow surface for the svisloch cascade of reservoirs *Land reclamation* **2** (64) p61–69

IOP Conf. Series: Materials Science and Engineering 1015 (2021) 012042 doi:10.1088/1757-899X/1015/1/012042

Geometric and static analysis of the historical truss of the medieval truss Church of St. Martin in Dolní Újezd

P Krušinský, J Gocál and M Holešová

University of Žilina, Faculty of Civil Engineering, Žilina, Univerzitna 8215/1, Slovak Republic

E-mail address: Peter.krusinsky@uniza.sk

Abstract. The late-Romanesque St. Martin´s Church in Dolní Újezd was probably built in the mid-13th century. The nave has a wooden joist ceiling whose beams serve as tie beams too. The Gothic vaults above the presbytery date from ´30s of the 15th century. The truss of the church is dendro-dated to 1420 - 1421. Some members of a truss structure are dendro-dated to 1823 - 1824, when the repair was undertaken. The examined truss is a rafter-collar tie structure with collar beams and pedestal struts. It has a characteristic shape and may be typologically classified as a structure of the earlier period. This type of a truss structure was typically built in this region and represents one of the features occurring in the particular social and culture circles at the time. Consequently, a numerical model of the roof structure was created in order to perform a static analysis of the roof structure in accordance with present standards. With regard to the structural analysis results, it can be said that the original design of the structure based on geometrical and proportional principles satisfy the reliability conditions defined by current European standards for structural design.

1. Introduction

Analysis of proportional conception is based on historical-structural, archival, and dendrochronological research of the object as a whole. The method of geometrical and proportional analysis of the concept of the roof is based on acquiring information from period literature and from previous roof analyses. The proportions of historic trusses were analyzed mainly at roof constructions from Slovakia and Czech republic. The object of research is analyzing the proportion of historical constructions and relation into the floor plan of building and layout of the building solution. The spatial arrangement of the historical building and structures is also built on the concept of proportional relationships. The roof height is proportionally and geometrically derived from the ground plan of the building. These principles were also identified in other historical trusses in Slovakia and Bohemia, which we analyzed (e.g. trusses in churches in Belá Dulice, Abramová, Okolicna, Klimkovice, Vlčovice, the church of St. Catherine in Banská Štiavnica and others [3, 4, 6]).

2. History of St. Martin´s Church in Dolní Újezd

The church (figure 1) was probably built in the mid-13th century. It has a late-Romanesque core - only the tower and west nave survived to this day. At the end of the 14th century was built a new presbytery on the eastern part in high-level crafts realization. The old presbytery was demolished and the new one was interconnected with a late-Romanesque nave, which can be seen in the irregular floorplan. At the time of the Hussite Wars, like many other churches in this region, the church was destroyed by the armies. Soon after was built today´s truss dated from 1420 - 1421d. Later, probably in the ´30s of the 15th century, the presbytery was newly vaulted with a brick vault and stone ribs.

IOP Conf. Series: Materials Science and Engineering 1015 (2021) 012042 doi:10.1088/1757-899X/1015/1/012042

Further modifications were made after 1533 - an addition of a northern antechapel, knocking out of windows, southern entrance and several small modifications that did not affect the overall appearance of the church, [1, 2] (figure 1).

Figure 1. The researched object – The St. Martin´s Church in Dolní Újezd.

Figure 2. The inner view of the truss.

3. Truss construction

The church has a gable roof with a single ridge above the nave and sanctuary with a polygonal roof-end above the polygonal end of the sanctuary. Above the sanctuary, there is a baroque polygonal Sanctus bell turret. The truss above the nave has a rafter-collar construction with two levels of collar beams and heel struts. The truss is longitudinally tied by a central frame that is divided into two parts by a Sanctus bell turret. The truss contains three types of ties. The first type consists of rafters, two levels of collar beams, and a central post. The central post of the first collar beam, into which it is halved, is jointed with the second collar beam using the straight halving; the top of a post is also halved into the rafters. The post is cross-braced by the heel struts into the first level of the collar beams. The heel struts are halved into the tie beam and rafters. The collar beams are dovetailed into rafters. The second type, unlike the first one, does not contain a central post with heel struts. The cross-bracing is made by a saltire cross dovetailed into the tie beam and rafters below the level of the second collar beam. The third type is similar to the second one, except to saltire crosses at the level of the first collar beam which is replaced by heel struts dovetailed into the collar beam and rafters. The truss is longitudinally braced using span pieces located on the upper and lower collar beams, with saltire crosses halved and dovetailed into the posts and span pieces (figures 2 and 3).

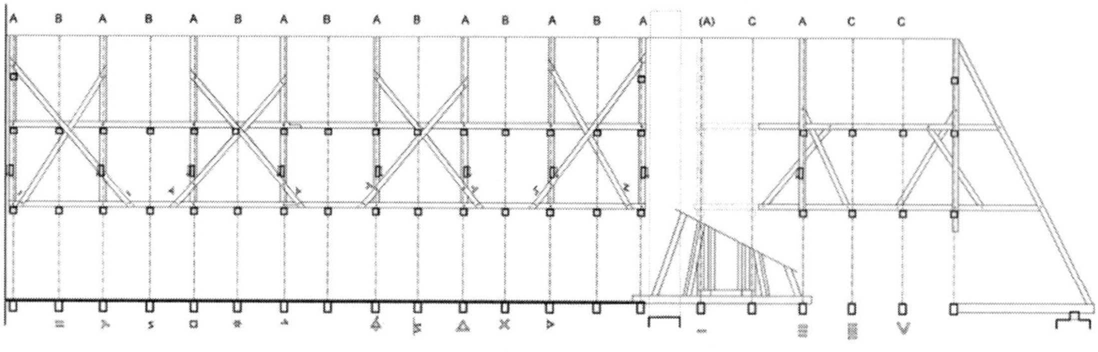

Figure 3. Longitudal truss section.

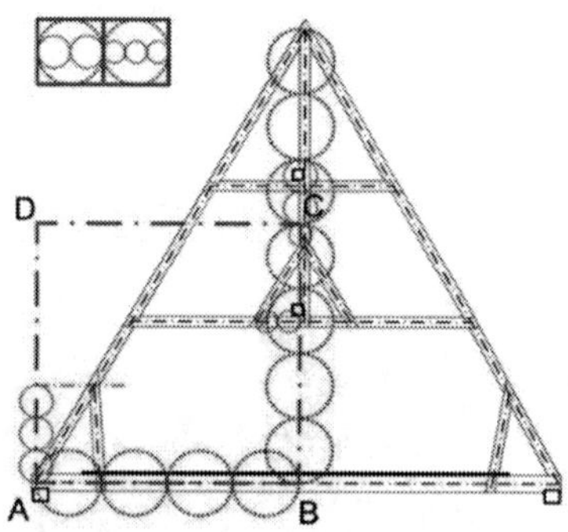

Figure 4. A cross-section.

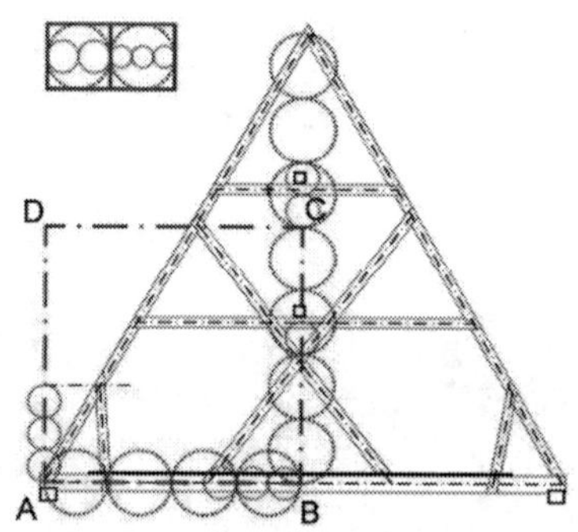

Figure 5. B cross-section.

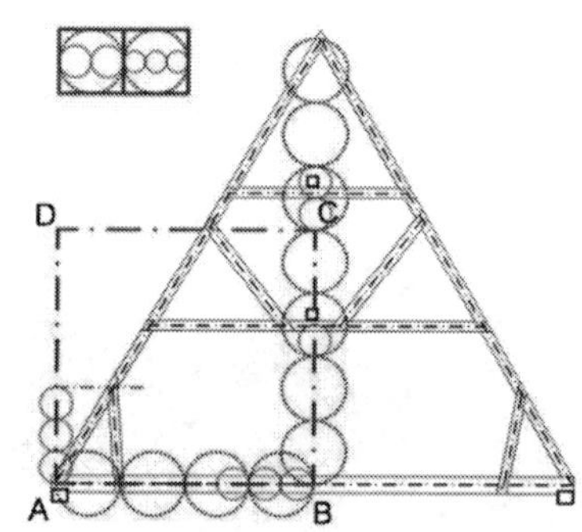

Figure 6. C cross-section.

4. Geometric concepts and proportional relationships within the truss

In the geometric analysis of historical trusses on sacral buildings, we encounter various ratios such as arithmetic, geometric, harmonic and with the Golden Ratio. When determining the length and width of the church, we can also find musical ratios.

We have found that a principle is very often used to determine the basic dimensions of the truss, which we denoted it as „*n* plus *k*", where $k = 1, 2, 3$. By this method, the height of the truss is derived from its width so that the width of the truss is divided into 2n of units x and the truss height is then $(n + k)$ of units x. In the St Martin´s Church in the village of Dolní Újezd was used this method for $k = 3$. The geometric analysis of the truss shows in the St Martin´s Church that the cross-sections of the truss above the nave are derived from the base square ABCD (figure 4). The length of the side AB of the square is nx, where in this case x is unit of measure and $n = 4$ („*n* plus 3"). Using the principle described before, the height of the truss is $(n + 3)\ x = 7\ x$. It is easy to realize that the height of the truss is 7/4 of the side AB of square ABCD, and the ratio of the truss width to its height is 8:7. The truss above the nave contains three different but quite similar in type of ties A,B and C. In the ties (figure 3.), the position of a lower collar beam is defined as 5/8 of the length of the side BC of the basic square ABCD. The upper collar beam is placed at the height 9/8 of the length of the side BC of the base square ABCD in all ties (figure 4). Hence, the ratio of the distance of a tie beam from a lower collar beam to the distance of both collar beams is 5:4. The all ties have the same positioned of a heel strut on the tie beam and on a rafter. The heel strut is attached to the tie beam at a distance x from the rafter heel (x represents 1/8 of the truss width) and is attached to the rafter at the height 3/8 of the length of the side BC of the square ABCD (figure 4).

Now we are dealing with the individual cross-sectional relationships of the truss above the ship and their differences. The cross-sectional A has a post that connects only the both collar beams and does not proceed to the tie beam. The heel struts of a post are horizontally positioned on a lower collar beam at a distance 1/6 of a length of the side AB. The heel struts of the post are positioned vertically on the post and the distance from the anchorage to the tie beam is 11/12 of the length of the side BC of the square ABCD (figure 4).
In the cross-sectional B was using a saltire cross. From the rafter heel, the strut is placed on the tie beam at a distance 5/8 of the length of the side AB. The strut mount on the rafter is where the side CD of the square ABCD intersect to the rafter (figure 5). The cross-sectional C is a modified the cross-sectional B so that the struts are shortened and are only down to the lower collar beam (figure 6).

5. Static analysis of the cross trusses

In order to verify the static behaviour of the truss, a computational model of the load-bearing structure was created in the FEM software SCIA Engineer [9]. The geometry of numerical model is related to the roof structure's geometrical analysis presented hereinbefore. Each of the cross trusses were modelled separately by means of planar models using beam elements. The softwood of grade C24 according to [10] was applied in the models, which corresponds to the applied type of wood. Geometrical schemes of the numerical models are presented in figure 7.

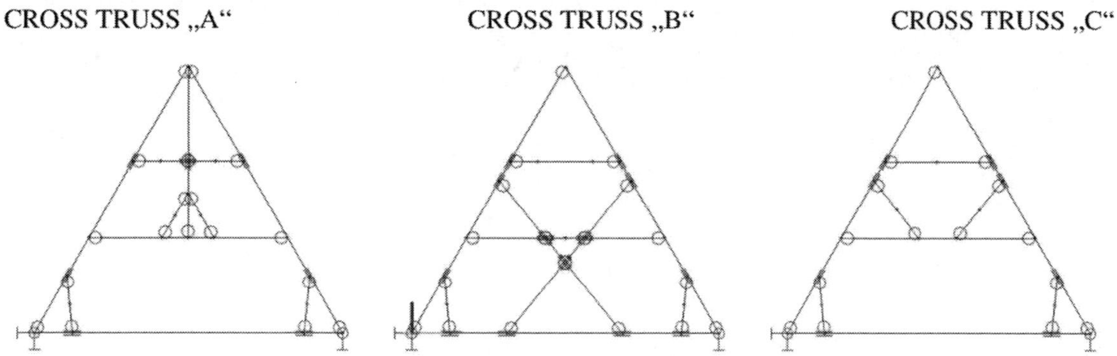

Figure 7. Geometrical schemes of the numerical models.

All the member connections are modelled as hinge joints with axial rigid connection and with capability of initial slip of 1 mm in the axial direction of member in order to consider an influence of gaps, cracks and geometry imperfections, occurring in historical carpentry joints. The effect of local weakening of the cross sections by carpentry joints is neglected in the models.
The individual load-bearing structures were loaded according to European standards [11, 12, 13] by permanent loads (self-weight and weight of roofing) and variable loads (wind actions only). With regard to the roof pitch angle (about 60°), the snow load was not applied on the roof. The self-weight load was generated by the FEM software. The combinations of load cases were generated according to the standard STN EN 1990 [11].

The results of numerical analysis are presented by the values of maximum (tension) and minimum (compression) normal stresses as well as displacements in the major structural members of the trusses, calculated for the decisive load combinations. Envelopes of the maximum and minimum normal stresses, respectively, in the major cross truss members are presented in figure 8. Envelopes of the vertical deflections caused by characteristic load combinations are presented in figure 9.

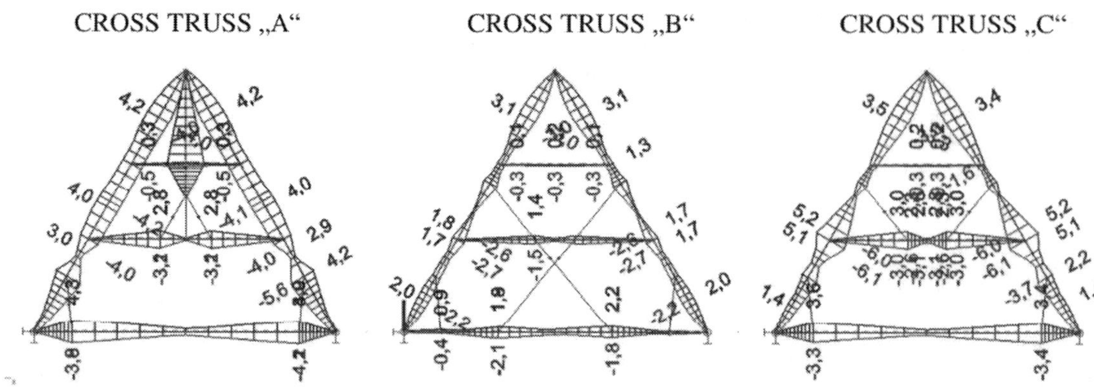

Figure 8. Envelopes of maximum and minimum normal stresses (MPa) in the major cross truss members.

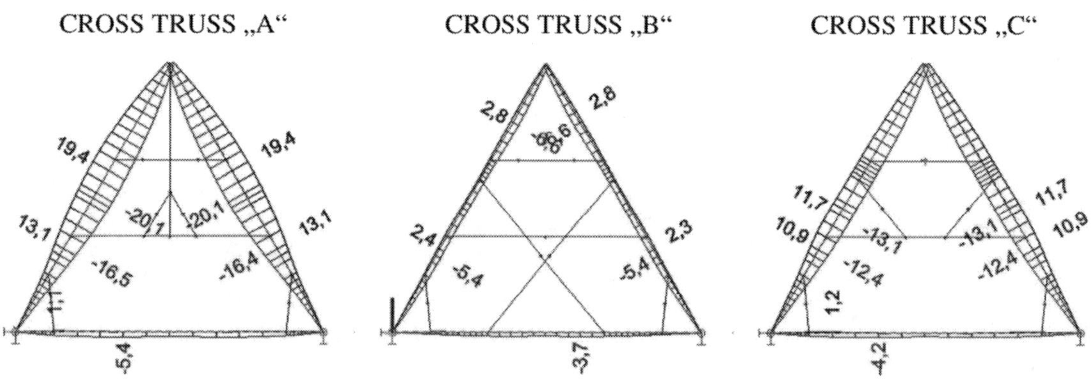

Figure 9. Vertical deformations (mm) in the major cross truss members.

Based on the results of numerical analysis it can be stated out that the normal stresses in all truss members do not exceed the design values of bending and compression strength equal to 16.60 MPa and 14.53 MPa, respectively, determined according to STN EN 1995-1-1 [14], assuming the timber strength class C24, modification factor $kmod = 0.9$ and partial safety factor $\gamma M = 1.3$. The possible buckling effects can be neglected, regarding the low share of compression forces due to normal forces (caused mainly by permanent loads). Similarly, the local vertical deformations of the major cross truss members do not exceed the standard [14] limit values (for example, in the case of rafters, it is equal to $L / 200 = 8220/200 = 41$ mm).

6. Conclusion

We discovered the proportional principle used in the truss of The St. Martin´s Church in Dolní Újezd. A truss with a similar proportional system and basic measure used is located in the village of the St. Catherine church in Banská Štiavnica [15], but its age (1655d) and typology are different. Therefore, the ratio of a half-width and a height of the truss is 4:7 [8]

The research results point to a rational approach to the design of roof trusses using geometric and proportional principles. Unlike now, an approach to the planning of the structure service life was significantly different, looked through a number of other generations. The results showed that the roof structure safely satisfy (with reserve about 58%) the reliability conditions defined by current European standards for structural design, both in terms of the ultimate limit states and serviceability limit states. The reserve extends the service life in relation to the potential long-term damage to the life of roof truss constructions and enables their rehabilitation.

Acknowledgements
This paper presents results of works supported by the VEGA - Grant Agency of the Slovak Republic under the project No. VEGA 1/0537/18.

References
[1] Líbal D 2001 *Katalog gotické architektury v České republice do husitských válek* (Praha: Unicornis)
[2] Václavík F 2000 Nové poznatky ze stavební historie kostela sv. Martina v Dolním Újezdě okr. Svitavy *Výroční zpráva za rok 2000 V Pardubicích: Památkový ústav v Pardubicích* pp 201-203
[3] Krušinský P, Capková E and Gocál J 2015 Comparison of two medieval trusses from the viewpoint of geometric and statistic *Advanced Materials Research* **1122** p 243-248
[4] Krušinský P Capková E Gocál J and Holešová M 2015 Geometric and static analysis of the historical trusses in roman catholic church of the holy Kozma and Damian in the Abramová village *Civil and environmental engineering: Scientific technical journal* **11** (2) pp 136-41
[5] Krušinský P, Capková E, Augustinková L and Korenková R 2016 Proportions - disposition relationship analysis of a historical truss in a rural house in Vápenná Village Czech Republic *Civil and Environmental Engineering* 12 (2) pp 111–5 (ISSN 2199-6512, DOI: https://doi.org/10.1515/cee-2016-0015)
[6] Gocál J, Krušinský P, Capková E and Kekeliak M 2014 Geometric and static analysis of the historical truss in village Belá Dulice In: *Advanced materials research* **969** p 199-207
[7] Struhár A 1977 *Geometric harmony of historical architecture in Slovakia* (Bratislava: Pallas)
[8] Makýš O 2014 *Včasnohistorické a rannostredoveké drevené a drevozemné pamiatky, hypotetické rekonštrukcie a konzervácia* (Renesans) p 189
[9] SciaEngineer 2010 Software for structural analysis. http://nemetschek-scia.com/sk
[10] Standard STN EN 338 2004 Structural timber. Strength classes (Bratislava: SÚTN)
[11] Standard STN EN 1990 2009 Eurocode. Basis of structural design (Bratislava: SÚTN)
[12] Standard STN EN 1991-1-1 Eurocode 1. Actions on structures. Part 1-1: General actions. Densities, self-weight, imposed loads for buildings 2007 (Bratislava: SÚTN)
[13] Standard STN EN 1991-1-4 Eurocode 1. Actions on structures. Part 1-4: General actions - Wind actions 2007 (Bratislava: SÚTN)
[14] Standard STN EN 1995-1-1 + A1 Eurocode 5. Design of timber structures. Part 1-1: General. Common rules and rules for buildings 2008 (Bratislava: SÚTN)
[15] Krušinský P, Holešová M and Ďurian K 2018 Proportional analysis of a transversal bond of the historic truss in the Gothic Roman-catholic Church of St Catherine in Banská Štiavnica dated to the mid-17th century *SGEM 2018 conference proceedings* **5** *Ancience Science*, Issue 2.3. Sofia: STEF92 Technology Ltd. p 179-84

IOP Conf. Series: Materials Science and Engineering 1015 (2021) 012043 doi:10.1088/1757-899X/1015/1/012043

Passage of low temperature piping through the external wall

J Leštach and R Ponechal

Department of Building Engineering and Urban Planning, University of Zilina, Faculty of Civil Engineering, Univerzitna 8215/1, 010 26 Zilina, Slovakia

Email address: jaroslav.lestach@uniza.sk

Abstract. The paper deals with the passage of the gas pipeline through the perimeter wall. The gas pipeline has a low temperature (the gas temperature drops significantly after the gas expansion in the regulator). When passing through the perimeter wall, the part of the wall near the internal surface will be significantly subcooled. Low temperatures in the building structure can cause local faults. The paper deals with the comparison of the course of temperatures and humidity of the passage of the gas pipeline in the protective tube with thermal insulation and without thermal insulation.

1. Introduction

Due to the type of piping and the perimeter structure, the passages of pipelines through the building envelope may have a location that meets the functional and safety needs of the pipeline. When designing buildings, when designing their architectural and structural solutions and especially in the stage of processing structural details, increased attention must be paid to the issue of possible thermal bridges. The materials used to make individual building structures have different physical properties, especially in terms of thermal insulation. Part of the problem of pipelines installation is taken over from abroad literatures. The main purpose is not to prevent the formation of thermal bridges. Due to the small number, considerable diversity and connection to the issue of several professions, these thermal bridges are not systematically solved. In contrast to possible thermal bridges in building envelope structures caused by material composition in the structure where thermal bridges are caused by transitions and installation lines, the determination of boundary conditions (medium temperature, periodicity, media properties, etc.) is also problematic, so this issue is not usually solved in project of building structure. According to Halahyja [1], the thermal bridge is a structural element with a different thermal conductivity than the whole structure. It usually has a higher thermal conductivity, the inner surface has a lower temperature.

Thermal bridges in the field of typical Slovak structures are dealt before by Sternova [2], similarly to other authors [3], [4], [5], [6], where they are solved only in connection with building structures, but not the passage of pipes through structures.

2. Boundary conditions

2.1. Temperature of the medium

The medium (mostly heating gas) has properties corresponding to its transport in the outdoor environment. The temperature is particularly important, as the medium is separated from the structure by a pipe and therefore no moisture is transferred by the transported gas. Table 1 shows approximate values, varies according to ambient temperature, installation depth, cable length, solar radiation, operating time, etc.

Table 1. Temperature of the medium as boundary condition in simulation

Material	Temperature of the medium	Notice
natural gas after regulation from pressure up to 0.4MPa	-15	winter season, average
natural gas without pressure regulation	0	winter season
natural gas without pressure regulation	15	summer season
air conditioning tube	5	winter season, recuperation only - no reheating

2.2. Pipeline position

Gas pipelines for public supply transport hydrocarbon heating gases (natural gas - methane; propane, butane). The gases are dehumidified during extraction; if the technological procedures for the installation of the pipeline are observed, the operation and functionality of the pipeline system will not be disrupted due to the action of residual moisture. Due to the safety requirements due to the possibility of the formation of an explosive mixture with air, the technical regulations (eg [8]) stipulate the requirements for the piping in such a way that no explosive concentration occurs. In particular, it is the piping through (un) ventilated spaces, if this cannot be observed, the piping is in a protective tube (tight piping leading to the ventilated spaces). One of the ways to prevent the formation of explosive concentrations is to run the pipeline through outdoor spaces - in the ground and on the facade of the building, respectively. its layers while ensuring constant ventilation of the grooves in which it is stored. The temperature of the heating gas (natural gas) entering the building (for heating purposes) varies according to its use, season and especially the pressure in the public part of the pipeline (before regulating the gas pressure) and fluctuates over a wide range. The temperature of the gas behind the regulation of the gas pressure is theoretically given by the equation of state of the gas (shown in the graph for methane), in reality it is significantly influenced by the transfer of heat from the surrounding environment.

The pipeline in the building is dealt with in Slovakia [8], which sets out the requirements for the passage of the pipeline through the perimeter wall in order to maintain safety with regard to the properties of natural gas and the building:

- the passage of the gas pipeline through the wall must not disturb the statics of the wall or the building;
- masonry and plaster must not act aggressively on the gas pipeline or protective pipe;
- when transferring a gas pipeline to a building, the gas pipeline must be placed in a protective pipe preventing gas from penetrating the building structure and at the same time protecting the gas pipeline from damage (specified in national standards).

3. The comparison of solutions

With regard to the main criteria that must be met in terms of safety, the solution of thermal protection and thermal bridges is solved marginally. The gas pipeline is usually made of a heat-resistant but thermally conductive material (steel, copper), which creates a thermal bridge in the structure. This thermal bridge can be manifested by a deterioration of the thermal insulation properties of the wall, an increase in humidity and damage to the building structure or a local deterioration of the climate in the building (humidity, mould). This article deals with the comparison of piping in a protective tube without thermal insulation and with thermal insulation.

To compare the influence of the temperature course, 2 variants are selected, meeting the requirements of technical regulations:

A) piping in a protective tube, without thermal insulation

B) piping in a protective tube, with thermal insulation.

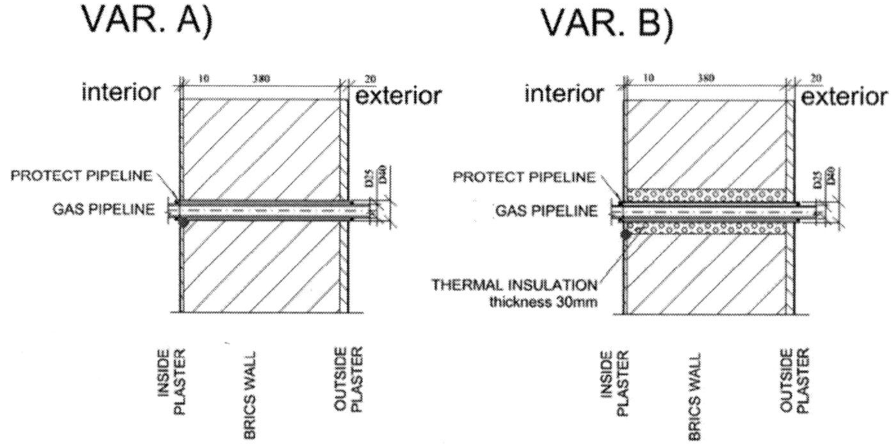

Figure 1. Variants of passage of pipeline

Table 2. Description of construction and material solution of the wall

Num.	Material	D [m]	Λ [W/m·K]	R [m²·K/W]
1	interior plaster	0.010	0.860	0.01
2	Porotherm Profi Dryfix 38T ceramic brick masonry	0.38	0.064	5.94
3	facade plaster	0.020	0.8	0.025
			Rt:	5.975

3. Conclusions and discussion

Pipe transitions through the outer wall create a thermal bridge. As with other thermal bridges, they must be considered. In case of risks for the construction, also evaluate and adjust.

As simulations have shown, the course of temperatures in the structure is significantly deformed along the entire pipeline. The low temperature was also reflected at the inlet of the pipe to the room, by a significant cooling below the dew point temperature. Due to the excellent thermal insulation properties of PUR foam, the temperature at the surface of the thermal insulation (VAR. B)) at the inlet of the pipe to the room is significantly higher than without thermal insulation (VAR. A)).

However, the thickness of the PUR foam affects the temperature (and thus the humidity) of the absorbent thermal insulation of the structure by moving the condensation zone to the non-absorbent layer, thereby limiting the wetting of the absorbent thermal insulation of the structure.

The course of relative humidity in the structure is influenced by the temperature of the structure and the diffusion resistance of the individual layers. The area where it is raised is therefore close to the pipeline. In particular, the following are problematic:

1) Over the internal surface of the structure - a condensation zone is created during VAR. A).
2) Relative humidity> 90% with protective tube for VAR. A) along the entire pipe.
3) Relative humidity> 90% with thermal insulation VAR. B) along most of the pipe (insulation) in the masonry. Condensed water vapour evaporates throughout the year.
4) Placement of pipes in thermal insulation VAR.B) requires major structural modifications in the structure - an enlarged hole for placement of insulation.
5) Filling the hole with polyurethane foam around the pipe will positively affect the course of temperatures near the pipe.
6) In the case of the implementation of masonry from organic absorbent materials (wood and wood products) in the condensation zone, it is necessary to re-evaluate the insulation thickness, or the overall solution.

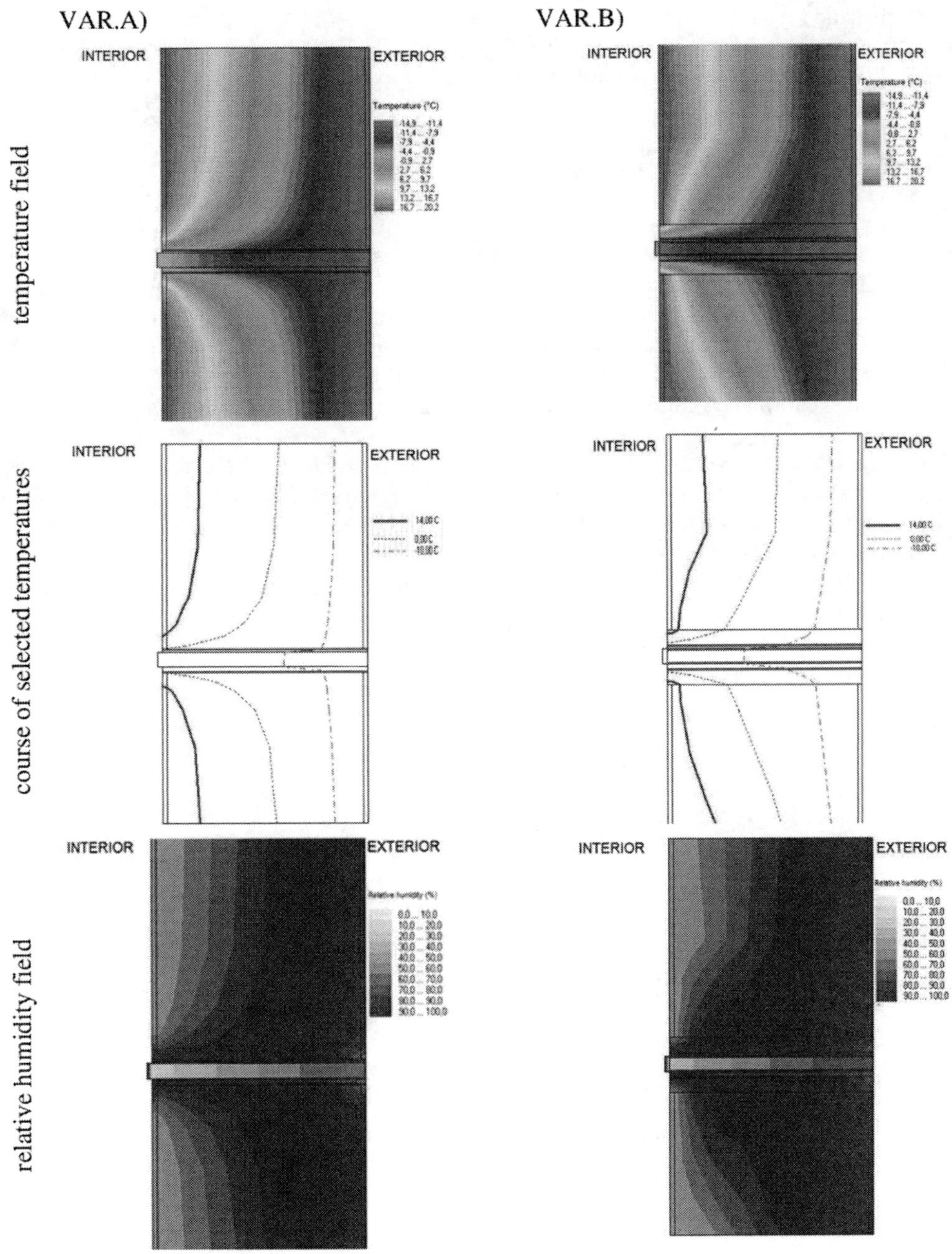

Figure 2. The comparison of temperature and relative humidity for variants A) and B), interior: θ_i=20°C, φ_i = 50%, exterior, θ_e= -15°C, φ_e = 84%, pipeline -15°C, steel pipe DN25, second protective pipe around the gas pipeline DN40, wrapping around the protective pipe of the gas pipeline (VAR.B) - PUR foam 30mm - λ_{iz} = 0.03 W·m^{-1} K^{-1}, Svoboda Cube 3D simulation program was used for the calculation [10].

Acknowledgments
This work was supported by the VEGA Project No 1/0673/20 Theoretical and experimental analysis of energy effective and environmental friendly building envelopes.

References
[1] Halahyja M 1985 *Building Thermal Technology, Acoustics and Lighting* (Bratislava: Alfa-SNTL)
[2] Sternova Z 2006 *Thermal Bridge Atlas* (Bratislava: JAGA group)
[3] Subrt R 2008 *Catalog of thermal bridges* (Praha: Grada)
[4] Subrt R 2016 *Catalog of thermal bonds III* (Praha: Isover)
[5] Martin K, Erkoreka A, Flores-Abascal I, Odriozola-Maritorena M and Sala JM 2011 Problems in the calculation of thermal bridges in dynamic conditions *Energy and Build.* **43** pp 529-35
[6] Berggren B and Wall M 2013 Calculation of thermal bridges in (Nordic) building envelopes – Risk of performance failure due to inconsistent use of methodology *Energy and Build.* **65** pp 331-39
[7] Tywoniak J 2012 *Low-energy houses 3. Zero, passive and others* (Praha: Grada Publishing)
[8] TPP 704 01: 2009 Gas offtake equipment for natural gas in buildings
[9] Lestach J 2019 *Technological Principles of Transitions and Conduction Realization of Installations in Low-Energy Buildings - Dissertation* (Zilina: University of Zilina)
[10] Cube 3D Software Manual, http://kcad.cz/cz/stavebni-fyzika/tepelna-technika-pro-specialisty/

The influence of the climatic features of the construction area on the level of energy-efficient thermal protection of of the office buildings

A A Frolova and E G Malyavina

Moscow State University of Civil Engineering, Yaroslavskoe shosse, 26, Moscow, 129337, Russia

E-mail address: FrolovaAA@mgsu.ru

Abstract. The choice of the level of the building thermal protection should be based on economic indicators of these buildings, taking into account the cost of maintaining a given room thermal microclimate. However, an economic calculation can only be performed after evaluating the impact of the thermal protection level on the building energy performance. The specific energy consumption of a building is affected by the size of the building, since in the buildings of different lengths and number of storeys, the ratio of ordinary rooms on the intermediate and the upper floors is different, as is the number of ordinary and corner rooms. The purpose of the article is to identify the energy consumption for heating, free and machine cooling of the buildings of various sizes and number of storeys in the areas which differ by the heating period duration and the solar radiation intensity, i.e. in Moscow and in Astrakhan. Three options of the building thermal protection have been considered herewith, differing from each other in the resistance to heat transfer of the outer wall and the coating. The area of the windows made 55% of the facade area. The calculation has been performed for several values of the outdoor air temperature during the year in each city.

1. Introduction

According to various estimates, the energy consumption for maintaining the thermal microclimate of buildings in Russia range from 30% to 40% of the energy of all extracted fuel [1]. That is why, reducing these energy consumption is an important task [2, 3]. A higher thermal protection of external enclosing structures should be considered as the most effective measure to save energy for maintaining thermal conditions in the building rooms [4].

When calculating the actual energy consumption of the systems that support the room specified thermal conditions, provision shall be made of taking into account the solar radiation heat inputs, that penetrate through the windows, as well as the heat coming from internal domestic or technological sources.

It is known [5-7] that the specific energy consumption of a building is influenced by the size of the building, since the ratio of ordinary rooms of intermediate and upper floors is different in the buildings of different lengths and storeys, as is the number of ordinary and corner rooms. Of great importance is the ratio of the size and the resistances to heat transfer of the windows and the exterior walls, due to the fact that in modern buildings the thermal insulation of the walls considerably exceeds the thermal performance of the windows, even of those having a bigger heat transfer resistance, in residential and most public buildings, according to the recent changes in the building rules SP 50.13330.2012.

The purpose of this work is to evaluate the energy performance of the buildings of various sizes and storeys in areas with different heating periods and solar radiation intensity: in Moscow and in Astrakhan.

2. Object of research

The study was performed by calculation. The calculation has been performed for the buildings of rectangular shape with the same width, equal to 20.2 m in external measurement. The length of the buildings varied from 13.6 m to 115.6 m. All end walls of the buildings are blind (without windows). The number of storeys of the buildings varied from 1 to 40 floors. The percentage of glazing of longitudinal walls is 0.55. The windows are dense enough to prevent infiltration. Individual characteristics of some buildings are shown in table 1.

Table 1. Main geometrical characteristics of the building.

Denomination of the value	The building option						
	1	2	3	4	5	6	7
Building length, m	20.4	40.2	20.4	61.2	40.2	61.2	115.6
Number of storeys	3	3	12	3	12	12	40
Building total area, m^2	1,236	2473	4,945	37,09	9,890	14,835	93,405
Area of exterior enclosing structures, m^2	1,362	2,252	4,212	3,141	6,534	8,855	44,705
Building volume, m^3	4,820	9,643	16,427	12,250	25,483	34,535	174,350

The buildings include four types of the rooms of the same size 6.8×10.1×3.9(h): row rooms of intermediate floors, row rooms of the upper floor, corner rooms on intermediate floors, corner rooms on the upper floor.

Three variants of the building thermal protection have been considered, differing from each other in the resistance to heat transfer of the outer wall and the coating. For option 1, the resistance to heat transfer of the external wall and the coating is close to the values, which are normalized by the formula (5.4) of SP 50.13330.2012 "Thermal protection of buildings" for sanitary and hygienic conditions. The option 3 of the thermal protection corresponds to the basic standards, based on the energy savings according to table 3 of the above mentioned SP. For option 2, the heat transfer resistances of the external walls and the coatings have been calculated using the formula (5.1) of the mentioned SP using a 0.63 reduction coefficient for the walls and 0.8 for the coating in relation to the option 3. The heat transfer resistance values, $m^{2\cdot O}C/W$, for external enclosing structures corresponding to options 1, 2 and 3 are shown in table 2.

The specific heat access to the premises has been considered from 9 a.m. to 18 p.m. and selected at 3 levels: 0 W/m^2, 40 W/m^2, 80 W/m^2. This value does not include the solar radiation coming through the windows.

Table 2. Heat transfer resistance of enclosing structures, $m^{2\cdot\circ}C/W$.

City	Denomination of the enclosing structure	Heat protection option 1	Heat protection option 2	Heat protection option 3
Moscow	Outer wall	1.226/1.2236	1.619/1.621	2.57/2.576
	Coating	1.379/1.371	2.74/2.7475	3.42/3.4226
	Window	0.66	0.66	0.66
Astrakhan	Outer wall	1.05/1.05667	1.40/1.40042	2.22/2.223676
	Coating	1.18	2.37/2.37091	2.96/2.966502
	Window	0.59	0.59	0.59

Note: The values of the heat transfer resistance required by the norms are shown above the line, and the calculated values are shown below the line.

During the heating season, i.e. when the outdoor temperature is +8 °C and below, a 24-hour operation of the heating system, which maintains the internal temperature of 20 °C round the clock, has been taken into account. It has been assumed, that the room cooling system worked only during the working hours, maintaining 22 °C in the room at this time.

Attention shall be drawn to the fact that the annual energy consumption considered only the building needs in heat and cold to maintain the specified room thermal microclimate. No losses due to inefficiency and additional energy consumption for the preparation of the required heat transfer agents for heating and cooling systems were considered. When calculating, it was assumed that the natural cooling shall be provided at an outdoor temperature not higher than +5 °C.

Table 3. Number of days in the year of the outdoor temperature monitoring.

Temperature range of the outdoor air (°C)	The interval average temperature (°C)	Duration (days)
	Moscow	
+22 ≤ t	+28	12
+18≤ t <+21.9	+20.5	29
+8≤ t <+17.9	+13	110
-2≤ t <+7.9	+3	117
-10≤ t <-2.1	-4	59
-20≤ t <-10.1	-15	31
t ≤-20.1	-25	8
	Astrakhan	
Above 30 °C	35	12
22 ≤ t <29.9	26	62
16≤ t < 21.9	19	62
8≤ t < 15.9	12	65
2 ≤ t <7.9	5	54
-4 ≤ t < 1.9	-1	62
-18≤ t <-4.1	-11	44
-26≤ t <-18.1	-21	4

It should be noted, that in Moscow the heating period takes 205 days and its average temperature is -2.2 °C. In Astrakhan the heating period is shorter i.e.169 days, and its average temperature is much higher: -0.8 °C.

To determine the energy consumption for maintaining a specified microclimate of the rooms having different heat transfer resistances of the external enclosing structures, provision has been made of a direct calculation of the office non-stationary thermal mode at different values of the outdoor air temperature, when the external air temperature was above +8 °C. No 24-hour maintenance of the room set temperature conditions and non-round-the-clock heat release explain the non-stationary character of the heat process. As a research tool, a program for calculating the non-stationary thermal regime of a room has been adopted; it is based on the calculation in finite differences with the construction of an implicit difference scheme using the heat balance method [8].

The calculations were performed at various outdoor air temperature values during the year: from the design temperature for heating to the design one for air conditioning in the warm season. The entire year was divided into several intervals in which the selected temperature values were the midpoints. The duration of the temperature monitoring was determined by [9] and is shown in the table 3.

IOP Conf. Series: Materials Science and Engineering 1015 (2021) 012044 doi:10.1088/1757-899X/1015/1/012044

3. The solar radiation thermal energy accounting method

It is known that the intensity of direct and scattered radiation varies from month to month. In addition, weather conditions change daily. However, the average monthly amounts of direct and scattered solar radiation intensity in multi-year terms are given by climatologists in reference books [9].

The total solar radiation, which inflows each month of the year on a horizontal surface, is the initial material for calculating the heat access to the rooms through translucent structures [10].

Actinometrists separately record in reference books [9] the monthly amounts of solar radiation in cloudy (actual cloud conditions) and clear (cloudless) sky. The monthly amounts of the solar radiation under actual cloud conditions shall be determined by averaging the measurement data under actual cloud conditions with any shape and number of clouds. However, the measurements are not taken into account when the sky is clear. Therefore, for each month, the sums of solar radiation should be put together for two states of the sky [11]. In this case, the monthly sum of the total solar radiation inflowing to the vertical surface takes into account the direct, scattered and reflected solar radiation according to the formula:

$$I_{\text{верт},m} = (S_{\text{я,гор},m} + S_{\text{обл,гор},m}) \cdot K_{\text{г.в},m} + (D_{\text{я,гор},m} + D_{\text{обл,гор},m})/2 + (I_{\text{я,гор},m} + I_{\text{обл,гор},m}) \cdot A_{\text{m}}/200 \quad (1)$$

where:

$S_{\text{я,гор},m}$, $S_{\text{обл,гор},m}$ – monthly amounts of direct solar radiation on clear days and under actual cloud conditions to the horizontal surface in the m month, MJ/m^2,

$D_{\text{я,гор},m}$, $D_{\text{обл,гор},m}$ – monthly amounts of scattered solar radiation on clear days and under actual cloud conditions on a horizontal surface in the m month, MJ/m^2,

$I_{\text{я,гор},m}$, $I_{\text{обл,гор},m}$ – monthly amounts of the summarized solar radiation on clear days and under actual cloud conditions on a horizontal surface in the m month, MJ/m^2,

$K_{\text{гв},j,m}$ – coefficient of conversion of direct solar radiation from the horizontal surface to the vertical one, that is oriented to the East and West sides, in the m month,

A_{m} – albedo of the underlying surface in the m month.

When considering the solar radiation heat inflows, the coefficients of total solar energy transmission by a 0.72 window double-chamber double-glazing, 0.8 bindings and circa 0.4 sun protection facilities have been taken into account.

4. Calculation results

The calculation investigated the need for heat and cold to maintain a given thermal mode in the building rooms within the year. Moreover, during the cold period of the year, the need for heating at night and cooling during the day was taken into account on certain days. The requirement to take into account the year-round maintenance of the specified temperature conditions in the rooms has been confirmed [12] by the fact that cold is required within a long period of time, when the outdoor air temperature is below the maximum allowable room temperature of 22 °C. At this time, increasing the resistance to heat transfer of walls and coverings plays a negative role, as it prevents the heat natural outflow from the room.

In Moscow, the heating period takes on average 205 days, i.e. the biggest part of the year, and in Astrakhan – 169 days, which is slightly less than half of the year. So, in Moscow, the main load is carried by the heating facilities and utilities, and in Astrakhan – by the cooling ones.

Calculations of the annual heat and cold consumption have been performed for the outdoor air temperatures indicated in the table 3. The intensity of solar radiation falling on the outer surface of the window was correlated with the value of the outdoor temperature. The consumed heat and cold have been estimated according to the primary fuel consumption. At the same time, the annual heat consumption of the building, MW·h, has been provided equal to the amount of the primary energy, MW·h. When recalculating the free cooling, the 6.95 electric energy use coefficient was taken into account, and in addition, the fuel efficiency coefficient at the power plant was equal to 0.3. When recalculating the machine cooling, the coefficient EER=3.31 has been entered, and the fuel use efficiency coefficient at the power plant has been provided equal to 0.3.

Table 4. Specific annual consumption of heat and cold by the building as a whole, kW·h/m^2.

The specific heat access	The specific needs for heat. free and machine cold	The building № 1 Heat protection option			The building № 2 Heat protection option			The building № 3 Heat protection option			The building № 4 Heat protection option		
		1	2	3	1	2	3	1	2	3	1	2	3
	Moscow												
0 W/m^2	Heat	76.95	55.36	43.65	66.30	49.43	41.04	60.20	48.00	38.29	59.44	43.46	36.58
	Free cold	9.06	12.79	17.51	13.78	10.63	20.77	12.09	15.31	20.06	14.14	17.95	21.17
	Machine cold	29.39	20.61	32.86	38.67	39.69	41.49	60.83	62.07	63.56	61.61	61.67	63.86
40 W/m^2	Heat	69.22	53.21	42.95	62.81	48.53	40.70	56.73	47.05	38.00	54.98	42.56	36.24
	Free cold	58.27	67.51	73.74	60.77	70.07	74.19	65.55	71.27	76.71	66.62	73.80	77.77
	Machine cold	82.90	84.84	87.09	92.49	94.10	95.90	134.88	136.30	137.79	135.61	135.90	138.10
80 W/m^2	Heat	69.04	53.21	42.95	62.73	48.53	40.70	56.68	47.05	38.00	54.92	42.56	36.24
	Free cold	115.03	124.52	130.68	116.17	124.69	131.39	122.44	128.22	133.65	123.50	130.92	134.71
	Machine cold	207.86	209.08	211.33	208.15	209.52	211.32	209.30	210.54	212.03	210.08	210.14	212.33
	Astrakhan												
0 W/m^2	Heat	66.03	49.06	39.41	55.23	41.55	34.70	52.96	43.22	34.79	51.63	39.05	33.13
	Free cold	25.02	32.04	37.68	30.12	36.46	40.49	30.46	35.42	40.44	31.82	37.93	41.43
	Machine cold	61.30	65.60	65.90	62.50	65.90	65.30	59.00	61.30	61.60	63.80	65.80	65.90
40 W/m^2	Heat	63.95	48.95	39.40	54.01	41.50	34.70	52.38	43.19	34.78	50.69	39.01	33.13
	Free cold	82.00	91.00	96.80	88.00	95.50	99.60	89.00	94.50	99.50	90.00	97.00	100.50
	Machine cold	137.11	138.07	138.35	137.65	138.89	138.34	129.39	130.73	131.03	137.39	138.19	138.28
80 W/m^2	Heat	63.95	48.95	39.40	54.01	41.50	34.70	52.38	43.19	34.78	50.69	39.01	33.13
	Free cold	141.10	150.10	155.90	147.10	154.60	159.00	148.00	153.60	158.60	149.00	156.00	160.00
	Machine cold	204.69	205.59	206.15	206.68	207.92	207.37	198.82	200.16	200.46	208.20	208.98	209.17

Table 5. Specific annual consumption of heat and cold by the building as a whole, kW·h/m^2.

The specific heat access	The specific needs for heat. free and machine cold	The building № 5 Heat protection option			The building № 6 Heat protection option			The building № 7 Heat protection option		
		1	2	3	1	2	3	1	2	3
	Moscow									
0 W/m^2	Heat	48.20	39.57	33.13	44.20	36.76	31.41	36.97	32.58	28.60
	Free cold	17.04	19.66	22.96	18.69	21.11	23.92	21.38	23.25	25.48
	Machine cold	62.50	63.22	64.37	63.04	63.60	64.65	63.93	64.44	65.09
40 W/m^2	Heat	46.11	39.05	32.97	42.57	36.39	31.29	36.36	32.44	28.56
	Free cold	71.88	76.04	79.73	73.99	77.64	80.74	77.71	80.04	82.38
	Machine cold	136.65	137.45	138.61	137.24	137.84	138.88	138.16	138.68	139.33
80 W/m^2	Heat	46.09	39.05	32.97	42.56	36.39	31.29	36.35	32.44	28.56
	Free cold	128.80	133.02	136.67	130.92	134.62	137.68	134.64	137.00	139.32
	Machine cold	210.98	211.69	212.85	211.53	212.07	213.12	212.40	212.91	213.56
	Astrakhan									
0 W/m^2	Heat	42.78	35.75	30.09	39.38	33.26	28.52	33.37	29.57	25.97
	Free cold	36.17	39.87	43.25	38.08	41.36	44.19	41.38	43.56	45.72
	Machine cold	64.70	66.00	65.80	64.90	66.00	65.70	65.30	66.00	65.60
40 W/m^2	Heat	42.44	35.74	30.09	39.13	33.25	28.52	33.31	29.57	25.97
	Free cold	94.94	98.90	102.30	97.00	100.40	103.30	100.40	102.70	104.80
	Machine cold	137.52	138.50	138.19	137.62	138.49	138.14	137.78	138.55	138.05
80 W/m^2	Heat	42.44	35.74	30.09	39.13	33.25	28.52	33.31	29.57	25.97
	Free cold	154.00	158.00	161.00	156.00	159.50	162.40	159.50	161.80	163.90
	Machine cold	206.62	207.55	207.44	207.84	208.67	208.45	208.97	209.71	209.29

The tables 4 and 5 show the specific needs for heat, free and machine cold per 1 m^2 of the floor space for buildings of various length and number of storeys in Moscow and Astrakhan.

As mentioned above, the calculation of the whole daily consumption shall take into account the heat of solar radiation that penetrates through the window, and the internal heat gain during the working hours, as given in the left part of the table 4. The fact that the heat demand at the internal heat gain of 40 W/m^2 and 80 W/m^2 of the floor are the same, specifies that all heat loss of the rooms is covered even with the heat gain of 40 W/m^2. The heat supply "excess" make a load of the room cooling. It is quite clear, that with an increasing heat protection (from option 1 to option 3), the heating load goes down, and the fact that this load falls with an increase in the length and the number of storeys of the building, may be explained by a smaller number of corner and upper rooms with regard to the number of ordinary ones.

It is interesting that the need for free and machine cold is growing from the option 1 to the option 3 of insulation of enclosing structure heat insulation, because the structures with a high heat transfer resistance actively prevent the heat outflow from the room. In a longer building, the loads on the room cooling under the free cooling mode increase in both cities, because of a long period of the free cooling time, when the outdoor air temperature is lower than the indoor air temperature of 22 °C. At the same time, an increasing share of ordinary rooms out of the total number of them plays a significant role.

The need for machine cold is formed during periods when the outdoor air temperature is lower than the indoor air temperature of 22 °C, and when the outdoor temperature is higher than the indoor one. During this second period, the load on the machine cold decreases with the increasing thermal protection. Despite the fact that this period is long in Astrakhan, the influence of the period with a lower outdoor temperature prevails in the formation of the machine cooling load, and the total annual load on the machine cooling increases with increasing the thermal protection. An increase in the size of a building leads to reduction the specific load on machine cooling.

The tables 6, 7 show the specific needs for primary energy per 1 m^2 of the floor space to maintain the specified thermal conditions in the buildings with different thermal protection. From the data in table 5, it follows that in both cities, with different sizes and storeys of buildings, it is energetically advantageous to provide the heat protection according to the option 3, that is, with basic heat transfer resistances in accordance with the building rules SP 50.13330.2012.

Table 6. Specific annual primary energy consumption by the building as a whole

The specific heat access	The building № 1 Heat protection option			The building № 2 Heat protection option			The building № 3 Heat protection option			The building № 4 Heat protection option		
	1	2	3	1	2	3	1	2	3	1	2	3
Moscow												
0 W/m^2	134.69	115.81	108.18	112.04	98.62	93.65	120.55	110.87	104.56	110.38	90.10	90.60
40 W/m^2	214.83	204.56	199.03	189.27	180.03	175.61	207.27	201.35	196.00	170.65	167.59	154.98
80 W/m^2	307.09	296.46	290.90	272.14	262.86	258.69	299.23	293.22	287.86	250.63	240.53	235.83
Astrakhan												
0 W/m^2	132.39	122.35	115.40	121.25	113.39	108.64	119.58	114.07	108.08	120.19	110.06	105.74
40 W/m^2	223.61	213.37	206.58	210.41	202.26	197.64	208.07	202.47	196.49	205.06	200.13	181.84
80 W/m^2	310.38	300.08	293.54	299.39	291.24	286.75	296.46	290.91	284.93	193.71	275.76	261.77

Table 7. Specific annual primary energy consumption by the building as a whole

The specific heat access	The building № 5 Heat protection option			The building № 6 Heat protection option			The building № 7 Heat protection option		
	1	2	3	1	2	3	1	2	3
Moscow									
0 W/m^2	112.20	105.36	101.38	109.40	103.52	100.33	104.14	101.02	98.59
40 W/m^2	200.99	196.45	193.01	198.89	194.83	192.01	195.12	192.68	190.40
80 W/m^2	292.91	288.33	284.88	290.79	286.71	283.88	286.98	284.55	282.26
Astrakhan									
0 W/m^2	117.03	112.78	108.40	114.64	110.93	107.14	110.42	108.19	105.16
40 W/m^2	208.06	203.96	199.50	205.73	202.11	198.31	201.52	199.47	196.33
80 W/m^2	296.18	292.05	287.60	294.84	291.23	287.55	287.55	291.56	289.48

5. Conclusions

As a result of the study it was found that the greatest thermal protection is the most appropriate in all cases in terms of the energy indicators in the cities under investigation. The solar radiation heat inflows an additional load on the building's cooling systems during the transitional and warm periods of the year. During the cold season these heat inputs reduce the load on the heating system. However. when the specific technological heat inflows to the room make 40 W/m^2 and more during the working hours. the need in the building cooling occurs rather frequently even during the heating period.

The obtained result confirms the economic calculations which compare the cost of the insulation material with the cost of the energy to maintain a given building mode. However the impact of the thermal protection on the economic component is not limited to these two cost indicators. The final decision should be taken after detailed economic calculations.

References

[1] Ardakanian O, Bhattacharya A and Culler D 2018 Non-intrusive occupancy monitoring for energy conservation in commercial buildings *Energy and Buildings* **179** pp 311-23 (DOI: 10.1016/j.enbuild.2018.09.033)

[2] Korniyenko S 2018 Complex analysis of energy efficiency in operated high-rise residential building: Case study *E35 Web of Conferences* **32** 02005

[3] Kwon M, Remoy H. Van Den Dobbelsteen A and Knaack U 2019 Personal control and environmental user satisfaction in office buildings: *Results of case studies in the Netherlands Building and Envir.* **149** pp 428-35 (DOI: 10.1016/j.buildenv.2018.12.021)

[4] Mingottia N, Chenvidyakarn T and Woods A W 2013 Combined impacts of climate and wall insulation on the energy benefit of an extra layer of glazing in the façade *Energy and Buildings* **58** pp 237–49

[5] Prozuments A, Borodinecs A, Odineca T and Nemova D 2020 Long-Term Buildings' Space Heating Estimation Method *Lecture notes in civil engineering* **70** pp 539-50

[6] Malyavina E and Frolova A 2019 Influence of solar radiation heat input into room on level of economically-efficient thermal protection of building *IOP Conf. Ser.: Mater. Sci. Eng.* **97** 03022

[7] Malyavina E and Frolova A 2014 The energetic and economic analysis of outdoor temperature to enable the transition to free cooling for conditioned rooms *9th Int. Conf. on Environmental Engineering*. ICEE 2014

[8] Berkovskiy B M and Nogotov E F 1976 *Difference methods of research of heat transfer problems* (Minsk: Science and technology)

[9] Scientific and applied reference book on the climate of the USSR. Episode 3. Multiyear data. (Hydrometeoizdat. St. Petersburg. 1989 – 1998)

[10] Bodrov V I Smyko A A 2014 *J. C.O.K.* **7** p 52 – 55

[11] Korkina E V, Gorbarenko E V, Gagarin V G and Shmarov I 2017 *J. Housing constr.* **6** pp 27–33

[12] Malyavina E and Frolova A 2019 The influence of the solar radiation heat input into the room on the level of energy-efficient thermal protection of the building *E3S Web of Conferences* **97** 03022

IOP Conf. Series: Materials Science and Engineering 1015 (2021) 012045 doi:10.1088/1757-899X/1015/1/012045

Identification of the collapse of the load-bearing facade cladding system of a building

D Papán, Z Papánová and M Ščotka

University of Žilina, Faculty of Civil Engineering, Department of Structural Mechanics and Applied Mathematics, Univerzitná 8215/1, 010 26 Žilina, Slovakia

Email address: daniel.papan@uniza.sk

Abstract. The aim of paper is to show how one structural fail can cause collapse of cladding system. This cladding system was one of most popular for use about 40 years ago for civic amenities buildings. The simplified theoretical approach for the collapse system reason is verified by modern FEM calculation. The results of investigation contain in situ inspections and measures on specimens. The main result of research is to amplify interest to older facades and its maintained and inspections.

1. Introduction

One of the most important long-term indicators in the design and planning for reliable buildings and their building components, as well as their "introduction to life" is maintaining of safety in their long-term use. In the event of non-compliance with technological procedures, due to climate change as well as age, each structure degrades and loses its useful properties and durability. At the time the construction is put into use, we can talk about the original state of the construction and its elements. Any change from this original state, that leads to a reduction in safety or service life, is considered a failure of the building structure or its elements. Accumulation of negative influences and factors may result in a serious disrepair or collapse of the structure. In these extreme cases, there is a general threat to the safety of the building or its main construction components, as well as the people in its vicinity

One of the monitored structural parts of buildings from the safety point of view is, among other things, the facade cladding system. The paper identifies a critical failure of the construction system of the overhanging facade at the Technical University in Zvolen, which led to its overall collapse and public danger.

2. Facade cladding systems on buildings

Currently the market offers various options for solving the final modification of the facades on individual buildings. In addition to a wide range of modern facade plasters and coatings (acrylic, silicate, silicone, mineral, cement, etc.), contact systems and various types of ventilated facade systems are also available (figure 1).

- Contact facade systems (adhesive) are characterized by gluing of the individual cladding elements directly onto the peripheral construction, masonry, plaster or thermal insulation. The base must be level and the weight or height to which the facing will extend must also be taken into account. Gluing is only possible under certain weather conditions [1].
- Ventilated facade systems (hanging facades) eliminate the shortcomings of adhesive cladding. The cladding is attached to the supporting structure by means of anchors, which always creates an air

chamber needed to vent the moisture penetrating from the interior. This construction solution protects the facade and the air chamber improves the thermal insulation properties of the wall.

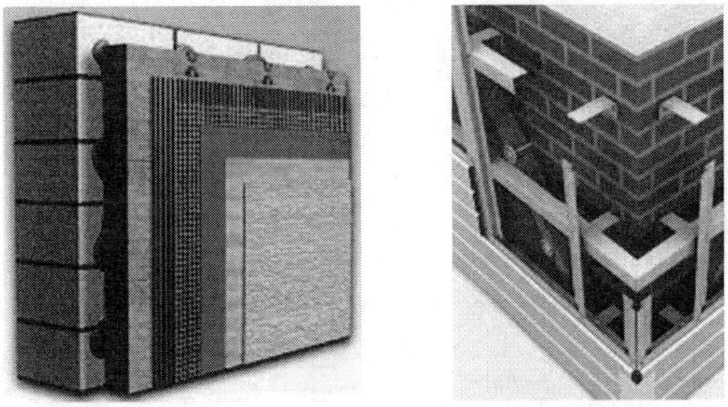

Figure 1. Example of contact and ventilated facade system of peripheral walls.

3. General description of the discussed object and specific facade system

The subject part of the discussed object´s building structure relates to the main building of the Technical University in Zvolen. The main building has an uneven floor plan with different heights of individual parts. A ventilated facade system was used for the side peripheral walls. The approximate age of the examined part of the building structure is 37 years. One of these sidewalls destructed in a short period of time, followed by a collapse. However, degradation processes of steel components did not show signs of excessive corrosion (figure 2).

Figure 2. University in Zvolen and collapse of the side wall facade system.

3.1. Description of the used facade cladding system

In terms of construction and technical solution, it is a self-supporting structure, which consists of an aluminium grate, anchored to hot-rolled profiles "U" 120. These profiles are anchored with mechanical anchors HILTI HSA M12×180 to the reinforced concrete load-bearing wall of the building. The components of the supporting grate have an atypical cross-section and are divided into vertical and horizontal aluminium profiles (figure 3). These components are connected by a screw connection using washers and aluminium extrusions. The single welded joint is implemented by connecting the L-shaped aluminium extrusions to the hot-rolled U 120 profiles to secure the vertical aluminium components. The individual travertine stone parts (facade blocks) are inserted into the aluminium grate. Their weight is approximately 16 kg and there are grooves milled around the perimeter for insertion and protection against falling out.

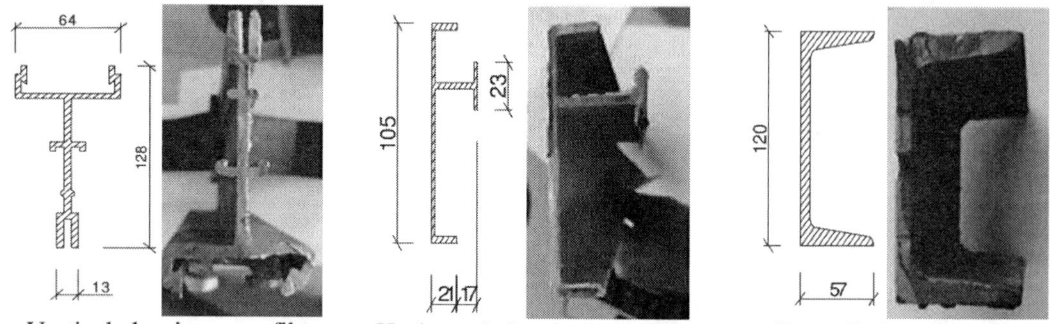

Vertical aluminum profile Horizontal aluminum profile Hot-rolled steel profile

Figure 3. Supporting components of the grate system for stone cladding.

The cladding was inserted to the height of 2.2 m in the lower part under the shading grate. The distance between two mechanical anchors on one profile is 2.75 m and the overhangs were 0.6 m long. The reconstructed system of hot-rolled profiles from which the dimensions and distances of individual elements were identified is on (figure 4).

Figure 4. Reconstruction of the load-bearing system of a fallen facade wall.

4. Critical failure analysis

The analysis of the accident was greatly simplified by a video recording, which indicated the mechanism of the system failure. This mechanism led to the failure of mechanical anchors, which were attached to the supporting "U" profiles to the existing building on which the facade was suspended. It was clear from the video that the anchors were broken first, which caused the fall of the entire facade system. Due to the impact force from the self-weight of the block, plastic joints began to form on the structure resulting in the formation of a kinematic mechanism, which collapsed unstable into a horizontal position. The consequence of collapse can be seen in (figure 5).

Figure 5. Collapsed structure of the stone cladding supporting system

Since there are several facade blocks such as those on the building, the upper parts of the load-bearing system were uncovered on the next block in order to examine the design solution of the not-yet- collapsed construction. After revealing, other failures of the support system were detected (figure 6):

- Mechanical anchor damaged by skidding
- Missing mechanical anchor
- Improper use of mounting washers

Figure 6. Structural failures of an exposed existing load-bearing structure (cut-off anchor, missing anchor, large number of washers)

One of the reasons for the incorrect design solution was an excessive unevenness of the reinforced concrete wall. In order to eliminate the inaccuracies created between the load-bearing grate of the hanging facade and the existing building, a considerable number of washers were used for mechanical anchors. This solution significantly affects the stress in these anchors and reduces their load-bearing capacity

4.1. Description of the used facade cladding system

As mentioned in above chapter, the entire collapse was initially caused by the failure of the mechanical anchoring of the supporting grate to the existing building. For this reason, the mechanical anchor itself can be considered a critical element of the structure and the cause of its collapse. In the calculation, only the actual weight of the facade was considered, except for climatic or temperature change loads. The self-load distribution was considered ideal [2], it means uniform for all mechanical anchors. The actual weight of the facade is given in table 1:

Table 1. Self-load of facade elements.

MEMBER	Length [m]	Area [m^2]	Number of pieces	Unit mass [kg]	Mass all members [kg]
stone facade block (thk. 30.1 mm)		79.26		67.6	5357.976
stone facade block (thk. 27.2 mm)		79.26		59.1	4684.266
horizontal aluminium profile	210			1.58	331.80
vertical aluminium profile	56			2.93	164.08
steel washers			108	0.61	65.88
hot-rolled profiles "U 120"	35.1			13.4	470.34
connecting material (aprox.)			1000	0.10	100
Sum (block thk. 30.1 mm)					6490.076
Sum (block thk. 27.2 mm)					5816.366

As the facade cladding has a large dispersion of thicknesses, the resulting vertical force will be determined for two variants:

- for the minimum measured thickness (ie 27.2 mm corresponding to the weight of the facade block $m_1 = 5816.366$ kg)
- for the maximum measured thickness (ie 30.1 mm corresponding to the weight of the facade block $m_2 = 6490.076$ kg)

These two limit weights, with sufficient reliability, represent the weight of all possible thickness variations that can occur on the facade. The part with a specified weight is suspended on 18 anchors, where the limit state designs of the force on the anchor were determined, i.e. $V_{sd\,(1)} = 3.54$ kN for variants with a stone block thickness of 30.1 mm and $V_{sd\,(2)} = 3,\ 17$ kN for variants with a cladding thickness of 27.2 mm.

After determining the load during collapse, the above-mentioned unsuitable design solution plays a significant role, when the anchor becomes a bracket, strained on the lever arm, which has a significant effect on the anchor´s cross-sectional stress. For this reason, a sensitivity analysis was performed, the parameter of which was the distance between the existing building and the anchored "U" profile at a constant load. Ideally, the anchor should be under shear load, without any free play. It was this ideal case that was the starting point for the calculation of the sensitivity analysis, which the values were compared with at an increasing free play.

4.2. *Sensitivity analysis of anchor resistance*

The calculation of anchors´ bearing capacity was carried out in accordance with ETAG 001 "METAL ANCHORS USE IN CONCRETE". Annex D – Design methods for anchorages. A schematic illustration of an anchor stressed on a lever arm is shown in figure 7.

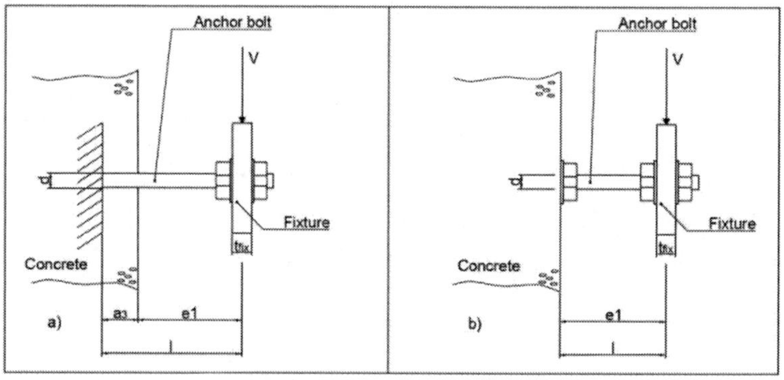

Figure 7. Mechanical anchor strain on the lever arm.

The first case (figure 7a) represents a situation without mounting washers and the second case (figure 7b) a situation when using mounting washers. Sensitivity analysis was performed in both cases, with image b) satisfying the resulting collapse. Computational models representing the use of mounting washers are shown in figure 8.

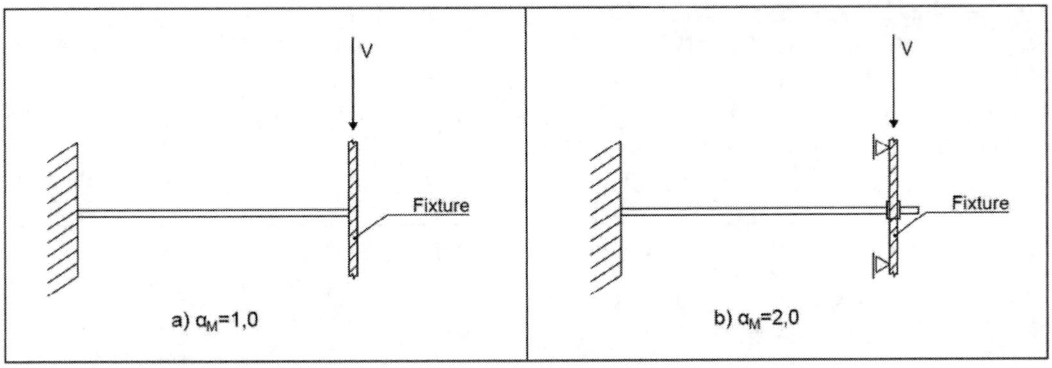

Figure 8. Computational models considering mounting washers.

The above-mentioned standard, on the basis of which the calculation of the load-bearing capacity of anchors was carried out, takes into account these mounting washer using the α_M coefficient, also shown in figure 8 for both cases. The total calculation according to the given standard is calculated as follows:

a) Shear load without lever arm [3]:
The characteristic resistance of the anchor in case of steel failure, $V_{Rk,s}$ is given by eq.(1)

$$V_{Rk,s} = 0.5 \cdot A_s \cdot f_{uk} \quad [N] \tag{1}$$

b) Shear load with lever arm [3]:
The characteristic resistance of an anchor, $V_{Rk,s}$ is given by equation (2)

$$V_{Rk,s} = \frac{\alpha_M \cdot M_{Rk,s}}{l} \quad [N] \tag{2}$$

where

α_M = see figure 8
l = lever arm
$M_{Rk,s}$ = $M^0_{Rk,s}(1 - N_{sd}/N_{Rd,s})$ [Nm]
$N_{Rd,s}$ = $N_{Rk,s} / \gamma_{Ms}$
$N_{rk,s}, \gamma_{Ms}$= to be taken from the relevant ETA
$M^0_{Rk,s}$ = characteristic bending resistance of an individual anchor.

After substituting specific values into the calculation, the resistance of the anchor was obtained as the ideal state.

$$V_{Rk,s} = 29.5 \text{ kN}$$

In the same way, the values of the resistance of the anchor at an increasing gap were obtained, see table 3.

Table 2. Input parameters of HILTI anchors for bearing capacity calculation.

MECHANICAL PROPERTIES - Anchor size		M6	M8	M10	M12	M16	M18
Nominal tensile	HSA, HSA-BW, HSA-F [N/mm^2]	650	580	650 650	700	650	700
strength $f_{u,thread}$	HSA-R2, HSA-R	650	560		580	600	625
Yield strength	HSA, HSA-BW, HSA-F [N/mm^2]	520	464	520 520	560	520	560
$f_{yk,thread}$	HSA-R2, HSA-R	520	448		464	480	500
Stressed cross-section A_s [mm^2]		20.1	36.6	58	84.3	157	245
Moment of resistance W [mm^2]		12.7	31.2	62.3	109.2	277.5	540.9
Char. bending	HSA, HSA-BW, HSA-F [N/m]	9.9	21.7	48.6	91.7	216.4	454.4
resistance	HSA-R2, HSA-R	9.9	21.0	48.6	76	199.8	405.7
CHARACTERISTIC RESISTANCE - Anchor size			M12			M16	
Eff. Anchorage dept h_{ef} [mm]		50	65	100	65	80	120
Tension N_{Rk}	HSA, HSA-BW HSA-R2, HSA-R [kN] HSA-F	17.9	26.5	35.0	26.5	36.1	50
		17.9	26.5	35.0	26.5	36.1	50
		17.9	26.5	35.0	26.5	36.1	50
Shear V_{Rk}	HSA, HSA-BW HSA-R2, HSA-R [kN] HSA-F	29.5	29.5	29.5	51	51	51
		29.3	29.3	29.3	56.5	56.5	56.5
		29.5	29.5	29.5	51	51	51
DESIGN RESISTANCE - Anchor size			M12			M16	
Eff. Anchorage dept h_{ef} [mm]		50	65	100	65	80	120
Tension N_{Rd}	HSA, HSA-BW HSA-R2, HSA-R [kN] HSA-F	11.9	17.6	23.3	17.6	24.1	33.3
		11.9	17.6	23.3	17.6	24.1	33.3
		11.9	17.6	23.3	17.6	24.1	33.3
Shear V_{Rd}	HSA, HSA-BW HSA-R2, HSA-R [kN] HSA-F	23.6	23.6	23.6	40.8	40.8	40.8
		23.4	23.4	23.4	45.2	45.2	45.2
		23.6	23.6	23.6	40.8	40.8	40.8

Table 3. Load capacity of the anchor strain as a lever arm.

LEVER ARM [mm]			1	2	4	8	15	30	50
without effective washer	α_M = 1.0	$V_{RK,s}$ [kN]	9.17	4.59	2.29	1.15	0.61	0.31	0.18
with effective washer	α_M = 2.0	$V_{RK,s}$ [kN]	18.35	9.17	4.59	2.29	1.22	0.61	0.37

4.3. FEM results verification

For FEM results verification was used computing environment "ANSYS Workbench". Geometrical model of anchor is available on the manufacture's websites ("HILTI") [4]. This verification has only principled character, because in computational model was used simplified material model. That is a reason why all result below describe reality only qualitative, but no quantitative. However, principle this simplification sufficiently fulfils the purpose of verification. The calculation was performed in three variants as was the case with the analytical solution. Length of lever arm was 50 mm, pure shear was considered without any space. In all variants was used same finite element mesh density and was used same element type. This is a reason why was possible mutual comparison. Loading and boundary conditions correspond with analytical solution.

Used material model: linear elastic - assumed linear relationship between stress and strain.

Used element type: SOLID186 - a higher order 3-D 20-node solid element that exhibits quadratic displacement behaviour. The element is defined by 20 nodes having three degrees of freedom per node: translations in the nodal x, y, and z directions. [7].

Results from FEM solution are in the following series of figures. Where are on the left side displacements and on the right sides are equivalent stresses due Von Mises theory. Figures 9-11 are in same order as it was in analytical solution: 1. pure shear, 2. lever arm without spacer and 3. lever arm with spacer.

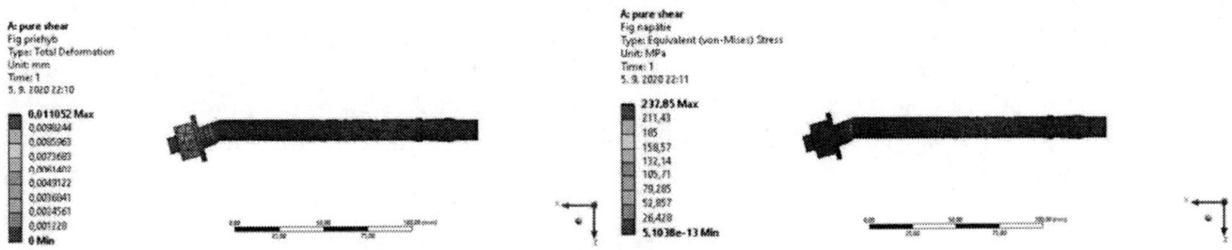

Figure 9. FEM solution of pure shear situation.

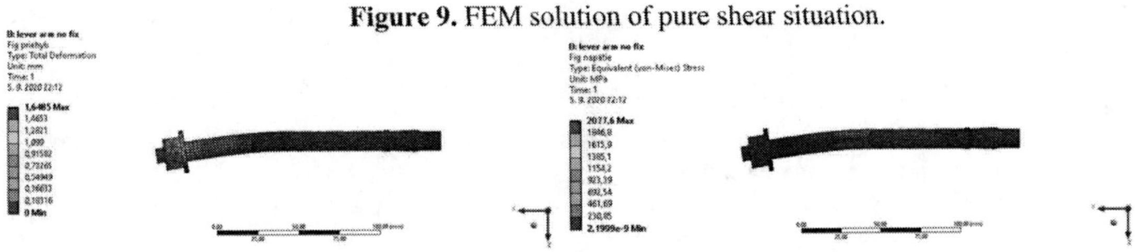

Figure 10. FEM solution of situation with lever arm without spacer.

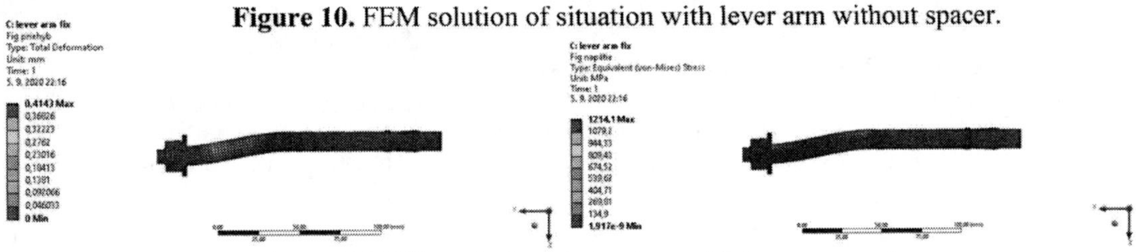

Figure 11. FEM solution of situation with lever arm and spacer.

From figures is the clear that the FEM solution is in principle as well as proportionally similar to the analytical solution from the previous chapters. This solution satisfactorily meets the initial assumptions of the FEM calculation.

Both of solution clearly show, that with increasing gap between anchored object and the substrate the resistance of the anchors rapidly decreases. Spacer partially help against resistance decreasing of anchor, but compared to the ideal shear condition, is it the reduction of resistance still quite large.

5. Conclusions

In general, the building's facade systems described in the paper are frequently used in Slovak Republic. This contribution obtained in previous chapters shows on real example how can be danger wrong technical realisation. Even when engineering design is not performed unsatisfactory. Civil engineers in 1983 designed the load-bearing system for facade stone cladding blocks but there were none to check the reality. That was the reason that anchoring did not works properly (ideal inertial forces redistribution). Many researchers and scientific authors are focused in this engineering area, not only main load bearing structures of buildings [8, 9]. That is because the facades, windows, doors, etc. are first stiff barriers for any static and dynamic load from environment. The reason of research investigation published in this paper is to increase interest on this self-load-bearing structures. In TU Zvolen was an accident of real facade system where only one factor "lever arm of anchor" caused collapse after 37 years of structure lifetime. Theoretical approach based on theory of elasticity considered in anchoring system manuals showed that the anchor without lever arm had large reserve of load-bearing capacity. But the real technical connection between steel anchor and reinforced concrete wall were performed "with lever arm". The load bearing capacity of such connection decrease several times rapidly depending on lever arm. This fact is confirmed by modern FEM calculation. Unfortunately, during the lifetime of investigated structure gradually the anchors loses the capacity and the system redistributed forces to stronger (better mounted) anchors. The last stage of redistribution was destruction. Today, no one will know if this fail was on civil structure constructers or accelerated completion due to the communist prestige of the then totalitarian leaders. From this case it is necessary to take the need to pay increased attention to diagnostics and inspections of exposed construction parts. All building's parts must be safe and maintained because it can also be danger for people moving around. The next process for the investigated building was to unmount the rest of facade parts built same way.

Acknowledgement
We kindly thank to Department of Structural Mechanics and Applied Mathematics for support in research activities.

References

[1] Electronic online source in Slovak https://www.stavebnik.sk/clanky/zateplovanie-obvodovych-stien-zateplovacim-systemom.html
[2] Standard STN EN 1991-1-1 Eurocode 1: Actions on structures. Part 1-1: General actions. Densities, self-weight, imposed loads for buildings
[3] ETAG 001, Edition 1997, *Guideline For European Technical Approval Of Metal Anchors For Use In Concrete*, European Organisation for Technical Approvals
[4] Hilti HSA EXPANSION ANCHOR, Technical Datasheet
[5] Design of Bonded Anchors, *Tech. Report, European Organisation For Technical Approvals*
[6] Bathe K J 2006 Finite Element Procedures (Cambridge: MA: Klaus-Jürgen Bathe)
[7] Ansys® Academic Research Mechanical, Release 20.1, Help System, Element Reference, ANSYS, Inc.
[8] Figuli L, Zvaková Z and Bedon C 2017 Design and analysis of blast loaded windows *Procedia Engineering*
[9] Herbut A, Rybak J 2017 Guidelines and recommendations for vibration control in the case of rapid impulse compaction, *Advances And Trends In Eng. Sciences And Technologies II*

IOP Conf. Series: Materials Science and Engineering 1015 (2021) 012046 doi:10.1088/1757-899X/1015/1/012046

Experimental studies of wind impact on religious buildings

O I Poddaeva, A N Fedosova and J S Gribach

NRU MGSU, 129337, Russian Federation, Moscow, Yaroslavskoe sh., 26

E-mail address: js-995@mail.ru

Abstract. This article is devoted to the description of experimental studies of wind impact on a religious temple in compliance with religious purposes. Studies of aerodynamic characteristics should be carried out using experimental methods for each Orthodox church individually. In the first chapter, the authors describe the specifics of the design of religious buildings, in the type of design of Orthodox churches. In the second chapter, brief information on the conduct of experimental studies of wind impact on the basis of NRU MGSU is launched. In the third chapter, a description of the experimental study of the wind effect on the Orthodox church, created in the Moscow region. Conclusions regarding this study.

1. Introduction

Design and construction of religious buildings and structures is one of the most complex and specific engineering tasks. The construction of objects of various religious denominations has a number of features in comparison with city-wide buildings.

First, it is necessary to determine the belonging of objects to a particular religion. For example, the specificity of designing Orthodox churches differs significantly from mosques or Catholic churches (figure 1).

a b c

Figure 1. Religious buildings: a - Cathedral of Christ the Savior, Moscow, b - White Mosque, Tatarstan, c - Catholic Church of the Immaculate Conception of the Virgin Mary, Moscow.

At the same time, the purpose and functional component of religious buildings should be noted: religious buildings of a particular religious affiliation include various buildings and related buildings, erected for specific purposes [1, 2].

The most widespread and large-scale type of religious buildings in Orthodoxy is the church (temple), intended for public worship. For example, in an Orthodox church, three parts are distinguished: the porch, the middle part of the temple, which is the main volume of the temple, and the altar (figure 2). It must be emphasized that the middle part of the temple, due to its intended purpose, must meet regulatory requirements to ensure the safety of people.

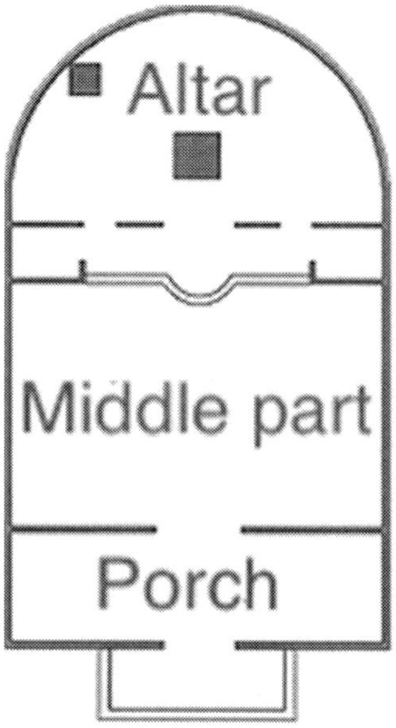

Figure 2. The main parts of the Orthodox church.

The most important role in the design of religious buildings is played by symbolism, which is closely related to their style, canonized requirements for the design of objects of a certain denomination and, consequently, with the design features of the object [3]. Based on the traditions of temple construction, as well as based on the provisions of regulatory documents related to the construction of religious buildings from an engineering point of view (for example [4]), designers should also take into account the requirements for ensuring the safety and reliability of building structures.

It is important to note that temple construction is a typical, which boils down to the need for a comprehensive structural analysis of building structures, which includes determining the wind load on the facades of objects. In the study of external aerodynamics, aerodynamic coefficients play an important role, depending on the geometric parameters of religious buildings. In the course of determining the distribution of aerodynamic forces over the outer surfaces of temples under wind action, engineers encountered inaccuracies or contradictions when using reference literature [5]. In this regard, studies of aerodynamic characteristics should be carried out using experimental methods for each Orthodox church individually [3].

2. Aerodynamic research at the National Research Moscow State University of Civil Engineering

A distinctive feature of NRU MGSU is the presence of a unique experimental base, which includes a wind tunnel of architectural and construction type, as well as a modern high-precision complex of measuring equipment. Aerodynamic tests are carried out on the basis of a unique scientific installation - the Large Research Gradient Wind Tunnel (figure 3).

Figure 3. Large research gradient wind tunnel of NRU MGSU.

NRU MGSU has developed a methodology for experimental studies of wind impact on construction sites. This method has been presented earlier in the following articles [6, 7]. Based on ensuring the requirements of the similarity theory [8, 9], the technique is the following algorithm:

1. Development of a three-dimensional geometric model of the object;
2. Analysis of meteorological and topographic data in the development area and assignment of wind characteristics;
3. Design and creation of a scaled-down model;
4. Tests in a wind tunnel of a scale model of the object. Determination of average and peak values of aerodynamic pressure coefficients Cp at control points on the surface of the object under study. The actual testing process is described in [6, 7];
5. Development of conclusions and recommendations based on the results of the study.

3. Aerodynamic research of a traditional Orthodox church

3.1. Description of the object under study and creation of a 3D model

In 2020, the Main Church of the Russian Armed Forces was opened in the Moscow region, the construction of which was timed to coincide with the 75th anniversary of Victory in the Great Patriotic War (figure 4a). This temple was designed in the monumental Russian-Byzantine style, which is a majestic cathedral with large golden domes. All dimensions of the temple are symbolic and refer to the most important figures associated with the history of the Great Patriotic War, the history of Russia and the Russian Armed Forces [10, 11]. A distinctive feature of this structure is its height (95 m), which makes the temple one of the highest Orthodox churches in Russia. The capacity of the inner church, taking into account its dimensions (65.4 m × 97.9 m), is up to 6 thousand people, which indicates that the building belongs to the category of a higher level of responsibility.

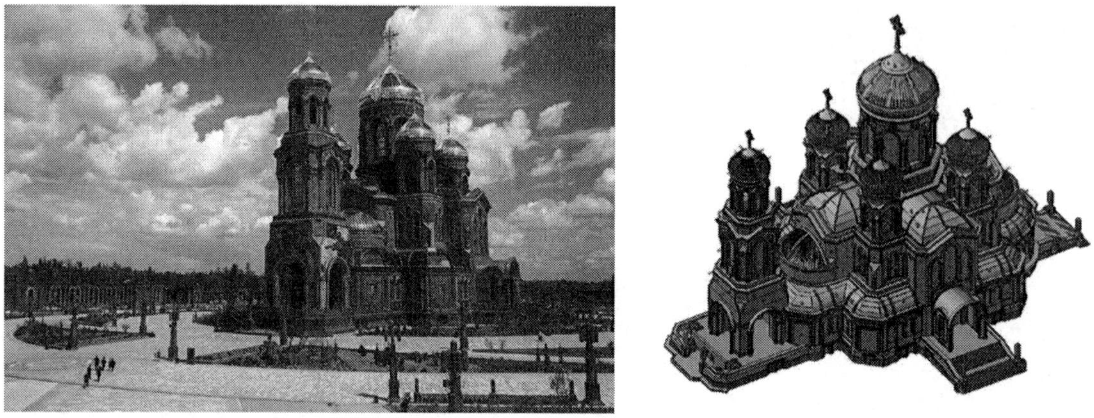

a b

Figure 4. The main temple of the Russian Armed Forces: a - view from the observation deck, b - 3D model.

Based on the initial data obtained, a 3D geometric model was developed, which was necessary for the study (figure 4b).

3.2. Analysis of meteo- and topographic data of the area

The projected construction site is located in the Moscow region, Odintsovo district, Kubinka. The climate of the region is moderately continental, with a clearly pronounced seasonality, but the degree of its continentality relative to other large European cities is much higher. The average annual temperature is + 5.8 ° C. The average annual wind speed is 2.3 m / s. The average annual air humidity is 77%. Figure 5 shows a wind rose.

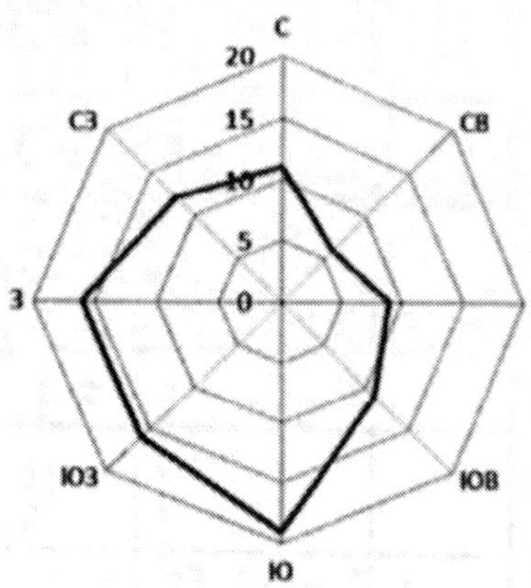

Figure 5. Wind rose in the construction area.

3.3. Designing a scaled-down model

For experimental studies, a 3D model of the investigated object was developed. Taking into account the dimensions of the wind tunnel test section, the scale of models 1: 100, which is the maximum possible from the conditions of blocking the flow, was chosen. The investigated model was installed on an automated rotary table located in the working area of the wind tunnel (figure 6). The model was made using 3D printing technology.

Figure 6. Model of the investigated object and installation of the pneumatic route.

Next, the model was prepared for experimental research; pressure collection points were placed on the surface of the model. From each point (hole), pressure is transmitted through copper and then through silicone tubes to differential pressure sensors (figure 6) [7].

3.4. Experimental results

During a series of physical tests on a scaled-down model of the temple, the average and peak values of the Cp coefficient were obtained at control points on the surface of the model. The research results are presented in a table 1 of the form:

Table 1. A table with the results of the study.

№ of point	0°	15°	...	75°	90°	...	345°
1	-1.66	-1.65	...	-0.40	-1.01	...	-1.38
...	...	...	...	...	...	...	...
186	-0.83	-0.72	...	0.20	0.33	...	-0.94

In the process of carrying out experimental studies. the average and peak values of the aerodynamic coefficient Cp at control points on the surface of the model were experimentally determined. while the average values are in the range from -2.94 to 2.39. the peak values are in the range from -3.32 to 2. 86. The maximum modulus values of aerodynamic coefficients are typical for the corner zones of facade structures. domes and facades of light drums.

4. Conclusions

In the course of the work. the most unfavorable wind directions were identified in terms of wind load - 75 and 90 (figure 7).

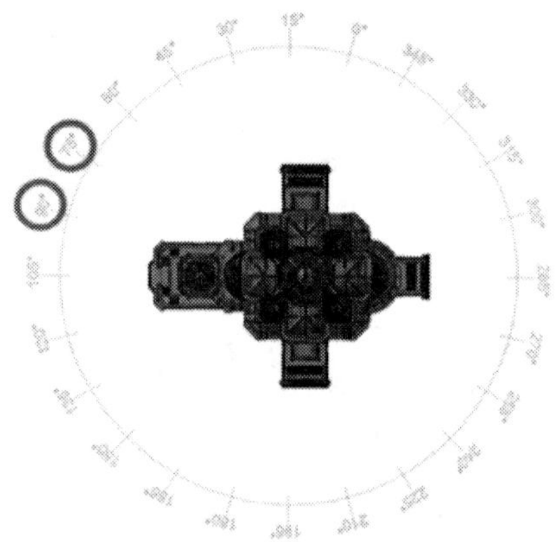

Figure 7. Wind load directions.

These directions require strengthening of the structural part in order to increase the reliability and safety of structures. Providing the required parameters in the design of the temple will allow for a long time to preserve its external and internal decoration and unique architecture [2. 5. 6].

Acknowledgments
This work was financially supported by the Ministry of Science and Higher Education of the Russian Federation (Project: Theoretical and experimental design of new composite materials to ensure safety during the operation of buildings and structures under conditions of technogenic and biogenic threats #FSWG-2020-0007). All tests were carried out using research equipment of The Head Regional Shared Research Facilities and Unique Scientific Installation Large Research Gradient Wind Tunnel of the Moscow State University of Civil Engineering.

References

[1] Abramyan S G etc. 2016 Construction of cultural facilities in Volgograd *International scientific review* **3** pp 55-7
[2] Cotorobey I C etc. 2017 Recovery renewable energy systems for romanian patrimonial religious buildings *Bulletin of the Transilvania University of Braşov* **10** p 59
[3] Kochev A G etc. 2016 Aerodinamics of natural ventilyation in unique buildings *Scientific review* **36** pp 36-8
[4] Standard SP 391.1325800.2017
[5] Kochev A G etc. 2015 Research into the aerodynamics of religious buildings *Modern problems of science and education* **2** (2)
[6] Poddaeva O I. Dunichkin I V and Curin P S 2016 Experimental study of wind loads on unique buildings and structures in Russia *MATEC Web of Conferences* **86** 02012
[7] Poddaeva O I etc. 2015 Experimental study of extreme wind loads on tall buildings and facilities *International Journal of Applied Engineering Research* **10** (21) pp 42797-9
[8] Elshaer A etc. 2016 LES evaluation of wind-induced responses for an isolated and a surrounded tall building *Engineering Structures* **115** pp 179-95
[9] Barskov K V et al. 2018 On the Applicability of Similarity Theory for the Stable Atmospheric Boundary Layer over Complex Terrain *Izvestiya. Atmospheric and Oceanic Physics* **54** pp 462–71
[10] Guryanov K V 2019 Memory of the people: the main Temple of the Armed forces of the Russian Federation *Bazis* **2** (6) pp 64-73
[11] Eloyan S N 2020 Architectural and artistic solution of the Patriarchal Cathedral of the Resurrection of Christ in the Park «Patriot» *Fine arts of the Urals. Siberia and the Far East* pp 141-7

IOP Conf. Series: Materials Science and Engineering 1015 (2021) 012047 doi:10.1088/1757-899X/1015/1/012047

Interaction of compressible barrette with surrounding soil and raft

Z G Ter-Martirosyan and Dam Huu Hung

Department of Soil Mechanics and Geotechnics, Moscow State University of Civil Engineering (National Research University), 26 Yaroslavskoye Shosse, Moscow, 129337, Russia

Email: hungdh1992@gmail.com

Abstract. The statement and analytical solution are handled simultaneously with the comparison with the results gained by the numerical method for addressing the problem of quantifying the stress-strain state (SSS) of a multi-layer soil cell consisting of a barrette interacting with the raft-plate. The solution to the assigned problem is considered by the authors for two cases, without and taking into account the compressibility of the underlying soil layer. In the first case, when the underlying soil layer is non-compressive, the formulas are formed in order to determine the SSS in the body of the barrette with the surrounding soil and the equivalent deformation modulus of the barrette-raft-surrounding soil system, so that engineers can absolutely estimate the bearing capacity of the underlying soil under the heel of barrettes. The second case, when the compressibility of the underlying soil layer is taken into account, the solution to the assigned problem is found by means of establishing and solving differential equations of the second order. As a result of the approach above, the formulas for determining the SSS in the body of the barrette are confirmed and the formulas for estimating the settlement of the raft and the equivalent deformation modulus for the barrette-raft-surrounding soil system are obtained. It is noted that taking into account the downward movement of the compressible barrette into the underlying soil layer lead to the decrease in the equivalent deformation modulus of the system.

1. Introduction

Upon considering the interaction of compressible barrette with surrounding multi-layer soil system of limited dimensions and the raft, a complex SSS arises in the barrette-raft-surrounding soil system, which determines the settlement of the system and the bearing capacity of the barrette. The load applied to the raft is distributed between the barrette, the surrounding soil, and the raft. Quantifying the distribution of the applied load is one of the most challenging problems in soil mechanics. Though the solution to the assigned problem the geometric parameters of the barrette-raft-surrounding soil system such as the laying deep of barrettes, the spacing among barrettes and dimensions of barrettes as well as the critical settlement of the underlying soil layer, the barrette-raft-surrounding soil system and the bearing capacity of barrettes are determined as well as estimated. In addition, by means of solving the assigned problem, we also can find the equivalent value of the deformation modulus to the barrette-raft-surrounding soil system.

IOP Conf. Series: Materials Science and Engineering 1015 (2021) 012047 doi:10.1088/1757-899X/1015/1/012047

2. Statement problem

Considering a barrette embedded in a soil cell, the applied load to the raft is distributed between the surrounding soil and the underlying soil layer. Let the following parameters be known: the dimensions of the barrette and the soil cell, its embedded length, the deformation modulus of the barrette, and the surrounding soil, underlying soil.

The following problems are needed to be solved:

The distribution of the applied load to the raft between the barrette and the surrounding soil;

The distribution of the applied load to the raft between the barrette and the underlying soil layer;

The settlement of the barrette-raft-surrounding soil system;

The equivalent value of deformation modulus of the barrette-raft-surrounding soil system

The solution to the assigned problem is conducted in two cases.

In the first case, when the underlying soil layer is possessed of vast rigid, i.e., $E_0 >> E_1$, the downward movement of the barrette's heel can be neglected ($S_R = 0;\ \sigma_R \neq 0$); in the second case, when the underlying soil layer is relatively compressive, i.e., the settlement of the barrette due to the punching of the heel into the underlying soil layer ($S_R \neq 0;\ \sigma_R \neq 0$). It is essential to be noted that, the stress at the surface of the barrette's heel is not allowed to exceed the critical load p^* proposed by L. Prandtl. i.e., $\sigma_R < p^*$.

3. Solving the problem in the condition of the settlement under the barrette's heel $S_R = 0$

According to the proposed calculation scheme below, the vertical displacements of the raft, the surrounding soil, and the barrette are equal, thereby it can be roughly assumed that the compression of the surrounding soil, in this case, behaves as in oedometer test. On the other hand, the vertical stress in the body of the barrette, and the shear stress at the contact surface between the barrette and the surrounding soil remain constant, as a result of the listed evidence, the condition of the equilibrium settlement of the raft S_r, the barrette S_b, and the surrounding soil S_s is fulfilled ($S_r = S_b = S_s$).

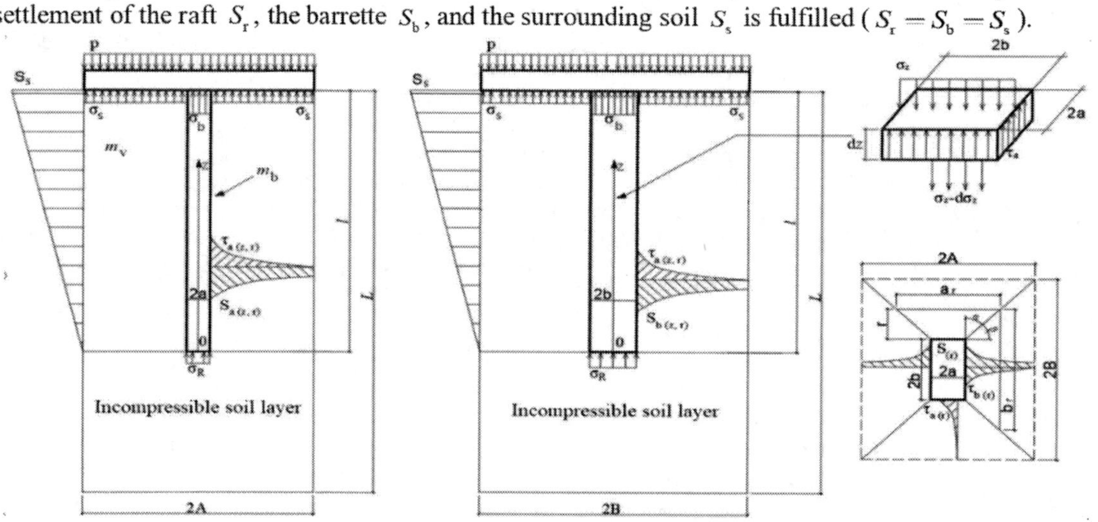

Figure 1. Calculation scheme of barrette interaction with surrounding and incompressible underlying soil layer.

From the calculation scheme (figure 1) we have an equilibrium condition

$$p = \sigma_N \omega + \sigma_s (1 - \omega) \tag{1}$$

in which σ_b is the stress in the body of the barrette, and σ_s is the stress under the raft and

$$\omega = \frac{ab}{AB} \tag{2}$$

The value of settlement of the compressible barrette corresponding to the settlement due to the compression of the soil cell can be written as follows

$$S_s = \sigma_s m_v L \tag{3}$$

m_v is the coefficient of volume compressibility of soil. The distribution of the stress under the raft σ_s along the depth z is shown in the calculation scheme (figure 1).

Considering the equilibrium condition $S_s = S_b$ from the calculation scheme (figure 1) we can get

$$\sigma_N = \sigma_s \frac{m_v}{m_b} \tag{4}$$

m_v и m_b are coefficients of volume compressibility of soil and barrette, respectively.

Substituting the expression (4) into (1), we obtain

$$\sigma_b = p \frac{m_v}{m_v \omega + m_b (1-\omega)} \tag{5}$$

By means of the equilibrium condition between the settlement of the soil under the raft and the settlement of the barrette $S_s = S_b$, we achieve $p\bar{m}L = S_b m_b L$, in which $\bar{m}$ is the equivalent value of the coefficient of volume compressibility of the barrette-raft-surrounding soil system. Substituting σ_b from expression (5) into the received expression above, we get

$$\bar{m} = p \frac{m_v}{m_v \omega + m_b \ 1-\omega} \tag{6}$$

$$S = p\bar{m}L \tag{7}$$

It is vitally essential to confirm the condition $\sigma_b < p^*$ under the barrette's heel, so that in this case the assigned problem is totally solved.

4. Solving the problem in the condition of the settlement under the barrette's heel $S_R \neq 0$

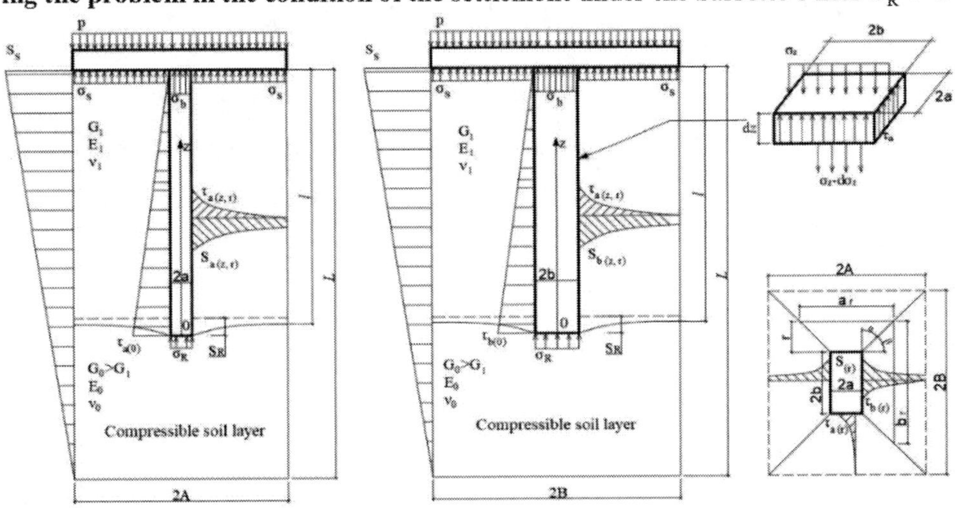

Figure 2. Calculation scheme of barrette interaction with surrounding and compressible underlying soil layer.

The settlement of the barrette and surrounding soil under the effect of the applied load to the raft occurs due to the shear deformation of the soil within the active zone around the barrette with the range from $2a \times 2b$ to $2A \times 2B$ along the embedded length (l) of the barrette and the length (L) of the soil cell. In this case, the increment of the arising settlement at the depth z can be determined by the following dependency:

$$dS(r,z) = -\gamma(r,z)\,dr \tag{8}$$

in which

$$\gamma(r,z) = \tau(r,z)/G \tag{9}$$

$$\tau_a(r,z) = \tau_a(z)\frac{(B-r_1)^2}{(B-b)^2},\quad \tau_b(r,z) = \tau_b(z)\frac{(A-r_2)^2}{(A-a)^2} \tag{10}$$

Substituting the gained value of the shear stresses into expression (8) we can find the distribution of the maximum value of the settlement on the side surfaces corresponding to the edges $2a$ and $2b$ of the barrette $S_{\text{side a}}$ and $S_{\text{side b}}$, respectively

$$S_{\text{side a}}(z) = \frac{-\tau_a(z)}{G}\int_b^B \frac{(B-r)^2}{(B-b)^2}\,dr = \tau_a(z)\frac{B-b}{3G} \tag{11}$$

$$S_{\text{side b}}(z) = \frac{-\tau_b(z)}{G}\int_a^A \frac{(A-r)^2}{(A-a)^2}\,dr = \tau_b(z)\frac{A-a}{3G} \tag{12}$$

The settlement of the barrette's heel is obtained from the known dependency of the vertical displacement of a rectangular rigid stamp taking into account the depth coefficient $K_1 < 1$ on the applied load, so we have

$$S_R = \frac{\sigma_R(1-v_0)a K_1 w}{G_0} = \sigma_R K \tag{13}$$

G_0 and v_0 are parameters of deformability of the underlying soil layer, and w is the coefficient considering the shape of the stamp (square, rectangle).

$$K = \frac{(1-v_0)a K_1 w}{G_0} \tag{14}$$

$$K_1 = \frac{1}{(1-v_0^2)w}\left\{\begin{aligned} &\frac{(1+v_0)(3-4v_0)}{4\pi(1-v_0)}\left[\ln\left[(n^2+1)^{1/2}+n\right]+n\ln\left[\left[(n^2+1)^{1/2}+n\right]/n\right]\right] \\ &+\frac{(1+v_0)(8v_0^2-12v_0+5)}{4\pi(1-v_0)}\left[\ln\left[\frac{(m^2n^2+m^2+16)^{1/2}+m.n}{(m^2n^2+m^2+16)^{1/2}-m.n}\right]+n\ln\left[\frac{(m^2n^2+m^2+16)^{1/2}+m}{(m^2n^2+m^2+16)^{1/2}-m}\right]\right] \\ &+\frac{2mn(1+v_0)}{\pi(1-v_0)(m^2n^2+m^2+16)^{1/2}}\left[(m^2+16)^{-1}+(m^2n^2+16)^{-1}\right] \\ &+\frac{2(1+v_0)(2v_0-1)^2}{\pi m(1-v_0)}\left[\arcsin\left[4n(m^2+16)^{1/2}(n^2+16)^{1/2}\right]+\arcsin\left[4n(m^2n^2+16)^{1/2}(n^2+1)^{1/2}\right]-\frac{\pi}{2}\right] \end{aligned}\right\} \tag{15}$$

where $m = 2a/l$ and $n = b/a$.

Determining the calculated value σ_R in the linear statement, the degree of its approximation to the limit state should be checked ($\sigma_R < \sigma_R^*$). This test can be carried out by the formula of L. Prandtl:

$$\sigma_R^* = (\gamma d + c_0 \operatorname{ctg}\varphi_0)\frac{1+\sin\varphi_0}{1-\sin\varphi_0}\exp(\pi \operatorname{tg}\varphi_0) - c_0 \operatorname{ctg}\varphi_0 \tag{16}$$

c_0 and φ_0 are strength parameters of the underlying soil layer (kPa) and (radian);
d is the depth at the level of barrette's sole (m);
γ is the unit weight of soil from the ground surface to depth d (kH/m^3).

Considering the interaction of a compressible barrette with the surrounding and underlying soils, embedded into the soil cell with the origin of coordinates $z = 0$ at the barrette's heel level, by dint of consideration the equilibrium conditions for the barrette's element dz, it is confirmed that the increment of the stress in the barrette's body $\mathrm{d}\sigma_b(z)$ is taken equal to the increment of the shear stress on the contact surfaces between the barrette and the surrounding soil

$$4ab\,\mathrm{d}\sigma_b(z) = (4\tau_a a + 4\tau_b b)\mathrm{d}z \tag{17}$$

Hence, it is followed that

$$\frac{\mathrm{d}\sigma_b(z)}{\mathrm{d}z} = \frac{\tau_a a + \tau_b b}{a \cdot b} \tag{18}$$

Analyzing the equilibrium condition of the settlement on the barrette's side surfaces $S_{\text{side a}}(z) = S_{\text{side b}}(z)$, we get the dependency of the shear stress $\tau_a(z)$ on $\tau_b(z)$

$$\tau_b(z) = \tau_a(z)\frac{B-b}{A-a} \tag{19}$$

The value $\tau_b(z)$ in expression (8) is superseded by the received from (19), thereby we get

$$\frac{\mathrm{d}\sigma_b(z)}{\mathrm{d}z} = \tau_a(z)\frac{(A-a)a + (B-b)b}{(A-a)ab} = \tau_a(z)I \tag{20}$$

where

$$I = \frac{(A-a)a + (B-b)b}{(A-a)ab} \tag{21}$$

Through the condition of the linear deformability of the compressible barrette, which is characterized by the value of deformation modulus of the constituting material, E_b (kPa), we obtain

$$\sigma_b(z) = \varepsilon_b(z)E_b = \frac{\mathrm{d}S_b(z)}{\mathrm{d}z}E_b \tag{22}$$

Differentiating the equation (22) with respect to z, we achieve:

$$\frac{\mathrm{d}\sigma_b(z)}{\mathrm{d}z} = \frac{\mathrm{d}^2 S_b(z)}{\mathrm{d}z^2}E_b \tag{23}$$

The new differential equation can be gained by putting $\tau_a(z) = \frac{3GS_{\text{side a}}(z)}{B-b}$ from (11) into (19), and equaling the received expression to (23)

$$\frac{\mathrm{d}^2 S_b(z)}{\mathrm{d}z^2} - \lambda^2 S_b(z) = 0 \tag{24}$$

in which $\lambda^2 = \frac{3GI}{E_b(B-b)}\left[\frac{1}{cm^2}\right]$ (25)

The expression received by solving the differential equation (24) can be written as follows:

$$S_b(z) = C_1 \,\mathrm{sh}\,\lambda z + C_2\,\mathrm{ch}\,\lambda z \tag{26}$$

C_1 and C_2 are the integration constants and determined from the boundary conditions when $z = 0$, $S_b(0) = S_R$, sh0 = 0, and ch0 = 1

$$\frac{\mathrm{d}S_b(z)}{\mathrm{d}z} = C_1\lambda\,\mathrm{ch}\,\lambda z + C_2\lambda\,\mathrm{sh}\,\lambda z = \varepsilon_b(z) \tag{27}$$

when $z = l$, $\varepsilon_b(l) = C_1\lambda \operatorname{ch}\lambda l + C_2\lambda \operatorname{sh}\lambda l = \sigma_N / E_b$.
With $C_2 = S_R$ we have

$$C_1 = \frac{\sigma_N}{E_b\lambda \operatorname{ch}\lambda l} - S_R \frac{\operatorname{sh}\lambda l}{\operatorname{ch}\lambda l} \tag{28}$$

Substituting all the obtained expression above into expression (26), we have the expression, describing the settlement of the body barrette depending on the depth z (m)

$$S_b(z) = C_1 \operatorname{sh}\lambda z + C_2 \operatorname{ch}\lambda z = \left[\frac{\sigma_N}{E_b\lambda \operatorname{ch}\lambda l} - S_R \frac{\operatorname{sh}\lambda l}{\operatorname{ch}\lambda l}\right] \operatorname{sh}\lambda z + S_R \operatorname{ch}\lambda z \tag{29}$$

$$S_b(z) = \left[\frac{\sigma_N}{E_b\lambda \operatorname{ch}\lambda l} - \sigma_R K \frac{\operatorname{sh}\lambda l}{\operatorname{ch}\lambda l}\right] \operatorname{sh}\lambda z + \sigma_R K \operatorname{ch}\lambda z \tag{30}$$

Conducting the differentiation of expression (30) with respect to z, we get

$$\frac{dS_b(z)}{dz} = \varepsilon_b(z) = \frac{\sigma_b(z)}{E_b} = \left[\frac{\sigma_N}{E_b \operatorname{ch}\lambda l} - \sigma_R K\lambda \frac{\operatorname{sh}\lambda l}{\operatorname{ch}\lambda l}\right] \operatorname{ch}\lambda z + \sigma_R K\lambda \operatorname{sh}\lambda z \tag{31}$$

$$\sigma_b(z=0) = \sigma_R = \frac{\sigma_N}{\operatorname{ch}\lambda l + K\lambda E_b \operatorname{sh}\lambda l} \tag{32}$$

Replacing the σ_R in expression (30) by the obtained σ_R from (32), we finally get the dependency of the compressible barrette's settlement on the depth z due to the application uniformly distributed load to the raft:

$$S_b(z) = \sigma_N \operatorname{sh}\lambda z \left[(E_b\lambda \operatorname{ch}\lambda l)^{-1} - \frac{K \operatorname{sh}\lambda l}{(\operatorname{ch}\lambda l)^2 + K\lambda E_b \operatorname{sh}(\lambda l)\operatorname{ch}(\lambda l)}\right] + \sigma_N \frac{K \operatorname{ch}\lambda z}{\operatorname{ch}\lambda l + K\lambda E_b \operatorname{sh}\lambda l} \tag{33}$$

$$S_b(z) = \sigma_N V_1 \operatorname{sh}\lambda z + \sigma_N V_2 \operatorname{ch}\lambda z \tag{34}$$

in which

$$V_1 = (E_b\lambda \operatorname{ch}\lambda l)^{-1} - \frac{K \operatorname{sh}\lambda l}{(\operatorname{ch}\lambda l)^2 + K\lambda E_b \operatorname{sh}(\lambda l)\operatorname{ch}(\lambda l)}; \quad V_2 = \frac{K}{\operatorname{ch}\lambda l + K\lambda E_b \operatorname{sh}\lambda l} \tag{35}$$

Considering the equilibrium condition between the settlement of the surrounding soil at the head of the barrette ($z = l$), we achieve the expression

$$S_R(l) + S_b = S_s \tag{36}$$

$$\sigma_R K + \int_0^l \frac{\sigma(z)}{E_b} dz = \sigma_s m_s l + \sigma_s m_0 (L - l) \tag{37}$$

$$\sigma_R K + \int_0^l [\sigma_N \lambda V_1 \operatorname{ch}\lambda z + \sigma_N \lambda V_2 \operatorname{sh}\lambda z] dz = \sigma_s [m_s l + m_0 (L - l)] \tag{39}$$

where m_s and m_0 are the values of the coefficients of volume compressibility of the surrounding soil in the range of the barrette's embedded length l and the underlying soil, respectively.
$S_R(l) = \sigma_R K$ is the settlement gained due to the punching of the barrette's heel into the underlying soil layer; S_b is the settlement of the barrette's body; S_s is the total settlement procured due to the compression of the surrounding soil around the barrette in the range of the barrette's embedded length l and the underlying soil layer.

Considering expression (20), we can draw out the distribution at depth z of the shear stress on the side surface corresponding edge $2a$ of the barrette.

$$\tau_a(z) = \frac{1}{I}\cdot\frac{d\sigma_z}{dz} = \frac{1}{I}\cdot\left[\frac{\sigma_N}{\mathrm{ch}\,\lambda l} - \sigma_R K\lambda E_b \frac{\mathrm{sh}\,\lambda l}{\mathrm{ch}\,\lambda l}\right]\lambda\,\mathrm{sh}\,\lambda z + \sigma_R \frac{K\lambda^2 E_b}{I}\mathrm{ch}\,\lambda z \tag{40}$$

From the equality of the settlement of the barrette's body and the underlying soil layer at the level of the barrette's heel $z = 0$, we obtain the following expression

$$\sigma_R K = \sigma_s m_0 (L-l) + \tau_a(0)\frac{B-b}{3G} \tag{41}$$

$$\sigma_R K = \sigma_s m_0 (L-l) + \sigma_N \frac{B-b}{3GI}\frac{K\lambda^2 E_b}{\mathrm{ch}\,\lambda l + K\lambda E_b\,\mathrm{sh}\,\lambda l} \tag{42}$$

After transforming three unknowns, a system of equations, which consists of expression (39), (42), and (1) is constituted as follows

$$\begin{cases} \sigma_R K + \sigma_N V_1\,\mathrm{sh}\,\lambda l + \sigma_N V_2(\mathrm{ch}\,\lambda l - 1) = \sigma_s m_s l + \sigma_s m_0 (L-l) \\ \sigma_N \omega + \sigma_s (1-\omega) = p \\ \sigma_R K = \sigma_s m_0 (L-l) + \sigma_N \dfrac{B-b}{3GI}\dfrac{K\lambda^2 E_b}{\mathrm{ch}\,\lambda l + K\lambda E_b\,\mathrm{sh}\,\lambda l} \end{cases} \tag{43}$$

Solving this system of equations we get the solution for the unknowns

$$\sigma_s = p\frac{V_3+V_4}{\omega m_s l + (V_3+V_4)(1-\omega)};\ \sigma_N = p\frac{m_s l}{\omega m_s l + (V_3+V_4)(1-\omega)};\ \sigma_R = \frac{p}{K}\cdot\frac{m_s l V_4 + (V_3+V_4)m_0(L-l)}{\omega m_s l + (V_3+V_4)(1-\omega)};$$

$$V_3 = V_1\,\mathrm{sh}\,\lambda l + V_2(\mathrm{ch}\,\lambda l - 1);\quad V_4 = \frac{B-b}{3GI}\cdot\frac{K\lambda^2 E_b}{\mathrm{ch}\,\lambda l + K\lambda E_b\,\mathrm{sh}\,\lambda l} \tag{44}$$

As a result, the expression for the compressive stress $\sigma_b(z)$ in the barrette's body, the settlement of the barrette, and the shear stresses on the barrette's side surfaces $\tau_a(z)$ and $\tau_b(z)$ are displayed as follows

$$S_b(z) = \sigma_N V_1\,\mathrm{sh}\,\lambda z + \sigma_N V_2\,\mathrm{ch}\,\lambda z;\ \sigma_b(z) = \sigma_N E_b \lambda V_1\,\mathrm{ch}\,\lambda z + \sigma_N E_b \lambda V_2\,\mathrm{sh}\,\lambda z \tag{45}$$

$$\tau_a(z) = \frac{1}{I}\cdot\frac{d\sigma_z}{dz} = \sigma_N\frac{\lambda^2 V_1 E_b}{I}\mathrm{sh}\,\lambda z + \sigma_N\frac{\lambda^2 V_2 E_b}{I}\mathrm{ch}\,\lambda z \tag{46}$$

$$\tau_b(z) = \tau_a(z)\frac{B-b}{A-a} = \frac{B-b}{A-a}\left[\sigma_N\frac{\lambda^2 V_1 E_b}{I}\mathrm{sh}\,\lambda z + \sigma_N\frac{\lambda^2 V_2 E_b}{I}\mathrm{ch}\,\lambda z\right] \tag{47}$$

where σ_N, σ_R, V_1, V_2, V_3, V_4 are described in expressions (35) and (44).

The settlement of the soil cell consisting of the barrette, the raft, the surrounding soil and underlying soil layer is constituted of two terms

$$S_{\text{soil cell}} = S_{\text{barrette-raft-sur.soil system}} + S_{\text{underlying soil layer}} \tag{48}$$

$$S_{\text{soil cell}} = \sigma_R K + \sigma_N V_1\,\mathrm{sh}\,\lambda l + \sigma_N V_2(\mathrm{ch}\,\lambda l - 1) + p m_0 (L-l) \tag{49}$$

Considering example:

$p = 500$ kN; $2a = 1.5$ m; $2b = 3$ m; $2A = 7.5$ m; $2B = 9$ m; $l = 30$ m; $L = 50$ m; $\alpha = \beta = 45^\circ$;

$K_1 = 0.722$; $G = 5000$ kPa; $G_0 = 30000$ kPa; $\omega = 1.22$; $\nu_1 = 0.35$; $\nu_0 = 0.3$; $E_b = 30000$ MPa;

$m_s = 4.615\cdot10^{-4}\ \text{kPa}^{-1}$; $m_0 = 9.905\cdot10^{-6}\ \text{kPa}^{-1}$.

Then we get:

$\sigma_R = 1341.75$ kPa; $\sigma_N = 7267.69$ kPa; $\sigma_S = 16.59$ kPa; $S_0 = 0.022$ m; $S_{30} = 0.026$ m;

$\tau_a(0) = \tau_b(0) = 94.36$ kPa; $S_{\text{soil cell}} = 12.5$ cm.

The result obtained by the numerical method using the PLAXIS 3D in linear formulation with the value of the uniformly distributed load $p = 500$ kPa

IOP Conf. Series: Materials Science and Engineering 1015 (2021) 012047 doi:10.1088/1757-899X/1015/1/012047

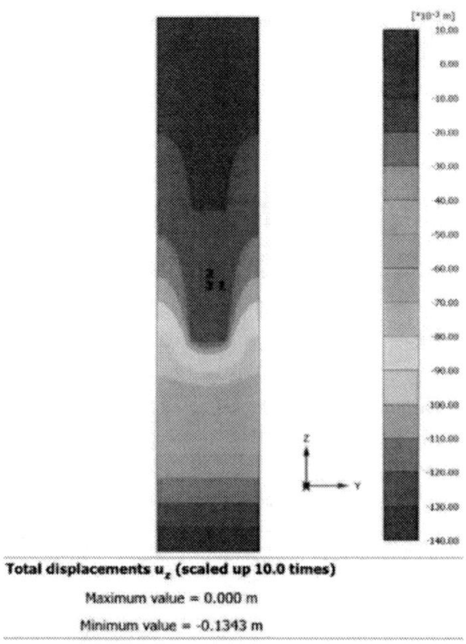

Figure 3. Iso-fields of the settlement of the soil cell consisting of a barrette, a raft, surrounding soil and underlying soil layer.

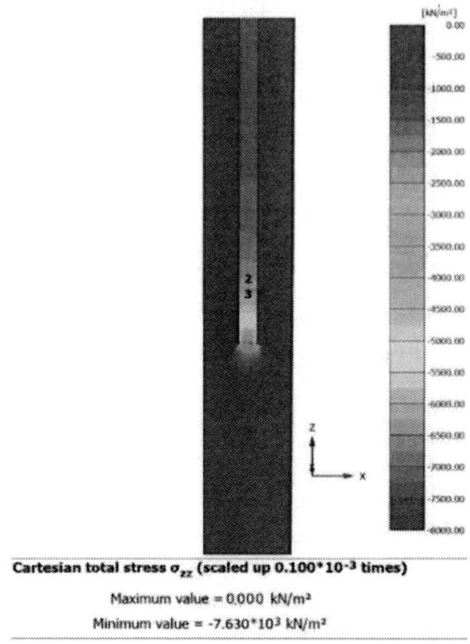

Figure 4. Iso-fields of the vertical stress of the soil cell consisting of a barrette, a raft, surrounding soil and underlying soil layer.

The average values of the stress under the barrette' heel, at the head of the barrette, and the settlement of the soil cell achieved by the numerical method are displayed as follows:

$$\sigma_{\mathrm{R}}^{\mathrm{plaxis}} = 1458\ \mathrm{kPa};\ \sigma_{\mathrm{N}}^{\mathrm{plaxis}} = 7342\ \mathrm{kPa};\ S_{\mathrm{soil\ cell}}^{\mathrm{plaxis}} = 13.43\ \mathrm{cm}.$$

The difference of the values shown above between analytical method and numerical method varies in the range from 1% to 8%.

The dependencies curve of $\sigma_{\mathrm{b}}(z)$, $S_{\mathrm{b}}(z)$, $\tau_{\mathrm{a}}(z)$, and $\tau_{\mathrm{b}}(z)$ on z (m) corresponding to 3 cases of the embedded length of the barrette and the depth of the soil cell 30-40; 40-50; 50-60m, respectively.

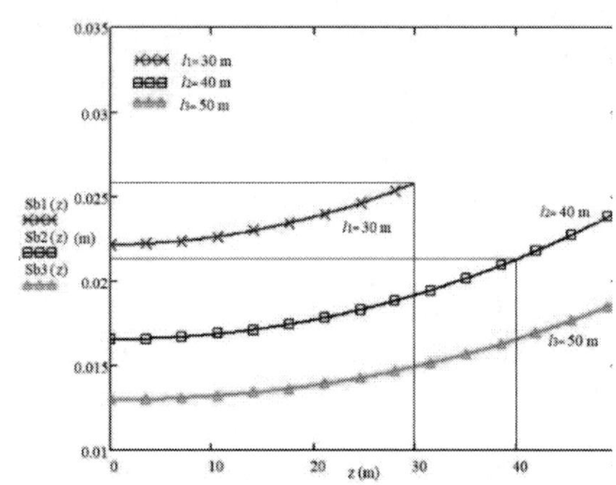

Figure. 5. Dependency curve of the settlement of the compressible barrette S_{b} on depth z (m).

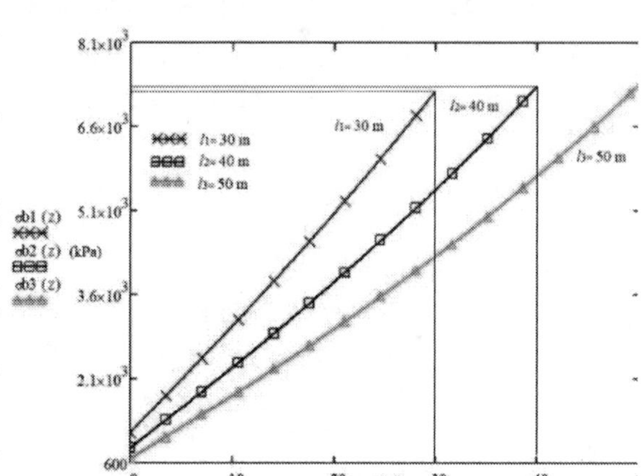

Figure. 6. Dependency curve of σ_{b} in the compressible barrette on depth z (m).

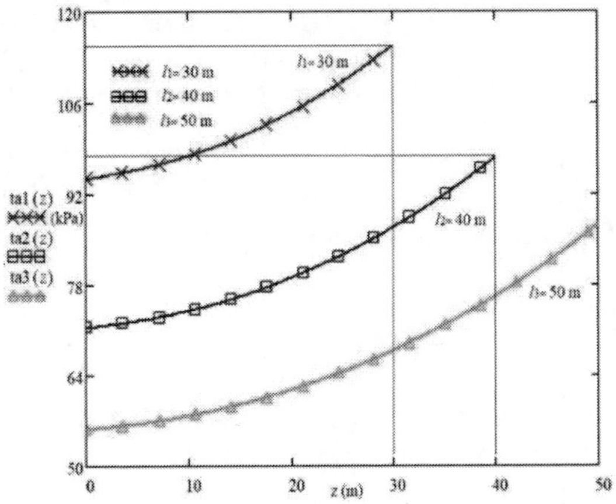

Figure. 7. Dependency curve of $\tau_a(z)$ on the side surface 2a of the compressible barrette on depth z (m).

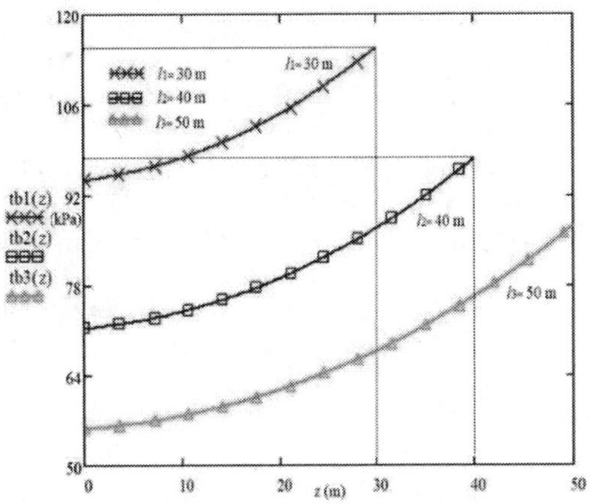

Figure. 8. Dependency curve of $\tau_b(z)$ on the side surface 2b of the compressible barrette on depth z (m).

5. Conclusions

- Upon the interaction of a compressible barrette with surrounding soils, underlying soil layer as well as the raft, a complex stress-strain-state (SSS) is formed, which completely depends on the geometric parameters of the barrette, as well as the physical and mechanical properties of the barrette, the surrounding soils, and the underlying soil layer.
- The solution of the assigned problems in a linear formulation is conducted by establishing and solving a second-order differential equation for the settlement of the barrette's body.
- It is drawn out that the uniformly distributed load applied to the raft transferred from the raft to the side surfaces and the barrette's heel is proportional to their geometric. mechanical parameters and this proportion changes with depth. The proportion stress at the barrette's heel accounts for approximately 15% of the stress at the pile head
- Upon comparison of the result obtained by means of calculating by analytical and numerical method is shown (no more than 8%), and the solution of the assigned problem can be recommended for performing calculation of the settlement and bearing capacity of the barrette-raft foundation.
- In order to increase the proportion of stress at the heel of the barrette and mobilize radically the bearing capacity of the underlying soil layer, the ratio of the barrette's dimension parameters to the soil cell should be adjusted appropriately.

References

[1] Ter-Martirosyan A Z, Ter-Martirosyan Z G, Trinh T V and Luzin 2015 *Vestnik MGSU [Proceedings of Moscow State University of Civil Engineering]* **5** pp 52-60.
[2] Zhijun Zhou, Duanduan Wangm, Lipeng Zhang and Weisi Ma 2015 *Journal of Traffic and Transportation Enginerring* **2** (6) pp 422-8.
[3] Xiag Fei, Yong Lai Zheng, Wei Wei Yang and Shu Xin Deng 2014 *Applied Mechanics and Materials.* **525** pp 541-5.
[4] Ter-Martirosyan Z G, Luzin I and Yakubov M 2019 *E3S Web of Conferences* **97** 04070.
[5] Zaven Ter-Martirosyan 2019 *IOP Conf. Ser.: Master. Sci. Eng.* **661** 012099.
[6] Nguen Z N 2006 *Proceedings of the 4th International scientific conference of young scientists, post-graduate and doctoral students «Building-forming living environment* pp 40-3.

[7] Ter-Martirosyan Z G, Angelo G O and Ter-martirosyan A Z 2019 *IOP Conf. Ser.: Master. Sci. Eng.* 698 022041
[8] Telichenko V I, Ter-Martirosyan A Z and Sidorov V V 2016 *Procedia Engineering* **165** pp 1359-66
[9] Telichenko V I and Ter-Martirosyan Z G 2009 *Vestnik MGSU* **4** 22-7
[10] Ter-Martirosyan A Z and Ter-Martirosyan Z G 2015 *Procedia Engineering* **111** pp 756-62
[11] Ter-Martirosyan A Z Ter-Martirosyan Z G and Sidorov V V 2016 *Soil Mechanics and Foundation Engineering* **3** pp 10-5
[12] Ter-Martirosjan Z G 2006 *Vestnik MGSU* **1** pp 38-49
[13] Ter-Martirosjan Z G and Nguen Z N 2007 *Vestnik grazhdanskih inzhenerov SPbGASU* **1**(10) pp 52-5
[14] Ter-Martirosyan Z G 2009 *Soil Mechanics* (Moscow, ASV Moscow).
[15] Ter-Martirosyan Z G Ter-Martirosyan A Z and Sidorov V V 2013 *Proc of the 18th Int conf on soil mechanics and geotechnical engineering* pp 2881-4
[16] Ter-Martirosyan A Z, Ter-Martirosyan Z G and Sidorov V V *Soil Mechanics and Foundation Engineering* **53** pp 166-73
[17] Ter-Martirosjan Z G Ter-Martirosjan A Z and Sidorov V V 2014 *Estestvennye i tehnicheskie nauki* **11-12** (78) pp 372-6
[18] Ter-Martirosyan Z G and Dam Huu Hung 2019 *J. Phys.: Conf. Ser.* 1425 012056

IOP Conf. Series: Materials Science and Engineering 1015 (2021) 012048 doi:10.1088/1757-899X/1015/1/012048

Predicting the settlement of foundation bases taking into account induced anisotropy

Z G Ter-Martirosyan, A Z Ter-Martirosyan and I N Luzin

Research and Educational Center «Geotechnics», Moscow State University of Civil Engineering (National Research University), 26 Yaroslavskoe highway, Moscow, Russia

E-mail assress: LuzinIN@mgsu.ru

Abstract. This article is devoted to the quantitative assessment of the stress-strain state of finite width foundations, taking into account their induced anisotropy. The article discusses the stress-strain state of a homogeneous base, which acquires the property of deformation inhomogeneity $E(x,y,z)$ under the influence of an external load, due in the general case to a nonlinear dependence of stresses and deformations of the form $\varepsilon_m = \sigma_m/K(\sigma_m, \tau_i)$ and $\gamma_i = \tau_i / G(\sigma_m, \tau_i)$, where ε_m and γ_i are volume and shear deformations, and σ_m are τ_i the average stress and intensity of shear stresses at each point of the base, K and G are modules of volume and shear deformation. As an example, we consider the stress-strain state of the foundation under the influence of a distributed load $p=const$ along the strip $b = 2a$ (plane deformation). To calculate the settlement with a nonlinear dependences $\varepsilon_m - \sigma_m$ and $\gamma_i - \tau_i$, it is proposed to use the experimental curves $K - \sigma_m$ and $G - \tau_i$. For the case of the absence of deformation inhomogeneity, when K and G are constant, an analytical solution of the problem is given for the vertical line $x=0$ under conditions of compression. For this case, a closed solution is obtained, according to which the settlement depends nonlinearly on the width of the loading band $b = 2a$.

1. Introduction

When calculating the settlement of foundations, it is often necessary to take into account the natural deformation inhomogeneity. This heterogeneity has been forming for a long time. At the same time, deformation heterogeneity [1,2] in the base soils can be observed in a homogeneous base under the action of a local load as a result of stress inhomogeneity $\sigma_{ij}(x,y,z)$. In this case, the deformation parameters K and G will depend on the arising inhomogeneous stress state - from σ_m and τ_i [3]. figure 1 shows a general view of the dependencies $\varepsilon_m - \sigma_m$ and $\gamma_i - \tau_i$.

Analysis of figure 1 shows that the deformation parameters significantly depend on the parameters of the induced inhomogeneity. In this article, an attempt is made to take into account the deformation inhomogeneity to determine the components of deformations. In this case, the nonlinear properties of soils are described based on known models.

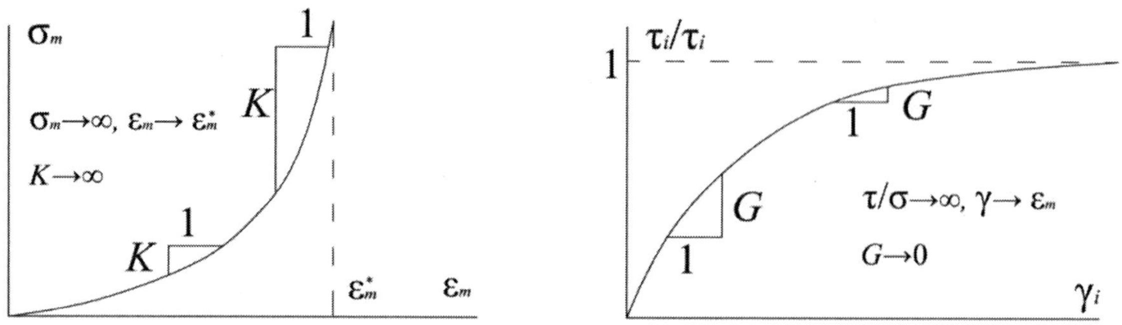

Figure 1. General view of the dependencies $\varepsilon_m - \sigma_m$ and $\gamma_i - \tau_i$.

2. Initial equations

To determine nonlinear volumetric deformations, the following dependence was taken [4,5]

$$\varepsilon_m(\sigma_m) = \varepsilon_m^*(1 - e^{-\alpha\sigma_m}) \tag{1}$$

where ε_m^* - volumetric deformation limit, ε_m - average volumetric deformation, α - nonlinearity parameter.

The secant deformation modulus can be determined by the formula

$$\frac{1}{K_s} = \frac{\varepsilon_m^*}{\sigma_m}(1 - e^{-\alpha\sigma_m}) \tag{2}$$

and at $\sigma_m \to \infty$ will be $K \to \infty$.

The tangent modulus of deformation can be obtained by differentiating equation (1) with respect to σ_m. Then we get

$$K_t = \frac{e^{-\alpha\sigma_m}}{\varepsilon_m^*\alpha} \tag{3}$$

It is accepted that the elastic-plastic properties of the soil during shear are described by the formula of S.P. Tymoshenko [6,7]

$$\gamma_i = \frac{\tau_i}{G^e}\frac{\tau_i^*}{\tau_i^* - \tau_i} \tag{4}$$

where γ_i - angular deformation, τ_i and τ_i^* - current and limit values of shear stresses and

$$\tau_i^* = \sigma_m \cdot tg\varphi_i + c_i \tag{5}$$

equation (4) implies that

$$G_s = G^e\left(1 - \frac{\tau_i}{\tau_i^*}\right) \tag{6}$$

From this it follows that for $\tau_i \to \tau_i^*$ is $G \to 0$ and $\gamma_i \to \infty$. Tangent shear modulus can be obtained by differentiating equation (4) with respect to τ_i. Then we get

$$G_t = G^e\left(1 - \tau_i / \tau_i^*\right)^2 \tag{7}$$

To describe the nonlinear relationship of characteristics, we use the Henki system of physical equations [8,9]

$$\begin{aligned} \varepsilon_x &= \chi(\sigma_x - \sigma_m) + \chi^*\sigma_m; \quad \gamma_{xy} = 2\chi\tau_{xy} \\ \varepsilon_y &= \chi(\sigma_y - \sigma_m) + \chi^*\sigma_m; \quad \gamma_{yz} = 2\chi\tau_{yz} \end{aligned} \tag{8}$$

$$\varepsilon_x = \chi(\sigma_z - \sigma_m) + \chi^* \sigma_m; \quad \gamma_{zx} = 2\chi\tau_{zx}$$

where

$$\chi = \frac{\gamma_i}{2\tau_i} = \frac{f(\tau_i, \sigma_m, \mu_\sigma)}{2\tau_i}; \chi^* = \frac{\varepsilon_m}{\sigma_m} = \frac{f^*(\tau_i, \sigma_m, \mu_\sigma)}{2\tau_i},$$

μ_σ - Lode-Nadai coefficient.

Note that with a linear relationship between stresses and strains, the parameters χ and χ^* take the form

$$\chi = \frac{\gamma_i}{2\tau_i} = \frac{1}{2G}; \chi^* = \frac{\varepsilon_m}{\sigma_m} = \frac{1}{K} \tag{9}$$

A distinctive feature of the Henki system of equations is that they allow representing linear deformation as a sum of volume and shear deformations.

To describe the inhomogeneous stress state of the foundation under the action of a local load, we use the Flamant formulas. For a distributed load $p=const$ across a strip of width $b=2a$, the Flamant equation system is as follows

$$\sigma_x = \frac{p}{\pi}\left[\text{arctg}\frac{a-x}{z} + \text{arctg}\frac{a+x}{z}\right] + \frac{2apz\left(x^2 - z^2 - a^2\right)}{\pi\left[\left(x^2 + z^2 - a^2\right)^2 + 4a^2z^2\right]}$$

$$\sigma_z = \frac{p}{\pi}\text{arctg}\left[arctg\frac{a-x}{z} + \text{arctg}\frac{a+x}{z}\right] - \frac{2apz\left(x^2 - z^2 - a^2\right)}{\pi\left[\left(x^2 + z^2 - a^2\right)^2 + 4a^2z^2\right]} \tag{10}$$

$$\sigma_m = \frac{2p(1+\nu)}{3\pi}\left[\text{arctg}\frac{a-x}{z} + \text{arctg}\frac{a+x}{z}\right], \quad \tau_{xz} = \frac{4axz^2}{\pi\left[(x-a)^2 + z^2\right]\cdot\left[(x+a)^2 + z^2\right]}$$

It should also be noted that under the conditions of the plane problem

$$\tau_i = (\sigma_1 - \sigma_3)/\sqrt{3}; \; \sigma_1 - \sigma_3 = 2\sqrt{\frac{\sigma_x - \sigma_z}{2} + \tau_{xz}^2}; \; \gamma_i = 2(\varepsilon_1 + \varepsilon_3)/\sqrt{3} \tag{11}$$

From the analysis of Henki equations taking into account equation (1) and (6) it is follows that the components of linear deformation $\varepsilon_x, \varepsilon_y$ and ε_z are shear and volumetric deformations. Indeed, if we introduce values χ and χ^* into the system of Henki's equations taking into account equations (7) and (11) for ε_z we obtain the following equation

$$\varepsilon_z = \frac{\sigma_z - \sigma_m}{G^e\left(1 - \tau_i / \tau_i^*\right)} + \varepsilon_m^*(1 - e^{-\alpha\sigma_m}) \tag{12}$$

From the analysis of equation (12) it follows that the components of linear deformation progress in different ways with an increase in the acting load p. The growth of deformation $\varepsilon_{z\gamma}$ is parabolic and tends to infinity, the development of deformation ε_{zv} is damped. The total deformation has a double curvature and tends to infinity at $\tau_i \to \tau_i^*$. Equation (12) taking into account equations (10) can be used for obtaining of color shadings of deformation and shear modules K and G, volumetric ε_{zv}, shear $\varepsilon_{z\gamma}$ and total ε_z deformations. The figures 2-6 show the results of solving equations (3), (7), (12) in PTC Mathcad.

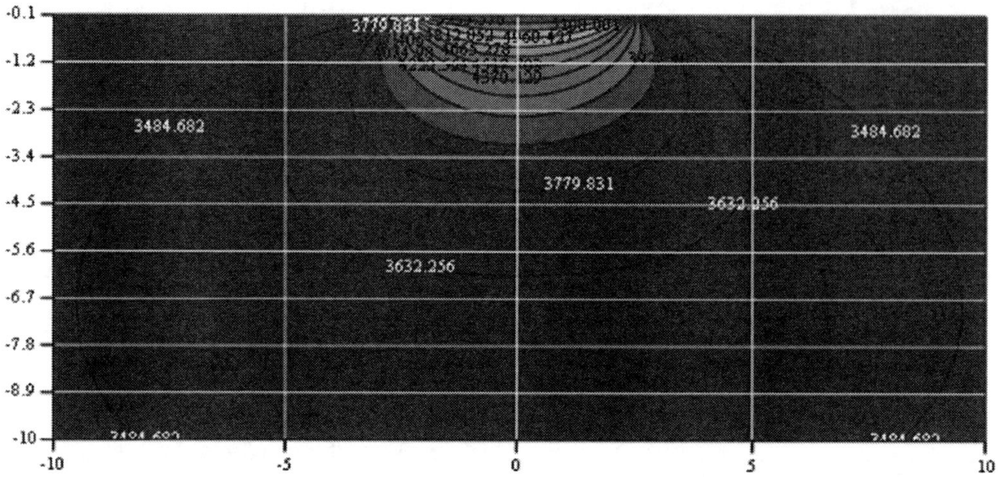

Figure 2. Color shadings of K_t by the formula (3).

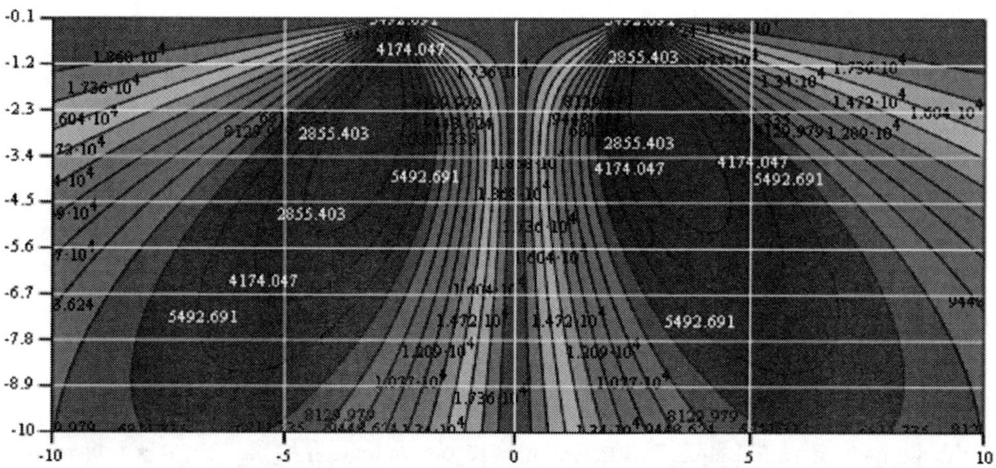

Figure 3. Color shadings of G_t by the formula (7).

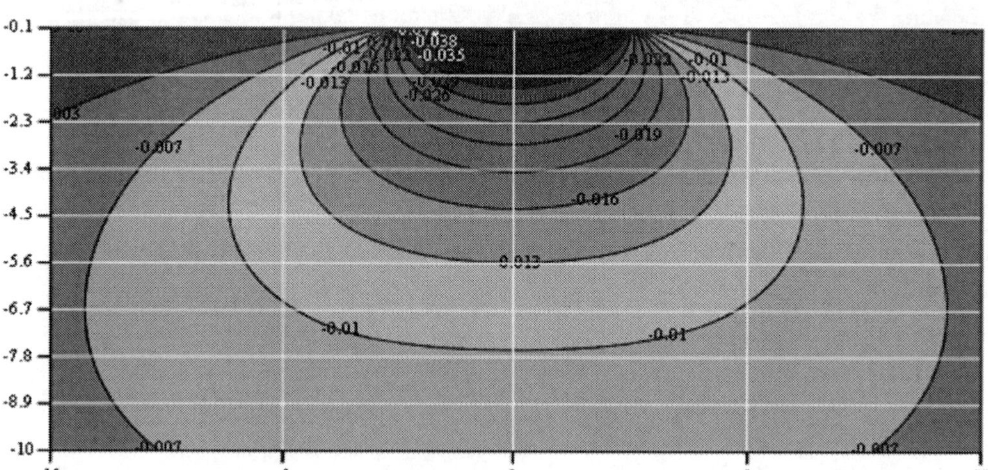

Figure 4. Color shadings of ε_{zv} by the formula (12).

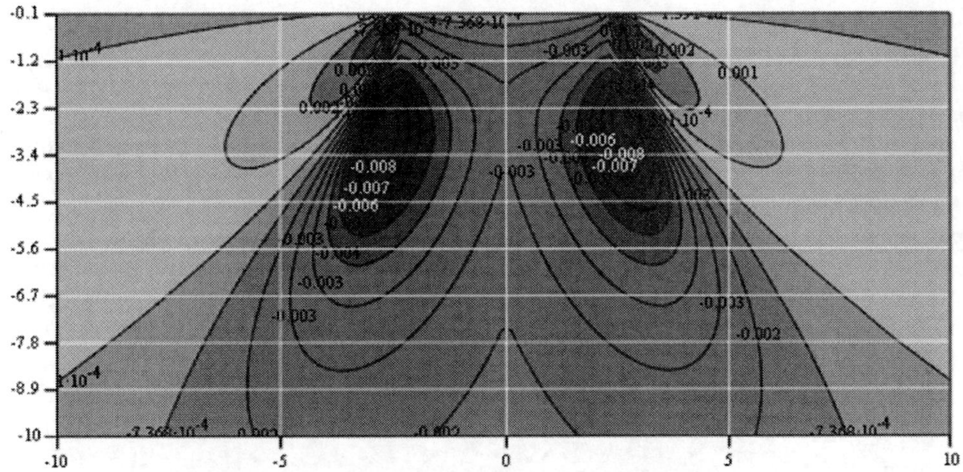

Figure 5. Color shadings of $\varepsilon_{z\gamma}$ by the formula (12).

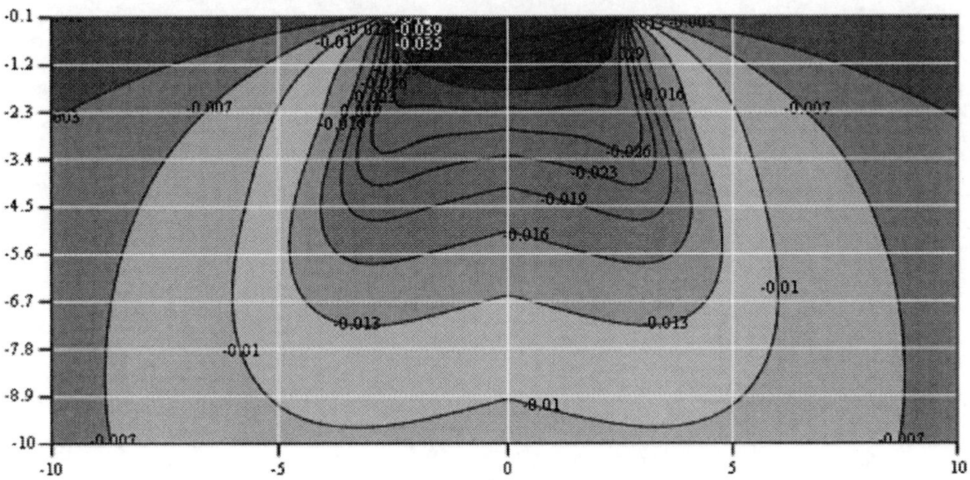

Figure 6. Color shadings of total ε_z by the formula (12).

3. Conclusions and discussion

This article presents the results of the analytical solution of the problem of deformation of the base taking into account the induced inhomogeneity. The obtained equations were used to construct color schemes of the deformation properties of soils and volumetric deformations, taking into account the inhomogeneous stress state of the soil mass. The results obtained can be used to refine the work of the foundations when a load is applied to them from foundations of a finite width.

References

[1] Chaun Gu, Jun Wang, Yuanqiang Cai, Lei Sun, Peng Wang and QuanYang Dong 2016 Deformation characteristics of overconsolidated clay sheared under constant and variable confining pressure *Soil and Foundations* **56** (3) pp 427–39

[2] Jocković S and Vukićević M 2017 Bounding surface model for overconsolidated clays with new state parameter formulation of hardening rule *Computers and Geotechnics* **83** pp 16–29

[3] Boldyrev G G, Arefiev D V and Gordeyev A V 2010 Determination of the deformation characteristics of soils by various laboratory methods (Opredelenie deformatsionnykh kharakteristik gruntov razlichnymi laboratornymi metodami) *Inzhenernye izyskaniya* **8** pp 16-23
[4] Ter-Martirosyan Z G 2009 *Soil mechanics (Mekhanika gruntov)* (Moscow: ASV)
[5] Kachanov L N 1969 *Basics of the plasticity theory (Osnovy teorii plastichnosti)* (Moscow: Nauka)
[6] Nadai A 1969 *Plasticity and destruction of solid bodies (Plastichnost i razrusheniye tverdykh tel)* (Moscow: Nauka)
[7] Timoshenko S P 1975 *Theory of elasticity (Teoriya elastichnosti)* (Moscow: Nauka)
[8] Vyalov S S 1978 *Rheological basis of soil mechanics (Reologicheskie osnovy mekhaniki gruntov)* (Moscow: Vyshaya shkola)
[9] Trofimov V T 2005 *Soil science (Gruntovedenie)* (Moscow: publishing of Moscow State University)

Numerical modeling of freezing of structural objects in transition zones of railways

S Hodás and A Pultznerová

University of Žilina, Faculty of Civil Engineering, Department of Railway Engineering, Univerzitná 8215/1, SK-010 26, Žilina, Slovak Republic

E-mail address: stanislav.hodas@uniza.sk

Abstract. The structure of the transition zone of the railway formation is composed of layered materials built into it, which must be resistant to adverse winter climatic conditions characteristic for the region. Among the most dangerous factors is the freezing of these materials and therefore the increased resistance to freezing in winter period is required. By this damage, change of their internal structures and material-technical characteristics can occur. For example, as a consequence of freezing, the structure may be affected by the consequent reduction of resistance of these layers to traffic load (axle loads, deformation modules of the individual layers, etc.). Special construction structures are also designed in the track and the paper is dedicated to transition zone, i.e. reinforced concrete tub as an intermediate element in changing the types of railway formation (the area between the slab track and the continuous track bed). In the experiments using the numerical models, the depth of freezing of the structural objects in the transition zones of the tracks is analyzed, i.e. the zero isotherm height position for 0 °C in the railway structure is determined and compared with respect to the bottom surface of the concrete tub as well as the FPL (frost protection layer) in the continuous track bed, subsequently the design of the objects may be modified.

1. Introduction

The transition between a slab track and a rail bed structure is one of the critical points on a railway track. In this section of a railway, a reinforced concrete tub is designed for the transition between them, which is intended to damp the dynamic effects arising from the pass of a train on this section. Within the research, the freezing of this transition zone (TZ) has been verified, i.e. the zero isotherm in the individual structural layers was monitored (curve or surface with a temperature of 0 °C). Freezing adversely affects to the materials built in the railway track and the layers above the zero isotherm must be frost resistant.

The paper describes research for climatic temperature load from frost at freezing index approx. I_F = -600 °C·day in [1], [2], using two thicknesses of protective gravel layer h_{FPL} = 450 mm and h_{FPL} = 600 mm for continuous rail bed behind the reinforced concrete tub. The I_F is the sum of the negative daily temperatures during the winter period, each period begins when the first four days have minus temperatures.

Photographs of the realized construction object of the transition zone on the double-track of the main corridor line Bratislava – Žilina in Slovakia for the speed V = 160 km/h are shown in figure 1.

Numerical modeling is beneficial for designers and achieves qualified results. Modeling of the transition zone TZ is realized by the Finite Element Method calculation [3].

Figure 1. Transition zone on double track $V = 160$ km/h.

2. Model of transition zone and climatic load

The numerical model must exactly correspond with the designed structure of the reinforced concrete tub and to the design of the track formation before and after this object (slab track and continuous track bed). The material-technical characteristics of structural layers verified in laboratory tests or standard characteristics are entered to the model.

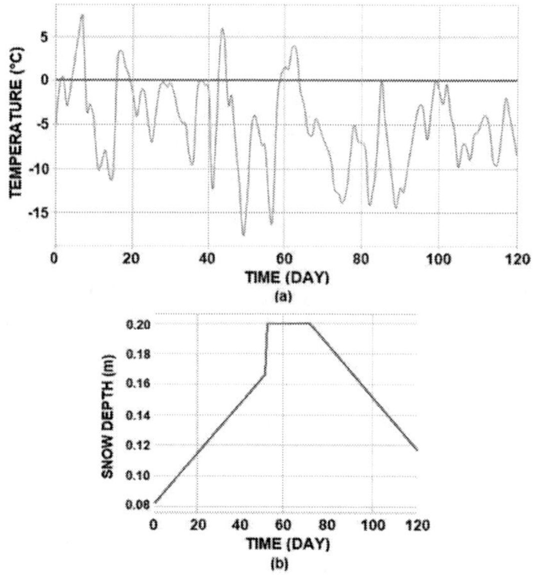

Figure 2. Climatic load: (a) mean daytime temperatures T_m and (b) snow load h_{snow}.

An important element is the initial setting of the initial temperatures in the particular structural parts of the TZ model and the track (also in the substructure). Numerical models of the transition zones are loaded by the climate regime in a 120 day freezing period with freezing index of I_F = -600 °C·day (winter freezing period), which is characterized by mean daily air temperatures T_m

(average measured 4 times over 24 hours). These railway track models were frost-loaded for 4 months, i.e. in the winter period from November to March of the consecutive year (in Central Europe approx. 15.11. - 15.03.), according to figure 2a. During this research, a frosty winter period with several freezing periods for the climate load of the model was searched, from a practical point of view, the period with I_F = -600 °C·day (town of Poprad in the High Tatras region) was selected.

Input technical characteristics of structural materials of particular layers of the railway track models in these experiments for the transition zone are considered according to table 1 [3] where the values are for thermal conductivity – λ, soil dry density – ρ, specific heat capacity – c and material humidity – w.

Table 1. Technical characteristics of structural layers.

Material	λ J/(day·m·C)	λ /(24×60×60) W/(m·K)	ρ kg/m^3	c J/(kg·C)	w %
Railway ballast	60000	0.69	1908	980	2
Reinforced concrete	136000	1.57	2400	1020	2
Subgrade	130000	1.50	2081	1585	17
Subsoil	105000	1.22	1646	1495	18

3. Numerical modeling of structural object freezing

The primary task is the preparation of a numerical model (design of a new or already existing transition zone of the track as part of testing). Numerical models for single and double track were created in research experiments. The paper and the figures show the model of the double-track of the main corridor line with the purpose of continuous presentation (calculations were done on both and their results are shown in table 2 to table 7).

The model must be created in 3D Environment Models, as it is done in our research using [3], in order the temperature transfer occurred throughout the 3D spatial model. In case of a railway track without objects, analysis of the characteristic cross-sections in 2D is sufficient (the longitudinal axis behaves in the same way). However, the transition zone is a complete three-dimensional structural object with spatial blocks, and therefore processing as a 3D model is advantageous as the frost penetrates all parts of the structural object in the *XYZ* space.

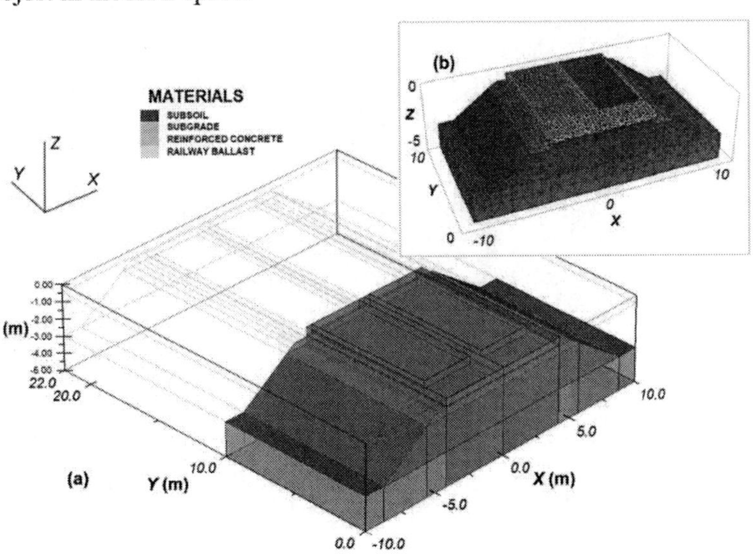

Figure 3. Numerical model of the structural object TZ: (a) definition of the surfaces and (b) 3D parts.

IOP Conf. Series: Materials Science and Engineering 1015 (2021) 012049 doi:10.1088/1757-899X/1015/1/012049

The numerical model of the transition zone with the design of particular spatial blocks and structural layers is shown in figure 3 (half of the length of the object was used for the calculation), where the finite element calculation network layout is also used. The length of the transition zone (the concrete tub is half of the length of the structure, but the model includes only for the first 10 m of the structure) is $l_{TZ} = V/2$ (m/s), for example, for the velocity on the railway corridor $V = 160$ km/h the length of the object is $d_{TZ} = 22.22$ m.

The design of the structural object of the transition zone is a complex issuer. In these cases, preparation for calculation of temperatures that are dependent on external influences is necessary, i.e. how the model was exposed to the freezing climatic exterior environment (the top surfaces in figure 3 that are affected by the climatic load). All surfaces and edges of 3D objects that make up the entire structural object must be defined.

3.1. Evaluation of freezing in the longitudinal sections YZ

This section is a longitudinal section of the structural element in 3D, as shown in figure 4a at $X = +2.10$ m, which is shown in the section in the track axis of a double-track line. A detailed graphical evaluation is shown in figure 4b (numerically is referred in the model databases) where the freezing of the zero isotherm for the 120th day of the experiment (maximum values) is marked. The displayed objects are dark, as the finite element net is dense, for color details it is necessary to use its close-up zoom to enlarge the profile. The particular temperature values are also color-coded in the legend.

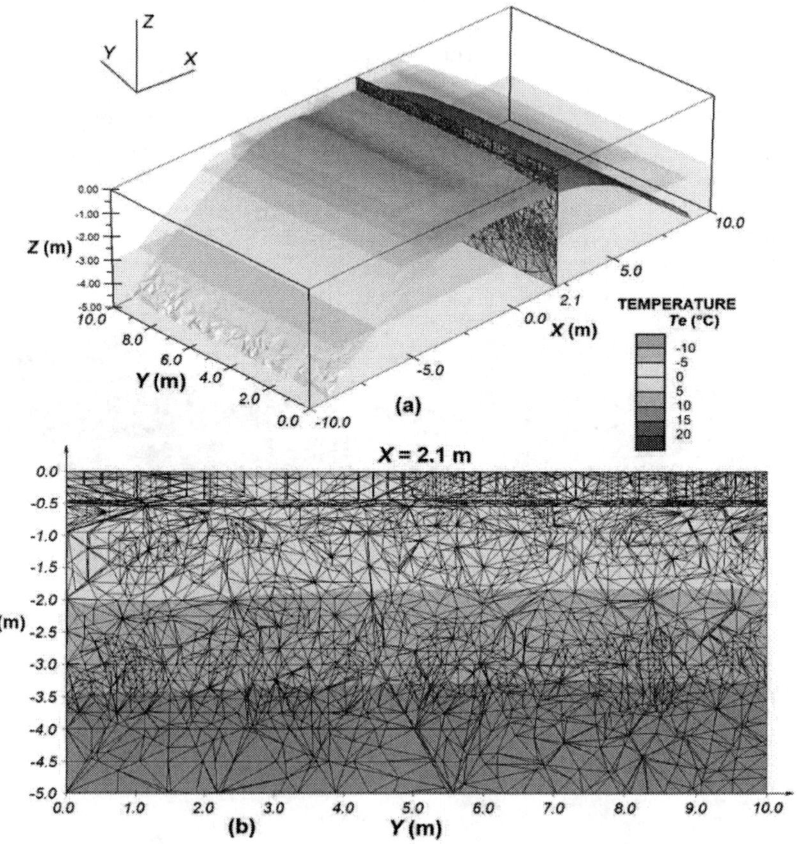

Figure 4. Temperature course in the *YZ* axis – 120th day of modeling: (a) 3D transition zone and (b) detail of longitudinal section.

The longitudinal section along the track axis with the temperature transition through the structure shows the parts where the reinforced concrete threshold of the tub (stationing 0.00 to 2.00 m) is located, the iron-concrete tub with rail bed (stationing 2.00 to 22.00 m) and two different railway formations (before stationing 0.00 a fixed track is built and after stationing 22.00 m is a continuous rail bed).

3.2. Evaluation of freezing in the longitudinal sections XZ

Sections *XZ* of the numerical model show the temperature transition and the zero isotherm in cross-sections, this is the most effective view of this graphical evaluation of the calculation results (figure 5 and figure 6). When interpreting the graphical results, it is evident that the zero isotherm is found at the depth h_{Fi} in table 2 and table 3 in the track axis for the 120th day of the numerical experiment. In figure 5b it is shown for stationing $Y = +6.00$ m, the result values are presented in table 2 and a section through the concrete threshold $Y = +1.00$ m is shown in figure 6.

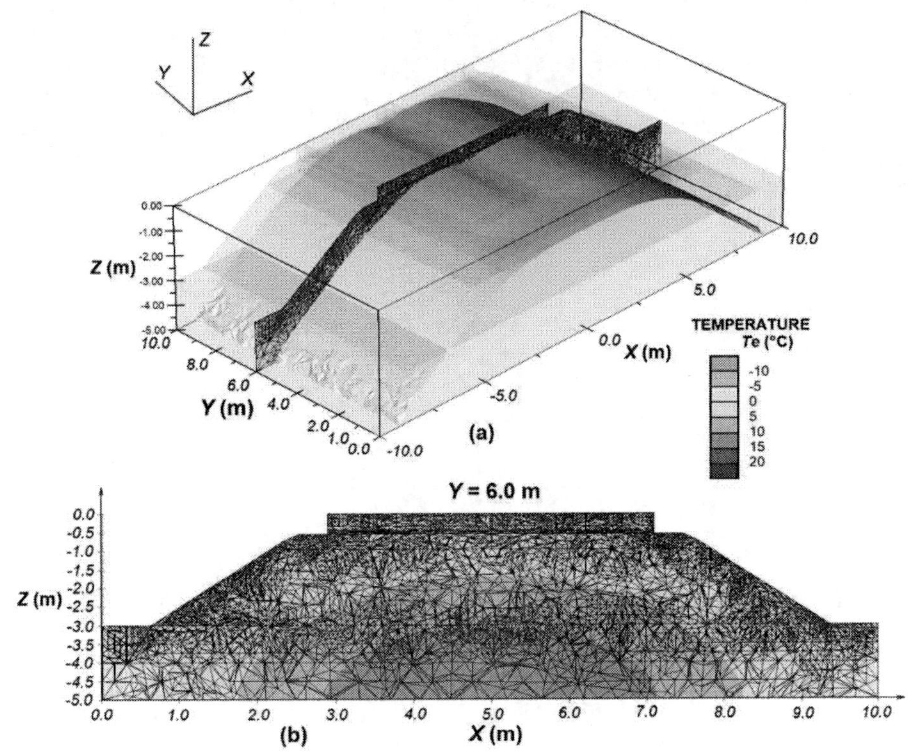

Figure 5. Temperature course *XZ* – 120th day of modeling: (a) 3D transition zone and (b) cross-section of the tub with railway bed.

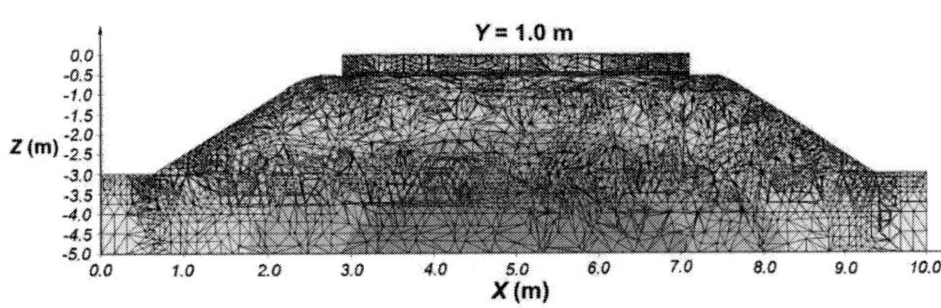

Figure 6. Temperature course *XZ*: cross-section of the reinforced concrete threshold.

IOP Conf. Series: Materials Science and Engineering 1015 (2021) 012049 doi:10.1088/1757-899X/1015/1/012049

The structural object can also be cut in the horizontal *XY* direction, for example in the longitudinal walls of the concrete tub, but the most important are the *XZ* cross sections.

The graphical and numerical results of the numerical model of the transition structure are processed in 60-minute intervals (or 30 minutes); 3D (or 2D) animation for each particular processing and model recalculation has been created.

4. Discussion – final evaluation of the experiments

Result values of the freezing experiment of the transition zone show that at the freezing index $I_F = -600$ °C·day (figure 5, table 2 and table 3) in the part of the concrete tub only with a railway bed, freezing under the concrete tub in the track axis does not occur (see cross section in $Y = +6.00$ m and longitudinal track section in $X = \pm\ 2.10$ m). Freezing in a part of the reinforced concrete threshold approached to the permissible limit of -0.95 m at the bottom of this threshold. However, below the marginal rail (L or R), the zero isotherm already exceeds this limit and interferes into one structural layer of the embankment, i.e. under the transitional reinforced concrete tub. Either it is necessary to design a new protective gravel sand layer FPL (as in the continuous track bed), or we will consider the structural layer partially frosty, but increased number of repairs and interventions in the geometry of track axes during track operation in the transition zone will be expected. The above applies to snow thickness $h_{snow} = 10$ to 20 cm at temperature conversion factor $n_F = 0.85$. For the marginal rails of both tracks of the double-track line (or both single-track rails L – left and R – right), there is also a tendency of freezing in the direction from the side slope as well as the margins of the reinforced concrete tub (figure 5, table 2).

Table 2. Depth of freezing h_F – cross sections *XZ*.

Stationing/cross sections (m)	h_F (m) L – rail	railway axis	R – rail
1.00	-1.10 ×	-0.97 ✓×	-0.93 ✓
6.00	-0.80 ✓	-0.62 ✓	-0.55 ✓
Bottom surface of the tub	-0.95	-0.95	-0.95

Table 3. Depth of freezing h_F – longitudinal profile *YZ* (track axis $X = \pm\ 2.10$ m).

Stationing (m)	h_F (m) railway axis
0.00	-0.95 ✓
1.00	-0.96 ✓×
6.00	-0.64 ✓
10.00 (detto 22.00)	-0.64 ✓
Bottom surface of the tub	-0.95

To supplement the data, experiments will also be described on the railway track with a continuous track bed for double-track and single-track track within freezing for the type of railway formation in the embankment with FPL with the protective layer with its thicknesses of $h_{FPL} = 450$ mm and $h_{FPL} = 600$ mm. The double-track freezes to depths according to the zero isotherm in table 4 and table 5, the climatic load is the same as in the transition zone at the freezing index $I_F = -600$ °C·day and the detailed division is 0.00 to 0.30 m depending on the snow thickness. At $h_{snow} = 0.00$ m we are talking about hard frost without snow cover with considerable freezing of layers; this is low probable situation. From a practical point of view, 0.10 to 0.20 m of snow is considered (the trains adjust the height of the running track by their own run), snow is considered an excellent thermal insulator.

Table 4. Depth of freezing h_F – railway bed (double track).

Track/conditions	h_F (m) L – rail	railway axis	R – rail
Double track h_{FPL} = 450 (mm) I_F = -600 (°C·day)· n_F = 0.85	-1.46 × -1.08 ✓ -0.55 ✓ -0.38 ✓	-1.25 × -0.78 ✓ -0.51 ✓ -0.32 ✓	-1.23 × -0.70 ✓ -0.45 ✓ -0.31 ✓
Bottom surface (FPL)	-1.15	-1.08	-1.01
Double track h_{FPL} = 600 (mm) I_F = -600 (°C·day) n_F = 0.85	-1.52 × -1.12 ✓ -0.55 ✓ -0.38 ✓	-1.35 × -0.89 ✓ -0.49 ✓ -0.31 ✓	-1.25 × -0.76 ✓ -0.44 ✓ -0.30 ✓
Bottom surface (FPL)	-1.28	-1.21	-1.14

Zero isotherm depths are investigated in the track axis and below particular track rails (L – left and R – right); pay attention to the different transverse slope of the lower surface of the FPL protective layer.

Another experiment of the numerical modeling was also solved on a single track (results are shown in table 6 and table 7). During the experiments it was found that the accumulated heat in the double-track railway body from the previous period (summer, autumn) is higher, because in the railway formation there is a larger volume of material [5], [6] and there is also somewhat less freezing depths that show table 4 to table 7 (values need to be compared only for the same thickness of h_{FPL} used). The values in the tables marked ✓ are within the limit and the values × exceed it; the required limit is the lower surface of the transition zone or protective layer (the last rows in the tables) to protect the other layers from frost destruction [7] and [8].

Table 5. Depth of freezing h_F– railway bed (single-track).

Track/conditions	h_F (m) L – rail	railway axis	R – rail
Single-track h_{FPL} = 450 (mm) I_F = -600 (°C·day) n_F = 0.85	-1.47 × -1.20 × -0.72 ✓ -0.44 ✓	-1.34 × -1.12 × -0.60 ✓ -0.38 ✓	detto L=R
Bottom surface (FPL)	-1.06	-1.00	-1.06
Single-track h_{FPL} = 600 (mm) I_F = -600 (°C·day) n_F = 0.85	-1.64 × -1.28 × -0.81 ✓ -0.46 ✓	-1.59 × -1.22 × -0.65 ✓ -0.41 ✓	detto L=R
Bottom surface (FPL)	-1.22	-1.15	-1.22

5. Conclusions

Civil engineers design various structural objects and their structures must meet different criteria required by standards. The railway formation, in our case the construction of transition zones (structure of a reinforced concrete tub with a rail bed), must comply with the transmission of dynamic loads (various shocks and vibrations frequencies), transmission of the load from the railway wheel to the particular structural layers (deformation modules) and it must not be forgotten of the resistance to frost penetration into the parts of this structure [9] and [1].

Repeated freezing of the structural elements under freezing climatic conditions (even several freezing periods during one winter) may disturb the material structure and reduce other structural features, such as the abovementioned load-bearing capacity, i.e. reduced modulus of deformation (different module standard values for individual structural layers). Thereafter, repeated maintenance of the spatial geometry of the track axis on the objects and its adjacent parts will be increased.

During experiments – numerical [7] and [5], but also on models in [10] and [11], the greatest depth of freezing is proved in the threshold area of the transition zone where the reinforcing concrete block is located. Here it seems necessary to design a protective layer of FPL. In the part of the concrete tub, where there is also a gravel railway bed, the structure is sufficient for the depth of freezing. The FPL layer should be with higher height at the outer rail (higher freezing from the side of the railway formation), for example, the design of the roof transverse slope of the lower surface of the FPL 4 to 5 %. Similarly as it is for the track with a continuous railway bed behind the transmission zone.

From the results of modelling track with only continuous track bed and FPL protective layer with the snow layer h_{snow} = 0.10 to 0.30 m it satisfies for double track until the snow layer falls below 0.10 m. At a snow thickness of 0.00 m, frost enters the structure clearly because there is no snow, which is a good thermal insulator (it would be a freezing winter without a snow cover).

The single track with the same conditions as previous experiments is worse. The design at the freezing index I_F = -600 ° C·day does not satisfy even with 0.10 m of snow. The experiments have clearly shown that the larger volume of material in the railway formation is more resistant to frost; the formation has more accumulated heat from the previous period. The double track is better from this point of view (solved with higher embankment, [6].

From this point of view, the initial temperatures in the layers and railway substructure are also important at the beginning of the experiments (on day 0), where temperatures of 10 to 15 ° C in the particular layers of the structure were considered (from examination of real models of the [11]). But if the temperature of winter starts for example [10] at a level of approx. 6 to 10 °C, mentioned zero isotherm depths will be deeper by approx. 4 to 8 cm at this freezing index on realized models of research made in previous years at the Department of Railway Engineering, Faculty of Civil Engineering, University of Žilina by authors [1], [11] and [6].

Finally, we can conclude, that building of the real models at the scale 1:1 has made possible to verify numerical modelling versus the true value of these models, and so in recent years, numerical models have been tested without laborious and costly constructions of models.

Acknowledgments

The presented parts of the paper were created within the framework of the research activities VEGA 1/0084/20 [4] by the Department of Railway Engineering at the Faculty of Civil Engineering of University of Žilina (DRE-FCE-UNIZA).

References

[1] Ižvolt L, Dobeš P and Hodás S 2018 Impact of the method of rail track routing on the thermal regime of subgrade structure - numerical modelling of non-traffic load, *Int. J. of Transport Development and Integration* **2** (3) pp 250-7

[2] Hodás S, Ižvolt L and Dobeš P 2016 Preliminary results and conclusions from mathematical modelling of thermal regime of railway track structure, *Int. J. of Computational Methods and Experimental Measurements* **4** (2) pp 69-79

[3] SV-HEAT 2D/3D 2020 Finite element - freeze/thaw modeling, *FlexPDE, SVOFFICE - Geotechnical modeling suite*, SVOFFICE™5/GE and SV-HEAT™GE, (SoilVision systems Ltd Canada)

[4] Numerical and experimental analysis of transition zones of objects of structures of railway superstructures and objects of formation substructure 2020 *Scientific research* (Dept. of Railway Engineering Faculty of Civil Engineering University of Žilina Slovakia)

[5] Hodás S and Pultznerová A 2017 Modelling of railway track temperature regime with real heat-technical values for different climatic characteristics. *Civil and Environmental Eng. Sci. Technical J.* **13** (2) pp 134-42

[6] Hodás S and Pultznerová A 2019 Freezing of the subballast layers of the railway formation – high embankment and double track. *Civil and Environmental Engineering Scientific Technical J.* **15** (1) pp 5-12

[7] Chen Ch and McDowell G R 2014 An investigation of the dynamic behaviour of track transition zones using discrete element modelling *Proc. of the Institution of Mechanical Engineers, Part F: J.of Rail and Rapid Transit* **230** (1) pp 117-128

[8] Zuada Coelho B and Hicks M A 2016 Numerical analysis of railway transition zones in soft soil. *Proc. of the Institution of Mech. Engineers, Part F: J. of Rail and Rapid Transit* **230** (6) pp 1601-13

[9] Wang H, Markine V and Liu X 2018 Experimental analysis of railway track settlement in transition zones. *Proceedings of the Institution of Mechanical Engineers, Part F: Journal of Rail and Rapid Transit* **232** (6) pp 1774-89

[10] Heydari-Noghabi H, Zakeri J A and Esmaeili M 2018 Field study using additional rails and an approach slab as a transition zone from slab track to the ballasted track *Proc. of the Institution of Mech. Engineers, Part F: J. of Rail and Rapid Transit* **232** (4) pp 970-8

[11] Hodás S and Ižvolt L 2014 Modeling of temperature regime of railway track structure and its comparison with the results of experimental measurements *COMPRAIL XIV Railway engineering design and optimization*, Rome, Italy, Edited by C.A. Brebbia, WESSEX Institute of Technology, WitPress, Southampton United Kingdom **14** pp 253-65

IOP Conf. Series: Materials Science and Engineering 1015 (2021) 012050 doi:10.1088/1757-899X/1015/1/012050

Shallow and deep foundations on permafrost soils at changing climate conditions

A V Stupivtsev and L I Cherkasova

Moscow State University of Civil Engineering, Yaroslavskoe Shosse, 26, Moscow, 129337, Russia

E-mail address: legolases@mail.ru

Abstract The purpose of the work is to analyze the possible consequences of global warming on the state of permafrost and the issues of predicting the degradation of foundations during construction and design in permafrost regions. The study determines the principles and constructive solutions of foundations in areas with permafrost. From the presented options, the most rational one is selected. The analysis of observational data on the degradation of permafrost is given, on the basis of which the forecast of further degradation is built. Based on the research results, a rational method for erecting foundations in permafrost areas was determined taking into account the predicted degradation of permafrost.

1. Introduction

In the 1950s, it was assumed that permafrost soils with a thickness of tens and hundreds of meters occupy 47% of the territory of the Russian Federation [1]. In modern editions, this figure exceeds 70% ZG Ter-Martirosyan Soil Mechanics for the Territories of the Russian Federation. [2]

For our country, the development of the northern territories is necessary, therefore, it is in Russia that much attention has been paid to the study of permafrost. The factor of climate change was not significant and was not considered in the works of past years, which already now can lead to environmental disasters. For example, at CHP-3 in Norilsk, in May 2020, a fuel tank was damaged and more than 20,000 cubic meters of fuel leaked into the local water area. One of the possible causes of the accident is local subsidence due to thawing of permafrost soil [3]. At present, the study of the permafrost zone of Russia is being carried out by Ya. A. Kronik[4], L.N. Khrustalev [5], V. P. Melnikov[6], S. A. Kudryavtsev[7], etc.

2. Materials and methods

Permafrost is considered to be a soil that is frozen for three years or more [8]. Features of construction in permafrost regions are associated with atypical physical and mechanical properties of soils arising from prolonged exposure to negative temperatures:

- when carrying out any work with frozen soil (testing, operation, construction) - it is necessary to control its temperature, since at different temperatures of one soil mass, its characteristics can differ significantly;
- practically waterproof frozen soil forms a water seal that raises the level of groundwater [9];
- permafrost soils are extremely sensitive to changes in temperature, at which the ratio of water in the solid and liquid state changes;

IOP Conf. Series: Materials Science and Engineering 1015 (2021) 012050 doi:10.1088/1757-899X/1015/1/012050

- in construction practice, problems of subsidence of the base and foundations arise as a result of errors in construction and operation;

- the development of permafrost soil in terms of labor intensity is close to the development of rocky soil and, in order to avoid laborious work, preheating of the soil is often used to transfer it to a thawed state.

These features lead to an increase in financial and labor costs when building on permafrost grounds.

Let's consider the existing principles of construction in conditions of permafrost foundations:

I - frozen soil is used as a base without preliminary defrosting (the frozen state of the soil remains during operation);

II - frozen soil is used as a base with preliminary thawing to the design depth (in special cases, thawing is allowed during operation) [10].

The choice of principle is based on economic feasibility. To reduce the cost of earthwork, it is preferable to choose principle I, which excludes the cost of preliminary thawing of the soil. This principle will be discussed below.

Construction on permafrost soils according to principle *I* does not allow the soil to thaw up to a certain design level, which creates the need for measures to heat the base soil. The following activities are widely used in construction practice:

- control of the amount of heat coming from the environment - construction on an previously untouched territory in permafrost regions changes the amount of heat entering the ground, in particular - the construction site is cleared of snow in winter, which increases the flow of solar radiation directly into the ground, thus the most violating the regime of seasonal changes in phases in the active layer; on the other hand, there is the possibility of creating or increasing the thickness of permafrost soil when clearing it from snow for several years, which is also associated with the ability of snow to isolate the soil from the environment, preventing it from freezing;

- ventilated underground - the building is raised above the ground mass, creating an air gap, allowing the soil to accumulate negative Celsius temperatures in winter and not to heat up from the sun's rays in summer, in addition, the soil does not heat up from the building itself;

- regulation of the moisture content of the base;

- shading of the construction site - also a possible option for creating or increasing the thickness of frozen soil, the method is acceptable, because after erection, the shadow will continue to be created from the erected structure itself [9].

- artificial maintenance of negative temperatures at the base in the summer.

In the conditions of the far north, the freezing depth can reach more than 1000 meters [1], thereby significantly increasing the cost of the underground part of the building. The foundation must lie at a depth below the active layer, subject to constant seasonal freezing-thawing, such a foundation, in accordance with the standards, refers to deep foundations [11].

Currently, global warming affects the territory of distribution and the thickness of permafrost. Under these conditions, the question of the influence of changes in the temperature of permafrost soil on the rheological properties of the foundation of an existing building, as well as taking these processes into account when designing new structures, becomes especially important.

For example, in conditions of global warming, the ratio of the thickness of frozen soils and the active layer will change. This fact will inevitably affect both the deformations of existing buildings and new construction.

According to the observations of soil temperature carried out by researchers, it is possible to reveal the dependence of the degradation of permafrost with an increase in the average monthly temperature in a given area. Consider the work done to monitor permafrost in Western Siberia. The study consisted in direct instrumental measurement of temperature at different depths of the soil. Four-year observation data are shown in figure 1 [12].

IOP Conf. Series: Materials Science and Engineering 1015 (2021) 012050 doi:10.1088/1757-899X/1015/1/012050

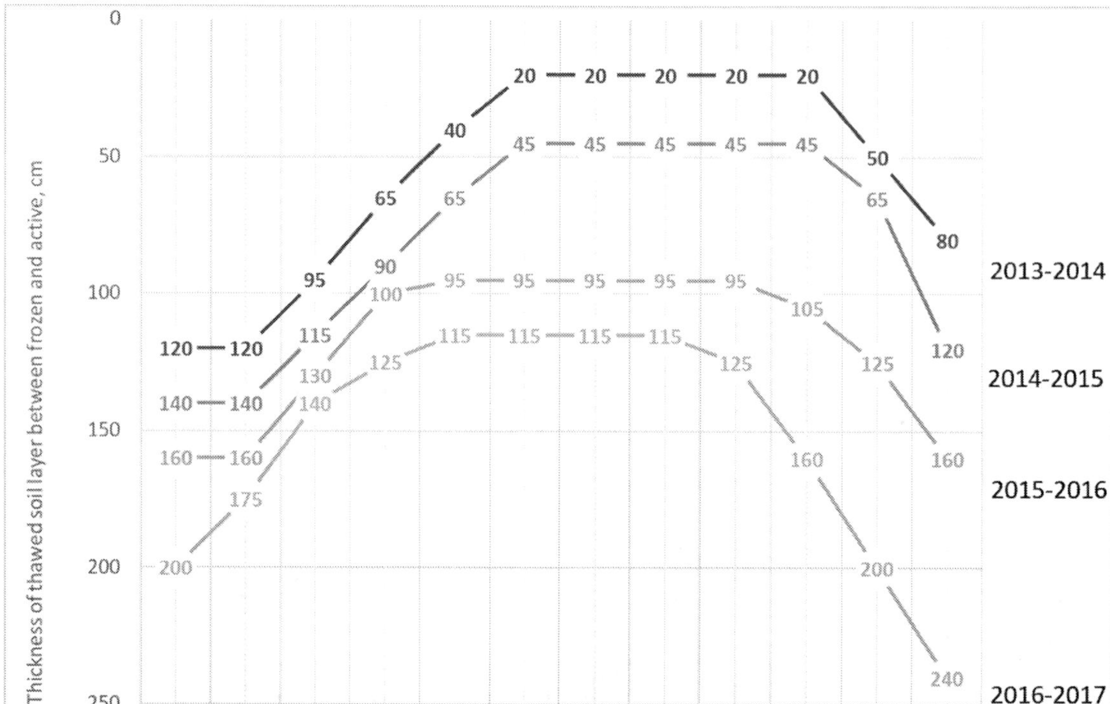

	Nov.	Dec.	Jan.	Feb.	March	Apr.	May	June	July	Aug.	Sep.	Oct.
2013-2014	120	120	95	65	40	20	20	20	20	20	50	80
2014-2015	140	140	115	90	65	45	45	45	45	45	65	120
2015-2016	160	160	130	100	95	95	95	95	95	105	125	160
2016-2017	200	175	140	125	115	115	115	115	125	160	200	240

Figure 1. Permafrost degradation rate in Western Siberia.

As can be seen from figure 1, every year there is an increase in the thickness of the thawed soil layer between the permafrost and active layers. The increase occurs due to the thickness of the layer of permafrost that has formed for many years and a decrease in seasonal freezing. In figure 3 for October is an illustrative example of such degradation.

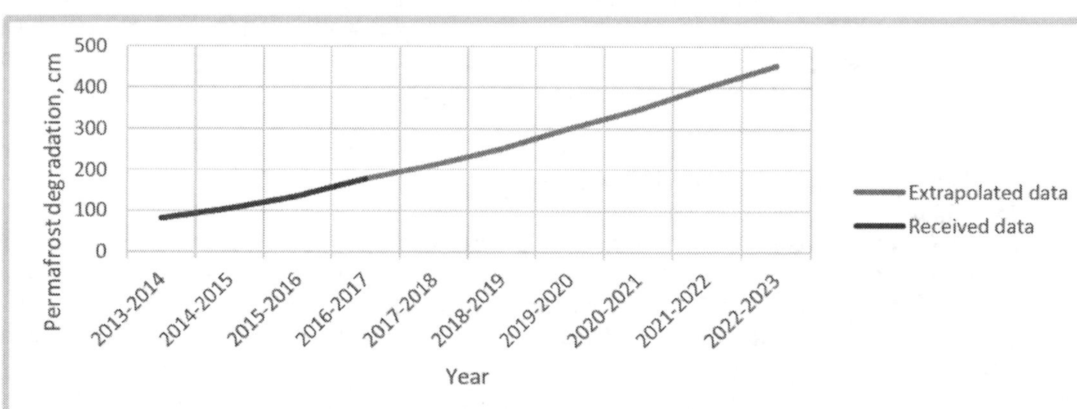

Figure 2. Predicted degradation of permafrost.

The data obtained were statistically processed. The forecast of further degradation was made by linear extrapolation for each month until 2023. Repetitive values of cold periods of time were removed from the obtained data series, suggesting that degradation stops in these months. For each year, the average value of the thickness of the thawed layer is taken, these values are plotted on the graph in figure 2. According to the results of statistical processing, the data are entered in table 1.

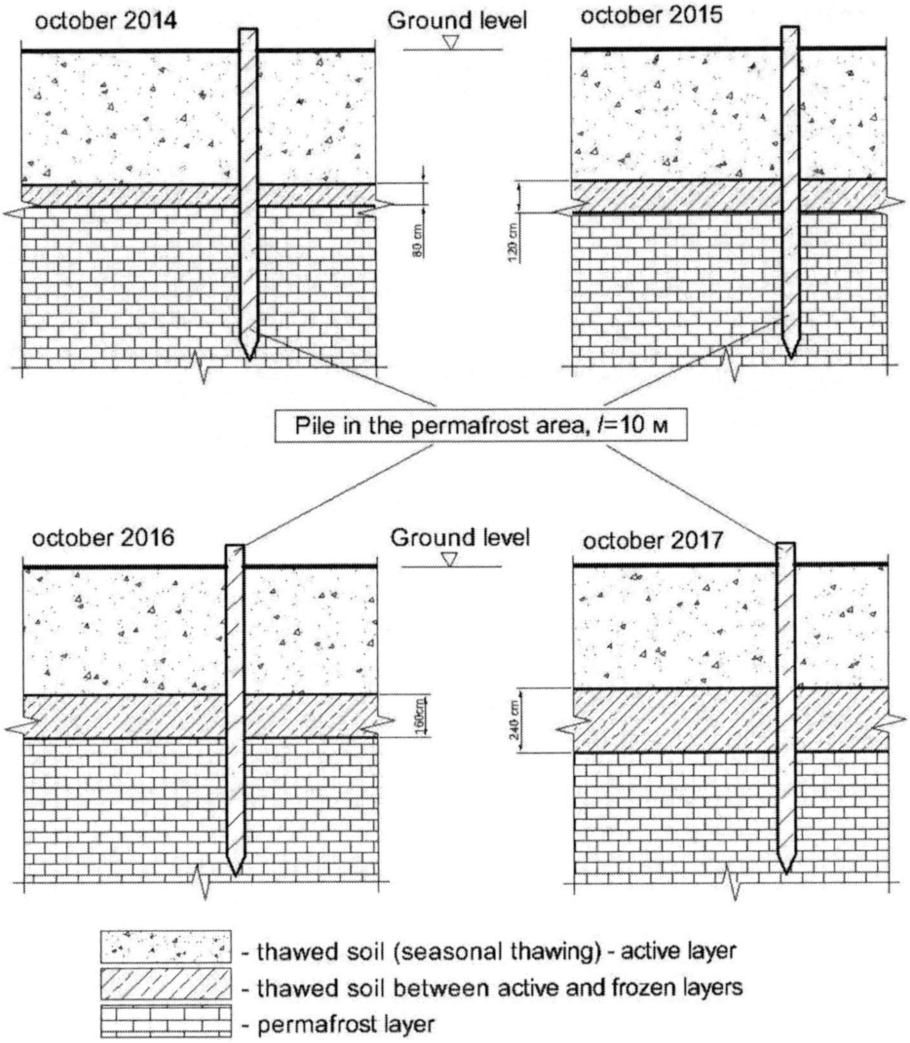

Figure 3. Permafrost degradation in October from 2014 to 2017.

From the data obtained, it can be seen that over 10 years the upper boundary of permafrost will drop by more than 5 meters. For example, figure 3, such de-gradation can cause a loss of stability of the pile, it will go from the "pile-rack" state to the "hanging pile", which is shown in figure 4.

Considering the long-term formation of the frozen layer, it is difficult to draw any conclusions based on only a four-year study. There is a need to draw up a map of the permafrost zone based on multiple point observations throughout the permafrost area with the dynamics of changes in permafrost over time. But the data already obtained demonstrate with sufficient confidence the relationship between the degradation of permafrost and global warming.

Table 1. Obtained and predicted thicknesses of thawed soil layer between permafrost and active layer.

	Thickness of thawed soil layer between permafrost and active for each year, cm									
Month	2013-2014	2014-2015	2015-2016	2016-2017	2017-2018	2018-2019	2019-2020	2020-2021	2021-2022	2022-2023
	Empirical data				Predicted data					
Nov.	120	140	160	200	220	246	272	298	324	350
Dec.	120	140	160	175	195	213.5	232	250.5		
Jan.	95	115	130	140	157.5	172.5				
Feb.	65	90	100	125						
March	40	65	*)							
Apr.										
May									348.8	407.5
June							261	318.2	375.3	432.5
July					179.2	234.8	290.5	346.2	401.8	457.5
Aug.			105	160	215	270	325	380	435	490
Sep.	50	65	125	200	237.5	288.5	339.5	390.5	441.5	492.5
Oct.	80	120	160	240	280	332	384	436	488	540
Average.	81.4	105.0	134.3	177.1	212.0	251.0	300.6	345.6	402.1	452.9

* data in empty cells are deleted as of no interest - degradation stops during these months.

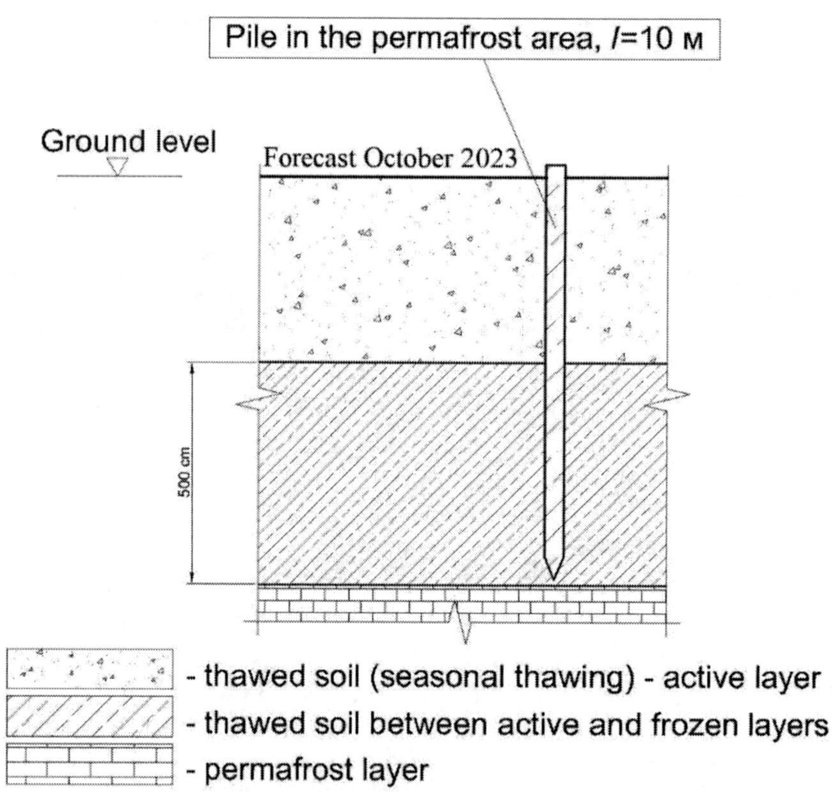

Figure 4. Predicted degradation of permafrost over 10 years.

In the areas of permafrost distribution, with climatic changes, in our opinion, one should also pay attention to changes in the characteristics of frost heaving.

Frost heaving occurs in the layer of the active layer, since the underlying layer, which is not subject to thawing, does not change its volume due to its constant presence in a frozen state. But with an increase in its volume under the influence of frost heaving, the active layer encounters greater resistance from the underlying frozen layers, as a result of which the pushing effect directly on the foundation installed in the heaving zone increases.

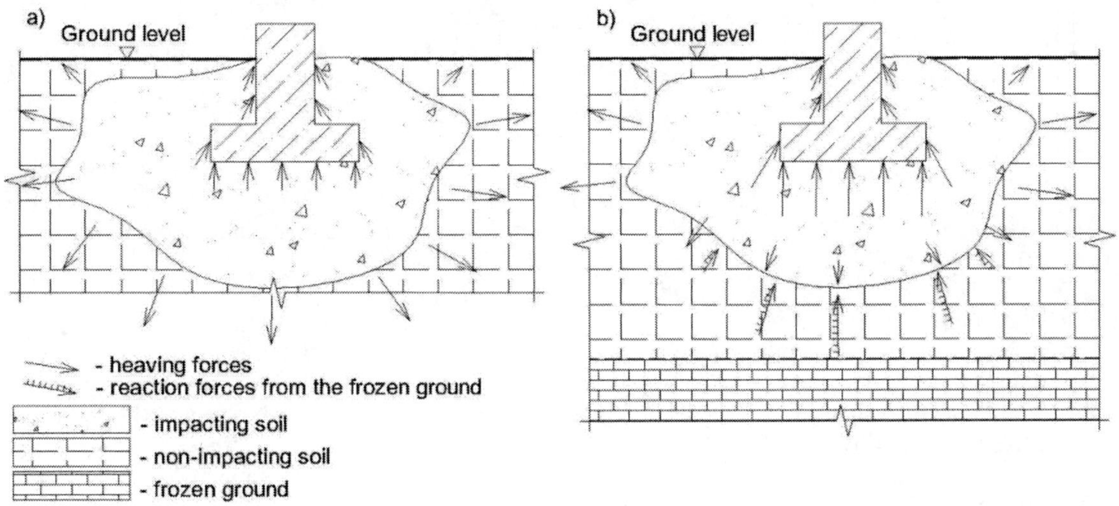

Figure 5. Forces of frost heaving, acting on the foundation: a) heaving during row construction; b) heaving with a frozen underlying layer.

With an increase in the thickness of the active layer, the influence of the forces caused by frost heaving on the foundation will increase. As a result, there is a need to strengthen measures to prevent deformation of the foundation:

- increasing the cross-section of reinforcing elements connecting parts of the foundation;
- anchoring of foundation elements to each other;
- an increase in the weight of the foundation.

The studies carried out show that even during construction according to the first principle - preservation of the soil in a frozen state, it is necessary to carry out rheological calculations that predict changes in the thickness of the active layer due to global warming.

At the design stage, geological surveys are carried out, which should include drawing up a map of isohypsum, which determines the thickness of the active layer and the upper surface of the permafrost soil. Carrying out these surveys, the task is also set to identify local permafrost or lenses of frozen soil, unsuitable for the construction of buildings and structures on them.

The base of the foundation should be supported on frozen soil with a certain predicted margin, taking into account the probability of thawing of the base under the erected structure by an amount greater than the normative seasonal thawing, since the thawed soil under the foundation for the first principle of construction on permafrost soil is unacceptable.

When considering the pile foundation, it is necessary to take into account the increase in seasonal heaving, leading to the rupture of the piles, as a result of the increase in the active layer. The rupture will occur in the active layer and this will be facilitated by the pinching of the pile by permafrost soil, as shown in figure 6.

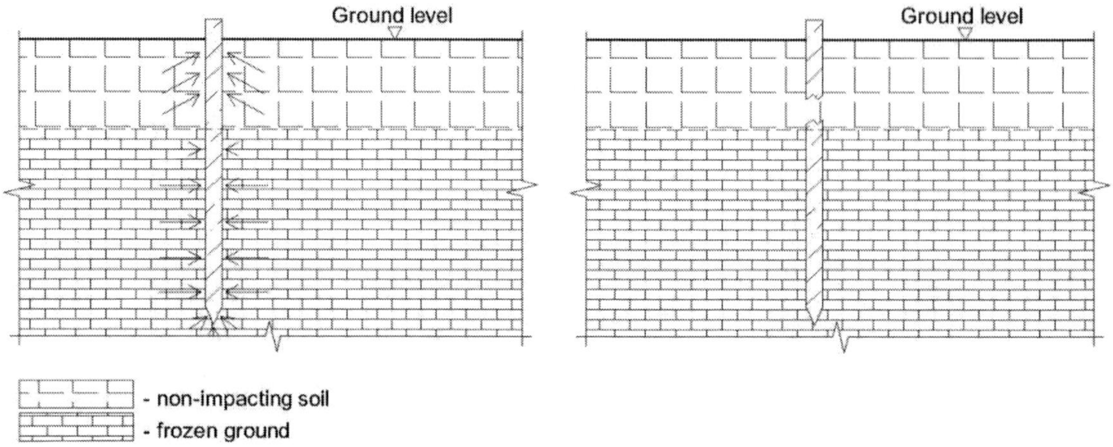

Figure 6. Destruction of the pile from the action of the forces of frost heaving and pinching in the permafrost layer of the base.

3. Conclusions

The influence of climatic conditions on permafrost soils affects primarily the depth of the active and thawed layers, as well as their ratio. An increase in temperatures in summer leads to an increase in the thickness of the thawed layer. With a change in the thickness and ratio of the active and thawed layers, the influence of the active layer on the state of the foundation structures increases. At the same time, the friction on the side surface and the frontal resistance of the base of the pile foundations can decrease. These processes can lead to uneven settlement of the base and the destruction of foundations.

When designing foundations, it is necessary to carry out rheological calculations, considering the general warming of the climate, and the development of methods for maintaining negative temperatures under existing buildings in these conditions.

It is necessary to consider the issues of changing the bearing capacity of pile foundations during operation in connection with a decrease in frontal friction and friction on the lateral surface of the piles. When developing new foundation structures, it is possible to consider the use of materials with reduced thermal conductivity.

References

[1] Tsytovich N A 1963 *Soil mechanics* (Moscow: State publishing house of literature on construction, architecture and building materials) p 636

[2] Ter-Martirosyan Z G 2005 *Soil Mechanics / Tutorial* (Moscow: Publishing house of the Association of building universities) p 488

[3] TASS report: Norilsk Nickel did not monitor permafrost in Norilsk: (https://tass.ru/ekonomika/8688585 date of access: 13.07.2020)

[4] Kronik Ya A 2019 *Analysis of data from monitoring the safety of hydraulic structures in the permafrost zone №2* (Moscow: Scientific and technical firm "Energoprogress") pp 44-54

[5] Khrustalev L.N. 2005 *Fundamentals of geotechnics in permafrost* (Moscow: Moscow University Publishing House)

[6] Melnikov V.P., Bruskov A.V., Vlasov A.N. and other 2010 *Cryogenic geosystems: Problems, research and modeling* (Novosibirsk: Academic Publishing. Geo.) p 392

[7] Kudryavtsev S.A., Paramonov V.N., Sakharov I.I. 2014 *Freezing and thawing of soil. Practical examples and finite element calculations. "Achievement of modern geotechnics"* (St. Petersburg: gr. Company Georeconstruction)

[8] Standard GOST 25100-11 *Soils. Classification.* 2018 (Moscow: Standartform)

[9] Sumgin M I, Kachurin S P, Tolstikhin N I and Tumel V.F. 1940 *General Permafrost Science*

(Moscow: Publishing House of the Academy of Sciences of the USSR) p 340
[10] Standard SP 25.13330.2012 *Bases and foundations on permafrost soils.* 2012 (Moscow: Ministry of Regional Development of Russia)
[11] Stabdard SP 22.13330.2016 *Foundations of buildings and structures.* 2017 (Moscow: Standard-tinform)
[12] Salinder E M 2018 Study of the degradation of permafrost in the Tazovsky region of the Yamalo-Nenets Autonomous Okrug *Collection of abstracts of the VI Regional Youth Conference named after VI Shpilman Problems of rational nature management and the history of geological prospecting in Western Siberia* (pos. Tazovsky: Limited Liability Company "Yugorsky Format" (Khan-ty-Mansiysk))
[13] Stupivtsev A V 2020 *Use of permafrost soil as a base for a shallow foundation* (Moscow: Publishing house MISI - MGSU) pp 430-5
[14] Ibragimov E.V., Kronik Ya.A., Paramonov V.N. 2019 *Forecast of the stress-strain state of a thermally stabilized base №6* (Moscow: Publishing House "Economy, Construction, Transport")

Long-term vibration laboratory tests of sandy soils

A Z Ter-Martirosyan and E S Sobolev

Department «Soil mechanics and geotechnics», Moscow State University of Civil Engineering (National Research University), 26 Yaroslavskoye Shosse, Moscow 129337, Russia

E-mail address: e.s.sobolev@mail.ru

Abstract. The presented work examines the features of testing at the laboratory of the Moscow State University of Civil Engineering "Research and Education Center" Geotechnics "on a triaxial device with the ability to create dynamic loads (vibration stabilometer). Information on the course of cyclic tests of sandy soils is given, their analysis is carried out and formulated the main conclusions about the effect of the duration of vibration loading on the final stress-strain state of the soil mass.

1. Introduction

During the design, construction and operation of industrial, civil, hydraulic engineering and transport structures in connection with the constant increase in the intensity of technogenic dynamic impact, construction in seismically active areas, in water areas and in other special situations, it is necessary to take into account the vibration effects on the foundation soils to assess their interaction with the underground part of structures, in order to determine possible additional deformations.

The experience in the construction and operation of the listed types of structures and high-rise buildings with a developed underground part shows that long-term cyclic and vibration effects lead to the accumulation of additional deformations in the soils of the bases and, ultimately, to additional precipitation and rolls of the foundations of structures, and sometimes to loss of stability, for example, during dynamic liquefaction of water-saturated sands.

2. Research method

To take into account the vibration creep of soils, the vibration creep coefficient (K_{vc}) is used, which reduces the value of the static modulus of total deformation. This approach to predicting the vibrational creep of soils can be called quasi-static, since it is carried out within the framework of conventional static methods for calculating the settlement of the foundations of structures.

The values of the vibration creep coefficient for the studied soils were calculated by the following formula:

$$K_{vc} = \frac{\Delta\varepsilon_s}{\left(\Delta\varepsilon_s + \Delta\varepsilon_g\right)} \tag{1}$$

where $\Delta\varepsilon_s$ and $\Delta\varepsilon_g$ - strain increments from static and dynamic loading in a given stress range.

The soil deformation modulus, taking into account the vibration creep deformation, should be determined by the formula:

$$E_{vc} = E_s \times K_{vc} \tag{2}$$

where E_s – Young modulus based on static test results, in kPa; K_{vc} – vibration creep coefficient.

One of the fundamental issues in the field of vibration creep assessment is the choice of test conditions: frequency, number of cycles and time. In most cases, the frequency is taken as the most typical for each site (for example, the vibration frequency of an electric machine, transport impact, etc.). The test is carried out until the full stabilization of deformations during vibration creep, which is assessed by the absence of increments of vertical deformation. However, given the laboriousness and high cost of such tests, it is necessary to develop an acceptable method for assessing the test time, as well as the need to conduct tests at different frequencies.

The mechanism of vibrational creep of dispersed soils is currently not fully understood. Many researchers [9, 11] believe that the vibration creep coefficient depends on the frequency of exposure, based on the fact that the conditions of interaction of particles change at different frequencies, and, consequently, their creep resistance. However, this theory needs additional experimental confirmation, since at the moment there are not enough test data at various frequencies.

In addition, consideration of the phenomenon of vibration creep from the point of view of energy [2, 3, 6] spent on overcoming internal friction leads to the conclusion that the manifestation of vibration creep will only be affected by the number of cycles with equal amplitude, regardless of frequency. However, in the case of soil, rheological processes associated with pore water and rigid bonds of various natures between particles are important, and these processes are always a function of time. Thus, the time between individual cycles will have a significant impact on the development of additional deformations due to vibration creep.

The purpose of this work is to study the parameters of vibration creep by analyzing the effect on vibration creep of the number of cycles and the frequency of exposure based on the experience of testing in the REC "Geotechnics" MSUCE soils for solving production problems.

3. Laboratory research

For testing, equipment manufactured by APS Antriebs-, Pruf- und Steuertechnik GmbH "Wille Geotechnik" (Germany) was used, which is a servo-hydraulic load frame with a maximum axial force of 63 kN, a triaxial compression chamber, a control unit for a servo-hydraulic drive, a data processing unit from pressure sensors and displacement, air pressure control unit, media separator, burette with differential volumetric deformation sensor and control computer. The general view of a three-axis device with the ability to create dynamic loads (vibration stabilometer) is shown in figure 1.

During testing in automatic mode (software GEOsys 8.7.8) [16], the specified loading path is realized.

The tests were carried out in type A chambers, designed for specimens 140 mm high and 70 mm in diameter. Top and bottom dies are made of stainless steel. Porous plates of pressed steel chips and filter paper were placed between the dies and the sample. Latex casings were used, 0.5 mm thick.

Samples of sandy soil of a given density ($\rho = 2.02$ g/cm^3), close to natural, were placed in the chamber of a triaxial device, after which they were completely saturated with water. After the completion of water saturation, a comprehensive pressure was applied to the sample, equal to the domestic one at a given depth, and the stage of consolidation was carried out under open drainage conditions. Compression pressure was 500 kPa. After stabilization of volumetric deformations, a deviatorial load was applied to the sample, equal to the additional pressure from the structure at the sample depth (150 kPa).

After the application of all acting loads and stabilization of deformations, the drainage was closed, and a vertical vibration load with a frequency of 20 and 50 Hz was applied to the sample. The number of cycles was 180×10^3 (at a frequency of 20 Hz) and 300×10^3 (at a frequency of 50 Hz), which represents about 2.5 hours of dynamic loading. The amplitude of the cyclic vertical loading was 30 kPa, which was 4.5% of the total vertical load.

Thus, in the course of the last stage of testing, the values of additional deformations of the sample were obtained due to the action of the load of a given frequency in the conditions of consolidated-undrained triaxial compression. After processing the test reports, the values of the vibration creep coefficient were obtained at the stage of vibration loading.

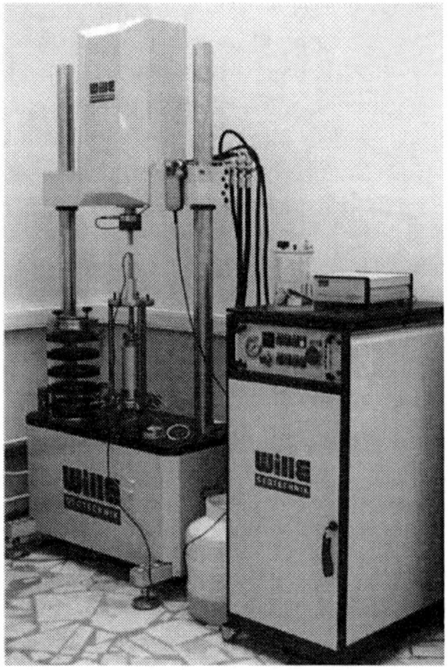

Figure 1. General view of the installation of triaxial compression with the possibility of vibration loading.

The results of the tests performed are presented in figures 2 and 3. The graphs show the dependences of the development of relative vertical deformations during the period of vibration loading on the number of cycles at frequencies of 20 and 50 Hz, respectively.

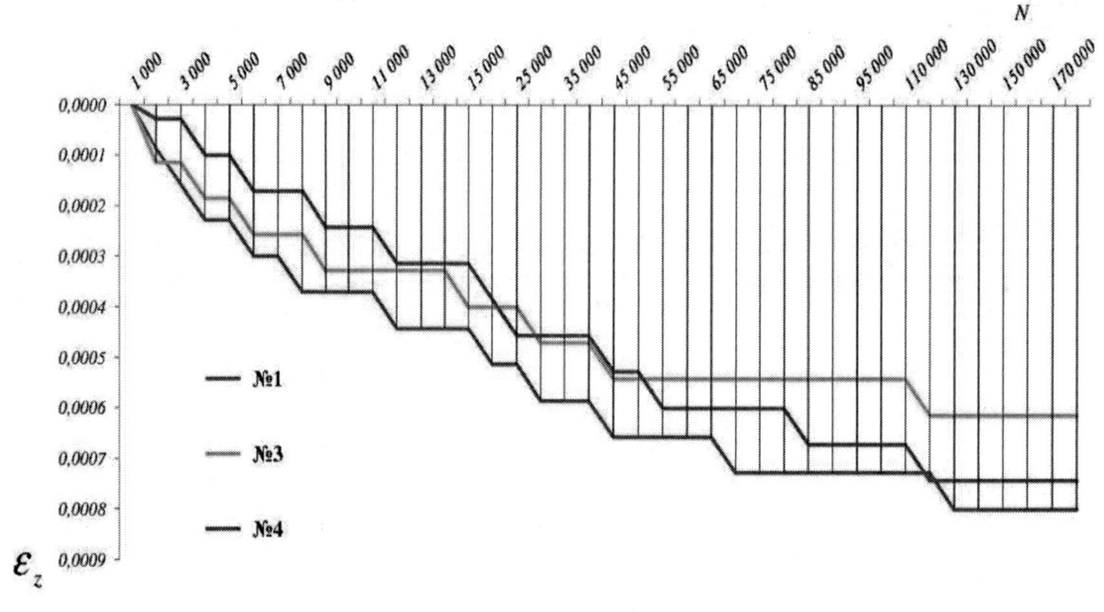

Figure 2. Dependence of the development of relative deformations on the number of vibration loading cycles with a frequency of 20 Hz.

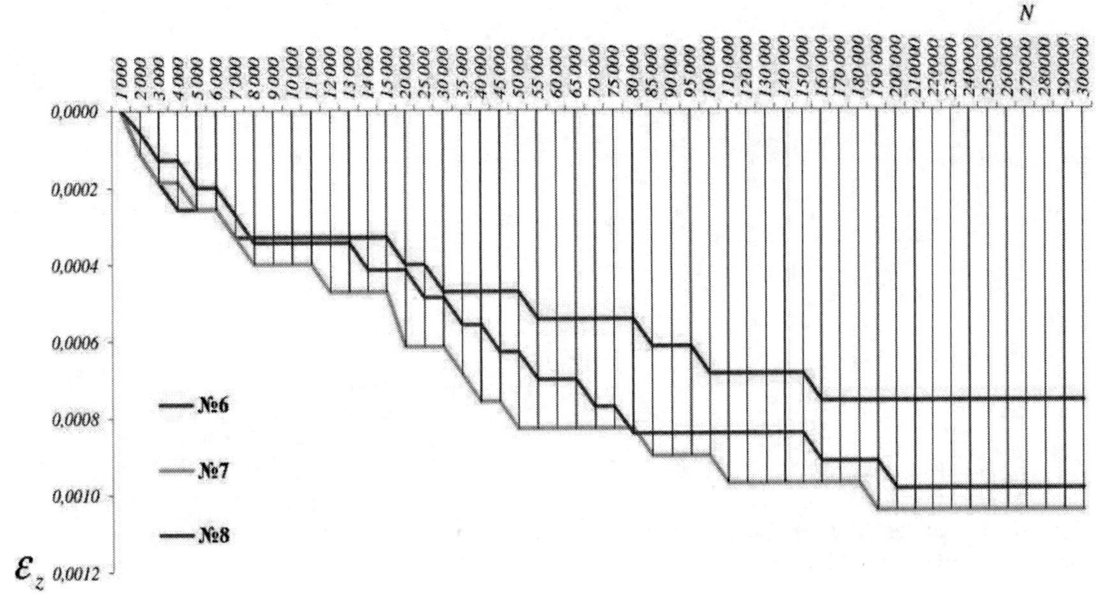

Figure 3. Dependence of the development of relative deformations on the number of cycles of vibration loading with a frequency of 50 Hz.

Based on the results of the tests carried out, trend lines were plotted using the least squares method, which made it possible to compare the results for frequencies of 20Hz and 50Hz (figures 4 and 5, respectively).

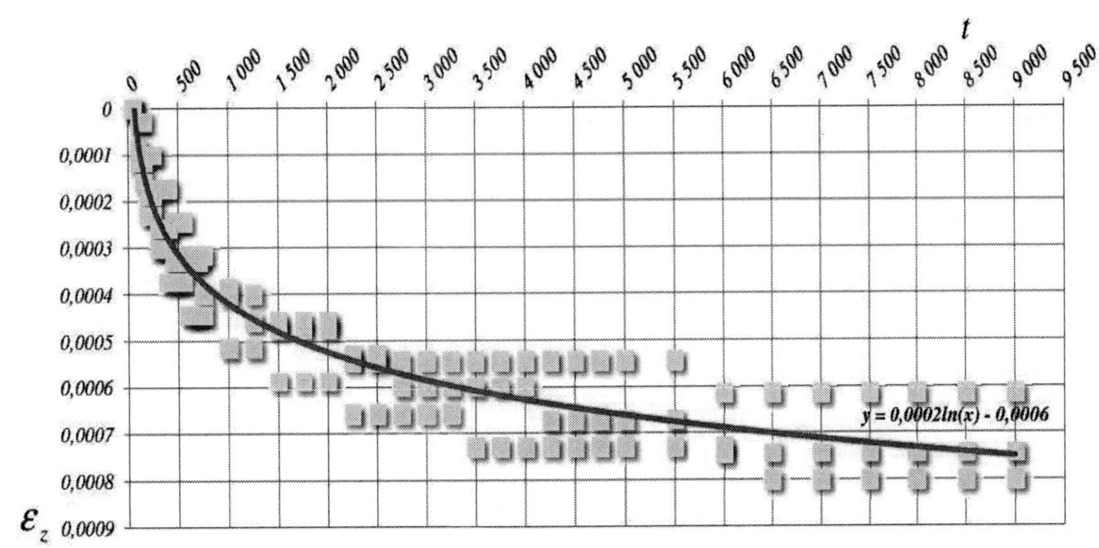

Figure 4. Dependence of relative deformations on loading time in seconds for testing four soil samples at a frequency of 20 Hz.

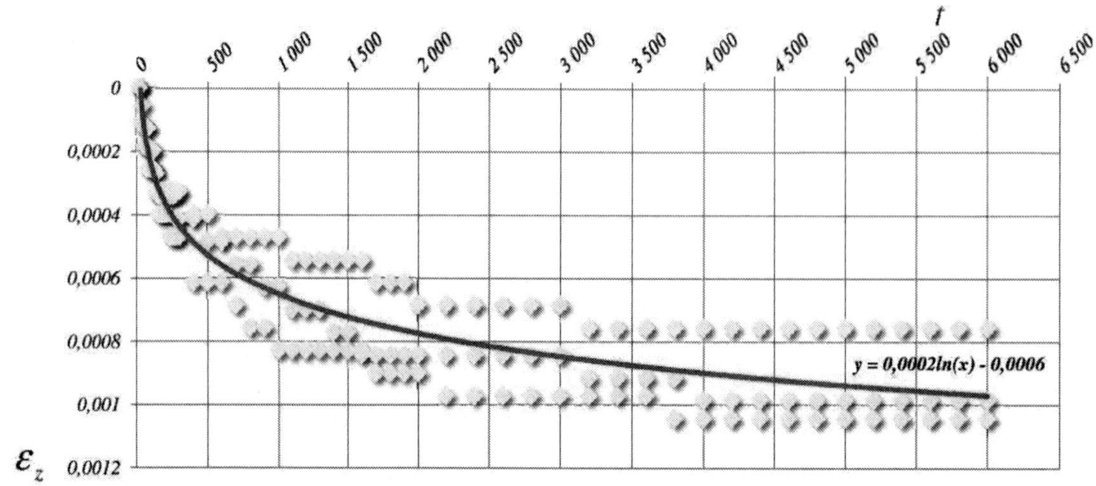

Figure 5. Dependence of relative deformations on loading time in seconds for testing four soil samples at a frequency of 50 Hz.

4. Conclusions

With the joint analysis of the obtained curves, the following general conclusions can be drawn:

- additional deformation during vibration creep increases with increasing frequency;
- regardless of the frequency and number of cycles in the same soil, deformation stabilization occurs in the same time;
- logarithmic equations of the trend line for the same soil differ exclusively in the free term. This allows us to conclude that the law of deformation development due to vibration creep will be identical for the same soil. Despite the fact that the logarithmic dependence does not allow obtaining complete stabilization of additional deformations, it can be assumed that it has a damping character, because at $t \to \infty$, $\dot{\varepsilon} \to 0$.

Based on a detailed analysis of the performed dynamic tests of sandy soils, the following main conclusions can be drawn:

- Stabilization of relative deformations during vibration tests of sandy soils under conditions of triaxial compression at a frequency of 20 Hz and an amplitude of deviatorial loading of 30 kPa (4.5% of the total vertical load) occurred on average after 4500 s.
- Stabilization of relative deformations during vibration tests of sandy soils under conditions of triaxial compression at a frequency of 50 Hz and an amplitude of deviatorial loading of 30 kPa (4.5% of the total vertical load) occurred on average after 4000 s.
- When carrying out dynamic tests of sandy soils in a vibro-stabilometer, the time of vibration loading plays a decisive role, since it is a factor that takes into account both the frequency and the number of cycles in the form

$$t = \frac{N}{f} \tag{3}$$

where t - time of vibration loading, s; N - number of vibration load cycles; f - vibration loading frequency, Hz.

- The vibration creep of each individual soil can be described by a general law that is unchanged for different frequencies in the form:

$$\varepsilon(t) = \alpha \times \ln(t) - \beta \tag{4}$$

where $\varepsilon(t)$ - additional relative deformations at the stage of vibration loading; α - rheological coefficient; β - constant value.

IOP Conf. Series: Materials Science and Engineering 1015 (2021) 012051 doi:10.1088/1757-899X/1015/1/012051

References

[1] Abramova T T 2016 Protection of soil masses from dynamic and seismic loads *Symvol Nauki* **4** pp 41-9

[2] Voznesensky E A 1999 *Dynamic instability of soils* (M: Publishing group URSS)

[3] Abramova T T and Voznesensky E A 2015 Modern methods of managing propeties of soils on the sites exposed to high dynamic loads *Geotechnics* **4** pp 6-25

[4] Standard GOST 12248-2010 Soils. Methods for laboratory determination of strength and deformability characteristics 2012

[5] Standard GOST 20522-2012 Soils. Methods for static processing of test results 2013

[6] Ishihara K 2006 *Behavior of soils during earthquakes* (S.-Pb: DIA)

[7] Sobolev E S, Ter-Martirosyan A Z and Morev D S 2019 Experimental studies of elastic shear waves speeds in soil-cement depending on density *Geotechnics* **11** (3) pp 6-12

[8] Technical report *Determination of the vibration creep coefficients of sandy soils at the facility:" Construction of the highway Drozhzhino - Bobrovo - Lopatino* at the address: MO, Leninsky municipal district, Bultnikovskoe rural settlement. - M .; MGSU, 2013.

[9] Ter-Martirosyan A Z, Mirny A Yu, Sidorov V V 2013 Laboratory testing of soils at the Moscow State University of Civil Engineering (MGSU-MISI) (M: Engineering Survey magazine) issue # 8

[10] Ter-Martirosyan A Z 2010 Residual deformations and stresses in the soil environment under the action of a cyclic load. Collection of scientific papers of the *XXIII International Interuniversity Scientific and Practical Conference of Young Scientists*, Doctoral and Postgraduate Students "Construction - Formation of the Living Environment", 14-21.04.2010 Moscow pp 815-9

[11] Ter-Martirosyan Z G 2009 *Soil mechanics* (Moscow: ASV)

[12] Ter-Martirosyan Z G, Ter-Martirosyan A Z 2009 Creep deformations of soils under cyclic and vibration effects *Proceedings of the XVIII Polish - Russian - Slovak seminar "Theoretical foundations of construction"* Moscow - Arkhangelsk 01-05.07.2009 (Warsaw) pp 473-80

[13] Ter-Martirosyan Z G, Ter-Martirosyan A Z 2008 Residual stresses in soils under cyclic loading *Proceedings of the XVII Polish - Russian - Slovak seminar "Theoretical foundations of construction"* Warsaw 02-06.06.2008 (Zilina) pp 278-83

[14] Ter-Martirosyan Z.G. 1990 *Rheological parameters of soils and calculations of the foundations of structures* (Moscow Stroyizdat)

[15] Ter-Martirosyan Z G, Nikolaev A P and Ter-Martirosyan A Z 2008 Residual deformations and stability of soil massifs under seismic impacts *Scientific and technical journal "Vestnik MGSU"* pp 41-7

[16] Operational documentation for a set of equipment for testing in triaxial compression 63kN APS GmbH "WilleGeotechnik" 2012 (Germany Göttengen-Rosdorf)

IOP Conf. Series: Materials Science and Engineering 1015 (2021) 012052 doi:10.1088/1757-899X/1015/1/012052

Engineering and geological features of technogenic deposits and cultural layers in the territory of Samara

D Vasilieva, M Baranova and A Maltsev

Department of Engineering Geology, Bases and Foundations, Samara State Technical University, 244 Molodogvardeyskaya St., Samara, Russia, 443100

E-mail address: vasilievadi@mail.ru

Abstract. The engineering and geological features of technogenic deposits in the territory of the city of Samara have been studied. Their features are revealed and a classification is carried out. The article presents the results of a petrographic study of samples taken at the archaeological site of Staraya Samara, Khlebnaya Square, carried out in 2019. The object represents cultural layers that were formed in the XVIII-XIX centuries at the site of the alleged location of the second Samara fortress. The studies were performed under a binocular microscope at an 8.75-fold magnification, in the passing light of a polarization microscope at a 72-fold magnification, and under a digital microscope at a 10-fold magnification. It has been determined that the pebble fraction is represented by quartzite and jasper flint, the sandy and silty fractions are angular quartz fragments with an admixture of undecomposed organic remains. Organic cultural layers, which reach 7-8 meters or more in the ancient part of the city, are especially powerful. Their presence is a limiting factor in modern urban construction.

1. Introduction

The study of man-made deposits and ancient cultural layers of settlements and cities provides a large amount of information about the first stages of settlement and use of the territory, about which few written sources have survived. Involvement of the methods of natural sciences, including soil science and geology, makes it possible to comprehensively consider the problems that people faced at the initial stages of the development of the territory, and understand the ways to solve them. The study of the history of changes in the geological environment under the influence of anthropogenic activity helps in making forecasts of its development and, possibly, in preventing geoecological problems.

2. Methods

Much attention is currently being paid to the study of technogenic deposits and cultural layers, but it is noted that their unified classification has not yet been created, which would take into account the different conditions and factors of their formation of technogenic deposits, the peculiarities of the inflow and transformation of technogenic material and the intensity of technogenic impact [1].

Technogenic deposits are a specific genetic group of modern continental formations, the origin of which is associated with practical human activity [2, 3]. The term "cultural layer" first appeared in the scientific works of archaeologists, who used it to refer to layers or formations of rocks and soils that have traces of ancient human activity and contain ancient artifacts [1]. The accumulation of cultural layers occurs constantly in the course of human life on a certain territory – during the production of earthworks, construction, landscaping, as well as during the accumulation of garbage.

In the Samara region, technogenic landscapes predominate, of which 75.93% account for agricultural landscapes; 6.72 % - urban landscapes, 1.32 % - industrial, including mining industrial landscapes [4]. On the territory of the city of Samara, the following technogenic deposits can be identified (table 1).

Table 1. Classification of technogenic sediments of the city of Samara.

Complex	Type	Kind	Variety
Bulk	Construction	Soils of modern buildings	Loam and sandy loam with inclusions of building material
		Road soils and road embankments	Carbonate crushed stone, gravel, sand
	Industrial	Waste of modern production	Slag, ash
	Household	Soils of modern and former cemeteries	Loam with inclusions of brickwork, stones, bones
		Cultivated soils of orchards and vegetable gardens	Loam, sandy loam, enriched with humus
		Mixture of household and household waste	Specially stacked or spontaneously accumulated
	Polygenic	Filled-in ravines	Loam, sand, organic deposits and variegated material
Sediments of artificial reservoirs	Sediments of a reservoir	Soils of floodplains of the Volga, Soka and Samara rivers	Alluvial sands

Technogenic sediments are found throughout the Samara city, while the most ancient part of the city is covered with a sedimentary sheath. Several zones can be distinguished on the urban territory:

1) unchanged relief, which is not characterized by the formation of technogenic sediments;

2) minimal changes in the relief, where the power of technogenic sediments does not exceed 1 m;

3) the maximum change in the relief, with the power of technogenic sediments over 5-6 m.

For the oldest part of the city Samara is characterized by the maximum change in relief and the greatest thickness of technogenic sediments.

The aim of the article is the study of the structure and engineering-geological peculiarities of technogenic sediments and cultural layers in the most ancient part of the city Samara, the examples of archaeological excavation at Khlebnaya Square and materials of a technical survey of the pavilion-cafe on the territory of the River Station.

Methods of scientific literature analysis, tabular and graphical methods were used in the research process. With the help of generally accepted methods researches under a binocular microscope (at magnification 8.75 times), in passing light of a polarization microscope (at magnification 72 times) and under a digital microscope at magnification 10 times were carried out.

3. Results and discussions

The territory of the research object "Old Samara. Wooden structures, erected no later than the middle of the XIX century ", located on the Khlebnaya square in the Samara district of the city Samara (figure 1). Archaeological excavations of this object were carried out many times, starting in 2013, which made it possible to obtain a large amount of data on the early stages of the settlement of Samara. Archaeological research in 2019 was carried out by JSC Transenergoproekt (Saratov) on the basis of an open sheet of V.V. Tikhonov. In the 2019 excavation, part of the slope of the second above-floodplain terrace of the Samara river was uncovered.

In geomorphological terms, the investigated territory represents the above-floodplain terraces of the Samara and Volga rivers of different ages and complex structures, with a significant slope of the terrain and gradually turning into the watershed surface [5].

IOP Conf. Series: Materials Science and Engineering 1015 (2021) 012052 doi:10.1088/1757-899X/1015/1/012052

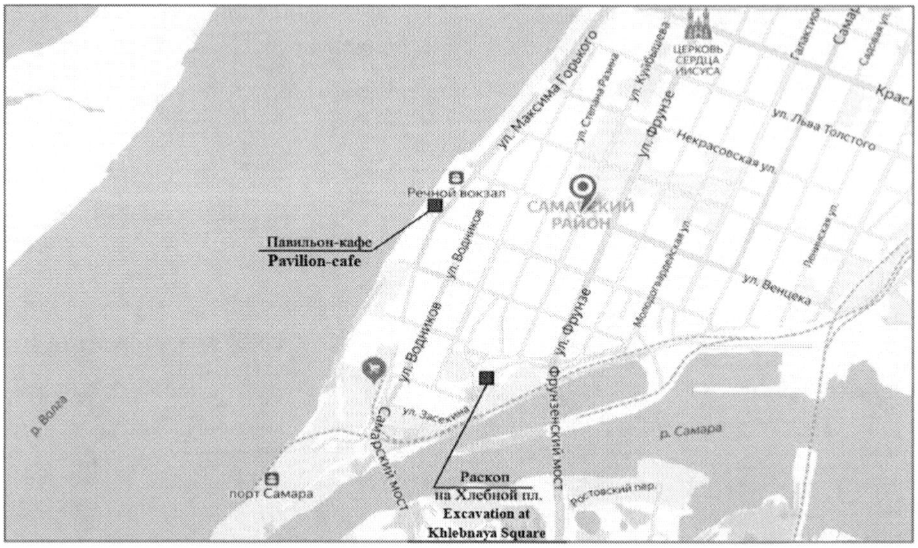

Figure 1. The layout of the objects of research in the historical center of Samara.

For this territory, before the development and leveling of the surface, the following exogenous geological processes were characteristic: ravines, planar erosion, landslides, flooding, etc. [6].

Gully formation is a very unfavorable exogenous geological process [7]. Most of the ravines belong to ancient erosional incisions, and only some of them are the result of modern erosion [8-10]. Perhaps it was to strengthen the construction site and prevent the activation of gully formation at the site of the fortress that wooden structures were erected, uncovered by archaeological excavations in 2013-2014 and 2017.

The 2019 excavation revealed the edge of the second terrace above the floodplain in Samara and a sharp drop in the relief to the first terrace. To smooth out the unevenness of the relief from about 1722 (the date was determined from the coins found in the cultural layers), the lowered areas of the terrain were backfilled with organic remains and construction debris. At present, the thickness of these layers reaches 7-8 meters and more (figure 2).

Figure 2. General view of the archaeological excavation site of Khlebnaya Square in 2019.

The cultural layers are located on the "mainland" - yellow ancient alluvial sand with a large number of ferruginous interlayers. At present, we can observe a sharp change in the continental layer up to 3 m high, which is completely planned by filling in organic cultural layers.

The pavilion-cafe object is located on the Volga river Embankment near the intersection of M.Gorky and Pionerskaya streets in the Samara district of Samara. The technical condition survey was carried out by employees of the Samara State Academy of Architecture and Civil Engineering (SamGASA) under the guidance of one of the co-authors of this article. In structural terms, the object was a one-story frame building made of metal elements, measuring 16 x 12 m in plan, operated only in the summer. The purpose of the survey is the possibility of reconstructing the building and replacing the light metal frame with capital brick walls for year-round operation of the cafe.

An analysis of the engineering-geological conditions of the pavilion-cafe site showed that, in geomorphological terms, the surveyed area is confined to the left-bank floodplain of the river Volga. Before the start of the Embankment construction, the absolute elevations of the natural relief were 30.1-34.0 m. As a result of planning works by the method of soil filling during the construction of the river station, the site currently has a flat surface with absolute elevations of 36.2-36.4 m with a weak slope in the northwest direction towards the river Volga.

Based on the results of engineering-geological surveys in the soil massif of the site of the building under study, 8 engineering-geological elements (IGE) were identified:

IGE-1 (tQ_{IV}) - bulk soil, packed, compacted, complex composition-a mixture of black soil (60%), clay, sand, crushed stone, broken bricks and glass, with a capacity of 0.5 to 2.4 m;

IGE-2 (tQ_{IV}) - bulk soil, similar to IGE-1, with a manure content of up to 30%, with a thickness of 1.8 to 3.2 m;

IGE-3 (tQ_{IV}) - black manure, decomposed, with a content of black soil and sand up to 20%, up to 2.0 m thick;

IGE-4 (tQ_{IV}) - brown-brown bulk sand, medium size, dense and medium density with inclusions of broken brick and glass, low moisture, up to 3.2 m thick;

IGE-5 (aQ_{IV}) - sand is dark gray, fine, medium density, moist, from a depth of 7.10 m from the ground surface, water-saturated, slightly humus, with a thickness of 2.8 to 4.2 m;

IGE-6 (aQ_{IV}) - gray sandy loam, plastic, slightly humus, up to 1.6 m thick;

IGE-7 (eP_{2kz}) - gray, highly weathered, loose dolomite, reduced strength, up to 3.4 m thick;

IGE-8 (eP_{2kz}) - grey dolomite, highly weathered, fractured, low-strength, with a thickness of 0.6 to 2.2 m.

The main indicators of soil properties are shown in table 2.

Table 2. Basic physical and mechanical properties soils the base.

Number IGE	Name IGE	Soil density ρ g/cm^3	Organic content, in unit fractions	Deformation modulus E MPa	Adhesive power c kPa	Internal friction angle φ deg.	Limit of compressive strength R_C MPa	Calculated resistance of the base soil R_0 kPa
1	Bulk soil	1.80	0.09	15	–	–	–	120
2	Bulk soil with a manure to 30 %	1.75	0.20	7	–	–	–	100
3	Manure with a content of black soil and sand up to 20%	1.50	0.40	3	–	–	–	50
4	Bulk sand	1.80	–	30	0	34	–	–
5	Sand fine	1.78	–	22	0	30	–	–
6	Plastic sandy loam	1.87	–	10	10	20	–	–
7	Dolomite strongly weathered of reduced strength	1.90	–	6	–	–	4	–
8	Dolomite highly weathered low-strength	2.20	–	35	–	–	12	–

The technical inspection of the object showed that the massif is represented by soils of different thickness and character of bedding, significantly differing in their physical and mechanical characteristics. Therefore, the base is characterized by heterogeneity and anisotropy of properties. Directly below sole of the existing foundations, the bearing layer is bulk soil. In addition, the base to a depth of 4-6 m is represented by various weak bulk soils (having low values of the design soil resistance in the range from 50 to 120 kPa), which are underlain by low-strength dolomites.

Most of the verification calculations showed that under the brick walls of the cafe building after its reconstruction, a strip foundation with a width of 1 m and depth laying down of the bottom plane of the foundation of 1.9 m, resting on bulk soils IGE-1, is sufficient. However, the presence of large amounts of manure and chernozem (i.e. organic matters) in the weak underlying layers of the IGE-2 and IGE-3 base made it possible to classify these soils as biogenic and consider their stress-strain state as unstabilized. The calculation for a weak underlying layer led to the fact that as the foundation for a small one-story cafe, a foundation plate was recommended for the entire building with dimensions in the plan of 18 × 14 m.

Therefore, the superstructure of existing and the construction of new multi-storey buildings in the historical center of Samara is associated with the need to take into account the complex geological conditions during construction. The presence of man-made and cultural layers, prone to heterogeneity and anisotropy of the properties and texture of the soil massif, in most cases leads to uneven settlements of the foundation under the load from the building. In addition, deformations of the base can be long in time and the dependence of settlement on time is not always predictable.

The study of engineering and geological features of construction on the territory of Samara depends on the physical and geographical conditions and geological structure, including stratigraphy, tectonics, karst occurrences and the level of groundwater.

Back in 1969, the research of doctor of geological and mineralogical sciences V.I. Rachitsky [12] made it possible to identify the engineering-geological area, confined to the high part of Samara from the street Polevay to the Khlebnaya Square. It covers the territory of distribution of quaternary alluvial deposits of the second above-floodplain ancient terrace at the confluence of the Volga and Samara rivers. The edges of the terrace are lowered by denudation processes and, especially on the Samara slope, are poorly expressed in the relief. This area, according to V.I. Rachitsky, composed of layered and cross-bedded ancient quaternary alluvial sands and sandy loams up to 30 meters thick. Sands and sandy loams were very well studied by V.I. Rachitsky from their physical and technical side (table 3).

Table 3. General characteristics of the properties of alluvial deposits.

Physical and mechanical properties of soils	Loams	Sandy loams	Sands
Soil moisture, w, %	17.25	13.50	8.60
Degree of soil moisture, S_r, in unit fractions	0.75	0.30	0.30
Density of mineral particles, ρ_s, g/cm^3	2.67	2.66	2.62
Soil density, ρ, g/cm^3	1.91	1.78	1.76
Porosity, n, %.	38.20	42.00	43.00
Porosity coefficient, e, in unit fractions	0.62	0.73	0.76
Humidity at the border of plasticity, w_p, %	16.75	16.60	-
Plasticity number, J_p, %	10.85	3.86	-
Calculated ground resistance of the base at a depth of 2 m, R_o, kPa	270.00	234.00	200.00

As follows from table 3 loam, sandy loam, and sand are partially suitable for construction.

The sand sequence is underlain by limestones and, more rarely, dolomites, and is partially suitable for construction. But in general, a specific engineering-geological region is less favorable, since in this region there are fears of encountering karst rocks of the Kazan tier. The groundwater level confined to the Kazan aquifer is located at absolute elevations from 30 to 40 m.

The authors carried out a petrographic study of the material of samples taken from the continental layer and organic cultural layers. The research was carried out under binoculars at an 8.75-fold magnification, in the passing light of a polarizing microscope at a 72-fold magnification, and under a digital microscope at a 10-fold magnification. Generally accepted methods were used [13-18].

Samples from the underlying "mainland" layer are represented by a weakly compacted bulk mass in the amount of 2 kg. The color of the whole mass is brown. The sandy-siltstone fraction predominates, in which there are single dense well-rounded pebbles.

Pebble fraction. The shape of the fragments is slightly flattened elliptical, ranging in size from 5 to 20 mm. The surface is well smoothed with shallow indentations, dark brown to black in color. In the cleavage, brown fragments have a conchoidal fracture, a light gray color with a faint creamy tint, and a fine-grained structure. On the action of 10% hydrochloric acid solution, boiling is observed. The Mohs hardness is 3, which indicates limestone. Black fragments in the cleavage also have a conchoidal fracture, cherry-red color, structure from fine-grained to latent-crystalline. The Mohs hardness is slightly above 6, indicating jasper-like flint (figure 3).

A) B)

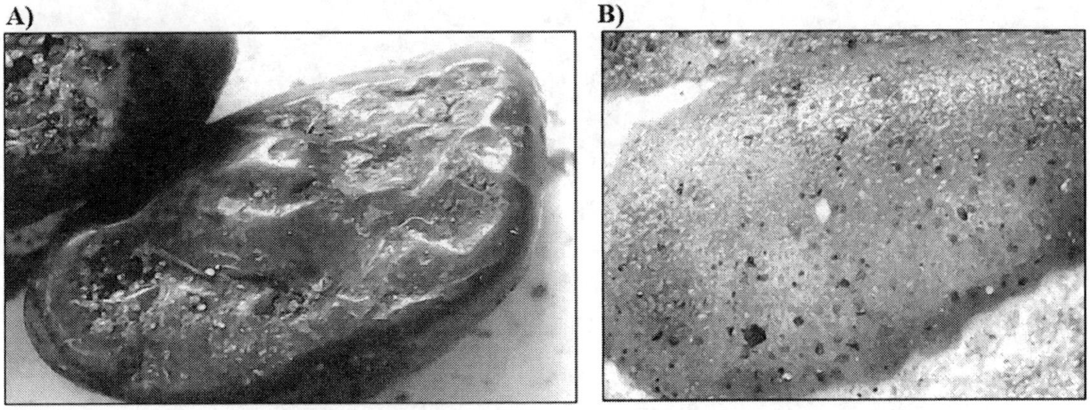

Figure 3. Photo of the pebble fraction (A-jasper flint, B-limestone).

Sand-siltstone fraction. The sand fraction has a size from 0.1 to 0.8 mm. The predominant grain is rounded with a strong glass luster, translucent in the passing light of a polarizing microscope, with a refractive index close to 1.54, which indicates quartz. In smaller quantities, there are dark brownish non-translucent angular fragments of the same size and large with a sliver-like elongated shape up to 3 mm in size, which probably indicates soil and plant formations (figure 4). The siltstone fraction has a size from 0.01 to 0.05 mm. Their shape is lamellar, angular. The composition is the same as in the sand fraction, but dark dusty grains predominate (figure 5).

Figure 4. Photo of the sand fraction.

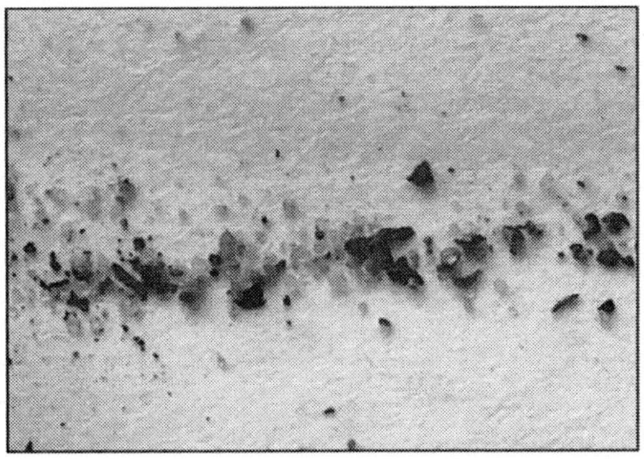

Figure 5. Photo of the siltstone fraction.

Samples from man-made deposits (organic cultural layers) are represented by a dense mass of dark brown color with not clearly defined layering. When splitting, the mass easily splits into lumpy and flattened fragments, in the cleavage of which there are intermittent and very thin layers of white color. Cracks are observed in these layers (figure 6). The mass structure is porous granular-fibrous. The pore configuration varies from incorrectly isometric to elongated in one direction (figures 6 and 7). Fine-grained inclusions are present in the total mass, but elongated wood remains with preserved microstructure predominate (figures 7 and 10). The contact of dense and uniformly porous layers is observed in individual samples. Cracks are present in dense layers, and oval formations are present in porous layers (figure 8).

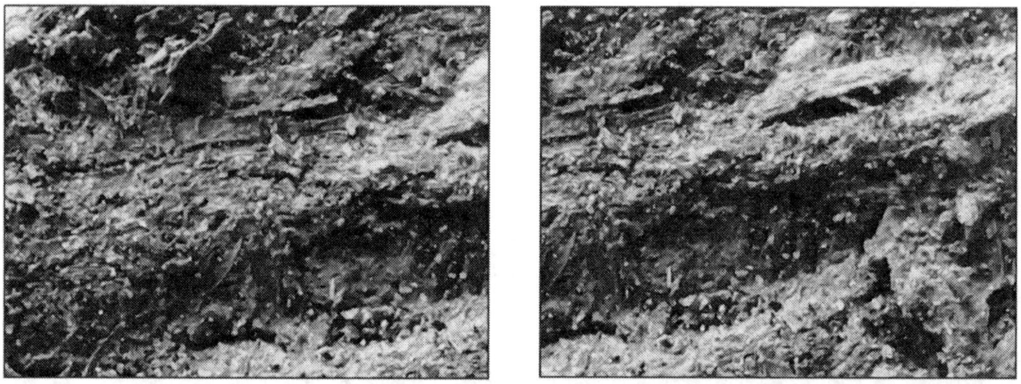

Figure 6. Chipped surface with cracks and white interlayers.

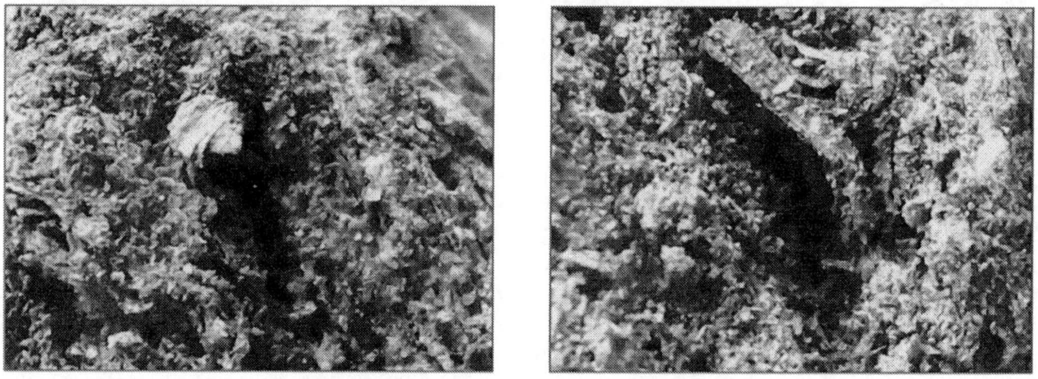

Figure 7. Treelike inclusions in a fine-grained mass.

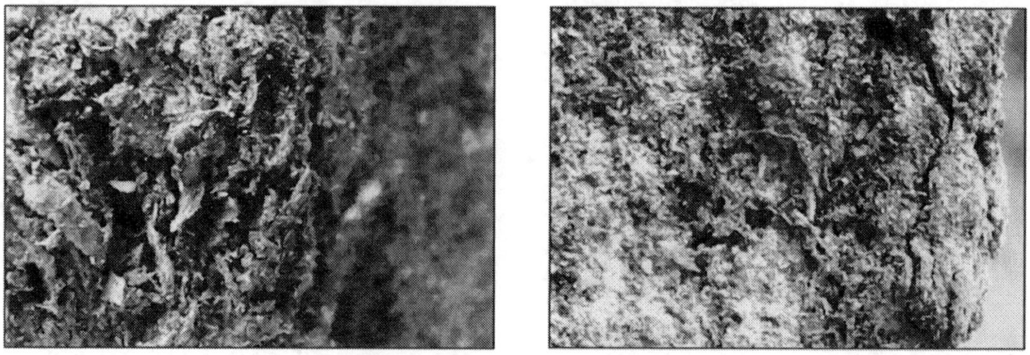

Figure 8. Contact of dense and porous layers.

These formations are covered with a white crust, which have a mossy surface, possibly fungal formations on oval inclusions of organic debris (figure 9). In the loose fraction, tree-like remains and individual quartz grains are clearly distinguished (figure 10).

Figure 9. White formations on oval inclusions.

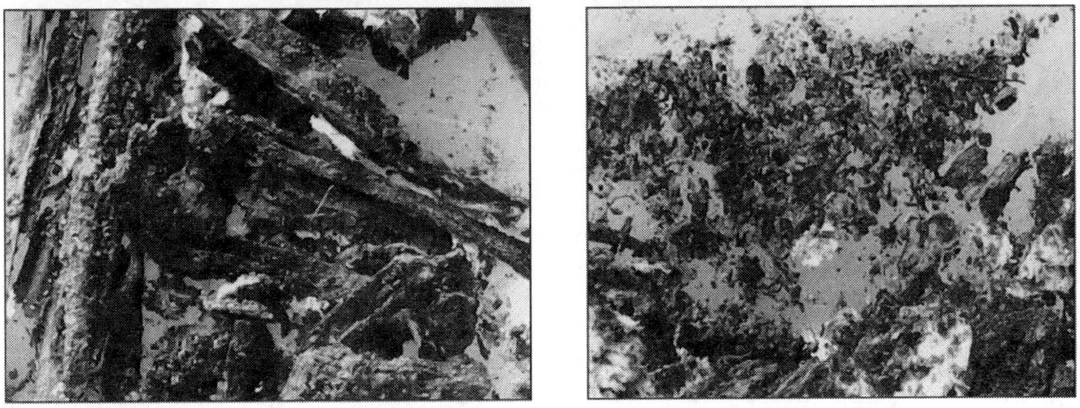

Figure 10. Plant residues and grains of transparent quartz in a loose fraction.

Thus, at the initial stage of settling the territory of Samara, the settlers faced significant difficulties due to the natural components of the landscape: high dissection and relief changes, loose sandy soils, active erosion processes. At the location of the modern Khlebnaya Square, these problems were solved with the help of the construction of wooden structures and the forming high-capacity of anthropogenic deposits of high power from organic materials and construction waste.

4. Conclusions

As a result of the study, the following conclusions can be drawn:

1. Technogenic deposits of Ancient Samara are characterized by a large thickness of organic cultural layers (7-8 meters or more), compacted, having a layer-by-layer occurrence, a porous structure, with remnants of wood and household waste.
2. Natural ancient alluvial deposits in the form of sand, sandy loam and gravel are located as the underlying rocks under the technogenic organic deposits.
3. Engineering and geological features of technogenic sediments can complicate the use of the territory for urban construction.

IOP Conf. Series: Materials Science and Engineering 1015 (2021) 012052 doi:10.1088/1757-899X/1015/1/012052

References

[1] Kazdym A A 2014 Technogenic soils and technogenic deposits, technogenic landscapes and the cultural layer - modern problems of classification and taxonomy *Soil science* **1** pp 54-70
[2] Afonin A P, Dudler I V and Ziangirov R S et al. 1990 Classification of technogenic soils *Engineering Geology* **1** pp 115-21
[3] Petrenko S A and Koff G L 1989 Engineering-geological structure and engineering-geological typification of Moscow In: *Engineering Geology and Hydrogeology of Moscow* (Moscow: Institute of the Lithosphere of the USSR Academy of Sciences) pp 22-45
[4] Report on the state and use of land in the Samara region in 2018. Federal Service for State Registration, Cadastre and Cartography Directorate of the Federal Service for State Registration, Cadastre and Cartography for the Samara Region 2019 (Samara)
[5] Baranova M N and Vasilyeva D I 2016 Geological and geomorphological zoning on the territory of Samara Traditions and innovations in construction and architecture Construction (Samara: Samara State University of Architecture and Civil Engineering) pp 189-92
[6] Vasilyeva D I 2015 Geological conditions for the construction of the Samara fortress of the 18th century on the Khlebnaya square of the city of Samara *Building a collection of articles "Traditions and innovations in construction and architecture"* ed. M I Balzannikova and etc. (Samara: Samara State University of Architecture and Civil Engineering) pp 252-7
[7] Vasilyeva D I, Baranova M N and Pivovarova L V 2018 Ravines as a geoecological factor in the use of land in settlements *Materials of the 7th international scientific-practical. Conf. "Bioecological local history: world, Russian and regional problems* (Samara: SGSPU) pp 107-10
[8] Kovalev S N 2001 *Gully erosion in urbanized areas Soil erosion and channel processes Release 13* (Moscow: Moscow State University) pp 55-84
[9] Lyubimov B P and Kovalev S N 2006 Zonal and regional types of ravines *Geomorphology* **1** pp 11-9
[10] Ibragimova S A and Kazantsev I V 2014 Characteristics of erosion processes on the territory of the Samara region *Bulletin of the Samara Scientific Center of the Russian Academy of Sciences* **16** (1-1) pp 243-6
[11] Scientific and technical report on the topic: "Inspection of the technical condition of the building of the pavilion - a cafe located on the Embankment of the river Volga in the Samara region, Samara " 2000 (Samara: SamGASA)
[12] Rachitsky V I 1969 *Engineering-geological zoning of the Kuibyshev region. Collection Geology, geochemistry and exploitation of oil fields* (Kuibyshev: KPTI)
[13] Platonov M V and Tugarova M A 2004 *Petrography of clastic and carbonate rocks*: textbook Allowance (St. Petersburg: SPbGU) p 72
[14] Baranova M N and Maltsev A V 2015 *Fundamentals of petrography of sedimentary rocks*: *Method instructions* (Samara: SGASU) p 32
[15] Kostyuk V P 2002 Fundamentals of mineralogy and petrography: a course of lectures (Samara: SGASU) p 156
[16] Kochurova R N 1977 *Fundamentals of Practical Petrography* (Leningrad: Leningrad State University) p 176
[17] Naumov V A 1981 *Optical determination of the components of sedimentary rocks* (Moscow: Nedra) p 203
[18] Tatarsky V B 1965 *Crystal optics and immersion method* (Moscow: Nedra) p 306

IOP Conf. Series: Materials Science and Engineering 1015 (2021) 012053 doi:10.1088/1757-899X/1015/1/012053

Experience in forecasting long-term settlements of energy facilities based on monitoring data analysis

A Z Ter-Martirosyan and V V Sidorov

Research and Educational Center «Geotechnics», Moscow State University of Civil Engineering (National Research University), 26 Yaroslavskoe highway, Moscow, Russia

E-mail address: vitsid@mail.ru

Abstract. The article discusses the problem of predicting the long-term settlement of reactor compartments of nuclear power plant with a long history of operation and ongoing creep processes. The features of determining the time of the primary consolidation and the application of the parameters of the secondary consolidation are considered. With regard to the existing buildings of the reactor compartments of a nuclear power plant, a forecast of long-term deformations of the base was made, from the present time to 2062, which is the end of the operation period. The analysis and comparison of the obtained results with the data of geotechnical monitoring were also carried out. Analytical forecasting showed a high accuracy in determining the end time of filtration consolidation, however, a comparison of the values of the obtained creep settlements to date and the rate of their development shows significant differences, which makes it necessary to take into account the real picture of deformation for further forecasting in terms of operation time. Possible technological reasons for discrepancies in the results of calculations and geotechnical monitoring are analyzed, as well as theoretical reasons like imperfections of the analytical method used and the possible application of modern numerical methods.

1. Introduction

Numerical methods are now widely used to predict deformations including creep. These methods give the most informative and convenient for processing and analysis results. However, such calculations require the use of special models containing specific soil parameters, which are difficult to obtain if their definition was not included in the geotechnical survey work program in advance [1,2].

For example, to perform such calculations in the PLAXIS software, the following specific parameters of the Soft soil creep model are required: modified swelling index k^*, modified compression index λ^*, modified creep index μ^* or set of alternative parameters – compression index c_c, swelling index c_s, creep index for secondary compression M. All that parameters can be determined by the testing curves analysis.

However, such data are not always provided for use, and besides, it is not always possible to perform a full-fledged numerical simulation (for example, in the absence of such works in the project). It is often necessary to perform calculations analytically using standard primary and secondary consolidation parameters[3,4]. In the described case, we are talking about predicting long-term deformations of structures that have been in operation for decades. In such conditions, it becomes necessary not only to make calculations, but also to compare them with the data of long-term observations [5]. This will allow you to correct the adopted parameters and more accurately predict subsequent deformations of the base [6-8].

The nuclear power facilities under consideration are located on the territory of Russia and represent the structures of reactor departments. The first and second power units were commissioned in 1981 and 1986.

The geotechnical conditions of the power unit operation site are well studied. From the surface, from a depth of 0.5 m to 7 m, there are technogenic, mainly sandy soils with various impurities and inhomogeneities. Under them lie Upper Quaternary glacial clayey deposits up to 11.8 m thick. In the same stratum, glacial-lacustrine and water-glacial deposits are found, represented by loamy interlayers and lenses. Below, there are mid-Quaternary water-glacial deposits, represented by sands of various sizes with a thickness of 0.5 m to 14.5 m. After that, there is a layer of mid-Quaternary moraine clays with a total thickness of up to 12.8 m. Under the Quaternary deposits, there are limestone layers with interlayers of Middle Carboniferous dolomite. In the lower parts of the boreholes, strata of heavy clays are also found. The hydrogeological conditions of the site can be characterized as difficult with three main aquifers.

Based on the systematic geodetic monitoring of settlements of the buildings, which has been carried out since 1977 for block 1 (RC-1) and since 1982 for block 2 (RC-2), it can be seen that the RC-1 building received a settlement of 483.1 mm (with an allowable value of 300 mm). For the RC-2 block, the settlement was 397.7 mm. The maximum relative settlements differences were also significantly exceeded.

Various technical materials on the object indicate various possible reasons for the development of such significant deformations of the base, among them the heterogeneity of the lithological composition of the base soils with different strength and deformation properties, the effect of prolonged dewatering on the site during the construction of an auxiliary building as part of a nuclear power plant, overestimation of the deformation modulus of the clay soils by the designers. Moreover, in the last 10 years, insignificant multidirectional values of vertical displacements of the plate have been recorded annually, which are associated with operational impacts from the station buildings, with seasonal changes in groundwater levels, and in general, in the coming years, further slow accumulation of settlements in the base of the reactor compartments is predicted with an average speed of 0.6-0.8 mm/year.

Reactor compartment No. 1 refers to the main production facilities and is part of the main production building. In the plan, it is a cylindrical monolithic reinforced concrete structure with a coating in the form of a flat spherical dome, rising to a level of 76 m. The diameter of the cylindrical shell is 47.4 m with a wall thickness of 1.2 m.

Reactor compartment No. 2 is a cylindrical structure with a diameter of 47.4 m and a height of 76.0 m to the top of the dome. The radius of the reactor compartment in the plan is 24.0 m, at a height of 12.30 - 23.7 m.

2. Methods for long-term deformations determining

2.1. Determination of the depth of the compressible strata

To assess the long-term vertical displacements of the base (settlement) of the RO-1 and RO-2 buildings, the well-known methods for determining the base deformations were used, described in PiNAE-5.10-87 "Foundations of reactor departments of nuclear power plants", SP 22.13330.2011 "Soil bases of buildings and structures".

The key point in calculating any type of foundation settlement is to determine the depth of the compressible stratum under the foundation of the structure, for which it is necessary to first construct diagrams of the distribution of natural and additional stresses in the soil base.

To construct the diagrams, data on the results of engineering and geological studies at the facility, as well as information on the bedding of soils within the structure under consideration, were used [9,10,11]. To calculate each base (under blocks 1 and 2), the most characteristic borehole in the contour of each

IOP Conf. Series: Materials Science and Engineering 1015 (2021) 012053 doi:10.1088/1757-899X/1015/1/012053

building was selected. The depth of the compressible strata under the foundation slab of RC-1 was 29.8 m, of RC-2 - 30.7 m. The graphical results are shown in figure 1.

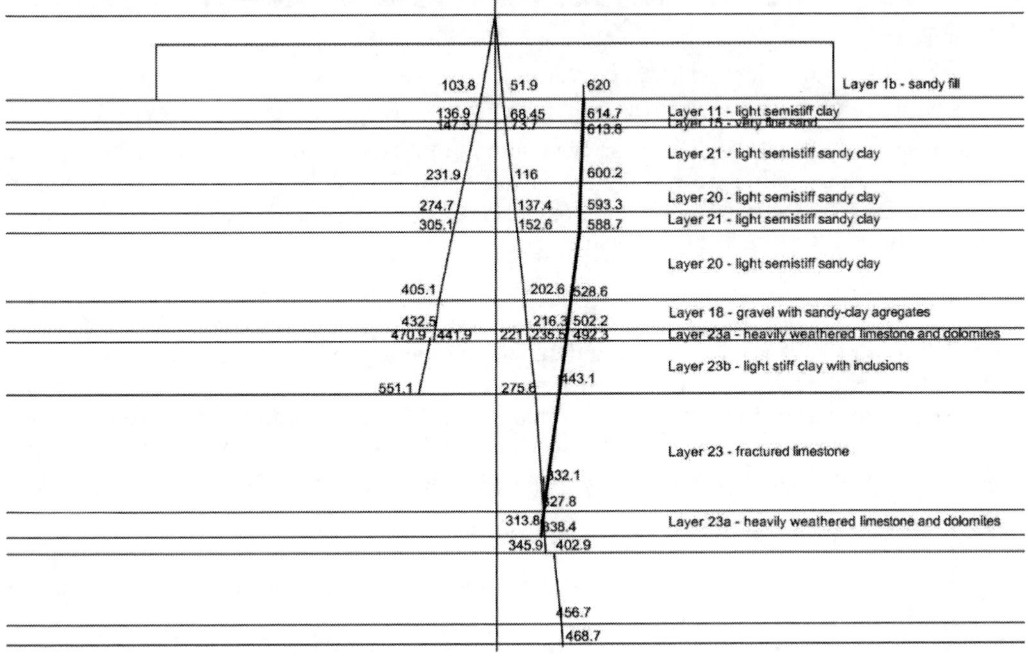

Figure 1. The scheme for determining the depth of the compressible strata under the bottom of the foundation of the RC-1 (values in kPa).

2.2. The settlement due to deformations of sandy layers and silty-clay non-creep soils

At the first stage, the settlement S_1 of the base was determined due to deformations of layers of sandy soils, as well as silty-clay soils that do not have pronounced creep properties.

It is calculated using the well-known formula (1)

$$S = \frac{\beta \cdot \sigma_{zp} \cdot h}{E} \tag{1}$$

σ_{zp}- average vertical additional stress in the elemental soil layer in the compressible strata;

β- lateral expansion failure factor;

h – layer thickness;

E – deformation modulus.

The settlement obtained for each elementary soil layer is summed up. The obtained value for the building of the reactor compartment No. 1 is 8.72 cm, for the reactor compartment No. 2 – 5.72 cm.

2.3. Settlement of the base due to deformation of layers of water-saturated silty-clayey soils

The second component of the total settlement of the structure over the entire lifetime of the project is the settlement of the base due to deformation of layers of water-saturated silty-clayey soils.

The settlement for such layers was determined in accordance with the expression (2):

$$S_w(t) = \sum_{i=1}^{n} \sigma_{zp,i} \cdot h_{cl,i} \cdot \left(m_{v1,i}(1 - \exp(-M_k)) + m_{v2,i} \cdot \ln \frac{t^1}{t_\phi} \right) \tag{2}$$

$\sigma_{zp,i}$ - average value of the additional vertical stress of the i-th soil layer;

$h_{cl,i}$ - thickness of the i-th saturated clayey soil layer;

$m_{v1,i}$ - relative compression index of the i-th soil layer due to additional pressure in soil base;

M_k - damping of the clayey soil layer coefficient;

$m_{v2,i}$ - relative secondary consolidation index of the i-th soil layer due to the additional pressure in soil base;

t^1 - secondary consolidation time (in years);

t_ϕ - primary consolidation time (in years).

It is also necessary to take into account the condition that determines the relationship between the filtration consolidation coefficient and the time during which 90% of the filtration consolidation process occurs. It is believed that during this time almost all significant precipitation occurs due to this process, the remaining 10% of the time is insignificant

$$e^{-Mk} = e^{\frac{-\pi^2 \cdot c_v}{h_{cl}^2}} = 0{,}1 \quad (3)$$

For soils showing creep, the primary and secondary consolidation factors have been determined under laboratory conditions (insert from Appendix 1).

The first part of expression (2) in brackets defines the settlement due to the primary (filtration) consolidation. The value of such a settlement can be calculated separately using the compression modulus of deformation of the layers under consideration using dependence (4)

$$m_{v1} = \frac{\beta}{E_k} \quad (4)$$

The settlement of water-saturated layers of clay soil due to primary consolidation for the base of the reactor compartment No. 1 was 10.6 cm, No. 2 – 11.8 cm.

2.4. Settlement of the base due to creep

The rest of the settlement (the second expression in brackets in (2)) will occur during creep over a long period of time in accordance with the coefficient of secondary consolidation and the corresponding coefficient of relative compressibility during creep.

Let us find the weighted average value of the secondary consolidation coefficient for a stratum of soils with creep with a thickness of 13.9 m (total thickness of creeping layers), taking into account its values in terms of cm²/year

$$\overline{c}_v = \frac{H}{\frac{1{,}6}{5751.3} + \frac{5{,}3}{8963.7} + \frac{7}{5026}} = 6150 \ (\text{cm}^2/\text{year}) \quad (5)$$

From expression (3), it is possible to express the passage time of 90% of filtration consolidation by taking the logarithm of both parts of the expression and transformations

$$t = \frac{h_{cl}^2 \cdot \ln 0.1}{-\pi^2 \cdot \overline{c}_v} = 18.32 \ (\text{years}) \quad (6)$$

Since forecasting is carried out for a long period of time (for decades), we will round this value to whole, i.e. up to 18 years (the calculation takes into account that the stratum has the ability to squeeze water out of the pores in two directions, up to the surface of the relief and down into the gravel deposits).

Then it is possible to determine the settlement of the RO-1 building in the period from the beginning of construction to the current state (that is, from 1977 to 2019), taking into account the creep.

Filtration consolidation ended in 1977 + 18 = 1995. Hence, the creep developed in 2019-1995 = 24 years.

Then the sediment from the creep process

$$S_{creep} = \sigma_{zp,i} \cdot h_{cl,i} \cdot m_{v2,i} \cdot \ln \frac{t^1}{t_\phi} \quad (7)$$

The given weighted average value of the compressibility factor for secondary consolidation is determined by the given formula

$$\overline{m}_{v2} = \overline{c}_{\alpha} \cdot \frac{1}{\overline{\sigma}} \cdot \ln \frac{80}{18} = 0.00000296 \quad (8)$$

80 years - station operation time, 18 years - the time taken to complete 90% of filtration consolidation

$$\overline{c}_{\alpha} = \frac{13,9}{\dfrac{1,6}{0.00124} + \dfrac{5,3}{0.0085} + \dfrac{7}{0.00149}} = 0.00114 \quad (9)$$

$$S_{creep} = 574,3 \cdot 13,9 \cdot 0,00000296 \cdot \ln \frac{24}{18} = 0,0068 \text{ (m)}.$$

3. Calculation results

3.1. Analysis and comparison of calculation results and monitoring data for RC-1

Thus, according to the calculations carried out, the total settlement by 2019 of the RC-1 soil base was: S = 0.0872 + 0.106 + 0.0068 = 0.2 (m) = 20 cm.

However, according to monitoring data (see figure 2), it can be seen that the settlement of the RO-1 slab is 41.63 cm by this time.

Moreover, the largest part of the settlement occurs before the end of filtration consolidation in 1995 (the curve shows a good correlation of this period with the calculations). Since the end of 90% of filtration consolidation until 2019, the building has accumulated a settlement equal to 3.63 cm. According to creep calculations, this value turns out to be much lower - only 0.68 cm. That is, the average settlement rate over a period of 24 years is 0.28 mm/year (if the average rate of settlement accumulation is determined from 3.63 cm, then it will be 1.5 mm/year). However, monitoring data from 2018 to 2019 (over 1.1 years) show an average rate of settlement increment of 0.88 mm/year. This indicates the impossibility of using the adopted creep parameters to predict the further course of the process in time, since they lead to a strongly underestimated creep settlement. Therefore, it was decided to adjust the coefficient of relative compressibility at the average rate of settlement accumulation, determined during the monitoring process.

In the technical materials on the conducted geodetic observations of the movements of the RO-1 building, carried out by specialists of NPO Gidrotekhproekt LLC in 2019, the following conclusions were drawn:

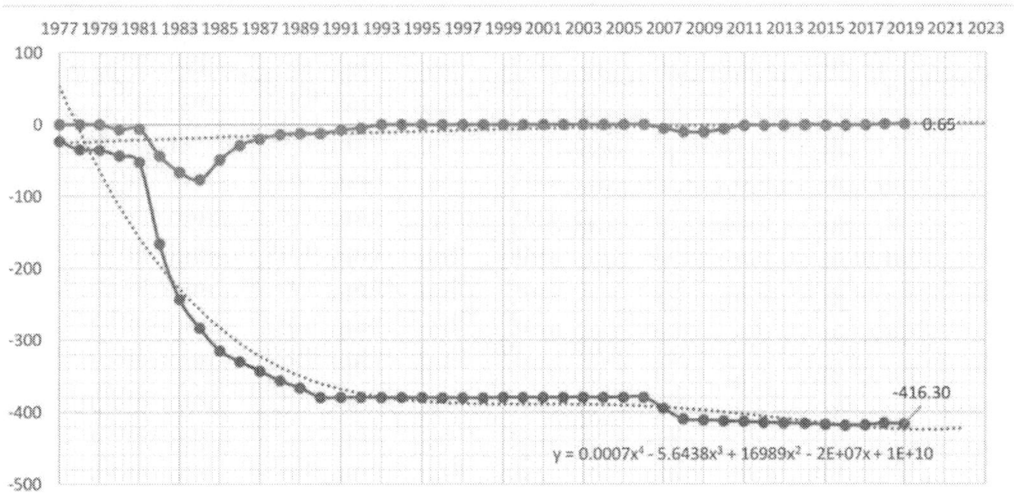

Figure 2. The curve of absolute settlement (blue line) and settlement rate (orange line) of the RC-1 foundation slab for the period 01.04.1977-29.07.2019 (NPO Gidrotekhproekt LLC).

1. The total settlement of the center of the foundation slab of the reactor compartment No. 1 for the period 01.04.1977 - 12.02.1991. is -379.1 mm.

The settlement of the plate center for the period June 25, 2018 - July 29, 2019 is -1.0 mm. The average annual precipitation rate for a period of 1.1 years is 0.88 mm/year.

The average settlement of the foundation slab for the period 12.02.1991 - 29.07.2019, with a duration of 28.6 years was -37.2 mm.

2. The total settlement of the center of the RO-1 foundation slab from the beginning of observations on 04/01/1977 to 07/29/2019 is -416.3 mm. This value was obtained by summing the settlement values over the observation periods 01.04.1977 - 12.02.1991 and 02/12/1991 - 06/25/2018 (-379.1) and (-37.2) mm.

3. The admissible values of the settlement of the reactor compartment's base are considered to be equal to 300 mm. Measured since 1977 to 2019 the average settlement of the RC-1 slab center exceeds the permissible value by 116.3 mm.

Figure 2 shows that the rate of settlement accumulation due to creep is close to the value of 0.792 mm/year (in the straight section from 2007-2017. Let us determine the value of the coefficient of relative compressibility during creep for a given speed. At this average rate of settlement for 24 years (from 1995 to 2019) will be 24*0.792 = 19 mm = 0.0190 m.

So the relative compressibility index due to creep is:

$0{,}0190 = 574{,}3 \cdot 13{,}9 \cdot m_{v2}^{\prime} \cdot \ln\frac{24}{18}$. Then $m_{v2}^{\prime} = 0{,}00000826$.

Further, the total settlement from the creep process was calculated for different periods of time:

$$S_{1995-2030} = 574.3 \cdot 13.9 \cdot 0.00000826 \cdot \ln\frac{35}{18} = 0.0438\,(\text{m});$$

$$S_{1995-2040} = 574.3 \cdot 13.9 \cdot 0.00000826 \cdot \ln\frac{45}{18} = 0.060\,(\text{m});$$

$$S_{1995-2050} = 574.3 \cdot 13.9 \cdot 0.00000826 \cdot \ln\frac{55}{18} = 0.0736\,(\text{m});$$

$$S_{1995-2057} = 574.3 \cdot 13.9 \cdot 0.00000826 \cdot \ln\frac{62}{18} = 0.0815\,(\text{m}).$$

Having received the full value of the creep settlement accumulated from 1995 to 2030, 2040, 2050 and 2057 (end of operation, t = 80 years), it is necessary to take into account that 3.63 cm has already been accumulated at the moment for the period from 1995-2019 according to monitoring data.

Therefore, was calculated the settlement increment that can be expected from now to 2030, 2040, 2050 and 2057:

By 2030: 4.38 -3.63 = 0.75 cm.

By 2040: 6.0-3.63 = 2.37 cm.

By 2050: 7.36-3.63 = 3.73 cm.

By the end of the service life in 2057: 8.15-3.63 = 4.52 cm.

You can also separately consider the pessimistic scenario, in which the rate of settlement accumulation is 0.88 mm/year (which was recorded for the period of 1.1 years from 2018 to 2019).

Having carried out similar calculations, it is possible to obtain the following characteristic values of the increment of the foundation settlement to the years under consideration:

By 2030: 4.88 -3.63 = 1.25 cm.

By 2040: 6.72-3.63 = 3.36 cm.

By 2050: 8.19-3.63 = 4.56 cm.

By the end of the service life in 2057: 9.07-3.63 = 5.44 cm.

Both obtained variants of forecasting the precipitation, taking into account different rates of development of precipitation, are shown in figure 3.

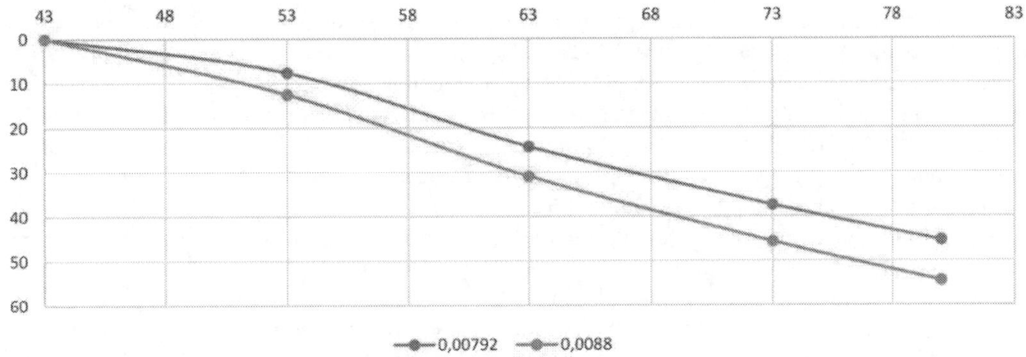

Figure 3. Curves of the increment of the RC-1 base settlement, starting from 2020 according to two scenarios (orange line due to rate of settlements of 0.88 mm/year, blue line due to rate of settlements of 0,79 mm/year), Axis X – time in years, axis Y – predicted settlement.

3.2. Analysis and comparison of calculation results and monitoring data for RC-2
Thus, according to the calculations carried out earlier, the total settlement by 2019 was:
S = 0.0572 + 0.118 + 0.0073 = 0.1825 (m) = 18.25 cm.

However, according to monitoring data (see Fig. 4), it can be seen that the settlement of the RC-2 slab is 33.58 cm by this time. Moreover, the largest part of the settlement occurs before the end of filtration consolidation in 1998 (the graph shows a good correlation of this period with the calculations). Since the end of 90% of filtration consolidation until 2019 (1998-2019), the building has accumulated a settlement equal to 335.8-327.56 = 8.24 mm = 0.00824 m. According to creep calculations, this value turns out to be less - 0.0073 cm That is, the average settlement rate over a period of 21 years is 0.35 mm / year (according to forecast calculations) and 0.39 mm / year (according to monitoring data). Therefore, to predict the further course of creep settlement, we will accept the two considered rates as optimistic and pessimistic scenarios.

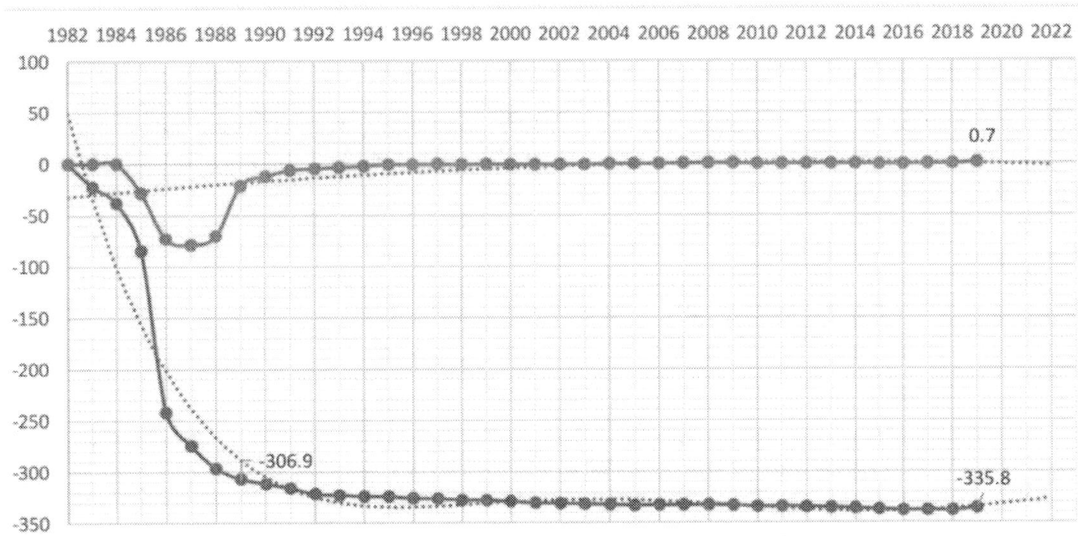

Figure 4. Curves of absolute settlement and settlement rate of the RC-2 foundation slab for the period 01.04.1977-29.07.2019 (NPO Gidrotekhproekt LLC).

The report on the conducted geodetic observations of the movements of the RC-2 building was made by specialists of NPO Gidrotekhproekt LLC in 2019.

Based on the observation results, the following conclusions were made.

1. The total settlement of the center of the foundation slab of the reactor compartment No. 2 for the period 04.11.1982 - 12.02.1991. is -316.00 mm.

Settlement of the plate center for the period of June 27, 2018 - 2.08.2019 is -0.7 mm.

The average annual precipitation rate is 0.6 mm / year.

The average settlement of the foundation slab for the period 12.02.1991 - 2.08.2019, lasting 28.5 years, was -19.8 mm.

2. The total settlement of the center of the RC-2 basement slab from the beginning of observations 01.04.1982 to 2.08.2019. is -335.8 mm. This value was obtained by summing the settlement values over the observation periods 04.11.1982 - 12.02.1991 and 12.02.1991 - 2.06.2019. (-316.00) and (-19.8) mm.

3. The admissible values of the settlement of the base of the reactor compartment are considered to be 300 mm. Measured since 1982 by 2019, the average settlement of the center of the RC-2 slab exceeds the permissible value by 35.8 mm.

For the rates obtained, the method described above was used to calculate the increments of settlement for two variants of the rate of its accumulation.

With a maximum settlement accumulation rate of 0.35 mm / year, the maximum value will be reached in 2062 with a value of 2.89 cm. Under a pessimistic scenario, with a rate of 0.39 mm / year (according to monitoring results), starting from 2020 and by the end of the service life, the increment in the settlement will be 3.38 cm.

4. Conclusions and discussion

According to the calculations performed, the following conclusions can be drawn.

1. The applied methodology for predicting the settlements accumulation in reactor compartments showed good agreement with the monitoring data in determining the time of the end of the primary consolidation (18 years for RC-1 and 16 years for RC-2), however, the settlement values obtained by this time do not converge with calculations. This is possible for various reasons, including the inaccuracy and inadequacy of the determination of the deformation characteristics of the considered soils, taken for analytical calculation, as well as the imperfection of the model adopted in the analytical method, given its relative simplicity, for example, in comparison with numerical methods.

2. Analysis of the geotechnical monitoring data over the past years indicates the practical stabilization of the sediment of the base of the reactor compartment vessels, which ended in 2008-2009. However, recent data indicate a slow and constant accumulation of deformations at a rate of up to 0.6-0.8 mm / year. This may be due to seasonal changes in groundwater levels, technological impacts during the operation of the station equipment.

3. In general, the predicted values of base deformations, taken for the calculation taking into account the data of geotechnical monitoring, look quite realistic in accordance with the long period of time remaining until the end of operation of the reactor compartments. However, 45.2 and 28.9 mm until 2057 and 2062, respectively, are significant values, given that the settlements of the departments has long exceeded the standard values established for this type of structure.

References

[1] Pen-Chen Wu, Wei-Qiang Feng and Jian-Hua Yin 2020 Numerical study of creep effects on settlements and load transfer mechanisms of soft soil improved by deep cement mixed soil columns under embankment load *Geotextiles and Geomembranes* **48** (3) pp 331-48

[2] Park C S, Ryu M Y 2019 Characteristics of Long-Term Settlement in the Soft Ground of Nakdong River by Numerical Analysis *J. of the Korean Geosynthetic Soc.* **18** (3) pp 55-67

[3] Kamao S 2019 A prediction method for long-term settlement of highly organic soft soil *Int. J. of Geomate* **17** (60) pp 162-9

[4] Karalar M, Caavusli M 2019 Examination of 3D long-term viscoplastic behaviour of a CFR dam

using special material models *Geomechanics and Engineering* **17** pp 119-31
[5] Kwon H W, Im J C, Chang J G and Kang S K 2018 A study of the long-term settlement characteristics of thick clay deposits using field monitoring *J. of the Korean Geosynthetic Soc.* **17** (4) pp 21-8
[6] Mirsayapov I T, Koroleva I V 2016 Forecast of long settlement of a high-rise building footing with the use of analytical diagrams of soils deformation *Civil Engineering* **12** pp 26-30
[7] Ter-Martirosyan Z G, Ter-Martirosyan A Z and Nguen H H 2013 Consolidation and creep of subfoundations having finite width *Bulletin of MSUCE* **4** pp 38-52
[8] Mirsayapov I and, Koroleva I V 2011 Prediction of deformations of foundation beds with a consideration of long-term nonlinear soil deformation *Soil Mechanics and Foundation Engineering* **48** (4) pp 148-57
[9] Nazli T O, Resat and Nihat S I 2020 Geo-engineering characterization and an approach to estimate the in-situ long-term settlement of a peat deposit at an industrial district *Engineering geology* **265** 105329
[10] Kapustin V V, Haustov V V, Kapustin V K 2017 Research in secondary consolidation of soils *Industrial and Civil Engineering* **5** pp 62-6
[11] Dasibekov A, Yunusov A A, Yunusova A A and Aiashova A 2013 Compaction of hereditary-aging heterogeneous earth foundations *Foundamental research* **8** (2) pp 323-31

IOP Conf. Series: Materials Science and Engineering 1015 (2021) 012054 doi:10.1088/1757-899X/1015/1/012054

Concept of management of the technolandscapes through the creation of innovative coal-watercoal clusters

T V Zommer and S N Chernyshev

Moscow State University of Civil Engineering (National Research University), 26 Yaroslavskoye Shosse, Moscow 129337, Russia

E-mail address: tzommer-1@yandex.ru

Abstract. Innovation should be at the heart of sustainable urban development. Innovations should become the basis for sustainable development of industrial and urban areas. A significant improvement in the quality of the environment can only be achieved through innovative technologies that should not degrade the environment. Coal mining at coal mining enterprises and coal burning at thermal power plants and in utility boilers are associated with a negative impact of economic activity on all geospheres. Therefore, it is important to ensure the rational use of natural resources and geo-ecological sustainability of industrial and urban areas through the introduction of innovative "green" technologies. In this regard, the authors of the article propose the concept of managing man-made and urban landscapes by creating innovative coal-watercoal clusters. The development of the concept of technolandscape management through the creation of innovative coal-watercoal clusters allows to increase the geoecological stability of industrial territories by significantly reducing the anthropogenic impact on the geosphere during coal mining and transportation. Development of scientific foundations, concepts and models of innovative coal-watercoal cluster includes the possibility of automatic control of innovative coal-watercoal technology, the concept of preparation and combustion of coal-watercoal slurry in housing and utilities boilers for the purposes of environmental management, energy and resources and reduce the negative anthropogenic impact on urbolandscape.

1. Introduction

Rational use of nature should be aimed primarily at the rational development of natural resources with the minimization of minerals extracted from nature and the prevention of possible negative consequences. However, the extraction, storage, processing, and burning of solid coal fuels at thermal power plants and in municipal boilers is associated with a negative impact of economic activity on all geospheres on a daily basis.

Modern technogenesis has features of an unmanageable process caused by almost unlimited exploitation of natural resources and increasing consumption. At the same time, technogenic processes are already comparable to geological processes that cause damage to the environment, change and create techno and urban landscapes commensurate with changes in the geological environment [1].

At the unmanaged stage of technogenesis, all geospheres and their functions undergo transformation under the influence of technology. V.T. Trofimov wrote about the technogenic transformation of ecological functions of abiotic spheres of the earth under the influence of economic and other human activities in the article [2].

IOP Conf. Series: Materials Science and Engineering 1015 (2021) 012054 doi:10.1088/1757-899X/1015/1/012054

The methodology for modeling biosphere compatibility, environmental and energy security of the "man – technology – environment" system was developed by V.A. Ilyichev – in General and in connection with the assessment of innovative technologies. In particular, in his articles, V.A. Ilyichev says that modern technologies should become the basis for the formation of an urban life support system compatible with the biosphere, and deduces the principles of reconstruction industrial and urban areas that ensure safety and comfortable living conditions [3, 4].

We believe that a significant improvement in the quality of the environment can only be achieved with the help of innovative "green" technologies, which should not worsen, but, if possible, improve the state of the existing environment [5].

Innovation should be at the heart of sustainable urban development, including in public utilities. According to the "Energy strategy of Russia for the period up to 2030", great attention should be paid to the development of innovative fuel technologies, including innovative water-coal technologies [6].

Therefore, the development of the concept of management of techno and urban landscapes through innovative coal-watercoal clusters in order to ensure rational use of natural resources and geoecological stability of territories is relevant [7].

Innovative watercoal fuel technologies have been developed around the world since the middle of the XX century. The principles and scientific and methodological foundations of rational nature management and the formation of a new technological order of sustainable development of mining enterprises and industrial territories were considered in a number of articles by academician K.N. Trubetskoy.

In particular, the scientific and methodological foundations of the strategy of a new stage of Geotechnology development in the context of the ecological crisis of modern technocratic civilization, the energy and material basis of which is the substance extracted from the lithosphere, were formulated in the form of five principles, which are mainly reduced to the conclusion of new requirements for geotechnologies related to the extraction of minerals from the lithosphere [8, 9, 10].

However, even today, almost all but a few countries in the world still burn solid coal fuels, causing damage to the environment. Therefore, we offer a simulation of controlled geoecospheric technogenesis using innovative technologies as a means of protection and automatic control of techno and urban landscapes.

In the article, we will focus in more detail on the development of scientific foundations for rational nature management with reducing the impact of economic activity on geospheres, energy and resource conservation by modeling an innovative coal-watercoal cluster of public utilities using innovative watercoal technology as a means of protection and automatic management of urban landscapes.

2. Materials and methods

In the article, the authors used systematization and analysis of data obtained from open sources, theoretical generalization used to identify shortcomings.

These methods allowed the authors to generate new ideas in the field of geoecological sustainability of industrial and urbanized territories in the development of scientific foundations, concepts and models of controlled technogenesis in the field of coal mining, processing it into a water-coal mixture with subsequent combustion in innovative clusters of public utilities.

3. Results

The main objectives of introducing new environmentally friendly innovative technologies in the energy sector should be to reduce the emission of pollutants into the atmosphere and improve energy efficiency when burning fuel.

The direction of the project under development corresponds to the priority strategy of scientific and technological development of the Russian Federation, which includes the transition to environmentally friendly and resource-saving energy, increasing the efficiency of production and deep processing of hydrocarbons, and the formation of new ways of processing hydrocarbons within the municipal economy.

IOP Conf. Series: Materials Science and Engineering 1015 (2021) 012054 doi:10.1088/1757-899X/1015/1/012054

Scientific and practical development of the model of an innovative coal-watercoal cluster of public utilities allows increasing the geoecological stability of urbanized territories by significantly reducing the anthropogenic impact on the Earth's geospheres during the production and combustion of innovative environmentally friendly coalwater fuel mixture in boilers instead of solid coal.

The project we are developing includes fundamental scientific research aimed at reducing the anthropogenic impact on the environment and improving the geoecological stability of urbanized territories based on the development of a model of a coal-watercoal cluster of public utilities, followed by its implementation at the pilot site.

The study identified the main objectives of the project, which includes developing the scientific basis, concepts and models of innovative coal-coalwater utility cluster based on existing and under construction boiler.

The tasks also include the search for the possibility of automatic control of innovative coal and watercoal technology, a schematic diagram of the preparation of watercoal fuel suspension using the principles of rational nature management, energy and resource conservation, and reducing the negative technogenic impact on geosfere.

At the moment, we are developing the scientific basis, concept and model of an innovative coal-watercoal cluster of housing and communal services based on existing and under construction boilers, including a comparative geoecological assessment of coal and innovative coalwater technologies taking into account their impact on geospheres, automatic management of innovative coalwater technology, a schematic diagram of the preparation of coalwater fuel suspension using energy and resource saving, reducing the negative technogenic impact on techno and urban landscapes.

As a result of our research, we have identified the main provisions of the concept in terms of coal-watercoal technogenesis (figure 1).

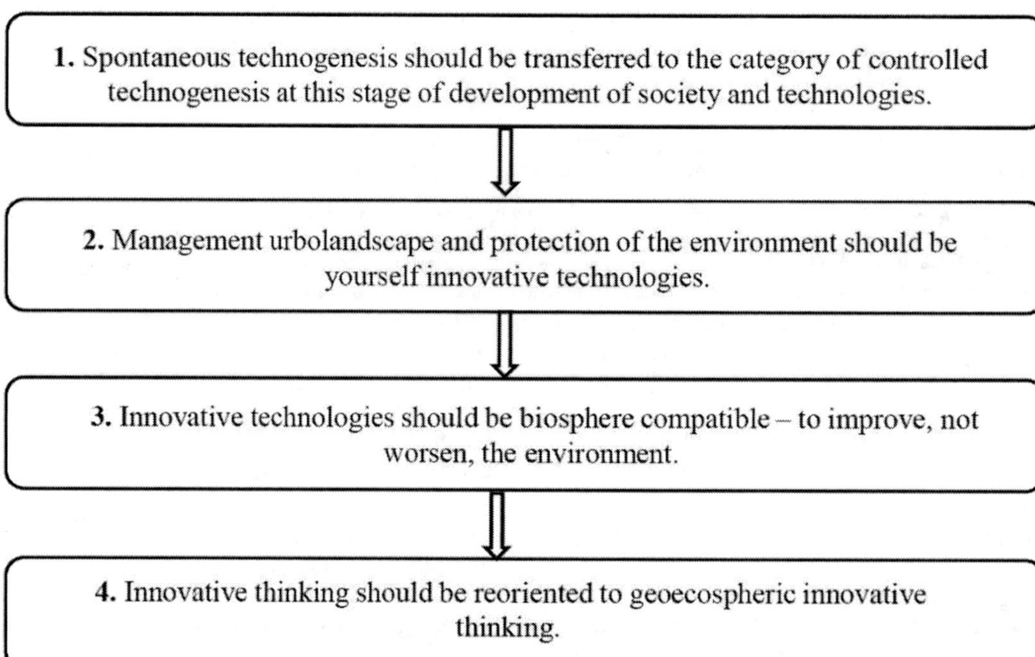

Figure 1. The main provisions of the concept in terms of coal-watercoal technogenesis.

IOP Conf. Series: Materials Science and Engineering 1015 (2021) 012054 doi:10.1088/1757-899X/1015/1/012054

Even earlier, in the article, we proposed to consider innovative "green" technologies as the basis for rational use of natural resources and management of technolandscapes on the example of the development and implementation of innovative coalwater technology [5].

Many people perceive innovative state policy only as a means of directing investments in high-tech production and high-tech innovative technologies, and as a result – in uncontrolled technogenesis as a product of technologies.

This approach to innovation needs to change. Taking this into account, we can conclude that new fuel technologies are needed at the moment, aimed not only at achieving an innovative effect, but also at reducing the man-made impact on all geospheres. In other words, we need innovative technologies that are aimed not only at technological progress, but also at improving the earth's ecology as a whole.

With the existing coal fuel technology, incomplete combustion of solid coal occurs, leading to mass emissions of dust, pollutants, which causes acid rain, thermal pollution of the atmosphere, and self-exaggeration of the greenhouse effect.

In the course of the work, a comparative geoecological assessment of coal and innovative coalwater technologies will be given, taking into account their impact on the Geosphere, based on a specially developed geoecological methodology.

The completion of the project includes the development of a schematic diagram for the implementation of an innovative coal and water coal cluster of public utilities (road map) with a set of measures.

Therefore, in comparison with existing boilers, the innovative coal-watercoal cluster of housing and communal services can be considered as a comprehensive solution to intersectoral problems of science and technology.

Tests of innovative coal fuel at the experimental site of municipal services have shown that the production and combustion of coal in the form of innovative watercoal suspension reduces nitrogen oxide emissions compared to coal combustion by 20-40%; mechanical underburning of coal in the composition of VUT is no more than 1%, which eliminates slag emissions into the atmosphere, environmentally safe fuel at all stages of production and use, even with accidental spills does not cause damage to the environment; ash from WUT incineration contains no more than 3% of unburned carbon and can be used as a concrete filler.

A fairly simple modernization of conventional coal-fired boilers in housing and utilities boiler houses with their transformation into special installations for burning innovative water-coal fuel leads to a reduction in fuel costs by 30%...60% compared to fuel oil and by 20...40% compared to burning solid coal, which is very significant [6, 7].

The formation of an innovative housing and utilities cluster can take place with the participation of existing sections of storm sewers and snowmaking plants. Currently, snow and storm water is discharged after treatment into the storm sewer, although it can be used in principle to prepare a water-coal suspension in a coal-watercoal cluster.

In this case, the solution to the oncoming problems leads to higher profitability of coal-watercoal suspension and coal-watercoal cluster housing in General, helps to achieve rational use of natural resources, energy and resource conservation and geoecological sustainability of industrial and urban territories.

In addition, it is important that in an innovative coal-water coal cluster, the raw material for self-prepared watercoal fuel can be coal screenings, as well as slurries, which are usually disposed of as waste. That is, a coal-watercoal cluster can work on coal fines and waste from the coal industry (figure 2).

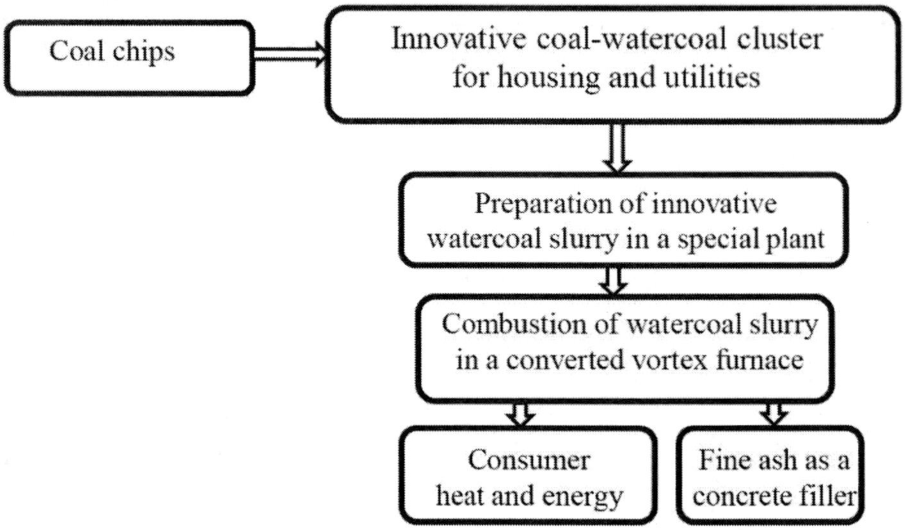

Figure 2. Schematic diagram of a coal-watercoal cluster housing and utilities.

The decrease in the number of mines will lead to the predicted reduction of the negative impact on the lithosphere, and the hydrosphere of techno-landscapes, including the reduction of disturbed areas, man-made cavities, heaps of inflow of mine waters, etc.

At the same time, almost one hundred percent combustion of coal with the mass introduction of innovative water-coal fuel technology will predictably contribute to a drastic improvement of the atmosphere of urbanized territories.

In our opinion, the geoecological stability of industrial and urbanized territories can be achieved only in the conditions of coal-watercoal geoecospheric technogenesis.

4. Conclusions

The development of scientific bases, concepts and models of controlled technogenesis in terms of coal technogenesis with rational use of strategic coal fuel reserves allows us to turn innovative technologies into means of protection and automatic management of techno and urban landscapes with a reduction in the impact of economic activities on geospheres.

In particular, the simulated innovative coal-water-coal cluster of housing and communal services makes it possible to use innovative coal-water-coal technology in economic activity as a means of protection and automatic control of urban landscapes with a reduction in the impact of economic activity on geospheres.

Scientific and practical development of a model of innovative coal-coal-water cluster housing allows to increase the geoecological sustainability of urban areas by significantly reducing the human impact on the Geosphere in the production and combustion in the boiler housing of the innovative clean coal-water fuel mixture, instead of hard coal.

In comparison with existing boilers, the innovative coal and water coal cluster of housing and communal services can be considered as a comprehensive solution to intersectoral problems of science and technology.

A fairly simple modernization of conventional coal-fired boilers in housing and utilities boiler houses with their transformation into special installations for burning innovative water-coal fuel leads to a reduction in fuel costs by 30%...60% compared to fuel oil and by 20...40% compared to burning solid coal, which is very significant.

The solution to the associated problems of Geoecology leads to higher profitability of coal-water slurry and coal-watercoal clusters in general, helps to achieve rational use of natural resources, energy and resource conservation and geoecological sustainability of industrial and urban territories.

The concept of management of techno and urbolandscape through the creation of innovative coal-coalwater clusters allows the transition from unmanaged technogenesis stage to the managed stage in the field of coal technogenesis, geoecological to improve the sustainability of industrial and urban areas by significantly reducing human impact on the geosphere.

Acknowledgments
The reported study was funded by RFBR, project number 19-35-90135.

References

[1] Lavrusevich A A 2010 The main features of technogenesis *Vestnik MGSU* **4** (2) pp 175-81
[2] Trofimov V T, Kharkina M A, Zhigalin A D and Baraboshkina T A 2019 Technogenic transformation of the ecological functions of the abiotic spheres of the earth under the influence of military activities *Moscow University Geology Bulletin* **74** (2) pp 121-30
[3] Ilyichev V, Emelyanov S, Kolchunov V and Bakaeva N 2015 About the dynamic model formation of the urban livelihood system compatible with the biosphere *Applied Mechanics and Materials* **725-726** pp 1224-30
[4] Ilyichev V. Kolchunov V, Bakaeva N and Emelianov S 2018 Principles of urban areas reconstruction ensuring safety and comfortable living conditions, *IOP Conference Series: Materials Science and Engineering* 032011
[5] Zommer T V and Chernyshev S N 2016 Innovative geotechnologies is the key to geoenvironmental sustainability of urban areas by reducing the load and control techno-landscapes on the example of innovative water-coal technologies, *MATEC Web of Conferences* **5** IPICSE 03011
[6] Morozov A G 2009 RU 82482 U1, *Installation for burning coal-water fuel*
[7] Zommer T V, Chernyshev S N and Zommer V L 2017 Management of the urban landscapes on the basis of geo-bio-positive coal-watercoal technologies *Economy and Entrepreneurship* **4** (2) pp 969-73
[8] Trubetskoy K N, Galchenko Y and Eremenko V A 2018 Fundamentals of converging mining technologies in integrated development of mineral resources of lithosphere *IOP Conference Series: Earth and Environmental Science* 012064
[9] Trubetskoy K N and Galchenko Y P 2017 Naturelike mining technologies: prospect of resolving global contradictions when developing mineral resources of the lithosphere *Herald of the Russian Academy of Sciences* **87** (4) pp 378-84
[10] Trubetskoy K N et al 2018 RU 2684647 C1 Method for controlling the geomechanical state of a rock mass

IOP Conf. Series: Materials Science and Engineering 1015 (2021) 012055 doi:10.1088/1757-899X/1015/1/012055

Preliminary results of geodetic diagnostics of transition areas of ballastless track

J Ižvoltová, J Chromčák and P Pisca

University of Žilina, Faculty of Civil Engineering, Department of Geodesy, Univerzitná 8215/1 Žilina, Slovak Republic

E-mail address: jana.izvoltova@uniza.sk

Abstract. Construction of ballastless track is usually used in high-speed railway, which is suffered by high traffic load. The main reason is to reduce the maintenance costs of the railway line, although the initial costs of its construction are incomparably higher than with the classic construction. From the point of view of long term deformation analysis of railway track, the critical point of such the type of the construction seems to be the place of its completion and transition to the classical the railway superstructure. In terms of dynamic effects, it is a place with a change in stiffness and therefore a special attention is paid to its diagnostics.

1. Introduction

A railway track with a classic construction is considered to be such a track whose rail grate is placed in a gravel bed. In modernized lines, this classic structure is often replaced by an unconventional structure in which the rail grate forms a monolithic structure, or is placed on a concrete or asphalt construction layer. Such a construction of the railway superstructure is called ballastless track and is characterized by the use of transverse sleepers in a modified shape [1]. The Faculty of Civil Engineering ensure of the University of *Žilina* carries out a comprehensive diagnostics of the spatial position of the ballastless track in the *Turecký vrch* tunnel and its surroundings, in order to detect the behaviour of such a structure during operational load. The diagnostics includes geodetic measurements, the task of which is to check the geometrical parameters of the railway line and its spatial position. For the reason to ensure the required accuracy and reliability of results, the precise geodetic technology was used such as robotic theodolites, 3D laser scanners, GNSS receivers and digital levels. The geodetic solution resulted in a deformation analysis of the given railway structure in order to define the height and positional displacements of both tracks. The main goal of the research task is to define the potential spatial changes of the track in its most critical sections, in the so-called transition areas of the track, represented by the transition between ballastless track and classical track. From the point of deformation analysis, the very interesting places of the new railway construction seems to be the transition area between tunnel and open railway line and places built on bridges. Because, the observation technology and evaluation methodology have been already presented in previous studies [1], [2], [3], this paper is devoted to the results publication especially those that deal with analysis of deformations arising in the transition sections of ballastless track.

2. Characteristics of railway locality

The *Turecký vrch* railway tunnel is the first tunnel in Slovakia designed and implemented according to technical specifications defined for unconventional lines. The total length of the tunnel is 1775 m and the tunnel tube of the excavated section has a length of 1738.5 m. The tunnel is followed by concave sections of the southern portal in the length of 25 m and the northern portal in the length of 10 m. The double-track is designed for a speed of 200 km/h, with opposite curves with a radius of 2000 m. Due to the complex directional conditions and bridge structures located immediately behind the northern portal, the ballastless structure is established not only in the tunnel itself, but on the northern side of the tunnel in the length of 800 m. The total length of the built ballastless structure is up to 4480 m of track. Due to the difference in the stiffness of the subsoil (tunnel bottom, bridge and railway bottom of the earth body), the system of ballastless structure is modified, which is reflected in the thickness of the concrete structure and also in its reinforcement [4].

3. Measurement of spatial track position

The diagnostics of the ballastless track in *Turecký vrch* locality was realized in the railway section km 102.360 000 – 102.530 000, of southern portal of the tunnel and in the section km 104.200 000 – 104.820 000 of its northern portal. The project of geodetic monitoring started in 2012 and continued firstly in half-yearly intervals and then in annual time measuring cycles, depending on the behaviour of the railway structure under traffic loading. In order to ensure high accuracy, a separate measurement of the position and height of the track was performed. The position of the track was determined by discrete and continuous measurements by done by robotic theodolites and the height was measured by digital levels. An essential part of geodetic measurements is the control of positional and height stability of the reference system, represented by railway benchmarks, which results in the definition of its accuracy, from which the a priori accuracy of the measurement is derived. Geodetic observation of particular points was realized on both of track's sides in the defined distances of the longitudinal direction (approx. every eight track node). The spatial changes of the track was related to the axis of the right rail string to maintain the accuracy and explicitness of the results and also because of the mutual comparison of the discrete values.

4. Analysis of positional track changes

Positional measurements were performed with universal robotic theodolites TRIMBLE S8, TRIMBLE VX, with a defined angular accuracy of ± 0.3 miligons and a distance accuracy of 1 + 1 ppm and 1 + 2 ppm. The reference system consists of railway benchmarks signalled by portable reflective foils and stakeout points stabilized by screws on the feet of electric poles. The stability of the reference points was determined by checking the relative position of the both types of marks. The control of the geometrical position of the particular observation cycles. Diagnostics of positional changes of the railway structure is performed by statistical hypothesis testing. The positional changes are tested by a simplified null hypothesis, defined by the following criteria:

- If the positional change is less as a-priori positional accuracy, then null hypothesis is confirmed and the motion of the structure is not proven
- If the positional change is less as twice a-priori positional accuracy, then null hypothesis cannot be confirmed and a positional change of the structure can be assumed
- If the positional change is greater as twice a-priori positional accuracy the positional deformation is statistically significant and the null hypothesis is rejected, which means that there is a probability of a positional change of the structure

By analysing the accuracy of measurements realized in particular observational cycles, the accuracy of determining the transitional track change is defined by value of standard deviation 0,022 m.

5. Analysis of height track changes

Height reference network consists of twenty benchmarks permanently stabilized in the concrete bases of electric poles. The process of checking the stability of the height reference network consisted of

IOP Conf. Series: Materials Science and Engineering 1015 (2021) 012055 doi:10.1088/1757-899X/1015/1/012055

estimating the parameters of the regression line, constructed by the differences of height changes (superelevations) between the first and the i-th measurement cycle, determination of outliers and estimating the a-priori accuracy of the network applying numerical solutions according to Marčák [5] and Rousseeuw [6]. When evaluating the results, the height shift of the track is considered significant, in case of exceeding the critical limits, which represent twice the a priori standard deviation of height.

6. Analysis of results

Due to the use of a special device for signalling the position of particular points, relative to the outer side of the rail, special attention was paid to the elimination of the pointing error. The process of evaluation of the positional measurements confirmed the expected value of pointing error in the transverse direction ± 2 mm and in the longitudinal direction ± 20 mm. For this reason, the transverse track displacement in the particular measuring cycles was related to the defined directional parameters of both tracks and the longitudinal track displacements were not defined due to the exceeded the value of the targeting uncertainty. The average value of the transversal shifts as well as the relevant accuracy characteristics of the measurements are introduced in the table 1. The value of the reliability intervals of the particular measurements was defined by using probability value α=0,05 and coefficient t_α=2 represented the quantile of Student distribution by assuming normal distribution of the measurement errors.

Table 1. Standard deviations of positional observation of ballastless track

Observation cycles	Southern portal		Northern portal	
	Track 1	Track 2	Track 1	Track 2
10.2012	4,8 mm	3,2 mm	4,4 mm	5,0 mm
04.2013	2,7 mm	3,1 mm	3,0 mm	5,9 mm
10.2013	3,0 mm	5,6 mm	3,6 mm	3,6 mm
05.2014	2.9 mm	3,2 mm	3,2 mm	3,8 mm
03.2015	7,5 mm	5,3 mm	3,2 mm	3,9 mm
03.2016	4,2 mm	3,8 mm	1,7 mm	3,8 mm
03.2017	2,9 mm	3,0 mm	4,0 mm	6,8 mm
04.2018	5,7 mm	4,1 mm	4,2 mm	4,0 mm
06.2019	4,6 mm	4,5 mm	3,8 mm	4,1 mm

The accuracy of height changes of ballastless track is defined by the value of unit height standard deviation obtained by precise levelling. The particular values of the unit standard deviations are introduced in table 2 for every measurement cycle.

There were four critical sections in the northern portal of the tunnel "Turecký vrch":

- Transition area between tunnel and river bridge in km 104.260 000 – 104.285 000,
- Railway bridge in km 104.280 000 – 104.300 000,
- Railway bridge in km 104.585 000 – 104. 600 000,
- Transition area between the slab track and the ballasted track in km 104.720 000 – 104.745 000.

A demonstration of time dependency of the positional and height changes of the particular points laying on the transition area between the ballasted track and the slab track in the observed area of the northern portal is in figures 1 and 2.

Table 2. Unit height standard deviations of ballastless track precise levelling

Observation cycles	Southern portal		Northern portal	
	Track 1	Track 2	Track 1	Track 2
10.2012	0,26 mm	0,30 mm	0,41 mm	0,41 mm
04.2013	0,42 mm	0,25 mm	0,32 mm	0,45mm
10.2013	0,34 mm	0,41 mm	0,33 mm	0,22 mm
05.2014	0,40 mm	0,41 mm	0,50 mm	0,18 mm
03.2015	0,25 mm	0,20 mm	0,42 mm	0,46 mm
03.2016	0,19 mm	0,21 mm	0,31 mm	0,34 mm
03.2017	0,22 mm	0,24 mm	0,38 mm	0,34 mm
04.2018	0,25 mm	0,20 mm	0,44 mm	0,37 mm
06.2019	0,35 mm	0,24 mm	0,38 mm	0,40 mm

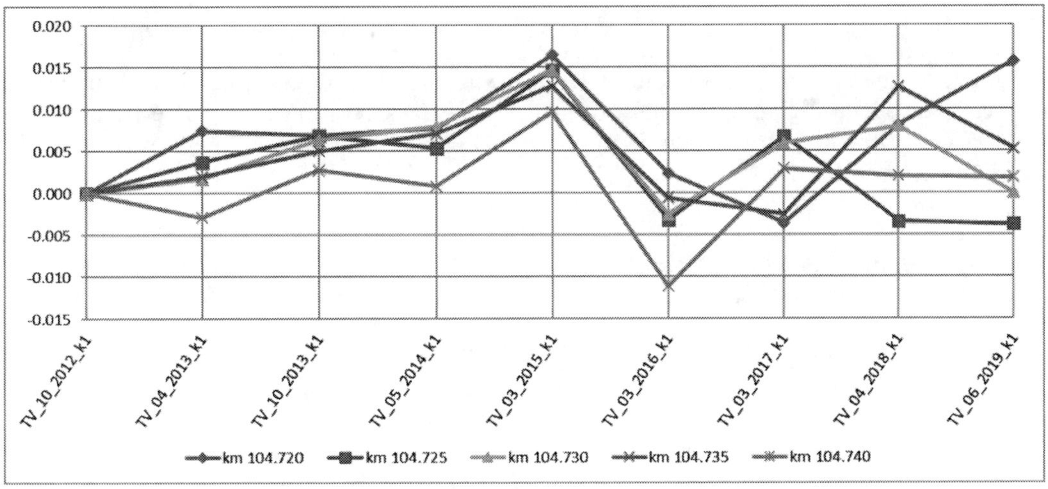

Figure 1. Time dependency of transversal shifts of transition area between ballastless and classical track.

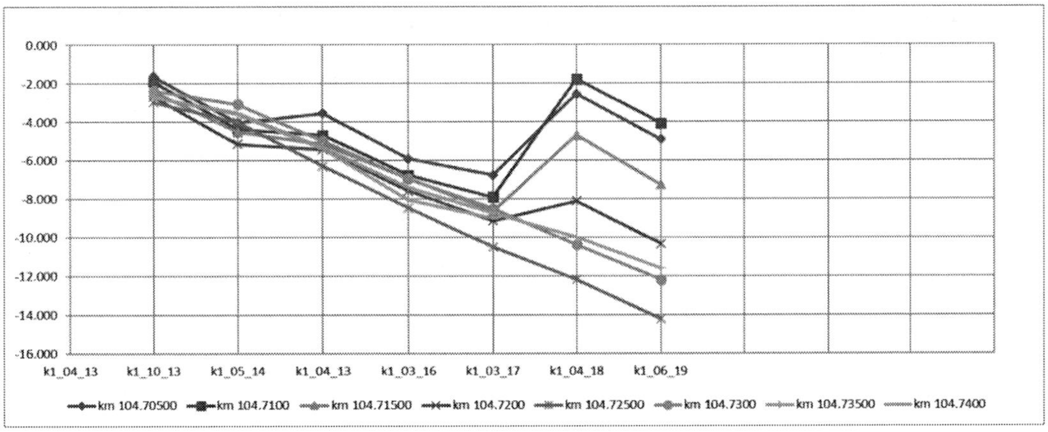

Figure 2. Time dependency of height changes of transition area between ballastless and classical track.

7. Conclusions

The main aim of monitoring the spatial position of the ballastless track in locality *Turecký vrch* is to determine its behaviour influenced by the traffic load. Since, nine observation cycles have been realized in one kilometre long area of both southern and northern portal of tunnel. The paper is devoted to the analysis of the accuracy and positional and height changes arise in critical sections of observed ballastless track, from which the special attention is devoted to the transition area between ballastless track construction and classical track. The track behaviour of this section is displayed in the figure 1 and 2. The measurement results processed so far show small positional a height changes of track, especially at points laying in the critical sections. The changes of spatial position of the other particular points did not confirm so far. The research task focused on the diagnostics of geometric parameters of ballastless track will continue by comparing discrete measurements with the point cloud applications obtained by 3D laser scanner and GNSS continual technique, to find out accuracy, reliability and efficiency of the use of the particular geodetic methods. The ambition of such a geodetic monitoring ballastless track is in contribution to the decision-making processes that are part of the regular inspections and maintenance of the railway line.

Acknowledgments

This article is the result of the implementation of the project VEGA 1/0275/17 "Application of numerical methods to define the changes of geometrical track position", supported by the Scientific Grant Agency of the Ministry of Education, Science, Research and Sport of the Slovak Republic and Slovak Academy of Sciences.

References

[1] Izvoltova, J, Pisca P, Villim A and Mancovic M 2014 Precision analysis of height measurements realized on ballastless track *Communications 4 EDIS 2014 University of Žilina*, ISSN 1335-4205) **16** pp 63-7
[2] Ižvoltová J and Koťka V 2016 Numerical adjustment of continual observations of ballastless railway line *16th International Multidisciplinary Scientific Geoconference* (SGEM 2016 Albena Bulgaria) **II** pp 473-8
[3] Ižvoltová J and Cesnek T 2017 Accuracy analysis of continual geodetic diagnostics of a railway line *International Scientific and Professional Conference on Geodesy, Cartography and Geoinformatics* 10.10.2017
[4] Ižvolt L, Hodás S and Šestáková J 2015 Railway buildings 1. Design, building, construction of railway lines and stations Handbook *EDIS 2015* p 561
[5] Marcak P and Kubacek L 1974 The Problem of the system of height reference in determining the settling of foundations and buildings *Studia Geophysica et Geodaetica* **18** pp 33-46
[6] Rousseeuw P J and Leroy A M 2003 *Robust regression and outlier detection* (New Jersey: Wiley & Sons) p 329

IOP Conf. Series: Materials Science and Engineering 1015 (2021) 012056 doi:10.1088/1757-899X/1015/1/012056

Map bases used for ecological stability assessment of land

J Chromčák and A Seidlová

University of Žilina, Faculty of Civil Engineering, Department of Railway Engineering, Univerzitná 8215/1, SK-010 26, Žilina, Slovak Republic

E-mail address: jakub.chromcak@uniza.sk

Abstract. The interest of land stability and its´ calculation due to the ecological stability and long term use of land as an entity is under pressure not only from ecological activists but the whole society at all. To calculate these kind of unknown values, which are interpreting many land possibilities and characters needed for many others calculations. That is the reason to find effective way to calculate the ecological stability index. The calculation of this index directly depends on character of area and its´ size. In Slovakia, there are many map bases with possible use for this purpose on whole territory. This paper is focused on the map base and their possibilities to be used.

1. Introduction

An ecological stability as an ecological systems´ stability to offset the external disturbances by their own functional mechanisms without an additional energy input (means for example without a human activity by using so called biological automation) has become part of many dialogues of political parties and hand in hand with the laws focused on this kind of interest, although an basic research of the stabilizing mechanisms will still remain under investigation for a long time. [4] An assessment of the landscapes´ ecological stability is considered to be the basis for the assessment of all conditions and assumptions of land use. This paper will focus on the possibilities of map selection for ecological stability assessing.

2. Land adaptations

The land adaptations are carried out in accordance with act 330/1991 of land adaptations, land ownership arrangements, land offices, land fund and about land associations. The content of the land adaptations is a rational spatial arrangement of the land ownership in a certain area and other immovable agricultural and forest property and property associated with it; carried out in the public interest accordance with the requirements and conditions of environmental protection and the creation of ecological stability system, landscape functionality and operational aspects of a modern agriculture and forestry hand in hand with the support of the rural development. [1]

Land adaptations include:

- an identification and a reorganization of the ownership and the conditions of use as well as related other material rights in the area of land adaptation and a redistribution of the land like merging, separation or other land readjustments;
- technical, biological, environmental, economic and legal measures related to the reorganization of the legal relations.

From a procedural point of view, land adaptation has 4 main stages. The main stages are proceedings for the commencement of land adaptation; proceeding of land adaptation; implementation of the land adaptation project; realisation of the project. The geodetic activities are necessary in the individual stages of the land adaptation. There are activities such as the establishment of a minor control, setting out of the perimeter of land adaptation area, a measurement of the focused area topography. In the other activities in the cooperation with the landscaping designers, there is focus on the ecological stability of the landscape. When the assessing the landscape ecological stability, unambiguous results can be achieved on the basis of a coefficient of ecological stability calculation (KES), which is based on a map on which the cultures′ individual areas can be defined.

3. Map basis

The degrees of the ecological stability can be ascertained from the projects of the regional territorial systems of the ecological stability in the scale of 1:50000 (or scale 1:25000) which is processed for the whole territory of Slovakia. [5] At the present, the products of GKÚ, such as orthophotomaps or opthophotomosaics can be used. Nowadays, the foresters and cartographers of Slovakia have connected to create an active orthophotomosaic, which can be used and is available to everyone for free online. In Slovakia, for the whole territory, there is possibility of digital cadastral map use. For the land adaptation, and to calculate the ecological stability coefficient, there is necessity to know not only the value of whole area, but the value of area hand in hand with the category of ecological stability. There is a requirement to divide the focused area due to its′ character (such as wood, water areas, meadow, urban area etc.), and to know the value of these areas. As it is obvious from the content of VKM, it is impossible to get to know the needed information, in this format it is impossible to use this kind of map.

Figure 1. Example of digital cadastral map on chosen area available online

The other possibility is use of digital base map, prepared for the whole territory of Slovakia as well. The character of area in meaning of use is better than in VKM. One example is, that one parcel (parcela) can obtain areas with other characters, but the ownership has priority before the character, the situation is different of digital base map. There is no focus on the ownership, and also better space to display the character of areas. The problem of the map is that the division of areas by the character is insufficient, and the question of map accuracy is also the theme needed to be focused on. The solution can be defining of new values of area character applicable for already created type of map. It means that there is way in reindexing of ecological quality coefficients due to used kind of characters, used in digital base map. Like it is possible to see on figure 2, the meadows or arable land are displayed the same, index of these two characters in regular used law is different. The question of reindexing of areas has to be theme of a professional discussion.

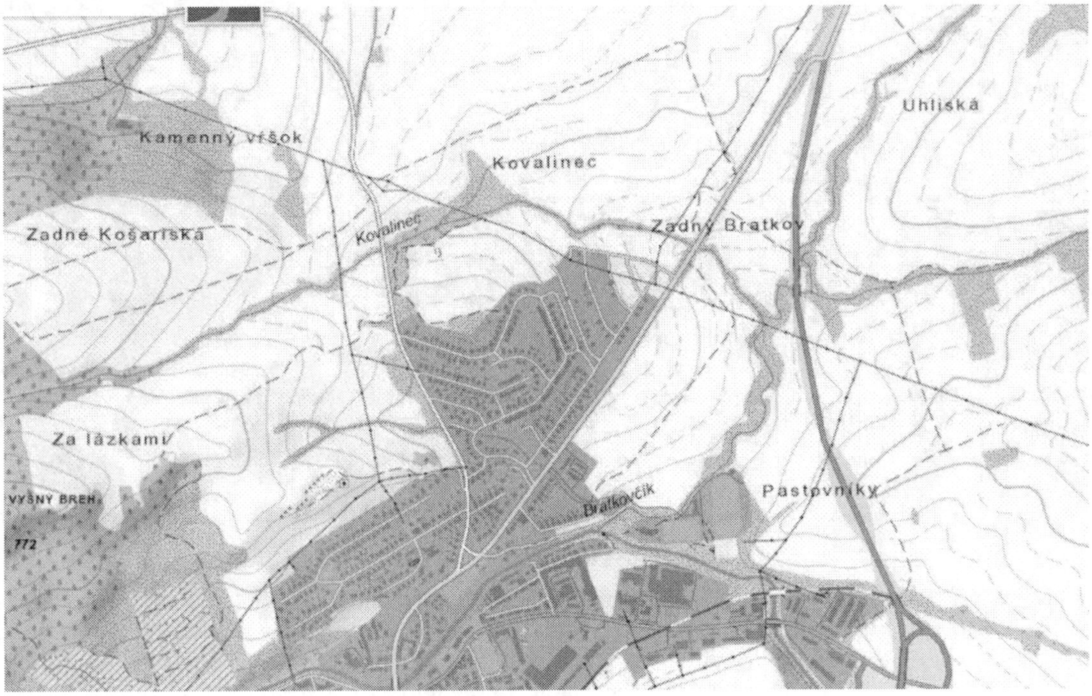

Figure 2. Example of digital base map available online

For the first time in national history, the National Forestry Center (NLC) in Zvolen and Geodetic and Cartographic Institute in Bratislava (GKÚ) have created the ortophotomosaic of the entire territory of Slovakia. This work, prepared in accordance with the rules of European Commission and available to the general public will significantly contribute to the streamlining of a public administration. As the NLC spokesman Ján Lichý informed for TASR, the creation schedule was divided in three years (2017-2019) so during each year in the growing season, data were collected and processed from one third of whole Slovak Republic territory (approximately 17,000 square kilometres). [2] For the first time in history, ortophotomosaics were processed with cyclical update of the agricultural production blocks register. It was ensured in time, thus eliminating the payment of support for ineligible land and subsequently preventing the possibilities of European funds shortening for the Slovak Republic. Another, equally important fact is that for the first time in forever, orthophotomosaics in a significant part of the territory were created from processed aerial laser scanning data. A new generation of the digital terrain relief model was used in the creation, which had a significant impact on achieving the accuracy of focused orthophotomosaics. Colour orthophotomosaics are available through a web application, hereinafter referred to the so-called web services and is available for free to all users, including commercial entities, also for download in the form of georeferenced raster images.

According to Lichý, approximately 80% of decisions made in public administration have a spatial aspect.

The products are used to create and update topographic map data, the so-called basic database for geographic information systems. Subsequently, all state organisations, state budget organizations especially, municipalities and higher territorial units are able to build their thematic geographical information systems over these reference data.

Figure 3. Example of digital orthophotomosaic available online [3]

Theoretically, there is possibility to compare the outputs with actual Google Maps data to check the accuracy. The difference is that the data given from the system are guaranteed and are able to be used for legal purposes, such as administrative proceedings for the purpose of issuing decisions. The system will also be used to control the payment of subsidies in the agricultural sector, as prevention against fraud- there is possibility for easier control of the land use not only for the owners but for everyone. Other uses are things, such as dispute prevention, control of ownership relations, preparation of project documentation for civil engineering constructions, preparation of project documentation, preparation of classification and delimitation of protected areas, simplification of construction procedures in cities, suburb areas, towns and villages, support of spatial planning, easier possibilities of map creation focused on tourism etc. For example, there was applied a formula to calculate the different data base on chosen sample. On each and every base type was applied the same formula in shape:

$$KES = \frac{\sum S_i * P_i}{P_Z} \tag{1}$$

where *KES* means coefficient of ecological stability of land, S_i is the value of ecological stability of the land type (e.g. garden index is 3, wood with index 4, rural area with index 1 or water areas with index 2) [6].

The value of the *KES* index for focused area was 1.31 for calculation from digital cadastral map, 1,64 for digital map and 2,18 for orthophotomosaic. The different values present the extreme necessity of precise area characterization, because the qualification of area can be incorrectly classified due to incorrect classification of given data.

4. Conclusion

The question of the ecological stability index increase on its´ importance due to the pressure on improving ecological stability of land as a complex. Next question is the question of index accuracy and its´ precision. There are many map bases possible to be used in this kind of calculations, the way of calculation and indexing of categories goes hand in hand with used base.

Acknowledgments

This article is the result of the implementation of the project VEGA 1/0275/17 "Application of numerical methods to define the changes of geometrical track position", supported by the Scientific Grant Agency of the Ministry of Education, Science, Research and Sport of the Slovak Republic and Slovak Academy of Sciences.

References

[1] Act 330/1991 coll. Of SNR from July 12 1991 on Land adaptations, land ownership arrangements, land offices, land fund and land communities

[2] Methodical instructions for performing geodetic activities for the project of land adaptation 2008 *Geodesy, Cartography and Cadastre Authority of the Slovak Republic*

[3] Geodetic and Cartographic Institute Bratislava, National forest centre in Zvolen; r.2017 - 2019

[4] Míchal I. 1982. Principy krajinářského hodnocení území. *Architektúra a urbanizmus* XVI/Z (Bratislava VEDA SAV) pp. 65–87

[5] Miklós L. 1986. Stabilita krajiny v ekologickom genereli SSR *Životné prostredie* **20** (2) (ÚKE SAV Bratislava) pp 87–93.

[6] Metodické pokyny na vypracovanie dokumentov RÚSES 2014 Slovenská agentúra Životného prostredia (SAŽP Banská Bystrica)

IOP Conf. Series: Materials Science and Engineering 1015 (2021) 012057 doi:10.1088/1757-899X/1015/1/012057

The role of underground urbanization in Samara city geological processes

D Vasilieva, A Maltsev and M Baranova

Department of Engineering Geology, Bases and Foundations, Samara State Technical University, 244 Molodogvardeyskaya St., Samara, Russia, 443100

E-mail address: vasilievadi@mail.ru

Annotation. In all largest cities due to lack of territory there is an intensification of underground space use - underground urbanization. It allows finding solutions to transport and environmental problems while preserving the unique appearance of the city's historical cents. However, the process of underground space development leads to activation of some exogenous processes, which may lead to negative consequences. More than one million people live in Samara city, subway line is built, there are subway objects of industrial and defense importance (bunkers) reaching considerable depth. Active underground construction leads to strengthening of geological processes (weathering, karst formation, suffosia, etc.), change of strength properties of rocks, violation of natural hydrological regime of the territory. By the example of Samara, stages of underground urbanization of the city were studied and its influence on geological processes was considered. Results of studying existing use of city underground space are described. Taking into account geological and geomorphological regionalization of the city, the areas, which underground space development is the most undesirable, were identified.

1. Introduction

At present, large cities are not only characterized by active development and growth of the city in width, but also the development of underground space - underground urbanization. According to international standards, the share of underground structures should be 20-25% of the total construction area. In the Russian Federation, this indicator is much smaller, but in the largest cities is gradually increasing, which allows the release of additional urban territory. Garages, warehouses, industrial and commercial enterprises, utility networks, administrative and cultural institutions, transport networks are being tried to be located underground. In some cities it is planned to create master plan of underground space development and create engineering-geological 3D-maps of districts and city blocks. Depth of the underground structures construction may reach 50 m. It is necessary to take into account existing environmental, especially geoecological features of the territory, including the possibility of activation of exogenous geological processes.

For the territory of Samara city, the problem of underground urbanization and its impact on engineering and geological conditions of construction has not been comprehensively studied before. In this article, for the first time, the city territory was zoned and the soil conditions of each district were analyzed. This study may be useful for planning further urban development, as it allows for taking into account information about the complexity of soil conditions.

2. Methods

Nowadays a lot of attention is paid to the issues of studying possibilities of using city underground space and consequences of underground urbanization [1-3].

The construction of underground structures on the territory of Samara began in the XVIII century, during the construction of the second Samara Fortress, where wood structures were erected to strengthen the slopes. This object is located on the territory of modern Bread Square in Samara and was studied during archaeological excavations in 2014, 2017 and 2019 (figure 1). It is shown that wooden structures were created to reduce landslide and erosion processes typical for the studied area at that time. At the same time, in order to level the surface around the fortress, powerful bulk layers (up to 7-8 meters high) were created, mainly from manure and construction debris [4-6].

A B

Figure 1. General view of the archaeological excavation of Bread Square 2017 (A) and 2019 (B).

The next significant stage of development of the city's underground space was connected with the beginning of laying of municipal networks - Samara water pipeline in the end of XIX century. The plan of Samara city of 1885 (figure 2) already shows water pipe network (project of engineer N.P. Zimin), and in 1886 Samara water pipeline was opened. By early 1917, the total length of Samara water pipeline was about 70 km. At present, according to the official website of SamaraVodokanal ME, the city's water supply networks are more than 1500 km long, and the sewerage networks are more than 1200 km long.

Significant volumes of underground construction were carried out during the Great Patriotic War, when in Samara (then Kuibyshev) a system of defense facilities (bunkers) was built [7, 8]. In 1941, the city of Kuibyshev received the status of "Western Capital", which led to the large-scale construction of defense facilities underground (Down Underground Command Points - DOC). They include: one of the most famous shelters in Russia - "Stalin's Bunker", as well as "Bunker Kalinin" or Red Army ZKP (on Kuibshev), "Bunker Zhukova" (on the 7th section), "Bunker Beria" or "Bunker NKVD" (near Bread Square), "Bunker Communications of the Soviet Navy". At present, the bunkers are in different states, for example, the Red Army ZKP is operational (command post of the Samara administration in case of emergency), Stalin's bunker is also operational and falls under the jurisdiction of the Ministry of Emergency Situations, but access to the bunkers is allowed, and most of the bunkers are abandoned, flooded by groundwater and destroyed.

In the middle of XX century in the conditions of confrontation and "Cold War" development of the city of Samara was carried out taking into account the need to create civil defense shelters. They can be divided into two types: blast shelters (equipped in the basement of houses) and bomb shelters factories. In Samara hundreds of blast shelters were created, most of which are now abandoned. During the Great Patriotic War, many businesses were evacuated to Kuibyshev. Bomb shelters built during and after the war were designed to survive in conditions of radioactive or biological contamination, they included shelters of plants "Progress", "Start", "ZiM" "GPZ № 4" and others. At present, most of them have been destroyed.

In 1980 in Kuibyshev the construction of the subway was started, and in 1987 the regular transportation of passengers began. At present the length of Samara Metro line is 11.6 km (total length of distillation tunnels and main tracks - 22.7 km), it consists of 10 stations (9 of them are underground and 1 aboveground). The deepest station of Samara Metro is "Gagarinskaya" (17.5 m), the deepest stations are "Sportivnaya", "Sovetskaya", "Pobeda" and "Bezymyanka" (about 8 m). At present, it is planned to continue construction of Samara Metro stations (figure 3).

Figure 2. Plan of Samara city with indication of full network of water pipes according to engineer N.P. Zimin's project.

The aim of the article is to study the features of underground urbanization on the territory of Samara and its impact on engineering and geological processes of the studied territory. In order to achieve the set goal, the following tasks were solved: first, to study main stages of underground construction development on the city territory; second, to conduct engineering-geological zoning and give characteristics of main exogenous geological processes; third, to reveal the most dangerous territories where exogenous geological hazards may activate during underground construction development.

The methods of scientific literature source analysis, field survey and description, tabular and graphical methods were used in writing the article.

3 Results and discussions

Samara is located on the watershed and valleys of Samara and Volga rivers. Back in 1969 researches of doctor of geological and mineralogical sciences V.I. Rachitsky [9] have given the chance to divide territory of Samara (Kuibyshev) on engineering-geological conditions of building into seven areas with instructions of their features (figure 3).

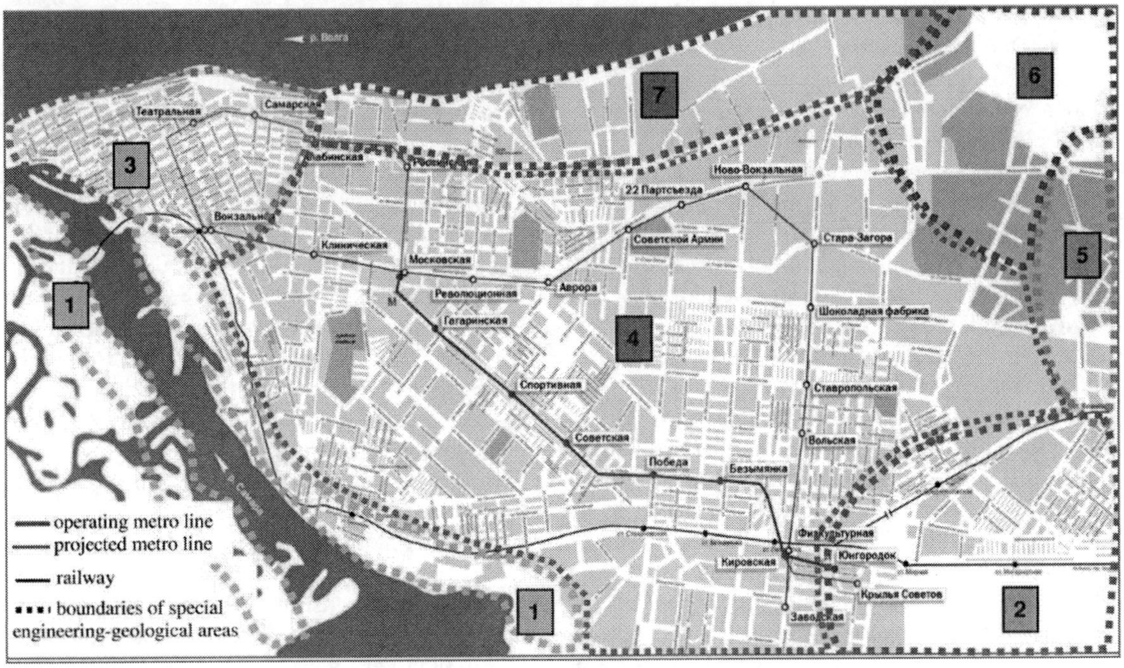

Figure 3. Plan of Samara city by indicating conditional borders of areas with different engineering and geological conditions of construction.

The first engineering and geological district covers the floodplain of the river. Samara. Due to covering of this region with water in floods and permanent covering of lower areas with water after construction of Saratov HPP, this region is not suitable for construction including underground one.

The second engineering-geological district is confined to the valley of Samara River, representing the 1st ancient (Khvalynskaya) terrace above the floodplain. It covers a part of Kirovsky District on the right bank, Zubchaninovka, Smyshlyaevka settlements and an airfield. On the left bank - Kryazh village, new buildings area Volgar, oil refinery and several other plants. This area is represented by alluvial deposits in the form of laminated and oblique loams, sands and loams, with a maximum capacity of 25 m. The depth of the mirror of groundwater, representing a single aquifer with waters of floodplain sediments within the absolute markings of 30-40 m. In general, within the limits of the second area the rocks are characterized by the allowable load from 1.5 to 2.5 kg/cm^2 and are to some extent subsidence. Evaluation of the subsidence in each individual construction site should be made additionally. For development of underground construction the second engineering-geological district may be suitable, but at present it is not relevant for this territory of the city.

The third engineering and geological district is dedicated to the high part of Samara city from Polevoy St. to Khlebnaya Square. It covers the territory of distribution of Quaternary alluvial deposits of the 2nd ancient terrace at the confluence of the Volga and Samara rivers. Terrace logs are lowered by denudation processes and, especially on the Samara slope, are poorly expressed in relief. According to V.I. Rachitsky, this area is composed of layered and oblique ancient Quaternary alluvial sands and loams up to 30 meters thick. The thickness of the sand is covered with limestone and less often dolomites, and are partially suitable for construction. However, on the whole, this particular engineering and geological region is less favorable, as there are fears of meeting the karsted rocks of the Kazan Stage. The level of underground water, confined to the Kazan aquifer, is at absolute levels of 30 to 40 meters. In this area there are already underground construction objects ("Bunker Beria", underground wooden structures of the second Samara fortress of the XVIII century), in addition, the construction of new metro stations is planned in the near future. The third engineering and geological district is characterized by old buildings, a large number of cultural heritage objects, the preservation of which is necessary to maintain the unique appearance of the historic city center. Underground construction in this area will help to solve the problem of territory deficit, but requires a comprehensive analysis of all potential consequences of underground construction. For example, the problem of possible activation of karst processes requires additional engineering-geological studies.

The fourth engineering-geological district covers separate sections devoted to the Volga and Samara rivers watershed and its slopes. From the surface it is composed of a solid cover of raw and delluvial loams with the capacity of 5 to 25 meters. Underlying rocks on a large area are limestones, dolomites with lenses of gypsum and anhydrite of the Kazan upper Perm, on a smaller area - red-colored clays of the Tatar layer. There is no permanent aquifer in loams. In general, this engineering and geological region is suitable for construction, especially where raw and sometimes deluvial loams are covered with clays. This district is also favorable for underground construction development, currently it is here that Samara metro stations are located.

The territory where the lines and metro stations are located covers separate sections dedicated to the watershed of the Volga and Samara rivers, where the engineering and geological situation is generally favorable for construction. Ground waters in this area are much lower than the depth of mine workings. From the surface of the watershed is formed by a continuous cover of loams with capacity from 5 to 25 meters. Underlying rocks on a large area are limestones and dolomites, which may be subjected to karst formation, which requires detailed exploration.

In the course of engineering-geological processes and phenomena during the metro construction in uncemented sedimentary rocks, shifts and subsidence may occur. Above underground excavations on the surface of the ground may form subsidence mulds with depth from 20 to 250 mm. Buildings and structures that are located in subsidence mulds can be deformed. On this basis, the lines of the Samara metro are laid mainly under the highways.

The fifth engineering and geological district is located in the area of outcrops and close to the Akchagylsky Neogene clay layer to the north of Smyshlyaevka settlement. The clays are mostly very plastic and are capable of landslide on slopes, especially under additional load, which should be taken into account when building in this area. In general, this district is suitable for underground construction by its characteristics but it is not relevant at present.

The sixth engineering-geological district covers the area of distribution of brick-red clays of the Tatarskaya Formation of Sokka, in some places covered with loams of insignificant capacity. This district is confined to watershed squares in the northeastern part of the city (Clinical Hospital district) up to Mekhzavod. The clays consist of rare but rather powerful limestone layers and dense siltstone. These interlayers in some areas are close to the surface and can be a very reliable basis for heavy structures. On the whole, the district is quite favorable for construction. This district is also suitable for underground construction development; it is not relevant here yet.

The seventh engineering-geological district occupies a significant area, stretching a wide strip along the left-bank slope of the Volga River valley from Postnikov ravine to the northern boundary of the city, bounded by the Sok River. Limestones and dolomites of the Upper Carboniferous, Lower and Upper

Perm come to the surface or lie close to it within the area. The rocks are heavily encrusted. When designing a construction in this area, it is always necessary to thoroughly explore the thickness by drilling to the depth of the influence of the projected construction pressure. If karst cavities are identified, they should be cemented or other measures should be taken to prevent their harmful impact on the construction. In this area development of underground construction is urgent because of territory deficit, but it requires additional technical and constructive solutions.

The seven engineering-geological areas mentioned above and noted in figure 3 have their own geological features. They are being developed, but construction in these areas is often conducted in difficult geotechnical conditions.

An example of influence of active underground construction on geological processes in district 7 is rise of ground water level due to leaks from communications that causes strengthening of karst processes resulting in formation of underground and surface cavities. As a result, gaps and funnels are formed on the surface. Similar result can be seen in districts 3, 4 as a result of intensification of suffosia and seepage processes. On the territory of Samara city there are known cases of cavities formation under buildings that lead to formation of cracks and even collapse of walls. At one of the metro stations (district 4), the surface is lowered due to the load on roads located above the metro line.

Thus, underground construction is now one of the most perspective directions of town-planning activity. However, it may lead to significant changes in the speed of exogenous geological processes, as well as in the hydrological regime of surface and ground waters [10].

4. Conclusions

As a result of the conducted research, the following conclusions can be drawn:

1. In city of Samara there is a high demand for increase in volumes of underground construction for the solution of transport problems and rational use of city territory, thus for a city difficult engineering-geological and hydrogeological conditions of construction of underground constructions are characteristic.
2. Engineering and geological features of Samara city territory are characterized by high spatial heterogeneity of Quaternary and indigenous (prequaternary) rocks.
3. When developing the underground space it is necessary not only to raise requirements to safety of work performance during construction of the facility and its reliability during operation, but also to take into account its geoecological conditions, including the possibility of activating exogenous geological processes.

References

[1] Petrov K S Gondusova A M Eskov V S and Batora A A 2020 Underground space development, its importance and advantages *Coll. of works of the All-Russian Scientific and Practical Conf. Actual directions of modern science, education and technologies* (Cheboksary: Expert and Methodological Center) pp 21-27

[2] Bratoshevskaya V V Beya N N and Gul'byakova M V 2017 Main tendencies of underground urbanization development *Coll. of articles of XIII International scientific-practical conf. Science and education: preserving the past, creating the future* (Penza: Science and Education) pp 246-248

[3] Korotayev V P 2009 Use of underground space in Moscow *Architecture and construction of Moscow* **543** (1) pp 39-44

[4] Vasilieva D I 2015 Geological conditions for the construction of the Samara fortress of the 18th century on the Khlebnaya square of the city of Samara *Traditions and innovations in construction and architecture. Construction* ed M I Balzannikova et al (Samara: Samara State University of Architecture and Civil Engineering) pp 252-7

[5] Baranova M N and Vasilieva D I 2016 Geological and geomorphological zoning on the territory of Samara *Traditions and innovations in construction and architecture. Construction* ed M I Balzannikova et al (Samara: Samara State University of Architecture and Civil

IOP Conf. Series: Materials Science and Engineering 1015 (2021) 012057 doi:10.1088/1757-899X/1015/1/012057

Engineering) pp 189-92

[6] Vasilieva D I Baranova M N and Shimanchik I P 2020 Petrographic study of the Samara fortress deposits (in Russian) *77th all-Russian scientific and technical conf. Traditions and innovations in construction and architecture* ed M V Shuvalov et al (Samara: Samara State Technical University) pp 276-82

[7] Pavlovich I L and Ratnik O V 2003 *Secrets and legends of Volga dungeons* (Samara: Scientific and Technical Center Publishing House) p 251

[8] Aleksandrov K 2009 Underground floors of Samara: Samara Diggers explore dungeons of Samara *Samara Fates* **9** pp 78-85

[9] Rachitsky V I 1969 *Engineering-geological zoning of the Kuibyshev region Collection Geology, geochemistry and exploitation of oil fields* (Kuibyshev: KPTI)

[10] Bukhman L M 2019 Geotechnical and Hydrological Characteristics of Urbanized Areas in the Development of Activities for Water Object Rehabilitation (Samara) *IOP Conf. Ser.: Earth Environ. Sci.* **272** p 22240

IOP Conf. Series: Materials Science and Engineering 1015 (2021) 012058 doi:10.1088/1757-899X/1015/1/012058

Deformation analysis of a construction pit in high-plasticity clay

R Bulko and J Mihálik

Geotechnics Department, Faculty of Civil Engineering, Univesity of Žilina

E-mail address: roman.bulko@uniza.sk

Abstract. In Slovakia, construction process already begins in less suitable locations. Such a location is also situated in Podhorie, small village near the city of Žilina. The subsoil is formed by high-plasticity clay with claystone in lower position. This article deals with the analysis of the influence of various input parameters as well as R_{inter} mesh discretization on the determination of safety factor – M_{SF}. The Mohr-Coulomb constitutive model was used in numerical model with respect to availability of input parameters. The ability to identify the parameter with the biggest impact on the resulting value of safety factor is an indisputable benefit. Various studies have been published on this topic. In our case, cohesion c had the biggest influence on the M_{SF} parameter when reducing shear parameters. In the future, the authors will compare the achieved results of numerical modelling with the actual deformations of the structure.

1. Introduction

Nowadays there is a high demand for building land in Slovakia. Nice flat sites are missing, having become the property of developers, or at high prices. Therefore, construction process already begins in localities that are less attractive and suitable for various reasons. Such a location is situated in Podhorie, small village near the city of Žilina, where the subsoil is formed by high-plasticity clay with claystone in lower position. In the past, high-plasticity clay was mined there for the production of solid fired bricks. Today, few people remember that there was a brickyard here and the construction of family houses began on the site in difficult geological conditions. The terrain is sloping 12-17° with frequent slope deformations. Therefore, the subsoil properties had to be verified. For this purpose, three static penetration tests were performed.

2. Geological conditions of area

The geological structure of the area consists mainly of claystone, which predominates over sandstones. In the past, several landslides have been documented in the area, mostly of a translational type [7]. These landslides have almost been occurred after high precipitation periods, or initiated by inappropriate human activities. The subsoil is formed by three main types of rocks as shown in figure 1.

Depth	Q_c Cone resistance	Q_{ist} sleeve friction	Penetration resistance $q_c(q_{st})$	Lokal friction f_s	Total resistance q_{tot}	Soils	E_{def}	φ'	I_c	I_D	c_u
[m]	[kgcm^{-2}]	[kgcm^{-2}]	[MPa]	[MPa]	[MPa]		[MPa]	[°]	[-]	[-]	[kPa]
0,2	4	16	**0,40**	0,080	0,48						
0,4	12	17	**1,20**	0,033	1,23						
0,6	10	26	**1,01**	0,107	1,12						
0,8	10	22	**1,01**	0,080	1,09	**F6/CI**	3,0	16,6	0,56		32,5
1	10	24	**1,01**	0,093	1,11						
1,2	13	33	**1,31**	0,133	1,45						
1,4	13	30	**1,31**	0,113	1,43						
1,6	27	31	**2,73**	0,027	2,75						
1,8	22	31	**2,23**	0,060	2,29						
2	11	24	**1,13**	0,087	1,21						
2,2	15	23	**1,53**	0,053	1,58						
2,4	11	24	**1,13**	0,087	1,21						
2,6	12	24	**1,24**	0,080	1,32	**F8/CH**	3,6	19,0	0,82		53,7
2,8	14	25	**1,44**	0,073	1,51						
3	13	26	**1,34**	0,087	1,43						
3,2	18	30	**1,84**	0,080	1,92						
3,4	24	30	**2,44**	0,040	2,48						
3,6	18	43	**1,86**	0,167	2,02						
3,8	107	124	**10,76**	0,113	10,87						
4	77	82	**7,76**	0,033	7,79						
4,2	90	161	**9,06**	0,473	9,53	**R5-R6**	39,4	30,6	2,34		248,7
4,4	105	122	**10,56**	0,113	10,67						
4,6	97	197	**9,77**	0,667	10,44						
4,8	139	216	**13,97**	0,513	14,48						
5											

LEGEND: Quarter Paleogen HPV

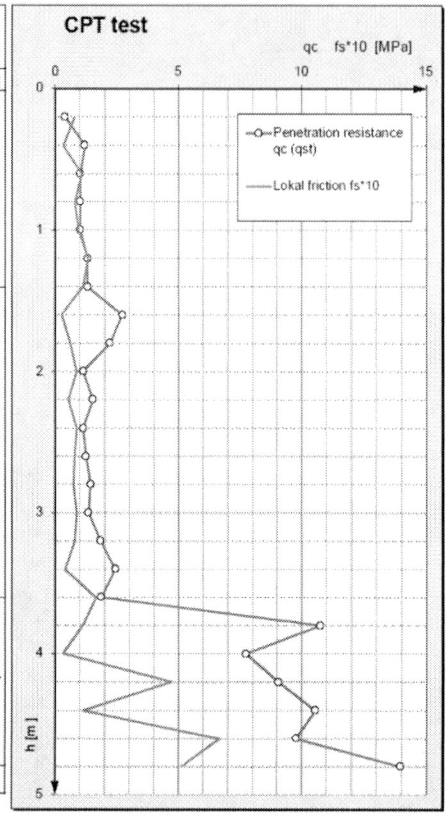

Figure 1. Results of penetration tests [1, 2, 3, 8].

3. Numerical model

The program PLAXIS 2D version 8.5 was used for numerical analysis. The construction pit is located in the cadastre of Podhorie, in Žilina district. Two types of rocks, clay and claystone, were included in the calculation. No groundwater level was indicated during the geological survey works. The main goal of deformation analysis by the help of numerical analysis was to determine how various factors may affect final safety factor of the construction pit. The construction pit consists of a soldier pile wall 7.0 m long, an IPE 220 profile with an axial distance of 2.0 m, and intermediate wooden profiles.

Table 1. The input parameters to Mohr-Coulomb constitutional model [8].

Rock type	γ [kN/m^3]	E_{ref} [kPa]	ν [-]	φ [°]	c [kPa]
Clay	20	3700	0,46	19,2	12
Claystone	20	43600	0,3	32,3	5

The final form of the model geometry is presented in Figure 2. We chose Mohr-Coulomb constitutional model with respect to the available input parameters. The input parameters to the constitutional model are listed in table 1. In the numerical model, the soldier pile wall is modelled by the structural element – plate with the values of stiffness as follows: $EA = 9.555 \times 10^{-5}$ kN/m, $EI = 8.55 \times 10^{-3}$ kNm^2/m. The individual phases of construction process are summarized in table 2.

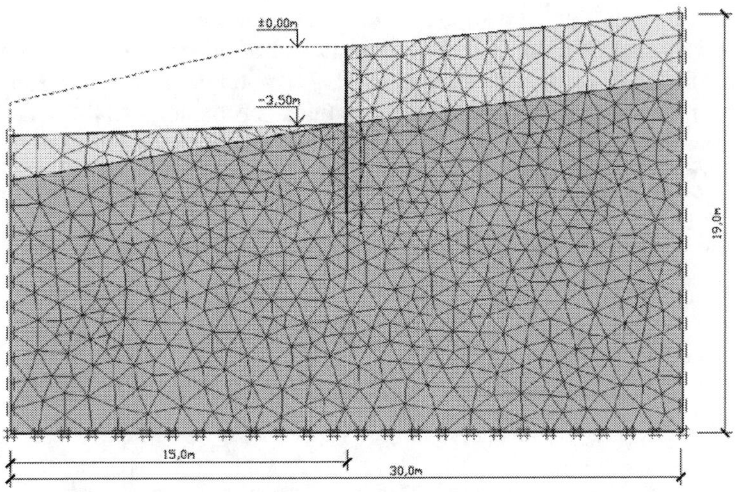

Figure 1. Finite elements mesh and geometry of the model.

Due to the inclination of the individual layers, the gravity loading method was chosen instead of the K0-procedure in phase 1 as a plastic analysis [5, 6]. In the second phase, the wall was activated and the soil was excavated into the final phase. The M_{SF} safety factor was calculated in the last phase by the method of phi-c reduction, when the strength parameters are reduced until the collapse of the structure.

Table 2. Individual construction phases.

Construction phases	Phase function	Phase type
1	Calculation of geostatic stress using gravity loading	Plastic
2	Activated support structure and excavation	Plastic
3	M_{SF} calculation	Phi-c reduction

4. Results of analysis

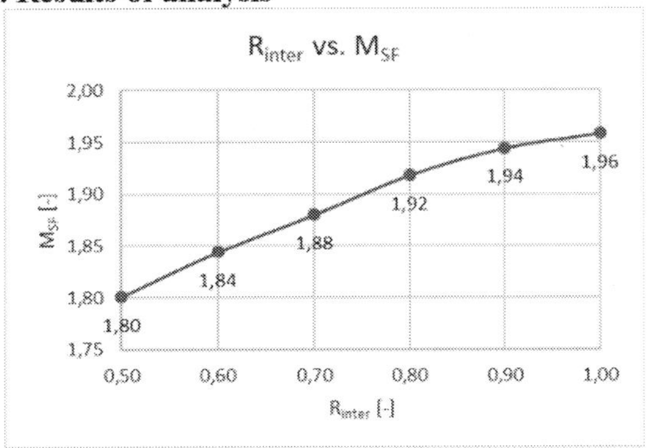

Figure 3a. Dependence between M_{SF} safety factor and parameter R_{inter}.

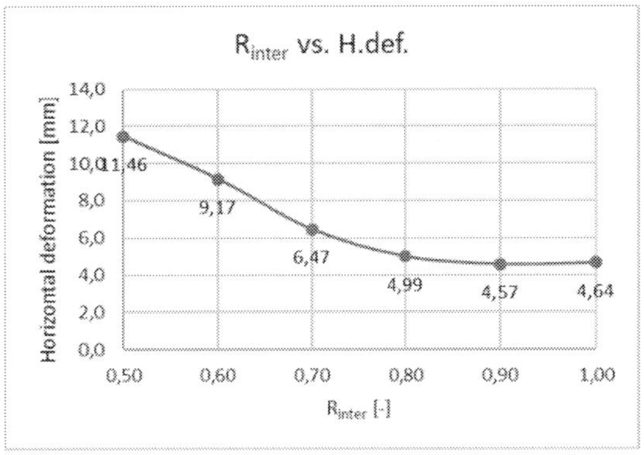

Figure 3b. Calculated horizontal deformations due to a change in the R_{inter} parameter.

In the parametric study we changed various parameters and determined their effect on the resulting safety factor. First, we analysed the effect of R_{inter} on the safety factor M_{SF} at mesh with discretisation into 1228 elements. We created six models in total. The parameter R_{inter} was entered in the range 0.5 - 1.0, while simulating a certain degree of discontinuity, respectively reducing the strength parameters at the soil-structure contact.

IOP Conf. Series: Materials Science and Engineering 1015 (2021) 012058 doi:10.1088/1757-899X/1015/1/012058

From the graph in figure 3 we can see a linear dependence between the examined parameters. Secondly, we also examined the horizontal deformations on the structure. The reduction of strength parameters due to smaller R_{inter} caused an increase in horizontal deformations. The appropriate choice of the R_{inter} parameter depends on the type of structure that is in contact with the soil. This issue has been addressed by several authors [4, 5]. Based on literature review of the articles published so far and our results, we decided to further use only $R_{inter} = 0.7$.

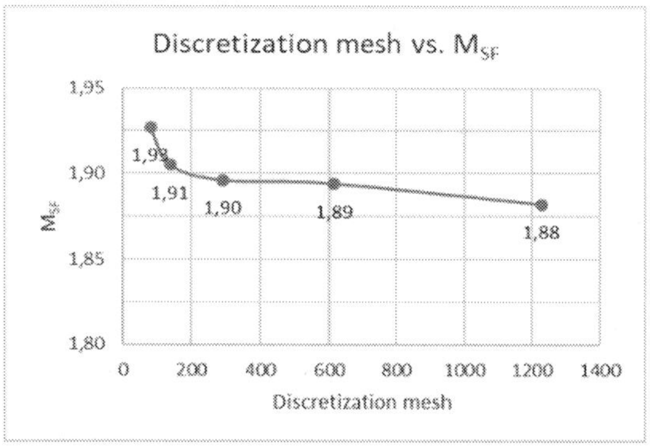

Figure 4a. Dependence between M_{SF} and mesh discretization.

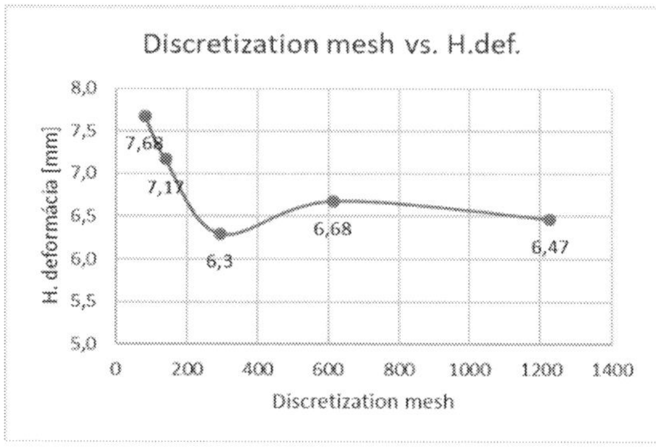

Figure 4b. Calculated horizontal deformations due to a mesh discretization.

Another parameter we analysed was the mesh density in our model, with the same input parameters. We created five models in total, choosing the number of 81, 139, 292, 614, 1228 elements. The results of our analysis are represented in the following graphs (figure 4a). A nonlinear dependence applies to these determined parameters. As the mesh discretization increases, the safety factor decreases. Respectively, with a higher number of elements, there should be a better approximation of the results. A sufficient degree of mesh discretization to obtain relevant results is presented in graph in figure 4a. From a certain number of elements, the M_{SF} safety factor does not change significantly as the number of elements in the network increases. If we look at the graph, we find that with some discretization of the mesh, the addition of other elements does not have a significant effect on the calculation. That is why we decided to keep $R_{inter} = 0.7$ and the number of 1228 elements in further analyses.

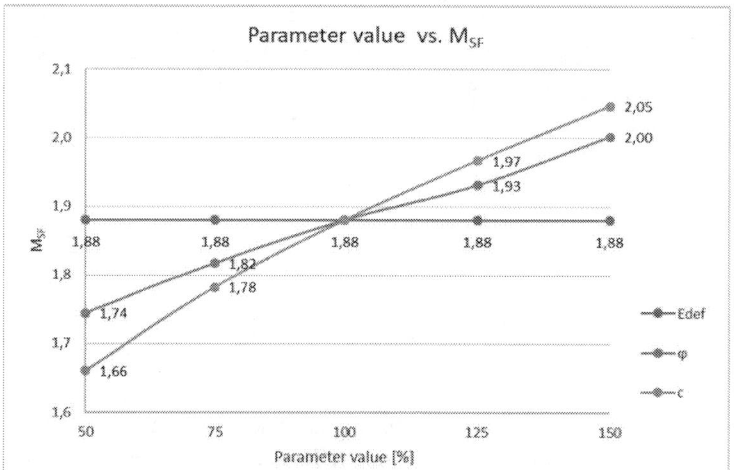

Figure 5. Dependence between safety factor and parameters values.

The next step of our analysis was to examine the influence of other input parameters on M_{SF}. We have created five models in total. We changed the deformation parameter E_{ref} and the parameters of shear strength φ and c. We entered 50 %, 75 %, 100 %, 125 % and 150 % of the parameter value into the calculation (figure 5).

By a parametric study for the Mohr-Coulomb constitutive model, we found that for our geotechnical problem, the value of cohesion c has the most significant impact and not the friction angle φ as we assumed. We assume that this is due to the higher value of c compared to φ. The mutual ratio of these two parameters is in favour of parameter c. In the case of the Mohr-Coulomb constitutive model for the final level of M_{SF} safety factor, we confirmed that the change in E_{ref} has no effect on our results.

Figure 6 shows the resulting slip surface, which is mostly present near to the surface. This is due to the lower rigidity of the soldier pile wall, which has the biggest displacement at the surface. For this reason, in these places there is also the least support of soil stability from pile wall and the software evaluates the slip surface at the surface behind the pile structure.

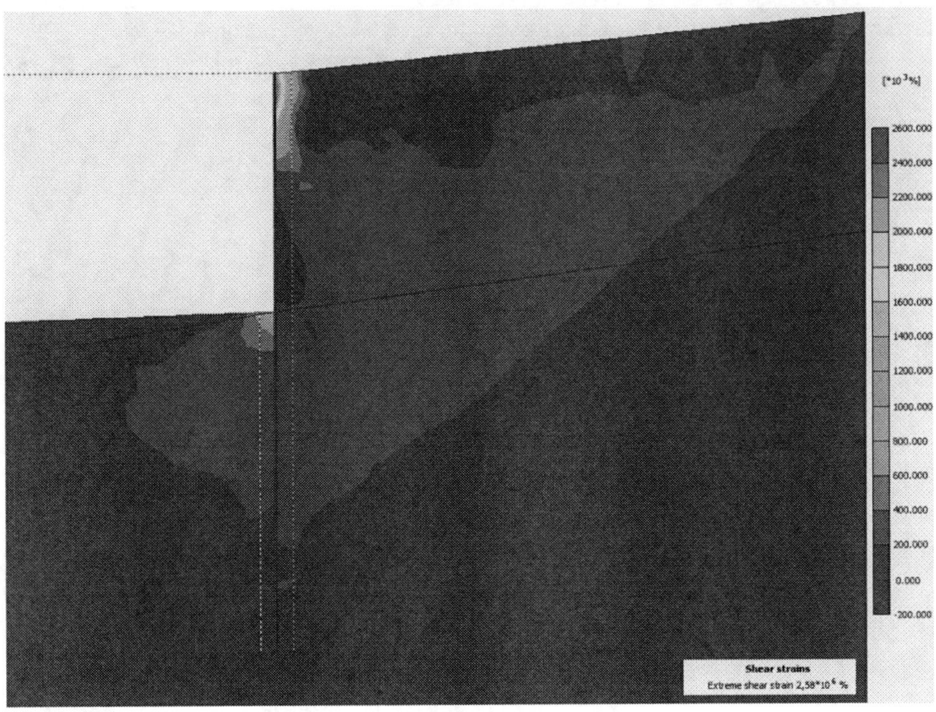

Figure 2. Slip surface.

5. Conclusions

The ability to identify the parameter with the biggest impact on the resulting value of safety factor in geotechnics is an indisputable benefit. This analysis contributes to better safety. This paper analyses the influence of individual parameters on the value of M_{SF}. Discretization of mesh had the least effect on M_{SF}, but only from a certain number of elements. R_{inter} has a significant impact on M_{SF} in terms of reduction of shear parameters at the soil-structure contact. Various studies have been published on this topic. In our case, cohesion c had the biggest influence on the M_{SF} parameter when reducing shear parameters. Before our parametric study, we thought that the friction angle would have a bigger effect than cohesion. We assume that this is caused by a small difference between the input parameters φ and c. To determine parameters for more advanced constitutional models such as e.g. Hardening soil, laboratory analysis and tests of high-plasticity clay are important. In future, authors will compare the results of numerical modelling with the actual deformations of the structure.

IOP Conf. Series: Materials Science and Engineering 1015 (2021) 012058 doi:10.1088/1757-899X/1015/1/012058

References

[1] Decký M, Drusa M et al 2009 *Navrhovanie a kontrola kvality zemných konštrukcií inžinierskych stavieb* (Zilina Krupa Print)

[2] Mayne P W 2007 Cone penetration testing state of practice Final Report NCHRP Project 20-05 Transportation Research Board (Washington)

[3] Robertson P K 1990 Soil Classification using the cone penetration test *Canadian Geotechnical Journal* **27** (1) pp 151-8

[4] Chalmovský J and Račanský V 2012 Vliv konstitučního modelu na vnější stabilitu pažící konstrukce v prostředí plně nasycených jílů *Geotechnika* **4** pp 19-25

[5] Brinkgreve R B J 2002 *Plaxis 2D – Version 8. Finite element code for soil and rock analysis user manual*(Delf University of Technology & Plaxis bv, The Nederlands)

[6] Potts M D and Zdrakovic L 1999 *Finite elements analysis in geotechnical engineering – theory* (Thomas Thelford Publishing)

[7] https://apl.geology.sk/gm50js/ (access: August 2020)

[8] Standard STN 72 1001 : 2010 : Klasifikácia zemín a skalných hornín.

Investigation of the diaphragm wall isolation of vibrations' transferred through the subsoil

M Wyjadłowski[1], **J Grosel**[1] **and Y Tyulyaeva**[2]

[1] Wroclaw University of Science and Technology, 27 Wybrzeże Wyspiańskiego, 50-370 Wrocław, Poland
[2] Florida International University, 11200 SW 8th Street, AHC-5 360 Miami, FL, USA

E-mail address: marek.wyjadlowski@pwr.edu.pl

Abstract Diaphragm walls are deep extended walls through granular or cohesive soils in shallow water table areas. This paper focus problems involved in excavating diaphragm walls including dynamic impact on the environment. Construction of diaphragm wall involves extracting a soil trench while simultaneously keeping heavy viscous slurry (bentonite) filling the excavation to provide lateral pressure that must be at least equal or greater than lateral soil pressure. Extraction of soil with machinery causes vibrations to be transmitted to the adjacent buildings. The research presented in the paper is also to assess the effectiveness of the concrete diaphragm wall in damping vibrations transmitted to the surrounding buildings. The first execution phase of the technology when a vertical finite-length trench element is excavated under a hydrostatic support from inside of a bentonite slurry is under consideration.

1. Introduction

The main advantage of diaphragm wall technique is to avoid a huge excavations in place where excavation is difficult or impossible to be performed. Terzaghi and Peck [1] defined "deep excavation" as those excavations that exceed a depth of 6 or 7 m. The deep excavations are required by needs of underground parking places (especially in the big cities), protection of existing objects in the excavation vicinities etc. The concept of vertical reinforced-concrete panels is both technically effective and economically justified also in order to reduce large loadings transmitted through the ground [2]. Diaphragm wall is also applicable construction in places where dewatering cannot be performed. This solutions are remedy solutions in granular soils with high ground water table, when low permeability layer underlies granular soil. Diaphragm walls are ended in underlying low permeability layer which consist of soil/rock keying into low permeability layer reduce ground water seepage below wall.

However, the construction of the diaphragm walls is always accompanied with a lot of difficulties and problems which are based mainly on how to estimate the deformation of the trench sides in the deepening phase. In the construction of such kind of deep excavation protection is the most serious problem encountered. Designing a retaining structure should be consistent with standards [e.g. 3, 4]. It is particularly important when the support works as a permanent structure. Estimating the predicted displacements at the design stage is essential for ensuring safety. The excavation protections are inspected by geodetic monitoring, which consists, first of all, in measuring displacements at points located in the adjacent engineering structures and in the excavation support top. These measurements enable evaluation of the size and range of excavation impact on surrounding structures, as well as displacements of excavation supporting system – usually its top [5, 6]. During the excavated trench execution, dynamic load on the surrounding buildings also occur and these dynamic impact should be carefully analysed.

2. Site preparation and excavation

First stage before starting the excavation a bentonite slurry plant must be prepared for mixing and providing bentonite to the trench or panels under excavation through a net of steel pipes. During excavation bentonite has to be provided to the trench while excavated soil mixed with bentonite are pumped again to the bentonite plant for recycling i.e. separation of soil particles and fractions mixed with bentonite during the process of excavation [7].

In the first phase of the technology when a vertical finite-length trench element is excavated under a hydrostatic support from inside of a bentonite slurry. During the trench excavation some dangerous effects can result from the variations of the groundwater table, the pore pressure, the soil collapse, the slurry level, weak soil lenses, suction forces, etc. [2, 8].

Figure 1. Cutter - excavation machine.

Figure 2. Monitored building.

At this stage it must be decided what kind of excavation machines must be used in order to achieve the basic requirements for a minimum disturbance of the soil around or surrounding the trench. [7]. Special clamshell also known as grabs/buckets/cutter are rectangular shaped and used to excavate vertical slots called panels/trench. Loose sand and gravels can be excavated by using grab while excavation of hard strata can be performed by using cutter. Digging mechanism may be cable or hydraulic operated. In practical geotechnical engineering many problems are caused by the response of soil subjected to the vibrations [9], so the more interest should be paid to the dynamic impact of excavation machine on adjacent buildings.

IOP Conf. Series: Materials Science and Engineering 1015 (2021) 012059 doi:10.1088/1757-899X/1015/1/012059

3. Monitoring process

Attempts to measure the influence of geotechnical works on existing infrastructure have been reported in the course of large pile driving, sheet pile driving or rapid impulse compaction [10]. Structural damage caused by vibrations can be limited by the use of construction and connections resistant to excitation based on passive, active or semi-active techniques of vibration reduction [11] and the use of barriers in the ground to prevent surface wave energy transfer [11].Vibrations caused by civil works, vehicle traffic, etc. are monitored according to respective guidelines and standards which specify admissible levels of vibrations depending on their frequency and the type of civil structure. The basis for the result analysis and the assessment of safe vibration levels is typically constituted by the following codes of practice: [12, 13, 14]. Signals generated during geotechnical works are of variable nature, which means that these vibrations require individual approach and detailed analyses to get as much information as possible. In most cases vibrations can be described as non-stationary stochastic process, i.e., the average value, variance and autocorrelation function varies along time [15].

Concrete trench counterparts or a row of piles is solution for vibration screening, this type of barrier is studied in the paper. The majority of the energy that affects nearby structures is carried by Rayleigh wave travelling on ground surface, away from source of vibration. This energy comprises approximately 67 per cent of the total energy [16]. A ground discontinuity makes wave propagation impeded. After incidence on an obstacle, the majority of the energy of the Rayleigh wave is reflected, with partial transmission through the barrier, and new body radiate outwards.

The research presented is to assess the effectiveness of the concrete diaphragm wall in damping vibrations transmitted to the surrounding sensitive buildings. In the studied case, a 18 m deep concrete trench was constructed. Excavation of hard strata was performed by using cutter. The work of the machine consists in short lowering the grab weighing 21.5 tons freely. This part of trench deepening is the source of highest vibrations amplitudes. Occurrence of big, hard and smooth surface stones that keep rolling under cutter wheels or stacking of these stones between the teeth of the cutter extends the process of the trench lowering. Crushing the boulders and stones causes also high vibration values.

The monitored structure was a sensitive building made in traditional technology (figure 2). The figure 3 shows the location of diaphragm wall section, building and measurement point. Vibration of the point at the corner of the buildings wall was monitored.

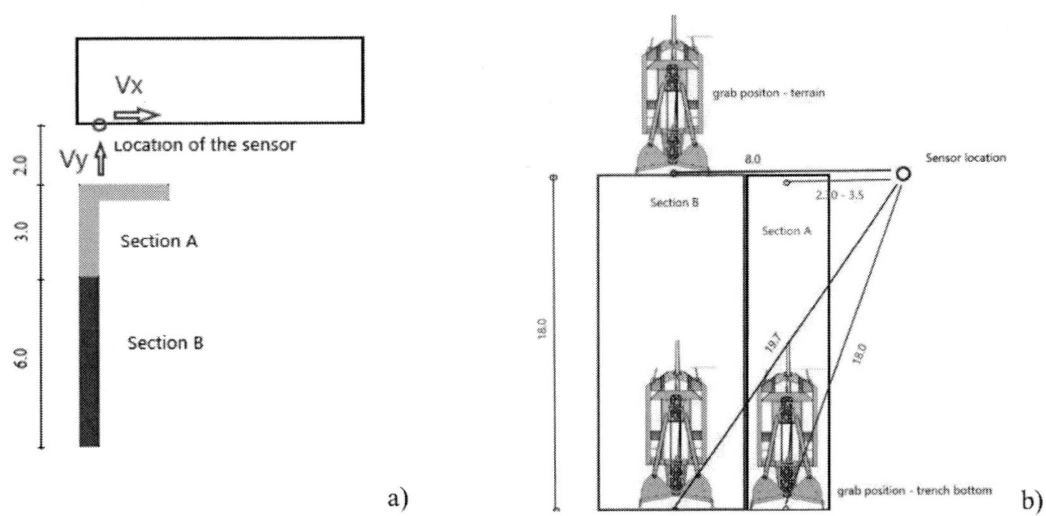

Figure 3. a) Scheme of diaphragm wall sections and location of the sensor, b)Trench cross section, extreme grab position.

In the first stage, the corner trench (Section A) of the excavation protection was made. The second stage of diaphragm wall (Section B) was completed after obtaining concrete strength of the corner

IOP Conf. Series: Materials Science and Engineering 1015 (2021) 012059 doi:10.1088/1757-899X/1015/1/012059

trench. The building wall was monitored during these phases and the distance from the source of vibrations (from a running cutter) to the sensor varied from 2.3 m to 19.7 m. According to the standard [12] the monitored building falls into first or second category, and as such requires dynamical monitoring.

4. Results of vibration measurements

The vibration parameters of the building during the execution of the closest to the building corner Section A were collected on the first day of monitoring (Section A). During the selected monitored time interval (6:00 – 7:52), the sensors recorded the highest amplitude V_x= 10 mm/s of observation from the whole working day (where x – direction parallel to the building front, y – direction perpendicular) . During the second day of diaphragm wall execution the sensors recorded the highest amplitude V_y= 2.50 mm/s. The graph showing velocity in time domain is presented in figure 4 and figure 5, where red lines represent V_x and blue lines represent V_y (see figure 3a).

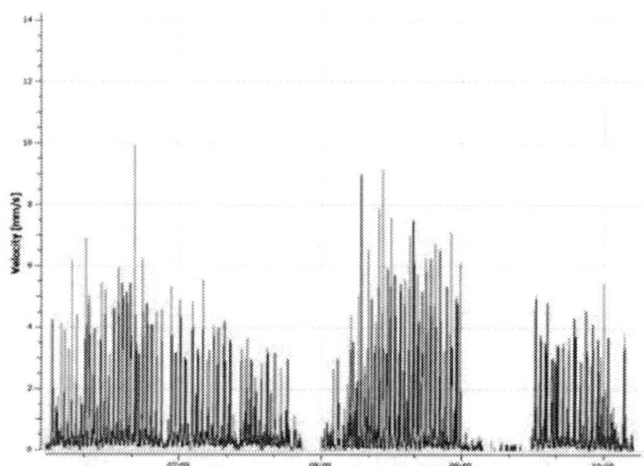

Figure 4. Velocity measurement during execution Section A.

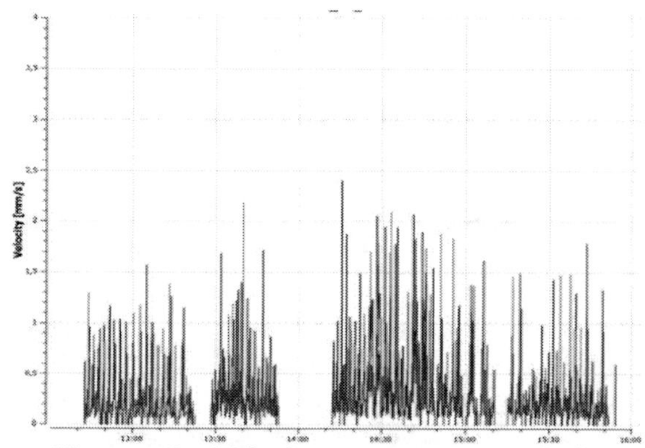

Figure 5. Velocity measurement during execution Section B.

The data for the two sections is set in the table 1. It is practical to count the time when the amplitude limit value is exceeded.

Table 1. Summary of the results.

Parameter	Section A		Section B	
	V_x (mm/s)	V_y (mm/s)	V_x (mm/s)	V_y (mm/s)
average value	0.41	0.32	0.20	0.24
median	0.22	0.17	0.03	0.17
skewness	5.37	5.19	2.72	3.96
maximum	9.92	8.15	1.37	2.40
number of occurrence velocity exceeding				
$V_{x\ or\ y} > 1$ mm/s	645	277	8	111
$V_{x\ or\ y} > 2$ mm/s	331	178	0	5
$V_{x\ or\ y} > 3$ mm/s	210	74		
$V_{x\ or\ y} > 4$ mm/s	115	22		
$V_{x\ or\ y} > 5$ mm/s	48	3		
$V_{x\ or\ y} > 6$ mm/s	21	2		
$V_{x\ or\ y} > 7$ mm/s	3	2		

Diagrams 6 and 7 show vibration parameters V_x, V_y for Section A and Section B. For the first section the most common vibration velocities are 0.0-1.0 mm/s, and the frequencies are in the range of 2–40

Hz. For the Section B the most common vibration velocities are 0-0.2 mm/s, and the frequencies are in the range of 2–30 Hz.

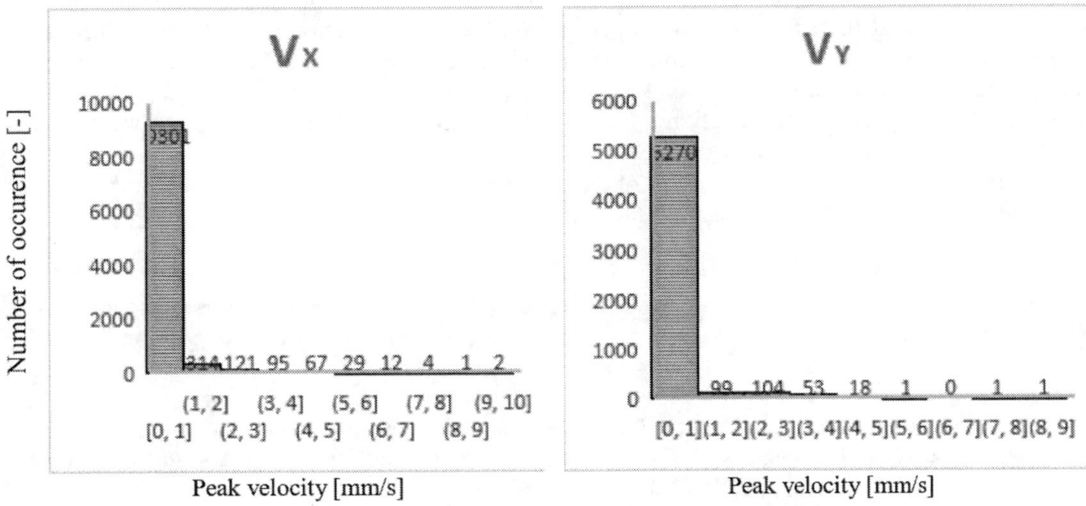

Figure 6. Histogram – frequency of occurrence of the vibration velocity for Section A.

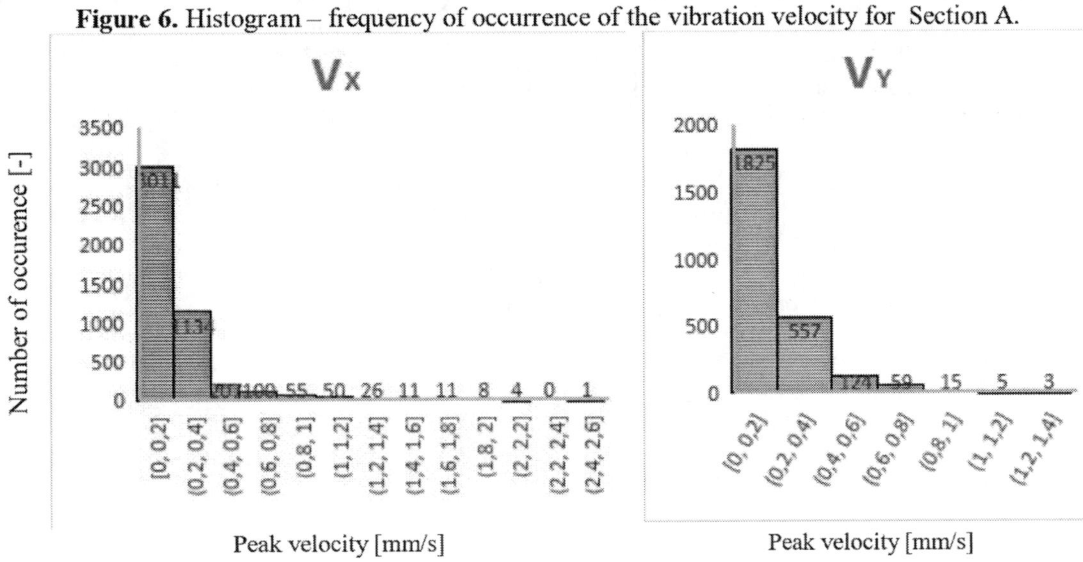

Figure 7. Histogram – frequency of occurrence of the vibration velocity for Section B.

5. Conclusions

It is evident that that the results are highly dependent on the ground conditions and vibration energy at the source.

The nature of vibrations generated by trench deepening has two sources: harmonic excitation, e.g. movement of heavy machine or low frequency cutter. Therefore conducted investigating provide varied results of vibration parameters in terms of each section of diaphragm wall.

The frequencies of individual peak particle velocity values presented in figures 6 and 7 show that data collected for Section A and Section B are divergent from one another. Both data sets are positively skewed. The mean value of the sample is raised by a few relatively high scores at the right-hand end of the distribution, and the most values are concentrated in the left-hand end of the distribution. This is more expressive for the measurements for Section A. This can be explained by the diverse distance sensor location from both trench section. The figure 3b shows the extreme position of the vibration source during trench deepening. The ratio of these distances for Section A and Section B is approximately equal 0.40 for initial grab position on the ground surface and 0.90 for deepest grab position at the trench bottom. It is assumed that the relationship between the amplitude of the source A_o and that measured A_r at the distance r is $A_r \approx A_o\, r$.

During the measurements for Section B, no velocity amplitudes greater than 3.0 mm/s were observed, while several hundred were recorded for the Section A.

The vibration attenuation effectiveness of fixed concrete trench is especially visible for the maximum amplitudes (AMF – amplitude mitigation factor)

$$\text{AMF} = V_{\text{BX}} / V_{\text{AX}} = 0.13 \text{ and AMF} = V_{\text{BY}} / V_{\text{AY}} = 0.30$$

The previously formed Section A gains function passive technique of vibration reduction as the barrier in the ground to prevent wave energy transfer while deepening further part Section B.

References

[1] Terzaghi K and Peck R B 1967 *Soil Mechanics in Engineering Practice* (Wiley Michigan)
[2] Brząkała W and Gorska K 2008 On safety of slurry-wall trenches, *Studia Geotechnica et Mechanica*, **30** (1-2) pp 199-206
[3] European Standard EN 1997-1 2009 *Eurocode 7 Geotechnical Design-Part 1: General Rules* (European Committee for Standardization: Brussels, Belgium)
[4] Siemińska-Lewandowska A and Mitew-Czajewska M 2008 Design of deep excavations according to Eurocode 7 *Studia Geotechnica et Mechanica* **30** (1-2) pp 207-214
[5] Rybak J, Ivannikov A, Kulikova E and Żyrek T 2017 Deep excavation in urban areas – defects of surrounding buildings at various stages of construction *9th Int. Scientific Conf. Building Defects* (Če ské Budějovice) ed I Juhásová Šenitková *MATEC Web of Conferences* **146** https://doi.org/10.1051/matecconf/201814602012
[6] Oliveira F and Fernandes I 2017 Influence of geotechnical works on neighboring structures *17th Int. Multidisciplinary Scientific GeoConference SGEM 2017 Science and technologies in geology, exploration and mining* **12** pp 993-1001
[7] Abdelhadi N, Abdelhadi M and Keinosuke G 1996 Problems Encountered in Diaphragm Wall Excavation *Reports of the Faculty of Engineering* **26** (46) (NAOSITE Nagasaki University)
[8] Polish Standard PN–EN 1538 2002 *Execution of special geotechnical works – diaphragm walls* (Warsaw)
[9] Massarsch K R 2000 Settlements and damage caused by construction-induced vibrations *Proc. of the Int. Workshop Wave 2000* (Bochum, Germany) pp 299-315
[10] Wojtowicz A Michałek J and Ubysz A 2019 Range of dynamic impact of geotechnical works on reinforced concrete structures *E3S Web Conf.* **97** 03026
[11] Herbut A 2020 Vibration mitigation efficiency of an inclined, curved, open trench *PLoS ONE* **15** (2): e0229010 https://doi.org/10.1371/journal.pone.0229010
[12] German Standard DIN 4150-3 1999 *Structural Vibration Part 3: Efects of Vibration on Structures* (Berlin)
[13] European Standard EN 1993-5 1993 *Eurocode 3. Design of steel structures. Part 5* (European Committee for Standardization: Brussels, Belgium)
[14] Polish Standard PN-B-02170:2016–12 2016 *Assessment of the Harmfulness of Vibrations Transmitted Through the Ground to Buildings* (Warsaw)
[15] Wyjadłowski M 2017 Methodology of dynamic monitoring of structures in the vicinity of hydrotechnical works – selected case studies *Studia Geotechnica et Mech.* **39**(4) pp 121-9
[16] Towhata I 2008 *Geotechnical earthquake engineering* (Berlin Heidelberg: Springer Verlag)

IOP Conf. Series: Materials Science and Engineering 1015 (2021) 012060 doi:10.1088/1757-899X/1015/1/012060

Long Short-Term Memory (LSTM) neural networks in predicting fair price level in the road construction industry

H Anysz[1] and Ł Brzozowski[2]

[1] Warsaw University of Technology, Civil Engineering Department, Al. Armii Ludowej 16, 00-637 Warsaw, Poland
[2] Warsaw University of Technology, Faculty of Mathematics and Information Science, ul. Koszykowa 75, 00-662 Warsaw, Poland

E-mail address: h.anysz@il.pw.edu.pl

Abstract. Running a tender procedure in the construction industry a client expects receiving reasonable prices in the tenders. However, the market competition and the contractors' will to win a contract to work out a profit, sometimes make the prices far from expected. They can be really low, almost impossible to be kept during the contract execution. Oppositely, the offered prices can be far above a client's expectation. If they are extremely high, a client has to decide, to accept one of them (even bid-rigging is suspected), or to cancel the procedure. Nevertheless, cancelling a procedure means the considerably postponed the start of the construction. The analysis of almost 400 completed tender procedures (in the Polish road construction industry) and the market trends proved that unexpectedly high prices do not always have to point the collusive behaviours of the offerors. As aforementioned analysis is based on past cases, there is a need to propose the predicting tool, to help clients making decisions in the new tender procedures: to accept the high price as a market-based, or to reject all of the offers as a bid-rigging is suspected. The above-mentioned analysis are based on the simple moving average tool. For predictions the autoregressive tool is proposed – long short-term memory (LSTM) neural networks.

1. Introduction

The unfair pricing (namely bid-rigging) – leading to a forced choice of the overpriced offer – happens all over the world (it is described e.g. in [1], [2], [3]). The Polish market is also not free from this unwanted phenomenon (some examples in the construction industry are described in [4]). The importance of the problem made national and international organisations publishing manuals, how to recognize and avoid bid-rigging in public procurements [5], [6], [7], [8]. If a client observes that the offered prices are much higher than expected, two decisions can be undertaken. Cancelling the procedure – then the expected results are postponed, as well as, the external financing of the project can be lost. Another possible decision is to accept one of the tenders, even if the unexpectedly high price is offered. Then, a client should be sure that the high level of the prices offered is based on other – than collusion or bid-rigging reasons. Otherwise, accepting the offer would be a loss of public funds (the amount exceeding the market price). There is not so long time to work-out this kind of decision. That is why, several bid rigging indicators has been found and published (e.g. in [5], [9], [10], [11]). One of the indicators (to be applied in the road construction industry), based on the relation of the tender price to the client's estimation is introduced in [12] (it was named PtE there – price to estimation). As the price level reflects the market trends (which are observed by the contractors), the simple moving average

(SMA) [13] of PtE is calculated on the 30 days basis. Then based on Bollinger bands concept [14], [15], lower and upper limit of reasonable, market bid prices are defined on the base of $SMA_{30}(PtE)$ values (see figure 1).

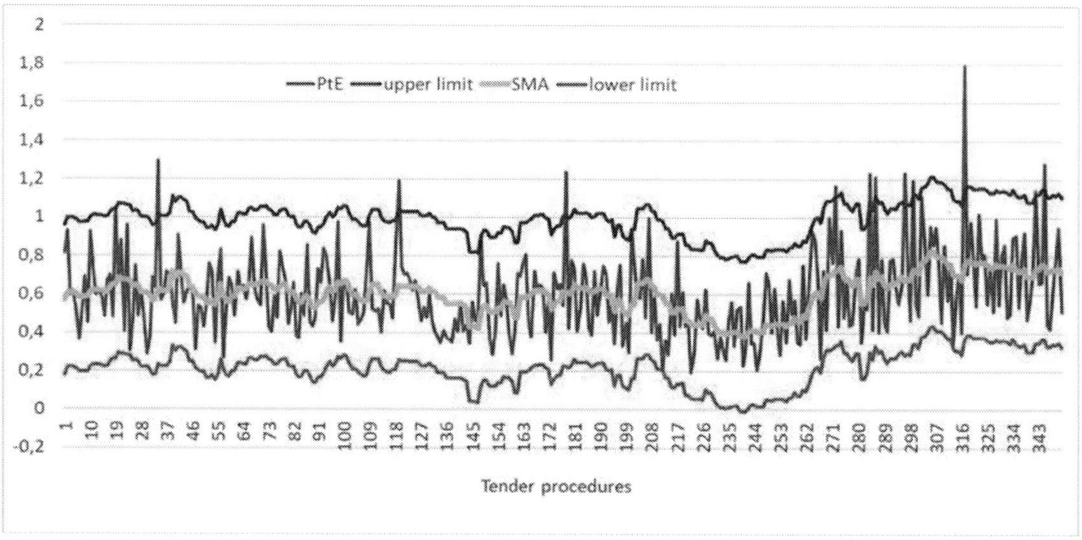

Figure 1. SMA and symmetric Bollinger bands (with PtE for 343 tender procedures) [12].

If the upper limit is not exceeded, even a certain bid price is high, the reason for that can arise from the market trends. The monitored period covers 3 years starting from June 1, 2014. Assuming a real case, when the choice of the best tender should be done by a client, the SMA curve, as well as, the upper and the lower limits are unknown, as the choice of the previous procedure was done several days ago. The maximum period between announcing the winner in the procedures recorded is 25 days. For details see table 1 and the histogram in figure 2. The total number of analysed procedures is 383, but 100 of them had a place simultaneously (2 or more in one day). Table 1 contains data concerning periods between 282 tender procedures (being single ones on certain days).

Table 1. Basic statistics of time between the tender procedures.

Number of procedures	Timespan			
	Minimum	Maximum	Mean average	Standard deviation
282	1	25	3,87	3,57

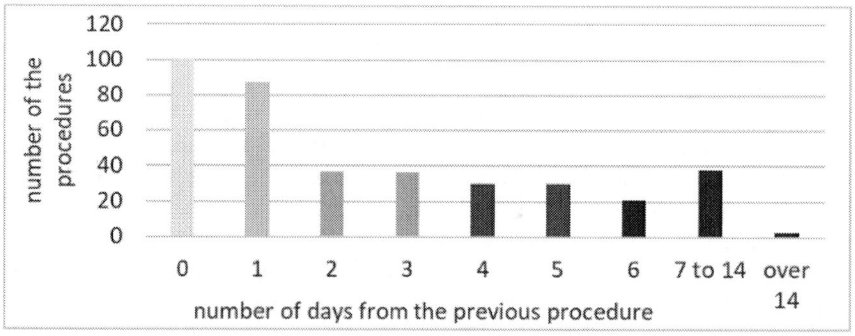

Figure 2. The distribution of the procedures with a certain time between them.

IOP Conf. Series: Materials Science and Engineering 1015 (2021) 012060 doi:10.1088/1757-899X/1015/1/012060

The client's estimation of lower and upper limits can't be done based only on the last recorded value of $SMA_{30}(PtE)$. The longer the period, the higher the possible error of $SMA_{30}(PtE)$ prediction (if the trends of it are not considered). The extreme daily changes of $SMA_{30}(PtE)$ for the recorded tender procedures are as follows: min -0,0186, max 0,0155. The possible prediction error is presented in figure 3.

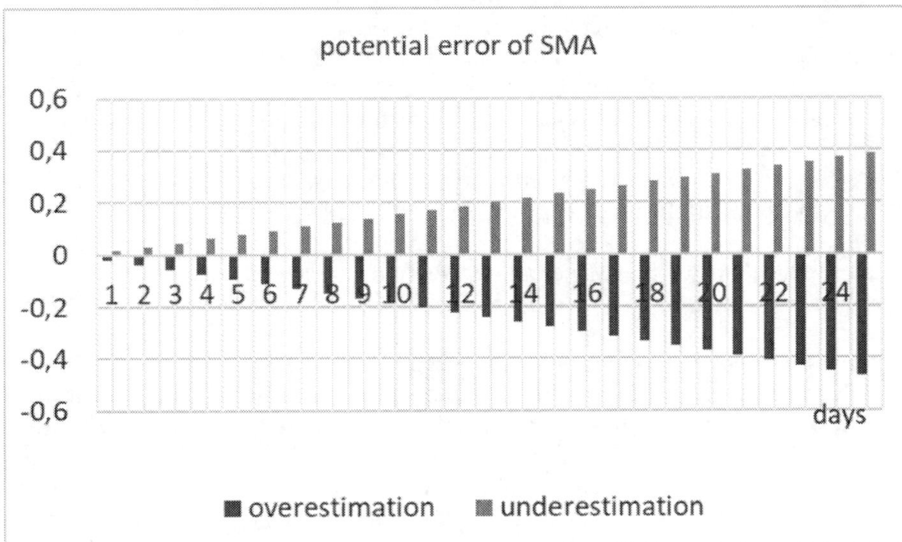

Figure 3. The potential error of $SMA_{30}(PtE)$ dependent on the number of days from the last recorded tender procedure.

In case of 12 days' time, the possible error can reach -0,2 to +0,2 values, while $SMA_{30}(PtE)$ values vary from 0,38 to 0,83. So, even for 12 days' time the error can be significant (over 50 %). Therefore, to provide clients with the tool allowing for predicting the market level of the bid prices (with possibly low error), the model based on long short-term memory neural networks (LSTM) is introduced.

2. LSTM model

2.1. *Data collected*

Several types of data are collected about tender procedures for building the sections of roads in Poland. The contracts over 10 million PLN were searched in the period starting from July 1, 2014 to June 30, 2017. Together with the values of the orders, the dates of clients' announcements (about the winners of the procedures) are collected. For over 400 contracts, the clients' estimations of the expected bid prices are available for 382 tender procedures. Then, the price to estimation ratio (PtE) is calculated for each case. Based on these data, the 30-days simple moving average is calculated. It consists of 343 cases starting from Aug 1, 2014.

2.2. *LSTM neural networks*

Unlike feedforward neural networks (applied e.g. in [16], [17]), Long Short-Term Memory (LSTM) neural networks introduced in [18] are a variation of recurrent neural networks. They are able to capture long-term dependencies in the data. The unique feature of LSTM networks is that each layer contains four gate sub-layers capable of removing and adding information to the data based on the context.

The main flow of the information in the LSTM is called the cell state, which conveys the data through the neuron with minor changes. As LSTM is a recurrent neural network, the updates to the data and the weights of the network happen in a loop, such that the refining of the weights is dependent mainly on the number of epochs. Parallelly to the cell state, the neuron conveys weights managing the information gain and loss of the cell state. The first sub-layer – *forget gate layer* – combines the input with the weights and a sigmoid activation, which results in a vector of values between 0 and 1 regulating how much of the information should be dropped (forgotten) from the data conveyed in the cell state. The network is trained to capture the relevant information from previous data which improves the prediction of new data. The next two sub-layers – *input gate layer* and *activation layer* – control which elements of the data in the cell state should be updated with new information. The *input gate layer* similarly to the *forget gate layer* uses a sigmoid activation to choose entries to be updated, and then the *activation layer* generates candidate values that should be added to the cell state. Having generated the vectors of elements to forget and to update, the network generates new data on the cell state as a linear combination of the previous data and the candidate data weighted with the outputs of the sigmoid functions in the sub-layers. Finally, the last sub-layer – *output layer* – generates new weights which will be used in the next epoch. A third sigmoid function is used to again choose what will be present in the output, and values of that vector are multiplied by the new values in the cell state after applying a chosen activation function.

The popular applications of LSTM neural networks are in language recognition (e.g. [19], [20]). However, predictions made with LSTM of the market processes and trends can also be found [21].

2.3. Input data preparation to LSTM neural networks

There were 343 announcements of the winners in the tender procedures in the 1064 days. There were 100 days with two announcements in the same days. So, there were 821 days when $SMA_{30}(PtE)$ couldn't be calculated, and the periods between the announcements vary. As the LSTM networks have to work on data with equal time spans between the variable values, it was decided to interpolate $SMA_{30}(PtE)$ values for each day without announcement. The following two method are applied: constant methods (where $SMA_{30}(PtE)$ remains unchanged from the lately recorded, up to the date of the next announcement of the winner) and linear method described below (see formula (1)).

$$SMA_{30}(PtE)_{\text{int}}^{(i)} = SMA_{30}(PtE)_r^{(k)} + \frac{SMA_{30}(PtE)_r^{(k+1)} - SMA_{30}(PtE)_r^{(k)}}{n} * i \qquad (1)$$
$$for \quad - = 1,2,\ldots,n-1$$

where:

int interpolated value
r real value
n number of days between *k* and *k*+1announcements

These processes allowed for making predictions of $SMA_{30}(PtE)$ with the use of LSTM neural networks. As the values of $SMA_{30}(PtE)$ vary from 0,38 to 0,83, there is no need to standardize them.

3. Results of LSTM application

3.1. Results achieved

The dataset consists of 73,5 % cases where the variable is not real, but interpolated. This implies the decision of calculating prediction errors only for days with announcements i.e. for days when the real value of the variable could be calculated. The first tries of forecasting $SMA_{30}(PtE)$ brought two important conclusions. The forecasting based on more than 15 days period does not bring an increase in the accuracy of predictions. Applying the dataset with linear interpolations provides much lower errors of predictions. All the results presented below are achieved for the dataset with linear interpolations, based on a 15-days historical period for predictions. Data concerning the last six months of analysed cases creates the validation dataset (approximately 1/6 of the total number of cases). So, the presented errors are calculated based on real cases from the validation dataset. The time horizon of the forecasts varies from 1 to 25 days. The forecast done for the 25th day comprise 25 values of predicted $SMA_{30}(PtE)$ from day 1 to day 25. The general rule is confirmed: the longer the forecast horizon, the bigger the error of prediction. For each time horizon LSTM networks are run separately, 50 times for each time horizon. There are different types of errors presented in table 2 concerning the farthest day of the forecast.

IOP Conf. Series: Materials Science and Engineering 1015 (2021) 012060 doi:10.1088/1757-899X/1015/1/012060

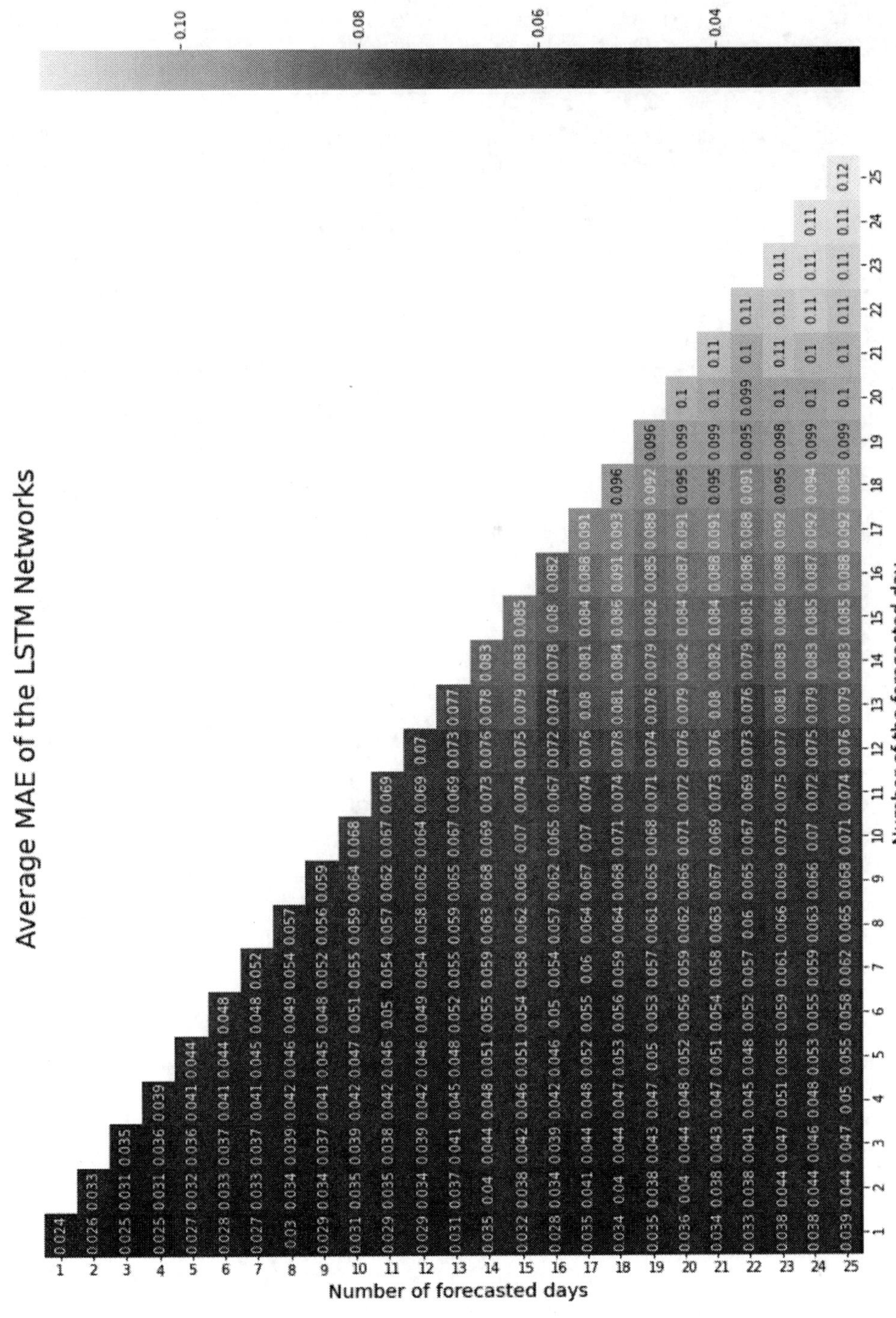

Figure 4. Heat map of MAE of $SMA_{30}(PtE)$ predictions.

Table 2. Errors of predictions for the last day of the forecast horizon.

Type of error	The time horizon of the forecasts in days 1	5	10	15	20	25
MSE	0.0011	0.0031	0.0062	0.0089	0.0128	0.0154
RMSE	0.033	0.056	0.079	0.093	0.113	0.124
MAPE in %	3.2	6.1	9.1	11.4	13.8	15.5
MAE	0.024	0.044	0.068	0.085	0.103	0.116

As the values in the dataset are not standardized the values of the errors RMSE and MAE are the same non-unit as the forecasted variable. Each from these types of errors can be presented as a "heat map". While MAE can be very easy to apply in real application of the invented prediction model, the MAE "heat map" is presented in figure 4. The results for the last day of predictions are compared to the highest potential error in figure 5.

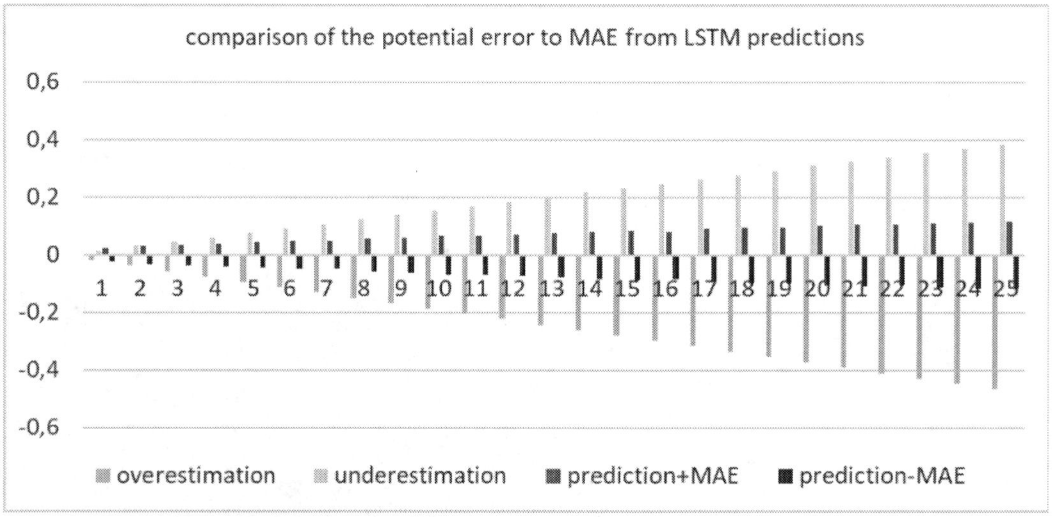

Figure 5. MAE of $SMA_{30}(PtE)$ predictions compared to the highest potential error that can be made without LSTM predictions.

The level of lowering of MAE seems very good, it is to remember that for the case without LSTM predictions the highest daily gradient of SMA is assumed (based on recorded data). To compare these approaches (considering the latest recorded SMA or calculating the predictions w LSTM network), the absolute errors (AE) are calculated for the validating dataset. There were 73 procedures in the first half of 2017 with the winner announced. The number of days between announcements is presented in table 3.

Table 3. Time span between announcements of the winner for the procedures in the first half of 2017 and number of procedures with a certain time span form the latest announcement

Time span	1	2	3	4	5	6	7	11	14
Number of procedures	25	12	10	5	7	8	3	2	2

Then, for each period category, the mean absolute error (MAE) is calculated for the case, where the latest SMA value is assumed as a current one. They can be compared to the MAE calculated based on predictions form LSTM networks. It is presented in figure 6.

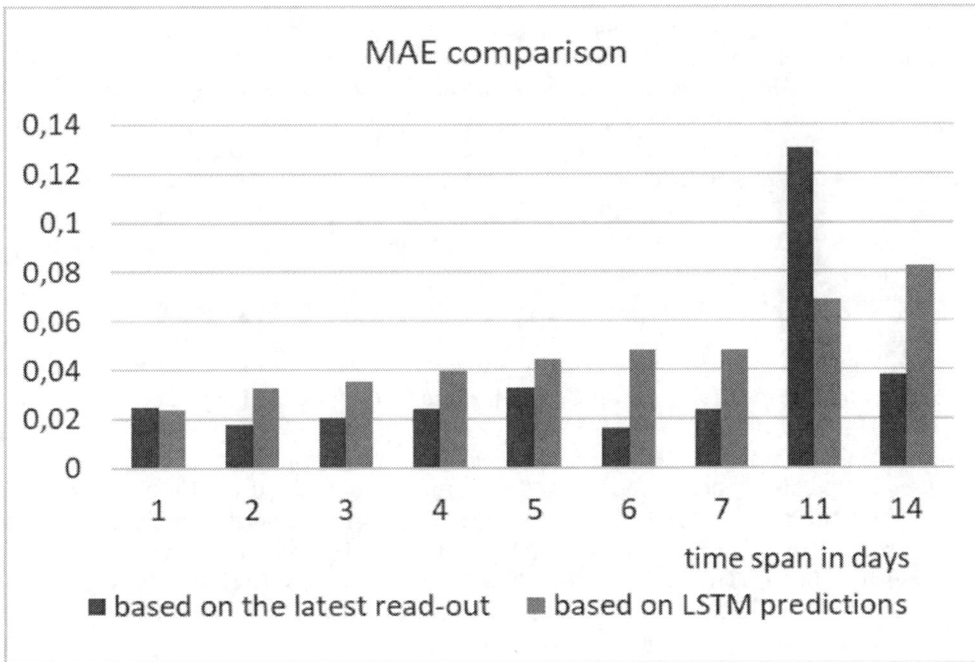

Figure 6. MAE comparison for the validating dataset (cases recorded in the first half of 2017).

Except for the 1-day period and for the 11-days period, the mean absolute error is lower, when the SMA is left unchanged from the latest announcements of the winner of the procedure. For the validating dataset – even the error values of the predictions are low – the short horizon of the forecast, and a rather stable trend of SMA, made prediction's accuracy not satisfactory. The total absolute error for these 73 procedures is 1,926 based on the latest SMA values, while for the predictions it is summed to 2,623.

4. Conclusions

The monitored price level in each tender procedure for the road construction in mid-2014 to mid-2017 differs. Sometimes the unexpected peaks of the price-to-estimation ratio are smoothed by the application of the simple moving average. It is necessary to estimate if the price level can be evaluated as market-based. It was suspected that even the period between the latest procedure and the new one is short, considering the latest value of SME may cause significant error in evaluating the price level in the new tender procedure. The errors produced by LSTM neural networks are low. When compared to the maximum recorded daily change of SMA, the absolute error can be lowered more than twice (see figure 5). Nevertheless, it was proved for the validating dataset, where the daily changes of SMA were not significant, that lower errors can be made if SMA is assumed on the latest recorded level (except for one-day span and 11-days span) i.e. without any advanced predictions.

References

[1] Zona J D 1986 *Bid-rigging and the Competitive Bidding Process: Theory and Evidence* Ph.D. Dissertation (State University)
[2] Gabrielli M F 2013 Detecting Collusion on Highway Procurement *Económica* **59** pp 1–31
[3] Oke A, Aigbavboa C and Mangena Z 2017 Prevention of collusion for innovative construction *Procedia Engineering* **196** pp 491–497
[4] Foremny A, Kulejewski J, Anysz H and Nicał A 2018 Collusion and bid rigging in the construction industry: case studies from Poland *Conference: Creative Construction Conference* (doi: 10.3311/CCC2018-107)

[5] Urząd Ochrony Konkurencji i Konsumentów 2017 *Zmowy przetargowe* Tender collusion
[6] Organisation for Economic Co-operation and Development (OECD) 2016 *Fighting bid rigging in public procurement: Report on implementing the OECD Recommendation*
[7] Organisation for Economic Co-operation and Development (OECD) 2012 *OECD recommendation of the council on fighting bid rigging in public procurement*
[8] Organisation for Economic Co-operation and Development (OECD) 2012 *Guidelines for fighting bid rigging in public procurement. Helping governments to obtain best value for money*
[9] Porter R H and Zona J D 1993 Detection of bid rigging in procurement auctions *J. of Political Economy* **101** (3) pp 518–538
[10] Harrington J 2006 *Behavioral screening and the detection of cartels* European Competition, Law Annual
[11] Ferwerda J, Deleanu I, and Unger B 2017 *Corruption in Public Procurement: Finding the Right Indicators European Journal on Criminal Policy&Research* **23** (2) pp 245–267 (doi: 10.1007/s10610-016-9312-3)
[12] Anysz H and Foremny A 2020 Client's expectations vs contractor's pricing *Fair Prices or Bid Rigging* (preprint accepted to DBMC 2020 conference) (doi: 10.13140/RG.2.2.22876.62085)
[13] Gençay R and Stegos T 1998 Moving average rules, volume and the predictability of security returns with feedforward networks *Journal of Forecasting* **17** pp 401-414 (doi: 10.1002/(SICI)1099-131X(1998090)17:5/6<481::AID-FOR709>3.0.CO;2-Q)
[14] Baiynd A -M 2011 *The Trading Book: A complete solution to mastering technical systems and trading psychology* (McGraw Hill)
[15] Lento C, Gradojevic N, and Write C S 2007 Investment information content in bollinger bands *Applied Financial Economoics Letters* **3** (4) pp 263-267 (doi: 10.1080/17446540701206576)
[16] Anysz H, Foremny A and Kulejewski J 2019 The comparison of ANN classifier to the neuro-fuzzy system for a collusion detection in the tender procedures in the road construction sector *IOP Conference Series: Materials Science and Engineering* **471** 112064 (doi: 0.1088/1757-899X/471/11/112064)
[17] Anysz H, Brzozowski Ł, Kretowicz W and Narloch P 2020 Feature importance of stabilised rammed earth components affecting the compressive strength calculated with explainable *Artificial Intelligence Tools* **13** (10) (doi: 10.3390/ma13102317)
[18] Hochreiter S and Schmidhuber J 1997 Long short-term memory *Neural Computation* **9** pp 1735-80 (doi: 10.1162/neco.1997.9.8.1735)
[19] Sundermeyer M, Schlüter R and Ney H 2012 LSTM neural networks for language modeling *13th INTERSPEECH Conference* pp 194-197
[20] Graves A, Mohamed A and Hinton G 2013 Speech recognition with deep recurrent neural networks *IEEE international conference on acoustics, speech and signal processing* pp 6645-49
[21] Tav N, Verenich I, La Rosa M and Dumas M 2017 Predictive business process monitoring with LSTM neural networks *Adv. Information Systems Eng. CAiSE 2017 Lecture Notes in Computer Science* ed Dubois E Pohl K **10253** (doi: 10.1007/978-3-319-59536-8_30)

IOP Conf. Series: Materials Science and Engineering 1015 (2021) 012061 doi:10.1088/1757-899X/1015/1/012061

Construction price assessment method based on inferential statistics

L Remek, J Mikolaj, L Mikulova and M Durinova

Department of Construction management, Univesity of Zilina, Zilina 01026, Univerzitna 1, Slovakia

E-mail address: lubos.remek@uniza.sk

Abstract. The paper presents a new objective price assessment method that can be used to accept, reject or negotiate prices for construction works proposed by the contractor of construction works. The method can be used as part of an acquisition of a new construction and in claim negotiation process both in private or a public sector. Price assessment is an investor's tool for the assessment of price eligibility. It should be focussed on evaluation of direct construction costs and indirect costs of the contractor in order to establish whether the costs declared by the contractor are justified. This creates a negotiation leverage in the creation of a contract or creation of legal basis for negotiations regarding claim reimbursement which can be recognized in accordance with special contractual provisions. Price assessment is a very time consuming task, the presented method uses statistical tools to assess the contractor's price proposal by detailed analysis of main items on the budget, which are then treated as a population sample. Population mean of difference in contractor prices and price catalogue prices are then used to establish a confidence interval which can be used for hypothesis testing, i.e. drawing conclusions and further negotiations.

1. Introduction

One of the main specificities of construction production is the uniqueness of the works that are carried out. Each construction is unique in terms of location of the construction site, conditions on the site [3], requirements for speed of execution [2], weather and other factors. This uniqueness is reflected in each part of the construction process including its price, which, moreover, is largely influenced by the situation on the market [6]. It is therefore difficult to establish a firm criterion according to which it is possible to decide what could be considered appropriate price of a construction. The main aspect that guarantees that the price stays within reasonable boundaries is competition. Public procurement procedures [5] take advantage of this fact by including "lowest bid" criterion as a main or sole criterion when selecting a contractor. While private procurement isn't regulated by law and private investors are free to choose contractors however they see fit, the "lowest bid" principle also naturally and reliably regulates prices also on the private markets. Specific conditions arise when changes in technical parameters of a construction are required. The related costs of these changes are known as claims.

The claim is a legitimate demand for additional reimbursement of costs (or additional time) as a result of changes made by an unforeseen circumstances and conditions that weren't laid down in the contract. For a claim to be eligible in the form of an addendum to the contract, it is first necessary to comply with the procedural rules of the requirement defined in the contract. These usually deal with technical justification of the claim rather than financial justification of items on the contractor's claim budget.

The subject of this paper is the objective price assessment method for items on the contractor's claim budget that can be used to negotiate prices for works submitted by the construction contractor. Price assessment is a tool for investors to assess the eligibility of declared direct construction costs and indirect costs of the contractor in order to create a legal basis for a claim for reimbursement, which can be recognized in accordance with special contractual provisions.

2. Price assessment method – methods, tool and logic

In civil engineering, the price analysis is followed by an assessment of technical justification, which evaluates the adequacy of the scope and nature of the construction work performed. Technical substantiation should be approved by the construction supervision and, in case of doubts, reviewed by the Institute of Forensic Engineering (in Slovakia) or other national authority or certified entity in the given country. Price assessment that follows uses four basic tools:

- Detailed price assessment focused on direct costs and indirect costs declared by the contractor
- Analysis of the existing price history.
- Price comparison with an independent third party estimate.
- Comparing competing offers from multiple suppliers.
- Use of an existing price catalogue.

The analysis of direct construction costs is based on the analysis of the unit price. It is the analysis of the items technical solution – materials, machinery and professions wage used for particular unit and their unit costs:

Direct material = Purchase price + price of transport to the construction site.
Wage costs = Direct wages increased by 15% as additional wages (increase for holidays, visits to doctor, vacation according to annual working time fund) + statutory deductions 35.2% of direct wages, as well as additional wages.
Machinery costs = Depreciations + operating cost (fuel, oil, repairs, etc.).

The analysis of indirect cost and profit calculation - the price of indirect costs is usually based as a percentage of direct construction costs and can be split into contractor's production and administrative overhead, profit, or other subjective surcharges. In the calculation formula, the contractor takes into account the costs of production processes on the construction site [3], the financing of the operation of the construction company and the market effects expressing the profit. The calculation formula sets the flat rate and percentage rates for determining overheads and profit. The form and structure of the calculation formula is not standardized, it can differ not only between construction companies and contracts, but also within the budget according to the type of item - different types of calculation formulas can be applied to calculated, uncalculated, material, external, etc. items.

Analysis of the existing price history is used mainly for items that were originally approved in the contract but unforeseen circumstances lead to increase of the item's number of units [1]. In these cases, we usually assume that unit costs remains the same. This however isn't always the case, if the technical and technological aspect changes, for instance earthwork increase that leads to the use of more powerful excavator, this approach is not valid anymore and detailed analysis is required. Historic prices from previous contract may also give a good estimate if new claimed contractor's price is appropriate, the historic prices are not binding however and other factors like construction site conditions can have an influence. Price comparison with an independent third party estimate is a good method if the third party is completely independent. A third party can give an estimate on prices claimed by the contractor, however, since the third party is usually a competition to the contractor on the market a conflict of interest can lead to objections from the contractor when referred to third party prices.

Comparing competing offers from multiple suppliers is the basic principle in public procurement, namely in the bidding process and in private procurement when multiple contractors are asked to bid for the contract. This method can't usually be used in claim management; it is usually not reasonable to search for a new contractor to provide extra construction works when original constructor has the capacity to carry them out.

Use of an existing price catalogue is a good method to assess the appropriateness of prices. The catalogue is usually ordered by a classification system. In the European union, NACE– Statistical classification of economic activities in the European Community and CPV– Common Procurement Vocabulary [4] or other national catalogue. In Slovakia the Classification of construction structures (TSKP) originated in 1977 is still the main classification system used today. Classification system ensure classification common to all normative documents, such as operational standards, price lists of construction works, etc. By assigning of a unique code to an item, the qualitative, delivery and other characteristics of the building structure or material are precisely defined. When performing price analysis, we usually consider a unit price analysis, which is a price analysis of one-unit quantity of a construction work. The requirement is the classification within the structure of classification system, which will achieve a clear definition of the subject of valuation. Quality is assumed to be at the level required by the relevant technical regulations and standards. However, the exemption of a unit price from the budget and subsequent individual assessment raises the problem of the possible absence of surcharges that are linked in the budget to the unit. These surcharges are usually items o material manipulation on the construction site, surcharges for soil stickiness, high elevation works risk surcharges etc.

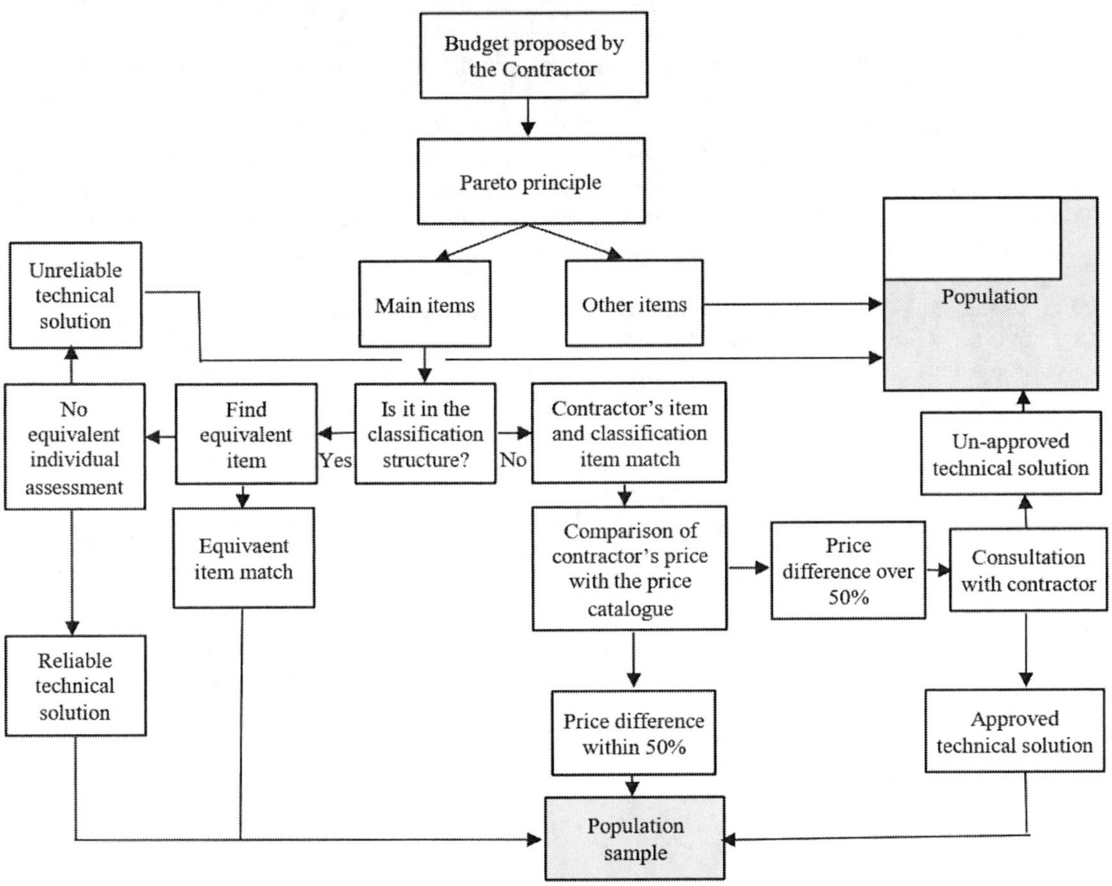

Figure 1. Price analysis and identification of items composing population and the population sample needed for inferential statistics.

Another principle used in price analysis is the Pareto principle, i.e. the 80/20 rule. In the civil engineering economics, about 80% of the total costs make up only about 20% of the items of the

IOP Conf. Series: Materials Science and Engineering 1015 (2021) 012061 doi:10.1088/1757-899X/1015/1/012061

construction works. The other approx. 80% of items are composing the residual 20% of the total costs. Thus, the detailed cost calculation is usually carried out only on these main items. Due to rationality, the price analysis is also carried out on these main items.

Figure 1 shows the overview where above mentioned methods and tools are applied and the constructor's budget is divided between items that composes the population sample and the population. This will be used in the inferential statistic to estimate statistical price difference of item's that weren't individually assessed and aren't included in the price catalogue.

Following the chart on figure 1, we can identify the steps of price analysis prior to the inferential statistic evaluation:

1. Identification and selection of main items on the contractor's budget
2. Identification of main items in the classification structure.
3. Individual assessment of main items outside of the classification structure.
4. Finding equivalent items within classification structure that matches the contractor's item – comparison of main items from the contractor's budget with those of technologically identical/similar.
5. For contractor's main items that don't have equivalent in the catalogue structure – analysis of the technical and technological solution from price catalogue items for the purpose of detailed price assessment and individual negotiation of item's price.
6. Comparison of the prices of all contractor's items in the classification structure or those that have an equivalent in the classification structure with those in the price catalogue.
7. Consultation of items with difference of over 50% contractor vs price catalogue.
8. Clarified items used as population sample to estimate overall price difference of all item's that weren't subject to detailed price assessment by inferential statistics.
9. Drawing of conclusions for the strengthening of the customer's position in the negotiations regarding unit prices of new items in order to protect public resources.

3. Price assessment of constructor's budget with the use of inferential statistic

If the investor follows the procedure described in previous chapter, he will have items that compose a population sample and the population. Compared to equivalent items in the price catalogue,

The price of contractor's items in the population sample will have a price difference against prices in the price catalogue. We will use this to estimate average price difference of all item's in the population. This process can be therefore split in two steps:

1. price assessment of items in the population sample, and
2. statistical analysis – estimating average price difference and confidence interval.

3.1. Price assessment of items in the population sample

Comparison of contractor's item with the price catalogue will produce a price difference. Since we want to apply the price difference on other items, we will work with percentages instead of financial units. Equation 1 and table 1 show how to calculate the percentile price difference of items in the population sample. Upper row for each item is the Contractor's price, lower row is the Price catalogue price.

$$PD = \frac{(CP - PCP) \cdot 100}{PCP} \tag{1}$$

PD Price difference [%]
CP Contractor's price [€]
PCP Price catalogue price [€]

IOP Conf. Series: Materials Science and Engineering 1015 (2021) 012061 doi:10.1088/1757-899X/1015/1/012061

Table 1. Example of calculated price difference of items in the population sample.

No.	Item code	Description	Unit	Quantity	Unit costs	Price	Price difference
99	460510201	Item 1	pcs	70.000	5.331	373.180	
			pcs	70.000	4.45	311.64	19.75%
		Difference			0.879	61.540	
101	460510202	Item 2	m	6.000	4.366	26.190	
			m	6.000	5.38	32.26	-18.79%
		Difference			-1.010	.	
103	460200174	Item 3	m	26.000	15.523	403.600	
			m	26.000	15.67	407.47	-0.95%
		Difference			-0.149	-3.872	
104	460560174	Item 4	m	26.000	3.318	86.280	
			m	26.000	3.24	84.24	2.42%
		Difference			0.078	2.040	
105	460200304	Item 5	m	44.000	29.474	1 296.850	
			m	44.000	27.05	1 190.11	8.97%
		Difference			2.426	106.738	
106	460560304	Item 6	m	44.000	7.090	311.950	
			m	44.000	6.50	286.18	9.00%
		Difference			0.586	25.774	
108	460120002	Item 7	m^3	3.360	19.335	64.970	
			m^3	3.360	11.96	40.20	61.61%
		Difference			7.371	24.771	
109	460490012	Item 8	m	70.000	0.544	38.080	
			m	70.000	0.58	40.32	-5.56%
		Difference			-0.032	-2.240	
111	584121111	Item 9	pcs	3.000	69.088	207.260	
			pcs	3.000	7.85	23.54	780.33%
		Difference			61.240	183.716	
113	210950203	Item 10	m	6.000	1.809	10.850	
			m	6.000	2.14	12.82	-15.32%
		Difference			-0.327	-1.966	
114	130001101	Item 11	m^3	1.000	19.611	19.610	
			m^3	1.000	17.19	17.19	14.08%
		Difference			2.421	2.420	
129	330510240	Item 12	m	1.000	3.019	3.020	
			m	1.000	25.30	25.30	-88.07%
		Difference			-22.281	-22.280	
Average weighted difference (weight=price)							65.63%
Average weighted difference (without local extremes)							9.41%
Contractor's price							2 841.84 €
Price according to price catalogue							2 471.27 €
Difference between contractor's price and price catalogue price							370.57 €

3.2. Statistical analysis – estimating average price difference and confidence interval

Now that we have the price difference of all items in the population sample expressed as percentages, we can proceed to perform a statistical analysis [7] [8]. The basic indicators that enter into more advanced statistical methods are the arithmetic mean of a population μ, standard deviation σ and standard error of the mean σ_μ. The arithmetic mean equation 2 follows:

IOP Conf. Series: Materials Science and Engineering 1015 (2021) 012061 doi:10.1088/1757-899X/1015/1/012061

$$\mu = \frac{\sum_{i=1}^{n} PD_i}{n} \tag{2}$$

μ arithmetic mean of a population [%]
PD_i price difference of a unit [%]
n population sample size (number of main items)

The standard deviation σ (equation 3), is a measure of statistical dispersion in probability theory and statistics. Simply put, it talks about how widely the values are distributed in a set, i.e. "how close the scores are to the mean.". Low value of standard deviation means that price differences of main items lie close together, high number means they are dispersed

$$\sigma = \left(\frac{\sum_{i=1}^{n} PD_i - \mu}{n-1} \right)^{1/2} \tag{3}$$

σ standard deviation

The standard error of the mean $\sigma_{\bar{x}}$ (equation4) is the standard deviation of a random variable, i.e. sampling mean obtained as a random sample from the base population set, this is called sampling distributions. The sampling distribution of a population mean is generated by repeated sampling and recording of the means obtained. Simply put, the standard error of the mean is a number that indicates how much the average of a random sample differs from the arithmetic mean of a population. We can imagine the mean error of the average as a random variable, which would be created by the repeated random selection of population samples groups from the population and the calculation of the average in each individual group.

$$\sigma_{\bar{x}} = \frac{\sigma}{n^{1/2}} \tag{4}$$

$\sigma_{\bar{x}}$ standard error of the mean

Table 2. Example of basic statistical indicators performed on the population sample.

Population sample n	Arithmetic Mean of Population n μ	Standard Deviation σ	Standard Error of the Mean $\sigma_{\bar{x}}$
12	63.96%	2.28	65.87

Final step is to apply inferential statistics and find the confidence interval for the arithmetic mean. Ideally, all items submitted by the contractor would be analysed in detail. This usually isn't rational for constructions with budgets that contains dozens of items, or is outright impossible for constructions that contains hundreds. However, since we can (and should) analyse in detail at least the main items of the building and determine their deviation from price catalogue prices, we acquire a population sample that we can statistically process (see previous chapter) and apply the result to the rest of the whole population (budget with all items), this procedure is called inference statistics. The essence of inference statistics is to determine the probability that the results obtained on a population sample will be valid for the whole population. In the case of this method, the sample is therefore a set of price differences of main items, that we assessed in detail. We assume that the contractor used his internal price catalogue and price settings procedures when valuing his construction works, these will be reflected in both the population sample and the population. We could assume that the average price difference in the sample and the base set will be therefore the same. But that doesn't have to be true. We can only say this with a certain probability. The degree of this probability is expressed by the confidence interval *CI* (equation 5). If we determine the average price deviation on the sample, the confidence interval will tell us in what interval falls the deviation of the whole population and with what probability, i.e. confidence.

$$CI = \mu \pm (z * \sigma_{\bar{x}}) \tag{5}$$

z Gaussian probability distribution coefficient

The coefficient z is based on the Gaussian probability distribution. It is possible to calculate it, or use coefficients from statistic tables for the probability percentage of interval we are interested.

Table 3. Gaussian probability distribution coefficients.

Confidence	85,00%	90,00%	95,00%	97,50%	99%	99,50%
z	1,036	1,282	1,645	1,960	2,326	2,576

For this case study, the arithmetic mean of population is 63,96 % price difference. Following equation 5 we can estimate confidence interval for 95%, 90% and 85% probability. This means that the price difference of the whole contract/claim budget will fall in these intervals with said probabilities.

Table 4. Example of confidence interval calculation for the presented case study.

Confidence	85,00%	90,00%	95,00%
Mean	63,96%	63,96%	63,96%
Confidence interval	11,92%	23,89%	35,96%
Upper limit of CI	69,92%	75,90%	81,94%
Lower limit of CI	57,99%	52,01%	45,98%

The confidence interval gives the upper limit ULCI and lower limit LLCI (see equations 6 and 7) from arithmetical mean of a population sample in which we expect the arithmetical mean of entire population will fall with certain probability.

$$ULCI = \mu + \frac{CI}{2} \quad (6)$$

$$LLCI = \mu - \frac{CI}{2} \quad (7)$$

As seen from table 4 and figure 2, the interval expands with rising probability. Thus we can say that we are 95% confident that the arithmetic mean of the population lies between 45,98% – 81,94% but we are only 85% confident that it lies between 57,99% – 69,92%.

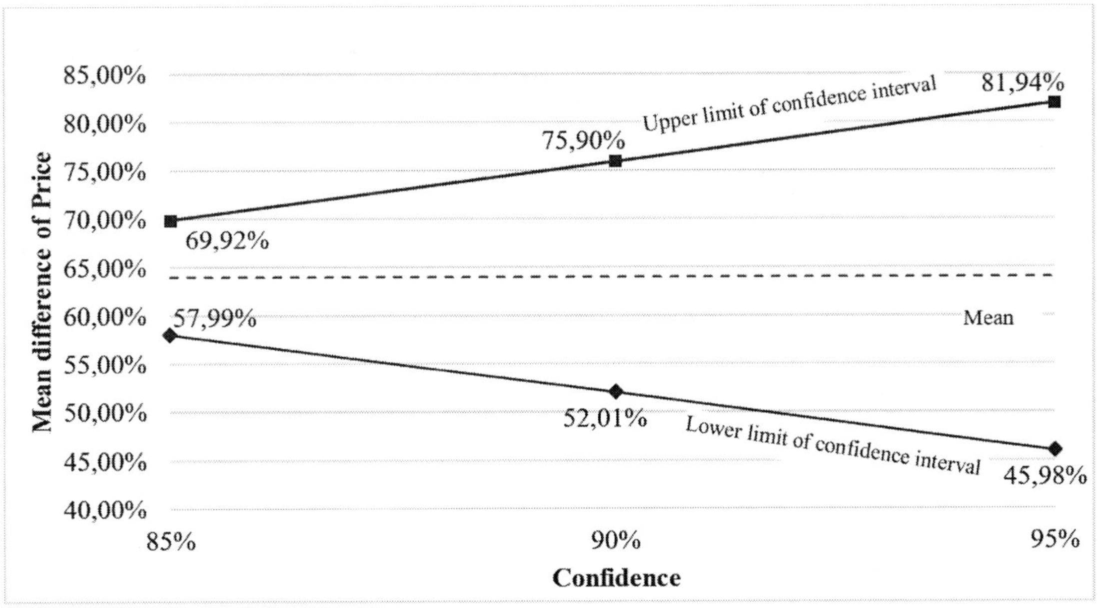

Figure 2. Graphical representation of confidence interval.

Chart such as one shown in figure 2 should be presented by the employer to the contractor. It should be used to negotiate deduction of contractor's claim for items that weren't part of the population sample. Simply stating that the average price difference of contractor items was a 63,96 % increase compared to price catalogue prices is a statement that doesn't give much space to negotiate a consensus between the contractor and the employer. Working with intervals may give a better framework for such discussion, and give the employer more space to propose deductions that don't look like an arbitrary percentage.

4. Conclusion and discussion

Assess the financial justification of contractor's budget remains the main concern for both private and public investors. The process is time consuming and requires a good knowledge of price setting which is usually lacking in the public sector. Comparison budget from third party or a comparison to the price catalogue remains the main method to assess financial justification of contractor's works. There are however severe limitations how this methods, the main being the sheer volume of budget counting sometimes hundreds of items. The described method can be used to give the investor objective data supporting his claim agenda and give him an advantage when negotiating price for additional construction works. The main items are to be negotiated separately. The price difference of main items from price catalogue (or comparisonn budget) can be used to estimate the average price difference that can be used for all other items. Two main question arise when considering this method:
What if our population sample is very slim and the price differences are extremely disparate? Due to the nature of inferential statistics, these disparities will be reflected in the statistical analysis, you will end up with very wide confidence intervals. We recommend to increase the population by adding item's that can be individually assessed by direct and indirect cost calculations. You may also consider to exempt the local extremes that range over 200% price difference, these items should also be considered for clarification by the contractor.
Which confidence interval should we use? Generally, the lower probability interval should be used if you expect the average priced difference be aggressively pushed downward by the contractor. Higher confidence should be preferred when you expect you can drive the price difference higher. Who will have the better leverage when negotiating the price will be determined by factor such as technical justification of claimed works, overstepping of budget specified in the contract, additional revenue possibilities from future contracts, etc. For instance, dubious technical justification of claimed works, high overstepping of the budget and further contract possibilities should give the investor upper edge when negotiating, thus you can push for the higher price difference in high confidence percentage.

5. References

[1] Kaczorek K and Anysz H J 2019 BIM 5D Czyli usprawnione zarządzanie kosztami przedsięwziecia budowlanego *Przewodnik Projektanta* **1** pp 20–2
[2] Kutera B and Anysz H 2016 The methodology of technical due diligence report preparation for an office, residential and industrial buildings, *5th Int. Sci. Conf. Integration, Partnership and Innovation in Construction Science and Education* (DOI: 10.1051/matecconf/20168607009)
[3] Anysz H 2017 The profit as in-company evaluation of the construction site effectiveness, *XXVI R-P-S Seminar Theoretical Foundation Of Civil Engineering* (DOI: 10.1051 /matecconf/201711700009)
[4] Official *Journal of the European Union*, COMMISSION REGULATION (EC) No 213/2008 amending Regulation (EC) No 2195/2002 of the European Parliament and of the Council on the Common Procurement Vocabulary (CPV) and Directives 2004/17/EC and 2004/18/EC of the European Parliament and of the Council on public procurement procedures, as regards the revision of the CPV, 2008
[5] Official *Journal of the European Union* 2014 DIRECTIVE 2014/24/EU OF THE EUROPEAN PARLIAMENT AND OF THE COUNCIL, on public procurement and repealing Directive 2004/18/EC, 2014

[6] Rosłon J H and Kulejewski J E 2019 A hybrid approach for solving multi-mode resource-constrained project scheduling problem in construction *Open Engineering* **9** (1) pp 7-13
[7] Aczel A D and Souderpandian J 2012 *Complete Business Statistics, Eight Edition*, ISBN 978-1935938187
[8] Anysz H and Rogala W 2018 The control of the mass of an aggregates deliveries and its impact on the effectiveness of earthworks execution *Sci. Rev. – Eng. and Environmental Sciences* DOI:10.22630/PNIKS.2018.27.3.27

IOP Conf. Series: Materials Science and Engineering 1015 (2021) 012062 doi:10.1088/1757-899X/1015/1/012062

Diagnostics data in the framework of railway tracks maintenance

J Šestáková and A Pultznerová

Department of Railway Engineering and Track Management, Faculty of Civil Engineering, University of Žilina, Univerzitná 8215/1, SK-01026 Žilina, Slovak Republic

E-mail address: janka.sestakova@uniza.sk

Abstract. Determining the requirement and location and deciding on the time and sources of repair action is a difficult process, influenced by requirements on cost minimization and operational constraints. The particular sections of the railway track react differently to traffic and non-traffic loads, which implies the need to determine the format of development of structural quality indicators and decision-making is based on a large amount of technical and economic information, the interconnection of which is determined by the infrastructure manager. The decision-making process on repair work in the railway infrastructure management system is therefore based on the analysis of diagnostic data, technical, technological, environmental, and economic information, or the personal professional experience of the staff of the infrastructure manager. Understanding the causes of the structure's behaviour in this process leads to the identification of its weaknesses and the ability to predict its behaviour, which allows for timely planning and accurate targeting of adequate repair work to minimize the occurrence of structural faults.

1. Introduction

During the life cycle of the railway track, or of its components, the infrastructure manager performs a set of technical, technological, administrative and management activities, collectively referred to as "maintenance" [1]. The objective of the realization of these activities is to maintain or achieve a state in which the railway track performs the required functions in the prescribed quality, comfort, and safety. From the above, it follows that the maintenance system also includes all activities performed to rehabilitate the quality of the railway track, i.e. diagnostics and repair work (routine maintenance, repairs, or reconstructions) and their interaction. Rehabilitation of the quality of the railway track creates a fundamental set of processes implemented with the aim of safe and reliable operation during the operation of the structure.

The extension of the service life of the structure within the operational-rehabilitation cycle is affected by:

- adequate information (diagnostic data) on the operational quality of the structure,
- comprehensive analysis of diagnostic data suitable for determining the causes of the operational quality of the structure,
- adherence to diagnostic strategies in relation to repair activities,
- appropriately chosen and used tools for repair work activities,
- adequate material and technical support for repair work activities,

- adherence to the recommended technological principles (complexity of activities) during the repair work activities.

2. Diagnostic data in the railway maintenance strategy

The main requirements of the infrastructure manager for diagnostics in relation to the planning and organization of repair work can be generally defined as the reduction of repair costs by introducing condition monitoring of the railway track in places where it is technically and economically possible to reduce repair activities, reduce repair costs with optimization, determining the life cycle costs used of the structures and their parts and repair activities, assisting to select the most appropriate type of repairs and to determine the optimal interval for repair work, maintaining the prescribed level of safety of railway operation. [2]

In the area of railway diagnostics, it is important to define the structures, their parts, or parameters that are decisive for maintaining the safety of railway traffic. Finding up-to-date information about them, in combination with "historical" data in the database of the results of previous diagnostic and repair activities, provides the basis for predicting the future state of the structure and for planning the future repair activities (figure 1).

Information on the quality of the railway track obtained by diagnostic methods should not be single use, only to inform about the current state of the structure. In the management of railway infrastructure, together with the history of already realized repair works, information databases are created, which allow behaviour monitoring of the structure in its history, predict the evolution of its quality and provide better conditions in time planning, financing, and repair technology.

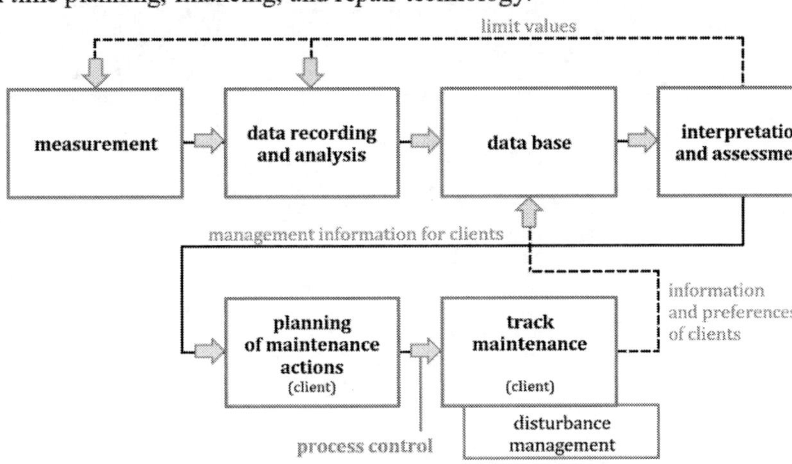

Figure 1. Diagnostics data and subsequent repair activities flowchart [2].

Achieving the complexity of the diagnostic system is conditioned by keeping the structure of primary diagnostic data. The structure of the data of diagnostics of the track geometry and structural elements of the track panel and the expected quality of indicators quality necessary for the prediction of the development of the quality of the structure and the planning and organization of repair work is evident from figure 2, in which the diagnostic methods are marked with the number: 1 – visually common, 2 – video systems with recording, 3 – point manual, 4 – continuous manual contact, 5 – continuously under load contact, 6 – continuously under load contactless.

The primary data also includes the most precisely defined information about the position and time of measurement (optimally synchronized in GPS). For an adequate evaluation of the quality of the railway track, it is important to supplement the diagnostic data with the data shown in figure 3: on the construction of the track, the history of repair work, and on the traffic and transport characteristics of the diagnosed section.

area of interest of diagnostics			quantity / characteristic	method 1	2	3	4	5	6	expected quality indicators (in the defined section)
geometric characteristics of the track	spatial position of the track		level of track			✓	✓			• quality code of spatial position of the track, expressing the influence of its deformations also on the change of force effects and traffic safety
			alignment of track			✓	✓			
	track alignment	structure	track gauge			✓	✓	✓	✓	• track quality index • SDV – standard deviation • QM – quality mark • OQM – overall quality mark • MT – mark of tamping • QN – quality number • number of local errors • AL – alert limit overrun • IL – intervention limit overrun or • IAL – immediate action limit overrun • the ratio of the length of sections with local error to the total length of evaluated section • graphical shape of track geometry errors • location of track geometry errors
			change of track gauge			✓	✓	✓	✓	
			nominal track gauge to mean track gauge over 100 m				✓	✓	✓	
			cross level			✓	✓	✓	✓	
			track twist			✓	✓	✓	✓	
		geometry	longitudinal level of rails			✓	✓	✓	✓	
			alignment of rails			✓	✓	✓	✓	
rolling stock movement	railway vehicle response		wheel/rail force						✓	• values calculated for the rolling stock with the most unfavourable parameters or for other rolling stock • Q vertical wheel/rail force • Y lateral wheel/rail force • Y/Q ratio • ΣY sum the lateral forces • a_y lateral acceleration in the vehicle body • a_z vertical acceleration in the vehicle body
			vehicle body acceleration						✓	
			acceleration of train floor						✓	
			acceleration of bogie						✓	
structural elements of track skeleton	rails and load-bearing structural elements in switches and crossings		rail section	✓		✓	✓	✓	✓	• number of defects • in the middle of the rail length • at the ends of the rail and • in the weld [categories A to D by [3]] • % of rails with defects (length) • number of rail wear defects • rail wear values • % of worn rails (length) • number of defects of the rail joints geometry • SDV of rails microgeometry
			shape deformation	✓	✓	✓	✓	✓	✓	
			rail head microgeometry	✓	✓		✓	✓	✓	
			breach of material integrity	✓	✓	✓	✓	✓	✓	
			material inhomogeneity			✓	✓	✓	✓	
			weld	✓	✓	✓	✓	✓	✓	
			rail joint	✓	✓	✓	✓	✓	✓	
	rail supports		shape deformation	✓	✓					• % of damaged rail supports with influence on fastening functionality • % of damaged rail supports without affecting on fastening functionality • length of sections with accumulation of damaged rail supports
			material failure	✓	✓					
			fastening failure	✓	✓					
			spacing	✓	✓					
	track fittings and rail fastenings		shape deformation	✓	✓					• % of loose fastenings, • % of missing fastening, • length of sections with accumulation of loose/missing component
			material failure	✓	✓					
			loose/missing component	✓	✓					

Figure 2. Track quality diagnostic data – geometry and structural elements of the track skeleton [4].

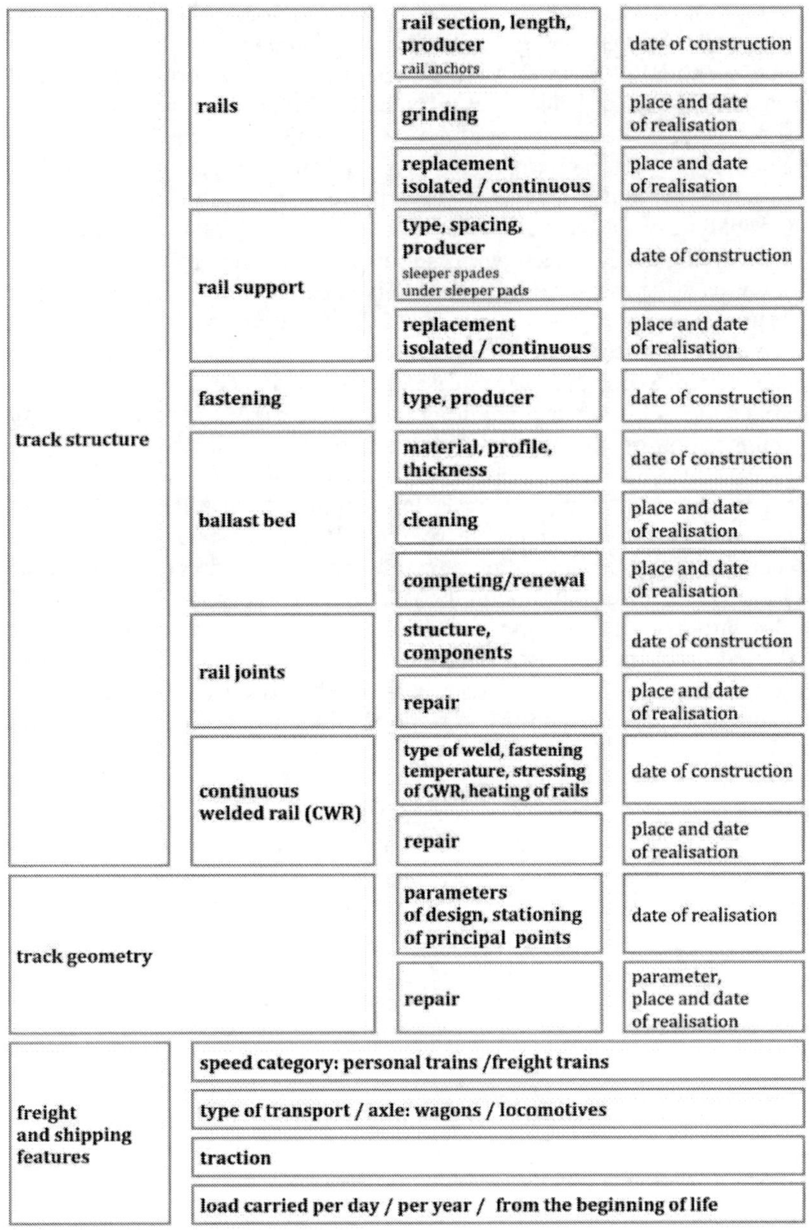

Figure 3. Track quality diagnostic data – basic section data [4].

The railway infrastructure also consists of other components, the quality of which affects the results of determining the quality of the track geometry, its development, and the planning of repair work. The quality of the track geometry is also affected by the parameters and condition of the rolling stock that moves along it and the conditions by which the railway track is affected by its surroundings and weather. Secondary diagnostic data are information on other variables of the track geometry (deficiency /*I*/ or excess /*E*/ track cant, horizontal track curvature /*k*/), railway substructure (ballast bed: condition description, % contamination, % material loss, presence of undesirable vegetation, substructure construction layers: geotechnical and hydro-technical parameters), types of rolling stock and state of its maintenance, contact geometry of wheelsets (contact angle, equivalent conicity /tg γe/), equipment and

IOP Conf. Series: Materials Science and Engineering 1015 (2021) 012062 doi:10.1088/1757-899X/1015/1/012062

railway superstructure structures, railway subgrade structures and equipment, signalling equipment and security technology (SST) and energy and electrical engineering (EE), climatic conditions of measurements and operation of the railway track, conditions of the surroundings of the railway track, the history of detected faults and deficiencies, eventually also the history of accidents.

Prediction of the expected development of the design quality requires organizing the data collection in the structure so that the data are comparable: unified in type, location and collection conditions, where, in terms of [5], [6] and [7], repeatability and reproducibility of measurements are required. Measurement and analysis of one set of diagnostic data allows only realization of the repair: if the value is exceeded, the repair activity is carried out, if not, it is not implemented.

With several consecutive measurements and their integrated analysis depending on time or other operational indicators, it is possible to detect quality deterioration trends that support the prediction of structural behaviour and thus the planning of structural rehabilitation before the values of diagnosed parameters reach the predetermined limit values and before faults occur.

At present, systems for continuous monitoring of selected parameters are being developed and used, thus obtaining information on the state of construction of the railway track, rolling stock, and the environment in real-time [8]. This information is combined with 'isolated' railway track interval diagnostics information, thus creating a basis for optimizing the planning and implementation of maintenance and repair activities in the predictive maintenance system. Such condition monitoring is important not only in terms of providing the information for decision-making in the predictive maintenance system but also for optimizing the activities of transport operations on the structure. Its properly evaluated information can help reduce operational reliability and safety risks, improve track performance, and reduce operating costs. As the current state is detected by real-time monitoring, the degradation of the monitored parameters and characteristics is operatively detected, and the railway infrastructure manager increases his ability to respond with adequate activities of routine maintenance and repair or reconstruction work [9].

Railway repair planning based on diagnostic data can be carried out in three ways. Manually – using the professional experience of an employee who visually evaluates data, e.g. in tables or graphs. This data can also be evaluated automatically. Some railway administrations use fully automated systems [10] to perform various optimizations, degradation modelling, and analysis to achieve optimal maintenance and repair. The latter method is less time consuming and allows a thorough analysis of all aspects and their development without the risk of human error, while manual methods, including visualization, are slow, inefficient, and affected by human error. Ensuring effective maintenance requires the continuous collection and analysis of data on the condition of maintained elements and data on financial costs incurred.

3. Models of relationships between repair work and diagnostic data

The planning and organization of repair work is influenced by the characteristics of the maintenance strategy, by which the method of using diagnostic data is determined. The results of the analysis of these data naturally affect the specification of the input information for the planning and organization of repair work, while they determine:

- location of the repair site within the construction of the railway track (railway superstructure, railway subgrade, etc.) and in the space (point, object, line),
- the method (work procedure) of organizing the repair,
- resources and characteristics of structure repair (material, people, machines, time, finances).

For effective financing, the repair work on the construction of railway tracks and structures (RTS) is carried out in coordinated (time, location) and comprehensively, including related work on SST and EE facilities.

The models of repair work activities express the relationship between the technology of the activity and the minimum range of diagnostic data from the group of primary indicators of the operational quality of the maintained section.

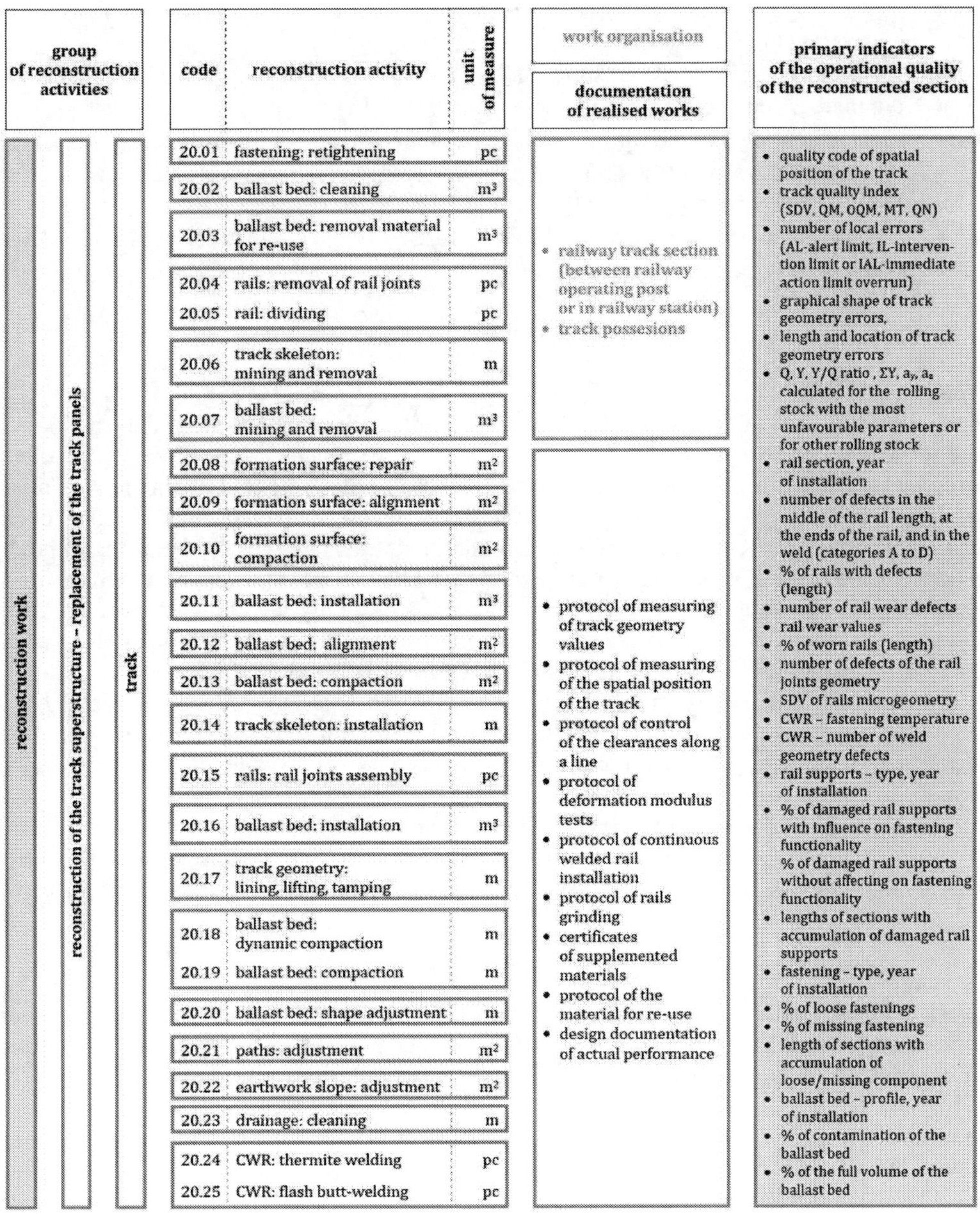

Figure 4. Relationships between railway superstructure reconstruction activities and diagnostics data – track (replacement of track panels) [4]

Relationships between activity and data were defined in [4] for maintenance, repair and reconstruction works of track superstructure as follows:

- routine maintenance (activities were not assigned by identification codes),
- repair works (activities are indicated by codes 10 to 17),
- repair works of track structure and geometry (10),
- repair works of switch (or crossing) structure and geometry (11),
- repair works of track (or switch or crossing) ballast bed by ballast cleaning machine (12),

IOP Conf. Series: Materials Science and Engineering 1015 (2021) 012062 doi:10.1088/1757-899X/1015/1/012062

- repair works of track (or switch or crossing) ballast bed by complete replacement with excavating (13),
- replacement of isolated rail supports in track (or switch or crossing) (14),
- continuous replacement of rail supports in track (15),
- replacement of isolated rails or welded rail in track (or switch or crossing) (16),
- reprofiling of rails in track (or switch or crossing) (17),
- reconstruction works (20 to 22)
- reconstruction of track railway superstructure by replacing of track panels (20) (figure 4),
- reconstruction of track railway superstructure by assembly of track skeleton in the track axis (21),
- reconstruction of switch (or crossing) railway superstructure (22).

4. Conclusions

The methods of diagnosing the operational quality of the railway track are different for individual railway infrastructure managers, even though the same parameters of railway infrastructure are measured and evaluated in principle, while the approaches to planning and organizing repair work are also different. In this area of activity of railway infrastructure managers, the regional, operational, technical, and economic conditions of railway administrations are even more manifested. These differences are the reason why it is not easy to create a universal or to apply a proven track repair planning system, although it is usually an open and constantly evolving system to which new technologies for the assessment and rehabilitation of operational quality are being adapted. A great influence on the maintenance systems has also been attributed to risk assessment, too. [11]

As the planning and organization of repair activities on a railway line relate several co-acting structures and structural elements of different quality and durability, it is not possible to create and use a single model for managing the activities of all types of repair work. In any case, management must be based on:

- correctly determined limit levels for repair action,
- properly implemented means for measuring and monitoring the quality of the structure, and
- reliable methods and tools for evaluating the current and predicting the future state of the structure.

Acknowledgments

This work was supported by the Ministry of Education, Science, Research and Sports of the Slovak Republic through the research project of scientific grant agency VEGA No. 1/0275/17: "Application of Numerical Methods in Defining the Change of the Geometric Position of a Track".

References

[1] Regulation No. TS 3-1 2011 *Works on the railway superstructure* (in Slovak) (Bratislava: Slovak Railways – ŽSR)
[2] Esveld C 2016 *Modern Railway Track.* [electronic] ed 3.8 (Zaltbommel: MRT-Productions)
[3] Regulation No. S 3-3 2005 *Rail Defects* (in Slovak) (Bratislava: Slovak Railways – ŽSR)
[4] Šestáková J 2017 *Diagnostics of Operational Quality of Railway Track and its Influence on Renewal Work* Planning (Habilitation thesis – in Slovak) (Žilina: University of Žilina) p 164
[5] STN EN 13848-2 2006 *Railway Applications – Track – Track Geometry Quality – Part 2: Measuring Systems – Track Recording Vehicles* (Bratislava: Slovak Office of Standards, Metrology and Testing)
[6] STN EN 13848-3 2009 *Railway Applications – Track – Track Geometry Quality – Part 3: Measuring Systems – Track Construction and Maintenance Machines* (Bratislava: Slovak Office of Standards, Metrology and Testing)

IOP Conf. Series: Materials Science and Engineering 1015 (2021) 012062 doi:10.1088/1757-899X/1015/1/012062

[7] STN EN 13848-4 2011 *Railway Applications – Track – Track Geometry Quality – Part 4: Measuring Systems – Manual and Lightweight Devices* (Bratislava: Slovak Office of Standards, Metrology and Testing)

[8] Roberts C and Goodall R M 2009 Strategies and Techniques for Safety and Performance Monitoring on Railways *Proc. 7th IFAC Symposium on Fault Detection, Supervision and Safety of Technical Processes (Barcelona)* **42** (8) pp 746-755

[9] Al-Douri Y K, Tretten P and Karim R 2016 Improvement of Railway Performance: A Study of Swedish Railway Infrastructure *J. Modern Transportation* **24** (1) pp 22-37

[10] Pace P and Jovanovic S 2011 Using Measurement for Decision Support *Int. Railway J.* **51** (7) pp 37-39

[11] Grenčík J, Poprocký R, Galliková J and Volna P 2018 Use of risk assessment methods in maintenance for more reliable rolling stock operation [electronic] *Machine modelling and simulations* 1st ed (London: Édition Diffusion Presse Sciences) pp 1-11

IOP Conf. Series: Materials Science and Engineering 1015 (2021) 012063 doi:10.1088/1757-899X/1015/1/012063

The cost of repairing damage caused by the inadequate demolition of walls in a multi-family building

M Sawicki and M Szóstak

Wroclaw University of Science and Technology, Faculty of Civil Engineering, 27 Wybrzeże Wyspiańskiego st. 50-370 Wroclaw, Poland

E-mail address: mariusz.szostak@pwr.edu.pl

Abstract. The article analyses the cost aspects related to the inappropriate demolition of interior walls in a multi-family building in Wroclaw. As a result of demolition and dismantle works on the ground floor of a multi-family building, which was erected using traditional masonry technology, the structural elements on the upper floors were damaged. The expertise report and performed tests confirmed the cause of damage, and the condition of the structure, resulting from the conducted works, was analysed. Recommendations for repairing the structure, and an analysis of the costs associated with the repair were also specified.

1. Introduction

A housing estate is an urban unit that is characterized by a unified residential development that consists of a set of residential buildings with an infrastructure, service and commercial facilities, and green areas [1]. In typical multi-family buildings, commercial premises, i.e. rooms intended for utility purposes, are usually located on the ground floor [2].

Commercial premises, depending on their function, have strict rules regarding their use [3]. A very common practice when changing the function of commercial premises is the enlargement of interior space. In order to achieve this, it is necessary to perform dismantling works of partition walls. Partition walls are internal, non-structural, and non-load bearing [4], and their main function is to divide premises into smaller rooms. As indicated in construction practice, a number of mistakes can be made during demolition and dismantle works. They result from the ignorance of the static behaviour of a structure, which in turn can lead to the occurrence of damage. There are many different methods for repairing damaged structural elements of walls, e.g. reinforcing masonry walls with FRP fibers [5] or using repair anchors [6] or the possibility of using post-tensioned concrete rocking walls [7]. In each case useful in the assessment of technical state of masonry walls are non-destructive methods [8]. Therefore, when planning demolition works, it is necessary to know the details of the design documentation and to determine the correct technology of conducting the works, while at the same time also taking into account the spatial rigidity and stability of the building [9].

The article presents an example of wall damage in a 20-year old multi-family building, which occurred as a result of the inadequate demolition of separating walls. The multi-family residential building was built in the year 2000 in Wroclaw. The building was erected using traditional masonry technology [10]. During the adaptation works, which involved a change in the function of the rooms and the demolition of the internal walls, numerous scratches and cracks were revealed both in the thin partition walls and in the thick walls separating apartments on the fourth and fifth floors (two-story apartments).

IOP Conf. Series: Materials Science and Engineering 1015 (2021) 012063 doi:10.1088/1757-899X/1015/1/012063

2. Research methodology

The article presents the impact of the incorrect demolition of the partition walls on the damage to the building's structural elements. Factors that may have affected the damage were analysed, and economic analysis of damage repairs was conducted with regards to the costs that were generated in one of the subsequent stages of the life cycle of the building, namely in the operational and maintenance phase. These phases include, among others, the costs of repairs and maintenance incurred during the operation of the facility [11].

In the article, a simple Life-Cycle Cost Analysis (LCCA) method was used to estimate the components of the cost of the renovation works, i.e. an engineering method of cost estimation (direct testing, component by component) without the discounting process [12].

3. Description of the structure and damage of the building

The investigated multi-family building was designed in 1999 and built in 2000. It is a five-storey building with two-storey flats on the fourth floor, which translated into a different division of the flats on floors 1 to 3 and also 4 to 5. This division of the flats on the last two floors required the use of additional partition walls with adequate sound insulation between the flats. These walls were made of aerated concrete and were 24 cm thick. The weight of the walls additionally loaded the ceilings, which meant that the floor slabs had to be strengthened in the central part of the building.

The walls and ceiling of the ground floor were made of monolithic concrete, while on the higher floors there were Filigran type ceilings with variously reinforced monolithic parts. The load-bearing and partition walls were made of silicate blocks and ceramic blocks with a thickness of 24 and 8(12) cm using traditional masonry technology. These walls suffered the most scratches during the operational phase due to rheological deflections of the ceilings. From the possessed documentation, it was not possible to determine whether the designer had determined the method and order of making the partition walls and floors. The documentation also lacked information about the conducted order of executing the above-mentioned works during the erection and modernization of the building.

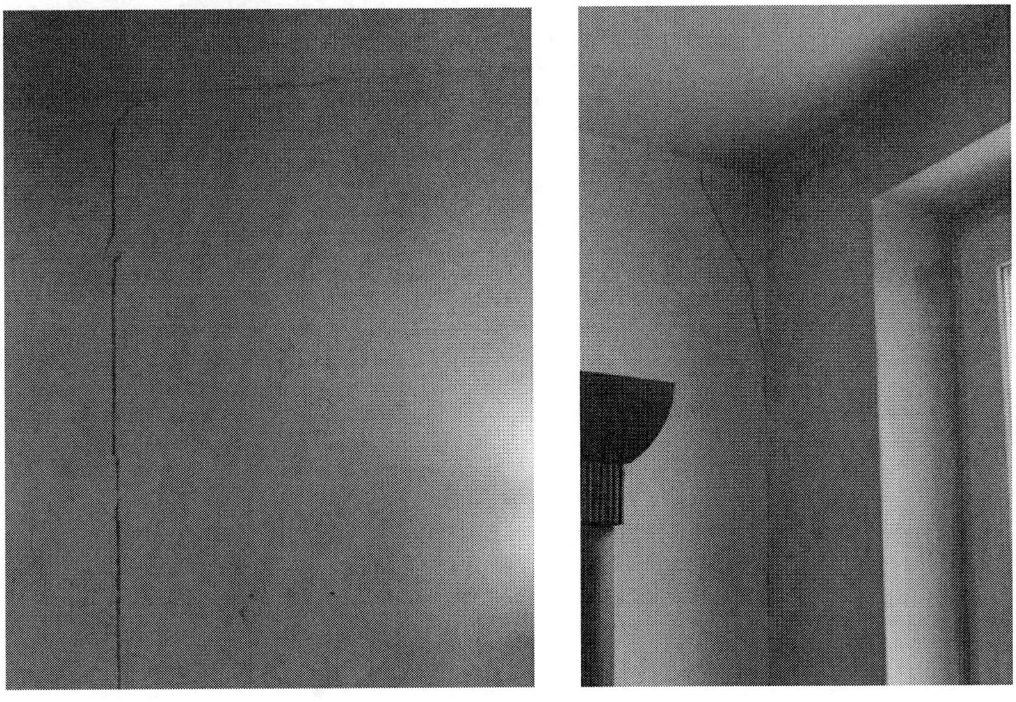

Figure 1. Damage in walls and ceilings.

In the investigated commercial premises located on the ground floor of the building, construction works started at the beginning of 2019 and consisted of the demolishing of parts of the internal partitions (using electric tools), making a new ventilation, and erecting new walls. After the start of the planned construction works, apartment tenants, whose flats were located next to the same staircase, reported many irregularities and damage, such as scratches, cracks, and a lot of loosening of the wall's surface, which forced the contractor to suspend the works. The facility manager commissioned the execution of technical expertise. During an on-site visit, no significant damage was found in the ceilings, which indicates that the ultimate limit state and safety of the structure were maintained. However, the serviceability limit state was not maintained, as was evidenced by numerous scratches and cracks in the walls. A lot of damage was observed within the walls in the form of scratches and cracks, and also in the form of gaps between the load-bearing walls and partition walls that were based on the ceilings (figure 1).

4. Analysis of research and observation results

Preliminary observations of the partition walls showed that the main cause of damage involved the deflections of ceilings that exceeded permissible values, and therefore ceiling deflection measurements were also conducted. The measurements were made using the geodetic method, and the obtained ceiling deflections on individual floors (for the limit value of 1/500 = 12.3mm [13] were as follows:

- ceiling above the ground floor - from 12.5 to 17.5 mm,
- ceiling above the first floor - approx. 13.5 mm,
- ceiling above the second floor - 14.3 mm,
- ceiling above the third floor - no consent was obtained for the measurements,
- ceiling above the 4th floor - up to 7.5 mm <12.3 mm.

A detailed analysis of the obtained measurement results showed that:

- the deflection of the ceiling above the ground floor exceeded the value of the permissible ceiling deflection (the permissible value: 12.3 mm). The ceiling above the ground floor had the largest deflection, which ranged from 12.5 to 17.5 mm. Such a large deflection is a consequence of the adopted method of supporting the ceiling, because the partition walls on the ground floor partly supported the ceiling before their demolition. It should be stated that the exceeding of the permissible deflection is not large and does not threaten the safety of the reinforced concrete structure, but it does pose a threat to the filling elements, i.e. partition walls. A similar situation occurred above the first and second floors (deflection was equal to 13.5 mm and 14.3 mm, respectively),
- the ceiling above the third floor was strengthened due to the different layout of the apartments on the last two floors of the building, and is heavily loaded with 24 cm thick aerated concrete walls. Measurements of the deflections of this ceiling would be interesting, but the consent of the apartment owners for measurements was not obtained,
- the ceiling above the top IV floor deformed relatively little (the deflection value was less than 7.5 mm). According to the authors, the small deflection value is caused by the fact that this is the highest storey and there are no higher floors from which loads could be transferred through the partition walls. Moreover, there are probably vaults created in the walls of the highest storey, which is indicated by a horizontal crack above the ceiling.

Based on the obtained measurements, it can be concluded that the exceeding of the permissible deflection is not large, although its effects can be seen to be significant. The building has a lot of damage (scratches, cracks) in the walls based on the ceilings. Almost 20 years of using the building, which was erected on quite unstable ground, as well as improperly carried out demolition and dismantling works, contributed to the rheological deflection of the ceilings and the leaning of them on the partition walls. As a result of changes in the load scheme, excessive stresses appeared on the walls, which is why scratches occurred most often between the walls and ceilings and between the partition and load-bearing walls.

Vibrations caused by inadequately performed demolition works in the premises on the ground floor also affected the numerous scratches and the chipping of plaster along the scratches. During the demolition of the partition walls on the ground floor with the use of electric tools, which caused large vibrations transmitted to the structure, the ceilings were not supported next to the walls that were dismantled. This caused the walls to be scratched on subsequent floors. The vibrations additionally caused the appearance of numerous scratches on the upper floors in the elements in which the state of stress was close to the limit values.

In addition, it is worth paying attention to the fact that there are balcony slabs of considerable size made on the gable wall of the investigated building (figure 2). The span of three balconies is from 1.40 to 2.00 m, and with their considerable length of about 8.0 m they may have an impact on additional damage to the apartment walls due to inadequate demolition and modernization works in the premises on the ground floor.

Figure 2. Façade of the investigated building.

According to the authors, the damage could be minimized if the partition walls were connected to the load-bearing walls in accordance with the principles of building art and also by being separated from the ceilings using flexible felt spacers, and if the building had properly executed foundations - with an appropriate soil compaction under the foundations during the construction phase.

To sum up, it can be stated that the investigated building did not show any signs of threat to the structure's safety (no damage of the load-bearing elements), and the measured ceiling deflections indicate a safe condition of the structure. However, the non-load bearing components and their connections with the load-bearing walls required repair.

5. Propositions of damage repairs

The building has a lot of damage (scratches, cracks) in the walls based on the ceilings, which was caused by the deflections of the ceilings. However, this damage does not threaten the safety of the structure. Some of the damage arose during the construction phase of the building, some during the operational phase, and a lot during the demolition and dismantling works in the ground floor premises. Damage, in particular scratches and cracks on the walls of residential buildings, are a typical execution error and a consequence of improper renovation works [14]. In the investigated building, vibrations that were accompanying the demolitions and which were transferred along the load-bearing walls and ceilings were the main reason for the occurrence of a significant number of scratches on the upper floors. Some of the scratches were not removed earlier, despite being created before the works on the ground floor were carried out.

The authors of the article proposed the following method of repairing the scratches and gaps in the walls and at the connections of the partition walls with the load-bearing walls:

- it is recommended to anchor vertical contacts between the partition walls and load-bearing walls using connectors at a height of 20 cm, 70 cm and 150 cm from the top of the ceiling,
- as the anchors - using steel expansion bolts screwed at an angle of 90o, or flat galvanized bars bent at an angle of 90o with a thickness of 4 mm mounted in carved recesses in 3 of the levels indicated above,
- widening the scratches, and after priming, filling them with putty reinforced with glass fibres, plaster or gypsum,
- forging damaged ceramic tiles in rooms and installing new ones.

6. Cost analysis of damage repairs

The authors and the insurer estimated the presented recommendations for repairing the damage with regards to cost. The following stages of works were proposed:

- the anchoring of the partition walls with the load-bearing walls in places where they are detached,
- the repairing of vertical, horizontal and diagonal scratches in partitions as described in section 4,
- the replacement of damaged glaze tiles in the kitchens and bathrooms,
- the execution of painting works for the entire area under renovation,
- the execution of preparatory and cleaning works including: removal of debris, removal of tiles, protection of floors and other necessary elements during works, disassembly of bathroom and kitchen installations, etc.

Above the premises on the ground floor in which the works were carried out, 13 apartments with areas ranging from 50-95 m^2 were located. The tenants of two apartments from the first and second floors with an area of about 80 m^2 in which the largest damage occurred applied to the insurance company for a valuation of the cost of repairing the damage. The insurer's initial valuation was PLN 3 500 – 4 500. The apartment owners questioned the amount of the valuation and scope of work, and commissioned to the authors of the paper to verify the costs of the works. Renovation works were carried out in the summer months, and the cost of renovating the flat within the required scope was equal to PLN 12 000. The valuation of the insurance company was three times lower and did not include all the works that were necessary and executed during the renovation.

The analysis of the estimated cost of repairing the damage was made with the following assumptions:

- the renovation concerns half of the premises above the renovated ground floor and covers half of the total floor area of the apartments. The total area of flats is 650 m^2, and therefore the area undergoing renovation works is equal to 325 m^2,
- the estimated cost of the renovation works was referred to 1 m^2 of the flat's area, taking into account the total cost of the full scope of the works, which was valued based on the bill of quantities, catalogues of inputs, and also current rates of labour, equipment and machinery using cost estimation software that is available on the Polish market,

IOP Conf. Series: Materials Science and Engineering 1015 (2021) 012063 doi:10.1088/1757-899X/1015/1/012063

- the scope of properly completed renovation works also included the placing of supports under the demolished walls and the manual dismantling of the partition walls,
- the conducted analysis also included the cost of correct compaction of soil under the foundations at the stage of erecting the building under its the entire surface. As demonstrated, incorrect soil compaction under the foundations during the construction phase had an indirect impact on the occurrence of damage.

Table 1 presents the analysis of the estimated damage repair costs.

Table 1. Analysis of the estimated costs of repairing damages and the increase in expenditure on the modernization of the commercial premises

Type of cost	Total cost [PLN]	Average cost per 1 m^2 of floor space [PLN/m^2]	Renovation cost per 325 m^2 of floor space [PLN]	Renovation cost per 325 m^2 of floor space + expertise report [PLN]
DAMAGE REPAIRS				
Valuation of the insurer - average value per 80 m^2 of premises according to the scope of works provided by insurer 1[1]	4 000	50	16 250	22 250
Author's valuation/valuation of renovation company 1[1]	12 000	150	48 750	54 750
Technical expertise report[2]	6 000	18.46		
INCREASE OF EXPENDITURE ON THE MODERNIZATION OF THE COMMERCIAL PREMISES - INVESTOR'S COST				
Correctly performed works: 1. manual demolition of walls ΔT=132 r-g, ΔR= 2 224 PLN 2. positioning of supports 10 pcs./week of work, ΔS=250 PLN	4 474	13.76		
Compaction of soil in the phase of erecting the building per 650 m^2 of the floor surface according to assumptions[3]	1 500	2.30		

* per 80 m^2 of premises, according to the scope of works of the insurer[1]
** per 325 m^2 of floor surface, according to assumptions[2]
*** per 650 m^2 of floor surface, according to assumptions[3]

The values presented in table 1 show that the correct manual dismantling of the walls and the executing of the support of ceilings generates an increase in the cost of renovation works on the ground floor by approx. PLN 4 500 and extends the work time by 132 man-hours, which, when employing a 6-person work crew, extends the work time by 2 days.

The actual time of completed works amounted to 30 working days. Correct execution of the renovation works, including the manual dismantling of the walls and the positioning of supports, would extend the work to 32 working days. Extending the working time by only 2 working days does not significantly affect the total implementation time.

The authors of the article did not achieve a detailed valuation of works carried out by the insurance company for individual residential premises, however, the proposed amount of PLN 3 500 to 4 500 for the renovation of an apartment of approx. 90 m^2 gives the total renovation cost of the damaged premises of PLN 22 500 (table 1). When referring to the real value of the necessary renovation costs that are equal to PLN 54 750, an underestimation of about 58.1% was obtained.

7. Summary and conclusions

As a result of the analysis of the case study, the following was indicated:

- in the investigated building, excessive stresses on the walls occurred as a result of changing the load scheme, which in turn resulted in the scratches,
- the formation of numerous scratches and chipping of parts of plaster along the scratches was affected by vibrations caused by inadequately performed demolition works in the premises on the ground floor,
- an increase in the cost of renovation works by about PLN 4 500 in the premises on the ground floor, and also an extension of the work execution time by 2 days, could have prevented the need for repair works in the premises above the ground floor for an amount 12 times higher.

In conclusion we also need to realise is that incorrect planning and the execution of the demolition and dismantling works of the internal partition walls in a building that was made using traditional masonry technology can cause a lot of damage (scratches and cracks) to the walls, what has been proved in this work.

References

[1] Siemiński W 2011 Housing estate as an urban-planning-social phenomenon *Człowiek i Środowisko* **35** (1–2) p 107–124 (in Polish)

[2] Act of 24 June 1994 on the ownership of premises (*Law Gazette*, No. **85**, it. 388 in Polish)

[3] Minister of Infrastructure Order of 12 April 2002 on technical requirements which should be met by buildings and their location (*Law Gazette*, No. **75**, it. 690, in Polish)

[4] Standard PN-EN 1991-1-1:2004 Eurocode 1 - Actions on structures, Part 1.1: General Actions - Densities, self-weight and imposed loads

[5] Hernan S-M, Alcaino P 2011 Repair of in-plane shear damaged masonry walls with external FRP. *Construction and Building Materials* **25** (3) p 1172-80 https://doi.org/10.1016/j.conbuildmat.2010.09.030

[6] Davidyuk A 2017 Technology of repairing brick masonry cladding of multi -layer external walls using helical repair anchors *MATEC Web of Conferences* **106** 03015 (https://doi.org/10.1051/matecconf/201710603015)

[7] Jafari A and Dugnani R 2018 Estimation of load-induced damage and repair cost in post-tensioned concrete rockin walls *Journal of Shanhai Jiaoton University (Science)* **23** p 122-131 (https://doi.org/10.1007/s12204-018-1917-x)

[8] Hoła A, Matkiewicz Z and Hoła J 2017 analysis of the moisture content of masonry walls in historical buildings using the basement of a medival town hall as an example *Procedia Engineering* **172** p 363-368 (https://doi.org/10.1016/j.proeng.2017.02.041)

[9] Drobiec Ł 2017 Causes of masonry damage. Part 2: Damage caused by poor performance and operat-ing errors *Izolacje* **22** (9) p 32–40 (in Polish)

[10] Stawiski B 2014 *Masonry structures. Repair and strengthening* (Polcen, in Polish)

[11] Plebankiewicz E, Zima K and Wieczorek D 2016 Life cycle cost modelling of buildings with consideration of the risk *Archives of Civil Engineering* **62** (2) p 149–166 (https://doi.org/https://doi.org/10.1515/ace-2015-0071)

[12] Grzyl B, Apollo M, Miszewska-Urbaska E and Kristowski A 2017 Management of exploitation in terms of life cycle costs of built structures *Acta Scientarium Polonorum Architectura* **16** (2) p 85–89 (https://doi.org/DOI: 10.22630/ASPA.2017.16.2.11)

[13] Standard PN-EN 1992-1-1:2008 Eurocode 2 - Design of concrete structures

[14] Plebankiewicz E, Zima K, Malara J and Biel S 2019 A procedure of repairing housing defects in development investments *MATEC Web of Conferences* **262** 07010 (https://doi.org/10.1051/matecconf/201926207010)

IOP Conf. Series: Materials Science and Engineering 1015 (2021) 012064 doi:10.1088/1757-899X/1015/1/012064

New flame-retardant composition for lowering contribution of expanded poly styrene to the propagation of fire

A M Mochalov[1], M V Akulova[2], J A Sokolova[3] and A G Sokolova[3]

[1] Ivanovo Fire and Rescue Academy of the State Fire Service of the Ministry of the Russian Federation for Civil Defense, Emergencies and Elimination of Consequences of Natural Disasters, 153040, Ivanovo, Stroitelei ave, 33, Russia
[2] Ivanovo State Polytechnic University, 153000, Ivanovo, Sheremetyevsky prospect, 21, Russia
[3] Moscow State University of Civil Engineering NRU MGSU, 129337, Moscow, Yaroslavskoye shosse, 26, Russia

E-mail address: as.falconi@yandex.ru

Abstract. Despite the major advances in ensuring fire safety of buildings and structures, fire risks still represent the serious issue not only on the territory of the Russian Federation but on the global basis. Presently, while erecting buildings and structures focused attention is given to lowering material consumption and reducing construction time. This is achieved by means of application of lightweight building envelopes with efficient insulators. For the most part, foam poly styrene plates are used as an insulator. Expanded or foam poly styrene has low coefficient of heat conductivity, high acoustic isolation, long-term durability, moisture resistance, and is characterized by ease of installation. However, application of foam poly styrene has constrains due to its high contribution to the propagation of fire.
In the present paper, the authors proposed new flame-retardant composition on the base of organo-siloxanes, which inhibits combustion initiating of foam poly styrene and constrains fire propagation when applied on the surface of foam poly styrene plates.

1. Introduction

Nowadays, when erecting buildings and structures thorough attention is paid to reduction of building materials consumption and cutting construction time. The desired outcome could be reached by means of lightweight envelope structures with efficient insulating materials. In the majority of cases, foam poly styrene plates are used as insulators. However, foam or expanded poly styrene is a combustible and highly flammable material forming burning melting drops [1, 2]. Contribution of foam poly styrene to the propagation of fire is assessed on two principal directions: fire hazard of a material itself (the effect of temperature and flame) and the hazard of thermal decomposition and oxidation products.

When having effect on human bodies at elevated temperatures and lowered oxygen concentration, various clinical implications could be observed that reflect, most notably, the state of oxygen deprivation of human tissues and organs.

On average, only 18% of people die of skin burns in fire condition, the remainder represents the victims of fumes poisoning combined with the effect of stress, heat and other kill effects. Statistics shows that if a relatively small fire breaks out in the premises saturated with combustion products of

IOP Conf. Series: Materials Science and Engineering 1015 (2021) 012064 doi:10.1088/1757-899X/1015/1/012064

polymer materials, there takes place uncontrollable life loss of the people inside the premises mainly due to volatile products poisoning.

All foam poly styrene plates are regarded as flammable and combustible building materials [3]. Thus, speed of flame propagation on the surface of foam poly styrene accounts for 1 meter per second that predetermines extremely high speed of flame propagation in the buildings and structures insulated with this material. Specific mass burning rate of foam poly styrene of the PSB grade is 2.19 kg/min·m^2. Due to high speed of flame propagation of foam poly styrene and increased specific heating value, the heat is released at the peak temperature of 1500^{0}C within relatively limited time.

Initiating combustion of foam poly styrene occurs at the temperature range 220÷380^{0}C while spontaneous combustion corresponds to the temperature range 460÷480^{0}C [4].

2. Materials and methods

In the present paper, the authors proposed new flame-retardant composition on the base of organo-siloxanes, which inhibits combustion initiating of foam poly styrene and constrains fire propagation when applied on the surface of foam poly styrene plates.

The paper methodology is based on the All-Union Standard GOST 30402-96 "Building materials. Ignitability Test Method", the section related to assessing ignitability of combustible buildings materials.

The object of study was presented by foam cellular plastics used in construction industry as insulators, enclosure structures of buildings such as foam poly styrene of the types RAVATHERM XPS STANDARD (Technical regulations 2244-002-00259620-2013, «NISKO Industry» plc, village Kryukovo, Moscow region) and PSB-S 15U (Technical regulations 2244-007-04001508-96, Joint-venture «TIGI KNAUF», Krasnogorsk, Moscow region) manufactured in compliance with the All-Union Standard GOST 15588 [5].

As the objects of research, the authors have chosen the compositions on the base of ethylene-siloxane, water-glass and sulfanol. The present substances have high thermal resistance and are used in various spheres of development and protection of buildings materials. Thus, organo-siloxanes and silicones present the most important polymer group, with the main chain formed by other elements in place of carbon [6]. Organo-siloxanes could be of linear, ring and branched structure. Their common features include increased flexibility and rotational freedom of atomic groups around bonds Si-C and Si-O. When rotating around bonds, the sections of diverse polarity of discrete siloxane groups of linear siloxane chains tend to arrange themselves in space in such an order that dipolar moments of different groups are compensated when possible [7-9]. In the result, linear organo-siloxanes form supramolecular structure. The present property is of utmost importance when applied in practice as it leads to the fact that the properties of organo-siloxanes are not much affected by the temperature.

In the present study, the following compositions have been used:
- composition No. 1 (5% organo-siloxane, 1% sulfanol solution);
- composition No. 2 (5% water-glass, 1% sulfanol solution);
- composition No. 3 (10% organo-siloxane, 1% sulfanol solution);
- composition No. 4 (10% water-glass, 1% sulfanol solution);
- composition No. 5 (2.5% organo-siloxane, 2.5 water-glass, 1% sulfanol solution);
- composition No. 6 (10% organo-siloxane, 30% aluminium flake, 5% soap stone, 30% water-glass);
- composition No. 7 (10% organo-siloxane, 30% aluminium flake, 5% soap stone, 30% water-glass, 5% carbamide).

3. Results and discussion

In the research, the authors have determined the parameters of material's ignitability at the given standard levels of radiant heat flux impact and ignition source flame on the specimen's surface. These parameters comprise critical area heat flow density (henceforth referred to as CAHFD) [10] and ignition time. Having implemented the tests, the authors applied CAHFD for subsumption of the material to a certain group.

Flame-retardant coats have been applied to the specimens (figure 1) on all sides.

Figure 1. The tested specimens.

Figure 2 is thus a representation of the general drawing of the machine designated for ignitability testing.

Figure 2. Ignitability testing machine.

The test results and their analysis are given in the tables 1 and 2.

IOP Conf. Series: Materials Science and Engineering 1015 (2021) 012064 doi:10.1088/1757-899X/1015/1/012064

Table 1. Dependence of ignition time of foam poly styrene on the type of coat. The tests results have been analyzed under Cochran's test.

Composition number applied for foam poly styrene's treatment	Ignition time, seconds	y^* (point estimation)	S_x (mean square deviation)	δ (half of the additional interval)	$y^* \pm \delta$ (interval estimation)
Non-treated	186 173 180	179.6667	6.506407	16.15285	179.6667 ± 16.15285
Composition 1	244 246 250	246.6667	3.05505	7.584487	246.6667 ± 7.584487
Composition 2	200 210 206	205.3333	5.033223	12.49551	205.3333 ± 12.49551
Composition 3	286 280 288	284.6667	4.163332	10.33591	284.6667 ± 10.33591
Composition 4	230 236 240	235.3333	5.033223	12.49551	235.3333 ± 12.49551
Composition 5	242 240 241	241	1	2.482606	241 ± 2.482606
Composition 6	241 247 242	243.3333	3.21455	7.980462	243.3333 ± 7.980462
Composition 7	231 236 230	232.3333	3.21455	7.980462	232.3333 ± 7.980462

where:

y^* - arithmetic mean for each series of tests (the formula 3.5 page 47 [1])
S_x - estimated by the formula 3.6 on the page 48 [1]
$y^* \pm \delta$ – estimated by the formula 3.31 on the page 65 [1].

When implemented few series of tests, check on compliance of results' resettability (check of homogeneity of variance) was done under Cochran's criterion (the formula 3.47 on the page 94 [1]):
- statistics calculated under the present criterion turned out to be equal 0.299528 at the significance point 0.5157, as $0.299528 < 0.5157$, then the condition of results' resettability is complied with (variances are homogeneous).

Check of significance of variances of mean values obtained in the result of discrete series of implemented tests was done under the Student's t-test (formula 3.50 on the page 100 [1]):

- with the help of the mentioned criterion, pair-wise comparison of the results of test 1-7 has been done with the results of "Non-treated" experiment.

The results are presented in the table 2.

IOP Conf. Series: Materials Science and Engineering 1015 (2021) 012064 doi:10.1088/1757-899X/1015/1/012064

Table 2. Test results' analysis (under Student's *t*-test).

Compared tests	Estimated statistics of *t*-criterion
0 – 1	16.1
0 – 2	5.4
0 – 3	23.5
0 – 4	11.7
0 – 5	16.1
0 – 6	15.2
0 – 7	12.6

Significance point t (significance level 0.05, number of degrees of freedom 4) = 2.776.

As the calculated values of t-criterion are higher than significance point, we can draw a conclusion that the observed differences between fixed parameters in the experiments 1-7 are statistically significant if compared to the zero-experiment.

Having analyzed the results of carried out research, the authors found out that the composition 3 (10% organo-siloxane, 1% sulfanol solution) is proven to be the most efficient one. As the table data demonstrate the average ignition time of foam poly styrene plates treated with the composition 3h as accounted for 219 seconds, i.e. it has increased by the factor 1.5 when compared with the check sample.

4. Conclusions

Thus, the authors have conducted research of the impact of new flame-retardant compositions on fire engineering characteristic of foam poly styrene. The specimens with applied coats are considered less hazardous. The composition containing 10% organo-siloxane and 1% sulfanol solution is proven to be the most efficient one. Time to reach maximum temperature has increased nearly by factor of 2, maximum temperature increment has lowered by 25.3%, when the reduction of specimen's mass loss is observed (-17.9%).

References

[1] Akulova M, Mochalov A, Lebedev D and Rodionov E 2017 On the safety of self-extinguishing polystyrene foam *Modern fire-safe materials and technologies: Proceedings of the International scientific and practical conference* pp 9-12

[2] URL: http://stkpenoplast.ru/Polystyrene/Properties.html/ 2019

[3] Serkov B and Firsova T 2019 *Buildings and structures*

[4] Stuart H, Ibarra C, Abdallah M, Boon R, Neels H and Covaci A 2008 *Environmental Int.* **34** (8) p 1170-75

[5] Guidelines for the use of expanded foam polystyrene panel systems in industrial buildings so as to minimise the risk of fire 2006 *Fire Eng. Research Report*, Department of Civil Engineering, University of Canterbury p 79

[6] All-Union Standard 15588-2014 Foam polystyrene heat-insulating plates Technical conditions

[7] Chestfield X, Volozhinsky L and Goldberg M 1968 *Chemistry Paint coating* p 640

[8] Voronkov M, Mileshkevich V and Yuzhelevsky V 1985 *Oligo organo siloxanes* p 413

[9] Jacobs L, Danen K, Kemmerel M and Keurentjes T 2007 *Polymer* **48** (13) pp 3771-80

[10] Hurlburt E, Krenitsky H and Bauer R 2009 *Proceedings of the ASME Summer Heat Transfer Conference* **3** pp 471-80

IOP Conf. Series: Materials Science and Engineering 1015 (2021) 012065 doi:10.1088/1757-899X/1015/1/012065

Relationship between water permeability and physical characteristics of polyolefins, vinyl polymers, and polycarbonates

T A Matseevich[1], T V Zhdanova[1], A V Matseevich[2] and A A Askadskii[1,2]

[1] Moscow State University of Civil Engineering (National Research University), 129337 Moscow, Russia
[2] A.N. Nesmeyanov Institute of Organoelement Compounds, Russian Academy of Sciences, 119991 Moscow, Russia

E-mail address: MatseevichTA@mgsu.ru

Abstract. Diagrams of compatibility of water permeability with glass transition temperature, density, and cohesion energy are constructed. The computer program "Cascade", INEOS RAS was used. Polyolefins, vinyl polymers, and polycarbonates were analyzed. Compatibility diagrams are constructed for the areas of lowest and highest water permeability for different classes of polymers. For polyolefins and vinyl polymers, the lowest water permeability is selected in the range from 0 to 100 barrer, and the highest water permeability is from 1000 to 2000 barrer. For polycarbonates, the intervals from 0 to 70 barrer and from 3000 to 7000 barrer are selected. These diagrams allow you to select polymers that meet the specified values of density, cohesion energy, glass transition temperature, and the onset of temperature of intense thermal degradation.

1. Introduction

Currently, due to the use of polymers in the production of various building materials, the requirements for their physical properties are much more complicated. There are problems that require polymers that possess a set of physical properties that simultaneously fall within the specified ranges of their characteristics. In this regard, the problem is raised about the possibility of combining the properties of polymers. The method described in monographs [1-2] for solving this problem consists in constructing "compatibility diagrams", which consists of the following: let's assume that for a large group of polymers, the values of the parameters of any two physical properties are known (from experiment or calculation). The diagram is drawn where each polymer corresponds to a specific point (figure 1), where the abscissus (X) axis represents the parameter values of the first property, and the ordinate (Y) axis represents the second property.

All these points are located inside an area that can be outlined. This area is called the "compatibility area" of the specified physical properties for this group of polymers. Among the groups of polymers are polyolefins, vinyl polymers, polystyrenes, polyethers and polyesters, polyamides, polyurethanes, polyetherketones etc. Note that the density of the distribution of points in this area can be quite uneven. Such diagrams allow us to evaluate the possibility of polymers with two specified properties without additional labor costs. In fact, let us define the ranges ($X_2 - X_1$) and ($Y_2 - Y_1$) of the property values that the polymers should possess. If the area defined by these ranges falls within the compatibility area on the diagram, and some points are located there, then such polymers may exist (case I). In the opposite case, obtaining the desired polymer among polymers of these classes has a low probability (case II).

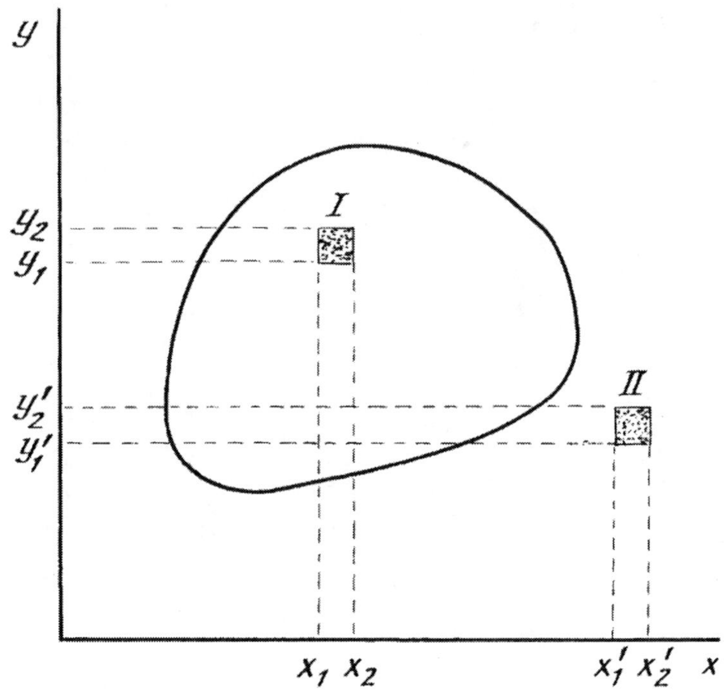

Figure 1. Schematic representation of the region of possible existence of two properties of polymers (see the text).

Either experimental data on various physical properties of polymers or calculated data can be used to create compatibility diagrams of this type. The combination of these data may be used. To calculate various physical properties of polymers, it is desirable to use the computer program "Cascade", developed at INEOS RAS, since it is used to consider the properties of polymers that have already been synthesized, or those that can be synthesized. This program was used in monographs [1-2] to compile compatibility areas for various properties of polymers. Among these properties, the following were considered: density, ρ; refractive index, n; glass transition temperature, T_g; ratio of glass transition temperature to melting temperature, T_g/T_m; the temperature of the onset of intensive thermal destruction, T_d; the Hildebrand parameter of solubility, δ (the energy density of cohesion); the stress-optical coefficient $C\sigma$.

As an example, see figure 2 shows the compatibility diagram of the glass transition temperature T_g and the onset temperature of intensive thermal degradation T_d. Let's look at a few features of these and other similar diagrams. Formulas that can be used to find a relationship between two specified property parameters are given in the monographs [1-3].

All compatibility areas have different densities. This means that there are compatibility subdomains in which the main number of polymers is located, and from these subdomains it is most easy to choose the chemical structure of the polymer that corresponds to the two specified properties.

All diagrams show clearly defined dependencies of one property on another. Each of these dependencies characterizes a certain number of polymers, such as polyesters, polyamides, polyimides, and so on. With such diagrams, it is possible to predict the possibility of synthesize of polymers that would have the necessary combination of one of several properties. If such diagrams are made using the Cascade computer program, in which the polymer is "synthesized" from the smallest base fragments, the polymer compatibility regions will consist of a huge number of polymers.

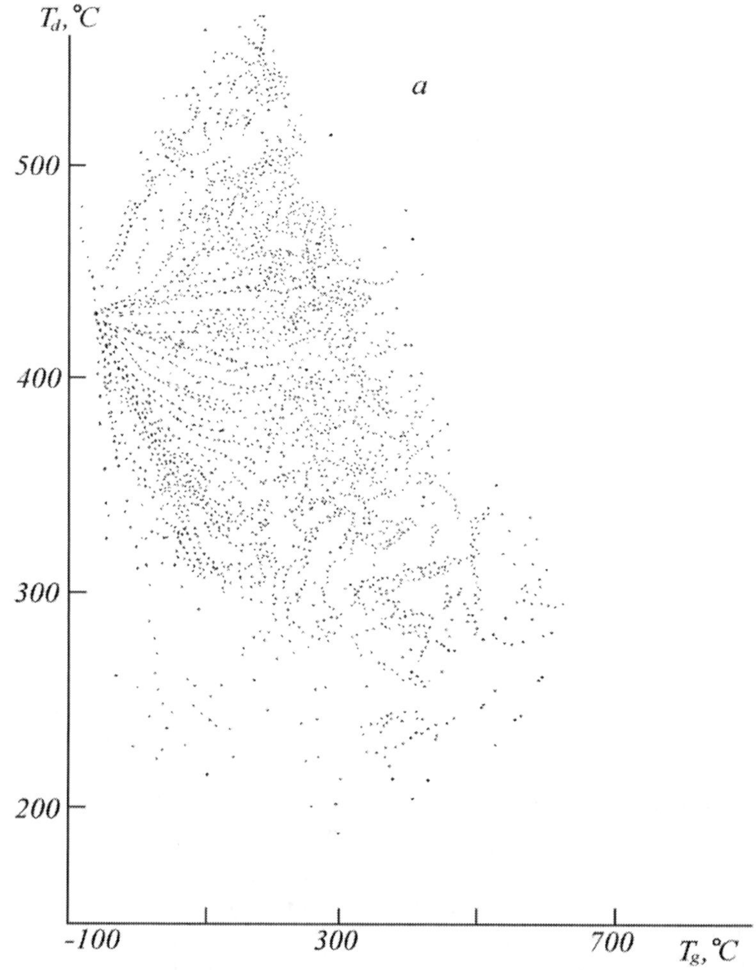

Figure 2. T_d and T_g compatibility diagrams (see text).

2. Experimental part

This work is devoted to obtaining diagrams of compatibility of water permeability of polymers with such physical characteristics as the glass transition temperature, the temperature of the onset of intensive thermal degradation, and the cohesion energy. Water permeability is important for polymer materials used in the construction industry for outdoor applications. The glass transition temperature allows us to determine whether the material is in the glassy state, i.e. whether it is high-modulus, and the cohesion energy is a fundamental factor for the dependence of water permeability on the chemical structure of the polymer. The temperature of the onset of intensive thermal degradation allows us to characterize the temperature of foaming of the polymer material, which leads to a decrease in combustibility

The description of the permeability P of polymers was performed using the general equation

$$P = P_0 e^{-\frac{\Delta E}{RT}} \tag{1}$$

where P_0 is the constant, ΔE is the activation energy of the permeability process, R is the universal gas constant, and T is the absolute temperature.

The value of ΔE was described by the ratio [4-6]

$$\Delta E = \frac{\sum_i \Delta E_i^{**}}{N_A \sum_i \Delta V_i} \tag{2}$$

where N_A is the Avogadro number, $\sum_i \Delta V_i$ is the van der Waals volume of a repeating polymer unit, $\sum_i \Delta E_i^{**}$ is the energy of the intermolecular interaction, consisting of the interaction energies of each atom and specific atomic groups that cause dipole-dipole interactions or hydrogen bonds.

Then

$$\ln P = \ln P_0 - \frac{\sum_i \Delta E_i^{**}}{N_A RT \sum_i \Delta V_i} \tag{3}$$

After the small transformations, we have

$$\frac{\sum_i \Delta E_i^{**}}{RT} - (N_A \sum_i \Delta V_i) \ln P_0 = -(N_A \sum_i \Delta V_i) \ln P \tag{4}$$

Atomic constants and parameters that characterize the energy of strong intermolecular interactions are available in [4-6]. The value $\ln P_0 = 3.002$ (barrer). The experimental values of the water permeability of polymers were borrowed from the works [7-14].

The glass transition temperature was described by the equation

$$T_g = \frac{\sum_i \Delta V_i}{\left(\sum_i a_i \Delta V_i + \sum_j b_j\right)_л + \left(\sum_i K_i \Delta V_i\right)_y} \tag{5}$$

where $\sum_i \Delta V_i$ is the van der Waals volume of a repeating fragment of a polymer network; $\left(\sum_i a_i \Delta V_i + \sum_j b_j\right)_l$ is the set of atomic constants for a repeating unit connecting the cross-linked points ; $\left(\sum_i K_i \Delta V_i\right)_{cr.p}$ is the set of atomic constants for the cross-linked points (for the physical meaning of the parameters a_i, b_j and K_i, see [1-2]).

The temperature of the onset of intensive thermal degradation is determined from the ratio

$$T_d = \frac{\sum_i \Delta V_i}{\sum_i K_i \Delta V_i} \tag{6}$$

where K_i are the atomic constants associated with the Morse potential parameters; $K_i = \frac{63}{2} \frac{R}{(ad_0 E)_i}$; a and d_0 are the Morse potential parameters, E is the energy of the chemical bond.

For copolymers, the ratio is as follows

$$T_g = \frac{\sum_{k=1}^{k=n} \alpha_k \left(\sum_i \Delta V_i\right)_k}{\sum_{k=1}^{k=n} \alpha_k \frac{\left(\sum_i \Delta V_i\right)_k}{T_{g,k}} + 0.03 \sum_{k=1}^{k=n} \alpha_k (1-\alpha_k)} \tag{7}$$

where α_k is the mole fraction of the k-th component, n is the number of components, $T_{g,k}$ is the glass transition temperature of the k-th component, $\sum_i \Delta V_i$ is the van der Waals volume of the k-th component.

The cohesion energy ΔE^* is determined by the ratio

$$\Delta E^* = \sum_i \Delta E_i^* \tag{8}$$

where ΔE_i^* is the contribution of each atom and the type of intermolecular interaction to ΔE^*.

The cohesion energy of a repeating polymer unit is described by the equation [1-2]

$$\sum_i \Delta E_i^* = \delta^2 N_A \sum_i \Delta V_i \tag{9}$$

where δ is the solubility parameter (Hildebrand parameter).

The density of polymers p is described by the ratio:

$$\rho = \frac{Mk}{N_A \sum_i \Delta V_i} \tag{10}$$

where M is the molecular mass of the repeating unit, and k is the coefficient of molecular packing (averaged value $k = 0.681$).

We get the relations connecting water permeability with the other characteristics considered above. This can be done because all the equations include the van der Waals volume of the repeating polymer unit. Water permeability and glass transition temperature of linear polymers are related by the ratio:

$$\ln P = 3.002 - \frac{\sum_i \Delta E_i^{**}}{N_A RT \cdot T_g \left(a_i \Delta V_i + \sum_j b_j \right)} \tag{11}$$

Water permeability and the temperature of the onset of intensive thermal degradation are related by the ratio:

$$\ln P = 3.002 - \frac{\sum_i \Delta E_i^{**}}{N_A RT \cdot T_d \sum_i K_i \Delta V_i} \tag{12}$$

Water permeability is related to density by the ratio:

$$\ln P = 3.002 - \frac{\rho \sum_i \Delta E_i^{**}}{RT \cdot Mk} \tag{13}$$

Water permeability is related to the cohesion energy by the ratio:

$$\ln P = 3.002 - \frac{\delta^2 \sum_i \Delta E_i^{**}}{RT \sum_i \Delta E_i^*} \tag{14}$$

The existing computer program Cascade, developed at INEOS RAS, allows for computer synthesis of polymers with a given water permeability. For polymer materials for construction purposes, polymers with very low water permeability are of interest, if they are exploited in outdoor conditions of action of a humid environment. In some other cases, materials with high water permeability are required. Therefore, in this work, we will limit ourselves to constructing compatibility diagrams for polymers of specified classes with both of narrow intervals of low and high water permeability.

Table. 1. Low water permeability of polyolefins and vinyl polymers (from 0 to 100 barrer).

№	Chemical structure	*P*, (barrer)	T_g, (K)	T_d, (K)	*E**, (J/mol)	*ρ*, (g/cm^3)
1	-CH2-CHCl-	10.5	341	534	16300	1.44
2	-CH2-CCl2-	0.002	268	473	20600	1.67
3	-CH2-CH(COOH)-	95.6	374	525	47200	1.35
4	-CH2-CF(Cl)-	22.3	491	557	26100	1.68
5	-CHCl-CF2-	42.7	315	577	16000	1.88
6	-CCl2-CF2-	0.00	262	508	20300	1.96
7	-CF2-CF(Cl)-	71.7	307	596	15800	2.04

Table 2. High water permeability of polyolefins and vinyl polymers (from 1000 to 2000 barrer).

№	Chemical structure	*P*, (barrer)	T_g, (K)	T_d, (K)	*E**, (J/mol)	*ρ*, (g/cm^3)
1	-CH2-CF2-	1117	218	746	17600	1.55
2	-CH2-C(CF3)2-	1761	291	673	25000	1.82
3	-CH2-CF(CF3)-	1793	300	694	21300	1.79
4	-CH2-CH(OH)-	1039	365	530	33000	1.20
5	-CF2-CH(CF3)	1151	294	706	11200	1.89

Figure 3 shows, as an example, diagrams of compatibility of water permeability with a number of other physical characteristics for polycarbonates "synthesized" on the basis of 5 base fragments.

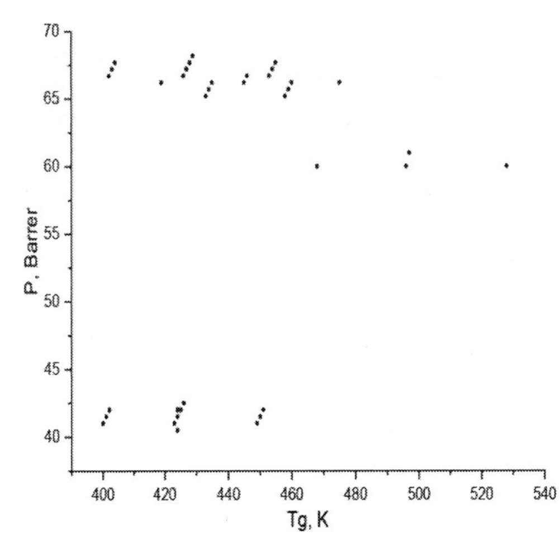

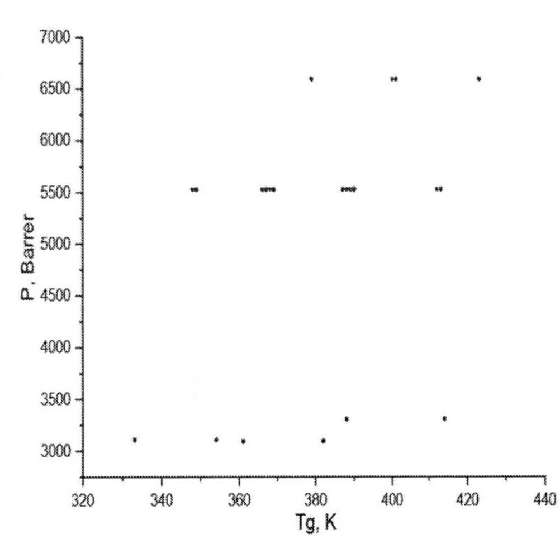

A *B*

Figure 3. Graph of *P* versus T_g for low and high water permeability of polycarbonates.

IOP Conf. Series: Materials Science and Engineering 1015 (2021) 012065 doi:10.1088/1757-899X/1015/1/012065

The range of low water permeability was from 0 to 70 barrer, and of high water permeability from 3000 to 7000 barrer. Under these conditions, 28 to 38 structures of repeating units were obtained. With the expansion of water permeability intervals and an increase in the number of base fragments, the number of structures reaches many tens of thousands. These diagrams allow you to select polymers that meet the specified values of water permeability, density, cohesion energy, glass formation temperature, and the onset temperature of intensive thermal degradation.

Acknowledgments

RF Ministry of Science and Higher Education (project Theoretical and experimental design engineering of new composite materials for safety exploitation of buildings and structures under conditions of technogenic and biogenic threats #FSWG-2020-0007 and theme of the state task of INEOS RAS [grant number 0085-2019-0004]).

References

[1] Askadskii A A 2003 *Computational Materials Science of Polymers* (Cambridge: Cambridge International Silence Publishing) p 695
[2] Askadskii A A and Kondrashchenko V I *2003 Komp'uternoe materialovedenie polymerov V 1. Atomno-molekulyarnyi uroven'* (Moscow, Nauchnyi Mir) p 543
[3] Askadskii A A 1996 *Physical properties of Polymers. Prediction and Control.* (Amsterdam: Gordon and Breach Publishers) p 350
[4] Askadskii A, Afans'ev E S, Matseevich T A, Popova M N, Kondrashchenko V and Qi Shichengn 2014 *Advanced Materials Research* **1033-1034** p 939
[5] Askadskii A A, Afanas'ev E S, Matseevich T A, Popova M N, Kovriga O V and Kondrashchenko V I 2015 The calculation scheme for estimation of the water permeability through polymers and copolymers *Polymer Science, Ser. A.* **57** (6) pp 924–945
[6] Askadskii A A, Popova M N and Matseevich T A 2017 *Vtorichnye polimernye materialy* (Moscow: ACU) p 492
[7] Park J and Crank G S 1968 *Diffusion in polymers* (London: Academic Press)
[8] Ho W S W and Sirkar K.K. 1992 *Membrane Handbook* (New York: Van Nostrand)
[9] Paul D R and Yampolskii Yu P 1994 *Polymeric gas separation membranes* (Florida: CRC Press Boca Raton)
[10] Reitlinger S A 1974 *Permeability of Polymeric Materials* (Moscow: Khimiya)
[11] Iordanskii A L, Startsev O V and Zaikov G E 2003 *Water transport in synthetic polymers* (New York: Nova)
[12] Vieth W R 1991 *Diffusion in and through polymers. Principles and applications* (Munich Hanser Publishers)
[13] Huang J, Cranford R J, Matsuura T and Roy C. 2003 *J. Membr. Sci.* **215** p 129
[14] Roberts A P, Henry B M, Sutton A P, Grovenor C R M., Briggs G A D, Miyamoto T, Kano M, Tsukahara Y, Yanaka M 2002 *J. Membr. Sci.* **208** p 75

IOP Conf. Series: Materials Science and Engineering 1015 (2021) 012066 doi:10.1088/1757-899X/1015/1/012066

Structure and properties of polyepoxyurethane-containing isocyanurate polymers

T A Matseevich[1], M D Kejmakh[2] and A A Askadskii[1,2]

[1] Moscow State University of Civil Engineering, 129337 Moscow, Russia
[2] A.N. Nesmeyanov Institute of Organoelement Compounds, Russian Academy of Sciences, 119991 Moscow, Russia

E-mail address: andrey@ineos.ac.ru

Abstract. The modification of epoxyurethane-containing isocyanate polymers of a network structure was performed due to variations in the curing temperature and the ratio of diol, diisocyanate components and epoxy resin. When the reaction systems are cured below 100°C, the polymers contain a large number of flexible chain structures and pass into the rubbery state at ~60°C. An increase in the processing temperature above 120°C leads to a significant change in the composition, which affects the properties of the resulting systems, the glass transition temperature of which increases in the limit to 225°C. An experimental change in the elastic modulus associated with an increase in the curing temperature from 80 to 200°C was found. At the processing temperature of 120°C, the Sol fraction content in the samples is decreased and the gel fraction is increased. Yield strength increases by 48%, modulus of elasticity by 107%, Brinell hardness by 40%, Shore hardness by 20 %. The glass transition temperature increases to 225°C.

1. Introduction

This work is devoted to preparation and studying the properties of modified epoxy urethane-containing isocyanurate network polymers. This work is preceded by previous studies [1-4] on this issue, including the study of the strength, deformation and thermal properties of such polymers synthesized on the basis of various oligoetherdiols, iso- and diisocyanates and epoxy resin. These properties vary significantly depending on the ratio of diol and isocyanate components. The properties also depend significantly on the reaction temperature and the concentration of the catalyst. The data obtained in [1-4] allowed us to evaluate the effect of the concentration of the polymer cross-linked points formed on the glass transition temperature and apply computational methods for quantitative description of this dependence.

In this paper, the influence of precursor curing temperatures on physical, strength and thermal properties is analyzed. Measurements were made in a wide range of temperatures. The paper uses the previously developed calculation scheme [5-7] for determining the glass transition temperature of copolymers to predict their physical characteristics.

2. Experimental part

The following systems were studied: 1) an epoxy resin cured with 2,4-toluylenediisocyanate (TDI) with a 3.5-fold excess of the latter; 2) polytetramethylene oxide (PTMO) cured with 2,4-toluylenediisocyanate (2.6-fold excess). The synthesis process of systems 1 and 2 was performed

IOP Conf. Series: Materials Science and Engineering 1015 (2021) 012066 doi:10.1088/1757-899X/1015/1/012066

simultaneously, but in different containers at a temperature of 75°C for 30-35 minutes. The reaction system consisting of a mixture of the products of both reactions in the presence of the catalytic hardener dimethylbenzylamine, equal to 0.2% in terms of the total mass of the system, was subjected to thermal curing. Temperature and time curing mode specified in table 1.

Table 1. Temperature-time mode of sample curing.

Samples of the copolymers	time, (min)						
	80°C	90°C	100°C	110°C	120°C	160°C	200°C
$SPEUI_{I}$	40	-	-	-	-	-	-
$SPEUI_{II}$	40	40	-	-	-	-	-
$SPEUI_{III}$	40	40	40	-	-	-	-
$SPEUI_{IV}$	40	40	40	40	-	-	-
$SPEUI_{V}$	60	60	60	-	60	-	-
$SPEUI_{VI}$	60	60	60	-	60	60	-
$SPEUI_{VII}$	60	60	60	-	60	60	60

Epoxy resin is a product of CHS-EPOXY 520 containing residual hydroxyl groups of phenol, as well as hydroxyl groups of oligomerized epoxy that can react with TDI. PTMO has a molecular mass of $M = 1000$. DMBA is used to accelerate the curing of epoxy resin and is a catalytic hardener. NMR spectra were recorded on a Bruker Avance 400 device with an operating frequency of 400 Hz (^{1}H); the signals of residual protons of deuterosolvators were used as an internal standard.

IR spectroscopy and titrimetric methods were used for continuous monitoring of the reaction in the TDI-epoxy, TDI-PTMO and epoxy-TDI + TDI-PTMO systems. IR spectra in the region of 440÷4000 cm^{-1} were registered on the Fourier spectrometer "NicoletMagna 750".

The choice of the optimal curing mode was determined by studying changes in the following properties of the system: strength at compression, modulus of elasticity (including at different temperatures), Brinell and Shore D hardness. Thermomechanical, differential-thermal and thermogravimetric methods of analysis were also used.

Measurements of stress-strain curves at compression were performed at different temperatures on the Dubov-Regel device (the device is equipped with special software for electronic registration of experimental data).

Standard methods for determining Shore D and Brinell (H_B) hardness were used. The measurement was performed on the Vostok-7 device (Shore D) and the TP-1 hardness tester. The load on the indenter was 5 kg (samples $SPEUI_I$-$SPEUI_{VII}$ (Shore D)) and 5 and 12 kg (H_B) samples $SPEUI_I$-$SPEUI_V$ and $SPEUI_{VI}$-$SPEUI_{VII}$ respectively.

According to thermomechanical analysis (TMA), the experimental glass transition temperature T_g was determined, on the basis of which the chemical composition of microphases was estimated. Methods of quantitative description of the dependence of T_g on the chemical composition of microphases using the Cascade computer program (INEOS RAS) were used [5-7]. TMA was performed on a thermomechanical analyzer "Q400 Instrument".

Thermogravimetric analysis (TGA) in combination with differential thermal analysis (DTA) was performed on a derivatograph "Derivatograth-C" in the temperature range of 20-700°C in air at a heating rate of 10 deg/min.

During the Sol analysis, fractions were isolated from samples at the curing depths, which were reached at temperatures of 80, 90, 100, 110, 120, 160 and 200°C. The Sol-fraction separation process was carried out in a continuous way in the Soxlett apparatus using acetone.

3. Results and discussion

Initially, when an excess amount of TDI 3.5 mol interacts with an epoxy resin at T ~75°C, isocyanateepoxyurethane is formed, which is the product of adding TDI to an oligomerized epoxy

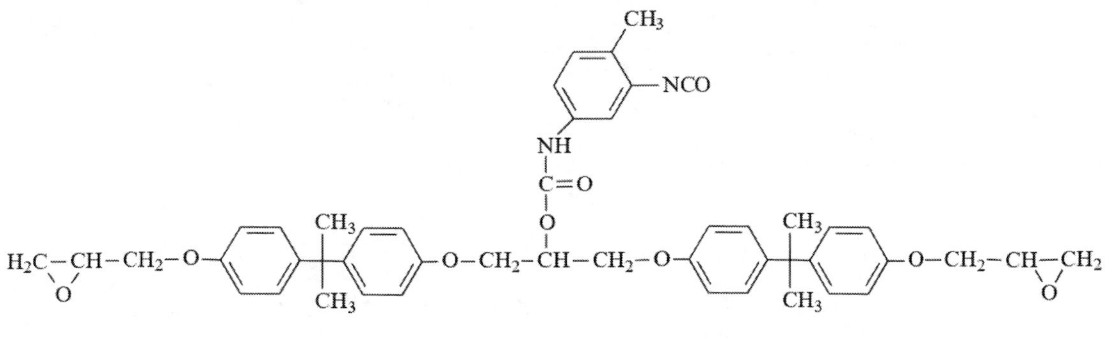

According to NMR data, the ratio of reacted TDI to non-reacted TDI is 1:3.

Based on the data of titrimetric analysis, it is calculated that the initial TDI remains in the reaction system in the amount of 2.3 mol. According to the same analysis, it was determined that the number of epoxy groups in the initial reaction system was ~14.02 wt. %. In the resulting product, the number of epoxy groups is ~11.8 wt.%, i.e. their consumption during the reaction does not exceed 16%. Quantitative processing of IR spectra shows that the main bands of the epoxy at 1511, 1247 and 830 cm^{-1} decrease by ~12.8, 23.6 and 11.2%, respectively, during the reaction 6.

As a result of the interaction of an excess amount of TDI (2.6 mol) with PTMO, we obtain the urethane for polymer of the following structure:

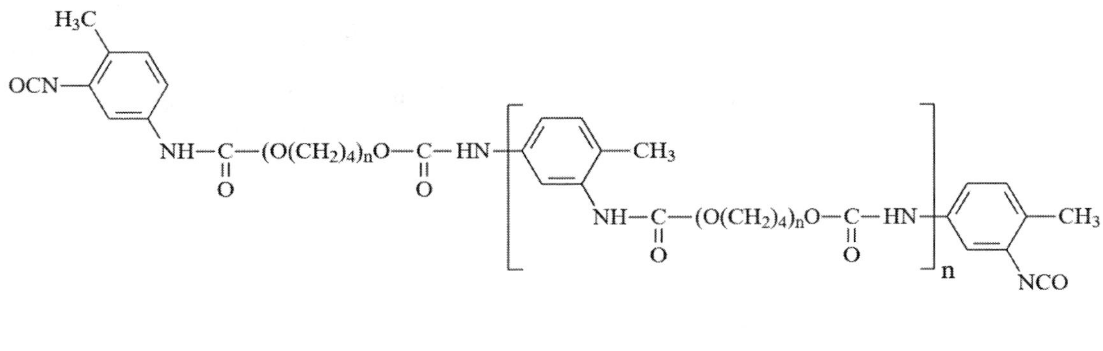

According to NMR data, the ratio of reacted TDI to non-reacted TDI is 1:2.

The TDI content in the resulting product (based on titrimetric data) is ~2.2 mol. IR spectroscopy data confirms a decrease in the peaks of isocyanate groups in the range of ~16%. This also corresponds to titrimetric data, according to which the TDI content is reduced by ~15%.

When further interaction between the urethane prepolymer and isocyanatomethane and add DMBA with increasing temperature epoxyurethaneisocyanurates formed of the following structure:

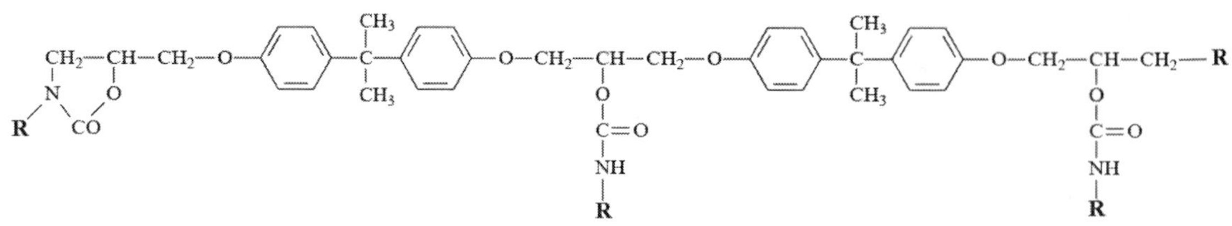

where **R** structures are shown below:

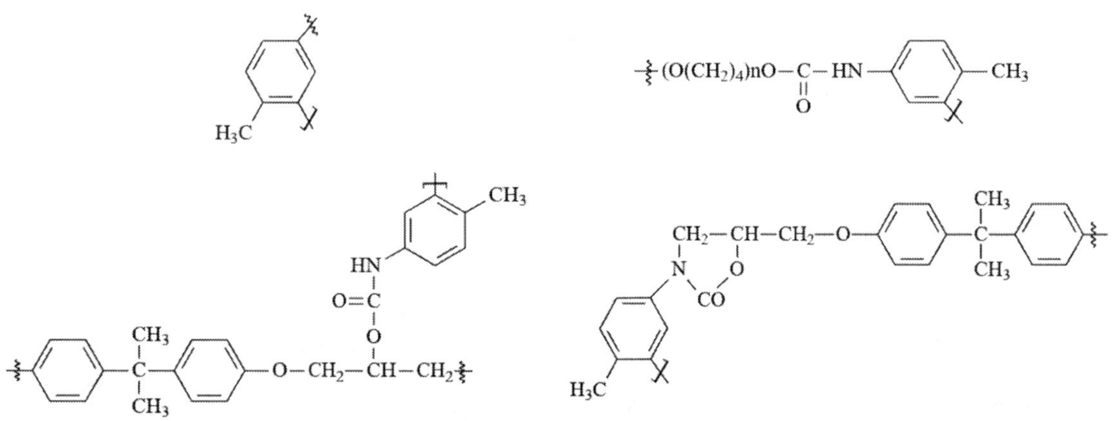

In the resulting system, initially containing 4.5 mol of TDI, the consumption of the latter, according to titrimetry data, exceeds 18% for more than 40 minutes. Thus, in the resulting epoxyurethanisocyanurate product, the number of TDI moles is reduced to 3.7. The data of titrimetric analysis are confirmed by the results of IR spectroscopy analysis, according to which the decrease in the characteristic peaks of isocyanate groups does not exceed 17%.

In general, the subsequent heating of the polymerizing system from 80°C and higher (see table 1) results in cyclotrimerization of residual NCO groups with the formation of isocyanurates [2-4]:

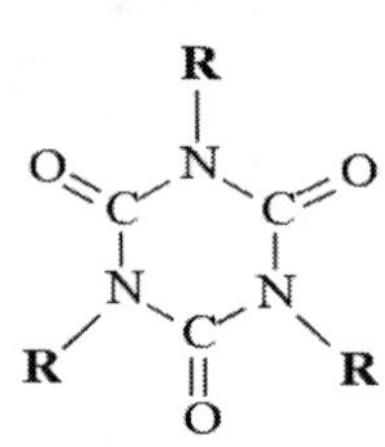

The chemical formulas of **R** structures are shown above.

The calculated and experimental data discussed below allow us to draw some conclusions about the choice of the final temperature of the curing process. Calculations of T_g of network polymers were performed using the method [6]. The structural features of the resulting systems were considered taking into account the chemical nature of the repeating units and their quantitative ratio. Based on the chemical structure of linear chains and cross-linked points (table 2), as well as experimental results of TMA and TGA, using the computer program "Cascade" obtained data on the quantitative ratios of the structures that make up each microphase.

Table 2. Chemical structure of repeating units of the copolymer.

A	$-[(CH_2)_4O]-$
B	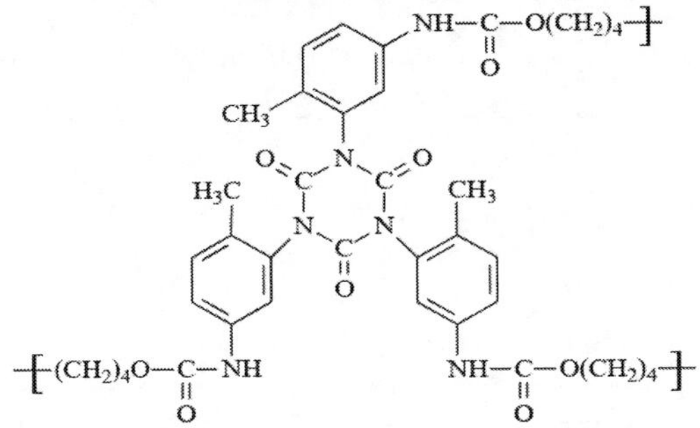
C	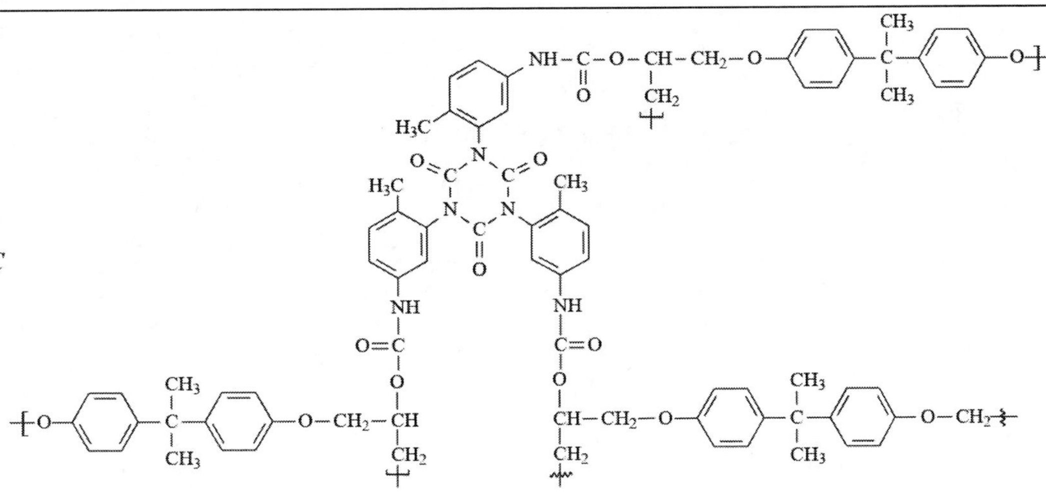
D	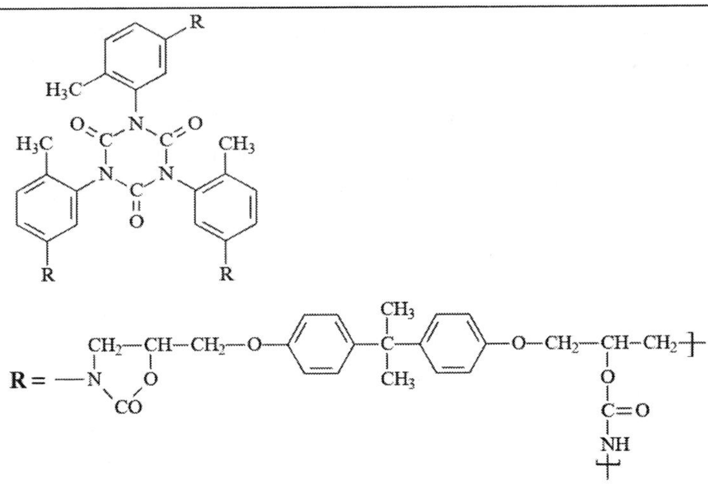

The calculated values of T_g and T_d of the studied samples are shown in table 3 (the table also shows the experimental values of T_g).

Table 3. Values of glass transition temperature and density of synthesized samples (calculations were made using the computer program "Cascade", INEOS RAS).

Samples	Relaxation transitions	M (g/mol)	$\sum_i \Delta V_i$ (nm^3)	ρ (kg/m^3)	T_g, K calculation	T_g, K Experimental values	T_d (K)
$SPEUI_I$	*I*	235	0.226	1160	332	328	653
	II	763	0.708	1220	432	433	651
$SPEUI_{II}$	*I*	235	0.226	1160	332	340	653
	II	448	0.420	1200	393	393	652
$SPEUI_{III}$	*I*	235	0.226	1160	332	338	653
	II	508	0.475	1200	403	407	651
$SPEUI_{IV}$	*I*	235	0.226	1160	332	338	653
	II	508	0.475	1200	403	407	651
$SPEUI_V$	*I*	263	0.251	1170	348	350	653
	II	1490	1.38	1230	478	475	656
$SPEUI_{VI}$	*I*	1360	1.26	1230	463	464	656
$SPEUI_{VII}$	*I*	1410	1.31	1220	497	498	641

The microphase cleavage for the $SPEUI_I$ sample takes place at $T \approx 328$K and $T \approx 433$K at the following ratio of structural fragments: $A = 26.4$ and 4%, $B = 15.8$ and 16%, $C = 28.9$ and 51%, $D = 28.9$ and 29%, respectively. For the $SPEUI_{VII}$ sample, the relaxation transition occurs only at $T \approx 498$K, which is 170K higher than for the $SPEUI_I$ sample. The composition of microphases is as follows: $C = 93\%$, $D = 7\%$.

It was found that the elastic modulus E decreases from an initial value of 700 MPa at 20°C to a value of 45 MPa at 105°C (figure 1, curve 1), which exceeds 93%. The most significant change in the composition of microphases associated with an increase in the final temperature of the curing process by 2.5 times (sample $SPEUI_{VII}$) is manifested in the practical loss of structures with increased flexibility (fragments C with a share of 93% and D with a share of 7% prevail in the composition of microphases). As for T_g, the highest value is observed here, equal to ~498K (table 3). In this case, the value of E reaches a value of 900 MPa at a temperature of 105°C with an initial value of 1450 MPa at 20°C (figure 1, curve 5). Here, the value E is reduced by 38 %. If in such a system (meaning the $SPEUI_{VII}$ sample) the concentration of cross-linked points increases, as indicated by the ratio of structural fragments, then the T_g of the system is higher by 170K than in the $SPEUI_I$ sample. The region in which the polymer $SPEUI_{VII}$ is located in this case corresponds to the glassy state.

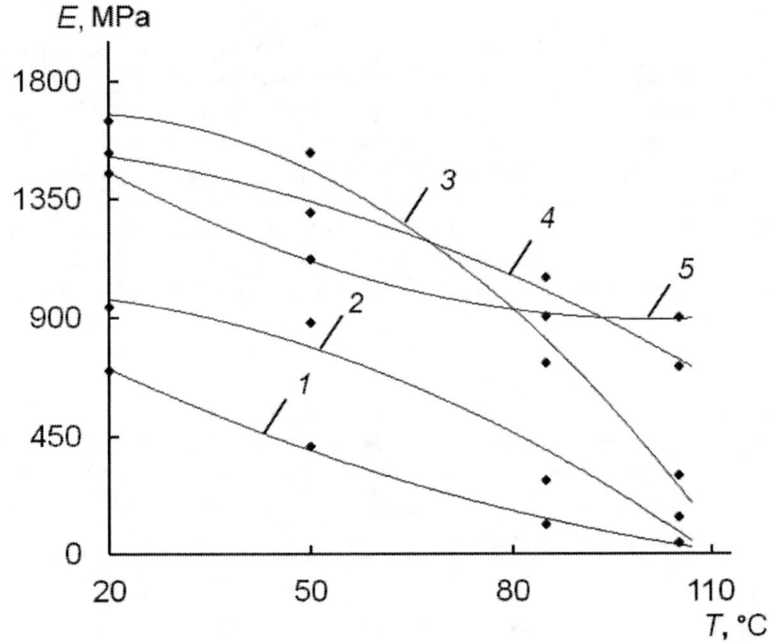

Figure 1. Dependence of the elastic modulus of copolymer samples on temperature: 1 – $SPEUI_{I}$, 2 – $SPEUI_{IV}$, 3 – $SPEUI_{V}$, 4 – $SPEUI_{VI}$, 5 – $SPEUI_{VII}$

The influence of the heat treatment temperature on the strength properties of samples can be seen in the obtained Brinell and Shore hardness dependencies (Shore D). The effect of changes in Brinell hardness with changes in the final temperatures of the curing process exceeds 40% (figure 2, curve 1), and shore hardness – 20% (curve 2).

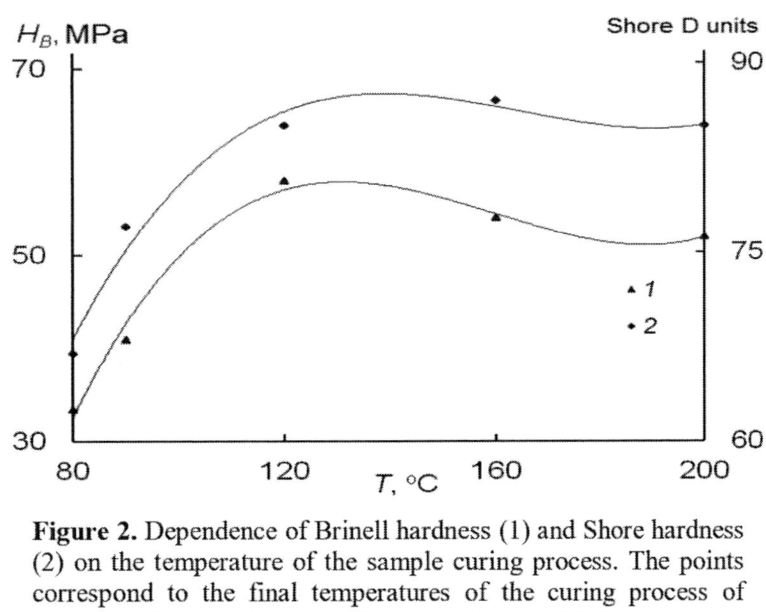

Figure 2. Dependence of Brinell hardness (1) and Shore hardness (2) on the temperature of the sample curing process. The points correspond to the final temperatures of the curing process of samples $SPEUI_{I}$, $SPEUI_{IV}$, $SPEUI_{V}$, $SPEUI_{VI}$, $SPEUI_{VII}$.

IOP Conf. Series: Materials Science and Engineering 1015 (2021) 012066 doi:10.1088/1757-899X/1015/1/012066

4. Conclusion

The experiments and calculations carried out in this work allowed us to obtain information about the effect of the curing temperature of the reaction medium consisting of epoxy resin, TDI, PTMO and DMBA on the strength and thermal properties of the resulting epoxyurethane-containing isocyanurate polymers of a network structure. The final curing temperature varied from 80 to 200°C. The dependence of the yield strength at temperature curing of the reaction mixture showed that at the curing temperature of 200°C, the value of σ_y acting increases by 48%, elastic modulus increased by 107%, Brinell hardness number increases by 40 %, and Shore hardness by 20%. The glass transition temperature increases to 225°C. As the analysis showed, such a change in strength properties with variations in the curing temperature is associated with a change in the chemical composition of microphases that form polymer systems.

Acknowledgments

This work was carried out with the financial support of the Ministry of science and higher education of the Russian Federation (project " Theoretical and experimental design engineering of new composite materials for safety exploitation of buildings and structures under conditions of technogenic and biogenic threats" FSWG-2020-0007 and the theme of the state task of INEOS RAS no. 0085-2019-0004).

References

[1] Askadskii A A, Goleneva L M, Konstantinov K V and Bychko K A 2001 *Russian Polym. News* **6** (2) p 6

[2] Askadskii A A, Goleneva L M, Afanas'ev E S and Petunova M D 2012 *Review J. of Chemistry* **2** (2) pp 105-152

[3] Askadskii A A, Goleneva L M, Afanas'ev E S and Petunova M D 2012 *Review J. of Chemistry* **2** (4) pp 263-314

[4] Askadskii A A 1998 *Russian Chem. Reviews* pp 681–712

[5] Askadskii A A 1996 *Physical properties of Polymers. Prediction and Control* (Amsterdam: Gordon and Breach Publishers)

[6] Askadskii A A 2003 *Computational Materials Science of Polymers* (Cambridge: Cambridge International Silence Publishing)

[7] Askadskii A A, Goleneva L M, Konstantinov K V and Bychko K A 2002 *Vysokomolek. Soedin. A.* **44** (4) p 567 [in Russian].

IOP Conf. Series: Materials Science and Engineering 1015 (2021) 012067 doi:10.1088/1757-899X/1015/1/012067

Optimization of technological parameters of reinforced concrete production

A G Chiknovoryan, S A Mizuryaev and A Yu Zhigulina

Samara State Technical University, Molodogvardeyskaya str. 194, 443001 Samara, Russia

E-mail address: umu-sgasu@mail.ru

Abstract. The article presents the results of research in the field of improving the technology of production of reinforced concrete products aimed at optimizing the composition of concrete; quality control of concrete for strength, taking into account its uniformity; adjustment of the composition of the concrete mix taking into account the characteristics of the aggregates; improvement of the modes of heat treatment of concrete; stabilization of the geometry of freshly molded products, obtaining a smooth and high-quality surface of the products and reduction of marriage associated with spontaneous damage to freshly molded products. It is shown that rational modes of heat treatment of concrete, allows you to abandon the traditional scheme of heat treatment of concrete with steam and provide the necessary conditions for obtaining concrete with design physico-mechanical properties.

1. Introduction

The production of reinforced concrete building envelopes and load-bearing structures in construction practice is a complex technological process and has its own characteristics. The search for solutions to optimal technologies and methods for the factory production of precast concrete structures is associated with the level of scientific and technological development of the country.

With the development of technology, technology is developing in parallel. A significant improvement in the technology of production of reinforced concrete and concrete structures occurred with the introduction of new modern technological processes [1-5], equipped with high-performance and automated equipment, for example, robots, manipulators, hoisting gears and etc. The modern direction for the development of precast concrete is the introduction of modernized and bench production lines.

The main range of products manufactured on such lines are multi-layer and single-layer external wall panels, internal wall panels, as well as floor slabs. Such lines provide high quality products and reduce the production time of prefabricated reinforced concrete structures due to the optimization of machines, equipment and special production solutions.

2. Technology research

From all that has been said, an obvious conclusion follows that the solution to the problem of ensuring the necessary technological properties of concrete mix and the design strength of hardened concrete and products is directly related to the optimization of concrete composition and the selection of rational technological equipment.

In the production of building materials, products and structures made of reinforced concrete, materials are widely used that can improve the physicomechanical properties of concrete and reinforced concrete products, as well as optimize the material and energy intensity of their production. Intensive development of modern building technologies as one of the main areas also includes increasing requirements for the quality of finished building products and the characteristics of the initial basic and auxiliary materials.

One of the ways to improve the molding properties of concrete mixtures is to improve its connectivity and reduce the water demand of concrete mixtures through the use of modern additives with multifunctional modifying properties. Complex modern additives meet the requirements of the modern production of building products and reinforced concrete structures, which have high plasticizing properties and, accordingly, accelerate the hardening of concrete and increase its strength.

So, for example, the technology of pallet circulation related to the semi-conveyor method of production allows the line to produce a wide range of panels with different sizes and use different rhythms of pallet movement within the same production line through the use of overtaking and dead ends, which is not feasible on classic conveyor lines .

This technology in comparison with traditional has several advantages. The main one is the automation of the technological process, which makes it possible to change the modes of preparatory operations and the molding process. This is due to the use of roller bearings and transmission machines, which allow you to move pallets between different speed parts of the line.

These projects are based on intelligent work processes and technologies that use the most economical equipment selection concept. This is expressed in the main technological features of the production line. These include, for example, the use of transverse lifting trolleys with a digital synchronization system, stationary cleaners from concrete residues from pallets that have rotating roller brushes and steel scrapers, rotating nozzles for applying lubricant to the molds, ejectors for even distribution of the coolant in concrete heat treatment chambers etc.

On the other hand, in the production of hollow core slabs, the most difficult, both from the technical and technological sides, is the formation of voids. The voids in the slab are necessary to reduce its weight and, accordingly, reduce the load on the building. However, this imposes additional requirements on the strength of concrete in the partitions and shelves of the slabs. The solution to this problem is associated with an increase in the strength of concrete slabs by optimizing the composition of concrete and the selection of rational technological equipment. Optimization of concrete compositions can be performed by reducing its water content while providing the necessary for effective compaction during molding workability and cohesion of the structure.

In the manufacture of hollow core slabs, a violation of their geometric dimensions is often observed (creeping of arches, faces, etc.). One of the reasons for these violations is the use of non-optimized concrete mixes. The device of voids in the plates allows to reduce their weight while maintaining the bearing capacity. However, this imposes additional requirements on the strength of concrete in sufficiently thin walls and partitions of hollow slabs. The required concrete strength in a hollow core slab is achieved by optimizing the consumption of binder and aggregates in concrete, reducing the water content of the concrete mixture, and also by choosing the optimal technological equipment.

Improving the technological modes of production of flat products on pallets circulation lines and modern bench lines requires solving a number of problems, including: to optimize the compositions and modes of heat treatment and their effect on the basic physical and mechanical properties of heavy concrete of flat products.

3. Results and discussions

Modern super- and hyperplasticizers are more effective additives for concrete and comply with current sanitary and epidemiological standards. Currently, they are increasingly being used in the manufacture of building products and structures. New RCA additives are developed taking into account the targeted effect on the properties of the concrete mix and hardened concrete, for example, by controlling the rheological properties of the mixture.

These additives are mass-produced by industry and they are actively used, for example, to obtain high-quality concrete surfaces in structures manufactured by continuous formless molding using specialized machines (extruders, splitformers, vibration casting machines, etc.) and on conveyor lines.

In the production of vibropressed pre-stressed molded concrete products on long heated stands by continuous formless molding, these additives have a stabilizing effect on the technological properties of moderately rigid and rigid concrete mixtures, as well as increase the density and strength of concrete and give products water-repellent and water-repellent properties.

Production research of bench production was carried out directly on the bench line of Volgotranszhelezobeton LLC (Samara) for the manufacture of hollow core slabs.

The conducted studies have confirmed the possibility of optimizing the composition of concrete for the production of hollow core slabs.

In the production line, the vibro-extruder of the Russian company Construction-Technological Machines (STM) was used as a molding machine. On this production line, the concrete mixture from the concrete mixer was discharged directly into the dispensing hopper (such as a “shot glass”), which was transported by crane to the molding machine. Further, the concrete mixture was reloaded from the distribution hopper into the storage hopper of the molding machine. Compaction of the concrete mix with the workability grade P1 (1 ... 4 cm) was carried out by the molding machine with two poly-frequency vibrators built into it. Vibroextruder molded products in one pass without the need for formwork (figure1).

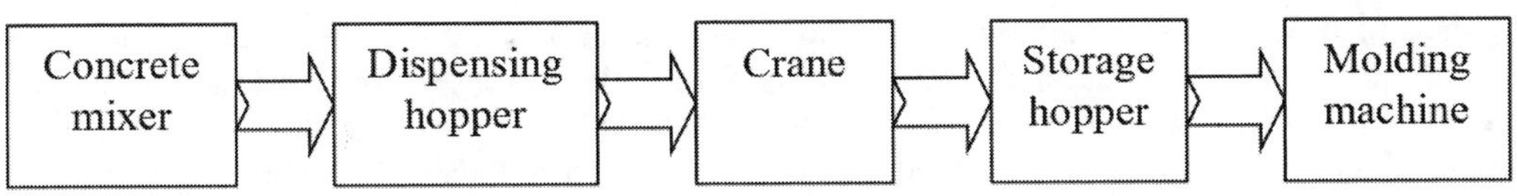

Figure 1. Scheme of the technological line for the production of plates.

Heat treatment of products on the stand was carried out using tubular registers built into the floor of the molding line, through which hot water was supplied, which ensured the temperature of isothermal exposure of concrete in products of about 50 ° C. The duration of the heat treatment of the products was 24 hours, including a temperature rise of 6 hours, isothermal aging of 12 hours, and cooling of 6 hours.

The concrete mixture was prepared using an additive of the “RCA” type in an amount of 0.16% in finished form by weight of the cement. Before use, the additive is dissolved in water in the ratio of one part of a concentrated solution of the additive to ten parts of water. To prepare the concrete mixture, Portland cement grade B42.5 in the amount of 365 kg/m^3, a mixture of fine (Mkr = 1.1) and coarse (Mkr = 3.4) sand in the amount of, respectively, 425 kg/m^3 and 505 kg/m^3 were used:, crushed stone (quarry “Ivanteevka”, Kazan) with a brand for crushability D800 and fractions 5 ... 20 mm in an amount of 1020 kg/m^3.

The main characteristics of concrete obtained by research: concrete strength class - B25; the actual average density of concrete is 2290 ... 2320 kg/m^3; the required concrete strength is 33 MPa with an average value of the coefficient of variation of concrete strength up to 6%; the transfer strength of concrete made it possible to release reinforcement tension on it after heat treatment.

In the manufacture of wall panels on conveyor lines, it all starts with the cleaning of pallets using a machine for cleaning pallets. On-board equipment is similarly cleaned using a special machine. The movement of pallets is carried out using roller tables: leading and guiding. Further, from the cleaning station, the pan moves to the lubrication station. After lubrication, marking is applied to the pallet with a plotter for the subsequent installation of a magnetic tool on it. Next, the pallet is sent to the post installation rig.

IOP Conf. Series: Materials Science and Engineering 1015 (2021) 012067 doi:10.1088/1757-899X/1015/1/012067

The installation of a magnetic tool is carried out automatically by the robot. After installing the rigging, the pallet goes to the post for installing additional formwork made of polystyrene foam. Next, the pallet moves to the reinforcement post, which is carried out by the robot. At the reinforcement station, elements are installed that ensure the formation of a protective layer of concrete in the product.

After that, the pallet goes to the installation post of embedded parts and sling loops. After the installation of the entire reinforcement, the pallet is sent to the molding station using the lateral movement devices. The concrete mix is laid on a pallet by a screw paver. The auger paver receives the concrete mix from the addressable feed plug, which drives up on separate suspension rail tracks from the concrete mixing unit.

After molding, the pallet goes to the vibratory compaction station of the concrete mixture, where on the vibrating platform the pallet is slightly raised to get off the axis of movement, and vibrated using a complex of vibrators. Then, after sealing, the pallet enters the surface smoothing station. Smoothing the surface is carried out using a vibrorail and a blade trowel. After smoothing the surface, the pallet passes to the post of heat treatment in a vertical cell chamber. Using a stacking device, the pallet is transported to a free cell in the chamber for heat treatment. After heat treatment, the pallet with the product enters the stripping station at the tilter, where the product is tilted together with the pallet for vertical removal, separated from the pallet and sent to the finished goods warehouse using an overhead crane.

Next, the product is sent with a bridge crane to the finishing station, and the pallet after stripping with the help of lateral movement devices is sent to the cleaning station. After finishing, the product is marked and sent to the warehouse using a self-propelled trolley. Also for the production of three-layer wall panels provides a post installation of insulation.Automatically through the central control system, the installation of on-board equipment, the movement of pallets along the posts of the processing line, the manufacture of a reinforcing cage for the product, the placement of molds with products in the compartments of the rack of the stacked heat treatment chamber are also carried out (figure 2).

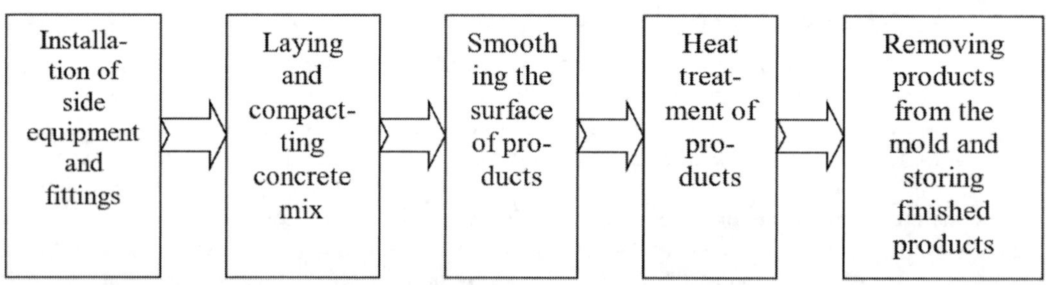

Figure 2. Panel production scheme.

One of the tasks on the lines of circulation of pallets is to optimize the modes of heat treatment of products in vertical cell chambers.

For heat treatment of concrete, as a rule, chambers for drying products are used, equipped with shelves, heat generators with gas burners and recirculation fans. This allows you to abandon the device costly boiler rooms, supplying steam for heat and moisture treatment of products. It should be noted that the cameras are loaded with pallets with products completely automatically using a lifting and transfer platform controlled by an automated computer system.

Cameras are vertical racks (compartments) with sections (cells) for placing pallets. Each section (cell) contains one pallet. Racks (compartments) are equipped with insulated walls to reduce heat loss. The dimensions of one section (cell) of the heat treatment chamber are about $11.2\times4.0\times0.6$ m.

For heat treatment, a low-temperature regime is used, at an isothermal holding temperature of about 50 ... 60 ° C.

Shows the calculation of the required number of sections (cells) in the heat treatment chamber with a line rhythm of about 35 minutes (see table 1).

Table 1. Modes of heat treatment of concrete in a vertical chamber

The strength of concrete from the design for 28 days, (%)	Temperature of isothermal exposure, (^{0}C)	The duration of the concrete heat treatment of, (h)	The number of sections (cells) in the heat treatment chamber, (pcs.)
70	60	32	48
70	50	40	60
50	60	16	24
50	50	20	30

The practice of improving the technological parameters of the production of reinforced concrete has revealed a number of problems that require further research aimed at optimizing the grain composition of aggregates and the content of the mortar component of the concrete mix and at minimizing the water-cement ratio due to plasticization while ensuring the necessary workability, structural cohesion, delamination and water separation of the concrete mixture.

4. Conclusions

The formation of precast concrete products using an additive of the type "RCA" leads to a decrease in water demand and an increase in the uniformity and cohesion of the concrete mixture and makes it possible to reduce the adherence of the mixture to technological equipment, which makes it possible to accelerate the set strength of concrete with the possibility of reducing cement consumption, improve the quality of the molded products, to stabilize the geometry of freshly molded products, to achieve a smooth and high-quality surface of products and to reduce defects associated with spontaneous damage to freshly molded products.

The use of the low-temperature regime of concrete heat treatment in vertical batch chambers equipped with heat generators with gas burners or on bench lines with hot water heating allows us to abandon the traditional heat-moisture treatment of concrete with steam and provide the necessary conditions for concrete hardening with design physico-mechanical properties.

References

[1] Cao G, Li N Z and Guo K 2017 Analytical study on the change of fluidity of fresh concrete containing mineral admixture with rest time *J. of Adv. Concrete Technology* **15** pp 713-23 (https://doi.org/10.3151/jact.15.713)

[2] Abbas S, Nehdi M L and Saleem M A 2016 Ultra-high performance concrete: Mechanical performance, durability, sustainability and implementation challenges *Int. J. of Concrete Structures and Materials* **10** (3) pp 271–95 (https://doi.org/10.1007/s40069-016-0157-4)

[3] Ma H and Li Z 2013 Realistic pore structure of Portland cement paste: Experimental study and numerical simulation. *Computers & Concrete* **11** (4) pp 317–36 (https://doi.org/10.12989/cac.2013.11.4.317)

[4] Bediako M, Kevern, J and Amankwah E 2015 Effect of curing environment on the strength properties of cement and cement extenders *Materials Sciences and Applications* **6** pp 33–9 (https://doi.org/10.4236/msa.2015.61005)

[5] Hilal A A, Thom N H and Dawson A R 2015 The use of additives to enhance properties of pre-formed foamed concrete *IACSIT Int. J. of Eng. and Technology* **7** (4) pp 286-7 (https://doi.org/10.7763/ijet.2015.v7.806)

IOP Conf. Series: Materials Science and Engineering 1015 (2021) 012068 doi:10.1088/1757-899X/1015/1/012068

Technical condition of apartment buildings in the central regions of Russia

M I Balzannikov

Samara State University of Economics, Samara, 141 Sovetskoi Armii St., 443090, Samara, Russia

E-mail address: mibsgasu@mail.ru

Abstract. The article notes the presence in Russian cities housing stock of the central regions of a large number of residential apartment buildings, built over 50 years ago and in operation. The bearing capacity of the structures of these houses has significantly decreased. In connection with this carried out a survey of technical condition of houses. For buildings that are for further operation invalid state, the procedure of recognition of their emergency. Analysis of data on the technical condition of apartment buildings showed that in the cities of Kazan, Samara and Volgograd, the physical deterioration of residential buildings is 25-28%, which is less than the average value in Russia, and in Nizhny Novgorod it is significantly higher and amounts to 38%. The data allow us to conclude that in the near future in Nizhny Novgorod one should expect a large number of residential buildings to go into emergency condition. Analysis of indicators of the state of emergency residential buildings in these cities shows that the percentage of the city of Samara operated homes, referred in 2013 to the state of emergency, the largest and is 10.1%, which is significantly higher than the average value in Russia. As one of the main reasons for the incomplete resettlement of emergency houses under the program for the period 2013-2018 indicates "the unknown whereabouts of the owner." The article concludes that it is possible to increase the efficiency of the resettlement program through operational work with the owners of residential premises of liquidated residential buildings.

1. Introduction

Currently, in the Russian Federation operates a large number of residential apartment buildings, built over 50 years ago. The main supporting structures of such buildings during the period of operation were subjected to various loads, including numerous dynamic effects and the influence of negative environmental factors [1, 2]. As a result of prolonged operation houses bearing capacity of structures could be significantly reduced [3, 4]. In connection with this very relevant is a thorough examination of the technical condition of apartment buildings, as well as the development of organizational and technical measures for the overhaul [5-7].

Events should provide a minimum negative impact on the environment [8, 9], and to objects having historic or architectural importance - the possibility of using an alternative design of retaining elements [10, 11]. For buildings that are inadmissible for further operation, it is necessary to apply the procedure for recognizing them as emergency and relocate residents. At the same time, it is important to timely prepare the resettlement housing stock for residents of houses that are subject to repair, reconstruction or demolition.

IOP Conf. Series: Materials Science and Engineering 1015 (2021) 012068 doi:10.1088/1757-899X/1015/1/012068

2. Materials and Methods

The generalized data on the technical condition of apartment buildings are analyzed using open statistical sources for large cities in the central regions of the Russian Federation. In particular, we consider the data about the state of residential buildings megacities located on the great Russian river Volga Nizhniy Novgorod, Kazan, Samara and Volgograd.

The technical condition of buildings is determined based on the results of their survey [12]. The main regulatory documents defining the procedure and composition of building inspections in Russia are: SP 13-102-2003 Code of rules for design and construction: rules for examining load-bearing building structures of buildings and structures and GOST 31937-2011 Interstate standard: Buildings and constructions. Rules of inspection and monitoring of the technical condition.

These documents determined that the buildings and structures can be operated, if they are on the results of state structures estimates are in the regulatory technical or operational. But if the physical deterioration of home designs exceeds 40%, and they are characterized by injuries and strains, confirming the exhaustion of their load-bearing capacity, the residential building should be evaluated for its reference to a crash.

3. Results

In Russia as a whole, the average value of the physical wear and tear of apartment buildings, information about which is placed in the information system [13], is currently 32%. In this case, the largest of the exploited wear homes (about 55%) have the buildings built in the period 1920-1940. For information about the serviceability of houses show that the average number of homes in the country, are in poor condition is 3.1%. At the same time, buildings of earlier construction correspond to a greater percentage of houses in emergency condition (table 1).

Table 1. Information about the technical condition of residential apartment buildings in Russia [13].

Period of putting residential buildings into operation	Average percentage of depreciation of residential buildings, %	Share of apartment buildings in emergency condition, %
1920-1930	55.9	11.9
1931-1941	55.0	13.5
1942-1952	49.2	7.1
1953-1963	43.4	5.0
1964-1974	36.6	1.9
1975-1985	29.8	1.4
1986-1996	23.2	1.1
1997-2007	13.7	0.3
2008-2018	3.8	0.1
2019-2020	1.1	0.0

To the largest cities in the central regions of Russia analysed parameters are shown in table 2. Of these, it follows that the average percentage of deterioration of residential apartment buildings in Kazan, Samara and Volgograd are practically equal at 25-28%, and in Nizhny Novgorod - is much higher: 38%. From this we can conclude that in the coming years in Nizhny Novgorod one should expect a significant increase in residential buildings that have passed into the status of emergency.

Comparison of indicators of the emergency state of residential buildings in the cities under consideration shows that at the moment in Kazan there are no residential buildings in the number of operated houses that were classified in emergency condition in 2013 (with the exception of one house).

IOP Conf. Series: Materials Science and Engineering 1015 (2021) 012068 doi:10.1088/1757-899X/1015/1/012068

However, in Samara, the percentage of such residential buildings - is the largest and is 10.1%, which is significantly higher than the average value in Russia (3.1%).

Table 2. Information on the technical condition of apartment buildings in the largest cities in the central regions of Russia [13].

City name	Average percentage of depreciation of residential buildings, %	Share of apartment buildings in emergency condition, %
Nizhny Novgorod	38.0	5.1
Kazan	25.4	0.0
Samara	28.3	10.1
Volgograd	25.4	8.4

To analyse the reasons for these significant differences, refer to table 3. It provides information about the resettlement of citizens during the period of municipal programs for the resettlement of citizens from emergency housing in the period from 2013 to 2017. Residential buildings were recognized as damaged as of January 1, 2013. The table shows that the total housing stock in Kazan and Samara is practically the same, while the number of homes resettled in Samara twice more than in Kazan. Thus, during the period under review, a large number of one- and two-apartment houses built in 1920-1930 were subject to resettlement in Samara. Figure 1 and having relatively small living space. In such homes uncomfortable and would be dangerous. It is obvious that many people temporarily moved to live to their close relatives. It is no coincidence that one of the main reasons for the incomplete resettlement of damaged houses of that period indicated: "the unknown whereabouts of the owner".

Table 3. Information on the resettlement of emergency housing stock in cities in the period 2013-2017 [14].

Indicator	Nizhny Novgorod	Kazan	Samara	Volgograd
Population, thousand people	1250	1250	1150	1010
Housing stock, thousand sq. m	38038	34632	34205	23008
Number of housing houses	10741	6300	12305	6374
Number of resettled houses	34	89	209	33
The area of the settled houses, sq. m	6200	50400	45500	12200
Average area of settled houses, sq. m	182	566	217	370
Number of incompletely settled houses	no data	1	111	11
Reasons for incomplete resettlement	-	Unforeseen circumstances (litigation, acceptance of inheritance, unknown whereabouts and other circumstances related to the personality of a citizen)		

The analysis of indicators of the technical condition of apartment buildings in the cities of the Samara region is carried out. At the moment, the average value of wear of houses varies in a small range (table 4). The exception is the city Pokhvistnevo data, wherein the analysed value is minimal and amounts to 13.5%, and the city gratifying for which the wear rate and the maximum is 39.0%.

Figure 1. General view of emergency residential buildings [14].

The relative number of apartment buildings in emergency condition in 2013 was recorded: the largest - in Syzran (19.9%), and the smallest - in Togliatti (only 0.1%). These Togliatti can be explained by the fact that the city is relatively young and active construction of residential apartment buildings in it began to develop only in the 70-ies of the last century due to the erection and commissioning of the Kuibyshev hydroelectric operation, enterprises of the chemical industry and the Togliatti automobile plant.

Table 4. Information about the technical condition of apartment buildings in the cities of the Samara region [13].

City name	Average percentage of depreciation of houses,%	Share of apartment buildings in emergency condition,%
Zhigulevsk	28.7	10.9
Kinel	27.3	12.9
Novokuibyshevsk	34.3	6.1
Oktyabrsk	21.2	11.0
Otradnyy	39.0	8.5
Pokhvistnevo	13.5	8.7
Samara	28.3	10.1
Sizran	30.3	19.9
Tolyatti	27.8	0.1
Chapaevsk	29.4	15.0

In the cities of the Samara region, the total number of houses for which the resettlement was not fully completed was 127. Most of them (excluding Samara) are in the city of Zhigulevsk (table 5).

Thus, a large percentage of emergency residential buildings in Samara to date can be attributed to a lot of dilapidated housing, as well as the result of incomplete implementation of the municipal programs of the previous period and the incomplete settling emergency apartment houses due to "unforeseen circumstances." In order to minimize such "unforeseen circumstances", it is necessary to begin organizational work to identify the whereabouts of the owners at the initial stage of the resettlement program.

In 2019, a new stage of liquidation of hazardous housing began. As part of the federal objectives of the project "Sustainable reductions unsuitable for housing accommodation", which is part of the national project "Housing and the urban environment", developed and adopted by the regional program of resettlement of citizens unfit for housing accommodation in the 2019-2025 years. The main objectives of the programs and the planned results are shown in table 6.

IOP Conf. Series: Materials Science and Engineering 1015 (2021) 012068 doi:10.1088/1757-899X/1015/1/012068

Table 5. Information on resettlement from emergency residential apartment buildings in the cities of the Samara region [13].

City name	Number of houses for which resettlement has been completed	Total area of settled houses, sq. (m)	Number of houses for which resettlement has not been completed
Zhigulevsk	116	47950	8
Kinel	2	1320	0
Novokuibyshevsk	39	16990	0
Oktyabrsk	38	6980	1
Otradnyy	49	18850	0
Pokhvistnevo	58	19180	0
Samara	209	45530	111
Sizran	94	34300	0
Tolyatti	0	no data	no data
Chapaevsk	32	21910	0

The new stage of the program is based on the assessment of the technical condition of apartment buildings, carried out in 2015-2016. and provides for the resettlement of citizens from the housing, a recognized emergency on January 1, 2017. As the table shows, the least amount of emergency housing after the first stage of resettlement (2013-2018.) to be relocated in Kazan: just 51 houses with total area of 15946 square m. It is planned to settle an order of magnitude more houses in the city of Samara. This situation is quite natural, since it follows from the general situation regarding the presence of residential apartment buildings in an emergency condition (tables 2 and 6).

Table 6. Key indicators of regional programs for resettlement of citizens from hazardous housing in the period 2019-2025 [13].

Indicator	Nizhny Novgorod	Kazan	Samara	Volgograd
Number of emergency houses in the constituent entity of the Russian Federation	1559	164	1304	629
included in the programs 2019-2025	901	85	1168	319
Settlement area of residential premises in the constituent entity of the Russian Federation, sq. m	26930	15946	392650	171440
Number of houses to be resettled in the city	192	51	500	253
The area of residential buildings to be resettled, sq. m	63206	15946	155160	146416
Number of resettled citizens, people	4900	1449	11710	9806
Average area of houses to be settled, sq. m	329	312	310	578
Average area of living quarters per person, sq. m	12,9	11,0	11,2	14,9
The proportion of the area resettled homes of the urban housing stock,%	0.17	0.05	0.45	0.64

The indicators characterizing the share of the areas of houses being settled in the total urban housing stock differ significantly for the cities under consideration. The average value of the indicator is 0.29%. The indicator is clearly insufficient for the megacities of the Volga region. Obvious need to take steps

IOP Conf. Series: Materials Science and Engineering 1015 (2021) 012068 doi:10.1088/1757-899X/1015/1/012068

to improve this integral index of at least three times. Otherwise, the number of accidental dwelling houses in the cities will not be reduced and continue to increase.

Conclusions

1. The experience in the development and implementation of federal and regional programs for the resettlement of citizens from emergency residential apartment buildings has shown their high need and prospects.

2. In order to prevent the incomplete implementation of the resettlement progress of resettlement programs and the elimination of emergency apartment houses due to "unforeseen circumstances," local authorities is recommended to start organizing work on identifying the seat of the owners at an early stage of implementation of the resettlement program.

3. To further increase the efficiency of emergency housing liquidation programs and achieve a stable dynamics of a decrease in the housing stock of million-plus cities in the central regions of Russia, related to an emergency state, it is necessary to significantly increase the volume of resettlement both in the number of apartment buildings and in their total living space.

References

[1] Chernoivan V, Novoseltsev V and Chernoivan N 2014 *Industrial and Civil Eng.* **4** pp 48-51
[2] Galitskova Yu and Balzannikov M 2018 *MATEC Web of Conferences* **196** 04025
[3] Alpatov V and Balzannikov M 2018 *IOP Conference series. Materials Science and Engineering* **456** 012011
[4] Zubkov V, Lukin A and Alpatov V 2018 *MATEC Web of Conferences* **196** 01005
[5] Bedov A, Znamenskiy V and Gabitov A 2016 *Evaluation of the technical condition, restoration and strengthening of foundations and building structures of operated buildings and structures. In 2 parts. Part 1. Inspection and assessment of the technical condition of the foundations and building structures of the buildings and structures in use* (Moscow: Publishing house ASV) p 702
[6] Shtengel V and Evdokimov B 2018 *V mire nerazrushayushchego kontrolya* **21** pp 26-31
[7] Alpatov V and Saharov A 2019 *IOP Conference Series: Materials Science and Engineering* **661** 012086
[8] Galitskova Y 2019 Vide. Tehnologija. Resursi - Environment, Technology, Resources *Proc. of the 12th Int. Scientific and Practical Conference* pp 62-64
[9] Galitskova Yu 2019 *IOP Conference Series: Materials Science and Engineering* **661** 012144.
[10] Balzannikova E 2015 Privolzhsky scientific journal 2 (34) pp 141-148
[11] Artemyeva T, Balzannikova E and Leonova A 2017 *Industrial and civil engineering* **3** pp 33-38
[12] Shulaikin A, Popov M, Kovalev V, Ivanov A, Artamonov Yu and Shurygin A 2016 *Modern equipment and technology* **4** (56) pp 25-27
[13] https://dom.gosuslugi.ru. Date of access: 07.03.2020
[14] https://www.reformagkh.ru. Date of access: 20.07.2020

IOP Conf. Series: Materials Science and Engineering 1015 (2021) 012069 doi:10.1088/1757-899X/1015/1/012069

Influence of type and quantity of magnesium additive on the hardening process of alumina cement-based cement stone

N Chumachenko and V Tyurnikov

Samara State Technical University, Molodogvardeyskaya st., 244, Samara, Russia, 443100

E-mail address: uvarovang@mail.ru

Abstract. The greatest effect on increasing the residual strength of cement stone based on alumina cements in the temperature range of 800-1200 °C is achieved when replacing part of the cement with thermally activated metakaolin which is associated with the formation of secondary mullite. To intensify the strengthening process of mullite formation, the influence of various magnesium additives-mineralizers has been studied. The effect of chemically pure magnesium oxide, caustic magnesite and magnesium oxide hydration product calcined at 390 - 450 °C were investigated experimentally. The content of additives in compositions based on high-purity high-alumina cement was accepted in the amount of 0.5, 1 and 1.5 %. Experiments have confirmed the positive effect of magnesium additives on the process of hardening of cement stone, both during hardening and after heat treatment. It was found that the type of magnesium additive affects the hardening process. The highest strength after hardening and drying is obtained by using caustic magnesite as a mineralizer additive in an amount of 0.5 %. Exposed to high temperature, the best results are recorded when using a chemically pure magnesium additive.

1. Introduction

Alumina and especially high-alumina cements have high heat-resistant properties. Alumina cement in comparison with Portland cement and aluminium phosphate binders has a higher fire and heat resistance. These properties determine the widespread use of this binder in the composition of heat-resistant concrete [1-3]. However, there are no ideal materials and these binders have one significant drawback - the cement stone based on them during the operation in the temperature range of 800-1200 °C significantly reduces its strength which is associated with the processes of destruction. Thus, according to [4], the residual compressive strength of samples made of high alumina cement after heating: up to 800 °C is 67.6 %; up to 1000 °C – 60.6 %; up to 1200 °C – 48.9 % of the strength of samples before heat treatment. According to [5], the residual strength after heating to 1200 °C is even lower – 14 MPa, which is 22% of the original. This process can be stabilized by introducing additives that allow the destruction products to be bound into a mullite that acts as a reinforcing frame. Cement stone mullitization which strengthens it is observed when an additive of refractory kaolinite clay is introduced into high-alumina cement [5]. In this case, mullite is formed from the clay itself, the so-called primary mullite and due to the interaction of the products of destruction of cement stone from high alumina cement and clay, the so-called secondary mullite.

The strength of cement stone at high temperatures depends on the type of mullite formed (primary or secondary) and its quantity. The process of formation of primary mullite from refractory kaolinite

IOP Conf. Series: Materials Science and Engineering 1015 (2021) 012069 doi:10.1088/1757-899X/1015/1/012069

clay occurs in the temperature range of 600-1100 °C [6]. Percentage of residual strength of cement stone from a combined binder high alumina cement + clay at temperatures above 1050 °C will, be directly proportional to the amount of clay in the binder up to a certain limit.

The greatest effect of hardening is achieved due to the formation of secondary mullite but its quantity is limited since secondary mullitization does not occur completely, as evidenced by the presence of free silica in the form of cristobalite [5].

The type of additive capable of interacting with degradation products with a hardening effect was theoretically justified and then experimentally confirmed. The best results were obtained when replacing part of the alumina cement with metakaolin which has a chemical activity. The effect of the greatest increase in strength was achieved by pre-thermal activation of metakaoline at 520-540 °C. This additive is recommended for all types of alumina cement although its effect is not the same [7, 8].

Further intensification of the strengthening process of mullite formation is the introduction of additives-mineralizers into the heat-resistant composition, along with metakaolin.

It was found that the process of mullite formation in kaolinite clays under the influence of mineralizing oxides decreases in the following series: $MgO > ZnO > Fe_2O_3 > CaO > K_2O > TiO_2$ [9]. It can be assumed that MgO is the optimal additive for mullite formation. However, the type of magnesium additive can be different and, accordingly, its activity will be different.

2. Methods and materials

In order to study the effect of the type and quantity of magnesium additive on the mullite formation process, experiments were carried out with the following components: chemically pure (c.p.) magnesium oxide, caustic magnesite and hydration product of magnesium oxide, calcined at 390-450 °C.

Lying high-purity high alumina cement (Highly pure high alumina cement) was used as alumina cement. Initially, its activity was 74 MPa, while at the time of the experiment it was 39 MPa. The composition includes metakaolin in the optimal ratio: 85 % Highly pure alumina cement + 15 % metakaolin [8].

The following series of samples of the corresponding compositions were formed:

- 1 series - compositions 1, 2, 3 with the addition of chemically pure magnesium oxide in the amount of 0.5; 1; 1.5 %, respectively;
- 2 series - compositions 4, 5, 6 with the addition of caustic magnesite in the amount of 0,5; 1; 1,5 %, respectively;
- 3 series - compositions 7, 8, 9 with the addition of magnesium oxide hydration product, calcined at 390 - 450 ° C in the amount of 0.5; 1; 1.5%, respectively;
- 4 series - compositions 10, 11, 12 with the addition of chemically pure magnesium oxide in an amount of 0.5; 1; 1.5 %, respectively (compositions of 4 series differ in that their heat treatment has a certain feature: when reaching 390 - 450 °C samples were kept at this temperature for 1 hour to dehydrate $Mg(OH)_2$, assuming that the most active phase of MgO is formed immediately after dehydration [10], and then burned according to a given scheme).

The amount of water was selected experimentally under the condition of obtaining mixtures with normal density. The obtained mixture was used for making by vibrating samples, which after hardening, drying and appropriate heat treatment at 1000 and 1200 °C were tested for strength. The samples density was also determined.

3. Results and discussions

Figure 1 and figure 2 show the kinetics of changes in strength and density of samples depending on the heat treatment temperature.

The strength data of the investigated compositions with different content of additives after drying at 110 °C shows that introduction of 0.5 - 1.5 % of caustic magnesite additive leads to strength increase 1.5 - 2 times in comparison with pure high alumina cement. In the temperature range of 110 - 1000 °C a different degree of intensity decrease in strength is observed. At 1000-1200 °C an increase in strength is observed, especially at compositions with chemically pure magnesium additive.

Experiments have confirmed the positive effect of magnesium additive on cement stone hardening process both during hardening and after heat treatment; they have allowed choosing the most optimal type and quantity of additive for different conditions of cement stone application.

The highest strength after hardening and drying was obtained using caustic magnesite as an additive-mineralizer in the amount of 0.5%.

When exposed to high temperature, the best results are recorded when using a chemically pure magnesium additive.

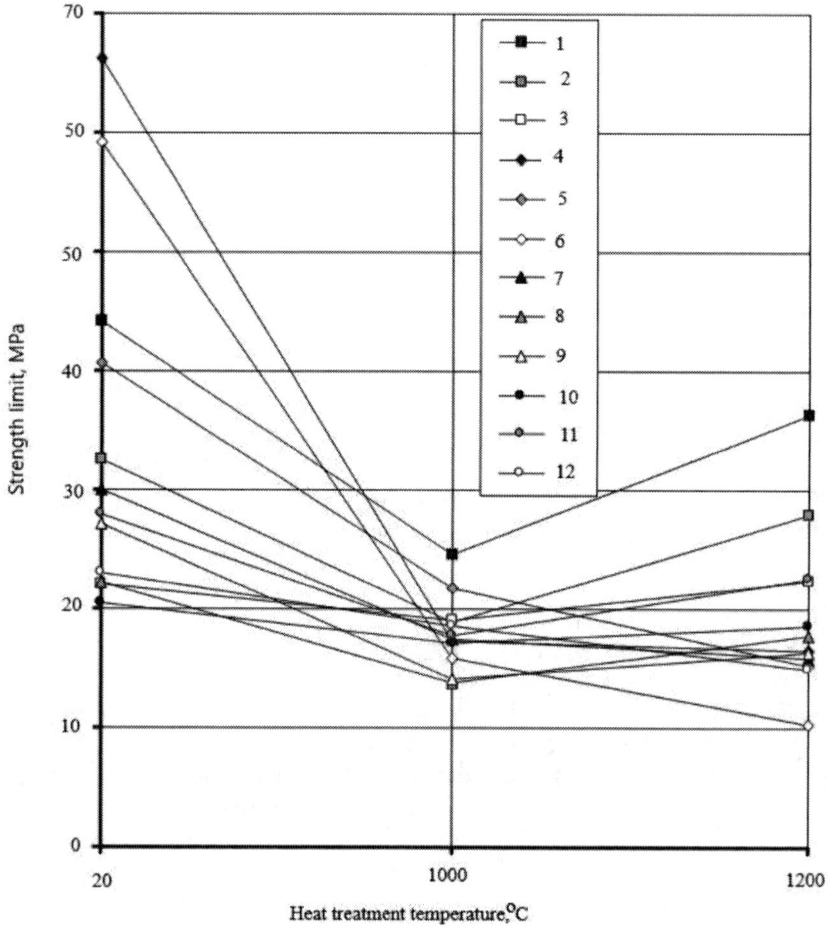

Figure 1 Dependence of the compressive strength of a composite heat-resistant binder (highly pure alumina cement + metakaolin) with magnesium additives-mineralizers on the temperature of heat treatment:

1, 2, 3 - addition of chemically pure magnesium oxide in the amount of 0.5; 1; 1.5 %;

4, 5, 6 - addition of caustic magnesite in the amount of 0.5; 1; 1.5 %;

7, 8, 9 - addition of the product of hydration of magnesium oxide at a temperature of 390-450 °C in the amount of 0.5; 1; 1.5 %;

10,11,12 - addition of chemically pure magnesium oxide, with exposure during heat treatment at a temperature of 390-450 °C for 1 hour, in an amount of 0.5; 1; 1.5 %.

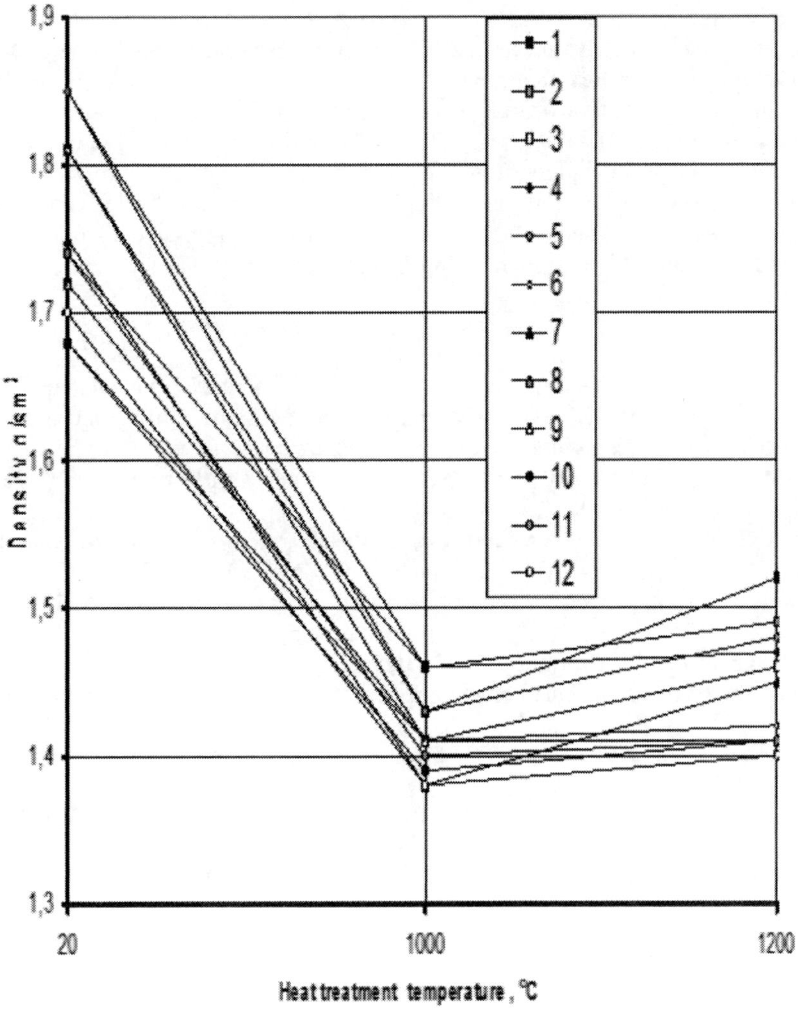

Figure 2 Dependence of the density of a composite heat-resistant binder (highly pure alumina cement + metakaolin) with magnesium additives - mineralizers on the temperature of heat treatment

1, 2, 3 - addition of chemically pure magnesium oxide in the amount of 0.5; 1; 1.5 %;

4, 5, 6 - addition of caustic magnesite in the amount of 0.5; 1; 1.5 %;

7, 8, 9 - addition of the product of hydration of magnesium oxide at a temperature of 390-450 °C in the amount of 0.5; 1; 1.5 %;

10,11,12 - an additive of chemically pure magnesium oxide, with exposure during heat treatment at a temperature of 390-450 °C for 1 hour, in an amount of 0.5; 1; 1.5 %.

IOP Conf. Series: Materials Science and Engineering 1015 (2021) 012069 doi:10.1088/1757-899X/1015/1/012069

4. Conclusions

Conducted experiments:
- confirmed the positive effect of magnesium additives on the process of hardening of cement stone, both during hardening and after heat treatment;
- confirmed the influence of the type of magnesium additive on the studied process;
- allowed to choose the most optimal type for different conditions of application of cement stone;
- allowed to determine the optimal amount of additives;
- the effect of almost complete return of the original activity when using "stale" cement was obtained.

The introduction of the developed compositions will significantly increase the strength of structures during operation. In conditions where such strength is not required it will lead to significant savings in expensive and scarce alumina cements.

References

[1] Standard GOST 969-2019 Alumina and high alumina cements. Technical specifications
[2] Kuznetsova T V and Talaber I 1988 *Alumina Cement* (Moscow: Stroyizdat) p 272
[3] Kuznetsova T V 1986 *Aluminate and sulphoaluminate cements* (Moscow: Stroyizdat) pp 38-40
[4] Pushkaryova E K, Glukhovsky V D, Krivenko P V and Chirkova VV 1983 *Fireproof binder* Patent № 998410 № 7 p 149
[5] Blizgareva T I 1992 *Heat-resistant fine-grained concrete on high-alumina cement with addition of refractory clay* (Dissertation Thesis Moscow)
[6] Strelov K K 1985 *Theoretical bases of refractory materials technology* (Moscow: Metallurgy) p 480
[7] Chumachenko N G and Tyurnikov V V 2019 Dynamics of destructive processes of cement stone on the basis of alumina cements at heat treatment *Refractories and technical ceramics* **1-2** pp 18-22
[8] Chumachenko N G and Tyurnikov V V Fireproof binder Patent (RF) № 2138456, MKI C 04 B 7/32 № 98100467/03
[9] Budnikov P P and Ginstling A M 1965 Reactions in mixtures of solid substances (Moscow: Stroyizdat) p 474
[10] Balkevich V L 1984 *Technical ceramics* (Moscow: Stroyizdat) p 138

IOP Conf. Series: Materials Science and Engineering 1015 (2021) 012070 doi:10.1088/1757-899X/1015/1/012070

Claydite dust - a unique technogenic raw material for heat-resistant concretes production

A Khlystov, D Isaev and V Suldin

Samara State Technical University, Molodogvardeyskaya st., 244, Samara, Russia, 443100

E-mail address: d.i.isaev@bk.ru

Abstract. The article presents various methods of heat-resistant materials synthesis on the basis of aluminosilicate wastes of expanded clay production. Researches have shown that the application of physical and chemical methods in the course of studying of expanded claydite dust composition, structure and properties makes it possible to define the reason of physical and thermal parameters increase of the heat-resistant binders prepared in its composition. The article investigates the influence of refractory mineral waste of petrochemistry - aluminochromium spent catalyst IM-2201 on the processes of increasing the strength and performance indicators of binders based on expanded clay dust. The possibility of the fire resistance increase of mixed binders synthesized on the basis of expanded claydite dust and aluminochromium waste has been established. Application of phosphate and liquid glass binders in the processes of heat-resistant binders synthesis on the basis of expanded claydite dust allowed to increase their performance.

1. Introduction

Industrial furnaces and other thermal units are used in many sectors of the national economy. These are metallurgical, chemical, oil refining, machine-building industries, enterprises producing firing building materials, as well as the heat and power complex. In most cases, the lining of these units uses piece ceramic refractories: fireclay, high-alumina, magnesite and chromomagnesite, corundum and zirconium. In the process of lining manufacture in the form of brickwork, the used refractory mortars do not correspond to the chemical composition of the basic lining material. Due to this reason the mortar is not sufficiently sintered. Therefore, as a result of thermal stresses due to sharp temperature changes, the destruction of refractories occurs at the point of contact. That is, the seams in brickwork are "narrow" place in any lining. The process of destruction starts here. That's why the recent use of heat-resistant concrete (in the form of large-format elements or blocks) in industrial furnaces has made it possible to minimize the seams and increase the lining resistance and durability.

The NIIZhBa (the head institute on concrete and reinforced concrete in Moscow) researches established the fact that for heat-resistant composites manufacturing portland cement, sodium liquid glass, silicate lump, alumina and high-alumina cements, periclase cement, phosphate cements, etc. can be used as binders. [1]

Previous studies at Samara State Technical University [2] have shown that industrial waste in the form of technogenic and nanotechnogenic raw materials is, for the most part, a valuable raw material that can be used to produce various materials useful for the economy, including heat-resistant concrete, mortars, torcret and refractory packaged masses.

IOP Conf. Series: Materials Science and Engineering 1015 (2021) 012070 doi:10.1088/1757-899X/1015/1/012070

2. Methods and materials

Mineral natural and artificial fired ceramic materials and industrial by-products can be used as fine refractory additives (chromite ore, broken fireclay, magnesite or ordinary bricks, andesite, pumice, granulated blast furnace slag, fuel slag, fly ash, spent chromium-alumina catalyst in the petrochemical industry IM-2201, etc.) According to the grain size composition, a requirement is imposed on fine-ground additives: the passage through sieve No. 008 should be at least 70-80%. [3-4]

At the expanded clay factories, a large amount of dusty wastes are formed with various dispersion, the chemical composition of which is close to the composition of either clay raw materials or burnt gravel of a similar product. But the efficiency of dust collecting devices operating at expanded clay plants is not perfect. Expanded clay wastes penetrate into all cracks of unsealed equipment and pollute the environment. So, when examining a number of expanded clay plants in Samara, operating on clay raw materials from the Smyshlyaevskoye deposit, the formation of four types of dusty waste was revealed. When using four furnaces, the total output of dusty waste exceeds more than one hundred cubic meters per day.

The harmful effect of aluminosilicate waste, which includes expanded clay dust, on the environment is associated with the content of aluminum, silicon, iron and their compounds, sulfates. This impact is very often beyond control and numerous factors indicate that it is increasing. This situation is associated with the development of old industries, the emergence of new ones, often without a reliable environmental forecast. [5]

Varieties of expanded clay wastes and their chemical compositions are shown in table 1. The table also shows data on the hydraulic activity of various varieties of expanded clay dust in relation to CaO. The rest of the physicochemical indicators are presented in table 2.

The high hydraulic activity of expanded clay dust taken under the junction of a two-drum furnace (115 mg CaO per 1 g of expanded clay dust) suggests its use in heat-resistant compositions using Portland cement. Figure 1 shows an X-ray diffraction pattern of a sample of expanded clay dust formed at the joint in a double-drum furnace, which has a high hydraulic activity in relation to CaO. X-ray studies were carried out using a modern diffractometer "ARLX'TRA".

The main crystalline phases in this sample of expanded clay dust are: quartz, aluminum oxide, hematite, cristobalite.

Table 1. Chemical compositions of clay production waste (Samara).

№	Mass content, (%)										Activity in mg relative to CaO
	SiO_2	Al_2O_3	Fe_2O_3	FeO	CaO	MgO	R_2O	TiO_2	SO_3	loss on ignition	
1	60.96	17.54	4.10	0.64	2.38	2.2	2.98	0.66	0.07	10.16	75
2	64.56	14.31	5.49	0.72	2.81	2.19	3.01	0.68	0.22	6.01	115
3	53.94	14.96	7.44	1.34	2.96	3.70	4.32	0.40	0.37	10.56	88
4	64.72	13.81	5.05	1.11	2.66	2.27	4.11	0.63	0.43	5.16	84,5

Note: No. 1-dust from cyclones of a two-drum furnace;
No. 2-dust formed at the joint in a two-drum furnace;
No. 3-dust from the dust-collecting chamber of a two-drum furnace;
No. 4-dust from the single-drum furnace's dust-collecting chamber;

Table 2. Physical and mechanical properties of expanded clay dust (1-4).

Indicator	Value
Bulk density, (kg / m^3)	750÷770
Fineness of grinding (pass h / c 0.14 %)	65÷70
Fire resistance, (°C)	1230÷1250

The nature of the *x*-ray shows the predominance of the amorphous component before the crystal phase. This circumstance explains the high hydraulic activity of expanded clay dust. One of the rational applications of expanded clay dust is its participation in the processes of synthesizing a heat-resistant binder based on Portland cement.

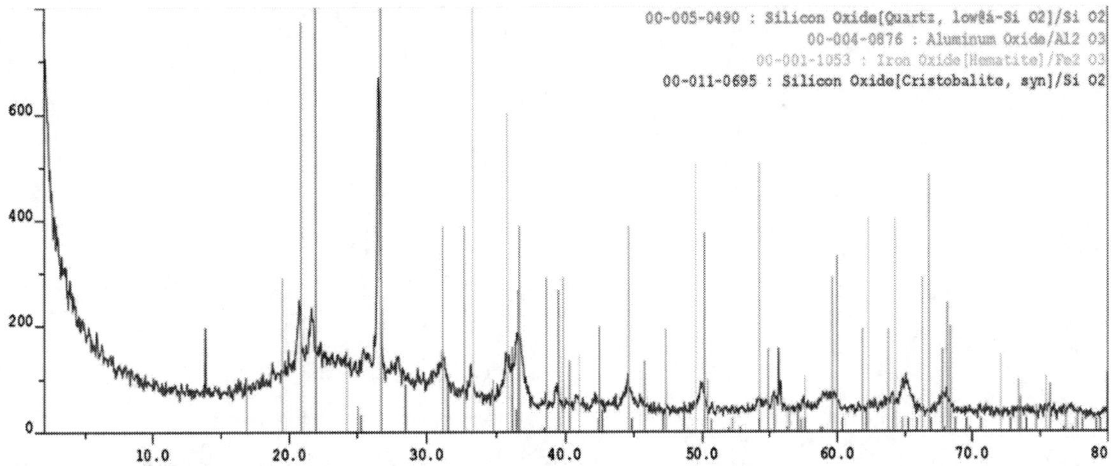

Figure 1. Radiograph of a sample of expanded clay dust.

During the research, the following materials were used to obtain binders: alumochromic waste - spent catalyst IM-2201 of the NovoKuibyshev petrochemical plant (Samara region), expanded clay dust - spillage from the two-drum furnace of the Samara expanded clay plant. The fillers of these concretes were class a firecracker bricks: firecracker crushed stone and firecracker sand obtained by crushing refractory scrap from a clay plant. The selection of concrete compositions was carried out experimentally, based on the conditions for obtaining mixtures with the maximum density.

3. Results and discussion

The analysis of physical and mechanical parameters of heat-resistant binders and concretes shows that it is possible to synthesize refractory composites with application temperatures up to 1100-1200 ° C on the basis of clay production waste. Such heat-resistant binders can be obtained not only by using hydraulic cements, but also by using sodium liquid glass, silicate blocks and phosphate binders.

However, the use of phosphate binders for the synthesis of the binder allowed increasing the fire resistance of the heat-resistant composition based on expanded clay dust to 1300-1350 ° C.

Orthophosphoric acid and alumochromophosphate binder were taken as phosphate binders. Metal phosphates are known to have high melting points [6]. Therefore, the course of reactions of the "oxide-phosphate binder" type allowed a number of low-melting substances in clay dust to be converted into refractory ones.

It is also possible to increase physical and thermal characteristics of heat-resistant composites (binders and concretes) based on expanded clay dust by synthesizing fine-ground refractory components of binders with the addition of alumochromium waste (spent petrochemicals catalyst IM-2201) and using liquid-glass or phosphate binders. Figure 2 shows graphical curves of changes in the fire resistance of mixed binders based on clay dust and alumina waste using sodium liquid glass. Analysis of curves shows that such an important technological parameter of heat-resistant binders as refractoriness can be easily regulated using high-temperature fine-ground additive in the form of alumochromium waste. The setting time of binders remains normal and, depending on the applied hardeners of liquid glass (sodium silicofluoride, alumina cement, the following values: the beginning - 45 minutes, the end of setting - 1 hour - 25 minutes). These parameters are provided by hardeners of liquid glass binders: sodium silicofluoride, introduced in an amount of 10% of the mass of liquid glass and alumina cement the content of which is one third of the mass of the fine-ground component.

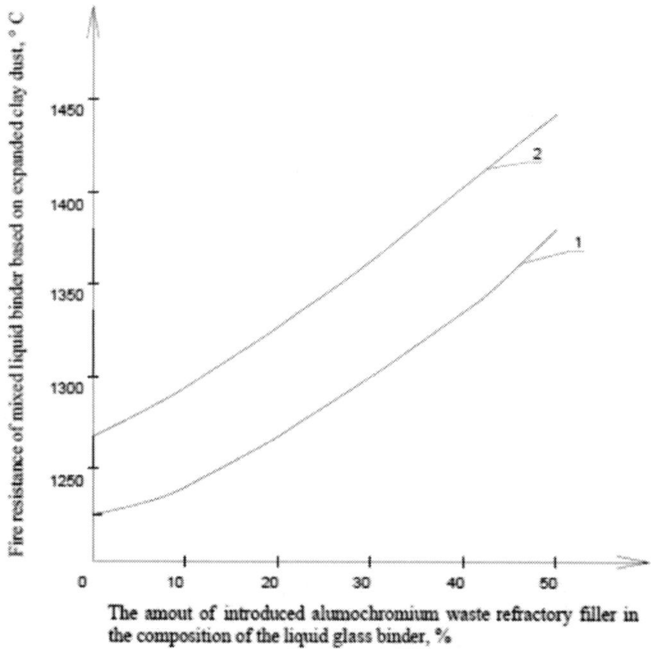

Figure 2. Fire resistance changes of mixed liquid-glass binders based on expanded clay dust:
1 – hardener - sodium silicofluoride,
2 – hardener - alumina cement

The compositions of heat-resistant concretes based on composite liquid-glass binders and their physical and thermal parameters are presented in table 3.

Table 3. Physical and thermal properties of heat-resistant concretes on mixed liquid-glass binders with the use of expanded clay dust.

№	Concrete composition, (kg/m^3)	Average density in dry state, (kg/m^3)	The limit of compressive strength MPa after heating, (°C)				Thermal stability, water heat transfer
			20 (3 days)	100	800	1200	
1	Expanded clay dust-180 Alumochromium waste - 260 Refractory rubble - 650 Fireclay sand-750 Liquid glass p=1.34 g/cm^3-360 Sodium silicofluoride-36	2045	18.6	28.6	26.5	25.9	22
2	Expanded clay dust-160 Alumochromium waste - 170 Alumina cement-110 Refractory rubble - 650 Fireclay sand-750 Liquid glass p=1.34 g/cm^3-380	2085	20.8	30.2	30.1	31.9 (1350 °C)	25

The tests results of mixed heat-resistant binders with the use of expanded clay dust and liquid glass (figure 2; table 3) showed that the replacement of sodium silicofluoride as a fusible substance with such a high-temperature compound as alumina cement significantly increased the initial strength indicators and performance. The fire resistance of the mixed heat-resistant binders exceeded 1450 °C, which makes it possible to obtain concretes based on them with an application temperature of 1300-1350 °C.

Figure 3 shows the curves of changes in the fire resistance of mixed binders based on expanded clay dust and alumochromium waste with the use of phosphate binders.

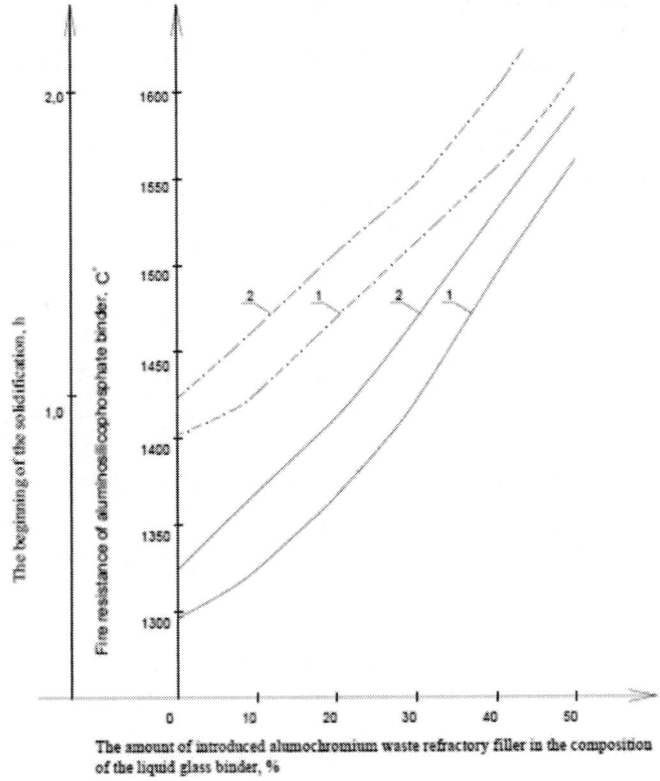

Figure 3. Changes in the refractoriness and binding properties of mixed cements of phosphate hardening: 1-stopper-H_3PO_4-70% concentration; 2-recluse-alumochromophosphate binder - p=1.52 g/cm^3; ———— - curves that characterize the fire resistance of the binder; — · — · — - curves that characterize the setting time of the binder.

Analysis of curves shows that the use of phosphate binders (H_3PO_4; alumochromophosphate binder) as a mixer of binders based on expanded clay dust allows to increase their fire resistance in a wide range, which is an important factor in the operation of these compositions. The setting time of phosphate-cured binders is slightly longer in comparison with liquid-glass binders, but still remains acceptable for the lining technology.

The heat-resistant concrete on phosphate composite binder with the use of claydite dust is presented in table 4. As shown by the test results of blended binders with the use of claydite dust replacement of liquid glass on phosphate mixers (H_3PO_4; alumochromophosphate binder) has improved performance of heat-resistant concrete and increased the operating temperature to 1400 °C.

Due to the fact that concrete mixtures based on orthophosphoric acid (H_3PO_4) slowly hardened in air and required heat treatment for acceleration, we developed a composition of a refractory filling mass. Expanded clay dust, as shown by experiments in its composition, contains a clay component, which ensures the plasticity of the filling mass. The samples of a given mass on orthophosphoric acid, which can be immediately removed from the mold, were made by filling.

The composition of the filling mass was as follows:

expanded clay dust from the cyclone - 453 kg/m^3;
fireclay sand - 1301 kg/m^3;
orthophosphoric acid ρ = 1.52 g/ m^3 - 190 kg/m^3.

The strength indicators of this refractory filling mass, which were determined after high-temperature firing (1300 °C), fluctuated within 37 ÷ 40 MPa.

IOP Conf. Series: Materials Science and Engineering 1015 (2021) 012070 doi:10.1088/1757-899X/1015/1/012070

The heat-resistant binder obtained in this way based on expanded clay dust and phosphate mixer was the basis for obtaining repair filling masses used directly to extend the service life of the lining of thermal units at factories where these wastes are generated. [7]

Table 4. Physical and thermal properties of heat-resistant concrete on phosphate binders with the use of claydite dust.

№	Concrete composition, kg/m^3	Average density in dry state, kg/ m^3	The limit of compressive strength MPa after heating, (°C)				Thermal resistance, water heat changes
			20 (1 day)	400	800	1300	
1	Expanded clay dust - 180 Alumochromium waste - 260 Fireclay crushed stone - 650 Fireclay sand - 750 Phosphoric acid (70% concentration) - 360	2100	5,5	28,6	33,2	30,5	25
2	Expanded clay dust - 160 Alumochromium waste - 280 Fireclay crushed stone - 650 Fireclay sand - 750 AChFS (ρ = 1.52 g/cm^3) - 380	2120	7,5	31,7	35,4	34,2 (1400)	28

4. Conclusions

Thus, the use of such chemical binders as liquid glass and phosphate binders allowed to create a wide range of heat-resistant concretes with relatively high physical and thermal characteristics on the basis of clay production waste. The resulting increased performance characteristics of heat-resistant concretes allow to be used for refractory lining works in many thermal units, including in contact with aggressive media, as well as in enterprises with refractory inorganic waste.

References

[1] Nekrasov K D 1974 *Heat-resistant concretes* (Moscow: Stroyzdat) p 176
[2] Arbuzova T B and et al. 1993 *Stroy materials from the industrial wastes* (Samara) p 96
[3] Manual for erection of Heat Units from Heat-Resistant Concrete 1983 (Moscow: Stroyzdat) p 65
[4] Khlistov A I 2017 *Heat Resistant concretes on the basis of industrial waste of Samara region* Monograph (Samara: SamSTU) p 171
[5] Arbuzova, T B 1991 *Utilization of the alumina-containing precipitation of the industrial wastes* (Samara: Saratov University Publisher Samara Branch) p 136
[6] Kopeikin V A 1975 *Physicotechnical properties of phosphate materials* (Moscow: Stroyizdat) pp 4-11
[7] Khlistov A I 2018 Heat-resistant compositions on the basis of waste of the enterprises of ceramic industry *Urban planning and architecture* **8** (4) (Samara) pp 56-60.

IOP Conf. Series: Materials Science and Engineering 1015 (2021) 012071 doi:10.1088/1757-899X/1015/1/012071

Developing physical and mathematical models of the water-air mode of gravity drainage network

V Orlov and S Zotkin

Moscow State University of Civil Engineering (National Research University), 129337 Yaroslavskoe shosse 26, Moscow, Russia

E-mail address: zotkinsp@mgsu.ru

Abstract. A physical and mathematical model of the water-air mode of free-flow drainage pipelines has been developed. The solution to the problem of physical modelling of pipeline network functioning is based on the study of the dynamics of changes in the hydraulic and aerodynamic parameters of the system, taking into account the phenomena of heat and mass transfer between liquid and air in the vaulted space of a free-flow pipeline. An algorithm for solving the problem is compiled on the basis of the criteria of hydrodynamic similarity, describing the phenomena of mass transfer as a result of forced air convection by using a mechanical ventilation system above the water surface. Based on the results of applying a special computer-aided calculation program, indicators of the required air exchange rate for effective removal of evaporation (steam) from the vaulted space of a pipeline were determined. As a result of solving the complex problem of studying the water-air mode of free-flow pipeline functioning, the authors compared and analysed the calculated values of the intensity of air exchange produced to remove steam, and to reduce harmful foul-smelling gaseous substances to reach the values of maximum permissible concentrations. The research results are aimed at preventing emissions of gases harmful to human health into the atmospheric air by constructive methods.

1. Introduction

An objective assessment of free-flow drainage system pipelines functioning is impossible without figuring out its physical model, including hydrodynamic and aerodynamic indicators [1]. Physical modeling is the initial stage of solving a complex problem associated with a comprehensive assessment of the water-air mode of pipelines operation [2, 3]. The model is to describe such complex phenomena as intensive release of foul-smelling toxic gases into the vaulted space of free-flow networks, and also make a detailed analysis of heat and mass transfer between liquid and air in the free air space of the pipeline [4]. A specific feature of the sewerage network is that chemical, biochemical and other processes actively take place in pipelines that affect the environmental component during transport and wastewater treatment [5, 6]. It is very difficult to fully cover these phenomena when compiling a model, therefore, when solving problems to study the water-air mode of free-flow pipelines functioning, certain assumptions were made, such as, for example, the condition of equilibrium at the "air-water" phase interface [7. 8].

A correctly built physical model is the basis for a mathematical model and becomes one when describing processes and phenomena using algebraic dependencies [9]. The mathematical model provides a multilateral quantitative analysis by changing the initial data, criteria and restrictions, reflecting the dynamics of changes in real physical processes [10]. Thus, the algorithm for solving mathematical modeling problems is reduced to a consecutive description of the transformation of the corresponding technological indicators used in the physical model.

The general requirement for the developed models, as a rule, boils down to making them available to specialists and unambiguously interpreted by them when solving a specific problem of assessing the sanitary safety of a designed or operated drainage network [11].

2. Materials and methods

The object (material) of the study is free-flow drainage pipelines, and the research method is an automated search for the optimal mode of their operation. The ultimate goal of research is to determine technical measures and indicators that provide the required sanitary and hygienic conditions for humans and the surrounding urban environment by neutralizing (removing) harmful foul-smelling gases and water vapour from the vaulted space of free-flow pipelines [12, 13].

The figure 1 shows a simplified physical model of a drainage network with a limited design area, for example, located near crowded places (shopping centres, entertainment and sports complexes, etc.), where the most significant issues are the localization of gas formation, caused by the presence of foul-smelling substances harmful to human health.

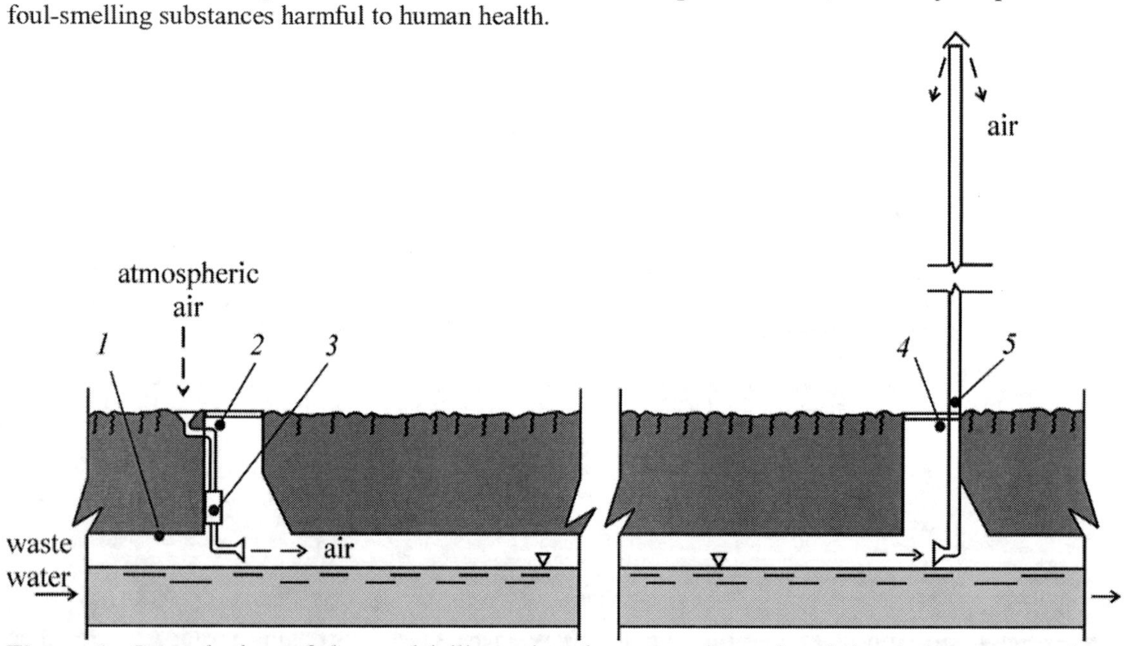

Figure 1. General view of the model illustrating the water-air mode of the drainage network operation: 1- free-flow pipeline, laid with a suitable slope; 2- initial sewer well; 3- autonomous ventilation unit with a diffuser oriented downstream; 4- final sewer well; 5-ventilation riser.

To solve the problem of physical modeling of the water-air mode of a free-flow pipeline network functioning, apart from studying the hydraulic and aerodynamic performance of the system, it is necessary to take into account the phenomena of mass transfer (gas-liquid reactions). It is very difficult to fully cover these phenomena when making up a model, therefore, when solving the problem at this stage of research, certain assumptions and simplifications were made, such as: the system is in equilibrium at the phase boundary (air-water). By this, it can be understood that the gas released from the surface of the water does not react with the transported waste water (there is no diffusion of gases, their adsorption and chemical reactions). But at the same time, the amount of air exchange should be determined to remove foul odors, as well as excess moisture entering the vaulted space of the pipeline as a result of wastewater evaporation. The elimination of these negative phenomena is one of the required tasks in the field of wastewater transport through free-flow

IOP Conf. Series: Materials Science and Engineering 1015 (2021) 012071 doi:10.1088/1757-899X/1015/1/012071

networks, since foul-smelling gases coming through the hatches of inspection wells reduce the people's life comfort and do damage to their health, and moisture in the vaulted pipelines, together with gas emissions, contributes to corrosion of pipe walls [14].

Input data for the formation of physical and mathematical models is based on such indicators as the diameter of the pipeline, its length, slope and filling. When coming up with models, the results of field measurements are presented, for example, the concentration of harmful gaseous substances in the vaulted space of the pipeline and the coefficient of aerodynamic (air) resistance. The input data also includes normative indicators, for example, the maximum permissible concentration of harmful gaseous substances in the atmosphere of the city. In addition, such input characteristics of atmospheric air as its temperature, kinematic viscosity coefficient and density are subject to processing.

The following target design indicators of the water-air regime must be analyzed and on that basis specialists make a decision on the efficiency of the drainage network: the calculated values of air exchange (I) and depression (II) of the air flow in the pipeline, the values of hydraulic and standard indicators (III). A specially developed computer program facilitated an operational analysis of the three presented design indicators [15].

I. As a basic indicator, assessed by a physical and then a mathematical model, the rate of air exchange in the free air space of a free-flow pipeline is regarded. When solving this problem at this stage of modeling, it is assumed that aerodynamic parameters are determined in conditions of still water in the pipeline with a wide range of fillings, limited by the relative hydraulic stability. In this case, the air exchange value is assigned based on the catalog characteristics of the fan installations used in industry. Ultimately, the value of the assigned air exchange is to be adjusted according to the specific results of calculating the required amount of air to remove foul-smelling gases and excess moisture, taking into account heat and mass shift between liquid and air in the sewer pipe, as well as the real time of neutralization of gaseous substances in the vaulted space.

The amount of air exchange QB (m^3/s) in a closed space to remove harmful foul-smelling gaseous impurities is determined by the formula (1)

$$QB = \frac{M}{C_{pdk} - C_0} \tag{1}$$

where M is the amount of foul-smelling gaseous substances released from water, mg/s; C_{pdk} stands for the maximum permissible concentration of the gas to be neutralized, (mg/m^3); C_0 corresponds to the concentration of the gas to be neutralized in the supply air, (mg/m^3).

To determine the values of air exchange in the vaulted space to remove excess moisture QBB (m^3/s) and the mass of water evaporating into the air WW (mg/s), it is necessary to simulate the process of heat and mass transfer between air and water in the vaulted part of the pipeline. In this case, it is necessary to start out from the fact that heat and mass transfer can be carried out in two ways, determined by the specific conditions of heat transfer, i.e. natural or forced convection.

Due to the fact that the developed model considers heat transfer as a result of forced convection (based on intensive ventilation), the main hydrodynamic criteria of the model are the Reynolds's criterion Re (2), the Prandtl's diffusion criterion Pr (3), the Nusselt's criterion Nu (4)

$$Re = \frac{V_b S^{0,5}}{\nu} \tag{2}$$

$$\Pr = \frac{\nu}{D} \tag{3}$$

$$Nu = 0{,}008 \cdot Re^{0,9} \cdot Pr^{0,43} \tag{4}$$

where V_b is the air flow rate, (m/s); S is the surface of heat and mass transfer (i.e. the area of the water surface in the pipeline with the corresponding filling), (m^2); v is the coefficient of kinematic air viscosity, (m^2/s); D is diffusion coefficient, (m^2/s)

To determine the diffusion coefficient D in the "air-steam" system at an appropriate temperature t and pressure H, it's possible to use the formula (5)

$$D = D_0 \left(\frac{t+T}{273} \right)^{1,89} \cdot \frac{760}{H} \quad (5)$$

where D_0 is the value of the diffusion coefficient (at air temperature t=0°C and pressure 760 mm Hg); T is the absolute air temperature in underwater space, K; H is pressure in vaulted space, (mm Hg); t corresponds to pressurized air temperature, 0°C.

According to Dalton's law, the mass of water evaporating into the air WW (mg/s) can be determined by the formula (6)

$$WW = 0,2778 \cdot \beta \cdot S \cdot (p_{nas} - p_{par}) \quad (6)$$

where 0.2778 is a conversion factor from (kg/h) to (mg/s); β is the coefficient of moisture exchange in the vaulted space, referred to the difference in pressure of steam (mg/m^{2}sPa); p_{nas} is pressure of steam above the water surface at the corresponding air temperature in the vaulted space, (Pa); p_{par} means partial steam pressures in the vaulted space, (Pa).

To determine the coefficient β, let's use the similarity equation (7)

$$\beta = Nu \cdot \frac{D}{S^{0.5}} \quad (7)$$

Knowing the values of the β coefficient and calculating the WW value, it is possible to determine the amount of air exchange to remove excess moisture QBB (m^3/s) in a closed space according to the formula (8)

$$QBB = \frac{WW}{d_y - d_n} \cdot \rho \quad (8)$$

where ρ is the density of the supplied air, (mg/m^3); d_y–d_n is the difference between the moisture content of the removed and supplied air, (g/kg) of dry air; d_y is the moisture content of the removed air, (g/kg) of dry air; d_n is the moisture content of the supplied air, (g/kg) of dry air.

The average air flow rate, taking into account its entrainment by the water flow VCR (m/s), can be determined by the formula (9)

$$VCR = (V^2 + V_b^2)^{0.5} \quad (9)$$

where V is the flow rate of waste water in the pipeline, (m/s).

Next, the real time TR (s) for the removal of gaseous substances and moisture is calculated in a section of length l between two wells when the rate of the air flows entrained by the water flow changes according to the formula (10)

$$TR = \frac{l}{VCR} \quad (10)$$

II. The second basic indicator for assessing the water-air regime is the value of the depression of the air flow in the vaulted space of the free-flow pipeline when setting up a pressurized ventilation system. To solve the presented problem, the depression can be considered as the pressure of the air medium in the calculated section of the pipeline between the wells (see figure 1). The discharge of pressurized air from the system is carried out at the end of the calculated section both directly onto the surface relief through an opening in the well hatch (first case) and through a ventilation riser of the corresponding height and diameter, which is set near the well hatch (second case).

The value of the depression *H1* (Pa) in the vaulted space of the pipeline is calculated by the formula (11)

$$HI = \frac{KAS \cdot P_{nn} \cdot l}{\omega_{nn}^3} \cdot QB^2 \quad (11)$$

where *KAS* is the coefficient of the aerodynamic resistance (N s^2/m^4), considered equal to 0.3-0.44 for ventilated spaces without obstacles in the path of the air flow; P_{nn} is the perimeter of the vaulted air space (m^2), determined by the formula (12)

$$P_{nn} = \pi d - \frac{\omega}{R} + a \quad (12)$$

where *d* is the diameter of the pipeline, (m); ω is the area of the free flow section in the shape of a segment bounded by a chord and the length of the wetted part of the circle, (m^2); *R* is the hydraulic radius acting as a function of filling, m; *a* is the chord length of the segment acting as a function of filling, (m); ω_{nn} is the cross-sectional area of vaulted air space, (m^2), determined by the formula (13)

$$\omega_{nn} = \frac{\pi d^2}{4} - \omega \quad (13)$$

The algorithm of the automated program includes the conversion of the depression value from Pa to MPa or m of water column.

III. The third calculated indicators of the water-air mode of the sewer network functioning include hydraulic and standard parameters: water flow rate *V* (m/s), air rate in the vaulted space V_b (m/s), water spending rate *Q* (m^3/s), air volume in the vaulted space of the pipe *WB* (m^3), the amount of gaseous substance *M* (mg/s) emitted per unit time, which is subject to subsequent neutralization to the maximum permissible value when ventilating the vaulted space, etc.

Below are the results of an automated calculation of an exemplary water-air mode of operation of a free-flow pipeline and the interpretation of the data obtained with regard to the required rate of air exchange to remove foul-smelling gases and excess moisture in the calculated section of the pipeline (see figure 1).

3. Results and discussions

The initial data on the object of modeling in a brief version is presented by means of the following digital indicators: the length of the network section is 500 m, the pipeline diameter is 0.4 m, the slope of the route is 0.0025, the assigned air exchange in the pipeline system is 0.00277 m^3/s, the concentration of harmful gaseous substances (for example, hydrogen sulfide), respectively, in the vaulted space 4.2 mg/m^3 (according to field measurements with instruments) and the maximum permissible concentration in the city environment is 0.008 mg/m^3 [16].

Taking into account the unevenness of water consumption during the day, and thus the different mode of heat and mass transfer, the simulation was carried out for three cases of filling the pipeline with water: 0.3 (minimum), 0.5 (calculated) and 0.7 (maximum) [17, 18].

The analysis of the results of the automated calculation was carried out on the basis of the most significant hydraulic and aerodynamic indicators, according to which there is an estimate of the sufficiency of the adopted air exchange for removing harmful gases from the pipeline in the vaulted space and removing excess moisture [19, 20].

The summary data on the results of modeling, taking into account the neutralization of the released hydrogen sulfide reaching $2.22 \cdot 10^{-5}$mg/s to its maximum permissible value in the atmospheric air of 0.008 mg/m^3 are shown in table 1.

IOP Conf. Series: Materials Science and Engineering 1015 (2021) 012071 doi:10.1088/1757-899X/1015/1/012071

Table 1. Summary results of automated calculation for various fillings *h/d* and assigned air exchange 0.00277 m^3/s.

Name of the calculated indicator	*h/d=0.3*	*h/d=0.5*	*h/d=0.7*
Waste water flow rate in the pipeline, m/s	0.678	0.786	0.849
Air velocity in the vaulted space of the pipeline in case of still water, m/s	0.0302	0.0436	0.0781
Air velocity in the vaulted space of the pipeline, in case of its entrainment by the water flow, m/s	0.679	0.787	0.853
Air volume in the vaulted space of the pipe, m^3	45.82	31.78	17.74
Total mass of gaseous substances entering the vaulted space of the pipeline from the water surface, mg	192.45	133.48	74.52
Duration of removal of gaseous substances from the vaulted space of the pipeline with the adopted air exchange in stagnant water conditions, *s* (h)	16542.2 (4.6)	11473.6 (3.2)	6405.0 (1.8)
The real time of moisture removal when changing the air flow rate, in case of its entrainment by the water flow, s (h)	736.4 (0.204)	635.1 (0.181)	586.1 (0.163)
Total depression of the air environment in case of its surpassing the atmospheric pressure, MPa (m water column)	0.09801 (10.001)	0.09802 (10.002)	0.09804 (10.004)

When the filling changes from 0.3 to 0.7, an increase in the air rate from 0.679 to 0.853 m/s (25.6%) is observed. At the same time, there is a sharp decrease in the total mass of gaseous substances entering the vaulted space of the pipeline from 192.45 to 74.52 mg due to a decrease in its volume from 45.82 to 17.74 m^3, which for both indicators is 30.72%.

For the same values of the range of fillings and the virtual condition of stagnant water in the pipeline (i.e., without taking into account the flow rate) and air rates in the range of 0.0302-0.853 m/s, as well as the duration of removal of gaseous substances from the vaulted space of the pipeline by ventilation will decrease from 4.6 to 1.8 h, i.e. by 39.1%.

In real conditions, the time for removing moisture from the vaulted space when changing the air flow rate, in case of being entrained by the water flow, will decrease from 0.204 to 0.163 h (21.1%). This indicates that moisture from the vaulted space will be removed from the sewer network faster than harmful gaseous substances, and in the range from 4.6/0.204 = 22.54 to 1.8 / 0.163 = 11.04 times, depending on the filling of the pipeline.

Regarding the air pressure (depression value) for the selection of the ventilation unit, it should be noted that the depression practically does not undergo changes due to low aerodynamic resistances in the vaulted space of the pipeline.

Thus, the air exchange value of 0.00277 m^3/s adopted in modeling ensures the removal of both hydrogen sulfide and moisture from the vaulted space of the pipeline (including the riser) to its maximum permissible value in the atmospheric air of 0.008 mg/m^3. However, the optimal operation of the ventilation unit will be secured by a relatively equal rate (closeness of values) for moisture and hydrogen sulfide removal. Since the time differences in the given example are significant, the modeling of the water-air mode of the pipeline functioning was carried out with a larger value of the assigned air exchange with automatically tracking the moment of equality of the two indicators. By using an automated program, calculated data were obtained with a new air exchange rate of 0.04 m^3/s.

To display the results of the automated calculation, figure 2 contains graphs characterizing the change in the range of time values for neutralizing foul odors in accordance with the formula $TZ=125.5(h/d)+603.18$ and for removing moisture from the pipeline by using the formula $TB=132.31(h/d)+124.86$.

In this case, the real time of moisture removal when changing the air flow rate, in case of being entrained by the water flow, does not change.

If we compare the results of time calculation with the previous air exchange (0.00277 m^3/s), when the results ranged from 22.54 to 11.04 times, then with the new value of air exchange the difference ranged from 0.3/0.1722 = 1.74 up to 0.1 / 0.0984 = 1.02. For h/d=0.8 the time values will practically coincide.

Thus, the value of air exchange selected by calculation will make it possible to successfully solve the problem of removing foul-smelling gases and moisture within a minimum period of time.

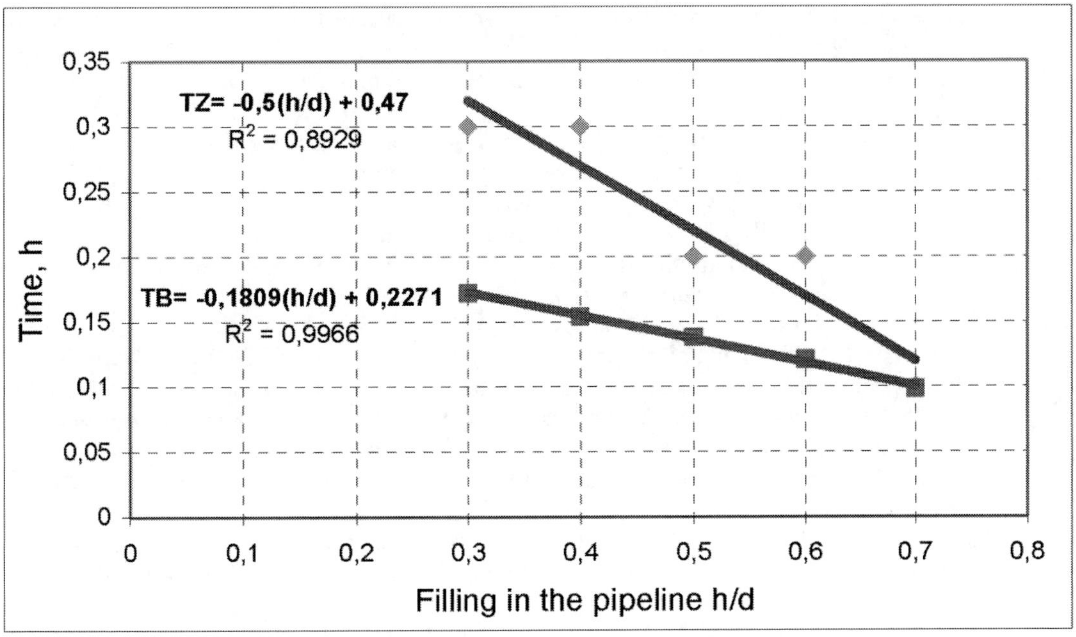

Figure 2. Graphs of the time dependence of the removal of odorous substances (the upper curve) and moisture (the lower curve) at the accepted air exchange.

4. Conclusions

1. Developing physical and mathematical models for the study of the water-air mode of free-flow pipelines functioning, as well as automated calculation, allow to quickly track the dynamics of changes in the parameters of the pipeline system in conditions of intensive release of gases harmful to human health.

2. In a specific example, taking into account the unevenness of water consumption through the day, the possibilities of automated calculation for determining the ranges of hydraulic and aerodynamic parameters of the pipeline system are considered.

3. As the basic parameters for modeling the water-air mode of non-pressurized pipeline functioning, the air exchange assigned by the user, as well as the duration of removing foul-smelling substances (for example, hydrogen sulfide) and moisture from the vaulted space of the pipeline are taken.

4. With the help of an automated complex, it is possible to adjust the amount of air exchange depending on technical circumstances and to minimize the cost indicators of the reconstruction and operation of pipeline networks equipped with ventilation units.

References

[1] Primin O G and Pupyrev E I 2013 Methods for improving the environmental safety of sewer pipelines *Ecology and Industry in Russia* **3** pp 13-17

[2] Fedorov S V, Vasiliev V M and Telyatnikova A M 2018 Development of a fundamental model of the sewer network *Journal of Bulletin of Civil Engineers* **2** (67) pp 168-174

[3] Chupin R V and Nguyen Tuan Anh 2015 Optimal reconstruction of sewerage networks *J. of Water*

Supply and Sanitary Engineering **2** pp 58-68
[4] Parker W J and Ryan H A 2011 Tracer study of headspace ventilation in a collector sewer *J. of the Air & Waste Management Association* **12** pp 581-592
[5] Gostelow P and Parsons S 2000 Sewage treatment works odourmeasurement *Water Sci. Technol* **41** (6) pp 33-40
[6] Kofman V Ya 2012 Hydrogen sulfide and methane in sewer networks *Water supply and sanitary engineering* **11** pp 72–78
[7] Stuetz R and Frechen F B 2001 *Odors in wastewater treatment* (London: Published by IWA Publishing SW1H QS) pp 436
[8] Pochwat K, Kida M and Ziembowicz S Odors in sewerage - a description of emissions and of technical abatement measures *Environments* **6** (89) pp 1-13
[9] Vasiliev V M and Malkov A V 2017Places of formation of aggressive gases in the sewer network *J. of Water Supply and Sanitary Engineering* **1** pp 66-74
[10] Malkov AV 2017 Calculation of the required gas exchange rate in the underwater space of the sewer networ k *Bulletin of civil engineers* **2** (61) pp 140-144
[11] Vasiliev V M, Morozov G V and Zhukov S V 2018 Problems of operation of sewerage networks and ways to solve them *Journal of Water Supply and Sanitary Engineering* **7** pp 44-50
[12] Rublevskaya O N 2013 Measures to prevent the spread of unpleasant odors at the facilities of the State Unitary Enterprise "Vodokanal of St. Petersburg" *Water supply and sanitary engineering* **10** pp 46 – 55 (in rus.)
[13] Wysocka I, Gębicki J and Namieśnik J 2019 Technologies for deodorization of malodorous gases *Environ. Sci. Pollut. R.* **26** p 9409–9434
[14] Vasiliev V M, Pankova G A and Stolbikhin Yu V 2013 Destruction of sewer tunnels and structures on them caused by microbiological corrosion *Water supply and sanitary engineering* **9** pp 55 – 61 (in rus.)
[15] Orlov V A, Zotkin S P, Storozhev A P, Gerasimov V A and Melnik O V 2020 Modeling the water-air mode of free-flow drainage networks functioning *Certificate of state registration of a computer program* 2020614973
[16] Standard GOST 32673-2014. Rules for the establishment of standards and control of emissions of foul-smelling substances into the atmosphere
[17] Vasilyev V and Stolbikhin Y 2015 Inspecting and monitoring the technical condition of sewage collectors *Trans Tech Publications Switzerland, Applied Mechanics and Materials* **725-726**, pp 1319-24
[18] Joyce J, Hunniford Ch and Plummer A 2013 Implementing vapor phase odor control on large diameter interceptor systems *Biosolids and Odor and Corrosion, Conference & Expo* pp 1-31
[19] Rodionov A I, Kuznetsov Yu P, Soloviev G S 2005 *Protection of the biosphere from industrial emissions. Basics of designing technological processes.* (Moscow: Chemistry) p 392
[20] Yushin V V, Popov V M and Kukin P P 2005 *Technique and technology for air protection* (Moscow: Higher school) p 391

IOP Conf. Series: Materials Science and Engineering 1015 (2021) 012072 doi:10.1088/1757-899X/1015/1/012072

Influence of mineral aggregate on asphalt pavement surface properties

E Remišová, N Holešová and M Brna

The University of Žilina, Department of Civil Engineering, Slovak Republic

E-mail address: eva.remisova@uniza.sk

Abstract. The road surface roughness is one of the factors influencing road safety. Therefore, the choice of materials and mixtures for the wearing course of pavement is important. The paper presents the results of research on the influence of the composition of asphalt mixtures on the surface parameters of PTV and MTD, which were analyzed according to the type of mixture, type of aggregate, and maximum aggregate size on laboratory-produced samples and samples with bitumen stripped surfaces. The skid resistance of SMA 11 and AC 16 mixtures were higher (PTV value 64.2 and 66.2) compared to AC11 and AC8 mixtures (PTV value 56.8 and 58.6) with all tested aggregates. After bitumen stripping from the surface, the PTV values of mixtures increased and the influence of the aggregate type and texture became more significant. To achieve better skid resistance properties, it is suitable to use coarse aggregate from andesite or granodiorite rocks (PTV value 67.0 or 65.8). Of the tested aggregates, limestone had the worst results (PTV value 63.3). Above all the portion of coarse aggregates, size distribution, and shape characteristics as shape index, angularity, determine the macrotexture of the surface.

1. Introduction

The extent to which a road can ensure functional requirements (safe pavement serviceability and pavement efficiency) depends on many factors, in particular its design and construction, environmental conditions, but also the amount and type of traffic load that uses the road [1, 2]. When assessing the quality of the road surface, the most important is the safety of road traffic, which is determined by the surface roughness by the macrotexture and microtexture of the surface. Insufficient road surface roughness can lead to skidding of the vehicle wheels as a common cause of an accident [3]. Many detailed analyses of accident sections performed on the territory of the Slovak Republic show that the reduction of roughness on a wet surface has a share of 25 % in a traffic accident.

Friction between the pavement and the wheel is influenced by a wide range of factors such as tyre parameters (pressure, tread pattern and depth), vehicle parameters (type, weight, speed, braking system), environmental characteristics (temperature, season, presence of water, surface contamination), the road horizontal and longitudinal alignment, and last but not least the surface texture and frictional characteristics (type and parameters of the mixture in the wearing course). The texture of the pavement surface, the type and parameters of the tyre, the presence of water on the road, and the speed of the vehicle have a major influence on the friction between the wheel and the pavement.

The aggregate size distribution and the void in the mineral aggregate in the wearing course of the pavement form the macrotexture of the surface. The high portion of bitumen binder and fine aggregate in asphalt mixture does not form the necessary surface texture. The critical values of air void content are about 2% and about 90% of the void filled with bitumen binder. Aggregates form the supporting skeleton of each building mixture. The maximum size and the particle size distribution of aggregate in the mixture are decisive for the macrotexture of the surface. The rock type of aggregate forms the microtexture of the pavement surface and the ability to maintain an aggregate surface texture during the effects of climatic conditions and traffic load. Therefore, the aim of the paper is determination the influence of the type of aggregate with the different minerals and different surface texture on the surface parameters of the asphalt mixture and the maximum aggregate size in the mixture and the type of mixture on the surface texture parameters.

2. Research methodology

The study of the influence of different mineral aggregate on asphalt surface properties conducted in laboratory-produced asphalt mixtures used in practice in the Slovak Republic. The asphalt concrete AC and stone mix asphalt SMA mixtures with maximum aggregate size 8 mm, 11 mm, and 16 mm were used. Figure 1 shows the particle size distribution of the aggregate mixture of tested asphalt mixtures AC 8, AC 11, AC 16, and SMA 11. Each mixture was produced with coarse and fine aggregates, limestone filler, and paving grade bitumen 50/70. The composition of asphalt mixtures, mass by weight in percentage, see in table 2. The used coarse aggregate, made from rocks from three different sources in terms of mineralogical composition content, were andesite (A), limestone (V), and granodiorite (G) rocks. The main physical characteristics of aggregate rock are in table 1. It was assumed that these aggregates would differ in micro and macro surface texture. The limestone fine aggregate used in all mixtures was the same. The mixtures SMA 11 additionally contained 0.3 % of natural cellulose fibers additive Viatop.

Table 1. Physical characteristics of aggregate rock [4].

Property	Aggregate		
	Andesite	Limestone	Granodiorite
Density (kg.m^{-3})	2 701	2 715	2 666
Particle bulk density (kg.m^{-3})	2 565	2 703.9	2 665
Porosity (%)	5.04	0.39	0.17
Water absorption (%)	0.94	0.11	0.22

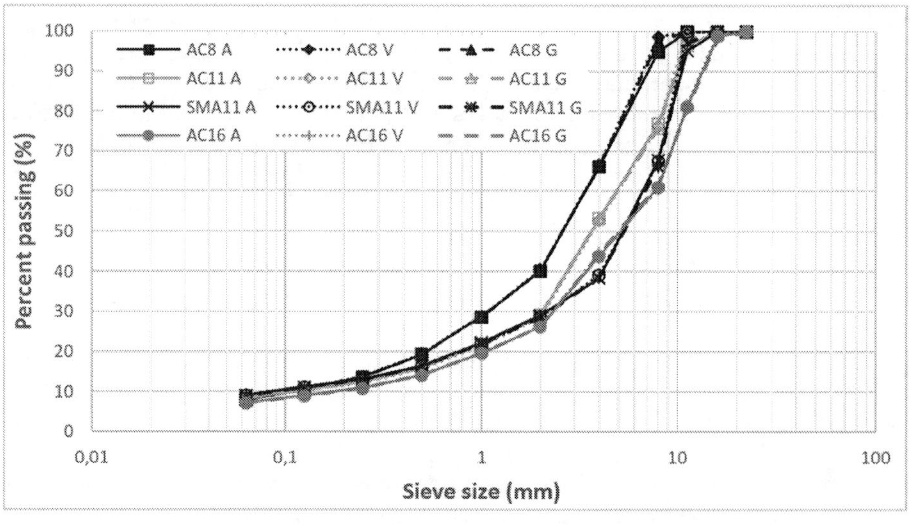

Figure 1. The particle size distribution of asphalt mixtures.

Asphalt mixtures have been designed in accordance with the requirements of European standards EN 13108-1 and EN 13108-5 and to meet the requirements specified in the national regulation KLAZ [7] (grading, bitumen content, air voids content, water sensitivity, resistance to permanent deformation). The decisive factor of the design was the grading of the aggregate mixture. The aim was to design the individual mixtures with different aggregate rocks that differed from each other as little as possible (figure 1). The proposed mixtures were produced in the laboratory of authors.

Table 2. Composition of asphalt mixtures.

Mixture	Type of coarse aggregate	Content in a mixture (%)						
		Aggregate						Bitumen
		Filler	0/2 (mm)	2/4 (mm)	4/8 (mm)	8/11 (mm)	8/16 (mm)	50/70
AC8	Andesite	1.9	35.9	26.4	30.2	-	-	5.6
	Limestone	1.9	35.9	24.5	32.1	-	-	5.6
	Granodiorite	1.9	35.9	25.5	31.1	-	-	5.6
AC11	Andesite	4.7	22.7	24.6	23.7	18.9	-	5.4
	Limestone	4.7	21.8	23.6	21.8	22.7	-	5.4
	Granodiorite	4.7	22.7	23.7	19.9	23.6	-	5.4
SMA11	Andesite	5.6	22.7	7.3	31.8	26.2	-	6.4
	Limestone	5.6	21.3	7.7	28.1	30.9	-	6.4
	Granodiorite	5.6	21.5	7.5	26.2	32.8	-	6.4
AC16	Andesite	3.8	20.8	18.0	17.1	-	35.0	5.3
	Limestone	3.8	20.8	18.0	14.2	-	37.9	5.3
	Granodiorite	3.8	20.8	18.0	15.2	14.2	22.7	5.3

The tests of asphalt texture parameters were performed on a specimen prepared by roller compactor with dimensions 300×400×40 mm. The measurements of skid resistance of a surface (PTV value) using the pendulum (according to EN 13036-4) and the measurements of texture depth (MTD value) using a volumetric patch technique (according to EN 13036-1) were used to determine asphalt surface texture. The pendulum test measures the skid resistance of a small area of 0.01 m^2. The representative value of the surface was determined as a mean of four individual PTV measurements (four determinations and areas on the surface) separately for the dry surface and wet surface. The outlier values of determination were not included in the result.

In practice, after the construction of the asphalt pavement surface, vehicle moving and climatic effects remove the bitumen film from the aggregate surface. Then the vehicle tire is in contact with the aggregate surface texture. After initial texture measurements, the bitumen film was removed from aggregate on the sample surface using a solvent. The determination of PTV values of all surfaces was performed under the same conditions.

3. Test results and discussion

The results of the PTV parameter (figure 2, left) showed differences between the single mixtures as well as between the coarse aggregate rocks. In terms of the type of mixture and the maximum aggregate size in the mixture, the highest values were obtained on the surfaces of mixtures of type SMA11 and AC16, the lowest values on the mixture AC11, which were comparable with the values of the mixture AC8.

It was assumed that mixtures with a larger contact area (AC8 and AC11) would have higher result values. However, it showed that the PTV value is influenced not only by the size of the contact area but also by the inter distribution of aggregate particles and the shape of particles, there may be a predominant "hysteresis" component. The pendulum impacts on larger protrusions on a surface (larger aggregate particle) when catching and braking than on samples with smaller aggregate particle size.

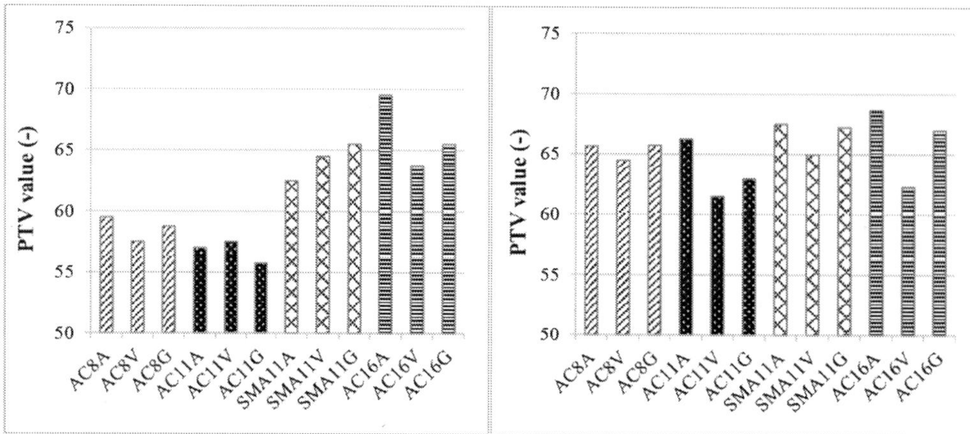

Figure 2. Results of skid resistance of asphalt mixture surfaces (samples after production – left, samples after removing the asphalt film from the aggregate surface – right).

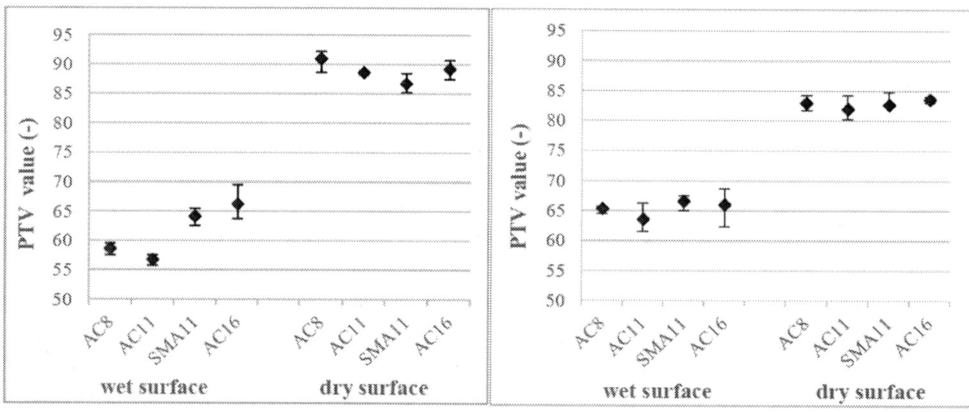

Figure 3. Skid resistance values PTV of different types of asphalt mixtures (samples after production – left, samples after removing the asphalt film from the aggregate surface – right).

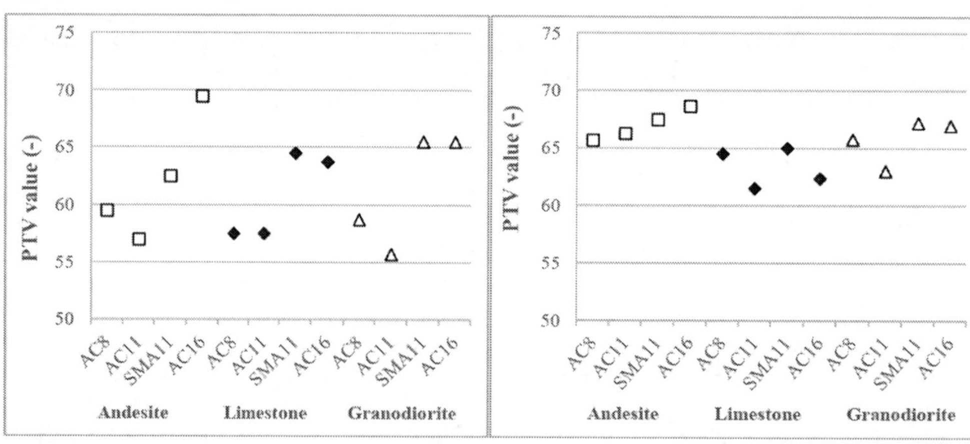

Figure 4. Skid resistance values PTV according to the type of rock of coarse aggregate (samples after production – left, samples after removing the asphalt film from the aggregate surface – right).

IOP Conf. Series: Materials Science and Engineering 1015 (2021) 012072 doi:10.1088/1757-899X/1015/1/012072

After removing the bitumen from the aggregate surface (figure 2, right), the PTV values of the AC8 and AC11 mixtures increased, the differences between the single types of mixtures increased partially (PTV in the range of 61.5 to 68.7), but the influence of the aggregate type became more significant. For all mixtures, it can be seen that in terms of skid resistance of the surface, the highest values were on the mixture with andesite aggregate and the lowest with limestone rock aggregate.

If we compare the mixtures with each other, regardless of the type of aggregate used (figure 3), the skid resistance of the dry surface of asphalt mixtures is equal (PTV in the range of 86.8 to 90.9). On a wet surface, the values are significantly lower. Mixtures SMA11 and AC16 had a higher skid resistance, which could be due to the good macrotexture of these compounds, which ensures good drainage of water from the surface and close contact of the tire on the surface. By removing the bitumen from the aggregate surface and the appearance of the aggregate texture, the differences between the mixtures were reduced, the skid resistance of wet surfaces increased, dry surfaces decreased.

From the results of the skid resistance of surfaces with stripped aggregate particles on the mixture surface (figure 4), we can see the positive effect of aggregate micro and macro texture that depends on aggregate rock. However, to determine the extent to which the aggregate texture affects the skid resistance of the surface, it is necessary to perform further measurements with several types of the aggregate of the same and different rock type and determine the texture parameters of the aggregate surface, e.g. scanning methods, scanner with high resolution 2.49 μm to determine the surface amplitude characteristics [10].

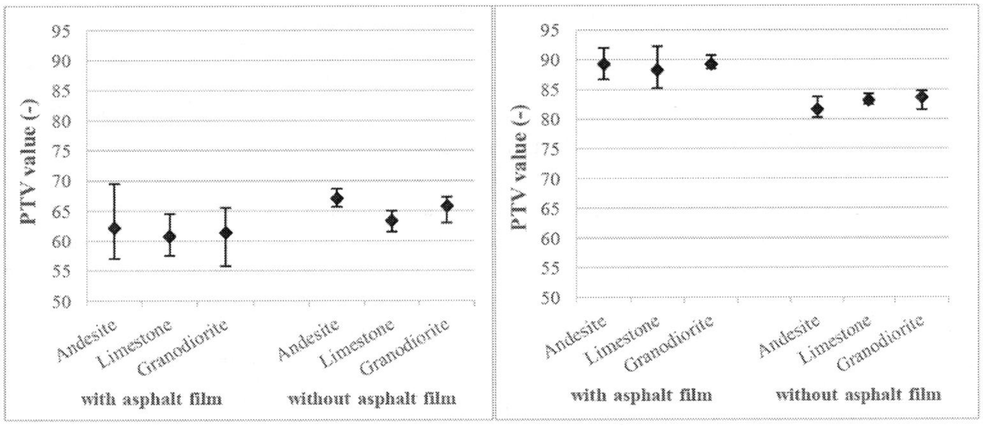

Figure 5. Average values of skid resistance PTV according to the type of aggregate rock (wet surface – left, dry surface – right).

Figure 5 shows the summary results of the tested surfaces from the coarse aggregate rock point of view. Small differences in average values can be observed on the coated samples. On dry surfaces of all mixtures, sufficient skid resistance (high PTV values) is ensured. The type of rock affects the skid resistance of the wet surface and contributes to road safety. The average PTV values according to the aggregate type clearly showed a positive increase, especially in the case of aggregates from andesite and granodiorite rocks, as well as limestone after bitumen stripping. The skid resistance of surface with andesite aggregate was PTV 67.0 and with the limestone aggregate 63.3 at an average of all tested mixtures.

The Mean Texture Depth using a volumetric patch technique is one of the methods that express a macro texture of surfaces. The measurements showed differences in the types of mixture and maximum aggregate size in the mixture (figure 6). It was confirmed that above all the portion of coarse aggregates, their size distribution, and shape characteristics as shape index, angularity, which also affect the air void content of the mixture, determine the macrotexture of the surface.

Finally, the macrotexture parameter was compared with the PTV values of the coated samples and stripping samples see figure 7. The correlation dependence of the parameters was higher on coated samples on which the bitumen film suppressed the aggregate microtexture. Skid resistance is mainly affected by the macrotexture.

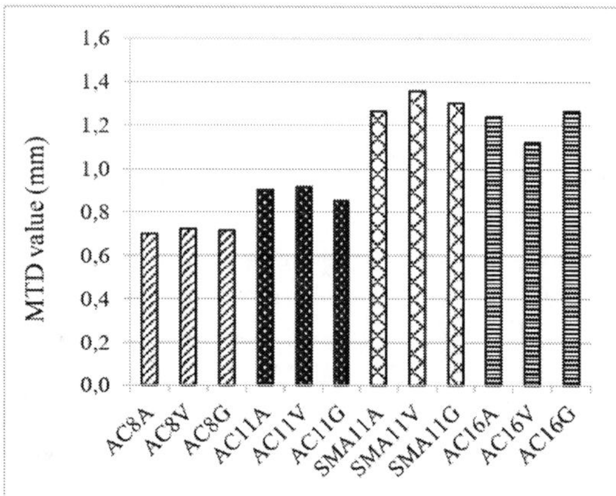

Figure 6. Average macrotexture depth of asphalt mixture surfaces.

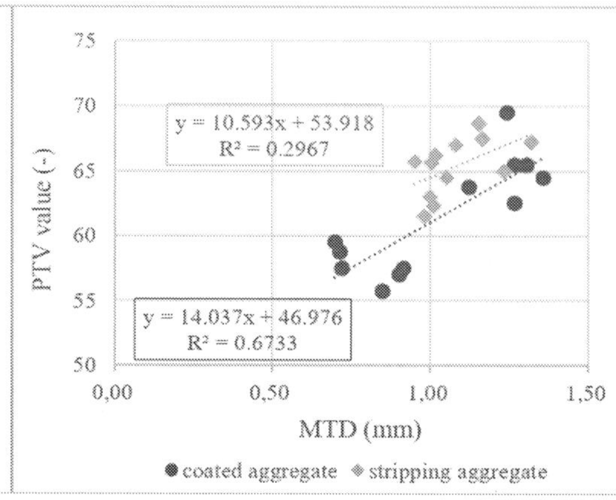

Figure 7. Comparison of MTD and PTV parameters.

4. Conclusions

The paper aimed to determine the effect of the mixture composition on the texture of the asphalt surface by expressing the roughness parameters PTV and MTD, which were analyzed according to the type of mixture, type of aggregate, and maximum aggregate size in the mixture.

The results obtained from the research showed that the coated samples, where the microtexture is absent, show that skid resistance depends on the macrotexture to some extent. The increasing values of macrotexture, the increase in the coefficient of friction. PTV values of stripping samples increased as expected, which is due to the effect of the aggregate microtexture, which affects friction. The effect of macrotexture has been eliminated to a minimum, the low values of the coefficients of determination as evidence, which may be due to the microtexture detection. This confirms the assumption that microtexture has a greater effect on the standardised value of skid resistance than macrotexture. In the case of dry conditions, the PTV values of stripping samples decreased compared to coated samples, which have higher adhesion forces, which are greatest on dry and smooth surfaces according to theory.

In terms of the type of aggregate rock, the skid resistance of asphalt mixture with coarse andesite aggregate was the highest, then granodiorite aggregate and the lowest value of PTV was on the surface with a mixture with limestone rock aggregate. To determine the extent to which the aggregate type affects the skid resistance of the surface, it is necessary to perform further measurements with several types of aggregate of the same rock and determine the texture parameters of the aggregate surface, e.g. scanning methods.

These measurements showed the results and dependencies on the produced and compacted mixtures. However, it is important the asphalt mixtures, used in road pavement, maintain these properties over the service life.

Acknowledgments

The paper is a result of the research supported by the research project VEGA 1/0537/17 „The influence of pavement surface morphology on pavement serviceability and emissions production"

IOP Conf. Series: Materials Science and Engineering 1015 (2021) 012072 doi:10.1088/1757-899X/1015/1/012072

References

[1] Celko J, Kovac M and Decky M 2011 Analysis of selected pavement serviceability parameters *Communications - Scientific Letters of the University of Žilina* **13** (3) (ISSN 1335-4205) pp 56-62
[2] Iwanski M and Chomicz-Kowalska A 2015 Evaluation of the pavement performance *Bulletin of the Polish Academy of Sciences-technical Sciences* **63** (1) pp 97-105
[3] Kovac M and Remisova E 2016 *Drsnosť asfaltových vozoviek* (Žilina: Žilinská univerzita) p 151
[4] Holzer R, Laho M, Wagner P and Bednarik M 2012 *Inžinierskogeologický atlas hornín* [online] (Bratislava: Štátny geologický ústav Dionýza Štúra)
[5] EN 13108-1 Bituminous mixtures. Material specifications. Part 1: Asphalt Concrete
[6] EN 13108-5 Bituminous mixtures. Material specifications. Part 5: Stone Mastic Asphalt
[7] KLAZ Katalógové listy asfaltových zmesí (Ministry of Transport and Construction of the Slovak Republic) p 40
[8] EN 13036-4 Road and airfield surface characteristics. Test methods. Part 4:Method for measurement of slip/skid resistance of a surface: The pendulum test
[9] EN 13036-1 Road and airfield surface characteristics. Test methods. Part 1:Measurement of pavement surface macrotexture depth using a volumetric patch technique
[10] Celko J, Kovac M and Kotek P 2016 Analysis of the pavement surface texture by 3D scanner *Transportation research procedia* **14** pp 2994-3003

IOP Conf. Series: Materials Science and Engineering 1015 (2021) 012073 doi:10.1088/1757-899X/1015/1/012073

Feasibility analysis for revitalisation of unused rail infrastructure

D Sitányiová and S Masarovičová

University of Žilina, Univerzitná 8215/1, 010 26, Žilina, Slovakia

E-mail address: dana.sitanyiova@uniza.sk

Abstract. Abandoned railway stations were constructed to serve as a railway stations but have fallen into disuse, because of different circumstances. Many rural stations and lines were closed with the introduction of new bus services, and also due to increased popularity of cars. Some of them were converted and now they have new functions. The purpose of this article is to introduce the concept of reinvention of rural stations in YOUMOBIL project with emphasis on feasibility of reinvention projects. Article gives an overview how is a typical feasibility study organised. Other parts of this document explain the specifics of YOUMOBIL feasibility studies, good practice of reinvention of disused railway infrastructure and example of YOUMOBIL reinvention pilot project in Brezno. YOUMOBIL makes a valuable contribution, helping rural communities get the most from their abandoned stations, turning stations and their surroundings into welcoming, thriving, and safe gateways and hubs. Working with the rail industry towards improvements, such as better pedestrian and cyclist access, supporting intermodality, running events, workshops or other activities encourages people to use public transport, and support the local community.

1. Introduction

There are many abandoned (or disused) railway stations or other rail structures in the world, which were constructed to serve for rail transport, but fell into disuse. There could be various circumstances when this occurred. A railway company may have gone bankrupt, the station was closed due to the failure of economic activity including insufficient number of passengers, or there were different operational reasons such as replacement of the rail line. Stations were also often relocated along the route of the line to new location usually nearer to the centre of population. Many rural stations and lines began to close in the past with the introduction of new bus services, and of course the increased popularity of the cars. Other lines and stations never lived up to the expectations of their operators, because they were badly sited, away from the towns and villages that were designed to serve and this also led to a rapid decline in passenger numbers.

Many former railway lines, railway buildings or rail carriages were converted and now they have new functions. In rural areas, especially in UK, former railway station buildings are often converted into private residences. Examples include the stations on the closed Didcot, Newbury or Southampton Railway in England. Example from Mayfield represents a conversion of the original ticket hall of Mayfield Station on the Cuckoo Steam Railway Line built in 1880 (figure 1) [1]. Also many former railway carriages have found a new life off the rails, being ingeniously reused as holiday homes, or as cafeterias and restaurants. Despite their derelict state, some of them have potential to be transformed into an incredible project. The property in figure 2 is located in the country of Central Otago region in

IOP Conf. Series: Materials Science and Engineering 1015 (2021) 012073 doi:10.1088/1757-899X/1015/1/012073

New Zealand. It is remote, with no access to city power or water supply, capturing rain water and generating all of its power from a small solar system.

Figure 1. Railway station converted into private house, Mayfield, UK, Source: [1].

Figure 2. Railway carriages converted into private house, New Zealand, Source: [2].

Former rail lines were repurposed as managed nature reserves, trails or other tourist attractions such as Bystrická Cycling Route which ranks among the longest tourist cycling routes in Slovakia. It was built on the former rail line on the narrow gauge historic forest railway (figure 3). There are also many abandoned and neglected buildings other than rail ones in rural areas. Countryside represents the picture of each region and country. Its image and structure were created historically and are continuously changing. In the past, countryside was predominantly inhabited by farmers. Overall picture of a countryside is now based on both nature and residential, farming and industrial buildings and infrastructure.

Figure 3. Cycling route on the former forest rail line, Slovakia, Source [3].

Today, people working in the cities and towns often move to the countryside and also both small and larger industries are being more and more established in these parts of country. Both the design and layout of residential, farming and industrial buildings, alongside with technical and green infrastructure, create a special rural environment that affects both social and environmental conditions of rural life. Rural buildings and their properties can be also reutilized in tourism, based on the protection and the valuable reutilization of popular rural architectural heritage with potential interest for tourism purposes.

2. YOUMOBIL reinvention projects in glance

YOUMOBIL is a project funded by Interreg Central Europe programme (2019 - 2021). There are partners from Slovakia, Italy, Germany, Croatia, Poland and Czech Republic in the project consortium, who face the same key development challenges including economic and labour market decline, peripheralisation and a deepening urban-rural divide. Project aims at enhancing the passenger transport system for young people living in rural areas and their access to the European and national transport networks. Poor mobility options other than the own car is among the most frequent reasons why young people choose to leave their native rural area and to migrate to larger cities or even beyond, intensifying the demographic change troubling most rural areas in Central Europe. Moreover, public transport and related infrastructures suffer from a rather poor image among Central European youth. YOUMOBIL partners therefore network with rural areas' local youth in piloting demand-driven novel service features, exploit the potentials of transport management through interfaces for mobile devices, and to explore how youth initiatives can reinvent disused rail infrastructure to enhance the attractiveness and image of public transport. This article is concentrated on reinvention of abandoned railway structures.

YOUMOBIL pilot projects support such reinvention and also social capital of when old railway infrastructure is converted to centres of rural community life. Recent experience has demonstrated that reinventing railway buildings and surrounding areas can create popular places with special character where people enjoy working, living, free time spending and visiting. Such reinventions represent opportunities both for conservation and development, to work together and transform the environment and public space for the communities that live and work there. Five rural or peripheral railway stations were selected for reinvention in partners' countries. Before technical development and implementation of reinvention project feasibility study (analysis) must be done to determine the viability of an idea how to reinvent unused railway infrastructure, such as ensuring project is legally and technically feasible as well as economically justifiable.

Stefan Zemp et al. [4] identified several generic functions of railway stations: linking catchment area and transport network, supporting transfer between modes of transport, facilitating commercial use of real estate, providing public space, and contributing to the identity of the surrounding area. Therefore, reinvention or redevelopment calls for the integration of many different objectives and common understanding among the multiple stakeholders with potentially conflicting interests. When divergent perspectives are concerned, reinvention not only poses technical challenges, but also significant social challenges. In YOUMOBIL project partners focused especially on two functions, providing public space and identifying surrounding area. Before preparation of feasibility study partners organised series of workshops with the aim to build consensus among the local youth and relevant stakeholders sharing the same aim of making rural stations as access points to the PT network more attractive through different actions driven by local youth.

Pilot sites were chosen together with local decision-makers, PT service providers or owners of the addressed sites, together with YOUMOBIL partners acting rather as facilitators and guides, and local stakeholders and youth initiatives acting as driving forces. Following good practices overview collected in form of the report, how local communities revitalised disused rail infrastructure for the well-being of the local society, YOUMOBIL partners facilitate concerted action of local stakeholders, foremost young people. In all partners' region adaptive reuse development is planned, where an existing railway station building is used, in whole or in part, or disused as a basis for a new facility. Activities were launched in the form of a series of local workshops during autumn 2019 and spring 2020, where visions were elaborated and selected and a strategy, and an action plan prepared. Focus was especially on local youth' interests.

The main objective of the series of workshops was to bring relevant people or groups together and provide a structure to make a good start on planning for revitalisation of selected sites in partners' regions. In total three workshops were organised. In visioning workshop different visions for future use were explored and critical factors identified. Inspiration was drawn from examples of best practice collected. Workshop participants defined a list of visions representing the possibilities for new use of facility at partners' sites. It was discussed which demand exists for this type of facility in a given site and surrounding rural region, and what type of spaces might best accommodate this demand understanding the community context. During strategy workshop participants together agreed on one vision and elaborate a strategy for its achievement, including identifying necessary actors and resources. During action workshop tasks assigned to different actors, and funding potentials were analysed. Once partners have found information about regional context, potential users' requirements and stakeholders' interests, it was then necessary to assess whether projects are also feasible from different perspectives.

3. Common criteria for feasibility analysis

Reinvention of unused railway infrastructure has its specifics; therefore, several attempts have been made in consortium to identify the critical success factors of such projects. Common criteria for organising feasibility study were defined. The purpose was to help partners to organise and elaborate feasibility studies, and set up common criteria how to do that. Partners were recommended to use modified feasibility tool called TELOS, an acronym for the five key areas that are needed to consider in an analysis: Technological (or Technical), Economic, Legal, Organizational (or Operational), Scheduling, described below (figure 4). The TELOS model was first presented in 2007 by James A. Hall [5] and later adopted across a huge range of settings, because it offers a simple way to consider the most important issues related to feasibility.

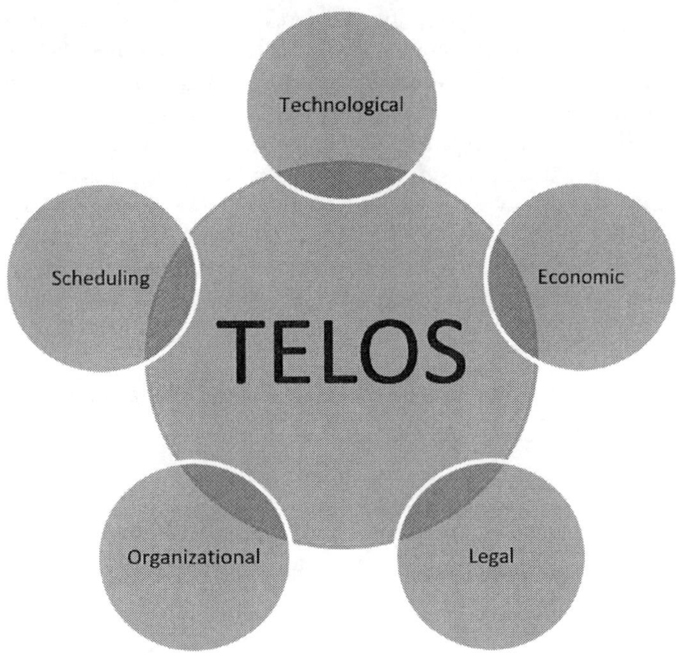

Figure 4. TELOS tool structure

Technical Feasibility focuses on the technical resources available to the organization and determine whether they meet enough capacity, converting the ideas developed during workshops into real project. Because YOUMOBIL revitalisation projects will be implemented by organisation outside consortium, a technical feasibility part of the study will assess only the basic details of how organisation responsible for project implementation should intend to bring project idea into reality. It should validate an idea, strategy and approach. The analyst must find out whether it is possible to develop the project given the current technical resources.

Economic Feasibility. Once the technical feasibility is established, it is important to consider the monetary factors also. Economic feasibility of the proposed project is carried out, involving typically a cost/benefits analysis determining the viability, cost, and benefits associated with a project before financial resources are allocated. There are two elements to the funding challenge: capital to redevelop the station, and revenue for the operation and maintenance. It is also necessary to ensure the sustainability of the project. In YOUMOBIL project it was recommended to develop also pro forma operating model for future to estimate operating revenues and expenses. In case of reinvented stations, it can be considered that revenue would come from 'earned' income through e. g. ticket sales for different events and services, space rental, and other ancillary income. It is anticipated that the remaining budget would be secured through a combination of rail operator, municipality or region support, grants and other contributions. There may however be an element of commercial activity within a plan.

Legal Feasibility. This assessment investigates whether any aspect of the proposed project conflicts with legal requirements like e.g. zoning laws. This assessment of the applicable legal framework includes the identification and analysis of pertinent laws and regulations that may affect the project.

Operational feasibility is mainly concerned with issues like how project will be operated, if it is developed and implemented. The essential questions that help in testing the operational feasibility of a project could be following. Are the users not happy with current state? Will then they welcome the change and the new project? Have the users can be involved in the planning and development of the project? Will the system effect the users in considerable way? It is also recommended to prepare operating plan, describing how such a facility would be owned, managed, and operated. As a project which accommodates significant civic activity, it makes sense that the ownership of facility would remain in public hands. However, other options exist for its operation including enter into an operating agreement with an existing or new established non-profit organization, hiring a commercial firm to operate, or operating by municipality or rail transport operator.

Scheduling Feasibility. This assessment is very important for project success; after all, a project will fail if not completed on time. In scheduling feasibility, an organization estimates how much time the project will take to complete.

YOUMOBIL feasibility study aims to objectively uncover the strengths and weaknesses of a proposed reinvention project, opportunities and threats present in the actual environment, the resources required to carry through, and the prospects for success. The common template was prepared to define the common standard for feasibility study (table 1).

Table 1. Template for standard feasibility study and Location Analysis for YOUMOBIL projects.

Name of reinvention project	Suitable name for the project.
Order party	Entity ordering the study.
Author/s of the study	Qualified external expert or company.
Date and place	Date and place of study preparation.
Introduction	The Introduction section of the feasibility study provides a general statement about the overall objectives and content of the document and information sources. It is short introduction describing the project of reinvention, its purpose and benefits, details on stakeholders and users, results expected. This section also introduces the project vision and its origins (workshops with local youth).
Site and location analysis	This section provides brief analysis of the project location. The information can be organized under the following heads: • Site (building), history and current site conditions • Location Overview • Location Assessment and Accessibility in local or broader context. Author visits the site, checks the accessibility and presence and availability of various services related to the project. At this level, the concern is also whether the proposal is legally feasible. It determines whether there are any conflicts with legal requirements.

Table 1. cont.

Alternative solutions	List of the alternatives to the proposed solution with their short description. This section describes the alternative solutions that have been considered during workshops and compares them to the proposed solution. Analysis of alternatives, highlighting the key differences between solutions.
Proposed solution	Description of proposed solution. Why the preferred solution was selected? Does the recommended approach satisfy the requirements of youth community, stakeholders, building owner, and PT operator? Is it a practical and viable solution? An overview of the proposed solution's functionality and features.
Impact	Impact of solution on the organization and young rural community including different benefits (with emphasis on non-monetary ones). Definition of values justifying the proposed solution.
SWOT analysis	A SWOT analysis likely involves many of the same elements as a feasibility study itself. The goal is the same – to determine the viability of the reinvention project. When looking at project Strengths, author should primarily consider what makes the project special and strong. This could include any unique characteristics of the project. Weaknesses – what aspects could potentially hold project back? (A lack of financing, experienced staff, or other attributes that could stop achievement of the project objectives) Opportunities – factors outside the project which could put project in a better position to succeed. Threats are the inverse of opportunities.
Resources	Identification of all the resources, technical, inventory, financial and human that will be needed to complete the project. E.g. determination of project's requirements of construction such as reconstruction of building, internal design, etc. Determination of project's requirements of skilled labour or managerial and financial labour. This section of the feasibility study also provides a comparison between the value of the proposed solution and its costs. Estimation of required investment, operational and maintenance costs.
Cost-benefit analysis	Cost Benefit Analysis (CBA) is mostly used to decide whether or not project should go ahead, or to decide between different solutions. It involves working out the total potential costs and benefits of a project, and then translating these into monetary terms. However many benefits/impacts are non-monetary. In other words, they cannot be numerically measured, or they may be too complex to measure, for example social or environmental benefits.
Timeline and project process	Framework for implementation of the project including milestones (Gantt chart recommended for visualization).
Operational analysis	Operational analysis including information on future reliability, maintainability, usability, sustainability, affordability, etc. Information of this section should be organized under following heads: operator decision, staffing issues, various operating models, maintenance issues, and other operational issues. This part is required if the project development will be a part of portfolio of the developer – facility owner or another entity. If the process is to reinvent unused railway infrastructure for community (perhaps also by community, or other entity), then a study determining how the new facility will operate in a way that is useful to its users. How well a proposed project solves the problems, and takes advantage of the opportunities identified during scope definition and SWOT analysis?
Conclusions and reflections	The findings of the feasibility study, a brief description of pros and cons for the reinvention project.

IOP Conf. Series: Materials Science and Engineering 1015 (2021) 012073 doi:10.1088/1757-899X/1015/1/012073

4. Reinvention pilot project in Brezno

Brezno is a small district town with a population around 21,000 located in the middle of Slovakia just under the mountains Low Tatras. It is the main access point to Horehronie region and two national parks. Area offers great opportunities in agriculture and tourism. However, the number of people living in Horehronie is still declining, including Brezno itself. Emigration plays a significant role in the overall decline of the population in region. The lack of jobs and the bad economic situation force especially young people to migrate outside Brezno region. The whole area is underdeveloped also in the sense of transport infrastructure and services. One of the main means of "how to get there" is a train. Brezno has two train stations. The main station is combined with the bus station, and located on the outskirt of the town. The train station that is a subject of YOUMOBIL reinvention project is a small station "Brezno mesto" very close to the city centre (figure 5).

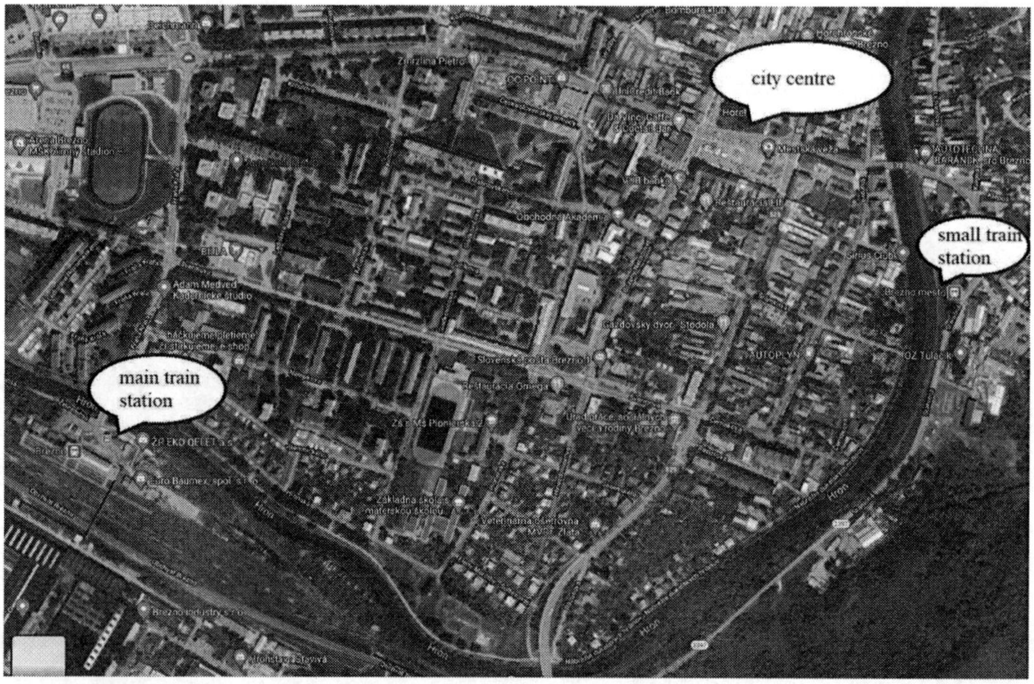

Figure 5. Brezno map.

Train station "Brezno mesto" is a property of Slovak Republic managed by Railways of the Slovak Republic, which is the main provider of local rail services. The station is no longer used and only serves as a boarding stop. Building and surrounding area were left as they were built and are in poor condition. Public space - the waiting room on the ground floor has not been used for a long time. First floor is used for rental. In exterior there's one platform with toilets for traveling public and space for buffet/pub. Whole space is neglected and in a need of refurbishment (figure 6-7) YOUMOBIL reinvention plans include complex reconstruction of this property with a library and gallery in the ground floor (space to be designed by local young artists). It will be also used as meeting room for Youth Parliament (figure 8). Another part of ground floor will be exhibition (small railway museum) accessible from the former ticket hall. The exterior of station will be transformed to more pleasant space for travelling and general public by creating community gardens, installation of street furniture and small playground for children. There will be also a bicycle stand (including e-bikes) as a part of a bike-sharing in the town installed.

Figure 6. Train station "Brezno mesto".

Figure 7. Platform in front of train station "Brezno mesto".

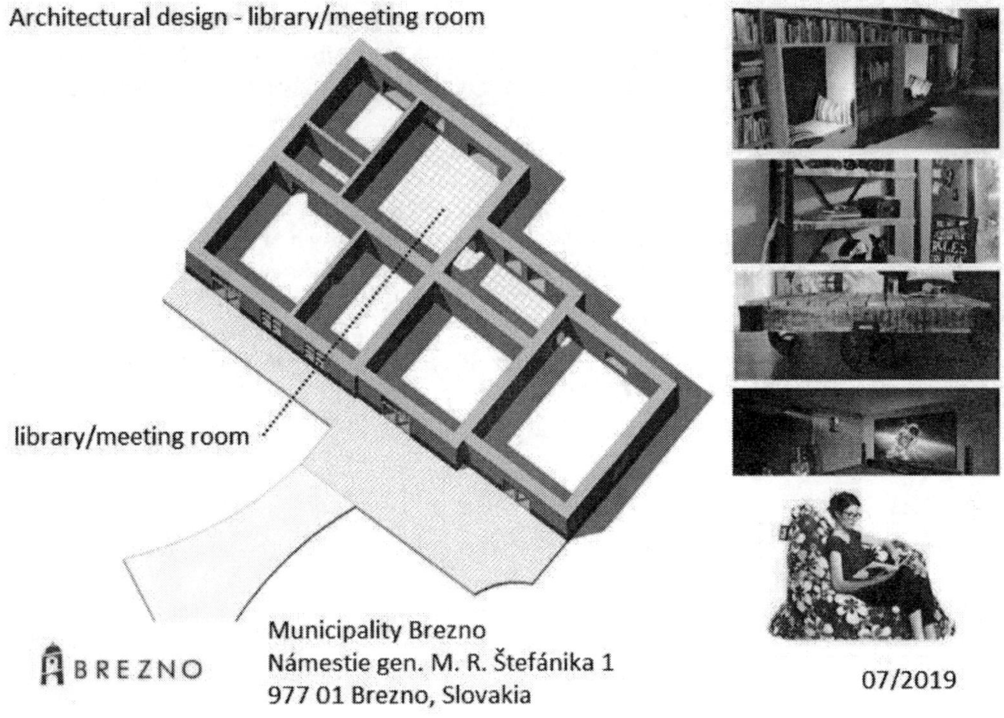

Figure 8. Architectural design – "Brezno mesto" train station.

Only a few passengers get off and get on at the stop. The aim of YUOMOBIL reinvention is to change this situation. Brezno municipality – a partner in the YOUMOBIL project wants to make people prefer this stop because of its geographical position close to the town centre and prefer train and bicycles as dominant means of transport.

By making this place more attractive, it will be used more, and people, including tourists, can also visit the sights and other services of the town, or meet with friends there. The associated partner of the project is Cluster Horehronie, which associates several organisations in tourist industry. The main role of the partner is to provide marketing activities and to work with young people participating in the project. Main beneficiaries of the project are residents of Brezno, whole region of Horehronie, young people, passengers and tourists.

IOP Conf. Series: Materials Science and Engineering 1015 (2021) 012073 doi:10.1088/1757-899X/1015/1/012073

5. Conclusions

There are plenty of examples of best practice of adaptive reuses of old railway infrastructure. It is an effective strategy for optimizing the operational and commercial performance of built assets. Adaptive reuse of buildings can be an attractive alternative to new construction in terms of sustainability and a circular economy. The reinvention of railway infrastructure can act as the catalyst to stimulate also revitalisation of whole area and also community life. Municipalities, building owners, architects, developers, builders and entrepreneurs who wish to become involved in rejuvenating and reconstructing a building must first make sure that the finished product will serve the need of the market, that it will be completely useful for its new purpose, and that it will be competitively priced. YOUMOBIL partners who will implement reinvention projects in pilot sites will have to prepare feasibility study to justify the "market" for the recycled project and its viability for future.

Acknowledgment

This article is published within the implementation of the project YOUMOBIL (CE1307), co-funded by the European Regional Development Fund under Interreg CENTRAL EUROPE.

References

[1] Knight F 2018 *Homes in former train stations – in pictures* [Photo] The Guardian https://www.theguardian.com/money/gallery/2018/nov/16/homes-in-former-train-stations-in pictures

[2] Old railway carriage converted into stunning off-grid home. By Living big in tiny house. August 8 2020, retrieved from https://www.livingbiginatinyhouse.com/railway-carriage-home-conversion/. Copyright 2018 by Living big in tiny house.

[3] Cycling route 005, Krásno n. Kys. – Vychylovka. By uknown author. August 8 2020, retrieved from https://www.turistika.cz/trasy/cyklotrasa-005-krasno-nad-kysucou-nova-bystrica-vychylovka-skanzen/foto?id=845391

[4] Zemp S, Stauffacher M, Lang D J and Scholz R W 2011 Generic functions of railway stations—A conceptual basis for the development of common system understanding and assessment criteria *Transport Policy* 18 (2) March 2011 pp 446-455

[5] Hall J A 2019 *Accounting Information Systems* (10th Edition. Cencage learning) p 816

IOP Conf. Series: Materials Science and Engineering 1015 (2021) 012074 doi:10.1088/1757-899X/1015/1/012074

Remote monitoring of road structures' condition as a way to increase the durability of motorway

T V Dormidontova and A V Filatova

Academy of Architecture and Civil Engineering of Samara State Technical University, 194, Molodogvardeyskaya ave., Samara, 443001, Russia

E-mail address: adisk63@yandex.ru

Abstract. The calculations of road structures' elements and the evaluation of reliability of data related to them are necessary for conducting justified and timely repair work, for forecasting the service life of road pavements, as well as for increasing the durability of motorway. Modernization of monitoring system, improvement of instrumental base structures, implementation of innovative methods in road construction allow carrying out more detailed research of road structures' temperature and humidity conditions, analyzing their deformation stability under real-life conditions, and as a result, increasing measurement accuracy and reliability. Scanning devices used in the process of remote monitoring are characterized by tightness, high reliability and maintainability, their data is automatically read and transmitted to the server [1]. The input data obtained as a result of monitoring road structures' condition are used to assess the impact of destructive factors of all elements of the system - the motor road, they are used to justify the application and selection of innovative materials and technologies. The obtained information allows imagining the processes that take place inside road structures and analyzing the peculiarities of external and internal factors' impact on the service life and durability of the road.

1. Introduction

The impacts that make it possible to evaluate the stability of road structure to the accumulation of residual and reversible deformations, as well as the estimate of irreversible displacements in cohesionless materials, are conventionally divided into three types: real dynamic load from moving vehicles; fixed load; tared shock loading. The wide capabilities of measuring sensors used in monitoring process and special software allow assessing deformations from any type of impact on road structures.

The methods of obtaining information in process of remote monitoring are characterized by long-term (for example, accumulation of residual deformations and stresses), accelerated (for example, monitoring carried out at special test sites or stands) and short-term (for example, evaluating the deflection of road structures) observations. As a rule, such devices as georadars (see figure 1) and shock loading units [2] are used in the process of diagnostics, inspection and monitoring of motorway.

2. Theory

No less important from practical view point is the study of the temperature regime of road structures' operation, the purpose of which may be identifying the absolute minimum and maximum temperature values for the operation of road structures; the daily range (temperature scatter) for the operation of materials in structural layers; identifying the periods of limited traffic load due to the high temperature of coating layer and the swampness of roadbed; evaluating the temperature drops - for example, in the coating (they have a direct impact on thermal cracking); estimating the impact of road structure elements' geometric parameters (different roadfill heights or excavation depths); checking the correspondence of design models to real-life conditions of road operation; identifying the duration (periods) of structures' operation under extreme conditions (maximum and minimum temperatures) and the duration of operation of roadbed under high humidity conditions; determining the number of drops of the road structure's operation temperature above and below zero degrees Celsius (see figure 2).

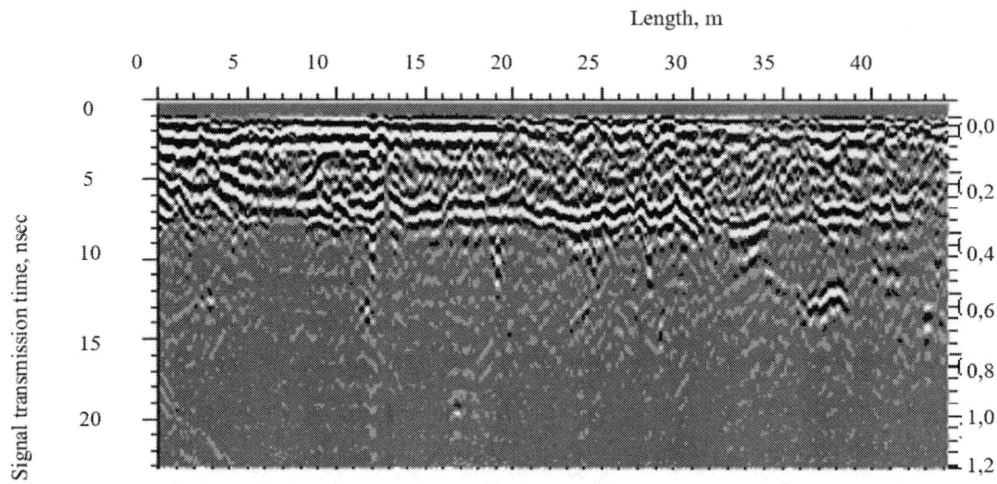

Figure 1. Determining the thickness of road pavement's structural layers, using a 3D georadar system.

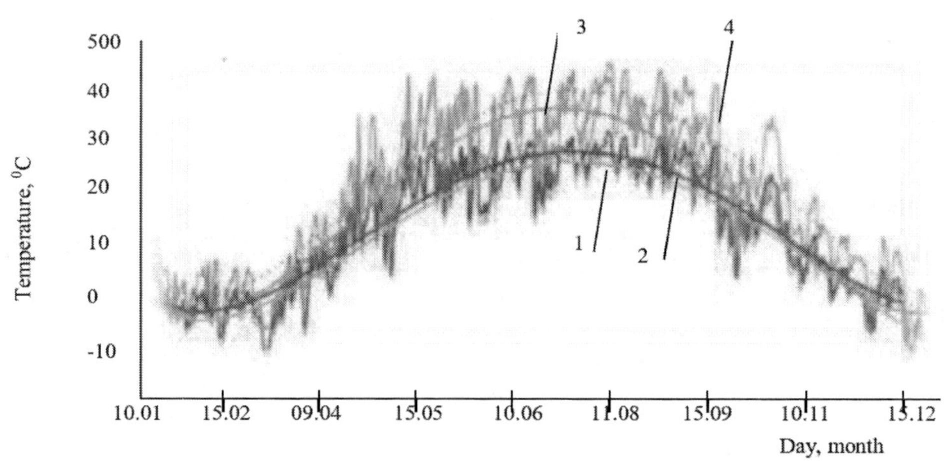

Figure 2. Checking the adequacy of design models and the roadbed:
1 - ambient temperature, 2 - calculated values in coating layer,
3 - temperature obtained by instruments used in monitoring process,
4 - temperature obtained according to formulas

The analysis of instruments applied in this sphere has shown that in many countries such units as «The Multi-Depth Deflectometer (MDD), Multistage Sensor (MS)» [3] are widely used. In process of this research the advantages and disadvantages of these devices have been evaluated, and they are described in the table 1.

Table 1. Advantages and disadvantages of MDD, MS units.

Advantages	Disadvantages
- presence of a corrugated tube, which prevents moisture ingress and partial collapse of the drilled hole;	- lack of temperature and humidity sensors;
- - the unit can be assembled prior to visiting the research and monitoring site; - it has got rigid fixing; - the data can be read at the monitoring site	- violation of the continuity of layers when mounting the devices; - displacement of the top layer, directly at the top point of the road structure, is not taken into account; - unreliable fastening of modules in cohesionless layers; - employees have to come to the site to read data, as a result, the working cost is high.

But, despite the listed disadvantages, the devices used in the research served as the basis for implementing an innovative system for remote monitoring of road structures' condition.

The instrumental base is used to obtain adequate input data, which can be further applied in the monitoring calculation base, and can be used to make recommendations on the service life of the motor road under study. Monitoring data are needed to make comparisons with development goals and criteria, and to assess the effectiveness and efficiency of measures taken.

3. Experiment

Wide use of innovative technologies is currently restrained by lack of data on the impact of these technologies on the operational condition of motorway [4]. Such data are necessary for an appropriate feasibility study of the effectiveness of new technologies' use. It is extremely important to develop comprehensive implementation programs that will allow monitoring the technologies at every stage, starting with trial design and construction.

Reliability should be taken as the criterion for assessing the technical condition of motorway. This criterion has a number of advantages; it is recommended for use in the monitoring calculation base. It is associated with time, can be used for any property of the object (strength, deformability, etc.), can be described mathematically, and can be used as the basis for forecasting the options of the condition of motorway, which is really vital for monitoring process.

Statistical characteristics of the parameters of motorway were used as input data in this research. The algorithms for calculating the statistical characteristics of the parameters were reduced to using linearization methods, enumeration with account of Monte Carlo technique. In the research the authors have determined the probability of destruction and the reliability of trouble-free operation of the pavement: during the period when the road was put into operation; after the expiration of the 1st and 15th years of operation of the structure.

RMS deviation of the required elastic modulus for dependence value was calculated using Formula 1

$$\sigma_{mp} = \overline{N}_{mp} \cdot E_{mp} \tag{1}$$

The minimal value of elastic modulus, at which, after opening the road for traffic, the road's destruction will reach 50%, was calculated using Formula 2

$$E_{\min} = \frac{\sqrt{E_{mp}^2 + \left(25 \cdot C_v^2 - 1\right) \cdot \left(E_{mp}^2 - 25 \cdot \sigma_{mp}^2\right)} - E_{mp}}{25 \cdot C_v^2 - 1} \tag{2}$$

The parameter σ_m was calculated using Formula 3

IOP Conf. Series: Materials Science and Engineering 1015 (2021) 012074 doi:10.1088/1757-899X/1015/1/012074

$$\sigma_m = C_V \cdot E_m \tag{3}$$

The reliability value of road pavement at the moment of putting the road into operation was found using Formula 4

$$Z = \frac{E_{mp} - E_m}{\sqrt{\sigma^2_{mp} + \sigma^2_m}} \tag{4}$$

Using the known tables, depending on the parameter *Z, W* (reliability value) was determined.

The probability of road pavement destruction in the beginning of the first year of operation was determined using Formula 5

$$P = 1 - W \tag{5}$$

E_{mp} - required modulus of elasticity
C_v - coefficient of variation of the total elastic modulus
$\bar{N}_{mp}$ - parameter of traffic intensity and composition

The probability of destruction and reliability after the expiration of the first year of operation of the road, and after 15 years of operation of the pavement expire, were determined in a similar way.

In the course of the research it has been found out that during the first months of road operation about 14 m^2 out of every 10 000 m^2 of the road structure will be destroyed (subsidence cracks, etc.) By the end of the first year of operation of the pavement, there will be about 4.1 m^2 of destruction per 1000 m^2 of coating. By the end of 15 years of operation, 18 m^2 of every 100 m^2 of coating will be destroyed.

Also, within the framework of this research, the reliability of road capacity was assessed. In this research the investigated and calculated parameters depended on road condition and type of traffic. The ensured vehicle speed was determined by permissible risk value, which is an objective characteristic used to calculate the speed of traffic flow, taking into account traffic safety.

4. Results and discussion

As a result of the research, the areas for applying the monitoring of road structures' condition have been determined, according to instrumental base data and calculations:

- maximum and minimum temperatures for selecting binder type;
- daily range of materials operation in structural layers for identifying possible co-use in adjacent layers of materials with different deformative properties from temperature effects;
- identifying periods and justified scheduling of measures to restrict traffic of vehicles with high axle load in hot summer time and in spring-autumn time when the roadbed (working layer) is swamp;
- identifying the areas where, due to this or that reason, asphalt-concrete layers of the coating have got different temperatures, which may lead to the formation of thermal cracks;
- identifying the sections of motorway (high road fill) that need additional thermal insulation layers;
- identifying the sections of motorway where additional measures are required to ensure drainage;
- checking the adequacy of the design models for forecasting temperature regimes of asphalt-concrete layers of the coating and humidity of the roadbed;
- identifying fatigue manifestations in the process of material aging caused by road operation under extreme conditions (extremely low and high temperatures, swampness of roadbed, etc.);
- determining the number of drops of the road structure layers' operation temperature above and below zero degrees Celsius, in order to forecast the frost resistance of the road pavement's structural layers.

In general, the use of data obtained as a result of monitoring opens up wide opportunities for improving the quality indicators of motorway. The wider the scope of such research, the more data will come from objects under study, the more objective picture of the operation of road structures under different climatic, geological and hydrogeological conditions can be obtained by different specialists in the field of road construction and design.

5. Conclusion

Except the purely practical benefits described above, monitoring will allow the accumulation of data for future adjustment of design models and methods for road structures' design, taking into account the variety of factors that have got an impact on motorway under their real-life operation conditions.

The information obtained as a result of monitoring, related to the rate of accumulating residual deformations in separate layers and to temperature-humidity regime, can serve as input data for the development and implementation of a new additional criterion describing the requirement for maximum permissible residual deformations of layers of road structures.

Data analysis has made it possible and will make it possible in future to understand the peculiar features of the impact of external and internal factors on the service life and durability of both the road section under study and of other motorway.

Acknowledgments
The authors of the article express their gratitude to the staff of the Department of Motorway and Geodetic Support of Construction of Samara State Technical University for their comprehensive scientific recommendations on the topic of this research.

References

[1] Chirva D V 2019 *Proc. Int. Conf. Current problems of science and technology* (Rostov-on-Don: DSTU) p 42
[2] Dormidontova T V and Domnin O V 2020 *Proc. Int. Conf. Traditions and innovations in construction and architecture* (Samara: SSTU) pp 126-31
[3] Mironchuk S A 2019 *Proc. Int. Conf. Current problems of science and technology* (Rostov-on-Don: DSTU) pp 60-1
[4] Filippov D V 2012 *Dorogi. Innovatsii v stroitel'stve* **20** p 74-8
[5] Nemchinov M V 2019 Car. *Road. Infrastructure* **1** (19) pp 7-20

IOP Conf. Series: Materials Science and Engineering 1015 (2021) 012075 doi:10.1088/1757-899X/1015/1/012075

Development of an algorithm for implementation of operation control automation systems using BIM technologies for a capital construction facility

O V Didkovskaya, O A Guzhova and M F KHairullin

Samara State Technical University, 194, Molodogvardeyskaya str., 443001, Russia

E-mail address: guzhova_oksana@inbox.ru

Abstract. The article is devoted to the problem of implementing BIM technologies for the operation of capital construction facilities. The research is relevant due to the problems that arose when transiting from traditional approaches to operation control of capital construction facilities to operation control based on the use of BIM technologies. The result of the research is the developed algorithm for implementing automation systems for operation control using BIM technologies. The implementation of the algorithm allows improving significantly the current operation systems of capital construction facilities. The algorithm distinguishes clearly the stages in their logical sequence, result, duration of each stage and the amount of costs.

1. Introduction

The information technologies are developing rapidly. They are changing the world around us including the techniques and methods used in different economic realm. The introduction of modern information and telecommunication technologies can significantly increase the productivity and efficiency of business entities. Nowadays, to build digital economy is one of the main tasks of the country's development.

At present, the world construction complex is undergoing a stage of radical transformation due to the rejection of traditional CAD technologies for designing buildings and structures in favor of a gradual transition to innovative operation control and implementation of investment and construction projects based on the use of modern information technologies. This fact makes the topic of the research relevant. Rapid development of information technologies, specialized software, the use of three-dimensional modeling that includes the creation of a unified information model of a construction facility (BIM-model) have become a key element of the construction development. BIM has become a part of operaion control tools in modern construction companies [1].

The term of BIM is given by V.V. Talapov in "Fundamentals of BIM. Introduction to information modeling" as "BIM is all numerically described, properly organized and managed information about a construction facility used both at the design and construction stage of the building and during its operation and even demolition" [2].

According to the definition given in applicable Construction Rules 333.1325800.2017 "Information modeling in construction. Rules for making an information model of construction projects at different stages of life cycle" , the information modeling of projects is the process of getting and using information on construction projects under construction and completed projects. It is done in order to coordinate the input data, organize joint production and storage of data, as well as their use for various purposes at all stages of the life cycle [3].

The issues of applying BIM technologies at the operation stage of capital construction facilities are the least studied in this field. Moreover, operation is the longest and most expensive period of a project's life cycle and it is here that you can get the greatest benefits from the introduction of BIM technologies. Information modeling forms the basis for all operatiom comtrol decisions throughout a project's life cycle [4].

The purpose of the research is to develop solutions for implementing BIM technologies in the operation of capital construction facilities and an automated system for operation control of facilities based on a BIM model.

The object of research is innovative technologies in the operation of real estate.

The subject of the research is the use of BIM technologies in order to optimize the operation of capital construction facilities.

2. Materials and methods

The methodological basis of the research is a systematic approach that allows considering the process of implementing BIM technologies as a complete project that contains many elements. Various general scientific and special methods were used during the research. The following research methods were used: analysis and synthesis, formal-logical (deduction, induction, justification and argumentation); abstract-logical, empirical, tabular, and others.

The use of the formalization method in the development of the algorithm for the implementation of automation systems for operation control using BIM technologies for capital construction facilities allowed to identify the stages of the algorithm and the logical sequence of their implementation and to form various classifications.

3. Results and discussions

The result of the research is a developed algorithm for implementing automation systems for operation control using BIM technologies for capital construction facilities.

At the beginning of the study, possible options, which should be evaluated for further development of a capital construction facility, were considered:

- to leave "as it";
- to sell the construction project;
- tomotivate the operating company to work more effectively;
- to increase funding;
- to change the operating company;
- to make a decision about implementing BIM technologies.

As it is seen from the list, only one of the options has the implementation of BIM. The most important thing when implementing BIM technologies for a capital construction facility is the owner's decision to organize its operation using information modeling. During the operation control process, property owners must obtain, integrate, edit and update much information related to various building elements, such as operating costs, warranties and specifications from various systems. BIM can effectively combine this primary data and provide convenient storage and retrieval of this data [5]. The main barriers to implementing BIM range from the lack of BIM structure, low awareness of the benefits of BIM, high initial investment costs to staff resistance to change [6].

Uncontrolled and increasing costs for the operation of a project, expenses opacity of the operating company, an increase in the number of accidents and failures, low quality service can be the reasons for using BIM technologies. Stage 1 of the developed algorithm is taking this decision.

Stage 2, the survey, starts after the owner takes the decision to operate a capital construction facility by using BIM technologies. To do this, it is needed to search for performers, consider competitive offers, evaluate the project budget, select a contractor and conclude a contract for implementing BIM technologies along with the entire complex of computer-controlled systems. The site survey is the basis

IOP Conf. Series: Materials Science and Engineering 1015 (2021) 012075 doi:10.1088/1757-899X/1015/1/012075

for the plan of implementing the operation automation systems using BIM technologies and is carried out in order to assess its condition and prepare a proposal for implementation.

The system for monitoring the technical condition of a construction facility should be evaluated by writing a conclusion report. The audit of business should be performed with a company's business process report. A company's business processes related to the operation should be described clearly and reliably. International standards ISO 9000 and 14000 series can be used as the audit methodology.

In general, an integrated report is made. On the results of the report, technical guidance on the implementation of BIM technologies is specified. It includes technical and economic feasibility study with a provisional financial plan.

Stage 3 is a plan development for implementing BIM technologies for the operation of a capital construction facility. After developing the implementation plan, the contractor must submit it to the customer (the owner of the building) and confirm its feasibility within the cost limits and deadlines set by the owner. As a rule, a plan for implementing BIM technologies is developed by licensed companies that are involved directly in the operation of information systems.

The study revealed that there are currently two approaches to the acquisition and implementation of information systems in Russia (table 1).

Table 1. Approaches to the acquisition and implementation of information systems.

Name of the approach	Recommended projects to use	Software
Boxed software	Small projects	IBM Maximo, EcoDomus, 1C Enterprise "Operation Management»
Customized software	Large and complex industrial complexes	RDC "Constractor", LLC "NPO SpecTech" (manufacturing group "TRIM"), ErmaSoft (manufacturing group "IGS-Zdaniye"), Sodis Lab (manufacturing group "Lement Pro"), LOY&HUTZ AG.

Each type of software is specific and requires the owner to make careful analysis and selection.

As a result, the owner should have a pack plan for implementing an automated operation control system. A consulting company or software developers should make this plan.

Stage 4 of the algorithm developed by the study authors is the implementation of BIM. On average, the information modeling implementation plan takes 4 months and costs up to 20 % of the project scope, while the implementation of BIM takes about 6-8 months. This conclusion is based on the experience of implementing such projects in Russia.

Stage 5 of the algorithm for implementing automation systems for operation control using BIM technologies for a construction facility is to control the results. To analyse the results is possible in a year from realizing the algorithm. It can be made by comparing the annual financial performance of the project's operation before the introduction of BIM and after the introduction of the new operation system. Based on the analysis, it is possible to draw appropriate conclusions and monitor the results of implementing BIM technology.

It should be noted that each construction project has its own characteristics. Construction projects differ in size, performed functions, technical condition, availability of technical documentation, qualification and motivation of personnel of the operating company, the amount of funding allocated, etc. As a result, there is no standard solution for automating the project operation tasks. In each case, you need your own project that takes into account the features of the project.

The stages discussed are shown in table 2.

IOP Conf. Series: Materials Science and Engineering 1015 (2021) 012075 doi:10.1088/1757-899X/1015/1/012075

Table 2. Developed algorithm for implementing BIM technologies for capital construction facilities.

Stages of the algorithm	Result of the stage implementation	Stage duration, months	Expenses, % of the project scope
Stage 1. The decision of the owner	Terms of reference for the survey	x	x
Stage 2. Survey	Integrated report and technical guidance for implementing BIM	2-3	10
Stage 3. Plan development for BIM implementation	Pack plan for BIM implementation	1	10
Stage 4. Implementation of BIM	Implemented BIM	6-8	70
Stage 5. Monitoring BIM results	Reasonable implementation results, conclusions	x	10

The authors considered a construction project with an operational model of good detailing. This model is operated by using a specialized software package. If an emergency occurs with engineering equipment at the facility, the operating company must identify the defective equipment and get all the necessary information about it: what kind of equipment it is, who installed it, where it is located, and how to turn it off. When working with an operational model, an employee refers to this model. The model shows what technological circuits the equipment is connected to, how to turn it off, turn on duplicate reserves. After that, they start the repairing. In order to generate the repair requirements correctly, complete information about the construction project and its engineering systems is required. This data is in BIM model that contains engineering documentation provided by the equipment manufacturer, various certificates, operating instructions, and so on.

Having found and identified the fault, the operating company starts making the activity plan and calculating the necessary funds to perform the work. With a small amount of work, the activity plan may not be developed and the fault will be fixed by the operating company in terms of ordinary working arrangements.

When the repair is completed, operating set of documents (register books, acts) is formed. It is reported in financial documents and in the operational model itself.

Thus, in the presented algorithm, each stage of implementing BIM technology for the operation control of capital construction facilities is considered in details. By using the algorithm, considerable time saving is achieved when the process is operated by a BIM technology model.

4. Conclusions

Operation of a construction facility using the building information model has significant advantages over traditional methods: the main indicators of operation become transparent and the response speed increases many times over. The research is of practical significance, as the developed algorithm for implementing automation systems for operation control using BIM technology for capital construction facilities allows improving the current system of capital construction facility's operation significantly. The results of the research indicate the need for active development and implementation of BIM technologies not only in the investment and construction sector, but also in operation control of capital construction facilities.

IOP Conf. Series: Materials Science and Engineering 1015 (2021) 012075 doi:10.1088/1757-899X/1015/1/012075

References

[1] Mesároš P and Mandičák T 2017 Exploitation and Benefits of BIM in Construction Project Management *IOP Conf. Series: Materials Science and Engineering* **245** (6) 062056

[2] Talapov V V 2017 *The Basics of BIM. Introduction to Information Modeling* (Saratov: Profobrazovanie) p 392

[3] Construction Rules 333.1325800.2017 Information Modeling in Construction. Rules for Creating an Information Model of Objects at Various Stages of the Life Cycle

[4] Kamenev A A, Abakumov R G and Yudov A E 2018 BIM Technologies in object and process management innovative economy: prospects for development and improvement (*Kursk: University Book*) 3 pp 59-64

[5] Siniak N, Źróbek S, Nikolaiev V and Shavrov S 2019 Building information modeling for housing renovation – example for Ukraine *Real Estate Management and Valuation* **27** (2) pp 97-107

[6] Wang Y, Wang X, Wang J, Yung P and Jun G 2013 Engagement of Facilities Management in Design Stage through BIM: Framework and a Case Study *Advances in Civil Engineering* **13** pp 1-8

IOP Conf. Series: Materials Science and Engineering 1015 (2021) 012076 doi:10.1088/1757-899X/1015/1/012076

Environment impact on the humidity change of beams from timber and glulam - experimental investigation

A Leśko

Warsaw University of Technology, Faculty of Civil Engineering, Division of Wood Structures, Al. Armii Ludowej 16, Warsaw 00-637, Poland

E-mail address: a.lesko@il.pw.edu.pl

Abstract. The paper relates to the humidity change in timber and glulam beams. Based on the literature moisture penetration into the wood phenomenon was briefly depict. The publication describes the carried out experiments: moisture measurement of glulam and timber beams in natural conditions. Results are presented in the form of graphs and photos on which cracks and warping of the samples is shown. Based on the literature and experimental results conclusions are drawn regarding the usage of glulam and timber beams in building industry.

1. Introduction

Timber structures are exposed to adverse impacts associated with the variability of the surrounding environment. Apart from biological and chemical factors, the influence of physical parameters of the environment such as relative humidity and air temperature are substantial. These parameters affect the physical and mechanical properties of wood, which successively leads to the creation of additional internal forces (in the form of stress) and affects the shape of beams or other timber elements. Additionally, the change in humidity causes shrinkage/swelling, which often results in numerous cracks in the whole mass of the solid. Therefore, the change of humidity is an important aspect which influences the stability and capacity of the structure as a whole load-bearing system.

Nowadays, timber structures are usually made of solid wood (rafter framings or other small scale structures) or glued laminated timber (e.g. construction of the warehouse roofing). These structures works in various conditions subjected to the constant changes of the environment (practically from the first day after sawing the timber). The design standards (e.g. [1]) take into account the influence of humidity on the timber structure by assuming a serviceability class consistent with the average, annual, ambient humidity. However, those regulations do not take into account (explicitly) changes, that occur naturally at the initial stage of the construction production process.

Studies available in the literature which relate to humidity changes in timber, are overwhelmingly focused on the extreme service conditions of the structures, e.g. in the case of bridges or swimming pools [2] where humidity is constantly heightened. However, in order to provide practical knowledge for the designers, who design buildings placed in less aggressive environment, it is necessary to carry out simple tests of the timber and glulam behaviour under “standard usage” conditions. Also consideration of structure production cycle in tests is relevant. Question is: how timber or glulam beam behaves "from wood sawing" (or delivery from stock for glulam) until material achieve equivalent humidity at the construction site.

IOP Conf. Series: Materials Science and Engineering 1015 (2021) 012076 doi:10.1088/1757-899X/1015/1/012076

2. Moisture penetration phenomenon in timber

Wood as an organic matter has hygroscopic properties, i.e. the ability to exchange humidity with the surrounding environment where absorbed water and water vapor are aggregated in liquefied form. Furthermore, it is possible to define three basic ways of damp storing [3]:

- free water - filling capillars;
- hygroscopic water - filling submicroscope spaces in cell walls;
- constitutional water - built into the chemical structure of wood.

Depending on the prevailing external conditions, hygroscopic material derive or remit moisture to/from the environment - this phenomenon is called sorption. Hygroscopic equilibrium is a state of balance in which the number of water vapour particles adsorbed by wood is equal to the number of particles desorbed into the environment [4]. This is achieved with different moisture contents of the wood and depends on temperature and relative humidity of the environment (Table 1). Theory called BET (from Braunner, Emmett and Taller theory) is defining which type of wood sorption phenomena is dominant for particular ranges of relative air humidity:

- mononuclear sorption (in the 0-21 (%) range) - coating of capillaries walls with a "single" layer of water molecules;
- polymolecular sorption (in the 21-60 (%) range) - coating of capillaries with up to five layers of water molecules;
- capillary condensation (over 60 (%)) - the phenomenon of water condensation after disturbing the balance on the surface of the capillaries.

Hygroscopic equilibrium state is desired. Achieved balance between environment and timber, allows for the inhibition of erosive processes of structures (such as cracks and delamination). In addition, in order to limit the potential "capillar burst", it is worth to limit the movement of humidity at the air/wood surface zone to the phenomenon of mononuclear sorption.

Table 1. Equivalent moisture content of timber [5].

relative air humidity (%)	air temperature (°C)																			
	0	5	15	20	25	30	35	40	45	50	55	60	65	70	75	80	85	90	95	100
100	29.0	28.7	28.5	28.4	28.2	28.1	28.0	28.0	27.9	27.9	27.4	27.0	26.6	26.1	25.6	25.0	24.6	24.1	23.6	23.0
95	24.6	24.4	24.2	24.0	23.6	23.6	23.2	23.0	22.8	22.6	22.4	22.0	21.5	21.1	20.6	20.0	19.4	18.8	18.2	17.5
90	21.0	20.8	20.6	20.4	20.0	19.7	19.4	19.1	18.9	18.7	18.3	18.0	17.5	17.0	16.5	16.0	14.9	14.3	13.8	13.3
85	19.2	18.8	18.5	18.3	17.8	17.5	17.2	16.8	16.5	16.2	15.9	15.5	15.0	14.5	14.0	13.4	12.9	12.4	12.0	11.2
80	17.5	17.2	17.0	16.7	16.0	15.7	15.3	15.0	14.7	14.4	14.0	13.6	13.1	12.6	12.1	11.7	11.2	10.7	10.3	10.0
75	15.3	15.2	15.0	14.9	14.4	14.1	13.8	13.6	13.3	13.0	12.5	12.0	11.6	11.0	10.7	10.3	9.9	9.6	9.2	8.8
70	13.6	13.4	13.2	13.1	12.8	12.6	12.4	12.1	11.8	11.5	11.1	10.8	10.4	10.0	9.6	9.2	8.7	8.4	8.1	7.8
95	12.3	12.2	12.0	11.8	11.4	11.2	11.0	10.8	10.6	10.4	10.2	9.8	9.5	9.0	8.7	8.3	7.9	7.5	7.0	6.8
60	11.3	11.0	10.8	10.6	10.4	10.3	10.1	10.0	9.7	9.5	9.2	8.9	8.5	8.2	7.8	7.5	7.1	6.8	6.5	6.2
55	10.2	10.1	10.0	9.9	9.7	9.5	9.3	9.1	8.9	8.7	8.4	8.1	7.8	7.4	7.1	6.8	6.4	6.1	5.8	5.6
50	9.6	9.4	9.2	9.1	8.8	8.6	8.4	8.2	8.0	7.8	7.6	7.4	7.0	6.7	6.5	6.2	5.9	5.7	5.5	5.2
45	8.8	8.7	8.6	8.5	8.1	7.9	7.7	7.5	7.3	7.1	6.9	6.7	6.3	6.0	5.8	5.6	5.3	5.1	4.9	4.3
40	8.2	8.1	8.0	7.8	7.4	7.2	7.0	6.8	6.6	6.4	6.2	5.9	5.6	5.4	5.2	5.0	4.8	4.6	4.3	4.1
35	7.2	7.0	6.9	6.8	6.6	6.5	6.4	6.2	6.0	5.8	5.6	5.3	5.0	4.8	4.7	4.4	4.1	3.9	3.7	3.5
30	6.3	6.2	6.1	6.0	5.8	5.7	5.6	5.4	5.2	5.0	4.8	4.6	4.4	4.2	4.1	3.9	3.5	3.3	3.1	2.9
25	5.4	5.3	5.2	5.1	4.9	4.8	4.7	4.6	4.5	4.3	4.1	3.9	3.8	3.7	3.5	3.4	3.1	2.8	2.6	2.4
20	4.7	4.6	4.5	4.4	4.2	4.1	4.0	3.9	3.8	3.6	3.5	3.3	3.2	3.1	2.9	2.8	2.6	2.5	2.3	2.1
15	3.9	3.8	3.7	3.6	3.4	3.3	3.2	3.1	3.0	2.9	2.8	2.7	2.6	2.5	2.3	2.2	2.0	1.8	1.7	1.6
10	2.7	2.7	2.7	2.7	2.6	2.6	2.5	2.5	2.5	2.4	2.3	2.2	2.1	2.0	1.9	1.8	1.7	1.6	1.4	1.2

In order to reduce erosion of wooden structures, glued laminated timber ("Glulam") is being used increasingly. Wood, which is the main raw material used to create that composite, is dried under strictly defined conditions (in a closed environment of the dryer). That process allows to control the water evaporation phenomena, which reduce defects of the timber caused by excessive or uneven dehydration. In addition, the multi-stage production process of glulam is based on mechanical and organoleptic sorting of the raw material, which eliminates discontinuities in the structure of the obtained material (so that during adsorption/desorption the glulam works freely, without local disturbances).

3. Experimental investigation

- *Aim of the measurements*

The aim of the investigation is to determine the magnitude and rate of changes in the moisture content of glulam (from the warehouse) and solid timber (raw) under normal use conditions.

- *Measurement specimens*

Six specimens (beams) were used for measurements:

- 3 beams from solid wood, pine, cross-sectional dimensions 10x20 +/-0,5 (cm), about 105 +/-2,5 (cm) length, raw (from own resources), initial moisture content not defined;
- 3 glulam beams, 10x20x100 +/-0,5 (cm), produced from spruce, („MUF" type glue), GL24h strength class, visual SI class, planed on four sides and chamfered along longer edges, humidity guaranteed by the manufacturer oscillates around 12,0 +/-2,0 (%);

(a)

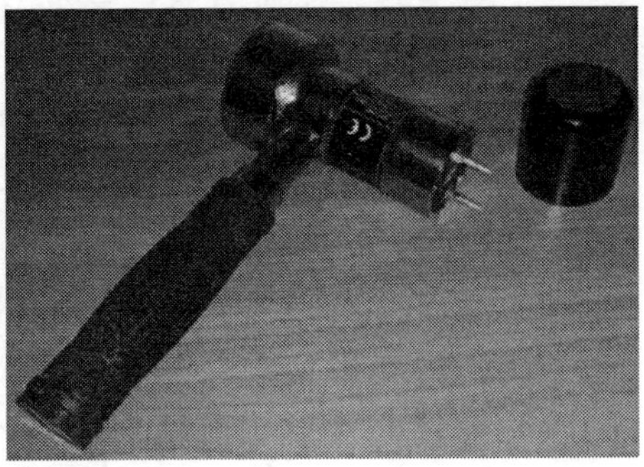
(b)

Figure 1. (a) View of the test stand. (b) Measuring device - „HIT3" hammer.

- *Measurements - stage I*

Timber beams after the sawing, were transported to the laboratory in natural conditions (without any protection). Glulam beams were purchased in the warehouse in a protective foil, which was a barrier mainly for surface contamination (it was not a tight humid insulation).

All beams and their walls were numbered (I-III beams are from glulam, IV-VI beams are from solid timber). Each beam was roughly, mechanically cleaned and placed in the laboratory about 20 (cm) above the ground so that a small part of the beam's edge touches the support (lower surface has free access to air) (figure 1 a). The laboratory had a natural atmosphere which was not controlled. During the measurements, the air humidity fluctuate between 26-46 (%) and temperature between 20.1-24.0 (°C).

Measurements were taken every 7 days. The measuring hammer "HIT3" was used for measurements (figure 1.b), calibrated in the „OUM" (District Office of Measures). Device specification:

- measuring range: 6-60 (%);
- measurement accuracy: +/- 1(%) (in the 6-15 (%) range), +/- 2(%) (in the 16-28 (%) range); about 10 (%) of the measured value (in the 16-28 (%) range);

- *Measurements - stage II*

The "HIT3" measuring hammer is equipped with needles which penetrate measured object into the depth of 5-7 (mm), through that the measurements in the first stage of the experiment can be considered as "superficial". In the second stage of research, the moisture content of the beams was measured according to section thickness (by destructive test).

IOP Conf. Series: Materials Science and Engineering 1015 (2021) 012076 doi:10.1088/1757-899X/1015/1/012076

After first stage of measurements, two selected beams (one from solid timber and one from glulam) were transported to the sawmill (which temporarily disturbed the ambient conditions). Then, on the stationary sawmill Pilous Forestor CTR 800, about 4-5 (mm) of the material was felled, without the use of coolant. After each cut, measurements with the usage of „HIT3" hammer were taken at 3 places of the beam: near the center and about 10-15 (cm) from ends of the specimen (figure 2). Ten cuts were made and each one reduced the width of the beam until the measurement was made at the symmetry surface of the original cross-section.

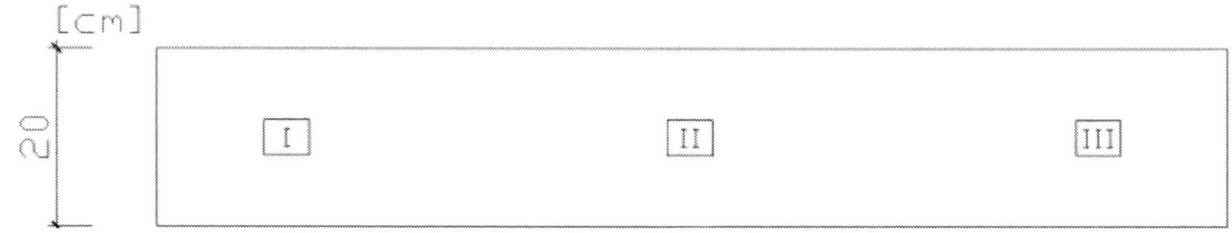

Figure 2. Measuring points of further layers of the beam.

4. Test results and data analysis

Moisture measurements were taken on each surface of the beam. Graphs below (figure 3 and figure 4) depict average values of moisture content level gained by beams, respectively by solid ones and from glulam.

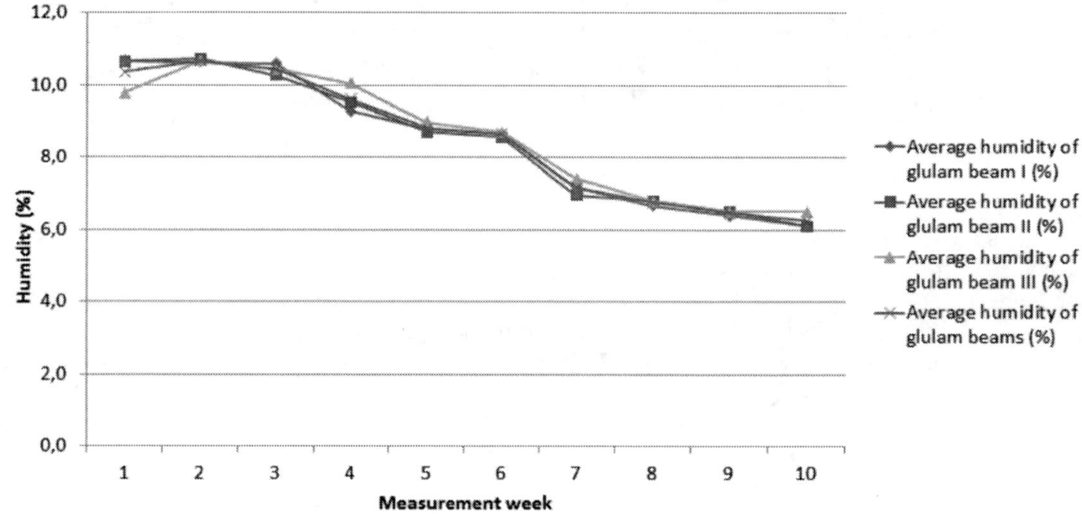

Figure 3. Average moisture content of glulam beams in each week of measurements.

During the tests, the moisture content of the individual glulam beams was dropping almost evenly (figure 3). The deviation of the results from individual beams towards the average from all measurements in a given week is smaller than the accuracy of the measuring device. Thereby obtained results can be considered as reliable. The average weekly decrease in humidity is 0.42 (%).

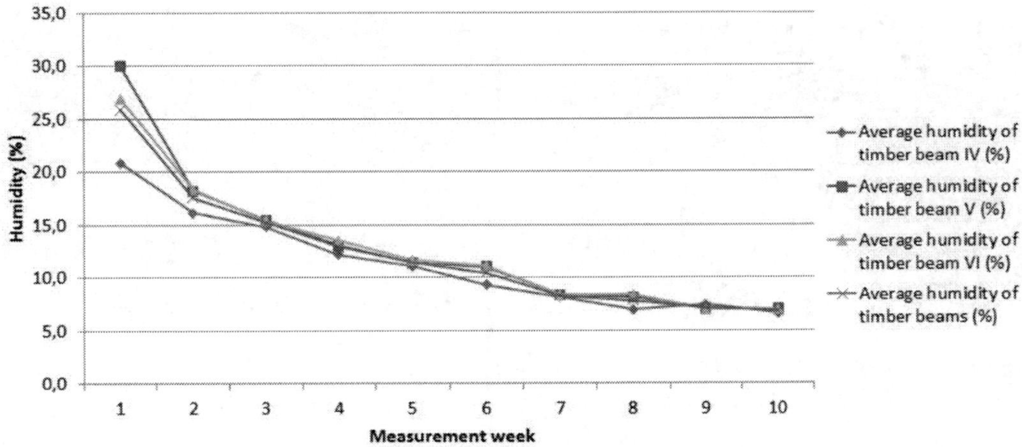

Figure 4. Average moisture content of solid timber beams in each week of measurements.

Beams made from solid timber, at the beginning of the experiment can be characterized as „unstable" (figure 4). Initial, high humidity decline rapidly (by about 40 (%) on the first week of measurements). During first 2-3 weeks of the experiment, moisture content of the beams normalized with a clear downward tendency. Following weeks of testing showed a steady decrease in humidity, where the measurements of each specimen did not deviate from the average by more than the accuracy of the measuring device.

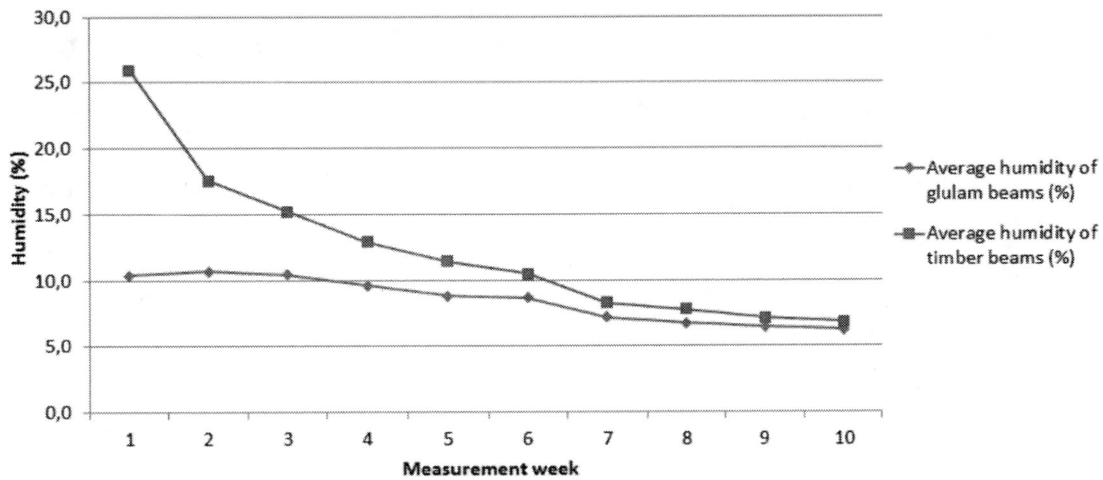

Figure 5. Average moisture content of solid and glulam beams.

On picture above (figure 5) we can see that the average moisture content of solid beams clearly drops down throughout the measurement period, stabilizing and approaching the average moisture content of glulam beams. Glulam beams behave more stable which can be caused by their earlier controlled drying in the production process. In the first phase of the measurements we can see that the average humidity of glulam beams increased slightly (by about 0.3 (%)) which may be caused by the inaccuracy of the measuring device (the accuracy of the hammer in the range of the prevailing environmental conditions is +/-1 (%)).

In order to carry out an organoleptic assessment of the behaviour of individual specimens, during the measurements photographic documentation was kept (figure 6 – figure 7).

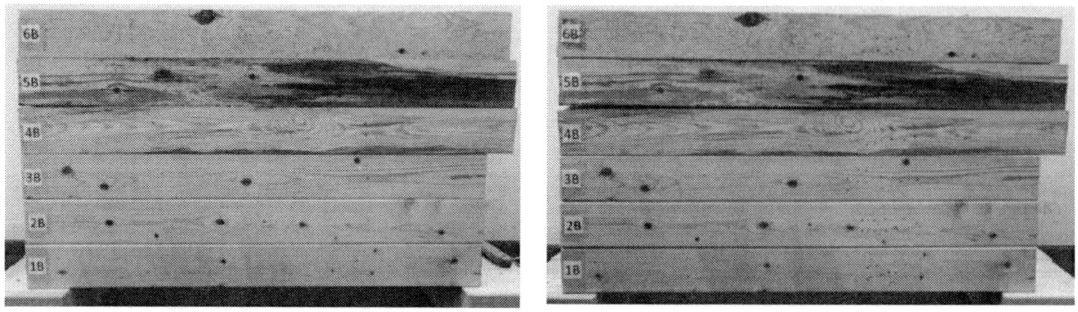

(a) (b)

Figure 6. (a) View of the beams before starting measurements, walls „B"
(b) View of the beams after the experiment, walls „B".

Comparing elements 1-3 (figure 6 a and b), there is a noticeable lack of "warping" of glued laminated beams. Each one retained the original shape despite an average change of humidity by more than 5 (%). Comparing elements 4-6 (figure 6 a and b), a "warping" of solid wood beams is visible (it should be noted that just after sawing all beams had a cuboid geometry). On both types of beams, on narrow walls, no cracks are visible (except for superficial, shallow cracks on beam no. 4, which has been significantly distorted).

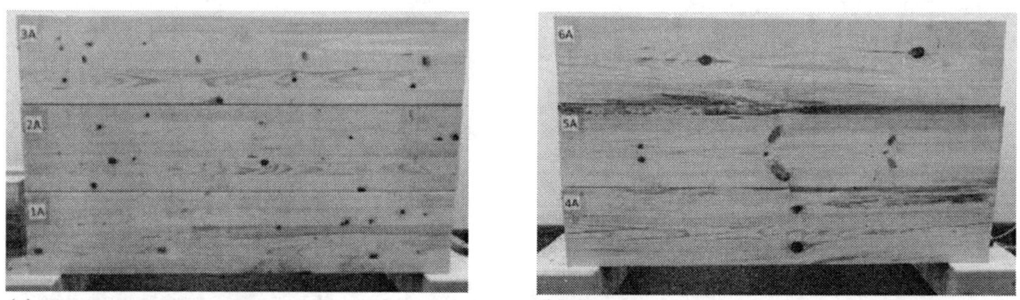

(a) (b)

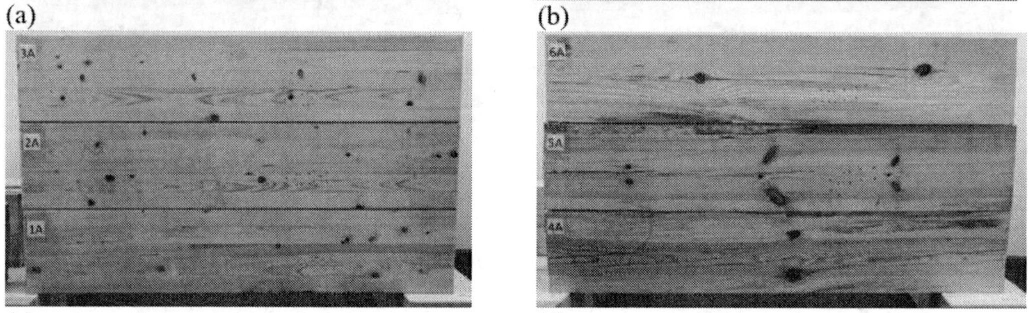

(c) (d)

Figure 7. (a) Glulam beams at the beginning of experiment. (b) Timber beams at the beginning of experiment.(c) Glulam beams at the end of the measurement. (d) Timber beams at the end of the measurement

Comparing wider surfaces of the glulam beams before starting the measurements (figure 7 a) with the same surfaces after the measurements (figure 7 c), there have been no major changes in the samples – only local, superficial, micro cracks appeared. On the next pictures (figure 7 b and d) we can see timber beams surfaces. On the picture (figure 7 b) cracks which appeared before the start of the tests are visible, practically just after the wood was sawn (in less than 1 day) cracks have appeared. Cracks in the middle part of the beams are conspicuous but shallow (less than 1 (cm)). At the ends of the elements, the cracks have a large opening (even 3 (mm)), but those are not propagated on the whole element. After moisture measurements, on solid beams additional cracks appeared (figure 7 d) (but only in the areas of attenuation of the beams - e.g. near knags). Existing cracks were elongated and their openness was slightly increased. In the middle parts of the solid beams, the cracks are superficial. At the ends of the beams, some of the scratches have a width of more than 4 (mm), but none of them causes complete delamination of the element. Sapstain appeared on beams number V and VI, which does not reduce the mechanical properties of the beams but affects the aesthetics and slows down the drying process of the element (especially at high humidity).

According to [6], we can distinguish six main directions cracks propagation (figure 9). The first letter gives the direction of the normal to the crack plane and the second letter indicates the direction of crack propagation. Because of the tension caused by water desorption (and the structure of the timber beams cutted out from the logs), during the experiment, most of the cracks propagate in "TL" direction, which is consistent with observations presented in other publications [6].

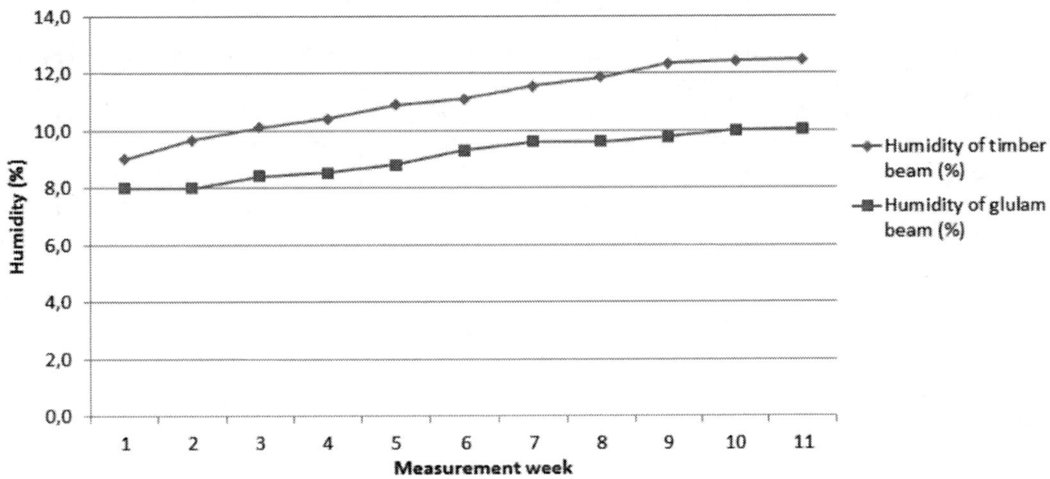

Figure 8. Graph of the average humidity level of solid and glulam beam towards to the subsequent measurement layers.

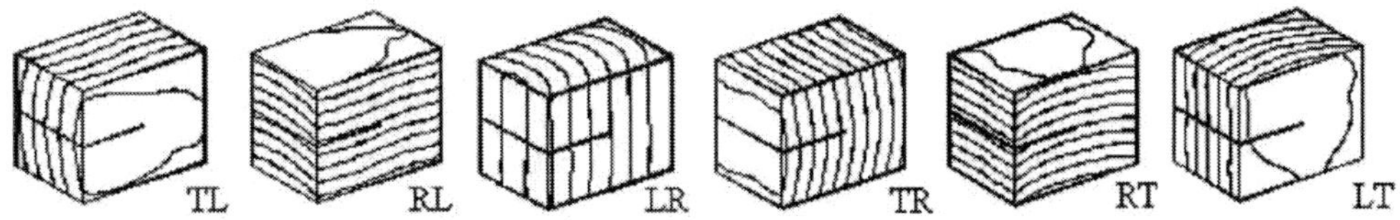

Figure 9. Six principal orientations of crack propagation in wood [6]

After the first stage of test, two beams were selected: one from solid wood and one from glulam. Then, the measurements were taken as described in in the chapter 3 of this paper. The graph (figure 8) shows the humidity of the beams as an average from 3 measurement points for each layer.

The diagram (figure 8) shows the change of humidity inside the specimens (from the outside-layer 1, to the inside of the beam - layer 11). The difference between the inner and outer part of the beam is about 2 (%) for a glulam beam and about 3.5 (%) for a solid beam. Both beams increase their moisture evenly towards to the central part of the measured specimen. The difference in humidity between the individual layers of solid and glulam beams is practically constant (apart from the two outer layers).

IOP Conf. Series: Materials Science and Engineering 1015 (2021) 012076 doi:10.1088/1757-899X/1015/1/012076

5. Summary and conclusions

On the basis of the obtained results, it can be concluded that:

- average humidity decrease of glulam beams over the entire measurement period was 0.42 (%/week);
- average humidity decrease of solid timber beams over the entire measurement period was 1.90 (%/week), and excluding the first part of the tests (the first 5 weeks of measurements) the decline was 0.48 (%/week);
- the rate of drying is important: solid timber beams attacked by sapstain dried out slowly in the initial stage of test and were not warped;
- glulam beams during humidity measurements behaved „more stable" than those from solid timber (decrease of humidity was steady and there were no cracks or warping);
- with a sufficiently long exposure time to weather conditions, the moisture content of glued laminated timber and solid timber beams is convergent;
- when cutting off subsequent measurement layers from timber and glulam beams, cracks caused by a change of humidity disappears (when cutting off a layer of about 10-15 (mm) thick, most cracks are already closed);
- most of the cracks caused by humidity decrease propagate in "TL" direction, which is consistent with observations presented in other papers;

Conducted research unambiguously indicate that the tested beams endeavor to equivalent humidity, which for an 26 (%) air humidity and temperature of about 20 (°C) is estimated at 5.0 (%) level. This value exceeds the range of measurement capabilities of conventional devices, so the tests were interrupted after 10 weeks (after that time, hammer "HIT3" was no longer able to make reliable measurements). Therefore, in order to perform measurements in the full range of the experiment it is necessary to use advanced measuring equipment or to perform destructive tests.

Studies have shown that glued laminated timber beams show much greater dimensional stability when the moisture content of the element changes. Additionally, the production process makes the glulam much more resistant to biological corrosion. Solid wood beams have warped at the initial stage of measurements (first week after sawing) where their moisture content dropped relatively quickly. It can be concluded that the care of solid, wooden elements in the first two weeks after sawing is crucial for their later suitability for construction purposes.

References

[1] PN-EN 1995-1-1

[2] Franke B, Franke S, Schiere M and Müller A 2016 Moisture diffusion in wood - experimental and numerical investigations *World Conf. Timber Eng.* (Vienna, Austria)

[3] Kozakiewicz P 2012 *Fizyka drewna w teorii i zadaniach* (Wydawnictwo SGGW)

[4] Adamson A 1963 *Chemia fizyczna powierzchni* (Wydawnictwo PWN, Warszawa, Polska)

[5] Krzysik F 1961 Wpływ wilgoci i wody na podstawowe własności drewna *Ochrona Zabytków* **2** (1–2) pp 50–65

[6] Ashby M and Easterling K 1985 The fracture and toughness of woods *Proc R Soc A* **398** pp 261–80

IOP Conf. Series: Materials Science and Engineering 1015 (2021) 012077 doi:10.1088/1757-899X/1015/1/012077

Research of qualitative properties of soils during construction works

Y M Galitskova

Academy of Architecture and Civil Engineering, Samara State Technical University, 194 Molodogvardeyskaya st., 443001, Samara, Russia

E-mail address: galickova@yandex.ru

Abstract. The development of humankind leads to an increase in the number and variety of sources of environmental pollution, to an increase in the volume of emissions of harmful substances. In urban conditions, environmental pollution occurs due to emissions of exhaust gases coming from various cars, leaks of fuels and lubricants coming from different equipment, etc. Analysis of the current state of urban areas shows the highest concentration of pollutants being present in soils. The article considers the problem of territories of construction sites contaminated with combustible substances coming from construction machinery, such as gasoline, kerosene and diesel fuel. These contaminants lead to the suppression and death of ornamental plants planted at the final stage of construction works. The authors describe in detail the methodology and results of laboratory studies of the process involving penetration of these contaminants into the soil. On the basis of the results obtained, recommendations are proposed for minimizing the negative impact of these factors on the soil and increasing the efficiency of work on the improvement of adjacent territories.

1. Introduction

Construction work has a negative impact on natural complexes. In the construction areas, one can observe high levels of air, water, soil pollution. This occurs at all stages of construction [1-3]: during design and survey work, during the construction of temporary access and adjacent roads, when arranging open pit mines for foundation, as well as directly in the process of construction of buildings and structures at the construction site. Each stage of construction is characterized by its own sources of negative impacts and characteristic pollutants. When performing any of the above works, the soils will be the components of the environment suffering the greatest from building mechanisms and materials, since they are the most inert environments with a low rate of self-cleaning process [4-6].

Under the influence of gravity, the upper layer of the soil is compacted, its porosity and moisture levels change, the soil may lose its ability to absorb and retain moisture, its water permeability decreases. Apart from changes in the physical properties of the soil, in the process of performing various works, soil contamination occurs due to bulk materials hitting the surface, as well as spills of fuel and liquid building materials, thereby causing changes in the chemical parameters of the upper soil layer [7-8]. Over time, contaminants that hit the surface penetrate into the soil assisted by rain and melt water. That is why it is not uncommon for the soil to be contaminated with heavy metals and various oil products [9-11]. With long periods of soil contamination and the achievement of high concentrations of oil products, irreversible changes occur in it, which further affect the biodiversity of the area and the vital activity of soil animals [12-13]. The thickness of the soil in which such changes take place depends on the type of soil, particle-size distribution of the original soil, as well as the precipitation depth and the amount of oil products that have hit the soil.

Responses to pollution from vegetation and wildlife are mixed. Small amounts of these pollutants may not cause any adverse effects if spilled into the ground. However, with significant volumes of pollution in the course of first few days, we observe the death of most soil animals for which the light fractions of oil are considered the most toxic.

At the end of construction work, landscaping improvement is carried out on the adjacent territories, i.e. playgrounds, flower beds and recreation areas are arranged. The possibility of soil contamination with oil products must be taken into account when carrying out such work. Since the processes of soil restoration in contaminated areas are slow, the ingress of oil products into the soil leads to a change in the photosynthetic functions of higher plants in the adjacent territories and slows down the germination or survival of annual and perennial plants planted for improvement [14]. Even low concentrations of oil products in soils lead to oppression of plants. The ingress of 1.1 l/m^2 of oil can lead to complete death. It can take up to 20 years to restore the soil so highly contaminated.

Many scientists are studying the process of penetration of oil products into the ground and the degree of its influence on vegetation [15-19]. As a rule, these are analytical predictions of the process involving penetration of substances into the soil, the spread of contaminants into the soil, the analysis of the dynamic characteristics of such spreads, as well as the determination of the amount of pollutants at different depths. It is also noted [18-20] that the degree of influence of pollution on vegetation depends on the quantitative indicators of the pollutant amounts in the soil. The purpose of this work is to conduct field studies of the process of penetration of oil products into the soil on a physical model. During the experiments, only the seepage of pollutants into the soil by gravity is investigated, while the lateral spread of oil products is excluded.

2. Methods

The results of any study involving soils depend on correct sampling and pretreatment. Therefore, at the preparatory stage of the research, soil samples were taken on the territory of the city in order to determine the particle-size distribution and type of soil. The sampling sites were outlined on a coordinate grid listing their numbers and coordinates. The sampling points were located at equal distances from each other. The results of the study show that the particle-size distribution of the soil in the urban area is close to the particle-size composition of river sand. In accordance with the theory of similarity, sand can be used for further laboratory research.

For the experiments, the most common fluids used in the operation of construction machinery at the construction site were selected: gasoline, kerosene, diesel fuel. A plastic pipe with a height of about 25 cm - the depth of penetration for the bulk of the root system of annual ornamental plants [20-21] was taken as a container for the experiments.

a) b)

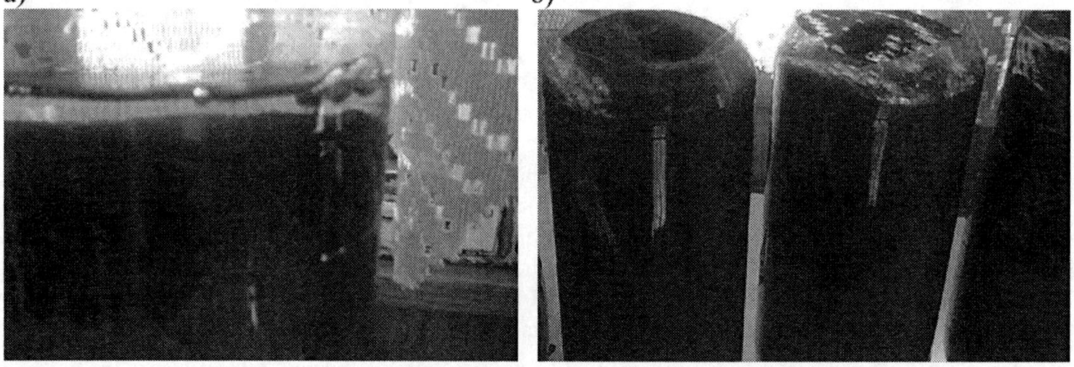

Figure 1. The process of penetration of contaminants into the soil (a – the first minutes of the experiment, b – the type of soil after 5 minutes).

The sequence of the experiments: The container was filled with sand. Continuous compaction was carried out during the filling. Then, using a measuring glass, the required volume of the test liquid was measured and poured onto the surface of the sand (figure 1). The volumes of 50, 100 and 150 ml took part in the experiment. The time of penetration of oil into the sand was recorded with a stopwatch. The depth of penetration was measured after 1, 3, 5, 10, 15, 30 minutes, then after 1, 2, 3, 6, 12, 18, 21 hours and one day after the moment of contamination.

3. Results and Discussion

Some research results are presented in the form of graphs reflecting the dependence of the penetration depth of the polluting liquid on the penetration time (figure 2-4).

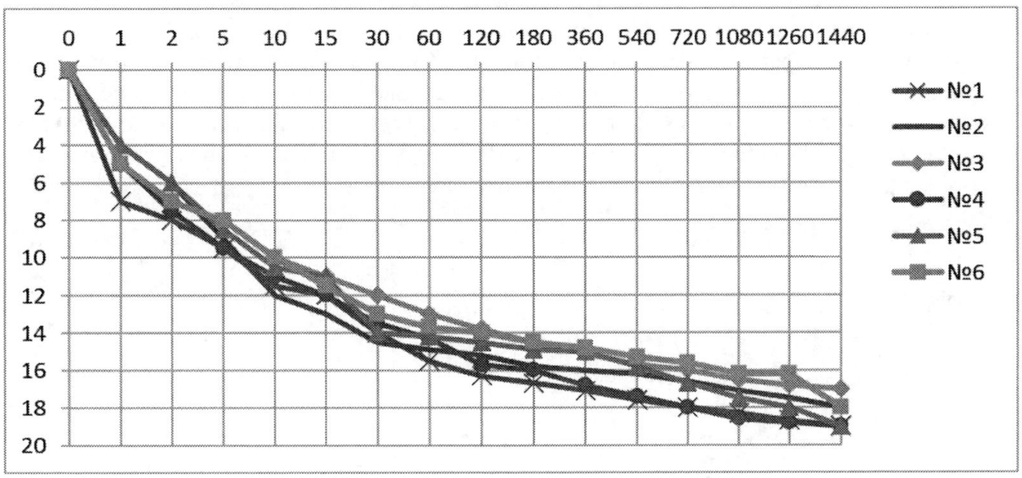

Figure 2. Results of a series of experiments on the penetration of kerosene (100 ml volume). Vertical – penetration depth in cm, horizontal – observation time in minutes.

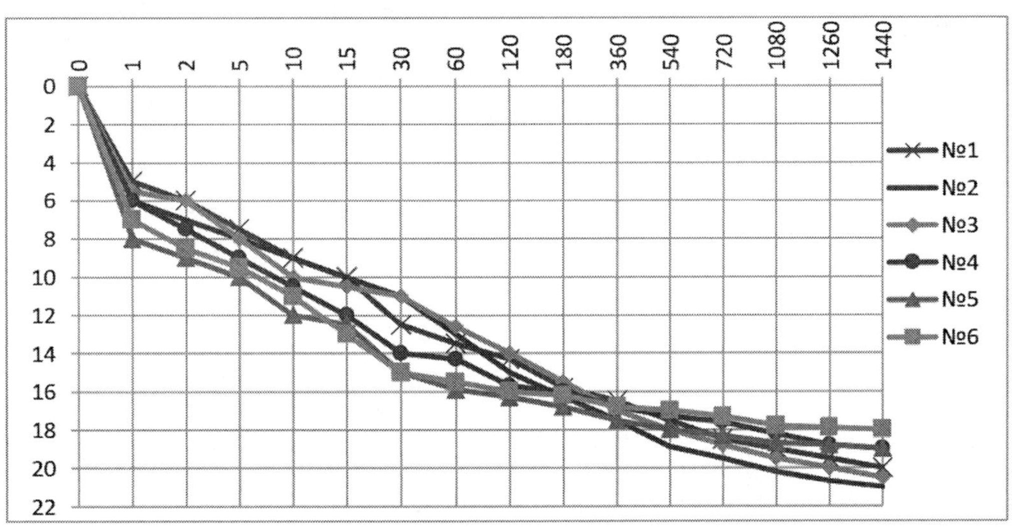

Figure 3. Results of a series of experiments on the penetration of gasoline (100 ml volume). Vertical – penetration depth in cm, horizontal – observation time in minutes.

IOP Conf. Series: Materials Science and Engineering 1015 (2021) 012077 doi:10.1088/1757-899X/1015/1/012077

The results of the studies have shown the unevenness with which the penetration of polluting liquids in the soil took place over time. The highest penetration rate is observed in the first hour after the ingress of contamination to the soil surface. With the smallest investigated volumes of contamination hitting the surface (50 ml), the soil becomes contaminated do the depth of 10 to 12 cm within an hour. If more liquid hits the soil (100 ml) the following is observed: diesel fuel penetrates 10-13 cm into the soil in an hour, kerosene penetrates 12-14 cm deep into the soil, and gasoline penetrates to a depth of 12-15 cm. The results of experiments with volumes of 150 ml showed that kerosene and gasoline reach a penetration depth of 21-23 cm after 30 minutes. Then there is a significant slowdown in speed penetration and the maximum penetration depth recorded in a day is 34-35 cm. It should be noted that the penetration of the liquid continued for several hours after all the liquid above the surface penetrated into the ground. Moreover, a larger amount of polluting liquid (up to 80%) is concentrated in the soil to a depth of 15-20 cm due to the liquid viscosity and the slowing penetration rate.

The graphs indicate that the higher the initial height of the liquid layer above the ground, the more intense the initial process of liquid penetration. Moreover, the dependence of the liquid viscosity and the intensity of penetration is observed. The viscosity of kerosene and diesel fuel is lower than that of gasoline, and the density is higher [22]; therefore, the graphs show that gasoline penetrates to greater depths and the process of gasoline penetration is more intensive.

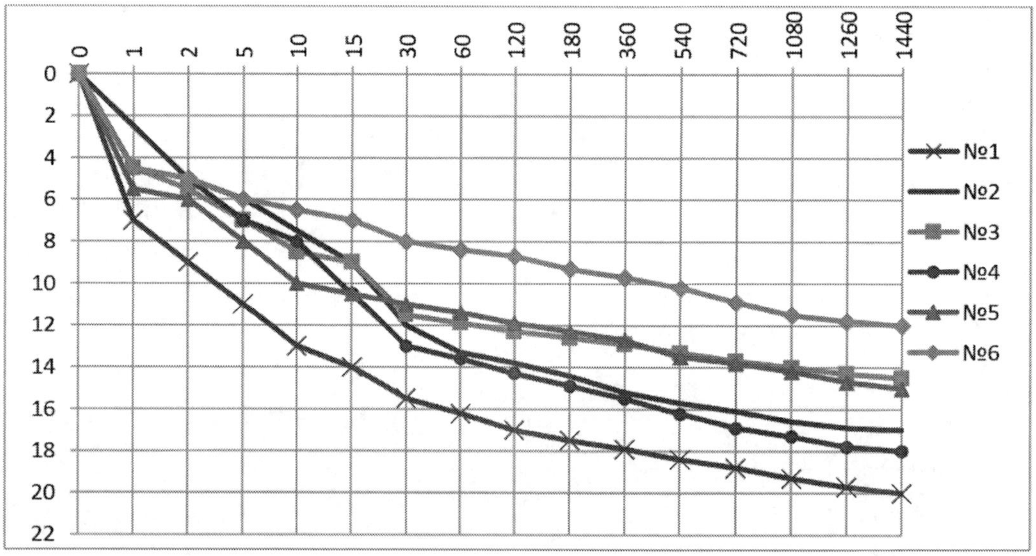

Figure 4. Results of a series of experiments on the penetration of diesel fuel (100 ml volume). Vertical – penetration depth in cm, horizontal – observation time in minutes.

4. Conclusions

The conducted studies of the process involving the penetration of oil products into the soil showed that the highest rate of penetration occurs in the first hours of the ingress of pollution to the soil surface. During construction work, it is necessary to immediately eliminate the spill of oil products to minimize the depth of pollution. This practice is typical for leaks on oil pipelines. On the territory of construction sites, the ingress of oil products in large quantities is extremely rare, therefore, uncontrolled contamination of the soil of construction sites with small volumes of oil liquids occurs.

As field studies have shown, by the time the improvement work begins, the soil can be contaminated to a depth of 25 cm and below. Consequently, in the places where ornamental planting is planned, it is necessary to remove at least 15 cm of the contaminated soil in order to prevent the plants death.

IOP Conf. Series: Materials Science and Engineering 1015 (2021) 012077 doi:10.1088/1757-899X/1015/1/012077

5. References

[1] Kozhevnikova E O 2017 The problem of dust pollution of urban areas during earthworks and ways to solve it *Alley of Science* **14** pp 676-88

[2] Goverdovskaya L G and Yushantsev A K 2015 Investigation of the processes of the impact of road construction work on the environment *Environmental safety of construction, Bulletin of SGASU. Urban planning and architecture* **1** (18)

[3] Islam M S, Ahmed M K, Al-Mamun M H and Eaton D W 2020 Human and ecological risks of metals in soils under different land-use types in an urban environment of Bangladesh *Pedosphere* **30** (2) pp 201-13

[4] Rodionovskaya I S, Trofimova T E and Sorokoumova T V 2016 Children's recreation in an urban environment *Scientific Review* **11** pp 112-16

[5] Xia W, Du Y, Li F, Wang F and Song D 2019 In-situ solidification/stabilization of heavy metals contaminated site soil using a dry jet mixing method and new hydroxyapatite based binder *Journal of Hazardous Materials* **369** pp 353-61

[6] Liu W, Liu S, Li X, Li Y and Mou Z 2020 Comparison of Chinese and Danish Soil Legislation Based on Soil Heavy Metal Values in Contaminated Sites: A Case Study in Sichuan *Soil and Sediment Contamination* **29** (3) pp 355-368

[7] Gainer A, Bresee K, Hogan N and Siciliano S D 2019 Advancing soil ecological risk assessments for petroleum hydrocarbon contaminated soils in Canada: Persistence, organic carbon normalization and relevance of species assemblages *Science of the Total Environment* **668** pp 400-10

[8] Al-Obaidy N, Al-Shueli A, Sattar H, Majeed Z N and Hamid A H 2019 An experimental study on geotechnical and electrical properties of an oil-contaminated soil at thi-qar governorate *International Review of Civil Engineering* **10** (3) pp 148-54

[9] Vesnikova A V 2018 Investigation of soil contaminated with residues of oil products *Chemistry. Ecology. Urbanism* **1** pp 337-41

[10] Oluremi J R and Adedokun S I 2019 Valorization of spent engine oil contaminated lateritic soil with high calcium waste wood ash *Journal of Engineering Research* **7** (1)

[11] Hewayde E, Abbas M and Kubba Z 2019 Influence of engine oil on geotechnical properties of cohesive soil *Int. J. of Engineering Research and Technology* **12**(1) pp 33-41

[12] Krüger A L, Snyman R and Odendaal J 2019 The impact of urban pollution on metal contamination of selected forest pockets in Cape Town *Environmental Science and Pollution Researc* (DOI: 10.1007/s11356-019-04679-0)

[13] Tabelin C B, Igarashi T, Villacorte-Tabelin M, Ito M and Hiroyoshi N 2018 Arsenic, selenium, boron, lead, cadmium, copper, and zinc in naturally contaminated rocks: A review of their sources, modes of enrichment, mechanisms of release, and mitigation strategies *Science of the Total Environment* **645** pp 1522-53

[14] Nazarov A V 2007 Influence of oil pollution of soil on plants *Bulletin of the Perm University* **5** (10) p 134-141

[15] Franchuk G M and Nikolayak M M 2007 Forecast of soil and groundwater pollution with oil products *Bulletin of BSTU im. V.G. Shukhova* **1** p 78-81

[16] Golovanov A I and Sychev S M 2008 Modeling the absorption of petroleum products into soils to substantiate ways to clean them from pollution *Melioration and water management* **6** pp 31-33

[17] Pankratova K G, Shchelkov V I and Sazonov Yu G 2007 Determination of the humus content in soil contamination with oil products *Fertility* **4** (37) pp 20-21

[18] Makarov O A and Kanzafarov E R 2016 Experience in assessing the resistance of urban soils to pollution and degradation *Ecology of urban areas* **4** pp 33-39

[19] Roeva N N, Voronich S S, Khlovaev A G, Zaitsev D A and Voronich NS 2018 Study of the dynamics of accumulation of mobile forms of heavy metals, benzo (a) pyrene and oil products in soils of urbanized areas *Ecology and Industry Of Russia* **22** (8) pp 39-43

IOP Conf. Series: Materials Science and Engineering 1015 (2021) 012077 doi:10.1088/1757-899X/1015/1/012077

[20] Khromovykh O V 2011 Influence of soil pollution with cobalt and manganese on the growth rates of some species of flowering ornamental plants *Problems of ecology and nature protection of a technogenic region* **1** pp 275-90
[21] Petukhova L V and Stepanova E N 2018 On the question of the intensity of the root systems of floral and ornamental plants *Bulletin of Tver State University. Series: biology and ecology* **1** pp 131-7
[22] Kuts Yu N, Korolchenko I A and Veksler G B 2012 Assessment of the conditions for the infiltration of oil products into the ground in case of corrosion damage to the shell of tanks *Izvestiya MGTU MAMI* **2** (14) pp 209-19

IOP Conf. Series: Materials Science and Engineering 1015 (2021) 012078 doi:10.1088/1757-899X/1015/1/012078

Requirements for engineering protection of structures and territories for the safe investment project implementation

P A Zhuravlev and A M Marukyan

Moscow State University of Civil Engineering, Yaroslavskoe shosse 26, Moscow, 129337, Russia

E-mail address: liza21@list.ru

Abstract. The article is devoted to the problems of engineering protection of buildings, structures and territories, when organizing territory, formalising principles (conditions), in accordance with which management and construction solutions for engineering protection arrangement are admitted at the planning and design stages. An integrated approach to the engineering protection arrangement is formulated on the basis of these principles (conditions). The approach requires preparation of an engineering protection project, including general layouts, detailed and specified plans, to develop various project solutions, optimize the design process, assess the prevented damage, justify the investment and estimate the consolidated approximate cost.

1. Introduction

The prosperity of the majority of Russia' population and successful development of the national economy are directly related to the cities and towns' environment. More than half of Russia's residents (64%) face with a shortage of public and business infrastructure within walking distance from home. The lack of trade, leisure-and-cultural infrastructure is typical for 70% of the territories of Russia's residential areas, formed both in the Soviet period and current condition. More than half of the residents of medium and large cities in Russia face with low transport accessibility problems, traffic jams and shortage of car parks.

About 20% of the Russia's residents live in cramped conditions and need more living space. A lack of parks, squares, low planting level near home, poor ecological situation, noise and atmospheric pollution are especially noted. As a consequence, formation of a high-quality urban environment should make a dent in the solution of the problems, which the majority of Russian residents face with every day.

Spatial development strategy of the Russian Federation for the period up to 2025, approved by the Russian Federation Government order dated 2.13.2019 No. 207-r, is developed to ensure sustainable and balanced spatial development of the Russian Federation, aimed at reducing interregional differences in the living standards of the population, accelerating the pace of economic growth and technological development, as well as ensuring the national security of the country.

As a consequence, formation of spatial development is acquiring a complex structure, covering the leading development areas (socio-economic, ecological, investment ones) [1-5].

Implementation of the main objectives of the national project "Housing and Urban Environment", which are to ensure an efficient land use, increase the level of housing construction, provide affordable housing, develop cities and towns in a holistic way, cardinally increase the comfort level of the urban

IOP Conf. Series: Materials Science and Engineering 1015 (2021) 012078 doi:10.1088/1757-899X/1015/1/012078

environment, increase the level of constant reduction of unlivable housing accommodation, leads to the increase in building construction.

In this regard, an important component of the integrated (spatial) organization of the territory is an engineering protection of buildings and territories, ensuring infrastructure construction, reasonable placement of construction objects in the ecosystem while preserving its natural relationships, protection of constructed or under construction objects from hazardous natural influences, as well as protection of the population from the risk of possible consequences of introduction into the ecosystem as a result of disruption of natural processes during construction.

2. Methods and materials

The objective of the work is to formulate principles (conditions), based on the current standards and legal requirements, in accordance with which management and construction solutions for engineering protection arrangement are admitted at the planning and design stages.

The Urban Planning Code of the Russian Federation defines the concept "activities for sustainability of the territories", according to which preparation and approval of documentation for the spatial planning of the capital construction objects for residential, industrial, public, business and other purposes and necessary for the functioning of such objects and ensuring the livelihoods of citizens communal, transport, social infrastructures, architectural and construction design, building construction, reconstruction of these objects are carried out in order to ensure the most efficient use of the territory.

Development of the territory planning projects is carried out to identify elements of the planning structure (district, microdistrict, quarter, common area, transport hub, linear facility, road network), set the boundaries of common and planned capital construction areas, characterize and prioritize the planned spatial development.

It should be noted, that the most important elements of the spatial planning project rationale are:

- a list of measures to protect the territory from natural and technogenic emergencies;
- a site plan, including vertical leveling, ground development and engineering protection of the territory.

Urban planning and technical solutions, taking into account conditions of safety and favorable human environment, limitating the negative impact of economic and other activities on the environment and ensuring the protection and rational allocation of natural resources, provide sustainability of a territory for the benefit of the people.

Modern design requirements take into account the assessment of the impact on the environment of the technogenic loads, but urban planning solutions, adopted without considering the dynamics of the territory properties, basing only on the social, urban planning, technical, economic and environmental requirements, lead to a contradiction with the natural-territorial resource [6-11].

A fairly common negative manifestation of the anthropogenic impact on the natural environment, that is wide spread, durable and large scale, is flooding of residential areas. The main causes of flooding are: water table rise, lack of storm sewers in settlements, as well as uncontrolled landscape disruption, increasing urbanization and development, accompanied by an increase in the area of waterproof coatings, which cause the maximum flow in cities [12-14].

In addition to the above impacts, the territory development is influenced by natural hazards (NH) and unfavorable weather conditions (UWC).

For example, according to the Federal State Statistics Service from 2010 to 2018 dangerous hydrological phenomena and heavy rains, snowfalls, heavy hail occupy more than one third of emergencies (33.73%) from all natural emergencies, that took place on the territory of the Russian Federation, and are 16.72% and 17.01% respectively, only large wildfires (28.85%), as well as frost and drought (19.53%) are less (they total 48.38%).

According to the data [13, 15-18] the territory of the Far Eastern, Southern Federal Districts, Irkutsk region had undergone the greatest impacts of moisture-saturated air masses. The regions with the greatest risk of heavy rainfalls, rains are concentrated in the areas, where high intensity of precipitation is combined with high density of population and economic entities.

Authors [19] report the results of the data analysis for the period from 1991 to 2017 on the changes in the frequency of dangerous and unfavorable hydrometeorological nature phenomena, distributed linearly with a clear increase in the frequency of phenomena as observations continue. The research notes an increasing trend of changes of dangerous hydrometeorological weather phenomena, affecting some sectors of the economy, as well as the population. A clear growing trend of damage to housing and utilities (including linear infrastructure) is noted among the considered sectors (agriculture, power industry, motor transport enterprises, housing and utilities).

Authors [20], applying the results of the data analysis on the natural hazards and unfavorable weather conditions, suggest indices for assessing the intensity of the impact on the social and economic systems of Russia. The first index shows the effect of recurrence of the NH and UWC in a specific area on the population density of this area. The second index shows the effect of average annual density of the NH and UWC on the gross regional product, thus being an assessment of the impact on the economic component.

In terms of the intensity of the impact of the NH and UWC on the socio-economic system the most unfavorable economic area is the North Caucasus region. Among the other Russian Federation's subjects, where the probability of the hazardous impact on the population is great, Moscow and Moscow region, Samara region and the Republic of Tatarstan are noted.

An example of the tragic impact of the NH and UWC on the regional socio-economic system is the 2019-s catastrophic flooding of settlements of the Irkutsk region, as a result of which 25 people died, more than 160 settlements and 10 000 houses, most of which cannot be reconstructed, were flooded. The economic damage is estimated at more than 35 billion rubles. Earlier in 2013 there were no less severe catastrophic floods in the Amur, Magadan regions, the Republic of Sakha (Yakutia), Primorsky and Khabarovsk regions, as well as in the Jewish Autonomous Region with a comparable economic damage.

The list of publications [21-26] should be noted, the authors of which shows, that anthropogenic impact enhances the negative effect of the unfavorable hydrological phenomena on the landslides and mudflows activity (for example, on the territory of Greater Sochi, right bank of the r. Volga near Ulyanovsk and Volgograd). Authors list the following main kinds of technogenic impacts: unauthorized storage of spoils on the slopes, cutting the slopes off during the construction of Olympic infrastructure, uncontrolled redistribution of the surface and underground runoff, development of the coastal (slope) territory. As a consequence, an increasing technogenic load on the slopes, low natural area stability (strength reduction of the ground) provoke soil deformations under the impact of hydrological phenomena, that leads to dangerous negative consequences for humans, economic facilities and the environment.

Thereby, the integrated development of territories depends on a large number of factors and requires taking into account:

- geographical, meteorological, seismic features of the Russia's territory;
- terrain and hydrographic conditions (reservoir configuration);
- design natural restrictions (flooding areas or areas with possible uneven subsidence of the earth's surface or ground shear after underground mining);
- heterogeneity and diversity of climatic, landscape, geocryological, geological, hydrogeological conditions;
- technogenic impacts.

3. Results

Any construction (of a facility or group of facilities) is preceded by site preparation, aimed at ensuring the necessary conditions for quality and timely erection of buildings and structures and including ground development and engineering protection.

An engineering protection is a set of measures, ensuring favorable conditions for construction and operation of settlements, setting-out and erection of buildings, laying of streets, utilities and other facilities of urban planning with the obligatory consideration of environmental requirements.

IOP Conf. Series: Materials Science and Engineering 1015 (2021) 012078 doi:10.1088/1757-899X/1015/1/012078

Engineering protection of the residential territory includes the following measures:

- vertical leveling of the earth's surface, providing the most expedient and economical conditions for placing buildings and structures, drainage of rain and melt water to places of discharge into water treatment facilities and reservoirs;
- creation of the necessary longitudinal slope of the streets and roads for car and pedestrian traffic, as well as laying of the underground engineering networks of gravity sewerage and drainage;
- territory protection from flooding during high water, lowering the watertable in the areas where it is needed, drainage of wetlands and anti-landslip and anti-mudflow measures, watering and irrigation of the territory in arid regions;
- control of ravine formation and erosion;
- recovery of the areas disturbed by human activity (waste dumps, spent quarries etc.).

Dangerous processes and phenomena, which may develop on the future construction territory, formed under the above factors, proves the necessity for their detailed research, forecasting and assessing risks, as well as the development of engineering protection measures from hazardous processes for both territories and buildings and structures.

Engineering protection of the territory always was an integral part of the infrastructure construction works, and at the same time, relatively recently, it began to stand out as a distinct large-scale section of the project documentation, having the city- and object-forming value.

The necessity for engineering protection of the territory from natural emergencies, caused by hazardous natural processes and phenomena, is determined in accordance with the Urban Planning Code in terms of urban development of the Russian Federation's subjects, cities and towns, taking into account the risk assessment of hazardous geological processes:

- in the general layout project, taking into account the variability of planning and technical solutions, for territories under development and reconstruction;
- in the construction, reconstruction and overhaul projects of buildings and structures, taking into account existing planning solutions and customer requirements, for built-up territories.

A complex of measures, preventing flooding of territories, should be developed when designing engineering protection of the territory against flooding according to the requirements for the functional use of this area and environmental protection or elimination of the negative effect of flooding.

The system of engineering protection against flooding should be geographically consolidated and should unite all local services of separate areas and objects. At the same time, it should be connected with general layouts and urban planning integrated plans.

Engineering protection measures should be designed comprehensively, taking into account possible deformations and impacts, the level of responsibility and cost of the protected areas, buildings and structures, their structural and operational characteristics.

Design of the urban and rural planning and development projects on the territory of the Russian Federation is carried out taking into account the requirements of SP 42.13330.2016. An engineering protection from floods, inundations, mudflows, landslips and avalanches should be included in the designing project, if necessary.

Surface water drainage should be carried out through closed rainwater sewerage with preliminary wastewater treatment. Application of the open sewerage facilities, which are ditches, cuvettes, trays, is permitted in the areas of one- and two-storey buildings and rural settlements, as well as in the parks with bridges or tubes at the intersection with streets, roads, driveways and sidewalks.

Lowering the watertable by closed drains in the area of capital construction should be provided in settlements with high watertable and wetlands.

Territories of settlements, located on coastal areas, must be protected from wind setup, inundation during high water and from groundwater flooding by ground bedding (inwash) or embankment.

Maximum preservation of the forest, planting of trees and shrubs, slope terracing, strengthening the banks of mudflow rivers, construction of barrages and weirs in the mudflow formation zones, dams, changing the mudflow direction, and tail races on the fan are necessary to be provided in hazardous mudflow areas to protect existing buildings.

IOP Conf. Series: Materials Science and Engineering 1015 (2021) 012078 doi:10.1088/1757-899X/1015/1/012078

Streamlining of surface runoff, interception of groundwater flows, protection of the natural buttress of the landslide massif from destruction, increasing the stability of the slope by mechanical and physico-chemical means, slope terracing, planting green spaces are necessary to be provided in cities and towns, located in landslide-prone areas.

Anti-landslip measures should be carried out basing on the comprehensive study of geological and hydrogeological area conditions.

When planning and building settlements, environmental safety and public health protection requirements should be fulfilled, measures for nature protection, rational application and reproduction of natural resources, improving the environment should be provided for. The regulations and standards, establishing the quality of atmospheric air, water, soil, as well as level of noise, vibration, radiation and other factors of natural and technogenic origin, are necessary to be complied.

According to SP 104.13330.2016 area protection from flooding should be carried out by the following ways:

- area embankment from the river, reservoir or other water facility. An embankment plan of the protected territory should be selected on the basis of a technical and economic comparison of options, taking into account the requirements of standards;
- grade elevation of the completely, temporarily or partly flooded areas should be artificially raised to the flood-free elevation for the building development;
- accumulation, regulation, removal of surface waste and drainage waters from the completely or temporarily flooded and lowland disturbed areas. At the same time, facilities, regulating surface runoff in the areas, protected from flooding, should be designed taking into account the estimated surface water discharge on these territories (rain and melt water, temporary and permanent streams) and according to the class of the protective facility.

When designing engineering protection of the permafrost residential areas, warming effect of the settlement and city development, reduction of evaporation from the surface of built-up areas and roads, increased snowdrift extent, significant thawing and watering effect of the thermal services and utility collectors, water pipes and sewerage, causing bedding and foundation deformations, should be considered.

Engineering protection systems should be designed taking into account features of environmental, sanitary and hygienic and antiparasitic requirements for each natural zone, as well as data of territorial integrated environmental protection schemes.

To protect territories from flooding, the following facilities should be applied:

- drainage systems;
- groundwater barriers and cutoffs;
- vertical leveling of the territory with the control of surface runoff, cleaning open courses and other elements of natural drainage, regulation of the water level in reservoirs.

Protective structures in flooded areas should be selected depending on the flooding kind (permanent, seasonal, episodic one) and the amount of damage it cause.

A wide range of measures and technical means are applied for engineering protection of buildings, structures and territories from hazardous processes, they are:

- regulation and control of surface runoff;
- bank strengthening and coastal slope improvement;
- consolidation and improvement of ravine areas;
- barrage water protection systems: diversion ditches, head and coastal drainages, groundwater cutoffs;
- horizontal drain system;
- local protection systems – drainages (pipe, flat, wall), waterproofing, groundwater barriers and cutoffs;
- vertical wells and special drains, i.e. combined, vacuum, ventilation, pneumatic ones;

IOP Conf. Series: Materials Science and Engineering 1015 (2021) 012078 doi:10.1088/1757-899X/1015/1/012078

- landslip protection measures: changes of terrain and watercourses, drainage, redistribution and strengthening of soil, construction of protective and regulatory structures;
- strengthening structures, i.e. retaining walls (on a natural or pile foundation), including reinforces ground structures, piles and pillars, anchors and dowels, buttresses, gabions;
- catchers: walls, nets, shafts, trenches, shelves with curb walls, posts;
- protection galleries;
- agroforestry, protective coatings and soil consolidation;
- mudflow control facilities and measures: mudflow protection structures (mudflow check-dams), mud flow ducts (check canals, mudflow chutes), mudflow-directing structures, slope stabilizing structures (retaining walls, drainages), channel stabilizing structures (debris dams);
- filling (grouting) of karst sediments during karst-suffusion processes (when designing tunnels);
- a set of measures to localize possible deformations of the surrounding buildings along the tunnels, routes: structural measures (for example, soil consolidation) and technological measures (advanced temporary support for mining methods, mechanized shields with face protection and sealing of the construction gap, slurry wall method, soil stabilization and others);
- technical measures of engineering protection of the above the tunnel under construction territory: compensatory filling of water into the ground after water table depression, artificial thawing of soil after its freezing, prevention of ground and surface water pollution with harmful substances during chemical soil consolidation, grouting or compensation injection of stabilizing compounds into the soil and other methods.

4. Conclusions

Organization and design of engineering protection of buildings, structures and territories, as spatial planning development factor, are carried out both at the stage of formation of spatial planning documentation and during architectural design stage taking into account the requirements and regulations for the project documentation development. Engineering protection includes measures and structures, designed and implemented at the site preparation, erection and operation stages of the capital construction. Engineering protection ensures safe operation of objects, territory and stay of citizens from all kinds of natural and technogenic hazards and dangerous geological processes.

In this regard, engineering protection of buildings, structures and territories is an important part of the integrated (spatial) territory organization, it provides infrastructure construction, reasonable placement of construction objects in the ecosystem, preserving its natural relationships, protection of constructed or under construction objects from hazardous natural impacts, as well as protection of the population from the risk of possible consequences of introduction into the ecosystem as a result of disruption of natural processes during construction.

Thus, based on the foregoing, it may be concluded, that engineering protection of buildings, structures and territories has a significant meaning in the infrastructure construction implementation and requires an integrated approach. Wherein a specific engineering protection project should be designed, including general layouts, detailed and specified plans, to develop various project solutions, optimize the design process, assess the prevented damage, justify the investment and estimate the consolidated approximate cost.

References

[1] Klochkov V V 2019 Interrelation of strategies of scientific and technical and spatial development of Russia *Rus.: trends and prospects for development* **14-1** pp 101-6

[2] Kolmakova E M, Kolmakova I D and Degtyareva N A 2018 Spatial development of the region in the context of the socio-economic growth strategy *Bullet. of the SUSHPU* 60 **3** (413) pp 30-37

[3] Kolmakova I D, Kolmakova E M, Silincev V S and Garipov R I 2016 Territorial planning of balanced socio-ecological and economic development *Managing the balanced development of territorial systems: issues of theory and practice Chelyabinsk* pp 250-91

[4] Kurushina E V and Petrov M B 2018 Criteria for success of spatial development projects based on interregional integration *Reg. Ec.* 1 **14** pp 176-89

[5] Andreev A V 2018 Prospects for spatial development of the Primorsky territory *Azimuth of Sci. Res: Economics and management* 2 (23) **7** pp 22-26

[6] Karpova N V 2019 Sustainable development of urban settlements : theoretical postulates and their practical implementation *Economy and ecology of territorial entities* **3** pp 64-70

[7] Korchak A V 2014 Problems, directions and ways of solving the problems of developing the underground space of megacities *Mining inf. and analyt. Bullet. (Sci. and Tech. J.)* **5** pp 89-98

[8] Abramyan S G 2009 Environmental support for the construction of linear-extended structures *Proc. of the MSUCE* **3** pp 114-9

[9] Kulikova E Y, Korchak A V and Levchenko A N 2004 Analysis of risk factors in the construction of urban underground structures *Mining inf. and analyt. Bullet. (Sci. and Tech. J.)* **11** pp 5-11

[10] Dushkova D O and Evseev A V 2011 Analysis of technogenic impact on geosystems of the European North of Russia *Arctic and North* **4** pp 1-34

[11] Zajchenko E N 2016 Features of green rationing and design in modern conditions *Mod. Tech.* **18** pp 71-9

[12] Faleev M I, Chernyh G S and Starostin A S 2014 Assessment of hazards and threats caused by catastrophic floods and proposals for protecting the population and territories from them *The Strat. of Civ. Prot.: Iss. and Res.* 1 **4** pp 18-32

[13] Avakyan A B and Istomina M N 2013 Natural and anthropogenic causes of floods *The Strat. of Civ. Prot.: Iss. and Res.* 1 **3** pp 269-81

[14] Belousova A P Assessment of the risks of flooding of territories with groundwater during floods (In Russian) *Water resources management* 1 **38** pp 30-38

[15] Timofeeva S S, Eglit V E and Morozova O V Flood monitoring in the Irkutsk region based on a retrospective analysis *Bullet. of the IrGTU* **9** (56) pp 82-89

[16] Kuz'mina Z V and Treshkin S E 2018 Climate changes in the Amur basin and their impact on ecosystems *Ecosystems: ecology and dynamics* **4** (2) pp 5-60

[17] Razumov V V, Razumova N V and Pchelkin V I 2015 Extent and risk of flooding in the Siberian region of Russia *The Sci. Inn. Tech.* **4** pp 103-44

[18] Belyakova A Y, Vashukevich E V, Ivan'o Y M and Petrova S A 2014 Trends in variability of natural events in the South of Eastern Siberia *Bullet. of the IrGTU* **10** (93) p 80-5

[19] Shamin S I and Sanina A T Analysis of the intensity of the impact of dangerous weather conditions on the socio-economic system *Proc. of VNIIGMI-MCD* **184** pp 54-66

[20] Korshunov A A, Rybanova A Y, Fokicheva A A and Shajmardanov M Z 2018 Analysis of the intensity of the impact of dangerous weather conditions on the socio-economic system *Sci. No. of the RSHU* **53** pp 18-33

[21] Krestin B M and Mal'neva I V Landslide and mudslide activity on the territory of Greater Sochi and its changes at the beginning of the XXI century *Geoecology* **1** pp 21–9

[22] Latypova V Z, Muhametshin F F and Gorshkova A T 2016 Landslide in Ulyanovsk 05.04.2016 as a classic example of a man- made disaster of the natural and man-made system of the right Bank of the Volga river *Ulyanovsk press* http://ulpressa.ru/2016/04/20/809050

[23] Mal'neva I V 2017 Natural disasters associated with dangerous geological processes and their prediction *The Earth's Life* **39** (1) pp 12-25

[24] Brylev V A and Korhova Y A 2010 Ancient and modern landslides of the Lower Volga region and factors of their formation *Geomorphology* **4** pp 37-47

[25] Konyushkov V V 2017 Engineering protection of the territory from slope processes taking into account natural conditions and man-made impacts *Bullet. of Civ. Eng.* **2** (61) pp 137-42

[26] Belov A A, Kiryushin A V and Maskajkin V N 2017 Engineering preparation of the city territory in case of flooding *Sci. Rev. Sect II. Sci. and Pr.* **1** pp 1-13

IOP Conf. Series: Materials Science and Engineering 1015 (2021) 012079 doi:10.1088/1757-899X/1015/1/012079

Principles of construction technical expertise (expert examination) informatization

N V Lazareva[1], A Y Zinoviev[2] and E M Kochenkova[1]

[1] Department of Technology, Organization and Management in Construction, Moscow State University of Civil Engineering, Yaroslavskoe shosse 26, Moscow, 129337, Russia

[2] Department of Construction of Thermal and Atomic Power Stations, Moscow State University of Civil Engineering, Yaroslavskoe shosse 26, Moscow, 129337, Russia

E-mail address: liza21@list.ru

Abstract. The article is devoted to principles of informatization of construction technical expertise (forensic, pre-trial and corporate ones). Information models of capital construction objects are considered in the article, in relation to the area of expertise they should include three components: standard (conceptual), basic (design), and real one. Modern information models, as a rule, include only the first two components, but it is the third component, that is crucial in the development of informatization of construction technical expertise. In this regard, the article defines the process of informatization of construction technical expertise and specifies its components: mediatization, computerization, intellectualization. Each informatization component has its own functionality, connected with its purpose, so the authors present their structuring and functional decomposition. The practical significance of the states, given in the article, is related to improving the efficiency and objectivity of construction technical expertise, as well as ensuring the quality of work, control of the cost and duration of contracts or services.

1. Introduction

There are conflict (disputable) situations between the investment-construction participants during all the life cycle of the capital construction object, and claim administration of business entities is aimed at resolving these disputes [1]. In some cases there is a need for an independent and objective construction technical expertise. It may be classified according to its institutional affiliation in a following way:

- *forensic* expertise, which is ordered by court;
- *pre-trial* expertise, which is initiated by investigating authorities as part of the offence investigation or by business entities in preparation for a trial to substantiate their position;
- *corporate* expertise, which is carried out by business entities for their own needs (for example, as a part of internal inspection or to remove questions or discrepancies in negotiations with contractors).

To implement these measures, structured full and electronic information about capital construction facility and all its changes throughout all the life cycle are needed [2-4]. Such set of interrelated information is called a construction information model [5-7].

If looking at the concept and methodology development of this area of expertise, capital construction information models are at their starting point. In our opinion, their progress is connected with the creation and operation of three components, which are standard (conceptual), basic (design), and real

IOP Conf. Series: Materials Science and Engineering 1015 (2021) 012079 doi:10.1088/1757-899X/1015/1/012079

one. The objective of this research is to form a methodological genesis of the organization of information support for construction technical expertises.

The standard (conceptual) element of the capital construction (CC) information model is embodied in the design specification and economical evaluation. The basic (design) element is design documentation. The real element is a parametric model of the real building or structure condition.

Modern information models, as a rule, include only the first two elements, and the third one is limited to information about the standard durability of structural elements [8, 9]. However, it is this element, that play a crucial role in the development of the construction technical expertise informatization.

Construction technical expertise informatization is a single organization and technical process, providing an objective and independent examination of the design solutions and economic calculations on the basis of formation and operation of the capital construction information models.

2. Methods and materials

CC information models are divided into automated (operate, basing on the computer-aided design systems) and manual (allow the storage of capital construction information both in electronic and paper form) ones.

Further in the article information models are interpreted as automated capital construction images.

In this regard, organization of the construction technical expertise informatization includes three components: mediatization, computerization and intellectualization. Each of the component processes has its own functionality, connected with its purpose [10-15].

Mediatization is a comprehensive process, associated with the increased availability of information, which leads to the quantitative and qualitative changes in the construction technical expertises and claim administration organization.

The following functions may be distinguished, basing on the definition:

- collection of information;
- storage of information;
- distribution of information.

The development of the "information society" and "digital economy" is impossible without constant evolutionary upgrade of the computer-aided design systems and technical devices. Thus, *computerization* is a technical component of the construction technical expertise informatization process, its functions are processing and display of information.

Intellectualization is the development of the information perception ability in terms of construction technical expertise, its intellectual potential improvement. It may be to provide decision-making and report (expert opinion) formation.

Informatization as an organizational system has a complex multilevel structure, which is constantly evolving due to objective reasons. However, taking into account the inertia of the construction technical expertise as field of activity, structuring and functional decomposition can be performed in a first approximation.

3. Results

In this regard, the basic mediatization structural elements (figure 1) may be determined on the basis of its definition, they are the object and users (consumers). The first element points at the source of information, which in turn is the third structural element of the construction technical expertise. The second element shows, that users need necessary and sufficient structured information about a capital construction object.

A real information model, in its turn, serves as a structural element of another component, which is construction technical expertise computerization (figure 2), intended in terms of functional decomposition to display information about a capital construction object.

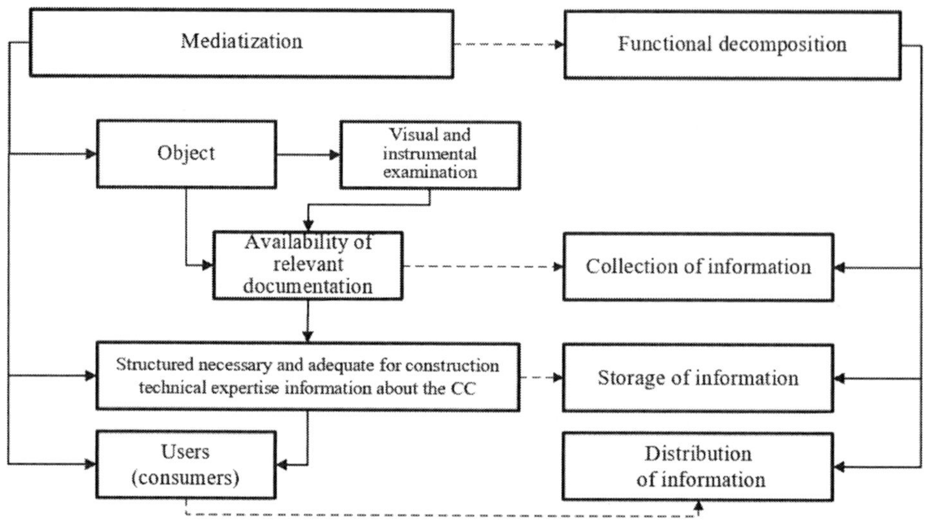

Figure 1. Functional decomposition and structure of the construction technical expertise mediatization.

To adequately display necessary and sufficient information about a building or structure an initial data needs to be structured and processes, and this requires computer-aided design systems and technical devices.

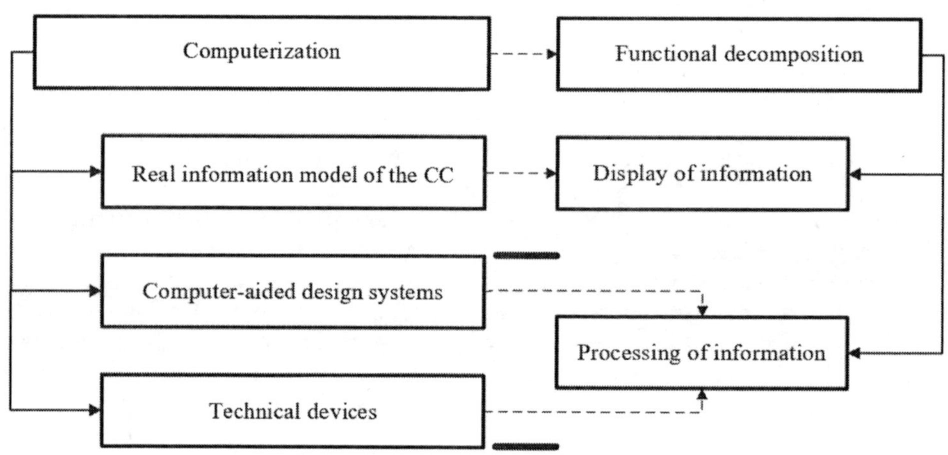

Figure 2. Functional decomposition and structure of the construction technical expertise computerization.

To support decision-making, intellectualization (figure 3) requires an appropriate visualization, related to the capital construction information model, and this may be a multi-D image of the building, as well as virtual and augmented reality. These statements may be relevant when conducting construction technical expertise in relation to the already built and operated capital construction projects, as well as their parts, which are hidden by other structures or finishing elements, so that there is no possibility to establish the availability and characteristics of the first ones without disruption of the last ones.

Another intellectualization aspect may be report (expert opinion) formation as a part of automated construction technical expertise, it requires standardization and unification of both procedures and documents through the development of industry and corporate regulations [16-20].

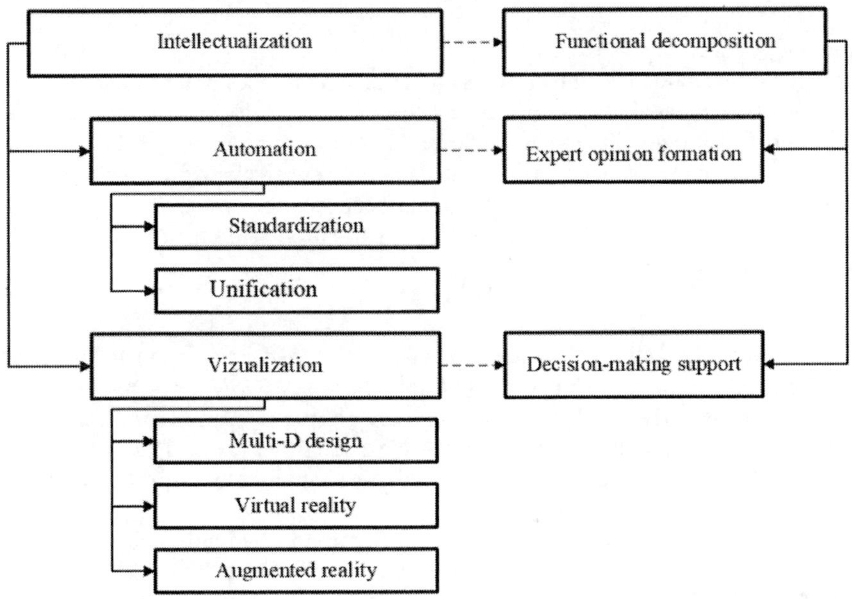

Figure 3. Functional decomposition and structure of the construction technical expertise intellectualization.

As it is seen from the above materials the informatization components are closely related to each other. The interaction of organizational elements of construction technical expertise is schematically shown in figure 4.

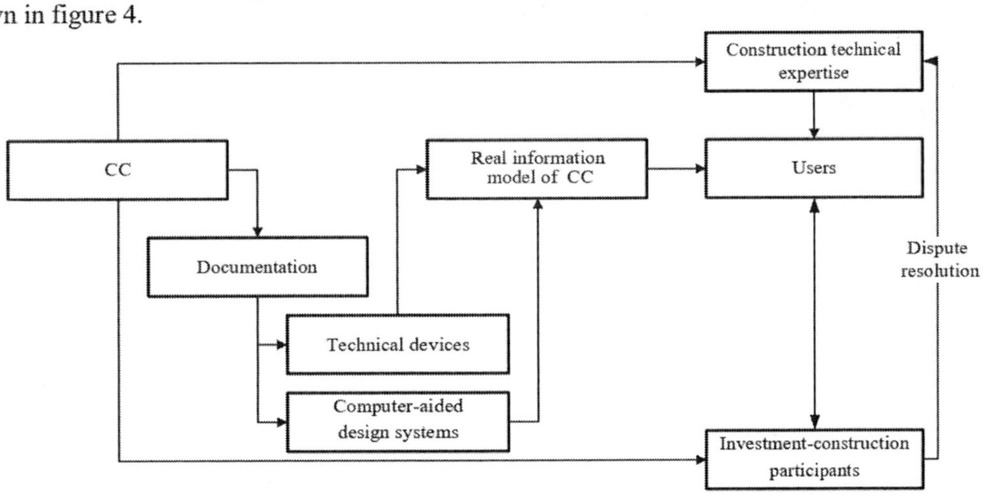

Figure 4. The scheme of interaction of organizational elements of construction technical expertise informatization.

It may be established here, that the initiating thing is the resolution of the dispute, which may arise between investment-construction participants. Although construction technical expertise is quite

IOP Conf. Series: Materials Science and Engineering 1015 (2021) 012079 doi:10.1088/1757-899X/1015/1/012079

expensive and long lasting, it is ordered to stop the dispute. The cost and duration of an expertise may be reduced by application of the capital construction information model (i.e. BIM-technology). Today such technologies are the driving force behind the whole investment-construction informatization process.

4. Conclusions

Undoubtedly, computer-aided design systems and technical devices will be developed, therefore, sooner or later construction technical expertise will be involved in the informatization process and will fall into the general information field. To give a single vector to the components of the informatization process of not only construction technical expertise, but also of the entire claim administration in the investment-construction, it is necessary to establish criteria and performance indicators for these subjects. Practical results of the construction technical expertise informatization will ensure the work quality, control of the cost and duration of contracts or services.

References

[1] Sborshchikov S B 2009 Teoreticheskiye zakonomernosti i osobennosti organizatsii vozdeystviy na investitsionno-stroitel'nuyu deyatel'nost' *Vestnik MGSU* **2** pp 183 – 7

[2] Shinkareva G N 2018 Model' inzhiniringovoy skhemy organizatsii stroitel'stva v perspektive zhiznennogo tsikla ob"yektov *Vestnik MGSU* 13 **10** (121) pp 1204–10

[3] Aleksanin A V 2015 Perspektivnyye napravleniya razvitiya organizatsii stroitel'stva *Nauchnoye obozreniye* **10-1** pp 378–81

[4] Zharov Y V 2018 Otsenka parametrov organizatsionno-tekhnologicheskikh resheniy na osnove neyrosetevykh modeley *Vestnik BGTU im. V.G. Shukhova* **2** pp 110-5

[5] Zharov Y V 2013 Uchet organizatsionnykh aspektov pri planirovanii stroitel'nogo proizvodstva v energetike *PGS Ind. and Civ. Eng.* **11** p 83-5

[6] Sborshchikov S B and Maslova LA 2019 Elementy informatsionno-analiticheskogo obespecheniya reinzhiniringa ob"yektov kapital'nogo stroitel'stva *Vestnik MGSU* 14 **7** pp 912–21

[7] Monastyrev P V, Yezerskiy V A. and Klychnikov R Y 2013 Imitatsionnaya model' optimizatsii parametrov termomodernizatsii zhilykh zdaniy v masshtabe goroda *Vestnik VolgGASU* **31-2** (50) pp 475-84

[8] Klyuyev V D and Levchenko A V 2014 Metodicheskiy podkhod k sozdaniyu informatsionno-analiticheskikh sistem stoimostnogo monitoringa v stroitel'stve *Nauchnoye obozreniye* **1** pp 214–8

[9] Lazareva N V 2015 Stoimostnoy inzhiniring kak osnova integratsii protsessov planirovaniya, finansirovaniya i tsenoobrazovaniya v investitsionno-stroitel'noy deyatel'nosti *Vestnik MGSU* **11** pp 178-85

[10] Silka D N 2018 Sistemotekhnika proektnogo finansirovaniia v stroitelstve aktualnye problemy i puti resheniia *Construction systems engineering. Cyber-physical construction systems* (Moscow: Collection of materials the VI IPICSE) pp 221–4

[11] Sborshchikov S B, Maslova L A and Zhuravlev P A 2019 Reinzhiniring v stroitel'stve *PGS* **7** pp 71–6

[12] Sborshchikov S B, Lazareva N V and Maslova L A 2019 Resursoobespecheniye reinzhiniringa ob"yektov kapital'nogo stroitel'stva *PGS* **12** pp 66–71

[13] Sborshchikov S B and Maslova L A Reinzhiniring ob"yektov kapital'nogo stroitel'stva i reinzhiniring tekhnologicheskikh protsessov *Vestnik MGSU* 14 **10** pp 1321–30

[14] Zhuravlev P A 2017 K voprosu ispol'zovaniya resursno-tekhnologicheskogo modelirovaniya pri formirovanii investitsionnykh program *Vestnik BGTU im. V.G. Shukhova* **7** p 198-201

[15] Zhuravlev P A and Kochenkova E M 2019 Stadiynost' resursno-tekhnologicheskogo modelirovaniya po ob"yektam kapital'nogo stroitel'stva pri formirovanii investitsionnykh programm *Vestnik BGTU im. V.G. Shukhova* **6** pp 76-80

[16] Zhuravlev P A 2015 Tsena stroitel'stva i etapy eye formirovaniya *Vestnik IRNITU* **9** (104) pp 174–8
[17] Lyapin A V and Lyapin V Y 2016 Sovremennyy podkhod k organizatsii smetnoy deyatel'nosti v stroitel'stve *Nauchnoye obozreniye* **8** pp 251–5
[18] Yermolayev Y Y 2013 Osobennosti opredeleniya fiksirovannoy stoimosti stroitel'stva v ramkakh gosudarstvennykh programm *Vestnik universiteta GUU* **11** pp 35–8
[19] Yermolayev Y Y 2013 Osobennosti opredeleniya fiksirovannoy stoimosti stroitel'stva v ramkakh gosudarstvennykh programm *The Hum. and Soc. Sci.* **3** pp 18-23

IOP Conf. Series: Materials Science and Engineering 1015 (2021) 012080 doi:10.1088/1757-899X/1015/1/012080

Organizational solutions for ensuring the quality of construction of nuclear power facilities

E Bachus, S Sborshikov and I Markova

Moscow State university of civil engineering (National research university) (NRU MGSU) 129337, 26, Yaroslavskoye Shosse, Moscow, Russia

E-mail address: markova@mgsu.ru

Abstract. The issue of quality assurance becomes extremely important in the implementation of investment and construction activities in the field of atomic energy use, since inconsistencies that may arise as a result of improper quality assurance of construction products during the construction of nuclear power facilities of the most representative nuclear facilities (NF) from the point of view of potential danger are not only highly undesirable events, but can also lead to significant environmental, social and political consequences. Solving quality assurance problems is an integral element of the development strategy of any of the industries, including the investment and construction sector. The implementation of such events at nuclear power facilities is currently a priority area of activity, given the increasing interest of foreign countries not only in nuclear power facilities of Russian design, but also in other nuclear facilities of the nuclear cluster.

1. Introduction

Globalization and increased competition have led to toughening of the requirements imposed by the consumer on the quality of products. Ensuring product quality is not only a competitive advantage, but also helps to reduce unforeseen resource costs and time from inconsistencies, which has a positive effect on the economic condition of the product manufacturer. [1-5]

At the same time, the requirements of investors for the quality of construction also have an upward trend, which is largely a result of the separation of the functions of the technical customer and the performance of certain types of work, and the legislative consolidation of the procedure for implementing contract relations and the distribution of responsibility for the quality of construction

2. Methods

Quality assurance in the implementation of investment and construction activities in the field of atomic energy use, being a priority area of activity not only for the developer, but also for other participants in the investment process, requires the formation and implementation of new organizational forms and schemes for the implementation of such activities, which do not contradict the current legislation in the field of urban planning activities and the use of atomic energy with the unconditional priority of nuclear and radiation safety [6, 8-11].

The emergence of new organizational forms and schemes for the implementation of investment and construction activities will cause not only a change in the distribution of functions, rights, duties, responsibilities, but also transform their very saturation and target orientation. A promising direction in

IOP Conf. Series: Materials Science and Engineering 1015 (2021) 012080 doi:10.1088/1757-899X/1015/1/012080

the field of construction management, including nuclear facilities, is the use of an engineering management scheme at the corporate level of construction.

This scheme is based on the separation of functions related to the direct organization of construction such as: preparation, planning, construction management, its material and technical support. Within the framework of such a scheme, powers are radically redistributed, an emphasis is placed on the functioning of a new subject of investment and construction activities - a construction organizer. In this regard, quality control as a component of the general system of organizing construction will require adjustments in relation to the above functional decomposition in order to ensure the solution of the main tasks - reducing the cost and construction time [7,12,13].

Considering the engineering management scheme as promising and, possibly, as the main one in the near future in relation not only to nuclear facilities, but also to a wider range of industrial and civil facilities, one should point to new organizational methods: strategic and operational-production controlling. the application of these methods can give some positive impetus to the development of a quality assurance organization. This statement is due to the fact that controlling from the point of view of its construction is complemented by such a component as the development and adoption of decisions (technical and managerial), and they, in turn, can be applied, among other things, to the tasks of ensuring the quality of construction of nuclear facilities.

3. Results and discussion

The analysis shows that organizational decisions can be aimed at:

1) transformation of the management system as a whole;
2) increasing the efficiency of private management functions (for example, ensuring the quality of construction).

In this regard, the quality assurance of construction is achieved by the development of a set of measures formed on the basis of studying the conditions and factors for the construction of a specific facility to meet the requirements of regulatory documents and should be carried out at all stages of the life cycle and contain control:

- design and construction solutions; incoming materials, products, structures, equipment (incoming control);
- production processes; structural units and elements;
- finished products and their transfer to the consumer;
- as well as the widespread use of statistical methods of accounting and analysis.

In the context of the life cycle concept, it is possible to distinguish such levels of construction quality as standard, actual and operational.

The normative level of construction quality is determined by the requirements of existing regulatory documents, technical regulations and is established at the R&D stage, based on the solution of socio-economic problems, the prospects for the development of investment and construction activities. The actual level is the level of quality of construction products achieved at the design and construction stages of a building or structure. [14-16]

The operational level of quality is expressed and supported by renovation and reconstruction works as part of the operation of capital construction facilities until their liquidation or decommissioning.

To confirm the quality of nuclear power facilities and ensure their safety in accordance with the provisions of the legislation on urban planning activities, as well as legislation in the field of the use of atomic energy, quality control of works is carried out, which is a multi-level system and includes not only control by the participants in the construction — the developer, a technical customer, a person carrying out the construction (general contractor or an engineering company — the organizer of the construction), persons performing types of work (contractors and subcontractors), but also from the operating organization, as well as the authorized body for managing the use of atomic energy.

In the light of the current state of affairs in construction enterprises, the assessment of the effectiveness of ensuring the quality of their products should include control, which is, at the corporate level, a subsystem of NF construction management. The component base of the subsystem is the

IOP Conf. Series: Materials Science and Engineering 1015 (2021) 012080 doi:10.1088/1757-899X/1015/1/012080

elements of incoming and outgoing flows, which are permanently transformed depending on changes in the course of construction production processes.

The elements of the driving stream are due to the performance of work and activities and the associated labour and resource costs for damping marriage, assessing the quality of construction products, and the elements of the outgoing flow are work and activities and associated labour and resource costs caused by the presence of defective construction products (internal and external).

The basis of this system is the control of the elements of incoming and outgoing flows, then in assessing the effectiveness of control of the quality of construction products, it is necessary to focus on this genesis.

Since the basis for assessing the effectiveness of this system is the control of incoming and outgoing elements, then in assessing the effectiveness of product quality control at an enterprise it is necessary to focus on this genesis and it can be stated that there is a need to form a common system of methods, criteria, indicators for evaluating and making organizational decisions for quality assurance. Construction (including nuclear power facilities) and regulatory and analytical support of this type of activity at the corporate level.

In development of the above provisions, a set of measures is proposed to improve the process of assessing and making organizational decisions to ensure the quality of construction of nuclear power facilities, based on the statement that the quality assurance subsystem should take into account: the specifics of construction production (as well as organizational forms and structures, methods used); compliance of the development directions of the quality assurance subsystem with the promising tasks of investment and construction activities, including in the field of nuclear energy; the specifics of interaction between participants in investment and construction activities.

Improving the organization of quality assurance in construction is possible in two directions.

The first direction is characterized by the content side, within which the quality assurance processes are organically linked to the content and justification of organizational decisions as corporate-level measures. [17,18]

Within the framework of the second direction, the procedural side of quality assurance in construction production is considered, based on the basic provisions and principles of cybernetics and associated with the development of effective technology, which is perceived as a systematically improving process that establishes the main stages of collecting and processing information for the implementation of the main goals of quality management within the existing organizational structures of investment and construction activities (so-called cyber-physical systems). [1-18]

Considering both the substantive and procedural components of the development of the organization for ensuring the quality of the construction of nuclear facilities, it seems appropriate to note the phenomenon of convergence, which, in technical and technological terms, leads to the emergence of new methods and methods of work, and they, in turn, will contribute to the qualitative transformation of activities and works ensuring the quality of construction, as well as structures implementing them. These transformations can be identified as technological reengineering, reengineering of business processes (management processes) and organizational reengineering.

The conformity of the elements of this subsystem, forms of interaction with other subsystems and control bodies should be determined by the quality assurance technology.

In this regard, the improvement of the organization of quality assurance in construction, especially in relation to its management processes, involves:

- substantiation of the standard scope of works (processes) for quality assurance;
- determination of rational methods for the implementation of quality assurance processes, taking into account the technical and information equipment, the level of qualifications of the performers;
- regulation of the implementation of quality assurance processes in the construction of nuclear power facilities;
- establishing the sequence of work and evaluating the effectiveness of achieving the set results.

IOP Conf. Series: Materials Science and Engineering 1015 (2021) 012080 doi:10.1088/1757-899X/1015/1/012080

4. Conclusions

In the course of the scientific search it was found that it can be a starting point for further research in the following areas (topics):

1. Quality engineering. It can be identified as the formation and organization of a set of interrelated activities aimed at creating and developing corporate quality assurance systems based on modern scientific achievements, international standards and domestic regulatory documents. In this regard, project quality management includes the processes and activities of the organization, which are determined by the quality policy, as well as the purpose of the project and the areas of responsibility of its participants so that construction products meet the established requirements and needs of the user of the facility.
2. Rationing of labour and resource costs as part of works to ensure the quality of construction, including nuclear facilities. Rationing of costs for ensuring product quality is the basis for the formation of a set of measures aimed at planning activities in one way or another related to the quality of construction products in order to achieve the best TP, establishing the amount of financial resources required to perform work to improve the quality of construction products, forecasting costs, necessary to manage the activities of the entire construction organization.
3. Planning and regulation of quality assurance measures at the stages of the life cycle of a nuclear power facility, including using BIM technologies. Due to the growing complexity of NF lifecycle management and increasing international competition, as well as the high level of requirements imposed by users on nuclear power facilities, it is necessary to apply end-to-end and integrated adaptive management at the lifecycle stages using BIM technologies and the use of cyber-physical systems.
4. Assessment of the efficiency of the corporate quality assurance system for the construction of nuclear power facilities. The key imperatives of the process of assessing the effectiveness of quality assurance systems for nuclear facilities, first of all, should include the requirements for reliability, adequacy and meaningfulness of parameters.

Automated systems for ensuring the quality of construction of technically complex and unique objects. Application in the construction of cyber-physical systems. The information and analytical system for ensuring the quality of nuclear facilities is an integrated system for designing and managing information on the quality of all nuclear power facilities under construction and in operation at the stages of their life cycle. As a result, a set of knowledge about the subject area is formed, structured to formalize the decision-making process, which determine the content of information systems and technologies, constitute the genesis of cyber-physical systems. The use, in turn, of such systems in the investment and construction sector indicates the need for integrated thinking, which involves making decisions based on a wider range of indicators, principles of complexity, multidimensionality, as well as imperatives of the concept of sustainable development.

It should also be noted that this automated system for ensuring the quality of nuclear facilities can become a component of the general information model of a capital construction object, which will reflect the real state of the building, structure, as well as its compliance with the project. By integrating a quality assurance system with BIM technologies, you can get access to promising areas in which these technologies are now developing, for example, virtual, augmented reality, modern planning methods. These facts can have a great potential for efficiency in relation to nuclear facilities not only in the considered subject area, but also in the areas of cost reduction and the duration of their construction.

References

[1] Sborshikov S B 2012 *Logistics of regulatory impacts in the investment and construction sector (theory, methodology, practice)* Dissertation for the degree of Doctor of Economics Russian Economic Academy GV Plekhanov Moscow

[2] Zhuravlev P A 2017 On the use of resource-technological modelling in the form of investment programs *Bulletin of the Irkutsk State Technical University* **7** pp 198-201

[3] Zhuravlev P A 2015 The price of construction and the stages of its formation *Bulletin of the*

Irkutsk State Technical University **9** (104) pp 174-8
[4] Ermolaev E E 2013 Management of the use value of construction sites *Humanities and social sciences (electronic journal)* **3** pp 18 - 23
[5] Alexanin A A 2009 Evaluation of the economic efficiency of the use of new technologies, materials and solutions in energy saving projects *Vestnik MGSU* **1** pp 125 - 9
[6] Zharov Ya V 2013 Accounting for organizational aspects in the planning of construction in the energy sector *Journal of Civil Service* **5** pp 69-71
[7] Bakhus E E 2016 To the question of improving the organizational and technological solutions to ensure the quality of construction of nuclear power facilities *Scientific review* **14** pp 20-3
[8] Lazareva N V 2015 Value engineering as the basis for the integration of planning, financing and pricing processes in investment and construction activities *Vestnik MGSU* **11** pp 178 – 85
[9] Alexanin A A 2015 Prospective directions of development of construction organization *Scientific review* **10** (1)
[10] Alexanin A A 2018 Organization of a logistics system for waste streams during the renovation of territories *FORM 2018. IOP Conference Series: Materials Science and Engineering*
[11] Lyapin A V and Lyapin V Yu 2016 Modern Approach to the Organization of Estimated Activity in Construction *Scientific review* **8** p 251-5
[12] Zhuravlev P Bachus E Markova I 2019 Multifactorial parametric model for assessing and making organizational decisions to ensure the quality of construction *Web of Conferences 22nd International Scientific Conference on Construction the Formation of Living Environment FORM 2019*
[13] Zhuravlev P, Bachus E and Markova I 2018 Directions of modernization of systems for ensuring the quality of construction of nuclear power facilities *MATEC Web of Conferences*
[14] Sborshikov S, Maslova L, Lazareva N and Shinksreva G 2019 Choice of the rational option of the organization and technological decision, taking into account the exposure of random factors *IOP Conference Series: Material Science and Eng. XXVIII R-P-S Seminar 2019*
[15] Shinkareva G 2018 Model of an engineering scheme for organizing construction in the perspective of the life cycle of objects *Vestnik MGSU* **9** (120) **13** pp 1090-105
[16] Sborshchikov S, Maslova L and Lazareva N 2019 Resource support for reengineering of capital construction facilities *Sc J Industrial and civil construction* **12** pp 66-71
[17] Sborshchikov S, Maslova L and Zhuravlev P A 2019 Reengineering in construction *Sc. J. Industrial and civil construction* **7** pp 71-6
[18] Zhuravlev P A and Kochenkova E M 2019 The staging of resource-technological modeling for capital construction objects in the formation of investment programs *Sc J Industrial and civil construction* **6** pp 76-80

IOP Conf. Series: Materials Science and Engineering 1015 (2021) 012081 doi:10.1088/1757-899X/1015/1/012081

Economic substantiation of construction of narrow profile hydraulic-fill structures

A A Mikhasek, M V Rodionov and R R Mingafin

Samara State Technical University, Academy of Architecture and Civil Engineering, Molodogvardeyskaya str. 194, 443001 Samara, Russia

E-mail address: rodionov_max@mail.ru

Abstract: Hydraulic-fill structures with free flow of pulp have a flattened transverse profile, especially in cases of hydraulic fill of the structure into water. The flattened profile of the structure demands significant financial and labor costs for its construction, reducing the efficiency of the hydraulic-fill structure. This paper deals with the existing methods for the construction of narrow profile hydraulic-fill structures, their economic indicators; the designs of hydraulic-fill structures using geosynthetic materials developed by the authors, which make it possible to increase their efficiency in comparison with previously known methods, especially when the structures are hydraulically filled into water.

1. Introduction

The construction of hydraulic structures using hydromechanization methods is one of the most highly productive and efficient methods of earth works [1, 2, 3]. During hydraulic fill of the structure the development, transportation and laying of soil are combined into a single technological process. Hydraulic depose of soil allows providing the required soil density in the road fill body, often without additional costs for mechanical compaction.

In hydraulic engineering hydraulic-fill structures have become widely used for the construction of dams, canals, coast protection and for the creation of artificial territories in water reservoirs and water ducts. The use of hydromechanization facilities in Russia started developing rapidly in the mid-30s - 40s of the XX century, during the construction of the Moscow and the Volgo-Donskoy named after Lenin canals. Later it was used at large hydroelectric power plants in many regions of the country (Gorkovskaya, Saratovskaya, Ivankovskaya, Uglichskaya, Rybinskaya, Tsimlyanskaya, Nizhnekamskaya hydroelectric power stations, etc.) [3]. Currently hydromechanization is still widely used in hydraulic engineering, however due to the fact that there's hardly any large-scale dam construction, its main areas of application are clearing riverbeds, coast protection and creation of artificial territories (figure 1, 2, 3, 4).

When creating hydraulic-fill soil structures, it is customary to distinguish between hydraulic filling into ground and into water, with freely and forcedly formed slopes. In the case of free flow of pulp hydraulic-fill structures have a flattened transverse profile, especially in cases of hydraulic fill of the structure into water.

The flattened profile of the structure leads to significant financial and labor costs for its construction, reducing the efficiency of the hydraulic-fill structure. There are various ways to increase the efficiency of hydraulic-fill structures. However they usually have got typical disadvantages: they are not applicable when the structure is hydraulically filled into water; they are labor consuming, material-intensive and expensive.

In this regard the search and development of technical solutions to improve the efficiency of the hydraulic-fill structure is a vital task.

Figure 1. Exit of pulp from hydraulic fill pipeline.

Figure 2. Hydraulic fill of dam.

Figure 3. Hydraulic fill of territories.

Figure 4. Hydraulic fill of coast protection (beach).

2. Materials and methods

In process of hydraulic fill of soil structures into ground forced formation of slopes is usually the most economical solution (see figure 1). Compression of profile is carried out by filling an earth banking, which is formed before the filling starts (primary banking) and limits the flow of pulp, as well as during the construction of the structure (accompanying earth banking). The steepness of the forcibly formed slopes of such structures ranges from 1: 2 to 1: 4, but in the case of free flow of pulp the steepness of their slopes can reach 1:50 [7]. The construction of hydraulic-fill structures with banking (compressed profile) into water is associated with a number of difficulties, i.e. when forming primary banking, which is built up as a rule using end-dumping method, it becomes necessary to fill the bank to a level exceeding the builder level of water by at least 35 cm [6], with slopes from 1: 1.5 (when using stone) to 1: 4 (when using sandy soils), which often leads to a significant decrease in the efficiency of hydromechanized work, especially if there are no reserves of stone in the area where the construction takes place. The hydraulic fill of the soil structure into water with freely formed slopes leads to the appearance of a flattened profile with the slopes laid at up to 1:10 (see figure 5). Thus, e.g. during the construction of coast protecting structures in narrow roadbeds, the use of hydraulic

IOP Conf. Series: Materials Science and Engineering 1015 (2021) 012081 doi:10.1088/1757-899X/1015/1/012081

filling method becomes technologically and economically inefficient, even if there are reserves of soil suitable for hydraulic fill at the construction site.

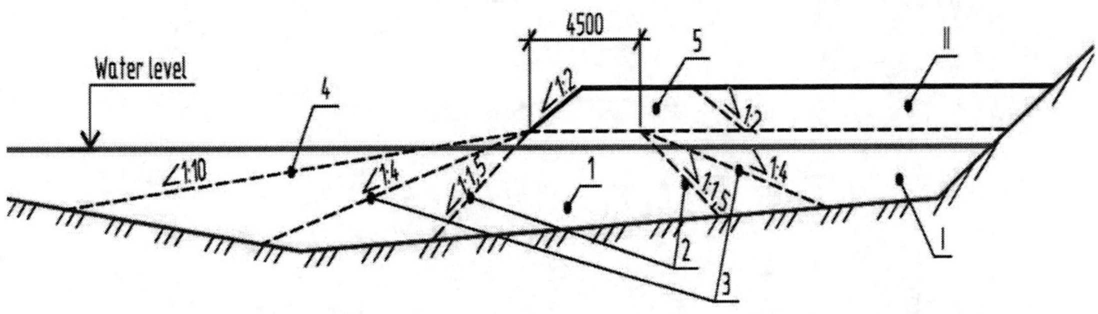

Figure 5. The methods of forming hydraulic filled slope into water: 1 - primary banking; 2 - slope with stone filling; 3 - slope with sand filling; 4 - slope with sand filled into water; 5 – accompanying earth banking; I, II – the stages of sand filling behind dams.

The authors faced this problem during the development of design documentation for the project "Coast protection of the Usta River near the village of Kirillovo, Krasnobakovsky District of Nizhny Novgorod Region". The purpose of the project was to protect the right-bank coastline, buildings, infrastructure and lands of Kirillovo village from destruction by the waters of the river Usta. The river bed is composed mainly of fine sands with interlayers of loamy soils on the right bank. Based on the topographic, geological and hydrological conditions of the area, as the first option for the construction of the coast protection structure a construction method was developed using rock placement into water at the foot of the slope, with the formation of a toe, the crest of which was located above the builder level of water in the river Usta. However stone quarries with required quality indicators were found only at a significant distance from the construction site, which led to high transport costs for its delivery and exceeding the permissible cost of construction of coast protection structures.

The geological conditions of the Usta riverbed prompted the development of technical solutions using local sand. The solution using hydromechanization seemed to be especially effective. Thus the installation of a dredger near the left bank made it possible to solve two problems at once - widening the riverbed and simultaneous formation of a stable slope near the right bank. However in the case of using the technology of soil hydraulic fill with a freely formed slope, with the construction depth of water at 4.8 m and laying the filled slope at 1:10, the slope with a length of 48 m was formed, practically covering the entire riverbed. For the above reasons the use of stone for the formation of the primary banking was not economically efficient. When searching for possible solutions geosynthetic materials were looked at. The first developed solution - using geosynthetic shells - has already been tested many times in various projects of coast protection structures and in the hydraulic fill of territories [4, 5]. In this case along the coast geosynthetic shells are stacked on top of each other and form a stable slope. The space inside and behind them is filled with alluvial sand (figure 6). This solution has made it possible to compress the profile of the hydraulic-filled structure significantly and to form a stable slope, while minimizing the amount of materials transported from offsite.

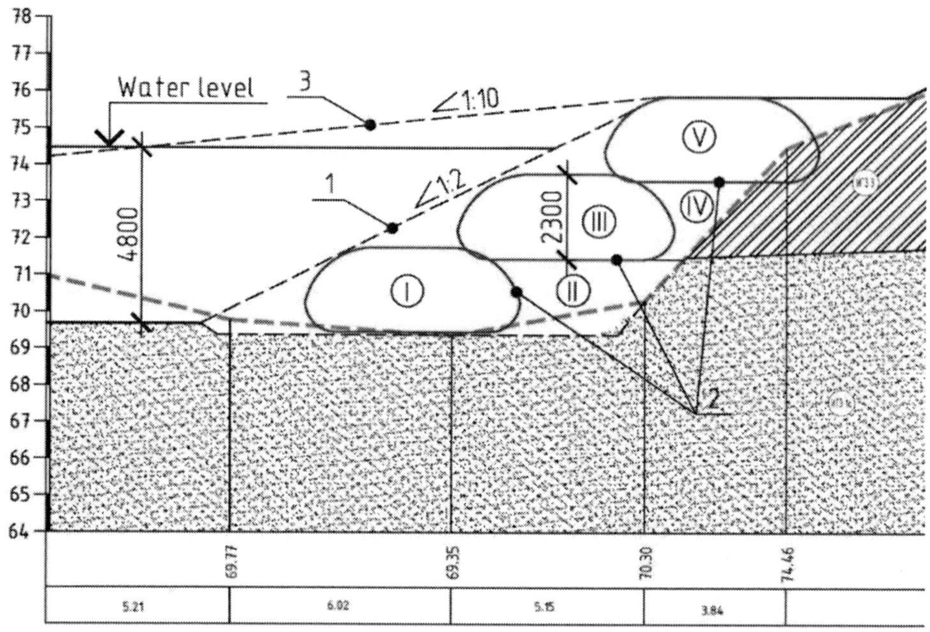

Figure 6. Narrow-profile hydraulic-fill coast protection structure with geosynthetic shells 1 - designed stable slope; 2 - geosynthetic shells; 3 - slope with free flow of the pulp into water; I-V - stages of hydraulic fill of the structure.

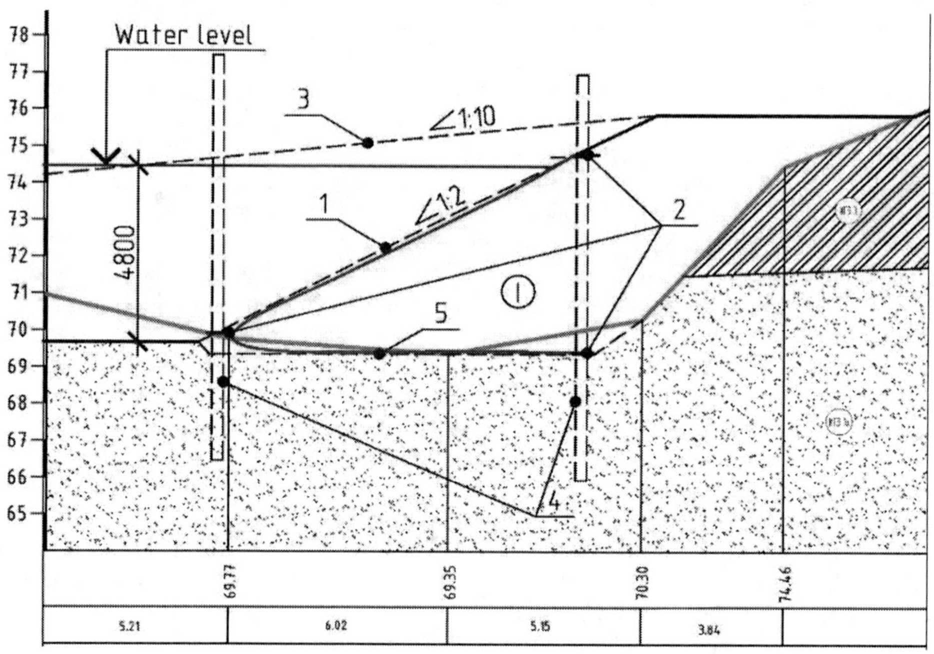

Figure 7. Narrow-profile hydraulic-fill coast protection structure with woven geotextile 1 - designed stable slope; 2 - joints for attaching geotextile to the guide; 3 - slope with free flow of pulp into water; 4 - guides; 5 - woven geotextile; I - stages of hydraulic fill of the structure.

IOP Conf. Series: Materials Science and Engineering 1015 (2021) 012081 doi:10.1088/1757-899X/1015/1/012081

As the second option for building a narrow-profile hydraulic-fill structure, a construction technology using woven geotextile has been developed, which has got no analogues at the moment. Woven geotextile is a water-permeable material made from durable synthetic threads by interweaving them at an angle of 90 degrees. Geotextiles must have the following characteristics: tensile strength at break of 100/100 kN/m, elongation at break of not more than 20%, open pore size of O_{90} - 100 microns, water permeability of at least 3 l/(m^{2}s). The essence of the invention is as follows. With the help of an excavator or a dredger, the bottom of the river at the foot of the slope is leveled, then metal guides (pipe, channel, etc.) are immersed at the foot of the existing and designed slopes. Rolls of woven geotextile of required length are cut and sewn together (taking into account the location of the guide, the formation of the necessary slope and the selected length of the grip). Either metal rings or woven loops are sewn on along the edges of the fabric and in the place of its bend. After the immersion of the guides, the rings (loops) fastened to the woven fabric are fixed on it, then the filling of the space between the coast and the woven fabric with sand pulp begins (figure 7). At the same time it is necessary to control the volume and location of pulp flow, in order to prevent the upfloat of the woven fabric, as well as to record the intensity of water drainage from the pulp through the fabric and the geometry of the formed slope. During the construction of such an alluvial structure, local sand is used, which, when reclaimed, turns into a slurry of 10% concentration. The slurry consumption during reclamation depends on the accepted brand of the dredger and ranges from 400 to 2000 m^3/h.

The described construction technology with the use of woven geotextile can also be applied for hydraulic fill of a structure into ground, eliminating the need to form the earth banking.

3. Results

The results of economic comparison of the existing and developed options for constructing the narrow-profile hydraulic-fill structure are presented in table 1.

Table 1. The economic comparison of the existing and developed options for constructing the narrow-profile hydraulic-fill structure

№	Construction method	Place of hydraulic fill	Construction cost[a], Euro
1	Banking	Ground	310,0
2	Geosynthetic shells	Ground	590,0
3	Woven textile	Ground	470,0
4	Banking	Water	1400,0
5	Geosynthetic shells	Water	750,0
6	Woven textile	Water	590,0

[a] The construction cost is given per one linear meter, the height being 5 meters, excluding the cost for hydraulic fill of the soil structure's body.

4. Conclusion

1 Hydromechanization techniques are still widely used in hydraulic engineering. The main areas of their application are associated with clearing water bodies and water drainage, as well as with creating artificial hydraulic fill territories.

2 The known technologies of soil hydraulic fill into water are not always effective, due to the need to create a flattened profile of the structure, especially when filling coast protection structures in not very wide roadbeds.

3 The technology developed by the authors for the hydraulic fill of the narrow-profile structure into water allows the creation of a hydraulic-fill soil structure of a given geometric shape. It is characterized by high productivity of work and low construction costs.

4 The economic comparison of the options has shown the efficiency of using geosynthetic materials when a narrow-profile structure is filled into water. The technology developed by the authors for creating the narrow-profile hydraulic-fill structure with use of woven geotextiles is 50% more economically

IOP Conf. Series: Materials Science and Engineering 1015 (2021) 012081 doi:10.1088/1757-899X/1015/1/012081

efficient than the technology of banking into water and 21% more economically efficient than technology with use of geosynthetic shells.

References

[1] Garnaga V V and Luchkova V I 2014. Methods for reconstruction of coastal areas in the Far East under flood conditions *The Far East: problems of the development of architecture and construction complex* **1** pp 58-63

[2] Panov S I 2008. Experimental research of soils at hydraulic-fill structures *Bulletin of All-Russian Scientific Research Institute of Hydraulic Engineering named after B.E. Vedeneev* **250** pp 82-91

[3] Tszhan R V and Kuzmin G P 2015 Hydraulic-fill structures in the permafrost zone of Russia: construction experience, problems and possibilities of using hydraulic borehole mining *Science and Education* **2** (78) pp 63-9

[4] Piyavsky S A, Rodionov M V and Kholopov I S 2012 Application of geosynthetic shells in hydraulic engineering *Bulletin of NRU MGSU* **6** pp 54-61

[5] Semenov D A and Kaloshina S V 2016 Construction of artificial islands using Geotube technology *Modern technologies in construction. Theory and practice* **2** pp 294-302

[6] SP 80.13330.2016 River hydraulic engineering facilities. Updated edition of SNiP 3.07.01-85

[7] ISCS 214-93. Standards for the design and performance of hydromechanized works in traffic facilities construction

Improving sintering by lowering the temperature of melt formation during firing

N Chumachenko, I Khafizov and D Kizilova

Samara State Technical University, Molodogvardeyskaya st., 244, Samara, Russia, 443100

E-mail address: uvarovang@mail.ru

Abstract. Liquid sintering for ceramic products is the main process providing necessary properties: strength characteristics, density and porosity. In order to reduce energy consumption for the production of ceramic products it is important to obtain a melt at a lower temperature with a wide sintering interval. Clay raw materials natural composition provides the melt formation at the temperatures above 710° C. The fusible silicate triple eutectics composition, providing the melt formation at the temperatures below 500°C is found. The components for complex fusion additive have been determined. The influence of the amount of complex fusion additive on the dynamics of melt charge change on the basis of Locosovo loam was studied. It was found out that the refractory additive in the amount of 6 to 18.5% extends the sintering interval, reduces the temperature of formation of the refractory melt and provides a decrease in firing temperature by 100°C. The optimum quantity of complex fusible additive providing improvement of sintering of ceramic materials at the expense of decrease in temperature of melt formation during firing is revealed.

1. Introduction

Sintering is the main process that ensures the formation of necessary properties of fired ceramic materials and products: ceramic bricks, tiles, ceramic tiles, sanitary products, etc. The process begins with solid-phase sintering, and, when the firing temperature increases, this happens due to liquid sintering. For ceramic products with a porous structure and high water absorption during firing, partial sintering is achieved due to the formation of 20-35% of the melt. If the melt is formed up to 50 %, then as a result of firing, completely caked products are formed. The degree of sintering is influenced by many factors: the properties of mineral raw materials and technological parameters of production.

On the basis of clay raw materials, which belong to the aluminosilicate systems, the formation of the primary melt occurs no earlier than 710 °C due to potassium aluminosilicate eutectic [1]. A lower-temperature melt without adjusting the composition of clay raw materials is impossible.

The scientific novelty is connected with the directed correction of clay raw materials with a specially selected complex additive that provides sintering at a lower temperature than in clays without additives.

It is important to obtain a melt at a lower temperature with a wide sintering interval in order to reduce energy consumption for the production of ceramic products.

2. Methods and materials

The main raw material used in the experiments was Locosovo loam of the Tyumen region, which is used for the production of ceramic bricks [2]. The loam has a light brown color with a grayish tinge.

IOP Conf. Series: Materials Science and Engineering 1015 (2021) 012082 doi:10.1088/1757-899X/1015/1/012082

Based on x-ray diffraction and differential thermal analyses, it was found that the loam contains a clay substance of the montmorillonite-hydrosluidic type. The raw material is moderately plastic, low-dispersed with a low content of large inclusions. According to the content of Al_2O_3, loam belongs to the group of acidic clay raw materials. The presence of lime lumps courses "lime blowing" in ceramic bricks [3 - 5]. To select the correction components, we analyzed the known triple silicate systems and determined the composition of the most fusible eutectic in the Na_2O - FeO - SiO_2 system [1] at the temperature less than 500 ^{0}C. The eutectic composition ensures the formation of the melt at the lowest temperature.To obtain a complex additive of this composition, pyrite calcine as an iron-containing component, silicate – block as an alkaline silicate, silica as a silica, and coal as an organic component necessary for iron recovery were selected. The ratio between the listed components was determined by calculation taking into account their chemical compositions with an orientation to the composition of the most fusible eutectic in the system SiO_2 - FeO - Na_2O. Six formulations were developed (figure 1). The content of the complex additive in the formulations varied from 6 (composition 2) to 35 % (composition 6). The amount of the resulting melt for experimental compositions was determined by the chemical composition of raw materials and the composition of triple aluminosilicate eutectic based on known state diagrams [1], using the calculation method [6]. When developing the calculation method, numerous studies of phase transformations of aluminosilicate raw materials performed by A. I. Augustinik, U. D. Kingery, V. F. Pavlov, S. P. Onatsky, A. A. Novopashin and V. V. Eremenko were taken into account [7-12].

3. Results and discussions

The dynamics of melt formation for the studied compositions according to the calculation method [1] is shown in figure 1. As it can be seen from figure 1, with the introduction of a complex additive, the amount of melt increases at a lower temperature, the sintering interval expands.

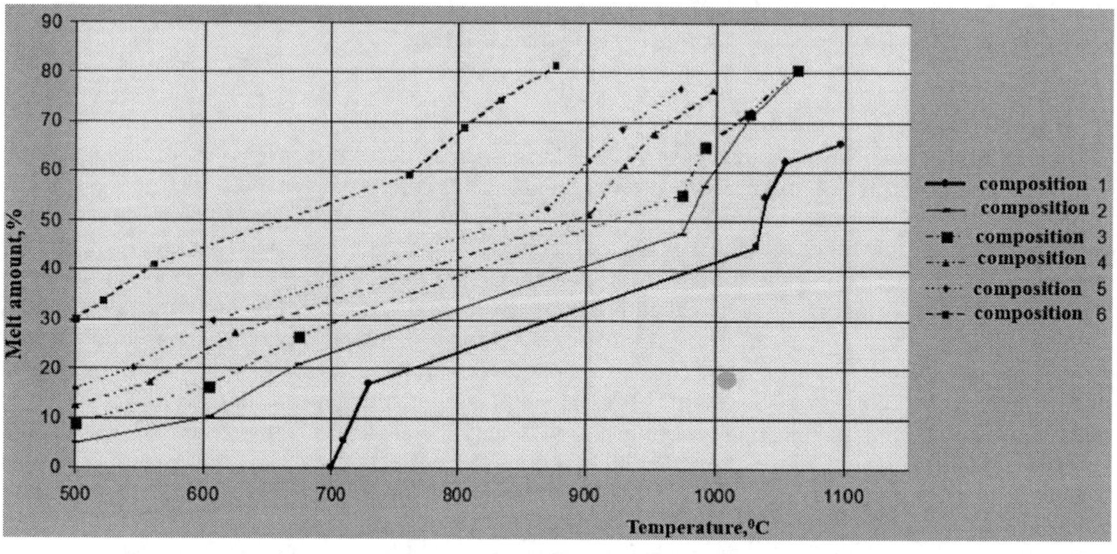

Figure 1. Dependence of the amount of melt (%) on the firing temperature (°C)
Description: No.1 - 100 % Locosovo loam (Reference Composition)
Composition No.2 - 94% Locosovo loam and 6 % complex fusible additive
Composition No.3 - 89.5% Locosovo loam and 10.5% complex fusible additive
Composition No. 4 – 85.5 % Locosovo loam and 14.5 % complex fusible additive
Composition No. 5 - 81.5% Locosovo loam and 18.5% complex fusible additive
Composition No. 6 - 65% Locosovo loam and 35 % complex fusible additive

In the Locosovo loam, sintering starts at 770 ° C and 35% of the melt can be formed at about 950 °C, provided the necessary degree of homogenisation is achieved. The introduction of a complex additive lowers the temperature of the lower limit formation required for melt sintering. Thus, when 6% of the additive is introduced, 20% of the eutectic melt is formed at 670 ° C. A further increase in the amount of additive leads to an even lower temperature that ensures the start of sintering. Compositions with an additive content of 6, 10.5, 14.5 and 18.5 % are characterized by a wide sintering interval. A further increase in the additive quantity reduces the sintering interval, so the additive composition with 35% additive content was not involved in further experiments. Compositions 1 to 5 with 0, 6, 10.5, 14.5 and 18.5 % additive content were used to study the impact of the percentage of complex fusion additive content on sintering. For the prepared charges, the molding humidity was determined; samples were formed by plastic method-cubes of 5x5x5 cm in size. After drying the samples were fired in one mode: the rate of temperature rise – 300 ° C/hour; firing at the maximum temperature for 1 hour; slow cooling in the oven. To determine the effect of the firing temperature, 3 series of experiments were conducted at the maximum firing temperature: 700, 800 and 900 ° C. The samples were tested for compressive strength after the firing.

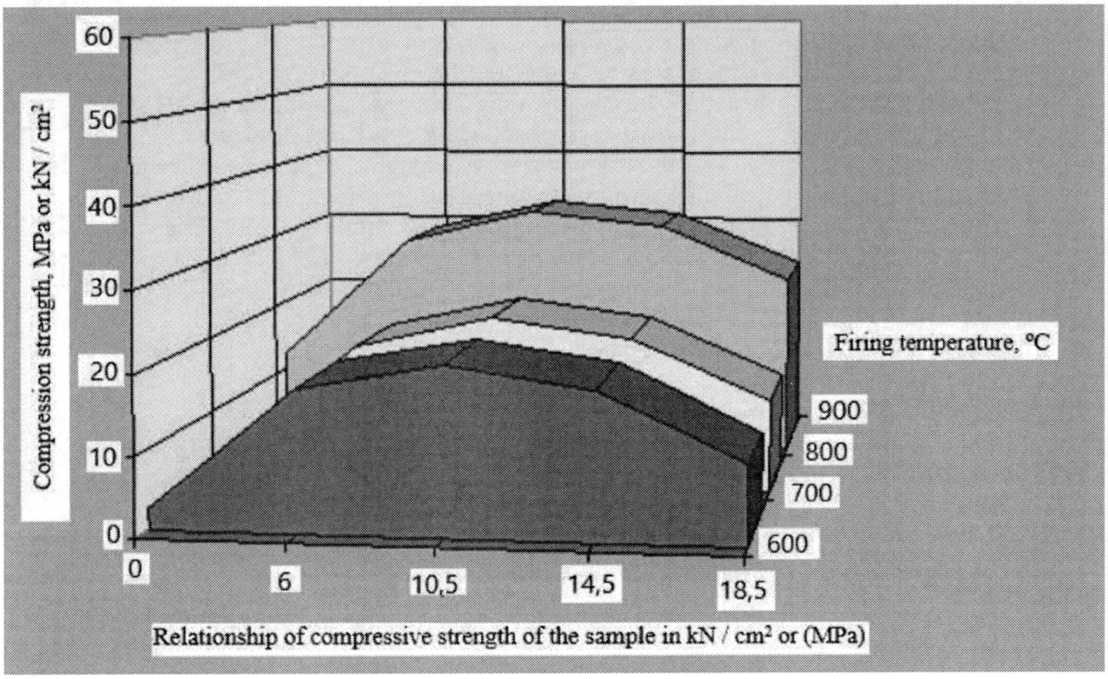

Figure 2. Dependence of the compressive strength of samples (in kN/cm^2) on the % amount of fusible additive (%) and the firing temperature (°C)

The authors carried out mathematical planning and statistical processing of the experimental results. A full quadratic regression equation with a confidence probability of 0.9 was obtained. Using this equation the dependencies of changes in the compressive strength of the specimens with changes in additive quantity and firing temperature shown in figure 2 were determined. Composition No. 3 which contains 10.5 % complex fusible additive has the greatest strength. As the temperature increases, the strength of this composition increases.

The experiment revealed the optimum amount of fusible additive.

IOP Conf. Series: Materials Science and Engineering 1015 (2021) 012082 doi:10.1088/1757-899X/1015/1/012082

4. Conclusions

1. The composition of a fusible silicate triple eutectic providing the formation of a melt at a temperature of less than 500° C was revealed.
2. The choice of components for obtaining a complex low - melting additive that provides the formation of low - melting eutectic in the SiO_2 - FeO-Na_2O system is justified: pyrite cinder - as an iron-containing component, silicate-block-as an alkaline silicate, silica - as a silica, coal - as an organic component necessary for iron reduction.
3. It is established that fusible additive in the amount of 6 to 18.5% extends the sintering interval, reduces the temperature of formation of a fusible melt and the firing temperature by 100 ° ^{0}C.
4. The optimal amount of complex fusible additive 10,5 % improves sintering of ceramic materials by reducing the temperature of melt formation during firing has been identified.

References

[1] Toropov N A, Burzakovsky V P, Lapin V V and et al 1972 *Diagrams of the state of silicate systems* issue **3** *Triple systems* (Leningrad: Nauka) p 488

[2] GOST 530 Bricks and ceramic stones. Technical specifications 2012

[3] Knigana G I, Vershinina E N and Tatsky L N 1985 *Laboratory work on the technology of construction ceramics and artificial porous aggregates* (Moscow: Higher school) p 223

[4] Technical analysis and control of ceramic production1986 (Stroizdat) p 272

[5] GOST 9169 Raw clay materials for the ceramic industry. Classification 1975 ГОСТ 9169 Сырье глинистое для керамической промышленности. Классификация

[6] Chumachenko N G 2014 The use of phase rule diagrams of aluminosilicate systems for calculating melt amount and constituents appearing in ceramic mixture under firing *23rd Russian-Polish-Slovak Seminar on Theoretical Foundation of Civ. Eng. TFoCE 2014 Wroclaw Szklarska Poreba Poland 25-29 VIII 2014 Procedia Engineering* **91** pp 381-5

[7] Augustinik A I 1975 *Ceramics* (Leningrad: Stroyizdat) p 592

[8] Kingery W D 1967 *Introduction to ceramics* (Moscow: Stroyizdat) p 499

[9] Pavlov V F 1977 *Physical and chemical bases of firing of construction ceramics products* (Moscow Stroyizdat) p 240

[10] Onatsky S P 1971 *Production of expanded clay* (Moscow: Stroyizdat) p 312

[11] Novopashin A A 1973 *Mineral part of the Volga shale* (Kuibyshev: Kuibyshev book publishing house) p 120

[12] Eremenko V V 1978 Investigation of raw material quality and calculation of optimal charge composition based on the state diagram of glass-like systems *Collection of materials of the all-Union seminar of workers of the expanded clay industry* pp 77-8

IOP Conf. Series: Materials Science and Engineering 1015 (2021) 012083 doi:10.1088/1757-899X/1015/1/012083

Economic justification for the use of narrow-profile structures for flood protection

M I Balzannikov[1], A A Mikhasek[2], M V Rodionov[2], R R Mingafin[2]

[1] Samara State University of Economics, Soviet army str. 141, 443090 Samara, Russia
[2] Samara State Technical University, Academy of Architecture and Civil Engineering, Molodogvardeyskaya str. 194, 443001 Samara, Russia

E-mail: andremixas@mail.ru

Abstract: The article notes that natural-anthropogenic emergencies often arise with the co-occurrence of factors of both man-caused and natural origin of very rare recurrence. In order to avoid such situations, it is necessary to provide for special arrangements and technical measures. The authors analyzed the conditions for preventing flooding of the territory built up with residential multi-storey buildings and social infrastructure. The territory has low surface elevation marks and is surrounded by water bodies: the Volga and Samara Rivers, as well as the oxbow channel path. At the same time, practically along the entire border of the territory, there are infrastructure engineering facilities and lines (roads, dams) with surface elevations that are significantly higher than the existing elevations of the built-up area. The authors present the results of the survey of these facilities, on the basis of which recommendations for the reconstruction of their individual sections are developed. When justifying the minimum elevation of protective lines, the parameters of a rise of water level in water bodies (at the times of discharge in the Samara River of rare recurrence) were taken into account. The conditions of originating wind waves and other disturbances of the free surface of the adjacent water area at maximum water levels and the damping effect of natural and artificial obstacles along the main line of wave acceleration were taken into account.

1. Introduction

In recent years, in the Russian Federation and many foreign countries, catastrophes caused by the forces of nature, in particular water forces, have become more frequent [1-6]. Depending on the conditions and causes of occurrence, the emergency situations involving destructive effects of the water environment can be grouped into three categories: natural, man-caused and natural-anthropogenic [7].

It is advisable to introduce into the first group the poorly predictable effects of the water environment, the severity of which does not depend on the results of human activity. These include, first of all, floods caused by tsunamis, typhoons, prolonged and intense precipitation, wind surges of sea water on the coast, mudflows, etc. The examples of such natural disasters are: the tsunami near Japan (2011) with a wave height of more than 40 m, the typhoon Morakot in the Philippines and Taiwan (2009). Floods caused by the lingering rainfall in Europe in June 2013 in Austria, Germany, Czech Republic, as well as in the territories of Northern China, Cambodia, Russia (in the Amur region, Khabarovsk territory), flooding in the city of Tulun, Irkutsk Region in June 2019, etc.

The causes of man-caused emergencies involving the destructive forces of water (second group) are mainly due to the human factors and occur in connection with engineering facilities. The examples of

IOP Conf. Series: Materials Science and Engineering 1015 (2021) 012083 doi:10.1088/1757-899X/1015/1/012083

emergencies in this group include: partial destruction of the Oroville Dam in the United States and the forced discharge of water through the crest of the dam, as well as accidents in Russia: at the Sayano-Shushenskaya HPP (2009), Zagorskaya PSP-2 (2013), erosion of the Moscow Canal dam in Tushino (January 2019), and also a large number of various accidents involving utility water supplies and collectors in cities and settlements. Almost all the accidents of this group involving the destructive forces of water could have been avoided by timely and regular monitoring of the technical condition of the infrastructure facilities and timely high-quality performance of preventive maintenance work [8-10].

A feature of the group of natural-anthropogenic emergencies involving the destructive forces of water is that they occur under the confluence of circumstances of both natural and anthropogenic nature. Here we are talking about the the factors manifested in cooperation. Separately, the forces of nature, for example, prolonged intense rainfall, or anthropogenic manifestations, such as the construction of a large facility, do not cause emergency situations. However, their combination in time and space can cause emergencies with catastrophic consequences and destructions [11]. Natural circumstances in such cases, as a rule, refer to rarely recurring events. For example, the destruction of a temporary dam on the Seiba River in the Krasnoyarsk region (October 2019). Examples from this group also include flooding of territories occupied by housing, infrastructure, industrial or other facilities during flood periods of rare recurrence. On the one hand, we have anthropogenic objects. On the other hand, there are natural phenomena that were not observed in the foreseeable previous period of time, but which, theoretically, with a very low probability, can still occur.

It should be noted that ideally such situations should not be allowed under any circumstances [12]. For this purpose, it is necessary to justify, develop and make special arrangements, as well as take particular technical measures in the form of complex technical regulations for already existing building codes.

2. Materials and methods

The authors study and solve the issues of ensuring safe operating conditions for waterworks facilities and the reliability of structures that ensure the protection of coastal residential areas in case of high levels of flood waters [13]. At the same time, engineering solutions are being developed in accordance with the regulatory documents and taking into account modern requirements in terms of environmental protection [14]. In particular, the authors investigated the operating conditions of the urban complex in the large residential area «Volgar» (future homes for 37 000 inhabitants) under construction, located in the southwestern part of the city of Samara (Russia) on the left bank of the estuary section of the Samara River (figure 1).

The area under study (118 hectares) is a floodplain with characteristic elements of microrelief, woods and shrub vegetation. At the same time, part of the territory is occupied by water bodies, such as lakes, eriks and scours, as well as abandoned artificial irrigation ditches. Its significant area has relatively low absolute land surface elevations of 30.00 ÷ 33.50 m with flooded or swampy areas. The groundwater level is notably high. In some areas, water is observed at a depth of 0.5 m from the soil surface [15]. At the same time, according to the construction design of a residential area, there is an active development of residential sectional multi-storey buildings, as well as public infrastructure, socio-cultural and retail facilities (figure 2).

The development area is surrounded by hydraulically interconnected water bodies: the rivers Volga and Samara, as well as the channel path of the Dubovy Erik oxbow. At the same time, practically along the entire border of the territory, there are infrastructure engineering facilities and lines with surface elevations that are significantly higher than the existing elevations of the built-up area. So, in the west and south of the territory under consideration there are earth dams of Narodnaya and Obuvnaya streets, in the north – Shosseynaya street is also located on the dam. In the east, the Moscow-Samara railway runs along a high dam with crest marks of up to 45.70 m. In the south, there is a site of the city's sewage treatment plant adjoined by an earth dam.

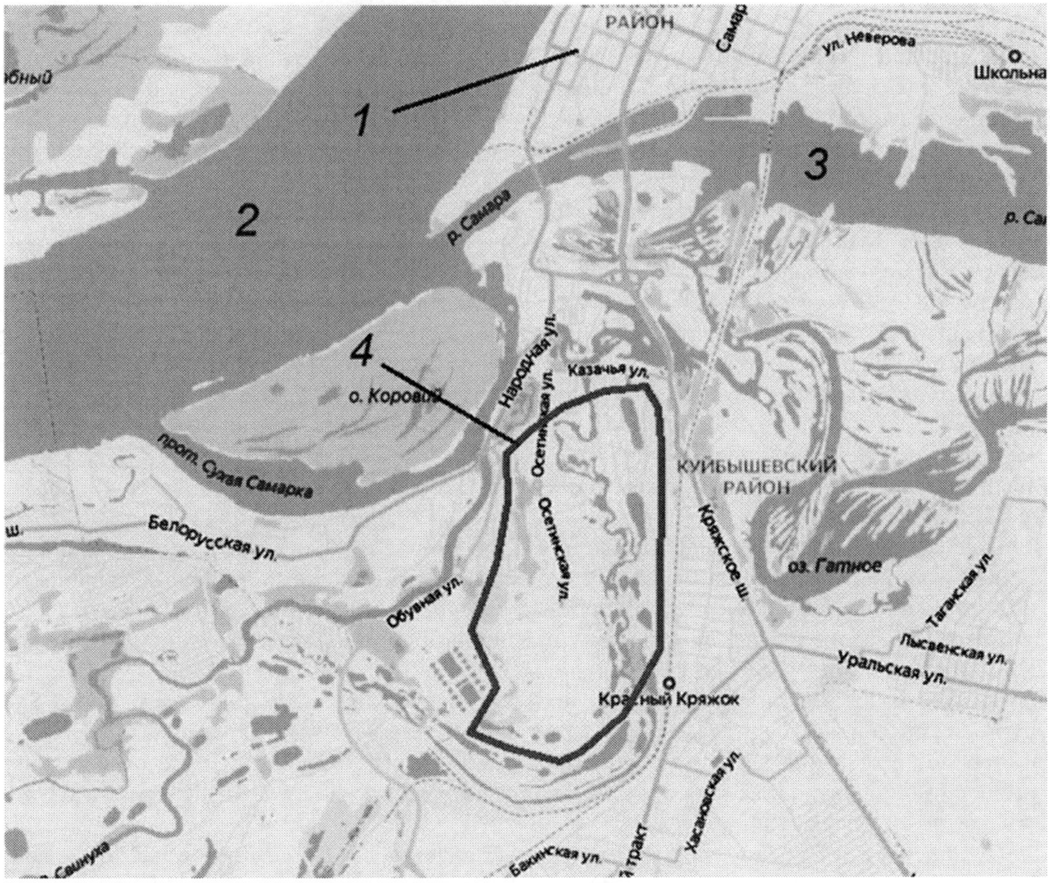

Figure 1. Layout of the residential area under construction: 1 – central part of the city of Samara, 2 – the Volga River, 3 – the Samara River, 4 – borders of the area under construction.

Figure 2. Residential buildings of the area under construction.

IOP Conf. Series: Materials Science and Engineering 1015 (2021) 012083 doi:10.1088/1757-899X/1015/1/012083

Thus, the study area with residential buildings has an already established protective contour in the form of an almost completely enclosed earth dam, which under certain hydrological conditions, protects the inner territory from direct flooding by high flood waters of the Volga and Samara Rivers.
The primary task of the study was to determine the degree of protection of the territory from flooding by the existing contour and to approve the optimal elevation mark of its crest as required by the current regulations.
The survey found that the maximum levels of the water body surrounding the area under consideration are related to the hydrological conditions of the Volga River in the operating schedules of the Saratov reservoir, and are determined by the modes of joint operation of the system of upstream hydroelectric complexes of the Volga-Kama cascade in the flood period (spring flood), and in the dry periods (summer, autumn and winter) - the backwater of the Saratov hydroelectric complex and the water releases of the Zhigulyovsk hydroelectric power station, depending, in turn, on the daily and weekly regulation of its capacity.

3. Results

To justify the required elevation of the crest of the protective outline dam, the following main factors were taken into account:

1) the calculated maximum permissible rise in the water level of open water bodies near the city of Samara was adopted based on the annual probability of an increase in the level depending on the class of protective structures for the main design case (probability P = 1%);
2) transformation of the maximum water flow passing through the downstream Saratov and upstream Kuibyshev reservoirs was taken into account;
3) influence of the operating conditions and hydropower modes of the downstream Saratov hydroelectric complex on the water level of the Saratov reservoir was taken into account;
4) the condition of inadmissibility of high flood waters overflowing through the crest of the dam protecting the built-up area was adopted, taking into account the swell of the wind wave and its wind setup on the upstream slope;
5) the conditions of originating wind waves and other disturbances of the free surface of the adjacent water area at maximum water levels and the damping effect of natural and artificial obstacles along the main line of wave acceleration were taken into account [12];
6) availability of a parapet wall on the crest of the protective dam, designed for wave effects, was taken into account, which made it possible to consider the top of the parapet wall a MWL of the protective structure.

The analysis of the initial data and the results of the calculations showed that the first and second factors have the greatest influence on making an informed decision when choosing the height of the crest of the protective contour dam, in other words, the maximum water level of the Volga River in the course of flood with 1% supply, taking into account the transformation of the maximum flow by the upstream reservoirs of the cascade. Moreover, the presence on the acceleration line of the flooded Koroviy Island (with quite sufficient tree vegetation) significantly suppresses the wind wave.

4. Discussion

The technical survey of the existing condition of the protective contour dams, the total length of which is 12.5 km, showed that in some sections of the roadway the elevation marks are lower than required and do not provide reliable protection of the territory of the residential area under construction from flooding. In this regard, engineering measures were recommended to increase the elevation of the contour crest within the two sections along Narodnaya street with a total length of 2240 m (1910 m and 330 m), within the section of Shosseynaya street (600 m), as well as on a partially destroyed dam that blocks the hydraulic connection of Dubovy Erik with the Tatyanka river, located between Krasny Kryazhok street and the territory of water treatment facilities.

It is recommended that the final elevation mark of the protective contour crest be set equal to at least 35.55 m. At the same time, the use of a solid parapet in lowered sections allows a decrease in the elevation mark of the dam crest to a height of 34.55 m.

Carrying out comprehensive measures for the reconstruction of engineering facilities of the protective contour with additional soil and (or) the installation of parapet structures on certain sections of the dam noted above will guarantee that no flooding of the residential area is happening and the negative impact of the water environment is prevented.

5. Conclusions

As a result of the study, the following conclusions were obtained:
In order to prevent natural-anthropogenic emergencies involving the destructive forces of the water environment under circumstances of very rare recurrence, it is necessary to develop, justify and make special arrangements, as well as take particular technical measures in the form of complex technical regulations approved by the interested organizations.

The authors examined the territory of a residential area under construction in the southwestern part of the city of Samara on the left bank of the mouth of the Samara River with an area of more than 118 hectares with low surface elevations. Engineering lines (roads, dams), located along the border of the territory and partially performing the function of its protection from flooding with low flood waters, have been identified.

Taking into account the complex of influencing factors, the minimum elevation of the crest of the existing borderline dam protecting the residential area was determined, at which the flooding of the territory of the built-up area with flood waters is excluded and the negative impact of the water environment on the territory facilities is prevented.

The state of the protective dam was determined, areas with unacceptably low elevations of the crest surface were identified, and practical recommendations were given for carrying out repair work on the dam and its reconstruction with the recommended increase of crest elevation. The values of m for spillways with a fusible link in the calculations for a step-curved edge with a 1:3 interval are recommended to be taken equal to at least 0.4.

References

[1] Gladkevich G I, Terskiy P N and Frolova N L 2012 *Water industry of Russia* **2** p 29-46
[2] SSFO State Hydrological Institute, St. Petersburg, Works on Hydrology p171 (2015)
[3] Peng G Lu, F Song Z and Zhang Z (2018) Key Technologies for an Urban Overland Flow Simulation System to Support What-If Analysis. *Journal of Water Resource and Protection,* **10** pp 699-724 doi: 10.4236/jwarp.2018.107040
[4] Kazaura W and Burra M (2017) Land Use Change and Traffic Impact Analysis in Planned Urban Areas in Tanzania: *The Case of Dar es Salaam City. Current Urban Studies* **5** pp 1-19 doi: 10.4236/cus.2017.51001
[5] John R (2020) Flooding in informal iettlements: iotentials and iimits for household adaptation in dar es Salaam City Tanzania *American Journal of Climate Change* **9** pp 68-86 doi: 10.4236/ajcc.2020.92006
[6] Lee Y (2015) Protecting the coastline from the effects of climate change: adaptive design for the coastal areas of Gangneung Korea. *Journal of Building Construction and Planning Research* **3** pp 107-15. doi: 10.4236/jbcpr.2015.32011.
[7] Balzannikov M I, Kruglikov V V and Mikhasek A A 2013 *Industrial and Civil Constr.* **11** p 61
[8] Zinoviev A T, Koshelev K B, Marusin K V and Kosheleva E D 2017 *Water industry of Russia* **2** pp 54-72
[9] Pulyaevsky A M 2015 *Far East: Problems of development of the architectural and construction complex* **1** pp 222-4
[10] Balzannikov M I and Mikhasek A A 2011 *Scientific Review* **7-2** pp 612-9
[11] Orlov E V and Kalinina M N 2018 *System Technologies* **29** pp 120-3

IOP Conf. Series: Materials Science and Engineering 1015 (2021) 012083 doi:10.1088/1757-899X/1015/1/012083

[12] Balzannikov M I, Rodionov M V and Litvinova M G 2016 *Procedia Engineering* **153** pp 77-82

[13] Balzannikov M I, Rodionov M V, Senitsky Yu E and Privolzhsky 2012 *Scientific Journal* **2** p 35-40

[14] Balzannikov M I, Rodionov M V and Seliverstov V A 2011 *SSUACE Bulletin Town Planning and Architecture* **1** pp 100-5

[15] Balzannikov M I, Kruglikov V V and Mikhasek A A 2013 *SSUACE Bulletin. Town Planning and Architecture* **2** pp 69-74

[16] Balzannikov M I, Rodionov M V and Melik-Pashaeva N E 2014 *Proceedings of higher educational institutions. Construction* **9-10** pp 48-55

IOP Conf. Series: Materials Science and Engineering 1015 (2021) 012084 doi:10.1088/1757-899X/1015/1/012084

Influence of window constructions on the creation of the working environment of buildings in the warm season

P Durica and P Barnak

Department of Civil Engineering and Urbanism, Faculty of Civil Engineering, University of Žilina, Univerzitná 8215/1, Žilina, 010 26, Slovakia

E-mail address: pavol.durica@uniza.sk

Abstract. The contribution is devoted to the needs of thermal and humid microclimate and aspects that affect a specific administrative character, ie in the work environment. It describes the measurement methods, measurement results and derives corresponding sub-plants for the warm climatic conditions of the year in the area of thermal comfort, determined on the basis of measurements in the selected office room.

1. Introduction

In this article we focus on the comparison of the heat-humidity microclimate in terms of its quality and the factors that affect it in summer and winter. We chose a room - an office in the Research Center UNIZA building, which is oriented to the west. The instruments used measured the physical components of the indoor environment, the effect of solar radiation through window structures and the temperature of the air from the cooling ceiling. Furthermore, the article will describe the measurement methods, the instrumentation used, and the resulting partial conclusions for the warm (summer) climatic period of the year.

2. Description of the building and the room to be evaluated

The subject of the evaluation is the building of the RC UNIZA, which was purposefully designed and built in an energy-passive standard (energy gains from the building, its environmental technology equipment and renewable energy sources should prevail over the energy losses related to its year-round operation) and managed by intelligent operations (central computer control system). It is described in more detail in the publication [1]. The building and its technical, technological and laboratory equipment are used not only for the purposes of administration and research workplaces, but also for experimental research focused on the design and operation of buildings with the lowest possible energy consumption for the workplaces of the three UNIZA faculties. Its general description and technical support can be found in [1], a brief description is given in [2]. The room we chose as the measuring point is located on the 2nd floor and is oriented to the west (figure 1).

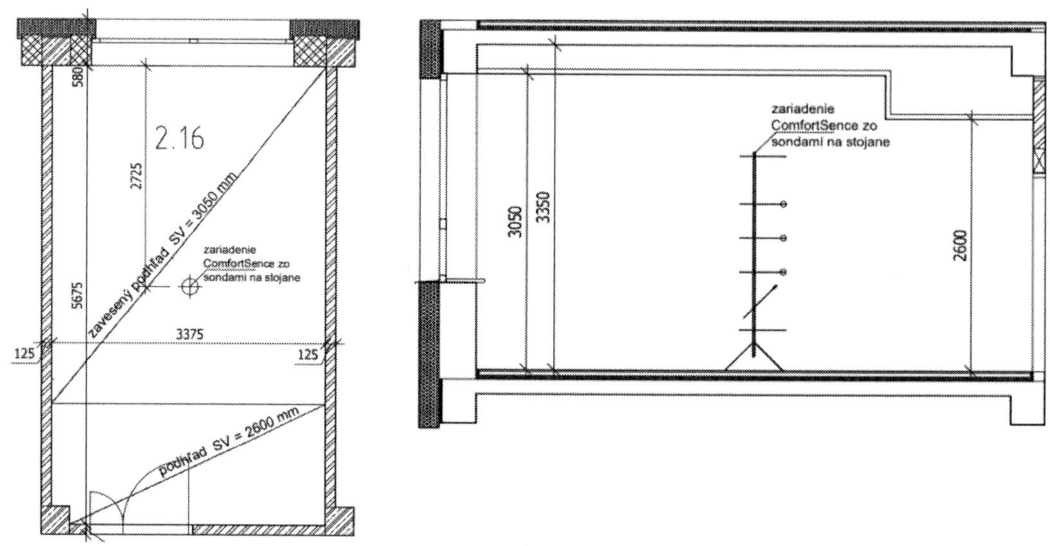

Figure 1. Floor plan and section of the room under consideration.

3. Indoor environment conditions in administrative buildings

The quality of the indoor environment in the working spaces of office buildings is required by the relevant legislation in most countries of the world. In Slovakia it is Act no. 355/2007 Coll. [3], which states that the indoor environment of buildings must meet the requirements for thermal-humidity microclimate, ventilation, heating, lighting, glare and other types of optical radiation. Persons who operate buildings for the public, whether natural or legal persons, are obliged by law to ensure the quality of indoor air in the building to avoid the risk of chemical, physical, biological and other harmful factors. More specific requirements for the indoor thermal-humidity microclimate are further discussed in the Decree of the Ministry of Health of the Slovak Republic no. 259/2008 Coll. [4]. Factors that can also affect the indoor microclimate are the production of heat by the body, the production of heat by devices and the clothes that people wear. Administrative work is included in activity class 1a (table 1) and the total thermal resistance of the garment is considered for standard clothing $R\text{cl} = 1{,}0$ duty. Optimal and permissible conditions of the thermal-humidity microclimate in the offices for the warm period of the year are given in table 2.

Table 1. Class of employment in offices according to [4]

Activity class	Total energy expenditure		Examples of activities
	q_M [W/m^2]	q_M [met]	
1a	66 - 80	1.13 – 1.38	Activity of a sitting room with minimal physical activity (administrative work, activity in classrooms); sitting activity associated with easy manual work of hands and arms (typing, working with a PC, simple sewing, laboratory work, assembling or sorting small light objects).

IOP Conf. Series: Materials Science and Engineering 1015 (2021) 012084 doi:10.1088/1757-899X/1015/1/012084

Table 2. Optimal and permissible conditions of the thermal-humidity microclimate for the cold season of the year [4].

Washing class	Operative temperature θ_0 [°C]		Permissible air velocity v_a [m/s]	Permissible relative air humidity φ [%]
	optimal	admissible		
1a	23 - 27	20 - 28	≤ 0.25	30 - 70

4. Description of measurements of thermal-humidity microclimate parameters

The measurement methodology and measuring instruments used were the same as in the cold period of the year [2] focused on monitoring the objective parameters characterizing the quality of the heat-humidity microclimate. A ConfortSense instrument with stationary and flexible probes (figure 2 on the left) placed on racks in the middle of the room at different height levels and on the windowsill [5] was used to measure indoor air temperature, operating temperature, relative indoor humidity and indoor humidity. The air velocity was measured using stationary probes placed on a stand. The measured values were recorded at five minute intervals. At the same time, measurements of solar radiation intensity immediately after glazing were performed using a pyranometer (figure 2 on the right) and internal surface temperatures using thermocouples placed on the window (identical to [2]). All probes were connected to the measuring station in a time step of five minutes, followed by entering the measured values into a computer. Measurements have been taking place since January 2018.

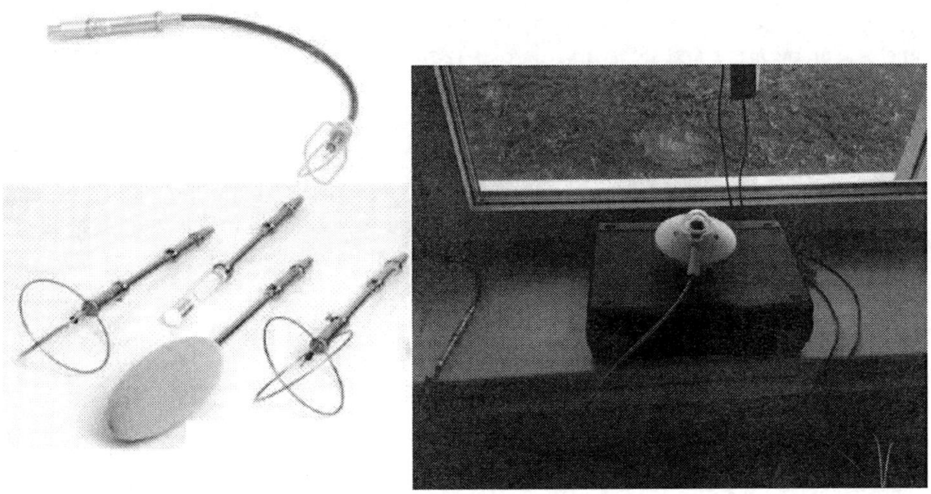

Figure 2. Measuring probes flexible and stationary (left) and pyranometer for measuring the intensity of solar radiation (right).

5. Measurement results

For the purposes of the paper, a section of one week with the highest temperatures in June 2019 with the range of outdoor air temperature measured just before the facade from 12,1 to 45.6 °C (average temperature 25.3 °C) is selected. Figure 3 shows, in addition to the outside air temperature, the indoor air temperature profiles together with the intensity of solar radiation, measured just in front of the façade in the vertical plane above the window under consideration and behind the window at the center of the windowsill in a horizontal position. The intensity of global solar radiation reached the maximum value of 743,9 W/m^2 in the summer on a clear sunny day, with an average value during measurements of 140.8 W/m^2. In the interior and due to the position of the pyranometer, it ranged only in the range from

IOP Conf. Series: Materials Science and Engineering 1015 (2021) 012084 doi:10.1088/1757-899X/1015/1/012084

0 to 94.6 W/m^2. The indoor air temperature, measured in the middle of the room, was in the range of the requirements of the optimal microclimate, even though the blinds were pulled off during the measurement. This was mainly due to the favorable effect of ceiling cooling from ground collectors.

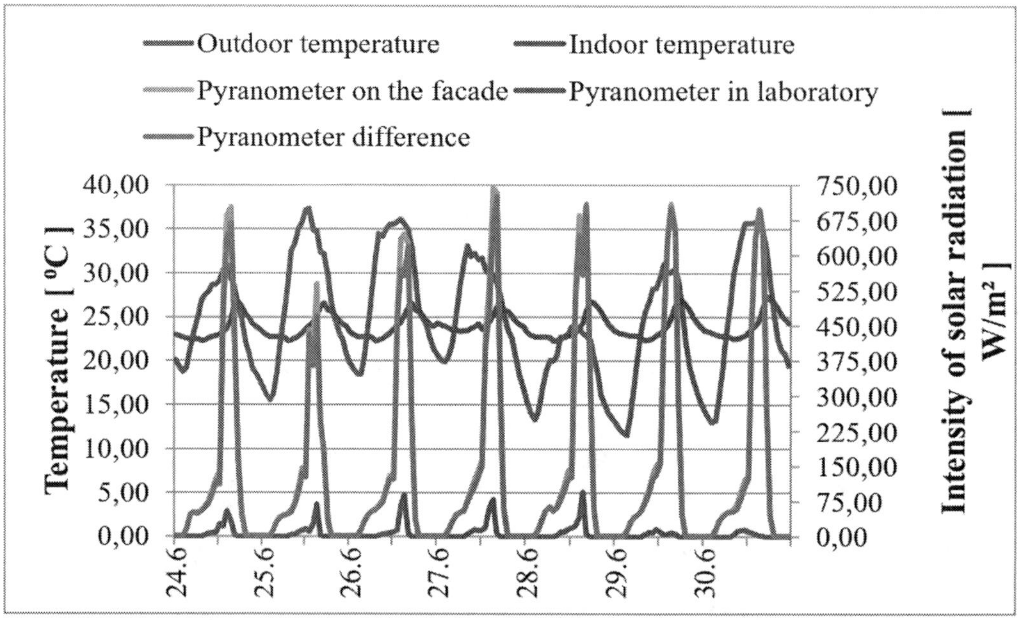

Figure 3. Temperature and intensity of solar radiation in front of the facade and behind the window in summer.

Figure 4 shows the intensity of sunlight and the movement of the sun across the sky in summer and the recorded temperature behind the window on the probes located on the windowsill (probe 1 - the left edge, probe 2 - the center, probe 3 - the right edge of the window). The monitored temperatures during the measurements were usually below the values of the air temperature in the middle of the room. This was influenced by their location under window friezes with partial shading.

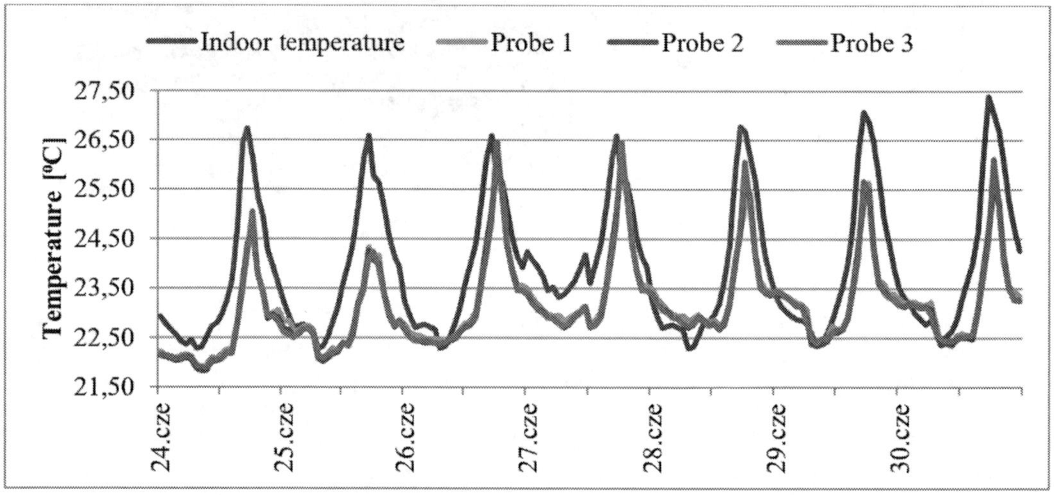

Figure 4. Temperature courses in the measured room in summer.

Figure 5 shows the temperature profiles from sensors located on the inside of the window structure in the summer. The lowest temperature was recorded from sensor 105 (horizontal cross-section between the sash), namely a temperature of 19.5 °C and, conversely, the highest temperature was measured by sensor 107 (lower right corner of the glazing), namely a temperature of 47.4 °C. At this measuring point, the temperature of the inner surface of the glazing was regularly higher than the temperature of the outside air at the peaks.

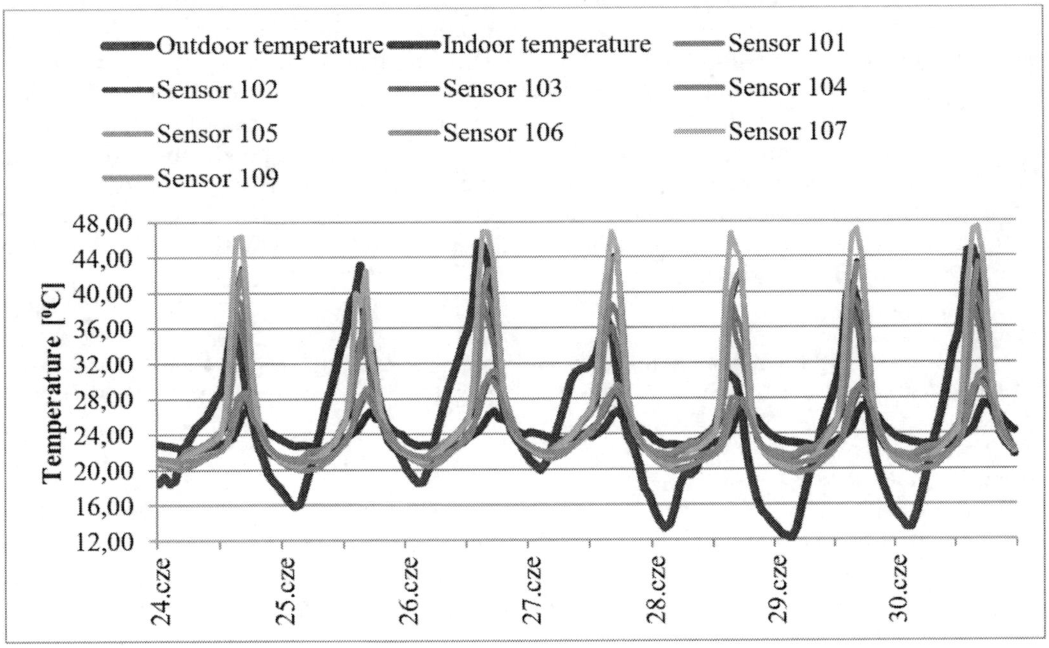

Figure 5. Course of temperatures on the inner surfaces of the window construction in summer.

Figure 6 shows the course of the air flow rate in the room. The figure shows that the values do not exceed the permitted standard values, even if the outside air temperature was higher than 30 °C.

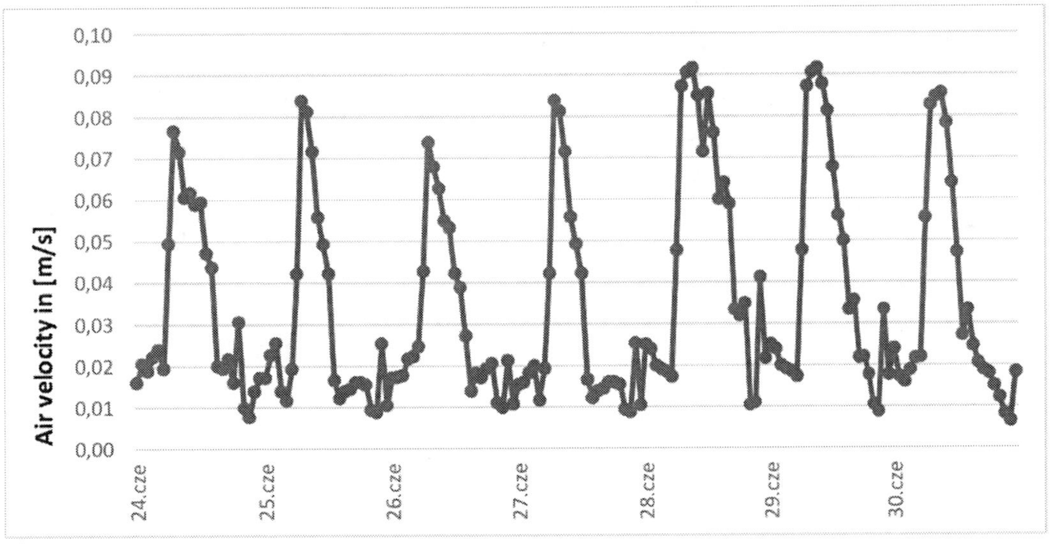

Figure 6 The course of air velocity in summer

IOP Conf. Series: Materials Science and Engineering 1015 (2021) 012084 doi:10.1088/1757-899X/1015/1/012084

6. Conclusions

The combination of the construction solution with the technological equipment on the building of the Research Center of the University of Žilina is to ensure the conditions of the indoor microclimate so that they meet the current standard requirements. Measurements that have taken place and are still taking place in the winter and summer period, we verify whether the indoor environment is not affected. In the winter, we focus on whether the underfloor heating system and its connection to geothermal wells in the underground is sufficient and whether it is not necessary to connect a secondary heating source. In the summer, we focus on whether the interior of the monitored room overheats in terms of surface temperatures and heat radiation to the environment. These measured values are then used for further evaluations needed to determine the indoor microclimate. The measured and evaluated quantities were in the optimal values during the evaluation, even though the shading element (exterior blinds) was completely pulled out. Optimum values have also been achieved due to the good orientation of the room (west) and the efficient cooling ceiling system which is connected to the ground collectors.

The evaluation of measurements and their analysis is currently ongoing and the positions of exterior blinds are also being varied. Subsequently, these variants will be evaluated for the shading efficiency of exterior blinds in different positions.

Acknowledgments
This publication was created thanks to support within the research project VEGA no. 1/0673/20: Theoretical and experimental analysis of energy effective and environmental friendly building envelopes.

References

[1] Durica P, Iringova A, Ponechal R, Rybarik J and Vertaľ M 2017 Energy and environmental design and evaluation of buildings *EDIS UNIZA* Žilina (ISBN 978-80-554-1296-2)
[2] Durica P and Barnak P 2019 Evaluation of the parameters of the internal environment of an administrative building in the winter period (paper presented at the conference *Thermal protection of buildings*)
[3] Act No. 355 of 21 June 2007 on the protection, promotion and development of public health and amending certain laws
[4] Decree of the Ministry of Health of the Slovak Republic no. 259 of 18 June 2008 on the details of the requirements for the indoor environment of buildings and the minimum requirements for lower-standard apartments and accommodation
[5] ComfortSense 2012: ComfortSense 264 http://www.dantecdynamics.com/comfortsense (access August 2020)
[6] Standard STN EN ISO 7730: 2006 (83 3563) Ergonomics of thermal environment. Analytical determination and interpretation of thermal comfort by calculating PMV and PPD indicators and local thermal comfort criteria

IOP Conf. Series: Materials Science and Engineering 1015 (2021) 012085 doi:10.1088/1757-899X/1015/1/012085

Synthesis of air-curing phosphate binders and heat-resistant composites based on them

A I Khlystov and V A Shirokov

Samara State Technical University, Molodogvardeiskaya St., 244, Samara, Russia, 443100

E-mail address: alex-x1950@yandex.ru

Abstract. The article discusses methods for synthesizing phosphate binders hardening in air at normal temperature. Mixed binders are offered to obtain heat-resistant composites. They have the ability to acquire stripping strength in air conditions. It is concluded that the proposed binders should be synthesized from powdered industrial waste or from pure oxides, providing a reasonable ratio of their activity degree.

1. Introduction

Nowadays, in lining constructions of most thermal devices and high-temperature installations both large block parts made of heat-resistant concrete and monolithic compositions in the form of solutions, mortars and other masses are used. This improvement of refractory lining technology allows increasing chemical resistance and durability of linings in general. The use of heat-resistant concretes in monolithic and precast variants in lining constructions of thermo technical plants allowed designers to create quite rational solutions for manufacturing of separate complex elements of thermal units, which cannot be created on piece refractories [1].

Heat-resistant concretes necessarily include such binders as portland cement with refractory additives, alumina cement, silicate liquid glass and its semi-finished products in the form of silicate - blocks, as well as phosphate compositions in the form of mixtures of powdered refractory fillers with orthophosphoric acid or liquid phosphate bonds.

Very scarce refractory components prepared on the basis of special ceramics (chamotte, mullite, magnesite, etc.) are used as fine aggregates and fillers in the composition of heat-resistant concrete. Technological production process of refractory fine aggregate and fillers is connected with such energy-consuming operations as hard ceramics crushing and grinding. In this connection, the use of local industrial refractory wastes and other technogenic raw materials in the composition of heat-resistant concrete is a very urgent task [2].

Multiple research and industrial tests have shown that the use of phosphate cements in the composition of heat-resistant concrete allows the development of very advanced refractory lining materials with improved physical, thermal and operational parameters [3].

Most phosphate compositions and heat-resistant concrete based on them require heat treatment up to 500-600°C for their curing. The curing process of plastic phosphate compositions can be controlled by adjusting the ratio between the components of the finely dispersed component and the chemical activity of the mixing liquid.

IOP Conf. Series: Materials Science and Engineering 1015 (2021) 012085 doi:10.1088/1757-899X/1015/1/012085

Recently, phosphate cements, which are chemical binders and allowing to create materials and products with high physical and mechanical properties, have become widely used in heat-resistant concrete technology [3].

It is possible to control the process of hardening of plastic phosphate cement masses by regulating the activity of the thin-walled component (e.g. increasing the fineness of grinding, heating temperature, ultrasound processing, etc.) and the activity of the mixing liquid (changing the acid concentration, quantity, the introduction of substances that accelerate or slow down the setting processes in air conditions).

2. Theoretical foundations for the synthesis of phosphate compositions

Powdered oxides of many metals and H_3PO_4 - orthophosphoric acid forming hydrophosphates of various degrees of substitution as a result of the reaction are used for the phosphate cements synthesis. Studying the processes of reactions of various powder components with H_3PO_4 of various concentrations, Kinzheri [4] noted that during hardening of phosphate compositions the heat formation as a result of an exothermic reaction is a very important factor.

Therefore, according to Hess's law, the heat effects of the Me_nO_m - H_3PO_4 reactions were calculated in order to determine the required amount of heat for the formation of phosphate masses under normal air conditions [5].

The heat effect of any chemical reaction "Q" is equal to the sum of the heat formation of the interacting substances of this reaction. When calculating the heat effect of reactions, the numerical values of the heat formation were taken from the reference book [6]. To perform the calculations, the assumption was applied that, as a result of the interaction of oxides with H_3PO_4, intermediate phosphates are formed. To reveal the influence of thermal effects on the hardening process of phosphate masses, the reduced reaction effect was taken, determined by the formula:

$$Q_{given} = \frac{Q}{M} \tag{1}$$

where – heat of reaction, found according to Hess's law, cal/g;

Q_{given} – conditional thermal effect of the reaction, cal/g;

M – gram-mole of oxide.

According to the values of the conditional thermal effect formed as a result of the interaction of powdery oxide with orthophosphoric acid, the hardening masses were divided into three groups based on the heat received:

1-group – phosphate compositions, curing at heat treatment (200 °C and more); Q_{given} – 30 cal/g

2- group – phosphate compositions, curing in normal air conditions (t = 20°C); 30 > Q_{given} > -200cal/g

3- group – phosphate compositions, which do not form monolith at the oxide reaction with H_3PO_4: Q_{given} < -200 cal/g.

Such structural and energy parameters of solids as ionic density and the energy of a single bond have a great influence on the course of chemical reactions. [7] In this regard, it becomes possible to examine l the hardening process of phosphate compositions and derive a new indicator determining the activity of powdered metal oxides [2].

Ionic density (P_i), determining the process of ion packing as a result of crystallization of the solid framework of substances, is calculated by the formula of prof. A.A. Novopashin [7]:

$$P_i = \frac{2.52 \cdot \rho_i}{M \cdot \sum n_i \cdot r_i^3} \tag{2}$$

where ρ_i – true oxide density, g/cm^3;

M – molecular weight;

n_i – the number of ions of a given type in a molecule;

r_i – ion radius in tenths of a nanometer.

IOP Conf. Series: Materials Science and Engineering 1015 (2021) 012085 doi:10.1088/1757-899X/1015/1/012085

The process of chemical interaction of solid metal oxides with H_3PO_4 completely depends on the packing density of the ions composing these oxides. However, the structure of a solid will be the more stable, the higher the value of the energy of a single bond between cations and anions. Thus, the ability of solid metal oxides to interact with H_3PO_4 and form compositions hardening with it can be estimated through their activity and presented as a dependence

$$K_a = f(P_i \cdot \varepsilon) \qquad (3)$$

where K_a – activity coefficient of oxides;
P_i – ionic density of oxides;
ε – the energy of a single bond between oxide cations with oxygen; kcal / mol, taken from Sun's experimental data [7].

It becomes possible to calculate the activity coefficient K_a of oxides by the formula:

$$K_a = \frac{44}{P_i \cdot \varepsilon} \qquad (4)$$

where P_i – ionic density of oxide;
ε – energy of a single bond between oxide cations with oxygen, kcal / mol [7].
44 – product value $P_i \cdot \varepsilon$ for phosphorus anhydride P_2O_5.

The results of mathematical calculations allowed classifying metal oxides by the following scheme:
K_a >2.42 – I group - phosphate compositions of the hardening structure do not form;
0.85< K_a <2.41 – II group - phosphate compositions hardening under normal air conditions (t = 20°C);
K_a <0.85 – III Group - phosphate compositions are formed during heat treatment (200°C and above).

Such classification allowed forming a new pattern of formation of solid structures in plastic phosphate compositions taking into account the modifying state of powdered oxide. This modified state of solid oxide allows fixing its ionic density. In this connection, it was found that the minimum amount of heat required to create air-curing phosphate compositions is 70 cal per 1 gram of solid oxide.

Therefore, it is possible to mix powdered metal oxides with each other, forming mixtures in which the total heat release will approach 70 cal per 1 g of total mass.

Naturally, the beginning of mass setting will be due to the heat generated by the reaction of active oxide with H_3PO_4. Phosphate composition in this case acquires initial strength, which is required for dismantling, transportation and individual elements assembly. Final structural formation of cement stone with stable phosphate component is observed during further heat treatment, which can be carried out during the first heating of the thermal unit due to the chemical reaction of the inactive part of the thin-walled additive with H_3PO_4.

As a result, it was found that using the obtained dependence of the effect of K_a of powdered oxides on the process of the appearance of hardening masses in phosphate masses, it is possible to synthesize cements with normal setting times.

It is obvious that the basis for the selection of raw components to obtain a normally solidifying phosphate plastic masses is the activity factor of K_a oxides. It allows to search for raw materials components, which are a natural mixture of oxides in the composition of industrial waste.

In the group of oxides having binder properties under normal conditions, there are the following mixtures Fe_2O_3 – H_3PO_4 and FeO – H_3PO_4. Among iron-containing industrial waste, the maximum amount of iron oxides is contained in the waste heaps of sulphuric acid production - pyrite cinder, formed after firing of raw materials in the form of pyrite.

The main components of pyrite cinders are oxides of ferrous and ferric iron. Their maximum content exceeds 70%. However, the chemical composition of pyrite cinders also includes such oxides as SiO_2; CaO; MgO and Al_2O_3.

For chemical compounds, which include many inorganic industrial wastes, the activity coefficient is recommended to be determined by the formula:

IOP Conf. Series: Materials Science and Engineering 1015 (2021) 012085 doi:10.1088/1757-899X/1015/1/012085

$$K_a = \frac{(K_a)1 \cdot A + (K_a)2 \cdot B + \cdots + (K_a)n \cdot N}{A + B + \cdots + N} \tag{5}$$

where K_a – industrial waste activity ratio;

A, B. … N – metal oxide powder content in waste, %.

Calculations show that taking into account the chemical composition of iron-containing waste - pyrite cinders, the activity coefficient is within $K_a = 1.17 \div 1.22$. Therefore, according to the above classification, inorganic chemical compounds with $K_a = 1.1\ 7 \div 1.22$ may exhibit binding properties in H_3PO_4 compositions in normal air conditions (t = 20°C).

In the process of experiments pyrite cinder of Voskresensky Chemical Plant (Moscow region) and 70% thermal orthophosphoric acid were used to produce air-curing phosphate cements. This cement was called iron-phosphate cement. Heavy heat-resistant concretes were filled with refractory pieces of chamotte brick type SHA or SHB, lightweight concretes were filled with expanded clay gravel of 5-20 mm fractions. Samples of concretes on iron-phosphate cement have a very high strength after hardening in the air for 7 days, which is preserved in the heating process to 1200 °C. Fire shrinkage of these concretes is within normal limits. The operating temperature of heavy heat-resistant concrete is within 1230-1240 °C, and the temperature of expanded clay concrete is 1000-1100 °C.

The results obtained show optimal physical, mechanical and thermal characteristics of the selected concretes. The main advantage is considered to be their ability to harden and increase the strength of time in air conditions. Thermal treatment for these compounds of concretes is excluded. This fact suggests that it is possible to produce large-size elements, as well as to use a concrete mixture in a monolithic form.

In order to increase the operating temperature of refractory concretes on phosphate cements using pyrite cinders, experiments were carried out to synthesize binder compositions with increased refractoriness for heat-resistant concretes. In this regard, the compositions of heat-resistant concretes were selected based on mixed aluminum-iron-phosphate and zircon-iron phosphate cements. To synthesize mixed alumina-iron-phosphate cement, alumina-chromium waste of petrochemistry was used as an alumina-containing component - a spent catalyst IM - 2201 and technical alumina. Zircon concentrate KC - 1 of the Volnogorsk deposit was used for zircon - iron phosphate cement. When the above raw materials are mixed in a mass of pyrite cinders with orthophosphoric acid, mixed plastic cement masses are formed, which have properties different from the masses on pure iron phosphate.

The refractoriness of the synthesized aluminum iron phosphate and zircon iron phosphate significantly increases after the introduction of alumina-containing raw materials and zircon into phosphate cements. The setting times of the synthesized phosphate cements increase, but even with 80-90% of the alumino-chromium waste IM-2201 of spent catalyst or zircon, the process of air curing of phosphate cements is observed.

The test results of the synthesized phosphate cements indicate that at an early age (1 day) they acquire the strength necessary for dismantling, transportation and installation of concrete elements prepared on their basis. This circumstance is explained by the formation of refractory compounds, such as aluminophosphates and zirconophosphates, which are capable of non-hardening when heated at high temperatures.

The results of samples testing of mixed cements in determining the refractoriness indicate that with an increase in the amount of a high-alumina product or zircon in the compositions of synthesized cements, their operating temperature increases: 1500 °C for aluminum iron phosphate and 1600 °C for zircon iron phosphate. Such an indicator of the durability of phosphate compositions as heat resistance is 30 or more water thermal cycles. Such indicators of heat-resistant compositions developed on the basis of waste are typical for many phosphate materials. Using the developed phosphate cements, the compositions of heavy heat-resistant concrete were selected. The test results shown in table 1 demonstrate that concrete mixtures based on chamotte and mullite aggregate harden under normal air conditions and being one day old they have strength indicators sufficient for dismantling and transporting products. In this regard, it becomes possible to transfer the heat treatment of concrete

IOP Conf. Series: Materials Science and Engineering 1015 (2021) 012085 doi:10.1088/1757-899X/1015/1/012085

products directly to a heating unit, in which any temperature conditions can be created during drying and first heating.

In heat-resistant concretes of all the developed compositions, the compressive strength after heating up to temperatures of 1400-1600 °C does not decrease and remains constant. The deformation temperatures of refractory concretes under load are also very high, i.e., the ability of concretes to non-harden when heated is retained due to the newly synthesized cements.

3. Results and discussions

Experimental data indicate a very high physical and thermal performance of the selected concrete. A great advantage of concretes is their ability to harden and increase strength in air conditions at a temperature of 20 °C, which affects the construction of large-sized elements and allows the use the concrete mixture in a monolithic version.

Table 1. Physicomechanical parameters of refractory concretes based on developed phosphate cements.

Concrete composition, (kg/m^3)	Average density in a dry state, (kg/m^3)	Compressive strength, MPa, after hardening and heating to temperature (°C)			Heat resistance, water heat cycle
		20 (1 day)	800	1400	
alumina-iron-phosphate concrete cinder – 90 Aluminum chromium waste – 350 Chamotte crushed stone (fraction 5-10mm) – 750 Chamotte sand– 650 H_3PO_4 -260	2082	2.1	44.7	45.2	28
zircon-iron phosphate concrete cinder – 90 Zircon KC 1– 350 Mullite crushed stone (fraction 5-10mm) – 780 Mullite sand– 680 H_3PO_4 -250	2124	2.2	48.1	50.8	32

The article also investigated nanotechnologie raw materials taken from the sludge of industrial waste of non-ferrous metallurgy. At the Samara metallurgical plant, sludge waste is formed in the form of secondary products, which are nanotechnogenic raw materials. Thus, the sludge of alkaline etching of aluminum is formed when processing aluminum-magnesium alloys with NaOH solution. The resulting high-plastic sludge (with *pH* level equals to 12.6) is mixed with carbonate sludge (chemical water treatment sludge) to neutralize the aluminum-alkaline sludge. As a result of this technology, alumocalcium sludge is formed at the plant treatment facilities, the chemical composition of which is shown in table 2.

Table 2. Chemical composition of calcium alumina sludge

Content, mass (%)								
SiO_2	Fe_2O_3	Al_2O_3	CaO	MgO	SO_3	R_2O	other mixtures	Σ
8.16	0.8	14.6	26.32	8.24	1.58	1.36	38.88	99.94

Physical and chemical studies conducted with aluminum-calcium, aluminum-alkaline and carbonate slurries have shown the presence of a large amount of the mineral-calcite $CaCO_3$ in the raw material.

IOP Conf. Series: Materials Science and Engineering 1015 (2021) 012085 doi:10.1088/1757-899X/1015/1/012085

The size of sludge particles, as shown by the studies [8], is 40-80 nm. This fact can be attributed to a slurry waste of Samara metallurgical plant to nanotechnologia raw materials.

Based on the sludge, liquid phosphate binders were synthesized by applying H_3PO_4 of a certain concentration. When these slurries come into contact with liquid orthophosphoric acid, an active chemical reaction occurs as a result of which almost all the solid substance is dissolved without a residue due to its nanoscale (while the resulting ligaments do not form a precipitate during long-term storage).

Using this technology, liquid phosphate binders are formed, namely, aluminum phosphate, calcium phosphate and aluminum-calcium phosphate binders due to the use of aluminum-alkaline, carbonate, and aluminum-quartz slurries, respectively.

Synthesized liquid phosphate binders due to their high chemical activity proved to be very effective mixing liquids in the development of air-hardening compositions of heat-resistant compositions: solutions, concretes and impregnating-coating masses.

Thus, the main factor determining the possibility of using oxides to produce phosphate binders is the thermal effect of the reaction Me_nO_m - H_3PO_4 (powdered metal oxide - orthophosphoric acid).

The activity coefficient of powdered oxides, calculated from their structural and energy parameters, is the basis for synthesizing compositions of phosphate cements that harden under normal conditions. According to this coefficient, metal oxides are distributed into III groups according to their degree of activity; thus, when interacting with H_3PO_4, the values of the oxide activity coefficient and the thermal effect found when interacting with H_3PO_4 allow for a very reliable synthesis of various phosphate cements to obtain concretes and solutions with the desired properties.

4. Conclusions

Heat-resistant heavy concrete based on iron-phosphate cement has high physical and mechanical properties. The operating temperature of such heat-resistant concrete with a fireclay component reaches 1200°C. The use of pyrite cinder in a mixture with high-alumina fine-ground powders and ground zircon made it possible to synthesize alumina-phosphate and zircon-phosphate cements and heat-resistant concretes that harden in air conditions (t = 20°C). Increased strength indicators of heat-resistant concrete are formed during drying and the first heating of the heat unit. Concrete elements of linings form a stable phosphate bond in their structure after such heat treatment. As a result, the physical and thermal performance indicators are significantly increased. This circumstance significantly increases the feasibility of using heat-resistant concrete of phosphate hardening instead of piece refractories in the linings of thermal units. Moreover, phosphate compositions (concretes and solutions) have increased chemical resistance, i.e. they are not wetted in oaluminum solutions and flux salts.

The use of acidic water-soluble alumophosphate binders, in particular alumocalcium phosphate, as a mixer of heat-resistant mortar and concrete mixtures, also allowed obtaining the effect of air-curing in phosphate composites.

References

[1] Krayukhin V. I 2014 *Heat-resistant concrete. Compositions. Exploitation. Restoration of basic properties after long-term operation* (Saratov) p 348

[2] Khlystov A I 2004 *Increasing the efficiency and improving the quality of refractory lining materials* (Samara) p 134

[3] Sudakas L H 2008 *Phosphate binders* (St Petersburg) p 260

[4] Sychev M M 1986 *Inorganic adhesives* (Leningrad : Chemistry) p 152

[5] Beskov S D 1962 *Techno-chemical calculations* (Moscow) p 468

[6] Karapetyants M Kh, Karapetyants M L 1968 *Basic thermodynamic constants of inorganic and organic substances* (Moscow Chemistry) p 395

[7] Novopashin A A 1973 *Mineral part of the Volga shale* (Kuibyshev) p 120

[8] Khlystov A I, Sokolova S V, Konnov M V 2012 Directed structural - chemical modification is one of the ways to improve the physical and thermal characteristics of aluminosilicate and high-alumina refractories *Refractories and technical ceramics* **11**. pp 35-39.

IOP Conf. Series: Materials Science and Engineering 1015 (2021) 012086 doi:10.1088/1757-899X/1015/1/012086

Influence of compaction on the deformation resistance of the sub-ballast layer

P Dobeš and L Ižvolt

University of Žilina, Faculty of Civil Engineering, Department of Railway Engeneering and Track Management (DRETM), Univerzitná 8215/1, 01026 Žilina, Slovakia

E-mail address: peter.dobes@uniza.sk

Abstract. In railway practice, the results of static load tests are used most frequently to check the quality of compaction. However, the static load tests are preferably carried out to demonstrate the achievement of the required deformation resistance on the sub-ballast layers. The paper deals with the analysis of measured values of deformation resistance of the sub-ballast layer on the modernised track Považská Teplá - Žilina, depending on the compaction (achieved degree of soil compaction) and the possibility of mutual correlation of these qualitative parameters. In the analysis, the values of deformation resistance and compaction quality (degree of soil compaction) were used as part of the quality diagnostics of the performed work on the part of the modernised section of the line Považská Teplá - Žilina.

1 Introduction

The ability of the sub-ballast layers to withstand traffic loads in railway engineering is termed deformation resistance. This term was introduced in Slovakia for the first time in [1]. Until then, the term bearing value of the sub-ballast layers, or track bed was used. The sub-ballast layers are sufficiently deformation resistant if they are able to absorb a defined load without harmful (permanent) deformations and without breaking them. If the sub-ballast upper surface or the subgrade surface loaded with increasing load, then in principle the deformations can be shown according to figure 1.

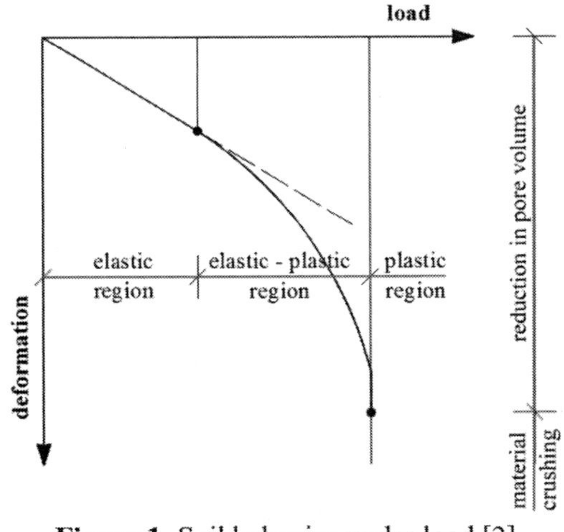

Figure 1. Soil behavior under load [2].

IOP Conf. Series: Materials Science and Engineering 1015 (2021) 012086 doi:10.1088/1757-899X/1015/1/012086

Initially, the load-induced deformations have a linear course and then a reverse (reversible), which can be characterized as elastic soil behavior. In another area, not only elastic, but also plastic (permanent) deformations occur due to the reduction in pore volume. When the breaking load (ultimate strength) is exceeded, deformations progressively develop under a constant load, which can be referred to as plastic behavior of the soil.

The deformation resistance of the sub-ballast layers or the subgrade surface can be expressed numerically eg. modulus of elasticity (static, dynamic, shear, volumetric), modulus of deformation (static, dynamic), reaction modulus, penetration modulus, CBR value or based on soil physical properties (moisture, plasticity index and consistency index). All these deformation characteristics indicate the relationship between load and deformation but differ significantly from each other in the boundary conditions of their detection [2]. For this reason, it is difficult to compare these deformation characteristics.

The following text of the paper will analyze the impact of the achieved degree of soil compaction on the deformation characteristic, which has the greatest application in railway engineering (in the diagnostic of the sub-ballast layers), namely the static modulus of deformation will be performed. The analysis will be performed on the basis of the measured values of the static modulus of deformation at the level of the sub-ballast upper surface determined during the diagnostic carried out in the modernised section of the line Považská Teplá - Žilina.

2 Characteristics of tested materials and applied method of diagnostic

This part of the paper will characterize the properties of built-in materials into the sub-ballast layer and test method for verification of deformation resistance and degree of soil compaction of the sub-ballast upper surface, which were used and identified on the assessed modernised section of the line Považská Teplá - Žilina.

2.1 Specification of material composition of the sub-ballast layer

The sub-ballast layer is the first structural layer of the sub-ballast layers located below the ballast bed (in the case of a conventional type of railway superstructure). Its main function is the spreading of the effects of line tonnage (traffic density) and load of the railway superstructure on the subgrade surface, eventually also the protection of the subgrade surface against adverse effects of water and frost. In some cases, the sub-ballast layer can only fulfill these functions if additional structural engineering measures are applied, such as:

- application of suitable geosynthetic materials and elements,
- improving or stabilisation the soil of the subgrade surface or changing the subsoil,
- application of waterproof materials in the groundwater protection area,
- incorporation of thermal insulation materials, etc.

The sub-ballast layer shall, by its grain size distribution and moisture, ensure sufficient compaction and deformation resistance of the structural layer and at the same time it shall comply with the filter criterion against the soil of the subgrade surface and the ballast gravel [3].

Within the diagnosed section of the modernised railway line Považská Teplá - Žilina, the sub-ballast layers were designed in accordance with TNŽ 73 6312 [1]. Structural and material composition of the individual structural layers resulted from the condition of ensuring a high-quality and safe railway for the train speed up to 160 km.h^{-1}. Based on these boundary conditions and in order to ensure the required deformation resistance at the level of sub-ballast upper surface $E_{sub} \geq 60$ MPa (railway line built on the original earthwork), a sub-ballast layer fr. 0/63 mm of variable thickness, reinforced by geogrids (depending on the measured value of the deformation resistance of the subgrade surface identified in the geotechnical survey - figure 2) was designed in the sub-ballast layers. In the case of a railway line on a newly constructed earthwork, constructed of frost-resistant and pervious soil (crushed aggregate fr. 0/63 mm), the required value of the deformation resistance at the level of the sub-ballast upper surface $E_{sub} \geq 80$ MPa was required. In this case, the sub-ballast layer was built of crushed aggregate fr. 0/63 mm thickness 0.30 m and placed on geotextile [4].

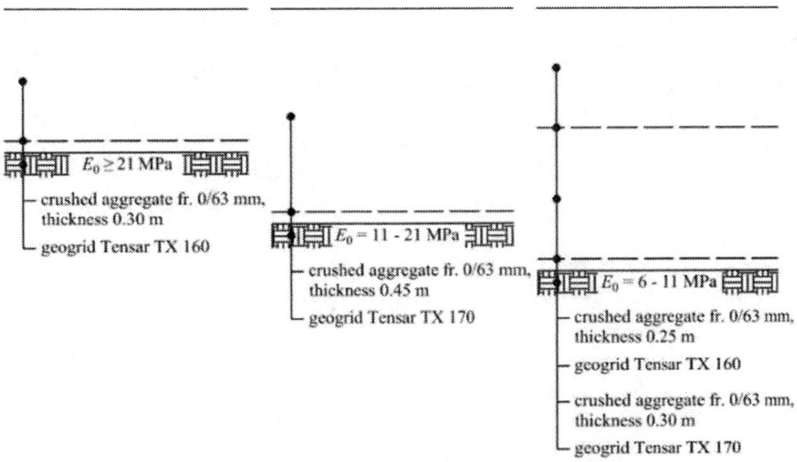

Figure 2. Design of construction and material arrangement of the sub-ballast layer depending on the deformation resistance of the subgrade surface [4].

As the values of deformation resistance $E_0 < 6$ MPa were diagnosed at the level of the subgrade surface during the modernisation of the railway line section in question, the design documentation had to be changed. In the line sections with very low deformation resistance of the subgrade surface (less than 6 MPa) the subgrade surface was stabilized with lime and the sub-ballast layer was carried out by the original procedure according to figure 2 (depending on the deformation resistance of the subgrade surface after stabilization).

2.2 Methodology of determination of deformation resistance and degree of soil compaction of the sub-ballast layer

Deformation resistance of the sub-ballast layers or in this case only the sub-ballast layer, was determined by a static load test carried out using the equipment shown in figure 3. The static load test is one of the oldest in situ tests, and it has been used informatively in our infrastructure since 1967 in railway substructure diagnostics.

Figure 3. – Realization of static load test at the level of the sub-ballast upper surface.

The Static plate load tests (PLT) at the level of the sub-ballast upper surface were performed according to the methodology given in STN 73 6190 [5]. The PLT principle is based on measuring the pressing of a rigid circular load plate of 300 mm radius into the diagnosed structural layer (in our case the sub-ballast layer) as a result of the action of a particular static load. Measurement of the deformation resistance of the sub-ballast layer was carried out in two load cycles with a maximum contact stress value of 0.20 MPa, while the loading / unloading of the rigid circular load plate is realized in steps of 1/4 maximum pressure. The decisive cycle for determining the static modulus of deformation is the second load cycle, from which the value of the rigid plate settlement is obtained.

Static modulus of deformation E_s or in this case, E_{sub} is calculated from the measured values based on the Boussinesque theory of resilient semi-space by adjusting the relation:

$$E_S = \frac{\pi}{2} \cdot \frac{p \cdot r}{y} \left(1 - \vartheta^2\right) \approx \frac{1.5 \cdot p \cdot r}{y} = 0.225 \cdot \frac{p}{y} \quad (1)$$

where: E_s – static modulus of deformation, (MPa),
p – contact stress under the load plate, (MPa)
r – load plate radius, (m),
y – total average settlement of the plate load, (m),
1.5 – constant for $\nu = 0.21$.

If the first load cycle is also used from the load diagram and the static deformation modulus is calculated from it, the measured values of deformation resistance can be used to demonstrate the degree of soil compaction of any structural layer of the sub-ballast layers or subsoil. The static modulus of deformation from the second load cycle E_{def2} is larger than the static modulus of deformation from the first load cycle E_{def1}, as the material of the test layer is compacted during the first load cycle. If the degree of soil compaction is very large (this corresponds to a large load plate settlement), then compaction was insufficient. According to STN 73 6133 [6], compaction is considered to be sufficient if for coarse-grained materials (in our case crushed aggregate fr. 0/63 mm) the degree of compaction is expressed as the ratio $E_{def2}/E_{def1} \leq 2.6$. As part of the diagnosed modernised section of the line, the compaction on the individual structural layers was carried out using a BOMAG vibratory roller, which also served as a counter-load in the performance of PLT.

The actual measurement of the static deformation modulus of the sub-ballast layer in the modernised (diagnosed) railway line section was in most cases performed every 200 m or the network of measuring points was condensed in cases of localization of areas with low deformation resistance.

3 Analysis of achieved results

Generally, the deformation resistance of any structural layer depends on the type and quality (physico-mechanical properties) of the layer material and the quality of its compaction (depends on the type of soil compaction equipment, compaction method and number of running gear). For this reason, we decided to identify the relationship between the degree of the soil compaction and the deformation resistance of the sub-ballast layer from the values measured by the diagnostics of the modernised section of the line Považská Teplá - Žilina. In the first phase of the analysis, the subject dependence was created for the crushed aggregate fr. 0/63 mm (sub-ballast layer) placed on the original earthwork - figure 4.

In total, 141 measured values were evaluated in the analysis. The deformation resistance values, expressed by the static modulus of deformation at the level of the sub-ballast upper surface (E_{sub}), varied between 60.8 MPa and 140.6 MPa and the degree of compaction values (E_{def2}/E_{def1}) between 1.13 and 2.6. figure 4 shows that it is not possible to create a dependence of the deformation resistance of the sub-ballast upper surface on the degree of soil compaction if the railway line is built on the original earthwork. It should be noted that in this case the value of deformation resistance $E_0 \geq$ 30 MPa was required at subgrade surface level, but in many cases this condition was not met.

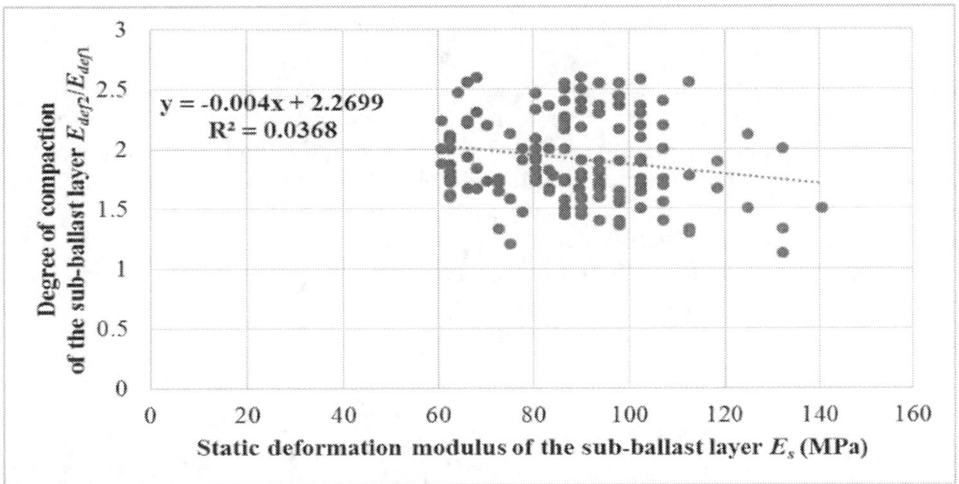

Figure 4. Dependence of degree of soil compaction and static deformation modulus of the sub-ballast layer built on the original earthwork of the railway line.

The deformation resistance of the subgrade surface was considerably uneven and the deformation resistance of the sub-ballast layers was ensured only through the sub-ballast layer reinforced by geogrids (see figure 2). In order to eliminate the influence of the reinforcing materials in the structure of the sub-ballast layer or insufficient deformation resistance of the subgrade surface to the examined dependence, analysis of measured values was performed for a part of the railway line built on a new earthwork (embankments). The dependence of the deformation resistance values determined at the level of the sub-ballast upper surface with the degree of soil compaction of the sub-ballast layer for the railway line built on new earthworks (embankments) is shown in figure 5.

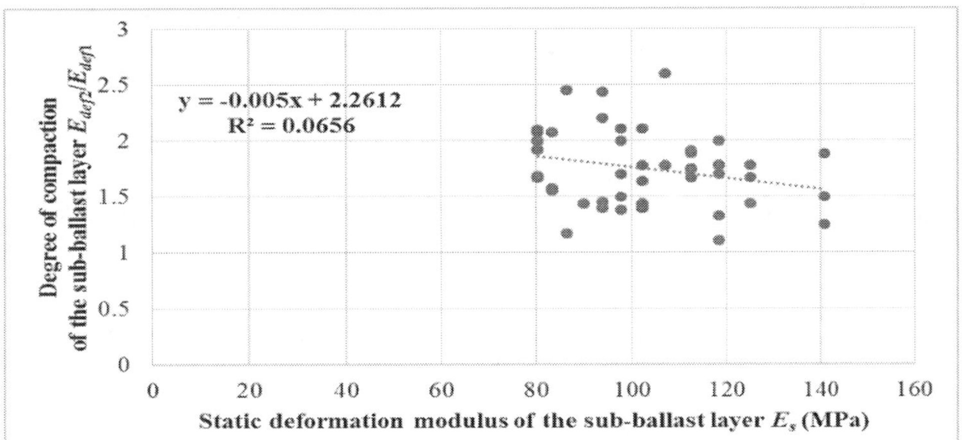

Figure 5. Dependence of degree of soil compaction and static deformation modulus of the sub-ballast layer built on the new earthworks (embankments) of the railway line.

In total, 50 measured values were evaluated in the analysis. The deformation resistance values, expressed by the static modulus of deformation at the level of the sub-ballast upper surface (E_{sub}), varied between 80.4 MPa and 140.6 MPa and the degree of compaction values (E_{def2}/E_{def1}) between 1.11 and 2.6. figure 5 shows that it is equally not possible to create a relevant dependence of the deformation resistance of the sub-ballast upper surface on the degree of soil compaction if the railway line is built on the new earthworks (embankments). It should be noted that in this case the value of deformation resistance $E_0 \geq 40$ MPa was required at subgrade surface level, which was in all cases met.

In the last phase of the analysis, the dependence of the monitored parameters (deformation resistance of the sub-ballast layer from the degree of its compaction) was created within individual coherent parts of a building (CPB 48 - CPB 52) or within station and interstation sections of the diagnosed railway line. A more detailed specification of individual CPB is given in the paper entitled *„Analysis of results of deformation characteristics measurements of embankments and transition zones of a modernised line Považská Teplá – Žilina“* [7] published in this Proceedings. In this analysis, the best determination coefficient for CPB 50 (interstation section between the railway stations Bytča and Dolný Hričov) was achieved - figure 6. The CPB 50 has a length of approximately 3.5 km, of which almost half of the CPB is built on a new earthwork.

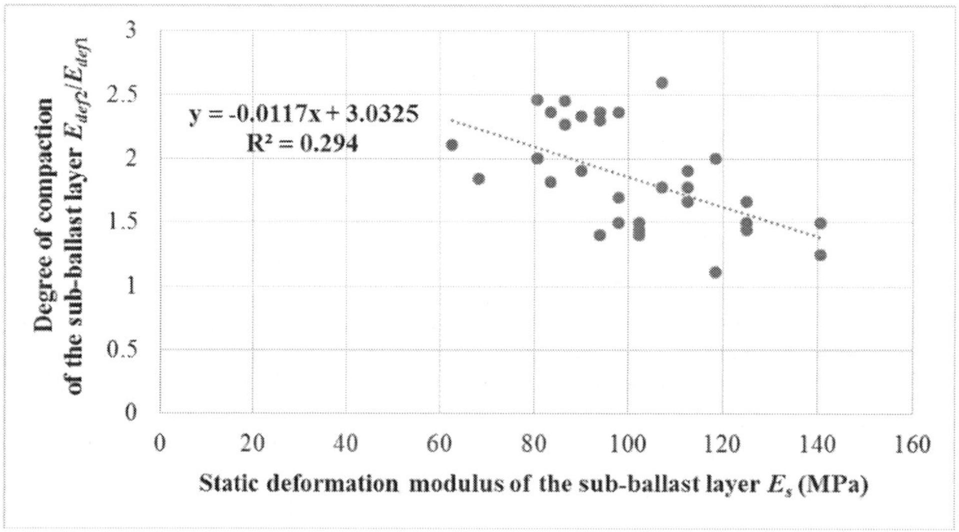

Figure 6. – Dependence of degree of soil compaction and static deformation modulus of the sub-ballast layer determined within CPB 50.

In total, 33 measured values were evaluated in the analysis. The deformation resistance values at the level of the sub-ballast upper surface (E_{sub}), varied between 62.5 MPa and 140.6 MPa and the degree of compaction values (E_{def2}/E_{def1}) between 1.11 and 2.6. figure 6 shows that the dependence of the deformation resistance of the sub-ballast upper surface on the degree of soil compaction in this case is better than in figure 4 or figure 5, but the determination coefficient is still too small. Within the other CPB, the determination coefficient $R^2 \leq 0.2$ was reached and therefore the dependence is not relevant.

4. Conclusions

The paper is linked to the issues published in [7] and [8]. The aim of these papers was to provide a comprehensive overview of the quality of construction work carried out in one of the diagnosed line sections (the modernised section of the line Považská Teplá – Žilina), which is part of the trans-European Corridor no. Va. In the paper the analysis was performed and the dependence between the deformation resistance of the sub-ballast layer and the degree of its compaction from the values obtained by static load tests was searched. In the first phase of the analysis, the dependence was monitored in railway line sections built on the original earthwork and subsequently in railway line sections built on the new earthworks (embankments). Within this phase, an unacceptable coefficient of determination $R^2 < 0.1$ was achieved (see figures 4 and 5).

Subsequently, the dependence was monitored within individual coherent parts of a building (CPB) or in interstation and station sections of the railway line in question. At this stage of the analysis, the highest coefficient of determination for the interstation CPB 50 was achieved, namely $R^2 = 0.294$ (figure 6), which is higher coefficient but still insufficient.

On the basis of the above, it can be stated that the obtained dependence between the deformation resistance of the sub-ballast layer and the degree of its compaction has a very low informative value, as the achieved coefficients of determination are very low. The deformation resistance of the sub-ballast layer depends not only on the compaction (type of soil compaction equipment, compaction method and number of running gear), but is fundamentally influenced by the deformation resistance of the subgrade surface, the quality (physico-mechanical properties) and thickness of the sub-ballast layer material and by applying geosynthetic to the sub-ballast layers (in our case to the sub-ballast layer). Given that all these influences changed from one place to another in the diagnosed section of Považská Teplá - Žilina, the values of deformation resistance had considerable variance as well as the achieved degree of compaction. It is therefore logical that, in this case of the diagnosed track section, an acceptable correlation could not be established.

Acknowledgments
The presented results are the partial results of solving the VEGA grant project 1/0084/20 "*Numerical and experimental analysis of transition areas of objects of structures of railway superstructures and objects of formation substructure*".

References
[1] Slovak Railway Technical Standard TNŽ 73 6312 The design of structural layers of subgrade structures". Directorate General of Railways of the Slovak Republic, Slovakia, 2005
[2] Ižvolt L 2008 *Railway substructure – stress, diagnostics, design, and implementation of body construction layers of railway subgrade* Scientific monograph, University of Žilina p
[3] Ižvolt L, Hodas S and Šestáková J 2015 *Railway Engineering 1 - design, buildings and constructions of railway lines and stations* Scientific monograph, University of Žilina p 561
[4] Technical reports of railway line Považská Teplá (out) – Žilina (out) modernization.
[5] Standard STN 73 6190 – Static plate load test of building constructions 1980 (Slovak Office of Standards, Metrology and Testing, Slovakia)
[6] Standard STN 73 6133 – Road Building. Roads embankments and subgrades 2017 (Slovak Office of Standards, Metrology and Testing, Slovakia)
[7] Dobeš P and Ižvolt L 2020 Analysis of results of deformation characteristics measurements of embankments and transitions zones of a modernised line Považská Teplá-Žilina *XII Int. Scientific Conf. "TRANSPORT PROBLEMS"* (in press).
[8] Ižvolt L and Dobeš P 2020 Analysis of measuring the deformation resistance of the subgrade surface of a modernised line Považská Teplá-Žilina *Civil and Environmental Engineering* **16** (1) pp 210-18

IOP Conf. Series: Materials Science and Engineering 1015 (2021) 012087 doi:10.1088/1757-899X/1015/1/012087

Research of the deformation of the shaping unit in an auger extruder caused by the extruding pressure in production of expanded clay

K S Galitskov, M A Nazarov and V A Borisov

Samara State Technical University,
244 Molodogvardeyskaya st., 443100, Samara, Russia

E-mail address: ksgal@yandex.ru

Abstract. The technological process of forming expanded clay granules in an auger extruder is considered. Based on the analysis of well-known studies, considering the influence of conditions for shaping a ceramic mass on the quality of the produced porous aggregate, the following factor was identified: clogging of the shaping grid leads to an increase in pressure on it, its deformation and the impossibility of further operation. The authors note that there are practically no theoretical studies in the field of determining the most rational modes of functioning of the shaping equipment. The simulation model of an auger extruder shaping unit was developed in the SolidWorks software environment, computational experiments on the influence of design parameters on shaping pressure were carried out, and recommendations for further technical application were formulated. The use of the results obtained will improve both the reliability of extruders with a perforated grid, as well as the quality of the final product with the specified quality parameters (bulk density and strength).

1. Introduction

The production of expanded clay gravel is a continuous technological process, which includes several main stages: preparation of clay raw materials; shaping of raw granules and their preliminary drying; burning and expansion; cooling of expanded clay. Each of the above technological limits is characterized by intermediate indicators, which, ultimately, form the qualitative characteristics of the finished product (bulk density, strength, uniformity coefficient, etc.).

Currently, issues related to the study of the technological process of burning expanded clay gravel in rotary kilns 40×2.5m and the development of recommendations aimed at energy-efficient control of this equipment in order to obtain the final product with the specified quality indicators (bulk density and strength), have been well studied [1-3].

The analysis of the results of works [1,2] shows that the authors, when considering the technological process of burning expanded clay gravel, applied the decomposition method, assuming that the rheological and physicochemical characteristics of the semi-finished product supplied for burning remain constant. However, as practice shows, it is at the stage of manufacturing raw granules in an auger extruder that the preliminary strength and the uniformity coefficient of the final product are formed. The constant change in the physicochemical properties of the raw materials, the lack of control over the processes occurring in the shaping devices, in particular, the mass flow rate, extruding pressure and the use of shaping grids with an increased hole diameter lead to a significant spread in the uniformity

IOP Conf. Series: Materials Science and Engineering 1015 (2021) 012087 doi:10.1088/1757-899X/1015/1/012087

coefficient of the resulting product. The authors [4] revealed the constructive drawback of existing granulating auger extruders – the presence of "dead" spots in front of the shaping nozzle, which leads to an increase in the shaping pressure even when using shaping grids with an optimal hole diameter. As a result, some pronounced uneven pressure distribution over the surface takes place, which causes deformation and the impossibility of further operation of the shaping grid. The article proposes a technique for studying the auger extruder shaping unit, formulates recommendations for its modernization, aimed at increasing its technological reliability and more rational use of the technical characteristics of the equipment.

2. Materials and methods

The results of experimental studies conducted by V. F. Veber [5] showed that the modes of shaping (speed of shaping, pressure and length of the channel of the auger extruder shaping head) significantly affect the quality characteristics (bulk density and strength) of expanded clay gravel. When the clay mass flows through the shaping channel of the auger extruder head, an inhomogeneous stress field occurs in its cross section, which leads to variations in the speed of clay movement in different flow channels [6]. The physical and mechanical properties of the clay mass depend on the velocity gradient $d\varepsilon/dt$; therefore, raw granules shaped at different speeds have different sensitivity to drying modes, which causes shrinkage cracks in the granules to appear. Burning granules with a layered structure, formed during shaping stage, leads to the occurrence of longitudinal cracks through which part of the gases escape and, as a consequence, the expansion of the granules worsens.

The analysis of the results of field experiments [6] showed that the pressure on the perforated grid depends on the perforation coefficient, the type of clay, the degree of its contamination and varies within 6-20 kg/cm^2. This is caused by the presence of solid inclusions 12-15 mm in size in the clay raw material, which leads to clogging of the shaping grids. Therefore, when about 25-50% of the grid holes are clogged, the extrusion pressure increases to the maximum attainable value of 20-22 kg/cm^2.

Thus, the goal of the study is to create a simulation model of the auger extruder shaping unit and to carry out computational experiments on the influence of design parameters (s is the thickness of the perforated grid; d is the diameter of the raw granules; C is the clogging of the shaping grid) on the shaping pressure in order to develop further technical recommendations.

The authors propose to use the SolidWorks software environment to solve the problem. When developing a model of the shaping unit, the results of field experiments [6-9], the main technological features and design characteristics of the extruder [10] were taken into account, namely: material - steel St 3; diameter – 500 mm; thickness – 16 mm; number of shaping channels (grid perforation coefficient) – 432 pcs; diameter of shaping channels – 10 mm; generated pressure – 6÷22 kg/cm^2; range of change in the degree of clogging of the shaping unit – 0÷80%; clay raw materials of the Smyshlyaevskoe deposit.

3. Results and discussions

Comparative analysis of extrusion pressure diagrams obtained during field experiments [6] with the results performed on the model (figure 1, figure 2) showed the adequacy of the developed computational model of the shaping unit and the possibility of further application for the purpose of experiments.

The revealed design drawback of the granulating auger extruder, described above, namely, the presence of "dead" spots in front of the shaping head, leads to an increase in the shaping pressure and the speed of shaping, and as a consequence, negatively affects the homogeneity of ceramic gravel [6]. The authors propose to solve the problem by constructive modification of the existing shaping unit by producing tapered lead-in parts of peripheral shaping channels, forming a complex geometric surface along the border of "dead" spots, which consists of cylindrical and conical parts (figure 3).

Therefore, when carrying out research on the deformation of the auger extruder shaping unit from the shaping pressure, we will require the development and determination of the most effective design of the tapered lead-in parts of the peripheral shaping channels. As a result, perforated grids with different design parameters of the lead-in parts of the peripheral shaping channels were modeled, and experiments

IOP Conf. Series: Materials Science and Engineering 1015 (2021) 012087 doi:10.1088/1757-899X/1015/1/012087

were carried out on the influence of the maximum allowable pressure of 22 kg/cm^2 on the deformation of the shaping unit (figure 4).

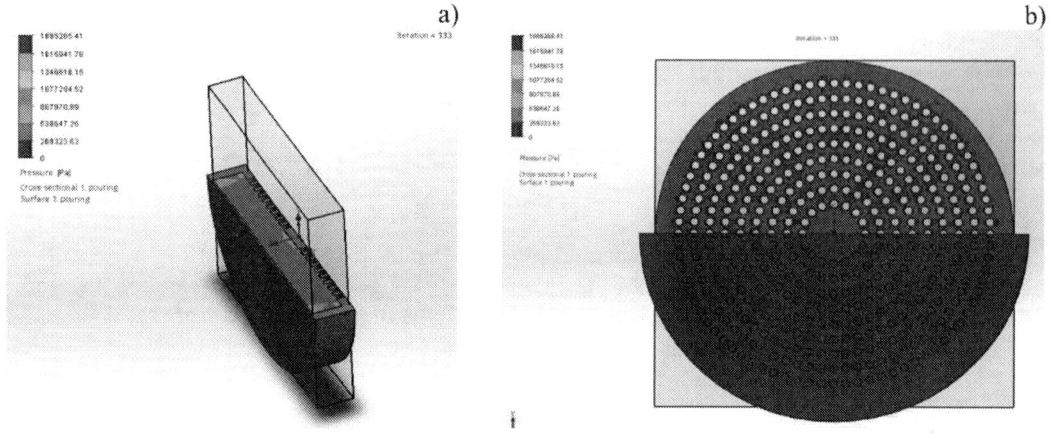

Figure 1. Pressure distribution in the auger extruder shaping unit at 0% clogging: a)inside; b) surface.

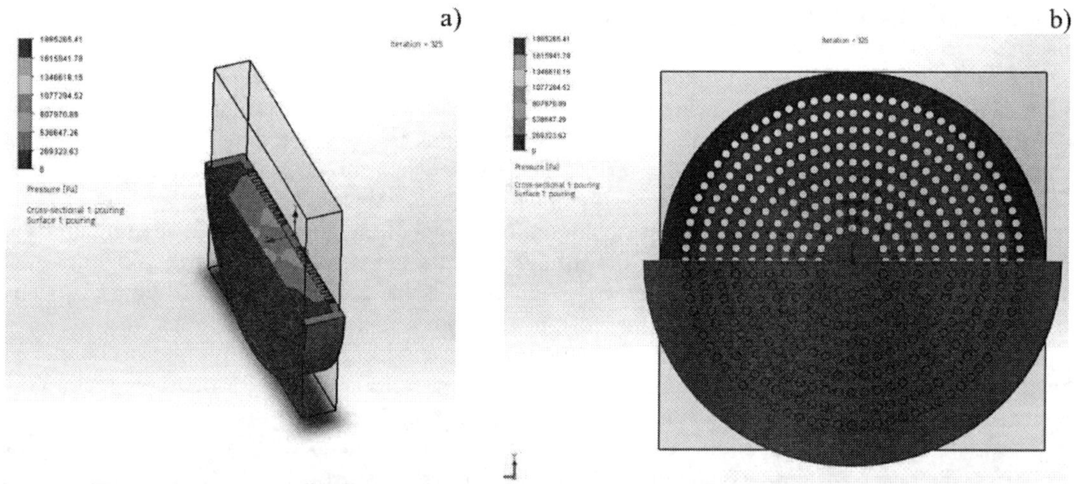

Figure 2. Pressure distribution in the auger extruder shaping unit at 80% clogging: a) inside; b) on the surface

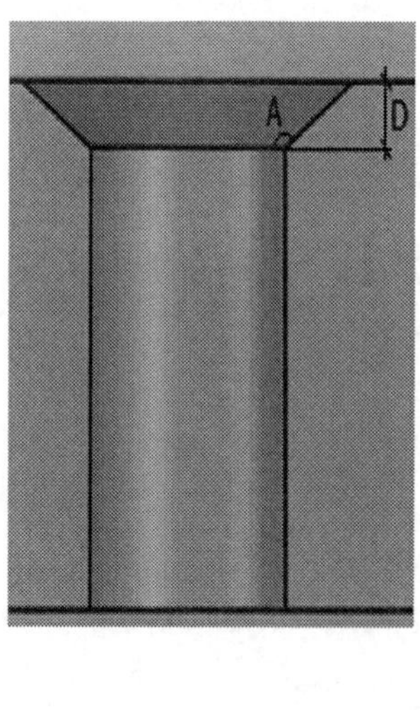

Figure 3. The tapered lead-in part of the peripheral shaping channel forming a complex geometric surface consisting of cylindrical and conical parts

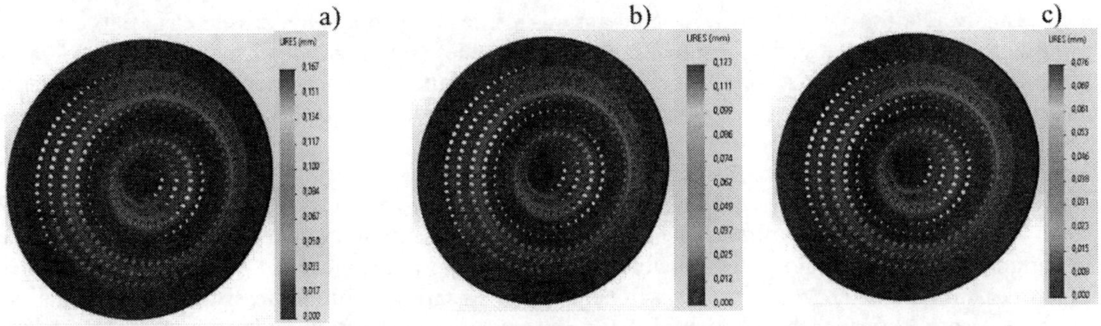

Figure 4. Diagram showing the change in the deformation of the shaping unit at a pressure of 22 kg/cm^2: a) without tapering lead-ins of peripheral shaping channels; b) design parameters for the tapered part – D=2 mm, A = 45°; c) design parameters for the tapered part – D=2 mm, A = 60°

The studies revealed that the design parameters of D = 2 mm and A = 60° (figure 4, c) are the most effective, but the resulting model does not reflect the law of pressure distribution inside the auger when the shaping nozzle is clogged and cannot be offered as a recommendation for further use. Therefore, additional studies of the developed model were carried out taking into account the clogging of the shaping nozzle with the effective design parameters of the lead-in parts (D = 2 mm, A = 60°) (figure 5, figure 6).

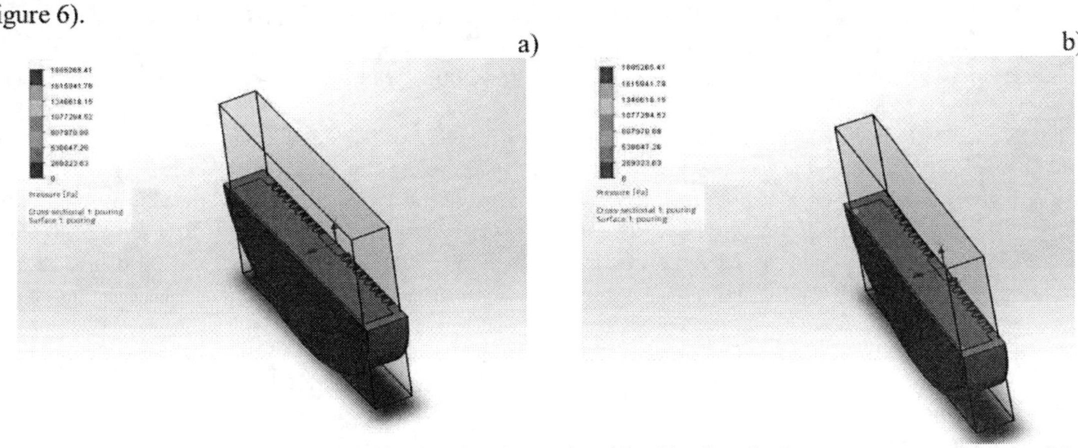

Figure 5. Pressure distribution inside the shaping unit with effective design parameters of the lead-in parts with clogged channels: a) 0%; b) 20%

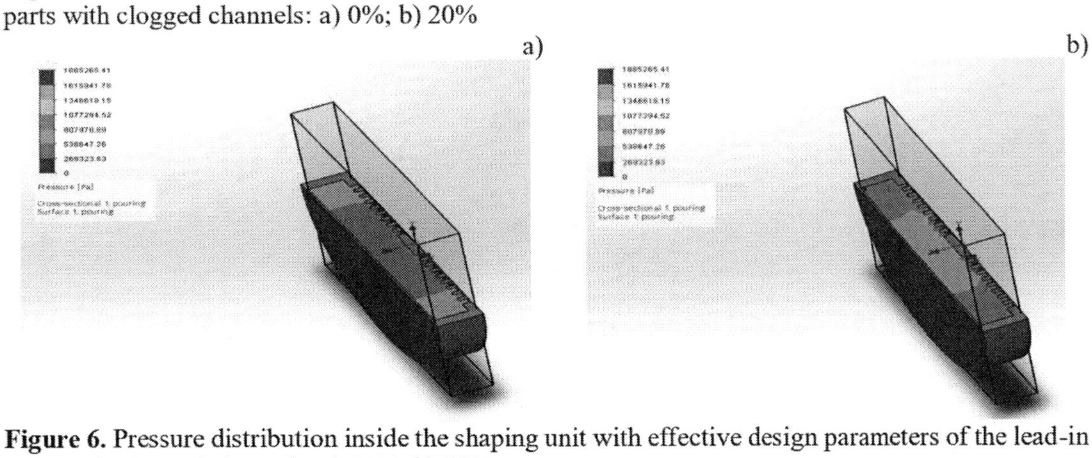

Figure 6. Pressure distribution inside the shaping unit with effective design parameters of the lead-in parts with clogged channels: a) 50%; b) 80%

IOP Conf. Series: Materials Science and Engineering 1015 (2021) 012087 doi:10.1088/1757-899X/1015/1/012087

The analysis of the pressure distribution diagrams inside the shaping unit, presented in figure 5, figure 6 shows that the use of effective design parameters of the lead-in parts (D = 2 mm, A = 60°) will allow not only to reduce, but also to stabilize the total pressure inside the auger extruder, which will lead to a more even compaction of the raw material over the entire surface of the shaping unit.

4. Conclusions

The developed model and the computational experiments carried out to study the deformation of the auger extruder shaping unit from the shaping pressure, considered in the article, are aimed at solving an important practical problem – the analytical design of newly projected and operating equipment. Moreover, the obtained research results will help to improve the reliability of existing extruders, as well as create preconditions for the creation of automatic pressure stabilization systems in order to obtain the final product with the specified quality parameters (bulk density and strength).

References

[1] Galitskov K S, Galitskov S Ya and Samohvalov O V 2018 Studies of the Dynamics of a Multidimensional System of Automatic Control of the Technological Process of Firing Expanded Clay *Bull. of the Samara State Tech. Univ. Series Technical Sci.* **2** (58) pp 7-15

[2] Galitskov S Ya, Galitskov K S and Samohvalov O V 2015 Modelling the Firing of ExpandedClay in a Kiln with a Controlled Rotation Speed as a Control Object *Scientific Review* **7** pp 227-37

[3] Galitskov K S, Samokhvalov O V and Fadeev A S 2017 Optimization of Burning Production Process of Ceramsite with Specified Density *Environment. Technology Resources Proc. of the 11-th Int. Scientific and Practical Conference* **11** (Rezekne: RA Izdevnieciba) pp 57-61

[4] Ravichev L V, Bespalov A V and Loginov V Ya 2003 Granulating Screw Press Russian patent № 2198787

[5] Veber V F 1972 Optimal Modes of Molding Raw Expanded Clay Granules on Screw Presses *Expanded clay and expanded clay concrete Proceedings of Scientific Conference* **6** pp 10-16

[6] Rybiev I A and Veber V F 1974 On Ways to Improve the Uniformity of Lightweight Aggregates *Expanded clay and expanded clay concrete* (*Proceedings of Scientific Conference* **6** pp 37-44

[7] Galitskov S Ya, Nazarov M A and Galitskov K S 2017 Parametric Identification of the Process of Preparing Ceramic Mixture as an Object of Control *Earth and Environmental Science IOP Conference Series*

[8] Handle F 2007 *Extrusion in Ceramics* (Berlin: Springer) p 468

[9] Tolkachev V Ya 2009 *Technology of High-Quality Extrusion of Clay Products* (Krasnoyarsk: Computer technologies) p 220

[10] Passport of the screw press SMK-435 1985 PO "Stromavtolinia"

IOP Conf. Series: Materials Science and Engineering 1015 (2021) 012088 doi:10.1088/1757-899X/1015/1/012088

On the technology of aluminosilicate and high-alumina phosphate hardening products

A Vlasov and E Vlasova

Samara State Technical University, Molodogvardeyskaya ul., 194, Samara 443001, Russian Federation

E-mail address: alvlass@yandex.ru

Abstract. The influence of alumocalcium sludge on the physical, mechanical and thermal properties of alumosilicate and high-alumina products of phosphate hardening intended for operation at temperatures up to 1500 °C. It was found that the use of alumocalcium sludge in the compositions of non-burning phosphate compositions helps to reduce the polymerization temperature and thus provides an earlier formation of rock-like phases that are not subject to hydration. Based on the obtained mathematical model, optimal non-burning compositions are developed that are not inferior to similar products in terms of physical, mechanical and thermal parameters, which provides an expansion of the range of applications of the raw material base.

1. Introduction

Significant progress has been made in the design and construction of heat engineering structures. However, even today, the design, construction and operation of structures operating in high temperatures have major disadvantages, which lead to a relatively rapid failure of them and to an increase in the cost of construction.

Damage to some structures of thermal units occurs due to physical and chemical processes occurring in the material, in other structures due to high temperature stresses or deformations, and in many cases from the combined influence of both factors.

The reason for the short service life of structures of thermal units is in most cases their execution from piece refractories on binders mainly of hydration hardening [1-4]. More promising is the creation of binders, which are based on other physical and chemical processes. Such binders, in particular, include compositions that harden as a result of reactions of certain materials with phosphoric acid or its acidic derivatives, i.e. binders of phosphate hardening [5-8]. However, based on the work of recent years, it is clear that the use of phosphate hardening materials leaves a large range of technical problems, which can be solved by nanotechnogenic waste from various industrial enterprises, including alumocalcium sludge - waste from non-ferrous metallurgy [9-12].

To improve the physical, mechanical and thermal properties of heat-resistant and refractory compositions, due to the General lack of natural raw materials, man-made waste from various industries is widely used, which also helps to combat environmental pollution, namely, the quantitative reduction of landfills and landfills for waste storage. It is proved that man-made waste can become a huge wealth from a heavy burden, and with a scientific approach to its processing, it can become a valuable, and sometimes even scarce raw material.

2. Materials and methods

Samara metallurgical plant (JSC «Alcoa SMZ») is one of the largest suppliers of man-made sludge raw materials. As a result of numerous studies, it was found that the slurries formed at the metallurgical plant can be used as refractory fillers, including alumocalcium slurries in the production of pressed phosphate-fireclay refractories.

The plant operates modern treatment facilities. At various installations of these structures, aluminum-alkaline sludge (alkaline etching sludge of aluminum products and parts), which has a pH > 12, is neutralized with another liquid sludge-containing waste, namely carbonate sludge, the main component of which is $CaCO_3$. By mixing almadellago slurry of the carbonate (i.e., by reacting the sludge with high pH) is formed aljumokalievye liquid slurry with pH = 7, which after passing the filter press (figure 1) in the form of «curd» sludge is disposed of on special polygons.

Figure1. Filter press installation.

Also, the dimensions of the alumocalcium sludge particles were tested using the Membrane-2 diffractometer at the B. p. Konstantinov St. Petersburg Institute of nuclear physics in 2010. The obtained results showed that the studied sludge can be attributed to nanoobjects of the 2nd level, since the nanoscale of the particles is less than 100 nm.

The chemical composition of the sludge is shown in figure 2.

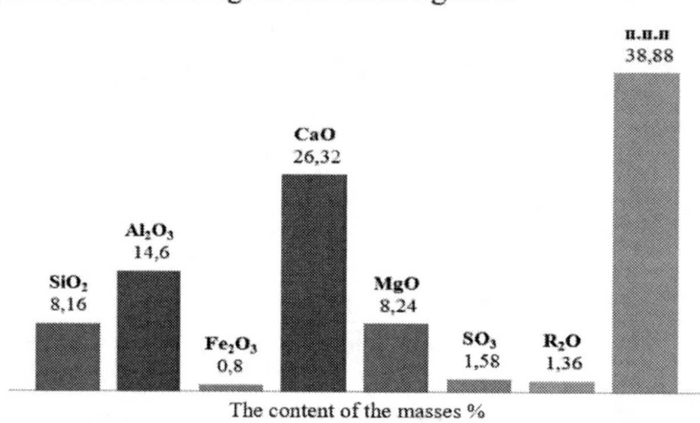

Figure 2. Chemical composition of alumocalcium sludge.

X-ray structural analysis of alumocalcium sludge showed that the main constituent minerals are calcite, quartz and hydroargillite (figure 3).

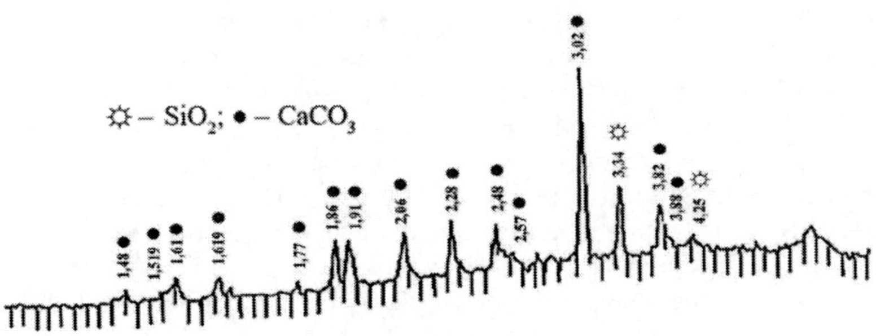

Figure 3. X-ray diffraction pattern of calcium alumina sludge.

In the tests were also used: fillers-fireclay sand (scrap), fr. 0-5 mm; mullite sand (scrap), fr. 0-5 mm; binder (sealing liquid) - orthophosphoric acid H_3PO_4 (concentration, %: 40, 50, 60, 70, 80).

In the study, using standard methods, General patterns were considered in changing the basic properties of phosphate-chamotte compositions obtained on the basis of orthophosphoric acid with the addition of nanotechnogenic raw materials – alumocalcium sludge. In this regard, at the first stage of the study, alumocalcium sludge was introduced into the charge compositions, while the charge composition consisted of 65 % fireclay sand fr. 0-5 mm and 35 % alumocalcium sludge. The resulting charge was closed in excess of 100 % orthophosphoric acid H_3PO_4 of different concentrations, which ranged from 40 to 80 %. Obtaining the required concentration was achieved using the relationship between the concentration of orthophosphoric acid and its density.

Samples for testing were made by semi-dry pressing from masses containing 12 % solution of orthophosphoric acid solution, a certain amount of moistened powdery mass consisting of fireclay sand and alumocalcium sludge was poured into the mold. The resulting mixture consists of solid particles of various shapes and sizes. Then the mass in the mold was compressed from opposite sides with two punches. Stepwise pressing was used to equalize the pressure and remove air from the pressed mass. After the end of pressing, the resulting product was pushed out of the mold and placed in a muffle furnace for heat treatment. The samples were heat-treated at three control temperature points: 400 °C, 800 °C and 1200 °C. After heat treatment, the samples were cooled and tested for compressive strength using a hydraulic press.

3. Results and discussions

As a result of the conducted experiments with compositions consisting of fireclay sand and alumocalcium sludge and closed with orthophosphoric acid of different concentrations, positive results were obtained on changes in physical and mechanical properties. Figure 4 shows the dependence of the compressive strength of fireclay-phosphate refractory compositions on the acid concentration and pressing pressure.

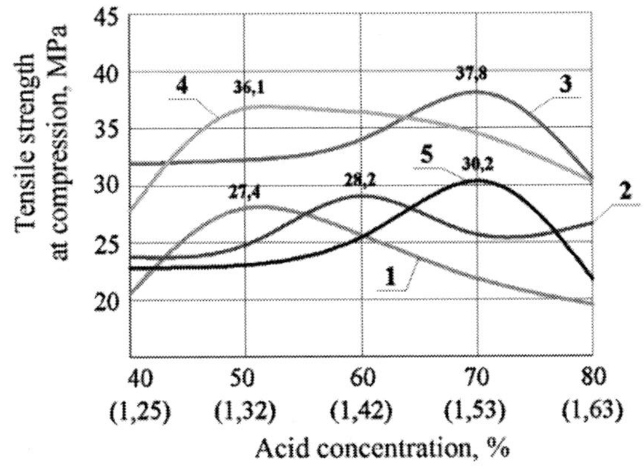

Figure 4. Dependence of the compressive strength of chamotte-phosphate refractory compositions depending on acid concentration and pressing pressure:
1 – sample pressing pressure 50 kgf/cm^2;
2 – sample pressing pressure 100 kgf/cm^2;
3 – sample pressing pressure 150 kgf /cm^2;
4 – sample pressing pressure 200 kgf/cm^2;
5 – the pressing pressure of a sample of 250 kgf/cm^2.

IOP Conf. Series: Materials Science and Engineering 1015 (2021) 012088 doi:10.1088/1757-899X/1015/1/012088

The graph shows the spread of experimental data. Compositions 3 and 4 are the best. The compressive strength of samples from fireclay-phosphate refractory compositions of these compositions after heat treatment at 400 °C is 37,8 MPa and 36,1 MPa, respectively. The constructed separate curves of the dependence of the strength of fireclay-phosphate refractory compositions on the concentration of acid and pressing pressure do not give an answer about the optimal parameters.

In order to reduce the number of experiments and determine the optimal values of the selected parameters, mathematical processing of experimental values was performed, for this purpose, factors with a variable component were included in the mathematical planning plan of the experiment – the pressing pressure taken in the range from 50 kgf/cm^2 to 205 kgf/cm^2 and the concentration of orthophosphoric acid in the range from 40 % to 80 % (table 1).

Table 1. Natural factors, their values and variation intervals.

Factor levels and their range of variation	Names and values of factors	
	the pressing pressure of	acid concentration
Top level (+2)	5	80
Top level (+1)	4	70
Basic (zero) level	3	60
Lower level (-1)	2	50
Lower level (-2)	1	40
The range of variation in	1	10
Unit	ton	%
Identification of factors	x_1	x_2

Also, for the two-factor experiment, a planning matrix was calculated, where the arithmetic mean values of the compressive strength after heating up to 400 °C were calculated.

In the regression analysis program «STATISTICA 10.0», a second-order regression equation was obtained from two factors. The regression equation of the «y» function, after calculating the coefficients found using the student's t-test, takes the following form:

$$R^{400}_{com.} = -25{,}9851 + 12{,}548x_1 + 1{,}3217x_2 - 2{,}07\,x_1^2 + 0{,}0157x_1x_2 - 0{,}0113\,x_2^2$$

The chosen mathematical model adequately describes the system under study with a confidence probability of 0,95. Using the obtained regression equation, the response surface isolines were constructed using the computational and graphical program «STATISTIKA» (figure 5).

The 3D isolines made it possible to accurately determine the optimal range of changes in parameters (pressing pressure and orthophosphoric acid concentration), providing maximum properties. At the same time, the optimal value of the pressing pressure was 160 kgf/cm^2 (3,2 t), and the acid concentration was 62 % (1,442 g/cm^3).

The study of a series of compositions based on phosphoric acid showed that already during heat treatment at 400 °C, the hardening of the structure of products occurs, and the mathematical model revealed optimal variable parameters that positively affect the physical and mechanical properties. To confirm the influence of alumocalcium sludge on the main properties of chamotte-phosphate compositions, the composition without additives (№1) and compositions with different amounts of alumocalcium sludge (№2 and №3) from 35 % to 65 % were studied (table 2).

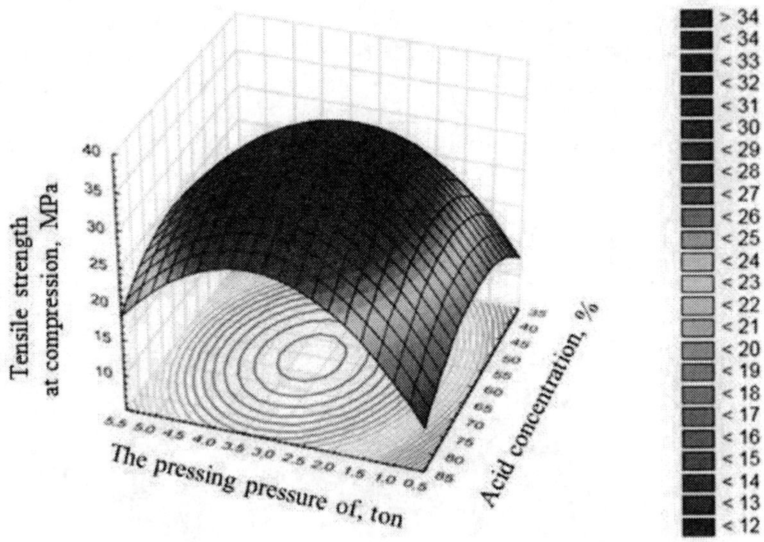

Figure 5. Response surface Isolines with optimal parameter range (3D surface for variable parameters).

Table 2. Influence of the amount of alumocalcium sludge and heat treatment on the physical and mechanical properties of chamotte-phosphate compositions.

№	Charge composition	Average density ρ, g/cm³ (numerator), compressive strength $R_{com.}$, MPa (denominator) of samples after heat treatment to temperature, (°C)		
		400	800	1200
1	- Fireclay sand (150 g) - H_3PO_4, ρ = 1,42 g/cm³, the concentration of the acid is 60 % - Alumocalcium sludge (50 g)	1.67 / 17.8	1.71 / 20.5	1.7 / 21.2
2	- Fireclay sand (100 g) - H_3PO_4, ρ = 1,42 g/cm³, the concentration of the acid is 60 % - Alumocalcium sludge (100 g)	2.11 / 36.8	2.12 / 39.6	1.97 / 41.4
3	- Fireclay sand (50 g) - H_3PO_4, ρ = 1,42 g/cm³, the concentration of the acid is 60 %	1.67 / 27.3	1.68 / 29.1	1.66 / 31.2

As can be seen from table 2, after heating to 800 °C and 1200 °C, all experimental samples show an increase in the compressive strength. At the same time, it is worth noting that the sample №2 initially selected for research has the best physical and mechanical properties, and in the sample №3, where the amount of sludge was brought to 65 %, there is an increased fire shrinkage and the content of the stopper up to 17 %. In the course of research, it was found that at an acid concentration of less than 60 %, not all components of the alumocalcium sludge react, which is clearly visible in freshly formed samples in the form of white grains that did not react with the acid. At a concentration of more than 60 %, a strong clumping is observed during mixing of the charge, which leads to an inhomogeneous distribution of sludge particles in the sample body and, accordingly, to a decrease in the compressive strength.

Similar studies were conducted with another aggregate, where fireclay sand was replaced with mullite. The initial components for the series of these compositions were: alumocalcium sludge; mullite sand; orthophosphoric acid with an optimal concentration of 60 % as a charge stopper.

The study of changes in compressive strength was also carried out in a wide temperature range from 400 °C to 1200 °C. To confirm the similarity of the results of physical and mechanical properties, a cycle of experiments was carried out during heat treatment at 400 °C. The data for the compressive strength is shown in figure 6.

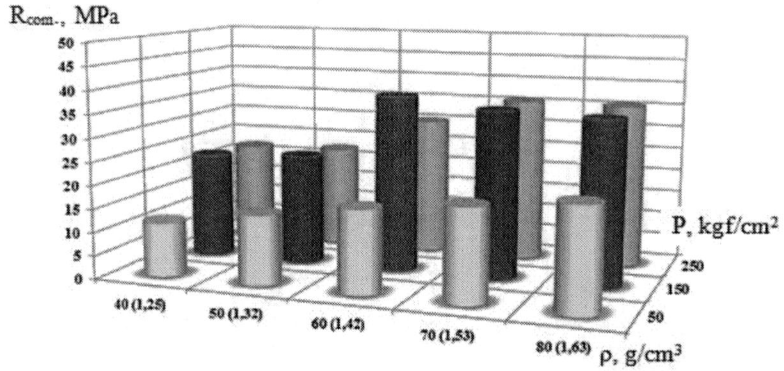

	40 (1,25)	50 (1,32)	60 (1,42)	70 (1,53)	80 (1,63)
50	11,7	15	17,8	20	22
150	23	24	38,2	36,1	35,4
250	22	22,4	30,1	35,7	35,5

Figure 6. Dependence of the compressive strength of mullite-phosphate refractory compositions from acid concentration and pressing pressure.

Much attention was paid to the mode of heat treatment of phosphate compositions (figure 7).

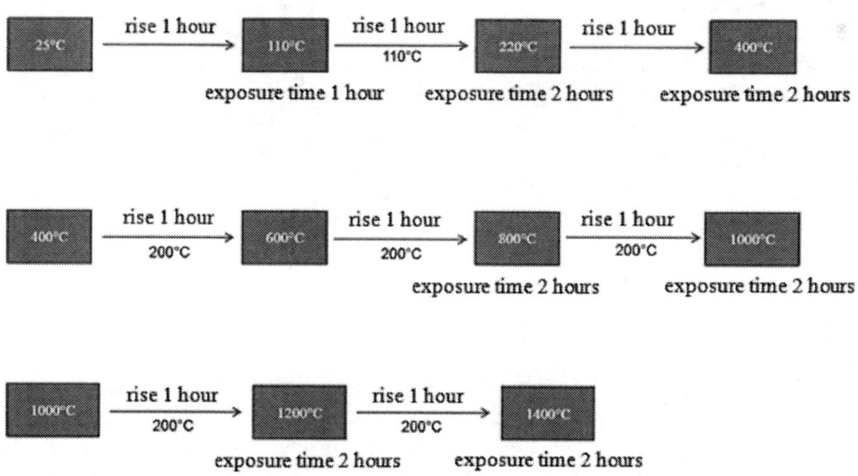

Figure 7. The Mode of heat treatment of samples.

Thus, it can be assumed that the regulation of the strength properties of pressed products is possible by optimizing the pressing pressure and regulating the fractional composition of mixtures, as well as by introducing the optimal amount of stopper.

IOP Conf. Series: Materials Science and Engineering 1015 (2021) 012088 doi:10.1088/1757-899X/1015/1/012088

4. Conclusions

As a result of research, it was found that the use of fireclay and mullite fillers in phosphate compositions based on acid of 60 % concentration allowed to obtain products with increased physical, mechanical and thermal parameters.

In the future study, instead of orthophosphoric acid, it is planned to use acidic water-soluble phosphate bundles (alumophosphate, alumocalcium phosphate, and alumochromophosphate bundles) obtained by partial neutralization of H_3PO_4 with aluminum, calcium, and chromium oxides.

References

[1] Kuznetsova T V 1988 *Aluminous cement* (Moscow: Stroiizdat) p 272
[2] Butt Yu M 1974 *Portland Cement* (Moscow: Stroiizdat) p 328
[3] Bazhenov Yu M 2006 *Modified high-quality concretes* (Moscow: Publishing House Assoc. construction universities) p 368
[4] Karklit A K 1974 *The refractories of the high alumina raw materials* (Moscow: Metallurgy) p 152
[5] Abyzov V A 2016 Energy-efficient technology of phosphate binders, adhesives and cellular concrete *Science of SUSU (Electronic resource) materials of the 68th scientific conferenc Sections of technical Sciences* pp 17-20.
[6] Sudakas L G 2008 *Phosphate Knitting Systems* (St. Petersburg: RIA Quintet) p 260
[7] Latypova L I 2012 Heat resistant phosphatic materials on the basis of high-alumina waste products *Bulletin of SUSU Series Construction and architecture* **15** (38) pp 69-71
[8] Sychev M M 1986 *Inorganic adhesives* (L.: Chemistry) p 152
[9] Khlystov A I 2013 The use of mineral sludge waste in the processes of synthesizing liquid phosphate ligaments *Bulletin of SUSU Series Construction and architecture* **13** (2) pp 43–46
[10] Chumachenko N G 2011 The use of high-alumina slimes in technological processes for the production of unburned refractories and clinker ceramic materials *Refractories and technical ceramics* **7-8** pp 47-50
[11] Korenkova S F 2004 *Basics and concepts of utilization of chemical precipitation of industrial waste in the construction industry* (Samara: Samarsk. state arch. building. Univ) p 203
[12] Khlystov A I 2017 Heat-resistant concretes based on industrial waste in the Samara region *monography* (Samara: ASI SamGTU) p 171

IOP Conf. Series: Materials Science and Engineering 1015 (2021) 012089 doi:10.1088/1757-899X/1015/1/012089

Analysis of various types of loads from a multi-layer structure of green roof systems

E A Korol and N S Shushunova

Moscow State University of Civil Engineering, 129337, 26, Yaroslavskoye Shosse, Moscow, Russia

E-mail address: nshushun@gmail.com

Abstract. For the sustainable development of green building technologies and the use of roof coverings with greening systems on buildings, it is necessary to take into account the complex of different types of loads from the multi-layer roof structure, as well as their effect on the building covering during operation processes. The introduction of new competitive technologies and methods of production of construction and installation works will ensure an increase in the operational quality of housing facilities. The article discusses the traditional technologies of roofing for operating roofs with greening systems, called «continuous» greening systems and new technologies for roofing with modular green roof systems, developed by the authors. The analysis and calculation of various types of loads for the given constructive and technological solutions of the operated roofing with greening systems is carried out.

1. Introduction

Today, more and more roofing systems have started to use greening systems. This is due to the fact that nowadays many property owners are concerned about the ecological image of their building and want to attract more tenants, clients and residents, so this problem is more urgent than ever, especially since it benefits the health of residents and looks aesthetically pleasing. Results show that the maximum roof surface UHI is reduced during the day by 1°C–3.8 °C by increasing green roof fractions from 30% to 90% [3].

However, the requirements for a green roof differ not only in the beauty of appearance, in the styles of green urbanism, but also have their own specific features when installing such coatings and taking into account various parameters in complex of technological processes. When installing greening systems on roofs, additional loads arise that must be taken into account based on local regulations and in relation to international standards and Building Codes.

Before designing a building, it is important to consider the building's construction features, climate conditions, including wind and seismic activity, the load bearing capacity of a building, the weight of saturated plants when planted but their weight at maturity, especially where shrubs and trees are proposed, as these are likely to be significantly heavier over time. The load assessment of green roof plants is provided in the table 1(in the relation to the international standards and Building Codes for green roof systems).

IOP Conf. Series: Materials Science and Engineering 1015 (2021) 012089 doi:10.1088/1757-899X/1015/1/012089

Table 1. Green roof vegetation weight loadings.

Green roof vegetation type	Weight loading (kg/m^2)
Low herbaceous (succulents and grasses)	10.2
Perennials and low shrubs up to 1.5 m	10.2-20.4
Turf	5.1
Shrubs up to 3 m	30.6
Small trees up to 6 m	40.8
Medium trees up to 10 m	61.2
Large trees up to 15 m	150

The study of the green roof system is based on the organizational and technological solutions of modern scientists in construction, including in the field of green technologies [1-6].

2. Methodology

2.1. Determination of types of loads for various roof greening systems
When designing and arranging the operated roofing of buildings and structures, the rules of the current standards for the design of buildings and structures must be observed. When installing a roof with landscaping systems, additional loads arise on the base of the coating, which must be taken into account at the design stage. There are only few standards for predicting the general loads on the roof and weight of a green roof system, such as SP 20.13330.2016 «Loads and impacts» (Russian standard), DIN EN 1991 «Einwirkungen auf Tragwerke Teil 1-1: Wichten, Eigengewicht und Nutzlasten im Hochbau, Teil 1-3: Schneelasten, Teil 1-4: Windlasten» («Effects on structures Part 1-1: Weights, dead weight and payloads in building construction, Part 1-3: Snow loads, Part 1-4: Wind loads», German standard) and ASTM E 2397/E2397M: 2019 «Standard Practice for Determination of Dead Loads and Live Loads Associated with Vegetative (Green) Roof Systems» (American standard) [7,8,17].

Depending on the duration of the loads, there are permanent (dead load) and temporary (long-term, transient or short-term and special loads) or live load on the roofing, while the calculated value of the load is determined as the product of its standard value by the load safety factor γf corresponding to the considered limiting state. Climatic loads: snow, wind, temperature and ice refer to short-term loads (live loads), which are also important to consider when designing roofing systems. The special loads include, for example, loads caused by fire; climatic (snow, wind, temperature and ice) loads, the action of which can lead to an emergency design situation, which is not covered in this article.

To study the loads on the roof, it is important to use in the investigation a systematic approach and a multi-variable structure [9-16]. The following options for the construction of green roof systems were chosen:

- roof with continuous greening systems (option 1, figure 1);
- roof with modular greening systems (modules with a diameter of 0.5 m) (option 2, figure 2).

The operated continuous green roof systems have shallow substrates on two types of roof, the first roof is on shipping containers, and is designed for a dead load of 250kg /m2 and live load of 100kg/m2 (figure 1). The extensive green roofs are the innovative and technical solutions for environmental sustainability for the urban areas in the city [18-20].

Figure 1. The operated continuous green roof systems in Melbourne (option 1).

Figure 2. The operated modular green roof systems, developed by the authors (option 2)

2.2. Calculation of loads for various roof greening systems

The calculated value of the load should be determined as the multiplication of its standard value and the safety factor for the load corresponding to the considered limiting state as follows:

$$L_c = L_n \cdot \gamma_f^* \quad (1)$$

where L_c - calculated value of the load, (kPa);
L_n - normative value of the load, (kPa);
γ_f^*- load safety factor.

By considering the averaged values of the loads from each structural element and systematizing the data obtained, the loads on roof with continuous greening systems and the loads on the roof with modular greening systems were obtained (tables 2 and 3).

Table 2. The load assessment of green roof with continuous greening systems.

Load type	ρ (t/m^3)	Normative value of the load, L_n, (kPa)	γ_f*	Calculated value of the load, L_c, (kPa)
Soil, $\delta = 0.1$ m	1.2	1.6	1.3	2.08
Drainage profiled membrane, $\delta = 0.008$ m	0.55	-	1.2	-
Insulation-mineral wool slabs, $\delta = 0.1$ m	0.035	0.15	1.3	0.195
Geotextile - 2 layers	150 g/m^2	-	1.2	-
Bitumen-polymer roll waterproofing - 2 layers	1.4	-	1.3	-
Primer	0.88	-	1.3	-
Reinforced cement-sand screed, $\delta = 30$ mm ($\gamma = 18$ kN/m^3)		0.54	1.3	0.7
Expanded clay gravel	0.4	0.23	1.3	0.3
Base covering, $\delta = 220$ mm	2.5	3.4	1.1	3.74
Summary:				7.015

*Load safety factor: a factor that takes into account, under normal operating conditions of structures, the possible deviation of loads in an unfavorable (higher or lower) direction from the standard values

IOP Conf. Series: Materials Science and Engineering 1015 (2021) 012089 doi:10.1088/1757-899X/1015/1/012089

Table 3. The load assessment of green roof with modular greening systems

Load type	Normative value of the load, L_n, (kPa)	$γ_f$*	Calculated value of the load, L_c, (kPa)
Solar panels	0.16		
Modules with wet soil, $δ = 0.4$ m	0.6	1.2	0.72
Drainage grid	1.5	1.2	1.8
Adjusting supports Buzon	-	1.2	-
Waterproofing	-	1.3	-
Base covering, $δ = 220$ mm	3.4	1.1	3.74
Summary:			6.26

We calculated short-term loads as snow loads on the roof in the conditions of the city of Moscow (IV snow region):

$$L_t(s) = C_w \cdot C_t \cdot m \cdot S_g = 2 \text{ (kPa)} \quad (2)$$

where

C_w - the coefficient taking into account the drift of snow from the coatings of buildings under the influence of wind or other factors, taken as 1, in accordance with [8];

C_t - thermal coefficient, taken as 1, in accordance with [8];

m - form factor, taking into account the transition from the weight of the snow cover of the earth to the snow load on the cover, taken as 1, in accordance with [8] (the slope of the roof covering is less than 30°);

S_g - the standard value of the weight of the snow cover per 1 m of the horizontal surface of the earth, taken as 2 in accordance with [8] (IV zone).

3. Results

Taking into account the basic combination of loads consisting of permanent (dead load) and long-term and short-term loads (live load), we get the following formula for calculation:

$$L = L_d + L_t(s) \quad (3)$$

where

L - the total value of the roof load;

L_d - calculated value of the load, L_c (table 2, 3);

L_t (s) - calculated short-term snow load.

The total value of the roof loads is calculated for the options: the total load on roof with continuous greening systems (L_1) and the total load on the roof with modular greening systems (L_2):

$L_1 = 7.015 + 2 = 9.015$ kPa;

$L_2 = 6.260 + 2 = 8.260$ kPa.

As we can see from this calculation, the total load on the roof with modular greening systems is almost 10% less than the total load on roof with continuous greening systems. Modular greening systems are lightweight structures, therefore they are best used on green roofs.

In the CAD program "LIRA" a check was made of the cross-section of the base covering, that it can withstand the given load (figure 3).

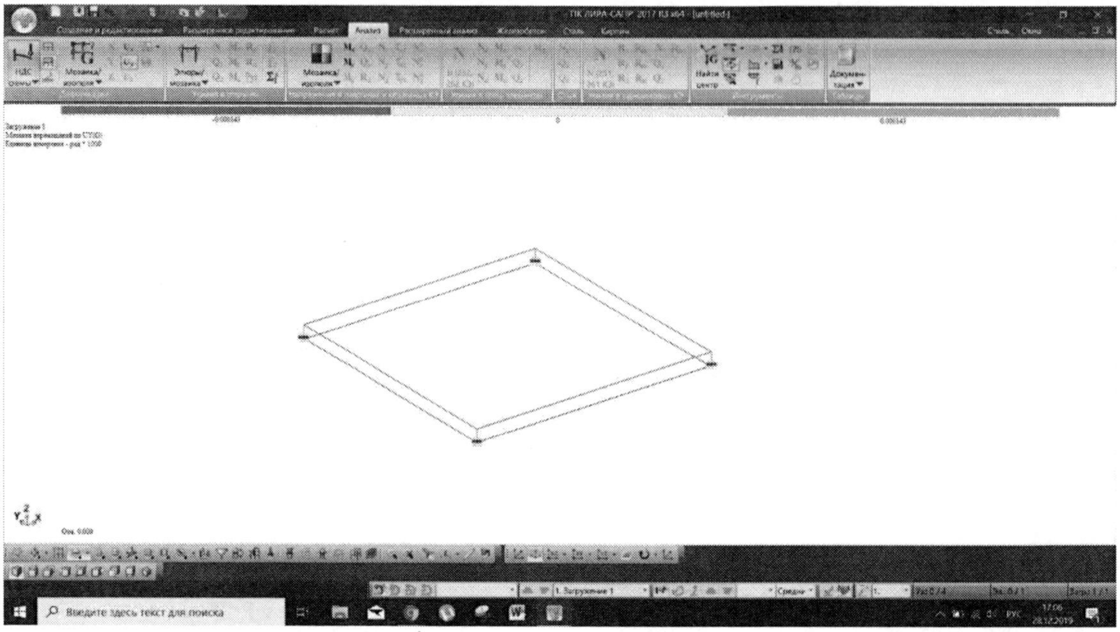

Figure 3. The cross-section view of the calculated base covering in CAD program "LIRA".

4. Conclusions

The basic loads associated with green roof systems were considered, based on several regulatory documents. The components that are encountered in green roof system as a multi-layer structure include: base covering, waterproofing drainage grid, insulation-mineral wool slabs and plants materials. The article analyzed the regulatory documents in this area of loads and impacts on the roof, the distinctive feature was found between them. However, these regulatory documents have some features, for example, American standard «Standard practice for determination of dead loads and live loads associated with vegetative (green) roof systems» does not address live loads associated with small architectural elements as a pavement, construction activities and loads associated with snow or wind. The Russian standard for loads and impacts on the roof of the greening system has not yet been developed at the moment, but there are prerequisites for its creation. Based on the database of important tools, the value of the roof loads for modular and continuous greening systems was considered and calculated, while it turned out that the modular greening system is easier on 10% and more convenient to use and for further operation they are better suited for the green roofs and green infrastructures in urban areas.

References

[1] Korol E and Shushunova N 2016 Research and Development for the International Standardization of Green Roof Systems *Procedia Engineering* **153** pp 287-91 https://doi.org/10.1016/j.proeng.2016.08.117

[2] Korol E, Shushunova N, Feoktistova O, Shushunova T and Rubtsov O 2018 Technical and economical factors in green roof using to reduce the aircraft noise *MATEC Web of Conferences* **170** https://doi.org/10.1051/matecconf/201817001081

[3] Imran H.M., J. Kala, A.W.M. Ng, S. Muthukumaran 2018 Effectiveness of green and cool roofs in mitigating urban heat island effects during a heatwave event in the city of Melbourne in southeast Australia *J. of Cleaner Production* **197** pp. 393-405 (https://doi.org/10.1016/j.jclepro.2018.06.179)

[4] Wuni I Y, Shen G Q P and Osei-Kyei R 2019 Scientometric review of global research trends on green buildings in construction journals from 1992 to 2018 *Energy and Buildings* **190** pp 69-85 (https://doi.org/10.1016/j.enbuild.2019.02.010)

[5] Ding Z, Fan Z, Tam V W Y, Bian Y and Moon S 2018 Green building evaluation system implementation *Building and Environment* **133** p 32-40 (https://doi.org/10.1016 /j.buildenv.2018.02.012)

[6] Darko A, Chan A P C, Huo X and Owusu-Manu D 2019 A scientometric analysis and visualization of global green building research *Building and Environment* **149** p 501-11 (https://doi.org/10.1016/j.buildenv.2018.12.059)

[7] Standard DIN EN 1991 Einwirkungen auf Tragwerke Teil 1-1: Wichten, Eigengewicht und Nutzlasten im Hochbau, Teil 1-3: Schneelasten, Teil 1-4: Windlasten

[8] Standard SP 20.13330.2016 Loads and impacts. Updated edition of SNiP 2.01.07-85

[9] Korol O A 2015 Issledovaniya i naukoyemkiye razrabotki v oblasti energoeffektivnogo stroitel'nogo proizvodstva *Stroitel'nyye materialy* **6** pp 13-15

[10] Borkovskaya V G 2014 Complex models of active control systems at the modern developing enterprises *Advanced Materials Research, Advanced Materials Research* **945-949** (22) pp 3012-15 DOI:10.4028 /www.scientific.net /AMR.945-949.3012

[11] Roe R, Bardenwerper W and Borkovskaya V 2018 Using a case study interactively to teach sustainability risk management *MATEC Web of Conferences* **251** p 06028 DOI: https://doi.org /10.1051/matecconf/201825106028

[12] Teng J, Mu X, Wang W, Xu C and Liu W 2019 Strategies for sustainable development of green buildings *Sustainable Cities and Society* **44** pp 215-26 (https://doi.org/10.1016 /j.scs.2018.09.038)

[13] Abu-Hijleh B and Jaheen N 2019 Energy and economic impact of the new Dubai municipality green building regulations and potential upgrades of the regulations *Energy Strategy Reviews* **24** pp 51-67 (https://doi.org/10.1016/j.esr.2019.01.004)

[14] On-line: https://www.grooflab.com/home-4-de (access August 2020)

[15] Sant'Anna D O, Dos Santos P H, Vianna N S and Romero M A 2018 Indoor environmental quality perception and users' satisfaction of conventional and green buildings in Brazil *Sustainable Cities and Society* **43** pp 95-110 (https://doi.org/10.1016/j.scs.2018.08.027)

[16] Amjad Khabaz 2018 Construction and design requirements of green buildings' roofs in Saudi Arabia depending on thermal conductivity principle *Construction and Building Materials* **186** pp 1119-31 (https://doi.org/10.1016/j.conbuildmat.2018.07.234)

[17] Standard ASTM E 2397/E2397M: 2019. Standard Practice for Determination of Dead Loads and Live Loads Associated with Vegetative (Green) Roof Systems (American Society for Testing and Materials)

[18] Chen Chi-Feng 2013 Performance evaluation and development strategies for green roofs in Taiwan: A review *Ecological engineering* **522** pp 51-8 DOI: 10.1016/j.ecoleng. 2012.12.083

[19] Bevilacqua, Mazzeo P, Bruno D and Arcuri R 2016 Experimental investigation of the thermal performances of an extensive green roof in the Mediterranean area *Energy and Buildings* **122** pp 63-79 DOI: 10.1016/j.enbuild.2016.03.062

[20] Korol O A and Kagan P B 2014 Calculation of integral indicators of investments in energy-saving measures at the stage of a feasibility study for the construction and reconstruction of real estate In the collection: Innovative and technical solutions for environmental sustainability in the construction and management of urban housing and communal services *Collection of materials of the VI International Scientific and Practical Conference* pp 100-7

IOP Conf. Series: Materials Science and Engineering 1015 (2021) 012090 doi:10.1088/1757-899X/1015/1/012090

The assessment of autonomy indexes of socially significant elements of urban infrastructure for creating a safe living environment

L A Shilova

National Research Moscow State University of Civil Engineering (MSUCE), Yaroslavskoye Shosse 26, Moscow, 129337, Russia

E-mail address: ShilovaLA@mgsu.ru

Abstract. The main task of modern society is to ensure and maintain a safe environment for human life. However, the large-scale introduction of information technologies leads to the fact that a technological process is constantly going on in residential buildings and structures, which can lead to the emergence of new previously unexplored sources of emergency situations, and the risks associated with them. In this regard, the article presents an approach to assessing the autonomy of construction projects in the event of an emergency. The autonomy of a construction facility will be understood as the ability of a construction facility to perform its target function during its design life without external intervention in the event of an emergency. At the same time, it is proposed to describe the autonomy of a construction object by the following indexes: reliability, stability and an index of natural and man-made safety.

1. Introduction

The main task of modern society is to ensure and maintain a safe environment for human life. This fact is confirmed by a significant number of publications in this area [1-4]. For example, the paper [1] confirms the fact that the world community is concerned about population growth, depletion of natural resources and the problems of global warming. Considering that control and reduction of the negative impact of construction on the environment has fundamental importance, the work focuses on the research project titled "Energy efficient LIghtweight Sustainable SAfe steel construction" (ELISSA). This paper also presents the life cycle analysis of the building developed as case demonstrator. At the same time, the problems of environmental impact were analyzed both at the stage of construction of the structure and at the stage of its dismantling.

The next work [2] is devoted to the development and technological research of an intelligent electrical building for creating a safe environment for human life. The authors of the study emphasize that the Intelligent Building is an important result of the use of computer technologies in the information age. This paper explains the basic concept of smart building, first describes the meaning and development goals of smart building with electricity, then analyzes the current situation of such buildings, identifies existing problems, and discusses key technologies of smart buildings. Finally, this paper examines the trends in the development of intellectual construction in general to ensure a comfortable living environment.

IOP Conf. Series: Materials Science and Engineering 1015 (2021) 012090 doi:10.1088/1757-899X/1015/1/012090

Also, in the process of creating a safe living environment, a lot of publications are devoted to green buildings and related to their construction factors (labor protection during the construction of buildings, safety, etc.) [5-11].

2. The relevance of research

The rapid development of science and technology leads to the active introduction of new technologies, especially in the industrial sector. At the same time, sometimes innovations led to significant environmental damage to many ecosystems, which, however, did not receive sufficient attention from society. Over time, humanity began to realize the depth of the possible consequences and make a conclusion about the need to develop measures aimed at solving the issue of the use of limited natural resources, calling for the use, for example, of renewable energy sources, for the introduction of environmentally friendly, nature-, energy- and material-saving technologies to support the stable development of social, cultural and ecological systems. This position is consolidated by the concept of sustainable development, which was put forward in 1987, and assumes such a development of mankind, in which the satisfaction of the needs of present generations is carried out without prejudice to the ability of future generations to meet their own needs.

If we consider this concept and its goals from the point of view of the development of the construction industry, then it is necessary to pay attention to the goal of sustainable development of cities and towns. To achieve this goal, leading research centers and scientists are conducting research aimed at introducing new technologies into the design, construction, operation, and disposal of construction projects [12-20].

Undoubtedly, in due time the development of the "smart home" concept was a significant and important achievement in this light. However, the large-scale introduction of smart home technologies leads to the fact that in residential buildings and structures a technological process is constantly taking place, which can lead to the emergence of new previously unexplored sources of emergency situations, and the risks associated with them.

In addition, at present time more attention is paid to the development of the information technology concept, which is focused on the introduction of computing resources into physical entities. Here, special attention is paid to cyber-physical systems that belong to the Fourth Industrial Revolution or the so-called Industry 4.0. The active introduction of cyber-physical systems in construction is also focused on creating a comfortable environment for human life. At the same time, the key technological trends at the heart of Industry 4.0 are: big data, cloud computing, the Internet of things, additive technologies, etc.

At the same time, in conditions of emergencies, socially significant elements of the urban infrastructure remain the most vulnerable. Wherein the concept of social infrastructure is quite extensive and includes a set of industries and enterprises that functionally ensure the normal life of the population. Based on this definition, it can be argued that housing, its construction, social and cultural facilities, housing and communal services, enterprises and organizations of healthcare systems, education, enterprises and organizations related to recreation and leisure, etc. can be classified as social infrastructure.

In addition, it should be noted that despite the fact that modern construction technologies allow creating sustainable and reliable construction projects, at the same time, in conditions of dense urban development, the safety of the living environment of the population, as a rule, is provided by the already existing infrastructure. In this light, the issue of ensuring the autonomy of the functioning of already existing objects, taking into account the definition of the properties that appear in these objects due to the introduction of new ones, including information technologies, becomes relevant.

In addition, the importance of solving the presented task of ensuring the autonomy of socially significant infrastructure elements is also due to the current situation with the COVID-19 pandemic, which began in 2019.

IOP Conf. Series: Materials Science and Engineering 1015 (2021) 012090 doi:10.1088/1757-899X/1015/1/012090

3. The basic principles for assessing the indexes of autonomy of socially significant elements of urban infrastructure

3.1. Basic terms and definitions

The term autonomy comes from an ancient Greek word that literally translated as "self-law" and it was used, as a rule, in psychology, philosophy and politics. However, over time, this term began to be used in various areas of human life. If we consider the term "autonomy" from the point of view of cybernetics, then it can be argued that it means that this or that part of it or a certain function is itself responsible for its regulation. This concept can be translated as "self-enforcement of the law."

The previous author's studies make it possible to formulate the concept of the building object autonomy and determine the indexes that allow it to be evaluated. Thus, the autonomy of a construction facility will be understood as the ability of a construction facility to perform its target function during the estimated service life without external intervention even in the event of an emergency.

At the same time, the *autonomy* of a building object can be described by the following indexes [21]:

Reliability (R) - the ability of a construction object to perform the required functions during the estimated service life [22];

Sustainability (T_{Sustb})- the state of the object, in which the actual functional and technical characteristics of the building (structure) and its elements correspond to the range of permissible values, and the nature of the dynamics of their change does not imply the possibility of inconsistency during the calculated period of time [23];

An index of the natural and technogenic *security (S)* of an object - the state of the object protection from accidents and disasters caused by complex influences - hazardous natural processes, damage and destruction of technical systems, errors and unauthorized influences of operators and personnel, in which conditions are created for its normal functioning and strict observance on it established modes.

3.2. Classification of emergencies

It is customary to classify emergencies by reasons of occurrence, speed of spread, scale of the covered territory, severity, departmental affiliation, etc. The consolidated classification scheme of emergencies is shown in figure 1.

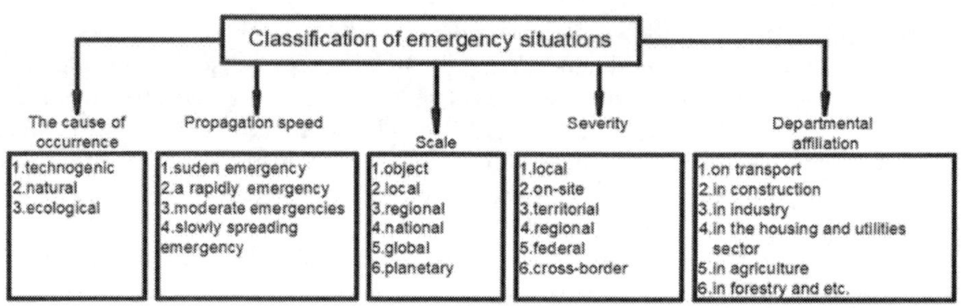

Figure 1. The consolidated scheme of classification of emergency situations

According to the sphere of occurrence of emergency situations, it can be technogenic, natural or environmental.

The source of technogenic emergency is a dangerous man-made incident: an accident or disaster at an industrial facility, transport, power and utility life support systems, treatment facilities, etc. The causes of natural hazards are: geophysical hazards (earthquakes, volcanic eruptions); dangerous geological phenomena (landslides, mudflows, avalanches, taluses, etc.); meteorological hazards (storms, hurricanes, tornadoes, showers, snow drifts, frosts, etc.); marine hazardous hydrological phenomena (typhoons, tsunamis, strong waves (5 points or more) or abnormal sea level fluctuations); hydrogeological hazards (low groundwater levels, high groundwater levels); natural fires (forest, peat, underground fires of fossil fuels), etc. Environmental emergencies can arise due to changes in the state and properties of the atmosphere, biosphere, or hydrosphere.

According to the speed of propagation, emergency situations are usually divided into sudden (explosive), fast (rapidly), moderate, and slow spreading.

According to the scale of possible consequences, emergency situations are divided into object (local), on-site, regional, national, global, planetary.

In addition, it is customary to classify emergency situations by departmental affiliation, depending on the industry in which the emergency occurred.

Within the framework of the study, emergency situations are proposed to be considered as internal and external. An example of emergency sources with such a division is presented in table 1.

Table 1. The internal and external sources of emergency situations.

The internal sources of emergency	The external sources of emergency
The level of technologies complexity used at the facility	Interruption of supply of energy carriers or technological products
The qualifications of users or personnel	Terrorism
Labor and technological discipline	Wars
Physical and moral deterioration of equipment	Epidemics
Design and engineering flaws	Natural sources

3.3. The approach to assessing the autonomy of construction objects

Each of the above indexes of autonomy is complex and can describe the state of a building object from different angles. In this regard, it becomes obvious that each index will consist of the parameter set. At the same time, according to the variety of construction objects types the number of parameters for each index may differ.

In general terms, the reliability of the construction object can be represented as follows

$$\mathbf{R} = [r_1\ r_2\ r_3\ \ldots\ r_m] \tag{1}$$

The sustainability of the construction object will take the following form

$$\mathbf{T}_{\text{Sustb.}} = [t_1\ t_2\ t_3\ \ldots\ t_m] \tag{2}$$

Natural and technogenic security will be described as follows

$$\mathbf{S} = [s_1\ s_2\ s_3\ \ldots\ s_m] \tag{3}$$

Thus, according to (1-3) the autonomy of a construction object can be represented as the following matrix

$$\mathbf{A}_n = \begin{bmatrix} r_{n1} & r_{n2} & \cdots & r_{nm} \\ s_{n1} & s_{n2} & \cdots & s_{nm} \\ t_{n1} & t_{n2} & \cdots & t_{nm} \end{bmatrix} \tag{4}$$

In this case, the index n means the standard or normal state of the parameter, and the number of parameters depends on the type of the object under consideration. To determine the level of autonomy of a construction object, it is necessary to collect a matrix with the actual values of the parameters under consideration, which in the general case may have the following form:

$$\mathbf{A}_f = \begin{bmatrix} r_{f1} & r_{f2} & \cdots & r_{fm} \\ s_{f1} & s_{f2} & \cdots & s_{fn} \\ t_{f1} & t_{f2} & \cdots & t_{fm} \end{bmatrix} \quad (5)$$

Next, it is necessary to determine the difference between the matrices (5) and (4), i.e. perform the operation of calculating the matrix $\mathbf{A}_c$, all the elements of which are equal to the pairwise difference of all the corresponding elements of the matrices $\mathbf{A}_n$ and $\mathbf{A}_f$, that is, each element of the matrix $\mathbf{A}_c$ is equal to: $a_{cij} = a_{fij} - a_{nij}$

$$\mathbf{A}_c = \begin{bmatrix} r_{f1} - r_{n1} & r_{f2} - r_{n2} & \cdots & r_{fm} - r_{nm} \\ s_{f1} - s_{n1} & s_{f2} - s_{n2} & \cdots & s_{fn} - s_{nm} \\ t_{f1} - t_{n1} & t_{f2} - t_{n2} & \cdots & t_{fn} - t_{nm} \end{bmatrix} \quad (6)$$

If the resulting matrix contains negative values, then the autonomy of the object under consideration is not ensured. If all parameters in the matrix take the value 0, then the autonomy of the object is fully ensured.

Let us illustrate with an example. The standard indexes of the autonomy of the construction object under consideration include the following parameters : $\mathbf{R}$=[1 1 1]; $\mathbf{T}_{\text{Sustb.}}$=[1 1 1 1] ; $\mathbf{S}$=[1 1 1 1]

Thus, the autonomy of the construction object under consideration can be represented in the form of the following matrix

$$\mathbf{A}_n = \begin{bmatrix} 1 & 1 & 1 & 0^* \\ 1 & 1 & 1 & 1 \\ 1 & 1 & 1 & 1 \end{bmatrix}$$

1 - means that the parameter has a normative value;
0* - dummy zero, which indicates that the index is described by fewer parameters than the rest.

The actual state of the building object is described by the following matrix

$$\mathbf{A}_f = \begin{bmatrix} 1 & 1 & 1 & 0^* \\ 1 & 1 & 1 & 0 \\ 1 & 1 & 1 & 1 \end{bmatrix}$$

Let us define the matrix $\mathbf{A}_c$, which will describe the calculated value of the autonomy of the object under consideration

$$\mathbf{A}_c = \begin{bmatrix} 0-1 & 1-1 & 1-1 & 0^* \\ 1-1 & 0-1 & 1-1 & 0-1 \\ 1-1 & 1-1 & 1-1 & 1-1 \end{bmatrix} = \begin{bmatrix} -1 & 0 & 0 & 0^* \\ 0 & -1 & 0 & -1 \\ 0 & 0 & 0 & 0 \end{bmatrix}$$

The obtained result shows that not all indexes have a normative value, and there are some parameters that require improvement in order to ensure the autonomous operation of the construction site. At the same time, given that construction objects are complex systems, which are inherent in failures and failures, it is advisable to distinguish more than two states of autonomy. This fact is due to the fact that autonomy can be provided or not, however, for example, individual functions performed by the object can be automated, and then we can say that the autonomy of the object is partially fulfilled.

4. Conclusions

Maintaining the indexes of autonomy within the framework of the normative values will speed up the process of creating a safe environment for the life of the population. At the same time, to achieve this goal, it is necessary to develop a classification of socially significant objects of urban infrastructure in order to empirically (based on statistical data on the functioning of complex systems) determine the number of parameters of the proposed indexes depending on the type of the construction object.

In addition, given that a technological process is constantly going on in residential buildings and structures, which can lead to the emergence of new previously unexplored sources of emergency situations, and the risks associated with them, special attention should be paid to studying these risks and identifying possible sources of emergency situations.

References

[1] Iuorio O, Napolano L, Fiorino L and Landolfo R 2018 Sustainability of modular lightweight steel building from design to deconstruction *United States Wei-Wen Yu International Specialty Conference on Cold-Formed Steel Structures* pp 515-27

[2] Zhang L-M, Liu B-C, Tang Q-H and Wu L-P 2014 The development and technological research of intelligent electrical building *6th China International Conference on Electricity Distribution* pp 88-92

[3] Lazar N and Chithra K 2020 A comprehensive literature review on development of Building Sustainability Assessment Systems *J. of Build. Eng* **32** 101450

[4] Hossaini N, Hewage K and Sadiq R 2015 Spatial life cycle sustainability assessment: A conceptual framework for net-zero buildings *Clean Technologies and Environmental Policy* **17** (8) pp 2243-53

[5] Ding W X, Zhao Y H and Ji X 2013 *Applied Mechanics and Materials* Analysis of the environmental benefits of green building **368-370** (1) p 1135-8

[6] Zhang X and Mohandes S R 2020 Occupational Health and Safety in green building construction projects: A holistic Z-numbers-based risk management framework *J. of Cleaner Production* **275** 122788

[7] Sedayu A, Setiono AR, Subaqin A and Gautama AG 2020 Improving the performance of construction project using green building principles *Asian J. of Civil Eng.* **21** (8) pp 1443-52

[8] Ge J, Zhao Y, Luo X and Lin M 2020 Study on the suitability of green building technology for affordable housing: A case study on Zhejiang Province, China *Journal of Cleaner Production* **275** 122685

[9] Liu Q and Ren J 2020 Research on the building energy efficiency design strategy of Chinese universities based on green performance analysis *Energy and Buildings* **224** 110242

[10] Xie B-C, Zhai J-X, Sun P-C and Ma J-J Assessment of energy and emission performance of a green scientific research building in Beijing, China *Energy and Buildings* **224** 110248

[11] Cianciarullo M I 2019 Green construction– reduction in environmental impact through alternative pipeline water crossing installation *J. of Cleaner Production* **223** p 1042-9

[12] AbouHamad M and Abu-Hamd M 2019 Framework for construction system selection based on life cycle cost and sustainability assessment *J. of Cleaner Production* **241** 118397

[13] Xu J, Shi Y, Xie Y and Zhao S 2019 A BIM-Based construction and demolition waste information management system for greenhouse gas quantification and reduction *J. of Cleaner Production* **229** pp 308-24

[14] Iacovidou E, Purnell P and Lim M K 2018 The use of smart technologies in enabling construction components reuse: A viable method or a problem creating solution *Journal of Environmental Management* **216** pp 214-23

IOP Conf. Series: Materials Science and Engineering 1015 (2021) 012090 doi:10.1088/1757-899X/1015/1/012090

[15] Mesaros P, Spisakova M and Mandicak T 2018 Sustainable design of construction through waste management using building information modelling *Int. Multidisciplinary Scientific GeoConference Surveying Geology and Mining Ecology Management, SGEM* **18** (6.4) pp 537-44
[16] Bilal M *et al.* 2016 Big data architecture for construction waste analytics (CWA): A conceptual framework *Journal of Building Engineering* **6** p144-56
[17] Guerra B C, Leite F and Faust K M 2020 4D-BIM to enhance construction waste reuse and recycle planning: Case studies on concrete and drywall waste streams *Waste Management* **116** 79-90
[18] Yung P and Wang X 2014 A 6D CAD model for the automatic assessment of building sustainability *International Journal of Advanced Robotic Systems* **11** (1) A131
[19] Park J W, Cha G W, Hong W H and Seo H C 2014 A study on the establishment of demolition waste DB system by BIM-based building materials *App. Mechanics and Materials* **522-524** pp 806-10
[20] Cheng J CP and Ma L YH 2013 A BIM-based system for demolition and renovation waste estimation and planning *Waste Management* **33** (6) 1539-51
[21] Volkov AA and Shilova LA 2020 Cyber-Physical Systems In Construction For Sustainable Urban Development *E3S Web of Conference* **143** 01019
[22] Standard GOST 27751-2014 Reliability of building structures and foundations. Fundamentals
[23] Shilova L A 2014 Information support for management of life supporting facilities taking into account the criteria of engineering and functional stability in the case of an emergency *Information resources of Russia* **6** (142) p 24-7

IOP Conf. Series: Materials Science and Engineering 1015 (2021) 012091 doi:10.1088/1757-899X/1015/1/012091

Compressive Properties of BFRP and HFRP Bars

K Protchenko, M Salha, M Urbański and P Narloch

Warsaw University of Technology, Warsaw University of Technology, Department of Civil Engineering, 16 Armii Ludowej Av., 00-637 Warsaw, Poland

E-mail address: k.protchenko@il.pw.edu.pl

Abstract: Durability of steel reinforced structures is one of the leading problems in construction field nowadays. Currently many studies are focused on the possibility of steel reinforcement replacement in concrete structures with a non-corrosive reinforcement, such as FRP (Fiber Reinforced Polymer) reinforcement. This paper presents and compare the results of compression testing of Basalt (BFRP) and a newly developed type of Hybrid FRP (HFRP) bars. The program examined 30 HFRP bars and 30 BFRP bars with a nominal diameter of 8mm and unbraced free-length varying from 50 to 220 mm. Compressive buckling load strength decreased as free length increased, and the modulus of elasticity under compression for diverse unbraced lengths bars slightly differed, but its value was similar to the modulus of elasticity at tensile. In addition, the test results showed that the ultimate compressive strength of non-buckled HFRP bars as a result of axial compression is about 46% of the ultimate strength. The relationships between the compressive buckling load strength and the unbraced length as well as the optimal unbraced length of the bars were determined. Current research is aiming to contribute the optimization of the transverse reinforcement design and the development of standard regulations in the area of elements with such compressed reinforcement.

1. Introduction

Inevitable obstacles, such as deterioration of building materials and their weak structural properties, have lead and forced engineers around the world to develop and implement new, efficient, and economical techniques on reinforcing and repairing structures. Steel bars are still the most commonly used building materials for reinforcing concrete structures, but researchers are working on an new renewable alternative. Fiber Reinforced Polymers (FRP) bars proved to be a potential alternative to steel bars providing high corrosion resistance, good resistance to chemicals, low weight, high strength to weight ratio, a fast and economical way of implementation [1-4].

Basalt FRP (BFRP) bars are made up of basalt fibers bounded together with matrix polymer. Good mechanical properties of constituents combined with cost-effective manufacturing process led to the production of BFRP bars to be used as internal reinforcement for concrete structures. BFRP bars are characterized also by high temperature resistance, good freeze-thaw performance, low elasticity modulus with respect to steel, chemical resistance and high tensile strength [4-6]. Hybrid FRP (HFRP) bars in context of this work might be understood as a combination of two or more fiber filaments together with the matrix polymer. An extensive research is being held in Warsaw University of Technology on an invented type of hybridization, where both basalt and carbon fibers were combined together creating a newly developed hybrid FRP bars. A detailed description on HFRP bars can be found in these companion papers [8-9]

IOP Conf. Series: Materials Science and Engineering 1015 (2021) 012091 doi:10.1088/1757-899X/1015/1/012091

Compression testing of FRP bars can specify the maximum static strength of bars with different constituents and formations. The use of FRP bars in the compression zone was not recommended by most of the norms and standards. Modified ASTM D695-15 [10] test method for rigid plastics recommends to test the sample by placing it between cylindrical heads in a compression device, where the tested sample should have equal diameter and a length twice than that of the bar. Crushing and unintentional rotation at both ends of the bar was not taken into account by this procedure, which affects the strength and deformation of the bar under compression.

In spite of standards restrictions, several compression tests on different types of FRP bars were made showing the importance of FRP bars for reinforced concrete (RC) structures. However, in FRP compression test, compressed elements with composite reinforcement should be designed by different methods than the compressed steel reinforcement. Different spacing of transverse reinforcement should be used [11].

Alajarmeh et al. [12] conducted a study using three high-modulus GFRP bars with different unbraced length and a nominal diameter of 9.5, 15.9, and 19.1 mm respectively. Some of the tests where done using ASTM- D695, where premature splitting was observed due to high-stress concentration at the ends. Some attempts were made to modify this standard procedure by inserting the bar ends in a steel rod. During the preparation of test specimens, hollow steel caps filled with cementations grout were used to restrain the top and bottom ends of the GFRP bars. The results showed that the increase in bar diameter increases the micro-fiber buckling and decreases the compressive-to-tensile strength ratio. Similarly, the failure mode changed from crushing to fiber buckling with the increase of L_f/d_b ratio, where L_f is the unbraced free length of the bar and d_b is its diameter.

Dietz et al. [13] conducted tests on 45 GFRP reinforcing bars, with a diameter of 15 mm and 50–380 mm free (unbraced) length to determine their ultimate strength and compressive modulus of elasticity. Individually, samples were placed into 135 mm long steel cylinders with an external diameter of 50 mm, intended for screwing into the UTM heads of the testing machine. An opening with 65 mm long and 17.5 mm in diameter was drilled in the center of each threaded cylinder so that any possible rotation of the sample ends would not lead to uncertainty of the results. GFRP samples with a ratio of free bar length to diameter $L_f/d_b < 7.33$ ($L_f < 110$ mm) were destroyed by crushing. The second failure mode for samples with $7.33 < L_f/d_b < 14$ of free length was a combination of crushing and buckling failures. For samples with free length of $L_f > 210$ mm ($L_f/d_b > 14$), the failure mode consisted of almost clean buckling of the free length bar. Based on the limited tests of the three samples, it was found that the modulus of elasticity under compression is approximately equal to the modulus of elasticity under tension. In addition, it was found that for unbuckled reinforcing bars with a free length of $L_f < 110$ mm, the ultimate compressive strength is about 50% of the final tensile strength.

Tavassoli et al. [14] conducted a study on nine circular concrete columns, reinforced with GFRP bars under simulated cyclic loading showing that the columns had a significant strength and ductility. The crushing of the compressed concrete was first to maintain followed by the longitudinal crushing of GFRP bars. Moreover, a part of the study was dedicated on testing 15 GFRP bars with 25mm diameter and various unbraced lengths from two different manufacturers. Free lengths specified based on the spiral pitch of the studied concrete columns, it varied between 50 to 275 mm, which gives the ratio of unbraced length to diameter of 2 to 11. Similar to Dietz method the samples were also placed in a steel cylindrical tube that holds GFRP bars in the machine, showing as a result that the compressive strength of GFRP bar is about 50% of its tensile strength, and its modulus of elasticity in compression is similar to the one received in tension [15].

Current research is focused on the investigation of compressive properties of BFRP and HFRP bars with nominal diameter of 8mm and unbraced free lengths between 50 to 220 mm showing the ways FRP bars acts under axial load depending on the free length and the type.

2. Hybridization concept and HFRP bars

Processing of hybrid bars is analogous to the process of producing of any other commercially available FRP bars, where a part of fibers is physically substituted with another fibers type [16]. The chosen constituents were: basalt and carbon fibers (low strength LS type) and epoxy resin. Carbon fibers (LS) were chosen because of its lower cost and its close extension parameters in comparison to basalt fibers [16].

Upon failure, carbon fibers in HFRP are the first to fail transferring the load to the basalt fibers. Carbon fibers can be characterized as rigid, however can be relatively expensive, at the same time basalt fibers are less expensive but do not have the carbon stiffness. For the aim of increasing stiffness, some basalt fibers are replaced with carbon fibers. The value of elasticity modulus is estimated taking into account the volume fraction of both fibers and the epoxy resin. Various combinations with different volume fractions have been checked during analytical consideration and numerical considerations, specifically the volume fractions were Carbon-to-Basalt [C:B]: 1:1; 1:2; 1:3; 1:4 and 1:9. The volume fraction of the used epoxy resin should be not less than 20% to ensure bonding. Analytical considerations, which were based on Voigt model, do not take into account the fiber distribution in the bar. Hence, it was decided to perform a numerical simulation with different configurations of fibers by finite element methods (FEM) in the software ANSYS® Academic Research Mechanical, Release 16.2 (Ansys Inc., Canonsburg, PA, USA) [17].

From the numerical analysis, it was found that the distribution of fibers does not play significant role as the volume fraction. Therefore, for the experimental phase HFRP bars were made up from carbon fibers, basalt fibers (with the volume ratio C:B - 1:4) and epoxy resin. Carbon fibers were placed in the core region because of the burning of carbon fibers in the near-surface region upon producing. The bars were characterized by higher stiffness and in general were having better mechanical properties [8, 9, 16]. The properties of HFRP bars and BFRP bars obtained experimentally are shown in the table 1.

Table 1. Mechanical properties of HFRP and BFRP bars of 8 mm diameter.

Type of bars	BFRP bars				HFRP bars			
Physical size	Maximum force	Tensile strength	Tensile strain at rupture	Modulus of elasticity	Maximum force	Tensile strength,	Tensile strain at rupture	Modulus of elasticity
Symbol	F_u	f_u	ε_u	E_L	F_u	f_u	ε_u	E_L
Unit	(kN)	(MPa)	(%)	(GPa)	(kN)	(MPa)	(%)	(GPa)
Average	60.03	1103.3	2.52	43.87	77.21	1277.92	1.73	73.89
Standard deviation σ	1.24	22.87	0.05	0.86	3.35	55.4	0.07	3.07
Variation	2.07%	2.07%	2.09%	1.95%	4.34%	4.34%	4.33%	4.15%

3. Research program

An experimental buckling load strength tests were done on BFRP and HFRP bars in accordance to standard ASTM D695-15 with some modifications. The bars were tested in ZD20 (WPM, Leipzig, Germany) testing machine with a maximum load of 200 kN.

3.1. Materials utilized for the work

The tests were conducted on 30 HFRP bars and 30 BFRP bars, 5 samples per different unbraced free length with an equivalent diameter of $d_b = 8.32$ mm for BFRP bars and $d_b = 8.77$ mm for HFRP bars.

The term "free length" – (L_f) means the distance between sample anchorages, where the value of the measuring length was taken times the diameter of the bar as $L_f = n{\cdot}d_b$, n values were taken differently for each type of bar (table 2).

Table 2. Number of tested samples

L_f – free length (mm)	L_f / d_b (–)	Number of samples tested BFRP bars	HFRP bars
50	6	5	5
85	10	5	5
120	14	5	5
170	20	5	5
185	22	5	5
220	26	5	5

3.2. Experimental program description

A relationship between buckling load strength and the non-anchored length of FRP bars is a necessity where the free length of FRP bars may vary from one column to another depending on the spacing of stirrups taken by the designer. Such relationship allows enhance transverse reinforcement spacing, which affects the possibility of having local buckling in longitudinal reinforcement of FRP.

BFRP and HFRP bars were tested taken from the same batch ensuring uniform material properties, total length of the samples of each of the specimens made of BFRP bars and HFRP bars respectively, was $L = L_a + L_f + L_a$. The equivalent bar diameter was determined as an average value according to procedure B1 of the ACI 440.3R standard [18]. L_a is the length of anchors placed at both ends of the bars with longitudinal dimension of 120 mm long steel sleeves and an external diameter of 42 mm.

Using a special adhesive with a hardener (ensured by developed resin mix with additives) the bars were centrally fixed to the sleeve, at first the bar was anchored in one sleeve after the connection was cured the other end of the tested specimen was anchored (figure 1).

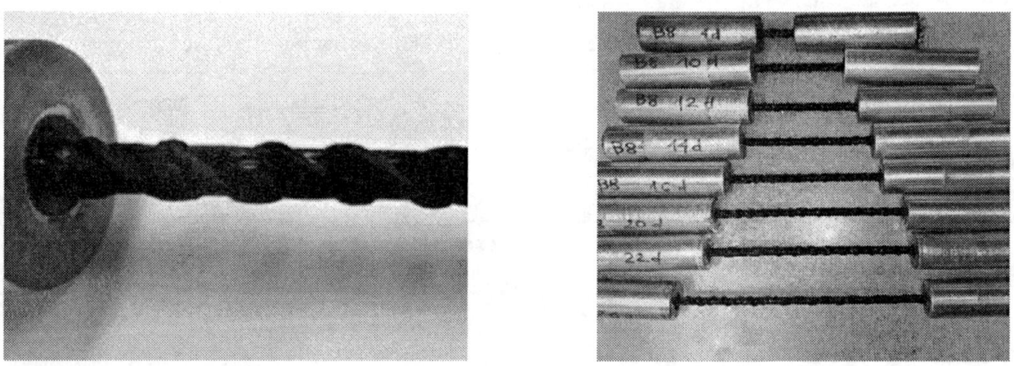

(a) (b)

Figure 1. Specimen preparation (a) anchoring at one end, (b) steel sleeves at both ends.

In the aim of absorbing and transferring compressive force, anchorages in the sleeves were applied preventing degradation of the bar ends. In addition, anchorages helped to prevent the movement of the bar during test. Avoiding unexpected crushing inside the sleeve, 16 mm diameter opening have been made along their entire length. Before installing strain gauges placed oppositely in the half free-length of the samples, the surface of the bars was ground and cleaned to eliminate errors and obtain approximate values. In the next stage two strain gauges (model RL20 Techno-Mechanik, Gdansk, Poland) were added for recording data, one with length of 30 mm and other of 20 mm and a gauge resistance of 120 Ω.

Upon measuring, the tested samples were placed in the center of the test machine ensuring clean compression load. To center the position of the bar properly, centering tube was used. Opening occurred in both sides of the tube along longitudinal axis was allowing the measurement and observation of the sample (figure 3).

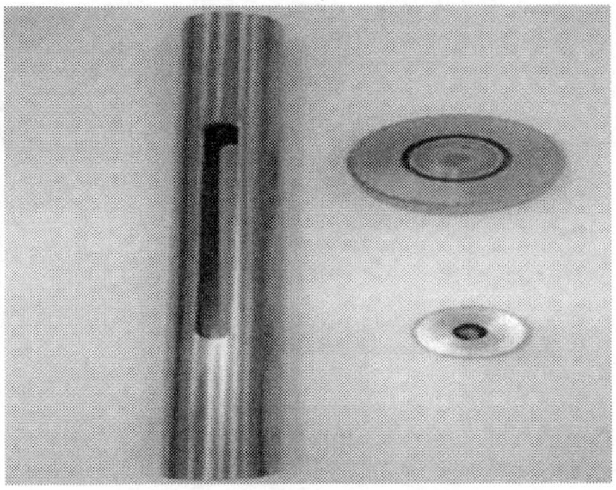
(a)

(b)

Figure 2. Compression testing device (a) compression test components (centering tube, fixing plate), (b) openings on the lateral surface of the centring tube

Avoiding accidental eccentrics, the anchored sample was placed at the bottom of a steel round centering plate allowing its rotation and self-centering operation. The load of the sleeve on the specimen is neglected since it was very low in comparison to the compressive load applied. In case of compressive tests and due to the high impact of transverse buckling load strength, 100 MPa/min load degree was taken so that the total demolish of the sample can occurs before seven minutes. When destruction occurs outside the free length zone another sample from the same batch was taken.

4. Results and discussion

The compressive strength of HFRP and BFRP bars was determined according to equation:

$$f_{f,c} = \frac{F_u}{A_{f,min}} \tag{1}$$

where:

F_u - is the maximum force registered during the test

$A_{f,min}$ - is the minimum bar diameter measured immediately before installing the strain gauges on the surface of the bar being tested.

4.1. Compressive (buckling load) strength of BFRP bars

During testing, several destruction modes were distinguished:

1. Mode 1: For specimens with an unbraced free length $L_f < 85$ mm ($10d_b$) failure occurred due to the shear of the bar for smaller free lengths. The load decreased rapidly after the bar was damaged and the damage occurred in the central part of the unbraced free length in this group of samples (figure 3a).
2. Mode 2: For specimens with the unbraced free length in the range from $10d_b$ to $12d_b$, the bars slightly buckled in the final buckled phase, then fiber groups accumulated split leading to a loss in load capacity (figure 3b).
3. Mode 3: For a specimen with unbraced free length greater than $12d_b$, buckling obtained at the final phase because of the loss in load capacity due to partial breaking of some fibers and detachment of the epoxy resin (figure 3c).

IOP Conf. Series: Materials Science and Engineering 1015 (2021) 012091 doi:10.1088/1757-899X/1015/1/012091

(a) (b) (c)

Figure 3. Failure modes of BFRP bars (a) crushing of the $6d_b$ free length specimen, (b) buckling with crushing a $12d_b$ free length specimen, (c) buckling of a $26d_b$ free length specimen

The obtained values of the average compressive (buckling load) strength of the bars with various values of free lengths presented in table 3.

Table 3. Compression and tension characteristics of BFRP bars.

L_f/d_b (–)	L_f (mm)	$f_{f,c}$ (MPa)	CoV (%)	$f_{f,c}/f_{f,t}$ (–)	E_{fc} (GPa)	E_{fc}/E_{ft} (–)
6	50	375.71	6.32	0.34	41.79	0.95
10	85	475.43	4.27	0.43	49.87	1.14
12	100	429.57	2.83	0.39	42.61	0.97
20	170	184.26	5.32	0.17	45.16	1.03
22	185	182.47	9.27	0.17	50.45	1.15
26	220	143.95	9.86	0.13	51.69	1.18

Column 1 shows the unbraced length L_f as times the equivalent diameter of the bar d_b. The compressive strength, $f_{f,c}$, was compared with the tensile strength, $f_{f,t}$, and the modulus of elasticity under compression, E_{fc}, with the tension modulus of elasticity E_{ft}, for BFRP bars. According to above results the group of bars which were damaged due to crushing (L_f < 10db), the stress-strain relationships observed were linear and close to each other for different samples in the same group (figure 4a).

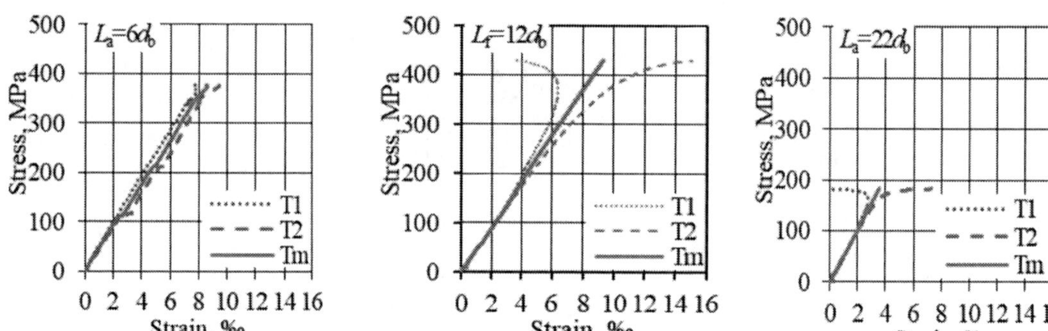

(a) (b) (c)

Figure 4. Stress-Strain relationship of BFRP bars (a) L_f =$6d_b$, failure by crushing, (b) L_f =$12d_b$, buckling with crushing (c) L_f =$22d_b$, buckling failure (T1, T2, T3 strain gauges reading, Tm - average strain)

For specimens with L_f in the range from $10d_b$ to $12d_b$, and as the free length is increasing an increase in strain difference was observed. This occurrence was associated with method of failure consisting from crushing with slight buckling of the bar (figure 4b).

For specimens with unbraced lengths $L_f > 12d_b$ full buckling of bars were significantly obtained leading to a reduction in compressive strength of the specimens as unbraced length increased (figure 4c).

Linearity of the modulus of elasticity, E_{fc}, was determined by dividing the difference of stresses by the difference of strains for the values of 0.5 and 0.2 of the breaking force. The tangent of the secant inclination passing through the two mentioned points is the determined modulus. Due to the anisotropic structure of basalt fibers, the modulus of elasticity under compression will be slightly different from the tension modulus of elasticity for BFRP bars.

4.2. Compressive strength of HFRP bars

Similarly, three different failure modes of HFRP bar samples were distinguished:

1. Mode 1: As in HFRP bars the destruction of the bars ($6d_b$) occurred as a result of splitting individual fibers and separating of epoxy resin. Bar failure is related to the shear of the bar for smaller free length (figure 5a).
2. Mode 2: In samples of intermediate length ($10d_b$, $14d_b$), the onset of failure occurred as a result of buckling. Sustain loading made that the fibers were simultaneously split and as a result the tested bars were crushed (figure 5b).
3. Mode 3: This mode of failure occurred in samples with unbraced free length of $22d_b$ and $26d_b$ as a result of buckling mixed with delamination of the individual fibers of the tested bars (figure 5c).

(a) (b) (c)

Figure 5. Failure mode of HFRP bars with a diameter of 8 mm (a) $L_f = 6d_b$, failure by crushing the sample, (b) $L_f = 14d_b$, buckling and crushing of the sample, (c) $L_f = 22d_b$, buckling of the sample.

HFRP bar with a free length of $6d_b$ (figure 5a) shows a damage due to crushing on a free length section, usually the damage occurs in the middle of the free length of bars.

As for bars with free length of $14d_b$ the damage occurs as a result of slight buckling followed by a crushing (figure 5b). Moreover, the failure mode of the bars exceeding the free length of $14d_b$ consisted of buckling only without any crushing while achieving of maximum compressive strength. The stress-strain diagram of bars with failure mode 1 were analogue and slight difference was observed concerning the values of modulus of elasticity obtained. It should be noted that for some samples with a free length above $20d_b$ at a maximum load, there were no signs of visible damage on the bar surface.

The stress-strain relationship for compressed bars was determined by stresses in accordance to eq.1 and strain was observed as mean strain value recorded during the test from two strain gauges attached on opposite sides in the middle of the sample. The buckling load strength of HFRP bars decreases with the increase of the unbraced free length of the bar. Buckling load strength (compressive strength) of tested HFRP bars was compared with the tensile strength of the bars with same nominal diameter of 8 mm. The compressive strength obtained was 46% of HFRP tensile strength (for $6d_b$ free-length), decreased to 15% as the anchor length increases (table 4, column 4). The average deformation values for the sample are shown in table 4, column 5. During compression of the deformation (ε_{uc}), in terms of absolute value, they reach from 1/3 to 2/3 of the tensile deformation (ε_{ut}), with no apparent effect of free length on the deformation value (table 4, column 7).

HFRP samples with free length of 6db and 10db, the stress-strain curves continued rising till crushing, the relationship were linear and characterized by high consistency (figure 6 a-b). The bar failure happens only on the free length of the samples with a sudden decrease in the load upon failure. As for bars with 14db free length the indications of the opposite strain gauges were increasingly divergent (figure 6 a-b).

Table 4. Compression and tension characteristics of HFRP bars

L_f	$f_{f,c}$	CoV	$f_{f,c}/f_{f,t}$	ε_{uc}	CoV	$\varepsilon_{uc}/\varepsilon_{ut}$	E_{fc}	$E_{fc,ave}$	$E_{fc,min}$	$E_{fc,min}/E_{fc,0.2}$
(mm)	(MPa)	(MPa)	(–)	(%)	(%)	(–)	(GPa)	(GPa)	(GPa)	(–)
50	583.60	28.13	0.46	9.01	0.42	0.52	73.37	73.68	73.46	0.99
85	516.43	23.55	0.40	6.06	0.30	0.35	78.67	82.75	81.08	0.95
120	439.12	20.99	0.34	6.95	0.31	0.40	83.89	82.41	81.65	1.00
170	289.48	23.85	0.23	7.73	0.62	0.45	78.18	78.22	77.75	1.00
185	296.18	25.74	0.23	12.49	1.15	0.72	71.11	73.02	72.07	0.98
220	195.68	18.86	0.15	7.31	0.72	0.42	81.26	82.34	81.55	0.99

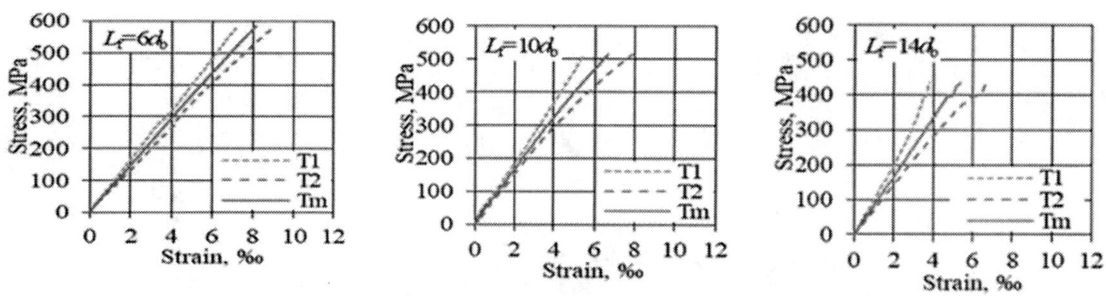

(a) (b) (c)

Figure 6. Stress–strain relationship (a): $L_f = 6d_b$ crushing of the specimen (b): $L_f = 10d_b$ crushing the sample (c): $L_f = 14d_b$ significant buckling followed by crushing.

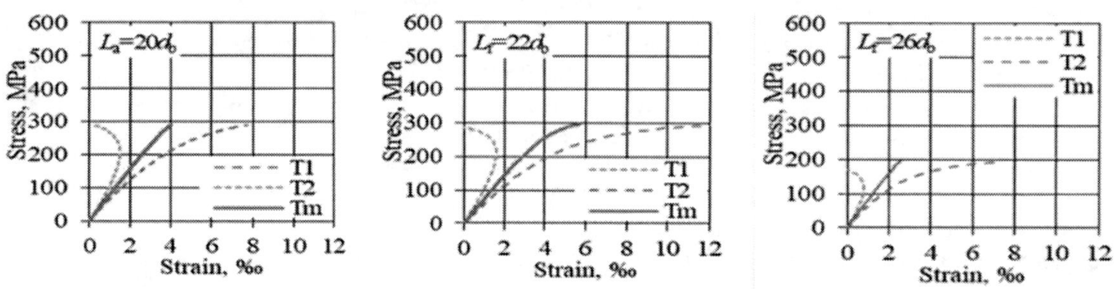

(a) (b) (c)

Figure 7. Stress–strain relationship (absolute value) (a) $L_f = 20d_b$ buckling without visible crushing, (b) $L_f = 22d_b$ buckling without crushing, (c) $L_f = 26d_b$ buckling without visible crushing.

In compression, the values of the modulus of elasticity obtained were approximately similar or slightly higher than those obtained under tension (table 4, column 12). Average values of modulus of elasticity shown in table 4, column 8. For specimens with a greater unbraced free length ($L_f > 14d_b$), stress-strain readings differs which in turn affects the estimation of the elastic modulus (figure 8, a-c).

Modulus of elasticity under compression can be calculated using several ways. First it can be determined based on B_2 procedure of the standard ACI 440.3R-12 used for tension bars, where the elastic modulus, E_{fc}, (table 4, column 8) is determined as a ratio of the difference in stress and strain for the values 0.5 and 0.2 of the ultimate force. In other words, it is the tangent of the initial inclination through the two stated points. Different way of elastic modulus calculation is presented in (table 4, column 9) is to calculate the average modulus of elasticity $E_{fc,ave}$ (table 4, column 9) based on subsequent readings of stress and strain in the range from 0.2 to 0.5 of the ultimate force. Results using this method of calculation are slightly higher than the previous one. The third way of elastic modulus calculation is to assess the minimum value of the tangent modulus of elasticity, $E_{fc,min}$ (table 4, column 10) using the same range of ultimate force.

The relationship between elastic modulus and the stress for HFRP bars with short and long non-anchored length shown in (figure 8, a-c) respectively. Moreover (figure 8, b) shows a comparison of the tangent tensile modulus of elasticity, E_{ft}, and the compressive elasticity modulus, E_{fc}, for the entire range of absolute stress values (from 0.1 to 1.0 ultimate stress) f_{ut} and f_{uc}. The tensile tangent modulus of elasticity increases as the load increase due to increase stiffness, when tensioning, however, the compression tangent modulus decreases with the increase in load related to mechanism of micro buckling of the bar. Therefore, minimum tangent modulus of elasticity, $E_{fc,min}$, is considered as the best value for defining the stiffness for the bars under compression. Hence, the compressive modulus of elasticity of HFRP bars can be calculated based on the modulus of elasticity in tension.

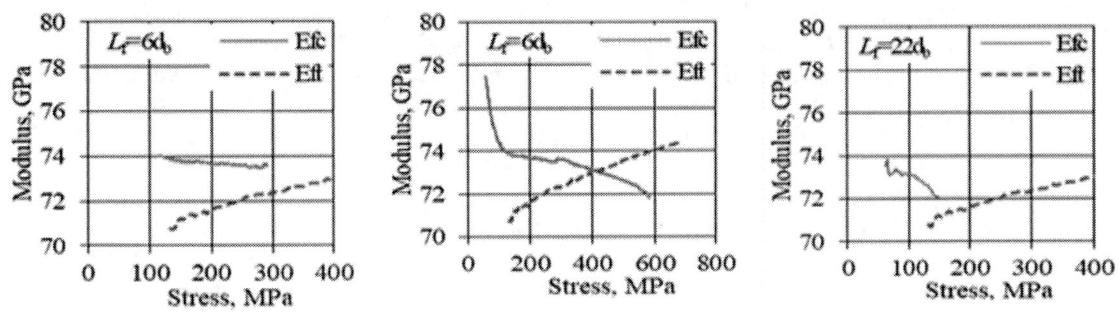

(a) (b) (c)

Figure 8. Comparison of the tangent modulus of elasticity at tensile E_{ft} in the range of $(0.1–1.0)f_{ut}$ with the tangent modulus of elasticity in compression (a) $L_f = 6d_b$, E_{fc} in the range of $(0.2–0.5)f_{uc}$, (b) $L_f = 6d_b$, E_{fc} in the range of $(0.1–1.0)f_{uc}$, (c) $L_f = 22d_b$, E_{fc} in the range of $(0.2–0.5)f_{uc}$ (absolute stress value).

5. Conclusions

The research on BFRP and HFRP bars with nominal diameter of 8 mm showed that the optimal free unbraced length of the bars can be established for which the ultimate compressive strength (buckling load strength) reaches its maximum value while meeting the non-buckling condition. Knowledge of the dependence on buckling load strength is of practical importance because it allows determining the optimal spacing of stirrups in RC compressed elements.

For BFRP bars, the compressive buckling load strength, $f_{f,c}$, depending on the free unbraced length has been exposed to several changes before reaching the optimal value according to the polynomial curve (for $L_f < 85$ mm) shown in crushing phase then the linear relationship (for $L_f < 125$ mm) achieved in the transition phase. As for HFRP bars, the optimal value was reached directly where the compressive buckling load strength decreases monotonically as free unbraced length increases. During the three obtained phases a linear relationship is obtained with correlation of $R_2 = 0.99$ allowing a simple selection of spacing in transverse reinforcement depending on the effort of compressed HFRP bars.

Moreover, the substitution of basalt fibers with carbon fibers (with the volume ratio C:B - 1:4 indicates a clear difference in the increased values of compressive strength (55%) and modulus of

IOP Conf. Series: Materials Science and Engineering 1015 (2021) 012091 doi:10.1088/1757-899X/1015/1/012091

elasticity E_{fc}; e.g., for bars with free length $6d_b$ compressive strength increased from 375.7 to 583.6 MPa and modulus of elasticity from 41.79 to 73.46 GPa in BFRP and HFRP respectively.

For non-buckled BFRP and HFRP bars the values of the modulus of elasticity under compression Ef,c are similar to those under tension, $E_{f,t}$. For bars with larger free lengths, the tangent modulus of elasticity under compression reached slightly higher values.

The minimum value of the tangent modulus of elasticity for the range of 0.2–0.5 ultimate load should be taken as the representative value of elasticity modulus considering that $E_{fc,min}$ is the best measure for stiffness for bars due to buckling load that should be taken into account while designing. Furthermore, according to the authors respectively, both BFRP and HFRP bars can be used as a compressive reinforcement but under special conditions. Further researches should be carried out on the remaining bar diameters for both HFRP and BFRP bars used to propose a general relationship for the design of reinforcement in the compression zone.

References

[1] Danraka M N, Mahmod H M, Oluwatosin O J and Student P 2017 *Intern J of Eng Science* **7** (6) p 13199

[2] Protchenko K, Włodarczyk M and Szmigiera E 2015 *Proc Eng* **111** pp 679–686 http://doi.org/10.1016/j.proeng.2015.07.132

[3] Garbacz A, Szmigiera E, Protchenko K and Urbański M 2018 *International Congress on Polymers in Concrete (ICPIC 2018): Polymers for Resilient and Sustainable Concrete Infrastructures, Springer* p 653-658 https://doi.org/10.1007/978-3-319-78175-4 83

[4] Protchenko K and Szmigiera E 2020 *Materials* **13** (5) p 1248 http://doi.org/10.3390/ma13051248

[5] Elgabbas F, Vincent P, Ahmed E A and Benmokrane B 2016 *Compos Part B-Eng* **91** 205-218. http://doi.org/10.1016/j.compositesb.2016.01.045

[6] Urbański M, Łapko A and Suprynowicz K 2016 *Solid State Phenomena* **240** p 55–60 http://doi.org/10.4028/www.scientific.net/SSP.240.55

[7] Protchenko K, Szmigiera E, Urbański M, Garbacz A, Narloch P and Lesniak P 2019 *IOP Conf Series: Mat Sci and Eng* **661** p 012081 http://doi.org/10.1088/1757-899X/661/1/012081)

[8] Protchenko K, Szmigiera E, Urbański M and Garbacz A 2018 *MATEC Web of Conf.* **196** p 04087 http://doi.org/10.1051/matecconf/201819604087

[9] Protchenko K, Dobosz J, Urbański M and Garbacz A 2016 *Czasopismo Inżynierii Lądowej, Środowiska i Architektury, JCEEA* **63 1/1** pp 149–56

[10] ASTM D695—15: Compressive Properties of Rigid Plastics; ASTM International: West Conshohocken, PA, USA, 2015

[11] Urbański M 2020 *Materials* **13** (8) p 1898 https://doi.org/10.3390/ma13081898

[12] Alajarmeh O S, Manalo A C, Benmokrane B, Vijay P V, Ferdous W and Mendis P 2019 *Constr Build Mater* **225** pp 1112–26 https://doi.org/10.1016/j.conbuildmat.2019.07.280

[13] Deitz D H, Harik I E and Gesund H H 2003 *J. Compos. Constr. ASCE* **7** pp 363–6 http://doi.org/10.1061/(ASCE)1090-0268(2003)7:4(363)

[14] Tavassoli A 2013 *Thesis: Behaviour of GFRP-reinforced concrete columns under combined axial load and flexure* (Toronto: University of Toronto) ON Canada p 201

[15] Khorramian K and Sadeghian P 2019 *J Test Eval* **49** pp 1–28 http://doi.org/10.1520/JTE20180873

[16] Szmigiera E, Protchenko K, Urbański M and Garbacz A 2019 *Arch of Civ Eng* **65** (1) pp 97–110 http://doi.org/10.2478/ace-2019-0007

[17] ANSYS® Academic Research Mechanical, Release 16.2, Help System, Coupled Field Analysis Guide; ANSYS, Inc.: Canonsburg, PA, USA, 2016

[18] 440.3R-12 Guide Test Methods for Fiber-Reinforced Polymers (FRPs) for Reinforcing or Strengthening Concrete Structures, ACI, 2012

IOP Conf. Series: Materials Science and Engineering 1015 (2021) 012092 doi:10.1088/1757-899X/1015/1/012092

Use of photovoltic systems in the construction of wooden houses in a sustainable standard

A Iringová and M Kovačic

University of Žilina, Faculty of Civil Engineering, Univerzitná 8215/1, 010 26 Žilina, Slovakia

E-mail address: iringova@uniza.sk

Abstract. The paper deals with the analysis of the efficiency and application of photovoltic systems in the roof plane and in the perimeter walls in the construction of wooden houses in a sustainable standard. The installation of photovoltic systems is a source of renewable energy in the construction of zero energy buildings with minimal emissions of environmental pollution. Semi-transparent photovoltics integrated in the building (BIPV) is one of the technologies that has the potential to increase the energy efficiency of the building and at the same time aesthetically complete the design of the building. This work deals with the construction and material solution of BIPV systems, including BIPV glazing products, tiles and modules and their application in the architectural solution of a model building.

1. Introduction

Achieving sustainable development has been a major goal of the international community since 1992, when the UN Conference on Environment and Development was held in Rio de Janeiro (Earth Summit). The summit document sets out the principles for addressing the sustainable development of the territory in terms of social, environmental and economic growth. In the Europe, the Strategy 2020, which is one of the main economic reforms of the European Union, defines requirements such as sustainability in the construction and operation of buildings, including their demolition. One of the main goals of the strategy, which the EU countries have committed themselves to meeting, is that from 31 December 2020, all new buildings will have energy consumption approaching zero. [1] Regulation of the European Parliament and of the Council of the EU no. 305/2011. This criterion can only be achieved through the use of active technologies, that could transform energy from renewable sources in buildings for its operational needs. Solar energy is the most available source for optimizing the thermal microclimate and operating technical equipment in buildings. If the potential for the use of solar radiation from exposed surfaces in the production of energy during the operation of buildings is increased, the amount of environmental emissions from non-renewable sources that escape into the air during the operation of heating and cooling systems will be significantly reduced. Sunlight is an unlimited source of energy. The latest European Union report on the State of Photovoltaics of 2019 [2] states that the average cost of generating direct current (DC) electricity produced by photovoltaic modules has fallen below € 0.02 (€) per kilowatt hour in many places around the world (kWh). The current problem of electricity production from PV modules is the high costs associated with its transport to the place of consumption. For this reason, building an integrated photovoltaic system (BIPV) at the point of consumption is ideal. BIPV systems are photovoltaic cells that are integrated into heat-exchange building envelopes, such as roofs or facades. Such a system performs two tasks in

IOP Conf. Series: Materials Science and Engineering 1015 (2021) 012092 doi:10.1088/1757-899X/1015/1/012092

the building at the same time. First, it is the final structural-material solution of the heat-exchange packaging of the building itself with the function of weather protection and they are early fulfilling the function of thermal, acoustic and fire insulation of the building. The second task of the BIPV system is to provide electricity for the operation of the building. They become the energy generator of the building [3,4]. The BIPV system supplies energy where the end user needs it. The system for storing the generated direct current is in a battery or in the public electricity network. It is ideal to store energy in batteries directly in the building, which can cover the night's electricity consumption in a building for heating, cooling or artificial lighting. In terms of environmental impact on the environment, the integration of PV panels into building envelopes is an ideal solution - the electricity generator is at the place of its consumption and their installation is without the need for additional occupation of agricultural land [5,6].

2. Use of photovoltic systems in the building envelope

We divide the use of photovoltic systems in the building in terms of their application on the heat exchange envelope, the method of their installation and the components used. We divide photovoltic energy systems in terms of their incorporation within the building envelope into applied PV panels - BAPV systems, and integrated systems in the construction of the heat exchange envelope - BIPV systems. BAPV systems are usually placed on a grate structure on the surface of the building envelope in grouped assemblies. With this installation system, the inclination, the area, the technology used and the orientation of the panels to the world side are decisive in terms of energy efficiency. As they are usually placed on facades and roofs additionally, they are a suitable solution for the reconstruction of energy systems of buildings during additional modernization of buildings. To increase their efficiency in heating buildings is possible. The use of BIPV systems is the optimal solution for new buildings. By incorporating integrated PV panels directly into the structure of the heat-exchanging building envelope or shielding lamellas, their effective area is usually significantly larger than in BAPV systems. They are most often applied to all-glass facades, which are oriented south, southeast, southwest in localities with a low degree of external shading by the surrounding urbanism or geo-relief. They are often located in the shading slats of sunshades, balustrades, awnings and the like. On roofs, they are most often applied to large-area skylights, glass roofs of atriums. With sloping roofs, they replace the classic roofing. When applying BIPV systems in general reconstruction of buildings - replacement of heat exchange envelope, it is possible with these technologies to achieve energy efficiency of a building with zero energy consumption (ZEB), or even plus energy buildings [7,8]. In terms of the elements used, they can be applied as - foil tiles, modules and glazing of solar cells. Application of PV systems in the heat exchange envelope of the building, see figure 1.

Figure 1. Classification of BIPVs BAPVs products. [9].

PV systems can also be implemented as photovoltaic thermal systems (PVT), which are with active or passive ventilation, behind the panels with subsequent removal of hot air to the heating system of the building. Cooling of PV modules is carried out using air or water. [10, 11, 12] In this way, the production of electrical and thermal energy in a building with higher efficiency is achieved [13, 14, 15]. For example, in a BIPV system with air ventilation in the gap of double-skin facades or roofs, a photovoltaic system is typically installed in front of the interior cladding of the heat exchange cladding. Fresh air in the summer naturally ventilates and cools the back of the BIPV system, increasing its efficiency in electricity generation. If the system uses this extracted hot air for heating purposes, it changes to a new configuration called the integrated photovoltaic heating system (BIPVT). Experimental assessment of the thermal performance of the BIPV ventilated wall is given in the study Chiu MS, Hou SP, Tzeng CT, Lai CM [16]. Examples of the use of different building facades with different focus on the distribution of energy production throughout the day are given in the literature [17,18]. The application of BIPV systems is not limited to a building, where, in addition to energy efficiency, it also performs a design function. It can also be used in other applications in vehicles - ships, cars, aircraft, where it contributes to their optimal performance in terms of energy consumption [19].

3. Methology of the procedure

The optimalization of the potential of BIPV and BAPV was based on studies and procedures reported in the literature [20, 21, 22].

There are currently four basic approaches available for assessing the potential of solar energy in photovoltaic systems. [22]. They are defined as theoretical, geographical, technical and economic potential. Theoretical potential is defined as all available solar radiation incident on the surface of the earth's surface, the distribution of which is without any geographical or technical restrictions.

Geographical potential is a fraction of the theoretical potential, which is usable depending on the inclination of the panels and their orientation to the sides of the world. For the capitals of the EU countries, it is shown in figure 2 on the basis of data from [23].

Figure 2 shows the average annual potential of geographical solar radiation in creating surfaces for the placement of PV modules in the capitals of all Member States of the European Union (EU) together with the capitals of Norway and Switzerland. The analysis and calculated quantities are based on hourly data on incident solar radiation from 2005 to 2016 from the Photovoltaic Geographic Information System (PVGIS)

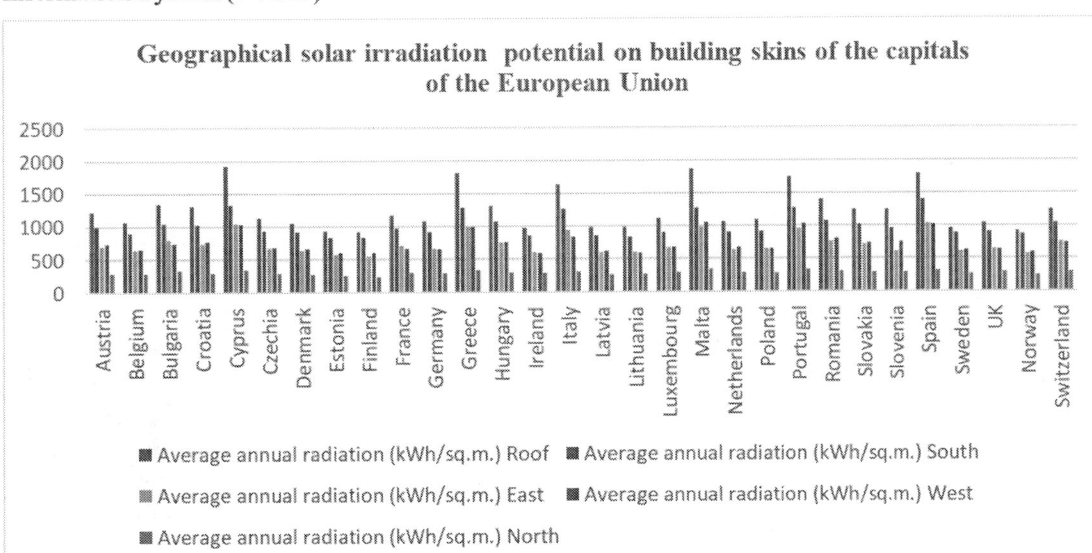

Figure 2. The average annual geographical irradiation potential on building skins of the capitals of the European Union member states (EU) with Norway and Switzerland.

IOP Conf. Series: Materials Science and Engineering 1015 (2021) 012092 doi:10.1088/1757-899X/1015/1/012092

Technical potential is defined as a fraction of the geographical potential that is technically usable depending on the efficiency of the photovoltaic module technology. Economic potential is part of the technical potential that is economically feasible.

2.1 Photovoltaic system technologies and their technical potential

The division of PV systems can be based on the composition of solar cells and the type of connection to the grid (separate - accumulation in a battery, connected to the public grid, or hybrid by a combination of the previous two systems. (GaAs), thin-film technologies, multi-cord cells and emerging PV technologies. With a broad classification, monocrystalline and multi-crystalline are the two basic forms of crystalline technology. Compared to all other types of photovoltaic technologies, crystalline silicon technology has the highest representation in the construction market. efficiency in the conversion of solar radiation into electricity see table1.

Table 1. Absolute efficiency values for different PV technologies [24].

PV cell technology (abbreviation)	Theoretical effectiveness of the PV cell	Laboratory result of PV cell efficacy	Laboratory result of PV module efficiency	Market efficiency for the PV module
MonoCrystalline silicon (C-Si)	29.8%	250%	22.9%	14-17%; 20%
Multi-crystalline silicon (MC-Si)	NA	20.4%	18.2%	12-15%
Copper Indium-Gallium Selenide (CIGS)	31.6%	19.6%	15.7%	11-13%
Telurid Cadmium (CdTe)	30.3%	16.7%	12.8%	11-12%
Gallium Arsenide (GaAs)	31.2%	28.3%	23.5%	NA
Multi-Junctional PV technology (Multi-J)	55.9%	43.5%	33.9%	25-30%

The efficiency of solar systems is directly dependent on the intensity of solar radiation incident on the surface of the structure, depending on its inclination and orientation to the world side. Figure 2 shows an analysis of the average values of solar radiation intensity in the capitals of the EU countries, Norway and Switzerland without the influence of external shading on the surfaces of buildings, depending on their orientation to the sides of the world. From the geographical potential of the intensity of solar radiation for the surfaces of the building, the technical potential of the produced electricity can be calculated. In order to calculate the technical potential, it is necessary to specify the technology and efficiency of the PV module, which is specified by the technology manufacturer. For selected systems, see Table 1. The average overall efficiency of BIPV systems varies depending on the technology used, configuration, site climate, ventilation, orientation and slope. Based on the experimental projects carried out so far, the technical potential of BIPV is between 10% and 22% [25]. Thus, if we take the average efficiency of 18% for BIPV panels - which is the average efficiency of commercialized BIPV panels on the market and not the system, the technical potential can be easily calculated by multiplying the 18% efficiency by BIPV x geographical potential from figure 2. Based on BIPV efficiency and data at figure 2 shows the technical potential of BIPV systems for European capitals in figure 3. In the future, it is expected that their efficiency will increase with the development of new PV module technologies. The expected life of a PV system is defined as the period during which the panels will generate electricity for at least 80% of their rated output. According to the manufacturers, the current BIPV systems have a warranty of about 30 years.

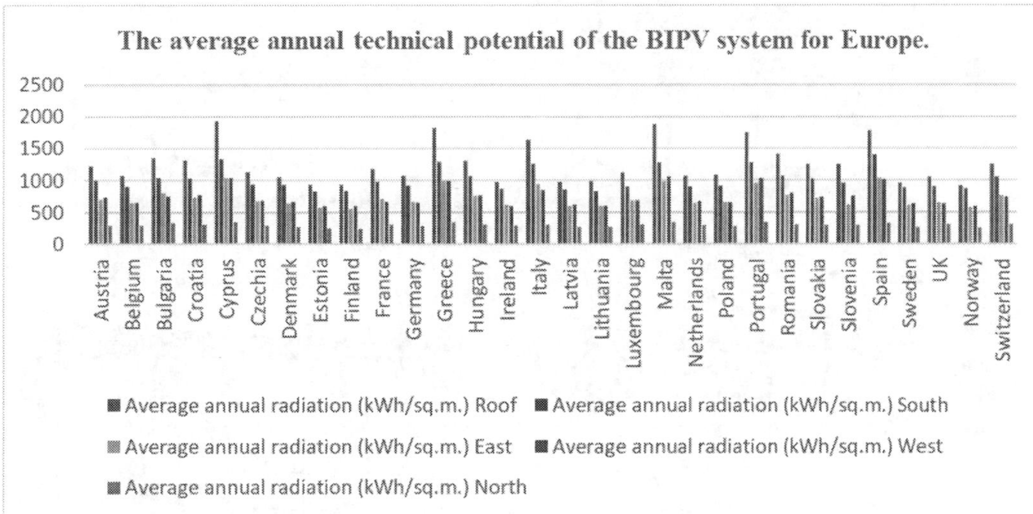

Figure 3. The average annual technical potential of the BIPV system of the capitals of the European Union member states (EU) with Norway and Switzerland.

4. Model solution of a PV system in a family wooden house in terms of their technical potential

The model solution compares the efficiency of the solution, of the location and technology of BIPV on the heat exchange envelope of the family wooden house in terms of their area, so that its energy consumption for the operation of the house throughout the year is ensured from the energy produced by PV. The family house has a gable wall with a glass wall facing south, its area is 28 m2. This wall is without external shielding and is ideal for placing a BIPV system. In terms of light comfort in the living room and visual contact of people with the exterior, the distribution of daylight is provided from the sides. To the west to the terrace are situated French windows with a height of 2.4 m, to the east in the dining room part of the window with a height of 1.5 m. The family house has a sloping roof, while the roof planes are oriented east - west. The second option is to place the BIPV system in the roof plane of the saddle roof with an orientation to the west. The geometry and layout of the apartment can be seen in figure 5

,

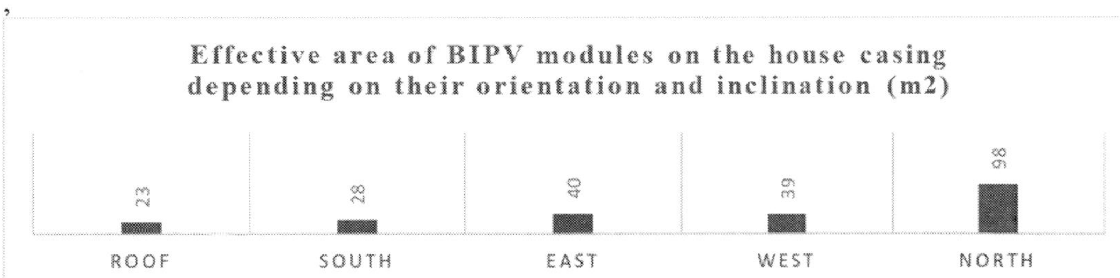

Figure 4. Area optimalization of BIPV modules depending on their orientation on the heat exchange envelope of the building with zero energy consumption from non-renewable sources.

IOP Conf. Series: Materials Science and Engineering 1015 (2021) 012092 doi:10.1088/1757-899X/1015/1/012092

Figure 4 shows the economic efficiency of placing BIPVs in walls depending on their orientation to the sides of the world and in a sloping roof.

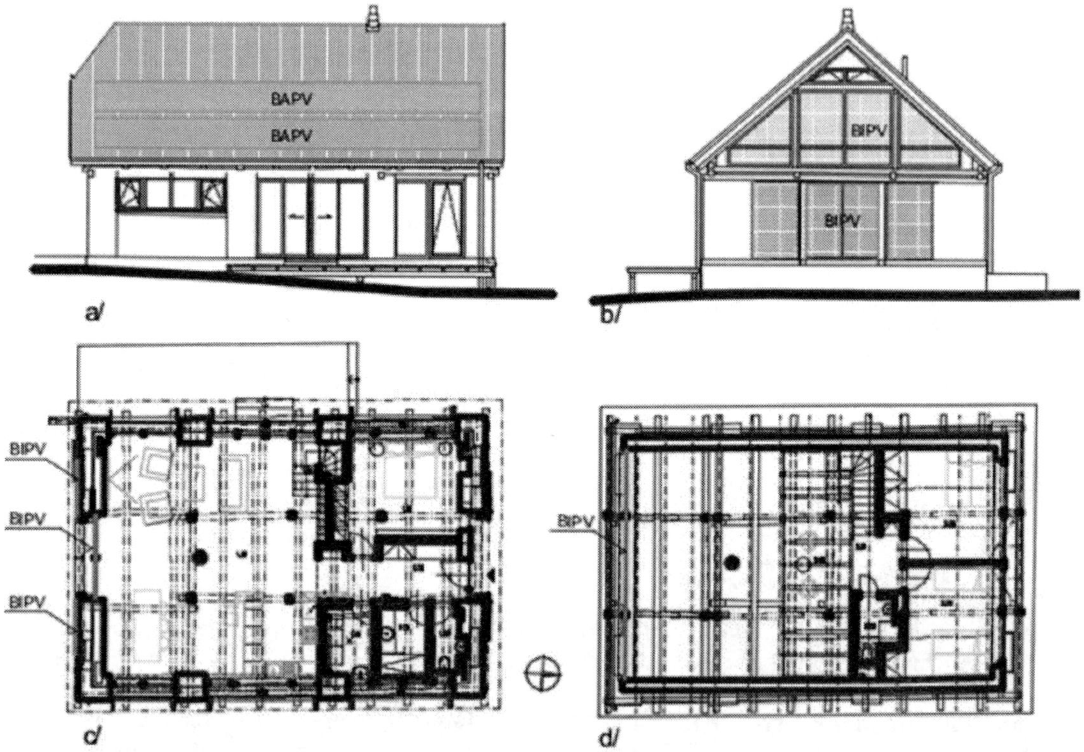

Figure 5. Example of a model solution for optimizing the location of a PV system in the envelope of a family house.

5. Results

In the case of family wooden houses, the optimal solution from the point of view of economic efficiency is the positioning of BIFV modules in the roof plane. In terms of equipment performance, it is ideal if the slope of the roof plane in our geographical location is in the range of 35-45 °. The second option, which is comparable to the variant of placing the BIPV on the roof, is the integration of BIPV modules into the southern gable wall in the model solution with regard to the boundary conditions of the area and the layout of the apartment. This solution will fundamentally affect the emotional quality of the living room space and the overall design of the building.

6. Conclusions

Current studies show that the lifespan of BIPV systems can be up to 50 years [26, 27], the return on the initial increased investment costs when installing them in the building is paid by their operation within 10 years. From the point of view of the service life of wooden houses, which are also realized for a service life of 50 years, the installation of BIVP on the casing of ZERO wooden houses is the optimal solution. PV systems in the building will provide at least 80% of their original projected electricity generation capacity during their lifetime. At present, more than 80% of BIPV systems in the world are installed on the roofs of buildings, the rest are PV systems mounted on facades [28]. BIPV products for facades are less widespread [29]. This is related to the optimalization of the microclimate in protected areas. If they are located in front of the windows, they fundamentally limit the distribution of daylight in protected areas and the visual contact of people with the exterior. The second major limitation in terms of their effectiveness in their application in the perimeter walls is the external

IOP Conf. Series: Materials Science and Engineering 1015 (2021) 012092 doi:10.1088/1757-899X/1015/1/012092

shading of the surrounding urbanism. In dense street buildings, their effectiveness is considerably limited.

Acknowledgment
This contribution was supported by the Scientific Grant Agency of the Ministry of Education of the Slovak Republic and the Slovak Academy of Sciences in the project VEGA 1/0248/19.

References
[1] Nariadenie Európskeho parlamentu a Rady EÚ č. 305/2011
[2] Jäger-Waldau A 2019 PV status report 2019. *Publications Office of the European Union;. Luxembourg*
[3] Gholami H, Røstvik HN and Müller-Eie D 2019 Holistic economic analysis of building integrated photovoltaics (BIPV) system: case studies evaluation. *Energy and Buildings* **203** p109461.
[4] Jing Yang R and Zou P X W 2015 Building integrated photovoltaics (BIPV): costs, bene-fits, risks, barriers and improvement strategy *Int. J. of Construction Management* **16** (1) pp 39-53
[5] Nguyen K C, Katzfey J J, Riedl J and Troccoli A 2017 Potential impacts of solar arrays on regional climate and on array efficiency *Int. J. of Climatology* **37** (11) pp 53-64
[6] Armstrong A, Waldron S, Whitaker J and Ostle N J 2014 Wind farm and solar park effects on plante soil carbon cycling: uncertain impacts of changes in ground-level microclimate *Global Change Biology* **20** (6) pp 1699-1706
[7] Sorgato M, Schneider K, Rüther R 2018 Technical and economic evaluation of thin-film CdTe building-integrated photovoltaics (BIPV) replacing façade and roof top materials in office buildings in a warm and sunny climate *Renewable Energy* **118** pp 84-98
[8] Gholami H, Røstvik H N and Müller-Eie D 2019 Analysis of solar radiation components on building skins for selected cities In: *Proceedings of 14th conference on advanced building Skins Bern* Switzerland: *Advanced Building Skins. ABS* pp 541-49.
[9] Shukla A K, Sudhakar K and Baredar P 2016 A comprehensive review on design of building integrated photovoltaic system in: *Energy and Buildings* **128** pp 99-110
[10] Gholami H, Khalilnejad A and Gharehpetian G 2015 Electro-thermal performance and environmental effects of optimal photovoltaic thermal system *Energy Conversion Management* **95** pp 326-33.
[11] Gholami H, Sarwat A, Hosseinian H and Khalilnejad A 2015 Evaluation of optimal dual axis concentrated photovoltaic thermal system with active ventilation using Frog Leap algorithm *Energy Conversion Management* **105** pp 782-90
[12] Mohammadi F, Gholami H and Menhaj M B 2016 Effect of ventilation on yearly photovoltaic performance. *1st international conference on new research achievements in electrical and computer engineering* **1**. IEEE p 6.1.
[13] Agrawal B, Tiwari G 2010 Life cycle cost assessment of building integrated photovoltaic thermal (BIPVT) systems *Energy and Buildings* **42** (9) pp 1472-81
[14] Ibrahim A, Fudholi A, Sopian K, Othman MY and Ruslan M H 2014 Efficiencies and improvement potential of building integrated photovoltaic thermal (BIPVT) system *Energy Conversion and Management* **77** pp 527-34
[15] Tripathy M, Joshi H and Panda S 2017 Energy payback time and life-cycle cost analysis of building integrated photovoltaic thermal system influenced by adverse effect of shadow *Applied Energy* **208** pp 376-89
[16] Chiu M-S, Hou S-P, Tzeng C-T and Lai C-M 2015 Experimental investigations on the thermal performance of the ventilated BIPV wall *Journal of Applied Sciences* **15** (3) p 613-8
[17] Brito M, Redweik P and Catita C 2013 Photovoltaics and zero energy buildings: the role of building facades. In: presentea *28th European Photovoltaic Solar Energy Conference and Exhibition* Paris

IOP Conf. Series: Materials Science and Engineering 1015 (2021) 012092 doi:10.1088/1757-899X/1015/1/012092

[18] Freitas S and Brito M C 2019 Solar façades for future cities *Renewable Energy Focus* **31** p 73-9
[19] Esmailian E, Gholami H, Røstvik H N and Menhaj M B 2019 A novel method for optimal performance of ships by simultaneous optimisation of hull-propulsion-BIPV systems *Energy Conversion and Management* **197** p 111879
[20] Kaltschmitt M and Wiese A 2013 *Erneuerbare Energieteager in Deutschland: potentiale und Kosten*. (Springer-Verlag)
[21] Quaschning V 2000 Systemtechnikeiner klimavertraglichen Elek-triziteatsversorgung in Deutschland für das 21. Jahrhundert. (VDI-Verlag)
[22] Mainzer K, Fath K, McKenna R, Stengel J, Fichtner W and Schultmann F 2014 A high-resolution determination of the technical potential for residential-roof-mounted photovoltaic systems in Germany *Solar Energy* **105** pp 715-31
[23] Eurostat 2019 Electricity price statistics Available: https://ec.europa.eu/eurostat/statistics-explained/index.php/Electricity_price_statistics#Electricity_prices_for_household_consumers
[24] Mehreen Gul, Yash Kotak and Tariq Muneer 2016 Review on recent trend of solar photovoltaic technology *Energy Exploration & Exploitation* **34** (4) pp 485-526 (doi: 10.1177/0144598716650552)
[25] Shukla A K, Sudhakar K and Baredar P 2017 Recent advancement in BIPV product technologies: a review *Energy and Buildings* **140** p 188-95
[26] Azadian F and Radzi M 2013 A general approach toward building integrated photo-voltaic systems and its implementation barriers: a review *Renewable Sustainable Energy Reviews* **22** p 527-38
[27] Ceron I, Caamano Martín E and Neila F J 2013 'State-of-the-art' of building integrated photovoltaic products *Renewable Energy* **58** pp 127-33
[28] Shukla AK, Sudhakar K and Baredar P 2017 Recent advancement in BIPV product technologies: a review *Energy and Buildings* 140 p 188-95
[29] Saretta E, Caputo P and Frontini F 2018 A review study about energy renovation of building facades with BIPV in urban environment *Sustainable Cities and Society* **44** pp 343-55

IOP Conf. Series: Materials Science and Engineering 1015 (2021) 012093 doi:10.1088/1757-899X/1015/1/012093

Improvement of Larsen sheet pile retaining wall structures with the use of coarse-porous concrete

A A Mikhasek M V Rodionov and A A Bezuglov

Samara State Technical University, Academy of Architecture and Civil Engineering, Molodogvardeyskaya str. 194, 443001 Samara, Russia

E-mail address: andremixas@mail.ru

Abstract: The construction of hydraulic structures is associated with a significant amount of construction and installation work. Reducing the cost of building such structures is a vital task. The solution to this problem is to improve the design of the retaining structure by means of combining the sheet pile (flexible structure) and the gravity wall (massive structure) into a single whole. In this case it is assumed that the sheet pile provides protection against negative atmospheric effects on the massive wall and ensures the stability of the structure against displacement. At the same time the gravity wall is made of "low" quality materials, e.g. using waste products - products of concrete crushing and / or coarse-porous concrete on crushed sedimentary rock. The gravity wall reduces the active pressure on the sheet pile wall from the backfill side, and this allows using composite and PVC based sheet piles. This article deals with the modeling of retaining walls with various types of composite sheet piles, types of soil, free height and thickness of the gravity element.

1. Introduction

The construction of hydraulic structures is associated with a significant amount of construction and installation work. Reducing the cost of building such structures is a vital task. It is proposed to solve this problem by improving the design of hydraulic structures by means of using retaining walls.

The solution to this problem is to improve the design of the retaining structure by means of combining the sheet pile (flexible structure) and the gravity wall (massive structure) into a single whole. In this case it is assumed that the sheet pile provides protection against negative atmospheric effects on the massive wall and ensures the stability of the structure against displacement. At the same time the gravity wall is made of "low" quality materials, e.g. using waste products - products of concrete crushing and / or coarse-porous concrete on crushed sedimentary rock. The gravity wall reduces the active pressure on the sheet pile wall from the backfill side, and this allows using composite and PVC based sheet piles, with the free height of the wall over 2 meters.

Designing the improved retaining wall consists in finding the optimal choice of the sheet pile wall material and of the gravity wall thickness, depending on the height of the retaining wall and the characteristics of backfill soil;

When designing, the results of calculations of the stress-strain state of the improved structure are used, which are made on the basis of commonly known and tested methods - contact models or continuum models, taking into account the requirements [1].

IOP Conf. Series: Materials Science and Engineering 1015 (2021) 012093 doi:10.1088/1757-899X/1015/1/012093

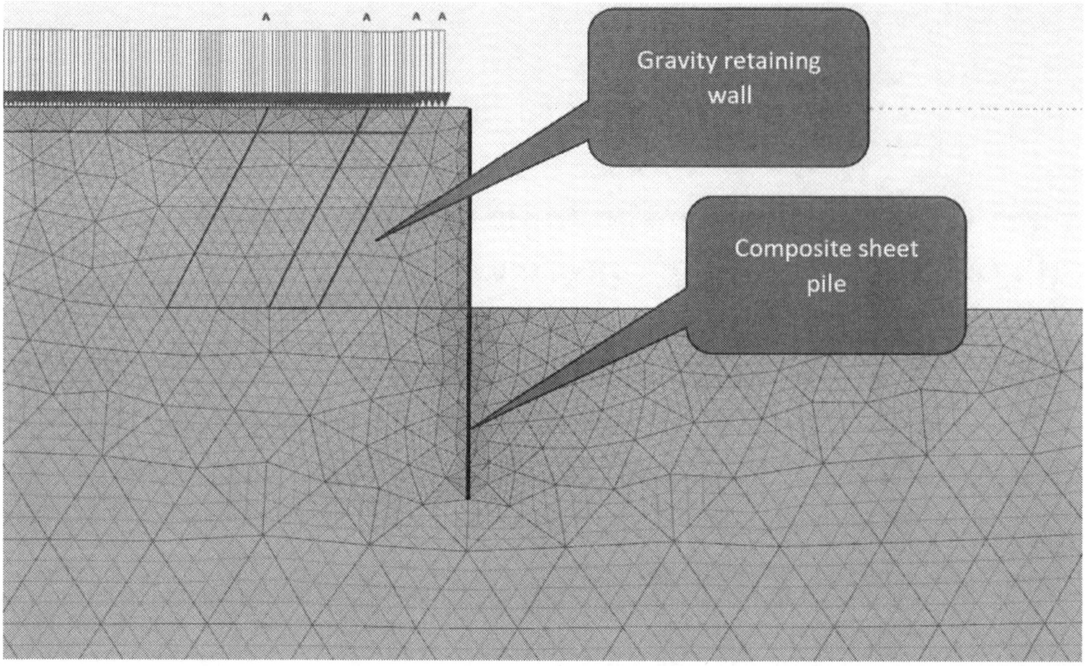

Figure 1. Improved structure's scheme.

This work has wide practical application, including on water bodies of the Russian Federation and beyond [2-4]. This research can be widely used in practice, for instance at water bodies in the Russian Federation, as well as in other countries [2-7].

2. Materials and methods

The general calculation in 2D-design has been made using PLAXIS 2D software package. PLAXIS 2D 2016 software is for complex calculations of the stress-strain state and stability of various purpose geotechnical objects, using finite element methods in a 2D problem. The software takes into account various features of geotechnical structures and construction processes, using debugged and theoretically justified calculation procedures. PLAXIS 2D is geotechnical software with an engineering user interface that offers flexible and unified geometry, realistic modeling of construction stages, error-adaptive and reliable calculation core and detailed processing of calculation results.

The calculation has been carried out by setting the stages of construction, in order to obtain the most accurate result. Each stage is a separate nonlinear calculation with iterative method for solving the stiffness matrix. The first stage is needed to obtain the initial stress values, that are present in the original earth body formed over many years. At this stage there are no displacements of model joints. At subsequent stages the recalculation of the deformations in the joints and the stress values in the elements is performed.

The software package uses methods for calculating the stability of slopes based on finite element method: the reduction method and the stress analysis method based on Coulomb's wedge theory.

The following boundary conditions have been accepted:

- free wall height H – from 2 to 6.0 m;
- gravity element width B – from 0.5 to 2.0 m;
- type of soil - sand and clay;
- type of sheet pile - GP-7 and ShK-200 UM;
- gravity element – coarse-porous concrete.

Soil characteristics in the calculation have been accepted as follows:

IOP Conf. Series: Materials Science and Engineering 1015 (2021) 012093 doi:10.1088/1757-899X/1015/1/012093

-Sand: $Y = 20.0$ kN/m³; $E = 28.0$ MPa; $c = 0.0$ kPa; $\varphi = 29.0$ °.
- Clay loam: $Y = 19.3$ kN/m³; $E = 11.0$ MPa; $c = 15.0$ kPa; $\varphi = 17.0$ °.
The calculated surface load for all cases has been set at $q = 24$ kN/m^2
The presence of coarse-porous concrete made it possible to exclude from the design scheme the impact of hydrostatic water pressure of backfill soil, since this kind of concrete works as drainage.
The task of modeling was to determine the depth of the sheet pile and the conditions for ensuring reliability, when minimum stability factor (F_{stab} (min)) was equal to 1.15.

3. Results

The calculations were made for 36 design cases, using the method of enumeration of indices H, B, type of soil and brand of sheet pile. The calculations determining the values of stress, deformation and the minimal factor of safety (the examples of calculations are presented on figures 2 and 3).

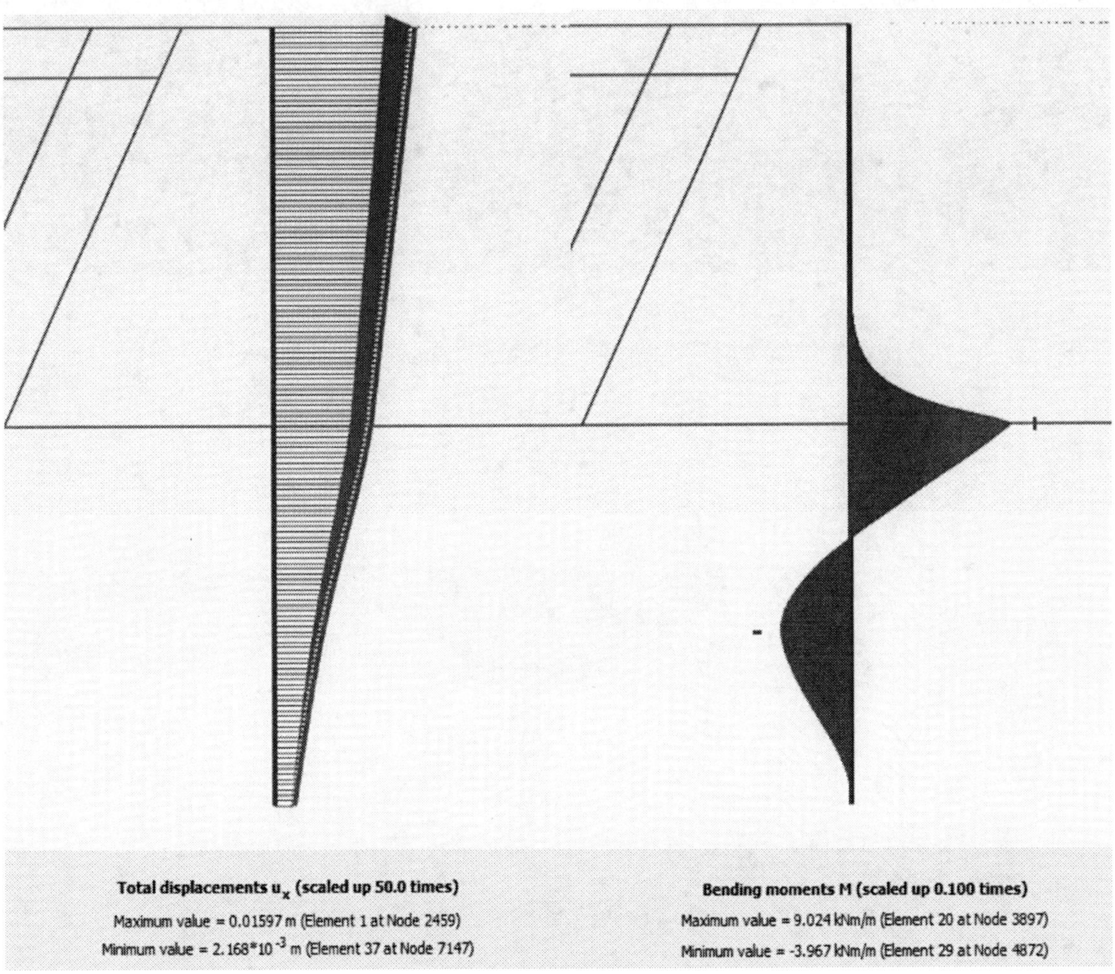

Figure 2. Examples of calculation results for sheet pile brand ShK 200 UM.

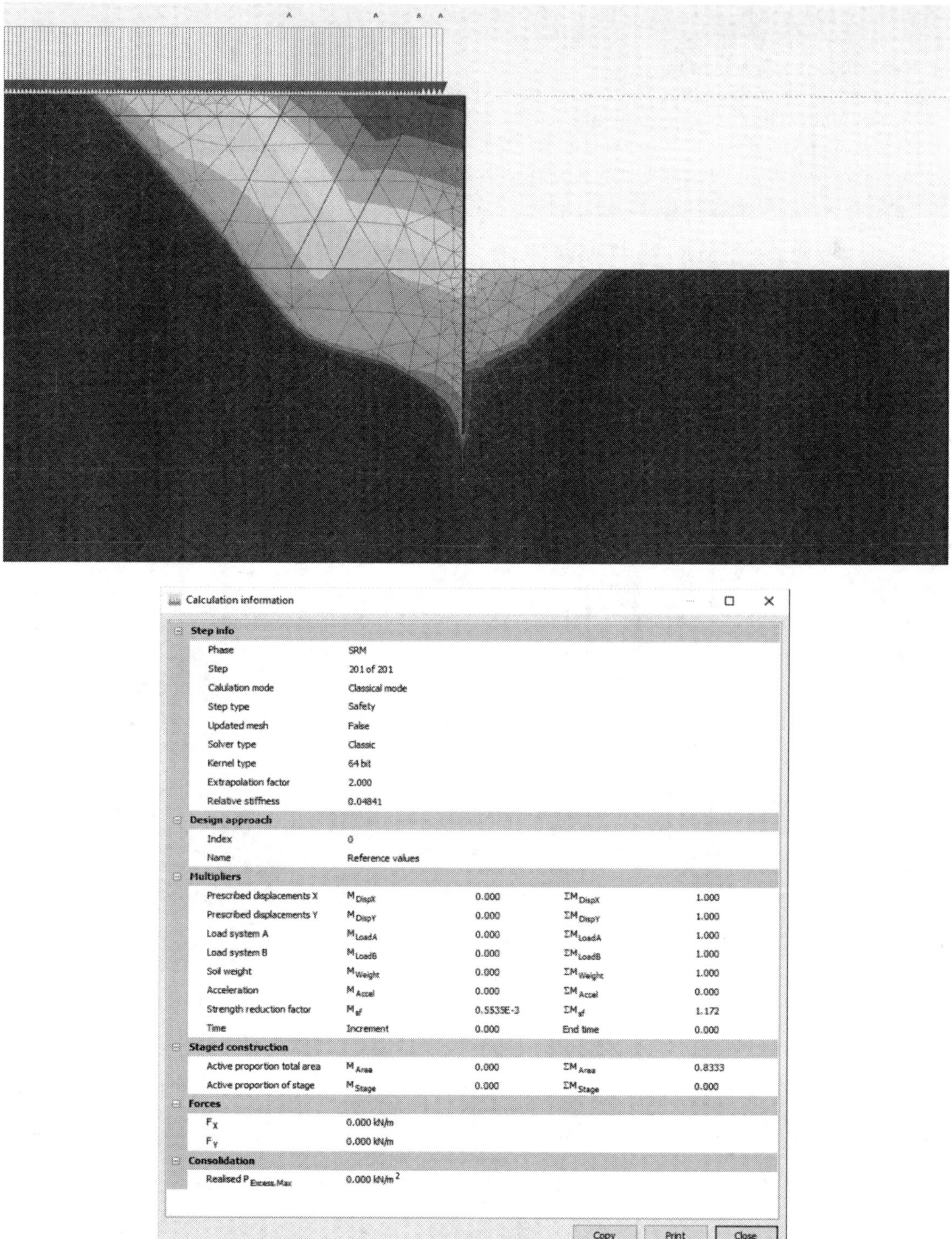

Figure 3. Examples of calculation results for sheet pile brand ShK 200 UM.

IOP Conf. Series: Materials Science and Engineering 1015 (2021) 012093 doi:10.1088/1757-899X/1015/1/012093

A summary table of the results of the calculation of the scheme of the combined structure of the sheet pile with the gravitational element made of large-porous concrete is given in table 1-3.The consolidated calculation results for the scheme of the complex structure consisting of sheet pile with gravity element made of coarse-porous concrete are presented in tables 1-3.

Table 1. Calculation results for the width of the element made of coarse-porous concrete – 0,5 m.

Free height, H (m)	Index	Obtained values for different kinds of soil and sheet pile materials			
		Sand		Clay	
		GP	ShK	GP	ShK
2.0	L (m)	3.9	3.9	3.2	3.3
	H_1 (m)	1.9	1.9	1.2	1.3
	d_x (mm)	17.44	15.97	20.15	21.59
	d_{xmax} (mm)	26.60	26.60	26.60	26.60
	M_y (kNm/m)	9.11	9.03	3.65	3.42
	F_{stab} (units)	1.168	1.172	1.159	1.157
4.0	L (m)	9.0	9.0	6.8	6.6
	H_1 (m)	5.0	5.0	2.8	2.6
	d_x (mm)	24.74	22.57	50.08	46.93
	d_{xmax} (mm)	53.30	53.30	53.30	53.30
	M_y (kNm/m)	43.51	47.73	9.21	8.87
	F_{stab} (units)	1.160	1.150	1.169	1.161
6.0	L (m)	14.8	14.8	10.4	10.1
	H_1 (m)	8.8	8.8	4.4	4.1
	d_x (mm)	69.21	63.87	74.24	77.66
	d_{xmax} (mm)	80.00	80.00	80.00	80.00
	M_y (kNm/m)	187.2	183.6	15.95	15.40
	F_{stab} (units)	1.177	1.178	1.192	1.166

[a] Where L – the total required length of the sheet pile, d_x – the horizontal displacements of the wall, M_y – the bending moment in the wall.

Table 2. Calculation results for the width of the element made of coarse-porous concrete – 1 m.

Free height, H (m)	Index	Obtained values for different kinds of soil and sheet pile materials			
		Sand		Clay	
		GP	ShK	GP	ShK
2.0	L (m)	3.5	3.5	2.8	2.8
	H_1 (m)	1.5	1.5	0.8	0.8
	d_x (mm)	12.32	10.96	18.12	16.69
	d_{xmax} (mm)	26.60	26.60	26.60	26.60
	M_y (kNm/m)	5.57	5.42	2.34	1.76
	F_{stab} (units)	1.169	1.167	1.166	1.174
4.0	L (m)	7.7	7.5	6.4	6.3
	H_1 (m)	3.7	3.5	2.4	2.3
	d_x (mm)	49.66	50.19	39.00	31.54
	d_{xmax} (mm)	53.30	53.30	53.30	53,30
	M_y (kNm/m)	18.08	17.91	6.68	6.61
	F_{stab} (units)	1.157	1.163	1.152	1.160

Free height, H (m)	Index	Obtained values for different kinds of soil and sheet pile materials			
		Sand		Clay	
		GP	ShK	GP	ShK
6.0	L (m)	12.2	12.2	9.7	9.7
	H_1 (m)	6.2	6.2	3.7	3.7
	dx (mm)	51.72	47.89	65.48	58.74
	dxmax (mm)	80.00	80.00	80.00	80.00
	M_y (kNm/m)	97.05	104.8	12.89	12.52
	F_{stab} (units)	1.160	1.158	1.158	1.165

Table 3. Calculation results for the width of the element made of coarse-porous concrete – 2 m.

Free height, H (m)	Index	Obtained values for different kinds of soil and sheet pile materials			
		Sand		Clay	
		GP	ShK	GP	ShK
2.0	L (m)	NA	NA	NA	NA
	H_1 (m)	0.0	0.0	0.0	0.0
	d_x (mm)	4.47	4.47	14.19	14.19
	d_{xmax} (mm)	26.60	26,60	26.60	26.60
	M_y (kNm/m)	-	-	-	-
	F_{stab} (units)	1.278	1.278	1.347	1.347
4.0	L (m)	6.3	6.3	5.0	5.0
	H_1 (m)	2.3	2,3	1.0	1.0
	d_x (mm)	29.54	27.78	27.26	27.17
	d_{xmax} (mm)	53.30	53.30	53.30	53.30
	M_y (kNm/m)	7.76	7.36	1.57	1.14
	F_{stab} (units)	1.163	1.166	1.206	1.211
6.0	L (m)	10.5	10.5	10.4	10.4
	H_1 (m)	4.5	4.5	4.4	4.4
	d_x (mm)	78.70	73.71	79.53	74.51
	d_{xmax} (mm)	80.00	80.00	80.00	80.00
	M_y (kNm/m)	13.39	12.32	13.65	12.18
	F_{stab} (units)	1.158	1.163	1.154	1,162

Basing on the analysis of the data in tables 1-3, the dependence of the value of sheet pile embedment on gravity element width, sheet pile free height, type of soil, etc. is obtained and presented on figures 4 and 5.

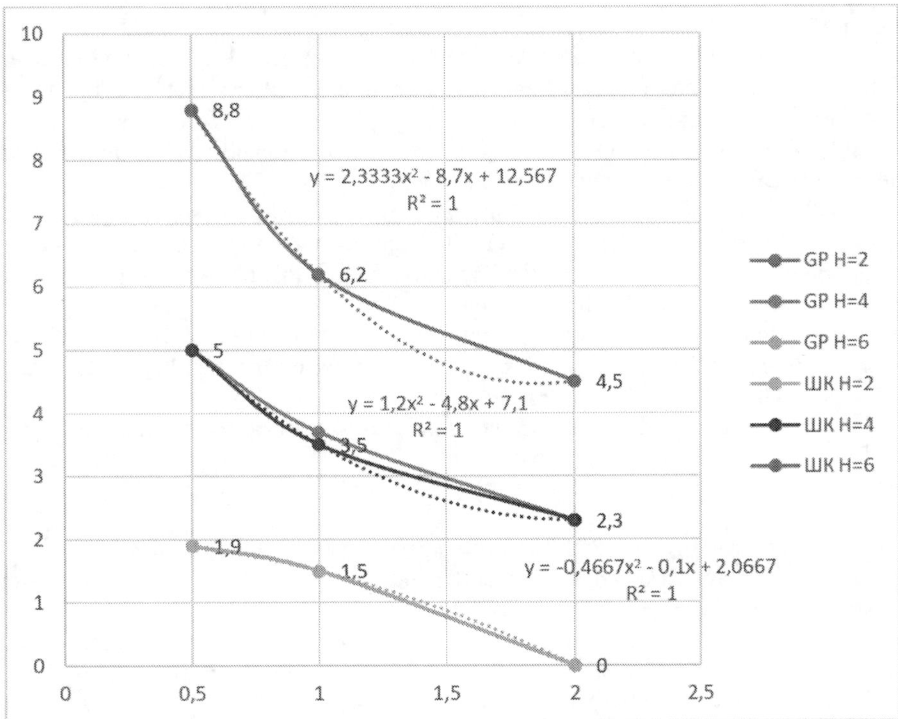

Figure 4. Dependence of sheet pile embedment depth H_1 for various brands of sheet pile, gravity element width B, free height H for sandy soil

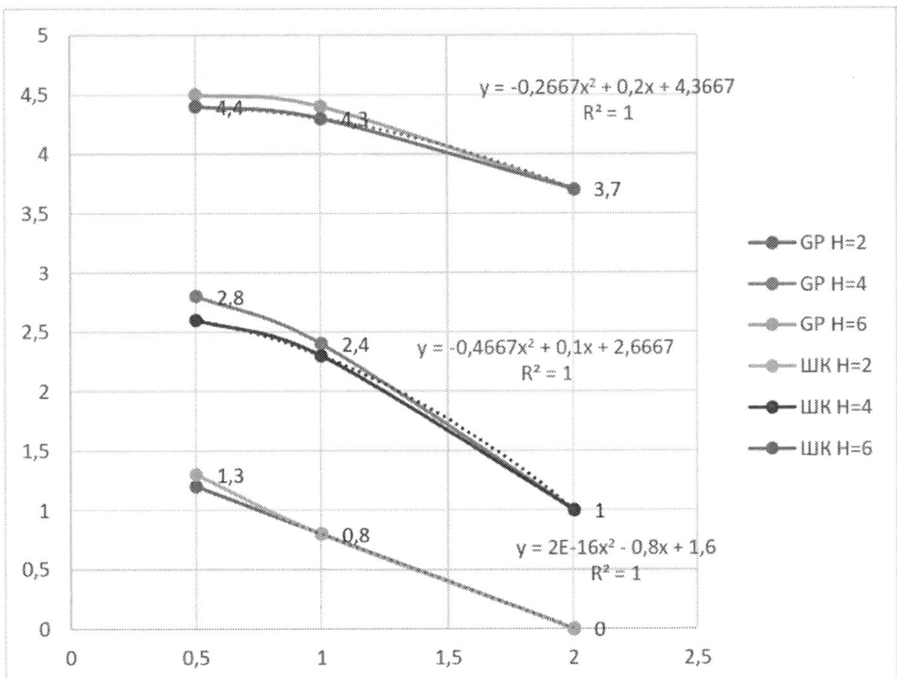

Figure 5. Dependence of sheet pile embedment depth H_1 for various brands of sheet pile, gravity element width B, free height H for clay

4. Conclusions

To reduce costs for hydraulic structures construction, it is recommended to use the improvement of the design of the retaining structure - combining the sheet pile (flexible structure) and the gravity wall (massive structure) into a single whole.

- The sheet pile provides protection against negative atmospheric effects on the massive wall and ensures the stability of the structure against displacement.
- The gravity wall is made of "low" quality materials, e.g. using waste products - products of concrete crushing and / or coarse-porous concrete on crushed sedimentary rock. The gravity wall reduces the active pressure on the sheet pile wall from the backfill side, and this allows using composite and PVC-based sheet piles with the free wall height of over 2 m;
-The type of sheet pile material ,when comparing different manufacturers of ShK and GP brands in the case of using a complex structure, has got practically no effect on the stability of the structure. In the future this factor can be excluded from consideration.
- In practice the sheet pile embedment depth into sandy soils must be greater than into clay which has adhesion.

References

[1] Code specification SP 381.1325800.2018 Retaining structures. Design rules.
[2] Mikhasek A A 2016 Using composite and synthetic materials for coast protection facilities construction *Scientific Review* **14** pp 67-72
[3] Mikhasek A A 2015 Peculiar features of coast protection facilities construction using composite sheet piles, *Traditions and innovations in construction and architecture. Construction: Collection of articles, ed. by M.I. Balzannikov, K.S. Galitskov, V.P. Popov. Samara State University of Architecture and Civil Engineering* Samara pp 210-2
[4] Mikhasek A A and Smyvalov A A 2015 Fundamentals of choosing a structure for a coast protection facility made of composite materials *Scientific Review* **14** pp 102-8
[5] Osório P, Odenbreit C and Vrouwenvelder T. 2010 Structural reliability analysis of quay walls with steel sheet piles *IANC MMX Congress Liverpool UK*
[6] Wall H and Wadsö L Corrosion rate measurements in steel sheet pile walls in a marine environment. 2013 *Lund University Publications. Marine structures* **33** pp 21-32
[7] Mikhasek A A 2017 Research on the technology of constructing a protection with joint use of no-fines concrete and bitumen *MATEC Web of Conferences* 117

IOP Conf. Series: Materials Science and Engineering 1015 (2021) 012094 doi:10.1088/1757-899X/1015/1/012094

Approaches to predicting the characteristics of fireproof materials

T Eremina, D Korolchenko and F Portnov

Moscow State University of Civil Engineering, Yaroslav shosse, 26, Moscow 129337, Russian Federation

E-mail address: wastingtimefilmart@gmail.com

Abstract. Planning an experiment is a new approach to research in which mathematical methods are given an active role. Based on the a priori information about the process under study, the researcher chooses some optimal research methodology. The results of the new approach to solving extreme problems depend to a large extent on the optimization of planning for a particular class of problems. The test plan presents a shortened plan of complete factor tests and can be considered acceptable due to the small correlation between affecting factors (AF) and output parameters (OP). The peculiarities of design of new composite materials to ensure safety in the operation of buildings and structures under conditions of man-caused and biogenic threats have been determined. The result of the article is the research using the method of forecasting the characteristics of fireproof materials.

1. Introduction

Many studies carried out in physics, chemistry and metallurgy come down to solving extreme problems aimed at finding optimal conditions for the processes or optimal choice of composition of multi-component systems. There are two different approaches to solving such problems. It may be required that the solution of extreme problems be preceded by a comprehensive study of both the process mechanism and the properties of substances. As a rule, systems to be optimized turn out to be so complex that they cannot be theoretically studied within a reasonable time. In most cases, extreme problems are solved experimentally with incomplete knowledge of the phenomena' mechanism.

Classical regression analysis is based on processing the results of so called "passive experiments". It is assumed that the researcher observes some uncontrolled spontaneously changing process or makes experiments in some arbitrary way, choosing experimental points in the factor space, based on intuition or any random circumstances.

Experience has shown that the classical regression analysis, despite a well developed theory, has not found any wide application for solving extreme problems in physics, chemistry and metallurgy. In solving such problems, we have to deal with a very large number of independent variables. In this case, the method becomes extremely cumbersome. Here, we have to deal with almost insurmountable difficulties associated, on the one hand, with the need to set a very large number of experiments and, on the other hand, with difficulties associated with the interpretation of the regression equation.

Planning an experiment is a new approach to research in which mathematical methods are given an active role. Based on the a priori information about the process under study, the researcher chooses some optimal strategy to manage the experiment. The research process is usually divided into separate stages. After each step, the researcher receives new information that allows them to change the research strategy [1-3].

2. Building mathematical model

In mathematical language, the task of planning the experiment is formulated as follows: at each stage of the study, the researcher has to choose the optimal, in some sense, location of points in the factor space in order to get some idea about the response surface. The choice of the optimal criterion is to a large extent arbitrary. Here, it is necessary to take into account both the setting of the problem by the experimenter and the real situation in which this problem has to be solved. In setting extreme experiments, it is natural to formulate the problem at the first stage of the study as follows: it is necessary to find the direction of movement to the area where the process conditions are optimal. To solve this problem, it is enough to investigate the response surface in a small area, limited to linear approximation. After reaching the area where the optimum is located the problem will be formulated differently. Here, the researcher needs to get a much better idea of the response surface by approximating it with polynomials of the second and sometimes even third order. In many cases, the research has to start with the so-called screening experiments, the purpose of which is to identify the dominant effects among a very large number of the potentially possible.

Experience shows that the success of the new approach to solving extreme tasks largely depends on how well we have been able to formulate an optimal planning criterion for solving a particular class of tasks.

The main idea of the new method - possibility of optimal experiment management with incomplete knowledge - is relative to those prerequisites on which the cybernetics is based. Appearance of cybernetics - science of control, including consideration of such complex objects as biological systems, only became possible after it was realized that optimal control in case of incomplete knowledge is possible [4-10].

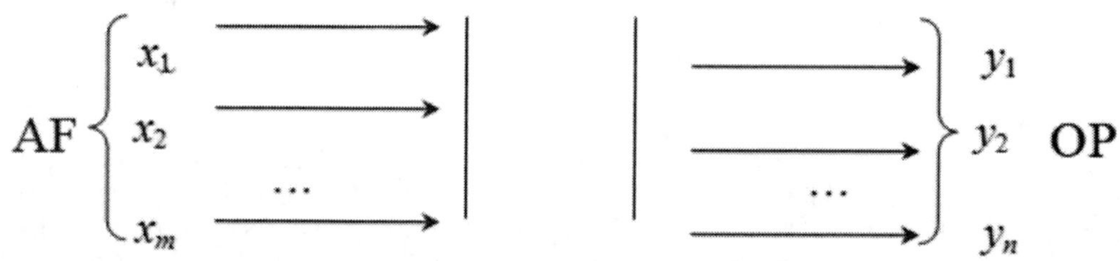

Figure 1. Presentation of the investigated complex system as a "black box" with m inputs and n outputs.

To build a mathematical model of a complex multifactor multiparameter system (figure 1), we need to perform the following actions:

1. make a list of m most significant affecting factors (AF) $[x_1,...,x_m]$ and n most informative output parameters (OP) $[y_1,..., y_n]$;

2. make a plan of active multifactorial tests in the form of matrix **X**, containing m columns (by number of AF) and n rows (by number of tests), the main requirements to which are:

a) the absence of correlation between AF (the coefficient of paired correlation r_{xkl} between factors x_k and x_l must be close to 0);

b) full coverage of the factor space (at least, the condition: $n > m$ must hold);

c) feasibility - i.e. compliance with the capabilities of the experimental base;

d) all experiments (AF combinations) in matrix **X** are equal.

3. carry out active tests, in the course of which to vary the AF combinations according to the plan (matrix **X**) and to determine (measure) the AF values, thus forming the matrix **Y** containing n rows (by

IOP Conf. Series: Materials Science and Engineering 1015 (2021) 012094 doi:10.1088/1757-899X/1015/1/012094

the number of tests) and n columns (by the number of OP). At the same time, the condition of unambiguousness of the result must be observed, i.e., when repeating the experiment (reproducing the same combination of AF), the spread of OP values must be insignificant;

4. perform mathematical processing of the results of active tests, which implies:

a) determining the relationship between OPs by finding paired correlation coefficients between OPs (the values of r_{ykl} between the y_k and y_l parameters must be close to 0, otherwise one of the OP y_k or y_l can be replaced by another OP);

b) estimating the correspondence of a sample of experimental values of each j-th OP$[y_{j1},\dots,y_{jN}]$ to the normal (Gaussian) distribution, in particular, by coefficients of asymmetry A'' and excess E_x (i.e., the condition $A_s = E_x = 0$ must be observed);

c) building an adequate mathematical model

$$y_j = f_j\left(x_1,\dots,x_m\right) \quad j\in[1,n] \tag{1}$$

which in this paper will take the form of a quasi-linear regression equation:

$$y_j \approx \sum\nolimits_{k=1}^{M_j} a_{jk} Z_{jk} \quad j\in[1,n] \tag{2}$$

where a_{jk} is the desired regression coefficient which is a component of vector $\mathbf{A}_j$,

z_{jk} is the k-th conditional factor, which is a component of the $\mathbf{Z}_j$ matrix and represents a function from BФ $x_1,\dots x_m$;

M_j is the number of regression coefficients or conditional factors ($M_j < N$).

d) using regression equations (2) for applied purposes:

- interpretation of OP dependence on AF;
- prediction of OP values in combinations of AF, different from those included in matrix **X**;
- estimation of the significance of the influence of AF on OP;
- construction of the working area on the set of AF, in which each j–th OP lies within permissible limits.

Conditional factors $\{z_{jk}\}$ are selected by the forced search method during the construction of the regression equation (2), and the vectors of the regression coefficients $A_1,\dots,A_m$ are calculated on the basis of the minimum dispersion condition of the regression equation (the least squares method)

$$D_j = (N-M_j)^{-1}\sum\nolimits_{i=1}^{N}(y^e_{ji} - y^b_{ji})^2 \to min, j\in[1,n] \tag{3}$$

where y_{ji}^e, y_{ji}^b – values of j-th OP, respectively obtained in the course of the i-th experiment and calculated from the regression equation (2) for the i-th combination of AF.

The adequacy of regression equations (2) can be estimated by the Fisher criterion, for which we need to calculate the value of

$$F_j = \frac{D_{jo}}{D_j} \quad j\in[1,n] \tag{4}$$

where D_{jo} – experiment dispersion calculated by expression

$$D_{jo} = (N-1)^{-1}\sum\nolimits_{i=1}^{N}(y^e_{ji} - y_{jav})^2 \tag{5}$$

$$y_{jav} = N^{-1}\sum\nolimits_{i=1}^{N} y^e_{ji} \quad j\in[1,n] \tag{6}$$

If the value of F_j calculated by expression (4) is greater than the tabular value at degrees of freedom N-1 and N-M_j, the equation of regression (2) is considered adequate with the corresponding confidence probability α and can be used to solve the above applied problems. It is possible to take with engineering accuracy $\alpha = 95\%$ at $F_j > 10$.

For the quasi-linear regression equation (2), condition (3) is reduced to matrix-vector expression:

$$\mathbf{A}_j = (\mathbf{Z}_j^{\mathrm{T}}\mathbf{Z}_j)^{-1}\mathbf{Z}_j^{\mathrm{T}}\mathbf{Z}_j \qquad j \in [1, n] \tag{7}$$

where $\mathbf{Z}_j$ – the conditional factor matrix, containing N rows (by the number of tests) and M_j columns (by the number of conditional factors);

^{T} - symbol of matrix transposition;

$\mathbf{Y}_j$ – N-dimensional vector of the values of the j-th OP.

Calculation of vector $\mathbf{A}_j$ as well as selection of conditional factors $[z_{j1},..., z_{jMj}]$ presents some difficulty overcome by using computer technologies.

It seems also expedient to use the multi-model principle, according to which the dependence of j-th OP on AF can be described not by one adequate equation, but by several such equations.

It is necessary to construct a mathematical model of a complex system in the form of quasi-linear regression equation (2), containing five OPs (n = 5), which are influenced by seven AF (m = 7). For this purpose was built a specialized plan of nine tests (N = 9) - matrix **X**, given in Table 1.

The test plan (Matrix **X**, Table 1) presents a shortened plan of complete factor tests. According to this plan, the system was tested, resulting in a matrix **Y**, shown in table 1.

Table 1. Tests plan and results

№ i	Matrix X							Matrix Y				
	x_1^*	x_2^*	x_3^*	x_4^*	x_5^*	x_6^*	x_7^*	Y_1^{**}	Y_2^{**}	Y_3^{**}	Y_4^{**}	Y_5^{**}
1	0.5	0.5	0.5	0.5	0.5	0.5	0.5	12	2	1	0	6.9
2	1.5	1.5	0.5	0.5	1.5	1.5	0.5	12	1	5	1	2.0
3	1.5	0.5	1.5	0.5	1.5	0.5	1.5	27	1	7	1	2.0
4	0.5	1.5	1.5	0.5	0.5	1.5	1.5	37	2	1	0	6.9
5	1.5	0.5	0.5	1.5	0.5	1.5	1.5	30	1	5	1	3.4
6	0.5	1.5	0.5	1.5	1.5	0.5	1.5	20	2	3	2	5.6
7	0.5	0.5	1.5	1.5	1.5	1.5	0.5	40	2	1	2	5.6
8	1.5	1.5	1.5	1.5	0.5	0.5	0.5	36	1	5	1	3.4
9	1.5	1.5	1.5	1.5	1.5	1.5	1.5	25	1	7	3	6.9
x_{max}	24	12	28	6	3	14	12	A_s=-0.202	0.178	-0.064	0.35	-0.187
x_{min}	10	7	21	1	1	5	5	E_x=-1.59	-2.17	-1.81	-1.11	-1.86

* - formulation components of a flame retardant,

** - y_1-intumescence ratio, y_2- adhesion, y_3- service life, y_4- hazard class, y_5- flushability of film.

Pre-processing of **X** and **Y** matrices allowed determining the following pair correlation coefficients: r_{x12}=r_{y12}=0.074 r_{y13} = -0.136; r_{y14} = 0.102; r_{y15} = 0.097; r_{y23} = -0.917; r_{y24} = -0.217, r_{y25} =0.692; r_{y34}=0.428; r_{y35} = -0.577; r_{y45}=0.107, r_{x12} = r_{x13} = r_{x14} = r_{x15} = r_{x16} = r_{x17} = r_{x23} = r_{x24} = r_{x25} = r_{x26} = r_{x27} = r_{x34} = r_{x35} = r_{x36} = r_{x37} = r_{x45} = r_{x56} = r_{x57} = r_{x67} = 0.1.

Thus, the test plan can be considered acceptable due to the small correlation between the AF. But among some OPs there is a significant correlation – y_2 and y_3 are closely connected, which allows one of them to be removed from consideration, thus simplifying the system analysis. The same conclusion can be made with respect to OP y_5, which is quite closely connected with OP y_2 and y_3. The values of coefficients A_s and E_x were found (table 1), which indicate some deviation of experimentally obtained values of OP from the normal law. Nevertheless, adequate quasi-linear regression equations (2) were obtained for all OP.

Table 2. Comparison of experimentally obtained values of the 1st system OP with the results of calculations using regression equations

№	y_1^e	y_{1a}^b	y_{1d}^b	y_{1b}^b	y_{1g}^b	y_2^e	y_{2a}^b	y_3^e	y_{3a}^b	y_{3d}^b	y_{3b}^b
1	12	16.8	14.7	11.2	11.8	2	2	1	1.63	1.33	0.95
2	12	6.98	14.6	11.9	11.8	1	1	5	4.90	5.40	4.93
3	27	27.0	27.1	26.2	25.8	1	1	7	6.52	5.87	6.99
4	37	35.3	39.9	40.1	36.8	2	2	1	2.17	1.49	1.02
5	30	30.2	28.3	31.5	29.4	1	1	5	4.77	5.20	4.93
6	20	20.8	20.2	20.1	19.7	2	2	3	3.38	3.41	3.01
7	40	39.5	37.0	35.7	40.0	2	2	1	1.77	1.44	1.02
8	35	35.3	34.2	34.6	35.8	1	1	5	5.31	5.34	5.16
9	25	25.0	24.3	25.8	26.4	1	1	7	6.52	6.76	6.99
	-	14.2	17.9	16.7	92.7	-	>10000	-	13.6	18.1	1004

№	y_4^e	y_{4a}^b	y_{4d}^b	y_{4b}^b	y_{4g}^b	y_{4e}^b	y_5^e	y_{5a}^b	y_{5d}^b	y_{5b}^b	y_{5g}^b
1	0	0.25	0.046	-0.017	0.018	0.054	6.9	5.62	7.01	7.07	6.90
2	1	0.76	0.739	0.732	1.01	0.984	2.0	1.93	1.83	1.73	1.72
3	1	1.13	0.982	1.05	1.03	1.01	2.0	1.93	1.47	1.92	2.20
4	0	0.36	0.316	0.145	0.225	-0.057	6.9	7.20	7.27	6.87	6.94
5	1	0.76	0.710	0.713	1.03	1.02	3.4	3.46	2.94	3.35	3.36
6	2	2.28	1.90	1.91	2.00	2.00	5.6	6.14	5.26	5.63	5.68
7	2	2.16	2.03	2.04	1.79	2.08	5.6	6.14	6.27	6.15	5.75
8	1	1.13	1.15	0.898	0.966	0.919	3.4	3.46	2.55	3.12	3.48
9	3	2.56	3.09	3.16	3.09	2.96	6.9	6.65	6.70	6.46	6.65
F_j	-	10.1	22.3	25.1	45.2	180.7	-	12.6	12.8	31.0	78.4

From the analysis of table it follows, that in particular. that the factors x_3 and x_4 have a positive effect on OP y_1 (i.e. with their increase OP y_1 also increases), and AF x_5 has a negative effect. As for OP y_2, it was found that it is only affected by AF x_1, and the effect is negative. The greatest positive influence on OP y_3 is produced by AF x_1, and the greatest negative influence is produced by AF x_6; the greatest positive influence on OP y_4 is produced by AF x_3, x_4 and x_5; the greatest negative influence on OP y_1 is produced by AF x_1, and the greatest positive influence is produced by OP x_2, x_5, x_6 and x_7.

As a result of the data confirmed by calculations: the optimal formulation of long-lasting weather-resistant fire retardant coating for metal structures has the following properties: life - 10 years, adhesion to the primed surface - 1 point, multiplier - 54 mm, the average density is 1.300 kg/m^3, the drying time is 24 hours, the resistance to the effects of variable temperatures is 15 cycles, the strength of the bend is 2 mm, the strength for the impact is 40 cm, the shelf life is 1.5 years, the fire retardant efficiency time is 45 minutes according to the NSA 236-97 (4 group) with an average coating thickness of 1 mm.

3. Conclusions

As a result of the study, methods were chosen to predict the characteristics of fire retardant materials to determine the main properties of fire retardant materials, including toxic effects on the human body.

Based on the calculations carried out using the mathematical planning method of the experiment, quantitative optimization of the composition of the compositions was made and the fire retardant formulation was developed, the characteristics of which are shown above.

A mathematical multifactor model has been developed, experimentally by the calculation method to quantify the relationship between the influence and output parameters, as shown in the results of calculations.

IOP Conf. Series: Materials Science and Engineering 1015 (2021) 012094 doi:10.1088/1757-899X/1015/1/012094

Further research may allow us to assess the performance of durability and fire retardant properties of coatings with the best performance of fire retardant efficiency.

Acknowledgments
This work was financially supported by the Ministry of Science and Higher Education of the Russian Federation (Project: Theoretical and experimental design of new composite materials to ensure safety during the operation of buildings and structures under conditions of technogenic and biogenic threats #FSWG-2020-0007).

References
[1] Telichenko V I and Roitman V M 2010 *Ensuring the resilience of buildings and structures with combined special impacts involving fire is a basic element of the integrated safety system*
[2] Roitman V M 2009 Resilience of buildings and structures against progressive collapse with combined special impacts involving fire *Moscow State University Herald* **2** p 37-59
[3] Roitman V M 2009 *Basics fire safety high-rise buildings* (Moscow: MSSU) p 107
[4] Nalimov V V 1960 *Application of mathematical statistics in substance analysis* (Moscow: Fizmatgiz) p 430
[5] Nalimov V V 1971 *Theory of Experiment* (Moscow: Nauka - Fizmatgiz Library of the Engineer) p 208
[6] Brodsky V Z 1976 *Introduction to factor planning of experiment* (Moscow: Nauka) p 423
[7] Born M and Kun H 1958 The dynamic theory of crystal lattices (Moscow: Foreign Literature Publishing House) p 488
[8] Akhnazarova S L and Kafarov V V 1985 Methods of optimization of experiment in chemical technology (Moscow: Vyshch.shk) p 327
[9] Nalimov V V and Chernova N A 1965 *Statistical methods for planning extreme experiments* (Moscow: Nauka) p 338
[10] Markova E V and Lisenkov A N 1945 *Combinatorial plans in problems of multifactor experiment* (Moscow: Nauka) p 345

IOP Conf. Series: Materials Science and Engineering 1015 (2021) 012095 doi:10.1088/1757-899X/1015/1/012095

Research of a new flame retardant material with set parameters of operational properties taking into account features of glued wooden structures

D Korolchenko[1] **and L Rodionova**[2]

[1] Moscow State University of Civil Engineering, Yaroslav shosse, 26, Moscow 129337, Russian Federation
[2] International Scientific Innovation Center of Construction and Fire Safety Ltd, Moscow, Dmitrovskoe shosse, 46 build.2, 127238, Russian Federation

E-mail address: da-vip@mail.ru

Abstract. The questions of peculiarities of using fireproof composition for glued wooden structures are considered. The results of testing the fire-protective efficiency of the composition, the influence on reducing the combustibility and flammability of wooden structures, compatibility with different glue components of glued wooden structures are given. The results of performance properties estimation are described

1. Introduction

In modern construction, despite the emergence of many different kinds of building materials, environmentally friendly wood still remains relevant. Currently, the use of wood in the construction of buildings and structures mainly comes down to the use of high-tech glued wooden structures.

By glued wooden constructions we mean constructions, whose main cross section consists of glued layers of boards joined together in length and glued in width. Gluing elements of wooden constructions with different polymeric adhesives expands engineering and architectural possibilities of wood application. Glued wooden structures can overlap the spans inaccessible to other materials, the distance between the supports of glued structures can reach 40 m, laminated - up to 55 m, bent-glued frames up to 70 m. Even entire stadiums, swimming pools and water parks are covered with glued wooden structure systems.[2-3].

However, any wooden structures are still among the combustible materials that spread fire quickly, which causes significant damage to buildings and structures and significantly reduces their fire safety in general. The study of fire resistance of glued wooden structures began in the 1970s due to the fact that glued and bent-glued wood began to be widely used as supporting structures in construction. [4] The issue of fire safety of such structures has become a priority and remains relevant today.

Recently, the demand for flame-retardant treatment of glued wooden constructions has especially increased, as their installation is carried out without confirmation of performance characteristics by certification bodies. Subsequent verification by fire authorities finds that it is necessary to ensure the compliance of glued wooden structures with fire safety requirements. Another difficulty is the fact that it is often impossible to identify glued wood samples and assess their fire resistance limit. Therefore, it is necessary to use fire protection coatings to reduce the fire hazard of glued wooden structures.

IOP Conf. Series: Materials Science and Engineering 1015 (2021) 012095 doi:10.1088/1757-899X/1015/1/012095

The first solutions concerning fire protection compositions for glued wooden structures were liquid-silicate pastes-slippings with asbestos, vermiculite or clay filling. In this way it was only possible to increase the fire resistance limit to 30 and 45 minutes. Nowadays, in order to increase the fire resistance limit of glued wooden structures, constructive fire protection has become the most widespread, namely: sheathing with mineral-based slab materials and fibrous fireproof mats with subsequent cladding. Essential drawbacks of this solution are the complexity of installation, high consumption and cost of passive fire protection, additional costs for cladding and decorative properties.

The use of the latest impregnation compositions and standard fire protection paints does not essentially affect the fire resistance limit of glued wooden structures, however, their use allows to increase the time of ignition and reduce the spread of flame over the structure.

2. Progress of work

Development of a complex fireproof coating on an organic basis for glued wooden structures allows to maintain operating properties under conditions of open atmosphere, temperature changes, high humidity and increase the reliability of the coating for the period up to 25 years.

The mechanism of fire protection of existing analogues of coatings is based on bloating, so it is necessary to have a fairly thick layer of coating and high consumption, to ensure a sufficient amount of foamcoke.

The action of the complex fire protection coating is based on the non-combustible properties due to liquid chlorinated organic phosphate. Coatings containing chlorinated phosphoric acid esters have high heat resistance, hydrolysis resistance, low viscosity, excellent resin compatibility.[7]

Due to the introduction of organic chlorinated phosphate, glued wood treated with a complex flame retardant coating has the properties of a non-combustible and non-flammable material. At the same time, the chemical structure of chlorinated flame retardant does not reduce the depth of penetration of the complex flame retardant coating into the treated wood and thus is securely fixed on it. Due to this unique development the complex flame retardant coating has low consumption and low required thickness of the required coating.

The efficiency of phosphorus compounds as flame retardants is linked to the following factors:

- the specific effect of phosphorus compounds on the combustion processes in the gas phase;
- the formation of a surface molten layer of polymetaphosphoric acid on the surface of glued wood, which is a barrier to heat transfer from the flame to the polymer and for the diffusion of oxygen to the surface, and pyrolysis products - into the flame.[8]

Copolymer of styrene with methacrylic acid is used as a film-forming substance in the formulation. Its polymeric matrix in interaction with liquid flame retardant leads to slowdown of the thermal destruction process, reduction of the amount of volatiles emitted, strengthening of carbonization and formation of carbonized residue resistant to oxidation, with high fire and heat protection properties. Fire retardant effect of silicon dioxide in the complex fireproof coating is due to the release of water vapor during decomposition in the flame, which leads to cooling the combustion zones and the formation of oxide film on the burning surface. Coverings filled with silicon dioxide have transparent matte texture that does not reveal the surface of glued wood.

Complex fireproof coating also has fungicidal properties due to the use in the formulation of additives based on oktilisothiazolinone and 3iodo-2-pinyl-butyl carbamate, and can be used on a variety of wooden surfaces and materials based on them, guaranteeing a long service life of coating, up to 25 years. In the course of development, the compatibility of the complex fireproof coating with the polymer adhesives available on the market for the production of glued structures has been established, the complex fireproof coating has low consumption and high technological efficiency in application.

Thus, as a result of the interaction of a mixture of flame retardants in the formulation of the flame retardant material, the processes of reducing the spread of flame and the formation of stable thermodynamic systems prevail over the reactions of thermal destruction, which reduces the rate of

IOP Conf. Series: Materials Science and Engineering 1015 (2021) 012095 doi:10.1088/1757-899X/1015/1/012095

release of volatile compounds and toxicity of pyrolysis products and increases the fire resistance limit of glued wooden structures. Also, due to the combination of flame retardants of different effect in the formulation, the flame retardant material has low consumption and the coating thickness required for protection, while having high decorative and aesthetic properties (figure 1).

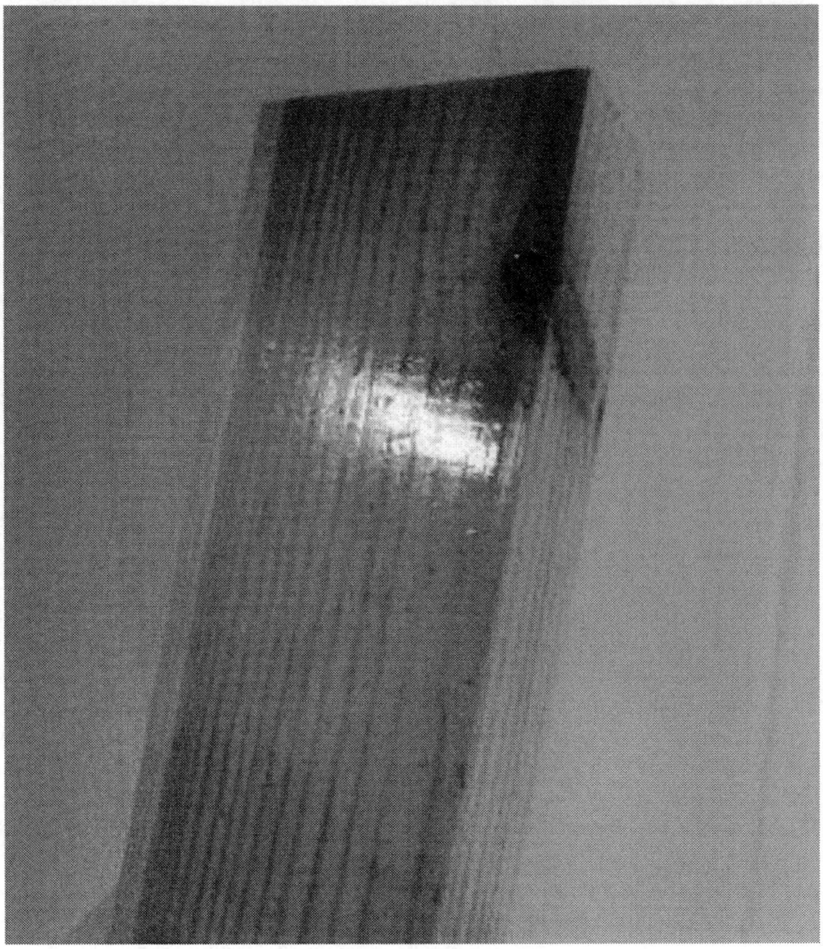

Figure 1. Appearance of new fireproof material.

3. Methods of testing

One of the main technical tasks was the development of a flame retardant material compatible with various adhesive components of a wood compound, as the polymer glue can influence the flame retardant properties, and a flame retardant material on an organic basis - on cohesion in an adhesive layer.

Compatibility of three types of glued wooden structures - laminated, solid and combined - with various adhesive layers was assessed. There are five types of polymer adhesives according to the type of resin component: phenol-formaldehyde, phenol resorcin-formaldehyde, aminoplastic, polyurethane, emulsion polymer-isocyanate (EPI-glues). [9] The latter corresponds to both domestic and some foreign standards for the production of glued woodblock. Emulsion polymer-isocyanate adhesive does not contain formaldehyde and belongs to the safest group of adhesives, it is universal and transparent.

Evaluation of the compatibility of the flame retardant material with various adhesive components was carried out visually, by fixing changes in the appearance of the glue joint and the structure of the

IOP Conf. Series: Materials Science and Engineering 1015 (2021) 012095 doi:10.1088/1757-899X/1015/1/012095

glued wooden structures in general. Compatibility of final adhesive films with fire protection material was also tested. According to the results of the compatibility testing and of trsting the effect of flame retardant material on the glue joint of the glued wooden structure it was concluded that the coating does not affect the strength properties of the glue, and the glue in its turn does not affect the fire retardant properties.

Fire tests of treated glued structures were carried out in accordance with GOST 30402-96 with the assessment of the critical surface density of the thermal flow (SFD) of the ignition process, flame propagation over the material surface with the determination of the flame propagation index (FPI) in accordance with GOST 12.1.044-89 p. 4.19.

When determining the flammability parameters, record was taken of the time and place of ignition, the nature of destruction of the sample under the influence of thermal radiation and flame, as well as of melting, swelling, delamination, cracking, swelling or shrinkage of the exposed surface. Based on the results of determining the ignition time for the samples under the influence of external heat flux of various intensities of 20, 30 and 40 kW/m^2, the values of critical surface heat flux density (CSHFD) were determined. CSHFD is characterized by the minimum value of the heat flux density, at which the samples burn with a steady flame.

When assessing the flame propagation index value, it was important to determine the time of flame passage by the front of each section of the sample surface, the temperature of the off-gases, the time of achieving maximum temperature values, the flame propagation velocity over the sample surface.

Determination of the flame retardant efficiency index of the flame retardant material was carried out during the mass loss tests of the sample in accordance with GOST R 53292-2009. As samples were used glued wooden structures made of laminated veneer lumber, treated with flame retardant material, with the consumption specified in Table 1.

Further, the operational properties of the coating were investigated. Samples of flame retardant material were tested for water absorption of the final film, moisture absorption of the loose film in a humid atmosphere was determined, and parallel testing was carried out in a climatic chamber for resistance to conditions of moderately cold climate.

To confirm the compliance with the declared area of application, the resistance of the flame retardant material to damp environment was tested experimentally. Water absorption of the film was determined without destruction by 24-hour liquid exposure of the sample glued wooden structure treated with flame retardant material.

In addition, moisture absorption of a loose film of the flame retardant material in a humid atmosphere was determined. The climate of swimming pools and water parks is characterized by the value of humidity of 50-60%. For our experiment we artificially created the humidity of 95% by filling the exciter with a saturated solution of sodium sulphide. The experiment was carried out by comparing the weight of a single film before the test and after conditioning for 24 hours in a humid atmosphere.

Resistance to environmental factors is also one of the main parameters required for a flame retardant material. Accelerated climatic testing of samples was carried out using a climatic chamber by means of formation of modes of temperature and humidity differences corresponding to conditions of a moderately cold climate during several cycles. In the course of tests changes in decorative and protective properties were recorded.

4. Test results

As a result of the test for the flame retardant material the flammability group - B1 was determined. Figure 2 demonstrates the appearance of samples after the test with flame retardant treatment (b) and without it (a).

According to the results of the test for the flame propagation index, the flame retardant material was identified as not spreading the flame over the surface. Figure 2 shows the samples of glued structures treated with the flame retardant material after the flame propagation test in accordance with GOST R 51032-97.

IOP Conf. Series: Materials Science and Engineering 1015 (2021) 012095 doi:10.1088/1757-899X/1015/1/012095

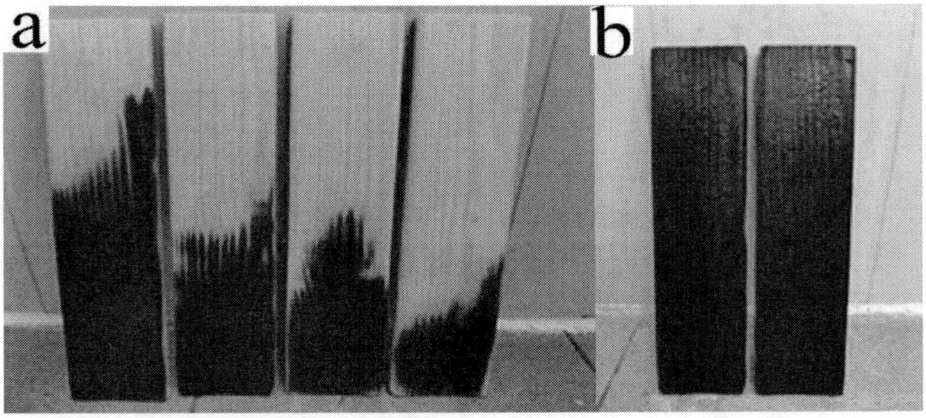

Figure 2. appearance of samples after the test: a - without flame retardant treatment; b - without flame retardant treatment

The weight loss test showed that the flame retardant material provides Group I fire protection efficiency, weight loss of the samples does not exceed 8.5%.

Table 1. Test results according to GOST R 53292-2009.

	Number of layers	Consumption (gr/m^2)	Sample mass loss (%)
Flame retardant material	3	340	8.3
	4	360	7.2
	3	335	8.5
	4	350	8.1
	3	375	8.1

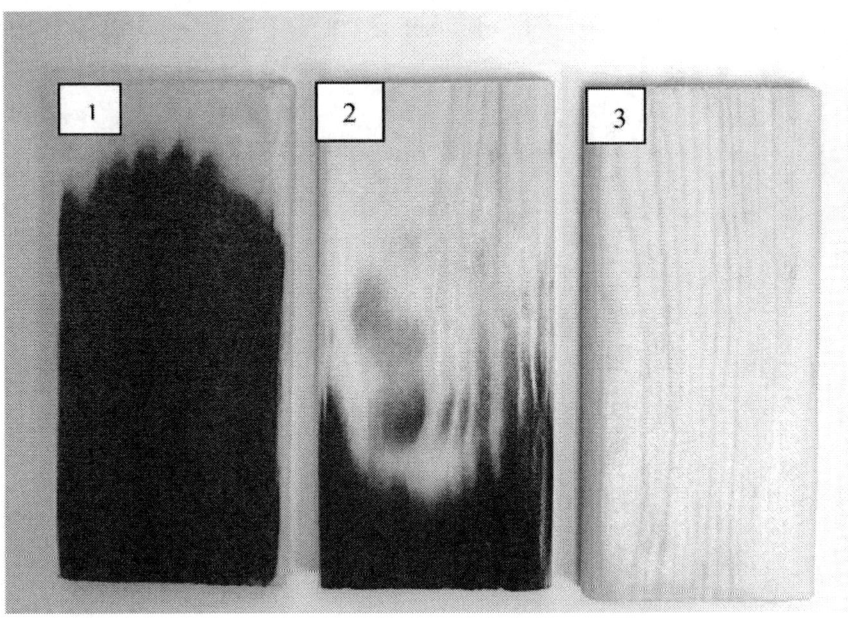

Figure 3. Appearance of fire retardant material: 1 - sample after testing; 2 - sample cleaned from ash residue; 3 - sample before testing.

IOP Conf. Series: Materials Science and Engineering 1015 (2021) 012095 doi:10.1088/1757-899X/1015/1/012095

The appearance of the coatings after the tests is shown in figure 3.

As a result, the value of water absorption coefficient equal to 0.08 kg/(m^2 h 0.5) characterizes the water absorption of the flame retardant film as low as compared to samples of competition flame retardant materials (figure 4).

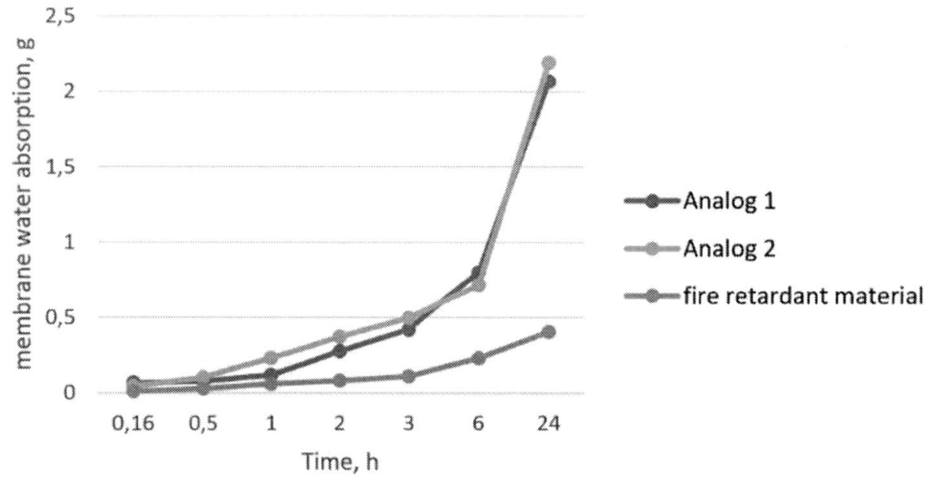

Figure 4. Weight gain of coating samples of Analog 1, Analog 2, new flame retardant material during film water absorption test

Dehumidification of loose film of the flame retardant material amounted to 10% in accordance with GOST 21513-76.

The accelerated climatic tests of the flame retardant material according to GOST 9.104 - 91 showed that in enclosed premises the service life of the material is not less than 30 years, and under the influence of temperature, humidity and UV radiation the protective properties are preserved for not less than 25 years. Figure 5 shows the state of samples with the CFR after accelerated climatic tests.

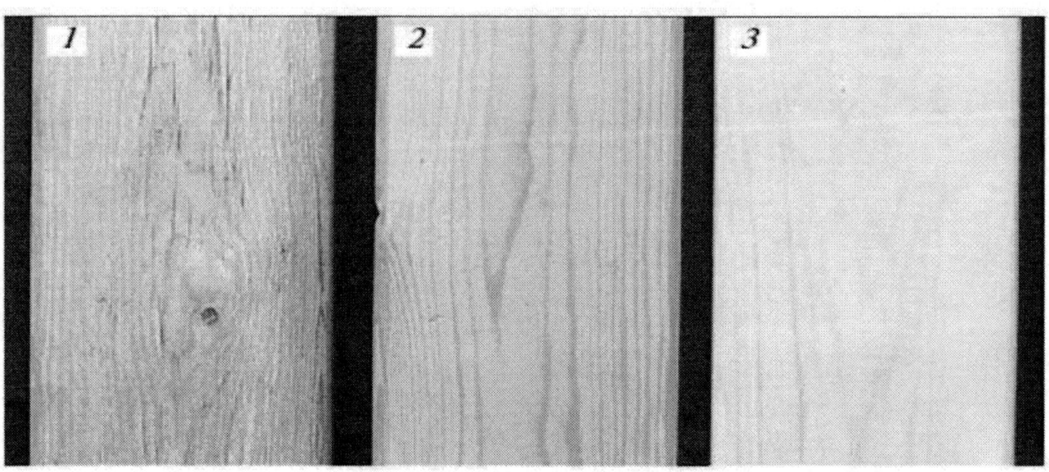

Figure 5. The state of the treated samples after 100 cycles of exposure in a climatic chamber 1,2 - tested samples, 3 - control samples.

IOP Conf. Series: Materials Science and Engineering 1015 (2021) 012095 doi:10.1088/1757-899X/1015/1/012095

Conclusions

The research of fire-technical characteristics of glued wooden structures samples treated with the new flame retardant material showed that protection of glued structures with the consumption of 350 g/m^2 makes the wood low-flammable. The new flame retardant material also reduces the spread of flame to group 1 and provides Group I fire protection efficiency. The results showed the possibility of using the new material for fire protection of supporting structures, finishing facades of buildings of all classes of fire resistance and fire hazard. The conducted researches of operational properties of the material have proved guaranteed service life of the fireproof material in conditions of open air for 25 years and indoors for 30 years.

Acknowledgments

This work was financially supported by the Ministry of Science and Higher Education of the Russian Federation (Project: Theoretical and experimental design of new composite materials to ensure safety during the operation of buildings and structures under conditions of technogenic and biogenic threats #FSWG-2020-0007).

References

[1] Lebedeva M A 2018 Place of glued wooden structures in modern construction and architecture *Young scientist* **50** p 46-48

[2] Serov E N , Labudin B V 2013 Kleenye derevyannye konstrukcii: sostoyanie i problemy razvitiya Lesnoy zhurnal **2**

[3] Kalugin A V 2011 Kleenye wooden structures in modern construction *Industrial and Civil Construction* **7** (2) pp 32–37

[4] Maximenko S A, Krivtsov Y V, Maximenko N A, Melnikov N O and Konstantinova N I 2013 Fire danger of new types of timber *Fire Safety* **2** pp 74-7

[5] Garashchenko N A and Beltsova T G 2005 Methods for fire protection of glue-wood structures. *Pozharovzryvobezopasnost* **5** URL: https://cyberleninka.ru/article/n/sposoby-ognezaschity-kleederevyannyh-konstruktsiy (date of access: 06/10/2020)

[6] Biryukov V G, Mishkov S N and Sobolev A V 2014 Fireproof glued materials on the basis of wood *Fires and Emergencies: prevention, liquidation* **3** pp 28-31

[7] Afanasyev S V and Korotkov R V 2012 Nitrogen-phosphorus-containing flame retardants of impregnating action for wood *Fire and Emergency Situations: prevention, elimination* **21** (6) pp 38-42

[8] Rustamov U I, Atashov A S and Mukhamedgaliev B A 2018 Investigation of combustion of fire-protective wood materials modified with polymeric flame retardants *Chemical Industry* **95** (4) pp 205-8

[9] Nikiforov E P 2019 Kleenaya wood: manufacturing technology, properties, advantages and disadvantages, areas of application *Modern Scientific Research and Innovations* **7**

IOP Conf. Series: Materials Science and Engineering 1015 (2021) 012096 doi:10.1088/1757-899X/1015/1/012096

Modal split analysis through the traffic volumes

M Drličiak, J Čelko and M Cingel

University of Žilina, Faculty of civil engineering, Department of highway engineering, Univerzitná 8215/1, 010 26 Žilina, Slovakia

E-mail addres: m.drliciak@rc.uniza.sk

Abstract. The article deals with the analysis of the modal split in Žilina self-region (ŽSK) during the morning and afternoon peak hours. The results are based on a mobility survey in the region. Households and inhabitants were divided into three main groups of the population. The first group characterized the inhabitants in the district towns, the second group in the other towns and the third covered the villages. The modal split was analyzed for all main groups and the most numerous subgroup - economically active people. In parallel with the analysis of mobility data, we analyzed data from the traffic surveys on several congested profiles. From the results, we determined the amount of traffic above the real road capacity limit. In the last part of the article, we present an example of combining data from the mobility and transport surveys. We defined the number of trips that need to be transferred to another transport system - the train.

1. Introduction

The mobility is generally defined as the ability to move or be moved freely and easily. The mobility (meaning as the number of trips per time unit) and travel time can be described as basic quantities of mobility. Their values represent the basic tool for SUMP (Sustainable Urban Mobility Plan) of functional urban areas (FUA). The data is preceded by several surveys, studies, and analyzes, which are usually imported into the transport model. In this article, we deal with the analysis of data from mobility surveys and traffic volumes surveys. The surveys were processed as a part of the input data for ŽSK. In principle, data from a mobility survey are mainly used to set up a demand model. The results are matrices of transport relations. The matrices show the number of trips between two points for a selected time interval. Theoretically, such data is calculated for the given time intervals to describe all the characteristics of the passenger load. But practically, it depends on the existing models for traffic [1]. The volumes survey data are used in the model calibration process. The article aims to demonstrate the possibility of linking data from both surveys. We focused on road profiles with the regularly exceeded saturation and with high concentrations of particulate matter (PM) [2]. These profiles are usually unsatisfactory also from measuring surface evenness [3]. The base-saturation flow rate value in HCM is derived from field data collected in developed countries. The adopted value in Slovakia is 1800 passenger car/h/lane. [4] We can determine the real capacity of roads from the measured data. Subsequently, we focused on expressing the number of vehicles above the capacity limit. The above volume limit was taken over into the demand model, which was set according to the data from the mobility survey. By changing the input quantities, such as e.g. fare, we can determine the necessary factors to change modal split for specific demand strata.

2. Mobility survey in Žilina self-governing region

Mobility survey (travel survey) includes information about the characteristics and determinants of travel activities such as trip purpose, start and end time, duration, cost, transport modes, location of origin, and destination [5].

The database for mobility characteristics is based on a mobility survey in Žilina region. The survey was made in 2016. The Zilina self-regionwith the area of 6,801 km² [6] lies on the north-western and northern part of Slovakia, it has borders with the Czech Republic and with Poland.

The actual data from the Slovak Statistical Office was used for specifying the credible sample. The survey was then performed in every village where the required number of households is greater than 10. Finally was asked 6,231 households with 18,382 inhabitants. The survey was conducted through questionnaires. The questions were formulated to describe the daily activity, transport habits, and trip purpose. Every interviewee was assigned to a predetermined population group.

The database contains 33,688 trip descriptions. Almost 50% of them contain the activity "JOB"(attraction or production). The economically active people share almost 67% of total inhabitants. The transport mode "car driver" was selected for 42% of work trips.

2.1. Peak time modal split

The congested roads reach the limit traffic volumes in the morning and afternoon peak hours. The main task of the article is to link traffic volumes data with mobility survey data. The evaluation of the modal split was analyzed for the morning time period (05:00 – 09:00) and for an afternoon time period (13:00 – 17:00).

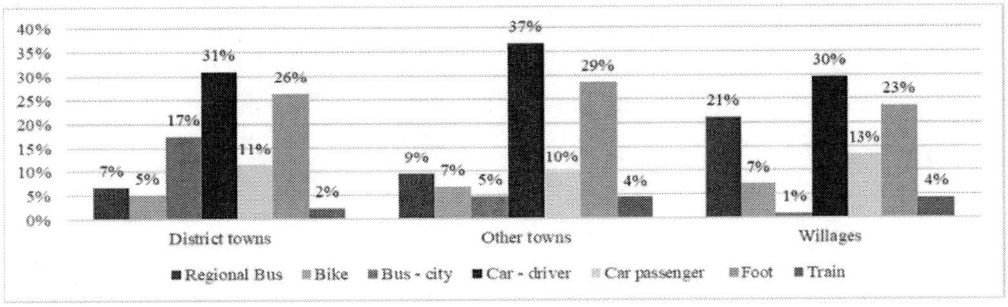

Figure 1. The morning modal split in district cities, other cities, and villages during the morning peak hours.

The equilibrium between the use of transport modes is the key to sustainable transport. The demand strata "economically active people - trips to the job" was selected to the next part of the analysis.

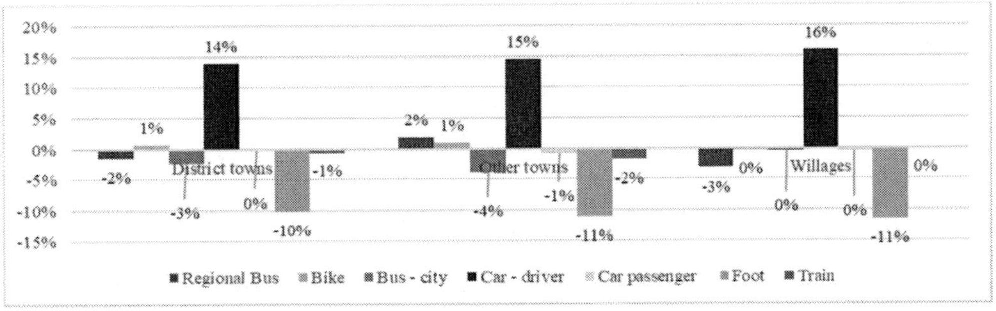

Figure 2. The modal split differences between total trip and demand strata "economically active people - trips to the job during morning peak".

IOP Conf. Series: Materials Science and Engineering 1015 (2021) 012096 doi:10.1088/1757-899X/1015/1/012096

The character of daily traffic is different depending on the distance from traffic-attractive zones. We divided the database set of morning and afternoon trips by location into three subcategories: district city, other city, villages. We decided to analyzed morning peak hours because the traffic load is more concentrated. The largest share (more than 30%) in the modal split is represented by trips marked as "car driver". The transport mode "Car passenger" is highest in villages. The results are presented in figure 1.

A transport mode "car driver" creates a significant proportion in modal split (figure 2). Approximately 45% of trips to work are realized by "car - driver" mode. The share of this mode is 15% higher compared to the modal split for all trips. The data from traffic count are analyzed in the next section.

3. Traffic survey

Profile surveys were carried out in the Žilina Regionin September and October 2018, on roads I., II. and III. classes in selected fifty-two profiles (a total of 104 traffic flows). Data from individual measurements were recorded for seven consecutive days from 00:00:00 to 23:59:59. For a better idea, in figure 3 the locations where the profile measurements were performed are marked.

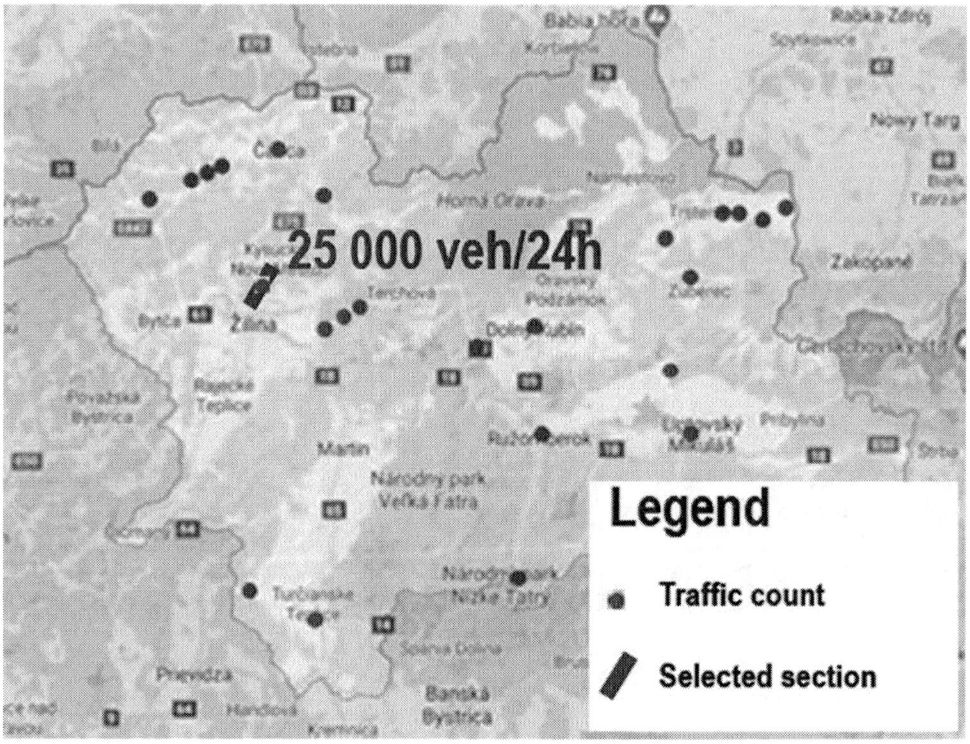

Figure 3. Traffic counts locality (red), application of results section (blue).

Gathered data was analyzed in Matlab. The example of the evaluated parameters of the overcapacity assignment traffic flow at road I/11 (figure 4).

The first graph (1-1) presents the relationship between speed and traffic volumes. The second (1-2) presents the relationship between traffic density and traffic volumes. These two data set were used for evaluation of the real capacity. The maximum value of the traffic volumes per hour was nearly 1800 veh/h. The third graph (2-1) complements the ratio between speed and density. The value of density increases as the speed decreases. The last graph presents the interrelation of traffic volumes, speed, and density.

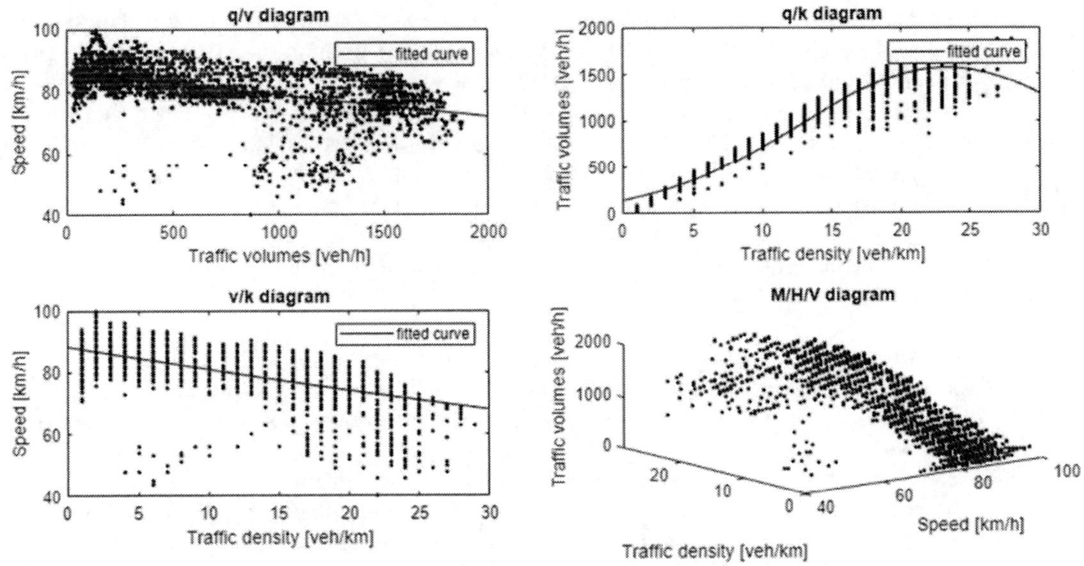

Figure 4. Traffic flow evaluation (traffic volumes, density, and speed).

3.1. Determination of the load capacity

The efficiency of the traffic system depends on the capacity of the traffic infrastructure. This capacity is defined as the "largest volume of traffic that traffic flow can reach at a given distance and traffic conditions at the cross-section determined for this flow" [7]. The capacity is determined by the density of the platoon of vehicles and the speed with which the platoon passes through the cross-section.[8]

The real capacity was determined as the maximum of fitted function (figure5) between density and traffic volumes.

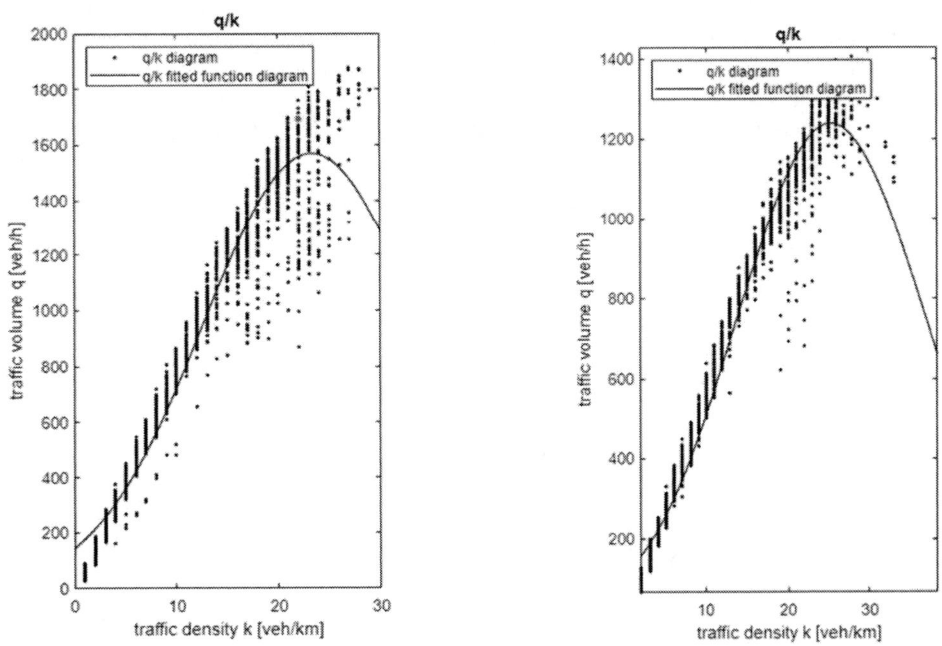

Figure 5. The compare of traffic volume/density ratio between two traffic flow.

IOP Conf. Series: Materials Science and Engineering 1015 (2021) 012096 doi:10.1088/1757-899X/1015/1/012096

Figure 5 presents the optimum capacities for two opposite lanes (two traffic flows). The optimum value was estimated by plotting the flow-efficiency diagram and finding the extremum point [9]. In this case, efficiency and related parameters should be estimated for various values of flow. The extremely fitted function was 1568 veh/h and 1293 veh/h. The results couldn´t be used for generalization of exact road capacity. The hourly traffic volumes were calculated as the sum of the four 15 min traffic volumes. The exact values should be evaluated for shorter intervals. Brilon and Zurlinden (2003) introduced 5 to15 min intervals as the best time intervals after investigating different values [10].

Our study collects the traffic count data at a specific place with data from a mobility survey between traffic-relevant areas.

4. Application of the survey results

We used data from both surveys to specific transport "funnel", which opens into congested place on the road I/11 in Slovakia. The maximum value of exceeding capacity was 350 veh/h from all measured days. The average maximum of exceedance was 202 veh/h.

We focused to import the value of overcapacity volumes to the demand model. Passenger transport demand is a function that covered the amount and type of transport that people will choose for specific factors (distance, price, and other quality conditions). Four-Step travel demand modelling is the traditional procedure utilized for transportation forecasts. It contains several submodels (trip generation, trip distribution, mode choice, and trip assignment). The mode choice model (modal split) is based on the utility functiontheory. Mode choice models determined the traveller's choice of which mode of transport to take, eg car, public transport, or whatever. The Utility functions are defined for each demand strata (stratum).

The sample for the database trips file was extracted from the mobility survey of Žilina Region. The factors (B) were calculated by software Biogeme which has been developed by Michel Bierlaire, Ecole Polytechnique Fédérale de Lausanne, Switzerland [10]. Biogeme is an open-source freeware designed for the maximum likelihood estimation of parametric models in general, with a special emphasis on discrete choice models. [11]

The comparison of current and proposal is presented in figure 6. The modal split was changed in favour of non-collapsing road transport when the saturation of traffic flow is under 1,0. We have reached the goal of reducing the number of trips by car by almost 7%. The region transport study evaluated sufficient capacity for train transport. The best assumption for reach the new modal split was reducing train travel costs to zero. This assumption changed the ratio of other modes. The total increase in train passengers was 14%.

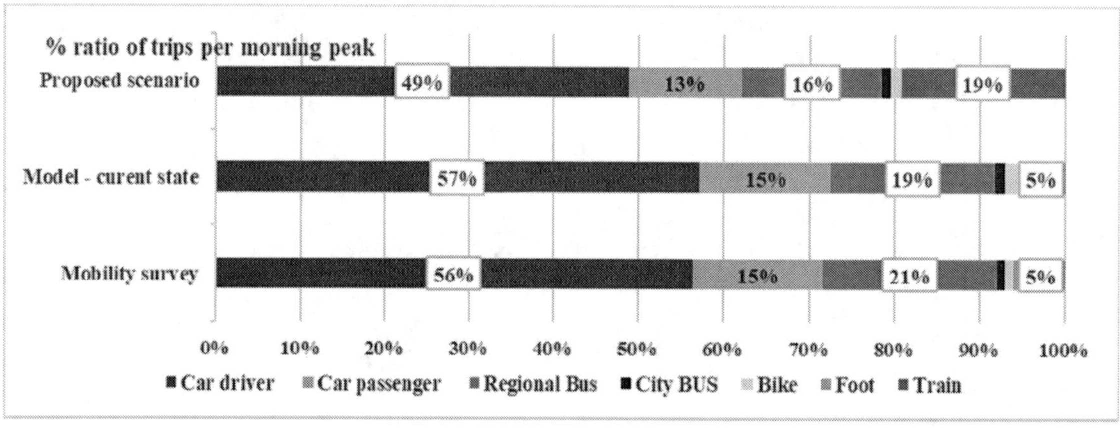

Figure 6. The compare of total modal split.

5. Conclusions

Travel time, travel cost mainly affect the choice of destination, route decision, and the choice of transport mode. We looked for opportunities for equilibrium between supply and demand for transport systems in the article. The purpose of the presented analysis was only regular trips. The most numerous trips are trips to work during the morning peak. The real route capacity was determined by density and speed. Data were gathered during 4 weeks at the regularly congested point.

The maximum value of exceeding the intensity was 350 veh/h from all measured days. The average maximum of exceedance was 202 veh/h. The utility theory was used to determine the changes in travel costs to achieve a capacity equilibrium. We analyzed scenarios with 100% subsidized train transport. The total increase in train passengers was 14%.In the next part, we want to evaluate the impact of traffic volumes on the environment by modal split changing [12, 13].

Acknowledgments

The presented results were partially obtained under the grant Vega No. 1/0428/19, Implementation of structural variables in the calculation of the modal split in the transport forecast (Implementácia štrukturálnych veličín pri výpočte deľby prepravnej práce v prognóze dopravy).

References

[1] Misharin A, Namiot D and Pokusaev O 2019 On Processing of Correspondence Matrices in Transport Systems *Int. Multi-Conference on Industrial Engineering and Modern Technologies, FarEastCon* October 2019, art. no. 8934188

[2] Jandacka D and Durcanska D 2019 Differentiation of particulate matter sources based on the chemical composition of PM10 in functional urban areas *Atmosphere* **10** (10) art. no. 583

[3] Leitner B, Decky M and Kovac M 2019 Road pavement longitudinal evenness quantification as stationary stochastic process *TRANSPORT* **34** (2) pp 195–203 (DOI: https://doi.org/10.3846/transport.2019.8577)

[4] TP102 - Výpočet kapacít pozemných komunikácií,

[5] Aschauer F, Rösel I, Hössinger R and et al. 2019 Time use, mobility and expenditure: an innovative survey design for understanding individuals' trade-off processes *Transportation* **46** pp 307–339 (https://doi.org/10.1007/s11116-018-9961-9)

[6] Statistical Office of the Slovak Republic (https://slovak.statistics.sk/)

[7] FGSV – Forschungsgesellschaft für Straßen- und Verkehrswesen (2001) *Handbuch für die Bemessung von Straßenverkehrsanlagen* (HBS) pp 3–191

[8] Friedrich B 2016 *The effect of autonomous vehicles on traffic* In: Autonomous Driving ed M Maurer J Gerdes B Lenz and H Winner (Berlin, Heidelberg: Springer) (https://doi.org/10.1007/978-3-662-48847-8_16)

[9] Optimum Capacity of Freeways: A Stochastic Approach. https://www.researchgate.net/publication/325287804_Optimum_Capacity_of_Freeways_A_Stochastic_Approach [accessed Aug 20 2020]

[10] Brilon W 2000 Traffic flow analysis beyond traditional methods *Proc 4th In. Symp on Highway Capacity*, TransportationResearch Circular E-C018,26–41 Washington, DC: TransportationResearch Board

[11] https://transp-or.epfl.ch/pythonbiogeme/ [accessed Aug 20 2020]

[12] Ďurčanská D, Decký M, Licbinsky R and Huzlik J 2013 Project SPENS - sustainable pavement for European new member states *Communications: scientific letters of the University of Žilina* **15** (2) pp 49-55

[13] Trojanová M, Decký M and Remišová E 2015 The implication of climatic changes to asphalt pavement design *24th Russian-Polish-Slovak Seminar on Theoretical Foundation of Civil Engineering TFoCE 2015* (Samara Russian Federation) pp 770-6 (DOI: 10.1016/j.proeng.2015.07.144)

IOP Conf. Series: Materials Science and Engineering 1015 (2021) 012097 doi:10.1088/1757-899X/1015/1/012097

The environmental impact of asphalt mixtures production for road infrastructure

Z Florkova[1], **S Sedivy**[2] **and J Pastorkova**[1]

[1] University of Zilina, Research Centre, Univerzitna 8215/1, Zilina, Slovakia
[2] University of Zilina, Faculty of civil engineering, Univerzitna 8215/1, Zilina, Slovakia

E-mail addres: zuzana.florkova@rc.uniza.sk

Abstract. Developed road infrastructure is a very significant factor which is influencing and conditioning the development of the economy and quality of transportation. Increasing demands for quality of road infrastructure lead to the construction of new roads. Moreover, the importance of keeping pavements in good condition increases proportionately with the increasing traffic. The construction of the road communications, their management and maintenance are material and energy consuming and the problem with the availability and exhaustion of resources is accretive. However, the development of road construction have to respect the protection of the environment, it have to be controlled and sustainable. Sustainable development ensures to meet subsistence for today's and future generations and respects the natural functions of ecosystems. The paper is focused on the analysis of the factors arising from the production of asphalt mixtures for road infrastructure by asphalt mixing plants and their subsequent impact on the environment.

1. Introduction

Asphalt mixture is one of the most commonly used materials in pavement constructions and it is a mixture of natural raw materials such as aggregate, filler and bitumen. Some of the additives may be added into the mixture for the reason of influence the performance of the product (e.g. adhesion agents, modifiers and fibers). Bitumen is the black, sticky component (binder) that bonds the aggregate together. Polymer modified bitumen's are increasingly used to modify the performance of the asphalt mixtures in special applications. Filler is used in asphalt mixture to fill out the smallest voids and to stabilize the binder [1].

The composition of the asphalt mixture has to meet the requirements of the appropriate technical standards. Depending on the asphalt production process and required mixing temperature there are different types of asphalt – Hot Mix Asphalt (HMA), Cold Mix Asphalt (CMA), and Warm Mix Asphalt (WMA).

Hot Mix Asphalt is a mixture of approximately 95% of well graded aggregates, together with filler and sometimes additives. Bitumen makes up the remaining less than 5% of the mixture [2]. These mixtures are generally produced at the temperature roughly between 120 and 190 °C. The production temperature of HMA depends on the used bitumen [3]. The production technology of HMA is the most commonly used technology and the main disadvantages are necessity of using high mixing temperatures and production of greenhouse gas emissions.

IOP Conf. Series: Materials Science and Engineering 1015 (2021) 012097 doi:10.1088/1757-899X/1015/1/012097

Cold Mix Asphalt uses bitumen emulsion as a binder and it is produced without heating the aggregate. Bitumen emulsions are heterogeneous two-phase systems consisting of bitumen and water. The bitumen is dispersed throughout the continuous water phase in the form of discrete globules, which are held in suspension by electrostatic charges stabilized by an emulsifier [1].

Warm Mix Asphalt mixtures are produced at temperature roughly between 100 and 140 °C, and thus they are also known as low temperature asphalts. These mixtures have properties and performance which are equivalent to conventional HMA [3].

The world production of asphalt mixtures is estimated to be about 1.500 million tons per year [1]. The statistics presented by The European Asphalt Pavement Association (EAPA) [4] have shown that approximately 300 million tons of hot and warm mix asphalt were produced in Europe (figure 1) in 2017 and to this day more than 90% of the 5.5 million kilo meters of roads in Europe are made of asphalt mixtures. The Figure 1 is showing the overview of total production of hot and warm mix asphalts in Europe from 2000 to 2017. Nowadays, in the Europe there are approximately 4.500 asphalt mixing plants, more than 10.000 companies are producing and/or paving asphalt, 180.000 people working in the asphalt industry and the industry has an estimated turnover of roughly 30 billion Euro [5].

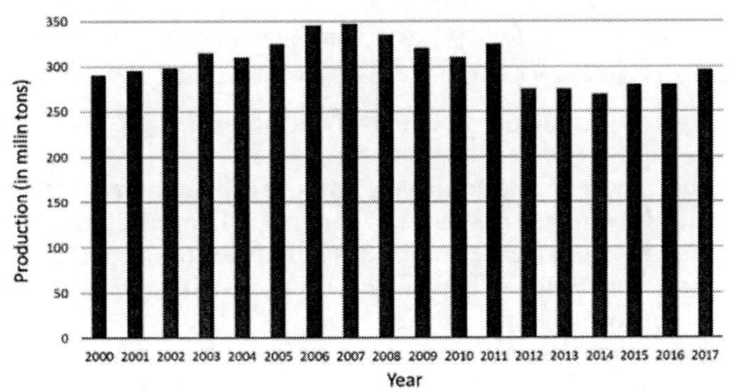

Figure 1. Total production of HMA and WMA in Europe from 2000 to 2017 (in million tons) [4].

2. Asphalt mixture production process

Production of asphalt mixtures is performed in a fully automated fixed or mobile mixing plant. Typical plant production capacities are 100-300 tons per hour [1]. There are two main types of plants - batch mix plants and drum mix plants. In batch mix plants (predominate in Europe) the mixing process is performed in a mixing unit (Fig. 2 - No.6) (called also as the pug-mill). In the case of drum mix plants is the mixing process carried out in the drum. The complete plant operation of asphalt mixture production is commanded and controlled by the electronic controlling system located in the control unit of the plant (figure 2 – No.10).

The batch mix plant´s production process consists of the following steps. The aggregate is loaded from storage piles and it is placed in to the cold aggregate supply system (figure 2 - No.1). Then, the aggregate is transferred in to the drum dryer (figure 2 - No.2) (typically gas-fired or oil-fired) by conveyer strand. The dust and flue gases are drawn off from the drum dryer by the ventilation unit into the dust catcher (figure 2 - No.7). In the next step, the aggregate is transported from the drum dryer into the vibrating screens (figure 2 - No.4) by the hot aggregate elevator (figure 2 - No.3), where it is separated on particular fractions and dropped into hot aggregate storages according to its size. From these storages is aggregate batched into the weighing system (figure 2 - No.5) and after that goes into a pug mill (mixing unit), where it is mixed and coated with bitumen. Bitumen is pumped from a heated bitumen supply system, weighed, and added into the pug mill. The specified amount of filler is added to the pug mill from filler supply system (figure 2 - No.8) at the same time. After the mixing time

IOP Conf. Series: Materials Science and Engineering 1015 (2021) 012097 doi:10.1088/1757-899X/1015/1/012097

(depending on plant type and mixture type) the hot mix is transported into a hot storage silo or it is dropped directly into a truck and hauled to the job site.

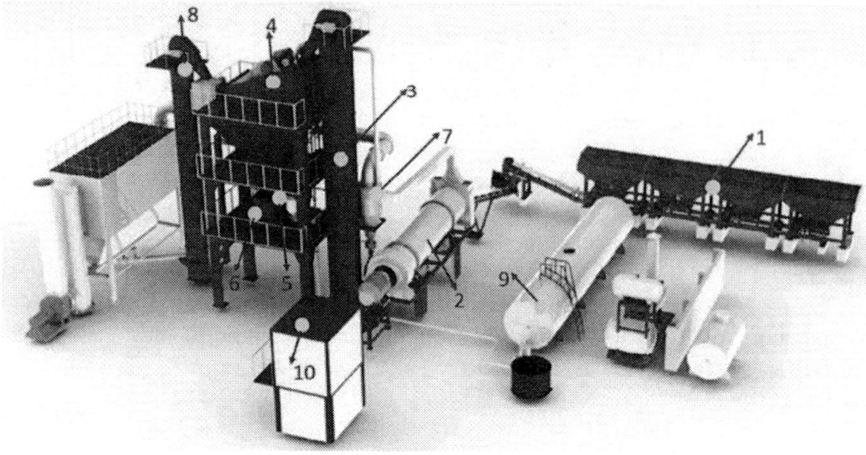

1 Cold aggregate supply system 2 Drum dryer 3 Hot aggregate elevator 4 Vibrating screen 5 Weighing system

6 Mixing tower 7 Dust catcher 8 Filer supply system 9 Bitumen supply system 10 Electronic controlling system

Figure 2. Batch mix plant scheme [6].

3. Analyse of possible environment impacts of asphalt mixtures production by mixing plant

A lot of studies are dealing with the possible environmental impacts of asphalt production [1] [7 - 11]. It should be noted that the consequences of the different types of impacts depend on the many factors and thus can be considerably different in many cases. The possible impacts are listed in the following paragraphs in accordance to importance.

Impact related to the normal operation of mixing plant – it may be a permanent impact (e.g. from the drum dryer) or temporary impact (e.g. due to the loading of storage piles);

Impact resulting from unplanned incidents - these impacts are difficult to quantify because they are unexpected in nature and may be the result of various circumstances;

Impact related to changes in operating conditions – e.g. changes in mixing temperature;

Impacts related to the commissioning and decommissioning of the mixing plant – e.g. impacts resulting from the planned filling, service or maintenance works.

Table 1. Possible pollutants emitted by asphalt mixing plant and their source [1].

Environmental impact (pollutants)	Source
Particulates - dust particles	Aggregate drying process, aggregate heating process, filler system, storage piles, material transport.
Gaseous organic and inorganic emissions	Heating system, fuel combustion process.
Noise	Particular parts of production, plant traffic.
Odour	Asphalt binder (bitumen), fuel tank.
Waste	Laboratory waste.
Sewerage water	Fuel tank.
Visual aspects	The design and layout of the plant.

These above mentioned impacts lead to the production of emissions of many different pollutants. The most common pollutants emitted from asphalt mixing plants are particularly dust particles,

IOP Conf. Series: Materials Science and Engineering 1015 (2021) 012097 doi:10.1088/1757-899X/1015/1/012097

gaseous organic and inorganic emissions. The noise, odour, waste, sewerage water and visual aspects are also important from the pollution point of view. Possible pollutants emitted by asphalt mixing plant and their sources are summarized in the table 1 and they are described in more details in following parts of the article.

3.1. Particulates - dust particles

Dust emissions are related mainly to the first phases of the asphalt production. The amount of dust depends on operating conditions. Drying and heating process of aggregate is the most important source of the dust emissions. Other sources of dust emissions are: material storage, material transport system (conveyor belts, elevators) and filler silos. Aggregates include certain percentage of fine fractions. These fine fractions can be emitted during the heating process by mechanical or thermal effects (separated by a hot gas flow in the drum). The produced dust consists of aggregate particles but may also contains particles of additives such as fibers. The dust can also contains binder particles when recycled material is used. The dust can contain also hydrocarbons as a result of the fuel combustion process in relation to the type of fuel. Moreover, the dust can also contains a small amount of heavy metals if waste oils are used. It is necessary to give attention to the dust that comes from some types of artificial aggregates which are used as a substitution for natural aggregates. The slags and ashes are also used as a substitutions. In this case, the special attention is needed. Silica dust may occurs if some types of aggregates are used.

The dust concentration depends on the type and moisture of the used aggregates, the method of treating materials in the drum, the amount and temperature of the waste gas, the rate of the waste gas in the drum, the shape of the extraction chimney and the performance of the plant.

Dust can be emitted from the complex filler system - starting with loading of storage silo from trucks, next with transportation of filler by hoses or screws (pneumatic or mechanical transport) and ending with adding the filler to the mixing unit. The filler system is a closed system with the additional filter accessory and under normal conditions small amount of emissions is emitted. Depending on the climatic conditions (dry weather in combination with wind), dust (especially fine fractions) may be emitted from storage piles of aggregates and also when the material is unloading from the trucks or when storage piles are filling. Dust can be whirled up by the traffic and can be standing in the air. Dust can be emitted from all parts of the system, where dry fine fractions are processed: conveyor system (conveyor belts, etc.), sieves (sorting), weighing, mixing units filling and dry mixing (without bitumen). Particular attention should be paid to the possible dust production during maintenance.

With regards to the information mentioned above, there is a lot of dust emissions sources produced by asphalt mixing plant and their production depends on many different factors, thus the many actions are necessary to be done to minimize the generation of these pollutants. The main prevention concerning dust emission problems should be to try to avoid its production, e.g. to keep the emitting processes in a closed systems with a negative air pressure or additional actions should be performed such as covering, moistening, hooding, and ducting. The main solutions for the reduction of emitted dust particles are to use effective dust extractors, ventilation and filtration systems.

3.2. Gaseous emissions

Gaseous emissions emitted from asphalt mixing plants can be divided to inorganic and organic emissions. The most common inorganic emissions are sulphur dioxide (SO_2), carbon monoxide (CO), nitrogen oxides (NO_x) and carbon dioxide (CO_2).

3.2.1. Inorganic emissions. SO_2 (Sulphur dioxide). The main source of sulphur dioxide SO_2 is the combustion process in the drying drum. The formation of SO_2 is influenced by the sulphur content in the fuel (mainly petroleum or brown coal). SO_2 is absorbed by some types of aggregates (limestone) and also by an alkaline dust layer in the dust filter. Other materials, such as slag, may also increase SO_2 emissions.

IOP Conf. Series: Materials Science and Engineering 1015 (2021) 012097 doi:10.1088/1757-899X/1015/1/012097

CO (Carbon monoxide). This emissions are mainly related to the combustion process in the drying drum as well. Fuel combustion is affected by direct contact of the pilot flame with the aggregate. The incorrect geometry of the drying drum, where the combustion space is too small, can also lead to deficient combustion of the fuel. In addition, CO emission is strongly influenced by the content of fine aggregate fractions and the water vapour content.

NO_x (Nitrogen oxide). The nitrogen oxide emissions are produced mainly from the burner in the drying drum. They are produced by the combustion of a fuel-air mixture by oxidation of atmospheric nitrogen with oxygen at high temperatures. The amount of these emissions depends on the nitrogen content of the fuel, the amount of excessed air in the drying drum, the pilot flame temperature and the type of the burner.

CO_2 (Carbon dioxide). The production of carbon dioxide emissions depends directly on the type of the used fuel and the energy consumption by the aggregate heating process and also on the type of treated bitumen and the type of heating system of asphalt tanks. The content of CO_2 emissions is determined by the production capacity of asphalt mixing plant and depends on the type of the used fuel as it is shown in table 2.

Table 2. Carbon dioxide emissions as a function of production capacity of asphalt mixing plant and the fuel used [1].

Specific heat consumption per ton of asphalt mixture	Light oil	Natural gas	Butane	Black coal	Brown coal
	kg CO_2 / ton of asphalt mixture				
70 kWh	18.65	11.63	15.27	23.05	23.61
85 kWh	22.65	14.13	18.54	27.99	28.67
100 kWh	26.64	16.62	21.81	32.93	33.73

In the common operational conditions of the plant are the values of SO_2 and NO_x emissions significantly lower than the regulation values. It is not possible to give specific requirements for CO and CO_2 emissions because of widely relation to the type of fuel used and the construction of burner installation. However all of these emission can be optimized by suitable plant operation.

All of the described inorganic emissions are specially related to the heating and drying processes in the drum. Generally, it should be alleged that reduction of these emissions are associated with the combustion processes optimization.

It is necessary to monitor the combustion process of the burner and also to ensure the maximum effectiveness of the combustion process. The one of the current solutions is to focus on production of Warm Mix Asphalt (WMA) mixtures. WMA mixtures are produced at lower temperatures compared to the commonly produced HMA and it leads to energy consumption and following emission reduction.

3.2.2. Organic emissions. Organic emissions consist of a large group of matters generally described as hydrocarbons. Their molecular structure is characterized by a combination of carbon and hydrogen atoms. In addition, these matters can also contain oxygen, nitrogen, sulphur and phosphorus. Hydrocarbon emissions are related to organic components and organic fuel used in the production process. These hydrocarbons are formed in the form of steam or reaction products. The most important source of hydrocarbon emissions is the deficient fuel combustion. The type of fuel, plan operating conditions and asphalt vapours in the mixing process lead to the different composition of the waste gases with respect to their organic components. Hydrocarbons are mainly emitted from the drying drum. These emissions can be reduced by periodic burner maintaining and fuel combustion volume optimizing. Another source of organic emissions is bitumen, which is heated in the production process. In places where the bitumen is heated to working temperature, vapours are generated.

Depending on the environmental impact of hydrocarbons and their impact on human health, hydrocarbons are divided into different categories. Polycyclic aromatic hydrocarbons (PAH) are most important because of their toxicity. Some of these polycyclic aromatic hydrocarbons (PAH) can be carcinogenic to human body under prolonged high-level exposures. The amount of these dangerous hydrocarbons in the bitumen is very small. Only a small amount of these emissions occurs at production temperatures which rarely exceed 200 °C. Available measured data [12] have shown, that PAH emissions are well below the national limit values, if at all are measured.

3.3. Noise
The main sources of noise in the asphalt mixing plant are drying drum, burner (especially air supply), ventilation unit, vertical conveying system and a hopper lift (especially when it is empty). Other associated sources of noise may be traffic noise from loaders as well as the traffic noise from trucks delivering raw materials and delivering the mixture. Operating noise level should not disturb houses located in the immediate vicinity of the asphalt mixing plant, especially out-off normal working hours. The overall solution of noise problems is to shield the emitting source, reduce the noise from the source or both.

3.4. Odour
Bitumen is the main source of odour. Odour is emitted during the filling of the bitumen tank and also when the mixing unit is emptied into the containers or trucks. The smell depends on the type of bitumen. The usage of recycled material can produce a smell during heating, especially if the recycled material has a high moisture content. During plant operation, all the practical steps should be taken to ensure that the odour won´t escape out of the plant area. There are several ways how to reduce odour, e.g. usage of products and fuels which produce less odour, reduction of the production temperature of the HMA mixtures or by usage of (chemical) additives which reduce the odour.

3.5. Sewerage water
Possible sources of groundwater pollution are: storage piles of artificial aggregate, leakage or mixing with natural soil, fuel tanks, hot oil for the heating system, pipelines and pumping stations and solvents (laboratory chemicals). In order to prevent groundwater contamination, preventive inspections (especially leakage inspections) should be carried out during operation of the plant. Bitumen tanks can be placed in a concrete base (or similar) to prevent leakage.

3.6. Waste
Asphalt mixture production is not a significant source of waste. Almost all of the basic raw materials are delivered in bulk so there is no packaging remained. Waste generated during the production process in the form of failure mixtures is in most of the cases either recycled or used in the construction of roads, e.g. in quarries. The source of the chemical waste may come from a laboratory, e.g. solvents used for bitumen analysis. All of the secondary waste should be recycled or stored according to regulations.

3.7. Visual aspect
The dominant features of the asphalt mixing plant are the high filler silos and asphalt mixture storage bins which should be of a neutral colour. Other visual aspects are the steam emitted from the storage tank, storage of aggregates or lighting of the plant area. The visual aspect should be maximized by appropriate landscaping, appropriate layout and fencing.

4. Discussion
A lot of environment impacts of asphalt mixtures production are described in the previous part of article. Some of these impacts have a minor or negligible effect on the pollution of environment. Despite this fact, it is still necessary to monitor them because of the prevention. For that reason, the measurements of emissions concentration and the pollution risk monitoring should be practiced regularly.

IOP Conf. Series: Materials Science and Engineering 1015 (2021) 012097 doi:10.1088/1757-899X/1015/1/012097

High energy consumption due to the high required temperature in the asphalt mixtures production (up to 190 °C) and the associated subsequent production of inorganic emission and greenhouse gas emissions appears as the biggest environmental impact of asphalt mixture production. High energy consumption and amount of CO_2 emissions depends on the type and condition of the production equipment (asphalt mixing plant), the settings of the combustion units and the type of fuel and other factors.

According to the measured data of the construction company [13] it can be stated that during the production of the asphalt mixture the CO_2 emissions were in the range of 13 to 30 kg/tone of the mixture (depending on the type of asphalt mixing plant and type of fuel). The least of CO_2 was emitted into the atmosphere during the combustion of liquefied gas, more by the combustion of natural gas and significantly more by the combustion of light fuel oil. It is very useful knowledge that will be used to decide on the type asphalt mixing plant. A positive result is if the amount of CO_2 is less than 20 kg/tone of the mixture. The similar results are described also in [1] (table 2).

High production temperature is important for reduction of the viscosity of bitumen and also for elimination of water from the aggregate surface [14]. Reduction of the production temperature leads to decreasing energy consumption. Therefore the reduction in the production temperature of the asphalt mixture production is currently under strong development due to environmental protection and high energy prices.

By usage of dry aggregates, it is also possible to reduce energy consumption during production. For that reason it is important to storage the aggregate in a place where the moisture content can be eliminated as much as possible. For example, only 1% higher aggregate´s moisture leads to increment of the energy consumptions about 14% [15].

As it is mentioned above, the one of the current solutions to reduce environmental impact of asphalt mixture production is to focus on production of Warm Mix Asphalt (WMA) mixtures. WMA is produced at a low temperatures compared to the commonly produced HMA and this leads to the reduction of energy consumption and following greenhouse gas emission. Studies [3] [16], also cited by [17] show, that the WMA production leads to the reduction of greenhouse gas emissions in the same rate as energy savings.

Results of the measurements [18] have shown that energy savings of asphalt mixing plant ranged from 20 to 35 % depending on the WMA technology, moisture content of the aggregate and the type and efficiency of the plant. The energy savings may be equivalent to approximately 1.5 to 2.0 litres of fuel per ton of material. The total energy savings are even greater with usage of recycled materials. The reduction in greenhouse gas emission is closely associated with the reduction in energy consumption i.e. 20 to 35 % reductions in CO_2, which is translated into approximately 4.1 to 5.5 kg of CO_2 per ton of mixture.

The emission parameters on several WMA projects were measured [19] including carbon dioxide (CO_2), carbon monoxide (CO), nitrogen oxides (NO_x), sulphur dioxide (SO_2) and particulates (dust particles). In general, the results from those measurements indicated lower emissions for WMA compared to conventional HMA.

The performance data of WMA mixtures have shown [3], [16], [20] that WMA will perform equally or possibly better than conventional HMA. WMA mixture exhibited similar or better performance including improved compaction, similar stiffness and rutting resistance, improved resistance to fatigue and thermal cracking, similar or less moisture damage and greater durability compared with HMA [19], [20].

Despite the all of these facts, that the production of WMA mixtures are the good way for reducing the environmental impact of the asphalt mixture production, the production of Warm Mix Asphalt (WMA) remained stable on a low level only about 2.4% of the total hot and warm mix asphalt production ranks to this asphalt. European leader in WMA production is France with a market share of 11.4%, followed by Norway (11.1%) and Denmark (8.5%) [4].

IOP Conf. Series: Materials Science and Engineering 1015 (2021) 012097 doi:10.1088/1757-899X/1015/1/012097

5. Conclusions

A general description of the types of asphalt mixing plants, operations and identification of environmental impact is the base for choosing the appropriate emission reduction methods. Knowledge about emissions and their concentration in the plant is important for preventive measures. For that reason the measurements of emissions concentration and the pollution risk monitoring should be practiced regularly.

The plan which clearly defines the complex long-term strategy for reducing the environmental impact should be established for each asphalt mixing plant. Different places (city/countryside), mobility (stationary/ mobile plant), etc. means different individual solutions. Solution applied in one plant, do not need to be applied in the other plant as well as not all emission reducing techniques are relevant to all types of asphalt mixing plants. The choice of specific measurement techniques and solutions which are appropriate to the particular situation will therefore depend on the location and many other factors.

Knowledge of the production process in combination with the input variables control can leads to a reduction of negative environmental impact along with other cost savings e.g. fuel consumption costs. The maintenance of the asphalt mixing plant, the control and handling with hazardous materials and recycling are also very significant in the process of environmental protection.

The sustainable development of road transportation which depends on the road infrastructure is the main goal of the technical policy in almost all of the countries. One of the solutions of the sustainable development problem of road transportation is also the saving of materials and energy related to construction or reconstruction of the roads and highways. In this context, we can argue that the development of road construction will be related to the use of environmental technologies. By the re-usage of road building materials, the recycling of materials, the using of industrial secondary products, the using of new binders and the using of less energy consumption technologies reduction of gaseous emissions and greenhouse gas emissions will be achieved.

Acknowledgments

This work was supported under the project of Operational Programme Research and Innovation: Research and development activities of the University of Zilina in the Industry of 21st century in the field of materials and nanotechnologies, No. 313011T426. The project is co-funding by European Regional Development Fund.

References

[1] Environmental Guidelines on Best Available Techniques (BAT) for the Production of Asphalt Paving Mixes 2007 *European asphalt pavement association (EAPA)* (Belgium Brussels)

[2] Asphalt pavements in tunnels 2008 *European asphalt pavement association (EAPA)* (Belgium, Brussels)

[3] The use of Warm Mix Asphalt 2014 *European asphalt pavement association (EAPA)* (Belgium Brussels)

[4] Asphalt in figures 2018 *European asphalt pavement association (EAPA)* (Belgium Brussels)

[5] The European asphalt pavement industry's position statement on the use of secondary materials, by-products and waste in asphalt mixtures 2017 *European asphalt pavement association (EAPA)* Belgium Brussels

[6] The Process of Asphalt Mixing Plant https://pistonclasico.com/wall/the-process-of-asphalt-mixing-plant, last accessed 2019/7/17

[7] Management of Potential Environmental Impact At Temporary Asphalt Plants 2011 *Work in progress 2nd Draft for discussion by Sabita Environmental Working Committee*

[8] The Environmental Impact of Asphalt Plants 2014 *NAPA National Pavements Assotiation*

[9] Preventing Pollution at Hot Mix Asphalt Plants A Guide to Environmental Compliance and Pollution Prevention for Asphalt Plants in Missouri 2004 *Missouri Department of Natural Resources Environmental Assistance Office 1-800-361-4827*

IOP Conf. Series: Materials Science and Engineering 1015 (2021) 012097 doi:10.1088/1757-899X/1015/1/012097

[10] Hot mix asphalt plants emission assessment report 2000 *Emissions Monitoring and Analysis Division Office of Air Quality Planning and Standards United States Environmental Protection Agency Research Triangle Park NC*
[11] Ozer H, Al-Qadi I and Harvey J 2016 Strategies for Improving the Sustainability of Asphalt Pavements. Federal Highway Administration *FHWA*
[12] Emissionen luftverunreinigender Stoffe aus Asphaltmischanlagen beim Einsatz von Ausbauasphalt – am Beispiel ausgewählter hessischer Anlagen. Schriftenreihe Umweltplanung, Arbeits- und Umweltschutz der Hessischen Landesanstalt für Umwelt 1995
[13] Budinsky V 2013 Energy intensity and greenhouse gas emissions in the production of asphalt mixtures *XVI Ivan Poliacek seminar with intern. participation (Bratislava)* pp 107-11
[14] Gabor P 2010 Environmental technologies in road construction and their topicality *Civil engineering*
[15] Bueche N, Dumont A G 2012 Energy in warm mix asphalt *5th Eurasphalt & Eurobitume Congress (Istanbul)*
[16] Croteau J M, Tessier B 2008 Warm mix asphalt paving technologies: a road builder's perspective. *Annual Conference of the Transportation Association of Canada (Toronto)*
[17] Tutu K A, Tuffour Y A 2016 Warm-mix asphalt and pavement sustainability: A review. *Open Journal of Civil Engineering* **6** 84
[18] Black and Green: Sustainable Asphalt, Now and Tomorrow. National Asphalt Paving Association (NAPA), Special Report Number 200
[19] Politano L 2012 Warm Mix Asphalt—A Greener Alternative to Hot Mix Asphalt. A Paper Submitted for the Transportation Association of Canada 2012 Technical Excellence Awards (Environmental Achievement Awards)
[20] D'Angelo J 2008 Warm-Mix Asphalt: European Practice. International Technology Scanning Program, Federal Highway Administration. Report No. FHWA-PL-08-007

IOP Conf. Series: Materials Science and Engineering 1015 (2021) 012098 doi:10.1088/1757-899X/1015/1/012098

Increasing the economic sustainability of winter road maintenance processes through the use of technological innovation

S Sedivy[1], L Mikulová[1], Z Florkova[2] and G Gaspar[3]

[1] Faculty of Civil Engineering, University of Zilina, Univerzitna 8215/1, Zilina, 010 08, Slovak Republic

[2] Research Centre, Division of Transportation Materials Research, University of Zilina, Univerzitna 8215/1, Zilina, 010 08, Slovak Republic

[3] TNtech, Ltd., Division of Applied Sciences, Lucna 1014/9, 014 01 Bytca, Slovak Republic

E-mail address: stefan.sedivy@uniza.sk

Abstract. Effective road management requires the constant generation of new approaches. It is our duty to focus on improving all technical and process areas of road management, especially in the context of environmental and economic sustainability. Otherwise, we will face the fact that neglected parts will slow down the progress of the whole system. This will be reflected in a reduction in the quality of the infrastructure and increased costs over its life cycle. The approach must be systematic and therefore it is necessary to address the elements that, at first sight, do not have a major impact on other elements contained in the whole structure of road infrastructure management. One of the key areas for improvement is the process of winter road maintenance. It must reflect global climate change, new approaches to the cost-effectiveness of finances spent by users and road managers and, last but not least, the need for modern and safe mobility of goods and people. Advanced technologies such as automated anti-icing spraying systems or predictive monitoring of road structure temperature development are one example. However, these innovations bring not only positive but also negative effects.

1. Introduction

The sustainability of road infrastructure must be seen as a comprehensive concept highlighted by the need to apply a multidisciplinary process. It is relatively easy to define, but very difficult to fill. And it is even more difficult to determine the extent to which the implemented solutions have contributed to progress in ensuring sustainability. Within standardly applied schemes, it is known that it is necessary to look for compromise solutions to a large extent. Solutions that, by interconnection, lead to specific environmental, economic and social effects. However, this approach is still exposed to new impetus. These need to be seen as a demand from road users as well as society as a whole, linked to natural technological progress, a more sensitive approach to the environment and an overall view of the future of mobility. On this basis, it will be necessary not only to deal with changes in the view of the concept of sustainability, but also to actively address the possibilities of implementing new approaches and elements that will have to be introduced into the entire road infrastructure environment.

One of the areas that is under great pressure in terms of sustainability is the technical maintenance of the road structure. Routine maintenance and repair processes still represent a large amount of expenditure within the communications managers' budget. It is no different in the conditions of the

Slovak Republic either. However, despite the growth trend in maintenance expenditure, it is not possible to call this state of road infrastructure satisfactory. Occupying the last ranks in the reports below should not be considered an acceptable state.

Table 1. Quality of roads – years 2018 and 2017.[a]

Country	2018		2017	
	Overall ranking	Score	Overall ranking	Score
Netherland	1	6.18	1	6.14
Germany	9	5.46	5	5.51
Poland	21	4.14	22	4.10
Slovakia	22	3.96	23	3.99
Czech republic	23	3.95	24	3.95
Hungary	24	3.89	21	4.15
Malta	27	3.24	26	3.23
Romania	28	2.96	28	2.70
EU		4.78		4.76

[a]Source: World Economic Forum Global Competitiveness Report. Rating based on a survey by the World Economic Forum, using a scale from 1 (extremely poor, among the worst in the world) to 7 (extremely good, among the best in the world). EU value is calculated as a simple average.

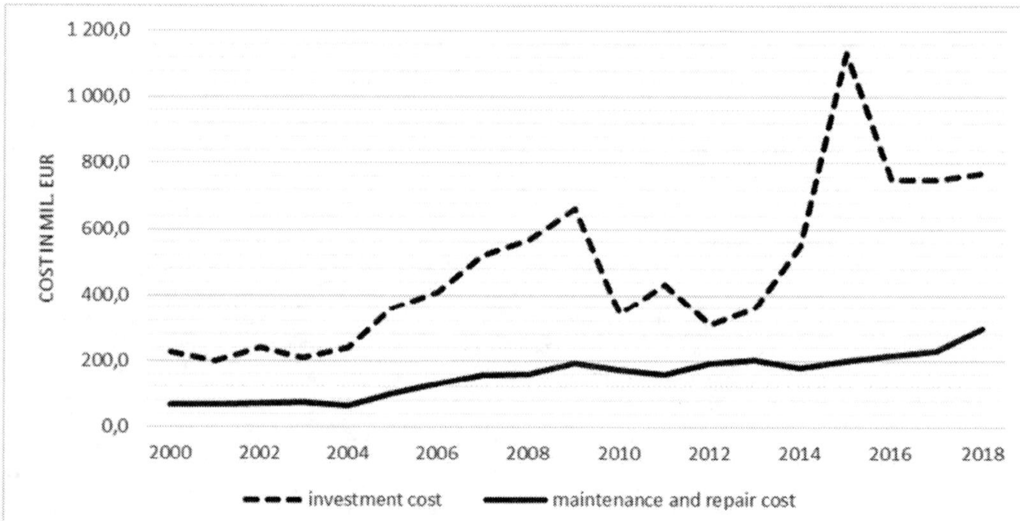

Figure 1. Investments and maintenance cost on road infrastructure in Slovak Republic (Source: own research from Statistical Office of the Slovak Republic data).

Table 2. Expenditures on road infrastructure maintenance (mil. EUR), current prices.

	2008	2009	2010	2011	2012	2013	2014	2015	2016	2017	Cost per 1km (2008 -2017) EUR
CZ	611.3	578.3	669.8	569.7	570.7	513.1	587.1	684.4	767.3	721.3	11262.12
HU	444.2	454.0	328.6	256.5	295.8	370.2	272.8	282.2	292.6	369.4	10599.18
SR	161.2	192.4	174.7	159.6	192.6	203.6	181.2	201.0	215.0	229.5	10580.87

[a]Source: own research from Statistical Office of the Slovak Republic and Eurostat data

IOP Conf. Series: Materials Science and Engineering 1015 (2021) 012098 doi:10.1088/1757-899X/1015/1/012098

2. Strategic overview about road sustainability and winter road maintenace

Models of road infrastructure sustainability are based on complexity. Strategies must think of all the elements that are present throughout the life cycle of the journey. The approach must be systematic, with the exception of the section on maintenance sustainability. Decision-making is carried out at the strategic, network and detailed project level. At the same time, louder tools are constantly emerging on the need to increase the rate and scope of innovation in individual parts of the model and its planning, decision-making and implementation processes. This is possible exclusively by incorporating new procedures and tools that emerge as suggestions from road users or road managers. However, it is not possible to forget about global demand either. These are mainly external factors, such as climate change or a changed approach to the value of money spent on the various phases of the communication life cycle. The common goal is to create an environment for transport that can be presented in the context of new platforms such as "Smart Cities", "Intelligent transport systems" or "Circular Economy". The space must have quite advanced technological elements that bring adaptive responses to changing climatic conditions. High emphasis must be placed on new, more environmentally friendly materials. Modern roads must include new infrastructural parts of the road, which e.g. they can actively redistribute traffic modes according to current needs (eg fast change of route for vehicles on a bicycle road).

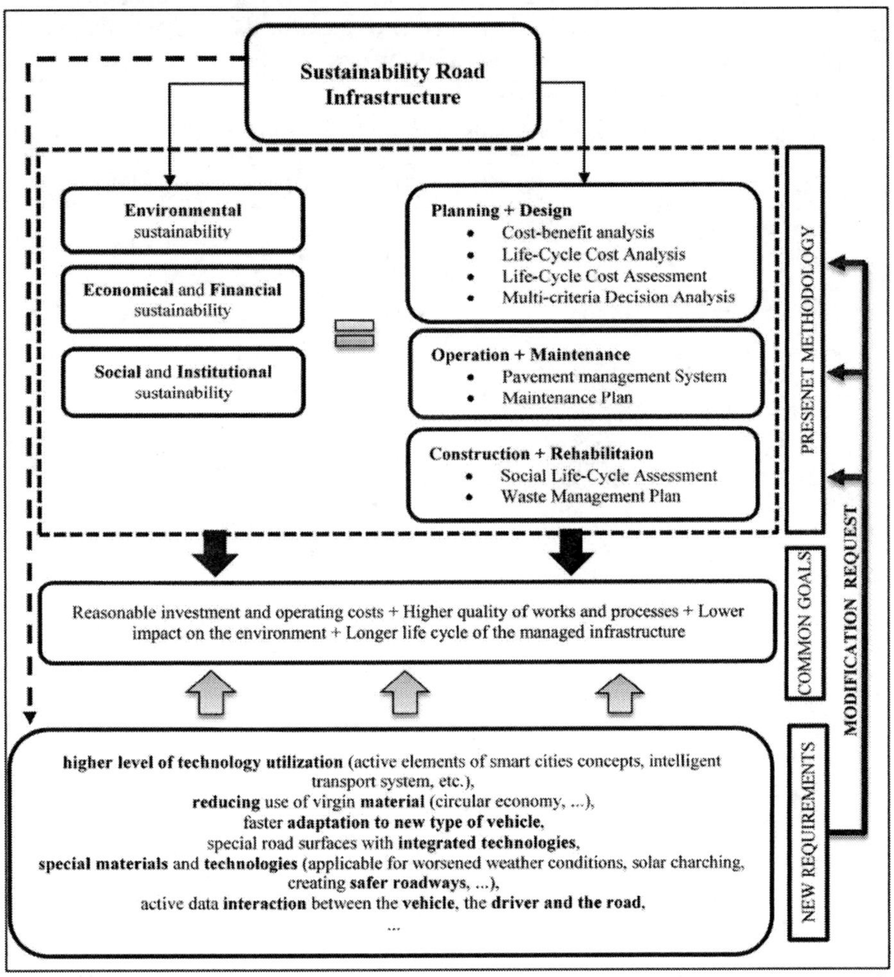

Figure 2. Simplified model of road management elements at context of sustainability (Source: own research).

IOP Conf. Series: Materials Science and Engineering 1015 (2021) 012098 doi:10.1088/1757-899X/1015/1/012098

Winter road maintenance (WRM) also has a place in sustainability models. Its quality, the way it is performed, is influenced by a considerable number of other factors within the sustainability model. The performance of winter road maintenance directly affects the technical condition of the road and the resulting service life, user satisfaction or the overall financial condition of administrators. In the next step, it is possible to present that the method of performing winter maintenance directly affects, of course, the costs associated with routine maintenance (road repairs). If winter maintenance is performed incorrectly, it can be expected that the quality of roads will decrease and investments in their repairs will increase in the short term.

Basically, it is possible to distinguish 4 main directions of WRM sustainability. They are technical, environmental, economic and procedural. It is appropriate if the individual directions go forward in parallel and in particular the trends are not preferred over the others. This is a prerequisite for ensuring the highest possible degree of sustainability in this area. The need to take economic sustainability into account is based primarily on two fundamental needs. The first is the cost-effectiveness of WRM performance. The second is the reduction of economic losses caused by the reduction of the road traffic (loss of travel time) and the reduction of the consequences of accidents caused by accidents in winter.

The performance of winter maintenance in Slovak conditions is demanding due to diverse geographical and climatic conditions. Similarly, it is possible to look at the economic side of the whole process. The table below shows the conditions from one of the Slovak self-governing regions. It shows the dependencies between the manifestations of the winter period and the performance in winter maintenance, including the financial expression.

Table 3. Costs and weather conditions one of the Slovak self-governing regions WRM[a].

Parameter	Year	19/20	18/19	17/18	16/17	15/16	14/15	13/14	12.13
Weather	number of ice days	17	26	23	52	21	13	16	16
	number of frost days	106	126	118	130	97	10	90	134
	number of precipitation days	106	96	110	102	99	89	79	133
	number of snow days	44	63	69	54	49	61	36	88
	number of maintenance service days	111	115	151	122	103	114	97	145
Maintenance volumes	plowing in km	20 111	47 952	84 185	47 836	35 929	30 079	19 989	99 524
	spreading in km	66 709	102 391	108 428	94 922	75 521	65 481	48 965	113 532
Costs	Total in mil. Eur	5	4	6	5	4	3	3	5
	cost per 1 km of maintained roads in Eur	2 386	2 279	2 903	2 549	2 006	1 816	1 479	2 494
	costs for 1 day of maintenance during the WRM period in Eur	30 039	28 964	36 928	32 639	25 470	23 061	18 784	31 167

[a]Source: own research from Zilina Slovak self-governing regions data

Financing of the development and maintenance of road infrastructure is not only long-term insufficient in Slovak conditions. Underfunded maintenance costs lead to negative impacts on the overall state of the infrastructure in the future and at the same time generally worsen the users' view of the activities of the communications manager. It is similar in the context of winter road maintenance. It can also be seen from the example above that the economic complexity of the WRM process is a relatively difficult thing to predict and therefore it often reaches a critical level. Inadequate road maintenance in the winter not only results in increased accident rates, but also increases fuel consumption, devalues the value of the vehicle and makes transport more expensive [2]. For this reason, the overall context of winter maintenance is currently receiving new impetus associated with several development trends [3]. This is despite the fact that the expected costs of climate change will fall [1]. These are the most important factors highlighting the need for innovation. Climate change is causing significant year-on-year fluctuations in the economic demands of winter maintenance. The combination of climate change, winter road maintenance processes and the context of economic impacts is becoming increasingly intense. As also stated [5] in his study, when a temperature increase of 1 °C in the studied locality also has a very significant impact on the costs associated with WRM. He further estimates that if the long-term atmospheric temperature rises by 2.5 °C, winter maintenance will not be necessary at all. Warming is expected to reduce the number of frost days in the long term, which could not only reduce road maintenance costs but also reduce accidents by up to 50% [11]. Several studies, such as [8, 10] draw attention to the fact that it will gradually be necessary to actively adapt the performance of winter maintenance to the specific manifestations of climate change. Another important factor that supports changes in approach is the policy of a low-carbon future. Different approaches within the WRM (driver behavior, scope and type of maintenance, ...) also affect fuel consumption and emission load [12, 13]. It is safe to say that winter maintenance can absorb a large number of new approaches and solutions. These can not only directly improve the condition of the road in terms of accessibility, environmental protection, improving the economic efficiency of expenditure, but also bring effects to ensure a longer life of the road structure. From this point of view, it is mainly the need for more gentle winter maintenance in the context of its geographical addressability and the adequacy of the intervention. Modern approaches include advanced interconnections of standard logistics maintenance processes with decision algorithms and simulations [4]. Another innovative direction is the application of new chemical mixtures based on agro-materials [6], which are not only more environmentally friendly, but also have financial competitiveness compared to traditional materials. Great attention is also paid to such advanced influences as the dependence between the type of asphalt, the type of salt used and the nature of the weather [7]. Relatively new research directions include investigating the effects of WRM on the life of bridges [9], [17]. These contexts can then be translated into economic expressions within the total cost of road infrastructure maintenance.

3. Selected solutions supporting increasing the quality of winter maintenance performance

The strategic goal of the research activities at the University of Zilina is to bring innovative elements to the road infrastructure environment with its own partial activities. The area of winter road maintenance is no exception. The aim is to make it possible to increase the quality of winter maintenance and at the same time to contribute to a more efficient performance of road management assets based on less wear and tear of the road body structure due to the action of chemical mixtures. The research team focuses primarily on solutions that are applicable in the context of the future thinking of users and communication administrators. This means that the emphasis is on solutions based on smart cities (local application for specific problems, connection to complex technological systems for urban zone management) and the context of intelligent infrastructure. The aim of the research is to bring new ways that would help increase the effectiveness of the communication manager's intervention. Specifically, it is the geographical addressability of the intervention (perform maintenance only where necessary), optimize the time horizon of the intervention (reduce the volume

IOP Conf. Series: Materials Science and Engineering 1015 (2021) 012098 doi:10.1088/1757-899X/1015/1/012098

of preventive interventions) and eliminate risks in the most critical places by automating the intervention (maintenance of communication without intervention).

3.1 New sensors for road temperature detection and monitoring
Effective weather forecast in the context of ice forecasting is not possible without measuring the temperature of the road structure. These data then enter into the prepared prediction models associated with the estimation of the time of ice formation on the road. It is the verification of the degree of reliability of this method of temperature measurement that is the basic step for future activities associated with the construction of quality road meteorological systems. These consist in the creation of a unique prediction system, the task of which will be to predict the need for winter maintenance at a specific point in the road network.

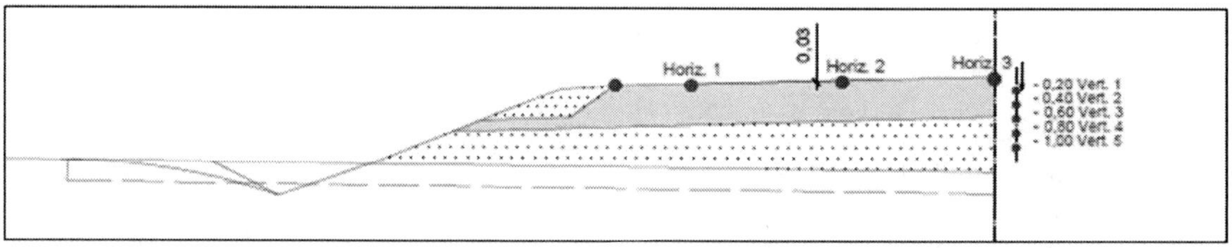

Figure 3. Scheme of installation of onw temperature sensors for weather prediction systems [a]Source: own graphics of construction solution.

It is a system of vertical and horizontal temperature sensors that are installed at different depths and distances within the road. Temperature sensors are designed to withstand not only the installation process, but also to be resistant to common right-hand pressures for a long time. Sensors are not installed in the context of measurements, which should monitor the development and behavior of the road structure in order to monitor the thermal mode of the road. They are installed in the context of the need to very sensitively monitor the development of that part of the road structure that directly affects the state of the road temperature on its surface. For this reason, the sensors used are also optimized for high temperature resolution up to 0.001 degrees Celsius.

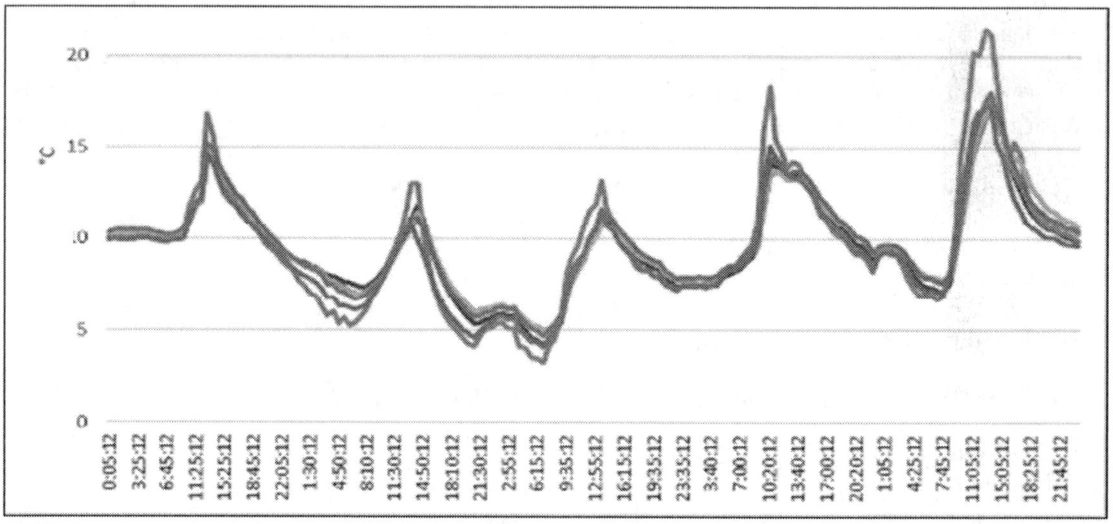

Figure 4. Overview of the sensitivity of sensors installed in the road to weather changes [a]Source: own research from University of Zilina data.

IOP Conf. Series: Materials Science and Engineering 1015 (2021) 012098 doi:10.1088/1757-899X/1015/1/012098

3.2 Autoamtized anti-icing spraying system

It is well known that increased slipperiness increases the risk of accidents. Deciding when and where to protect communication against the ice is based on meteorological measurements and prediction. However, these predictive models have some uncertainties. This undermines the effectiveness of administrators' intervention, which in turn translates into increased costs that may not be spent efficiently [16]. As stated in [14], data from meteorological stations, together with weather forecast models and data from meteorological radars and satellites, are processed, presented and used to predict the condition and temperature of the road surface. The result is information support for the right maintenance decision and for efficient and timely road care in winter. Gradually, however, there is a need to innovate older approaches that integrate not only meteorological conditions, traffic data, road surface condition or temperature [15]. It is necessary to gradually address the symptoms of risks associated with different situations of roads (sections with less sunlight, sections near reservoirs,...). It is also necessary to ask for new meteorological approaches that can be used to predict the occurrence of sudden weather conditions. The University of Zilina is developing a comprehensive system consisting of a new meteorological forecasting model, which will also be connected with a reaction system preventing the occurrence of negative situations on the road. One of these systems is an automated spraying system. These are not completely new in the world. However, they still cannot be considered perfect and must be constantly improved and tested. Their primary task is to prevent and reduce the formation of a layer of ice on the road surface at a critical time interval ending with the arrival of maintenance vehicles. It should be borne in mind that its purpose is not to remove thick layers of ice, to prevent the accumulation of a layer of snow on the road or to keep the road dry. It is understood as a supplement in the process of winter service, resp. winter maintenance in order to respond operatively to the possibility of reducing the operability and safety of a critical section of the road.

Figure 5. Automatized anti-icing spraying systems of University of Zilina. Source: authors' photos

The entire installation also includes a software application that not only informs the road administrator about the status of the road section, but also helps in deciding whether to start interventions or processes. The program allows you to manage a centralized early warning network in front of the ice through its own alert algorithm.

IOP Conf. Series: Materials Science and Engineering 1015 (2021) 012098 doi:10.1088/1757-899X/1015/1/012098

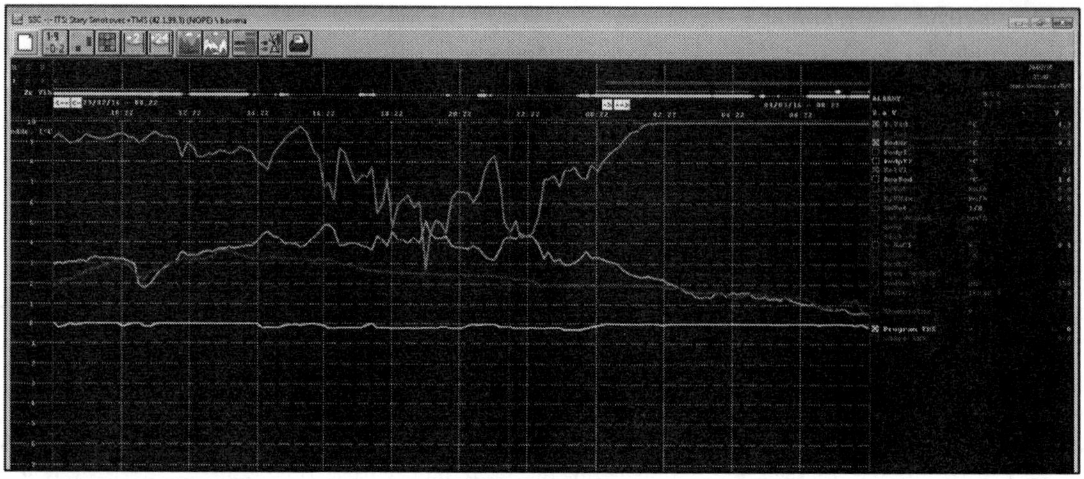

Figure 6. Software application as part of Automatized anti-icing spraying systems
Source: authors' photo

4. Economic benefit of technological innovations on the maintenance process

As mentioned in the introductory chapters, winter maintenance directly affects the overall economic condition of road maintenance. The common goal should be to optimize the technological background and processes in order to achieve the greatest possible economic, social benefits and benefits in the field of environmental protection. At this point, process and technological equipment innovations are needed. However, these not only generate benefits, but are also associated with certain costs. An overall comprehensive assessment of the impact of their implementation is relatively difficult to define. The whole process of winter maintenance creates a relatively large set of costs and benefits. These are also influenced by a wide range of factors. Factors are divided into uncontrollable (climatic soils) and controllable (WRM performance strategies, quality of technological equipment, scope of WRM, type of material used in relation to the environment, etc.). The overall approach to WRM then brings additional benefits. For safety, the environment, user costs and benefits (corrosion, travel time), costs associated with ensuring the quality of communication after winter maintenance.

The basic method can be presented as follows:

$$Q_{WRM} = Q_{WRMold} \cdot x \cdot k + Q_{WRMinn} \cdot x \cdot (1-k) \tag{1}$$

Q_{WRM} - quality of winter road maintenace
Q_{WRMold} - quality of winter road maintenace - unchanged state
Q_{WRMinn} - quality of winter road maintenace – innovated parts
k - percentage share of original winter road maintenance processes from the total volume of WRM

This method is explicitly conducted in the context of professional estimation, which expresses how innovations (technological, process) affect the resulting quality of WRM. The second method is based on exactness and is based on the methodology of cost-benefit analysis. For our questions, it is possible to present the statement as follows through a modified NPV.

$$NPV_{WRM} = \sum_{i=1}^{T} \frac{CF_T}{(1+i)^T} - cost_{inn} \tag{2}$$

where

$cost_{inn}$ - innovation cost
T - time
i - discount rate

IOP Conf. Series: Materials Science and Engineering 1015 (2021) 012098 doi:10.1088/1757-899X/1015/1/012098

This method is relatively accurate, but only if the calculation is detailed and includes all the costs and benefits associated with implementing the innovation. The following table shows the impact of the above technological innovations on selected parts of winter road maintenance in simple terms positive/negative/neutral.

Table 4. Effect of technolgical innovation implementation to Winter Road Maintenance.[a]

Paramaters or Specifics	New sensors for road temperature detection and monitoring as part of complex alarm systems	Automatized antiicing spraying systems
Road Administrators' investment costs		
Costs of new technologies	Negative • new technologies represent additional investment costs, including cost for new staff • needs to have money for services and repair	Negative • new technologies represent additional investment costs, including cost for new staff • needs to have money for services and repair
Road Administrators' operational costs		
Cost of sanding material and salt	Positive • if the Road Administrator does not know the detailed road surface temperature, it is necessary to take precautionary interventions. When using predictive technologies, he sees when and whether maintenance is necessary at all.	Positive • in case of early intervention against the ice, it is possible to use less material. Ice is still very weak or is yet to form. Negative • necessary to have sufficient supplies of chemical mixtures at the place of installation of the technology, as well as clean water for cleaning the system
Costs of mechanism maintenance	Positive • lower volume of interventions, reduce the cost of maintenance vehicles	Positive • lower volume of interventions, reduce the cost of maintenance vehicles Negative • there are costs associated with refilling chemical tanks
Staff costs	Neutral • Standard technological innovations bring savings in personnel costs. In this case, however, this does not apply due to the low scope of innovation on the road network. • every technology requires human supervision	Positive • Standard technological innovations bring savings in personnel costs. In this case, however, this does not apply due to the low scope of innovation on the road network. • every technology requires human supervision
Road maintenance costs	Positive • Higher quality of winter maintenance combined with adequate salting contributes to lower damage to the structure of the road structure. From this it is possible to predict lower financial needs for repairs or cleaning of ditches from spreading material	Positive • Higher quality of winter maintenance combined with adequate salting contributes to lower damage to the structure of the road structure. From this it is possible to predict lower financial needs for repairs or cleaning of ditches from spreading material

Environment quality		
Contamination of air, surrounding water, soil and vegetation	Positive • increasing the efficiency of road administrator interventions within the WRM reduces the volume of produced emissions of maintenance vehicles • the reduced volume of chemicals used reduces the negative impact on the environment	Neutral • Due to the point technology, which solves only very risky sections of the road, the positives are only minimal from this point of view and it is rather possible to classify it as a neutral effect.
Selected user costs		
Travel time saving	Positive • A road that is maintained according to the expected conditions is more comfortable than a road where maintenance work is only being carried out	Positive • if the road administrator eliminates risk places on the road network (eg ascents, shaded places) and there are no delays in these parts
Safety	Positive • the road manager can have information about the planned change of weather conditions and subsequently the conditions on the road surface - he can perform the intervention in advance and thus increase safety)	Positive • the road administrator eliminates risk places on the road network (eg ascents, shaded places
Corrosion of vehicles	Positive • (Interventions are targeted and there is no unnecessary performance, the road is less exposed to chemicals, which subsequently degrade vehicles)	Positive • (Interventions are targeted and there is no unnecessary performance, the road is less exposed to chemicals, which subsequently degrade vehicles)
Impact on the economy of the country and the region	Positive • Higher quality of peoples and goods mobility	Positive • Higher quality of peoples and goods mobility

[a] Source:own research

5. Conclusions and future work

Road sustainability is a complicated model depending on a wide range of factors. At first glance, it may seem that the performance of winter maintenance only slightly affects its quality. However, in the context of climate change and the need for more efficient expenditure on road infrastructure management, it is necessary to consider introducing innovations in this area as well. The economic sustainability of WRM comes to the fore. Innovations that can improve meteorological forecasting models and action technology platforms for reducing road slip are one such tool. They not only have a positive effect on driving safety and the environment, but also contribute to increasing the economic sustainability of the WRM. Future challenges will include bringing exact approaches that can quantify this benefit. This step is necessary to allow administrators to compare the benefits of each innovation.

Acknowledgments

This paper was supported with project of basic research: "Expanding the base of theoretical hypotheses and initiating assumptions to ensure scientific progress in methods of monitoring hydrometeors in the lower troposphere", which is funded by the R&D Incentives contract.

References

[1] Matthews L, Andrey J and Picketts I 2017 Planning for Winter Road Maintenance in the Context of Climate Change *Weather Climate And Society* **9** (3) pp 521-32, DOI: 10.1175/WCAS-D-16-0103.1

[2] Ratkevicius T and Laurinavicius 2017 Assessment Model Of Levels For Winter Road, Maintenance, *Baltic J.l Of Road And Bridge Eng.* **12** (2) pp 127-34, DOI: 10.3846/bjrbe.2017.15

[3] Pilli-Sihvola E, Aapaoja A, Leviakangas P, Kinnunen T, Hautala R and Takahashi N 2015 Evolving winter road maintenance ecosystems in Finland and Hokkaido *IET Intelligent Transport Systems* **9** (6) pp 633-38 (Japan) DOI: 10.1049/iet-its.2014.0220

[4] Tang WJ, Zhang GH Liao MJ 2015 A Framework of Winter Road Maintenance Optimization, *11th Int. Conf. On Natural Computation* pp1057-61

[5] Lorentzen T 2020 Climate change and winter road maintenance *Climatic Change* **161** (1) DOI: 10.1007/s10584-020-02662-0

[6] Nazari MH, Shi XM 2019 Developing Renewable Agro-Based Anti-Icers for Sustainable Winter Road Maintenance Operations *J. of Materials in Civil Eng.* **31** (12) DOI: 10.1061/(ASCE)MT.1943-5533.0002963

[7] Hossain SMK, Fu LP, Hosseini F, Muresan M, Donnelly T and Kabir S 2016 Optimum winter road maintenance: effect of pavement types on snow melting performance of road salts *Canadian J. Of Civil Eng.* **43** (9) pp 802-11, DOI: 10.1139/cjce-2016-0010

[8] Venalainen A and Kangas M, 2003 Estimation of winter road maintenance costs using climate data *Meteorological Applications* **10** (1) pp 69-73 DOI: 10.1017/S1350482703005073

[9] Stemberk P, da Silva WRL, Sykorova J and Bartova J 2013 Fuzzy modeling of combined effect of winter road maintenance and cyclic loading on concrete slab bridge *Advances In Eng. Software* **62-63** pp 97-108 DOI: 10.1016/j.advengsoft.2013.04.008,

[10] Norem H 2009 Selection of Strategies for Winter Maintenance of Roads Based on Climatic Parameters *J of Cold Regions Eng.* **23** (4) pp 113-35, DOI: 10.1061/(ASCE)0887-381X(2009)23:4(113)

[11] Andersson AK and Chapman L 2011 The impact of climate change on winter road maintenance and traffic accidents in West Midlands UK *Accident Analysis And Prevention* **43** (1) pp 284-9 DOI: 10.1016/j.aap.2010.08.025

[12] Vignisdottir HR, Ebrahimi B, Booto GK, O'Born R, Brattebo H, Wallbaum H and Bohne RA 2020 Life cycle assessment of winter road Maintenance *Int. J. of Life Cycle Assessment* **25** (3) pp 646-61, DOI: 10.1007/s11367-019-01682-y,

[13] Vignisdottir HR, Ebrahimi B, Booto GK, O'Born R, Brattebo H, Wallbaum H and Bohne RA 2019, A review of environmental impacts of winter road maintenance *Cold Regions Science and Technology* **158** pp 143-53 DOI: 10.1016/j.coldregions.2018.10.013

[14] Kocianova A 2015 The intelligent winter road maintenance management in Slovak conditions, *24RSP (TFOCE 2015) Procedia Engineering* **111** pp 410-19 DOI: 10.1016/j.proeng.2015.07.109

[15] Mahoney WP and Myers WL 2003 Predicting weather and road conditions - Integrated decision-support tool for winter road-maintenance operations *Highway Maintenance Safety, Support And Services: Maintenance*, Book Series: Transportation Research Record Series **1824** pp 98-105

[16] Riehm M and Nordi L 2012 Optimization of winter road maintenance energy costs in Sweden: a critique of site specific frost warning techniques *Meteorological Applications* **19** (4) pp 443-53 DOI: 10.1002/met.275

[17] Bujnakova P, Jost J, Farbak M and Kotes P 2019 Experimental study of the modulus of elasticity of concrete at different ambient temperature *IOP Conference Series: Materials Science and Eng.* **549**

IOP Conf. Series: Materials Science and Engineering 1015 (2021) 012099 doi:10.1088/1757-899X/1015/1/012099

Justification of the possibility of laying cable communication lines at the side of roads

M V Rodionov, T V Dormidontova and S E Lavrov

Samara State Technical University, Academy of Architecture and Civil Engineering, Molodogvardeyskaya str. 194, 443001 Samara, Russia

E-mail address: rodionov_max@mail.ru

Abstract. One of the ways to improve road safety is to equip motor roads with automated traffic control systems. Such systems allow recording violations of traffic rules, responding to emergencies on the roads, providing automation of fare collection systems and driver information services, etc. To ensure the operation of automated traffic control systems, it is necessary to construct long cable communication lines. The best place for laying them can be the side of the road, however, at present in Russia there is not enough experience in construction that confirms the safety of such a solution. The authors present the results of research devoted to assessing the feasibility and safety of laying cable communication lines at the side of roads using micro-trenches.

1. Introduction

Ensuring road safety is a vital social-economic and demographic task in many countries, including the Russian Federation. It is possible to improve road safety by equipping most motor roads with automated traffic control systems (ATCS). Such systems consist of various subsystems, which are controlled from a single traffic control center [1]. ATCS subsystems allow continuous video monitoring of the traffic situation, in order to record and promptly respond to traffic violations and other emergencies that occur on the roads; allow informing drivers about road conditions; solving commercial problems, such as automating the fare collection system and managing advertising information.

To ensure the operation of ATCS along motor roads, the construction of long cable communication lines is required. Current regulatory requirements specify that communication cables should be laid on the right-of-way. In this regard technical difficulties arise, as well as problems with the allotment and occupation of land property, even for temporary use, and problems in subsequent operation. The most effective solution is to lay cable communication lines (CCL) at the side of the road, as it has got the following advantages: reducing the time and cost of design, construction and operation; year-round and round-the-clock availability for operation and maintenance; minimum terms for carrying out repair and reconstruction work; facilitating the procedure for approval and land allocation; increasing the reliability of work due to the relative evenness of the right-of-way centerline; practically no intersections with other communications; reduction of damage arising from excavation work and the impact of rodents. At the same time there is not enough experience in the Russian Federation that confirms the safety of such a solution. In this regard the authors, basing on the analysis of world experience in laying CCL at motor roads and on applied numerical simulation, have evaluated the possibility and safety of laying cable communication lines at the side of the road.

2. Materials and methods

The analysis of world experience of constructing CCL at motor roads [2-8] has proved that to lay them micro-trenches (width = 4 ÷ 8, depth = 20 ÷ 30 cm), mini-trenches (width = 8 ÷ 20, depth = 30 ÷ 60 cm) and macro-trenches (width = 20 ÷ 30 cm, depth = up to 100 cm) are used. Within the earth roadbed the communication lines are located at the traffic way or at the side of the road, inside asphalt-concrete coating (micro- and mini-trenches) and inside earth layers (mini- and macro-trenches) (see figure 1). As a rule backfilling (filling) of trenches for laying cable communication lines is performed using materials with properties similar to design data.

a b c

Figure 1. Types of trenches used when constructing cable communication lines at motor roads [5]: a. Micro-trenches; b. Mini-trenches; c. Macro-trenches.

To assess the impact of laying CCL using mini-trench technologies at the side of the motor road, verification calculations of sections of motor roads have been made along the traffic way adjacent to the side of the road where it is planned to lay cable communication lines (see figure 2). The calculations have been made by means of modeling the stress-strain state of the motor road's roadbed, the change of which may occur as a result of laying cable communication lines and may cause deterioration in the road's operational and traffic-operational characteristics.

The stress-strain state has been modelled for three design cases.

Design case № 1: Existing operational conditions before laying CCL. The width of the traffic way, the structure of the road pavement and of the side of the road, as well as the physical and mechanical properties of their layers are accepted according to the engineering surveys for the public motor road of federal significance M-5 "Ural" km 1034 – the city of Togliatti. The temporary traffic load is taken according to "GOST 32960-2014. Motor Roads for Public Use. Design Loads, Design Load Distribution". In the calculations stability is 49.3 kN/m^2, and in the calculations for deformations it is 12.4 kN/m^2.

Design case № 2: Construction period, at the moment of building the trench up to design parameters. The difference between design cases № 2 and №1 is the presence of an open trench at the side of the road, and the decrease in temporary traffic load to the values typical for the construction period - 24 kN/m^2. During calculations various parameters of the trench have been modeled: according to depth (from 0.3 m to 0.6 m, with a step of 0.15 m), according to width (from 0.05 m to 0.15 m, with a step of 0.05 m), as well as according to the planned position relative to asphalt-concrete coating edge (from zero to 1.5 m, with a step of 0.5 m).

Design case № 3: Operational period. At this stage the stress-strain state of the road is assessed, taking into account the building and subsequent backfilling of the trench located at its side, as well as the presence of a package of micro-pipes at the bottom of the trench. The assessment has been carried out for the parameters of the trench described in design case № 2, taking into account its filling with excavated soil, and the temporary traffic load is taken as similar to design case № 1. The parameters of the micro-pipe package are as follows: width = 160 mm, height = 17.7 mm, shell thickness = 0.85 mm, weight = 975 kg / km, tension load when laying it = over 9200 N.

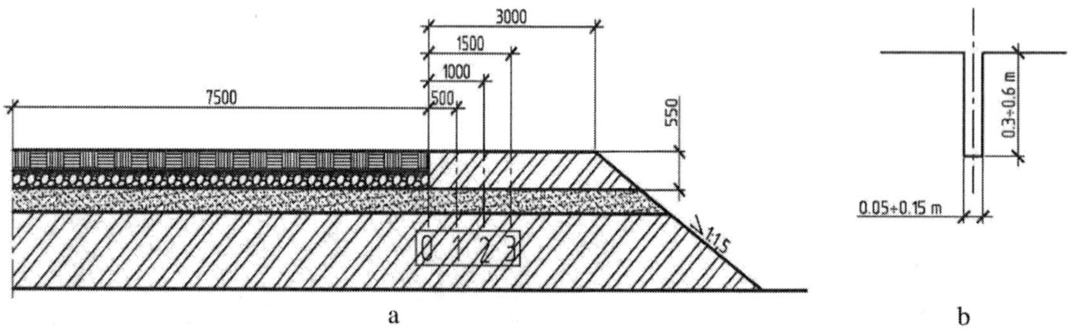

Figure 2. Geometric diagram of the design structure: a - geometric diagram of the earth roadbed; 0-3 accepted options for the location of the trench, b - geometric diagram of the accepted trench options.

The calculations have been made using Plaxis 2016 2D software package, which allows performing an advanced calculation of the stress-strain state and the stability of geotechnical objects used for various purposes, applying finite element method in a 2D problem. Design models are made of 2D elements. The calculation has been made by setting the stages of the CCL life cycle, in order to obtain the most accurate result. Each stage is a separate nonlinear analysis with iterative method for solving the stiffness matrix. The first stage is necessary to obtain the initial stress values in the original earth body and in the road pavement, which have developed over years. At this stage there is no movement of the model joints. At the following stages the values of deformations in the joints and of stress in the elements are recalculated. So, within the framework of the calculations 37 design models have been considered, examples of the graphical presentation of the calculation results are shown in Figures 3-5.

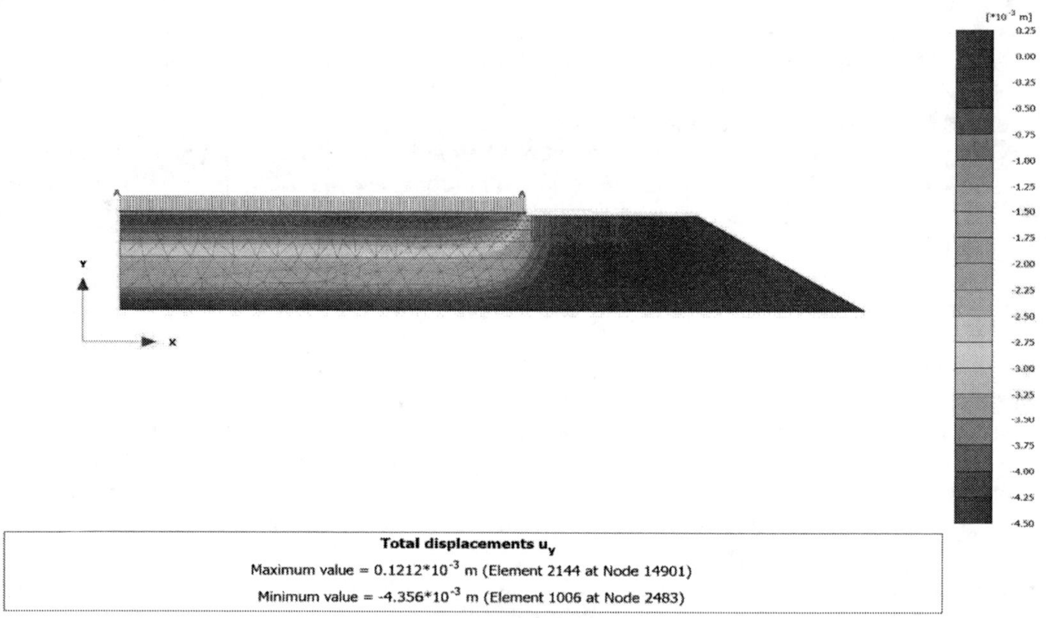

Figure 3. Defining placement.

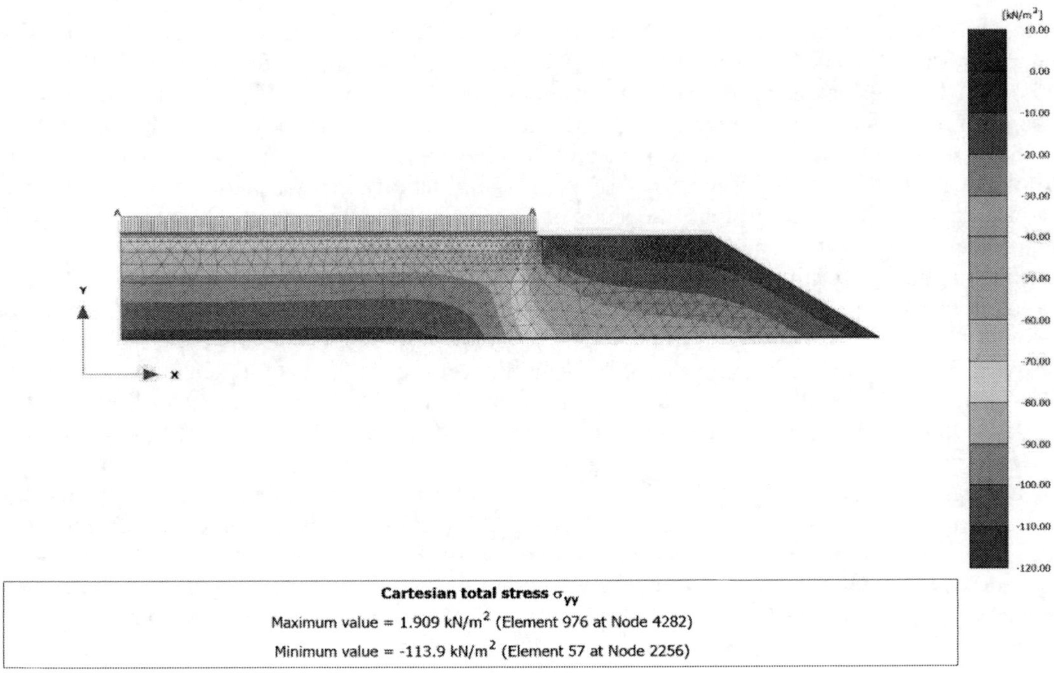

Figure 4. Determining stress values

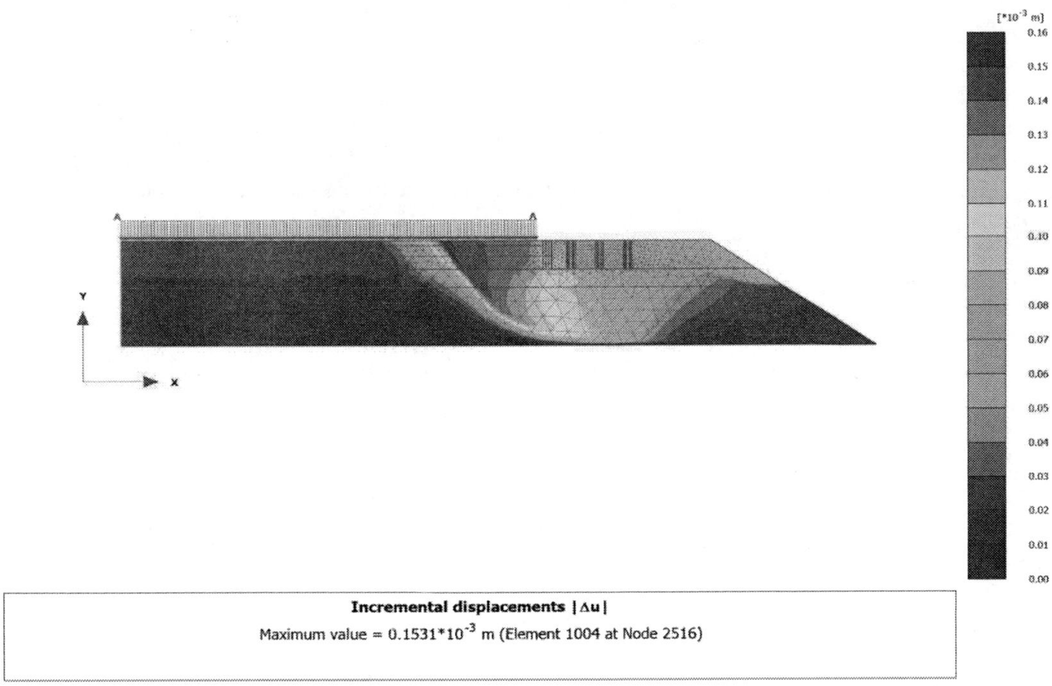

Figure 5. Determining stability factor (further – F_{stab}).

IOP Conf. Series: Materials Science and Engineering 1015 (2021) 012099 doi:10.1088/1757-899X/1015/1/012099

3. Results

The results of verification calculations have been assessed for compliance with "GOST R 50597-2017. Motor Roads and Streets. The Requirements to the Level of Maintenance Satisfied the Traffic Safety. Methods of Testing.", namely: the traffic ways next to the side of the road should not have any defects in the form of an S-wave more than 3.0 cm deep, and in the form of a track - more than 2.0 cm deep; the side of the road in which the CL is laid should not have any defects in the form of the height of the side of the road reduced by more than 4.0 cm; loss of stability of the motor road is evaluated by determining the stability factor, thus, the slope or its morphological element is considered stable if its stability factor is more than one (F_{stab}> 1).

Basing on the verification calculations results, the following conclusions have been made:

- the deformation values of the soils in the body of the motor road and in the layers of the road pavement for the initial design case (Design case № 1) are 4.36 mm, during the period of maintenance (Design case № 2) they are between 1.3 and 1.4 mm (except for the case of trench location at the edge of the traffic way, when the walls of the trench collapse), and during the operation of CCL (Design case № 3) they are between 4.3 and 4.4 mm, which is similar to the indicators for the initial design case (Design case № 1).

- the stress values of the soils in the body of the motor road and in the layers of the road pavement for the initial design case (Design case № 1) are 113.9 kN/m^2, during the period of maintenance (Design case № 2) they are 62.0 ÷ 78.0 kN/m^2 (except for the case of the trench location at the edge of the traffic way, when the walls of the trench collapse), and during the operation of CCL (Design case № 3) they are between 113.0 ÷ 114.0 kN/m^2, which is similar to the indicators for the initial design case (Design case № 1);

- when the trench is located at the edge of asphalt concrete coating (the location of the trench is "0") at the depth of 300 mm or more, the stability of the soils in the body of the motor road and in the layers of the road pavement cannot be ensured (F_{stab}<1);

- the stability factor values of the soils in the body of the motor road and in the layers of the road pavement for the initial design case (Design case № 1) are F_{stab} = 1.805, during the period of maintenance (Design case № 2) they are between F_{stab} = 1.1 ÷ 3.2 (except for the case of traffic load application at the edge of the trench, when its walls collapse), and during the operation of communication cables (Design case № 3) they are between F_{stab} = 1.5 ÷ 1.9, which is similar to the indicators of the initial design case (Design case № 1). The difference of the calculation results for Design case № 2 from the other cases is largely due to the fact that during maintenance period the traffic load from vehicles on the traffic way adjacent to the side of the road was accepted as lower than during operation period.

4. Conclusions

The performed analysis of the international experience of constructing CCL at motor roads and the verification calculations of motor roads' sections along the traffic way adjacent to the side of the road, as well as of the side of the road where it is planned to lay the communication cables, have shown that it's possible to build up a safe trench, if the trench is reconstructed with materials identical in characteristics, but not worse in quality than those used when constructing the motor road. In general it can be concluded that laying the cables at the side of the road does not affect the stress-strain state of the road, however it is recommended to apply traffic load at a certain distance from the edge of the mini-trench, which is justified by a calculation in each specific case, based on the design features of the motor road. Also proceeding from the condition that the drainage of the pavement should be maintained, the micro-pipes should be laid above the drainage layers (not deeper than 0.4 - 0.5 m), and in case there is a separating layer of geotextile the depth of location should be over it.

IOP Conf. Series: Materials Science and Engineering 1015 (2021) 012099 doi:10.1088/1757-899X/1015/1/012099

References

[1] Andreev V, Andreev R, Popov B and Popov V. 2015. Laying optical cables in the right of way of the motor roads, *First mile* **1** pp 30-5

[2] Department for Culture, Media and Sport 2011 *Micro-trenching and street works: An advice note for local authorities and communications providers* (London)

[3] International Telecommunication Union 2003 *Construction, installation and protection of cables and other elements of outside plant* (Geneva.)

[4] ITU-T Recommendation L.49 2003 *Micro-trench installation technique* Construction, installation and protection of cables and other elements of outside plant Series L (International Telecommunication Union Geneva)

[5] Arbeitspapier fur Bauleistungen zur Glasfaserkabelverlegung – Sonderverfahren Mikro / Mini-Trenching der Focusgruppe "Mikro- / Mini- Trenching" der Projektgruppe "Branchenubergreigen - de Zusammenarbeit" des IT-Gipfelprozesses 2012 (Berlin, Germany)

[6] Duraline Technical Report 2016 *Recommended Procedures for Conduit Microtrenching and Restoration* Technical Bulletin: DCEB-10002

[7] Hashemian L, Rezaei M, and Bayat A 2017 Field and laboratory investigations on pavement backfilling material for micro-trenching in cold regions *Int J of Pavement Research and Technology* **10** (4) pp 333–42

[8] Vaseli H, Hashemian L, Bayat A, Gay L, Williams I, and Melzer J. 2017 Evaluation of fiber optic installation methods, a case study on micro-trenching in Alberta, Canada. *FACETS* **2** (2) pp. 642–59

IOP Conf. Series: Materials Science and Engineering 1015 (2021) 012100 doi:10.1088/1757-899X/1015/1/012100

The influence of the pavement surface texture on the Pendulum Test Value

M Kováč and M Brna

Department of Highway Engineering, Faculty of Civil Engineering, University of Žilina, Univerzitná 8215/1, 010 26 Žilina, Slovakia

E-mail address: matus.kovac@uniza.sk

Abstract. The article deals with the problem of finding a relation between surface texture and skid resistance represented by the Pendulum Test Value. For the texture measurements was used a new device that can capture surface irregularities up to the microtexture level. Besides Pendulum Test Values were also evaluated standard roughness parameters as well as the Mean Profile Depth widely used for pavement surface macrotexture evaluation. The texture parameters were evaluated for both macro- and micro-texture of all tested surfaces. Two groups of surfaces were measured. The first group, represented by reference surfaces with known parameters, was used for evaluation of the reliability of new device as well as for determining the relation between microtexture and pendulum test value. The second group was represented by asphalt mixture specimens designed to create various surfaces with different parameters of both macrotexture and microtexture. The obtained results show that there is no unambiguous direct relation between the size of irregularities and skid resistance. That means that high values of both macro and microtexture won't have to necessarily mean a high level of friction. The proper size and proportion of both parts in their optimal combination are needed.

1. Introduction

Accurate and precise diagnostics of pavement serviceability parameters plays a key role in the aim to keep road safety at the required level for the whole time of their live cycle [1]. Skid resistance is one of the most important parameters for evaluating the quality of road pavement in terms of road safety. Although many features influence the friction on the contact between the wheel and the pavement surface (vehicle speed, tire tread depth and pattern, tire inflation, presence of water film and its depth, temperature, etc.), there is no doubt that pavement surface texture plays a key role in point of view a road pavement surface diagnostics. Pavement surface texture, besides skid resistance, influences also many other important things concerned the quality of life, such as noise production and propagation as well as particulate matter emissions [2]. Even though the main source of emissions is road traffic [3], the wearing of the pavement surface depended on its texture has a significant contribution [4]. All this has a significant influence on air quality and whole environment and climate changes and vice versa these changes should be taken into account when designing a road pavement construction [5]. The knowledge about how the mixture composition affects the pavement surface texture is important in terms of wearing course design because it can positively influence, besides other mentioned factors, also the skid resistance and the ability of the surface to keep the high level of it in long term, what is very important for increasing of road safety and decreasing the risk of car accidents.

There have been many methods and devices designed for pavement texture evaluation. The most common are the methods for contactless texture measurement by the use of laser sensors. There are

IOP Conf. Series: Materials Science and Engineering 1015 (2021) 012100 doi:10.1088/1757-899X/1015/1/012100

also other optical methods, for instance, stereophotogrammetry, but these are affected by boundary conditions such as light uniformity during measurement, image noise or multicolored road pavement surface. For comprehensive and objective evaluation of skid resistance are still mostly used devices measuring a coefficient of friction despite the problems with repeatability and reliability of many of them. On the other hand, these devices are still the only ones that can measure macro- and micro-texture at the same time and the traffic speed. In addition to that, there still hasn't been created a satisfying and objective model or formula for predicting or estimating the skid resistance based on contactless texture measurements only. The main reason is the non-sufficient resolution of currently used devices without the ability to detect surface irregularities up to microtexture level, which seems to be crucial in terms of overall friction level.

The goal of the article is to bring a little more light to the relationship between surface texture and skid resistance. For the purpose to determine the rate of influence of the texture on the coefficient of friction was tested a wide spectrum of surfaces, where besides Pendulum Test Values (PTV) were also evaluated standard roughness parameters as well as the Mean Profile Depth (MPD) widely used for pavement surface macrotexture evaluation.

2. Measuring device and testing surfaces

2.1. Static road Scanner

For the precise and accurate measuring of the tested surfaces was used the Static road Scanner (SRS). The device was designed at the University of Zilina in order to measure both macro and microtexture of the pavement surface and for this purpose uses two different optical methods. The first one uses two red lasers scanning in opposite directions with the resolution up to 15 μm and the second one uses the blue structured light with a resolution up to 2.49 μm in the X and Y-axis. The resolution in the Z-axis is approximately orders of magnitude higher than in the X and Y axes. The maximum captured size of a scanned surface at the one test represents the area of 120×100 mm, whereas the goal is to scan the surface area corresponding to the path of the rubber slider of British Pendulum Tester. The output of the scanning process is a text-file with X, Y, Z coordinates that are used as an input to the algorithm created by authors [6] in MATLAB® software. The algorithm allows calculating standard roughness, spacing, and waviness surface texture parameters [7], including the parameter - Mean Profile Depth (MPD) used for characterization of road pavement texture by use of surface profiles [8]. The algorithm provides a possibility to calculate all texture parameters for irregularities at both scales, i.e. for macrotexture as well as for microtexture.

2.2. Testing surfaces

Two groups of surfaces were measured. The first group was represented by reference surfaces with known sizes of irregularities (grit size). The reference surfaces were used for evaluation of the reliability of new device as well as for determining the relation between microtexture and pendulum test value. The second group was represented by asphalt mixture specimens designed to create various surfaces with different parameters of both macrotexture and microtexture.

2.2.1. Reference surfaces. These surfaces were represented by sandpapers with given standardized grit sizes. The texture measurements and pendulum tests were performed on more than 50 sandpapers with more than 20 different grit sizes from P7000 to P16. The grit size of sandpaper is stated as a number that is inversely related to the particle size. The grit size P7000 represents a surface with irregularities corresponding to the particle diameter (grain size) of 3 μm. The grit size P16 represents a surface with irregularities corresponding to the particle size of 1324 μm. Considering the homogeneity of surfaces of factory-produced sandpapers, the surface scanning area was limited to 30×10 mm.

2.2.2. Asphalt mixture specimens. The specimens were designed and produced to create a wide spectrum of surfaces with different macrotexture and microtexture. The different macrotexture was created by the use of different maximum grain size of aggregates in the mixture. The asphalt mixtures with the smallest macrotexture level were AC8, asphalt concrete with maximum grain size (8 mm). The mixtures AC11 and SMA 11 had the same maximum grain size (11 mm), but different particle size distribution. The mixtures with the roughest macrotexture were AC16 with maximum grain size (16 mm). Mixtures with different microtexture were created by using a different type of aggregates. The roughest microtexture level was obtained by using a Granodiorite (Gra) and the finest one by a Limestone (Lim). The middle level of microtexture was obtained by an Andesite (And) [9]. According to the above-mentioned conditions were created 12 specimens with different macro and microtexture of their surfaces [10].

3. Measuring on reference surfaces

The reliability of the new device was tested on reference surfaces with known values of geometric parameters. One of the goals was the examination of the influence of different measurement conditions on the accuracy of measurement results. Different measurement conditions were represented by different colors of sandpapers (blue, grey, black and red), different light conditions (without covering device – normal conditions, or with covered device for establishing complete darkness).

Skid resistance was measured on all reference surfaces by British Pendulum Tester [11]. The result of a pendulum test is a Pendulum Test Value (PTV).

3.1. Texture measurements and measuring device reliability verification

The reliability of the device was evaluated by comparison of measured roughness parameters with the declared grain size of sandpapers. Examples of relations between values of measured texture parameters and grain size of sandpapers are shown in figure 1 and figure 2. For better visualisation of obtained results are relations showed on semi-log graphs.

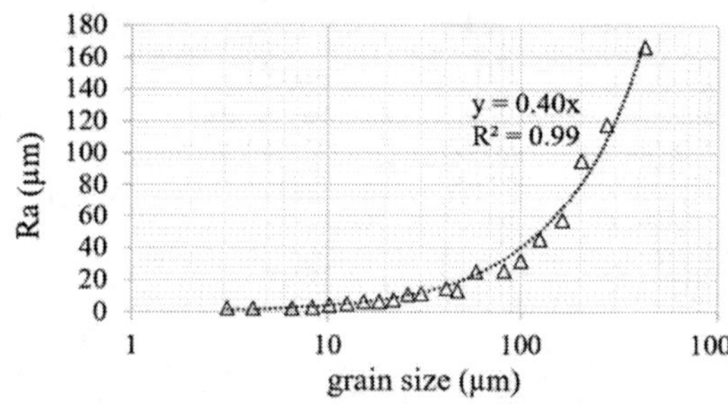

Figure 1. Comparison of measured values of an arithmetical mean deviation (Ra) of the assessed profiles with the grain size of the sandpapers.

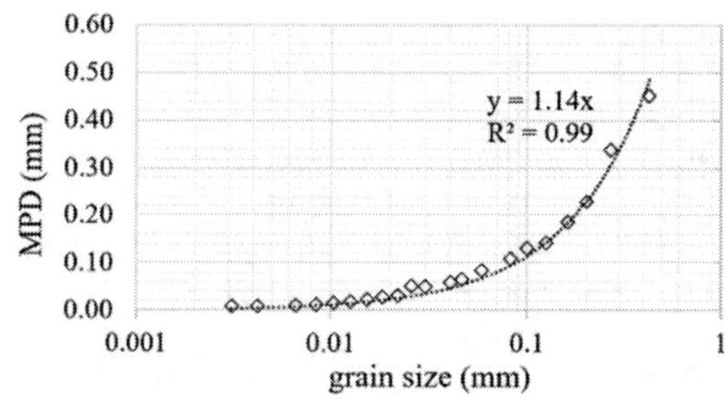

Figure 2. Comparison of measured values of a Mean Profile Depth (MPD) values with the grain size of the sandpapers.

The results obtained on reference surfaces demonstrated the accuracy and reliability of the device. Almost the same results were obtained at all different measurement conditions. Based on these promising results were prepared experiments on asphalt mixture specimens.

3.2. The Pendulum test

After texture measurements were on all reference surfaces performed the Pendulum Tests at wet conditions according to the procedure specified in the standard. Considering the believed fact that microtexture represents the deviation from a true planar surface with the characteristic dimensions along the surface of less than 0.5 mm and peak-to-peak amplitudes are in the range of 0.001 mm to 0.5 mm, the test was performed with the goal to determine the relation between microtexture and skid resistance. All sandpapers were firmly glued to the completely planar base to exclude any influence of macrotexture. Only two sandpapers (P16 and P24) in the test exceed the size of irregularities characteristic for microtexture. For a better picture, however, values obtained on these surfaces were also taken into account. Pendulum Test Values obtained on sandpapers characterised by grain sizes are shown in figure 3. Comparison of obtained Pendulum Test Values and measured MPD values on the same sandpapers are shown in figure 4. For better visualisation of obtained results are relations showed on semi-log graphs. The dashed line in the chart represents the transition between macro and microtexture.

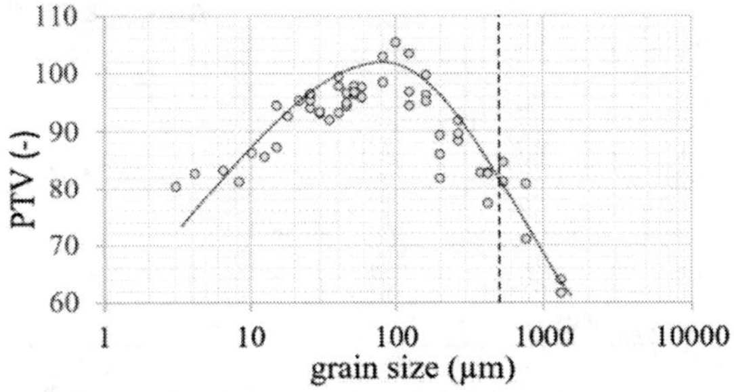

Figure 3. Pendulum Test Values measured on reference surfaces represented by the grain size of sandpapers.

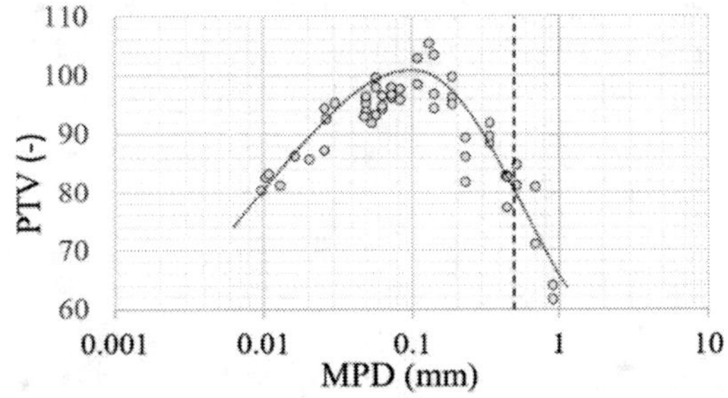

Figure 4. Comparison of Pendulum Test Values and MPD values measured on reference surfaces by the new device.

As can be seen in figure 3 and figure 4, contrary to the expected result, there is no only increasing trend of Pendulum Test Values within the whole spectrum of increasing grain sizes or MPD values. PTV increases up to grain size of approximately 125 μm and from there begins a steep decrease of Pendulum Test Values. Differences between obtained Pendulum Test Values reached up to 40 %. The highest values of PTV were obtained at surfaces with the size of irregularities in range of 80 to 125 μm. Relatively high Pendulum Test Values were also measured on surfaces with grain sizes in the range of 50 to 150 μm. As expected, on surfaces with very small values of texture parameters (3 - 20 μm) were obtained relatively small PTV. However, surprisingly low Pendulum Test Values were measured on surfaces considered as more harsh, i.e. with surface irregularities bigger than 150 μm. All this suggests that it will not be possible to make a prediction of skid resistance based only on simple profile hight parameters, since high values of texture don't have to necessarily mean a high level of friction.

Clear increasing tendency where can be find a relatively close correlation between microtexture and skid resistance using mentioned parameters is when considering surfaces with the size of irregularities up to 125 μm. For the best fit was used a power function of the regression curve shown in figure 5 and figure 6.

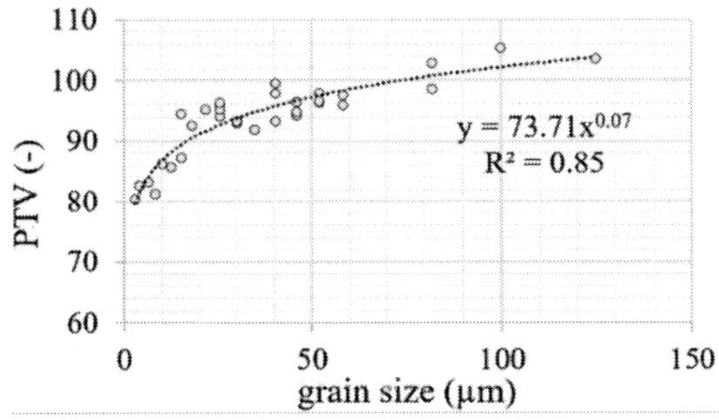

Figure 5. The relation between Pendulum Test Values and the grain size of sandpapers with the limited size of surface irregularities to 125 μm.

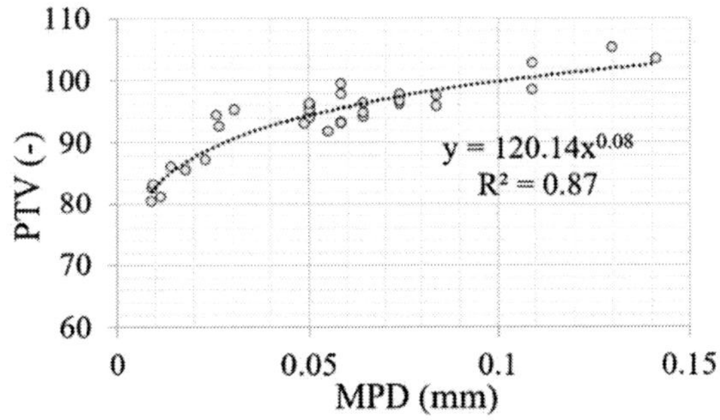

Figure 6. The relation between Pendulum Test Values and MPD values with the limited size of surface irregularities up to 125 μm.

IOP Conf. Series: Materials Science and Engineering 1015 (2021) 012100 doi:10.1088/1757-899X/1015/1/012100

4. Measuring on surfaces of asphalt mixture specimens

On 12 surfaces of asphalt mixture specimens with different macro and microtexture, created according to conditions mentioned above, were performed the 3D scanning by SRS and the Pendulum Test. All tests were performed on the brand new specimen surfaces where the aggregates were coated by a bitumen film, as well as on the same surfaces with revealed microtexture of the aggregates after the removal of bitumen film by perchloroethylene. The scanning of macrotexture was performed by use of red laser sensors at the area corresponding to the rubber slider path of the pendulum. The microtexture was measured only on few (4 – 8) aggregate grains due to the small depth of focus of structured light sensor. The Pendulum tests were performed on both, dry and wet surfaces, at two places in two opposite directions at each specimen. Obtained results of all measurements are shown in table 1.

Table 1. Results of texture measurements and pendulum test.

Mixture	COATED				REVEALED			
	MPD_{macro} (mm)	MPD_{micro} (mm)	PTV_{DRY} (-)	PTV_{WET} (-)	MPD_{macro} (mm)	MPD_{micro} (mm)	PTV_{DRY} (-)	PTV_{WET} (-)
AC8-And	0.85	0.092	92	61	1.07	0.098	81	66
AC8-Lim	0.88	0.066	92	58	1.15	0.077	82	65
AC8-Gra	1.23	0.075	86	60	1.35	0.123	83	66
AC11-And	1.69	0.044	88	58	1.29	0.114	80	67
AC11-Lim	1.58	0.059	85	58	1.22	0.099	84	62
AC11-Gra	1.41	0.055	88	56	1.34	0.112	81	64
SMA11-And	2.08	0.107	86	63	1.55	0.117	81	68
SMA11-Lim	2.01	0.090	85	65	1.41	0.123	82	64
SMA11-Gra	1.80	0.099	88	66	1.42	0.125	84	68
AC16-And	1.86	0.109	89	70	1.38	0.148	83	68
AC16-Lim	1.75	0.155	87	65	1.26	0.150	83	63
AC16-Gra	1.62	0.142	86	67	1.33	0.111	84	68

4.1. Texture measurements results

The texture of measured surfaces was evaluated by MPD value for both macro and microtexture. Results of macrotexture measurements for coated and revealed surfaces of asphalt mixture specimens are shown in figure 7 and figure 8.

As can be seen in figure 7, the macrotexture level of the coated surfaces of specimens corresponds to the maximum grain size of the asphalt mixture. The smallest MPD values are obtained at specimens denoted as AC8 for all used aggregate type. As expected, second in order with higher MPD values are specimens AC11. However, surprisingly highest MPD values were obtained at stone mastic asphalt specimens denoted as SMA 11, even though specimens AC16 has a bigger maximum grain size. This can be explained by the fact that the stone mastic asphalt is a gap graded hot mix asphalt and contains only a small percentage of aggregate particles in the mid-size range what can influence the average profile hight and then the overall MPD value.

Results of macrotexture measurements obtained on revealed surfaces after the removal of the bitumen film are presented in figure 8. It can be seen the same order of the specimens according to MPD values depended on grading and maximum grain size, however, the differences between mixtures are much less than at coated specimens. The MPD increased only for specimens AC8 and for all other mixtures MPD values decreased. This could be caused by sufficient removal of the bitumen film only from the surface of the aggregates and at the same time by insufficient film removal from valleys and cavities.

MPD values of microtexture level of specimen surfaces are shown in figure 9 and figure 10.

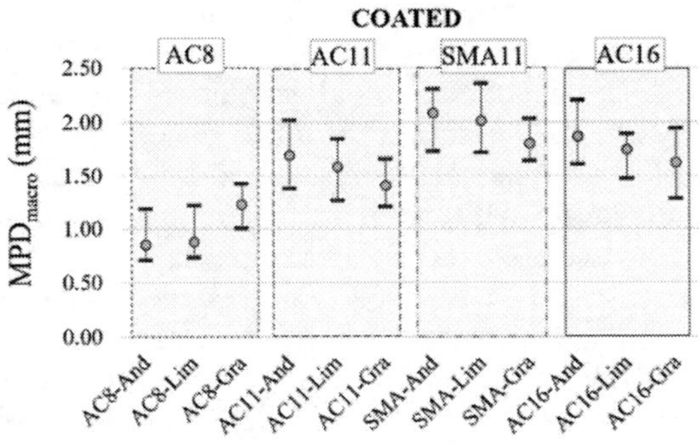

Figure 7. MPD values of macrotexture measured on surfaces of mixtures with bitumen film on the aggregate grains.

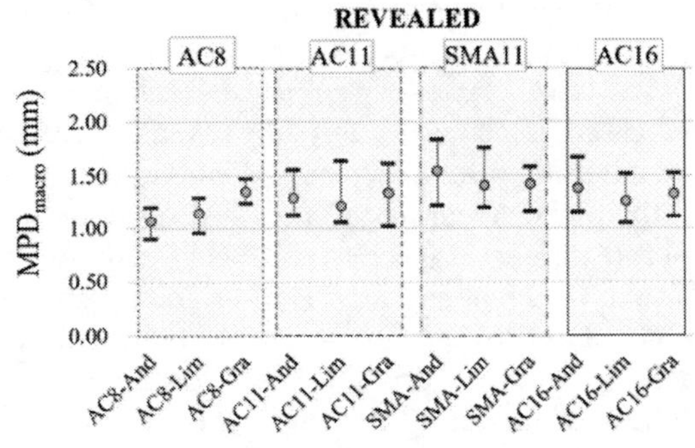

Figure 8. MPD values of macrotexture measured on surfaces of mixtures after removal of bitumen film from the aggregate grains.

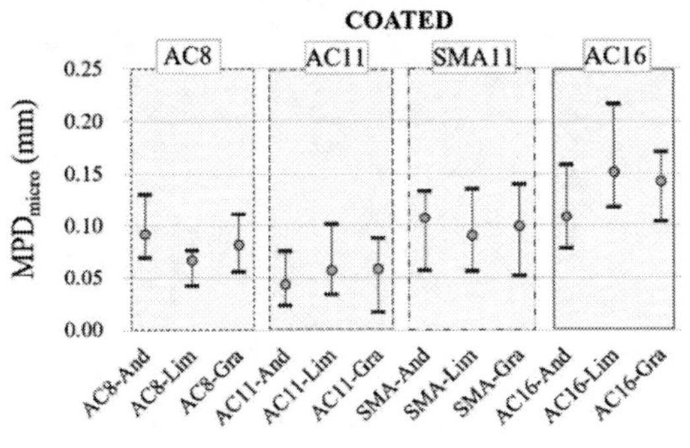

Figure 9. MPD values of microtexture measured on surfaces of mixtures with bitumen film on the aggregate grains.

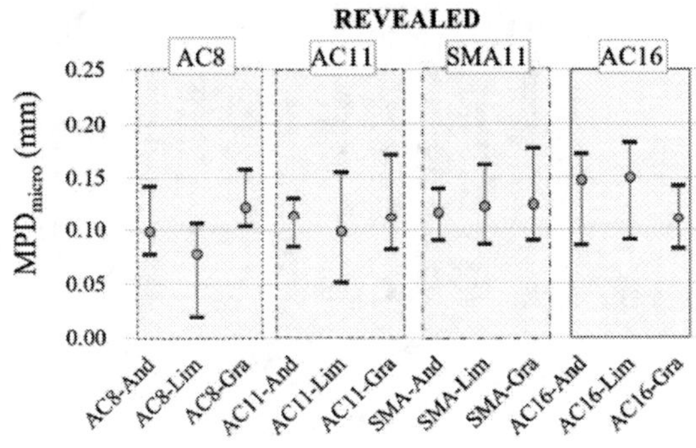

Figure 10. MPD values of microtexture measured on surfaces of mixtures after removal of bitumen film from the aggregate grains.

The premise was that at coated surfaces will be obtained approximately the same and low levels of microtexture at all specimens, regardless of types of aggregates. The results didn't confirm the expectations and microtexture levels vary significantly. The second premise was that after removing the bitumen film from grain surfaces the microtexture level will increase at all specimens and the levels will be in the order: the least for the Limestone, the middle for Andesite and the highest for Granodiorite. As was expected, the average microtexture level increased at all revealed surfaces, however, the measured values show a very high level of dispersion. Also, the expected order of microtexture levels for aggregate types was confirmed only at mixtures AC8 and AC11. In addition to that, the high dispersion set the examined surfaces of different aggregates approximately at the same level. This could be caused by calculating of microtexture parameters based on raw measured data without filtering the longer wavelengths (although measured by the sensor with low depth of field) where were taken into account bigger surface irregularities than are representative for microtexture level. Also, the amount of measured data for this purpose turned out to be insufficient.

4.2. Pendulum test results

Results of measured Pendulum Test Values on coated and revealed surfaces of all specimens at dry conditions are shown in figure 11. As expected, the Pendulum Test Values measured on dry surfaces were significantly higher than on wet surfaces for both, coated and revealed aggregate grains. At coated surfaces with bitumen film were obtained higher PTV at all dry specimens than on surfaces with revealed aggregate grains. This is due to adhesion forces which are highest at dry an smooth surfaces. The highest PTV were obtained on mixtures AC8 with fine grading influencing the area size of contact with a rubber slider. Pendulum test values on rest of mixtures reached a little lower but very similar level. After revealing the bitumen film from grain surfaces the Pendulum Test Values decreased, which was caused by revealing the microtexture, the surfaces were not so smooth anymore and adhesion forces decreased. Pendulum Test Values are still relatively high and at a similar level for all mixtures. According to obtained results, it is hard to claim what type of aggregates and hence the microtexture level has the most beneficial influence on the overall level of skid resistance.

A quite different situation is after getting surfaces wet (figure 12). As expected, there were obtained significantly lower values of PTV on all surfaces in comparison with dry surfaces. Contrary to the measurements on dry surfaces, Pendulum Test Values measured on wet and coated surfaces are highest on mixtures with the highest level of macrotexture, i.e. SMA11 and AC16. Considering the missing contribution of microtexture, this can be caused by better drainage properties and a higher decrease in the kinetic energy of rubber slider after bumping into the bigger particles. After the microtexture revealing by the bitumen film removal from grain surfaces, the Pendulum Test Values are higher almost at all specimens.

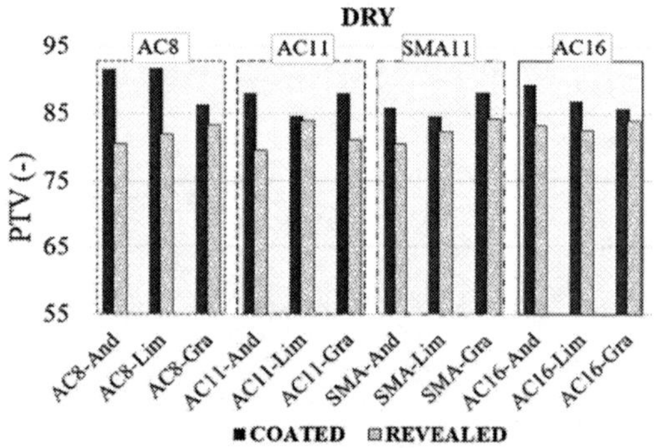

Figure 11. Pendulum Test Values measured on dry surfaces of mixtures with bitumen film on aggregates (COATED) and after its removal.

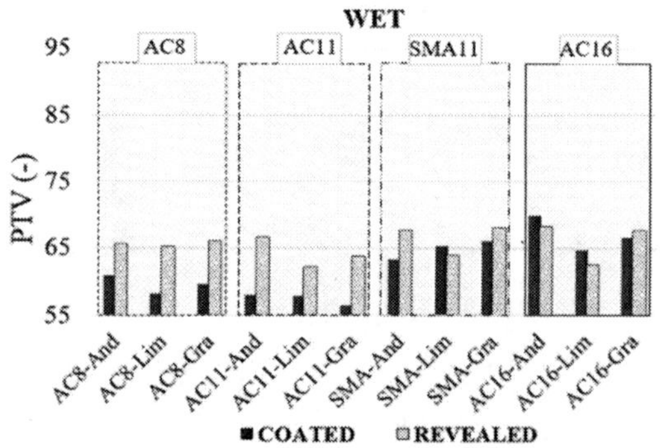

Figure 12. Pendulum Test Values measured on wet surfaces of mixtures with bitumen film on aggregates (COATED) and after its removal.

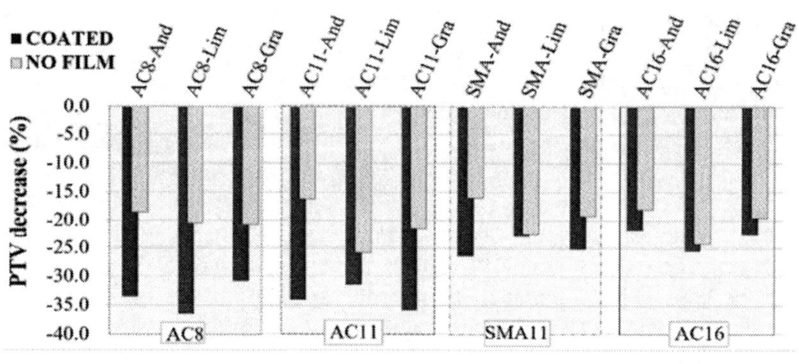

Figure 13. Decrease of Pendulum Test Values after getting surfaces wet.

The ratio of the decrease of Pendulum Test Values after getting surfaces wet is shown in figure 13. As can be seen in figure 13, as expected, there is a decrease of Pendulum Test Values after getting surfaces wet on all surfaces no matter the surfaces are coated or revealed. However, the decrease in PTV on specimens without bitumen film is significantly lower than on coated specimens, especially on mixtures with fine grading (AC8 and AC11, mixtures with low macrotexture level). The decrease of PTV after getting surfaces wet on the coated surfaces can reach up to 37 % (AC8-Lim, AC11-Gra), while the decrease at the same mixtures with revealed microtexture was only about 21 %. The microtexture and the bigger contact area at these mixtures complement the low macrotexture. Lower differences between the decreases at coated and revealed mixtures SMA11 and AC16 show the bigger influence of macrotexture at these specimens.

Comparing different types of aggregates, the lowest decrease within individual mixtures show specimens with Andesite and the highest show specimens with Limestone. Although the differences between results compared on revealed surfaces are not very significant there can be seen an inalienable benefit of microtexture on skid resistance at wet conditions.

The ratio of the decrease of Pendulum Test Values, or the increase respectively, according to measurements on dry and wet surfaces after removing a bitumen film from the surface of the aggregate grains and hence the revealing a microtexture is shown in figure 14.

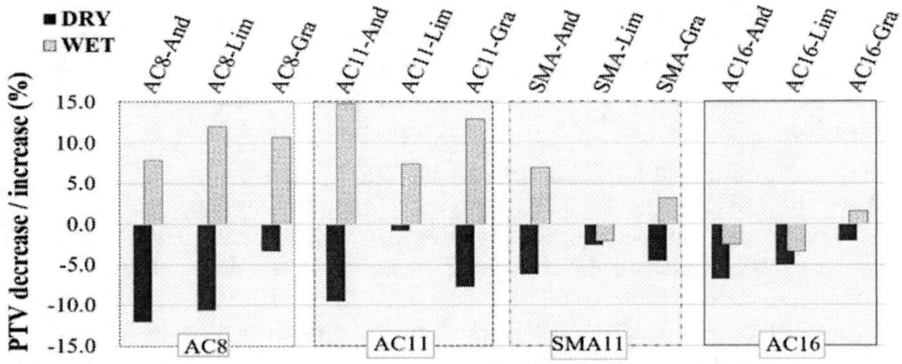

Figure 14. Decrease/Increase of Pendulum Test Values after removing a bitumen film from the surface of the aggregates and hence revealing a microtexture.

As can be seen in figure 14, at all specimens was noticed a decrease of PTV measured on dry surfaces after revealing a microtexture. As mentioned above, this was due to the decreasing of adhesion forces which are biggest at dry and smooth surfaces. The biggest decrease of PTV on dry surfaces was recorded on most mixtures with fine grading and biggest contact area (AC8 and AC11) where were obtained the highest values at coated surfaces. Nevertheless, the overall Pendulum Test Values stayed at a very high level. Contrary to the measurements on dry surfaces, after the microtexture revealing the Pendulum Test Values on wet surfaces increased at all specimens of mixtures AC8 and AC11, at two specimens of mixtures SMA11 and one specimen of mixture AC16. This is proof of the important role of microtexture to provide sufficient friction at wet conditions. An unexpected decrease in PTV on wet specimens SMA11-Lim, AC16-And, AC16-Lim show the bigger influence of macrotexture then the revealed microtexture at the smaller contact area than on mixtures with fine grading.

5. Relations between MPD and PTV

To determine the rate of influence of texture on skid resistance were examined correlations between MPD and PTV measured on both, dry and wet surfaces. Correlations between these variables for macrotexture are shown in figures 15 and 16.

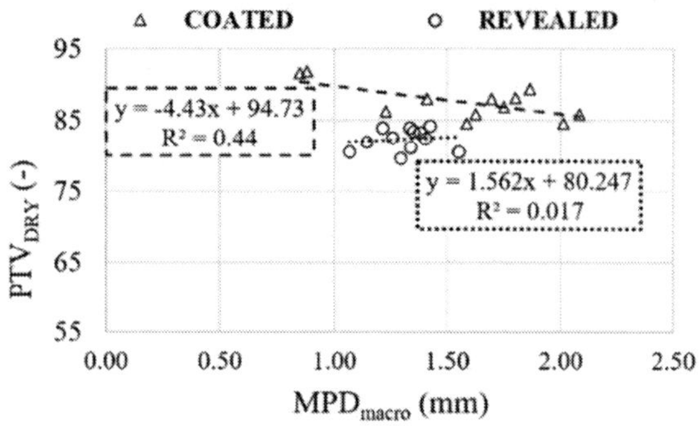

Figure 15. Relations between MPD values of macrotexture and Pendulum Test Values measured on dry surfaces.

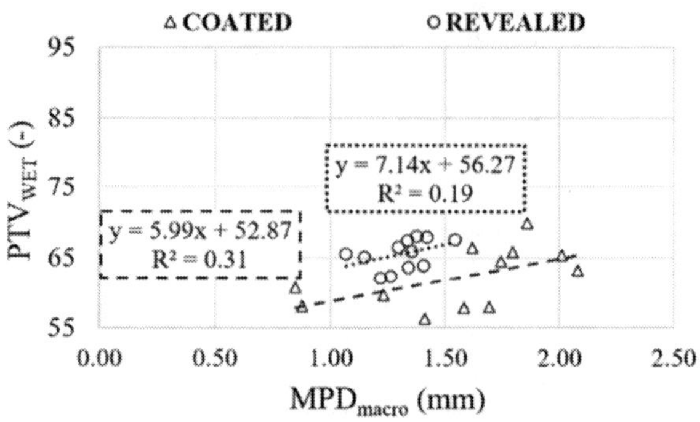

Figure 16. Relations between MPD values of macrotexture and Pendulum Test Values measured on wet surfaces.

As can be seen, the strength of an association between macrotexture characterised by parameter MPD and skid resistance characterised by PTV is very low for all examined cases.
The highest value R^2 was obtained from the relation between MPD and PTV measured on dry and coated surfaces what was explained above by the influence of high adhesion and bigger contact area at fine grade mixtures. However, it is still not satisfying strength of the association between these variables. There wasn't found any significant relation between variables measured at dry and revealed surfaces. The main reason is the small range of macrotexture levels over the spectrum of the surfaces as well as the fact that the differences between obtained Pendulum Test Values are also not very high.
Relations between MPD values of macrotexture and Pendulum Test Values measured on wet surfaces are also not very significant. A certain upward trend can be seen on coated surfaces, which can be explained by better water drainage from the surfaces of mixtures with higher macrotexture level. However, the R^2 value is still very low. There also wasn't found any significant relation between variables measured at wet and revealed surfaces due to the same reason as at dry surfaces i.e. the small range of measured values.
Examined correlations between MPD and PTV measured on both, dry and wet surfaces according to microtexture level are shown in figures 17 and 18.

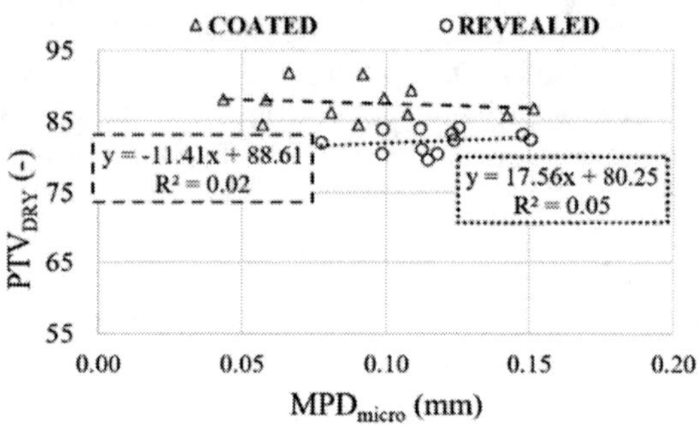

Figure 17. Relations between MPD values of microtexture and Pendulum Test Values measured on dry surfaces.

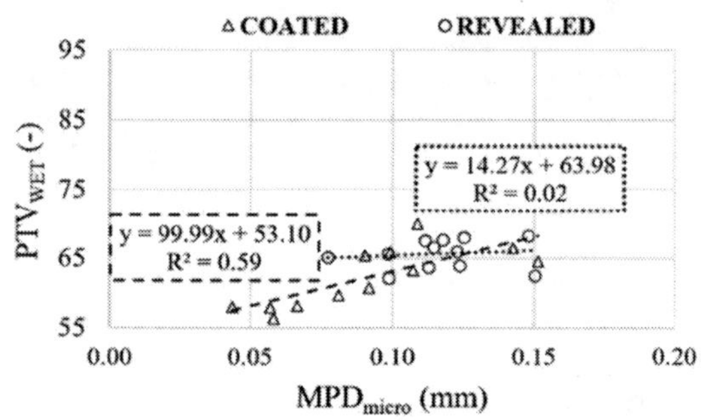

Figure 18. Relations between MPD values of microtexture and Pendulum Test Values measured on wet surfaces.

As can be seen, the strength of an association between microtexture characterised by parameter MPD and skid resistance characterised by PTV is in most cases even lower than at macrotexture evaluation. The main reason at dry surfaces is that even though there is a relatively wide range of measured values of microtexture obtained Pendulum Test Values were approximately at the same level. The average values of microtexture after revealing was in the range of 80 – 150 μm, which

IOP Conf. Series: Materials Science and Engineering 1015 (2021) 012100 doi:10.1088/1757-899X/1015/1/012100

corresponds to the range of grain sizes of reference surfaces where were also obtained Pendulum Test Values not with big variance. That means that change of microtexture depth in such range didn't cause any significant change of PTV.

Relations between MPD values of microtexture and Pendulum Test Values measured on wet surfaces are also not very significant. Practically, there wasn't found any relation between considered MPD values for microtexture and PTV on revealed surfaces. The relatively high value of R^2 obtained for the relation between MPD and PTV we consider as a coincidence due to a very high level of dispersion of measured values at individual specimens.

6. Conclusions

The main goal of the article was to analyze the influence of surface texture on skid resistance represented by Pendulum Test Value. For this purpose were analysed results of texture measurement by use of a new device designed at the University of Zilina. Texture measurements were performed on two groups of surfaces. The first group was represented by reference surfaces with known sizes of irregularities. The second one was represented by specimens of asphalt mixtures designed and produced to create surfaces with different macrotexture and microtexture. The different macrotexture was created by the use of different maximum grain size of aggregates in the mixture. Mixtures with different microtexture were created by using a different type of aggregates. For the evaluating of the relation between texture and skid resistance were compared parameters MPD and Pendulum Test Value.

According to obtained results, we can state the following conclusions and recommendations for further research:

- The new device, Static Road Scanner is reliable a can be used for measurement of pavement surface texture. However, measurements on specimens by the use of structured light showed the need to change the methodology of microtexture measurement and evaluation.
- Measurements on reference surfaces (without macrotexture) showed that there is no only increasing trend of Pendulum Test Values within the whole spectrum of increasing values of microtexture.
- The highest values of PTV were obtained at surfaces with the size of irregularities in range of 80 to 125 μm. A steep decrease of Pendulum Test Values was observed on surfaces considered as more harsh, i.e. with surface irregularities bigger than 150 μm.
- Clear increasing tendency where can be find a relatively close correlation between microtexture and skid resistance was found at surfaces with the size of irregularities up to 125 μm.
- Measurements on created asphalt mixture specimens confirmed some expected results such as level of macrotexture corresponding to aggregate grading, a decrease of PTV on wet surfaces, an increase of PTV after revealing the bitumen film. However, a lot of results didn't end up as was anticipated. Especially microtexture evaluation didn't confirm expected levels of microtexture for different types of aggregates at all specimens and showed a very high dispersion of measured values.
- The strength of an association between macrotexture characterised by parameter MPD and skid resistance characterised by PTV was very low for all examined cases. The main reason was unreliable microtexture data and a small range of different levels of both macro and microtexture achieved on created specimens.

All this suggests that it will not be possible to make a prediction of skid resistance based only on simple texture parameters, since as showed up that high values of texture don't have to necessarily mean a high level of friction. The proper size and proportion of both parts in their optimal combination are needed. In further research, we will focus on the evaluation of 3D areal (volume and feature) parameters on more surfaces with a wider range of macrotexture and microtexture.

IOP Conf. Series: Materials Science and Engineering 1015 (2021) 012100 doi:10.1088/1757-899X/1015/1/012100

Acknowledgements
The paper is a result of the research supported by the research project VEGA 1/0537/17 „*The influence of pavement surface morphology on pavement serviceability and emissions production*".

References

[1] Mikolaj J, Remek L and Kozel M 2016 Life cycle extension of a pavement structure. Communications *Scientific Letters of the University of Zilina* **18** pp 83-89
[2] Ďurčanská D, Decký M, Licbinsky R and Huzlik J 2013 Project SPENS - sustainable pavement for European new member states. Communications *Scientific Letters of the University of Zilina* **15** pp 49-55
[3] Drličiak M and Celko J 2018 Air TRITIA transport model *27th Russian-Polish-Slovak Seminar. Theoretical Foundation of Civil Engineering MATEC Web of Conferences* **196** p 1485-93
[4] Jandacka D and Durcanska D 2019 Differentiation of particulate matter sources based on the chemical composition of pm10 in functional urban areas *Atmosphere* **10** pp 583
[5] Trojanová M, Decký M and Remišová E 2015 The implication of climatic changes to asphalt pavement design *24th Russian-Polish-Slovak Seminar on Theoretical Foundation of Civil Engineering Procedia Engineering* **111** pp 770-776
[6] Kotek P, Kováč M and Decký M 2015 The comparison of standard methods for pavement texture evaluation with unconventional approach using 3D scanning *15th Int Multidisciplinary Scientific Geoconference and EXPO - SGEM* pp 1167-74
[7] Standard STN EN ISO 4287:1998 Geometrical Product Specifications (GPS). Surface texture: Profile method - Terms. definitions and surface texture parameters
[8] Standard STN EN 13473-1:2002 Characterization of pavement texture by use of surface profiles - Part 1: Determination of Mean Profile Depth
[9] Florkova Z, Kovac M and Komacka J 2016 Comparison of aggregate microtexture values obtained from different quarries. *16th International Multidisciplinary Scientific GeoConference - SGEM* pp 111-118
[10] Zummer L 2019 *The analysis of the influence of the pavement surface texture on the coefficient of friction* Diploma Thesis University of Zilina (Zilina, Slovakia)
[11] Standard STN EN 13036-4:2004 Road and airfield surface characteristics – Test methods – Part 4: Method for measurement of slip/skid resistance of a surface. The pendulum test

IOP Conf. Series: Materials Science and Engineering 1015 (2021) 012101 doi:10.1088/1757-899X/1015/1/012101

Issues of traffic surveys within the TRANS TRITIA project

P Danišovič[1], M Pitoňák[2], M Trnka[3] and R Ondrejka[3]

[1] University of Žilina, Faculty of Civil Engineering, Department of Construction Management, Univerzitná 8215/1, 010 26 Žilina, Slovakia
[2] University of Žilina, Faculty of Civil Engineering, Centre of Excellence in Transport Engineering, Univerzitná 8215/1, 010 26 Žilina, Slovakia
[3] Transport Research Institute, JSC., Veľký Diel 3323, 010 08 Žilina, Slovakia

E-mail address: peter.danisovic@uniza.sk

Abstract. Road infrastructure is increasingly loaded by national and international freight transport, and so the European Union is trying to force Member States to try to shift some road freight transport to another mode of transport to create a large and efficient intermodal freight network. In this paper, we present a research project called TRANS TRITIA solved within the resources of the Interreg Central Europe program, which falls under priority axis no. 4 "Cooperation on transport to better connect CENTRAL EUROPE" and the specific program objective SO 4.2 "To improve coordination among freight transport stakeholders for increasing multimodal environmentally-friendly freight solutions. We will focus on its objectives by which wants to contribute to the efficient utilization of intermodal freight transport, to make transport more environmentally friendly. The project partners from Slovakia, the Czech Republic and Poland, together with representatives of the relevant regional governments, focused on improving the interconnection of Central European countries in the field of freight transport. In the second part of the paper, we will discuss in more detail the issue of traffic surveys, which served to obtain important data for the transport model of the TRITIA area.

1. Introduction – about project TRANS TRITIA

The project focuses on cross-border, transnational and interregional cooperation to strengthen economic and social cohesion in order to achieve the objectives defined in the Europe 2020 strategies and the EU White Paper on Transport. White Paper entitled "Roadmap to a Single European Transport Area - Towards a competitive and resource efficient transport system". The objectives of the document point primarily to minimizing the impact of transport on the environment. In particular, Europe's dependence on oil imports should be reduced, while CO_2 emissions from transport should fall to 60% by 2050 in the context of growing transport and mobility support. To achieve these goals, the intention is to transfer 30% of road freight transport over 300 km by 2030 to other modes of transport, such as for rail or shipping, and by 2050 it should be more than 50% [1].

The main objective of the project is to improve information, planning and coordination between regional authorities, transport network managers and freight transport participants due to more efficient use of the development of transport infrastructure with the support of environmentally friendly transport modes in the TRITIA region. The EGTC TRITIA is a European grouping of territorial cooperation TRITIA, which unites 3 regions of Poland, Slovakia and the Czech Republic (figure 1). It is a cross-border territory, which includes the Žilina self-governing region (SK), the Moravian-Silesian region (CZ) and the Silesian Voivodeship (PL). TRITIA is one of the most populous cross-border areas in Europe, with important transport corridors. The most used are the transport corridors of road infrastructure, less railway infrastructure and the least insufficiently built water infrastructure. The region covers an area of approximately 24.5 thousand km^2 and a population of approx. 6.5 mil. inhabitants [2].

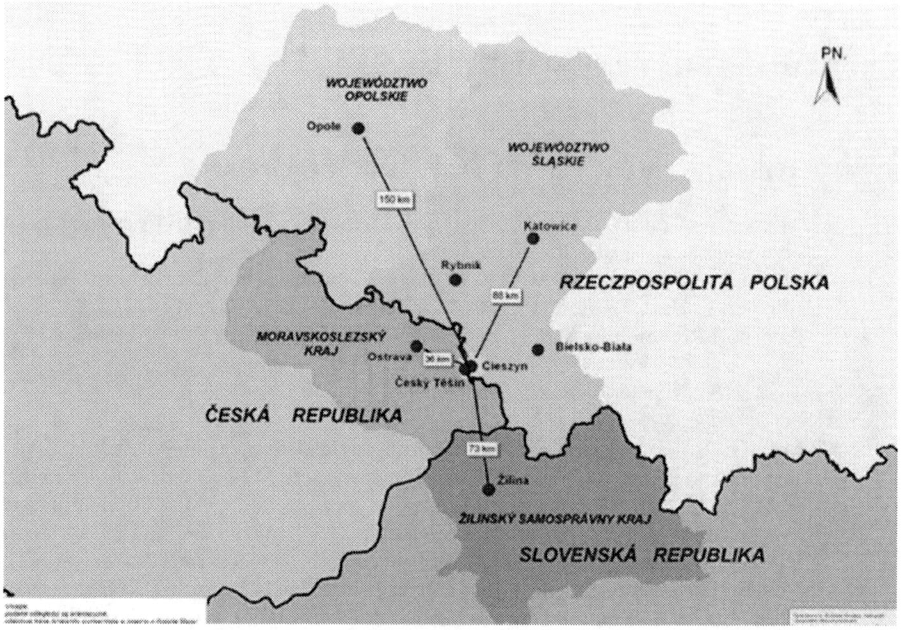

Figure 1. TRITIA region + Opole voivodeship.

The project consists of 2 parts:
1. Multimodal freight transport strategy and action plans for implementation in the TRITIA area.
2. Transport model describing the current cross-border situation between individual regions with a view to 2030.

The project consists of 5 work packages:
1. WP M: Project Management (4 activities and 21 partial outputs),
2. WP C: Communication / publicity of the project (7 activities, 2 main outputs and 18 partial outputs)
3. WP T1: Preparation of the TRITIA strategy for regional multimodal freight transport and action plans (2 activities, 2 main outputs and 17 partial outputs)
4. WP T2: Multimodal transport across the territory of TRITIA (3 activities, 3 main outputs and 12 partial outputs)
5. WP T3: TRITIA transport model (2 activities, 1 main output and 9 partial outputs)

2. Traffic surveys

One of the most important parts of the project is the elaboration of a multimodal freight transport strategy in the monitored regions, while the main tool for verifying the proposed measures is the elaboration of a cross-border multimodal model describing in particular the state and development of freight transport. The main task of the model is to identify the potential of shifting road transport over 300 km from roads to alternative modes of transport (rail, inland waterway) in the TRITIA area of interest with a forecast of its development until 2030. In addition to transport relations in the TRITIA area, the transport model should also take into account wider transport relations across its borders. For the above reasons, the model is highly demanding on the input traffic-engineering data for the needs of model creation and its calibration. To obtain most of the input data, 3 types of surveys were carried out within the project:

- The questionnaire traffic survey on border crossings (SK-CZ, CZ-PL, SK-PL).
- Profile traffic survey on identified road infrastructure profiles.
- Transport demand survey between operators of freight transport and manufacturing enterprises [3].

In this paper, we focus on the Questionnaire Traffic Surveys, as the main source of data for international transit freight transport, in which there is a presumption that it could be implemented in another mode of transport.

2.1. Questionnaire Traffic Surveys

These surveys were the main element of the project, on the basis of which the most important transport relations in the TRITIA area were determined. They were implemented at road border crossings intended for freight transport:

1. SK/CZ, I/10-I/35 Makov – Bíla Bumbálka
2. SK/CZ, I/11 Svrčinovec – Mosty u Jablunkova
3. SK/PL, D3-S1 Skalité – Zwardoń
4. SK/PL, I/59-7 Trstená – Chyzne
5. CZ/PL I/57-41 Bartultovice-Vysoká – Trzebina
6. CZ/PL I/67-78 Bohumín – Chalupki
7. CZ/PL D1-A1 Bohumín – Gorzycki Laziska
8. CZ/PL I/48-52 Český Těšín – Czieszyn

As one of the project partners from Slovakia, we carried out surveys at the first four SK-CZ and SK-PL border crossings (figure 2).

Figure 2. Performance of questionnaire traffic surveys.

The organization of such surveys required a long time preparation in terms of projects of temporary traffic signs and subsequent permits from the competent authorities and, last but not least, the assistance of the Slovak Police, without which it would not be possible (unjustified) to stop the lorries at border crossings (near them). In total, this 12-hour survey from 6:00 am to 6:00 pm in 1 day at 4 locations was carried out by 40 interviewers, who were divided according to the expected intensity of freight traffic. The survey itself was preceded by the training of interviewers, which explained the required data from the counting sheet (figure 3).

Border crossing: Name:
Date: Hour:
Direction: No. of standpoint:
Serial number of counting sheet:

N	Time	Vehicle category		Origin	Destination	Route – according to the coding key	Frequency Scheduled		Frequency Not-scheduled	
1		MGVT	HGV				<	1-2*W	1*Y	2-5*Y
		HGVT	ST				3-4*W	5-7*W	6-10*Y	>10*Y
2		MGVT	HGV				<	1-2*W	1*Y	2-5*Y
		HGVT	ST				3-4*W	5-7*W	6-10*Y	>10*Y
3		MGVT	HGV				<	1-2*W	1*Y	2-5*Y
		HGVT	ST				3-4*W	5-7*W	6-10*Y	>10*Y

Figure 3. Counting sheet.

The transit time, vehicle category, source and destination of the transport, route and frequency of transport were recorded. During the survey, prepared map base and encryption keys were used (figure 4).

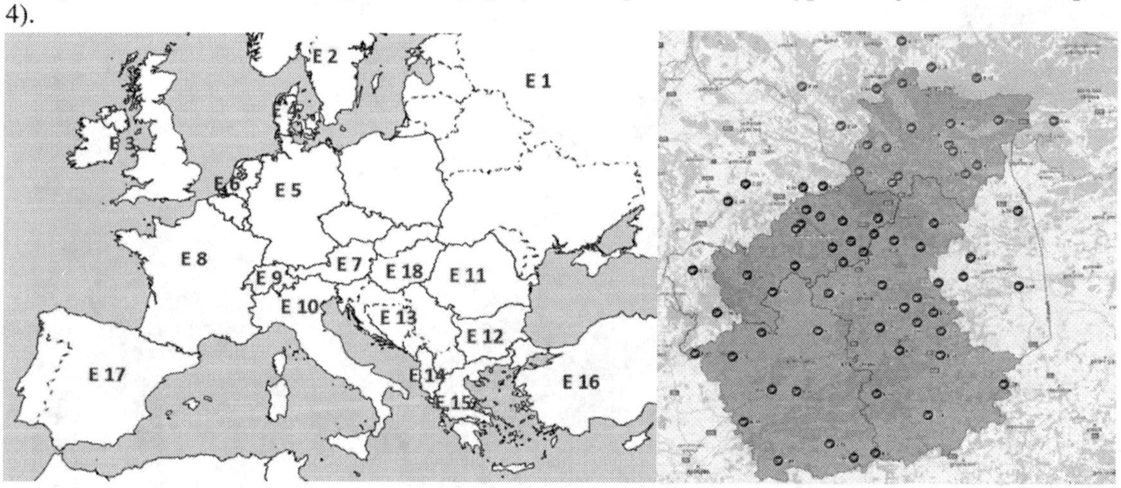

Figure 4. Example of map base and encryption key [4].

Despite several problems (dissatisfaction/complaints of residents near the counting point, delays of police officers, traffic jams due to a not working truck on the 1st class road in Svrčinovec city, etc.) we managed to capture a high proportion of passing trucks. Together, we managed to inquire 3696 drivers at 4 Slovak border crossings (in 12 hours). This huge amount of data was thoroughly analyzed and, together with other data from other surveys, were used in the creation of the transport model of the TRITIA area and its calibration.

3. TRITIA traffic model

The need to develop the TRITIA transport model stems from the current situation of the regions. Searching for solutions to remove cross-border problems between different transport modes, which are designed to increase efficiency and promote the development of individual regions. The increasing demand for transport points to infrastructure bottlenecks in individual regions and also in cross-border connections. The aim of the transport model is to point out the need for the development of transport infrastructure in individual regions and also in cross-border interconnections and to support the unification of procedures in the countries concerned. For this reason, two main objectives were defined during the elaboration of the transport model:

1. Quantification of the usable potential for the redistribution of traffic from road freight to alternative modes of transport (rail and inland waterway) in relation to the objectives defined in the White Paper (shift of road freight over 300 km to other modes of transport by at least 30%).
2. Defining bottlenecks in transport infrastructure and proposing typological measures to increase their capacity with the aim of increasing the volume of usable potential of shifting the traffic load to more environmentally friendly transport modes [5].

The construction of the TRITIA transport model (figure 5) takes into account the purpose and goals of the project described above. For this reason, the TRITIA transport model consists of two submodels:

- Submodel describing intra-zoning and inter-zoning transport relations;
- Submodel of the surrounding area, including inter-zoning roads and international transport [6].

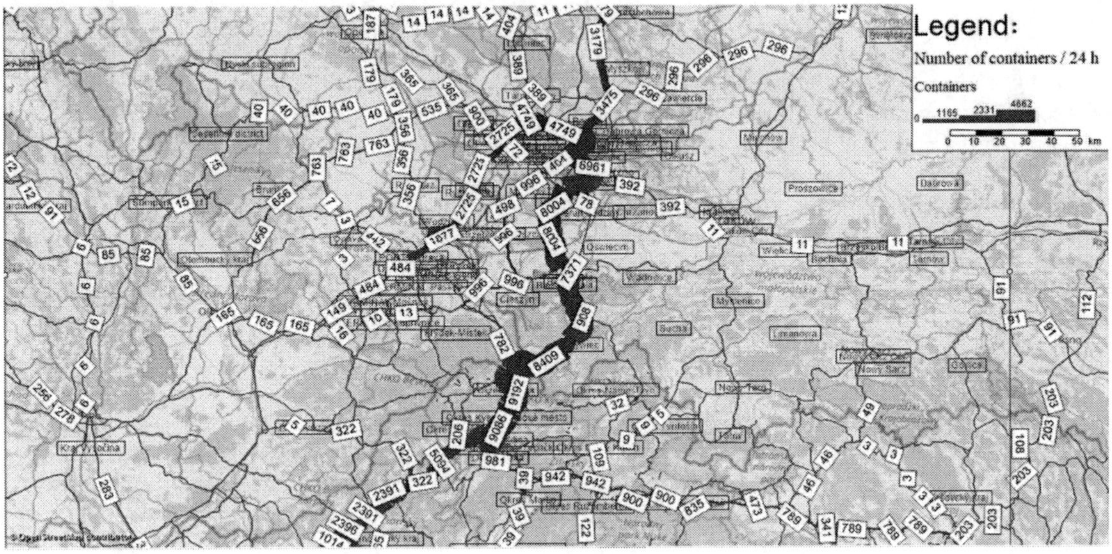

Figure 5. Model of traffic load redistribution to the multimodal transport network of the TRITIA area for zero state (2030) [5].

The outputs of the TRITIA multimodal potential model after the redistribution of transport load show that out of the total volume of transport performance in road freight transport of 12,546,256 container kilometers per year (total potential), almost half of this load is transferred to rail transport and about 4% to inland waterway transport. In the case of rail and inland waterway infrastructure, its capacity constraints were not taken into account when modeling the shift of traffic load. The remaining part of the modeled transport load (46.7 %) remains on the road infrastructure, where it is transported by heavy trucks.

The purpose of the transport model was to identify the potential shift of long-distance road transport over 300 km to alternative transport modes in the time horizon until 2030. The results of the zero scenario and alternative scenarios are that from the potential shift identified in road transport it is possible to shift approx. 40 – 50 % on rail transport and 2 – 4 % on inland waterway transport. At the same time, these values point to the fact that by 2030 it is possible to shift more than 30 % of road transport over 300 km, which would mean that the objectives set out in the White Paper - Roadmap to a Single European Transport Area could be met.

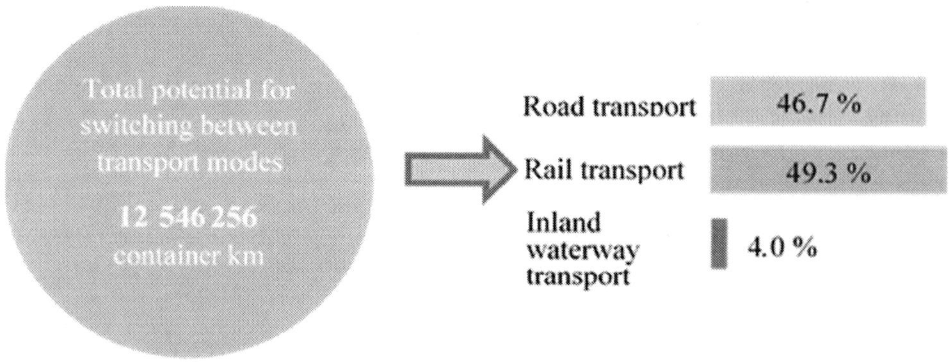

Figure 6. Transfer of the total potential of the transport load to individual transport modes for zero state (2030).

4. Conclusions

Transport, as a key element of the economy, requires ensuring its sustainability in view of the new challenges we are currently facing and which we expect in the near future. Transport is important for EU countries, so the effectiveness of actions requires strong international cooperation. Such an example is the cooperation of the countries Slovakia, Poland and Czech Republic within the TRITIA area with the aim of supporting such infrastructure solutions that will be efficient, environmentally acceptable and economically sustainable for ensuring the transport of goods. Transport infrastructure creates mobility, so without finding and without the support of adequate transport systems, it will not be possible to achieve major changes in transport.

The EU is trying to support the development of alternative modes of transport (rail and inland waterway), which should reduce the quotient of road freight transport. The shift of road freight transport to rail transport can cause problems on rail infrastructure that is not yet sufficiently prepared for this and needs to be taken into account in further strategic planning. The transport model TRITIA is an infrastructure model for identifying potential shift from road transport to alternative transport modes. The results of the model are focused on the analysis of the infrastructure, resp. assessments of the impact of the price for the use of infrastructure on the change of the redistribution between transport modes. In any case, it has been confirmed that there is sufficient potential in road transport to shift road freight transport. If this is not really the case and quality infrastructure is available, then it is necessary to address systemic measures with respect to the organization in individual modes of transport [7].

Acknowledgments

This contribution is the result of the project implementation: "TRANS TRITIA CE960" supported by the Interreg CENTRAL EUROPE funded by European Regional Development Fund of European Union.

References

[1] Application Form CE960 TRANS TRITIA
[2] Florkova Z, Duris L, Veselovsky M, Sedivy S and Kovalova D 2018 Three-dimensional mobile mapping system and its use in road engineering *MATEC Web of Conferences* **196**
[3] Project TRANS TRITIA, D.T3.1.1 Report; Methodology of development of traffic surveys (https://www.interreg-central.eu/Content.Node/Methodology-of-development-of-traffic-surveys.pdf)

IOP Conf. Series: Materials Science and Engineering 1015 (2021) 012101 doi:10.1088/1757-899X/1015/1/012101

[4] Project TRANS TRITIA, D.T3.1.2 Report, Preparation and performance of annual traffic surveys (https://www.interreg-central.eu/Content.Node/Preparation-and-performance-of-traffic-surveys-FINAL.pdf)
[5] Project TRANS TRITIA, D.T3.1.3 Report, Evaluation of traffic surveys (https://www.interreg-central.eu/Content.Node/Evaluation-of-traffic-surveys.pdf)
[6] Project TRANS TRITIA, D.T3.2.2 Report, Zero scenario of TRITIA traffic model (https://www.interreg-central.eu/Content.Node/Report-on-the-zero-scenario-of-TRITIA-traffic-model.pdf)
[7] Project TRANS TRITIA, D.T3.2.4 Report, Implementation plan for TRITIA region (https://www.interreg-central.eu/Content.Node/Implementation-plan-for-TRITIA-region.pdf)

IOP Conf. Series: Materials Science and Engineering 1015 (2021) 012102 doi:10.1088/1757-899X/1015/1/012102

Requirements and practice from measuring the deformation resistance of the formation of the railway subgrade of the Slovak railways modernized line Plevník-Drienové-Bytča-Žilina

M Mečár and A Pultznerová

University of Žilina, Faculty of Civil Engineering, Department of Railway Engeneering and Track Management (DRETM), Univerzitná 8215/1, 01026 Žilina, Slovakia

E-mail address: martin.mecar@uniza.sk

Abstract. Deformation resistance of the formation of the subgrade is the main indicator of the quality of the designed structural layers of the subgrade formation. Deformation resistance is determined in Slovak Railways (ŽSR) by a static plate load test and is expressed by the modulus of deformation E_s. The static load test not only verifies the correctness of the design of the type of structural layers of the body of the railway subgrade, but it is also an indicator of the quality of the performed construction works. On modernized ŽSR lines, is designed for subgrade type No. 3 and in the most unfavorable geotechnical conditions also the type of structural layers of the formation of the railway subgrade No. 6. The required value of deformation resistance at the level of the formation of the subgrade is designed with regard to the speed zone of the railway line. ŽSR corridor lines are designed for speeds up to 160 km/h, so they belong to the RP4 speed zone. Such lines also include a section of the ŽSR modernized Plevník-Drienové-Bytča-Žilina line.

1. Introduction

Since 2002, ŽSR has begun extensive modernization of its railway lines. The priority of modernization is the main lines, which are included in the strategic European transport routes TEN-T and railway corridors on the basis of international agreements AGC and AGTC. Three important railway corridors pass through the territory of the Slovak Republic, while the double-track line Považská Teplá-Žilina is a part of corridor no. V. (Venice-Trieste/Koper-Ljubljana-Budapest-Uzhhorod-Lviv). With the integration of the ŽSR network into European transport routes, ŽSR has taken over the obligation to respect international agreements and the resulting technical requirements that are imposed on these lines.

The modernization of railway lines focuses on the modification of stations and trainstops, the building of level crossings and especially on increasing the line speed, where the highest speed of trains reaches 160 km/h after the modernization. The line speed is the basic parameter used to design the type of structural layers of the subgrade body and the required deformation resistance at the level of the subgrade formation.

2. Railway line Plevník–Drienové–Bytča–Žilina

Construction of the modernized line ŽSR Plevní-Drienové - Bytča is an independent investment project of the modernization of the line Púchov-Žilina to a speed of up to 160 km/h, II. stage Považská Teplá (outside) - Žilina (outside). The investment project passes through the districts of Považská Bystrica, Bytča and Žilina. This area is part of the main development axis of Slovakia, through which the corridors of transport and technical infrastructure of supraregional importance meaning. In the draft regulations for territorial development, the project is included as a public benefit measure of the superior transport infrastructure. The structure consists of interstation track sections and railway station structures, which represent separate coherent parts of the project (CPP):

- CPP 48 Považská Teplá – Bytča with trainstop Plevník-Drienové and Predmier,
- CPP 49 Bytča – figure 1,
- CPP 50 Bytča – Dolný Hričov,
- CPP 51 Dolný Hričov,
- CPP 52 Dolný Hričov – Žilina with trainstop Horný Hričov.

Individual CPP were realized as separate buildings. Each of them was designed to be independently feasible and to be able to operate independently also with certain limitations and specifications.

The proposal for the modernization of the railway line makes maximum use of the existing railway formation, taking into account compliance with the required speed parameters, basic conditions for capability and area modification. The technical conditions for the implementation of directional adjustments of the track curves required the necessary deviations in order to achieve its required speed parameters. The necessary track alignments, caused by the required speed parameters, with regard to the existing railway track, have been done in:

- track section Považská Teplá – Bytča, km 176.7 – 177.9 in length 1 200 m,
- track section Bytča – Dolný Hričov, km 187.0 – 189.7 in length 700 m,
- railway station Dolný Hričov, km 190.7 – 191.7 in length 1000 m and km 191.9 – 192.8 in length 900 m,
- track section Dolný Hričov – Žilina, km 195.2 – 196.5 in length 1 300 m.

Figure 1. Railway station Bytča after modernization.

The structure realisation was based on the conditions of maintaining operation in inter-station sections in the exclusion of one of the two railway tracks. The order of implementation of individual CPP was chosen on the basis of close cooperation of the construction contractor (TEBS Association) with the client of the building (ŽSR), the railway administrator of the railways (ŽSR, OR Žilina), building designer (Reming Consult) and building supervision (DEC International), on the basis of approved work procedures and construction organization plan in specified exclusions of railway operation.

The construction procedures and durations of construction took into account the time horizons set by the Ministry of Transport, Posts and Telecommunications of the Slovak Republic due to the efficient drawing of finance. The total length of the modernized section is 22.702 km, 50 613 m of new rail tracks were laid, 44 new rail switches and crossing structures, 9 railway bridges were reconstructed and 7 new railway bridges were built, 5 new underpasses for passengers and the public and 13 113 m of noise bariers [1].

3. Deformation modulus measurements of the subgrade formation

In the conditions of ŽSR, since 1971, the deformation resistance of the structural layers of the railway substructure has been expressed by the value of the deformation modulus determined by a static load test [2]. Determination of deformation characteristics by static plate load test (PLT) with a circular plate is based on the general theory of elastic half-space. In the Slovak Republic, the following normative documents apply to the static plate load test: STN 73 6133 [3], STN 73 6190 [4] and for railway subgrade structures the ŽSR TS4 regulation [5]. The deformation resistance of the plain of the railway subgrade body is characterized by its modulus of deformation determined by a static load test using the methodology specified in the regulation [5]. According to this regulation, the resulting of static modulus of deformation E_{si} is determined from the second of two load cycles - figure 2. One load cycle has a specified 4 load stages, after reaching the maximum specific pressure, the same stepped relief of the plate follows until reaching the zero applied pressure. The graphical course of the static PLT is shown in figure 2. The value of 0.2 MPa was used as the maximum contact stress for both load cycles, which is the standard stress used on ŽSR lines for the formation of the subgrade. The measured quantity was the static modulus of deformation E_{si}, which was determined from modernized line ŽSR Plevní-Drienové - Bytča is an independent building of the relation:

$$E_{si} = \frac{1{,}5 p \cdot r}{y} \tag{1}$$

where: E_{si} – static modulus of deformation, [MPa],
p – contact stress under the load plate, [MPa],
r – load plate radius, [m],
y – total average settlement of the plate load, [m],

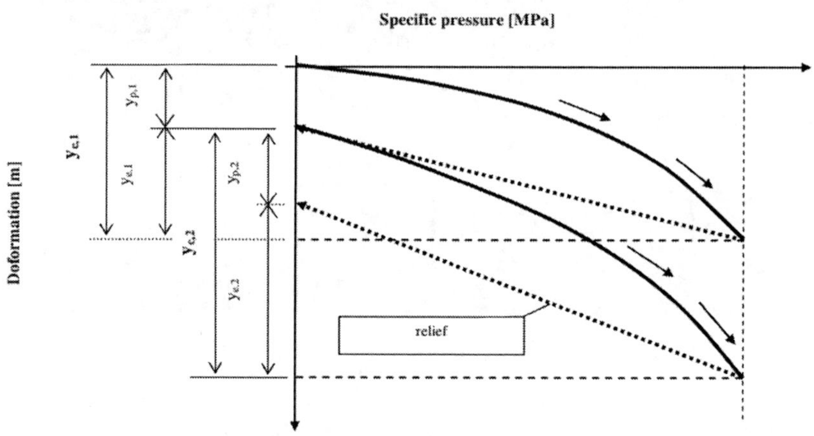

Figure 2. The course of PLT according to the methodology of ŽSR TS4.

IOP Conf. Series: Materials Science and Engineering 1015 (2021) 012102 doi:10.1088/1757-899X/1015/1/012102

Static load tests were performed by employees of the Department of Railway Engineering and Track Management, Faculty of Civil Engineering, University of Žilina. The equipment of the Accredited Laboratory of the Faculty of Civil Engineering, the static load set FROWAG with single-point deformation sensing with a digital indicator IDU25 Mitutoyo and a load plate with a diameter of 0.300 m were used for the measurement.

Static load test at the level of the surface of the formation of the railway subgrade belongs to the basic measurements in situ of the inspection and acceptance process of building. The measurement provides accurate results in a relatively short period of time, other field quality control tests, e.g. compaction degree, last disproportionately longer. This is an important parameter especially in the conditions of ŽSR structures, because the deficiencies found by the control test must be removed immediately in the possesions, for example additionally compacting the tested layer, or replace it with a more suitable material, etc. The static load test at the level of the railway subgrade formation surface is the main acceptance test of the quality control of the railway subgrade structure, it is the basis for taking over the structure works and the permit for the commencement of structure works of the railway superstructure. The prescribed number of static load tests is according to [5], one PLT every 200 m. The minimum required values of the static modulus of deformation of the formation of the subgrade E_s on main railway tracks are specified in the project documentation, namely at least 80 MPa on newly built relocation embankments and at least 60 MPa on existing railways.

4. Evaluation of measurements

On the modernized section of the Plevník-Drienové-Bytča-Žilina railway line, a total of 277 static load tests were performed at the level of the formation of the subgrade, of which 243 load tests in the main railways tracks and 34 static load tests in other tracks in the modernized railway stations Bytča and Dolný Hričov [7]. The results of the average values of the modulus of deformation at the level of the formation of the railway subgrade on particular CPP and tracks are shown in figure 3.

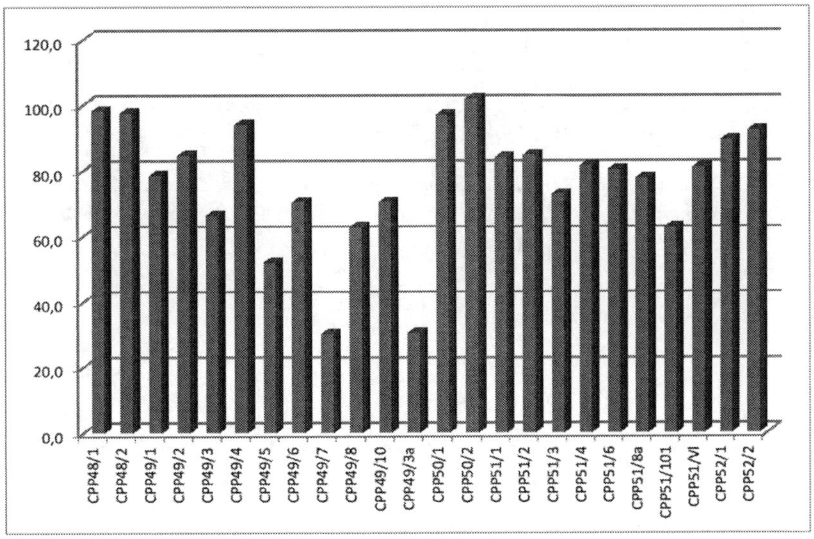

Figure 3. – Overview of the results of the achieved average values of deformation modules in the level of the railway subgrade formation.

As mentioned in the introduction, the modernization of the ŽSR corridor lines focuses mainly on increasing the line speed to 160 km/h.

This line speed applies to the main railway tracks, therefore in the further assessment of the quality of the structure of the modernized line, the measurement results are related only to the main railway

IOP Conf. Series: Materials Science and Engineering 1015 (2021) 012102 doi:10.1088/1757-899X/1015/1/012102

tracks. Table 1 shows the average values from the deformation modulus measurements identified in the main railway tracks.

Table 1. Average values of deformation modulus in main railway tracks.

	CPP 48		CPP 49		CPP 50		CPP 51		CPP 52	
Track	1	2	1	2	1	2	1	2	1	2
$E_{s\ average}$ [MPa]	98.1	97.5	78.3	84.5	96.8	101.7	83.8	84.6	89.2	92.2
$E_{s\ reglets}$ [MPa]	113.4	104.3	-	-	98.5	113.8	91.3	91.4	100.2	101.4
$E_{s\ exist}$ [MPa]	93.9	95.9	-	-	96.2	99.2	75.6	79.9	85.1	88.8

In table 1, in addition to the average values of the measured deformation modulus ($E_{s\ average}$), the average values of the deformation modulus measured on the sections with the existing railway formation ($E_{s\ exist}$) and on the sections with the newly built railway formation ($E_{s\ reglets}$), i.e. on relocations of the railway track due to a change of alignment (to achieve its required speed parameters). Graphic course of deformation modules of track No. 1 and 2 in the level of the formation of the railway subgrade of the modernized line Plevník-Drienové - Bytča - Žilina is shown in figures 4 and 5.

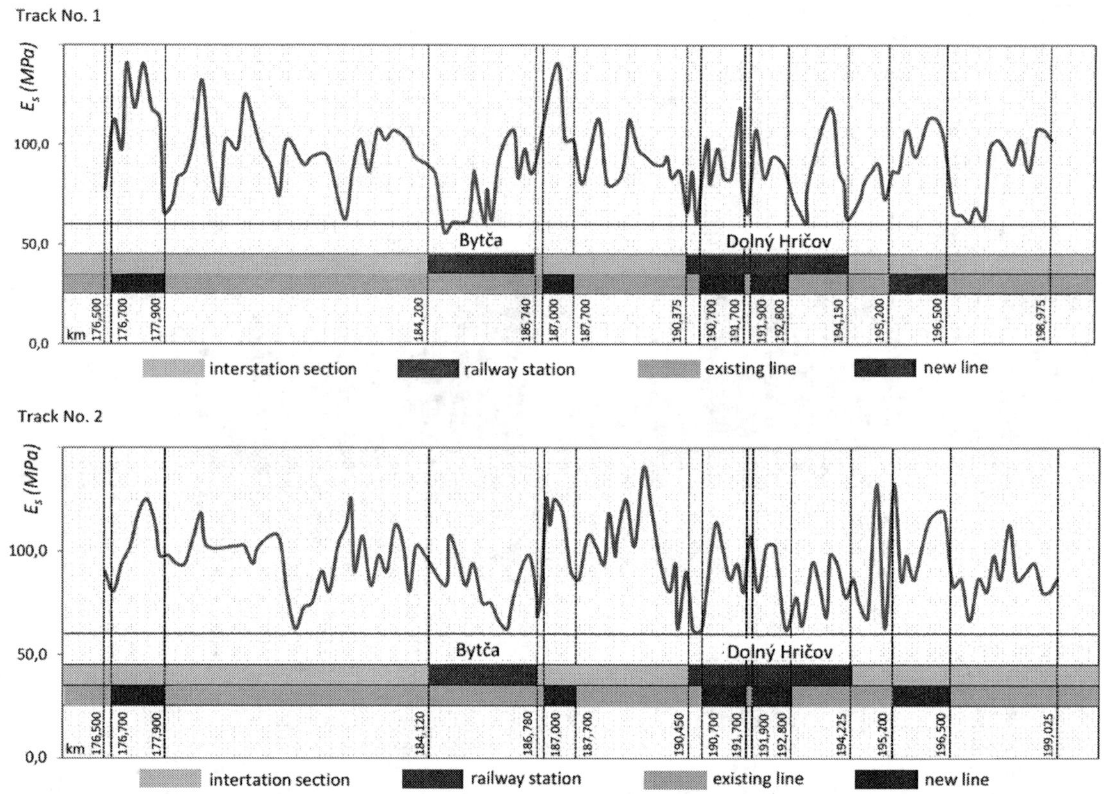

Figure 4. Graphic course of achieved values of deformation modulus of the formation of the railway subgrade surface.

From the measured values of the deformation modulus of the formation of the railway subgrade surface, it is clear that on a newly built railway formation achieves higher values of the modulus of deformation than the modulus of deformation of the formation of the railway subgrade on the existing body. Similarly, the modulus of deformation of the formation of the railway subgrade surface of the modernized railway stations Bytča - CPP 49 and Dolný Hričov - CPP 50 reaches lower values than on the interstation sections - CPP 48, CPP 50 and CPP 52.

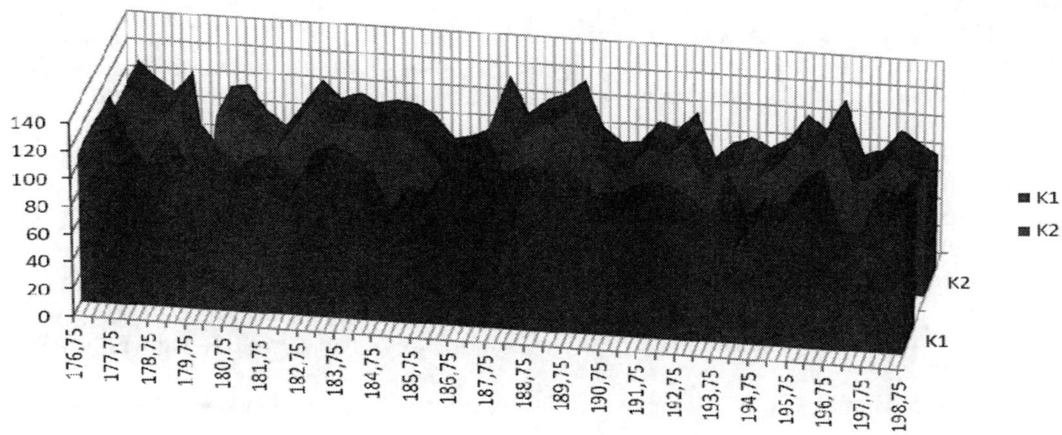

Figure 5. Comparisons of values of deformation modules in the level of railway subgrade formation surface track. No. 1 and track No. 2.

4. Conclusions

At ŽSR, the deformation resistance of the railway subgrade formation surface is determined by the minimum value of the deformation modulus [6]. The design of the structural layers of the body of the railway subgrade is based on the required technical requirements and on the results of the geotechnical survey. The technical requirements are uniquely determined; we propose the required value of the deformation modulus at the level of the surface of the formation of the railway subgrade with regard to the speed zone of the railway line. ŽSR corridor lines are designed for speeds up to 160 km / h, they belong to the RP4 speed zone, for which the minimum values of the deformation modulus are considered.

However, the results of the geotechnical exploration introduce a certain degree of uncertainty into this design, because another important parameter of the design of the structure of the structural layers of the railway subgrade formation is the modulus of deformation of the ground plane determined by a static load test made in borrov and drill hole. The selected number of survey tests may not reveal problem areas of the existing state and these will be revealed only by uncovering the ground plan during the actual realisation. In such a case, it is important that the construction work and the chosen technology respond flexibly to this situation and that appropriate measures are proposed. This was also the case for the structure of the modernized line Plevník-Drienové - Bytča - Žilina, where the project proposed a type of structural layers of the formation of the railway subgrade No. 3 with different thickness of the base layer and different types of geosynthetics. As far as the situation allowed, these were operatively adjusted according to the existing state of the ground plan, and in the most critical places, the type of structural layers of the formation of the railway subgrade was changed to type No. 6, i.e. on the type of structural layers of the formation of the railway subgrade with a stabilized ground plane. From the results of measurements of the deformation resistance of the surface of the formation of the railway subgrade, it is clear that the cooperation of the building contractor, construction designer and construction supervisor was effective and the specified design parameters and the overall quality of the realized building were observed with sufficient reserve.

IOP Conf. Series: Materials Science and Engineering 1015 (2021) 012102 doi:10.1088/1757-899X/1015/1/012102

Acknowledgments
The presented results are the partial results of solving the VEGA grant project 1/0084/20 "*Numerical and experimental analysis of transition areas of objects of structures of railway superstructures and objects of formation substructure*".

References
[1] Martiš J 2017 Modernization of the train line Považská Teplá – Žilina, *Engineering buildings* **1**
[2] Ižvolt L 2008 Railway substructure – stress, diagnostics, design, and implementation of body construction layers of railway subgrade Scientific monograph, University of Žilina (in Slovak)
[3] Standard STN 73 6133 – Road Building. Roads embankments and subgrades. Standard, Slovak Office of Standards, Metrology and Testing 2017 Slovakia (in Slovak)
[4] Standard STN 73 6190 – Static plate load test of building constructions. Standard, Slovak Office of Standards, Metrology and Testing 1980 Slovakia (in Slovak)
[5] Regulation ŽSR TS4 Train subgrade 2018
[6] Slovak Railway Technical Standard TNŽ 73 6312 The design of structural layers of subgrade structures 2005 Directorate General of Railways of the Slovak Republic Slovakia (in Slovak).
[7] Technical reports of railway line Považská Teplá (out) – Žilina (out) modernization

IOP Conf. Series: Materials Science and Engineering 1015 (2021) 012103 doi:10.1088/1757-899X/1015/1/012103

Economic aspects of pavement surface degradation in terms of noise pollution in their surroundings

M Trojanová[1], M Decký[2], A Ivanovičová[1] and K Hodásová[1]

[1] TheUniversity of Žilina, Department of Construction Management, Slovak Republic
[2] TheUniversity of Žilina, Department of Civil Engineering, Slovak Republic

E-mail address: maria.trojanova@uniza.sk

Abstract. The morphology of the road surface is changing in time which has a consequence in increasing noise levels in areas near the roads. This fact was determined by measurements on two road sections. The measurements were done using the Statistical Pass-By method. Different types of asphalt surfaces were used on the road sections. The Statistical Pass-by Index (SPBI) was quantified from the measurements, which were carried out over the years. Functions of road surface noise evolution in time were calculated from quantified values of SPBI for both measured road surfaces.

1. Introduction

For the purpose of objectification of noise from road transport, currently, the decisive codified regulations are Act no. 355/2007 Coll. [2] and Decree of the Ministry of Health of the Slovak Republic no. 549/2007 Coll. [3]. Act o. 355/2007 [2] in Section 27 Noise, infrasound and vibration in the environment in relation to the described issue states the following facts:

- a natural person-entrepreneur and a legal person using or operating noise, infrasound or vibration sources, a road manager [1], a waterway operator, a rail operator, an airport operator, and operators of other objects producing noise (hereinafter referred to as " the operator of sources of noise, infrasound or vibration ") shall ensure that the exposure of the occupants and their environment is as low as possible and does not exceed the permissible values for day, night and night provided for in the implementing regulation;
- in the design, construction or substantial reconstruction of traffic structures and infrastructure, noise in related external or internal environments shall not exceed the permissible values at the assumed traffic load,
- in the design, construction or substantial renovation of buildings, it is necessary to ensure the protection of the indoor environment of buildings from noise from the external environment
- municipalities are authorized to objectify the exposure of the population and their environment to noise and vibration in accordance with the requirements laid down in the implementing regulation. Only those skilled in the business who hold an accreditation certificate can objectify the exposure of residents and their noise, infrasound and vibration environments.

The Decree of the Ministry of Health of the Slovak Republic 549/2007 Coll. [3], of December 1, 2007, prescribes the details of permissible values for determining noise quantities and requirements for the objectification of environmental noise. The Decree determines individual noise limits for:

- the external environment,
- the indoor environment of buildings.

IOP Conf. Series: Materials Science and Engineering 1015 (2021) 012103 doi:10.1088/1757-899X/1015/1/012103

Section 3 [3] states that the protection of health against noise is ensured if the assessed values of the determining quantities of noise are not higher than the permissible values (table 1). In relation to the issue addressed, the decisive provisions are presented in Section 6, which states:

- noise shall be assessed at the locations covered by the permissible values as follows:
  - outside noise outside buildings shall be assessed at the height of 1.5 m ± 0.2 m above the ground,
  - external noise in front of the perimeter wall shall be assessed at a distance of 1.5 m ± 0.5 m from the wall and at the height of 1.5 m ± 0.2 m above the floor of the relevant story,
- the measured value shall be increased by the value of the expanded measurement uncertainty determined in accordance with the metrological practice and, if necessary, in further corrections in accordance with the Annex. If necessary, it is determined for the relevant reference time interval. The obtained assessment value is compared with the permissible value.

Table 1. Permissible values of determining quantities in the external environment according to the Decree of the Ministry of Health of the Slovak Republic 549/2007 Coll. [3].

Area category	Description of the protected area or an external environment	Ref. time interval	Admissible values[a] (dB)				
			Traffic noise				Noise from other sources
			Road and water transport [c] $L_{Aeq,p}$	Railways [c] $L_{Aeq,p}$	Air transport $L_{Aeq,p}$	Air transport $L_{ASmax,p}$	$L_{Aeq,p}$
I.	Areas with special noise protection, e.g. spas, spa and health resorts	day	45	45	50	-	45
		evening	45	45	50	-	45
		night	40	40	40	60	40
II.	Areas in front of windows of residential rooms of residential houses, area in front of windows of the protected rooms of school buildings, medical facilities and other protected buildings, [d] outdoor area in residential and recreational areas	day	50	50	55	-	50
		evening	50	50	55	-	50
		night	45	45	45	65	45
III.	Areas as in category II near motorways, roads of I. and II. class, local public transport roads, railways, and airports, city centres	day	60	60	60	-	50
		evening	60	60	60	-	50
		night	50	55	50	75	45
IV.	Non-residential and unprotected outdoor areas, manufacturing zones, industrial parks, factory premises	day	70	70	70	-	70
		evening	70	70	70	-	70
		night	70	70	70	95	70

[a] The permissible values apply to dry road surfaces and terrain not covered in snow,
[b] Inland transport is transport on the road, including tramways,
[c] Stops for local public transport, bus, rail, waterway, and taxis intended for embarkation and disembarkation shall be assessed as part of land and waterway transport,
[d] The permissible values in front of the facade of non-residential buildings shall be applied at the time of their use; schools during class times.

The 2012 dissertation [10] presented an overview of permissible road noise limits for EU countries. The overview of the permissible values of the determining quantities of noise in the Member States of the European Union shown in figure 1 [11] was prepared for a similar type of built-up areas and affected areas (new constructions in the area) and for transregional transport (motorways, expressways, bypasses, railway lines). The data on permissible values for determining noise quantities were obtained from foreign websites, foreign literature and the competent authorities of the Member States. The methodological tool for drawing up such an overview of limits was the document [7] (Inspection of the implementation of the Directive 2002/49 / EC on environmental noise), issued in 2010.

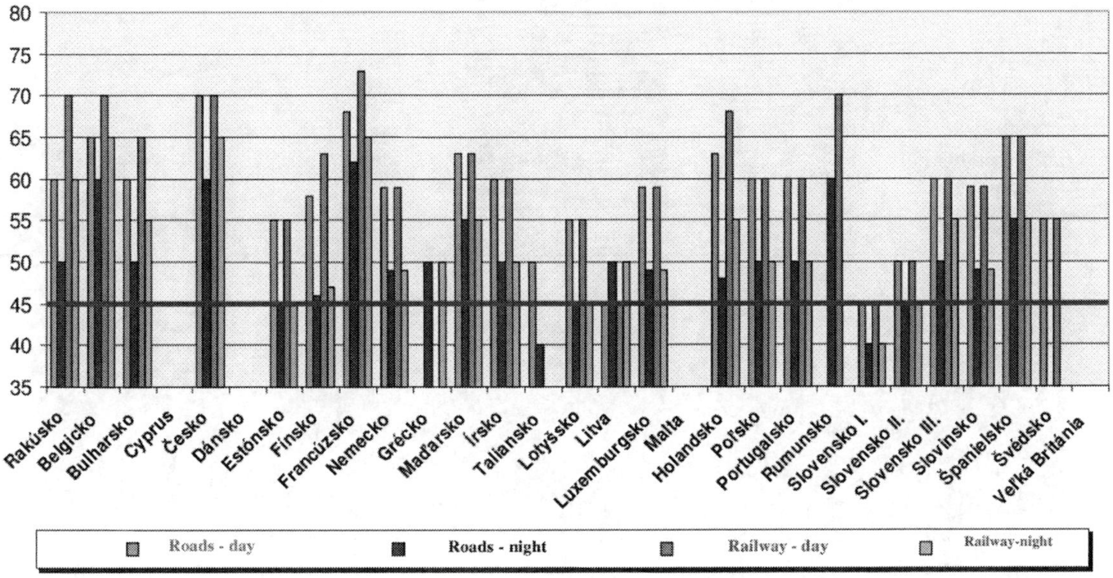

Figure 1. Permissible values of noise emissions from road and rail transport in EU countries EU [6].

Specific levels of noise and particle pollutions [8, 9] from road traffic are determined mainly by the intensity and composition of traffic flow [4, 5, 16], the state of road surface morphology [10, 12, 15] and vehicle-road interaction [6, 13]. Noise limits presented in table 1 and figure 1 do not take into account the increase in the noise load around the roads induced by a change in the morphology of the road surface. And this is exactly the problem discussed in this article.

2. Measurement of the influence of road surfaces on traffic noise.

Traffic noise measurements were performed by using Statistical Pass-By-Method (SPB) that is described in the ISO 11819-1. It provides a method to determine an index that can be used to compare the noise emission impact of different road surfaces by measuring vehicle pass-bys at the roadside. The SPB method is based on the measurement of the maximum A-weighted sound pressure levels of a statistically significant number of individual vehicle pass-by (table 2) together with the vehicle speeds.

Specific measurements of A sound levels L_A were conducted by sound analyser NOR-121 (Norsonic, Norway - ID 31211) accuracy Class 0. The sound analyser was calibrated prior to each set of measurements by microphone calibrator NOR N-1251 (Norsonic, Norway - ID 31069), accuracy Class 1. Applied measuring string of noise emissions and pollution had separate verification for own sound level meter, measuring microphone Nor-1225 (Norsonic, Norway - ID 48045), one-third-octave filters, acoustic calibrator. Velocity measurements were realised by speed radar gun Bushnell Speedster (table 1) and meteorology conditions by Wireless weather station - WS 888 set.

IOP Conf. Series: Materials Science and Engineering 1015 (2021) 012103 doi:10.1088/1757-899X/1015/1/012103

Table2. Vehicle categories in STN EN ISO ISO 11819-1 and results of measurements made 4.11.2008 and 23.6.2020 at measuring station by the road I/61 in Dolný Hricov (figure 2).

Category	Typical examples		Number of pass-by				
			minimum	4.11.2008		23.6.2020	
				No.	$L_{category}$	No.	$L_{category}$
1) cars	2 axles, 4 wheels passenger cars		100	102	76.6	175	80.9
2a) dual-axle heavy vehicles	2 axles, more than 4 wheels, light trucks, buses		30 (2a+2b = 80)	35	79.6	35	85.0
2b) multi-axle heavy vehicles	heavy trucks more than 2 axles		30 (2a+2b = 80)	48	84.5	54	86.6
SPBI (dB) corrected to reference temperature 20°C				82.7 dB		78.7 dB	

Figure 2. Views of the measuring station on the BUS stop by the road I/61 in Dolný Hričov on 8.6.2008 and 23.6.2020 with an indication of measuring distances.

This regression line was then used to determine the average maximum A-weighted sound pressure level L_{veh} at the reference speed. The L_{veh} of the three vehicle categories ($L_{veh,1,regression} \equiv L_1$, $L_{veh,2a,regression} \equiv L_{2a}$, $L_{veh,2b,regression} \equiv L_{2b}$)can be combined to give a single index called SPBI (Statistical Pass-By Index) which is indicative of the influence of the road surface on the noise emission of a mixed vehicle collective

$$SPBI = 10 \times \log\left[W_1 \times 10^{\frac{L_1}{10}} + W_{2a} \times \left(\frac{v_1}{v_{2a}} \right) \times 10^{\frac{L_{2a}}{10}} + W_{2b} \times \left(\frac{v_1}{v_{2b}} \right) \times 10^{\frac{L_{2b}}{10}} \right] \quad (1)$$

where:

$SPBI$	- the Statistical Pass-by Index, for a standard mix of light and heavy vehicles	(dB)
L_1, L_{2a}, L_{2b}	- the Vehicles Sound Levels for vehicle categories 1, 2a, 2b (figure 1)	(dB)
W_1, W_{2a}, W_{2b}	- the weighting factors, which are equivalent to the assumed proportions of vehicle categories in the traffic (table 2)	(-)
v_1, v_{2a}, v_{2b}	- the reference speeds of individual vehicle categories (table2)	(km·h^{-1})

The passing vehicles were classified into one of three vehicle categories and one of three reference speeds was chosen according to the average operating speed of the road (table 2).

Table 3. Reference speeds and weighting factors (W_x) in the different road aped categories.

Vehicle category		Road speed category					
		Low		Medium		High	
Name	No.	Ref. speed (km/h)	W_x	Ref. speed (km/h)	W_x	Ref. speed (km/h)	W_x
Cars	1	50	0.900	80	0.800	110	0.700
Dual-axle heavyvehicles	2a	50	0.075	70	0.100	85	0.075
Multi-axle heavyvehicles	2b	50	0.025	70	0.100	85	0.225

A regression line of the maximal A-weighted sound pressure level versus the logarithm of speed was calculated for the pass-bys of every category (Figure 3 to 5).

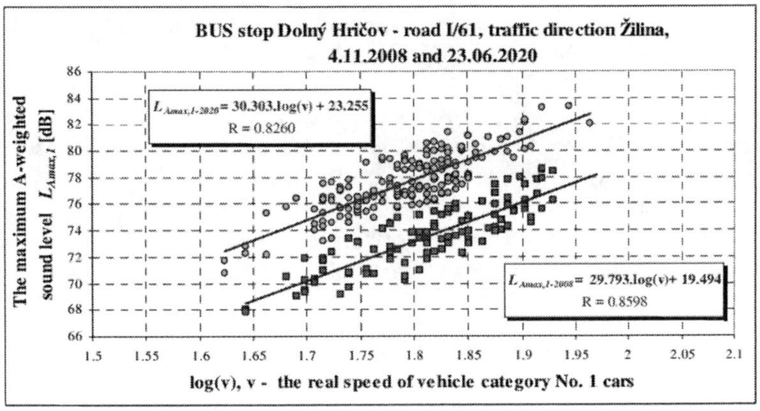

Figure 3. Correlation dependences of the maximum A-weighted sound level on the logarithm of the real speed vehicles of category 1 cars.

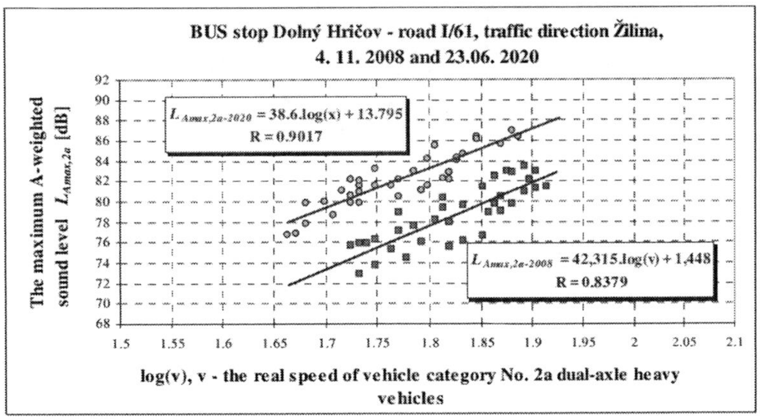

Figure 4. Correlation dependences of the maximum A-weighted sound level on the logarithm of the real speed vehicles of category 2a dual-axle heavy vehicles.

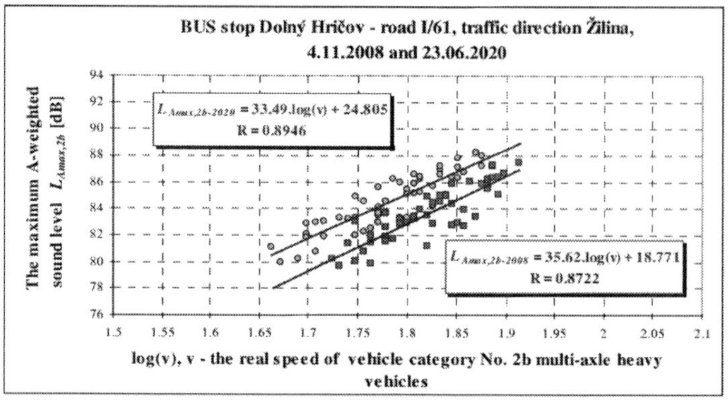

Figure 5. Correlation dependences of the maximum A-weighted sound level on the logarithm of the real speed vehicles of category 2b multi-axle heavy vehicles.

3. Valuation of noise costs

The monetary valuation of the negative impacts that arise from the use of the road expresses society-wide costs. Impacts caused by the use of the road by its users are referred to as user costs. External costs value the negative impacts that arise during the construction and operation of the road and have an impact on society and the environment. The principle of design quantification and valuation of the costs of road traffic noise is elaborated in Fig.6 Principle of quantification and valuation of noise costs.

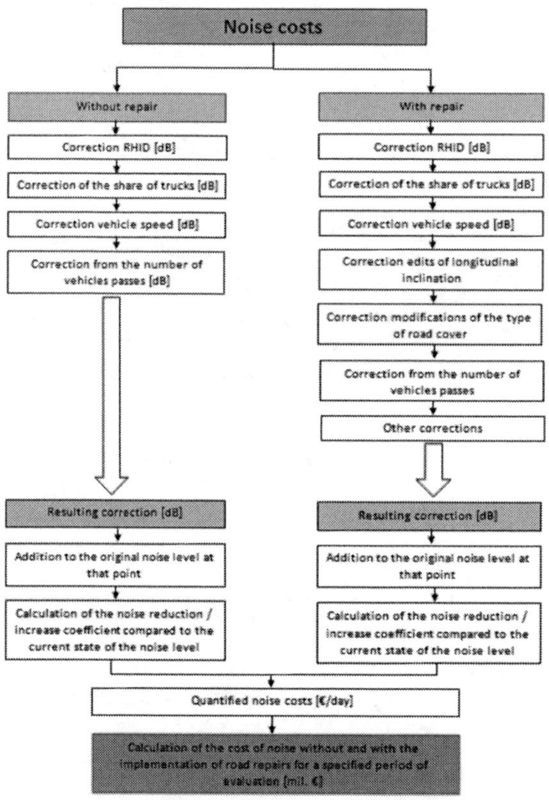

Figure 6. Principle of quantification and valuation of noise costs [14].

Quantification and valuation of society-wide costs are relatively complicated in practice. The society-wide costs in a given year are calculated according to the formula (2)

$$CN_t = \sum_{t=tu}^{T} UN_t + \sum_{t=0}^{T} EN_t \tag{2}$$

where:

CN_t	- the Society-wide costs in the considered year	(€)
UN_t	- the User costs in the considered year	(€)
EN_t	- the External costs in the year under consideration	(€)
t	- the year considered	(years)
t_u	- the start of road use	(years)
T	- thenumber of years of the road life cycle	(years).

On the selected section of the road I/61, the asphalt cover was replaced due to unsatisfactory parameters of the road cover. Table 4 shows the quantified noise costs for the site of road I / 61 without road repair and with road repair. Valuation of costs for noise from road traffic on this road was processed in the excel program. [14]

Table 4. Cost of noise without repair and with repair.

Without repair		With repair	
Low intensity	High intensity	Low intensity	High intensity
11 778.39	4 184.79	7 852.26	2 789.86
11 778.39	4 184.79	7 852.26	2 789.86
11 778.39	4 184.79	8 314.16	2 953.97
11 778.39	4 184.79	8 314.16	2 953.97
11 778.39	4 184.79	11 085.55	3 938.63
11 778.39	4 184.79	11 085.55	3 938.63
11 778.39	4 184.79	11 085.55	3 938.63
11 778.39	4 184.79	11 085.55	3 938.63
11 778.39	4 184.79	11 085.55	3 938.63
11 778.39	4 184.79	11 085.55	3 938.63
11 778.39	4 184.79	11 085.55	3 938.63
129 562.32	46 032.71	109 931.66	39 058.06

Conclusions

The article presents comparisons of traffic noise measurements performed according to requirements STN EN ISO 1819-1:2002 Acoustics. Measurement of the influence of road surfaces on traffic noise. Part 1: Statistical Pass-By method [17] from the years 2008 and 2020. Measurements of SPBI were realised 5 months (4.11.2008) and 12 years (23.6.2020) after the rehabilitation of wearing course of asphalt pavement of road I/61 in Dolný Hričov. Degradation of the surface morphology of asphalt concrete wearing source during 12 years operation caused the SPBI index the increase 4 dB (table 2).

The total cost of noise without road repairs is € 129,562.32. In the case of road repairs, the total cost of noise for the evaluation period of 10 years would reach € 109,931.66. The cost savings are € 19,630.65. For comparison, if the road were located in an urban location, the cost of noise would reach (low traffic intensity) the value of € 2,953,206.61 without repair of the road and € 2,505,751.07 with the repair. The cost savings would be € 447,455.55.

Acknowledges

The paper is a result of the research supported by the project KOR / 7479 / 2019.

IOP Conf. Series: Materials Science and Engineering 1015 (2021) 012103 doi:10.1088/1757-899X/1015/1/012103

References

[1] Act no. 135/1961 Coll. on Roads (Road Act), as amended

[2] Act no. 355/2007 Coll. on the protection, promotion and development of public health and on amendments to certain acts

[3] Decree of the Ministry of Health of the Slovak Republic no. 549/2007 Coll., laying down details on permissible values of noise, infrasound and vibration and requirements for the objectification of noise, infrasound and vibration in the environment

[4] Drliciak M and Celko J 2018 Air TRITIA transport model *XXVII Russian-Polish-Slovak Seminar MATEC Web Conf.* **96**

[5] Drliciak M, Celko J and Cingel M 2019 The Economically Active People in the Transport Process *XXVIII Russian-Polish-Slovak Seminar on Theoretical Foundation of Civil Engineering IOP Conference Series: Materials Science and Engineering* **661**

[6] Durcanska D, Decky M, Drliciak M, Jandacka D and Papan D 2019 *Road Transport and the Environment* (University textbook, EDIS – Editing centre ZU 2019) p 278

[7] Final Report on Task 1. Review of the Implementation of Directive 2002/49/EC on Environmental Noise, May 2010

[8] Jandacka D and Durcanska D 2014 Particle pollution in urban and non-urban *4th International Multidisciplinary Scientific Geoconference and EXPO* SGEM **2** pp 467-74

[9] Jandacka D, Durcanska D and Kovalova D 2019 Concentrations of traffic related pollutants in the vicinity of different types of urban crossroads *Communications - Scientific Letters of the University of Zilina* **21** (1) pp 49-58

[10] Kotek P and Kovac M Comparison of valuation of skid resistance of pavements by two device with standard methods *24th Russian-Polish-Slovak Seminar on Theoretical Foundation of Civil Engineering Procedia Engineering* **111** pp 436-43

[11] Krokker A 2012 *Environmentálna kapacita ciest z aspektu hlukových imisií* (Dissertation work, Faculty of Civil Engineering, University of Žilina) p 115

[12] Leitner B, Decky M and Kovac M 2019 Road pavement longitudinal evenness quantification as stationary stochastic process *Transport* **34** pp 195-203

[13] Lumnitzer E, Hricová B, Bednárová L and Pacana A 2018 Development of materials obtained from recycled cars and their subsequent use in noise reduction *Progress in Rubber, Plastics and Recycling Technology* **24** pp 221-9

[14] Margorinova M 2019 *Analýza celospoločenských nákladov projektu výstavby a prevádzky pozemnej komunikácie* Dissertation work (Faculty of Civil Engineering, University of Zilina)

[15] Mikolaj J, Remek L and Macula M 2017 Asphalt concrete overlay optimization based on pavement performance models *Advances in Materials Science and Engineering* **2017**

[16] Petru J, Dolezel J and Krivda V 2017 Assessment of the transport routes of oversized and excessive loads in relation to the passage through roundabout *Building up Efficient and Sustainable Transport Infrastructure 2017 BESTInfra* **236**

[17] STN EN ISO 1819-1:2002 Acoustics. Measurement of the influence of road surfaces on traffic noise. Part 1: Statistical Pass-By method (ISO 11819-1:1997)

IOP Conf. Series: Materials Science and Engineering 1015 (2021) 012104 doi:10.1088/1757-899X/1015/1/012104

Sustainability obstacles of free bike-sharing model

A Kocianova[1] **and L Slebodnik**[2]

[1] Department of Highway Engineering, Faculty of Civil Engineering, University of Žilina, Univerzitná 8215/1, Žilina, 010 26, Slovakia
[2] Municipality of Žilina, Nám. Obetí komunizmu 1, Žilina, 011 31, Slovakia

E-mail address: andrea.kocianova@uniza.sk

Abstract. Cycling is one of the options for sustainable urban mobility. Although this transport mode is not so widely used in Slovak cities, the bike-sharing project in Žilina with 80 thousand inhabitants has been remarkably successful. The system comprises of 120 shared bikes and 20 bike stations located mainly in the city core and wider centre to cover important transport destinations. During the first-year usage, a great interest of citizens has been recorded also thanks to the one-hour-free-of-charge service. That leads to an average rental number of 35 thousand bicycles per month (with a maximum of 47 thousand) and a high usage rate of the system (on average, 9.5 rides per bike per day). A more detailed analysis, including the Covid-19 pandemic's impact on reducing bicycle rental, is presented in the article. In addition to the positive experience with the bike-sharing system, the disadvantages of the one-hour-free service model are discussed. In addition to the reduced service availability for citizens with real transport needs or higher users' disrespect to the system, the technical operation costs are about 40 % higher due to an excessive wear system, poorer technical condition, and shorter lifespan of bicycles, or increase in a certain form of vandalism due to the bicycle joyriding. Ultimately, the overall loss is much higher because of no income, which threatens the sustainability of this system.

1. Introduction

There are several options to ensure sustainable urban mobility and to reduce the environmental impact and energy consumption related to the mobility of citizens. One of them is the creation of infrastructure and conditions for cycling, for instance, a bicycle sharing system. Bike-sharing service provides an opportunity for residents as well as visitors to use other alternative modes of urban transport in the city. This mode ensures relatively fast, flexible, and especially emission-free transport. Furthermore, this also motivates people to move, which improves health as can be seen from the conclusions of twelve major bike-sharing systems in Europe [1]. In addition, bike-sharing can be motivational for the use of a bicycle as a commonly available mode of transport in the city on a private basis. Also, preferring a bicycle over a car improves public space and the quality of the urban environment, and reduces demands on the parking spaces. A well-designed and operated bike-sharing system can brings also economic benefits. Individual benefits and public good benefits in an economic context are demonstrated for example on the Dublin bike-share scheme [2]. However, for a successful and long-term sustainable bike-sharing system in the city, it is important to have a well-established service pricing policy.

The article describes the bike-sharing system in the Slovak city of Žilina with 80 thousand inhabitants and its enormous usage of this system. One of the main reasons for its widespread use is the fact that the service is free-of-charge for up to one hour. The effects of this free service model on the sustainability of the bike-sharing system are discussed in this article.

2. Brief description of the bike-sharing system in Žilina

The bike-sharing system in Žilina (BikeKIA) has started at the end of March 2019. It is a part of the international bike-sharing system Nextbike operated by ARRIVA Slovakia. This system comprises of 120 shared bikes and 20 bike-sharing stations located mainly in the city centre and wider central parts of the city to cover the important strategic points, such as centres of residential areas, university area, university dormitories, or health and shopping centres (see figure 1). The maximum distance between stations is 4.5 km, most are within 500 meters. Except for two stations (No. 18 and 20), they are all on a flat plane terrain.

The bike-sharing station network is connected via the Internet. Each bicycle is equipped with a SMART Lock, GPS locator, and GSM communication. Renting a bicycle is based on a registration system with payment card verification. This registration is possible via the web or an application, and then the user opens the bicycle from the stand using a smartphone. Visitors and residents of Žilina can rent a bike free of charge for one hour, without a registration fee or deposit. It is possible thanks to the support of the non-profit Kia Motors Slovakia Foundation and the City of Žilina. Based on the memorandum, they will finance the operation of the service for 5 years without any revenue from the system.

A pricing scheme for the BikeKIA system is quite simple. It contains only basic tariff (for the first 60 minutes for free), service fee, and technical fees as shown in table 1.

Table 1. The pricing scheme for the BikeKIA system.

Basic tariff	the first 60 min FOR FREE	The bike must be returned to an official bike station within 1 hour of biking FOR FREE. Then it is possible to rent another bike straight away – for 1 hour FOR FREE again.
Service fee	20 EUR	The fee for every additional hour if the bike is not returned within 1 hour.
	100 EUR	24 hours max.
Technical fee	75 EUR	The bike was stolen because of negligence.
	75 EUR	The damage was done to the bike / rental station / bike stand.
	25 EUR	The bike was not returned to an official station and returned or locked incorrectly.
	10 EUR	The use of the bike does not comply with the General Terms and Conditions.

3. Utilization of the BikeKIA system

The analysis of bike-sharing utilization in this contribution is conducted based on the raw data from the BikeKIA system supplied by the operator (start time, end time, duration, rental place, return place, and bike number). The period analyzed in this paper is the first-year operation between April to November 2019.

The location and utilization of bike-sharing stations are shown in figure 1. The size of the circle represents the number of bike rentals from a given station. The most frequently used stations are in the city centre and wider city centre (No. 1, 4, 5, 6 and 8), next to the university dormitories (No. 2 and 7), inside the largest residential area (No. 3), next to the shopping centres (No. 9, 11 and 13), and Žilina reservoir (No. 10) with several opportunities for sports and relaxation. The least frequently used stations are on the hill at the university area and the forest park (No. 18 and 20) and also in the city center near frequently used stations (No. 17 and 19).

The difference between the number of bicycle rentals and returns on frequented bike-sharing stations is between 1-7% (see figure 1, right), which indicates a sufficient redistribution of shared bikes and variety of bike-share rides during a day.

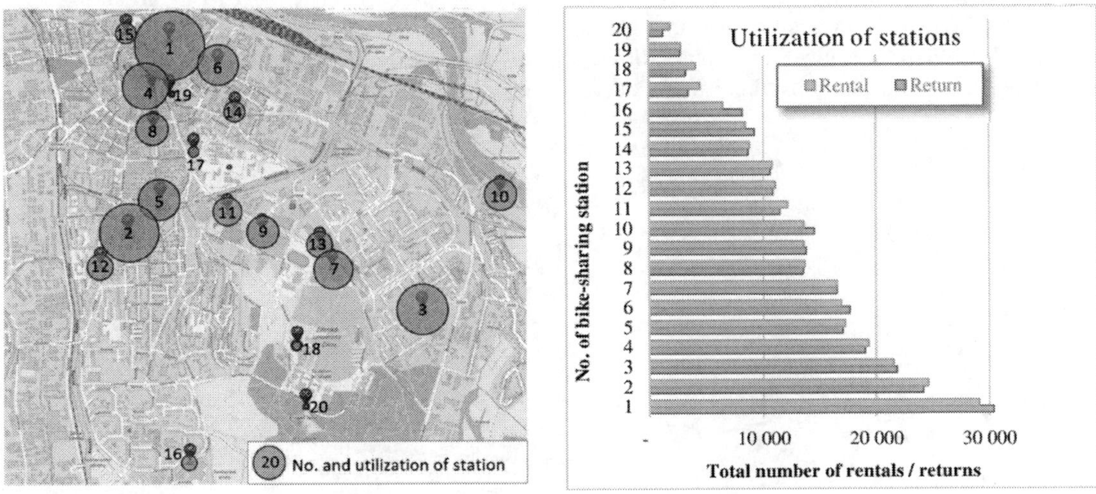

Figure 1. Utilization of rental stations.

The usage of the BikeKIA system is shown in the chart in figure 2. The total number of bicycle rentals (bike-share rides) was about 280 thousand over the period with an average of about 35 thousand per month. The highest number of bicycle rentals was achieved in the first month of full operation, the lowest in November due to the weather. A lower number of rentals were also recorded during the summer holidays and in September when there are fewer university students in the city.

To demonstrate the BikeKIA system utilization, the usage rate of the system (an average bike-share rides per day per bicycle) was determined. It ranges from 6 to 13 rides per day per bike depending on the month with an average value of almost 9.5 (see figure 2). On some days in the first period of the service operation, the usage rate reached the value of 20 rides per day per bicycle.

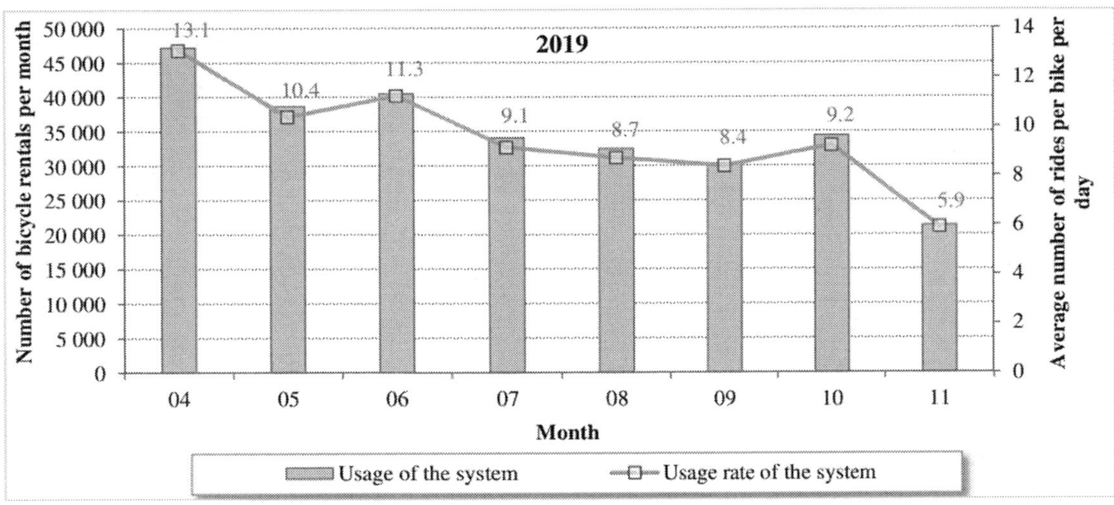

Figure 2. Utilization of the BikeKIA system.

Covid-19 pandemic's impact on reducing the usage of the system is presented in figure 3. The first two months, March and April, are omitted because the system was not in operation at that time due to coronavirus and state of emergency with strong restrictions in Slovakia. In May 2020, when several companies had a home-office mode, and the university, all schools, and many shops were closed, there was a decrease in the total number of bike-share rentals in comparison with 2019 of about 60 %. In June 2020, when the university and most of the schools were closed, there was a decrease of about 40%. In July 2020, during the summer holidays and the standard mode without more significant restrictions due to coronavirus, there was a decrease of only about 14%.

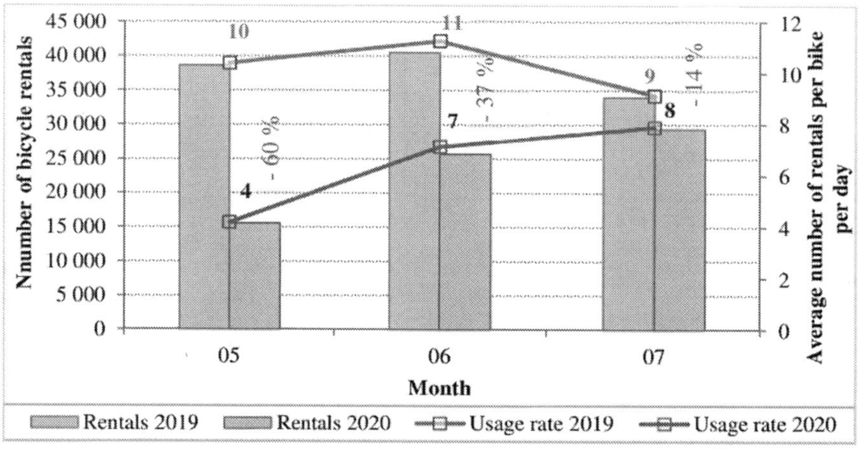

Figure 3. Covid-19 pandemic's impact on the usage of the BikeKIA system.

Presented data indicate a huge usage of shared bicycles in the city of Žilina. This can be confirmed in comparison with systems in some selected Slovak and Czech cities (see table 2). It is possible to see up to 3-4 times higher usage rate of the BikeKIA system compared to other systems.
A high utilization rate of the BikeKIA system is also evident in comparison with the average number of rides per day per bike of bike-sharing systems in large cities such as Barcelona (6.5), Washington DC (3.8), Miami Beach (3.2) or Melbourne (0.7) [3].

Table 2. Comparison of usage rates of bike-sharing systems (2019) [a]

City	Number of inhabitants in thousand	Number of stations	Number of bicycles	Average number of rentals per day	Usage rate of the system
Žilina	81.1	20	120	1 142	**9.5**
Bratislava	424.4	74	400	1 020	**2.6**
Kladno	69.0	50	100	180	**1.8**
Ostrava	289.6	210	600	1 893	**3.2**
Prostějov	43.7	37	100	212	**2.1**
Havířov	72.1	20	80	265	**3.3**

[a] Source: processed based on data in [5, 6, 7]

The main factors influencing a high utilization of the BikeKIA system include:

- free-of-charge bicycle rental for up to one hour, even several times a day,
- easy registration without any registration fee or deposit,
- inhabitants with a relatively low average age (36 years), quite a lot of young people - students and university students,
- flat plane terrain (just only two stations are on the hill),
- optimal allocation of bike-sharing stations mainly in the city core and wider center with covering strategic points,
- sufficient variety of bike-share rides ensuring natural redistribution of the shared bicycles during a day,
- existing cycle routes, which are relatively safe,

- short distances between stations,
- reliable and simple system,
- good quality of operational service,
- good marketing.

4. Rental duration analysis

Since the BikeKIA system is free for 60 minutes, the bicycle rental duration was also analyzed. The results in figure 4 show that most rentals take less than 20 minutes (73.1 % to 86.5 % by month, with an average of 78.0 %). The proportion of rentals exceeding an hour is only 0.4 %, which is related to the fact of free-of-charge rental for one hour. The proportion of rentals exceeding 20 minutes (20-60 minutes) is more than 20 %. The average rental duration is around 16 minutes.

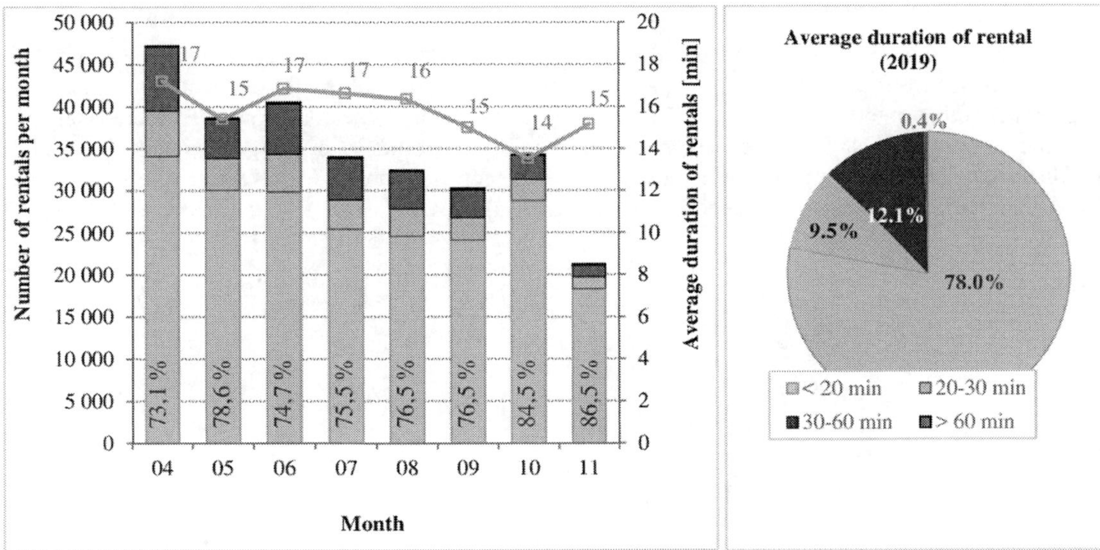

Figure 4. The average duration of bicycle rentals.

A more detailed analysis of raw data has been revealed, that too long free-of-charge rental time supports an unnecessary prolongation of rentals and usage of shared bikes for non-transport purposes instead of effective interstation rides. These too long rentals mainly include "too long rides" (extended or interrupted rides because they are free) and "rental rides" (rental station and return station is the same). Experience has shown that these "rental rides" often include the usage of shared bikes mainly by young users for fun (usage for joyrides).

The proportion of "rental rides" from all bike rentals during the observed period of 2019 is represented by an average value of 11.6 % (7.1 % - 15.3 % depending on the month). Most "rental rides" were recorded at the Žilina reservoir station (No. 10; about 42 %), and at the most frequently used stations (No. 1, 2, and 3; about 15 %).

To determine the proportion of unnecessary long rides, the rental-return matrix of rental duration limits was determined. For the setting of these limits, average travel times (depending on distance and average speed 10 km/h) and the Modes of a set of data (rental durations) for all trips between stations were calculated. As an example, processed data for the two most frequently used trips are shown in figure 5. Rental duration limits have always been set higher than the mode of rental duration and average travel time (by about 30 % higher than the average travel time depending on distance and longitudinal slope). Rental durations exceeding the set limits were considered as "too long rides".

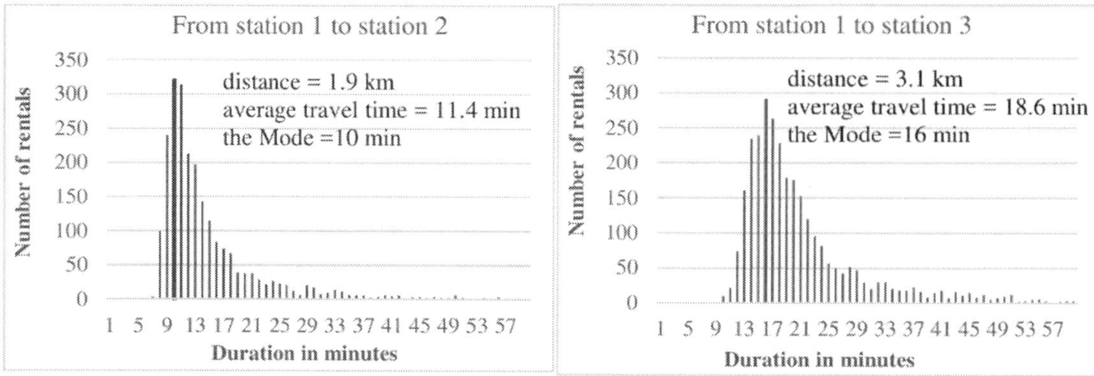

Figure 5. Processed data of rental durations for two most frequently used trips.

Results for the observed period of 2019 show that only 65 % of all rentals were carried out for travel purposes from the origin (rental station) to the destination (return station) without unnecessary delays (effective interstation rides). The remaining percentage means using the shared bike for purposes other than just effectively moving from point A to point B (too long rides and rental rides). Further analysis revealed that these rentals are much longer than rentals for effective interstation rides. More than half of these rentals exceed 20 minutes (see table 3).

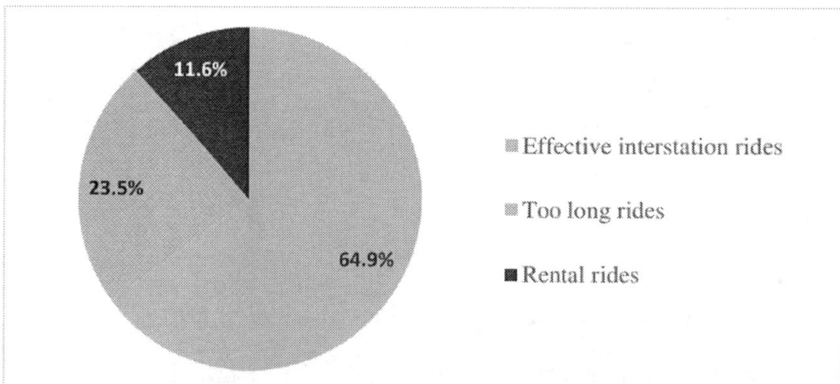

Figure 6. The proportion of bicycle rentals by type of rides.

Table 3. Comparison of rental durations by type of bike-share ride.

Rental duration	The proportion of the rentals in % Effective interstation rides	Too long rides	Rental rides
< 20 min	96.6	42.4	47.7
20-30 min	3.2	24.6	14.0
30-60 min	0.2	32.1	37.0
> 60 min	0.0	0.9	1.2
Total	100.0	100.0	100.0

5. Effects of the free-of-charge service of the BikeKIA system

Excessive usage of the BikeKIA system thanks to the free-of-charge service model up to an hour significantly affects the sustainability of this system, especially in two areas:

- quality of the service for users,

IOP Conf. Series: Materials Science and Engineering 1015 (2021) 012104 doi:10.1088/1757-899X/1015/1/012104

- financial sustainability of the system.

They are threatened mainly by:

- excessive wear of the system,
- worse technical condition and shorter lifespan of bicycles,
- increased requirements for redistribution,
- increased number of unnecessarily long rentals,
- increased certain form of vandalism and careless handling of bicycles due to their usage mainly by young users for fun (usage for joyrides),
- higher users' disrespect to the system.

In terms of the service quality, these aspects have the effect of reducing the functionality and reliability of the system, user safety as well as the availability of this service for users with real transport needs. Availability of the service is given by the availability of bicycles, which is reduced due to more frequent repairs and charging bicycles, or using bicycles for joyrides mainly by young users, unnecessary rides, or too long rides.

In terms of the financial sustainability of the system, the above-mentioned aspects result in higher requirements for repairing, maintenance, and cleaning of bicycles which is associated with an increased need and amount of work and materials (spare parts). For such an excessive used system, material costs have proved to be almost double and the need for additional working time for technicians almost 30 % higher compared to the standard used system. These increase the technical operating costs of the system by up to 40 %.

6. Conclusions

The article presents a successful bike-sharing project in Žilina with data demonstrating the enormous usage of this system. One of the most important reasons for such a large usage of the system is the free-of-charge service - for up an hour, several times a day, and without registration fees or deposit. This aspect, together with a well-designed and operating system thus offers a new opportunity for free and ecological mobility of residents and visitors to the city. On the other hand, this threatens its long-term sustainability, especially in two areas: quality of the service for users and the financial sustainability of the system. In terms of the service quality, there are effects of reducing the functionality and reliability of the system, user safety as well as the availability of this service for users with real transport needs. In terms of the financial sustainability of the system, there are impacts on higher technical operating costs of the system, which can be up to 40 %. Ultimately, the overall loss is much higher because of no income.

The data analysis in the BikeKIA system shows that too long rental time for free unacceptably supports an unnecessary prolongation of rentals and the use of shared bikes mainly by young users for joyrides instead of an effective interstation rides This results not only in reducing the availability of bicycles but also increased financial demands due to a certain form of vandalism and careless handling of bicycles.

To ensure the use of a bike-sharing system for effective transport, while enabling free rental service, it is necessary to change the rental conditions to eliminate as much as possible undesirable rentals. One of the options is a significant reduction in rental time for free and the collection of a registration fee or deposit. The rental time for free should be considered based on detailed data and financial analysis. Another contribution to reducing vandalism and negligent treatment of bicycles can be an increase in the fine, improving the intelligent monitoring of bicycles and rental stations, and the support for other users in this monitoring.

Acknowledgments

The presented results were partially obtained under the grant Vega No. 1/0428/19, Implementation of structural variables in the calculation of the modal split in the transport forecast (Implementácia štrukturálnych veličín pri výpočte deľby prepravnej práce v prognóze dopravy)

References

[1] Otero I, Nieuwenhuijsen M J and Rojas-Rueda D 2018 Health impacts of bike sharing systems in Europe *Environment International* **115** pp 387-94

[2] Bullock C H, Brereton F and Bailey S 2017 The economic contribution of public bike-share to the sustainability and efficient functioning of cities *Sustainable Cities and Society* **28** pp 76-87

[3] Fishman E, Washington S and Haworth N 2013 Bike share: a synthesis of the literature *Transport Reviews* **33** pp 148-65

[4] Petru J and Krivda V 2017 Utilization filled spandrel bridge structure for interchange cycle transport *17th International multidisciplinary scientific geoconference SGEM 2017 Geoinformatics* Albena Bulgaria 06/07 2017 pp 867-74

[5] www.cyklodoprava.sk/wp-content/uploads/2019/10/slovnaftbajk.pdf

[6] www.akademiemobility.cz/aktuality/1112/predstavujeme-system-sdileni-kol-nextbike

[7] https://nasregion.cz/kladensko/sdilena-kola-se-kladenakum-libi/

IOP Conf. Series: Materials Science and Engineering 1015 (2021) 012105 doi:10.1088/1757-899X/1015/1/012105

Impact of environmental actions on steel structures

M Jančula[1], J Vičan[1] and A Spiewak[2]
[1] University of Žilina, Faculty of Civil Engineering, Department of Structures and Bridges, Slovakia
[2] MAAD Sp. z o. o., Częstochowa, Poland

E-mail address: miroslav.jancula@uniza.sk

Abstract. The durability of steel constructions is strongly influenced by damage due to environmental actions, which need to be consider in design and maintenance of new and also existing structures. Corrosion aggressiveness of the atmospheric environment is classified by the corrosion rate, what it represents thickness losses of structural steel. That leads to a smaller resistant area, producing a decrease in the structural performance in terms of resistance, stiffness and ductility. The object of this contribution is to present a partial report of experimental research of corrosion processes influence on steel bridges in Slovakia and Poland.

1. Introduction

Several factors influence the service life of metal structures. Atmospheric corrosion is recognized to be one of the major risks, which decrease the resistance of constructions, resulting in huge economic and social losses [1]. Corrosion life prediction provides a key part for the optimum selection of materials or coatings for those structures. Studies made over the last year's show that the world's climate is changing. Besides global warming, there are also changes in other parameters which affect aggressiveness of environment.

Researchers in many countries all around the world have accomplish exposure tests to investigate the effects of the environment on corrosion rates what leads to classify category of corrosivity of atmosphere from C1 to CX by European standard STN EN ISO 9223 [2]. This standard allows also other way to determine category by dose-response function from the data related to the environment like rainfall, pollution, humidity or average temperature. These information lead to the formation of corrosion maps for different materials [3, 4] to show the environmental effect of the atmosphere on engineering structures.

2. Observation of corrosion process in Žilina

Roadway and railway bridges together with other structures, like tunnels, are the most critical components of transport infrastructures not only in Slovakia but also in all European countries. Also other types of engineering structures connected with traffic are affected, like stations or parking houses [5]. Real bridge structures are affected by environmental conditions. The environmental actions in time cause the degradation of bridge members what leads to a decreasing cross-section area, or in other words significantly affect safety of structure [6, 7]. Monitoring process of corrosion degradation during time on existing bridge structures is challenging, on the other hand with test sample placed at construction it is much easier.

For the study of the corrosion effects Žilina self-governing region was selected, where 12 bridges on the road or railway were selected. Those structures overpass various types of obstacles like a river

IOP Conf. Series: Materials Science and Engineering 1015 (2021) 012105 doi:10.1088/1757-899X/1015/1/012105

or a highway and different structural design types as a girder, truss or composite bridge. Samples of structural steel with dimensions of 150×100×3 mm were placed on the bridges at the position of the angle of 45° figure 1. Each sample was measured and weighed with accuracy of 0.01 mm and 0.01 g.

Figure 1. Test samples "Y" placed at bridge near Nová Bystrica – Vychylovka.

Test samples were exposed to aggressive environment and their degradation due to pollution and chloride ions from traffic was measured annually. Every sample after period of exposure was cleaned and compared with initial values at the beginning of experiment. Figure 2 shows a test sample before and after cleaning.

a) b)

Figure 2. Test samples a) before cleaning, b) after cleaning.

Corrosion rate r_{corr} which is thickness loss in micrometres per year can be influenced by many effects. Within in-situ measurement, it was decided to determine the impact of repeated cleaning - figure 3. Test samples were exposed to aggressive environment during last three years. Discussed samples "Y" are placed on concrete roadway bridge, it is a connection between Orava and Kysuce districts, location with significant snowfall during winter. The first set was cleaned three-times after first, second and third year (Y1, Y2), the second set of samples was cleaned once after second year (Y3 – Y5) and the last third set was cleaned also once but after three years (Y6 – Y8). Chart shows results that the repeated cleaning of steel structural members has negative impact on thickness losses.

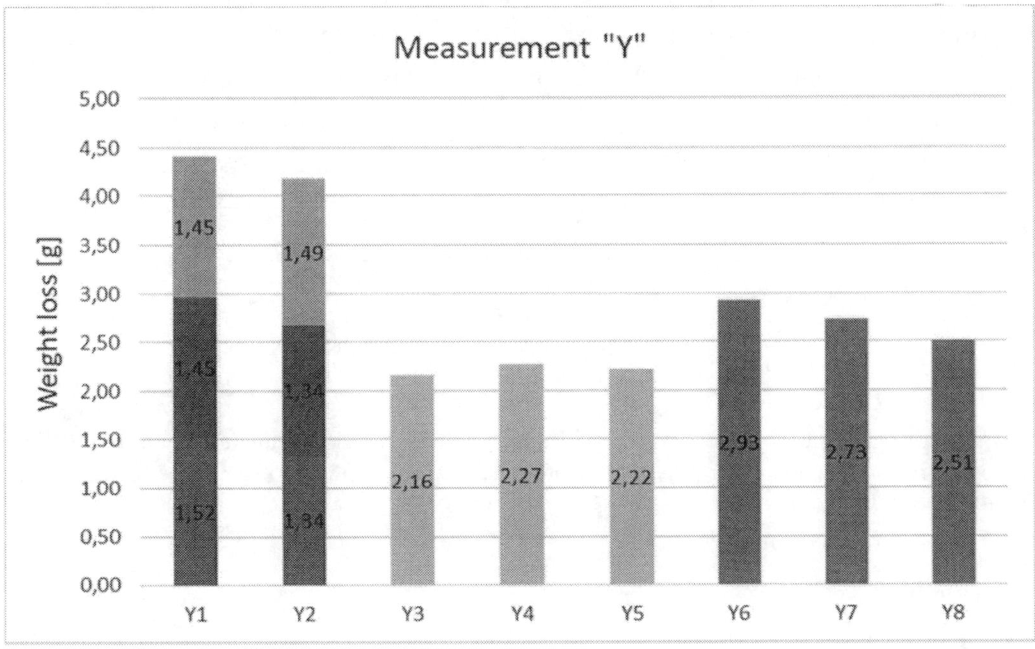

Figure 3. Impact of repeated cleaning of samples during three years exposure.

As it can be seen in figure 3, samples Y1,Y2 have far greater weight loses after three years of exposure compared to samples Y6, Y7, Y8, which can indicate that repeated cleaning of corrosion products cause bigger mass loses. However, the results on different bridges are not identical. From total twelve sets of samples placed on bridges, three locations in Žilina region show similar loses between 3-times and once cleaned samples after third year of experiment. These discrepancies will be monitored in next years of our long-term experiment of environmental action effects on steel bridges.

3. Observation of corrosion process at Silesia

The basic function of bridges is, above all, to ensure reliable traffic, so that the bridge does not become a limiting element of performance and permeability of the road [8]. The administrators of bridge structures with quality and qualified maintenance, repairs and reconstructions ensure fulfilment of this function. For these purposes it is necessary to know not only the current state of the structure, but also how it will degrade in the future [9]. The main factors influencing the condition of bridges is increasing traffic intensities and especially the degradation processes of structural elements, which are caused by environmental action from the surrounding environment [10]. Polluted air has a significant impact on the degradation of all materials used in the transport infrastructure. As pollution is increasing, the polluted air becomes considerably more aggressive with a bigger influence on the structural parts of bridges.

In Poland, five classes of winter maintenance are classified, in which the roads are divided by traffic, and 3 of these classes are located in the Silesia region where a study of the influence of chloride ions, such as winter maintenance products, on corrosion processes. In selected locations, the samples of structural steel with dimensions 150x100x2 mm were placed on 11 bridges. A total of 12 samples were placed for the horizontal (upper side of the lower flange), as well as the vertical (beam web).

After third year of in-situ experimental measurement three samples were taken from the horizontal and vertical positions and after cleaning they were weighed. The measured data were processed, evaluated and result are presented in table 1. where higher standard deviation signalize that even bridges which are in the same type of winter maintenance don't have similar degradation of test samples due to impact of environmental load combined with chloride ions.

Table 1. Results after the third year of exposition on bridges.

Type of WRM	Place of installation	Weight loss $g/m^2/year$	
		mean	standard dev.
WRM I	web of I-beam	218.94	29.22
	bottom flange of I-beam	265.07	264.97
WRM II	web of I-beam	97.56	29.01
	bottom flange of I-beam	116.37	18.62
WRM III	web of I-beam	190.81	110.97
	bottom flange of I-beam	226.67	158.63

Comparison of results after three years exposition at aggressive environment are shown in figure 4, where outcomes are divided to 3 types of winter maintenance and 2 types of positioning, web of I beam and bottom flange of girder.

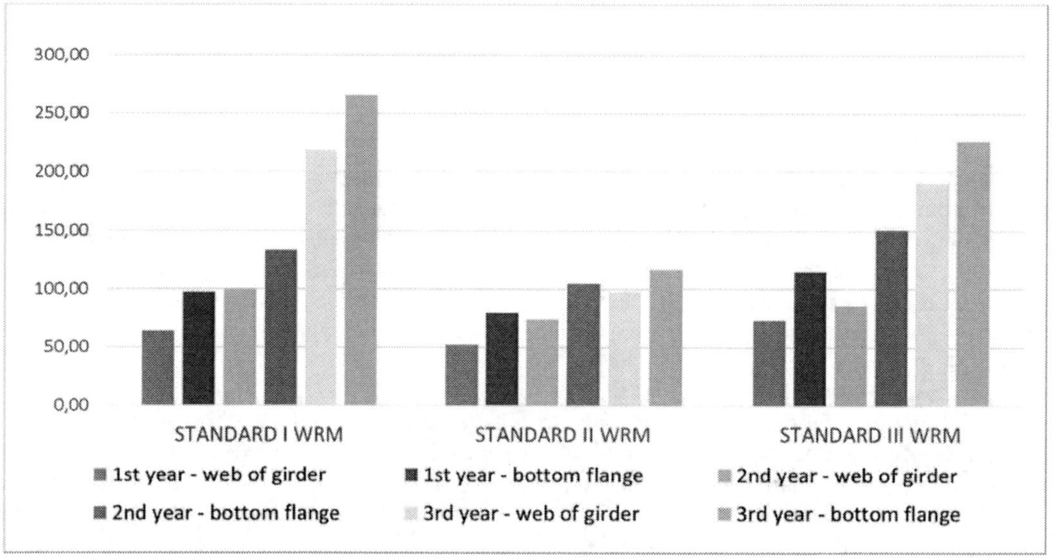

Figure 4. Comparison of results after three years of exposure.

The same approach as on bridges were used for samples placed in reference stations, where samples are both horizontally and vertically positioned. Collected data from stations after third year of measurement are in table 2, where Poraj, Katowice and Ustroń represent three districts at Silesia region. Results of horizontal and vertical positions are together presented in table 2, what have impact on the higher standard deviation.

Table 2. Results after the third year of exposition at reference stations.

Location	Weight loss ($g/m^2/year$)	
	mean	standard dev
Poraj	173.06	64.69
Katowice	103.11	20.76
Ustroń	114.45	62.78

Significant increment after first year does not continue as it can be seen in figure 5, but it only grows slowly. This results can be affected by several impacts like average temperature and rainfall during years, also protection of sample by corrosion itself, where greater weight losses after initial time of corrosion processes are exchanged by slower progress of material degradation.

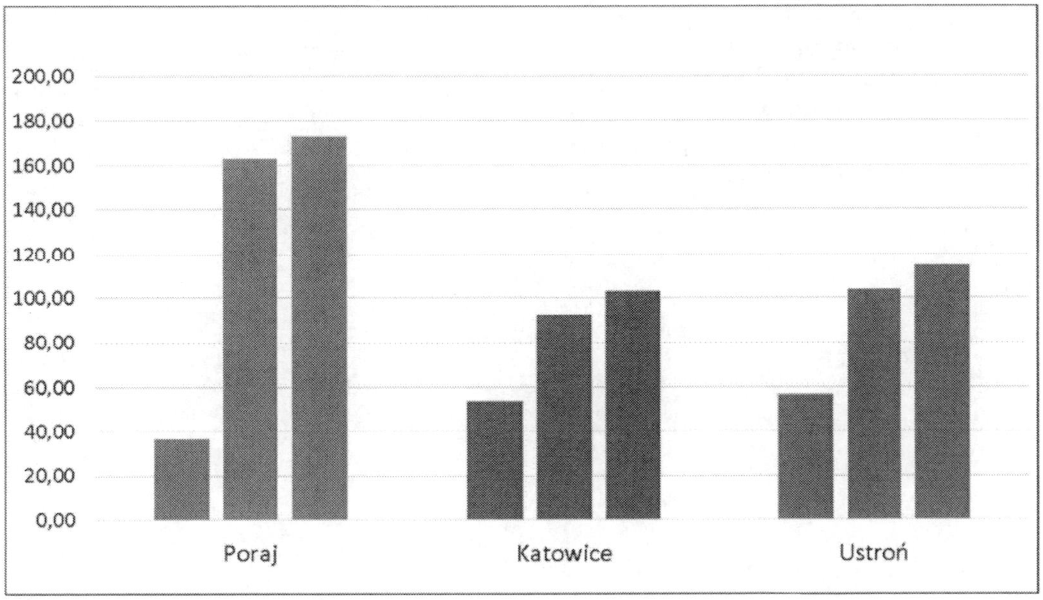

Figure 5. Comparison of results from the reference stations after three years.

4. Conclusions

Even though efforts during the design and construction of bridge structures, various damages and failures ensue within design lifetime. Without proper maintenance, small defects can develop into failures significantly affecting bridge load-carrying capacity, traffic safety and bridge remaining lifetime. Process of analyses considering environmental degradation processes and their future development is complicated and problematic

Corrosion has a stochastic character and it dependents on factors like average temperature, humidity or rainfall, also on material characteristics, but even on structural design of the bridge. The results obtainable in this article approve the randomness of corrosion. More relevant data for the estimation of the influence of corrosion on any structure can be obtained only from long-term measurements. Therefore, the experimental investigation of impacts of aggressive environment should continue.

Acknowledgments

This research was supported by the Research Project No. 1/0413/18 and No. 1/0336/18 of the Slovak Grant Agency.

References

[1] Bujňák J, Gocál J and Hlinka R 2016 Assessment of railway steel bridge structures *Procedia Engineering* **156** pp 75-82
[2] STN EN ISO 9223, Korózia kovov a zliatin. Korózna agresivita atmosfér. klasifikácia, stanoveniea a dohad, SUTN, 2012
[3] Strieška M and Koteš P 2018 Corrosion map of zinc in Slovakia *Pollack Periodica* **13** (2) pp 129-36

[4] Koteš P, Ivašková M and Brodňan M 2015 Air Pollution as an important factor in construction materials deterioration in Slovak Republic *Procedia Engineering* **108** pp 131-8
[5] Kvočák V, Vargová R, Beke P and Terpáková E 2013 Effects of atmospheric corrosion on the car park roof structure *Komunikacie* **15** (1) pp 80-7
[6] Vičan J, Gocál J, Meliš B, Koteš P and Kotula P 2008 Real behaviour and remaining lifetime of bridge structures *Komunikacie* **10** pp 30-37
[7] Gocál J, Odrobiňák J 2020 On the influence of corrosion on the load-carrying capacity of old riveted bridges *Materials* **13** (3) 1 February 2020 Article number 717
[8] Vičan J and Koteš P 2018 Hodnotenie existujúcich mostných objektov (Zylina: EDIS)
[9] Bujňák J 2005 Kovové mosty spravovanie, údžba, rekonštrukcia *EDIS* 80-8070-361-2
[10] Leygraf C, Wallinder I O, Tidblad J and Graedel T 2016 *Atmospheric corrosion* (New York NY USA: Wiley & Sons)

AUTHOR INDEX